Gas Measurement Technology in Theory and Practice

Gerhard Wiegleb

Gas Measurement Technology in Theory and Practice

Measuring Instruments, Sensors, Applications

Springer

Gerhard Wiegleb
Department of Electrical Engineering
University of Applied Sciences
Dortmund, Germany

ISBN 978-3-658-37231-6 ISBN 978-3-658-37232-3 (eBook)
https://doi.org/10.1007/978-3-658-37232-3

This book is a translation of the original German edition „Gasmesstechnik in Theorie und Praxis“ by Wiegleb, Gerhard, published by Springer Fachmedien Wiesbaden GmbH in 2016. The translation was done with the help of artificial intelligence (machine translation by the service DeepL.com). A subsequent human revision was done primarily in terms of content, so that the book will read stylistically differently from a conventional translation. Springer Nature works continuously to further the development of tools for the production of books and on the related technologies to support the authors.

This Springer imprint is published by the registered company Springer Fachmedien Wiesbaden GmbH, part of Springer Nature.
The registered company address is: Abraham-Lincoln-Str. 46, 65189 Wiesbaden, Germany

This book is dedicated to my wife Petra Anna, who has supported me sacrificially and lovingly for over 45 years. Without her this book would not exist.

Foreword

I would like to comply with the author's wish to write a foreword to this comprehensive work on gas measurement technology with a few words. I do this all the more gladly because no such useful work was available to me during my 36 years of professional activity as a developer and head of development at Hartmann & Braun AG (now ABB Automation GmbH). In this book you will find important basics of all currently realized measuring principles as well as many interesting applications of gas analysis technology. Developers, product engineers, salesmen, and users of gas analysis devices and systems need all this for their daily work.

The basics needed to understand the measurement methods discussed here are interdisciplinary. The whole variety of *MINT subjects* is needed to enter the field of gas measurement technology. Students who enjoy such a broad education should therefore be advised to take a look at this book. They will certainly enjoy the rich offer of smart technical solutions.

Everyone is currently talking about automobile exhaust gas analysis, which is also dealt with in this book. In the future, mobile measuring equipment will be needed for measurements on the road instead of on the chassis dynamometer. Automobile exhaust gas analysis is just one example of the importance of gas measurement technology as a whole for maintaining an environment worth living in.

Last but not least, I hope this book will be of great interest to students and professionals in industry and research. I would like to thank the author for the invitation to write this foreword.

Kronberg/Taunus, Germany — Werner Schaefer
November 2015

Preface

This book on gas measurement technology is the result of almost 40 years of work in this varied and interesting field. Already as a young physics student in 1978 I had the opportunity to work for some months in the laboratories of Dr. K. F. Luft,[1] at the Mining Research in Essen-Kray, as an intern. At that time, Dr. Luft had just retired, and so I received the first introductions to gas measurement technology from my then supervisor, Mr. Dankwart Mohrmann, which had a lasting influence on me. Since then, gas measurement technology has never left me. My final thesis at the University of Essen, in the field of laser spectroscopy, was therefore also oriented in this direction. The extraordinarily good supervision by Dr. Schomburg enabled me to work scientifically in this field.

My first job in industry, at Leybold-Heraeus Gmbh in Hanau, was also characterized by very free work as a development engineer. The close cooperation with Mr. Albert Randow and many others in this working group led to the development of several new gas measuring instruments and many publications and corresponding patents.

The move to KOSTAL GmbH in Lüdenscheid then brought completely new experience in the field of automotive sensor technology. With the development of a new type of Halbeiter gas sensor for determining ventilation quality in motor vehicles, I was able to continue my activities in the field of gas measurement technology there. Prof. Dr. Heitbaum and Prof. Dr. F.-J. Schmitte were valuable supporters and helpful colleagues during this time.

The start of my work as head of development for the entire analytical technology at Hartmann & Braun AG in Frankfurt-Praunheim was the highlight of my 15 years of industrial activity. During this time, I gained insights into all common methods of gas measurement technology. Here, I met a large number of experts in gas measurement technology, of whom I would like to mention Walter Fabinski, Dr. Werner Schaefer, and Dr. Michael Zöchbauer in particular.

[1] Karl-Friederich Luft (1909–1999) German physicist and pioneer of physical gas measurement technology.

Furthermore, I had the opportunity to welcome Dr. Luft at the Praunheim plant during the series introduction of the URAS 4. Together with Dr. Schaefer and Dr. Luft, I wrote a historical paper in 1993, in which the developments of the past 50 years in the field of IR gas measurement technology were described.

Visit of Dr. K. F. Luft (*front left*) at the Hartmann & Braun factory in Frankfurt-Praunheim, on the occasion of the production launch of the fourth generation of URAS devices.

Since my appointment to the University of Applied Schiences Dortmund, the miniaturization of sensor-based processes in gas measurement technology has always been at the forefront of research activities. However, this R&D focus has also been carried out in many other working groups at universities, research institutes, and companies. In the context of the description of the different measuring methods, this book refers to these international developments at the appropriate points.

In this respect, the cooperation with Sensors Inc. in Saline, Michigan, USA, in the years 1999 to 2009 was particularly noteworthy: Together with my US colleagues Robert Zummer, Gideon Eden, and Juma Achoki, a mobile NO_x-measuring device (*SEMTECH*) was developed for the first time, which today is part of the standard measurement of on-*road exhaust gas analysis* (automotive industry) worldwide.

Although the book has a total of about 1300 printed pages, it was not possible to present all data and information in full in this work. The author has therefore created an Internet access where further data and current information/developments are available:

http://www.gasmesstechnik-wiegleb.de

Despite careful correction, errors in the present work cannot be ruled out. I would therefore be pleased to receive information about possible errors. Furthermore, suggestions for improvement are always welcome, which will be taken into account in a new edition.

Dortmund, Germany Gerhard Wiegleb

Acknowledgments

A major focus of this book is the wide range of applications in gas measurement technology. At this point, I would like to thank the 40 authors who have enriched Chaps. 14, 15, 16, 17 and 18 with their competent and up-to-date application reports in a form which is probably unique in the field of gas measurement technology.

Valuable support was also given to me by expert engineers and scientists during the work on individual chapters and on certain sub-areas of gas measurement technology.

My special thanks:

- Mr. Gabriele Dietrich, Dr.-Ing. Holger Födisch, and Mrs. Anika Sauer from Dr. Födisch Umweltmesstechnik AG in Markranstädt, who have worked together on Chap. 13 on dust measurement technology
- My colleague, Mr. Christian Fried (University of Applied Sciences-Dortmund), for relieving me of teaching duties during the time I was particularly busy with the book project
- Dr. Klaus Kaltenmaier of Gasmet Technologies GmbH in Karlsruhe for important information on gas drying (Sect. 14.1)
- My colleague Prof. Dr. Ulrich Hahn (University of Applied Sciences-Dortmund) for reviewing Chap. 2
- Dr.-Ing. Frank Hammer of LAMTEC Meß- und Regeltechnik für Feuerungen GmbH & Co KG in Walldorf for reviewing the section on solid electrolyte sensors (Sect. 4.2)
- Dr. Wolfgang Jessel (formerly Dräger Safety AG, Lübeck) for reviewing Sect. 18.1
- Dr. Olaf Kiesewetter and Jürgen Müller from UST Umweltsensortechnik GmbH in Geschwenda for reviewing the sub-chapter on semiconductor gas sensor technology
- Dr. Uwe Lawrenz of ZIROX GmbH in Greifswald for important information on humidity measurement with solid electrolyte sensors and for reviewing this sub-chapter
- Dr.-Ing. Joachim Ritter of Dr.-Ing. Ritter Apparatebau GmbH in Bochum for reviewing parts of Chap. 11
- Dr. Werner Schaefer (formerly Hartmann & Braun AG Frankfurt am Main) for the foreword

- Dr. Guenter Schierjott and Dr. Phillip Schierjott of Wösthoff GmbH Bochum for valuable information on gas mixing systems based on piston pumps
- Dr. Peter Schley from SmartSim GmbH in Essen for important information on the physics of gases and especially on the special features of natural gas
- Dr. Martin Schmäh of IAS GmbH Oberursel for notes and corrections to the dynamic calibration methods (Sect. 12.3)
- Dr. Wernecke from Feuchtemesstechnik GmbH in Potsdam for information on the physics of humidity
- My son Robert Wiegleb (AIM Infrarot-Module GmbH Heilbronn) for checking important mathematical correlations of IR optics (Chap. 7)
- My son Sebastian Wiegleb (Wi.Tec-Sensorik GmbH Wesel) for important information on sensor signal processing
- Mr. Thomas Wortelmann from *G.A.S.* Gesellschaft für analytische Sensorsysteme mbH in Dortmund for information and searching the section on *ion mobility spectroscopy*.

Furthermore, I would like to thank Springer-Vieweg Verlag Wiesbaden for the publication of this work. At this point I would like to mention in particular the editing by Ms. Andrea Broßler and Mr. Rainhard Dapper, as well as the project coordination and production by Ms. Walburga Himmel, Ms. Gabriele McLemore, and Ms. Yvonne Schlatter.

Contents

List of Contributors

Martin Bender WITT-Gasetechnik GmbH & Co KG, Witten, Germany

Thomas Beyer SICK AG, Reute, Germany

Roland Bianchin DURAG Holding AG, Hamburg, Germany

Steffen Biermann Micro-Hybrid Electronic GmbH, Hermsdorf, Germany

Gabriele Dietrich Dr. Födisch Umweltmesstechnik AG, Markranstädt, Germany

Julian Edler SICK AG, Reute, Germany

Thomas Eisenmann DURAG Holding AG, Hamburg, Germany

Walter Fabinski Formerly ABB Automation Frankfurt, Kriftel, Germany

Peer Fietzek Kongsberg Maritime Germany GmbH, Hamburg, Germany

Holger Födisch Dr. Födisch Umweltmesstechnik AG, Markranstädt, Germany

Oliver Franken Sensors Europe GmbH, Erkrath, Germany

Axel-Ulrich Grunewald Messkonzept GmbH, Frankfurt am Main, Germany

Frank Hammer LAMTEC Meß- und Regeltechnik für Feuerungen GmbH & Co KG, Walldorf, Germany

Thomas Heckler WIKA Alexander Wiegand SE & Co KG, Klingenberg, Germany

Alexander Hein WIKA Alexander Wiegand SE & Co KG, Klingenberg, Germany

Frank Herrmann Weinmann EMT GmbH, Hamburg, Germany

Wolfgang Jessel Dräger Safety AG, Reinfeld, Germany

Joachim Kastner Weidmüller Interface GmbH & Co KG, Detmold, Germany

Olaf Kiesewetter UST Umweltsensortechnik GmbH, Geschwenda, Germany

Rainer Krage GfG GmbH, Dortmund, Germany

Roland Kurte WIKA Alexander Wiegand SE & Co KG GmbH, Dortmund, Germany

Tobias Lehmann Vocational College for Technology Moers, Moers, Germany

André Magi Micro-Hybrid Electronic GmbH, Hermsdorf, Germany

Matthias May UST Umweltsensortechnik GmbH, Geschwenda, Germany

Holger Müller BlueSens gas sensor GmbH, Herten, Germany

Jürgen Müller UST Umweltsensortechnik GmbH, Geschwenda, Germany

Ernst Murnleitner AWITE Bioenergie GmbH, Langenbach, Germany

Peter Paplewski Bruker AXS GmbH, Karlsruhe, Germany

Jürgen Reinmann Envea Germany, Bad Homburg, Germany

Stefan Römisch Dr. Födisch Umweltmesstechnik AG, Markranstädt, Germany

Anika Sauer Dr. Födisch Umweltmesstechnik AG, Markranstädt, Germany

Rolf Schiffler SICK AG, Reute, Germany

Peter Schley SmartSim GmbH, Essen, Germany

Udo Schmale BlueSens gas sensor GmbH, Herten, Germany

David Triebel DURAG data systems GmbH, Hamburg, Germany

Michael Unruh ExTox Gasmess-Systeme GmbH, Unna, Germany

Michael Wagner Dr. Födisch Umweltmesstechnik AG, Markranstädt, Germany

Thomas Weyrauch ABB Automation GmbH, Frankfurt am Main, Germany

Gerhard Wiegleb University of Applied Sciences Dortmund, Dortmund, Germany

Wi.Tec-Sensorik GmbH, Wesel, Germany

Patrick Zimmerman C-Lock Inc, Rapid City, SD, USA

Scott Zimmerman C-Lock Inc, Rapid City, SD, USA

Introduction

1

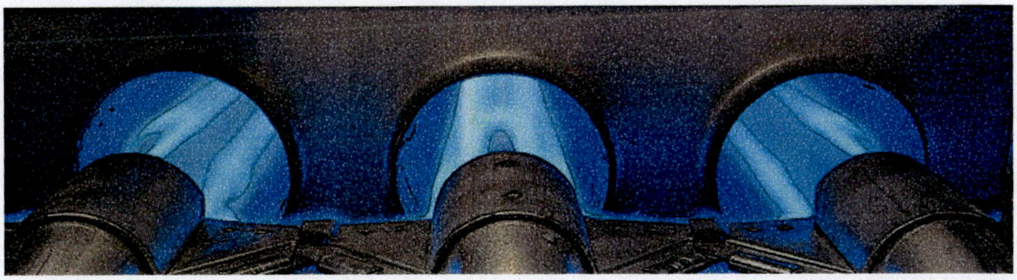

Combustion nozzles of a gas burner. (Source: fotolia)

Scientific knowledge about the existence and properties of gases was established very late in comparison to other physical theories. This was certainly because gases cannot be seen or touched. Most gases are also odourless, so humans cannot grasp gaseous substances with their natural senses. While Isaac Newton[1] published the fundamentals of mechanics in 1687, which are still valid today, there was hardly any knowledge about gases at that time, which is still valid today. The word *gas* first appeared around 1610 in the work of the Flemish doctor Johann Baptist van Helmont.[2] At that time, he was investigating substances that are released during alcoholic fermentation. We now know that this was carbon dioxide (CO_2). He used the Greek word *chaos* to describe these volatile substances. In the Dutch debate, this then became the word *gas* (Jessel 2001). By that time, chemical experiments carried out by various scientists had already identified various gases.

[1] Sir Isaac Newton (1642–1727) English naturalist and civil servant.

[2] Johan Baptist van Helmont (1580–1644) Flemish doctor, naturalist and chemist.

G. Wiegleb, *Gas Measurement Technology in Theory and Practice*,
https://doi.org/10.1007/978-3-658-37232-3_1

Fig. 1.1 Historical experiment by Otto von Guericke to prove the air pressure

However, most of the experiments during this period were carried out with air. The most famous experiment of this kind was probably the *Magdeburg hemisphere* from 1650.

The mayor and scientist at the time, Otto von Guericke,[3] had been experimenting with air for several years. For the experiment he used two hemispheres, which he sealed at the junction and pumped (evacuated) the air inside. He then had 8 horses pull on both sides of the hemispheres without the horses managing to separate the hemispheres (Fig. 1.1). The external air pressure that pressed the halves of the sphere together was simply too great. Otto von Guericke was able to impressively demonstrate the effect of the air pressure with this experiment. At the same time, he also invented the air pump and built the first water barometer. In 1664 von Guericke developed the manometer.

People have always been fascinated by the processes that take place during combustion. A first scientific description of the processes that take place during combustion was provided by Stahl[4] in 1722. According to this theory, a substance which he called *phlogiston* escapes during combustion. From today's perspective, the phlogiston could be described as a product of energy and carbon dioxide. This *phlogiston theory* was used to describe most of the combustion processes known until then, and also certain reduction processes (ore with charcoal). Also the German natural scientist Johann Christian Wiegleb,[5] who is considered to be the founder of modern Chemistry in Germany, became

[3] Otto von Guericke (1602–1686) German politician, lawyer, physicist and inventor.
[4] Georg Ernst Stahl (1659–1734) German alchemist, chemist, physician and metallurgist.
[5] Johann Christian Wiegleb (1732–1800) German natural scientist and pharmacist.

a supporter of this theory (Klosa 2009). However, the increase in weight during the oxidation of iron could not be explained by the phlogiston theory (Golze 2008). With the discovery of oxygen by Lavoisier,[6] this theory was then transferred to the *oxidation theory which is* still valid today.

In the following period, more and more gaseous substances were discovered and the physical properties of the gases were described more and more precisely. Many important scientists were involved in this gas theory, which was developed over a period of 100 years. The first systematic investigations into gas measurement technology were carried out by *Robert Bunsen*[7] as early as 1837. He is therefore considered the founder of scientific gas analysis (Neumann 1901). In Kassel, Bunsen examined, for example, the chemical processes that take place in a blast furnace process and published a treatise on the analysis of combustion gases in 1839 (Bunsen 1839). During his work Bunsen found out that 75% of the calorific value of the coal used is lost. He also found that only 20% of the carbon monoxide (CO) was used for the reduction process and the majority (80%) escaped from the blast furnace (blast furnace gas).

In 1846, Bunsen received an invitation from the Danish government to accompany an expedition to Iceland. On Iceland, he investigated the *Great Geyser* and found hydrogen, hydrogen sulphide and carbon dioxide in the escaping gases.

In 1857, Bunsen published the world's first textbook on gas measurement technology (Bunsen 1857), which appeared in the same publishing house as the present work.

In the age of industrialisation, gaseous substances also became increasingly important for practical applications. As early as 1786, the first attempts were made in England and Germany to install gas lighting in interiors. The illuminating gas made of hard coal was discovered by Joachim Becher[8] as early as 1682. Gas lighting in factory halls has been known since 1802. By 1860 there were already 350 gas works in Germany. The use of gas for chemical processes, lighting ticks, medical use and for energy production took a rapid upswing during this time. However, measurement technology for the various gases and their concentration in the ambient air lagged far behind this development. As a result, many serious accidents occurred again and again in the following period. The first detection methods for the dangerous carbon dioxide had been known to winegmakers and brewers for a long time. Rooms in which enriched CO_2 was present, were only entered with a burning candle in the hand. If the candle began to flicker or even went out, this meant that the CO_2content in the air was too high, and the room (usually a cellar room) was not entered. A relatively simple but also unreliable method.

Canaries were used in mining until the 1950s to warn of mine gas (Fig. 1.2). When hard coal is mined, mine gas is released, which mainly contains methane, carbon dioxide and also carbon monoxide. The birds react very sensitively to an increase in the gas

[6] Antoine Laurent de Lavoisier (1743–1794) French chemist, attorney, principal tenant of customs.

[7] Robert Wilhelm Eberhard Bunsen (1811–1899) German chemist.

[8] Johann Joachim Becher (1635–1682) German scholar, economist and alchemist.

Fig. 1.2 Portable bird cage with a canary, for use underground in a hard coal mine. The oxygen cylinder (handle) is used to revive the bird in case of a gas warning. It was also used as a carrying handle. (Source: © Bettmann/CORBIS Canary Used for Detecting Gas in Mines)

concentration and could therefore warn of a danger. However, this method was also unreliable and was subsequently replaced by electrically registering gas warning devices.

The first electrically operating gas measuring instruments used the different thermal conductivities of the gases to determine the gas concentration of carbon dioxide or hydrogen in the ambient air. These measuring instruments are well known from submarines for measuring the CO_2concentration during a diving trip. This method was also used to measure the hydrogen content (H_2) in a Zeppelin, and to warn of dangerous gas leaks.

Nowadays there are many technical possibilities to detect gases and vapours in the ambient air or in an industrial process. Gas measuring instruments can detect concentration ranges from 100 vol.% up to the ppt[9] range. The motivation for this gas measurement technology is very different. In addition to the first mentioned applications in safety engineering, there is also an increasing demand for gas detectors in quality assurance. These include the measurement of pollutant concentrations in the ambient air (immission measurement technology) and the release of pollutants from combustion processes into the ambient air (emission measurement technology/exhaust gas analysis, Fig. 1.3). The limit values to be monitored are specified by the legislator and must be complied with according to certain rules (e.g. TA-Luft[10]), which are precisely defined. If these limit values are exceeded, this has legal consequences. The measuring instruments used therefore meet a very high quality standard. Similarly, high demands are also made in process measurement technology. However, the operators of these plants have a different motivation. They want to control the processes as optimally as possible in order to reduce the use of energy and

[9]ppt = parts per trillion 10^{-12}.

[10]Technical Instructions on Air Quality Control (1st version of 8 September 1964).

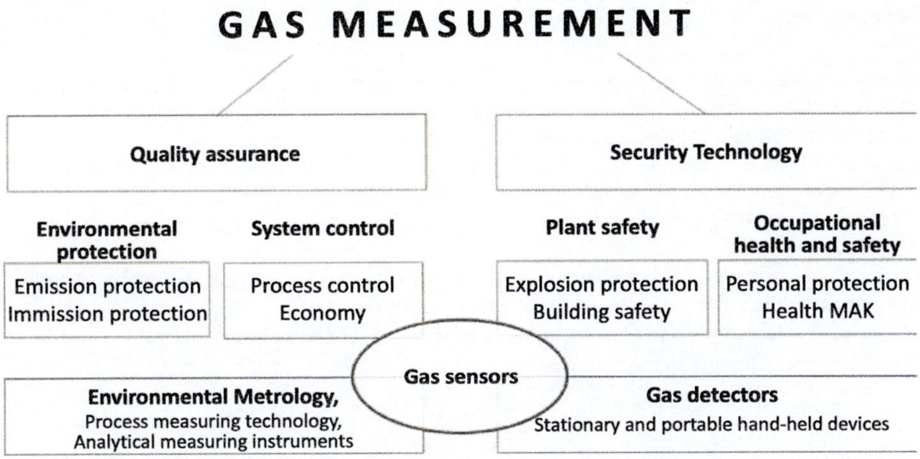

Fig. 1.3 Division of the gas measurement technology into the different application areas

raw materials. This saves time and money. In addition, some of these plants are also subject to legislation that regulates the emission of pollutants into the environment. One example of this is Sulphur hexafluoride (SF_6), which is used in electrical switchgear. Other gases which are under special observation by the authorities are the so-called CFCs, which are used in air conditioning systems and refrigerators.

Another large area of application for gas measurement technology is in medical technology. This area has gained more and more importance, especially in the last 20 years. Here, however, we are still at the beginning of a significant development. In addition to the well-known anaesthetic gas analysis, gas measurement technology now also plays a role in the diagnosis and treatment of diseases. In the breathing air of patients there are, for example, gaseous trace substances which provide information on certain diseases (Hering et al. 1993). After it became known that dogs can *smell* certain cancers, there is a desire for reliable gas detection devices for this application. In 1998, researchers Robert Furchgott,[11] Ferid Murad[12] and Louis Ignarro,[13] received the Nobel Prize for Medicine for their discovery of the effect of nitric oxide (NO) on the human body. The group found out that NO is important for the blood supply of organs and its role as a messenger in the organism. This knowledge about NO opens up new possibilities in the treatment of vascular diseases and the resulting organ damage. This application also requires NO-measuring devices that add low NO-concentrations to the patients' breathing air.

[11] Robert Francis Furchgott (1916–2009) US biochemist.

[12] Ferid Murad (1936–) American doctor and pharmacologist.

[13] Louis José Ignarro (1941–) American scientist.

References

Bunsen, R.: Analyse der Verbrennungsgase. Dinglers Polytechnisches Journal. **71**, 321 (1839)

Bunsen, R.: Gasometrische Methoden. Vieweg-Verlag Braunschweig (1857)

Golze, D.: Phlogiston vs. Sauerstoff. Uni Leipzig (2008)

Hering, P., Fuß, M., Haisch, M., Wiegleb, G.: Verfahren und Vorrichtung zur Bestimmung derIsotopenverhältnisse in Gasen. DE 4224146.414.7.1993, Deutsches Patent

Jessel, W.: Gase-Dämpfe-Gasmesstechnik. Ein Kompendium für die Praxis. Dräger AG, Lübeck (2001)

Klosa, M. A.: Johann Christian Wiegleb (1732–1800) eine Ergobiographie der Aufklärung. Wissenschaftliche Verlagsgemeinschaft Stuttgart (2009)

Neumann, B.: Gasanalyse und Gasvolumetrie. Verlag von S. Hirzel Leipzig 1901.

2 Physical Properties of Gases

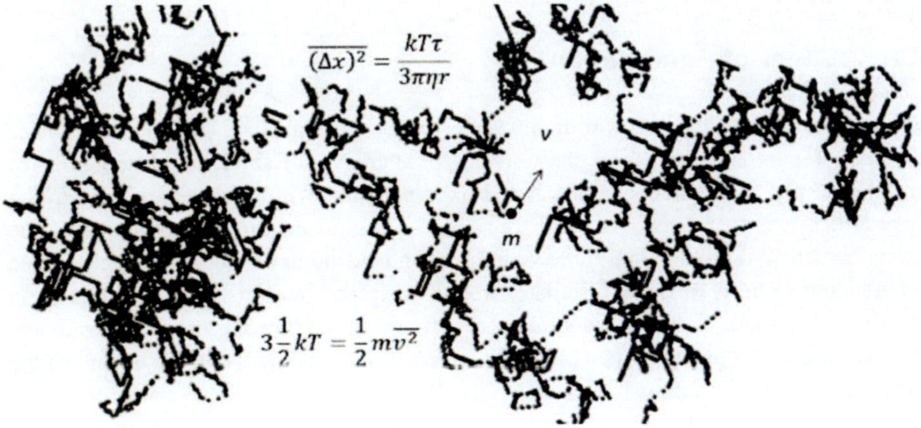

Brownian movement

Definition of Gases and Vapors

A gas or gaseous substance is a substance that is neither a solid nor a liquid at room temperature (20°C) and normal atmospheric pressure (1013 hPa). The gaseous state is therefore closely related to temperature and pressure. The physical properties of gases form an important basis for applications in gas measurement technology. Depending on the structure of the gaseous substance, a distinction is made between three different forms. In addition to the noble gases occurring in atomic form and the monatomic molecular forms (e.g., N_2), there is a multitude of so-called multiatomic gases (e.g., CO_2). A maximum of 200 substances can be described as *gas* under the term given above. In total there are only 12 elementary gases (6 noble gases and 6 monatomic gases). The vapors are a special

G. Wiegleb, *Gas Measurement Technology in Theory and Practice*,
https://doi.org/10.1007/978-3-658-37232-3_2

Table 2.1 Classification of gases according to their atomic structure

Noble gases	Singleatomic molecules	Multiatomic molecules
Helium, He	Fluorine, F_2	Carbon monoxide, CO
Neon, Ne	Chlorine, Cl_2	Carbon dioxide, CO_2
Argon, Ar	Oxygen, O_2	Nitric oxide, NO
Krypton, Kr	Ozone, O_3	Nitrogen dioxide, NO_2
Xenon, Xe	Nitrogen, N_2	Hydrocarbons, C_nH_m
Radon, Ra	Hydrogen, H_2	Sulfur hexafluoride, SF_6 Laughing gas, N_2O

feature. These are substances which are present in liquid form at room temperature (20 °C) and normal pressure (1013 hPa), but which nevertheless gasify to a certain extent and then become vapor. Vapors physically behave like gases. The best known substance in this context is water vapor (Table 2.1).

2.1 States of Aggregation

The transition from the solid or liquid state to the gaseous state is called the phase transition. Figure 2.1 shows the 3 phases (states of aggregation), solid, liquid, and gaseous. In the solid phase, the atoms are located at fixed positions within a group of atoms. The outer dimensions (contours) of a solid are rigid and do not adapt to the environment. The atoms are held in this position by attractive forces and the atoms cannot move freely. There is only the possibility to swing (oscillate) around this place. With rising temperature these oscillations increase. If the temperature continues to rise, the binding forces are overcome and the solid passes into the liquid phase. In this phase the binding forces are lower and the atoms can then move almost freely.

Evaporation

The range (volume) in which the liquid is located is determined by an outer container. Due to the high particle density, which is similar to that of a solid, the particles collide with other particles after a short time, which then disturb this movement (momentum transfer). If the temperature is further increased, the average particle velocity increases and the particles then gradually change into the gas phase. The transition from the liquid phase to the gas phase is called or evaporation (Fig. 2.2). In this phase the particle density is much lower. If, for example, 1 kg of water (approx. 1 L) is evaporated, then at a boiling temperature of 100° C and a pressure of 1013 hPa, approx. 1700 L of water vapor is obtained. The density in the gas phase is thus lower by a factor than in the liquid phase. This already shows that the particles in the gas phase are much more mobile and can therefore move freely in space (fewer collisions). The transition from the liquid phase to the gaseous phase at the phase boundary can only take place by supplying energy.

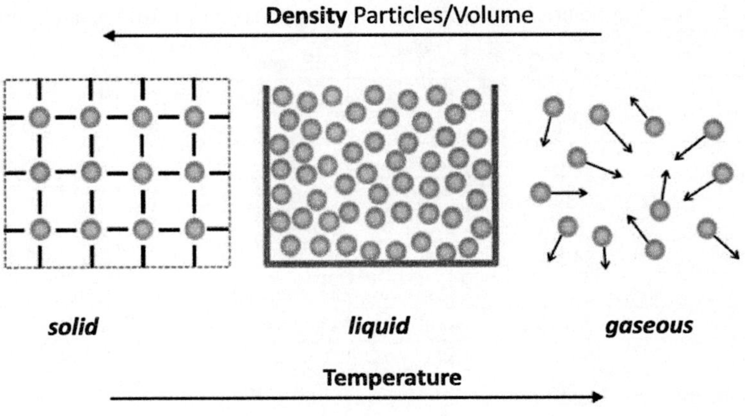

Fig. 2.1 Particle model of the states of aggregation

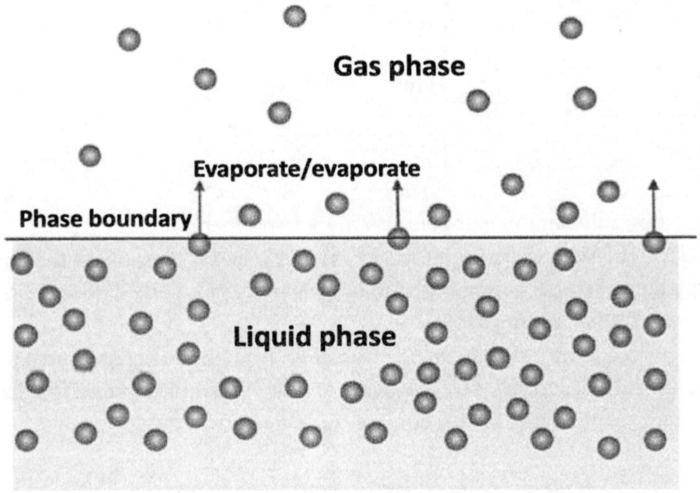

Fig. 2.2 Transition of the molecules from the liquid phase to the gas phase at the phase boundary by evaporation or vaporization

Heat of Vaporization

In order to realize the transition to the individual phases (solid→liquid→gaseous), a higher temperature T is required, which is achieved by supplying energy. For the transition from the liquid to the gaseous state of aggregation, this is the so-called heat of vaporization Q_{sd}, which is the product of the specific heat of vaporization r (Table 2.2) and the mass m:

$$Q_{sd} = r \cdot m \tag{2.1}$$

Table 2.2 Boiling temperature T_S and specific heat of evaporation r for 1013 hPa (Kuchling 2011)

Substance	$T_S/\,^\circ\mathrm{C}$	$r/\frac{\mathrm{kJ}}{\mathrm{kg}}$	Substance	$T_S/\,^\circ\mathrm{C}$	$r/\frac{\mathrm{kJ}}{\mathrm{kg}}$
Argon	−186	163	Naphthalene	218	314
Benzene	80.1	394	Neon	−246.1	105
Bromine	58.8	183	Octane	126	299
Butane	−0.65	385	Ozone	−113	316
Chlorine	−34.1	290	Pentane	36.1	360
Diethyl ether	34.5	384	Propane	−42.1	426
Ethane	−88.6	489	Propanaol-1	97.2	750
Ethanol	78.3	840	Mercury	356.6	285
Ethylene	−104	483	Oxygen	−183	213
Helium	−269	20.6	Sulphur dioxide	−10	389
Heptane	98.4	318	Carbon disulfide	46.3	352
Carbon dioxide	−78.5	574	Hydrogen sulfide	−60.4	548
Carbon monoxide	−192	216	Nitrogen	−195.8	201
Krypton	−153	108	Nitric oxide	−88.5	376
Methane	−162	510	Nitrogen dioxide	−151.8	461
Methanol	64.6	1100	Water	100	2257
Methyl acetate	57	406	Hydrogen	−252.8	461
Methyl chloride	−23.8	428	Xenon	−108.2	96

Figure 2.3 shows the temperature curve T by adding thermal energy Q. A linear rise in temperature T can be seen in the liquid phase, which remains constant at the boiling point T_S until all the liquid has been transferred to the gas phase. Only then does the temperature T rise again with further energy input.

If the gas is then cooled down again, the transition from the gas phase to the liquid phase takes place (Fig. 2.2). The energy Q_{sd} required for the evaporation process is then released again. This energy is then also called condensation heat.

$$\textit{Heat of evaporation} = \textit{Heat of condensation}$$

It is also possible to go directly from the solid phase to the gas phase. This process is called sublimation. The energy required for this process results from the heat of fusion and the heat of evaporation.

$$Q_{\mathrm{sb}} = m \cdot (r + s) \tag{2.2}$$

The direct transition from the gas phase to the solid state is called desublimation. Also in this case, all energy is released again.

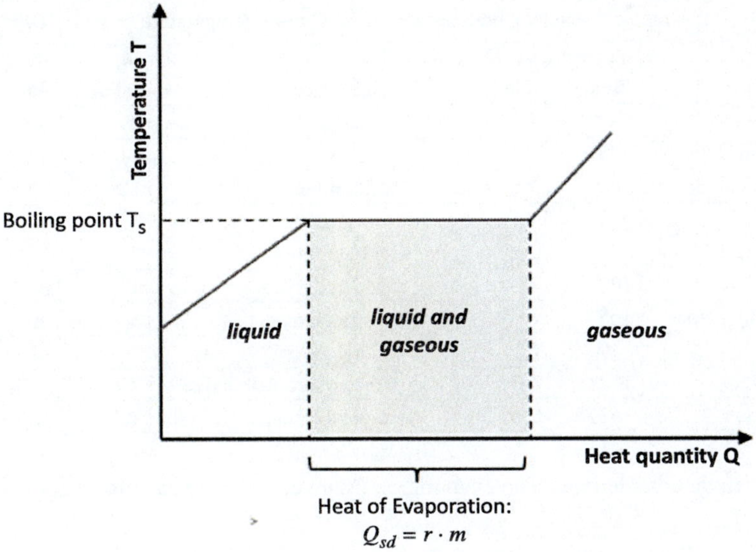

Fig. 2.3 Energy input during the transition from the liquid phase to the gas phase (= evaporation)

$$Heat\ of\ sublimation = Heat\ of\ desublimation$$

Evaporation Rate

The evaporation rate is a relatively comparative dimension. It indicates the factor by which the evaporation rate of a substance differs in comparison to a reference substance (Table 2.3). In general, diethyl ether is used as a reference substance. The evaporation rate of this reference substance is then by definition =1. The determination of this figure is carried out empirically and is described in more detail in DIN 53170. A simple test is to dribble a small amount of liquid (0.5 ml) onto a piece of filter paper and then measure the time until complete evaporation. The result is then divided by the time value for the reference substance and the evaporation figure is obtained as the result.

In gas measurement technology, this number plays an important role as it gives an indication of how quickly an explosive atmosphere can form in a closed room. If, for example, there is an acetone layer on an area of $1 m^2$ at 25°C, an explosive volume of 2 m is formed after only $1 min^3$ (Olenik et al. 1983).

A functional relationship can also be derived empirically by plotting the evaporation number against the vapour pressure of the respective substance at room temperature (Fig. 2.4). In a double logarithmic curve (Fig. 2.5) an empirical dependence on this physical

Table 2.3 Vapor pressure and evaporation rates at 20°C room temperature (Jessel 2001)

Substance	p_D [hPa]	Evaporation rate	Substance	p_D [hPa]	Evaporation rate
Diethyl ether	587	1	Ethanol	59	8.3
Dichloromethane	475	1.8	*n*-butyl acetate	11	12
Carbon disulfide	400	1.8	Chlorobenzene	12	12.5
Acetone	233	2.1	*o*-xylene	6.7	13.5
Ethyl acetate	97	2.9	*n*-butanol	7	33
Benzene	100	3	Methyl glycol	11	34
Methylethylketone	105	6	Cyclohexane	5	40
Toluene	29	6.1	Ethyl glycol	5	43
Methanol	128	6.3	Tetrahydronaphthalene	2.7	190
1.4-Dioxane	41	7.3	Cyclohexanol	1.2	400

quantity can then be derived. The evaporation figure can also be calculated approximately as follows:

$$\text{Evaporation number} = 170 \cdot p_D^{-0.804} \tag{2.3}$$

Vapour Pressure Curve
Above the phase boundary between liquid and gas, a gas mixture is formed, which consists, for example, of ambient air and water vapor.

This gas mixture can also be described by the respective pressures of the gases, which in the case of a mixture consist of partial pressures or partial pressures. The partial pressure of the substance that passes from the liquid phase into the gas phase depends very much on the temperature and is called saturation vapor pressure. As the temperature rises, the partial pressure or vapor pressure also increases and reaches exactly the ambient pressure at the boiling point. The temperature curve of the vapor pressure $p_D(T)$ is substance-specific and is determined experimentally for each substance. The values can be found in the corresponding tables. For applications in gas measurement technology, the description of the vapor pressure using the so-called Magnus[1]-equation. The temperature T in this equation is given in °C.

[1] Heinrich Gustav Magnus (1802–1870) German physicist and chemist.

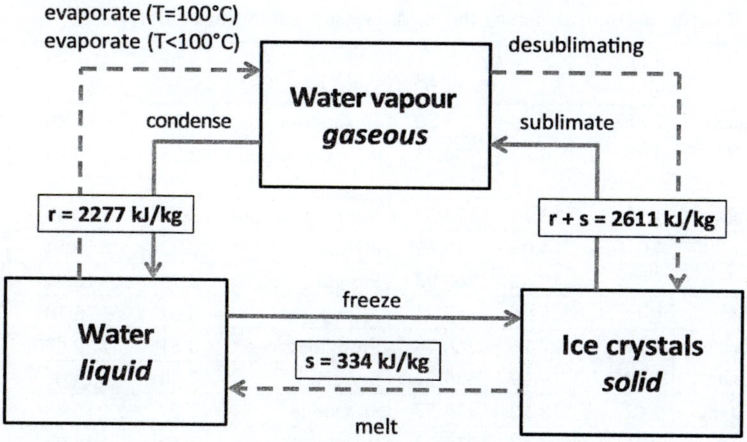

Fig. 2.4 Transformation energy for phase transitions of water (Lauer and Bendix 2006, p. 36)

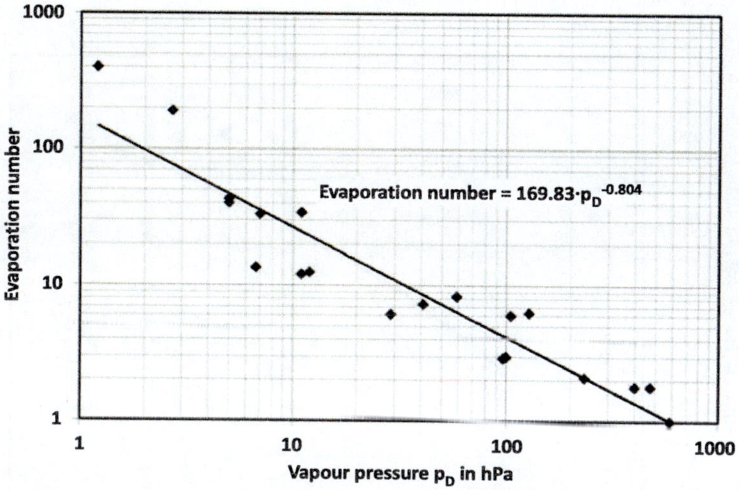

Fig. 2.5 Dependence of evaporation figure on vapour pressure at 20°C

$$p_D(T) = a \cdot \exp\left[\frac{b \cdot T}{c + T}\right] \tag{2.4}$$

The constants a, b, and c are summarized in Table 2.4 for the most important substances. The coefficient a represents the vapor pressure of the respective substance at $T = 0^\circ\mathrm{C}$ again, since $\exp(0) = 1$. Figure 2.6 shows the vapor pressure curve for water in the temperature range from -50 to 100°C.

Table 2.4 Coefficients for calculating the vapor pressure curve (Jessel 2001)

Substance	*a* in hPa	*b*	*c* in °C	Substance	*a* in hPa	*b*	*c* in °C
i-Amyl acetate	1.27	17.25	223.50	Hexanon-2	3.57	15.80	230.00
n-Amyl acetate	0.97	14.83	234.73	*n*-Hexylamine	2.95	16.22	243.73
Amyl formate	3.79	13.13	174.57	Methoxyethanol	3.23	15.90	233.06
Butanol-2	4.01	24.03	315.65	*n*-Nonan	0.92	16.14	192.36
i-Butanol	3.03	19.80	260.03	Picolin	3.38	16.71	253.42
n-Butanol	1.47	28.26	363.91	3-Propanol	6.61	26.10	378.60
n-Butyl acetate	3.29	17.26	260.78	Propylbenzene	0.61	9.690	95.190
Butyl formate	9.40	19.99	339.21	Styrene	2.24	19.05	296.34
Chlorobenzene	3.42	18.12	274.57	*m*-Xylene	2.27	16.45	232.38
Cumene	1.03	23.39	315.78	*o*-Xylene	1.69	18.26	252.86
Acetic acid	4.18	15.95	224.65	*p*-Xylene	2.41	15.68	220.73
Ethylbenzene	2.84	20.32	312.56	Water	6.11	17.08	234.17

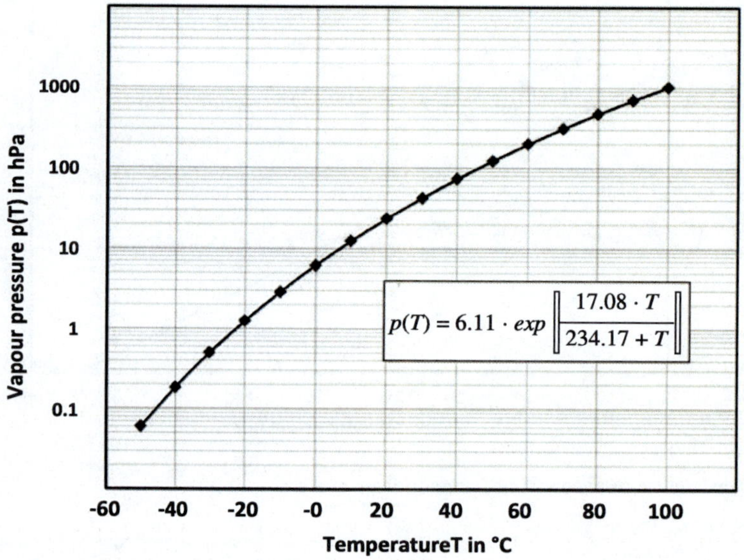

Fig. 2.6 Vapor pressure curve of water

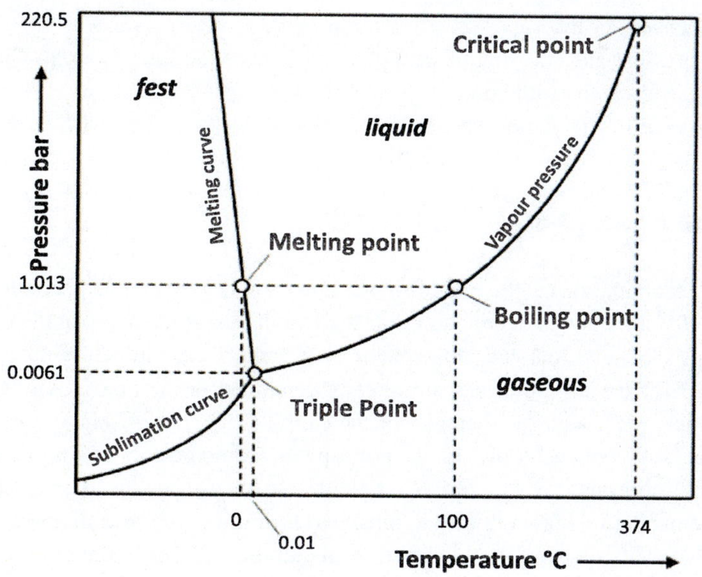

Fig. 2.7 Phase diagram for water with non-linear scaling of the axes in the *pT*-diagram

Triple Point
The transition of a substance from the liquid phase to the gas phase is not affected by the presence of other substances in the gas phase. According to the law of Dalton[2] the total pressure p_{ges} of a mixture of ideal gases is equal to the sum of the partial pressures of the individual components.

$$p_{ges} = p_1 + p_2 + \ldots + p_n \tag{2.5}$$

Nevertheless, the solidification point and the boiling point are dependent on temperature and pressure. In Fig. 2.7 the three phases for water are summarized in a *pT* diagram.

The transition from the liquid to the gaseous phase occurs at a pressure of 1013 hPa and a temperature of 100 °C. If the pressure rises, the boiling point increases along the vapor pressure curve (therefore also called boiling curve). If the pressure exceeds 220 bar, a critical point is reached at a temperature of 374°C. Below this temperature, a gas changes to the liquid phase at the appropriate pressure. Above this temperature, this is no longer possible and the medium continues to compress with increasing pressure. The melting curve characterizes the transition from the liquid to the solid phase and vice versa. Compared to the vapor pressure curve, it goes up very steeply. The sublimation curve is

[2] John Dalton (1766–1844) English chemist

quasi the extension of the vapor pressure curve. All three curves meet at the triple point. Only at this pressure and temperature do 3 phases of a pure substance coexist (Gibbs' phase rule). For pure water the triple point is $T = 273.16$ K (=0.01°C) and $p = 610.6$ Pa. This point is also used for the definition (fixed point) of the temperature unit Kelvin.

2.2 The Atmosphere

The largest accumulation of gaseous substances on earth is in our atmosphere. It has a total mass of $5 \cdot 10^{18}$ kg, but which only accounts for 1 millionth of the mass, of the entire earth. Meteorologists believe that the atmosphere is located at an altitude h up to 100 km, although 90% of the air molecules are already contained in the first 20 km. Comparing this layer thickness h with the diameter of the earth of $d = 12,742$ km and transfers this ratio to a ball with a diameter of 1 m, the atmosphere would be only 157 mm thick. For us humans, the troposphere is especially important, which reaches up to about 10 km and contains the habitat of flora and fauna (biosphere). Furthermore, the stratosphere plays a decisive role for life on earth, as this layer, at an altitude of 35 km, contains the maximum of the known ozone layer (Fig. 2.8). This ozone layer absorbs the short-wave UV radiation of the sun and thus protects life on our planet. The concentration of water vapor is highest near the ground, as higher and higher temperatures usually prevail there and lead to evaporation of water. As a result of condensation and cloud formation, the water vapor content decreases with increasing altitude h and is almost zero at altitudes $h > 10$ km.

Composition of the Atmosphere

The main components of the atmosphere are nitrogen (N_2), oxygen (O_2), and argon (Ar). These three substances already account for 99.96% of the total atmosphere. The remaining 0.04% is distributed among the trace gases (Table 2.5). The most important trace gas is carbon dioxide, with a current content of 0.039%. The exact composition of the atmosphere for gas applications was defined in 2005 in the standard DIN EN ISO 6976 (Cerbe 2008). As the water vapor content of the atmosphere is subject to very strong fluctuations, the concentration proportions of the other components are always based on dry air.

If the entire atmosphere were considered at a normal pressure of 1013 hPa and 0°C (273.15 K), the layer thickness would be much smaller. In this case, the total height would be only about 8 km, that is, smaller than the highest mountains in the Himalayas (Table 2.6). The trace gas CO_2 would then be no higher than a room, namely 2.5 m.

Change in CO_2- and Methane Concentration

In particular, the gas components of the atmosphere that change in concentration play an important role in the current debate on the greenhouse effect and the climate change

Fig. 2.8 Distribution of ozone concentration and water vapor content in the atmosphere (Klein and Werner 1993, p. 9)

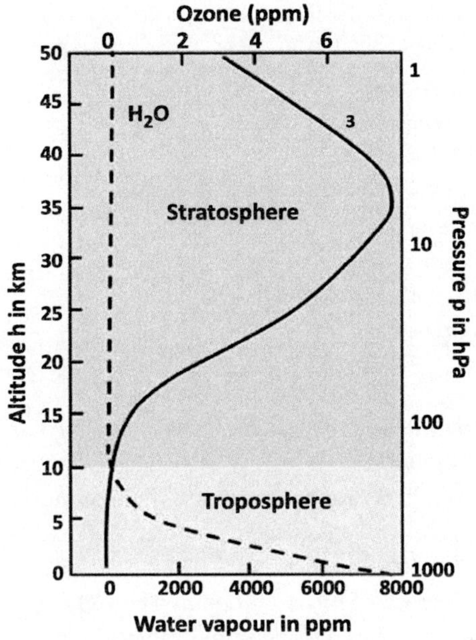

Table 2.5 Composition of the steam-free atmosphere (Roedel and Wagner 2011). The *italic* and **bold** have a variable concentration in the air

Component	Molar mass	Volume shares
Nitrogen N_2	28.013	78.09%
Oxygen O_2	32.0	20.95%
Argon Ar	39.95	0.93%
Steam H_2O	18.02	***0–5%***
Carbon dioxide CO_2	44.01	***0.039%***
Neon Ne	20.18	18.21 ppm
Helium He	4.0	5.24 ppm
Methane CH_4	16.04	***1.88 ppm***
Krypton Kr	83.80	1.14 ppm
Hydrogen H_2	2.02	0.5 ppm
Laughing gas N_2O	56.03	***0.3 ppm***
Carbon monoxide	28.01	***0.2 ppm***
Xenon	131.3	0.087 ppm
Ozone O_3	48.0	***0–0.1 ppm***

expected from it. Since 1958, the carbon dioxide content in the atmosphere on Mount Mauna Loa in Hawaii (USA) has been continuously measured to determine the change in this trace gas. Already after the first years it was found that the CO_2-content in the atmosphere is continuously increasing. This rising concentration curve was described as

Table 2.6 Column height of various gases in an isobaric atmosphere under normal conditions (0 °C and 1013 hPa), with a total amount of approximately 8 km

Substance	Column height (100%- gases)
Nitrogen	Approx. 6250 m
Oxygen	Approx. 1670 m
Argon	Approx. 74 m
Steam	Approx. 35 m
Carbon dioxide	Approx. 2.5 m
Noble gases (without argon)	Approx. 20 cm
Ozone	Approx. 3.5 cm

the *Keeling*[3] *curve* and today forms the basis for all climate models. The increase in CO_2-concentration is mainly associated with the combustion of fossil fuels (coal, oil, and natural gas). In addition, an increase in this concentration is assumed to be due to outgassing of the world's oceans, which store very large quantities of CO_2. The figure shows the increase in CO_2-concentration since 1958. In 1958, this value was approximately 314 ppm whereas today (2015) it is 400 ppm. The CO_2 content therefore increased by 86 ppm during this period. The CO_2-measured values can be described very well by a quadratic equation (trend line) (x= year). According to this trend, a CO_2 concentration of over 670 ppm would be expected in 2100.

What is striking about the Keeling curve is that the increase in CO_2 per year has doubled from 1 ppm/year to 2 ppm/year. The reason for this disproportionate increase is the sharp rise in CO_2 emissions in recent decades. China and India in particular have caused this development, while in most industrialized countries CO_2 emissions are stagnating or even declining.

Since 1958 the global temperature has increased by about 0.6°C. These two values result in a gradient of $8.8 \cdot 10^{-3}$°C ppm CO_2. According to the climate models currently available, this temperature increase is to be caused exclusively by the rise in CO from 314 ppm to 400 ppm. If this is the case, a further increase should result in even higher temperatures. The CO_2 increase of the last 60 years can be described empirically very well by a quadratic equation, using the respective year for x:

$$CO_2 \text{ in ppm} = 0.0118 \cdot x^2 - 45.376 \cdot x + 43922 \tag{2.6}$$

A doubling of the CO_2 content, compared to pre-industrial times (290 ppm), would therefore not result in a reduction of CO_2−emisions! in the year 2077. A further increase

[3] Charles David Keeling (1928–2005) US-amerikanischer Klimaforscher.

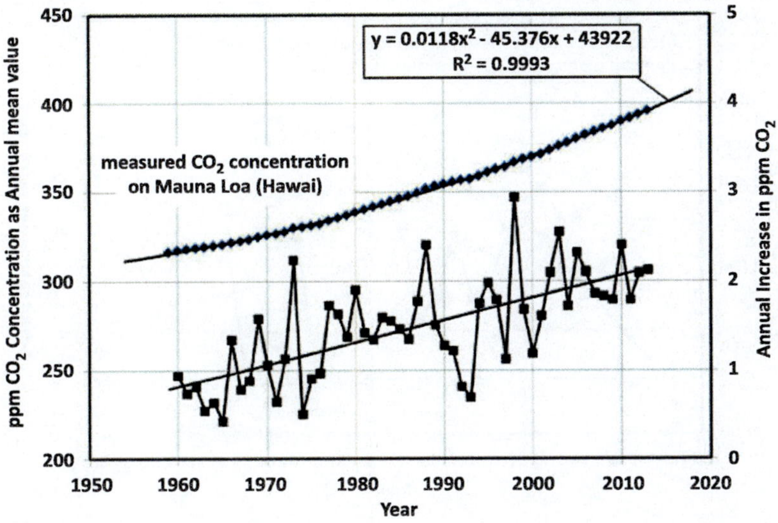

Fig. 2.9 Keeling curve of the CO_2 -concentration on Mauna Loa and the annual increase in ppm CO_2

of 180 ppm over the next 60 years would have led to an increase in temperature ΔT of max. 1.6 °C.

$$\Delta T = 180\text{ppm} \cdot 8.8 \cdot 10^{-3}\,^{\circ}\text{C/ppm} \tag{2.7}$$

This simple calculation is consistent with some of the climate researchers' models. However, the range of predictions is also very wide due to the large uncertainties of the climate models. The German Weather Service (DWD) even gives a bandwidth of 0 °C (no change!) to +3°C for the period from 2014 to 2077. If one were to assume, however, that the rise in temperature over the last 100 years is not exclusively attributable to CO_2 the actual temperature increase in the future could, however, be much lower than generally feared.

In the meantime, 20 additional gases are continuously recorded and evaluated on Mauna Loa. All data is available via Internet[4] and can be used for own calculations. Figures 2.9 and 2.10 shown in this chapter are based on this data source.

Another important gas that is repeatedly mentioned in connection with the greenhouse effect is the methane CH_4. The current methane content is 1.85 ppm and is thus significantly lower than the carbon dioxide content. The methane is mainly caused by the sharp increase in mass livestock farming, leaks from biogas plants, outgassing from dams, and from the cultivation of rice. Furthermore, a release of methane from permafrost areas is

[4] http://www.esrl.noaa.gov/gmd/obop/mlo/index.html

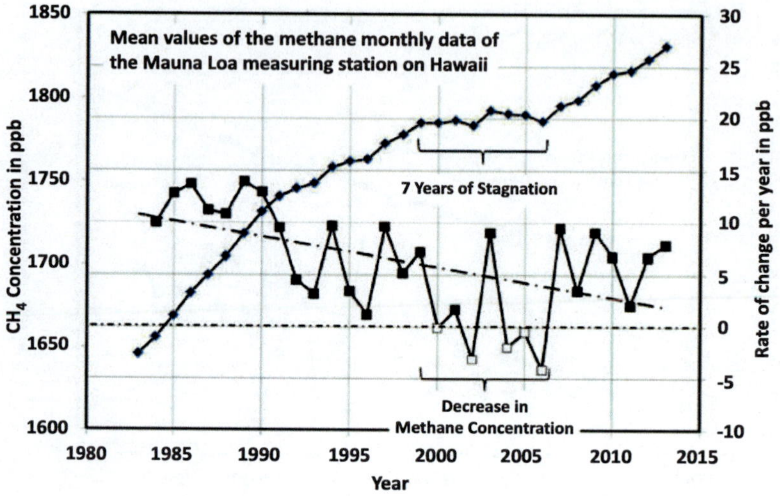

Fig. 2.10 Measured values for the increase in methane concentration on Mauna Loa

reported, which should also lead to a further increase in the global methane concentration. In fact, scientists observed a stagnating methane concentration on Mauna Loa between 1999 and 2006. Since 2006, it has been rising again, but the rate of increase at 7 ppb is only half as high as in the years 1984–1990, although there are apparently more and more emission sources. This behavior was not expected and cannot be explained conclusively at present. It is very difficult to measure the chemical and physical processes in the atmosphere, as the values only apply to one location at a time. Therefore, metrological attempts are increasingly being made to use satellites for a global recording of concentration values and their changes.

Satellite Measurements

Especially the observation of ozone concentrations in the stratosphere is difficult without satellite measurements. The transport of gas measuring instruments into the stratosphere with a weather balloon allows only a punctual measurement of the altitude profile (see Fig. 2.8). However, since the ozone (O_3) in the atmosphere has only a limited lifetime and decomposes again into molecular oxygen after a short time (minutes to hours), continuous ozone measurement can only be realized with remote measurement.

Figure 2.11 shows such a measurement by NASA. In this case, the ozone concentration values are shown as Dobson units, which reflect an integral value through the entire atmosphere (see Sect. 9.3). The different concentration distributions over the geographical locations can then be visualized with a colored display.

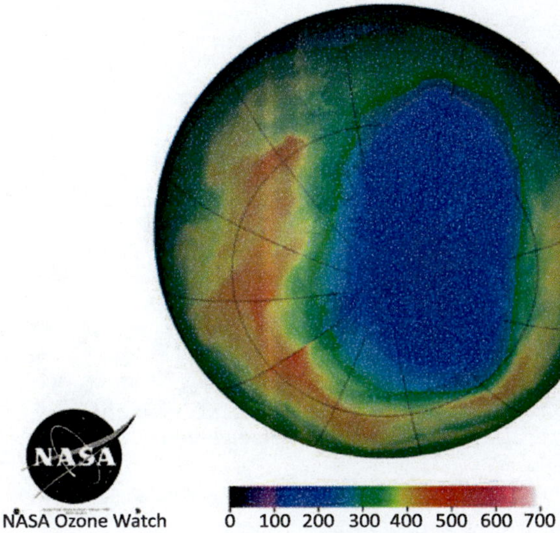

Fig. 2.11 Measurements of ozone concentration in the stratosphere taken by satellite. The results are shown in Dobson units as a false-color display. (Source: NASA Ozone Watch)

In the picture shown (Fig. 2.11) the Antarctic (South Pole) can be seen. Above the Antarctic the ozone concentration is thinned out (<200 Dobson units) and is only half of the *normal* ozone concentration of about 400 Dobson units. This colored representation, which can be chosen at will, suggests the impression of an opening, which in common parlance is simply called an *ozone hole* designated. In fact, these are only areas where ozone concentrations are lower. There are various reasons for the thinning of the ozone layer over the polar caps. Since ozone formation in the atmosphere is only possible through the sun's UV radiation, solar radiation plays the greatest role in this process. At the polar caps, the specific radiation (W/m^2) due to the geometrical arrangement of the earth to the sun. Due to this fact, less ozone is naturally formed there. The O_3formation process is still influenced by air pollutants, the so-called CFCs...[5]affected. For these reasons, the UNO decided in 1989 to ban CFCs worldwide. Since 1995, industrialized countries have been subject to a general ban on the production and use of CFCs.

Air Pressure

If there are gas molecules with the mass m in the gravitational field of the earth, then these will be $F_G = m \cdot g$ dressed. The resulting pressure on a piece of the earth's surface Awould thus be:

[5]Fluor-Chlor-Kohlenwasserstoff Verbindungen.

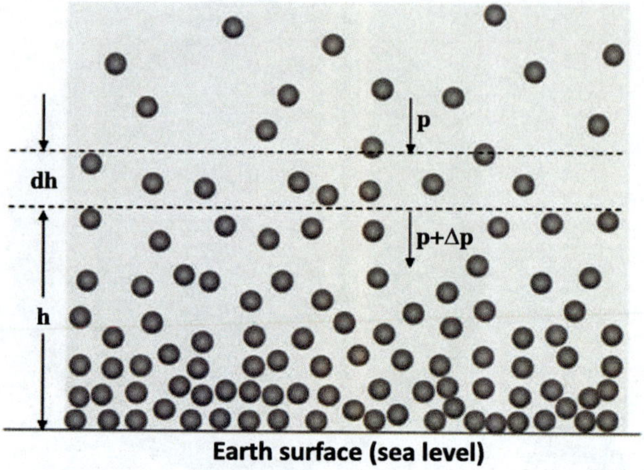

Fig. 2.12 Distribution of molecules under the influence of gravity and density change

$$p = \frac{m \cdot g}{A} \tag{2.8}$$

Describing the mass m by the density $\varrho = m/V$ or rather $m = \varrho \cdot V$ and now sets for the volume $V = A \cdot h$ one receives heavy pressure:

$$p = \varrho \cdot g \cdot h \tag{2.9}$$

However, since the gas density depends on the pressure, the density changes with height h. This compressibility is determined by the *Boyle*[6]*–Mariotte Act*[7] described:

$$p \cdot V = \text{constant} \quad @T = \text{const} \tag{2.10}$$

The figure shows the distribution of the gas molecules in the Earth's gravitational field. On the ground there are many molecules (→ high density) while with increasing height h the density decreases (Fig. 2.12).

According to *Boyle–Mariotte,* this density change decreases proportionally with height h at constant temperature T (Meschede 2010). Equation 2.9 can therefore only be applied in a thin layer. If the height changes h at dh the more the pressure changes by $dp = -\varrho \cdot g \cdot dh$. The density ϱ can then be replaced by $p/\varrho = p/\varrho_0$. The following equation is then obtained:

[6]Sir Robert Boyle (1627–1691) britischer Physiker.

[7]Edme Mariotte (1620–1684) französischer Physiker.

$$\frac{dp}{dh} = -g \cdot \frac{\varrho_0}{p_0} \cdot p \tag{2.11}$$

This then results in the so-called barometric altitude formula:

$$p(h) = p_0 e^{-\frac{\varrho_0 \cdot g \cdot h}{p_0}} \tag{2.12}$$

With $\varrho_0 = 1.293 \text{kg/m}^3$ and $p_0 = 1013$ hPa and $g = 9.81 \text{m/s}^2$ you get the simplified form of the barometric altitude formula:

$$p(h) = p_0 e^{-\frac{h}{7.99 \text{km}}} \tag{2.13}$$

However, this formula is only an approximation, as this calculation assumes a constant temperature $T = 0°C$. In fact, the temperature changes very much with altitude, so that appropriate corrections must be made to this equation. These considerations led to the international altitude formula. This equation applies to the entire troposphere up to an altitude of $h = 11$ km validity (Fig. 2.13):

$$p(h) = 1013 \text{hPa} \cdot \left(1 - \frac{6.5 \cdot h}{288 \text{km}}\right)^{5.255} \tag{2.14}$$

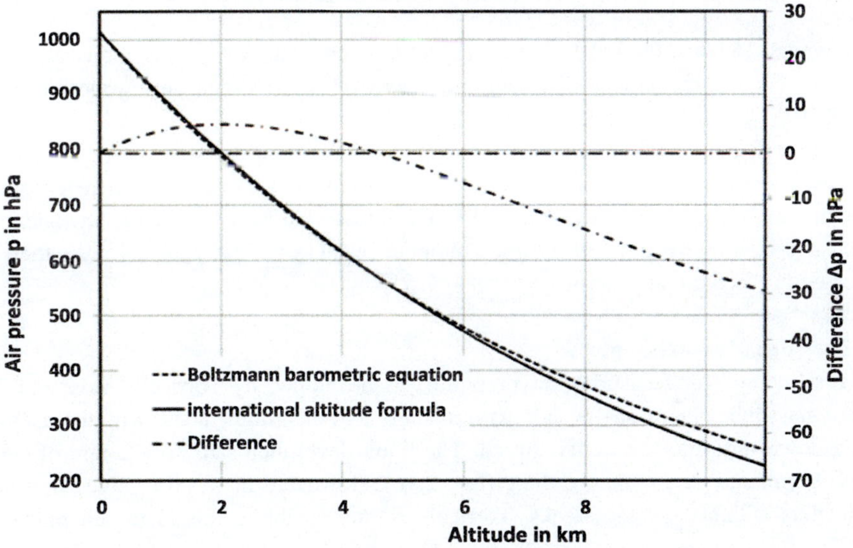

Fig. 2.13 Comparison between the barometric altitude formula (Eq. 2.13) and the international altitude formula (Eq. 2.14). Larger deviations (Δp) occur above $h = 8$ km

For density, this equation can be transformed accordingly and the following expression is obtained:

$$\varrho(h) = 1.2255\frac{\text{kg}}{\text{m}^3} \cdot \left(1 - \frac{6.5 \cdot h}{288\text{km}}\right)^{4.255} \quad (2.15)$$

With the international altitude formula it is also possible to calculate the altitude from the current air pressure by converting to h:

$$h = \left[1 - \left(\frac{p(h)}{1{,}013}\right)^{\frac{1}{5.255}}\right] \cdot \frac{288\text{km}}{6.5} \quad (2.16)$$

For the calculation of height differences (Δh) the current air pressure p is displayed at the beginning of the measurement instead of 1013 hPa. Usually this is the first measured value (reference altitude). This method is used especially for mountain hikes to record the altitude differences during the hike. In the following measurement a trip from South Tyrol to the Sauerland/Westphalia was recorded. At the starting point ($h = 950$ m) there was a pressure of 914 hPa. With the current start pressure the altitude difference was determined and added to the known start altitude (950 m):

$$h = 0.95\text{km} + \left[1 - \left(\frac{p(h)}{914}\right)^{\frac{1}{5.255}}\right] \cdot \frac{288\text{km}}{6.5} \quad (2.17)$$

Both the first 15 h and the last 15 h the pressure measurement system was at the respective locations, without movement. The pressure p was largely constant during these periods (Fig. 2.14).

If the pressure p is observed over a longer period of time, there are sometimes large changes. These pressure changes Δp are weather related and are related to the meteorological high- and low-pressure areas (Fig. 2.15). Gas measuring instruments are dependent on these air pressure changes and the measurement results of a gas analysis must then be corrected for accurate evaluation.

Air Pressure Measurement

The first method for measuring air pressure was developed by Torricelli[8] invented. He filled a glass tube closed on one side with mercury and then immersed it with the opening facing downward into a mercury trough. The liquid level then sank to a height of about $h = 760$ mm and then remained there (Fig. 2.16). However, this level changed by a few millimeters within days and weeks. Torricelli rightly attributes this to the changing air

[8]Evangelista Torricelli (1608–1647) Italian physicist.

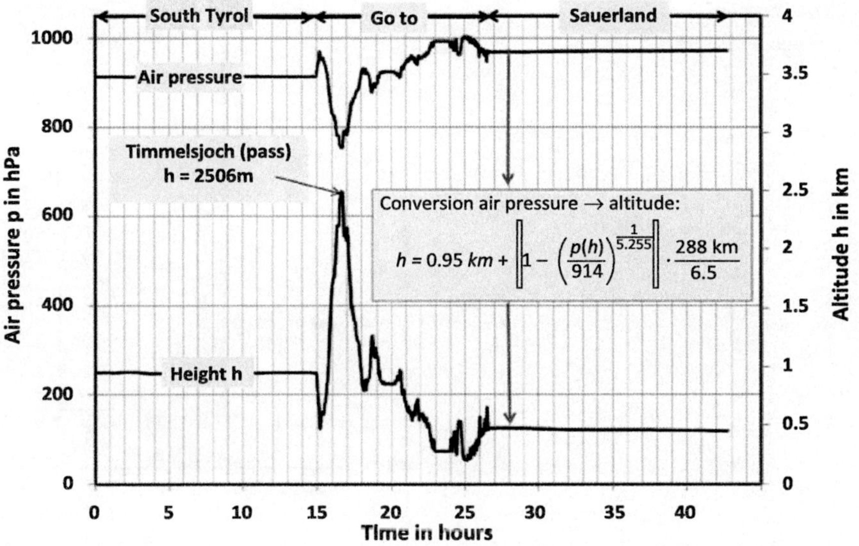

Fig. 2.14 Air pressure measurement and conversion to altitude h with the international height formula (Eq. 2.17)

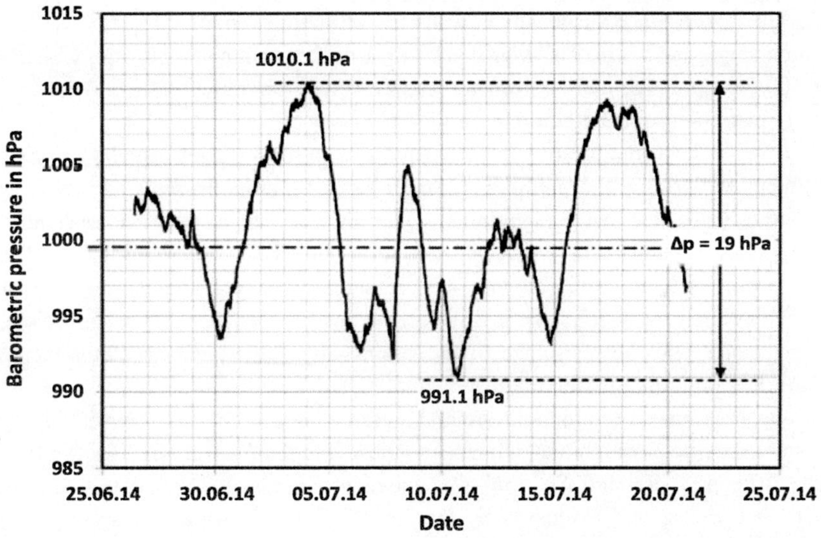

Fig. 2.15 Natural air pressure changes at one location (Dortmund) within 1 month (June 2014)

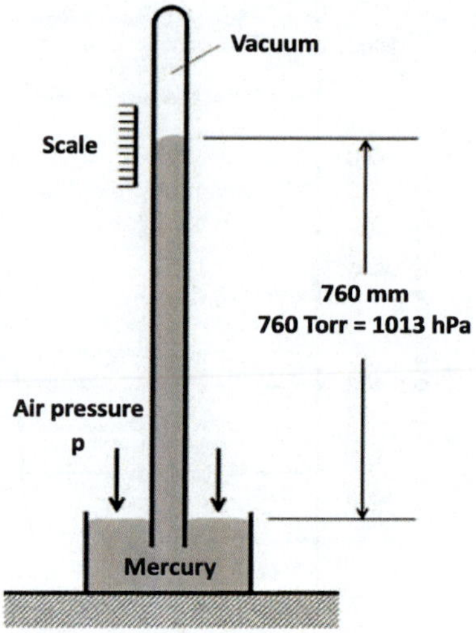

Fig. 2.16 Torricelli barometer for measuring air pressure

pressure. This is because the air pressure presses on the mercury column and thus prevents leakage. This height h from 760 mm was later, in honour of Torricelli, assigned the pressure unit Torr (1 mm Hg corresponds to 1 Torr).

This experiment was later repeated by other scientists, using water as a medium (e.g., Otto von Guericke). Since the density of water is 13.546 times smaller than that of mercury, the column height is 10.3 m. This was much too large for practical applications, so that for almost 300 years the Torricelli barometer was the most important and also most accurate measuring instrument for determining air pressure.

Today, only electronic sensors are used for this purpose. They usually consist of a silicon microchip with a thin membrane. Strain gauges (Dehnungsmessstreifen = DMS) are applied to the edges of the membrane, which increase their electrical resistance with increasing strain. The pressure space underneath the membrane is closed and a reference pressure is applied (Fig. 2.17). If the two pressures are equal, that is, air pressure = reference pressure, the diaphragm is in the rest position. If the air pressure rises, the diaphragm bends and the resistance of the strain gauge also increases. This method is called absolute pressure measurement.

A similar structure is used for a differential pressure measurement. However, an opening is now created in the closing plate, which is connected to the pressure p_2 (Fig. 2.18). In the case of a differential pressure $\Delta p = p_1 - p_2$, the diaphragm is also bent in this case, which can then be detected and evaluated by the strain gauge. For practical applications the pressure sensors are then provided with gas connections to connect gas hoses.

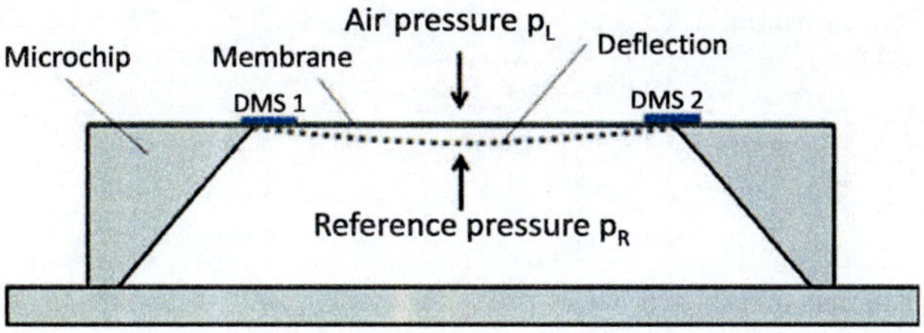

Fig. 2.17 Micromechanical absolute pressure sensor

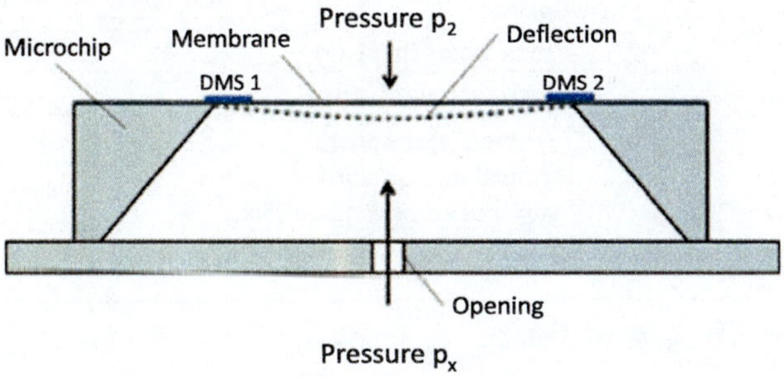

Fig. 2.18 Micromechanical differential pressure sensor

As a printing unit, Pascal is used today as an internationally valid unit (SI). The definition for 1 Pa is 1 Newton/m^2:The printing unit used today is Pascal[9] is used as the internationally valid unit (SI). The definition for 1 Pascal (Pa) is 1 Newton/m^2:

$$1\text{Pa} = \frac{1\text{N}}{\text{m}^2} = \frac{1\text{kg}}{\text{m} \cdot \text{s}^2} \tag{2.18}$$

In gas measurement technology, work is usually carried out at atmospheric pressure ($p \approx 100,000$Pa). In order to simplify the conversion to the millibar (mbar), which has been the standard procedure up to now, the air pressure is given in hectopascals (hPa). Then 1 hPa are exactly 1 mbar.

In some areas of medicine and technology the *old units* are still used. Therefore it is important to be able to convert these units. Table 2.7 shows the still commonly used non-SI units and their conversion to Pascal.

[9]Blaise Pascal (1623–1662) French mathematician and physicist.

Table 2.7 Conversions from non-Si units to Pascal

Non-SI unit	Pascal
1 at[a]= 735.6 Torr	98.07 kPa
1 mWs = 0.1 at	9.807 kPa
1 mmWS	9.907 Pa
1 mmHg	133.3 Pa
1 bar = 750 Torr	1000 hPa
1 mbar	1 hPa
1 Torr	133.3 Pa
1 atm[b]= 760 Torr	1013 hPa
1 psi[c]	6.895 kPa
1 inch water (in H_2O)	249.1 Pa
1 inch mercury [inHg]	3.386 kPa

[a]Technical atmosphere
[b]Physical atmosphere
[c]Psi = pound per square inch

2.3 Kinetic Theory of Gases

Particle Model of Ideal Gases

In the context of a theoretical description of gases, the gas atoms or molecules are regarded as small spheres with the mass *m* which can move completely freely in space at the velocity *v*.. Only the force of gravity (F_G) acts on these particles. When these particles collide, an elastic transfer of momentum occurs according to the laws of classical mechanics. The gaseous state is characterized by the fact that the distances *x* between the molecules are significantly larger than the diameter *d* of the molecules (Fig. 2.19). With this rather simple model, most physical effects in the gas phase can be explained.

Mean Free Path Length

When a gas molecule with diameter d moves in a straight line, it inevitably collides with another molecule. The frequency of these collisions per second, at a molecule number density $n = N/V$, can be calculated from the mean number of impactsWhen a gas molecule with the diameter *d* moves in a straight line, it will inevitably collide with another molecule.

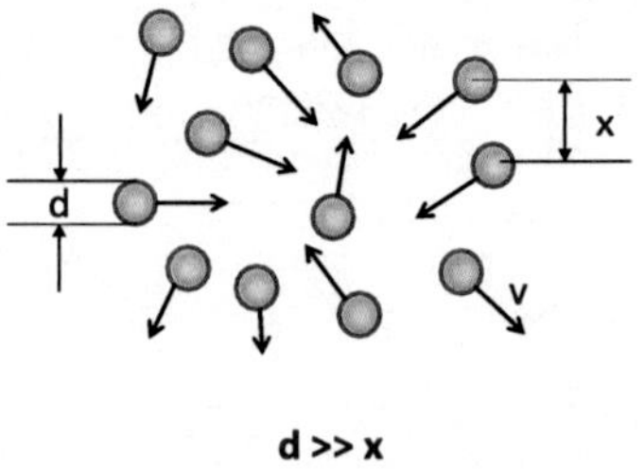

Fig. 2.19 Particle model of gases as freely moving mass spheres in space

The frequency of these collisions per second, at a molecule number density $n = N/V$can be calculated from the mean number of impacts $\bar{z}$:

$$\bar{z} = \pi\sqrt{2}d^2\bar{v}n \tag{2.19}$$

with the pressure p, the temperature T and the Boltzmann constant k the following expression results:

$$\bar{z} = \pi\sqrt{2}d^2\bar{v}\frac{p}{kT} \tag{2.20}$$

Under normal conditions (0°C, 1013 hPa), this numerical value is between 10^9 and 10^{10}s^{-1} for most gases. If there are different atoms or molecules in a gas mixture, the radii r_1 and r_2 of the different impact partners are introduced into Eq. 2.19 and we obtain the following expression:

$$\bar{z} = \pi\sqrt{2}(r_1 + r_2)^2\bar{v}n \tag{2.21}$$

If the number of collisions per time is given by equation, it is quite easy to calculate the distance between two collisions. This distance is also called the mean free path length $\bar{l}$:

$$\bar{l} = \frac{k \cdot T}{\sqrt{2}\pi(r_1 + r_2)^2 p} \tag{2.22}$$

For the above equations the diameter d or the radius r which requires atoms/molecules. Real gas molecules do not behave like hard balls. The effective diameter d or the radius r depends on the temperature T. In 1894 Sutherland[10] found an empirical equation that could describe this behavior:

$$r(T) = r_\infty\sqrt{\frac{T_V}{T} + 1} \tag{2.23}$$

T_V is the Sutherland constant or the doubling temperature. r_∞ is the molecular radius at an infinitely high temperature. Figure 2.20 shows the behavior of the radius from temperature for different gases. With increasing temperature the curves run to a limit value r_∞ between 1 and $2 \cdot 10^{-10}$ m is located.

So one determines the radii r with the Sutherland equation (Eq. 2.23) and can then calculate the mean free path length $\bar{l}$ using the equation. Figure 2.21 shows for hydrogen

[10] William Sutherland (1859–1911) Australian physicist and chemist.

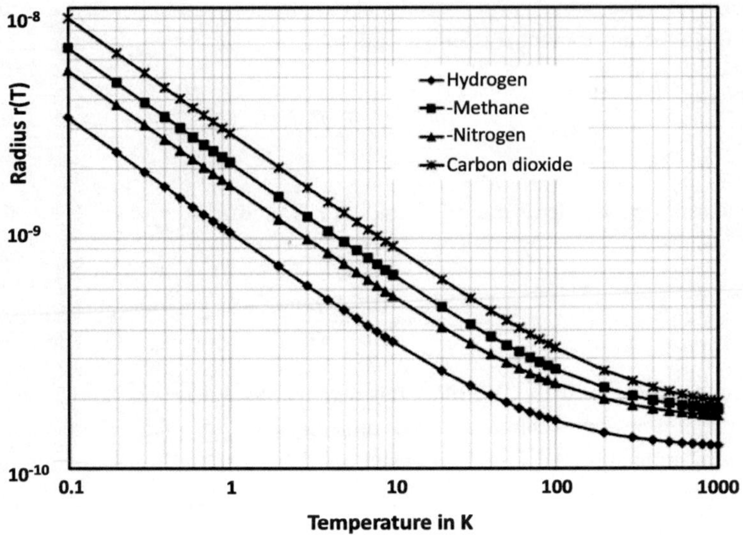

Fig. 2.20 Molecular radii of different substances as a function of temperature

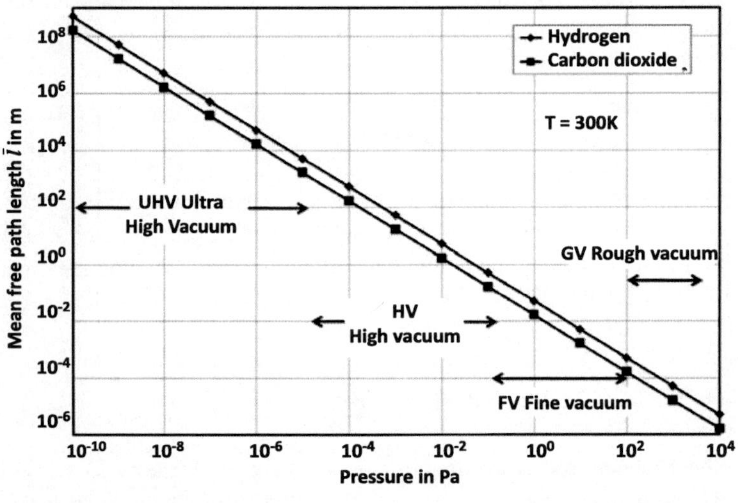

Fig. 2.21 Division of the vacuum into different pressure ranges with the corresponding mean free path lengths according to Eq. 2.22

and carbon dioxide at $T = 300$ K The two curves are close together, although the size of the two gases differs significantly. At atmospheric pressure 1000 hPa the mean free path is 10^{-7}m. The diagram also shows the different vacuum ranges. In ultra-high vacuum, the particle

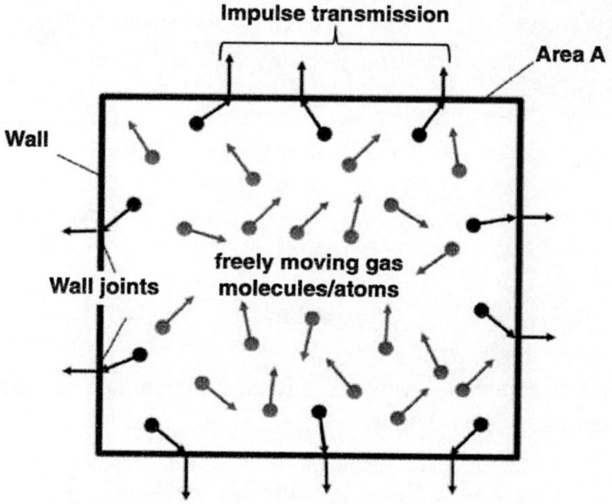

Fig. 2.22 Freely moving atoms/molecules in a closed chamber. The collision of the particles with the chamber wall creates a momentum transfer and thus a force effect F

density is so low that the molecules hardly touch each other. The impact frequency in this range is $0.0001\,s^{-1}$, that is, the particles only meet every 2–3 h on average.

Calculation of the Pressure

If these freely moving particles are located in an enclosed space, they regularly collide with the speed v on one of the 6 walls (Fig. 2.22). Each of the 6 walls has the same area A.

As the particles move in all directions, 1/6 of the particles hit each wall. Per time interval dt, however, only those particles hit the wall that are located at a distance:

$$ds = v \cdot dt \tag{2.24}$$

in front of the wall. This results in the volume dV in which these particles are contained:

$$dV = A \cdot v \cdot dt \tag{2.25}$$

With the particle density n you can now calculate the number of blows to the wall per time:

$$\text{Number of shocks} = \frac{1}{6} \cdot n \cdot A \cdot v \cdot dt \tag{2.26}$$

At each impact with the wall, in the form of a reflection, the impulse p_i is absorbed by the wall.

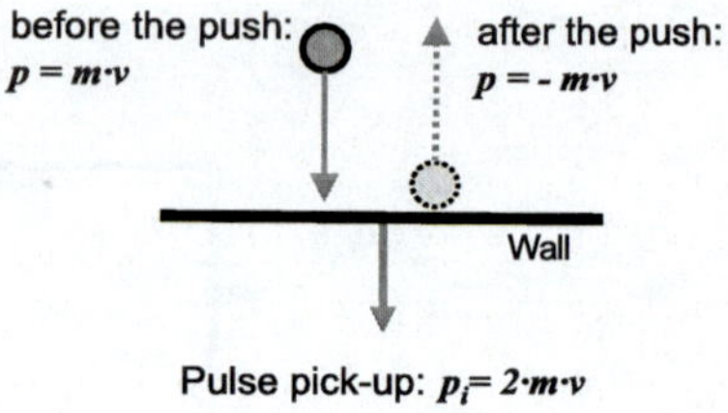

Fig. 2.23 Collision of a particle with the chamber wall and the resulting momentum transfer

$$p_i = 2mv \tag{2.27}$$

Impulse transmission per time dt however is a force F which then acts directly on the wall (Fig. 2.23). The pressure p is by definition:

$$\text{Pressure } p = \frac{\text{Force}}{\text{Area}} = \frac{F}{A} \tag{2.28}$$

If we now use Eqs. 2.26 and 2.27 in Eq. 2.28, we obtain the following expression for the pressure p:

$$p = \frac{F}{A} = \frac{\frac{2mv}{dt}\left(\frac{1}{6}Anvdt\right)}{A} \tag{2.29}$$

or rather

$$p = \frac{1}{3} n \cdot m \cdot \overline{v^2} \tag{2.30}$$

Equation 2.30 is also called the *fundamental equation of kinetic theory of gases* and has already been described by *Bernoulli.*[11] In this case, the mean value of the velocity squares is used as velocity v^2, since the particles do not all have the same velocity. The advantage of this equation is that the average velocity of the particles (atoms or molecules) can only be determined by macroscopic pressure measurement and knowledge of the gas density $\varrho = nm$:

[11] Daniel Bernoulli (1700–1782) Swiss mathematicians and physicists.

$$\overline{v} = \sqrt{\frac{3 \cdot p}{\varrho}} \tag{2.31}$$

Example

We want the average speed $\overline{v}$ of nitrogen molecules at 0 °C and 1013 hPa. The density ϱ of nitrogen under these conditions is 1.2505 kg/m^3. This results in the average speed as follows:

$$\overline{v} = \sqrt{\frac{3 \cdot 1.013 \cdot 10^5 \frac{\text{kg}}{\text{m} \cdot \text{s}^2}}{1.2505 \frac{\text{kg}}{\text{m}^3}}} = 493 \frac{\text{m}}{\text{s}} \tag{2.32}$$

However, this velocity is only an average value that results from the multitude of individual velocities of the particles. The next question therefore deals with the distribution of the velocities.

Maxwell's Velocity Distribution

The movement and thus also the velocity of the molecules can be described with statistical laws. The individual velocities can therefore deviate considerably from the mean value. Based on probability theory, *Maxwell*[12] derived a formula with which the distribution function $f(v)$ of the velocities in a gas can be calculated. An exact derivation of this important formula can be found in Richter (2010).

$$f(v) = \sqrt{\frac{2}{\pi} \left(\frac{m}{k \cdot T}\right)^3} \cdot v^2 \cdot \mathrm{e}^{-\frac{m \cdot v^2}{2k \cdot T}} \tag{2.33}$$

with $k=$ Boltzmann constant ($1.381 \cdot 10^{-23}$J/K), $T=$ temperature in Kelvin, $m=$ molecular mass, and $v=$ molecular velocity.

Figure 2.24 shows a maximum speed for each temperature T a maximum speed v_{max}. The speed is also known as the *most likely speed*. This value is obtained from the first derivative of the distribution function $f(v)$. If the first derivative is set to zero, this velocity can then be calculated:

[12] James Clerk Maxwell (1831–1879) Scottish physicist.

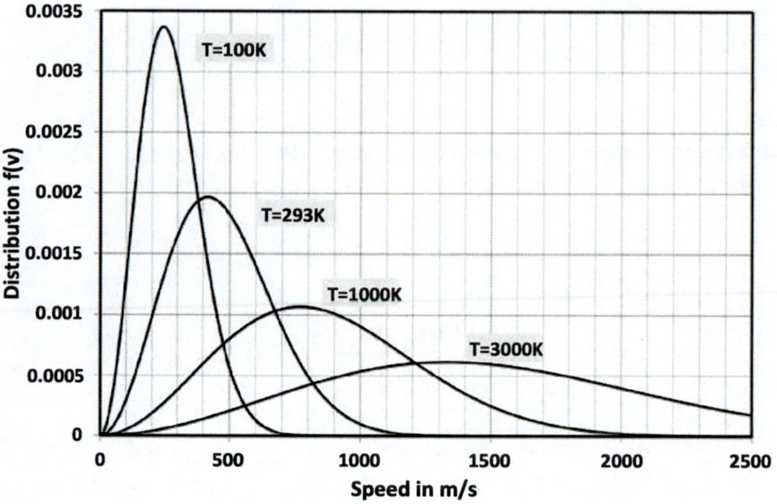

Fig. 2.24 Speed distribution $f(v)$ of nitrogen (N_2) at different gas temperatures T

$$\widehat{v} = \sqrt{\frac{2k \cdot T}{m}} \tag{2.34}$$

We had already used the mean square velocity in Eq. 2.30 to calculate the pressure (Table 2.8). If one sets this formula with the expression $p = nkT$, we obtain:

$$\sqrt{\overline{v^2}} = \sqrt{\frac{3k \cdot T}{m}} = 1.225\widehat{v} \tag{2.35}$$

The mean value of all velocities results from the arithmetic mean of all velocities:

$$\overline{v} = \sqrt{\frac{8k \cdot T}{\pi m}} = 1.128\widehat{v} \tag{2.36}$$

These 3 different speeds differ only very little from each other (Table 2.9). In Fig. 2.25 the individual positions have been entered to scale in the diagram.

The mean value of the velocities naturally depends very much on the type of gas. Very light atoms, such as Helium, have a much higher velocity at the same temperature T compared to very heavy molecules, such as sulphur hexafluoride (Fig. 2.26).

From the velocity v of the molecules it is now possible to infer the energy E. If one considers only the movement in space, the average kinetic energy can be calculated as follows:

Table 2.8 Molecular radii and doubling temperatures (Richter 2010)

Gases	$r_\infty/10^{-10}$ m	T_V/K
Hydrogen	1.21	76
Helium	0.97	79
Methane	1.66	164
Ammonia	1.23	503
Water	1.34	600
Nitrogen	1.60	112
Oxygen	1.48	132
Hydrogen chloride	1.54	360
Argon	1.43	169
Carbon dioxide	1.73	273

Table 2.9 Overview of the different speeds according to Kuchling (2011)

Molecular velocity	$\sqrt{\overline{v^2}}$	$\overline{v}$	$\widehat{v}$
Mean square velocity	1	1.085	1.225
Mean value of the speeds	0.921	1	1.128
Most probable speed	0.816	0.886	1

$$\overline{E_{\text{kin}}} = \frac{m}{2}\overline{v^2} \tag{2.37}$$

with $\overline{v^2} = \frac{3 \cdot k \cdot T}{m}$ one gets the following expression for the average kinetic energy:

$$\overline{E_{\text{kin}}} = \frac{3}{2}k \cdot T \tag{2.38}$$

The average kinetic energy of a gas thus increases linearly with temperature *T*. At the absolute zero point of temperature (= 0 K) the molecules are thus in the rest position (v = 0) and the energy is therefore also zero. Since the particles can move freely in space, there is an energy fraction[13] per degree of freedom of:

$$E_F = \frac{k \cdot T}{2} \tag{2.39}$$

For a pure translational movement there are then 3 degrees of freedom in space (*x*, *y*, *z*).

Uniform Distribution Theorem

So far we have looked at the particle model in the form that they are small spheres. Strictly speaking, however, this model is only valid for noble gases that consist of only one atom.

[13] Possibility of movement of the atoms/molecules.

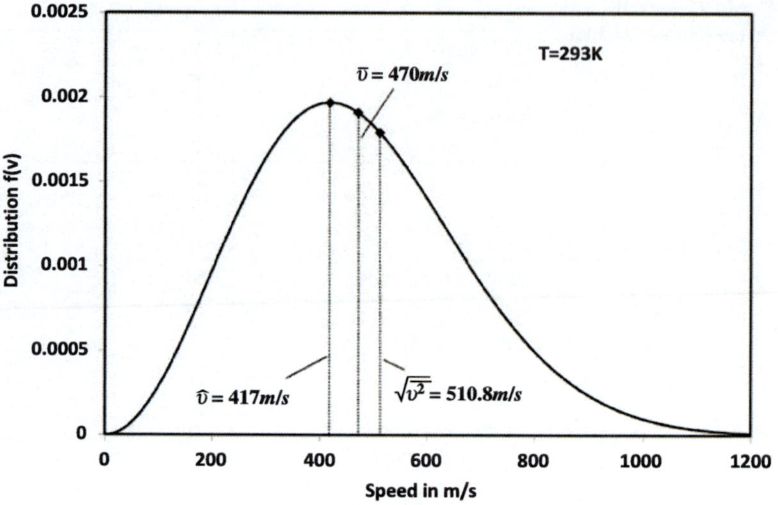

Fig. 2.25 Position of the different velocities for nitrogen for $T = 20^{\circ}\mathrm{C}$

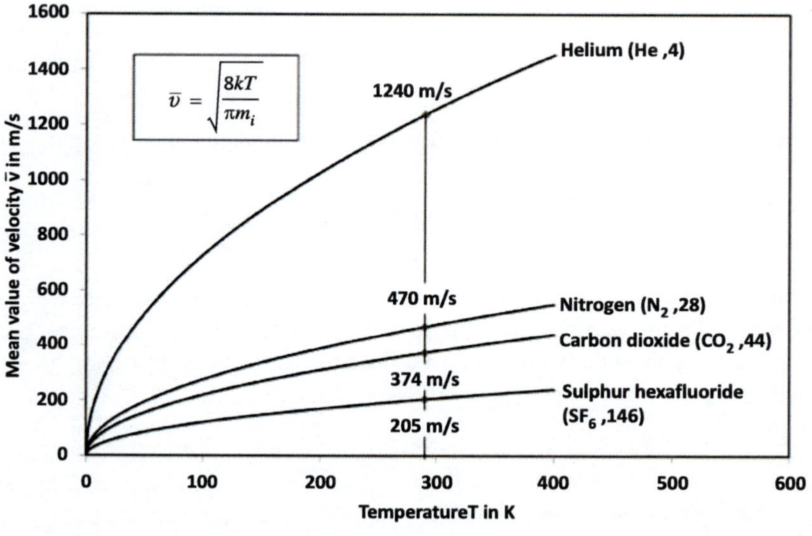

Fig. 2.26 Average velocities for different gases as a function of temperature T

Molecules always consist of several atoms, so that the structure changes accordingly. Molecules also have the possibility to absorb energy in the form of rotational energy around different axes. At very high temperatures it is also possible to absorb energy in the form of vibrations. With hydrogen this happens at temperatures>2000 K (Fig. 2.27). At temperatures above 3200 K, the hydrogen already decays (Giancoli 2010). According to

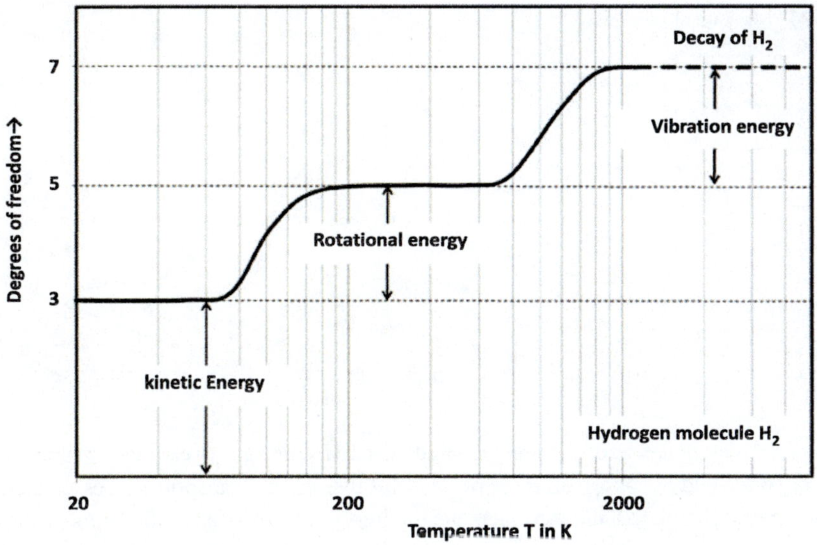

Fig. 2.27 Degrees of freedom of hydrogen as a function of temperature *T*

Table 2.10 Degrees of freedom for different types of gas

Substance	Degrees of freedom *f*		
	Translation	Rotation	Total
1 Atomic gas	3	–	3
2 Nuclear gas	3	2	5
3 Atomic gas	3	3	6

Clausius[14] and Maxwell, the energy of a molecule is distributed evenly over all degrees of freedom, so that an energy fraction E_F is required for each degree of freedom. Depending on the structure of the molecule, 3–6 degrees of freedom can be available (Table 2.10). This property is known as the *principle of equal distribution* or the *principle of equipartition*. The total energy can be calculated as follows (Hahn 2007)

$$\sum E = (f_{\text{Translation}} + f_{\text{Rotation}} + f_{\text{Oscillation}}) \cdot \frac{1}{2} kT \tag{2.40}$$

2.4 Transport Operations

Due to the high mobility of atoms and molecules in the gas phase, these particles can change location relatively quickly and transfer both substances and energy. These transfer mechanisms can take place not only in the gas space, but are also possible at the interfaces

[14] Rudolf Julius Emanuel Clausius (1822–1888) German physicist.

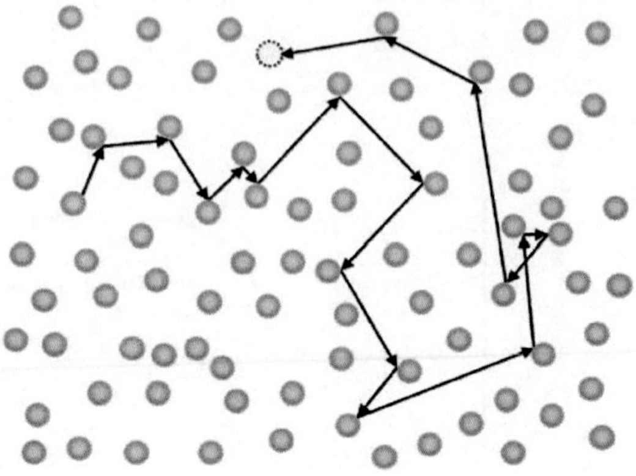

Fig. 2.28 Movement of a gas molecule in space

between the gas phase and a liquid or solid. Furthermore, gases can also penetrate into liquid or solids and be dissolved there or pass through them. Transport processes therefore play a major role in gas measurement technology and are dealt with in the following chapter.

Molecular Motion

If you look at an accumulation of gas molecules in space, all gas molecules move in different directions (x, y, z) with different velocities corresponding to the *Maxwell distribution.* Figure 2.28 shows the path of an individual gas molecule moving in zigzag paths. This kind of movement was already observed by Brown[15] under a microscope in 1828 when he examined flower pollen. He discovered that these movements are statistically distributed and are not predictable. This type of movement is also known as *Brown's molecular movement. Perrin*[16] developed a theory on this and was awarded the Nobel Prize in 1929.

Thermal Conductivity

The heat conduction of gases is a transport phenomenon of the transfer of thermal energy through gas molecules. For this purpose two plane-parallel plates are considered, which are located at a distance $s = x$ from each other. One plate has the temperature T_1 while the other plate has the temperature T_2. The temperatures are different ($T_1 > T_2$). The gas molecules, in between, first come into contact with the left plate and experience an energy transfer through the contact, which is expressed by an increased speed v and thus also an increased kinetic energy ($E = 0.5m \cdot v^2$).

[15]Robert Brown (1773–1858) Scottish botanist

[16]Jean Baptiste Perrin (1870–1942) French physicist.

These molecules now collide with other molecules in space and therefore transfer the energy further in the direction of the right plate with each collision. The molecules that are close to this plate collide with it and then release the energy. This one-dimensional case can be described by the empirical *Fourier equation:*

$$\frac{\mathrm{d}Q}{\mathrm{d}t} = -\lambda \cdot \mathrm{grad}T \cdot \mathrm{d}A = -\lambda \cdot \frac{\mathrm{d}T}{\mathrm{d}s} \tag{2.41}$$

In this equation, the thermal conductivity λ is introduced as the substance size of the gases. λ has the unit $\mathrm{Wm^{-1}\,K^{-1}}$. By solving the differential Eq. 2.41 the following expression is obtained:

$$Q = -\lambda \cdot \frac{A \cdot t \cdot \Delta T}{s} \tag{2.42}$$

respectively

$$P = \frac{Q}{t} = -\lambda \cdot \frac{A \cdot \Delta T}{s} \tag{2.43}$$

Using a solution approach based on the kinetic gas theory, the thermal conductivity can also be expressed by the viscosity η and specific heat capacity c_V:

$$\lambda = \frac{1}{2}\eta \cdot c_V \tag{2.44}$$

Taking into account the degrees of freedom of different gases, Eq. 2.44 is modified as follows (Meschede 2010):

$$\lambda = \frac{1}{2}\alpha \cdot \eta \cdot c_V \tag{2.45}$$

The constant α has the value ≈2.4 for diatomic gases, ≈1.9 for monatomic gases and ≈1.6for triatomic gases. Since both c_V and η are independent of pressure, the thermal conductivity λ is also independent of pressure (Fig. 2.29). Only when the mean free path *l* comes into the order of magnitude of the plate distance *s*, λ decreases proportionally (→*Prirani effect*[17]). This pressure dependence (Fig. 2.30) is used in the field of vacuum measurement technology to determine the pressure.

The thermal conductivity is measured experimentally with so-called *catharometers*. Figure 2.31 shows such an experimental setup. It consists of a cylindrical measuring chamber in which a thin platinum wire is coaxially tensioned. Since the wire heats up and expands thermally during the measurement, it must be mechanically pre-tensioned to

[17] Pirani, Marcello Stefano (1880–1968) German physicist.

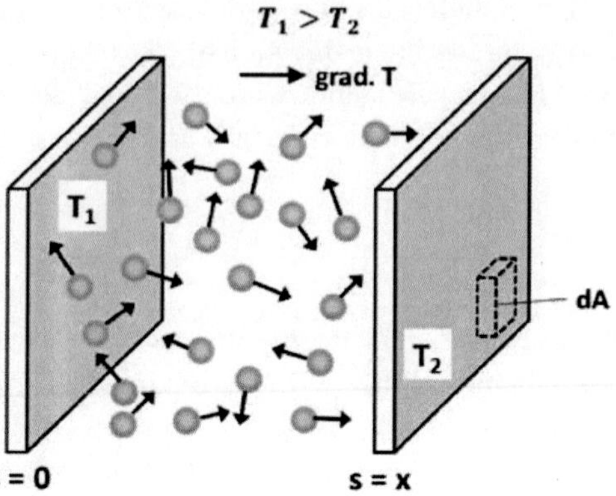

Fig. 2.29 Model of energy transfer through the heat conduction of gas molecules

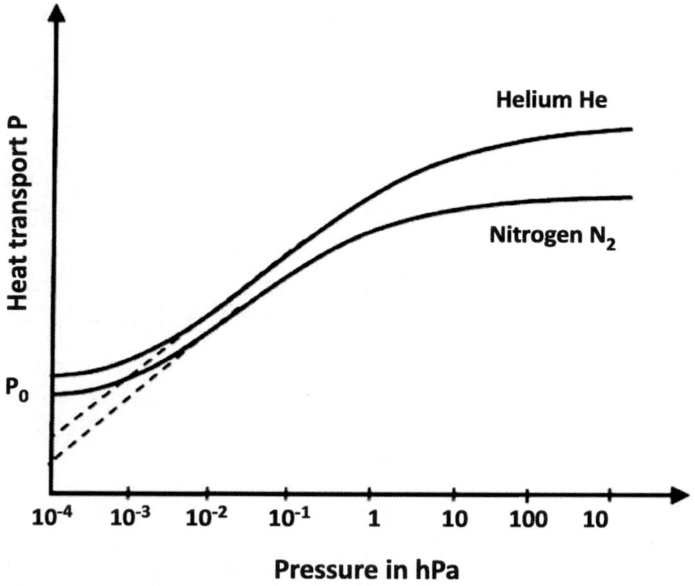

Fig. 2.30 Heat transfer (Pirani effect) in the vacuum range between 10^{-3} hPa and 10 hPa for helium and nitrogen

ensure a coaxial geometry in all cases. The temperature T_U of the outer wall of the measuring cell must be controlled at a constant value. This requirement can be met by a circulator that pumps a liquid (e.g. Water) around the measuring cell. In addition, the entire structure is also thermally insulated. The power balance then results from the following equation:

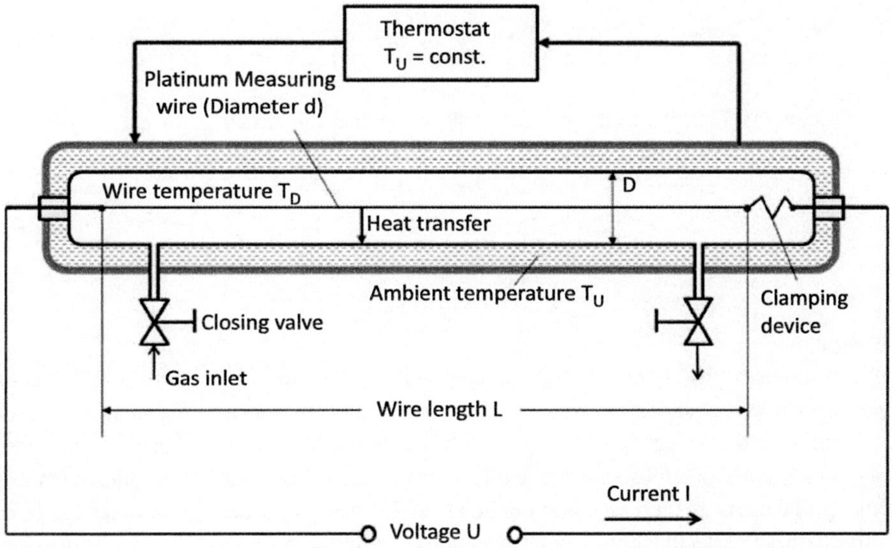

Fig. 2.31 Experimental setup (catharometer) for determining the thermal conductivity of gases. The measuring chamber is located in a water bath, which is permanently exchanged by the thermostat and is therefore at a constant temperature T_U

$$P = \lambda \cdot \pi \cdot L \cdot \Delta T \cdot \ln\left(\frac{D}{d}\right) \tag{2.46}$$

With $P = U \cdot I$ and $\Delta T = T_D - T_U$, the following expression results for the experimental value of thermal conductivity λ_{exp}:

$$\lambda_{\text{exp}} = \frac{U \cdot I}{\pi \cdot L \cdot (T_D - T_U) \cdot \ln(D/d)} \tag{2.47}$$

The wire temperature T_D can be calculated from the electrical resistance of the wire $R(T_D)$ as follows

$$R(T_D) = \frac{U}{I} \tag{2.48}$$

The resistance value $R(T_D)$ depends on the temperature T and can be described for small temperature changes by a linear approach:

$$R(T_D) = R_0 \cdot [1 + \alpha \cdot (T_D - T_U)] \tag{2.49}$$

By reshaping, one then obtains:

$$T_D = T_U + \frac{1}{\alpha} \cdot \left[\frac{R(T_D)}{R_0} - 1 \right] \tag{2.50}$$

If we now use Eqs. 2.50 and 2.48 in Eq. 2.47, we obtain the equation for determining the experimental thermal conductivity:

$$\lambda_{\text{exp}} = \frac{U \cdot I}{\pi \cdot L \cdot \left[\frac{1}{\alpha} \cdot \left(\frac{U/I}{R_0} - 1 \right) \right] \cdot \ln(D/d)} \tag{2.51}$$

Diffusion

Diffusion refers to the equalization of concentration differences by a mass flow. Diffusion is based on the statistical movement of gas molecules in space and can therefore be calculated with the kinetic gas theory. This velocity distribution has already been determined with the *Maxwell distribution* and is therefore an important basis for understanding diffusion. If there is an increased number of gas molecules (= concentration c_1) at one point in a room, these molecules move in all directions according to *Maxwell's velocity distribution.* In doing so, they collide with other molecules and thus spread out in the whole space (Table 2.11). This superimposed diffusion current J goes in the direction of the areas with a low particle density (= concentration as particles per volume $= c_2$).

This diffusion current J is directly proportional to the concentration gradient grad$c$, which for the one-dimensional case can be[18] described by *Fick's* first *law:*

$$J = -D \cdot A \cdot \frac{\mathrm{d}c}{\mathrm{d}x} \tag{2.52}$$

or

$$J_{1 \to 2} = D \cdot A \cdot \frac{c_1 - c_2}{\Delta x} \tag{2.53}$$

Where A is the area through which the diffusion current J passes and D is the substance-specific diffusion coefficient (Table 2.12). This coefficient can be calculated from the mean velocity $\overline{v}$ and the mean free path length $\overline{l}$:

[18] Adolf Fick (1829–1901) German physiologist.

Table 2.11 Thermal conductivity of different gases. λ_0 at 0 °C. λ_{rel} at 100 °C (Hengstenberg et al. 1980)

Gas	$\lambda_0[mWm^{-1}K^{-1}]$	$\alpha[10^2K^{-1}]^a$	$\lambda_{rel} = 100^*\lambda_{Gas}/\lambda_{Luft}$
Argon Ar	16.3	0.31	69
Helium He	143	0.18	554
Neon ne	46.1	0.24	182
Krypton Kr	87.8	0.32	37
Xenon Xe	5.15		
Hydrogen H_2	172	0.27	694
Nitrogen N_2	24.0	0.28	99.7
Oxygen O_2	24.5	0.3	101.3
Air	24.1	0.28	100
Carbon dioxide CO_2	14.3	0.48	71
Carbon monoxide CO	23.1	0.32	96.8
Nitrogen monoxide NO	23.2	–	102
Nitrous oxide N_2O	15.3	0.54	75
Ammonia NH_3	21.6	0.54	108
Chlorine Cl_2	7.7	–	34
Sulphur dioxide SO_2	8.4	0.6	44
Sulphur hexafluoride SF_6	14.1[b]		
Methane CH_4	30.2	0.48	141
Ethane C_2H_6	18.3	0.73	101
Propane C_3H_8	15.1	0.8	87
Butane C_4H_{10}	13.5	0.81	78
Pentane C_5H_{12}	12.2	0.75	73
Hexane C_6H_{14}	12	0.68	64
Water vapour H_2O			78

[a]At 30 °C

[b]Temperature coefficient in the range from 0 to 100 °C. $\alpha = \frac{1}{\lambda_0}\frac{d\lambda}{dT}$

$$D = \frac{1}{3}\bar{v} \cdot \bar{l} \tag{2.54}$$

If the expressions $\bar{v}$ for (Eq. 2.36) and $\bar{l}$ (Eq. 2.22) are used in Eq. 2.54, the following relationship is obtained

$$D = \frac{1}{3} \cdot \sqrt{\frac{8}{\pi} \cdot \frac{R \cdot T}{m}} \cdot \frac{k \cdot T}{\sqrt{2} \cdot \pi \cdot p \cdot (r_1 + r_2)^2} \tag{2.55}$$

If the constant values are now combined to a factor *f*, the following expression results for *D*:

Table 2.12 Diffusion coefficients for gases and vapours in air at 20 °C and 1013 hPa (Jessel 2001)

Substance	D in cm²/s	Substance	D in cm²/s	Substance	D in cm²/s
Hydrogen	0.69	Formic acid	0.150	Diethyl ether	0.0918
Ammonia	0.246	Carbon dioxide	0.1437	Ethyl acetate	0.0861
Steam	0.241	Ethanol	0.1181	Benzene	0.0859
Oxygen	0.231	Acetone	0.1049	*n*-Pentane	0.0842
Acetylene	0.22	Dichloromethane	0.1037	Toluene	0.0763
Methane	0.22	*i*-Propanol	0.1013	*n*-Hexane	0.0732
Nitrogen	0.211	Carbon disulfide	0.1013	*n*-Heptane	0.0664
Phosgene	0.175	Diethylamine	0.1	*n*-Octane	0.0616

$$D = f \cdot \frac{T^{3/2}}{p \cdot r^2} \cdot \sqrt{\frac{1}{m}} \tag{2.56}$$

Accordingly, the diffusion coefficient D is dependent on temperature, pressure and the size of the molecules. Small molecules, such as Helium and hydrogen, therefore have a large diffusion coefficient. However, this equation only describes the diffusion of one type of gas, so that in this case a pure self-diffusion is assumed. For a gas mixture this equation must be modified accordingly. For a binary gas mixture one receives then:

$$D_{1,2} = f \frac{T^{3/2}}{p \cdot (r_1 + r_2)^2} \cdot \sqrt{\frac{1}{m_1} + \frac{1}{m_2}} \tag{2.57}$$

For the temperature dependence $D(T)$ of the diffusion coefficient, the following expression can then be used to convert table values for $D_1(T_1)$ real applications at a temperature $T2$

$$D_2 = D_1 \cdot \sqrt{\left(\frac{T_2}{T_1}\right)^3} \tag{2.58}$$

Another interesting question is how fast a gas can propagate in the air (Fig. 2.32). This temporal propagation behaviour of gases is particularly important in the field of safety engineering (gas detectors) in order to calculate the propagation of "gas clouds".

Example (Giancoli 2010)

We look at an opened bottle filled with ammonia solution (Fig. 2.33). At the bottle opening, ammonia (NH_3) with a concentration c_1 escapes. Now the question is: How long does it take for a person at 10 cm to smell the ammonia?

Since the odour threshold of ammonia with 0.03–0.05 ppm (Altmann et al. 2015) is very small compared to the escaping concentration c_1, one can assume that $c_2 \approx 0$. The diffusion

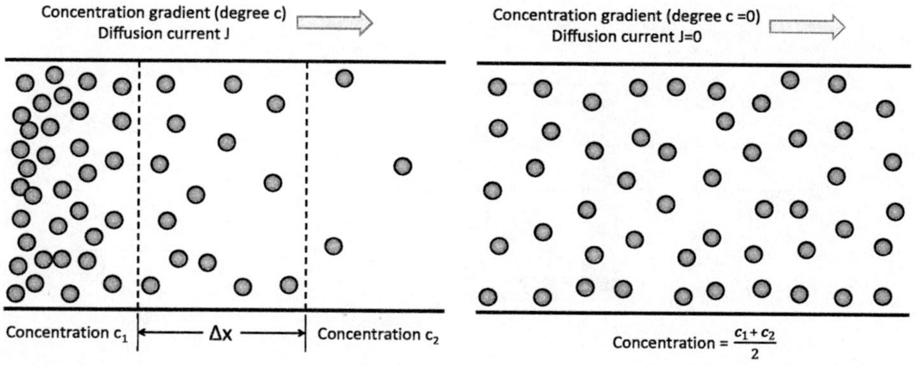

Fig. 2.32 Concentration profile before diffusion and after completion of the equalization process

Fig. 2.33 Experiment for the transmission of a "gas cloud" from the source (bottle) to the sensor (nose)

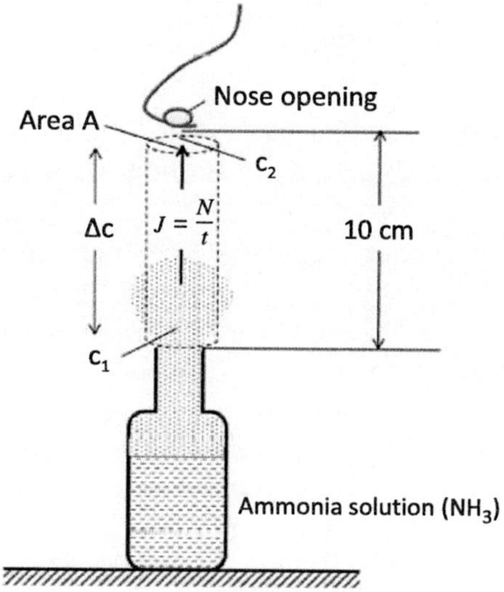

current J can be equated with the number of molecules N that flow through the area A in time t. From this follows:

$$t = \frac{N}{J} = \frac{N}{D \cdot A} \cdot \frac{\Delta x}{\Delta c} \tag{2.59}$$

The average concentration $\overline{c}$ that is between the bottle opening and the nose can be estimated by $\overline{c} \approx N/V$. The volume results from $V = A \cdot \Delta x$. For N we can now insert the expression into the equation and get $N = \overline{c} \cdot A \cdot \Delta x$:

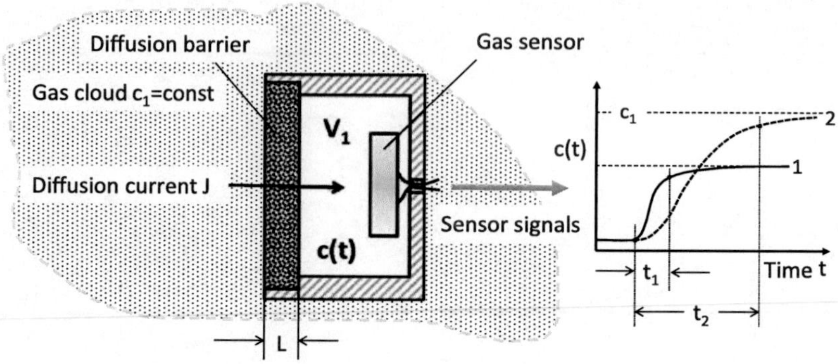

Fig. 2.34 Measuring chamber in which a gas sensor for detecting substances in the ambient air (gas cloud) is located. Chemical gas sensors (1) behave differently than physical gas sensors

$$t \approx \frac{\overline{c} \cdot A \cdot \Delta x}{D \cdot A} \cdot \frac{\Delta x}{\Delta c} = \frac{\overline{c}}{\Delta c} \cdot \frac{(\Delta x)^2}{D} \tag{2.60}$$

Since the concentration c_2 in the area of the nose is virtually zero, you can use for $\overline{c}/\Delta c \approx \frac{1}{2}$ or$\overline{c} \approx \Delta c/2$ With $D_{NH_3} = 0.246\text{cm}^2/\text{s}$ you can calculate the time:

$$t \approx \frac{1}{2} \frac{(0.1\text{m})^2}{\left(2.46 \cdot 10^{-5}\text{m}^2/\text{s}\right)} \approx 200\,\text{s} \tag{2.61}$$

This time of 200 s appears first of all as very large, since one knows from experience that the propagation of, for example, odours in space is faster. This is obviously due to the fact that this temporal process is not determined by diffusion, but by other effects. Even within closed rooms air movement never is null. By door-grooves or window-seals, there will always be some flow within room, which will transport these gases and vapours forced from A to B. The movement of people who are in the room will also cause considerable turbulence in the room.

The situation is different if a small volume V_1 (e.g. the measuring cell of a gas sensor) is sealed with a diffusion barrier (Fig. 2.34). This barrier can, for example, consist of a membrane or a sintered disc (metal or plastic). Diffusion barriers are open-pored and allow gases to diffuse through them, while direct inflows are prevented. Although the temporal behaviour of the gas sensors is impeded by this type of gassing, the protective effect of the diffusion barrier against dust or splash water is essential for practical applications.

If the gas sensor in the measuring cell is based on a physical principle (no chemical reactions), the diffusion current causes an accumulation in$c(t)$ the cell and the concentration gradient $\Delta c/\Delta L$ decreases. However, the smaller the gradient, the lower the diffusion current J and thus the gas exchange across the diffusion barrier. The gas sensor records

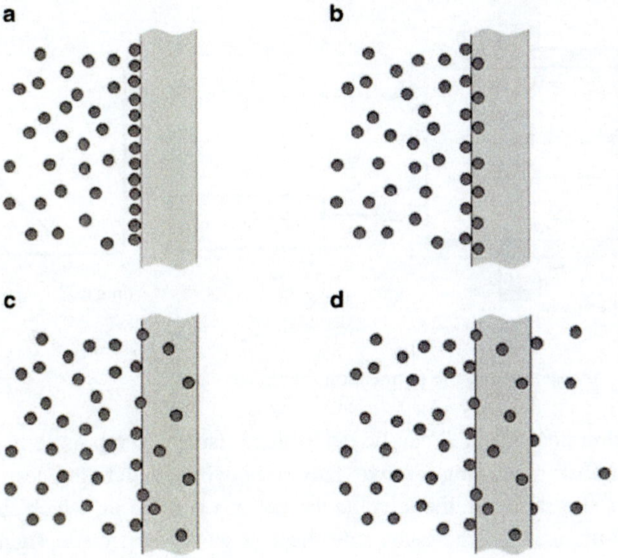

Fig. 2.35 Partial steps of the permeation of gas through a solid. (**a**) Adsorption, (**b**) Absorption, (**c**) Diffusion, (**d**) Desorption

the course of $c(t)$, whereby only after a time t_2 *does* the sensor signal (2) come close (90% value) to the concentration value c_I in the surrounding gas cloud. In a chemical gas sensor, the sample gas is converted by a reaction so that it disappears from the measuring cell or assumes a valuc bclow c_I. The gradient thus remains always larger in comparison to a physical gas sensor, so that the time response $t_1 < t_2$ is also better.

Permeation

Gaseous substances also have the possibility to migrate through solids (Fig. 2.35). This process is called permeation. In this process, the gas molecules are first adsorbed on the surface of the solid. Then the gas penetrates into the layer (absorption) and migrates through the substance (diffusion). On the other hand, the gas molecules are released from the surface again (desorption) and return to the gas space. Simplified, this permeation can be represented as the product of the solubility S in the solid and the diffusion coefficient D. P or P_0 is the permeation number. The temperature dependence of P is described by an *Arrhenius equation* with the activation energy E_a:

$$P = S \cdot D = P_0 \cdot e^{\left(-\frac{E_a}{k \cdot T}\right)} \tag{2.62}$$

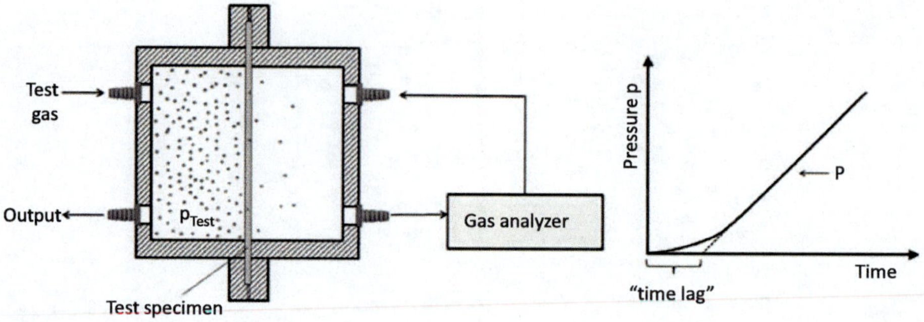

Fig. 2.36 Setup for determining the permeation number *P*

The permeation coefficient *P* can be determined using a setup as shown in Fig. 2.36. One has a chamber consisting of two halves between which the test specimen (e.-g. Membrane) is stretched. On the left side the test gas is filled in, which should permeate through the membrane. On the right side the test gas is expected. There are different methods to evaluate this passage. On the one hand, one can record the pressure increase in the right half. The permeation number *P* can then be calculated from the slope. On the other hand, one can also record the gas concentration with a gas analyzer and thus draw conclusions about the permeation number *P*. The time lag (time lag), which elapses until the pressure curve rises linearly, can be used to calculate the diffusion coefficient (Burger 2001).

The gas volume *V which* has passed through the permeation can be calculated with the following equation (Jessel 2001):

$$\frac{V}{t} = P \cdot \frac{A}{d} \cdot (p_1 - p_2) \tag{2.63}$$

The permeation depends very much on the chemical structure of the material and can be adjusted to the respective gas for specific purposes. Especially for gas separation, selective membranes are sought that only allow a certain gas to pass through (Fig. 2.37). Such a membrane can be used to enrich oxygen from the ambient air, for example, so that it can then be used for medical applications.

Polymer membranes consisting of very long molecules are used for these applications and partly have molar masses of up to 200.000 g/mol. Figure 2.38 shows a typical polymer chain that was synthesized specifically for gas separation applications (Burger 2001).

With the help of these processes, gas separations can also be carried out in large-scale plants. One example is the separation of methane and carbon dioxide from biogas plants. The biogas process supplies a gas mixture of approximately equal proportions (53% CH_4, 47% CO_2). However, energy recovery is only possible with methane. Therefore the methane must be separated from the carbon dioxide. So-called composite membranes

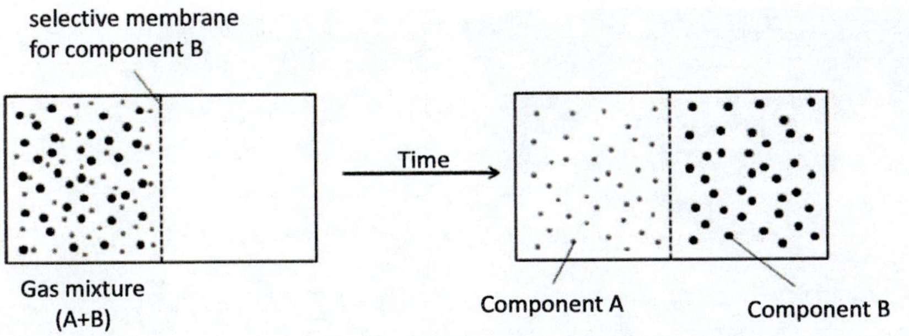

Fig. 2.37 Ideal mode of operation of a gas separation by a selective membrane

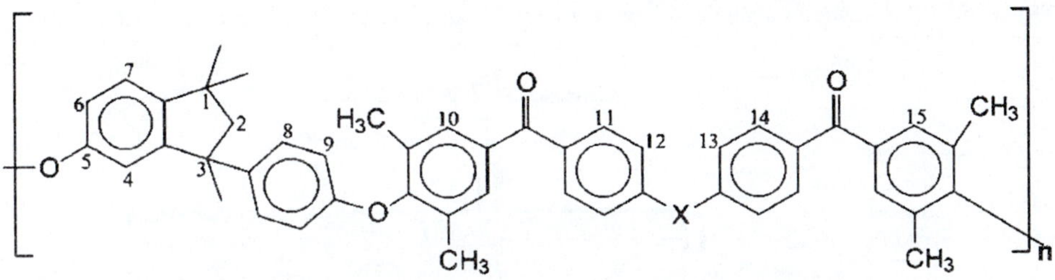

Fig. 2.38 Structure of a polymer molecule for gas separation (Burger 2001)

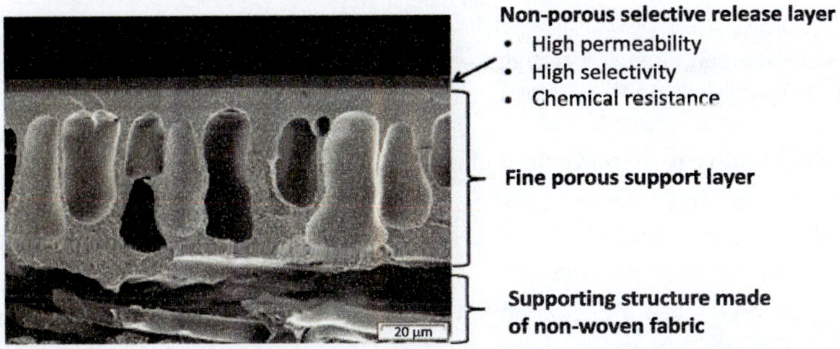

Fig. 2.39 Sectional view (REM image) through a composite membrane with different support layers. (With kind permission of GMT Membrantechnik GmbH)

consisting of several layers are used for this purpose. Figure 2.39 shows a SEM image of such a membrane. The selective and pore-free carrier layer is located on the outside. Underneath this is a fine porous support layer, which is primarily intended to ensure mechanical stability. The lower supporting structure made of non-woven fabric has the same function.

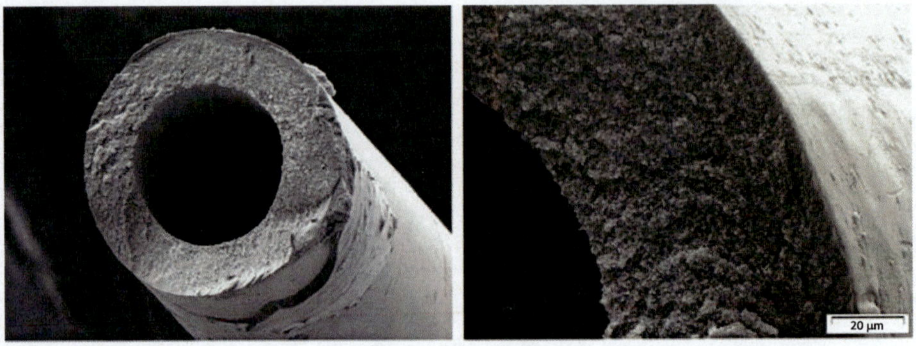

Fig. 2.40 SEM image of a single hollow fiber membrane. (Source: Evonik Industries AG Marl)

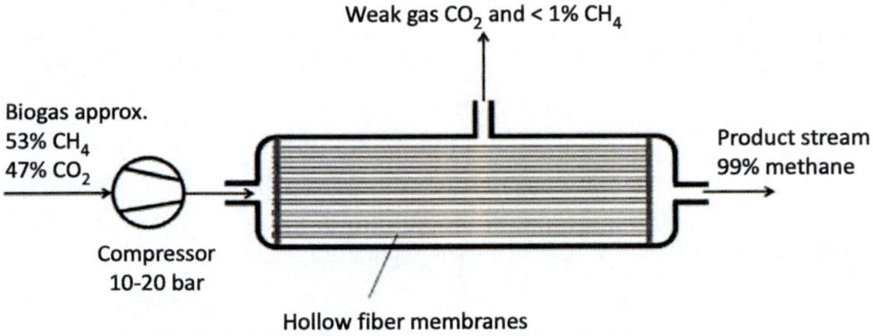

Fig. 2.41 Cross-section through a cartridge with a large number of thin hollow fiber membranes. (According to Evonik Industries AG Marl)

Another possibility is to use hollow fiber membranes with a diameter of only some µm (Fig. 2.40). If these fibers are bundled into a small space, a cartridge with a very large surface is obtained (Fig. 2.41). If these cartridges are cascaded to a complete plant, biogas can be upgraded per hour via 1000 m^3.

The base material can be, for example, polyimides with a permeation number of $P_{CO_2} = 0.2$ Barrer[19] where the ratio is $P_{CO_2}/P_{CH_4} = 64$ (Scott 1995). Even better selectivities can be achieved with other high performance polymers, but these are not disclosed by the manufacturers.

Solubility of Gases in Liquids

Gas molecules that are present at an interface with a liquid (e.g. water) are absorbed to a certain extent by the liquid. These gas molecules go into solution and are then part of the

[19]Converted to SI units, 1 barrer corresponds to $=7.5 \cdot 10^{-8}[m^3(STP)] \cdot [m] \cdot [m^2] \cdot [s]^{-1} \cdot [Pa]^{-1}$ STP =standard temperature and pressure.

Table 2.13 Coefficient of solubility α at 1013 hPa and 20 °C in litres of gas per litre of water (Jessel 2001)

Gas	Water solubility	Gas	Water solubility
Ammonia	685.7	Nitric oxide	0.0480
Hydrogen chloride	448.0	Methane	0.0330
Sulphur dioxide	39.40	Oxygen	0.0310
Chlorine	2.260	Carbon monoxide	0.0230
Acetylene	1.040	Hydrogen	0.0180
Carbon dioxide	0.879	Nitrogen	0.0160
Nitrous oxide	0.665	Sulphur hexafluoride	0.0056

liquid. The proportion depends very much on the temperature and the gas pressure. With increasing pressure, the number of gas molecules that are dissolved in the liquid also increases. The temperature, on the other hand, has an opposite effect. With rising temperature the number of gas molecules in the liquid phase decreases. This relationship is known as *Henry's law.*[20] The solubility L_i depends on the partial pressure p_i of the respective molecules in the gas phase. The index i indicates that gas mixtures with several components ($i = 1, 2, 3, \ldots$) are also subject to this law.

$$L_i = \alpha_i \cdot p_i \tag{2.64}$$

The proportionality factor α_i is called the solubility coefficient and depends on the type of gas, the temperature T and the solvent (liquid).

The values from Table 2.13 show that the solubility of the different gases varies greatly. While ammonia is very soluble in water and forms the well-known ammonia solution, nitrogen is a very poorly soluble gas. Nevertheless, nitrogen in the air we breathe is a major problem for deep-sea divers. For example, a diver, who wants to finish a dive at 40 m depth is not allowed to surface immediately. Due to the high pressure of 5 bar, a part of the nitrogen in the blood dissolves, which would lead to nitrogen bubbles in the blood if the diver surfaced immediately. Therefore the diver must slowly get *used* to the atmospheric pressure so that the nitrogen can escape from the blood without bubbles forming (Fig. 2.42).

The temperature dependence of the solubility coefficient is shown in Fig. 2.43. It can be described by a fourth order polynomial and therefore decreases disproportionately with increasing temperature.

[20]William Henry (1774–1836) English physician and chemist.

Fig. 2.42 Transition of the gas molecules at the boundary layer into the liquid phase

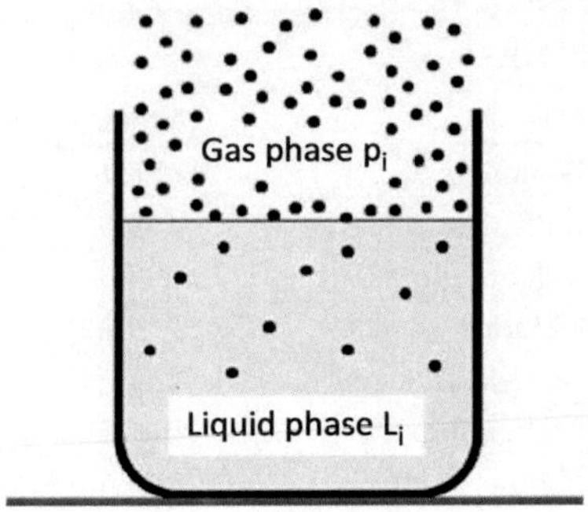

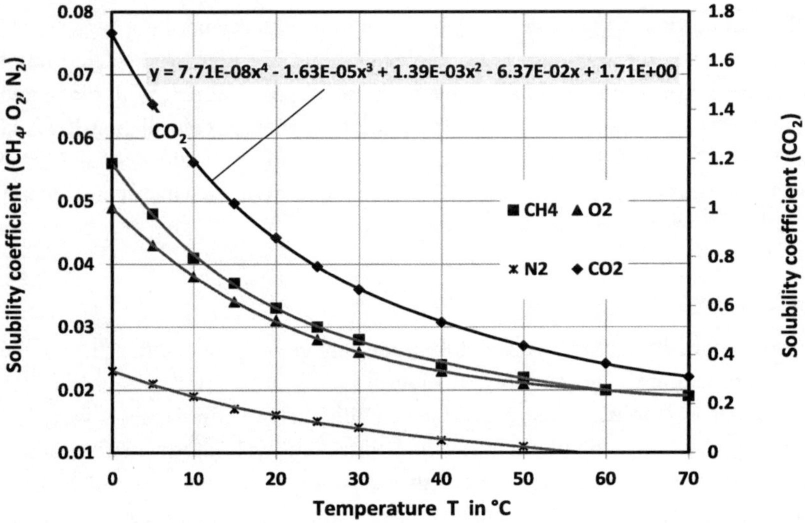

Fig. 2.43 Temperature behaviour of solubility in water for different gases. (Data from Jessel 2001)

The reverse process of the solubility of gases in a liquid is outgassing. This process can be described analogously. If there is a partial pressure above a liquid layer which is less than p_i, the gas passes from the liquid phase back into the gas phase. The simplest possibility is to pass clean gas (e.g. purified air) through a *bubble bottle.* With a glass frit at the transition into the liquid, many very small gas bubbles are produced, which together form a large surface for a transition of the dissolved gases from the liquid phase into the gas phase. So in each gas bubble molecules are collected according to *Henry's law,* which then saturate with the gas on their way up (concentration c_i). The substance can then be quantitatively determined from the liquid phase with a downstream gas analyzer. This method is known as *stripping* (Fig. 2.44). This method can also be used to detect liquid substances dissolved in water (e.g. alcohol).

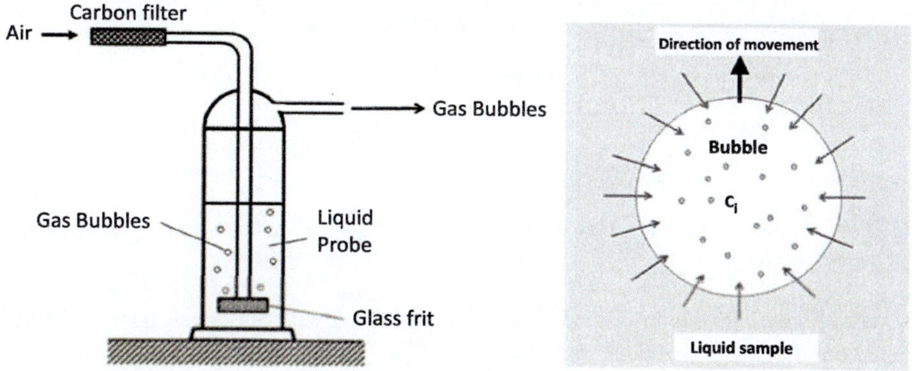

Fig. 2.44 Stripping the dissolved gases with a gas washing bottle

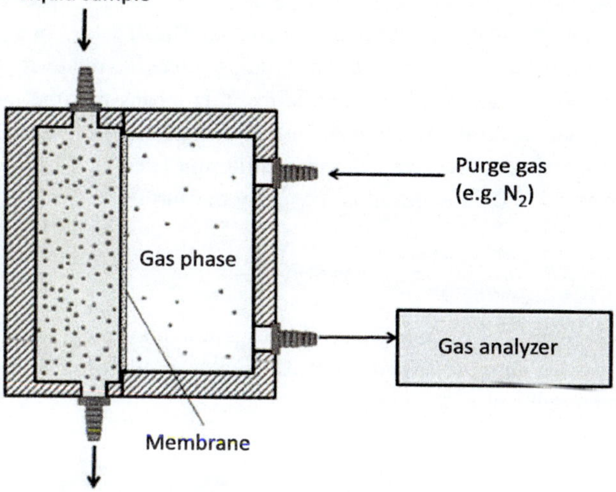

Fig. 2.45 Flow arrangement for stripping dissolved gases or evaporable liquids

A different method is used in process measurement technology. It uses the permeation of gases through a membrane. This can also be done from the liquid phase if the membrane is arranged as a transition between the two phases. The liquid sample flows past a membrane for this purpose. The dissolved gas molecules come into contact with the membrane and then have the opportunity to enter the gas phase via permeation through the membrane. If the gas phase is permanently purged, a stationary concentration gradient is established (→ continuous operation, see Fig. 2.45). Alternatively, the right-hand chamber can also be closed so that the gas accumulates there over time for better detection.

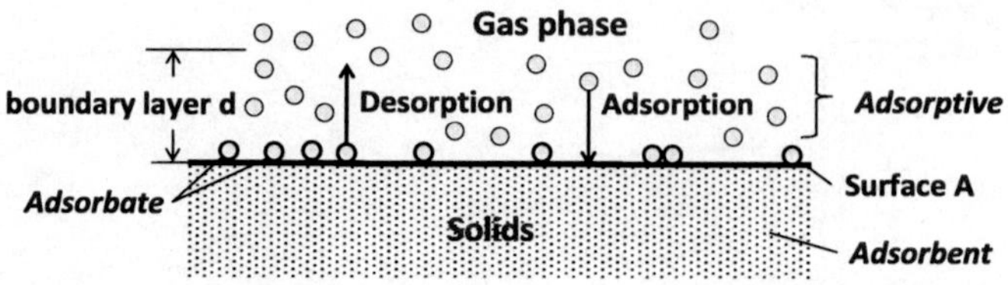

Fig. 2.46 Dynamic equilibrium between adsorption and desorption at the interface between a solid and the gas phase

At certain time intervals the chamber must then be filled with purge gas again and again (→intermittierender operation).

Gases on Surfaces

Under certain conditions gases bind or stick to a surface. This binding is caused by physical forces acting between the gas molecules and the atoms of the solid surface (Fig. 2.46). This process is called *adsorption* and is caused by the *van der Waals forces* in a purely physical bond. In this case it is called *physisorption*. The binding forces are relatively low and can be overcome by thermal activation. In concrete terms, this means that the molecules move away from the surface again by an increase in temperature (higher kinetic energy). This process is called *desorption*. Accordingly, an equilibrium is created between the particle flows that are adsorbed ($\dot{n}_{Ad}$) and desorbed ($\dot{n}_{Des}$) by the surface.

$$\dot{n}_{Ad} = \dot{n}_{Des} \tag{2.65}$$

According to *Langmuir,*[21] the molecules bind as a monomolecular layer on the surface ($\Theta = 1$), whereby the number of molecules increases proportionally with the partial pressure p of the component in the gas phase.

$$\dot{n}_{Ad} = k_1 \cdot (1 - \Theta) \cdot p \tag{2.66}$$

The quantity k_1 is a velocity constant that can be derived from the kinetic theory of gases.

Desorption, on the other hand, depends on the area Θ covered by the gas molecules and a hypothetical pressure p_{Ad} in the sorbed phase.

$$\dot{n}_{Des} = k_2 \cdot \Theta \cdot p_{Ad} \cdot \exp\left(\frac{-E_{Ad}}{R \cdot T}\right) \tag{2.67}$$

E_{Ad} is the energy required to bind the molecules from the gas phase to the surface (= adsorption energy). If one now uses both expressions in Eq. 2.65 and defines a factor K:

[21] Irving Langmuir (1881–1957) American chemist and physicist.

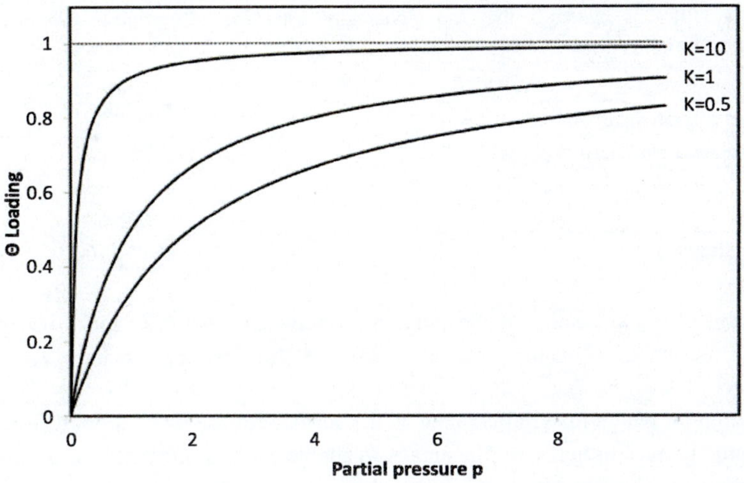

Fig. 2.47 Sorption isotherms according to Langmuir for different K-factors according to Eq. 2.69

$$K = \frac{k_1}{k_2} \cdot \exp\left(\frac{E_{Ad}}{R \cdot T}\right) \tag{2.68}$$

the equation for Langmuir's sorption isotherm is obtained with the degree of coverage or the load Θ

$$\Theta = \frac{K \cdot p}{1 + K \cdot p} \tag{2.69}$$

Figure 2.47 shows these isotherms for different *K-factors.* With increasing partial pressure p the isotherms approach the value 1, which is equivalent to a complete occupation of all possible places on the surface.

In fact, the gas molecules can also bind to the surface in several layers one above the other. This fact has been investigated by several scientists and is described by the so-called *Brunauer-Emmet-Teller equation*[22,23,24] or *BET equation for* short:

$$\Theta = \frac{C \cdot p_r}{(1 - p_r) \cdot [1 + (C - 1) \cdot p_r]} \tag{2.70}$$

[22] Stephen Brunauer (1903–1986) Hungarian chemist.

[23] Paul Hugh Emmett (1900–1985) American physicochemist.

[24] Edward Teller (1908–2003) Hungarian physicist.

Table 2.14 Different materials that are technically used for adsorption purposes, with the corresponding surface-to-mass ratios

Adsorbent	Surface area/mass in m^2/g
Wide pored activated carbon	300–2500
Activated carbon with narrow pores	750–850
Silica gel	300–350
γ-Alumina	200–500
Molecular sieves	500–1100

The constant C is a measure for the interaction between *adsorbent* and *adsorptive* (see Fig. 2.46). For large C *values,* the adsorption of the first layer occurs very quickly (monolayer).

Depending on which material is taken as the surface for the interaction, different gases can be bound there. Furthermore, the area A available for this interaction is a measure for the total amount of *adsorbate* that can be absorbed. In practice, therefore, attempts are made to *artificially* increase the surface area of materials or to use materials which already have a large surface area due to their lattice structure. Typical materials that are used for these purposes are listed in Table 2.14.

For example, activated carbon is very well suited to absorb air pollution. This property is used on a large scale in paint shops to treat the exhaust air (Fig. 2.48). For this purpose, large containers are filled with activated carbon and the loaded air flows through them to absorb the pollutants. The cleaned air then exits at the outlet of the filter (Fig. 2.49). As the load increases, the filter element will be able to absorb fewer and fewer pollutants until saturation is reached (Fig. 2.50). After this saturation, the *breakthrough* occurs and the pollutants now exit unfiltered. This point in time must be detected by using appropriate gas measuring devices in order to switch to another filter element at an early stage. During the time the second filter element is loaded, the first element can be regenerated.

A special feature of the binding of gases to surfaces is the so-called *chemisorption*. In contrast to the physisorption described so far, where only physical forces caused the binding of the gas molecules to the surface, chemisorption has further effects. In chemisorption, reactions take place at the surface, which cause an electron transfer between the adsorbent and the adsorbate. The electron transfer is characteristic of chemical reaction. The adsorbates therefore change their material properties or even their composition and reaction products are released into the gas phase. The chemisorbed bond forms a monolayer on the surface and thus hinders the further adsorption of possible gas molecules. The main differences between these two forms of attachment are summarized in Table 2.15.

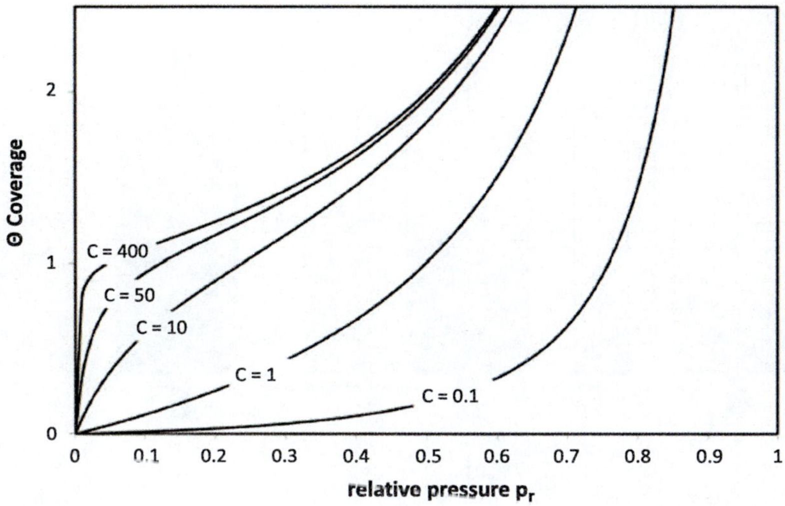

Fig. 2.48 Sorption isotherms according to Brunauer, Emmett and Teller for different *C-values* according to Eq. 2.70

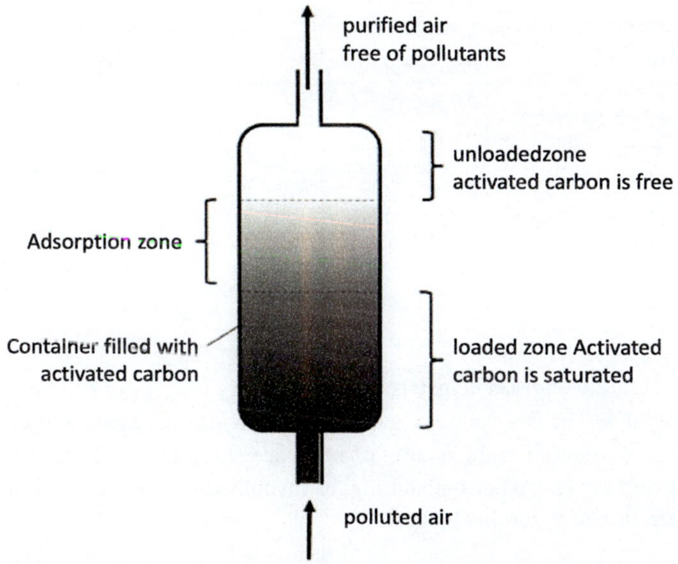

Fig. 2.49 Filter element with activated carbon for cleaning air

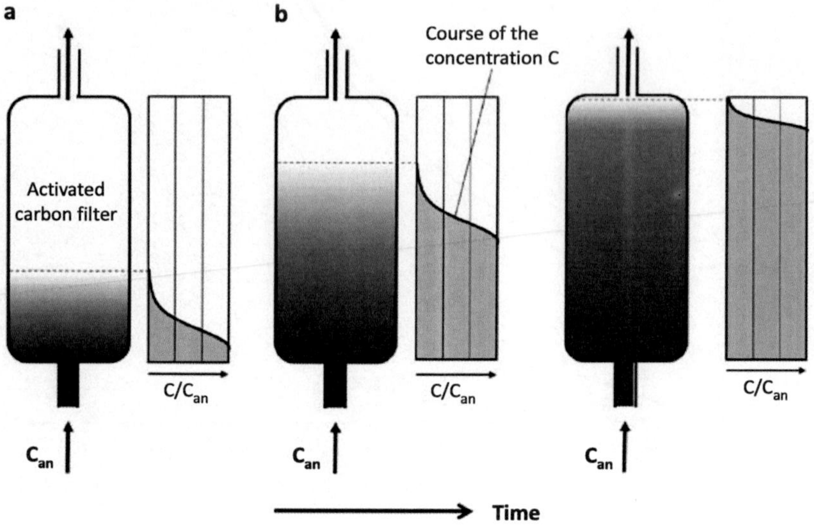

Fig. 2.50 Course of the loading zone at different times. (**a**) Start of loading. (**b**) The adsorption zone moves through the activated carbon filter. (**c**) The filter breaks through

Table 2.15 Differences between physisorption and chemisorption of gases on surfaces

Parameters	Physisorption	Chemisorption
Adsorbent	All solids	Specific solids
Adsorbate	All gases	Chemically active gases
Temperature	Low temperatures	High temperatures
Adsorption heat	Small	High
Assignment	Multiple occupancy	Monolayer
Reversibility	Fully reversible	Often irreversible

Solubility of Gases in Metals
Although the solubility of gases in metals is generally very low, it can still be detected with appropriate methods. The absorption of gases plays a major role, especially in metallurgy, in order to be able to describe the quality of materials (Fromm and Gebhardt 1976). The dissolution process can take place both during the manufacturing process (e.g. molten steel) and during operation (e.g. turbine blades).

First the gas comes into contact with the solid surface and is adsorbed there. After this adsorption, the gas now has the possibility to diffuse into the solid and then to settle in the crystal structure. This process is called absorption. Absorption occurs much better at higher temperatures and high pressures, whereas adsorption becomes less effective at higher temperatures. A state of equilibrium is formed here, which can be very complex in detail.

Chemical reactions, which can take place both on the surface and inside the solid, are also possible. The amount of m_{Gas} that can be absorbed by a metal, for example from the gas phase, can be calculated as follows:

$$m_{Gas} = k \cdot \sqrt{p} \tag{2.71}$$

The constant k depends on the reaction partners and is only valid for a constant temperature. At different temperatures the absorption changes very much and is then described by the following equation (Kubaschewski 1938):

$$m_{Gas} = \alpha \cdot e^{-\frac{1}{T}} \cdot \sqrt{p} \tag{2.72}$$

The absorption also depends on the size of the molecule and its chemical activity. Hydrogen, for example, as a small molecule with a high chemical activity, it is well suited to be embedded in the metal lattice structure. This type of compound is called metal hydride and is of great importance for the storage of gaseous hydrogen. Depending on the metal used, hydrogen can be absorbed up to 600 times the volume of the metal. More hydrogen can therefore be stored in metal hydride than in liquid form for the same volume. In addition to this form of hydrogen absorption, however, unwanted inclusions also occur. If, for example, mechanical components are exposed to a high hydrogen concentration in a technical process, this leads to embrittlement of the metal. This changes its material properties permanently.

Convection

Convection is a combination of heat transfer and mass transport (Fig. 2.51). For this purpose, we consider a heating plate that emits a heat output dQ/dt to the ambient air (Fig. 2.52). The gas absorbs this amount of heat and thus changes its temperature from T_1 to T_2. With this temperature increase, the gas density also changes. With increasing temperature (in K) the gas density decreases accordingly:

$$\varrho_2 = \varrho_1 \cdot \frac{T_1}{T_2} \tag{2.73}$$

The heated gas now has a lower density and will now flow upwards by buoyancy, against gravity. This flow is called *natural convection* or *free convection.*

The calculation of the heat transfer can be represented by the following equation:

$$\frac{dQ}{dt} = \alpha_K \cdot (T_H - T_1) \tag{2.74}$$

The proportional factor α_K is called the heat transfer coefficient and has the unit $W/m^2 \cdot K$. This coefficient depends on many factors (especially on the orientation of the heating plate

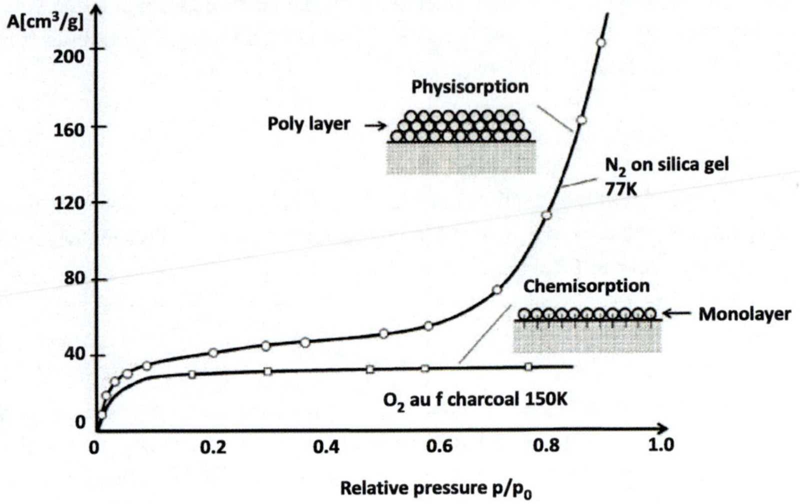

Fig. 2.51 Course of the adsorption isotherms for physisorption and chemisorption (Dörfler 2002)

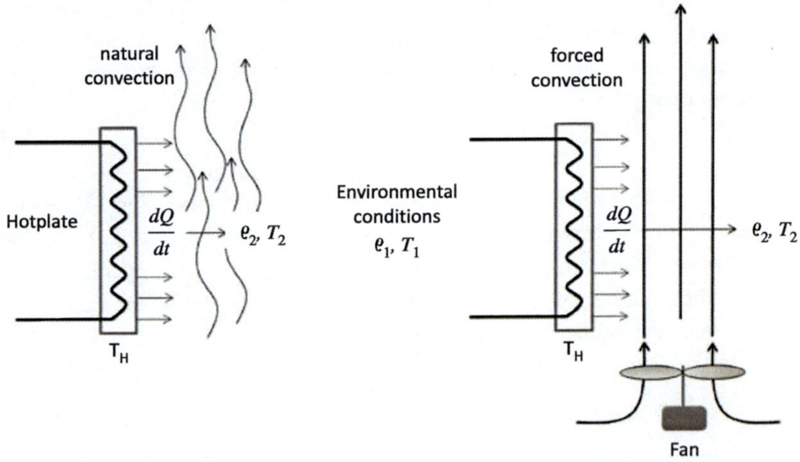

Fig. 2.52 Different forms of convection. Due to the change in density caused by temperature, the heated gases rise to the top and thus lead to a natural flow against gravity. If a fan is used to support this flow, it is called forced convection

to the direction of gravity) and can only be determined empirically (Hahn 2007). It is typically in a range from 3.5 to 35 $W/m^2 \cdot K$.

This convection can be increased by a forced flow from outside (e.g. fan). In this case one speaks of forced convection. Here too, the heat transfer coefficient α_K can only be determined experimentally. Depending on the external flow, this value can be between 35–300 W/m^2 · K.

Since the external factors are very large in the experimental determination of α_K, and can therefore significantly influence the result, this approach quickly reaches its limits. In the years 1909–1915 Nußelt[25] developed a theory in which he applied Reynolds' similarity theory to this problem. With this approach he was able to achieve a better result (Marek and Nitsche 2012). The *Nußelt number* (Nu), named after him, sets the characteristic length L of a body in relation to the thickness of the thermal boundary layer δ and also indicates by which factor convection is stronger than pure heat conduction in the gas (air).

$$\mathrm{Nu} := \frac{\alpha_K \cdot L}{\lambda_{\mathrm{Gas}}} \tag{2.75}$$

With this key figure, different but similar superstructures can be compared. Prerequisite for this comparison are identical Nußelt numbers. This consideration applies to both forced and free convection.

2.5 Gas Laws

The distances between individual atoms or molecules, which are in the gas phase, are under normal conditions $x \approx 3 \cdot 10^{-9}$ m. The molecule diameters are at$d \approx 3 \cdot 10^{-10}$ m. Gaseous substances can therefore be compressed by a factor ≈1000 before they have a similar density to solids or liquids. Gases can therefore be compressed very strongly by increasing the pressure. *Boyle*[26] recognized this connection as early as 1664, when he carried out the experiment shown in Fig. 2.53.

He found out that the enclosed air volume V is reduced by the pressure p of the mercury column and the product of the height h *of* the mercury column and the enclosed volume V is constant. To do this, he gradually filled in more and more mercury on the right side and then found that the volume decreases in proportion to the amount of mercury. *Mariotte*[27] carried out similar experiments and in 1676 the relationship known today as the *Boyle-Mariotte Law* was formulated$T =$ const. for:

$$\frac{p_1}{p_2} = \frac{V_2}{V_1} \quad \text{or} \quad p \cdot V = \text{const.} \tag{2.76}$$

[25] Ernst Kraft Wilhelm Nußelt (1882–1957) German engineer and physicist.

[26] Sir Robert Boyle (1627–1691) British physicist and chemist.

[27] Edme Mariotte (1620–1684) French physicist.

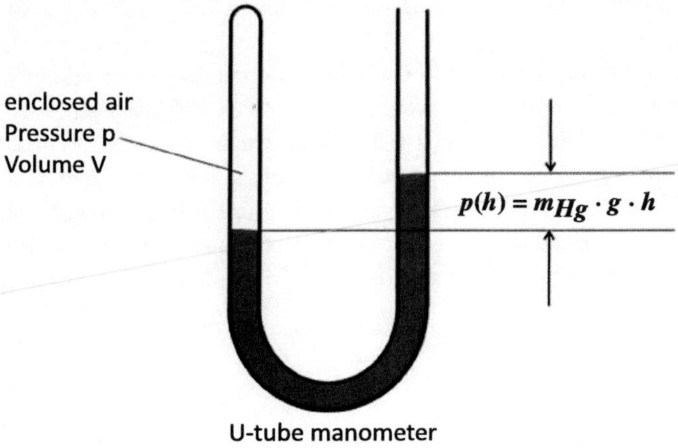

Fig. 2.53 Boyle's experiment to prove the compressibility of air

Example Calculation
A gas cylinder contains nitrogen at a pressure of $p_{\text{Fl}} = 129$ bar. The volume (capacity) of the gas cylinder is $V_{\text{Fl}} = 10$ L. How much V_U can be taken from the gas cylinder under ambient atmospheric conditions ($p_U = 985$ hPa)?

$$\frac{p_{\text{Fl}}}{p_U} = \frac{V_U}{V_{\text{Fl}}} \rightarrow V_U = \frac{p_{\text{Fl}} \cdot V_{\text{Fl}}}{p_U} \tag{2.77}$$

With the corresponding numerical values one then receives:

$$V_U = \frac{12.9\text{MPa} \cdot 10\text{L}}{0.0985\text{MPa}} = 1309.6\text{L}$$

As the cylinder content is 10 L, and this residual quantity of nitrogen then remains in the cylinder, the gas quantity removed is $V_U \approx 1300\text{L}$.

Figure 2.54 shows this relationship graphically using a concrete example.

However, this relationship only applies if the temperature T is constant. Boyle and Mariotte were unable to quantify this temperature influence because the thermometer had not yet been invented. It was not until 126 years later that *Gay-Lussac*[28] experimentally investigated and formulated this relationship. If one changes the temperature T of the gas at a constant pressure p, the volume V changes according to the following relationship (*first law of Gay-Lussac*)

[28] Joseph Louis Gay-Lussac (1778–1850) French physicist and chemist.

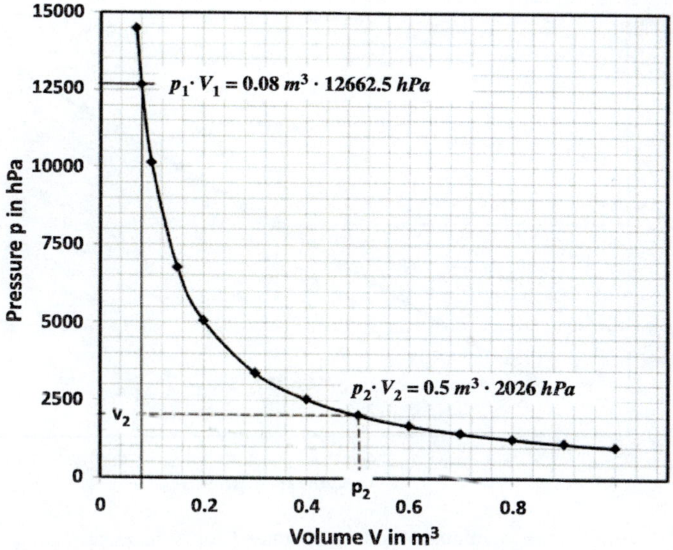

Fig. 2.54 Pressure and volume profile of an isothermal change of state with example calculation

$$\frac{V_1}{V_2} = \frac{T_1}{T_2} \quad \text{or} \quad \frac{V}{T} = \text{const.} \tag{2.78}$$

According to this relationship, the volume V thus reduces linearly with the temperature T and would have a volume of zero at 0 K (= − 273.15°C). If the external gas volume V is reduced, however, an increasing reduction in volume would have to leave as minimum volume the intrinsic volume of the atoms and molecules. This shows that this law obviously only works under certain ideal conditions and must be modified in extreme cases, such as low temperatures and high pressures (see Chapter Real gases). The intersection point at −273.15°C today is called absolute zero, where all atoms and molecules are at rest (no kinetic energy). Furthermore this point serves as zero point for the temperature scale in Kelvin (K not °K). The temperature in K is therefore also called absolute temperature. Figure 2.55 shows the *first law of Gay-Lussac* with a temperature axis in Kelvin.

The slope of this curve has the value according to Eq. 2.78:

$$\gamma = \frac{1}{273.15\text{K}} = 0.003661\,\text{K}^{-1} \tag{2.79}$$

The constant γ is the volume expansion coefficient for gases and varies slightly depending on the type of gas. Table 2.16 shows the γ values for the most important gases. The deviation from the γ value according to Eq. 2.79 characterises the non-ideal behaviour of the respective gases.

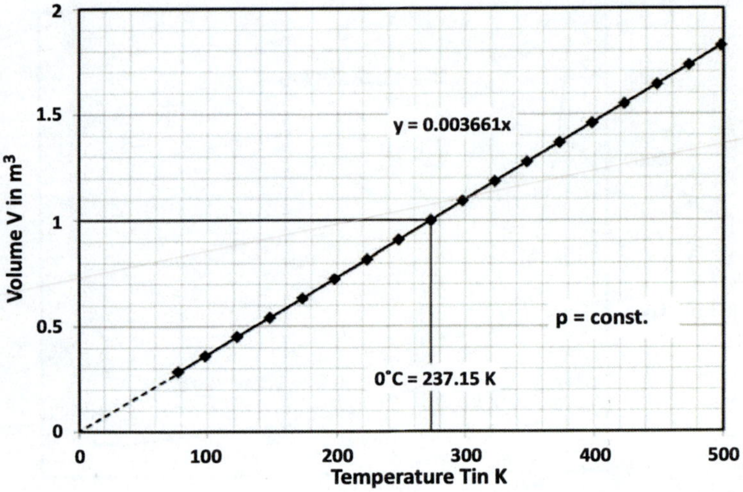

Fig. 2.55 Relationship between volume V and temperature T for an ideal gas

Table 2.16 Volume expansion coefficient γ of gaseous substances for $T = 0\ldots100^\circ C$ at $p = 1013$ hPa (Kuchling 2011)

Gas	$\gamma[10^{-5} \cdot K^{-1}]$	Gas	$\gamma[10^{-5} \cdot K^{-1}]$
Ammonia	377	Air	367
Argon	368	Methane	368
Chlorine	383	Neon	366
Hydrogen chloride	372	Oxygen	367
Ethane	375	Sulphur dioxide	385
Ethine	373	Nitrogen	367
Helium	366	Nitric oxide	368
Carbon dioxide	373	Hydrogen	366
Carbon monoxide	367	Steam	394
Krypton	369	Xenon	373

If the volume is now kept constant and the pressure increase is measured with an increase in temperature, a similar result is obtained. This connection is called the *2nd law of Gay-Lussac*:

$$\frac{p_1}{p_2} = \frac{T_1}{T_2} \quad \text{or} \quad \frac{p}{T} = \text{const.} \tag{2.80}$$

As can be seen in the concrete example in Fig. 2.56, the gradient in this case $\gamma = 0.003661/$K is also the same as in the *first law of Gay-Lussac*.

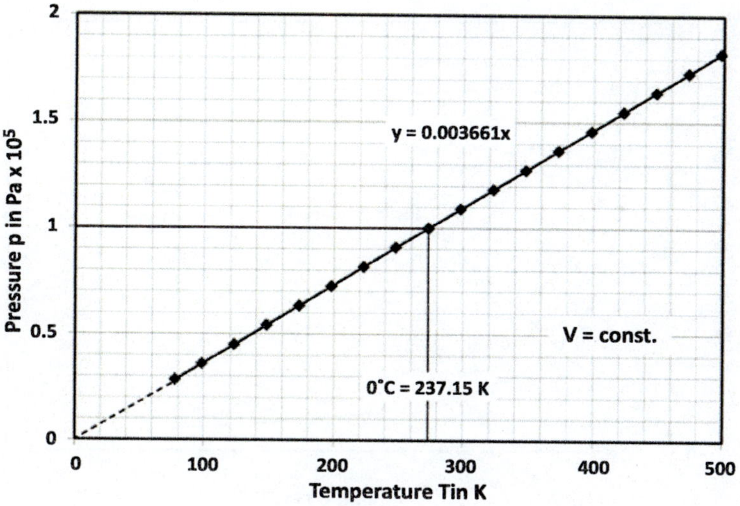

Fig. 2.56 Pressure increase in a closed volume *V* with a temperature increase according to Eq. 2.80

The gas laws of Gay-Lussac and Boyle-Mariotte can be combined to form a comprehensive equation of state for ideal gases. This relationship now states that for a given mass (quantity) of a gas, the product of pressure and volume divided by the absolute temperature is always constant. The general gas equation is therefore

$$\frac{p_1 \cdot V_1}{T_1} = \frac{p_2 \cdot V_2}{T_2} \quad \text{oder} \quad \frac{p \cdot V}{T} = \text{const.} \tag{2.81}$$

Figure 2.57 shows this relationship graphically. The pressure change $p(V)$ increases with rising temperature T according to Eq. 2.81.

In this equation of state for ideal gases, the gas laws described so far are included as special cases (Table 2.17). In this context a distinction is made between:

- Changes of state at constant pressure (= isobaric)
- Changes of state at constant volume (= isochoric)
- Changes of state at constant temperature (= isothermal).

By definition, the equation of state can also be expressed by the gas quantity m, since the following relationship applies:

$$\frac{p \cdot V}{T} \sim m \tag{2.82}$$

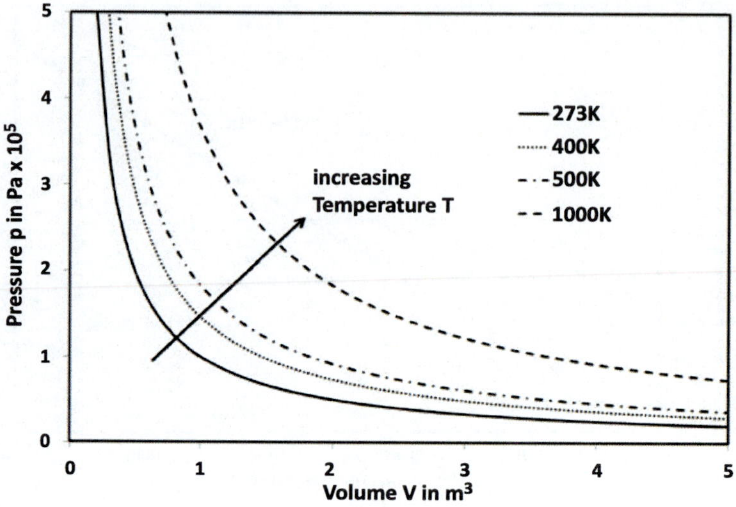

Fig. 2.57 Course of the isotherms according to the general gas equation

Table 2.17 Possible changes of state of ideal gases

Designation	Isobar	Isochor	Isotherm
Condition	$p = \text{const.}$	$V = \text{const.}$	$T = \text{const.}$
Equation	$\frac{V_1}{V_2} = \frac{T_1}{T_2}$	$\frac{p_1}{p_2} = \frac{T_1}{T_2}$	$\frac{p_1}{p_2} = \frac{V_2}{V_1}$
Act	Gay-Lussac		Boyle Marmot

By introducing a proportional factor *Ri, the* following equation of state is then obtained for ideal gases:

$$p \cdot V = m \cdot R_i \cdot T \tag{2.83}$$

In this case, the proportional factor *Ri* is the specific gas constant whose numerical value depends on the type of gas. Table 2.18 shows the different gas constants of the most important gases.

Gas Density

The density ϱ of a gas depends on both the pressure p and the temperature T of the gas, since both can affect the volume V. The term standard density was therefore introduced as a reference value. The standard density applies under normal conditions, that is, at $p_N = 1013.25$ hPa and $T_N = 273.15\text{K} = 0^\circ\text{C}$ (Table 2.19).

Important for technical applications is of course the conversion of the density under real conditions to the standard density and vice versa. With the relation for the gas density

Table 2.18 Specific gas constant for different gases

Gas	$R_i[\mathrm{J} \cdot \mathrm{K}^{-1} \cdot \mathrm{kg}^{-1}]$	Gas	$R_i[\mathrm{J} \cdot \mathrm{K}^{-1} \cdot \mathrm{kg}^{-1}]$
Ammonia	481	Methane	518
Argon	208	Methyl oxide	161
Butane	137	Neon	412
Chlorine	115	Ozone	173
Hydrogen chloride	226	Phosgene	82
Nitrous oxide	188	Propane	185
Ethane	273	Propene	194
Ethene	294	Oxygen	260
Ethine	316	Sulphur dioxide	127
Helium	2078	Hydrogen sulfide	241
Carbon dioxide	188	Nitrogen	297
Carbon monoxide	287	Nitric oxide	277
Krypton	99	Hydrogen	4127
Air	287	Xenon	63

Table 2.19 Standard density of gaseous substances

Gas	$\varrho_N[\mathrm{kg} \cdot \mathrm{m}^3]$	Gas	$\varrho_N[\mathrm{kg} \cdot \mathrm{m}^3]$
Ammonia	0.7714	Air	1.2923
Argon	1.784	Methane	0.7174
Butane	2.703	Neon	0.9002
Chlorine	3.214	Propane	2.0096
Hydrogen chloride	1.6422	Oxygen	1.42895
Dimethyl ether	2.1098	City gas	≈0.6
Helium	0.1785	Nitrogen	1.2505
Carbon dioxide	1.9769	Water vapor (100 °C)	0.768
Carbon monoxide	1.250	Hydrogen	0.08989
Krypton	3.744	Xenon	5.897

$$\varrho = \frac{m}{V} \rightarrow m = \varrho \cdot V \tag{2.84}$$

the following relationship is obtained for different gas volumes (V_1 and V_2) and the same mass m:

$$m = \varrho_1 \cdot V_1 = \varrho_2 \cdot V_2 \rightarrow \frac{V_1}{V_2} = \frac{\varrho_2}{\varrho_1} \tag{2.85}$$

The equation of state for ideal gases can then be formulated as follows:

$$\frac{p_1 \cdot \varrho_2}{T_1} = \frac{p_2 \cdot \varrho_1}{T_2} \tag{2.86}$$

The standard density is calculated by converting Eq. 2.86 and inserting the real values (T_r, p_r, ϱ_r):

$$\varrho_N = \varrho_r \cdot \left[\frac{1013,25\text{hPa}}{273.15\text{K}}\right] \cdot \left[\frac{T_r}{p_r}\right] \tag{2.87}$$

Gas Mixtures

Both the gas density and the special gas constant are gas-specific variables. In a gas mixture, therefore, a weighted average value is given, which refers to the volume or mass. For the mean gas density ϱ_M one then obtains for *n-components*:

$$\varrho_M = \frac{\varrho_1 V_1 + \varrho_2 V_2 + \ldots + \varrho_n V_n}{V_1 + V_2 + \ldots + V_n} \tag{2.88}$$

Analogous to this expression, the mean special gas constant Ri_M can then also be calculated:

$$R_{iM} = \frac{R_1 m_1 + R_2 m_2 + \ldots R_n m_n}{m_1 + m_2 + \ldots m_n} \tag{2.89}$$

Buoyancy in Gases

The laws for buoyancy in gases are the same as for buoyancy in liquids. According to the *Archimedean principle,* every body experiences an upwardly directed buoyancy force F_A, which is equal to the weight of the surrounding liquid or gas that it displaces. As the density, and thus the weight force of liquids (water) is much greater than that of gases (air), the buoyancy force in air is usually neglected. The volume of[29] an average adult human being is about 0.08 m^3. The buoyancy (F_A) experienced by a human being can be calculated as follows:

[29]The average density of a human is ($\varrho \approx 1000\text{kg/m}^3$similar to water). With a body weight of 80 kg this then corresponds to a volume of approximately 0.08 m^3.

$$F_A = V_{\text{Gas}} \cdot \varrho_{\text{Gas}} \cdot g \tag{2.90}$$

$$F_A = 0.08\text{m}^3 \cdot \frac{1.2923\text{kg}}{\text{m}^3} \cdot \frac{9.81\text{m}}{\text{s}^2}$$

$$F_A = 0.8\text{N}$$

This buoyancy corresponds to approximately 103 g in body mass. Based on the current example (80 kg), this would be only 0.80 kg% change in weight due to the buoyancy.

With large volumes (e.g. gas balloon) this lift force can increase strongly. In order to be able to calculate the load capacity of a gas balloon with the volume V_{Balloon}, one needs not only the buoyancy force F_A but also information about the load and weight of the envelope + basket (Fig. 2.58). The equation for the resulting ascent force ΔF is then calculated as follows:

$$\Delta F = \left(\varrho_{\text{Air}} \cdot V_{\text{Balloon}} - \varrho_{\text{Gas}} \cdot V_{\text{Balloon}} - m_{\text{Last}}\right) \cdot g \tag{2.91}$$

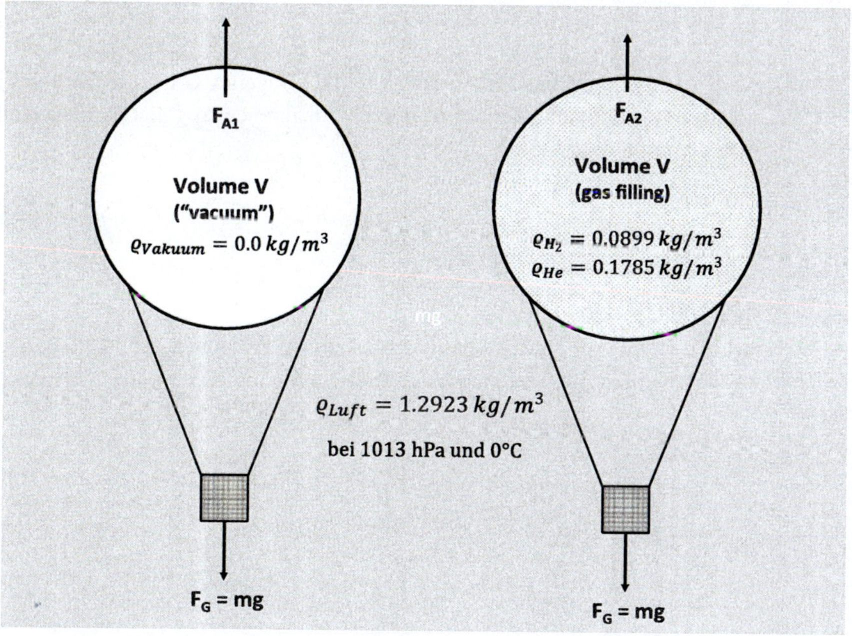

Fig. 2.58 Lift of a gas balloon in the atmosphere

Example

Let us assume that the balloon envelope including basket weighs 100 kg. The filling gas should be vacuum ($\varrho_{\mathrm{Vaccum}} = 0$), hydrogen ($\varrho_{H_2} = 0.09\mathrm{kg/m^3}$), helium ($\varrho_{\mathrm{Air}} = 0.18\mathrm{kg/m^3}$) and hot air from 100 °C $\left(\varrho_{\mathrm{Air}_{100\,^\circ\mathrm{C}}} = 0.9\mathrm{kg/m^3}\right)$. The gas balloon should have a diameter of 7 m. This results in a volume of:

$V_{\mathrm{Balloon}} = \frac{1}{6} \cdot \pi \cdot (7\mathrm{m})^3 = 179.5\mathrm{m}^3$

Using this value in Eq. 2.91, the following values are obtained for the ascending force ΔF:

$$\begin{aligned}
\Delta F(\mathrm{Vaccum}) &= 1295\mathrm{N} \\
\Delta F(\mathrm{Hydrogen}) &= 1136\mathrm{N} \\
\Delta F(\mathrm{Helium}) &= 9776\mathrm{N} \\
\Delta F(\mathrm{Hot\ air} 100\,^\circ\mathrm{C}) &= -290\mathrm{N}
\end{aligned}$$

Filling with vacuum fails for practical reasons, as the envelope would collapse. A hydrogen filling comes very close to the vacuum. With helium you only get 86% of the ascending force. A hot air balloon of this size would not take off at all. It would have to get much bigger.

Even in natural gas pipelines used for energy supply, buoyancy occurs if the pipes are not laid horizontally. This buoyancy leads to a pressure difference Δp_A that depends on the difference in density and height:

$$\Delta p_A = (\varrho_{\mathrm{Air}} - \varrho_{\mathrm{Gas}}) \cdot (h_1 - h_2) \cdot g \tag{2.92}$$

Example (Cerbe 2008, p. 137)

The vertical part of a supply pipeline for natural gas starts at 1.2 m and ends 20.6 m above the cellar floor. To calculate the pressure difference due to buoyancy, the densities are needed:

$$\varrho_{\mathrm{Gas}} = 0.783 \frac{\mathrm{kg}}{\mathrm{m}^2} \quad \text{and} \quad \varrho_{\mathrm{Air}} = 1.2929 \frac{\mathrm{kg}}{\mathrm{m}^3}$$

$$\Delta p_A = \left(1.2929 \frac{\mathrm{kg}}{\mathrm{m}^3} - 0.783 \frac{\mathrm{kg}}{\mathrm{m}^2}\right) \cdot (20.6\mathrm{m} - 1.2\mathrm{m}) \cdot 9.81 \frac{\mathrm{m}}{\mathrm{s}^2} \tag{2.93}$$

$$\Delta p_A = 97.04 \frac{\mathrm{kg}}{\mathrm{m} \cdot \mathrm{s}^2} = 97.04\mathrm{Pa} = 0.9704\mathrm{mbar} \tag{2.94}$$

Molar Sizes

Up to now, we have considered the quantity of gas either as mass m or as volume V. But behind the macroscopic mass m is hidden the mass of the individual atoms or molecules multiplied by the number N of particles that are in a certain volume V. Under normal conditions (1013.25 hPa and 273.15 K) each gas contains the same number n_0 of molecules. In one m^3 there are:

$$n_0 = 2.68678 \cdot 10^{25} \mathrm{m}^{-3} \quad (2.95)$$

This numerical value is called the *Loschmidt constant.*[30] In the International System of Units (SI), the unit mole (mol) is defined for the amount of substance n and indicates the number of particles contained in 12 g of the carbon isotope C-12. This number is called the *Avogadro constant* and is

$$\frac{N}{n} = N_A = 6.02214 \cdot 10^{23} \mathrm{mol}^{-1} \quad (2.96)$$

For certain applications it is quite helpful to relate the volume V or the mass m to the amount of substance n. This results in the following quantities:

$$\text{molar Mass } M = \frac{m}{n} \quad (2.97)$$

$$\text{molar Volume } V_m = \frac{V}{n} \quad (2.98)$$

The molar mass M is given in kg/mol or g/mol and its numerical value is identical with the relative atomic mass or molecular mass. Furthermore, the following relationship applies to the molar volume V_m under normal conditions:

$$\text{molar Normvolume } V_{mn} = 22.414 \frac{1}{\mathrm{mol}} = 22.414 \frac{\mathrm{m}^3}{\mathrm{kmol}} \quad (2.99)$$

For certain fields of application (e.g. physical chemistry and analytical technology) the equation of state is not calculated with the mass m but with the amount of substance n. With Eq. 2.83, the mass can then be expressed by $m = M \cdot n$ and the equation of state then has the following form:

[30] Josef Loschmidt (1821–1895) Austrian physicist and chemist.

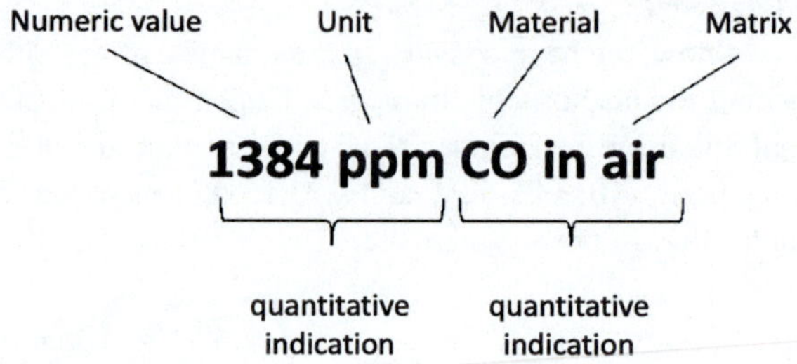

Fig. 2.59 Units of measurement in gas measurement

$$p \cdot V = n \cdot M \cdot R_i \cdot T \tag{2.100}$$

The product $M \cdot R_i$ is called universal gas constant R and is independent of the substance. With this universal gas constant R the equation of state changes as follows:

$$p \cdot V = n \cdot R \cdot T \tag{2.101}$$

The universal gas constant R can be calculated as follows, if Eq. 2.83 is related to the amount of substance 1 mol under standard conditions:

$$\begin{aligned} R = M \cdot R_i = \frac{p_n \cdot V}{T_n} = \frac{p_n \cdot V_{mn}}{T_n} &= \frac{101.325\text{Pa} \cdot 22.414 \cdot 10^{-3}\text{m}^{-3}}{273.15\text{K} \cdot \text{mol}} \\ = 8.3145 \frac{\text{J}}{\text{mol} \cdot \text{K}} \end{aligned} \tag{2.102}$$

Concentration Data

The molar size gives us an indication of how many particles are in a given volume. Under normal conditions these are approximately particles $6 \cdot 10^{23}$ in 22.4 L gas volume. In gas mixtures there are different components whose proportion or content is an important metrological parameter. In gas measurement technology, therefore, the concentration c is also important in addition to the substance being measured. A distinction is therefore made between the data in Fig. 2.59.

Various units have been developed for the quantitative information, which differ significantly depending on the application. Some examples are listed in Table 2.20.

It follows:

Table 2.20 Common concentration data in analytical technology (Wiegleb 2010)

Share	Volume information	Example	Weight data
10^{-2}	Percent [%]	Alcohol in beer	g/L, g/m^3
10^{-3}	Per mil [%]	Blood alcohol	mg/L, mg/m^3
10^{-6}	Parts per million [ppm]	Emission concentrations	µg/L,µ g/m^3
10^{-9}	Parts per billion [ppb]	Ambient air concentrations	ng/L, ng/m^3
10^{-12}	Parts per trillion [ppt]	Trace analysis and detection sensitivity of animals (insects)	pg/L, pg/m^3

$$\begin{aligned} \mathbf{1 vol.\text{-}\%} &= \mathbf{10.000 ppm} &&= \mathbf{10.000.000 ppb} \\ 0,1 vol.\text{-}\% &= 1000 ppm &&= 1.000.000 ppb \\ 0,01 vol.\text{-}\% &= 100 ppm &&= 100.000 ppb \\ 0,001 vol.\text{-}\% &= 10 ppm &&= 10.000 ppb \\ 0,0001 vol.\text{-}\% &= 1 ppm &&= 1000 ppb \end{aligned}$$

Conversion Between Volumes and Weights

This conversion is very important, because the indicating measuring instruments are usually calibrated to volume data (vol.-% or vol.-ppm = vpm[31]) and weight data are needed in different areas (e.g. power station).

Assuming that one mole of a gas under normal conditions (273 K and 1013 hPa) occupies a (mole) volume of 22.4 L, one can conclude about the volume concentration. In applications[32] where measurements are made exclusively in the ambient air at 20 °C, the molar volume changes accordingly to 24.1 L.

$$\begin{aligned} 22.4\text{L} &\cong \text{molar Mass [g]} \\ 1\text{L} &\cong \text{molar Mass [g]}/22.4\text{L} \end{aligned}$$

If this volume (1 L) is in one cubic meter of air, this corresponds to a volume concentration of 1000 ppm.

[31]The ppm and vpm values are identical in the gas analysis

[32]For example, at the MAC value determination.

$$1000\text{ppm} \cong \text{molar Mass }[\text{g}]/22.4\text{L pro m}^3$$

or

$$1\text{ppm} \cong \text{molar Mass }[\text{mg}]/22.4\text{L}$$

The following conversion formulas can be derived from this:

Conversion from mg/m³ to ppm

$$c[\text{ppm}] = \frac{\text{Molar volume}}{\text{molar Mass}} \cdot c\left[\text{mg/m}^3\right] \quad (2.103)$$

Conversion from ppm to mg/m³

$$c\left[\text{mg/m}^3\right] = \frac{\text{molar Mass}}{\text{Molar volume}} \cdot c[\text{ppm}] \quad (2.104)$$

Under real conditions, the general gas equation must be used to determine the molar volume.

$$\frac{p_r V_r}{T_r} = \frac{p_0 \cdot 22.4\text{L}}{273\text{K}} \quad (2.105)$$

with p_r= real pressure, V_r= volume, T_r= absolute temperature.

Important relative atomic masses: nitrogen (N) = 14, sulphur (S) = 32, oxygen (O) = 16, carbon (C) = 12, hydrogen (H) =1.

Example
A flue gas contains 135 ppm SO_2 and 169 mg/m³ NO_2. Concentrations in mg/m³ and ppm shall be calculated under normal conditions and under real conditions if the flue gas temperature is 66 °C at a pressure of 1179 hPa.

1. Conversion to mg/m³:

The molar mass of SO_2 is:32g/mol (S) + 2 · 16g/mol (O) = 64 g/mol.

This value is now used in Eq. 2.104 and you get.

$$c\left[\mathrm{mg/m^3}\right] = \frac{64\mathrm{g/mol}}{22.4\mathrm{L/mol}} \cdot 135 \cdot 10^{-6} = 385.7\mathrm{mg/m^3}$$

2. Conversion to ppm:

The molar mass of NO_2 is:14g/mol (N) + 2 · 16g/mol (O) = 46 g/mol.
This value is now used in Eq. 2.103 and you get.

$$c[\mathrm{ppm}] = \frac{22.4\mathrm{L/mol}}{46\mathrm{g/mol}} \cdot 169 \cdot 10^{-6} = 82.3\mathrm{ppm}$$

3. Under normal conditions the molar volume changes as follows:

$$V_r = \frac{1013\mathrm{hPa} \cdot 22.4\mathrm{L} \cdot 339\mathrm{K}}{1179\mathrm{hPa} \cdot 273\mathrm{K}} = 23.9\mathrm{L}$$

The SO_2 concentration in mg/m^3 therefore changes as follows:

$$c\left[\mathrm{mg/m^3}\right] = \frac{64\mathrm{g/mol}}{23.9\mathrm{L/mol}} \cdot 135 \cdot 10^{-6} = 361.5\mathrm{mg/m^3}$$

The following then applies to the NO_2 concentration:

$$c\left[\mathrm{mg/m^3}\right] = \frac{22.4}{23.9} \cdot 169 = 158.4\mathrm{mg/m^3}$$

The concentrations in ppm do not change under normal conditions.

In the natural gas industry, conversions of concentration data for gas mixtures are of great importance, as the pressure ranges can vary from 1 bar to 1000 bar. The real gas behaviour must therefore be considered for these applications. These relationships are described in detail in DIN 1492.

Heat Capacity

The state of a gaseous substance can be clearly defined by the pressure p, the volume V and the temperature T. The equation of state for ideal gases (Eq. 2.81) formally describes this relationship. Of particular interest are above all the processes associated with an exchange

Table 2.21 Specific heat capacities of gases at 20 °C

Gas	c_p	c_V	κ	Gas	c_p	c_V	κ
Ammonia	2.160	1.655	1.305	Carbon monoxide	1.042	0.744	1.40
Argon	0.523	0.317	1.648	Air	1.005	0.717	1.402
Hydrogen bromide	0.360	0.254	1.42	Methane	2.219	1.696	1.308
Chlorine	0.745	0.552	1.35	Neon	1.030	0.628	1.64
Chorethane	1.151	0.967	1.19	Ozone	0.795	0.568	1.40
Chloromethane	0.762	0.593	1.285	Propane	1.595	1.412	1.13
Hydrogen chloride	0.803	0.578	1.39	Oxygen	0.917	0.656	1.398
Ethane	1.729	1.455	1.188	Sulphur dioxide	0.640	0.504	1.27
Ethene	1.549	1.249	1.24	Hydrogen sulfide	1.047	0.799	1.31
Ethine	1.683	1.368	1.23	Nitrogen	1.038	0.741	1.401
Generator gas	1.05	0.75	1.40	Nitric oxide	0.883	0.690	1.28
Helium	5.23	3.21	1.63	Nitrogen dioxide	0.996	0.717	1.39
Hydrogen iodide	0.226	0.161	1.40	Hydrogen	14.32	10.17	1.41
Carbon dioxide	0.837	0.647	1.293	Xenon	0.159	0.095	1.67

of energy. In principle, it applies that in all processes no more energy can ever be produced than is supplied.[33]

The total energy (internal energy U) in a gas (system) is composed of the thermal energy Q supplied and the work W performed on the system:

$$U = Q + W \tag{2.106}$$

This internal energy can also be described as the sum of the kinetic energy E_{kin} over all molecules (N) in the system:

$$U = N \cdot \left[\frac{m_M}{2}\overline{v^2}\right] = \frac{3}{2} \cdot N \cdot k \cdot T \tag{2.107}$$

However, this equation cannot be used for practical applications, so it is better to switch to the following practical form:

$$U = c_V \cdot m \cdot T \tag{2.108}$$

The factor c_V is the specific heat capacity of the gas at a constant volume (Table 2.21). If a certain amount of heat Q *is* supplied to the system, only pressure and temperature change if

[33] Impossibility of a perpetuum mobile (first species).

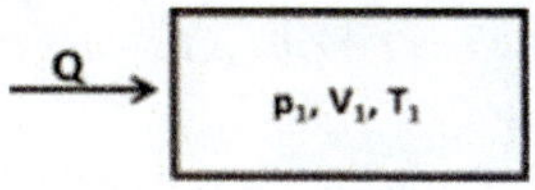

Fig. 2.60 Closed system, with supply of thermal energy Q

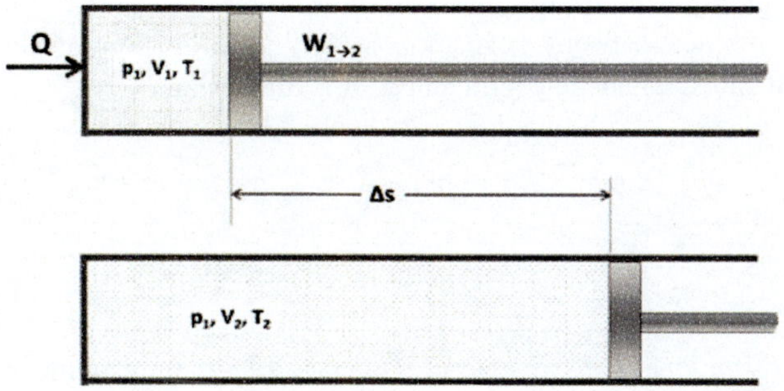

Fig. 2.61 Expansion of a gas volume in a cylinder

the system cannot change its volume (Fig. 2.60). The following applies to the amount of heat supplied

$$Q = c_V \cdot m \cdot \Delta T \tag{2.109}$$

If heat energy is supplied from outside and the pressure p *is* to remain constant, the volume V must change. The work W transferred by such a system can be explained clearly by an expansion of the gas volume in a cylinder (Fig. 2.61).

The pressure p in the piston presses on the cylinder, which travels a distance due to the expansion Δs. For the work performed, the following applies $W = F \cdot \Delta s$ with $F = p \cdot A$. By inserting the compressed volume $\Delta V = A \cdot \Delta s$, the expression for the work[34] done is obtained:

$$W = -p \cdot \Delta V \tag{2.110}$$

For the heat quantity is now valid with (Eq. 2.106):

$$Q = U - W = Q = c_p \cdot m \cdot \Delta T = c_V \cdot m \cdot \Delta T + p \cdot \Delta V \tag{2.111}$$

c_p in this case is the specific heat capacity for constant pressure conditions.

With $p \cdot \Delta V = m \cdot R_i \cdot \Delta T$ we get then:

[34]By convention, the gas (system) gives off work on expansion ($\Delta V > 0$), which is counted negatively, so W must be counted negatively.

$$c_p \cdot m \cdot \Delta T = c_V \cdot m \cdot \Delta T + m \cdot R_i \cdot \Delta T \tag{2.112}$$

Dividing this equation by $m \cdot \Delta T$ results in

$$c_p = c_V + R_i \quad \text{or} \quad c_p - c_V = R_i \tag{2.113}$$

The unit of specific heat capacity is $\mathrm{kJ \cdot kg^{-1} \cdot K^{-1}}$. The quotient of both specific heat capacities is also called adiabatic coefficient $\varkappa$. It is dimensionless.

$$\varkappa = \frac{c_p}{c_V} \tag{2.114}$$

Real Gases

Under extreme conditions, such as high pressures and low temperatures, the equation of state for ideal gases finds its limits, since the molecular volume and the intermolecular forces can no longer be neglected. *Van der Waals*[35] therefore extended this equation of state in 1873 with corresponding coefficients (*a* and *b*) to better describe the real gas behaviour. The equation developed from this is known as the *van der Waals equation* and has the following expression

$$\left(p + \frac{a}{V^2}\right) \cdot (V - b) = R \cdot T \tag{2.115}$$

The coefficient *a* characterizes the forces of attraction (cohesion) between the molecules when they come closer and closer together as a result of increasing pressure (high density). In addition to the pressure *p*, this results in a further pressure component actinga/V^2 in the interior of the gas. This part is therefore also called internal pressure or internal pressure. The constant *b* describes the so-called *covolume of* the molecules. This means the minimum volume of the molecules that they can take up. It is four times the molecules' own volume (Richter 2010).

Figure 2.62 shows the behaviour of carbon dioxide as an ideal gas and as a real gas. The van der Waals equation was changed to *p*, so that the pressure *p* can be represented as a function of the volume *V* in a *pV diagram.*

$$p_1(V) = \frac{R \cdot T}{V - b} - \frac{a}{V^2} \tag{2.116}$$

[35] Johannes Diderik van der Waals (1837–1923) Dutch physicist.

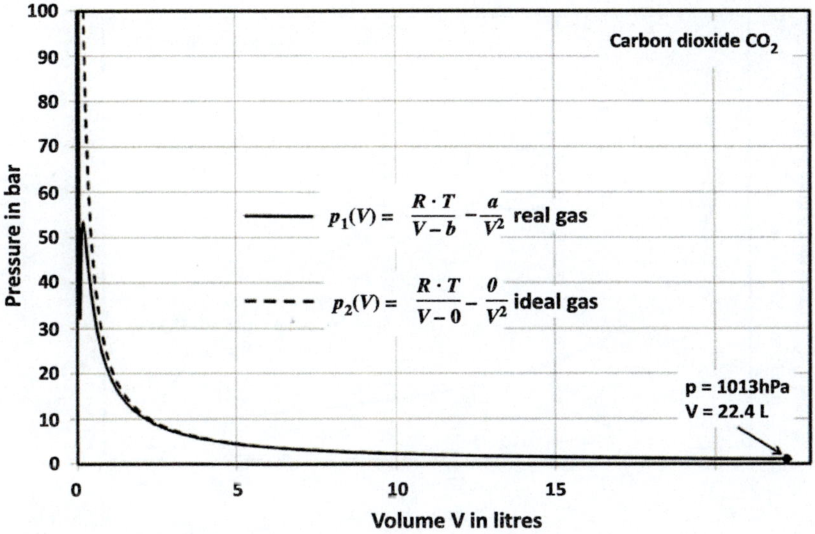

Fig. 2.62 Course of the isotherm according to the *van der Waals equation* Eq. 2.116. The *dashed course* shows the ideal gas. In this representation (scaling) hardly any differences between both courses are to be recognized

With $a = 0$ and $b = 0$ you get the expression for the ideal behaviour:

$$p_2(V) = \frac{R \cdot T}{V} \tag{2.117}$$

Below 1 MPa (10 bar) there are only minor differences between the two curves. Only above 10 MPa larger differences can be seen, which then describe the real gas behaviour (Fig. 2.63).

Especially the pressure range above 20 bar is of special interest for the explanation of the real gas behaviour. Figure 2.64 shows the real gas behavior of carbon dioxide at three different temperatures. If the gas is compressed at a temperature of $T < T_K$, the pressure increases up to a certain point (B). At this point the condensation process starts and the gas is liquefied. At a temperature of 0 °C and a pressure of 47 bar the liquefaction of carbon dioxide starts. The pressure remains constant until the liquefaction process at point A is completed. The transition from B to A is also known as the Maxwell Straight. Since carbon dioxide is only present in liquid form from point A onwards, the pressure p increases extremely during further compression, since liquids are almost incompressible. At higher temperatures, points A and B move closer and closer together until they fuse (degenerate) at the critical temperature. The pressure present here is called critical pressure p_K and the volume is called critical volume V_K. Mathematically, the critical point on the tangent of the van der Waals equation is $T = T_k$. The first two derivations are therefore valid:

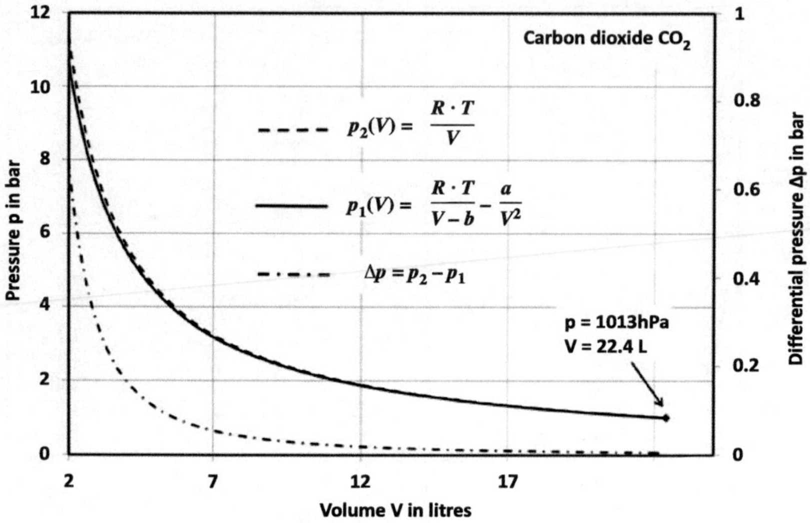

Fig. 2.63 Course of the isotherms for a print from $p < 12$ bar. The difference Δp between the two curves increases with the pressure and then leads to considerable deviations

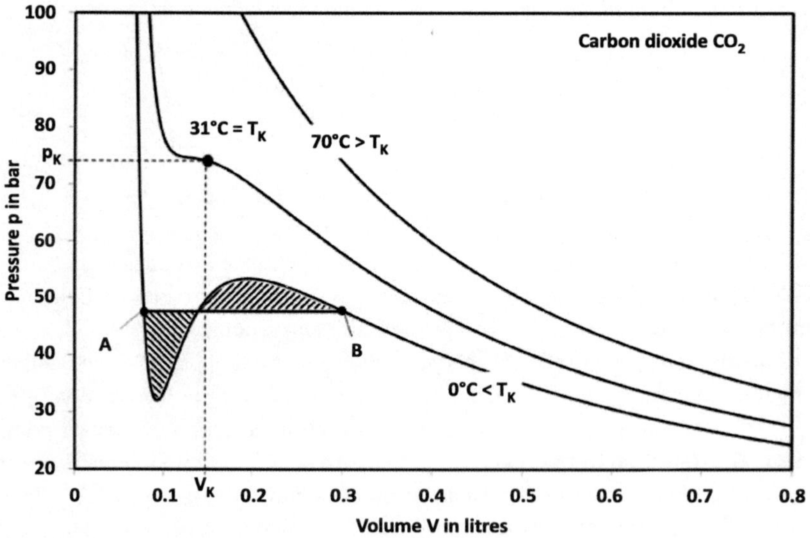

Fig. 2.64 Course of the isotherms at the critical temperature and above and below

Table 2.22 Van der Waals coefficients of some gases (Jessel 2001)

	Constant a (internal pressure)		Constant b (molecular volume)	
Gas	$N \cdot m^4/mol^2$	$100kPa \cdot L^2/mol^2$	m^3/mol	L/mol
Hydrogen	0.019	0.19	$23 \cdot 10^{-6}$	0.023
Nitrogen	0.132	1.32	$40 \cdot 10^{-6}$	0.040
Oxygen	0.137	1.37	$30 \cdot 10^{-6}$	0.030
Methane	0.229	2.29	$43 \cdot 10^{-6}$	0.043
Carbon dioxide	0.365	3.65	$43 \cdot 10^{-6}$	0.043
Ammonia	0.423	4.23	$37 \cdot 10^{-6}$	0.037
Steam	0.554	5.54	$30 \cdot 10^{-6}$	0.030
Chlorine	0.659	6.59	$56 \cdot 10^{-6}$	0.056
Sulphur dioxide	0.682	6.83	$56 \cdot 10^{-6}$	0.056

$$\left(\frac{\delta p}{\delta V}\right)_{T_K} = 0 \quad \text{and} \quad \left(\frac{\delta^2 p}{\delta V^2}\right)_{T_K} = 0 \tag{2.118}$$

It follows from this then:

$$a = 3 \cdot p_K \cdot V_K \quad \text{and} \quad b = \frac{V_K}{3} \tag{2.119}$$

Above the temperature T_K, liquefaction of the gases is no longer possible, regardless of the pressure.

The *van der Waals coefficients* (Table 2.22) are determined experimentally from the data of the critical point. For this purpose, the expressions for a and b given in Eq. 2.119 are used. The values for the critical pressure p_K and the critical volume V_K are determined experimentally with a corresponding arrangement. Figure 2.65 shows an apparatus with which these experiments are carried out (Richter 2010).

Virial Equation

In technical applications, a different type of correction is often used. In this case, the equation of state for ideal gases is modified with a *real gas factor Z*.

$$p \cdot V_r = Z \cdot R_i \cdot T \qquad \textit{for } 1\text{kg} \tag{2.120}$$

$$p \cdot V_{mr} = Z \cdot R \cdot T \qquad \text{for } 1\text{kmol} \tag{2.121}$$

The index r stands for *real* and the index m for *molar*. The real gas factor is of course for ideal gases $=1$. For real gases it is described by so-called *virial coefficients:*

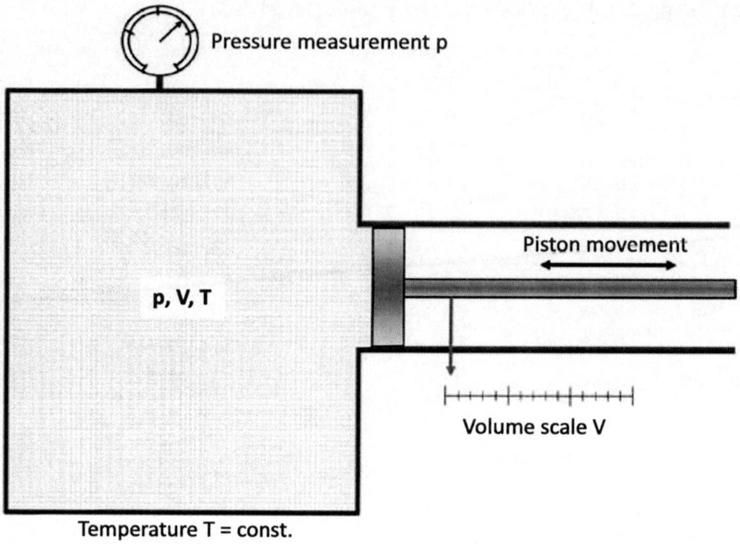

Fig. 2.65 Experimental set-up for the determination of the *van der Waals constants a* and *b*

$$Z = 1 + \frac{B(T)}{V} + \frac{C(T)}{V^2} + \frac{D(T)}{V^3} + \ldots \tag{2.122}$$

The real gas factor Z is dependent on the physical conditions, so that a reference to the standard state is useful:

$$p_n \cdot V_{nr} = Z_n \cdot R_i \cdot T_n \quad \text{for 1 kg} \tag{2.123}$$

$$p_n \cdot V_{mnr} = Z_n \cdot R \cdot T_n \quad \text{for 1 kmol} \tag{2.124}$$

The ratio between the *real gas factor* Z for any given state and that for the standard state is called the *compressibility number K*.

$$K = \frac{Z}{Z_n} = \frac{p \cdot V_r \cdot T_n}{p_n \cdot V_{nr} \cdot T} = \frac{p \cdot \varrho_{nr} \cdot T_n}{p_n \cdot \varrho_r \cdot T} \tag{2.125}$$

For gas density under real conditions then applies:

$$\varrho_r = \varrho_{nr} \cdot \left[\frac{p \cdot T_n}{p_n \cdot T}\right] \cdot \frac{1}{K} \tag{2.126}$$

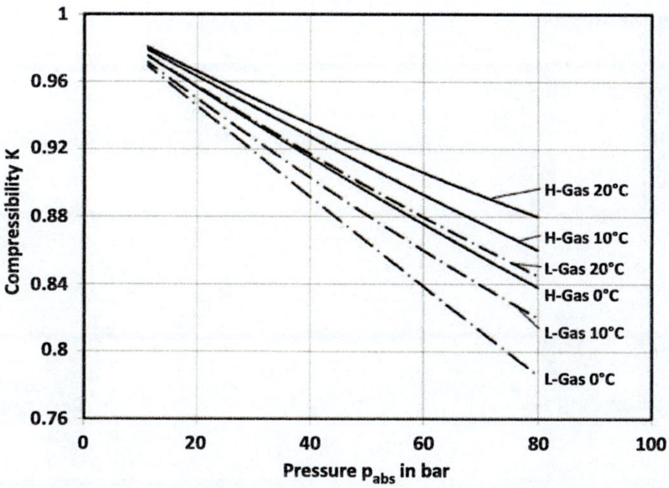

Fig. 2.66 Compressibility coefficient K for the natural gas type H (H gas: vol. %>90 methane) and L (L gas: vol. %≈82 methane, high N_2 content (>10 vol. %)) calculated according to the GERG 88 calculation method (Kunz et al. 2007)

With gas mixtures, the real gas behaviour can only be described by a complex analysis of the individual gas components. For natural gas there is a standardised method (AGA8 method), which is described in detail in DIN ISO 12212 (2005).

As an approximation, the following value applies to natural gas at 12 °C and 70 bar (Cerbe 2008):

$$K \approx 1 - \frac{p_{abs}}{450 \text{bar}} \tag{2.127}$$

And for[36] coke oven gas:

$$K \approx 1 + \frac{p_{abs}}{6200 \text{ bar}} \tag{2.128}$$

Figure 2.66 shows the dependence of the compressibility factor K on the pressure p at different temperatures T. With increasing pressure p, K decreases and behaves according to Eq. 2.127. However, in extremely high pressure ranges, above 150 bar, the compressibility coefficient K then increases again (Schley 2011).

[36]For example: 55 vol.% H_2, 5 vol.% CO, 25 vol.% CH_4, 2 vol.% C_nH_m, 2 vol.% CO_2, 10 vol.% N_2, 1 vol.% O_2.

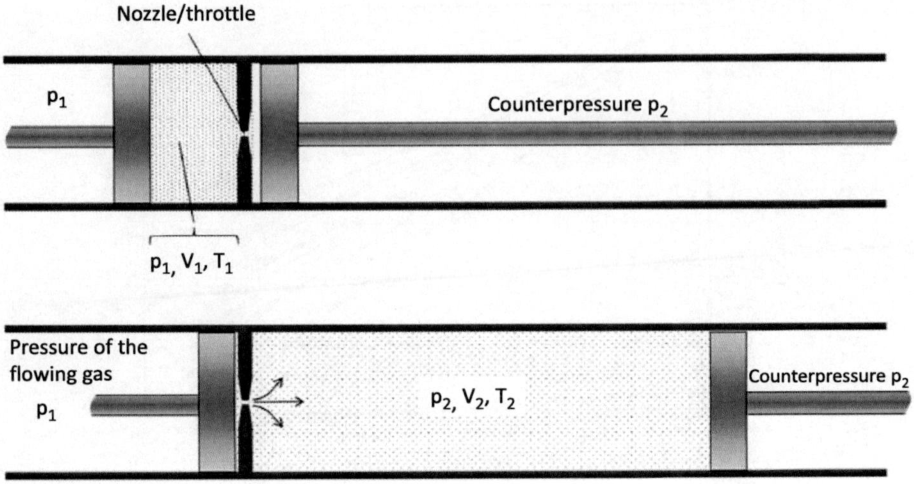

Fig. 2.67 Experiment to prove the Joule-Thomson effect

Joule Thomson Effect

If a gas under high pressure p_1 is expanded through a nozzle to pressure p_2, a change in temperature ΔT can be detected. Joule[37] and Thomson[38] carried out an experiment on this, which is shown in Fig. 2.67. The compressed gas with the pressure p_1, the volume V_1 and the temperature T_1 flows through the nozzle into the second chamber where the gas expands. For the volume $V_2 > V_1$ and for the pressure $p_1 > p_2$. The following relationship applies to ideal gases:

$$V_1 \cdot p_1 = V_2 \cdot p_2 = \text{const.} \tag{2.129}$$

For the temperature T is:

$$T_1 = T_2 \quad \text{or} \quad \Delta T = 0 \tag{2.130}$$

Real gases behave differently, and so for many gases there is a measurable temperature difference afterΔT the balancing process. This behaviour is called *Joule-Thomson effect* and can be explained by the *van der Waals equation* (Meschede 2010).

[37] James Prescott Joule (1818–1889) British physicist.

[38] William Thomson (later Lord Kelvin) (1824–1907) British physicist.

The temperature change ΔT is proportional to the pressure difference $\Delta p = p_2 - p_1$ and depends on a substance-specific constant μ (*Joule-Thomson coefficient*).

$$\mu = \frac{\Delta T}{\Delta p} \tag{2.131}$$

The numerical values for μ depend on the gas type, temperature T_1 and pressure p_1. Under standard conditions, the literature values are enclosed:

Oxygen, nitrogen	$\mu \approx +2.5\text{K/MPa}$
Carbon dioxide	$\mu \approx +7.5\text{K/MPa}$
Hydrogen	$\mu \approx -33.3\text{K/MPa}$

Cooling only occurs when there is $\mu > 0$. Under these conditions we can observe a significant warming of hydrogen, which can lead to an ignition or explosion if this gas expands uncontrolled.

Furthermore, cooling only occurs if the initial temperature T_1 is lower than the inversion temperature T_i. If this condition is not met, relaxation leads to heating.

The inversion temperature T_i can be calculated from the parameters *a* and *b* of *van der Waal's equation of state*

$$T_i \approx \frac{2 \cdot a}{R \cdot b} = \frac{27}{4} \cdot T_K \quad \text{or} \quad T_i \approx 6.75 \cdot T_K \tag{2.132}$$

T_K is the critical temperature of the respective gas and is 126.2 K for oxygen. The inversion temperature is then approximately 600 °C.

A practical application in gas measurement technology is the expansion above a pressure reducer. The cylinder pressure can be up to 200 bar. If a volume flow d*V*/d*t* is now released to the environment (atmospheric pressure=p_2) via a pressure reducer, the temperature in the fitting drops by ΔT. If this volume flows permanently, the pressure reducer will cool down further and further until finally, due to the humidity, icing occurs. At small volume flows $\dot{V}_n$, the heat output d*Q*/d*t* is transferred from the environment without icing up. At high volume flows this is no longer sufficient and the gas must be preheated additionally.

The required gas preheating is given by the following equation (Cerbe 2008):

$$\dot{Q}_E = \dot{V}_n \cdot \varrho_n \cdot c_p \cdot [(p_1 - p_2) \cdot \mu + (T_2 - T_1)] \tag{2.133}$$

Example

From a CO_2 gas cylinder ($p_1 = 100$ bar) 10 L gas/minute shall be delivered to an ambient pressure of$p_2 = 1$ bar. The gas cylinder has a temperature of $T_1 = 5°$ C and should be

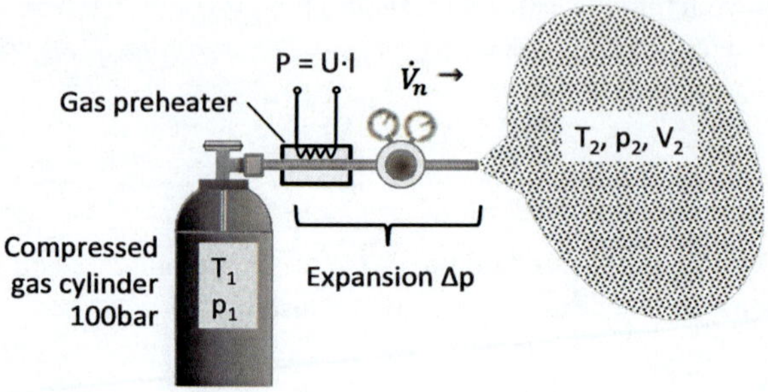

Fig. 2.68 Extraction of gases from compressed gas cylinders with the necessary preheating to prevent icing at the valves and at the pressure reducer

adjusted to a temperature T_2 (room temperature 20 °C). What heating power is required for this?

With Eq. 2133 this results in

$$\dot{Q}_E = 1.67 \cdot 10^{-4} \frac{\text{m}^3}{\text{s}} \cdot 1.976 \frac{\text{kg}}{\text{m}^3} \cdot 0.837 \frac{\text{kJ}}{\text{kg} \cdot \text{K}} \cdot \left[(9.9\text{MPa}) \cdot 7.5 \frac{\text{K}}{\text{MPa}} + 15\text{K} \right] \quad (2.134)$$

$$\dot{Q}_E = 24.65 \frac{\text{J}}{\text{s}} \approx 25\text{W}$$

In the case of large plants for the transmission of natural gas, for example, 100.000 m^3/h is reduced from 60 bar to 5 bar. The required preheating capacity is then already in the range of 1.715 MW.

Linde Process for Air Liquefaction

Based on the *Joule-Thomson effect, Linde*[39] succeeded in 1895 in developing a refrigerating machine with which he was able to generate very low temperatures. This machine generated a pressure of 200 bar in a multi-stage compression. At an expansion valve the air was then expanded to a pressure of 20 bar. The *Joule-Thomson coefficient* is the main component of air (O_2 + N_2) $\mu \approx$ + 2.5K/MPa. The pressure difference Δp of 180 bar thus leads to a temperature difference ΔT = 45K. However, this temperature difference was not yet sufficient to liquefy the air (Fig. 2.68). With a simple but ingenious idea, Linde solved the problem by using the cooled gas to cool the compressed gas (Fig. 2.69). As a result, the expanding gas already had a lower temperature, so it cooled down further and further. Linde used a countercurrent cooler and pumped the air in a circuit

[39]Carl Paul Gottfried Linde, Knight of Linde (1842–1934) since 1897, was a German engineer, inventor and founder of Linde AG.

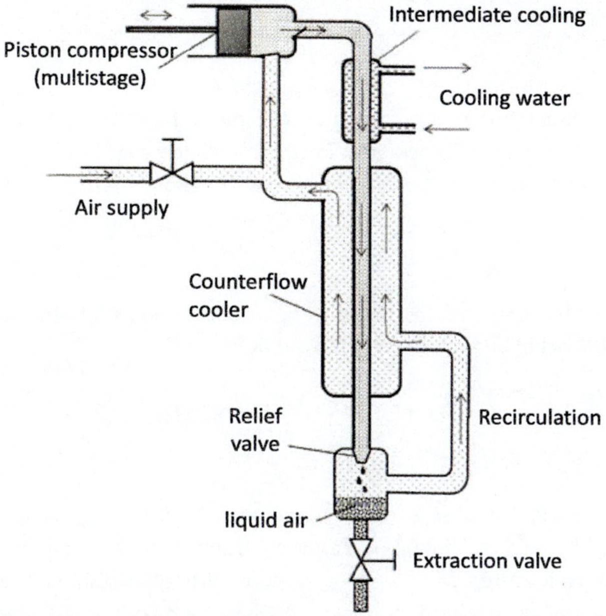

Fig. 2.69 Apparatus for air liquefaction according to Linde

until it liquefied at a temperature of about −191°C. The air was then cooled down to a temperature of about C.

The main components of liquid air are nitrogen, oxygen and argon. These three components are carried out by countercurrent distillation in a rectification column (Veranneman 2000). Since the boiling points of the gases (oxygen −183°C, argon - 186°C, nitrogen −196°C) are very close to each other, this distillation must be carried out in several stages.

2.6 Combustion Processes in Gases

Combustion is a chemical reaction (oxidation) in which a fuel (gas, liquid or solid) reacts with oxygen in a flame, thereby releasing thermal energy (Q). This energy is also known as chemically bound energy.

Calorific Value

The amount of heat Q released depends on the mass m *of* the fuel and the specific calorific value H_i.

$$Q = m \cdot H_i \tag{2.135}$$

To make this combustion process as optimal as possible, it is necessary to mix the fuel and the combustion air homogeneously. Furthermore, the mass ratios between the two materials must correspond to the stoichiometric[40] ratio resulting from the chemical reaction.

Example

If, for example, hard coal (assumed to be 100% carbon C) burns with oxygen, we have the following chemical reaction equation:

$$C + O_2 \rightarrow CO_2 + \text{Energy } (Q) \tag{2.136}$$

With the relative atomic masses is then obtained: 12g (C) + 32g (O_2) = 44g (CO_2).

Accordingly, 12 g fuel (hard coal) is produced from an optimal combustion 44 g CO_2.

As hard coal is a solid, *good mixing* is very difficult to achieve and incomplete combustion is inevitable, which then releases carbon monoxide (CO). Furthermore, hard coal, and brown coal in particular, contains sulphur compounds that burn in the flame to form sulphur dioxide (SO_2). This sulphur dioxide has to be removed from the flue gas of large power plants by means of complex process engineering processes. The *waste product of* these desulphurisation plants is gypsum, which is used in the construction industry. A further problem with solid fuels is the ash (incombustible solid matter), which is present as dust both in the combustion chamber and in the exhaust gas. While the ash in the combustion chamber can be removed relatively easily, the dust must be removed from the exhaust gas with electrostatic filter systems. Gaseous fuels (e.g. Natural gas) do not have these disadvantages and are therefore considered environmentally friendly.

Natural gas is offered in two different qualities, which differ in their composition and thus in their calorific value (Table 2.23). Natural gas of group H[41] usually comes from the CIS states and from the North Sea of the natural gas fields of Norway, the Netherlands and Denmark. Group H natural gas has a methane content of between 87 and 99.1% by volume. H-gas is characterized by a low nitrogen (N_2) and carbon dioxide (CO_2) content. The calorific value is usually between 11.0 and 12.4 kWh/m^3. Group L[42] natural gas generally has a methane content of between 79.8 and 87% by volume. The nitrogen (N_2) and carbon dioxide (CO_2) content of L-gas is slightly higher than that of H-gas. The calorific value usually varies between 9.2 and 10.5 kWh/m^3.

[40] Stoichiometry =Chemical Calculation.

[41] H for high = high calorific value.

[42] L for low =low calorific value.

Table 2.23 Typical compositions (reference values) of some typical fuel gases (Cerbe 2008)

	Gas composition in % by volume									Calorific value
Fuel gas	H_2	CO	CH_4	C_2H_6	C_3H_8	C_4H_{10}	CO_2	N_2	O_2	
Blast furnace gas	4.1	21.4	–	–	–	–	22.0	52.5	–	
Coke oven gas	54.5	5.5	25.3	–	2.3[a]	–	2.3	9.6	0.5	
Natural gas L	–	–	81.8	2.8	0.4	0.2	0.8	14.0	–	
Natural gas H	–	–	93.0	3.0	1.3	0.6	1.0	1.1	–	

[a] Sum of the residual hydrocarbons.

Another advantage of natural gas is that the combustion energy released is much greater in relation to CO_2 emissions than with hard coal or lignite. This advantage can be determined quantitatively with a simple calculation:

Example: Natural Gas Versus Hard Coal
The specifications are the calorific value of hard coal $H_i \approx$ 30MJ/kg and the calorific value of natural gas $H_i \approx$ 50 MJ/kg, which is only calculated as methane (for example, for gas from Russia). The stoichiometric equations are as follows:

$$16\text{g } (CH_4) + 64\text{g } (O_2) = 44\text{g } (CO_2) + 36\text{g } (H_2O) + 0.6\text{MJ} \tag{2.137}$$

$$12\text{g } (C) + 32\text{g } (O_2) = 44\text{g } (CO_2) + 0.36\text{MJ} \tag{2.138}$$

So when natural gas (methane) is burned, an amount of energy equal to 13.64 kJ is released per gram of CO_2. When coal is burned, the figure is only 8.2 kJ. For the same amount of heat, the combustion of natural gas (methane) 40 results in CO_2 emissions that are reduced by 50%.

As efforts are being made worldwide to reduce CO_2 emissions from combustion processes, the use of gaseous fuels such as natural gas would be very helpful. Unfortunately, this enormous technical advantage has so far not been taken into account in the German Energiewende.

The exhaust emission values of vehicles are also of particular interest in the automotive industry. In addition to optimum combustion, which leads to better performance with lower fuel consumption, aspects of air quality (environmental protection) naturally also play a major role. Motor vehicle exhaust gases are usually analysed for components such as carbon dioxide CO_2, carbon monoxide CO, hydrocarbons HC, nitrogen oxides NO_x and particulate matter. This exhaust gas measurement technology is now subject to EU-wide standardisation and must be complied with by all EU states. Only vehicles that meet this standard may be put into circulation.

The emission of CO_2 has become the measure of all things when assessing *environmental friendliness* today. On closer inspection, however, this CO_2 value depends exclusively on the known fuel consumption. The following example shows how fuel consumption (e.g. L/100 km) is converted into the emission value *g CO_2 per 1 km*.

Example: CO_2 Emission per km

A vehicle runs on H natural gas (CNG[43]) and has a consumption of 4.9 kg/100 km. Assuming that more than 90% of the natural gas is methane, the following combustion process can be assumed:

$CH_4 + 2O_2 \rightarrow CO_2 + 2H_2O + \text{Energy}$

With the molar masses for the individual substances, the following equation is obtained for the masses:

$$16\text{g}\ (CH_4) + 64\text{g}\ (O_2) = 44\text{g}\ (CO_2) + 36\text{g}\ (H_2O)$$

With a consumption of 4.9 kg per 100 km, this is 49 g per km. The CO_2 emissions are then:

$$CO_2\text{-Output per km} = \frac{\frac{49\text{g CH}_4}{\text{km}} \cdot 44\text{g CO}_2}{16\text{g CH}_4} = 134.75\text{g CO}_2/\text{km}$$

This value corresponds very well with the data (133 g) from the data sheet of the vehicle.

Other fuels (petrol, diesel, LPG[44]) require the average molar masses of these mixtures. In the European Union, gasoline (petrol) consists of the following groups of substances:

- *35 vol.% Aromatics*
- *18 vol.% Alkenes*
- *1 vol.% Benzene*
- *50 ppm Sulphur*
- *46 vol.% Alkanes*,

the average molar mass is approx. 100 g/mol.

Combustion Plants

The combustion of gases usually takes place in a combustion chamber, which also contains a heat exchanger that transfers the heat of combustion Q for further use (Q_N). In private combustion plants (Fig. 2.70), this is usually the service water heating (bathing, showers) and the heating heat (central heating, floor heating). In power plants, this heat is used to generate steam to drive the turbines and thus the generators for conversion into electrical

[43]CNG = Compressed Natural Gas.

[44]LPG =Liquid Petrol Gas (colloquial language LPG).

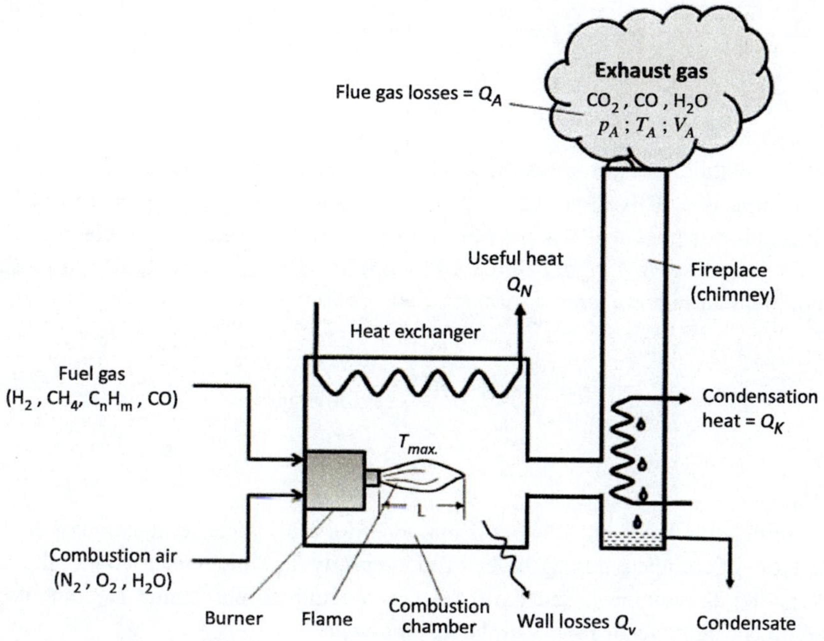

Fig. 2.70 Combustion plant with gaseous fuels and downstream condensation to use the heat of condensation *QK* of the water vapour

energy. In some cases, the residual heat is extracted after the turbines for use as district heating. In the winter months, this provides inexpensive heating energy from buildings located near the power plant (district heating network). This dual use is also known as combined heat and power (CHP). Power plants that use this advantage achieve efficiencies of 80%.

Oxygen Demand

For optimum combustion, exactly as much oxygen O_2 is required as has already been calculated in Eq. 2.137. Accordingly, two moles of O_2 are required per mole of CH_4. Since the volume per mol is the same for all gaseous substances[45] at the same pressure and temperature, exactly 2 m^3 O_2 is required for the combustion of 1 m^3 CH_4. The following reaction equation can be given for any hydrocarbons (C_nH_m):

[45] 22.4 L under normal conditions.

$$C_nH_m + \left(n + \frac{m}{4}\right) \cdot O_2 \rightarrow n \cdot CO_2 + \frac{m}{2} \cdot H_2O \tag{2.139}$$

Air Demand

With this calculation basis, it is now possible to determine the necessary air content for optimum combustion. This air volume is also known as the minimum air requirement L_{min}. This calculation assumes that the combustion air and the fuel gas do not contain water vapour (dry gas). Since in 1 m^3 dry air 0.2095 m^3 $\approx$0.21m^3 oxygen contained in this results in the minimum air requirement:

$$L_{\text{min}} = \frac{O_{\text{min}}}{0.21} \quad \text{in} \quad \frac{\text{m}^3\ \text{Air}}{\text{m}^3\ \text{Flow gas}} \tag{2.140}$$

Excess Air

A good mixing with the combustion air is important for a complete combustion of the fuel gas. Therefore, if complete mixing is achieved in exactly the stoichiometric ratio, the ratio of the supplied air volume L and the theoretically required air volume L_{min} should be 1. This ratio is called the air ratio λ or the *lambda value*:

$$\lambda = \frac{L}{L_{\text{min}}} \tag{2.141}$$

In practical cases, however, the supplied combustion air quantity L is not identical with the required combustion air quantity. Due to this imbalance we have the air ratio $\lambda \neq 1$.

The air ratio λ has a great influence on the combustion. If, for example, less oxygen than required is made available for combustion, incomplete combustion will inevitably occur. In particular, the carbon monoxide (CO) content and the proportion of hydrocarbons burned (HC) then increases sharply. Furthermore, additional soot formation is to be expected in this area. Figure 2.71 shows these ratios, as a function of the λ value. Since the efficiency η drops steeply $\lambda < 1$ in this range, the λ value must always be greater than 1. In practice, an optimum lambda value of $\lambda_{\text{opt}} = 1.02$.[46] For λ values greater than 1 (excess air), the exhaust gas heat losses increase because the excess air is also heated without any benefit. The efficiency η therefore also drops in this area, although not as much as in the case of a lack of air.

[46] Hammer 2014.

Fig. 2.71 Changes in the emission values for various vehicle exhaust components as a function of the λ value (Baunach et al. 2006)

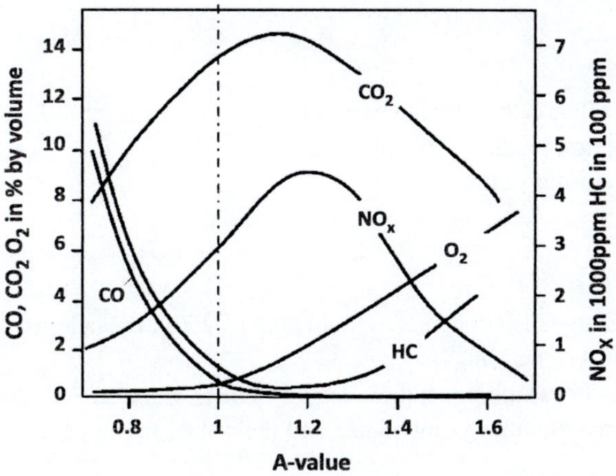

Table 2.24 Combustion products $\lambda = 1$ for humid flue gas (Cerbe 2008)

Size	Blast furnace gas	Coke oven gas	Natural gas L	Natural gas H
Air demand L_{min}	0.61	4.60	8.41	9.85
Exhaust gas flow V_F	1.48	5.09	9.44	10.88
Vol. % CO_2	29.3	7.9	9.5	9.7
Vol. % H_2O	2.8	22.4	18.5	18.7
Vol. % N_2	67.9	69.7	72.0	71.6

Exhaust Gas Quantity

In the case of complete combustion, the exhaust gas contains CO_2, H_2O, N_2 and O_2 (Table 2.24). The total moist exhaust gas volume V_F is therefore the sum of the individual components:

$$V_F = V_{CO_2} + V_{H_2O} + V_{O_2} + V_{N_2} \tag{2.142}$$

Calorific Value

Especially when natural gas is burned, a high proportion of water vapour is produced in the exhaust gas. According to Eq. 2.137, 16 g natural gas results in 36 g water vapour, and 1 kg natural gas results in 2.25 kg water vapour. This water vapour now contains an additional energy component Q_K, the heat of condensation ($r_{H_2O} = 2447 \text{kJ/kg}$).[47] According to Eq. 2.1 this energy share can be calculated:

[47] Cerbe 2008.

$$Q_K = r_{H_2O} \cdot m = 2447 \frac{\text{kJ}}{\text{kg}} \cdot 2.25\text{kg} = 5.506\text{MJ/kg} \tag{2.143}$$

The calorific value *Hs* differs from the calorific value H_i by this amount and is therefore approximately 10% higher than the calorific value. The following therefore applies to the calorific value:

$$H_s = H_i + m \cdot r \tag{2.144}$$

In so-called condensing boilcrs (Fig. 2.70), this additional energy can be used to generate heat in buildings.

In gas mixtures, the calorific value H_i (mixture) is calculated from the calorific values of the individual components and their mass fractions μ

$$\mu_1 = \frac{m_1}{m_{Mi}} = \frac{\text{Mass of the component 1}}{\text{Mass of the gas mixture}} \tag{2.145}$$

$$H_{i_{Mi}} = \mu_1 \cdot H_{i1} + \mu_2 \cdot H_{i2} + \ldots + \mu_n \cdot H_{in} \tag{2.146}$$

Natural gas can contain more than 10 components, depending on the place of origin. The concentration values of the individual gas components vary considerably (Table 2.25). To calculate the calorific value, the concentration values of the individual components must be known. Usually this analysis is carried out with a gas chromatograph (GC see Chapter 5). Until a few years ago calorimeters were also used for this purpose, but they are very expensive to operate. In 2002, a new measuring method (Schley et al. 2002) was presented, with which it was possible to achieve results comparable to those obtained with calorimeters and GC analysers by means of correlation analysis. Especially for fast control processes of natural gas mixtures this sensor-based method has become generally accepted.

Wobbe Index

The heat energy supply of a burner depends on the supplied energy. The decisive factor here is the energy flow $\dot{Q}_B$, which results from the volume flow $\dot{V}_B$ of the fuel gas (e.g. Natural gas) and the calorific value *Hs:*

$$\dot{Q}_B = \dot{V}_B \cdot H_s \tag{2.147}$$

This energy flow is of great importance for the gas industry, as the economic gas billing is based on this value (Fig. 2.72). Furthermore, it also results in the interchangeability of fuel gases (e.g. Change in the quality of the natural gas supplied) for certain applications

Table 2.25 Examples of the composition of different types of natural gas in % by volume

Gas type	RG1	RG2	RG3	RG4	RG5	RG6	RG7	RG8	RG9	RG10
CH_4	84.21	86.60	85.78	98.24	86.368	95.61	83.37	88.96	82.26	86.05
N_2	9.640	2.980	1.370	0.825	5.250	1.343	10.15	1.235	11.83	11.00
CO_2	1.560	1.160	1.100	0.081	1.314	0.322	1.450	0.855	4.420	1.510
C_2H_6	3.510	7.470	9.964	0.578	5.591	2.125	3.830	6.795	0.759	0.743
C_3H_8	0.640	1.270	1.340	0.183	1.000	0.414	0.732	1.442	0.293	0.303
C_4H_{10}	0.229	0.361	0.319	0.062	0.315	0.129	0.258	0.512	0.386	0.200
C_5H_{12}	0.095	0.070	0.050	0.014	0.072	0.026	0.057	0.096	0.050	0.100
C_6	0.039	0.041	0.034	0.004	0.038	0.012	0.040	0.057		0.100
C_7	0.016	0.024	0.021	0.002	0.021	0.007	0.040	0.033		
C_8	0.005	0.004	0.003	0.001	0.003	0.001	0.017	0.004		
He	0.054	0.020	0.011	0.013	0.030	0.015	0.050	0.012		

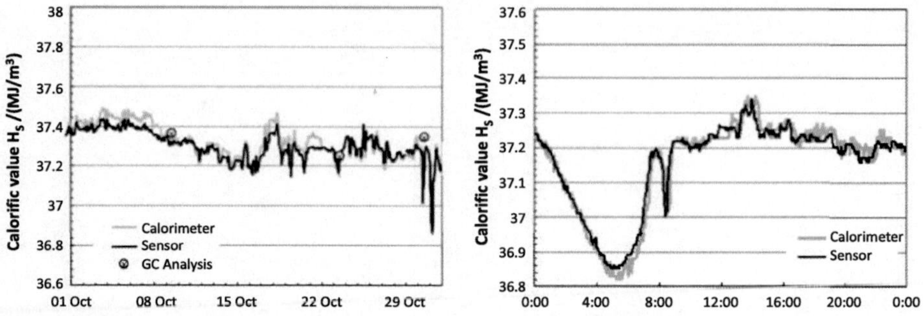

Fig. 2.72 Comparison measurements between a sensor-based calorimeter (Q1 Elster-Instrumet GmbH, Dortmund) and conventional calorimeters or gas chromatographs (selective control measurement) on a natural gas pipeline

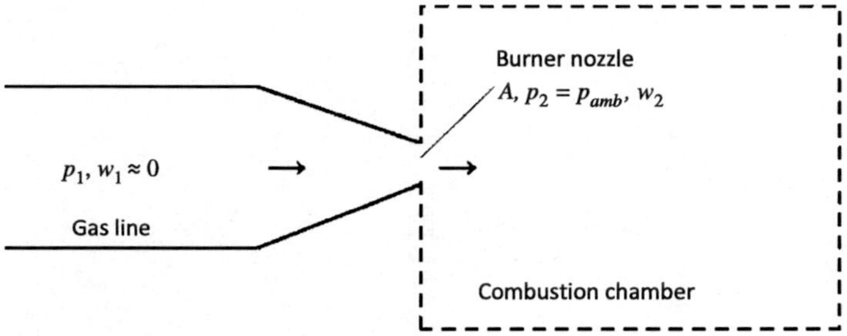

Fig. 2.73 Definitions for determining the Wobbe Index (Schley 2011)

(e.g. Gas boiler). In 1927 a parameter was introduced which was named after the inventor *Wobbe.*[48]

The following applies to the volume flow $\dot{V}_B$ according to the continuity equation (Eq. 2.157) with the cross-sectional area A and the average flow velocity w (Fig. 2.73)

$$\dot{V}_B = A \cdot w \tag{2.148}$$

The flow velocity w can further be expressed by the Bernoulli equation (Eq. 2.172) if the height, density and pressure are the same and the flow velocity $w_1 \approx 0$ is in front of the burner nozzle:

[48]Geoffredo Wobbe, Italian physicist and gas engineer

Table 2.26 Wobbe index of different fuel gases (Herdin 2012)

Substance	W_S[MJ/m³]	W_i[MJ/m³]
Hydrogen H_2	48.34	40.90
Methane CH_4	53.45	48.17
Propane C_3H_8	81.18	74.74
Verbund Erdgas Nord	51.55	46.54
Russian natural gas	53.21	47.97
Biogas (65% by volume CH_4)	28.44	

$$\frac{w_2^2}{2} = \frac{\Delta p}{\varrho_g} \rightarrow w_2 = \sqrt{\frac{2\Delta p}{\varrho_g}} \tag{2.149}$$

Since the maximum speed w_2 is reduced due to losses, a so-called nozzle coefficient α must be introduced at this point. With $w = \alpha \cdot w_2$ you get:

$$\dot{Q}_B = A \cdot w \cdot H_s = A \cdot \alpha \cdot \sqrt{\frac{2 \cdot \Delta p}{\varrho_g}} \cdot H_s \tag{2.150}$$

If one now leads the relative density $d = \frac{\varrho_g}{\varrho_{Air}}$ so one receives:

$$\dot{Q}_B = A \cdot \alpha \cdot \sqrt{\frac{2 \cdot \Delta p}{\varrho_{Air}}} \cdot \frac{H_s}{\sqrt{d}} \tag{2.151}$$

Since the expression$A \cdot \alpha \cdot \sqrt{\frac{2 \cdot \Delta p}{\varrho_{Air}}} = $const. is the upper Wobbe index (Table 2.26) can be determined as follows:

$$W_S = \frac{H_S}{\sqrt{d}} \tag{2.152}$$

Figure 2.74 shows the relationship between the Wobbe index and the calorific value for different types of gas. For the lower Wobbe index, heating value H_i= then applies:

$$W_i = \frac{H_i}{\sqrt{d}} \tag{2.153}$$

Methane Number

The methane number MZ is a term from the field of gas engine technology and describes the so-called knock resistance. If the engine is operated in the stoichiometric combustion range, this range is restricted by the knock limits. The maximum output varies depending on the methane number. It is therefore attempted to run the engine within these limits, at

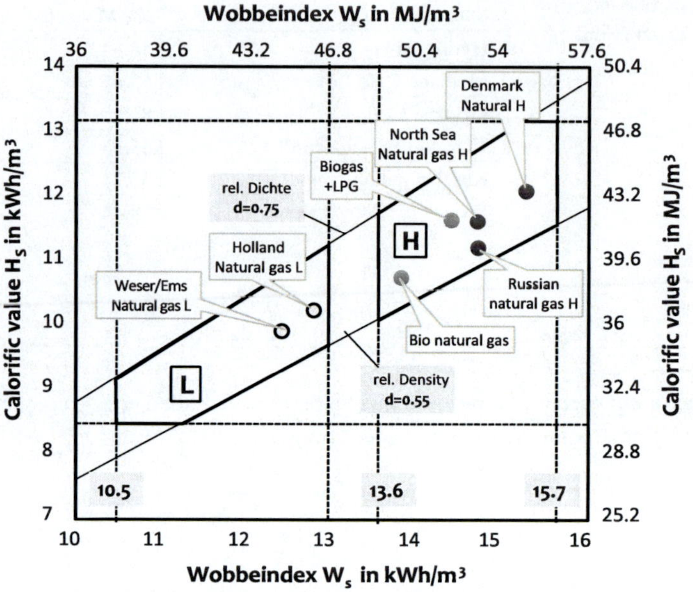

Fig. 2.74 Relationship between the Wobbe index *W* and the calorific value *H*. Specification of the gas quality according to DVGW Code of Practice G260 and presentation of the limits for H_S, W_S and *d*

maximum possible power. The determination of the methane number is therefore very important for the operation of gas engines (Hoppe and Wolf 1999). Knock resistance is also a problem with gasoline engines that run on liquid fuels (gasoline). In this area, knock resistance is described by the octane number.

As with the determination of the octane number, the methane number is also determined empirically. A test engine is used which is operated with different mixtures of methane (MZ = 100, antiknock) and hydrogen (MZ = 0, knocking) and this binary gas mixture is compared with a fuel gas of unknown methane number (Table 2.27). The methane concentration, at which the CH_4-H_2-compound shows the same knocking behaviour as the unknown fuel gas, then indicates the methane number. There are several possibilities to determine the methane number:

- Calculated from a detailed gas composition using the *AVL method* (Christoph et al. 1972)
- Calculated using various correlations between calorific value/density/CO_2 content.
- Metrologically by IR absorption measurement of hydrocarbons up to C_4H_{10} and corresponding correlations with a methane number controller (Hoppe and Wolf 2000).

Table 2.27 Methane numbers of important gases (Herdin 2012)

Fuel	Designation	Methane number	L_{min}[Nm^{3}L/Nm^3]	H_i[kWh/Nm^3]
H_2	Hydrogen	0	2.379	2.996
CH_4	Methane	100	9.537	9.971
C_2H_4	Ethylene	15	14.39	16.521
C_2H_6	Ethane	43.7	16.85	17.89
C_3H_8	Propane	33	24.24	26.00
C_4H_{10}	Butane	10	32.26	34.34
CO	Carbon monoxide	75	2.381	3.51

2.7 Gas Flows

If external forces *F* act on gas molecules or on a gas volume *V*, it is accelerated and moves in the direction of the force *F*. The laws of this movement are derived from Newton's fundamental law of dynamics:

$$F = m \cdot a \qquad \text{or} \qquad F = \varrho \cdot V \cdot a \tag{2.154}$$

The external force *F* can be specified, for example by a pump, fan or a pressure difference. The resulting flow can be viewed both macroscopically (volume) and microscopically (atoms, molecules). Experimentally, one usually starts from the macroscopically measurable quantities, while the physical effects are explained on the microscopic level.

Continuity Equation

To illustrate this gas movement, we consider a friction-free gas flow through a pipeline. The gas volume V_1 moves with a velocity w_1 in a pipe with the diameter d_1. By tapering the pipe (Fig. 2.75), a new pipe diameter d_2 is obtained in which the same gas volume $V_2 = V_1$ now moves at a greater velocity w_2.

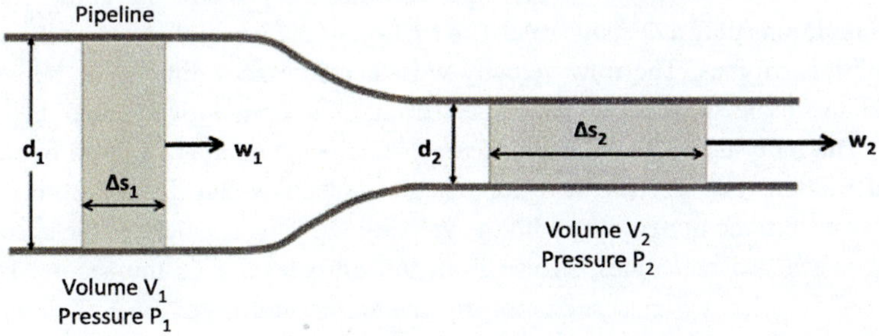

Fig. 2.75 Gas flow through different pipe cross-sections

The surface A of the pipe is given by

$$A_1 = \frac{\pi \cdot d_1^2}{4} \quad \text{or} \quad A_2 = \frac{\pi \cdot d_2^2}{4} \tag{2.155}$$

If we now assume that the distance Δs is covered in time Δt, we obtain for the volume flow:

$$\dot{V} = \frac{A \cdot \Delta s}{\Delta t} \tag{2.156}$$

Since the velocity is a result of $w = \frac{\Delta s}{\Delta t}$, the following expression can be given for the individual volume flows:

$$A_1 \cdot w_1 = A_2 \cdot w_2 \quad \text{or} \quad d_1^2 \cdot w_1 = d_2^2 \cdot w_2 \tag{2.157}$$

This then follows for the velocity w_2:

$$w_2 = w_1 \cdot \left[\frac{d_1}{d_2}\right]^2 \tag{2.158}$$

This equation is called the *continuity equation* or flow equation. However, it is only valid under the following conditions:

- The medium is incompressible, which means that the density remains constant.
- The flow through the pipe runs without friction.
- The flow is stationary in both sections.

If the pressure and/or temperature of a gas flow changes, these changes must be taken into account with the general gas equation (Eq. 2.101).

Viscosity

Especially when the diameter d of the pipe becomes very small, the above mentioned boundary conditions play a decisive role. The friction of the gas molecules on the pipe wall creates additional forces. The flow velocity w is therefore zero directly at the wall. This means that the molecules are at rest, apart from their own movement directed in all directions. The individual layers have different velocities, which are caused by the forces of internal friction (Fig. 2.76). For example, the molecules can diffuse from the lower boundary layer into the upper layer with the velocity w_1. This decelerates the speed. In the other case, when gas molecules diffuse from the third layer into the second layer, the second layer experiences an acceleration by the faster molecules (w_3). This change in velocity then leads to a force F. Newton's friction law then applies (Richter 2010):

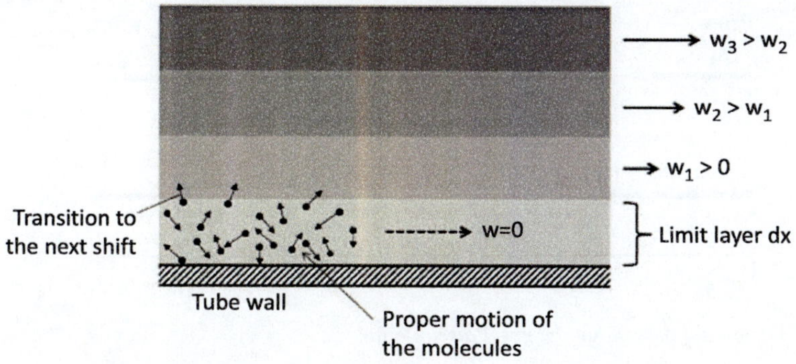

Fig. 2.76 Boundary layer model for gas flows in pipelines

Table 2.28 Dynamic viscosity η and kinematic viscosity ν at 0 °C and 1013 hPa (Kuchling 2011)

Gas	η inμ Pa s·	ν in mm²/s	Gas	η inμ Pa s·	ν in mm²/s
Acetylene	9.5	8.1	Carbon monoxide	16.6	13.3
Ammonia	9.3	12.1	Krypton	23.3	6.22
Argon	21.2	11.9	Air	17.2	13.3
Hydrogen bromide	17.0	4.67	Methane	10.2	14.2
i-Butane	6.9	2.58	Neon	29.8	33.1
Chlorine	12.3	3.83	Propane	7.5	3.70
Hydrogen chloride	13.1	7.99	Oxygen	19.2	13.4
Ethane	8.6	6.34	Sulphur dioxide	11.6	3.96
Ethylene	9.4	7.46	Nitric oxide	17.9	13.4
Helium	18.7	105	Nitrogen	16.5	13.2
Hydrogen iodide	17.3	2.99	Hydrogen	8.42	93.7
Carbon dioxide	13.7	6.93	Xenon	21.1	3.58

$$F = \eta \cdot A \cdot \frac{dw}{dx} \tag{2.159}$$

A is the contact area of the gas layers, *w* the velocities of the individual gas layers and *x* is the coordinate perpendicular to the direction of flow. η is the coefficient of friction or dynamic viscosity (Table 2.28). The resolution according to η provides:

$$\eta = \frac{F \cdot x}{A \cdot w} \tag{2.160}$$

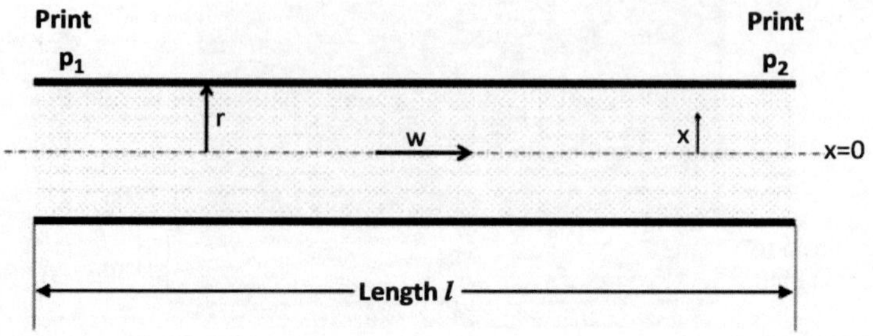

Fig. 2.77 Geometric data on the *Hagen-Poiseuille law*

The SI unit for η is therefore N · s/m² or Pa · s (Pascal second). The dynamic viscosity can also be expressed by the mean velocity $\overline{w}$, the mean free path $\bar{l}$ and the ϱ density:

$$\eta = \frac{1}{3} \overline{w} \cdot \bar{l} \cdot \varrho \tag{2.161}$$

The kinematic viscosity is the quotient of the dynamic viscosity and the density of the gas:

$$\nu = \frac{\text{Dynamic viscosity } \eta}{\text{Density } \varrho} \tag{2.162}$$

Hagen-Poiseuille Law

If the flow is through a pipe, there is a pressure difference Δp between the inlet (p_1) and the outlet (p_2) due to internal friction (Fig. 2.77). This frictional force can then be expressed by the relationship *pressure* = *force/surface area* as follows:

$$F_R = (p_1 - p_2) \cdot \pi \cdot x^2 \tag{2.163}$$

Through further transformations (Richter 2010) the *Hagen-Poiseuille law*[49,50] can then be derived:

---

[49] Gotthilf Heinrich Ludwig Hagen (1797–1884) German engineer.

[50] Jean Léonard Marie Poiseuille (1797–1869) French physiologist.

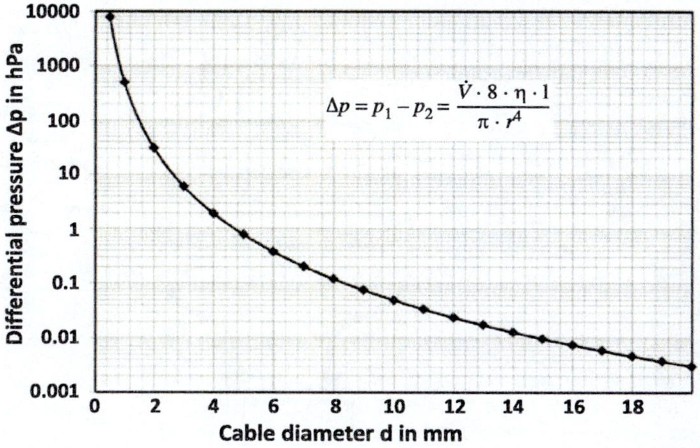

Fig. 2.78 Differential pressure Δp according to Hagen-Poiseuille's law for a 1-meter measuring section and a volume flow of 1 L/min

$$\dot{V} = \frac{\pi}{8 \cdot \eta \cdot l}(p_1 - p_2) \cdot r^4 \tag{2.164}$$

By rearranging the equation Δp, one then obtains an expression for the pressure drop in a gas pipe:

$$\Delta p = p_1 - p_2 = \frac{\dot{V} \cdot 8 \cdot \eta \cdot l}{\pi \cdot r^4} \tag{2.165}$$

Determination of the Internal Friction (Dynamic Viscosity)

The coefficient of internal friction η is of great importance for flow calculations. This quantity can also be determined experimentally (Fig. 2.78). This is done by using the pressure dropΔp that occurs in a capillary during a gas flow. The experimental setup is shown in Fig. 2.79. The capillary of length l is located in a liquid bath which is brought to a constant temperature by a thermostat. This temperature T is required as a measured variable for the calculation of η. The pressure drop across the capillary is measured with a U-tube manometer, while the volume flow $\dot{V}$ is determined with a bubble flow meter (see Sect. 11.3). The gas supply is provided by a pressurized gas cylinder with a corresponding pressure reducer, which regulates the cylinder pressure from $p_{Fl} \approx 100$ bar to the pressure $p_1 \approx 2$ bar.

With the measurable variables of the bath temperature T, differential pressure Δp, inner radius of the capillary r, volume flow $\dot{V}$ and the length of the capillary l, the internal friction can be calculated by transforming Eq. 2.165:

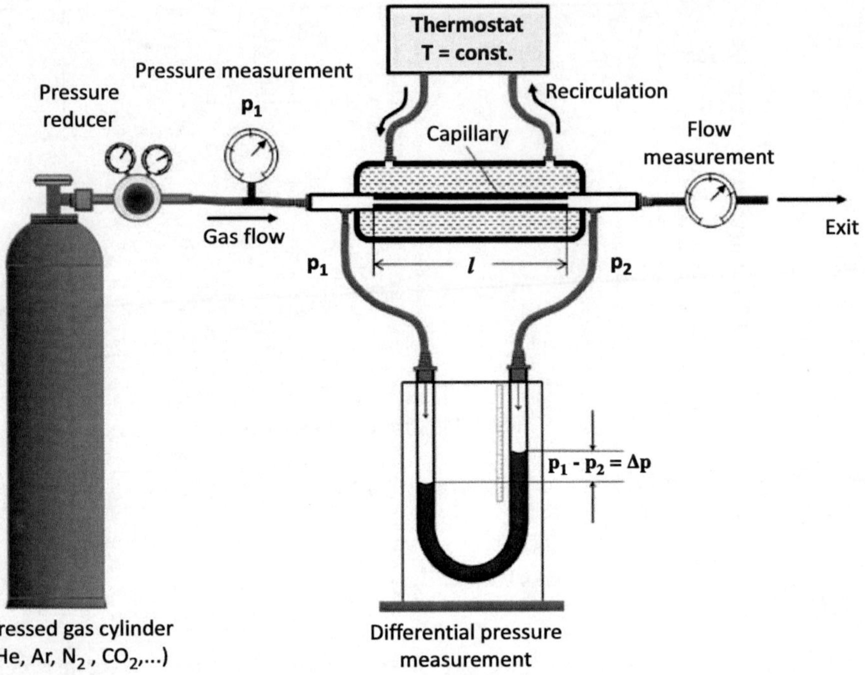

Fig. 2.79 Experimental setup for determining internal friction η

$$\eta = \frac{\pi \cdot r^4 \cdot \Delta p}{8 \cdot \dot{V} \cdot l} \tag{2.166}$$

Pipe Friction

Flows in real pipes are subject to additional laws, which are not covered by the *Hagen-Poiseuille law*. Especially the friction of the gases on the inner wall of the pipes leads to an energy conversion and thus to pressure losses. The equation of *Darcy*[51] takes this friction into account by introducing a pipe friction coefficient λ.

$$p_R = \lambda \cdot \frac{L}{d} \cdot \frac{\varrho}{2} \cdot w^2 \tag{2.167}$$

The pipe friction coefficient λ is determined experimentally and depends on the *Reynolds number Re*. For laminar flows λ results from:

[51] Henry Darcy (1803–1858) French engineer.

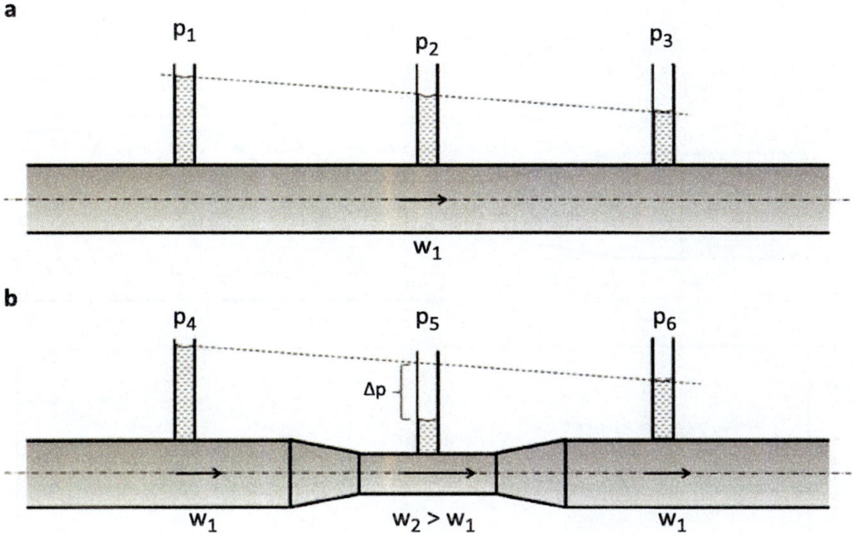

Fig. 2.80 Pressure conditions for a pipe flow with and without tapering of the pipe cross-section

$$\lambda = \frac{64}{Re} \tag{2.168}$$

λ is, depending on the Reynolds number, between 0.1 and 0.02 for laminar flows and between 0.08 and 0.008 for turbulent flows (Cerbe 2008).

Bernoulli's Equation

Pipelines are naturally not constructed in the same way at all points. There are branches, tapers, extensions and differences in height Δh have to be overcome. All these changes have a serious influence on the flow velocity w and the pressure p in the pipeline. Figure 2.80 shows such a taper in comparison with a straight pipe. The pressure in a pipeline therefore decreases linearly with length l. In Fig. 2.80a, pressure gauges are placed at three different points, which indicate a linear pressure curve according to Hagen-Poiseuille's law. If one now introduces a constriction into this pipeline (Fig. 2.80b), the conditions change significantly. At the constriction, the velocity increases due to the continuity equation, so that obviously a significantly lower pressure is established here. This additional pressure reduction can be explained by the law of conservation of energy.

Figure 2.81 shows the flow conditions in a pipeline with a tapered cross-sectional area from A_1 to A_2. Afterwards the height changes from h_1 to h_2. The mass flow m is the same in both cases. For the entire flow in the pipeline, the energy law (energy conservation) must be

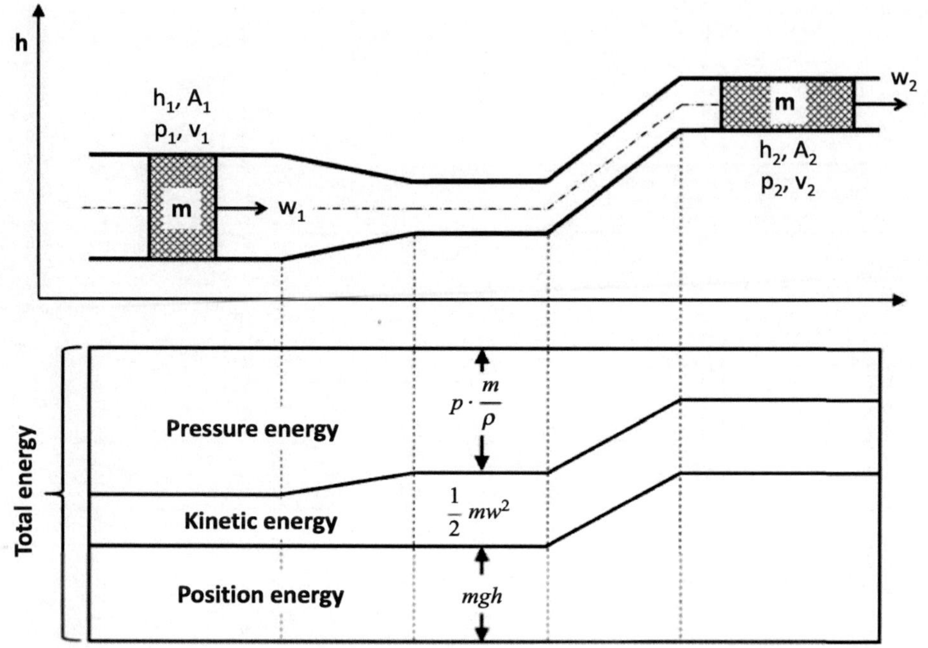

Fig. 2.81 Energy distribution in a pipeline according to Bernoulli's equation (after Böswirth 2004)

fulfilled for each point. Therefore the following relationship can be derived (Böswirth 2004):

$$\textit{Kinetic energy} + \textit{Position energy} + \textit{Pressure energy} = \textit{Total energy}$$

or expressed formally:

$$\frac{m \cdot w^2}{2} + m \cdot g \cdot h + p \cdot \frac{m}{\varrho} = E_{\text{ges}} = \text{const.} \tag{2.169}$$

If all energy components must always remain constant in total, then an increase in one component will reduce another. At the point where the flow area $_{A1}$ decreases to $_{A2}$, the kinetic component increases. This leads to the fact that the proportion of pressure energy is reduced if the position energy remains constant. If the positional energy (h_2) changes, I reduce the pressure energy at constant flow velocity ($w \approx$ kinetic energy) as well. With this equation, which was established by Bernoulli, all movement and flow processes in pipelines can be described. Equation 2.169 can also be represented as a pressure equation and as an elevation equation:

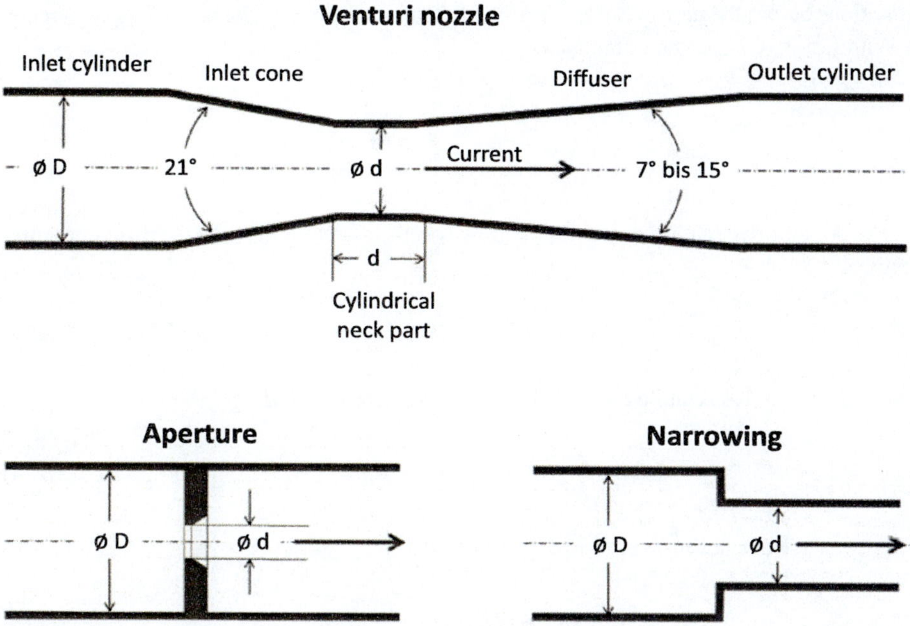

Fig. 2.82 Examples of throttling pipe flows with different devices. For metrological purposes, the Venturi nozzle and certain orifice plate geometries are mainly used

$$\text{Pressure equation}: \varrho\frac{w^2}{2} + \varrho \cdot g \cdot h + p = p_{ges} = \text{const.} \tag{2.170}$$

$$\text{Height equation}: \frac{w^2}{2 \cdot g} + h + \frac{p}{\varrho \cdot g} = h_{ges} \tag{2.171}$$

Naturally, the relationships described above only apply to friction-free flows. By inevitable losses, part of energy will be lost by friction so that, for example, pressure conditions will change. So at very long pipelines (natural gas-pipelines) pressure must be increased again and again by compressor-units.

Flows Through Constrictions (Orifices, Nozzles, Etc.)

If there are obstacles or narrow places in a flow channel or pipe, the gas flow is disturbed. Due to this disturbance/throttling, the flow conditions change and the pressure and flow velocity will also change. The simplest case is a transition from a large pipe diameter *D* to a smaller pipe diameter *d*. This transition is called a constriction (see Fig. 2.82). With the already derived Bernoulli equation, these conditions can be calculated. We assume that the pipe flow is horizontal and therefore the hydrostatic component is constant. The flow conditions before the narrowing are described with index 1 and in the narrowing with index 2. With Eq. 2.169 one then obtains

$$\frac{p_1}{\varrho}+\frac{w_1^2}{2}=\frac{p_2}{\varrho}+\frac{w_2^2}{2} \tag{2.172}$$

The continuity equation describes the flow speeds at both points and the different cross-sections:

$$w_1 \cdot A_D = w_2 \cdot A_d \tag{2.173}$$

The areas A_D and A_d can be described by the diameters D and d:

$$A_D=\frac{\pi}{4}\cdot D^2 \qquad \text{and} \qquad A_d=\frac{\pi}{4}\cdot d^2 \tag{2.174}$$

These expressions are now used in Eq. 2.173:

$$w_1\cdot\frac{\pi}{4}\cdot D^2 = w_2\cdot\frac{\pi}{4}\cdot d^2 \tag{2.175}$$

Now this equation is converted to w_1 and used in Eq. 2.172:

$$w_1 = w_2\frac{d^2}{D^2} \qquad \text{it follows that} \qquad \frac{p_1}{\varrho}+\frac{w_2^2}{2}\frac{d^4}{D^4}=\frac{p_2}{\varrho}+\frac{w_2^2}{2} \tag{2.176}$$

For the velocity w_2, the volume flow $\dot{V}_{\text{th}}$ and the mass flow $\dot{m}_{\text{th}}$ then obtains the following expressions by transformation as $p_1 > p_2$ and $w_1 < w_2$:

$$w_2=\sqrt{\frac{2\Delta p}{\varrho\left[1-\frac{d^4}{D^4}\right]}} \tag{2.177}$$

$$\dot{V}_{\text{th}} = w_2\cdot A_d = \frac{\pi}{4}d^2\cdot\sqrt{\frac{2\Delta p}{\varrho\left[1-\frac{d^4}{D^4}\right]}} \tag{2.178}$$

$$\dot{m}_{\text{th}} = \varrho\cdot\dot{V}_{\text{th}} = \frac{\pi}{4}d^2\cdot\sqrt{\frac{2\Delta p\varrho}{1-\frac{d^4}{D^4}}} \tag{2.179}$$

These three equations describe the flow behaviour at throttle points that behave ideally. In real cases, however, many factors play a role which must be taken into account for an exact calculation. In particular, the pipe wall roughness, compressibility of the gases and the flow profile must be mentioned here (Bohl and Elmendorf 2005). By introducing a flow coefficient C and an expansion coefficient, one obtainsε the corrected flow equations for real gas flows at throttling points:

Volume Flow

$$\dot{V} = \frac{\pi d^2 C \varepsilon_1}{\sqrt[4]{1-\left[\frac{d}{D}\right]^4}} \cdot \sqrt{\frac{2 \cdot \Delta p}{\varrho_1}} = \frac{\pi d^2 C \varepsilon_2}{\sqrt[4]{1-\left[\frac{d}{D}\right]^4}} \cdot \sqrt{\frac{2 \cdot \Delta p}{\varrho_2}} \tag{2.180}$$

Mass Flow

$$\dot{m} = \frac{\pi d^2 C \varepsilon_1}{\sqrt[4]{1-\left[\frac{d}{D}\right]^4}} \cdot \sqrt{2 \cdot p \cdot \varrho_1} = \frac{\pi d^2 C \varepsilon_1}{\sqrt[4]{1-\left[\frac{d}{D}\right]^4}} \cdot \sqrt{2 \cdot p \cdot \varrho_2} \tag{2.181}$$

The expansion coefficient downstream of the throttling point (downstream) can be determined as follows:

$$\varepsilon_2 = \varepsilon_1 \cdot \sqrt{1 + \frac{\Delta p}{p_2}} \tag{2.182}$$

The values for C and ε the values applicable to the different types of throttling units can be found in the relevant standards.[52]

Critical Nozzle Flows

If the diameter d or the area A_d at the throttle point is selected very small, a very high flow velocity w_2 occurs in this area (Fig. 2.83). If the speed of sound $w_2 = c$ is reached at this point, a so-called critical range is reached.

According to Saint-Venant[53] and Wantzel,[54] the exit velocity w_2 can be calculated as follows (Bohl and Elmendorf 2005)

---

[52] DIN EN ISO 5167-1.

[53] Adhémar Jean Claude Barré de Saint-Venant (1797–1886) French engineer, mathematician and physicist.

[54] Pierre-Laurant Wantzel (1814–1848) French engineer and mathematician.

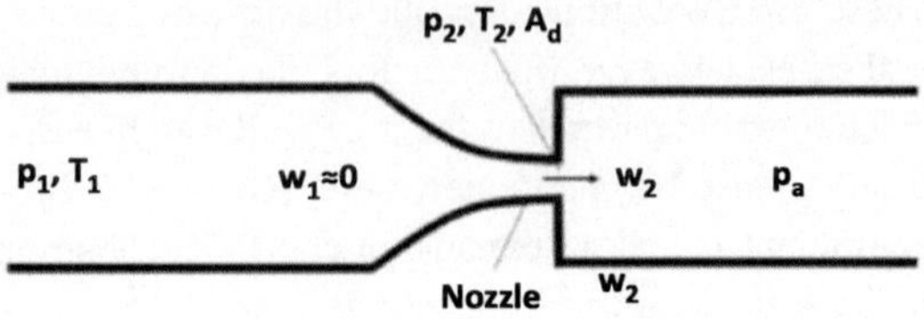

Fig. 2.83 Flow conditions at a nozzle in a pipe system

$$w_2 = \sqrt{2 \cdot \frac{\varkappa}{\varkappa - 1} \cdot \frac{p_1}{\varrho_1} \cdot \left[1 - \left(\frac{p_2}{p_1}\right)^{\frac{\varkappa - 1}{\varkappa}}\right]} \tag{2.183}$$

The mass flow can then be calculated using the following equation:

$$\dot{m} = A_d \cdot \sqrt{\frac{\varkappa}{\varkappa - 1}\left[\left(\frac{p_2}{p_1}\right)^{\frac{2}{\varkappa}} - \left(\frac{p_2}{p_1}\right)^{\frac{\varkappa + 1}{\varkappa}}\right]} \cdot \sqrt{2 \cdot p_1 \cdot \varrho_1} \tag{2.184}$$

By introducing an outlet function ψ, a simplified expression for the mass flow is then obtained:

$$\dot{m} = A_d \cdot \psi \cdot \sqrt{2 \cdot p_1 \cdot \varrho_1} \tag{2.185}$$

With

$$\psi = \sqrt{\frac{\varkappa}{\varkappa - 1}\left[\left(\frac{p_2}{p_1}\right)^{\frac{2}{\varkappa}} - \left(\frac{p_2}{p_1}\right)^{\frac{\varkappa + 1}{\varkappa}}\right]} \tag{2.186}$$

The maximum of the outlet function ψ then results for a critical pressure ratio:

$$\left(\frac{p_2}{p_1}\right)_{\text{crit}} = \left(\frac{2}{\varkappa + 1}\right)^{\frac{\varkappa}{\varkappa - 1}} \tag{2.187}$$

For air the critical pressure ratio is enclosed with then:$\varkappa = 1,4$

$$\left(\frac{p_2}{p_1}\right)_{\text{crit}} = 0,528 \tag{2.188}$$

If the pressure p_1 *is* further increased, the mass flow changes as follows (Grollius 2012).

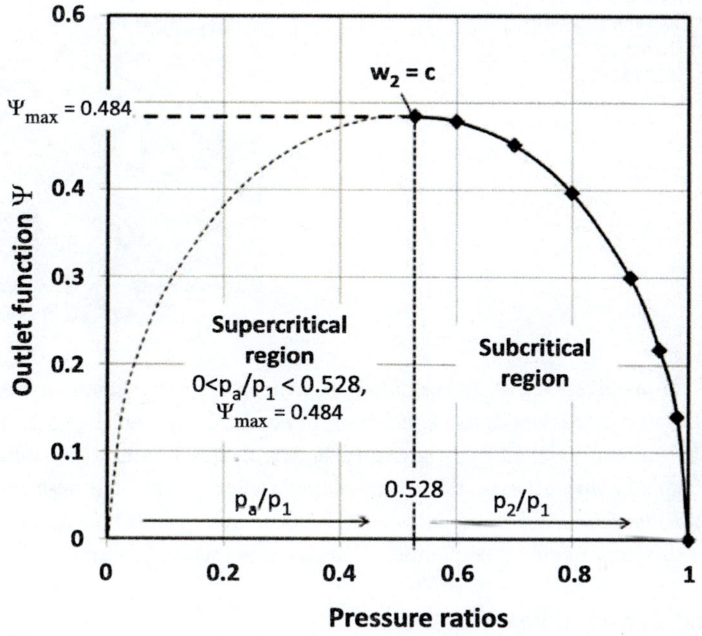

Fig. 2.84 Outflow characteristic for a nozzle with an air flow and $\varkappa = 1.4$ (Grollius 2012). The supercritical range between 0 and 0.528 is blocked for practical reasons. The outlet function is constant above the maximum and thus the mass flow

$$\dot{m} = A_d \cdot \psi_{max} \cdot p_1 \cdot \sqrt{\frac{2}{T_1 \cdot R_L}} \rightarrow \dot{m} = k \cdot \alpha \cdot \frac{p_1}{\sqrt{T_1}} \tag{2.189}$$

In the simplified form of Eq. 2.189, the factor α describes the geometric conditions. Figure 2.84 shows the course of the outlet function for different pressure conditions. The course of the mass flow has an analogous course, since both variables are linked with each other. At a pressure ratio of $p_2/p_1 = 1$, the same pressures are present on both sides of the nozzle and the mass flow is then logically zero. If the pressure behind the nozzle drops, this flow increases until it reaches a maximum at a pressure ratio of 0.528 (for dry air). According to Eq. 2.186, mass flux then should decrease again (dotted line), which however cant be true for practical reasons. By experiments one can prove that the mass flow above this point remains constant. So this behaviour is independent of pressure p_2 at outlet of nozzle (supercritical area). In this context, the falling range is thus not possible. This phenomenon is also called *blocking of* this area. Figure 2.85 shows examples of critical nozzles.

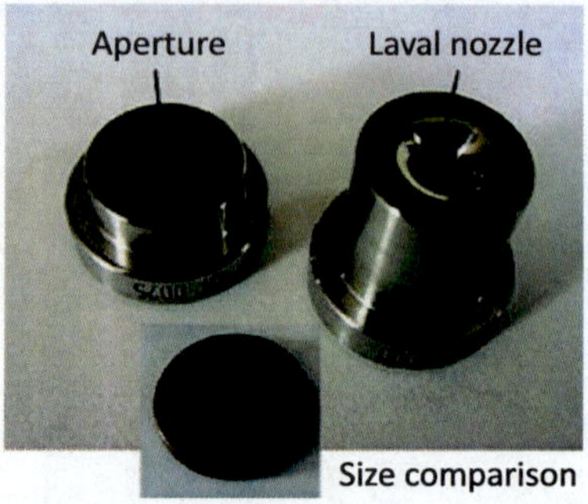

Fig. 2.85 Critical nozzles with different inner diameters. (Source: Ing.-Büro IMBUS, Potsdam)

In the equations given above, the calculations with ideal gases, which are not available as gas mixtures, are listed. Under real conditions, of course, the properties of the gases must be taken into account even at high pressures and with gas mixtures. Especially in the natural gas industry this behaviour is of great importance. A detailed description of the critical nozzle flow can be found in Wendt (2000). Schley (2001) has described the behaviour of different natural gas qualities in detail in his work.

Laminar and Turbulent Pipe Flow

Figure 2.86 shows two different types of flow in a pipeline. The gas (e.g. Air) is to be sucked in through the funnel-shaped opening (left side) and passed on in the pipe. Depending on the average flow velocity w_m, either a laminar or a turbulent flow is formed. In general, laminar flow conditions exist at low flow velocities and turbulent flow conditions at high flow velocities. In a laminar flow, the individual gas layers lie on top of each other and form a parabolic flow profile after a certain inlet distance due to the internal friction (viscosity). The velocity distribution in a laminar pipe flow then results from Hagen-Poiseuille's law:

$$w(r) = \frac{k}{4 \cdot \eta} \cdot \left(\frac{d^2}{4} - r^2\right) = \frac{p_1 - p_2}{4 \cdot \eta \cdot l} \cdot \left(\frac{d^2}{4} - r^2\right) \tag{2.190}$$

The maximum velocity w_{max} is therefore in the middle of the pipe flow at $r = 0$ and has the value

$$w_{max} = \frac{p_1 - p_2}{4 \cdot \eta \cdot l} \cdot \left(\frac{d^2}{4}\right) \tag{2.191}$$

The mean velocity w_m in the pipe can be derived by integration and is then

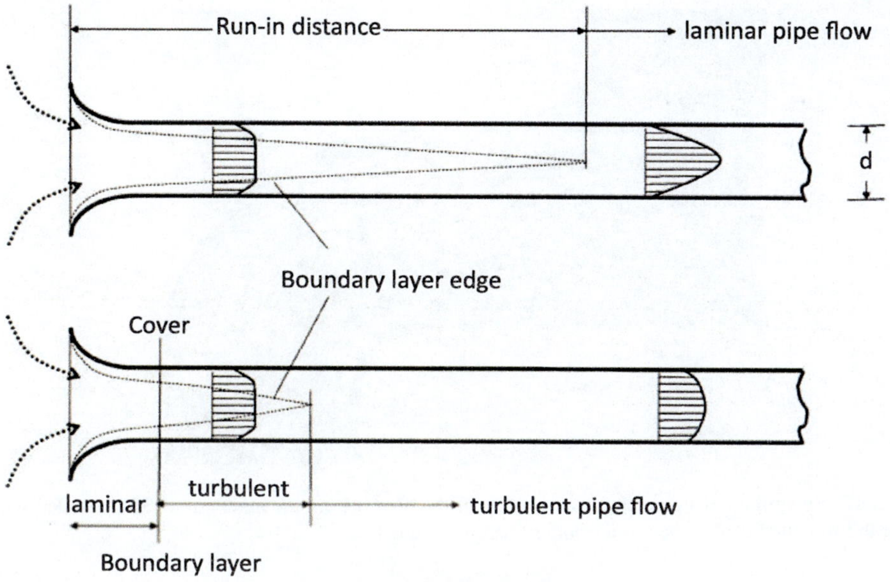

Fig. 2.86 Inlet section for a laminar or turbulent pipe flow

$$w_m = \frac{1}{2} w_{\max} = \frac{p_1 - p_2}{32 \cdot \eta \cdot l} \cdot d^2 \tag{2.192}$$

The pressure drop Δp for a laminar pipe flow can be calculated by changing Eq. 2.192 as follows:

$$\Delta p = p_1 - p_2 = \frac{32 \cdot w_m \cdot \eta \cdot l}{d^2} \tag{2.193}$$

According to *Reynolds*[55] '*law of similarity*, a critical value can be determined from which a laminar flow becomes unstable and changes into a turbulent flow. This *Reynolds number Re* depends on the mean velocity and a characteristic length of the body. For a pipe flow this is the diameter *d*.

[55] Osborne Reynolds (1842–1912) English physicist and engineer.

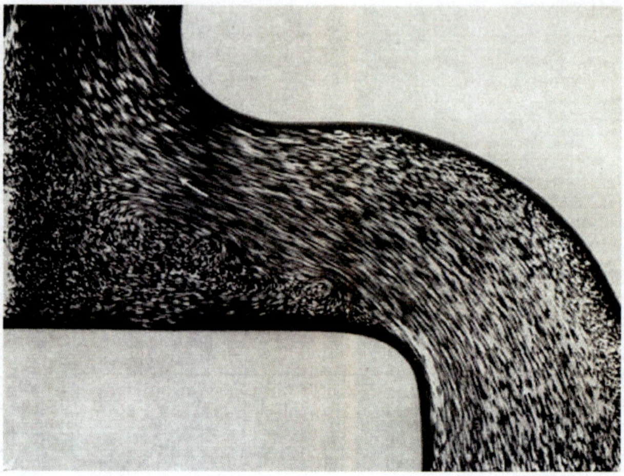

Fig. 2.87 Streamline in a pipe elbow and break off the flow behind the inner curve. The turbulences mark the so-called dead water area (motionless area) (Eck 1941)

$$Re = \frac{w_m \cdot d}{\nu} \leq Re_{\text{crit}} = 2320 \tag{2.194}$$

However, a stationary flow profile $w(r)$ only develops after a certain inlet distance L_{ein}. The running-in distance depends, for example, on whether the inlet is sharp-edged or guided, as shown in Fig. 2.87. In principle, however, at every point where there is a fault (widening, tapering, elbow, valve, etc.) there is also a fault in the flow profile, so that a new running-in section is then required. This influence is particularly important when placing flow sensors. The inlet section can be estimated as follows (Böswirth 2004):

$$\text{Laminar Flow} \qquad L_{\text{ein}} = (0.06 \text{ to } 0.065) \cdot Re \cdot d = 80 \cdot d \tag{2.195}$$

$$\text{Turbulent Flow } L_{\text{ein}} = (10 \text{ to } 30) \cdot d \tag{2.196}$$

Example: Gas Flow in Thin Tubes

In gas measurement technology, the gas sensors or analysers are usually operated with a volume flow of $\dot{V} \approx 1\text{L}/\min$. The sample gas is transferred by a pump with this volume flow. The average flow velocity w can be calculated for a typical inner diameter of the hose with d = 4 mm:

$$w = \frac{\dot{V}}{A} = \frac{1.67 \cdot 10^{-5}\text{m}^3/\text{s}}{\pi \cdot \frac{0.004^2}{4}\text{m}^2} = 1.329\text{m/s}$$

The Reynolds number Re is then calculated with the kinematic viscosity for$\nu = 13.3\mathrm{mm}^2/\mathrm{s}$ air at 0 °C and 1013 hPa, as follows

$$Re = \frac{w \cdot d}{\nu} = \frac{1329\,\frac{\mathrm{mm}}{\mathrm{s}} \cdot 4\mathrm{mm}}{13.3\,\frac{\mathrm{mm}^2}{\mathrm{s}}} \approx 400$$

The Reynolds number is considered to be clearly below the critical value 2320, so that in this case a laminar flow can be assumed. The inlet section is then $L_{\mathrm{ein}} = 80 \cdot 4\mathrm{mm} = 320\mathrm{mm}$.

The calculated flow velocity can also be used to calculate the time t with which the sample gas is transferred from a sampling point in the process to an analyser (e.g. in a remote control room). With a hose length of $s = 10\mathrm{m}$ would be the transmission time then:

$$t = \frac{s}{w} = \frac{10\mathrm{m}}{1.329\mathrm{m/s}} = 7.5\mathrm{s}$$

Pipe Network (Gas Grid)

Gaseous substances can be transported relatively easily over long distances through pipes (pipeline). This mass transport is described by the volume flow $\dot{V}$ and is therefore an important parameter and calculation basis for economic trade via the pipeline network (Fig. 2.88). This gas flow is driven by pumps or compressors that maintain the pressure in the pipeline over long distances (Fig. 2.89). In order to transport the largest possible quantity of gas (= mass) through the pipeline network, the pressure is increased, to up to 100 bar. The real gas behaviour of the gases to be transferred thus plays a decisive role in all further calculation methods.

In the high pressure range, due to the compressibility of the gases, both the pressure p and the flow velocity v in the pipe change as a function of the length L.

For gases that are pumped through a pipe at a low pressure, an incompressible medium can also be assumed for short lengths (Fig. 2.90b). The pressure drop is then linear, according to the Hagen-Poiseuille law.

For highly compressed gases, the flow velocity v increases with increasing pipe length L due to expansion of the gas, since at the same time the pressure p decreases non-linearly (Fig. 2.90a). In this context, one speaks of an expansion flow (Bohl and Elmendorf 2005).

For horizontally laid pipes, the pressure drop can be expressed by the *square pressure drop formula* as follows (Mischner et al. 2011):

$$p_1^2 - p_2^2 = 4.6401353 \cdot \lambda \cdot \frac{L}{d} \cdot \varrho_n \cdot T \cdot K \cdot \dot{V}^2 \quad (2.197)$$

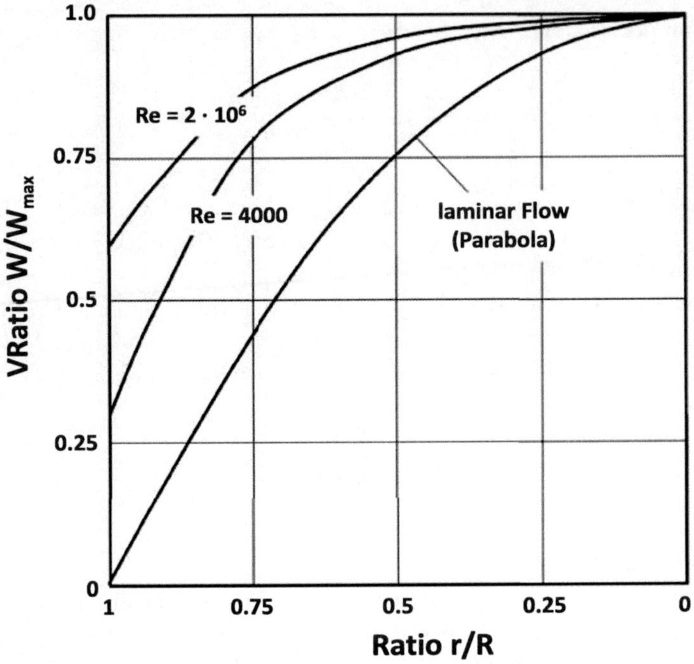

Fig. 2.88 Distribution of flow velocity in a turbulent or laminar pipe flow

p_1 is the pressure at the beginning of the pipeline and p_2 at the end. K is the dimensionless compression ratio.

The most important application for the transmission of gases takes place in power engineering. More than 150 years ago, local gas networks were already established in cities in Germany in order to realize the supply of so-called town gas. For 50 years this poisonous gas has been replaced by natural gas which is transported over long distances (North Sea, Russia, etc.) to the cities. Berlin was the last city in Germany that was also completely converted to natural gas only between 1990 and 1996.

The gas grid in Germany is structured into different areas, which can be seen in Fig. 2.91. The gas is fed into the German grid via large high-pressure pipelines from the gas fields across national borders. From here, it is forwarded to the various local gas suppliers. In the network, there is also mixing of the different natural gases that come from the different sources. In addition, a small proportion from local biogas plants is also fed into the network today. In the future, hydrogen and methane, which is produced by converting excess wind power, will also be fed into the natural gas grid.

The distribution within the local grid depends on various factors, which are characterized by the system costs, power reserves and security of supply. Figure 2.92 shows two typical structures as they are found in many cities in Germany. Version A., with

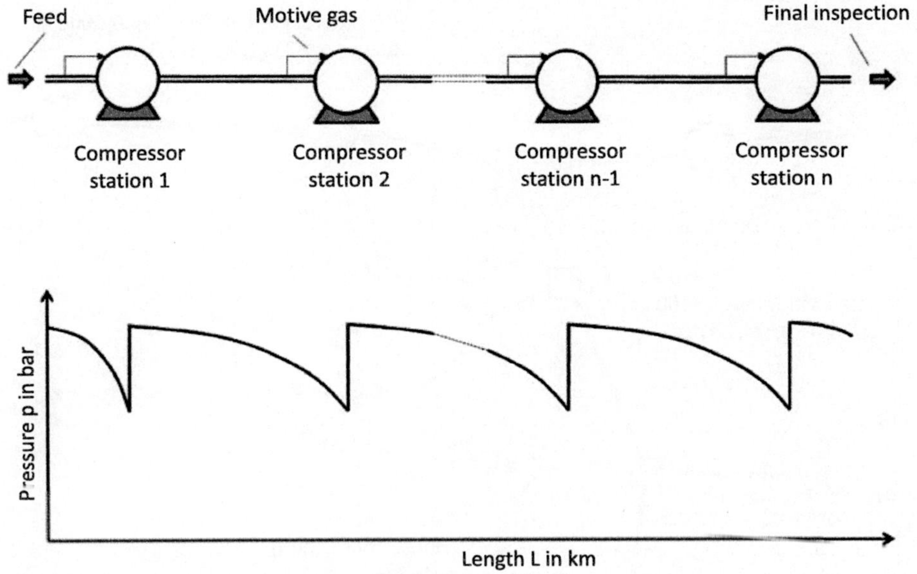

Fig. 2.89 Pressure curve in a natural gas pipeline during a transmission from >1000 km

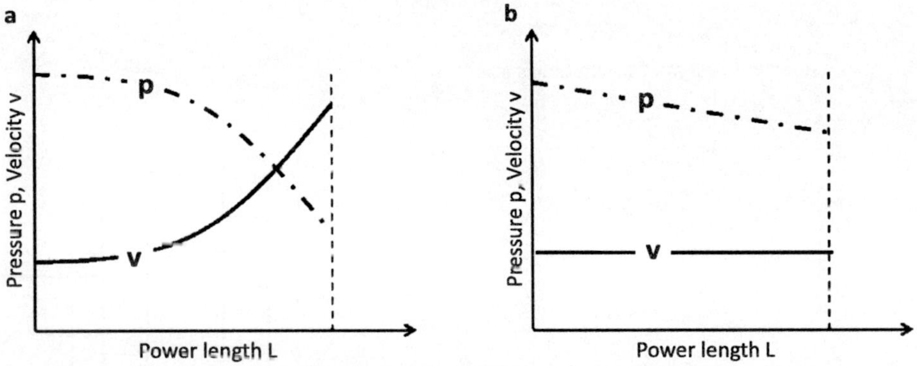

Fig. 2.90 Pressure p and flow velocity v for different pipe flows. (**a**) Compressible pipe flow. (**b**) Incompressible pipe flow

a branched radiation pipe network, is very cost-effective. However, this structure has significant limitations with regard to reserves and security of supply, whereas the very expensive version B does not have these disadvantages.

The distribution of gas volumes within these networks can be determined by appropriate calculations. In analogy to electrical engineering, these network analyses can be performed with the same laws. These are in particular *Ohm's law* and *Kirchhoff's rules*.

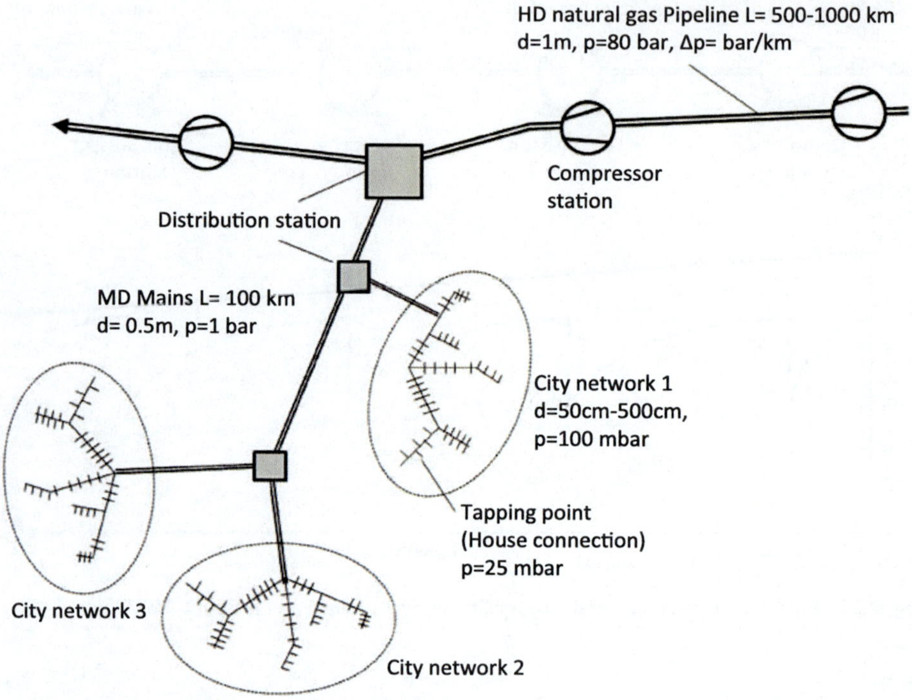

Fig. 2.91 Structural design of a natural gas grid

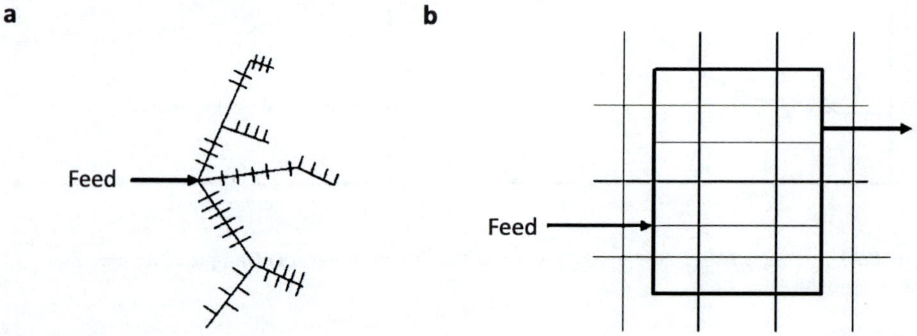

Fig. 2.92 Different possibilities of gas distribution in an urban gas grid. (**a**) Branched ray pipe network. (**b**) Meshed ring pipe network

Figure 2.93 shows a node in a gas grid. According to the first Kirchhoff rule, the supplied volume flows must result in zero. In this case the consumption is calculated as outgoing volume flow. Each individual node in the network is given its own number (here

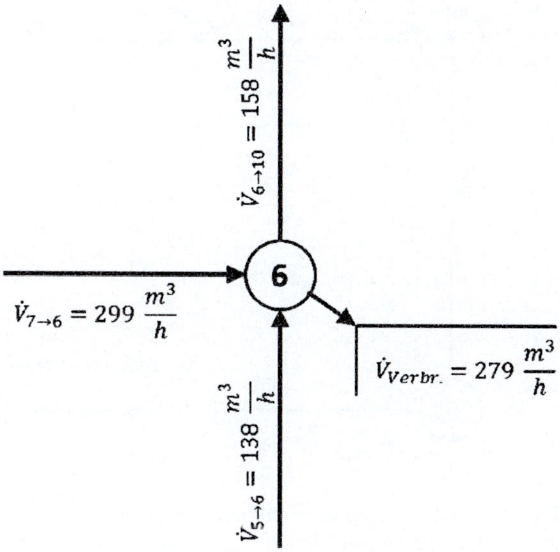

Fig. 2.93 Definition and labelling of a node in a gas network according to the first Kirchhoff law

no. 6), which also serves as an index for the different volume flows and directions (e.g. $V_{6\,\to\,10}$).

The second Kirchhoff law applies when several nodes are linked to form a mesh. The sum of all pressure drops Δp_s must also be zero. For this purpose, it is necessary to take into account the respective flow direction, from which the *direction of* the pressure drop then results. As agreed, these directions are evaluated as positive clockwise and negative counterclockwise (Fig. 2.94).

Strictly speaking, these calculation bases only apply in the case that there are no storage effects (buffer volumes) in the pipelines (stationary operating conditions). Under these conditions, *Kirchhoff's laws* can be used to calculate the pressure and volume flow conditions. In this case, an iterative approximation method from *Hardy Cross* is used.[56] Which is also suitable for complex systems.

First, the volume flows in the individual nodes are determined in this procedure. Then the resulting pressure losses are compared and the volume flows are corrected until the mesh condition is met.

For the volume flow correction, the following results are then obtained (Cerbe 2008)

$$\Delta\dot{V} = -\frac{\sum\Delta p_s}{2\cdot\sum\frac{\Delta p_s}{\dot{V}_s}} \tag{2.198}$$

[56]Hardy Cross (1885–1959) American engineer.

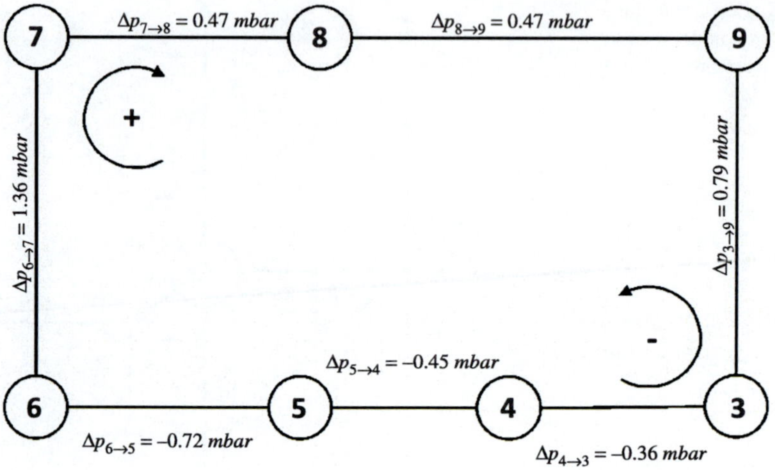

Fig. 2.94 Pressure losses with directional losses (6 → 5) of the pressure drop Δp within a mesh according to the second Kirchhoff law. The feed into the mesh is via the node no. 6

Table 2.29 Comparison of the laws of electrical engineering with fluid mechanics, neglecting the compressibility of gases

Electrical engineering	Fluid mechanics
Current I	Volume flow $\dot{V}^2$
Voltage U	Pressure p
Resistance R	Pneumatic resistance ξ
Capacity C	Buffer volume V
Ohm's law:	
$I = \frac{U}{R}$	$\dot{V}^2 = \frac{p}{\xi}$
1. Kirchhoff's node rule: The sum of all currents in a node is zero:	
$\sum I_i = 0$	$\sum \dot{V}_i = 0$
2. Kirchoff's mesh rule: The sum of all tensions in a closed mesh is zero:	
$\sum U_i = 0$	$\sum \Delta p_s = 0$

Also in this case, the signs for the respective flow direction must be observed. Corresponding example calculations on this topic can be found in Cerbe (2008).

Flow Resistances

The analogy of fluidic variables to electrotechnical variables shown in Table 2.29 can also be applied to the calculation of resistance networks. With Darcy's equation and the continuity equation one obtains (Cerbe 2008)

$$\Delta p_R = \lambda \cdot \frac{L}{d} \cdot \frac{\varrho}{2} \cdot \frac{1}{A^2} \cdot \dot{V}^2 \tag{2.199}$$

The flow resistance *Ri* for pipes can therefore be described by the following term:

$$R_i = \lambda \cdot \frac{L}{d} \cdot \frac{\varrho}{2} \cdot \frac{1}{A^2} \tag{2.200}$$

For pipe installation elements (elbows, bends, fittings, ...) this equation can be modified with the ζ value:

$$R_i = \zeta \cdot \frac{\varrho}{2} \cdot \frac{1}{A^2} \tag{2.201}$$

The total pressure loss, which results from a series connection of several flow resistances, is additive:

$$\Delta p_{\text{ges}} = \Delta p_1 + \Delta p_2 + \Delta p_3 + \ldots + \Delta p_n \tag{2.202}$$

$$\Delta p_{\text{ges}} = R_{\text{ges}} \cdot \dot{V}^2 = (R_1 + R_2 + R_3 + \ldots + R_n) \cdot \dot{V}^2 \tag{2.203}$$

$$\Delta p_{\text{ges}} = \frac{\varrho}{2} \cdot \dot{V}^2 \cdot \sum\nolimits_{i=1}^{n} \left(\lambda_i \cdot \frac{l_i}{d_i} \cdot \frac{1}{A_i^2} + \frac{\zeta_i}{A_i^2} \right) \tag{2.204}$$

If these flow resistances are connected in parallel, the volume flow is divided according to the individual flow resistances, whereby the pressure drops Δp are equal. It therefore applies:

$$\Delta p_1 = \Delta p_2 = \Delta p_2 = \Delta p_n \tag{2.205}$$

$$\Delta p = R_i \dot{V}_i^2 = R_{\text{ges}} \dot{V}^2 \rightarrow \dot{V}_i = \sqrt{\frac{\Delta p}{R_i}} \tag{2.206}$$

$$\dot{V} = \dot{V}_1 + \dot{V}_2 + \dot{V}_3 + \ldots + \dot{V}_i \tag{2.207}$$

If we now insert the terms for the volume flow from Eq. 2.207 into Eq. 2.206, we obtain (Fig. 2.95)

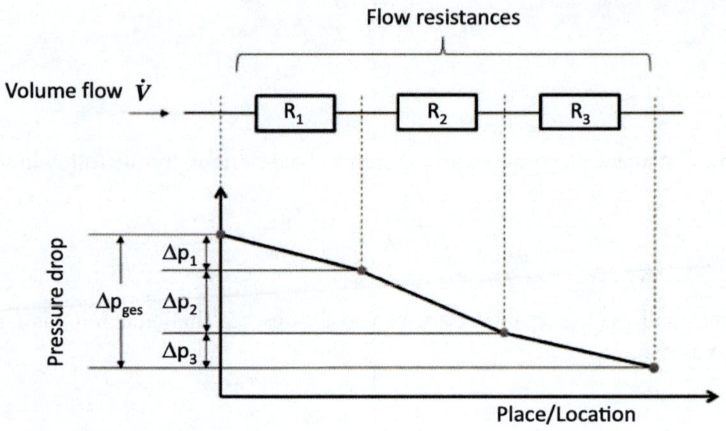

Fig. 2.95 Series connection of flow resistors and the corresponding pressure drop (corresponds to the voltage drop in electrical engineering)

$$\sqrt{\frac{\Delta p}{R_{\text{ges}}}} = \sqrt{\frac{\Delta p}{R_1}} + \sqrt{\frac{\Delta p}{R_2}} + \ldots + \sqrt{\frac{\Delta p}{R_n}} \tag{2.208}$$

$$\frac{1}{\sqrt{R_{\text{ges}}}} = \frac{1}{\sqrt{R_1}} + \frac{1}{\sqrt{R_2}} + \ldots + \frac{1}{\sqrt{R_n}} = \sum_{i=1}^{n} \frac{1}{\sqrt{R_i}} \tag{2.209}$$

The following printout is then obtained for the pressure drop when individual flow resistances are connected in parallel (Fig. 2.96)

$$\Delta p = \frac{\dot{V}^2}{\left[\sum_{n=1}^{n} \frac{1}{\sqrt{R_i}}\right]^2} \tag{2.210}$$

For the individual flow resistances, depending on the design of the pipe installation elements, the corresponding key figures must then also be integrated. These key figures are usually provided by the manufacturers of the components.

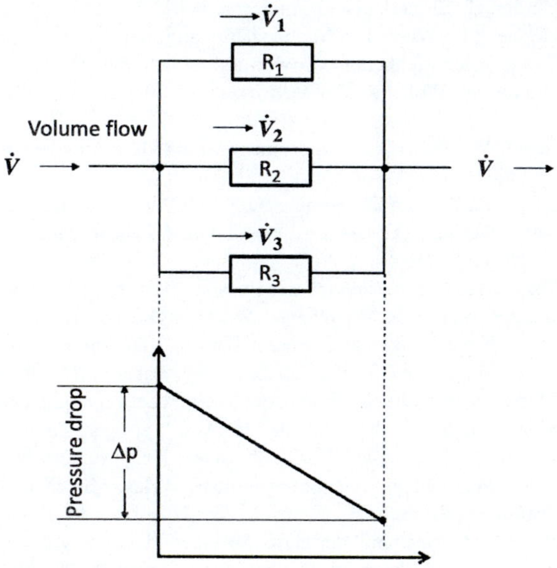

Fig. 2.96 Parallel connection of flow resistors

References

Altmann, L. et al.: Geruch und Reizwirkung von Ammoniak gemessen an Probanden. Poster Nr. 170 der BGFA-Bochum. (2015). http://www.ipa.ruhr-uni-bochum.de/image/poster/170.pdf

Baunach, T., Schänzlin, K., Diehl, L.: Sauberes Abgas durch Keramiksensoren. Phys. J. **38**(5), 33–38 (2006)

Bohl, W., Elmendorf, W.: Technische Strömungslehre. 13. Aufl. Vogel Buchverlag, Würzburg (2005)

Böswirth, L.: Technische Strömungslehre. Vieweg Verlag, Wiesbaden (2004)

Burger, M.: Polyetherketone mit Phenylindangruppen als Membranen zur Gastrennung. Dissertation, TU-München (2001)

Cerbe, G.: Grundlagen der Gastechnik. 7. Aufl. Hanser-Verlag, München (2008)

Christoph, K., Cartellieri, W., Pfeiffer, U.: Die Bewertung der Klopffestigkeit von Kraftgasen mittels Methanzahl und deren Anwendung bei Gasmotoren. MTZ Motortechnische Zeitschrift. **33**(10), 391–403 (1972)

Dörfler, H.-D.: Grenzflächen und kolloid-disperse Systeme. Springer (2002)

Eck, B.: Technische Strömungslehre. Springer, Berlin (1941)

Fromm, E., Gebhardt, E.: Gase und Kohlenstoff in Metallen. Springer (1976)

Giancoli, D.C.: Physik Lehr- und Übungsbuch. 3. Aufl. Pearson Studium, München (2010)

Grollius, H.-W.: Grundlagen der Pneumatik. Hanser Verlag, Leipzig (2012)

Hahn, U.: Physik für Ingenieure. Oldenbourg, München (2007)

Hammer, F.: Sensorische Verbrennungsoptimierung in Gasfeuerungsanlagen. gwf-Gas|Erdgas. **2014**(4), 376–380 (2014)

Hengstenberg, J., Sturm, D., Winkler, O.: Messen, Steuern und Regeln in der Chemischen Technik. Bd. II Betriebsmesstechnik (Physikalische Analytik). Springer, Heidelberg (1980)

Herdin, G.: Grundlagen Gasmotoren. PGES-Skriptum (2012)
Hoppe, M., Wolf, D.: Methanzahl-Controller zur Steuerung von Gasmotoren bei wechselnden Gasbeschaffenheiten. Gaswärme-International. **48**, 419–422 (1999)
Hoppe, M., Wolf, D.: Der Methanzahl-Controller im Praxiseinsatz. Gwf Gas/Erdgas. **141**, 778–781 (2000)
Jessel, W.: Gase-Dämpfe-Gasmesstechnik, Ein Kompendium für die Praxis. Dräger AG, Lübeck (2001)
Klein, V., Werner, C.: Fernmessung von Luftverunreinigungen. Springer, Heidelberg (1993)
Kubaschewski, O.: Die Löslichkeit von Gasen in Metallen. Zeitschrift für Elektrochemie. **44**(2), 152–167 (1938)
Kuchling, H.: Taschenbuch der Physik. 20. Aufl. Hanser, München (2011)
Kunz, O., Klimeck, R., Wagner, W., Jaeschke, M.: The GERG-2004 Wide-Range Equation of State for Natural Gases and Other Mixtures. VDI GmbH, Düsseldorf (2007)
Lauer, W., Bendix, J.: Klimatologie. Westermann, Braunschweig (2006)
Marek, R., Nitsche, K.: Praxis der Wärmeübertragung. Hanser, Leipzig (2012)
Meschede, D.: Gerthsen Physik. 24. Aufl. Springer, Heidelberg (2010)
Mischner, J., Fasold, H.-G., Kadner, K..: gas2energy.net Systemplanerische Grundlagen in Gasversorgung. Oldenbourg Industrieverlag, München (2011)
Olenik, H., Rentsch, H., Wettstein, W.: Handbuch für Explosionsschutz. Giradet, Essen (1983)
Richter, D.: Mechanik der Gase. Springer, Heidelberg (2010)
Roedel, W., Wagner, T.: Physik unserer Umwelt. Die Atmosphäre. 4. Aufl. Springer, Heidelberg (2011)
Schley, P.: Thermodynamische Stoffgrößen von Erdgasen zur Beschreibung einer kritischen Düsenströmung. Fortschritt-Berichte VDI Reihe 7, Nr. 418. VDI-Verlag, Düsseldorf (2001)
Schley, P.: Vorlesungsskript Gasmesstechnik. Ruhr Universität Bochum (2011)
Schley, P., Kastner, J., Wiegleb, G.: New Measurement Instrument to Determine Natural Gas Quality. Proceedings of the GTI-Conference „Natural Gas Technologies", Orlando, FL, September 2002
Scott, K.: Handbook of Industrial Membranes. Elsevier, Oxford (1995)
Veranneman, G.: Technische Gase. Herstellung – Verteilung – Anwendung. Verlag Moderne Industrie, Landsberg (2000)
Wendt, G.: Einsetzbarkeit kritisch durchströmter Düsen mit REYNOLDS-Zahlen kleiner 10^5 für die Durchflussmessung von Gasen (Dissertation). PTB-Bericht PTB-MA-69 Wirtschaftsverlag NW, Bremerhaven (2000)
Wiegleb, G.: Umweltmesstechnik. Skriptunterlagen Fachhochschule Dortmund (2010)

Related References

Cerbe, G., Hoffmann H.-J.: Einführung in die Wärmelehre. 2. Aufl. Hanser-Verlag, München (1970)
DIN EN ISO 12213-1 Erdgas – Berechnung von Realgasfaktoren (2010)
DIN EN ISO 14912 Gasanalyse – Umrechnung von Zusammensetzungsangaben für Gasgemische (2003/2006)
DIN EN ISO 5167-1 Durchflussmessung von Fluiden mit Drosselgeräten. Teil 1: Blenden, Düsen und Venturirohre in voll durchströmten Leitungen mit Kreisquerschnitt
EN ISO 6976 Erdgas – Berechnung von Brenn-Heizwert, Dichte, relative Zusammensetzung und Wobbeindex aus der Zusammensetzung. Europäisches Komitee für Normung (1999)
Henzler, M., Göpel, W.: Oberflächenphysik des Festkörpers. Teubner (1991)

Kast, W.: Adsorption Aus der Gasphase. Weinheim, VCH (2008)

Kienle, H., Bäder, E.: Aktivkohle und ihre industrielle Anwendung. Ferdinand Enke, Stuttgart (1980)

Müller-Syring, G.: Permeation von Gasen durch Kunststoffrohrleitungen. Energie, Wasser-Praxis DVGW Jahresrevue 12/2006, 70–73 (2006)

Smithells, J.C.: Gases and Metals. Chapman & Hall Ltd., London (1937)

Wedler, G.: Adsorption – Eine Einführung in die Physisorption und Chemisorption. Verlag Chemie (1970)

3 Physical Gas Sensors

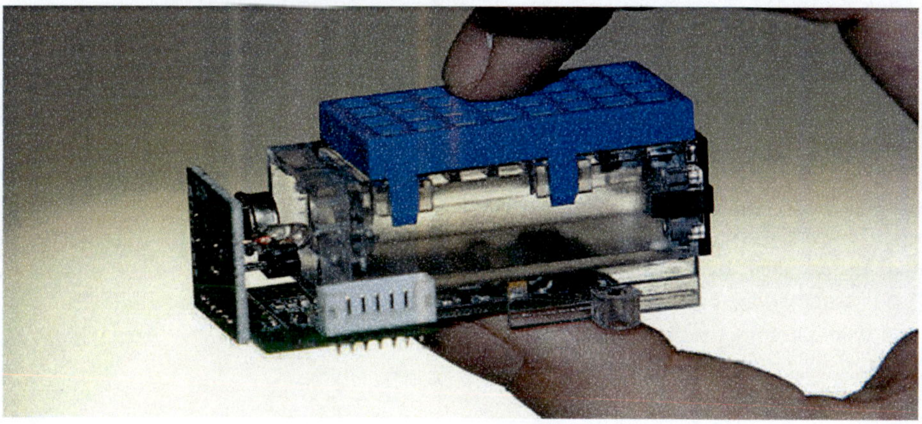

Gas Sensor. Source: TriAss/Fachhochschule Dortmund

The voltage supply of gas sensors is usually in the low voltage range (12–24 VDC). Both analog values (0–1 V or 4–20 mA) and digital values (I2C, Modbus, CAN bus, etc.) are common output signals (Fig. 3.1). In the meantime, however, intelligent gas sensors are increasingly being used, which have an integrated microcontroller that enables a wide range of compensation and calculation options. These are pressure and temperature compensation, linearization function of the characteristic curve and calculation of the drift behavior.

The gases to be investigated are not changed in physical gas sensors and return to the process or ambient air after measurement (Fig. 3.2).

G. Wiegleb, *Gas Measurement Technology in Theory and Practice*,
https://doi.org/10.1007/978-3-658-37232-3_3

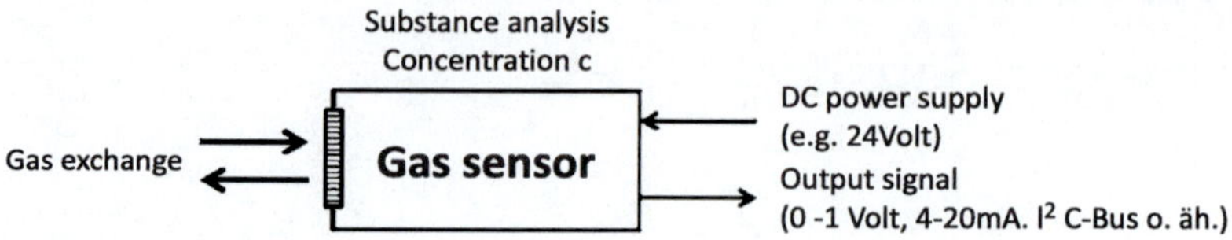

Fig. 3.1 Interfaces of a gas sensor

Fig. 3.2 Infrared gas sensor with gas exchange (diffusion) via a sintered metal frit (Source: Dortmund University of Applied Sciences)

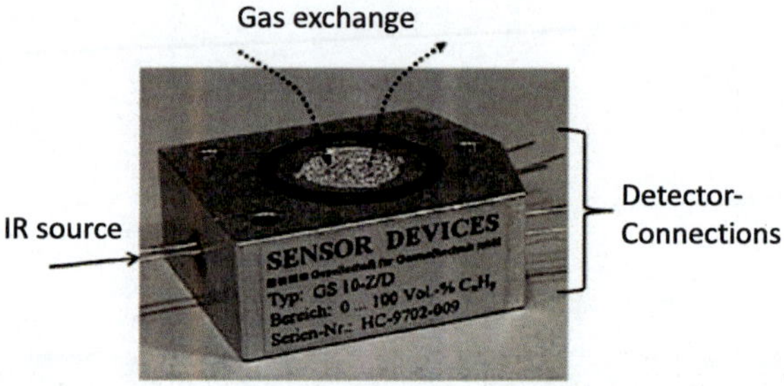

Chemosensor

Another detection method is to convert the gas by a defined chemical or biochemical reaction. During this conversion, electrochemical effects are used which lead to a measurable signal voltage or current. Since these sensors are current-generating cells, they are also called amperometric methods in this context. However, gases can also be detected by auxiliary chemical reactions if the gas concentration of a certain component is very low. This is usually done in the liquid phase by color reactions, which can then be detected very sensitively photometrically (Hengstenberg et al. 1980; Richly 1993).

Gas Analyzer

If these gas sensors are combined with additional components to form a complete measuring system, a gas analyzer or gas analysis device is obtained. An important difference to a gas sensor is the much more complex construction. In most cases, a gas analyzer also integrates extensive operating and display elements in addition to the pure sensor measurement function (Fig. 3.3). The energy supply is usually provided via the 230 VAC network, as the electrical power consumption can be via 100 W. Battery-powered gas analyzers are therefore the exception. These devices are usually also considerably larger and heavier than the actual gas sensors. For most applications, the 19″ -technology has become established as the housing form (Fig. 3.4).

Typical peripheral components of a gas analyzer are:

- Integrated gas delivery (sample gas pump) with gas conditioning (e.g. dust filter, gas cooler, condensate trap, flow measurement, etc.)

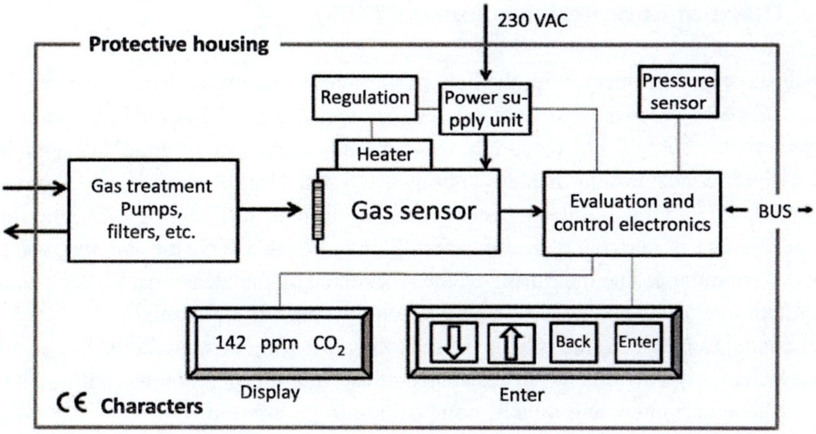

Fig. 3.3 Structure of a gas analyzer integrated with extensive peripheral components in a protective housing

Fig. 3.4 Gas analyzer (GS-IRA-8000) with display, operating functions, fine filter in the front panel and a flow control (variable area flow meter). The entire assembly is located in one 19″ housing. (Source: GS-Messtechnik GmbH Ratingen)

- Thermostatization of the gas sensor to prevent condensation and reduce the temperature error if necessary
- Power supply unit for energy supply or rechargeable battery for mobile use
- Operating and display units
- Additional sensors for compensation (e.g. cross-sensitivities, pressure influence, temperature, etc.)
- Powerful microcontroller for signal processing and data communication
- Possibilities of an automated calibration (solenoid valves for switching the gases)
- Protective housing to ward off environmental influences (splash water, dust, explosive atmosphere, etc.)

3.1 Thermal Conductivity Sensors (TCS)

The analysis of gas mixtures, with the help of thermal conductivity cells, has been known for over 100 years and is therefore considered the oldest physical method in the field of gas measurement technology. The reason for this is mainly the relatively simple design, which allows a direct concentration display without additional electronic effort (e.g. amplifier). For this purpose, a temperature-dependent measuring element (heating wire, thermistor, thin-film resistor) is integrated in a measuring chamber in which the gas mixture to be analyzed is conducted. The measuring element is heated by the measuring current, so that a constant temperature is achieved in a state of equilibrium. If a gas mixture with a higher thermal conductivity enters the measuring cell, the measuring element cools down and the resistance changes. This change in resistance allows the thermal conductivity of the gas mixture to be determined, and thus its composition to be inferred.

The first technical applications of thermal conductivity measuring instruments were in purity control and leakage monitoring of filling gases (hydrogen/helium) in airships (Zeppelin). In the first submarines, this measuring method was also used to control underwater air quality, which referred to the CO_2 content as a controlled variable (Fig. 3.5).

Today, gas analysis using the thermal conductivity method is mainly used in chemical process engineering (e.g. chlorine gas analysis, purity control in air separation, etc.) and gas chromatography. In the mostly technical applications of thermal conductivity sensors (TCS), there are only two (binary) or three (ternary) gas components. Due to new technological approaches (silicon micromechanics), miniaturized measuring cells have been used to an increasing extent since 1985, leading to a renaissance of this measuring technology.

Basics

The first exact measurements of the absolute thermal conductivity λ of gases were already carried out by Schleiermacher[1] in 1888. In the following period, extensive theoretical treatises on the thermal conductivity of gases were developed, which today provide us with a comprehensive picture of this material size.

The thermal conductivity λ of a gas or gas mixture is defined by the energy transfer per time unit (→heat flow=$\mathrm{d}Q/\mathrm{d}t$), related to an area element $\mathrm{d}A$ over the temperature gradient $\mathrm{d}T/\mathrm{d}s$. Figure 3.6 shows the basic design of a thermal conductivity measuring cell. The heat source is located on the left side at $s = 0$, and transfers the heat to the gas molecules. The gas molecules in turn transfer the heat to other gas molecules by impact and finally to the right side of the measuring cell at $s = x$. The geometric conditions are such that the transfer path x is only a few mm.

[1] August Ludwig Eduard Friedrich Schleiermacher (1857–1953) German physicist.

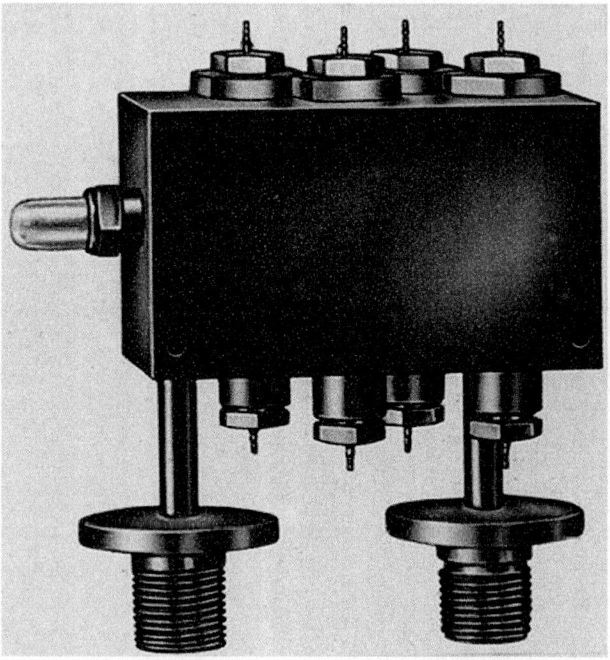

Fig. 3.5 Historical photograph (≈1930) of a thermal conductivity meter from Hartmann & Braun, Frankfurt a. M. with 4 platinum measuring resistors and closed reference chamber *(left screw connection)*. The two gas inlets are located at the bottom. The entire construction consisted of a solid metal block, which was not thermostatted

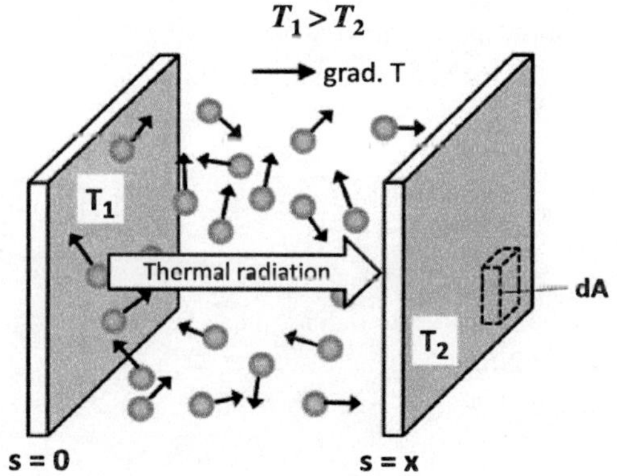

Fig. 3.6 Heat transfer by heat conduction (collisions of the gas molecules with each other) with superimposed heat radiation for a planar geometry

The calculation of the heat transfer is based on the empirical law of Fourier (Herwig and Moschallski 2006):

$$\frac{dQ}{dt} = \lambda \cdot \text{grad}T \cdot dA \tag{3.1}$$

The physical unit of thermal conductivity λ is W m^{-1}K^{-1}. In the different literature and tables you can also find information like [μW cm^{-1}K^{-1}], or [mW m^{-1}K^{-1}]. Another possibility of representation is the relative thermal conductivity to air, which is standardized as “1”. In the following diagrams and tables, the units of the corresponding original sources have been used.

For a stationary heat conduction one obtains for a sensor geometry according to Fig. 3.6:

$$\dot{Q} = \lambda \cdot \frac{A}{s} \cdot (T_1 - T_2) \tag{3.2}$$

Although the entire heat transfer takes place to a large extent via thermal conduction, convection and thermal radiation cannot be neglected. The complete heat transfer therefore results from:

$$\sum \dot{Q}_{\text{ges}} = \dot{Q}_{\text{Management}} + \dot{Q}_{\text{Radiation}} + \dot{Q}_{\text{Convection}} + \dot{Q}_{\text{Connections}} \tag{3.3}$$

Example

For a setup according to Fig. 3.3, with an area of $A = 1\ \text{cm}^2$, a distance s = 5 mm and the temperatures $T_1 = 40^\circ\text{C}$ or $T_2 = 140^\circ$, the two heat transfer mechanisms “conduction and radiation” are to be compared with each other. According to Eq. 3.2, the following value is obtained for heat transport by conduction in air ($\lambda = 24\ \text{mW/m} \cdot \text{K}$):

$$\dot{Q}_{\text{WL}} = 24 \frac{\text{mW}}{\text{m} \cdot \text{K}} \cdot \frac{(0.01\ \text{m})^2}{0.005\ \text{m}} \cdot (413\ \text{K} - 313\ \text{K}) = 48\ \text{mW} \tag{3.4}$$

Heat transfer by radiation can be calculated using the *Stefan-Boltzmann law* (Tipler and Mosca 2014):

$$\dot{Q}_{\text{Str}} = \sigma \cdot \varepsilon \cdot A \cdot \left(T_1^4 - T_2^4\right) \tag{3.5}$$

By inserting the values given above, as well as the Stefan-Boltzmann constant $\sigma = 5.67 \cdot 10^{-8}\ \text{W/m}^2 \cdot \text{K}^4$and the emissivity $\varepsilon \approx 0.055$(valid for a rough aluminium surface) one obtains the proportion of heat transfer by radiation:

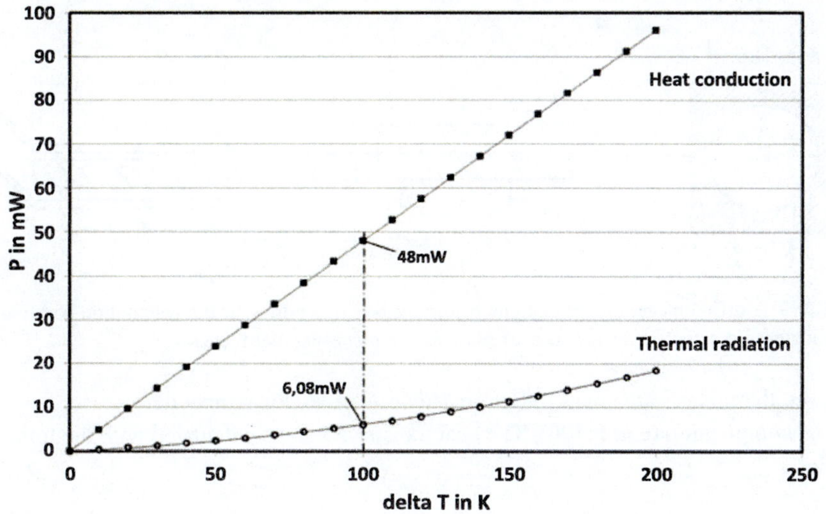

Fig. 3.7 Distribution of the different heat transfer capacities ($P = \dot{Q}$) depending on the temperature difference (ΔT)

$$\dot{Q}_{Str} = 5.67 \cdot 10^{-8} \frac{W}{m^2K^4} \cdot 0.055 \cdot (0.01\ m)^2 \cdot (413^4\,K^4 - 313^4\ K^4) = 6.08\ mW \quad (3.6)$$

The radiation share is therefore in a range of ≈10 % of the total heat transfer (Fig. 3.7). Since the temperature difference ΔT, due to the different thermal conductivities, changes by only a few °C, radiation and convection (if present) can be included as constant variables. Also, the heat transfer through the electrical connecting wires can be considered as a constant part.

Geometry of the Measuring Cell

With Eq. 3.1, different geometric designs can now be calculated. The calculation of plane-parallel plates (Fig. 3.8c) has already been described in Eq. 3.2. In this setup, the measuring element is considered, for example, as a thin-film sensor on a substrate (microchip with area A), which is located at a distance d from another area (e.g. cell wall). The gas mixture is introduced between the two surfaces. The necessary gas exchange must of course not lead to an additional heat transport by forced convection (flow), which would disturb the heat conduction sensitively. This requirement naturally applies to all cell geometries.

The use of thermistors, which are available in the form of a small bead with a typical diameter of d < 1 mm, leads to a concentric structure (Fig. 3.8a) of the cell geometry (Fig. 3.8a). Thermistors have a very large negative temperature coefficient (NTC) compared to platinum or nickel. This property allows TCS to be produced with high sensitivity.

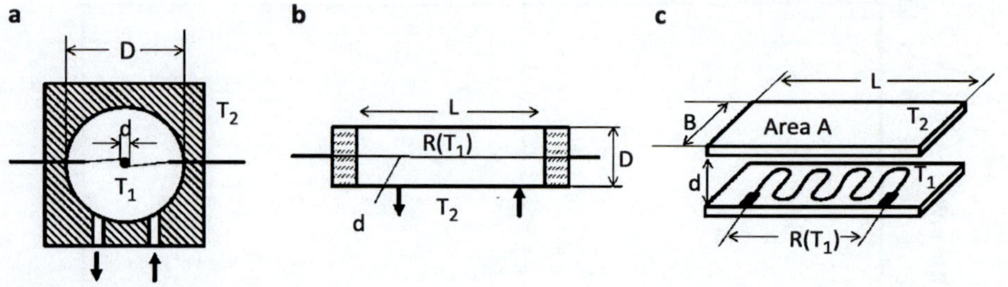

Fig. 3.8 The different geometric designs of the measuring cells with the calculation formulas for heat transfer. **a** concentric design, **b** coaxial design, **c** plane-parallel plates

However, thermistors are not as long-term stable as a metallic temperature sensor due to the material composition (e.g. Pt100). The heat transfer *P* can be calculated with the following equation:

$$\dot{Q}=P=\lambda\cdot 8\pi\cdot D\cdot d\cdot\left(\frac{T_1-T_2}{D-d}\right) \tag{3.7}$$

The simplest construction consists of a taut hot wire (e.g. Platinum) in a tubular measuring cell (3.8b). This coaxial structure was frequently used in the past and is also used for the absolute determination of thermal conductivity. Today, this cell geometry is hardly used anymore, as the production is very complex and therefore cost intensive. Furthermore, it is very difficult to flush the measuring cell without causing disturbing effects by forced convection. The calculation formula is the same as for the structures already described:

$$P=\lambda\cdot 2\pi\cdot l\cdot (T_1-T_2)/\ln\left(\frac{D}{d}\right) \tag{3.8}$$

The calculation formulas for the different geometric structures can also be presented in a simplified form, in which the geometric conditions are expressed by a geometry factor (cell factor) *G:*

$$P=\lambda\cdot G\cdot \Delta T \tag{3.9}$$

For an indeterminate cell geometry that differs from the three basic types, the geometry factor *G* can also be determined experimentally. For this purpose, Eq. 3.9 is converted to *G* and the temperature difference ΔT is measured for a known thermal conductivity λ of the gas and the electrical power ($P = U \cdot I$).

$$G = \frac{P}{\lambda \cdot \Delta T} \quad (3.10)$$

The unit for the geometry factor G is therefore meter [m].

Example

One measuring chamber has an inner diameter of 5 mm and a length of 20 mm, in which a Pt-100 measuring resistor with the diameter 1 mm is concentrically arranged. According to Eq. 3.8 there is a geometry factor of:

$$G = 2 \cdot \pi \cdot 0.02\ \text{m}/\ln(5\ \text{mm}/1\ \text{mm}) \approx 0.08\ \text{m} \quad (3.11)$$

Miniaturized gas sensors (e.g. MTC 2601, Neroxis SA, Switzerland) have measuring chamber sizes of $V < 0.1\ \text{cm}^3$and a geometry factor G of 3.9 mm (= 0.0039 m).

Temperature Dependence of Thermal Conductivity

The thermal conductivity λ can be described by the kinetic gas theory as follows:

$$\lambda = \frac{1}{2} \cdot f \cdot \eta \cdot c_V \quad (3.12)$$

If we now use Eq. 2.161 for the dynamic viscosity η and take the mean value 2 for $f = 1.6$ up to 2.4 (depending on the molecular structure), we obtain

$$\lambda \approx \frac{1}{3} \bar{v} \cdot l \cdot \varrho \cdot c_V \quad (3.13)$$

Now the mean free path length is replaced by Eq. 2.22, the mean velocity by Eq. 2.36 and the density by Eq. 2.87:

$$\lambda = \frac{1}{3} \sqrt{\frac{8 \cdot k \cdot T}{\pi \cdot m}} \cdot \frac{k \cdot \rho_N \cdot \left(\frac{273K}{1013\text{hPa}}\right) \cdot c_V}{\sqrt{2} \cdot \pi \cdot (r_1 + r_2)^2} \quad (3.14)$$

The temperature dependence of the thermal conductivity λ can be approximately described by a root function:

$$\lambda \approx \sqrt{T} \quad (3.15)$$

Figure 3.9 shows this relationship for the temperature range from T = 273 K to 373 K (=0°C-100 °C). The values were normalized to "1" for 273 K (0 °C). The slope of this curve

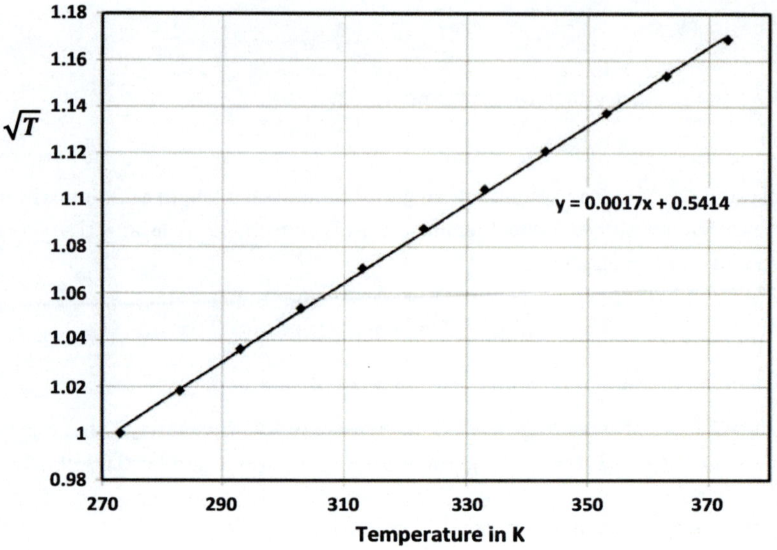

Fig. 3.9 Theoretical course of temperature dependence according to Eq. 3.15

is 0.0017 K^{-1} . Real values show a similar dependency, although the gradient can be very different. Figure 3.10 shows the curves for helium and carbon dioxide. Helium, as a monatomic gas, behaves almost ideally according to Eq. 3.15. The slope is 0.0022 K^{-1} . The gradient for carbon dioxide is more than twice as large at 0.0054 K^{-1} . This property, the different temperature sensitivity, can be used for gas analysis in complex gas mixtures (Göldner et al. 1987, and Grunewald 1999).

For large temperature ranges, which can go from −100 to 1000 °C, the thermal conductivity can be described by a fourth degree polynomial. The coefficients *A* to *E* can be found in the corresponding tables (e.g. VDI-Heat Atlas):

$$\frac{\lambda}{\mathrm{W/m \cdot K}} = A + B \cdot \frac{T}{K} + C \cdot \left[\frac{T}{K}\right]^2 + D \cdot \left[\frac{T}{K}\right]^3 + E \cdot \left[\frac{T}{K}\right]^4 \tag{3.16}$$

The temperature coefficient α from Table 3.1 was determined as follows:

$$\alpha = \left[\frac{\lambda_{200^\circ\mathrm{C}}}{\lambda_{0^\circ\mathrm{C}}} - 1\right] / 200^\circ\mathrm{C} \tag{3.17}$$

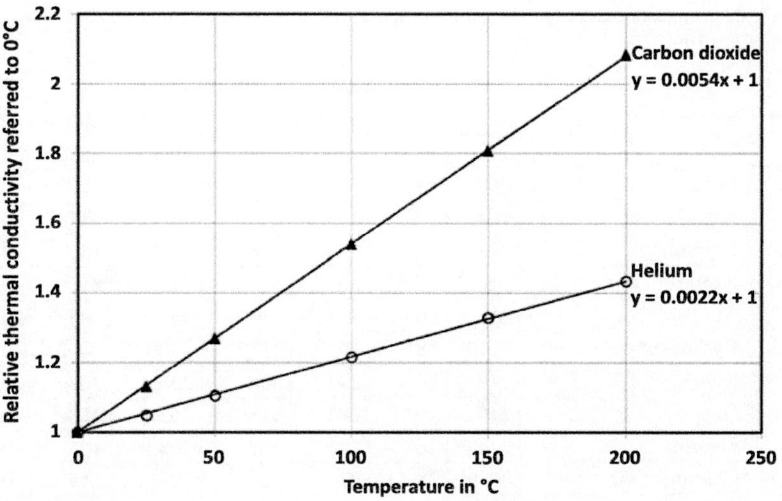

Fig. 3.10 Temperature dependence of thermal conductivity for carbon dioxide and helium. The absolute values were normalized to "1" at a temperature of 0°C (Data basis: Landolt-Börnstein 1968)

Pressure Dependence

According to the kinetic gas theory, the thermal conductivity λ is independent of pressure for single-atomic gases whose molecules consist of rigid spheres. In fact, this dependence is very low in the range from 0.2 to 2 bar and is only 1% per bar for most gases with non-polar molecules. For gases with large molecules the pressure dependence can be a few % per bar.

Only when the mean free path length, at a very low pressure, reaches the order of magnitude of the cell geometry, can a clear pressure dependence be observed. This effect is used, for example, for vacuum measurements (Pirani effect) in rough vacuum. Figure 3.11 shows the output signal of a measuring bridge with a Pirani sensor (TCS). Especially, the pressure range from 0.1 to 10 hPa can be used for vacuum measurement.

Atmospheric pressure changes (±100 hPa) can usually be completely neglected. Only with extremely miniaturized setups (micromechanics, cell geometry d < 1 mm), can atmospheric pressure dependence be demonstrated.

However, measurement in the vacuum range also depends on the type of gas. Figure 3.12 shows the measured value display of a Pirani pressure gauge for different pressures. The curve for air (nitrogen) is taken as the base curve to which the gauge is calibrated. The deviations for hydrogen and argon are particularly large at high pressures, whereas in the vacuum range (<1 hPa) only a constant deviation (offset) can be determined.

The thermal conductivity shows a rather unusual behavior at high pressures. In addition to the polarity of the gases, the temperature also plays a major role. Near the critical point this behavior becomes particularly clear. There is no generally valid theoretical explanation

Table 3.1 Thermal conductivity of various gases at different temperatures ($p \approx 1$ bar) (Source: D'Ans and Lax 1992)

Formula	Substance	Thermal conductivity λ inμ W/cm · K						
		0 °C	25 °C	50 °C	100 °C	150 °C	200 °C	α [K^{-1}]
He	Helium	1430	1500	1580	1740	1900	2050	0.0022
Ne	Neon	460	490	–	570	–	670	0.0023
Ar	Argon	164	177	189	213	235	255	0.0028
Kr	Krypton	87.8	95.1	102	116	129	142	0.0031
Xe	Xenon	51	55.5	60	69	77.7	81.5	0.0030
N_2/O_2	Air	241	260	276	314	354	385	0.0030
H_2	Hydrogen	1710	1810	1910	2110	2300	2490	0.0023
N_2	Nitrogen	240	259	277	308	338	367	0.0026
O_2	Oxygen	245	264	283	318	353	386	0.0029
F_2	Fluorine	243	264	286	327	368	406	0.0034
Cl_2	Chlorine	80	88.4	97	114	131	149	0.0043
Br_2	Bromine	43	47	50	58	66	74	0.0036
CO	Carbon monoxide	231	249	267	304	335	368	0.0030
CO_2	Carbon dioxide	145	164	184	223	262	302	0.0054
H_2O	Steam	182	197	–	246	296	332	0.0041
H_2S	Hydrogen sulfide	130	140	–	190	–	250	0.0046
NH_3	Ammonia	216	241	270	332	400	472	0.0059
NO	Nitric oxide	239	257	275	310	346	380	0.0029
NO_2	Nitrogen dioxide	–	140	–	190	–	250	0.0032
N_2	Nitrous oxide	153	173	193	236	280	322	0.0055
CS_2	Carbon disulfide	78	81	–	120	–	160	0.0053
SO_2	Sulphur dioxide	86.2	99.4	113	138	165	191	0.0061
SF_6	Sulphur hexafluoride	120	130	–	190	–	250	0.0054
CCl_4	Tetrachloromethane	128	143	–	189	–	250	0.0048
CH_4	Methane	303	337	371	442	520	605	0.0050
C_2H_6	Ethane	183	212	245	316	385	465	0.0077
C_3H_8	Propane	151	180	210	272	340	425	0.0091
C_4H_{10}	Butane	135	163	185	245	310	380	0.0091
C_5H_{12}	Pentane	130	150	173	228	294	340	0.0081
C_6H_{14}	Hexane	120	130	146	194	–	331	0.0088
C_2H_2	Acetylene/ethine	181	215	246	303	–	420	0.0066
C_6H_6	Benzene	89	107	125	165	214	260	0.0096
CH_4O	Methanol	137	157	195	236	276	340	0.0074
C_2H_2O	Ethanol	130	154	180	220	273	350	0.0085
C_3H_6O	Acetone	98	130	146	176	227	270	0.0088
CH_2O	Formaldehyde	140	160	–	220	–	320	0.0064
CH_5N	Methylamine	130	160	–	250	–	410	0.0108

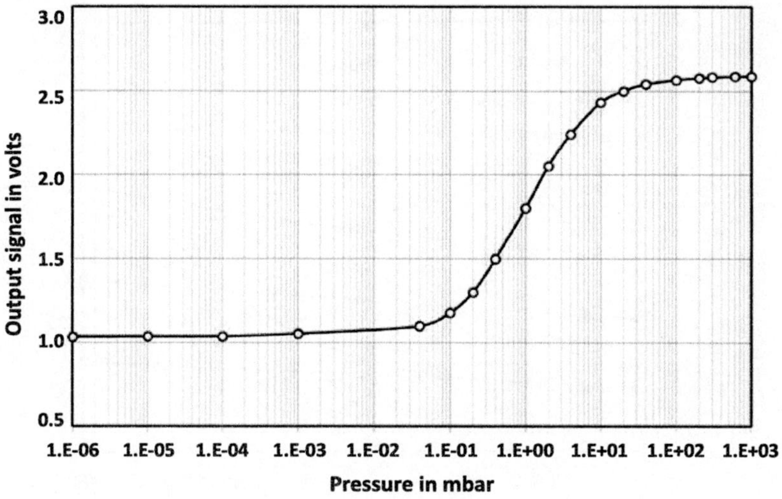

Fig. 3.11 Pressure dependence of a TC sensor (type MTC 2601) in the vacuum range ≫ 1 bar or 1000 hPa. (Based on the data of Neroxis SA Switzerland)

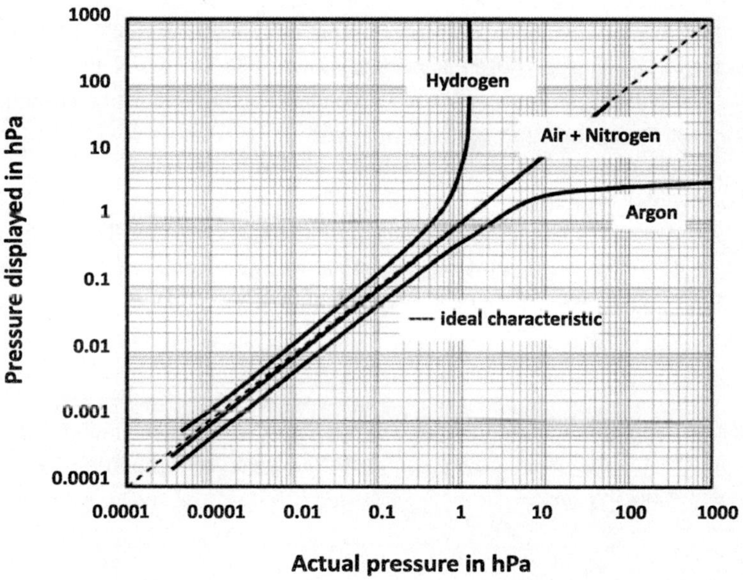

Fig. 3.12 Measuring error of pressure measurement in the vacuum range for different gases

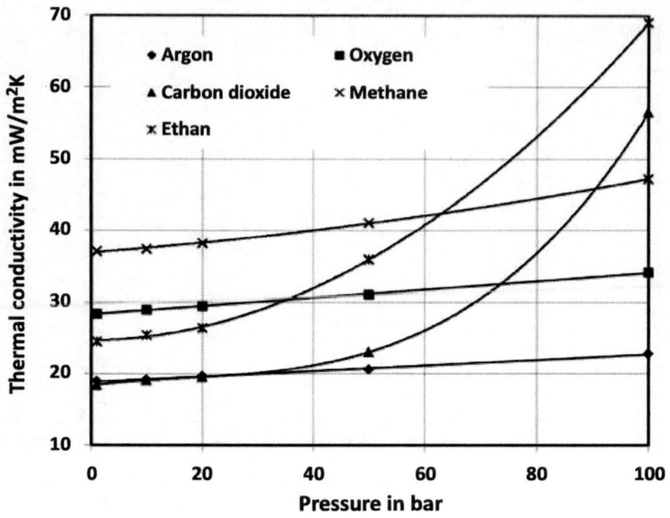

Fig. 3.13 Dependence of thermal conductivity on pressure (1–100 bar) for various gases at 50 °C. (Database: Landolt-Börnstein 1968)

for this, so that this area is dependent on experimental results. In Fig. 3.13, the pressure dependence up to 100 bar for 5 different gases is shown at 50 °C. Especially carbon dioxide and ethane show a disproportionate pressure dependence, which leads to intersections with other gases (oxygen, methane).

Gas Type Dependence

The type of gas has a very large influence on the thermal conductivity. The most important influencing factor is the molar mass M of the respective substance. If, for example, in Eq. 3.12, the specific heat capacity c_V *is* set to the molar heat capacity c_V according to

$$C_V = M \cdot c_V \tag{3.18}$$

then you get an expression for the thermal conductivity λ in which the molar mass M is the denominator. The thermal conductivity λ is therefore inversely proportional to the molar mass M of the respective gas.

$$\lambda = f \cdot \eta \cdot \frac{C_V}{M} \qquad \text{d.h.}\lambda \approx \frac{1}{\text{M}} \tag{3.19}$$

If the measured thermal conductivity is plotted against the molar mass M, a decreasing curve with increasing molar mass M is obtained according to Eq. 3.14. It can be clearly seen

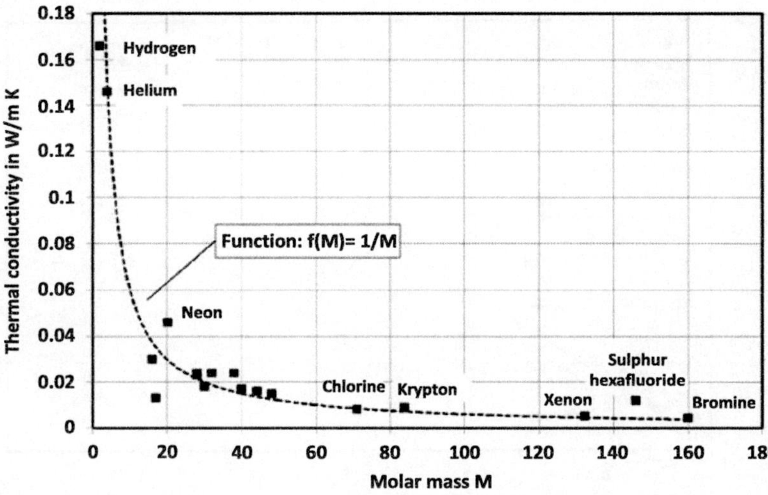

Fig. 3.14 Dependence of thermal conductivity for 18 gases between the maximum value (hydrogen) and the minimum value (bromine). The values were taken from the VDI heat atlas at 0 °C

that not all gases lie on the $1/M$ curve. The thermal conductivity λ is therefore also influenced by other molecular structures (e.g. dipole moment).

If one takes the values from Fig. 3.14 and enters these data over $1/M$, a linear relationship should be established. Figure 3.15 shows this relationship. In fact, this correlation is now almost linear. The deviations for the gases methane, ammonia and neon are also clearly visible in this diagram.

Gas Mixtures

The thermal conductivity of any gas mixture is composed of the additive concentrations ($c_1, c_2, \ldots, c_n$) and thermal conductivities ($\lambda_1, \lambda_2, \ldots, \lambda_n$) of the individual components in the mixture. The concentrations c_n are usually given as mole fractions or molar contents. In practice, the concentrations are available as volume %, so that c_n can also be given as follows:

$$c_n = \frac{c[\text{Vol.-\%}]}{100\text{Vol.-\%}} \tag{3.20}$$

The resulting thermal conductivity λ_M can then be calculated using the mixing formula:

$$\lambda_M = \lambda_1 c_1 + \lambda_2 c_2 + \ldots + \lambda_n c_n \tag{3.21}$$

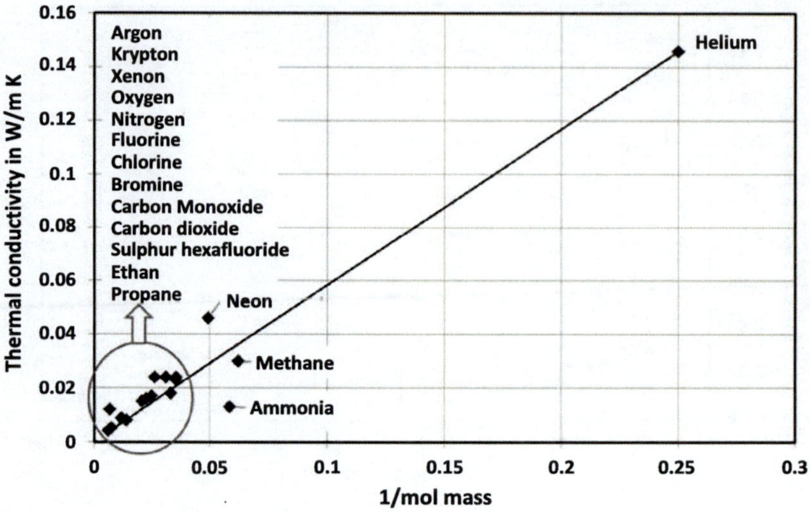

Fig. 3.15 Thermal conductivity λ applied via 1/*M*. The database is the same as in Fig. 3.14

For quantitative gas analyses, the thermal conductivity method can usually only be used for binary gas mixtures, since changes in concentration in a multi-component mixture cause the clear assignment to be lost. Ternary gas mixtures (e.g. air, hydrogen, carbon dioxide) can also be analyzed by means of additional information (CO_2 measurement with an NDIR analyzer).

Unfortunately, this linear approach is only valid for a few gas mixtures. Depending on the gas combination, the deviations from Eq. 3.21 can be considerable. To be able to calculate the thermal conductivity λ for a known gas mixture nevertheless, a correction factor *r* is introduced. However, this factor cannot be calculated exactly, but must be determined experimentally by tests.

$$r = \frac{\lambda_{\text{Mess}}}{\lambda_{\text{Theory}}} \tag{3.22}$$

This correction factor *r* is then used to determine the real thermal conductivity of the gas mixture. Information on the correction factor *r can* be found in the D'Ans-Lax and Landolt Börnstein.

$$\lambda_M = r \cdot (c_1\lambda_1 + c_2\lambda_2) \qquad \text{or.}\lambda_M = r \cdot [c_1\lambda_1 + (1 - c_1)\lambda_2] \tag{3.23}$$

However, the correction factor also changes with the mixing ratio c_1/c_2. The correction factor is therefore given for 5 different ratios. Table 3.2 shows some important gas mixtures with the corresponding correction factors.

Table 3.2 Correction factor *r* for different binary gas mixtures (According to D'Ans-Lax)

Mix	*T* in ° C	*r* for $c_1 \cdot 10^2$				
		10	25	50	75	90
Ar-H_2	101	0.99	0.97	0.97	0.96	0.98
Ar-he	38	0.77	0.67	0.68	0.76	0.88
Ar-O_2	29	0.88	0.75	0.65	0.62	0.71
CO-CO_2	38	0.96	0.93	0.94	0.96	0.98
CO-N_2	101	1.00	1.00	1.00	1.00	1.00
CO_2-H_2	25	0.87	0.77	0.63	0.6	0.71
CO_2-N_2	50	0.99	0.97	0.96	0.96	0.98
H_2-He	50	0.99	0.97	0.96	0.96	0.98
H_2NH_3	25	0.81	0.72	0.75	0.83	0.91
He-Kr	25	0.86	0.77	0.79	0.85	0.94
He-Ne	18	0.6	0.52	0.55	0.70	0.85
Kr-N_2	18	0.93	0.89	0.91	0.94	0.98
Kr-He	30	0.96	0.93	0.88	0.87	0.92
N_2-H_20	18	0.98	0.97	0.96	0.97	0.98
N_2-NH_3	65	1.04	1.07	1.09	1.08	1.04
Xe-H_2	75	1.05	1.11	1.14	1.12	1.08
Xe-N_2	45	0.84	0.69	0.56	0.49	0.49
Xe-O_2	45	0.90	0.84	0.78	0.81	0.89
CH_4-C_3H_8	45	0.91	0.86	0.81	0.78	0.88
CO_2-C_2H_8	75	0.98	0.95	0.93	0.94	0.96
Air CH_4	95	1.00	1.00	1.00	1.00	1.00
Ar-H_2	22	1.00	1.00	1.00	1.00	1.00

Figure 3.16 shows the correction factors for the different concentration ratios for 7 gas mixtures. The factor can be either >1 or<1. The mixture, nitrogen-ammonia and air-water vapor have, for example, an *r-factor*>1 . For gas mixtures of noble gases and hydrogen the value is <1. Obviously, gas mixtures with polar gases lead to a value of $r > 1$, while mixtures with non-polar gases and clearly different molecular weights have a value of $r < 1$. The mixture of nitrogen and carbon monoxide has a factor of $r = 1$ throughout, since both gases have the same molecular weight (Wiegleb 1980). The influence of the dipole character on the *r-factor* was already known by Schmick in 1928.

Since the data basis for the *r-factor is* limited to 5 values only, one has to extrapolate for other values in between. Usually a linear approach is sufficient for this purpose. Alternatively, the nonlinear relationship can also be described by a higher-order polynomial. Figure 3.17 shows the course for a mixture of helium in neon and water vapor in air. The fourth-order polynomials fit quite well with the actual curves, so that the intermediate values can be calculated with sufficient accuracy. However, it must be pointed out that these curves can change with temperature. In the relevant tables, therefore, pairs of values for different temperatures are also given. Unfortunately, these values are not sufficiently

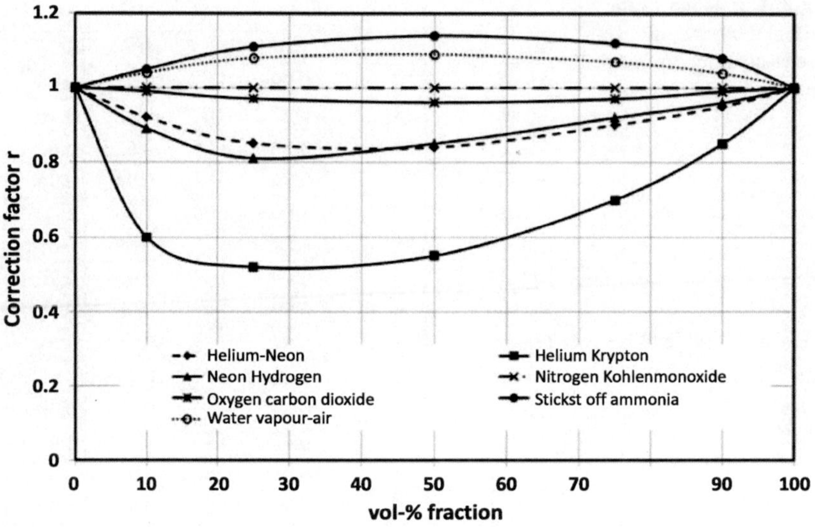

Fig. 3.16 Display of the *r-factor* for different mixing ratios according to D'Ans-Lax

available for all gas mixtures and all temperatures, so in practical cases one must carry out own investigations. Manufacturers of measuring instruments calibrate the gas measuring instruments only for a special gas mixture, so that these influencing factors are automatically included in the factory calibration. However, this connection is problematic if it is a gas mixture with more than two components.

Example

In an incubator,[2] both carbon dioxide (5% by volume) and water vapor (37 °C dew point) are added to the air. The water vapour content is predetermined by a water container which, at a chamber temperature of 37 °C, results in a relative humidity of almost 37%. The CO_2 concentration, on the other hand, is controlled by adding carbon dioxide. This is therefore a 3-component mixture consisting of air (N_2/O_2), water vapor and carbon dioxide. However, since the water vapor is always saturated at 37 °C, it can be considered constant. WL sensors are usually used for this control, which have a sufficient accuracy. At a sensor temperature (*T1*) of 100 °C the high purity gases (nitrogen was chosen for air) have the following thermal conductivities:

$$\lambda_{N_2} = 30.8 \frac{\text{mW}}{\text{K} \cdot \text{m}}, \lambda_{CO_2} = 22.3 \frac{\text{mW}}{\text{K} \cdot \text{m}}, \lambda_{H_2O} = 24.6 \frac{\text{mW}}{\text{K} \cdot \text{m}}$$

[2]Incubator in biotechnology for cell and tissue culture.

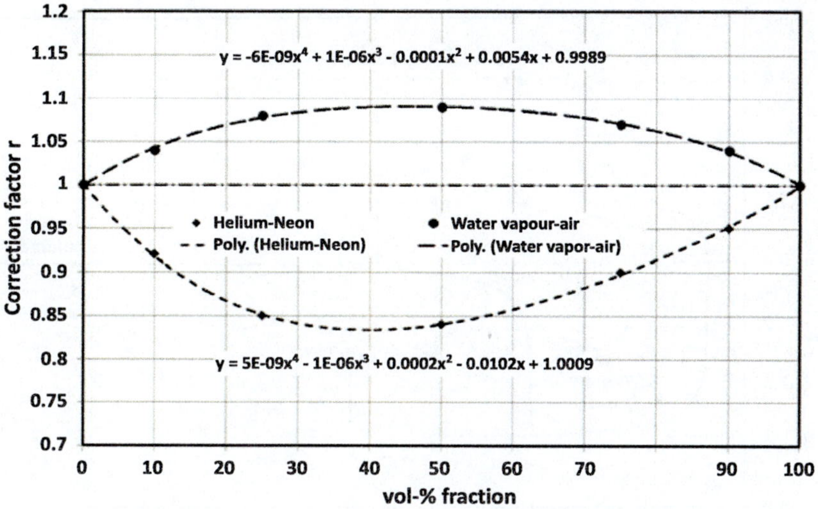

Fig. 3.17 Approximation of the measured *r-factors* by a polynomial of order 4 $r(c)$

According to the mixing rule, for 5 vol.% CO_2 and 4.5 vol.% H_2O (measured humidity of approximately 70% rel. H), a thermal conductivity of:

$$\lambda_{N_2+CO_2} = 30,375\frac{\text{mW}}{\text{K}\cdot\text{m}} \rightarrow \lambda_{N_2} - \lambda_{N_2+CO_2} = \Delta\lambda_{N_2+CO_2} = 0.425\frac{\text{mW}}{\text{K}\cdot\text{m}}$$
$$\lambda_{N_2+H_2O} = 30,521\frac{\text{mW}}{\text{K}\cdot\text{m}} \rightarrow \lambda_{N_2} - \lambda_{N_2+CO_2} = \Delta\lambda_{N_2+H_2O} = 0.279\frac{\text{mW}}{\text{K}\cdot\text{m}}$$

According to these calculations, a TC gas sensor should provide a positive signal for both components, whereby the CO_2 signal should be approximately 50% higher than the water vapor signal. In fact, a signal is measured which is the same for both mixtures, but points in different directions. Figure 3.18 shows the measurement result in relation to the zero gas (air or N_2). For the CO_2 mixture, a positive voltage signal of approximately 3 V is obtained, while the H_2O mixture generates a negative voltage signal of approximately 2.8 V (Hildebrandt 2013).

If one now uses the results from Fig. 3.17, an *r-factor* of 1.018 can be determined for a gas mixture of approximately 4.5 vol.% CO_2 in N_2. The thermal conductivity of the nitrogen-H_2O mixture then changes as follows

$$r\cdot\lambda_{N_2+H_2O} = 1.018\cdot 30.521\frac{\text{mW}}{\text{K}\cdot\text{m}} = 31.07\frac{\text{mW}}{\text{K}\cdot\text{m}}$$

This value results in a change of:

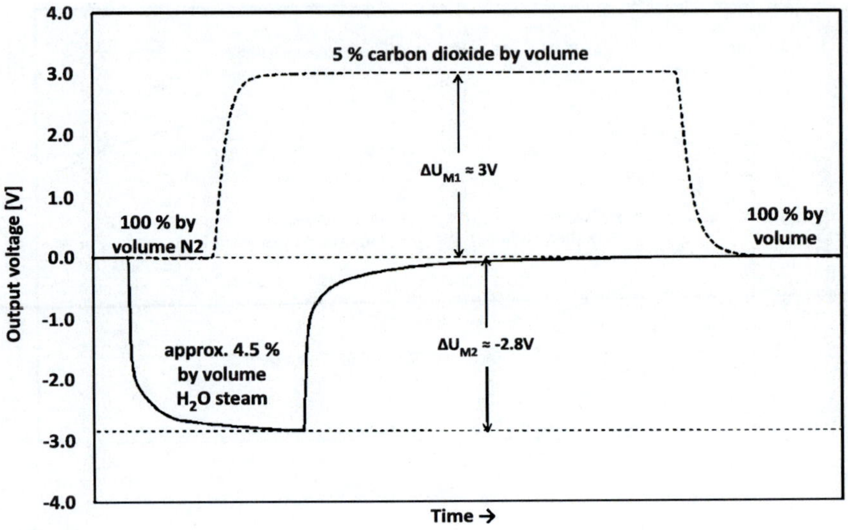

Fig. 3.18 A TC gas sensor was successively exposed to a mixture of nitrogen-CO_2 and nitrogen-H_2O and the signal curves were recorded

$$\Delta\lambda_{N_2+H_2O} = -0.27\frac{\mathrm{mW}}{\mathrm{K}\cdot\mathrm{m}}$$

This value now points in the same direction (negative) as the measured values and also the order of magnitude is the same. In the described incubator application, this water vapor error is considered a constant value by adjusting the zero point with the air-water vapor mixture.

Another possibility is to calculate the thermal conductivity λ_M by an approach (*WMS equation*) of *Wassilijewa*, *Mason* and *Saxena.* In this approach, the dynamic viscosity η and the molar mass *M* are considered (Poling et al. 2001):

$$\lambda_M = \frac{\lambda_1}{1+\frac{c_2}{c_1}\cdot F_1} + \frac{\lambda_2}{1+\frac{c_1}{c_2}\cdot F_2} \tag{3.24}$$

The two factors F_1 and F_2, result from the following relationship:

$$F_1 = \frac{\left[1+\left(\frac{\eta_1}{\eta_2}\right)^{1/2}\cdot\left(\frac{M_2}{M_1}\right)^{1/4}\right]^2}{\sqrt{8\left(1+\frac{M_1}{M_2}\right)}} \quad \text{and} \quad F_2 = \frac{\left[1+\left(\frac{\eta_2}{\eta_1}\right)^{1/2}\cdot\left(\frac{M_1}{M_2}\right)^{1/4}\right]^2}{\sqrt{8\left(1+\frac{M_2}{M_1}\right)}} \tag{3.25}$$

Example: CO_2 in N_2

Now the thermal conductivity of a gas mixture of 0–100 vol-% CO_2 in N_2 shall be calculated according to the *WMS equation.* Given:

$$M_{CO_2} = 44.00\ \text{g/mol} \qquad \eta_{CO_2} = 18.47\,\mu\text{Pa}\cdot\text{s} \qquad \lambda_{CO_2} = 22.87\ \text{W/K}\cdot\text{m}$$
$$M_{N_2} = 28.01\ \text{g/mol} \qquad \eta_{N_2} = 21.10\ \mu\text{Pa}\cdot\text{s} \qquad \lambda_{N_2} = 18.47\ \text{W/K}\cdot\text{m}$$

With these values the two factors F_1 and F_2 can now be calculated:

$$F_1 = \frac{\left[1 + \left(\frac{18.47}{21.1}\right)^{\frac{1}{2}} \cdot \left(\frac{28.01}{44}\right)^{\frac{1}{4}}\right]^2}{\sqrt{8\left(1 + \frac{44}{28.01}\right)}} = 0.7429 \tag{3.26}$$

$$F_2 = \frac{\left[1 + \left(\frac{21.1}{18.47}\right)^{\frac{1}{2}} \cdot \left(\frac{44}{28.01}\right)^{\frac{1}{4}}\right]^2}{\sqrt{8\left(1 + \frac{28.01}{44}\right)}} = 1.3677 \tag{3.27}$$

If these two factors are now added to Eq. 3.24, the equation for determining the thermal conductivity of the gas mixture for the concentration range 0–100 vol.% is obtained. In Fig. 3.19, this connection is represented graphically. For comparison, the uncorrected, linear mixture formula was also drawn in. The correction with the r-factor is also shown in the diagram (= experimental values). The *WMS equation* comes very close to the experimentally determined values (*r-factor*) and can therefore be used for all gas mixtures for which the *r-factor* is not known.

Sensor Design

In the measuring chamber of a thermal conductivity sensor, there is a temperature sensor which is heated by a measuring current I_M. Normally this temperature T_1 is approximately 100 °C above the block temperature (T_2). If the sensor block is temperature controlled (e.g. on 50 °C), the temperature T_1 is 150 °C. Between the temperature sensor and the inner wall of the measuring chamber the heat conduction through the measuring gas then takes place. The gas exchange is mostly done by diffusion of the gas from outside into the measuring chamber. Forced flows should be avoided at all costs in order not to disturb the measuring effect of the heat transfer. Aluminum or stainless steel is selected as the material for the sensor block. Aluminum has the advantage that it is mechanically easier to process and has very good heat conduction. This property is advantageous for thermostatization the sensor block. Stainless steel is preferred especially for corrosive gas mixtures (chlorine gas +humidity) as it is much more resistant than Aluminum. Figure 3.20 shows a sensor construction which explains the basic functionalities.

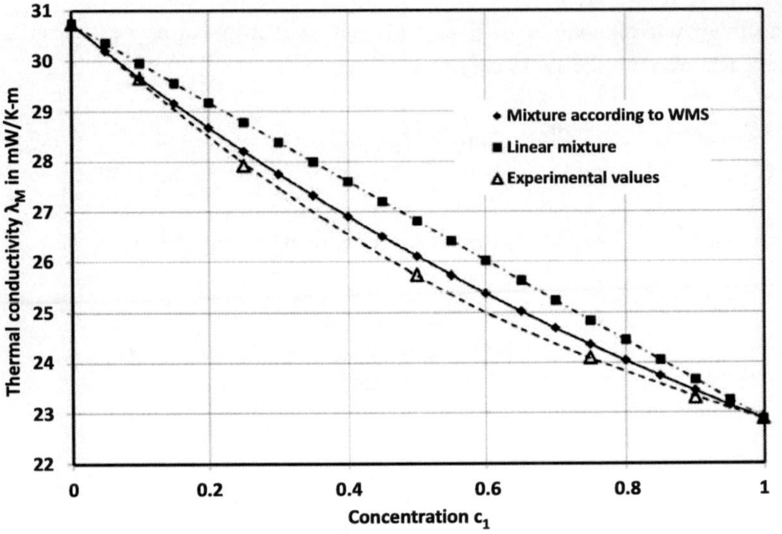

Fig. 3.19 Calculated thermal conductivities of a gas mixture of 0–100 vol% carbon dioxide in nitrogen by different methods. (100 vol.% 1=)

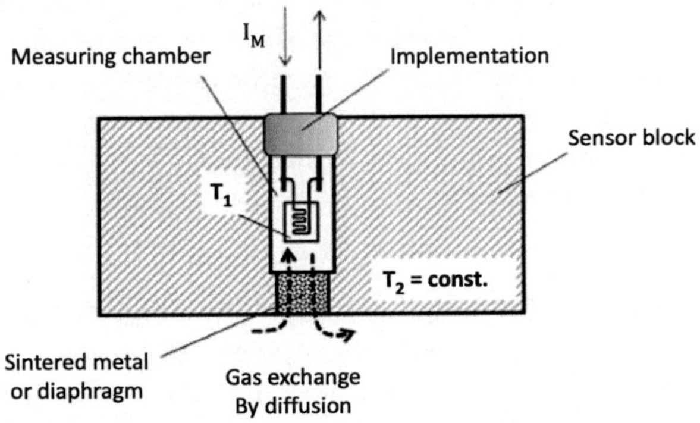

Fig. 3.20 Basic structure of a thermal conductivity sensor (TCS)

Today, thin-film sensors based on platinum (Pt 100) or nickel (Ni 100) are mostly used as sensor elements (Fig. 3.21). The change in resistance, as a function of the temperature *T* (in °C), of a Pt 100 is calculated according to DIN IEC 751:

$$R(T) = 100\ \Omega \cdot \left(1 + 3.90802 \cdot 10^{-3}\,{}^{\circ}\mathrm{C}^{-1} \cdot T - 5.802 \cdot 10^{-7}\,{}^{\circ}\mathrm{C}^{-2} \cdot T^2\right) \tag{3.28}$$

This correlation applies to a temperature range from 0 to 850 °C. For small temperature ranges (0–100 °C) a linear approach can be chosen:

Fig. 3.21 Measuring resistor (Pt 100) mounted on a pressure glass feedthrough for use in a TCS. (Source: Fachhochschule-Dortmund)

$$R(T) = 100\ \Omega \cdot \left(1 + 3.85 \cdot 10^{-3\,\circ}\mathrm{C}^{-1} \cdot T\right) \tag{3.29}$$

A temperature sensor based on a Ni-100 can also be described by a second degree polynomial in a temperature range from −20 to130° C (Karbach 1997):

$$R(T) = 100\ \Omega \cdot \left(1 + 5.4506 \cdot 10^{-3\,\circ}\mathrm{C}^{-1} \cdot T + 7.2742 \cdot 10^{-6\,\circ}\mathrm{C}^{2} \cdot T^{2}\right) \tag{3.30}$$

Nickel-based temperature sensors provide a greater change in resistance, but have stability problems at higher temperatures $T > 200^{\circ}$C. Furthermore, the thin nickel layer also absorbs hydrogen and thus changes its electrical resistance. For this reason, this material is only suitable for applications where no hydrogen is present. Alternatively, the nickel layer can be additionally coated with a gas-impermeable protective layer (e.g. SiO_2).

The measuring resistor is heated to temperature T_1 by the measuring current I_M. The electrical power loss P is 90% ($\rightarrow \eta = 0.9$) due to heat conduction through the gas into the sensor block. The remaining 10% is dissipated by parasitic effects (radiation, convection etc.). In a stationary state, a constant temperature gradient from the sensor resistance to the chamber wall is also formed $\Delta T = T_1 - T_2$. It then applies:

$$I_M^2 \cdot R_1(1 + \alpha \cdot T_1) \cdot 0.9 = \lambda \cdot G \cdot (T_1 - T_2) \tag{3.31}$$

$$\lambda = 0.9 \cdot \frac{I_M^2 \cdot R_1(1 + \alpha \cdot T_1)}{G \cdot (T_1 - T_2)} \tag{3.32}$$

If you convert Eq. 3.31 to T_1, you get the following expression with $\eta \approx 1$:

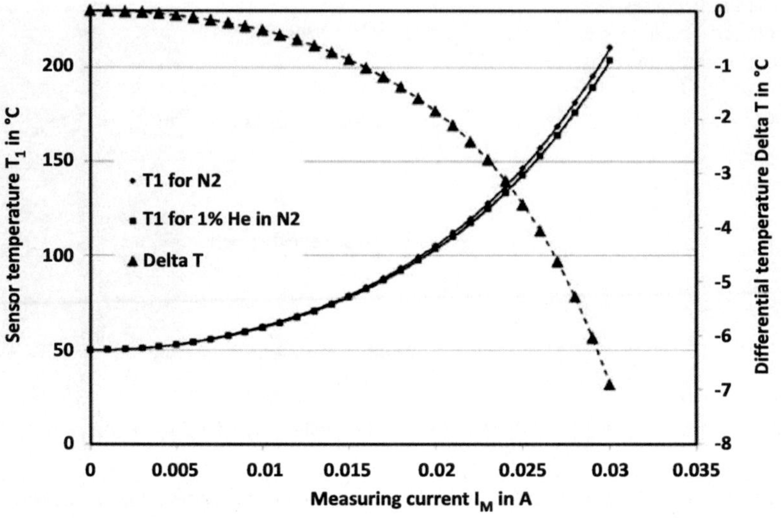

Fig. 3.22 Change of the sensor temperature with increasing measuring current for a given geometry ($G = 0.03$ m). For comparison, the curve was calculated for nitrogen and a 0.5 vol% hydrogen in nitrogen. The values for temperature difference (Delta T) between the two curves were also plotted

$$T_1 = \frac{I_M^2 \cdot R_1 + \lambda \cdot G \cdot T_2}{\lambda G - I_M^2 \cdot R_1 \cdot \alpha} \tag{3.33}$$

With real assumptions for $G = 0.03$ m, $\lambda_{N_2} = 0.034 \mathrm{Wm}^{-1} K^{-1}$, $\alpha = 0.0039$ K^{-1}, $T_2 = 50^{\circ}$C, $R_1 = 100$ Ω, the temperature T_1 was calculated and shown in Fig. 3.22. The course can be approximated very well with a third degree polynomial, which is also consistent with the theory (Hengstenberg et al. 1980). For comparison, a curve was plotted for a gas mixture of 0.5 vol% hydrogen in nitrogen, with a calculated thermal conductivity of $\lambda_{H_2 \text{ in } N_2} \approx 0.0035$ Wm^{-1} K^{-1}.The temperature difference *Delta T* increases disproportionately with increased measuring current. At a sensor temperature of 150 °C (corresponds to a temperature increase of $\Delta T = 100$ K) the difference is approximately -4 K (Fig. 3.23). The sensor resistance thus cools down due to the better thermal conductivity of the hydrogen mixture. Assuming that this is the entire gas measuring range (MB =0to 0.5 vol.% H_2 in N_2), this measuring range must be accurate to 1%. Related to the temperature T_1 this means a stability of ±0.02 K.

For evaluation, the TCS is connected in a Wheatstone measuring bridge (Fig. 3.24). In this case, the bridge current is divided into both bridge branches. If there is only one active element in the measuring bridge, the current I_{Br2}through the second bridge branch is kept as low as possible. For this purpose, resistors R_3 and R_4 are selected to have a very high resistance (×1000) compared to R_1 and R_2. The entire current I_{Br} . then only passes through

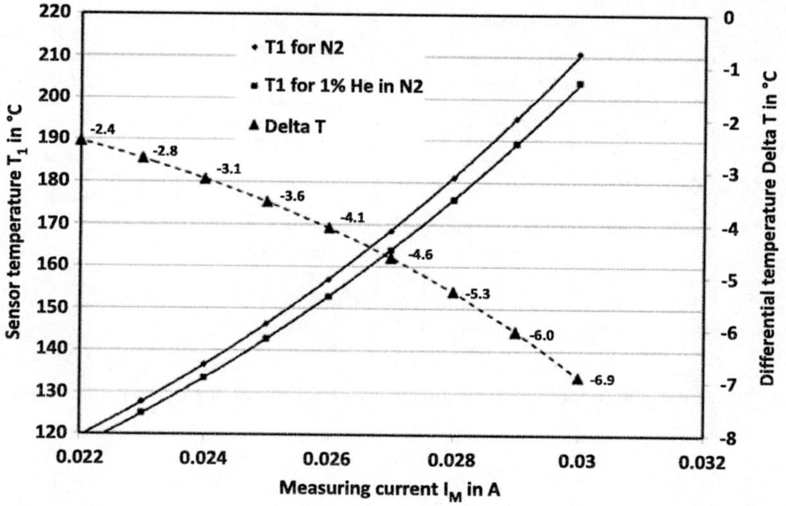

Fig. 3.23 Results from Fig. 3.22 with a different scale to show the small temperature differences between the two different gas curves. The delta *T-values* for the discrete data points were also shown in the diagram

the active bridge branch. The current I_{Br1} corresponds to the measuring current I_M. The bridge voltage ΔU_{Br} can then be calculated as follows:

$$\Delta U_{Br} = U_{Br}\left[\frac{R_2}{R_2 + R_1(1 + \alpha \cdot \Delta T)} - \frac{R_4}{R_4 + R_3}\right] \tag{3.34}$$

With $R_3 = R_4$, Eq.. 3.34 can also be written in the simplified form:

$$\Delta U_{Br} = U_{Br}\left[\frac{R_2}{R_2 + R_1(1 + \alpha \cdot \Delta T)} - \frac{1}{2}\right] \tag{3.35}$$

With this equation the measuring voltage can now be calculated. Assuming that the bridge adjustment was done with nitrogen, the value for R_2 must correspond to the value of R_1 at 150 °C. With a TC of $\alpha = 0.00385\,K^{-1}$, the value of 157.75Ω for a Pt-100 measuring resistor[3] results.

[3] 100 Ω at 0 °C according to DIN IEC 751.

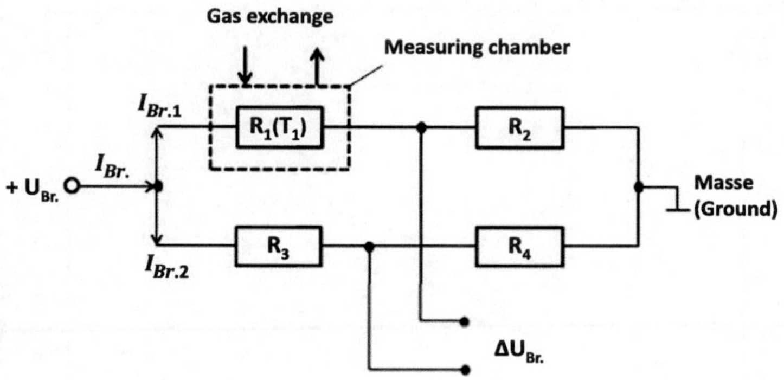

Fig. 3.24 Measuring bridge for evaluating the change in resistance due to the thermal conductivity in the measuring cell

$$\Delta U_{\mathrm{Br}} = 5\ \mathrm{V}\left[\frac{157.75\Omega}{157.75\ \Omega + 100\ \Omega\left(1 + 0.00385\ \mathrm{K}^{1} \cdot \left[150^{\circ}\mathrm{C} - 4^{\circ}\mathrm{C}\right]\right)} - \frac{1}{2}\right]$$
$$= 12.3\ \mathrm{mV} \qquad (3.36)$$

With a supply voltage of the bridge of 5 V, one thus obtains a measurement voltage of 12.3 mV and a resulting stability requirement of ≈100 μV. With a highly integrated PGA,[4] this measurement task can be solved without any problems.

However, a setup as shown in Fig. 3.24 has the disadvantage that unavoidable temperature changes are transferred to the measurement signal and thus lead to distortions. The calculations have shown that temperature changes in the 1/1000 K range can already lead to measurable and thus disturbing effects. Thermostatization of the sensor block is also possible with great effort only in the ≈1/10 Kelvin range.

However, the Wheatstone bridge can compensate for this error influence by using a second setup as shown in Fig. 3.20 with a closed reference chamber. This reference chamber is filled with air, nitrogen or another reference gas, and hermetically sealed (Fig. 3.25). Room temperature changes are also detected by this reference sensor and can be compensated accordingly in the bridge circuit. Both resistors (1 + 2) are located in one pointer of the measuring bridge. If both resistors change equally, for example, due to an increase in ambient temperature, this does not lead to a change in the bridge voltage ΔU_{Br}. However, the influence of heat transfer due to changes in gas concentration in the measuring chamber is only detected by this resistance. The reference channel remains unaffected and the measuring signal ΔU_{Br}can then be clearly assigned to the gas concentration.

[4]Programmable Gain Instrumentation Amplifier.

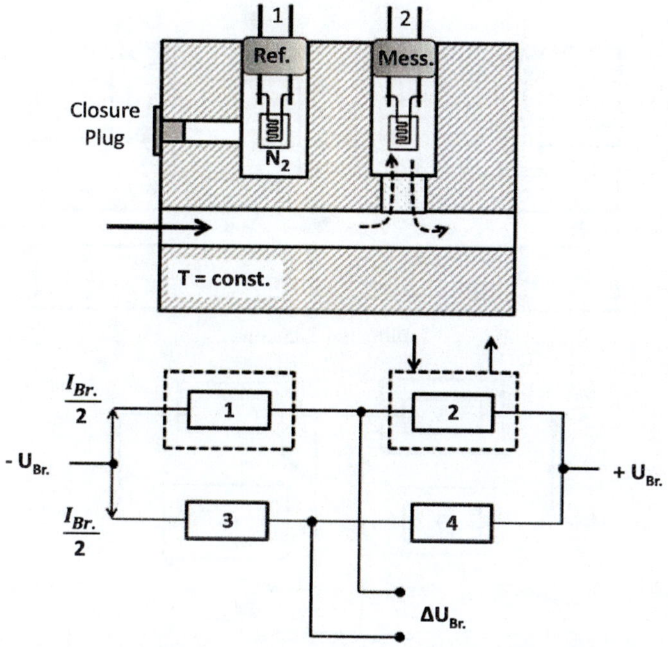

Fig. 3.25 Sensor block with integrated reference measurement (e.g. Nitrogen)

The bridge voltage is calculated in the same way as Eq. 3.34, but with the difference that R_1 and R_2 are temperature-dependent resistors. If the temperature change ΔT is caused by external influences (e.g. room temperature fluctuations), this effect is almost completely eliminated from Eq. 3.37.

$$\Delta U_{\mathrm{Br}} = U_{\mathrm{Br}} \left[\frac{R_2(1 + \alpha \cdot \Delta T)}{R_2(1 + \alpha \cdot \Delta T) + R_1(1 + \alpha \cdot \Delta T)} - \frac{1}{2} \right] \tag{3.37}$$

It is important in this context that the gas filling of the reference chamber remains stable over a long period of time (several years), otherwise the measurement signal would drift away. Bonded connections are therefore only conditionally suitable for this task, so that soldered or welded connections/sealings must be used. However, this also leads to a considerable effort in sensor installation.

In certain sensor setups, there is also a Wheatstone full bridge in which all four resistors are active (Fig. 3.26). The advantage of these superstructures is that the measuring signal is twice as large, since two identical measuring sensors are available. Also, the compensation of environmental influences is improved by two reference cells. The two measuring cells are offset from each other in the measuring bridge in different bridge branches. This is necessary to obtain the double measuring effect. In this case, the resistors 2 + 3 are flowed

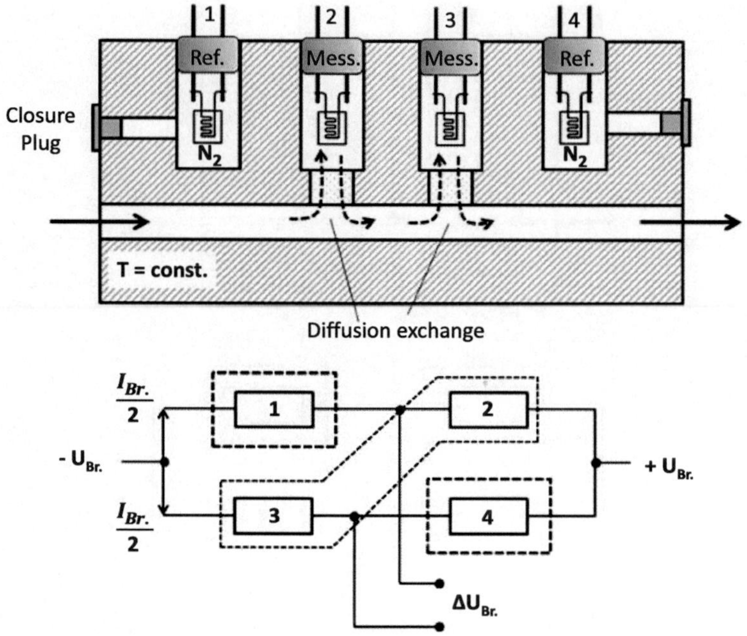

Fig. 3.26 Arrangement of the 2 measuring and 2 reference resistors in the sensor block and in the bridge circuit

through with the sample gas ($\rightarrow \Delta T^*$), while the resistors 1 + 4 are filled with a reference gas. The bridge voltage is then calculated using the following equation:

$$\Delta U_{\mathrm{Br}} = U_{\mathrm{Br}}\left[\frac{R_2(1+\alpha\cdot\Delta T^*)}{R_2(1+\alpha\cdot\Delta T)+R_1(1+\alpha\cdot\Delta T)} \frac{R_4(1+\alpha\cdot\Delta T)}{R_4(1+\alpha\cdot\Delta T)+R_3(1+\alpha\cdot T^*)}\right] \tag{3.38}$$

Another option, which is often used, is to also flow through the reference chambers (see Figs. 3.27 and 3.28). In this case, however, a continuously flowing reference gas is required. This option is often used to determine changes in a gas mixture or to monitor the purity of a gas. This is required, for example, for the production of gases (O_2 and N_2) using the Linde process (air separation).

Miniaturized TCS

Since the response time of TCS depends on the chamber volume to be flushed, an attempt was made to keep this volume as small as possible. Especially in gas chromatography (GC) this approach was used to reduce the response times to $t_{90} < 1$ s. In process measurement technology, on the other hand, glass-encapsulated platinum resistors were used, which were integrated in appropriately large measuring chambers. The response

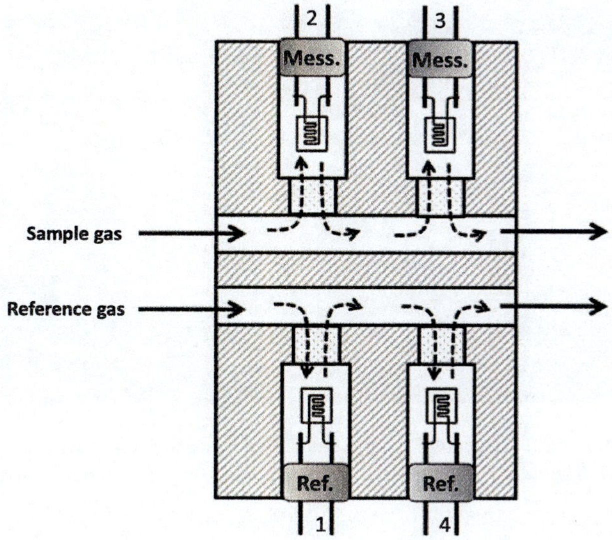

Fig. 3.27 Sensor design with four active sensor elements. If the two gas flows (measuring gas and reference gas) are identical, the bridge voltage is zero. If the two gases differ in their composition, the measuring signal changes

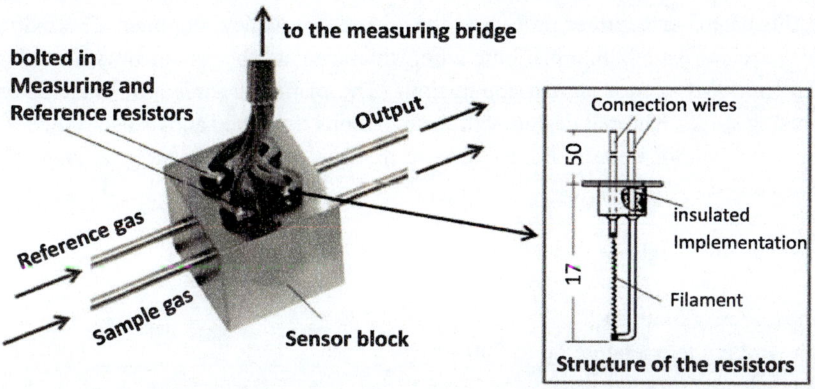

Fig. 3.28 Thermal conductivity sensor with 2 measuring and 2 reference resistors in one measuring bridge. The reference path is also flushed. The resistors consist of individual platinum wire filaments. (Source: Gow-Mac Instrument Co. Bethlehem USA)

times were then at $t_{90} > 10$ s. A first approach to the miniaturization of TCS, which were suitable for process gas analysis, was demonstrated by Schunck in 1979 and is still used today in a gas measuring device (*Conthos*, LFE GmbH Bruchköbel). The measuring cell consists of several ceramic parts, which are fused together by a high-temperature glass solder connection. Inside the microcell (Fig. 3.29) there are two wire loops that are bent

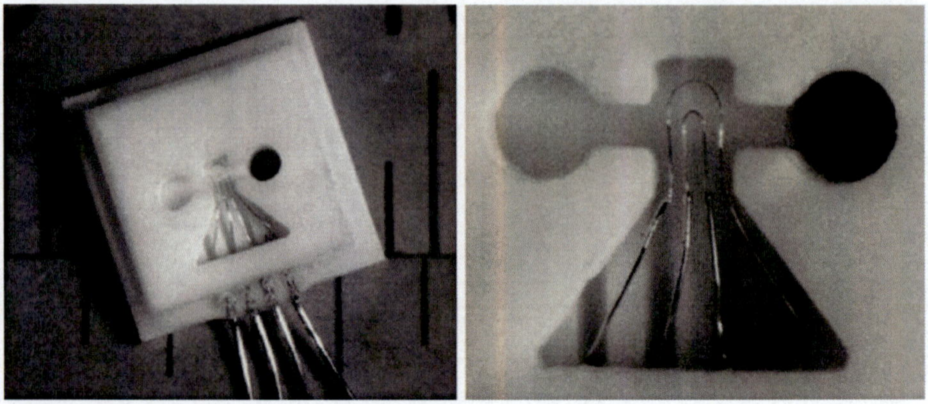

Fig. 3.29 Miniaturized TCS according to Schunck in corrosion-resistant design (ceramic-glass solder connection). The entire structure has a dimension of only 5 × 5 mm. (Source: LFE GmbH Bruchköbel)

into a U-shape. The inner loop (3 → 4) is heated by a current, while the second outer loop (1 → 2) detects the temperature in the temperature gradient of the inner loop. By changing the thermal conductivity in the microcell, the gradient also changes, so that the temperature of the outer loop changes. This change is converted into a voltage signal by a Wheatstone bridge (Fig. 3.30) and further processed in a carrier frequency amplifier. The individual resistors each consist of a thin platinum wire, which can also be provided with a thin glaze layer (silicon dioxide) for certain applications. The platinum wires are contacted with a gold brazing solder. The cell design with the two loops described above also provides flow independence within certain limits, as a cross-flow over the sensor arrangement is compensated.

The main advantages of this TCS are:

- Short response time <1.5 s
- High working temperature up to 200 °C
- Independence from gas flow
- Highly corrosion-resistant measuring cell
- Compensation of gas convection in the measuring cell
- High long-term stability for zero point and sensitivity.

The main applications for this TCS are in chemical process engineering, where corrosive components (e.g. moist hydrogen chloride) are to be expected, and therefore a cell temperature of 100–200 °C is required. The fast response behaviour is required above all for measurement tasks in the iron and steel industry (fast gas control) and medical technology.

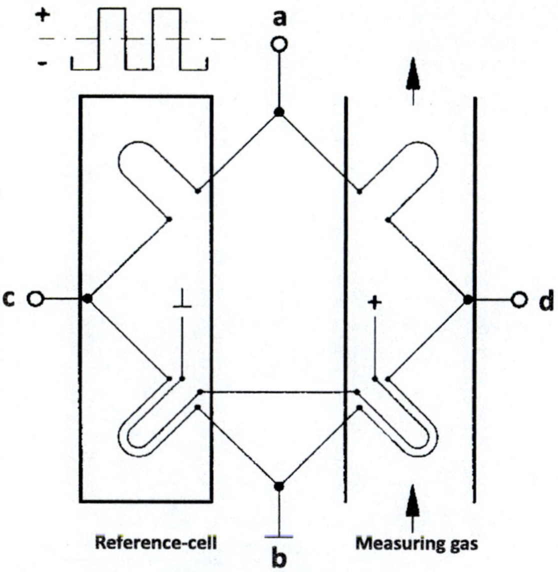

Fig. 3.30 Measuring bridge connection of the *Conthos-TCS* with indirect heating of the measuring elements

The construction according to Schunck has clear metrological advantages, but the manual manufacturing process is very complex and therefore also expensive. In 1984, a further design was presented (Wiegleb 1985a) in which at least the measuring resistors are manufactured by a foil etching process. The measuring cell consists of a miniaturized sensor with which an extremely small chamber volume (approximately 2 µL) was realized. By this measure the physical time constant $t_{90} < 1$ s could be drastically reduced. The sensor construction consisted of a thin (5 µm) nickel foil, which was structured to a measuring resistor by a photolithographic etching process. By using two measuring resistors integrated near each other, it was also possible to reduce the flow dependence (Wiegleb 1987a; Wiegleb 1987b). This effect can be explained by the thermal coupling of the measuring resistors. Figure 3.31 shows a greatly enlarged image of this miniaturized sensor element.

Micromechanical WL-Sensors

Since 1990, miniaturized WL sensors, which are manufactured with the help of micromechanical production processes, have been increasingly used. Figures 3.32 and 3.33 show such a sensor, which was developed at Hartmann & Braun AG (today ABB) in Frankfurt a. M. Due to the miniaturized design of the sensor and the associated small chamber volume ($V < 0.2$ mm^3),[5] a very fast measured value display is obtained from$\tau < 100$ ms. This property is of great importance especially for control processes.

[5]TCS208F (www.grwsensors.de)

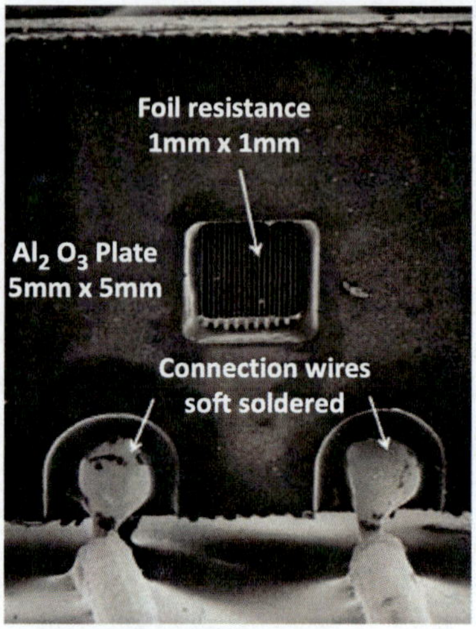

Fig. 3.31 Miniaturized TCS in foil etching technology. (Wiegleb 1985b)

Fig. 3.32 Micromechanical thermal conductivity sensor type TCS 208F

Figure 3.34 shows the sensor element in cross-section. The construction consists of three parts, namely the Si base plate, the sensor chip and the Si cover plate. The sensor chip has a trapezoidal breakout in the center, which is produced by an anisotropic etching process. This produces an extremely thin (≫1 μm) silicon nitride membrane which serves as a support for the measuring resistors (13). The measuring resistors (nickel) are protected with an additional passivation layer of silicon dioxide (SiO_2), to be able to measure corrosive gases and hydrogen. The measuring resistors are connected to the bonding wires via conductor tracks, which in turn are connected to the contacts of the pressure glass feedthrough.

The temperature of the measuring resistors can be controlled relatively easily to a constant value, which is usually at 180 °C. The power consumption is then, depending

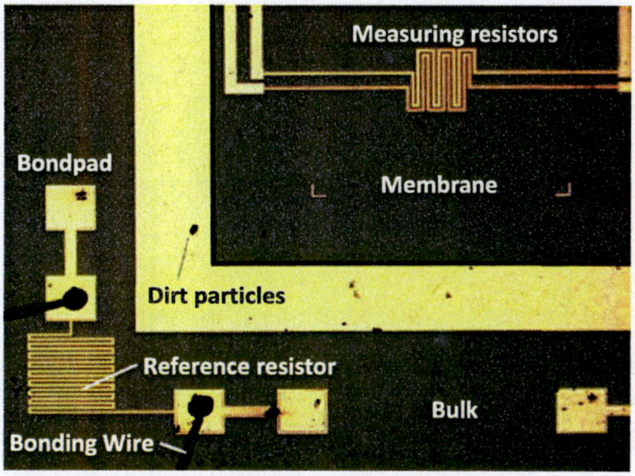

Fig. 3.33 Partial view of the sensor structure of the TCS 208F. In the center there are the two measuring resistors which are integrated in a meander-like manner. The reference resistors are integrated on the bulk structure of the microchip. (Source: Fachhochschule-Dortmund)

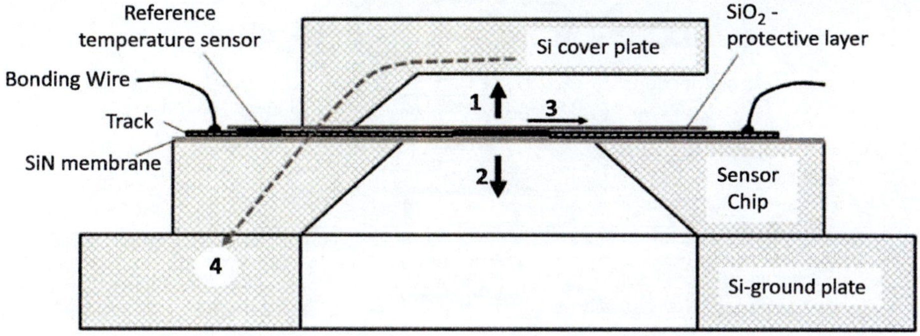

Fig. 3.34 Cross-section through the thermal conductivity sensor TCS 208F. The external dimensions of the entire structure are approximately 3 mm × 3 mm × 1 mm

on the type of gas, maximum P = 30 mW. The supplied electrical power P is then a direct measure of the heat conduction λ_M of the gas mixture, and thus of the concentration c_X in the microcell. The heat transfer components are also shown in Fig. 3.34. The main part is the heat conduction through the gas, to the lid (1) or base (2). The heat conduction (1) to the cover plate is transferred directly to the environment by the well heat-conducting silicon components (4). It can therefore be assumed that the temperature of the Si cover plate corresponds to the ambient temperature. The lateral heat conduction (3) via the membrane and convection can be neglected. This also applies to radiation to a certain extent. Since the gas exchange in the microcell only takes place by diffusion, the proportion of forced convection can also be neglected. The Si cover plate additionally reduces this part of the heat transfer.

In the version shown in Fig. 3.35, the thermal conductivity sensor is extremely sensitive to mechanical influences. Especially the thin bonding wires must be protected. For this

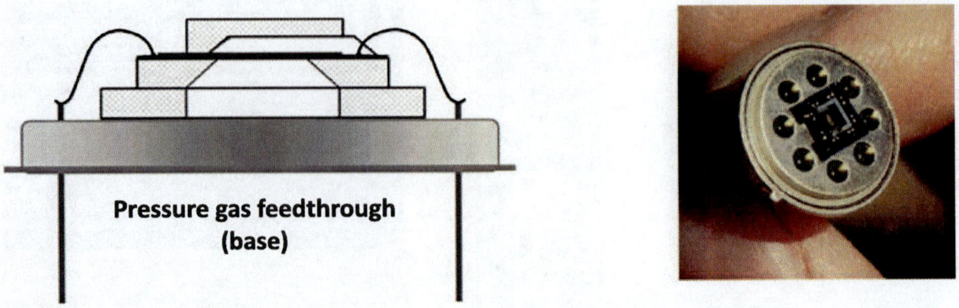

Fig. 3.35 Ready mounted thermal conductivity sensor on a pressure glass feedthrough. The outer diameter of the feedthrough is 13 mm

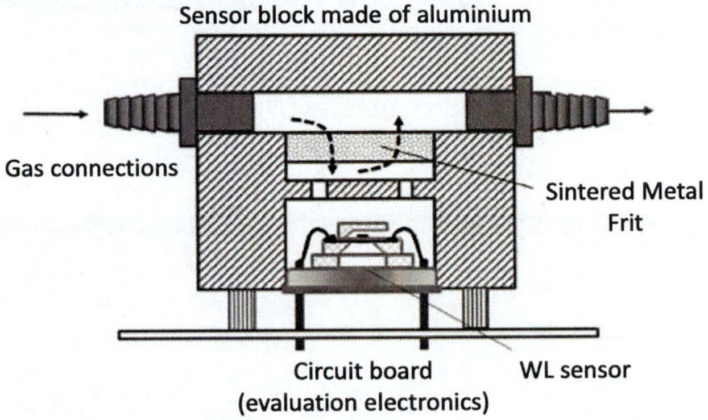

Fig. 3.36 Cross-section through a sensor structure with integrated WL sensor

purpose, the sensor is installed in a housing, in which the gas connections are also located (Fig. 3.36). The flow in the measuring chamber, in which the WL sensor is located, is usually achieved by pure diffusion via a sintered metal frit or a thin gas-permeable membrane. A bypass flow, without the protective cover, is also possible and is preferred especially for extremely fast indicating sensor designs. It is also possible to heat the sensor (thermostat) to reduce the temperature error or prevent condensation. For applications in process engineering and process analytics, the sensor block is usually made of stainless steel.

The company Messkonzept GmbH from Frankfurt a. M., has developed various detectors and measuring devices based on this sensor technology. An OEM assembly in which the TCS 208F sensor element was integrated is shown in Fig. 3.37. The mechanical structure consists of a stainless steel block (1.4571). The gas connections are located directly on the sensor block and are firmly connected to it by a laser welded joint. This provides a high mechanical stability and a gas-tight connection.

The signal processing is located above the sensor block. A heating resistor can be seen on the side, with which the sensor block can be heated to a constant temperature

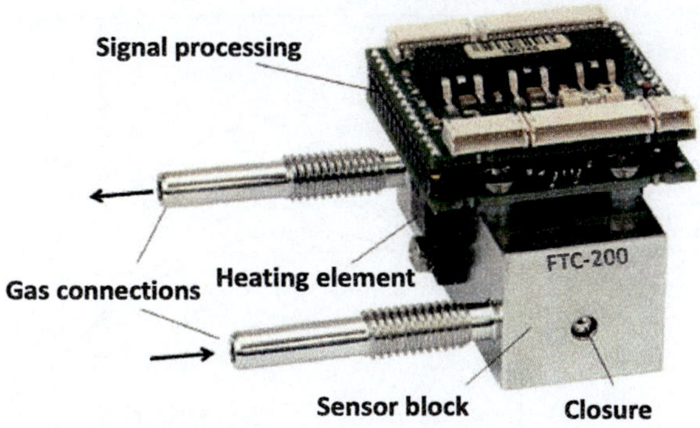

Fig. 3.37 TC sensor integrated in a stainless steel sensor block. The gas connections are also made of stainless steel and are welded to the sensor block. (Source: Messkonzept GmbH Frankfurt a. M)

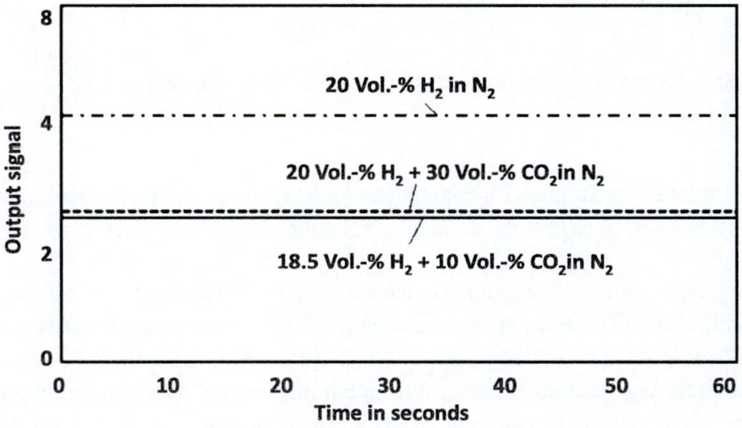

Fig. 3.38 Signal curve of a TC sensor with constant chip temperature for three different gas mixtures

(e.g. 50 °C). The required control electronics are also located above the sensor block (second circuit board).

FTTCA[6] Procedure

This method is based on the temperature behavior of thermal conductivity. Especially for gas mixtures with more than 2 components, this method has clear advantages over the measuring methods with constant temperature. The method was patented by Grunewald in 1991, and further developed by Grienauer (2006, 2012).

Figure 3.38 shows the signal curve of a TC sensor at a constant chip temperature. The three gas mixtures differ significantly from each other in their composition. Nevertheless,

[6] Fourier-Transformed Thermal Conductivity Analysis.

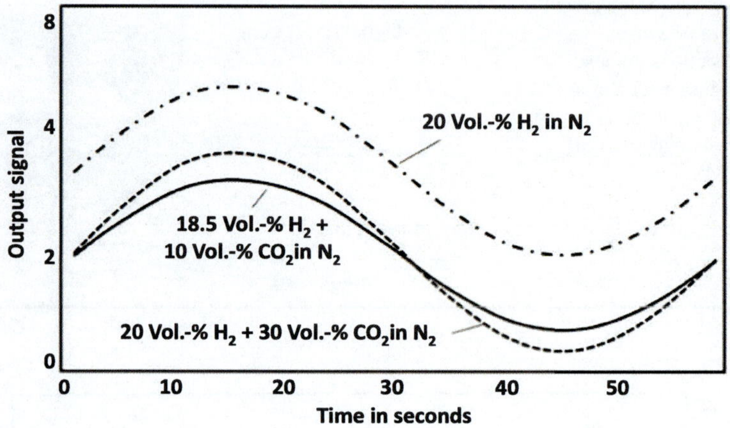

Fig. 3.39 Signal curve of a TC sensor with variable (modulated) chip temperature for the same gas mixtures as in Fig. 3.38

the two gas mixtures with the CO_2 content cannot be distinguished from each other, since both gas mixtures have almost the same thermal conductivity at this temperature λ.

If the temperature is now modulated with a very low frequency, the picture changes fundamentally. The two gas mixtures containing CO_2 can now be distinguished from one another. The amplitude of the temperature modulation reveals the different behavior of the gases at different temperatures reached during the modulation. With this additional information, increased selectivity can then be achieved by appropriate evaluation (Grienauer 2012) (Fig. 3.39).

The temperature dependency of the different gases and the response to temperature modulation are shown in Fig. 3.40. Helium (He) and carbon dioxide (CO_2), have different temperature sensitivities, which lead to different amplitudes $f(T)$ when modulated (ΔT). With the help of a Fourier analysis, the individual components in a gas mixture can be detected (Grienauer 2006).

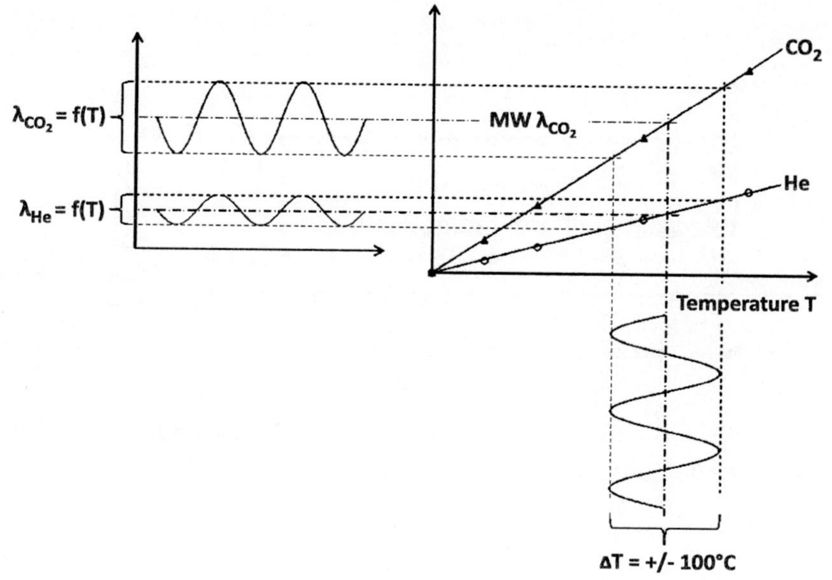

Fig. 3.40 Effect of temperature modulation for different gases (He, CO_2)

3.2 Mass Sensitive Sensors

The molar mass M and the resulting properties of the gas molecules can also be used for a gas measurement. These properties are proven by the speed of sound c, the gas density ϱ and the deposition (Δm) of gases on surfaces. Although these physical properties are not selective for a certain gas, they can be used for the gas analytical determination of concentration ratios in gas mixtures. Based on the different effects, gas sensors have been developed, which show excellent measuring characteristics in special applications.

Sound Velocity

The propagation of acoustic waves in a gaseous medium can be used for gas analytical purposes by evaluating the phase shift $\Delta\varphi$ of the wave. Since this type of evaluation is not selective for a specific gas, the method can only be used in binary gas mixtures. Figure 3.41 shows such a structure.

The sound is coupled, for example, into a measuring cell by a piezoelectric transducer (loudspeaker) and received by a microphone on the opposite side. The phase position φ_0 of the transmission frequency f_0 and the phase position φ_1 of the received signal, can be converted into an output signal c_n proportional to the concentration by means of

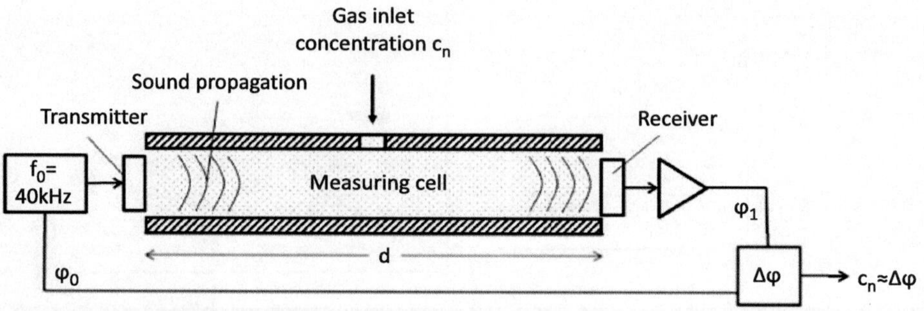

Fig. 3.41 Ultrasonic detector according to Noble et al. (1964)

appropriate evaluation electronics. The phase shift results from the transmitting frequency f_0, the distance d and the speed of sound c:

$$\Delta\varphi = \frac{180^\circ \cdot f_0 \cdot d}{c} \tag{3.39}$$

For ideal gases the *Laplace equation* applies at low frequencies:

$$c = \sqrt{\frac{1}{K \cdot \varrho}} = \sqrt{\frac{p \cdot c_p}{\varrho \cdot c_V}} = \sqrt{\frac{\varkappa \cdot R \cdot T}{M}} \tag{3.40}$$

With $\varkappa = c_p/c_V$, R=universal gas constant, T=temperature in K, M=molecular weight, K=compressibility, ϱ=density.

For a binary gas mixture, consisting of the carrier gas (TG) and the substance to be measured (X), with the mole fraction ξ, one obtains approximately the following equation to determine the phase shift $\Delta\varphi$:

$$\Delta\varphi \approx \frac{180 \cdot d \cdot f_0 \cdot \xi}{\sqrt{RT}} \cdot \sqrt{\left(\frac{M}{\varkappa}\right)_{\mathrm{TG}}} \cdot \left[\frac{M_X}{M_{\mathrm{TG}}} - 1\right] \tag{3.41}$$

According to Eq. 3.41, the speed of sound c in a gas is independent of the pressure p, and is only determined by the substance size M (molecular weight) and the temperature T. Figure 3.42 shows the velocity of sound in air at an absolute pressure of 1 bar and 5 bar, for the temperature range 0–100 °C. The values are taken from the VDI Heat Atlas. Both curves lie almost identically one above the other. The deviations are only −0.5 m/s. The percentage deviation is therefore <0.03 % / bar. Atmospheric pressure changes, in the range of ±100 hPa, therefore do not lead to a significant change in the speed of sound.

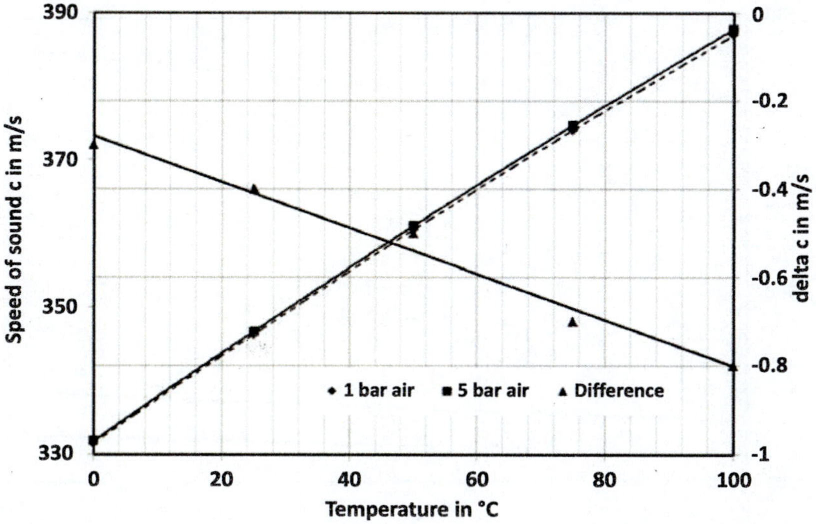

Fig. 3.42 Sound velocity c of air as a function of temperature T at two different pressures (1 bar and 5 bar)

The speed of sound can also be calculated using Eq. 3.40, if the data in the relevant tables are not available. The temperature curve for carbon dioxide (CO_2) was taken from the VDI Heat Atlas and calculated. The deviations were $\pm$ 1% over the entire temperature range from 0 °C to 100 °C. This deviation cannot be shown in Fig. 3.43, because the curves are congruent at this scale. The temperature dependence of the sound velocity was only calculated. The deviation at 0 °C is +8% in relation to a table value (Kuchling).

Another possibility is to measure the transit time Δt to calculate the speed of sound c. The structure is identical to that shown in Fig. 3.41. For this purpose, the sound source emits individual sound pulses which are received by the microphone with a time delay. The speed of sound c is then calculated:

$$c = \frac{d}{\Delta t} \tag{3.42}$$

For a measurement path of d = 10 cm, Eq. 3.42 then gives a transit time of $\Delta t_1 = 288.8$ μ sfor air (1 bar, 25 °C), and $\Delta t_2 = 372.2$ μs for carbon dioxide. This method was used to develop gas measuring instruments for measuring the CO_2 concentration in exhaust gases (Greiser and Heller 1981). The difference in transit time between 100% air and 20% CO_2 by volume is ≈ 17 μs .

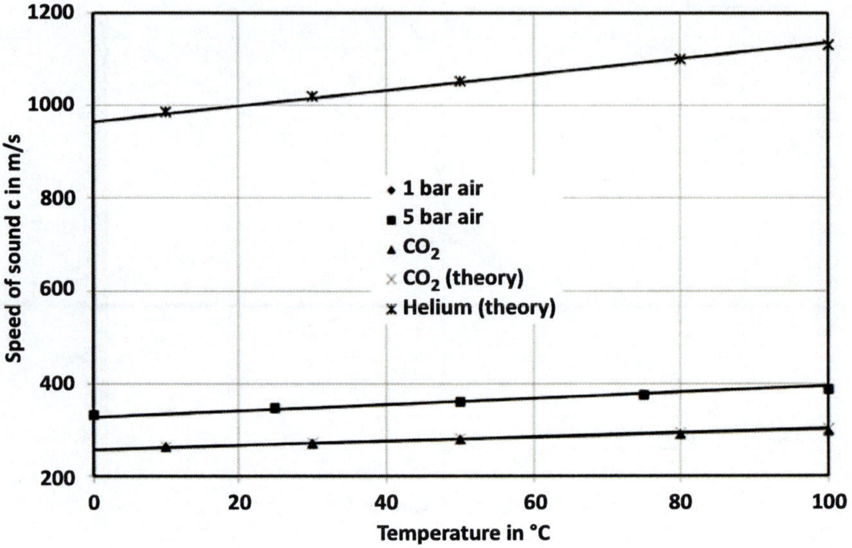

Fig. 3.43 Calculated and measured values (VDI Heat Atlas) for the sound velocity of different gases

Sound Amplitude

In addition to the change in sound velocity (Δc) and the phase shift ($\Delta\varphi$), the amplitude change (ΔA) of a sound wave can also be used to determine the gas density (Jochum 1986). In Fig. 3.44, the gas density meter of the company LFE[7] is shown. It consists of two single chambers which are separated by a membrane. The membrane is excited to oscillate by an external sinus generator and the amplitude A_{S1} generated is recorded by a sound receiver (microphone). The sound transmission in the measuring chamber changes the amplitude ($\Delta A = A_{S1} - A_{S2}$) as the density ϱ_2 in the measuring chamber increases. The relationship between the change in density $\Delta\varrho$ and the change in amplitude $\Delta A$and provides a linear output signal (Haran 1983). However, the density is also dependent on other influencing effects, such as the chamber pressure and temperature. Furthermore, the diaphragm can change over time (e.g. due to contamination), so that the amplitude A_0 changes. Therefore, a second identical sound receiver is integrated in the reference chamber, which provides a control signal to keep the amplitude A_0 constant. An increased measuring stability is therefore achieved with this reference chamber that it is in contact with the ambient air, and thus controls barometric pressure changes to the vibration amplitude.

The influence of temperature is kept within narrow limits by thermostatting the entire setup, so that this setup allows very accurate density determination. The influence of external mechanical influences (shock and vibration) is reduced by electronic filtering of

[7]Laboratory for Industrial Research GmbH & Co, Bruchköbel.

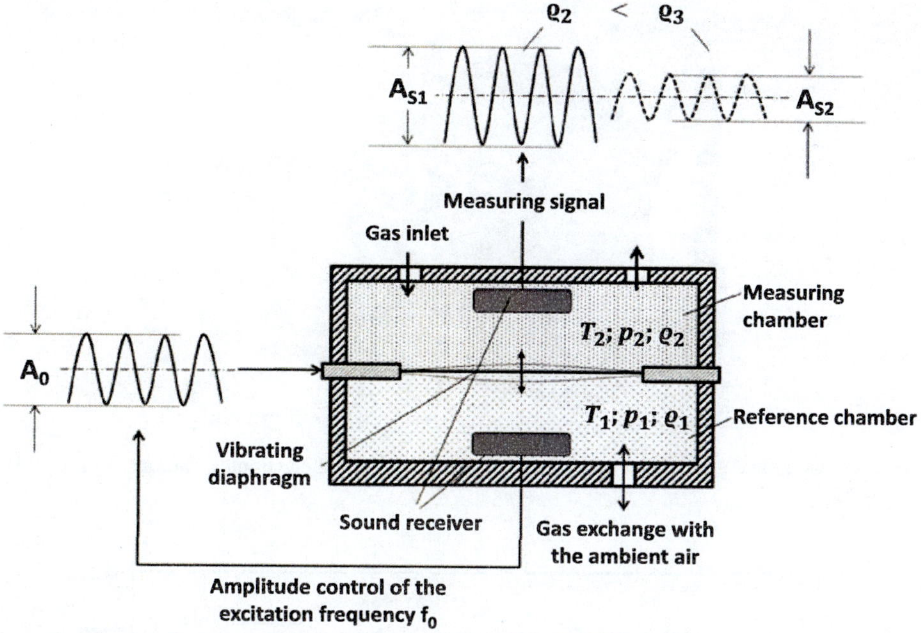

Fig. 3.44 Basic structure of the LFE gas density meter

the mechanical suspension. The main application for this gas density meter is in the gas industry for determining standard density and calorific value/Wobbe index.

Ultrasonic Sound Propagation

Another interesting application of sound propagation is leakage monitoring in larger plants by ultrasound. Since the gas pipes to be monitored are usually under high pressure, the escaping gas generates a sound wave at a leakage point. This sound wave can be detected even at a greater distance (up to 40 m) with appropriately sensitive microphones. The greatest challenge, however, is to reliably differentiate the sonic noise caused by the gas leak from the background noise. This requires appropriate evaluation algorithms which should guarantee a clear assignment. Therefore, usually only a frequency range from 25 to 100 kHz is investigated, since background noise is often found in the lower frequency range. Some companies also use mathematical methods of artificial neural networks (ANN[8]) to separate the background noise from the wanted signal. The dynamic range of the sound wave that is detected is 40–120 dB (standard) or 58–104 dB (Figs. 3.45 and 3.46).

[8] Artificial Neural Network.

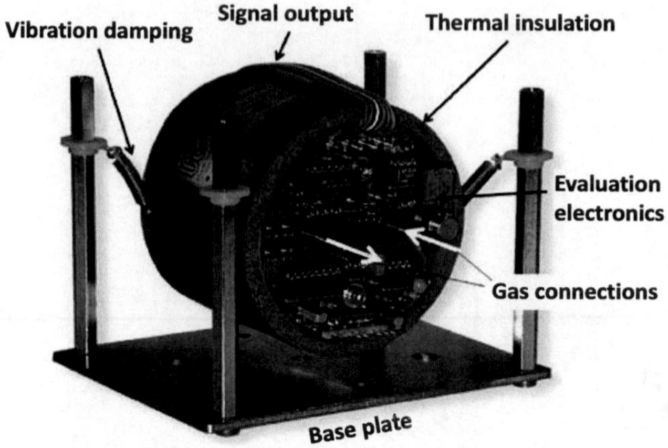

Fig. 3.45 Complete gas density meter with integrated evaluation electronics. (Source: LFE GmbH Bruchköbel)

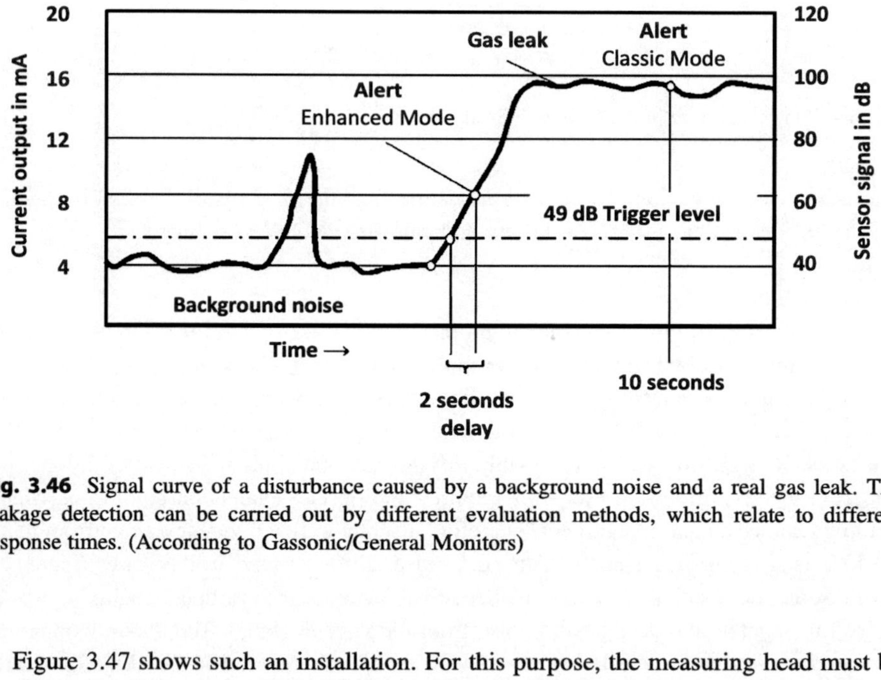

Fig. 3.46 Signal curve of a disturbance caused by a background noise and a real gas leak. The leakage detection can be carried out by different evaluation methods, which relate to different response times. (According to Gassonic/General Monitors)

Figure 3.47 shows such an installation. For this purpose, the measuring head must be installed at a position from which the entire danger area can be covered as far as possible. If this is not possible, several of these sound receivers must be distributed over the site.

The advantage of this measuring method is the fact that the measuring head does not require direct contact with the measuring medium (gas). Also, the wind direction does not

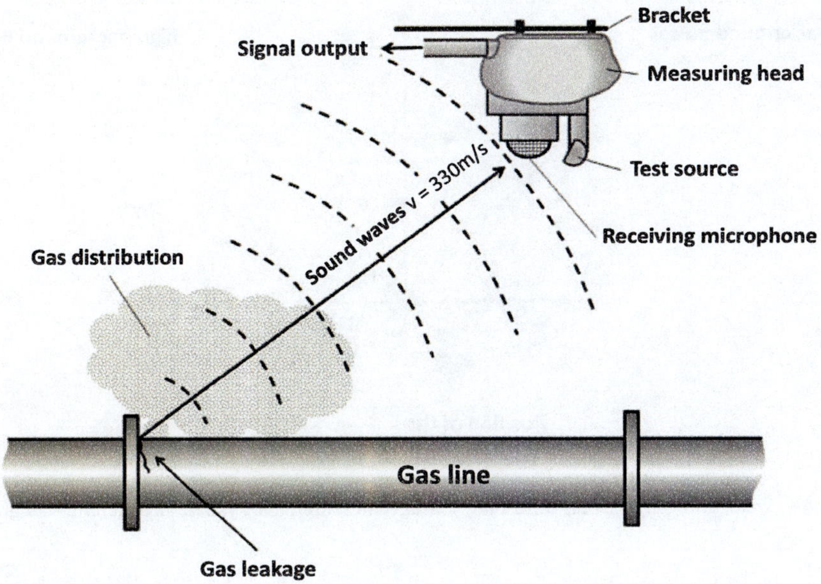

Fig. 3.47 Installation arrangement of an ultrasonic sensor for detecting a gas leakage in a high-pressure pipe

play a decisive role, since the sound, in contrast to a gas cloud, propagates evenly in all directions. Only the receiving properties of the microphone limit the detection range. Figure 3.48 shows the sensitivity distribution for the different directions of observation. With this method, at least one 180° coverage is possible. However, the detection range can be limited if a system is to be monitored where there is regularly strong background noise. The detection range is then halved in these cases. This situation can be found, for example, for turbo compressors and for an installation on an offshore weather deck. The *quiet* areas are near *onshore* boreholes, or gas storage facilities.

Quartz Resonance Sensors (QMB)

For some years now, so-called quartz resonance sensors (BAW[9] or also QMB[10]) have been used to detect trace substances in the air. As a basis, oscillating crystals are used, which are cut out of a quartz crystal with a certain orientation of the lattice structure. This *AT cut* leads to a stable frequency f_0, which is not affected by temperature changes. This type of oscillating quartz crystal is already used in large quantities as an electronic component for the time base of computers. For gas analytical measurements, these crystals are provided with a sensitive layer to which certain gaseous species can bind. This structure

[9] Bulk Acoustic Wave (=volume oscillator).

[10] Quartz microbalance.

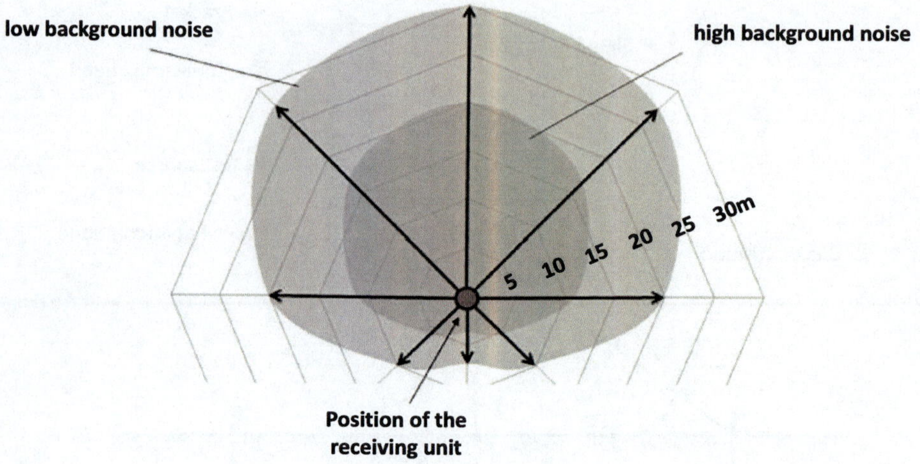

Fig. 3.48 Detection range of an ultrasonic leakage measurement system at different background noise

is shown in Fig. 3.49. Due to the associated increase in mass (m + Δm), the resonance frequency ($f_0 + \Delta f$) changes. This frequency change can then be used to draw conclusions about the gas or gas concentration. The frequency change Δf (Hz), can be calculated with the *Sauerbrey equation* (Kurtz et al. 2010), if the frequency f_0 *is* given in megahertz, Δm in grams, and the area A in cm^2:

$$\Delta f = -2.3 \cdot 10^6 \cdot f_0^2 \frac{\Delta \mathrm{m}}{\mathrm{A}} \tag{3.43}$$

Example

A QMB sensor has an active area of 1 cm^2 and an $f_0 = 10$ base frequency MHz. What is the smallest increase in mass Δm that can still be detected?

With Eq. 3.43, and switching to Δmone obtains:

$$\Delta \mathrm{m} = 1\ \mathrm{Hz} \cdot 1\ \mathrm{cm}^2 \cdot \frac{10^{-6}}{10^2 \cdot 2.3} = 0.43\ \mathrm{ng}$$

The detection limit of only 0.43 ng shows that even very small amounts of substances can be detected with this method. An important application area is therefore the use in so-called electronic noses. Several of these QMB sensors with different coatings are used in these setups. The materials used for the coatings are mostly polymers which have similar

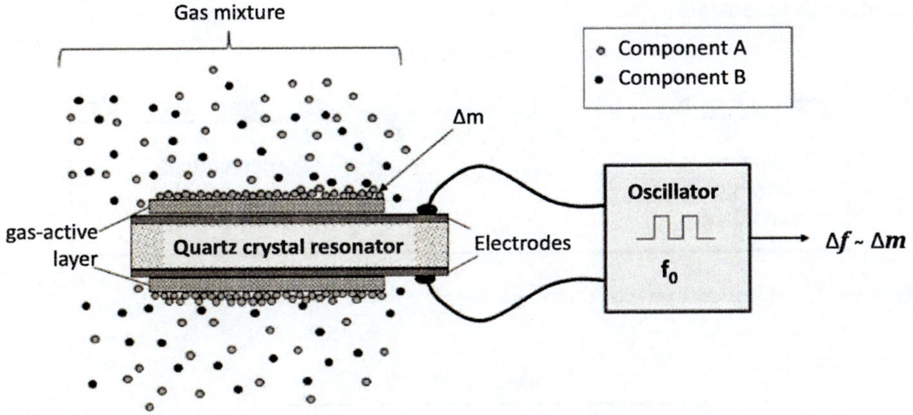

Fig. 3.49 Structure of a QMB gas sensor with a coating on both sides of the resonator with a gas-active layer (e.g. polymers), which are known from gas chromatography

properties to the coatings of the separation columns in a gas chromatograph. For example, identical materials are also used for both applications. Due to the different coatings, each sensor has a different characteristic and is therefore sensitive to different groups of substances. With these *sensor arrays* it is possible to draw conclusions about very complex compositions by pattern recognition. Different smells, aromas and fragrances can be recognized, which are otherwise only distinguished by the human nose. Furthermore, these electronic noses can also detect differences in the quality of food (Mashayekhi 2005).

Surface Acoustic Waveguide (SAW)

Another possibility is to use surface acoustic waves (=SAW) to detect gaseous substances. The SAW element consists of a substrate of quartz, lithium niobate or a piezoceramic. On one side, there is a transmitter which generates a surface acoustic wave in the ceramic which travels in the direction of the receiver. There the mechanical wave is then converted back into an electrical signal. The excitation frequency lies between 10 MHz and 1 GHz. When driven with 350 MHz, the surface wave has a wavelength of 8 μm. On the way from the transmitter to the receiver there is a gas-sensitive coating, as with a QMB sensor (see Fig. 3.50). In this case too, identical layers can be used as in gas chromatography. If this layer is now occupied by the gases to be detected, the surface waves are damped and a smaller amplitude reaches the receiver. Since this attenuation is quite low, a reference measurement is used (see Fig. 3.51), which is arranged in a planar technique, parallel to the described setup (Bender 2004).

Gas Density Measurement

When determining the mass flow, in the form of mass flow $\dot{m}$ or volume flow $\dot{V}$, an exact knowledge of the gas density is required. Furthermore, the gas density ϱ_{Gas} also plays an

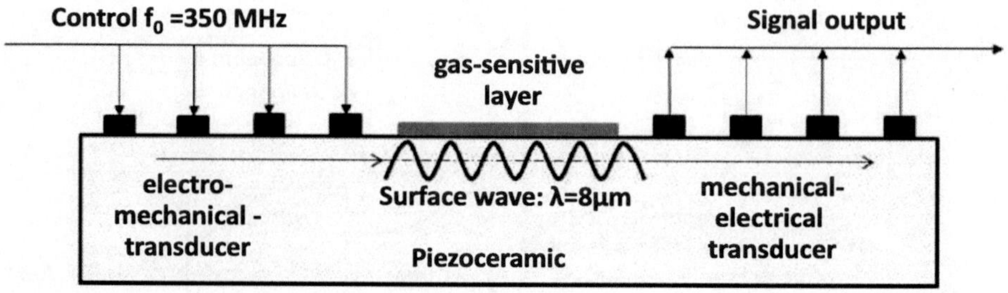

Fig. 3.50 Cross-sectional view of a SAW gas sensor

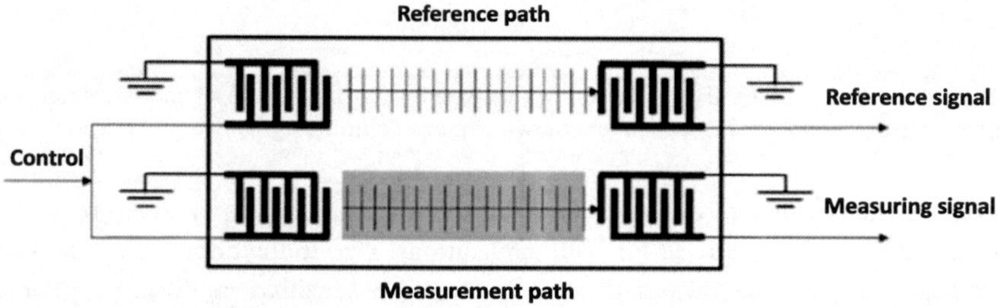

Fig. 3.51 SAW gas sensor with integrated reference path to compensate for temperature influences. The measuring path contains the gas-sensitive layer that attenuates the wave

important role in determining the energy flow $\dot{Q}_B$. The energy flow $\dot{Q}_B$ can be calculated from the product of the volume flow and the calorific value:

$$\dot{Q}_B = \dot{V}_B \cdot H_s \tag{3.44}$$

The volume is again related to the pressure p and the temperature T or the gas density ϱ_{Gas}. For billing purposes, the volume flow is therefore related to the standard density ϱ_N, which can be calculated as follows:

$$\varrho_N = \varrho_r \cdot \left[\frac{1013.25 \text{ hPa}}{273.15 \text{ K}}\right] \cdot \left[\frac{T_r}{p_r}\right] \tag{3.45}$$

In addition to these applications, gas density sensors can also be used for gas analytical purposes. The gas density of a mixture results from

$$\varrho_M = \frac{\varrho_1 V_1 + \varrho_2 V_2 + \ldots + \varrho_n V_n}{V_1 + V_2 + \ldots + V_n} \tag{3.46}$$

In binary gas mixtures, the gas density can therefore be used to draw conclusions about the substance quantity ratios in the mixture. Although this method is non-selective (as with thermal conductivity sensors), it is nevertheless successfully used for certain measurements (Fig. 3.52).

Standard Density Scale

The oldest measuring method for determining density, which is still used today, is the so-called gas density balance. Figure 3.53 shows a historical setup with which the gas density was determined by buoyancy forces.

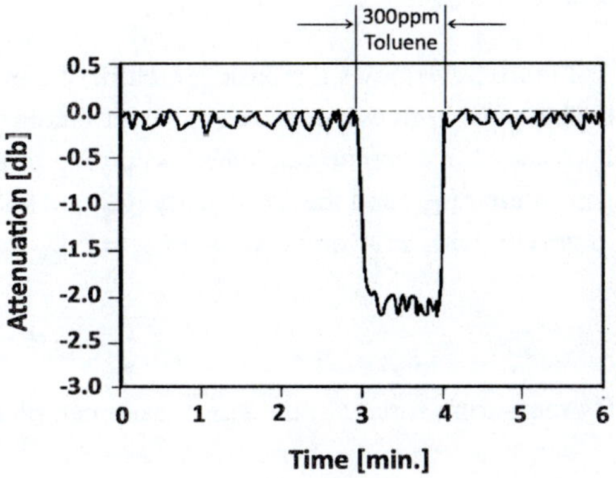

Fig. 3.52 Measurement of 300 ppm Toluene in air with a SAW gas sensor (according to Bender 2004)

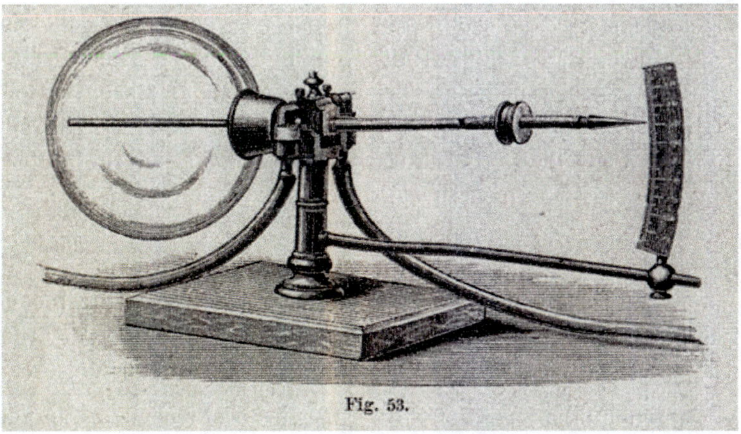

Fig. 3.53 Historical representation of a mechanical gas density balance (Brunck 1919)

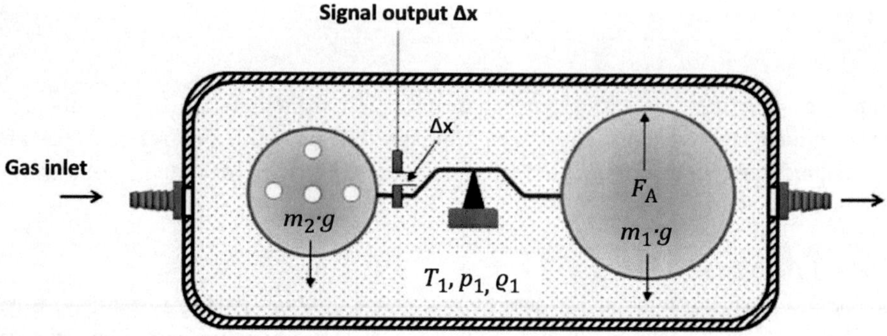

Fig. 3.54 Standard density balance for determining the gas density according to the buoyancy method

Figure 3.54 shows the basic structure of a current measuring system. This method uses the buoyancy forces that arise in the surrounding gaseous medium at a defined volume. The buoyancy force can be calculated according to the *Archimedean principle,* if the displaced gas volume V_{Gas} and the gas density ϱ_{Gas} are known. The constant g is the acceleration due to gravity with an average value of $g = 9.81\ \mathrm{m/s^2}$.

$$F_A = V_{Gas} \cdot \varrho_{Gas} \cdot g \tag{3.47}$$

By measuring force F_A, the gas density can be calculated as follows:

$$\varrho_{Gas} = \frac{F_A}{V_{Gas} \cdot g} \tag{3.48}$$

Since the forces are very small, a beam balance is used to measure them. On both sides of the beam balance there is a glass ball, the right ball is filled with nitrogen and melted. It serves as buoyancy body. The left ball is open so that the gas to be measured can also enter the inside of the ball. This ball therefore experiences no buoyancy. The surfaces of the balls are identical, whereby the surface of the left ball consists of the inner and outer surface. The diameter is therefore smaller. The reason for this surface equality is due to possible adsorption effects on the sphere surface. With identical surfaces, this undesirable effect of weight increase Δm is completely compensated. The weights of the two balls are balanced so that they are also in equilibrium. Thus, the following relationship applies in the zero position of the balance:

$$\Delta m + m_1 \cdot g = \Delta m + m_2 \cdot g \tag{3.49}$$

Fig. 3.55 Versions (models 11–625 and 11–140) of the gas density detector from Gow-Mac Instrument Co. Bethlehem USA

The entire weighing system is installed in a gas-tight sealed chamber, through which the gas to be measured flows completely. The flow conditions in the chamber are selected in such a way that the measurement is not affected. The deflection Δx of the balance beam is sensed by an inductive displacement sensor, compensated and converted into a current proportional to the standard density (4 20 mA) by downstream electronics (Wernekinck 2009).

Gas Density Detector

The gas density detector presented in the following was originally developed for use in gas chromatography but can also be used for other gas analytical applications (Heise et al. 1981). As early as 1956, Martin and James described the basic principles of this method as a *gas density balance*. Nerheim (1963) further developed the method for gas measurement technology. Today, the setup is still produced in almost unchanged form by the company Gow-Mac USA (see Fig. 3.55).

Figure 3.56 shows the schematic structure of the measurement procedure. It consists of a reference branch and a sample branch. The reference gas, with a known gas density ϱ_R, flows into the chamber system from the left and is initially distributed evenly over the upper and lower path. The two partial flows are equal ($\varnothing_1 = \varnothing_2$). In both the upper and the lower gas path, there is a flow sensor (S_1 and S_2) to detect the partial flows. The gas sample to be

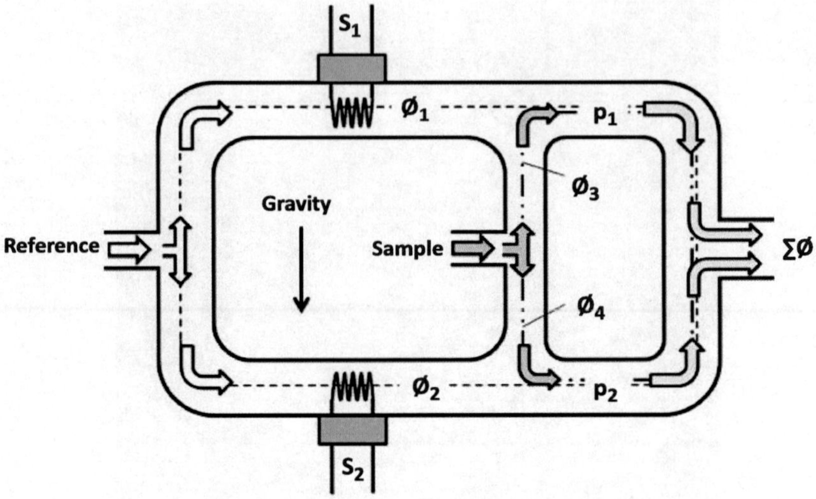

Fig. 3.56 Gas density detector according to Nerheim (1963) for measuring density differences between a sample and a reference gas

measured with the density ϱ_R is coupled in at the point provided for this purpose, and also splits into two identical partial flows ($\varnothing_3 = \varnothing_4$). All partial flows are combined at the outlet ($\sum\varnothing$). If the gas density of the sample is greater than the gas density of the reference gas, the volume flow $\varnothing_2$ is *slowed down* by the pressure increase *p2* and is detected by the flow sensor S_2. It then applies:

$$\Delta p = p_1 - p_2 = (\varrho_R - \varrho_M) \cdot g \cdot h \tag{3.50}$$

Both flow sensor elements are located in a Wheatstone measuring bridge, whose detuning is a measure of this gas density veining. In the opposite case, if the gas density of the sample is lower than the gas density of the reference gas, a lower flow occurs in the upper branch ($\varnothing_1 < \varnothing_2$). The advantage of this arrangement is that the flow elements are only in contact with the known reference gas. Corrosive gases can therefore also be measured with this setup.

Vibration Density Meter

With this method, an element that can vibrate (tuning fork, quartz crystal or similar) is electrically excited so that it performs a harmonic vibration. If this element is in a vacuum, i.e. no damping gases are present, the natural frequency f_0 of the element can be calculated as follows:

$$f_0 = \frac{1}{2\pi}\sqrt{\frac{c}{m_0}} \tag{3.51}$$

where m_0 is the mass of the vibrating element, and c is a constant resulting from the spring stiffness. If this oscillating element is now surrounded by a gas, microscopic collisions between the oscillator and the gas occur. The natural frequency f then changes as follows:

$$f = \frac{1}{2\pi}\sqrt{\frac{c}{m_0 + \varrho \cdot V}} \tag{3.52}$$

If both equations are divided up, the ratio of the frequencies is obtained:

$$\left(\frac{f}{f_0}\right)^2 = \frac{1}{1 + \varrho \cdot V/m_0} \tag{3.53}$$

this follows with the apparatus constant $K = m_0/V$:

$$\left(\frac{f}{f_0}\right)^2 = \frac{1}{1 + \varrho/K} \quad \text{and} \quad \frac{f}{f_0} = \sqrt{\frac{1}{1 + \varrho/K}} \tag{3.54}$$

The gas density ϱ can then be calculated by transforming Eq.3.54:

$$\varrho = K \cdot \left[\frac{f^2}{f_0^2} - 1\right] \tag{3.55}$$

The difference frequency $\Delta f = f_0 - f$ is also obtained by transforming Eq. 3.54:

$$\Delta f = f_0\left(1 - \sqrt{\frac{1}{1 + \varrho/K}}\right) \tag{3.56}$$

Equation 3.56 shows that the measured value Δf increases proportionally with the fundamental frequency (Fig. 3.57). It is therefore advisable to select the largest possible base frequency f_0. The factor K also affects the sensitivity. The lesser K is, the greater the sensitivity. K is proportional to the mass m_0, i.e. miniaturized structures have a greater sensitivity here. The development trend in gas density measurement is therefore towards miniaturized setups. Figure 3.58 shows the principle of a gas density sensor. The entire sensor construction consists of two identical oscillating crystals, each of which is housed in a separate chamber. The reference element is in a closed and evacuated chamber. This element therefore oscillates with the fundamental frequency f_0. The second element is in the

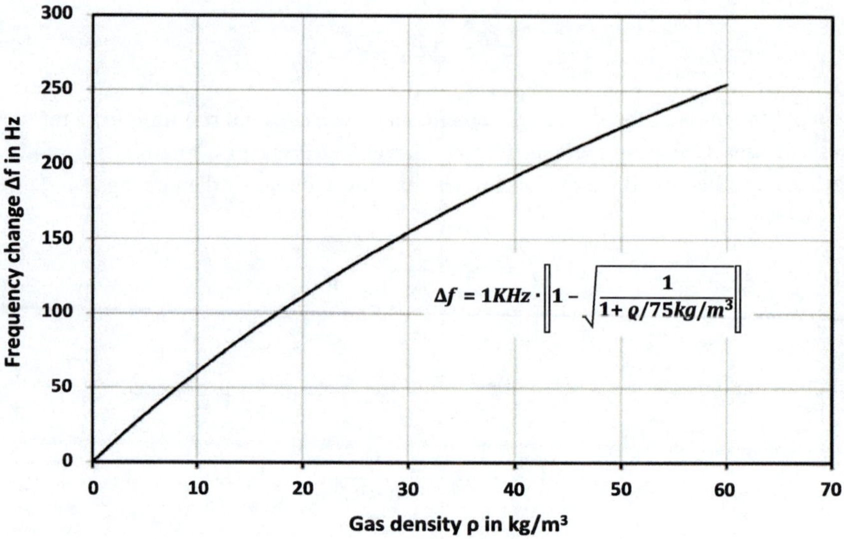

Fig. 3.57 Calculated characteristic curve according to Eq. 3.56

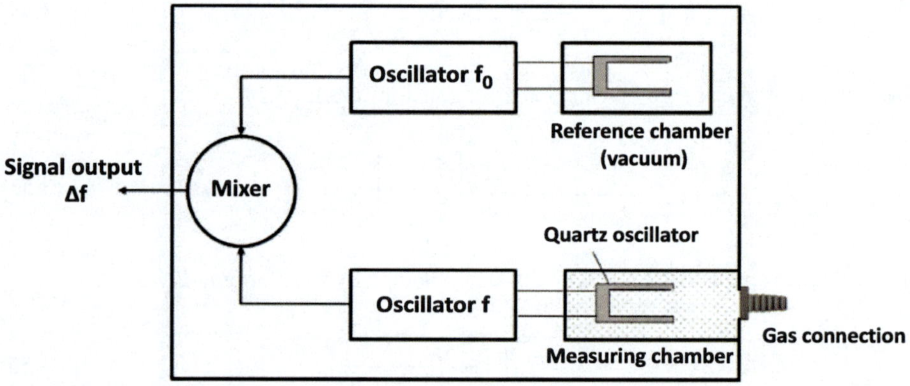

Fig. 3.58 Basic design of a gas sensor based on gas density measurement with two oscillator circuits for compensation purposes

measuring chamber and is surrounded by the gas to be measured. According to Eq. 3.52, the frequency f changes according to the gas density ϱ in the chamber. Both measuring results are connected to an electronic mixing stage in which the difference frequency Δf is determined. This difference frequency increases proportionally with the gas density ϱ in the measuring chamber.

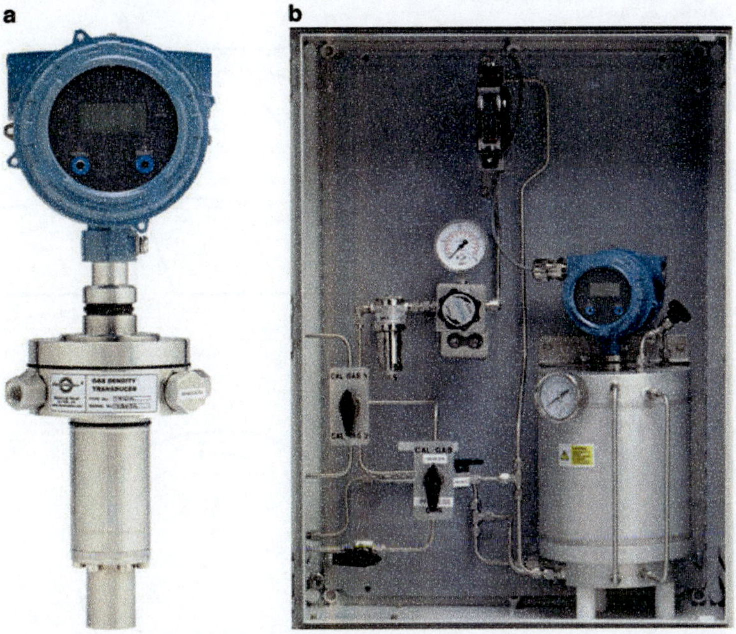

Fig. 3.59 Industrial gas density transmitter Type: *GDM Gas Density Meter* as single device (**a**) and complete standard density measuring device (**b**). (Source: Emerson/Micro Motion)

The described sensor *(Gas Density Sensor 8774)*[11] is used to control the gas atmosphere of sulfur hexafluoride (SF_6) in high-voltage switchgear. The density of SF_6 at room temperature (20 °C) and a pressure of p = 1013 hPa, $\varrho = 6.4\ \text{kg/m}^3$. The measuring range of the sensor is specified in the data sheet from 0 to 60 kg/m^3 (− 60 g/liter), which corresponds to a 100% gas at 8.87 bar. The frequency change Δf for the sensor is 243 Hz, for a density measuring range of 60 kg/m^3 . Per Hertz these are then $\Delta\varrho \approx 0.25\ \text{kg/Hz}$. This value also corresponds to the metrological resolution of this measuring method. With Eq. 3.56 and a given basic frequency $f = 1$ kHz, this results in a *K value* of 75 kg/m^3 . With this value the characteristic curve can then be explicitly calculated. Figure 3.55 shows the calculated characteristic curve. The curve corresponds to the experimental values from the data sheet (Fig. 3.59).

Fluidic Oscillators

In the 1980s, a special technique for the acoustic detection of gases was developed at the Karlsruhe Nuclear Research Centre (Kernforschungszentrum Karlsruhe). A so-called fluidic oscillator was set up there which recorded the gas density of a binary gas mixture to

[11] Trafag AG Sensors & Controls, Bublikon Switzerland.

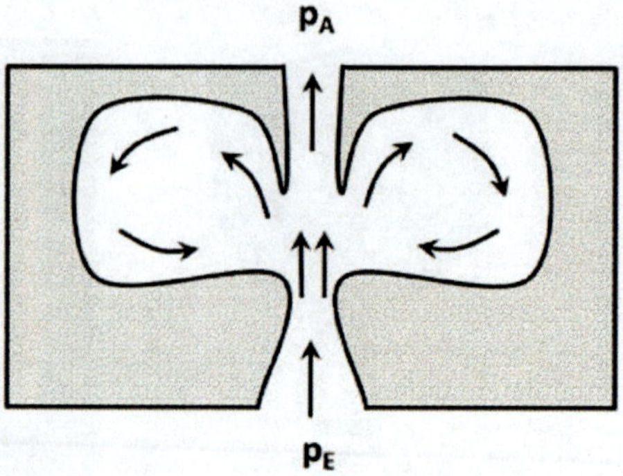

Fig. 3.60 Fluidic, acoustic oscillator with internal feedback (Zipser and Labude 1991a)

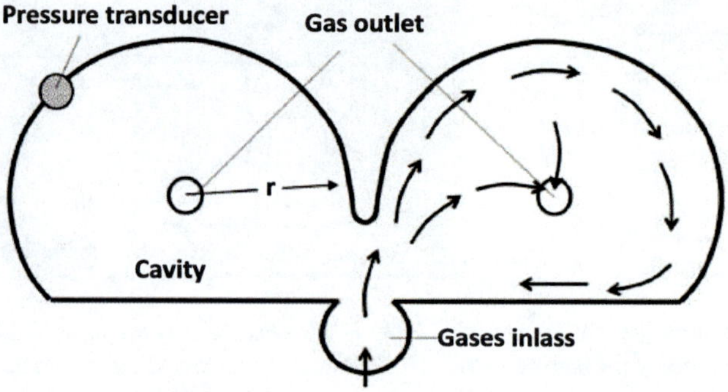

Fig. 3.61 Cavity resonator with the inner radius *r*. The gas outlets are each located in the centre of the circular path. Phase 1 (start) begins in one of the two circular areas. In this example, the pressure measurement is carried out in the *left* chamber (Source: according to Kreiter GLP-mbH Wiesbaden)

calculate the concentration ratios. The first application for this gas measurement technique was the measurement of uranium hexafluoride (UF_6) in hydrogen (H_2). The oscillator consisted of two chambers which were connected to each other and through which the gas to be measured flowed. Figure 3.60 shows such an oscillator with internal recirculation. A comprehensive theoretical treatise can be found in Zipser and Labude (1991b).

The oscillation can be explained most simply by the three phases that occur during an oscillation period. These phases are shown in Figs. 3.61 and 3.62.

The gas enters the cavity resonator via the entrance and is then forced into one of the two halves by the kinetic energy of the gas jet. The centrifugal force presses the gas against the wall and therefore keeps it close to the wall. Part of the gas will leave the cavity resonator at the outlet point. The other part enters a linear movement after the rotational movement and meets the incoming gas there. Through this impulse transfer, the gas jet is directed to the other side and "hangs" there due to the *Coanda*[12] *effect.* When this circular path in the left

[12]Henri Marie Coandă (1886–1972) Romanian physicist.

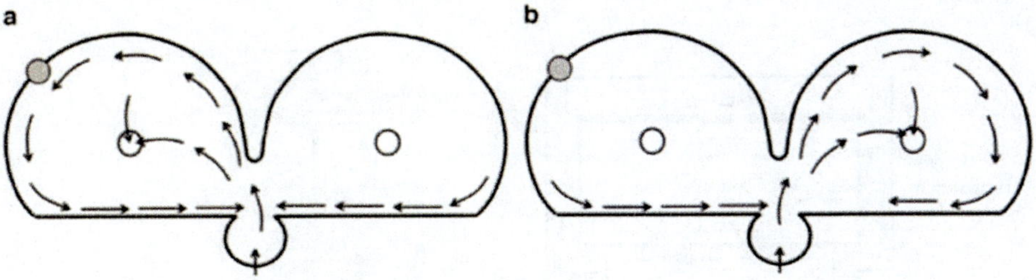

Fig. 3.62 Phase 2 (**a**) and 3 (**b**) of the oscillation process

chamber is also completed, the subsequent linear movement again hits the gas jet and deflects it again. Now the gas jet is directed into the right chamber and the oscillation process is started again (→first oscillation period).

The oscillation frequency f for an acoustic oscillator result from a standing wave of the wavelength λ, with $L = \lambda/4$. There is a relationship $c = \lambda \cdot f$ between the speed of sound c and the frequency f, so that the following expression is obtained:

$$f = \frac{c}{4L} \tag{3.57}$$

For the fluidic oscillator, the characteristic length L corresponds to half the circumference of the chamber ($L = \pi \cdot r$). Using the *Laplace equation* for the speed of sound c, the frequency f is obtained for pure gases:

$$f = \frac{1}{4 \cdot \pi \cdot r} \cdot \sqrt{\frac{\varkappa \cdot R \cdot T}{M}} \tag{3.58}$$

In a binary gas mixture, the frequency $f(\xi)$ then changes as a function of the material properties (M=molecular weight, heat capacity c_p, c_V), and the gas concentration ξ (mole fraction) of the first component as follows:

$$f(\xi) = \frac{1}{4 \cdot \pi \cdot r} \cdot \sqrt{\frac{R \cdot T}{\xi \cdot M_1 + (1-\xi) \cdot M_2} \cdot \frac{\xi \cdot c_{p,1} + (1-\xi) \cdot c_{p,2}}{\xi \cdot c_{V,1} + (1-\xi) \cdot c_{V,2}}} \tag{3.59}$$

If the temperature T is kept constant, the frequency measurement can be converted directly to the gas concentration ξ. In practice, the frequency f as a function of the gas concentration is measured experimentally and stored as a characteristic curve.

Figure 3.63 shows the sensor design according to Mayer (1980). The fluidic oscillator consists of several thin plates (Fig. 3.64), which are placed on top of each other and form the oscillator through the various cutouts. The plate material can be chosen almost arbitrarily, although for practical reasons stainless steel is usually used. However, glass and oxide ceramics are also used in special applications. The sound frequency f is sensed by

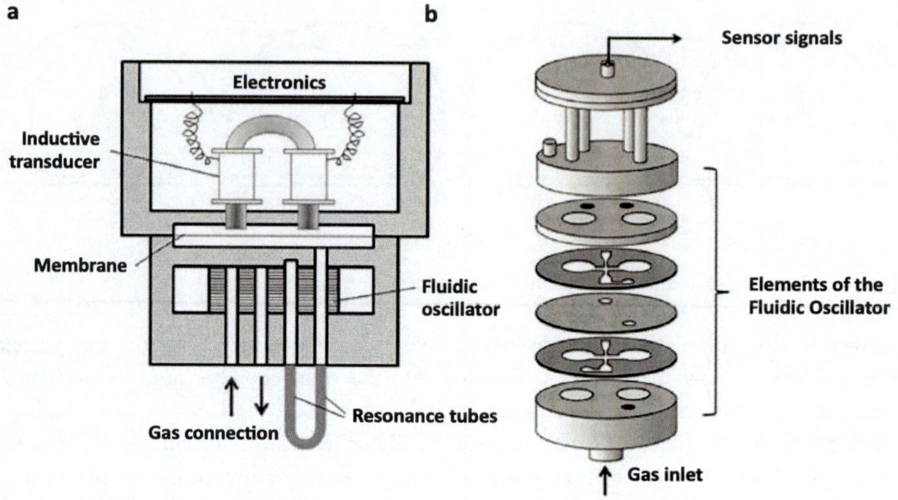

Fig. 3.63 Acoustic gas sensor according to Mayer (1980). The resonance tubes serve for a possible adaptation of the characteristic length L by an external feedback. In **b** is an exploded view of the fluidic elements (according to Zipser)

Fig. 3.64 Single fluid element of the company GLP-mbH Wiesbaden

a corrosion-resistant diaphragm which, like a microphone, converts the sound waves into a voltage. An inductive transducer is used for this purpose, which transmits the signals to the evaluation electronics. Since only the frequency is measured here, this type of measurement is comparatively insensitive to interference.

Coriolis Gas Density Sensor (MassSense^{TM})

A special gas density sensor is shown in Fig. 3.65. The extremely miniaturized design was built in silicon micromechanics and consists of a silicon tube capable of oscillating. This microstructure is produced by anisotropic etching techniques. The movable tube, with an

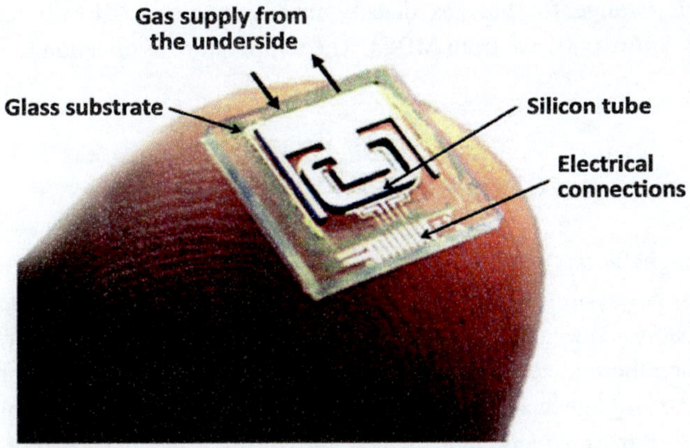

Fig. 3.65 Miniaturized Coriolis gas density sensor (MassSense™) from Integrated Sensing Systems USA. (Source: Dr.-Ing. Ritter Apperatebau GmbH Bochum)

internal volume of only 200 nL, is excited to oscillate at a frequency *f* of above 20 kHz. The gas flows through the silicon tube at a constant velocity, whereby the gas density can change. The resulting or changing *Coriolis force*[13] leads to a measurable change in the oscillation frequency *f*. A temperature sensor is also integrated on the microstructure, so that the temperature of the gas is also recorded and can be compensated by calculation (Fig. 3.66).

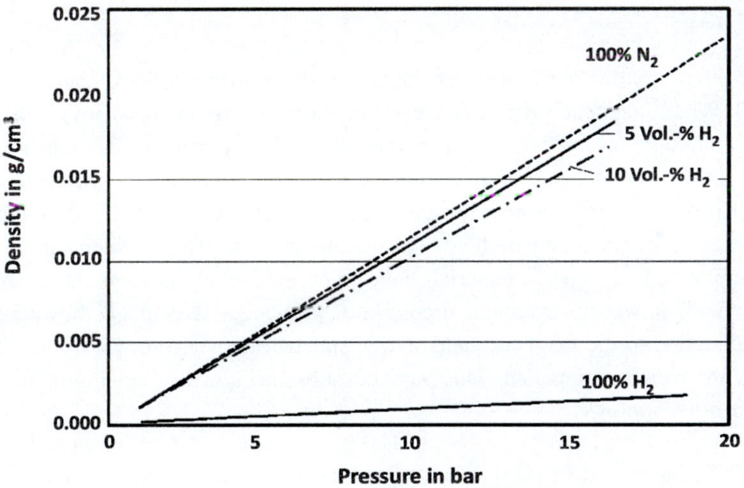

Fig. 3.66 Gas density measurement with the Coriolis sensor in a pressure range from 0–20 bar

[13] Gaspard Gustave de Coriolis (1792–1843) French mathematician and physicist

The measuring range for the gas density measurement is 0 to 30 kg/m^3, with an accuracy of 0.1 kg/m^3($\approx$0, 3% from MBE). The temperature application range is between −20 and 60 °C.

3.3 Paramagnetic Oxygen Sensor

The first paramagnetic oxygen sensors were developed in Germany in the 1930s (Lehrer 1926), and were based on the thermal properties of magnetic materials. The discovery of this effect probably came about more by chance than by investigating the influence of magnetic fields on thermal conductivity. In the case of oxygen, a considerable dependence of the measuring voltage was observed, which was not due to the change in thermal conductivity but to the magnetic field. A new measuring effect had been discovered. The first setups for oxygen measurement therefore consisted of thermal conductivity sensors in which the measuring cell was in a strong magnetic field. In the 1940/50s, further processes were developed on this basis and described theoretically for the first time (Krupp 1953). Figure 3.67 shows such a paramagnetic oxygen sensor, which was used as a series device in industrial measurement technology after 1950.

Interesting in this presentation is also the temperature control, which was realized by a mercury thermometer with integrated electrodes. When the temperature rises, the mercury column also rises and then touches the contact, which switched off the heating via a relay. As a result, the structure cools down and the mercury column sinks, opening the contact and switching the heating back on (2-point controller).

Basics

The magnetic properties of gases are extremely small compared to ferromagnetic materials (e.g. iron). Weak magnetic properties are called paramagnetism and diamagnetism. Paramagnetic materials have a permanent, magnetic dipole moment m. Without an external magnetic field, the molecules are disordered, so that no magnetic influences can be detected. However, if an external magnetic field H is applied, molecules align themselves and are attracted by the magnetic field. This results in force effects which can be used for gas measurement. Diamagnetic materials have no permanent dipole moment. In an external magnetic field, however, magnetic dipoles are produced, which are, however, in the opposite direction to the magnetic field. These substances therefore weaken the magnetic field and are therefore repelled. Diamagnetic substances are of no importance in gas measurement technology.

Fig. 3.67 Historical image (≈1950) of a paramagnetic oxygen sensor (type MAGNOS 2, Hartmann & Braun AG, Frankfurt a. M)

The magnetic properties of paramagnetic materials are described via the magnetic permeability μ. The relative permeability μ_r results from the magnetic field constant $\mu_0 = 4\pi \cdot 10^{-7}$ Vs/Am:

$$\mu_r = \frac{\mu}{\mu_0} \tag{3.60}$$

Another important parameter is magnetic susceptibility χ_m, which is defined as follows:

$$\chi_m = \mu_r - 1 \tag{3.61}$$

For paramagnetic substances we use $\mu_r > 1$ and $\chi_m > 1$, while for diamagnetic substances $\mu_r < 1$and $\chi_m < 1$ applies. The magnetic flux density (induction) B results from:

$$B = \mu_0 \cdot \mu_r \cdot H \tag{3.62}$$

Another useful quantity is the magnetization M, which is defined as the magnetic dipole moment per unit volume:

Table 3.3 Molar susceptibility χ_{mol} and relative susceptibility χ_{rel} of various gases (Vonderschmidt 2012)

Gas	$\chi_{mol} \cdot 10^{-9}\,m^3/mol$	$\chi_{rel}[O_2 = 100\%]$
Oxygen	43.34	100
Nitric oxide	18.36	42.4
Nitrogen dioxide	1.88	4.33
Nitrous oxide	−0.24	−0.55
Helium	−0.025	−0.06
Hydrogen	−0.05	−0.12
Neon	−0.087	−0.13
Nitrogen	−0.15	−0.35
Steam	−0.16	−0.37
Carbon monoxide	−0.12	−0.28
Carbon dioxide	−0.26	−0.60
Argon	−0.24	−0.55

$$M = \frac{m}{V} \tag{3.63}$$

The magnetization is proportional to the magnetic flux density B, and inversely proportional to the temperature T (in K). This relationship is also known as the *Curie*[14] *law:*

$$M = C \cdot \frac{B}{T} \tag{3.64}$$

The proportional factor C is the so-called Curie constant. With increasing temperature, the materials lose their magnetic properties. This relationship can also be used for gas measurement.

Table 3.3 shows the susceptibilities χ_{mol} of various gases. In this list, oxygen has by far the highest susceptibility, followed by nitric oxide (NO). Paramagnetic measuring methods are therefore preferably used for the measurement of oxygen, since interference (cross-sensitivities) caused by the other gases can almost be excluded.

In a gas mixture (concentration c), the susceptibility χ_M changes according to the mixing rule:

$$\chi_M = \chi_1 \cdot c_1 + \chi_2 \cdot c_2 + \ldots + \chi_n \cdot c_n \tag{3.65}$$

[14] Pierre Curie (1859–1906) French physicist

Example

Oxygen content is often measured in combustion processes and then lies between 0–20 vol % (measuring range). The NO content can be e.g. 2000 ppm (= 0.2 vol.%). For the oxygen measuring range of 0–20 vol.%, the susceptibility[15] therefore changes from 0 to$43.34 \cdot 0.2 = 8,668$. This value therefore corresponds to an O_2 concentration of 20 vol.%, i.e. 0.4334/vol.%-O_2. The NO concentration of 0.2 vol.% has a susceptibility of $18,36 \cdot 0,002 = 0,03672$. With a sensitivity of the O_2 sensor of 0.4334/vol.%-O_2, this value results in an O_2 indication of 0.085 vol.%. Related to the entire measuring range of 20 vol.% O_2, the error is thus <0.5 % of the measuring range end value.

This cross-sensitivity can therefore be neglected in most cases, so that one can speak of a selective measuring method in paramagnetic oxygen measurement. In chemical process engineering, on the other hand, high NO concentrations in the volume percent range can occur, which then have a correspondingly greater error influence. Nitrogen monoxide NO, can therefore also be monitored in high concentrations using the paramagnetic measuring method.

Due to the different mechanisms of action, three basic measuring methods for the detection of oxygen on the basis of paramagnetism have been developed, which will now be described in the following.

Thermomagnetic Method

As already mentioned, the first paramagnetic oxygen sensors consisted of a measuring cell, which was similar in design to a thermal conductivity sensor. Figure 3.68 shows such a structure. The measuring resistor is located in a measuring cell, and is heated by the bridge current to a temperature of e.g. $T = 150^{\circ}C$. In this cell, the heat power is now transferred to the outer wall by the surrounding gas. These effects have already been described in detail in Sect. 3.1. If a permanent magnet is now additionally used to generate a magnetic field H in the measuring cell, oxygen molecules are attracted by the magnetic field. The O_2-molecules move in the direction of the highest magnetic field strength, which is present in the middle between the two pole shoes. This is where the heated measuring resistor is located, and transfers the heat to the O_2-molecules, which lose some of their magnetic properties due to this increase in temperature (→ *Curie law*) and are displaced by the *cold* O_2-molecules that are pressing after them. This process creates an additional convection in the chamber, which is caused by the magnetic properties of the O_2 molecules. This magnetic convection is also called *magnetic wind*. This convection causes the measuring resistor to cool down additionally, and thus changes its electrical resistance $R(T)$. This resistance change ΔR is proportional to the oxygen concentration c_{O_2} in the measuring chamber:

[15] The unit was left out here.

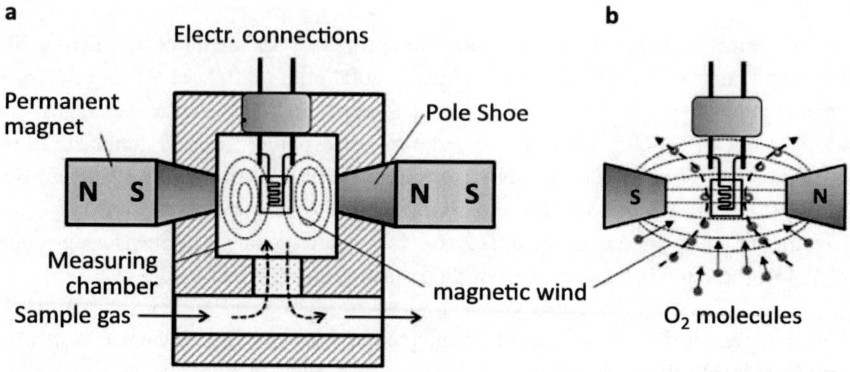

Fig. 3.68 a Measuring cell with integrated magnets, b Generation of the magnetic wind by local heating at the measuring resistor

$$\Delta R \sim \alpha \cdot c_{O_2} \tag{3.66}$$

The proportional factor α is a constant resulting from the sensor design (magnetic flux B, temperature T etc.). The change in resistance ΔR can in turn be converted into a voltage signal ΔU using a Wheatstone measuring bridge.

As with the thermal conductivity sensors, one must ensure that no forced convection occurs in the measuring cell with these thermomagnetic sensor structures. Too strong a current lead to additional cooling, and the effect of the magnetic wind is considerably disturbed. To prevent this, the chamber is only flowed over by diffusion, which leads to a response time in the range of $t_{90} > 10$ seconds.

Figure 3.69 shows the design of a MAGNOS type oxygen sensor. So-called ring resistors (W_1 and W_2), which are located rotationally symmetrically between pole shoes, were used as resistance materials here. These ring resistors consist of thin platinum wires embedded in glass. Other manufacturers use glass-encapsulated thermistors here, which serve a similar purpose. In the reference chamber is the resistor W_2, in a cell with the same geometry, but without the magnetic field. With this reference chamber, influences of the ambient temperature T, the thermal conductivity λ of the gas mixture and the position dependence (x, y, z) are compensated. Especially the position dependence is an important influencing factor, because the natural convection and the magnetic convection (magnetic wind) are always superimposed.

The described effect of the magnetic wind is a very small effect. Nevertheless, it could be made visible with an optical method (Krupp 1953). For this purpose, the structure

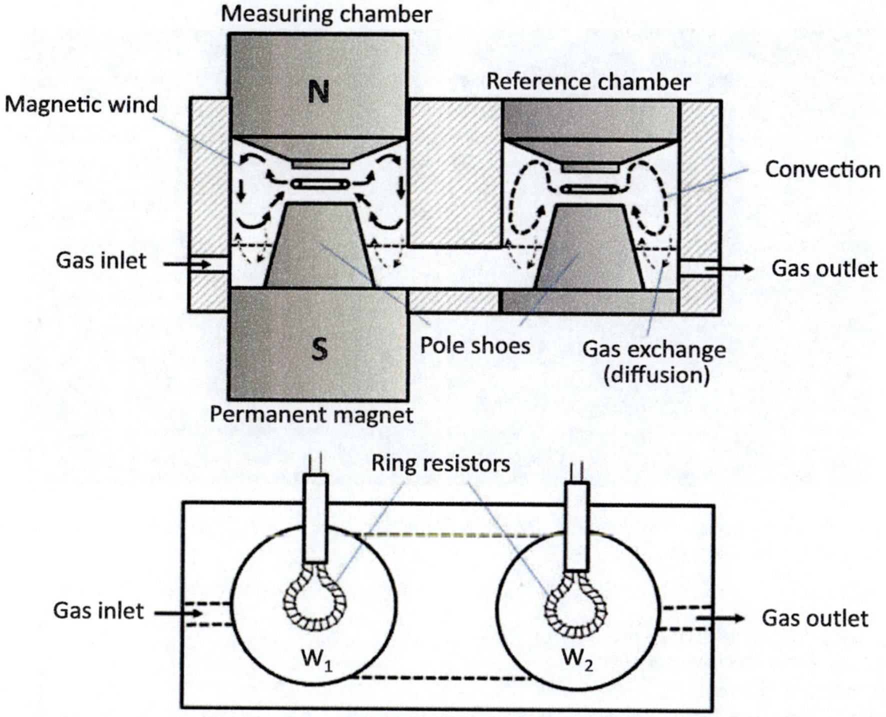

Fig. 3.69 Cross-sectional view through a thermomagnetic oxygen sensor (MAGNOS). The gas exchange with the measuring or reference cell takes place by diffusion and thus limits the time response of the display

described above was photographed with the *Toepler's*[16] *Schlieren method.* The results are shown in Fig. 3.70. One can clearly see the symmetrical distribution of natural convection (Fig. 3.70a) in the reference cell (without *B-field*), which conducts the heated gas upwards. Due to the magnetic wind, a superposition with the natural convection (Fig. 3.70b) takes place, which leads to a vectorial alignment of the resulting quantity.

The vectorial superposition can be determined by the different directions of force. The natural convection acts against the gravitational force, while the repulsion of the *cold* oxygen molecules leads out of the gradient of the magnetic field. The superposition can be seen in Fig. 3.71 and corresponds to the experimental results in Fig. 3.70.

Ring Chamber Method

Another interesting variant of the thermomagnetic measuring method was developed by *Lehrer and Ebbinghaus in* 1943 (Lehrer and Ebbinghaus 1950). It consists of an annular

[16]August Joseph Ignaz Toepler (1836–1912) German physicist.

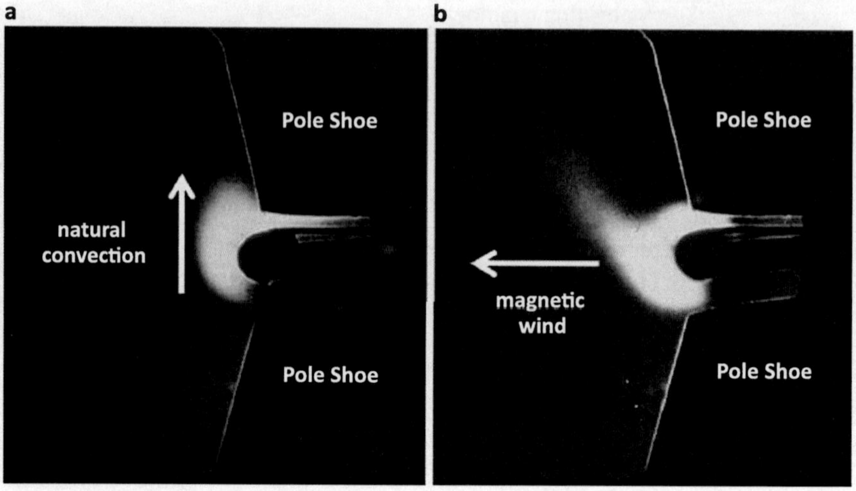

Fig. 3.70 Representation of natural convection in the reference chamber (**a**) and magnetic wind in the measuring chamber (**b**)

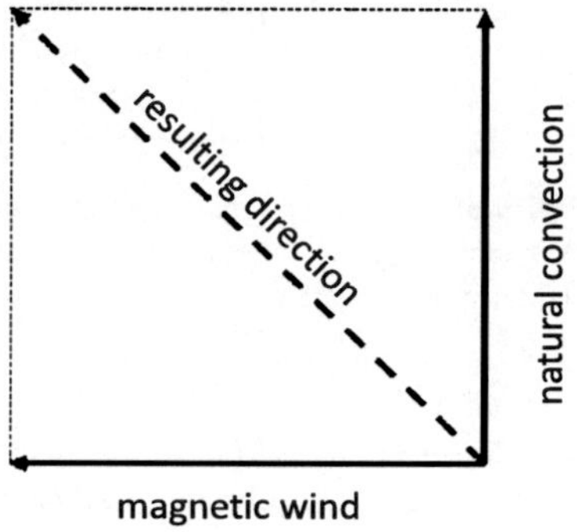

Fig. 3.71 Determination of the resulting direction by vectorial superposition of natural convection (against gravity) and magnetic wind (against the magnetic field)

chamber through which the gas to be analyzed flows. Since this chamber is symmetrically constructed, there is no cross flow under normal conditions (horizontal installation). If a permanent magnet is placed at the entrance of the cross connection, oxygen molecules in the gas mixture are attracted by the magnet. In the area of the inhomogeneous magnetic field there is a platinum wire winding on the outside of the cross-connection (thin glass tube). The two halves of the wire winding are part of a Wheatstone measuring bridge. The bridge current heats up the wire winding to the temperature T_2, and the oxygen molecules located in this area are pushed out of the magnetic field again.

The *cold* oxygen molecules (T_1) located in the vicinity can thus flow in. This initially creates a pressure difference Δp which can be calculated approximately (Lehrer and Ebbinghaus 1950):

$$\Delta p = \frac{1}{2} H^2 \cdot \left[\frac{1}{T_2^2} - \frac{1}{T_1^2} \right] \cdot \varrho_{O_2} \cdot C \cdot \frac{p\,273K}{1013\text{hPa}} \tag{3.67}$$

This pressure difference Δp induces a transverse flow (magnetic wind). The transported volume is calculated with the viscosity η and the resistance coefficient k of the annular chamber as follows:

$$V = \frac{\Delta p}{k \cdot \eta} \tag{3.68}$$

The flow velocity w, with which, for example, a gas mixture of 1 vol.-O_2 in N_2 flows through a 6 mm pipe is$w \approx 0,\ 8$ mm/s (Krupp 1953). The wire winding experiences an asymmetrical temperature distribution due to the flow, so that the temperature difference ΔT in the measuring bridge is converted into a voltage ΔU_{Br}. In this case too, there is a proportional relationship between the oxygen concentration c_{O_2} in the gas mixture and the voltageΔU_{Br}:

$$\Delta U_{\text{Br}} \sim \alpha \cdot c_{O_2} \tag{3.69}$$

If the temperature increase ($T_1 - T_2$) of the wire winding is set, for example, at 100 °C, a sensitivity of $\alpha = 6$ mV/vol% O_2 is obtained. However, this sensitivity is only valid for a concentration range up to c_{O_2} 25 vol.% O_2. For higher concentrations the sensitivity decreases, so that the characteristic curve must be linearized electronically (Fig. 3.72).

Position Dependence

Due to the high temperature ($T \approx 100$ °C) in the tube, a natural convection is also formed, which is superimposed on the magnetic wind. In order to keep this influence low, the structure is installed in a horizontal position and usually also permanently installed (e.-g. wall mounting). However, if the superstructure is tilted at an angle α, a cross-flow (*chimney effect*) is inevitable, which is independent of the oxygen concentration (Fig. 3.73). Ebbinghaus used this effect to realize strongly suppressed measuring ranges of 20–21 vol. O_2 or 95–100 vol. % O_2 (Ebbinghaus 1953).

For the adjustment of a measuring range of 20–21% by volume O_2, the assembly is first flowed with 20% by volume O_2. Then turn the angle α until the $\Delta U_{\text{Br}} = 0$. In this position, the natural convection (B.) due to the *chimney effect* is just as large as the effect by the magnetic wind (A.). In this position there is *no wind*. Now the oxygen concentration is increased to 21% by volume and the end point is set. Suppressed measuring ranges are

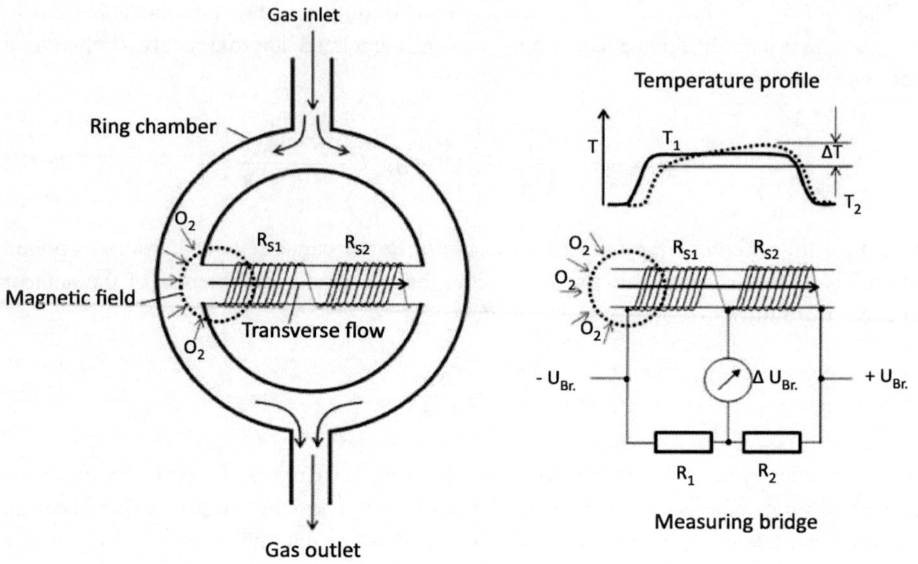

Fig. 3.72 Ring chamber structure for the detection of oxygen according to Lehrer and Ebbinghaus (1950)

Fig. 3.73 Tilted annular chamber for zero suppression. The magnetic wind (*A*.) is diametrically opposed to natural convection (*B*)

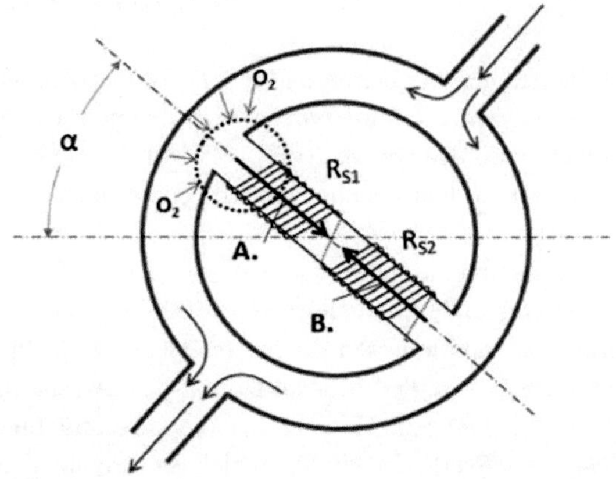

used, for example, to examine the plant growth (20–21 vol.%) or to carry out a purity control (98–100 vol.%) of oxygen.

Magnetomechanical Method

This method has been known since 1940 and is based on a magnetic rotary balance (Pauling et al. 1946). The construction consists of two thin-walled glass spheres ($d \approx 3$ mm), which are fused with a glass rod of approximately 4 mm length. The

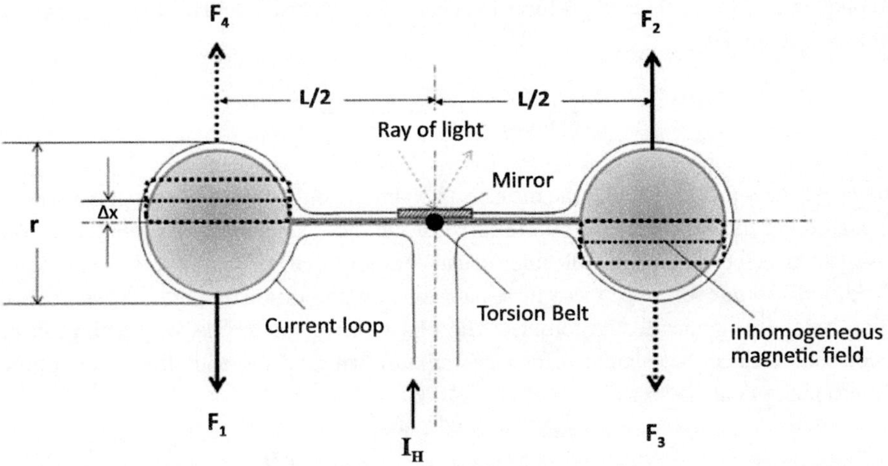

Fig. 3.74 Structure of the dumbbell with the different forces that lead to a resulting torque. The current loop is used to compensate for this rotational movement

susceptibility χ_K of the glass spheres results from the glass material and the gas filling (e.g. N_2). The mass *m of* the glass spheres is a few mg. This arrangement looks like a *dumbbell, which* is why it is sometimes referred to as the *dumbbell process*. Both balls are in an inhomogeneous magnetic field *H,* and are held in the middle between the balls by a torsion band (Fig. 3.74). There is also a current loop around the two balls, through which an adjustable current I_H flows. In the initial position, the balls are offset by an amount Δx from the center of the magnetic field *H*. In this position, the torsion band is pretensioned. There is no oxygen in the measuring chamber yet. To hold the dumbbell in position, a basic current I_H *is* already flowing, which builds up an opposing magnetic field *H.* The magnetic field *H* can be calculated as follows:

$$H = \frac{I_H}{2r} \tag{3.70}$$

Since the loop is in an external magnetic field, a force *F is* generated which is proportional to the magnetic field *H*. The proportional factor, δ, results from the geometrical arrangement and the field strength of the permanent magnet:

$$F_3 = F_4 = \delta \cdot H \tag{3.71}$$

This *H-field* generates a torque, which is opposite to the torsion band torque M_{Torsion}. Both torques then cancel each other out in the rest position. This measure ensures that the dumbbell is always at a defined position and cannot be adjusted at will be due to shocks and vibrations. It applies:

$$M_{\text{Torsion}} = F_3 \cdot \frac{L}{2} + F_4 \cdot \frac{L}{2} \tag{3.72}$$

If oxygen is now introduced into the measuring chamber, the O_2-molecules will try to reach the position of the glass spheres by the force of the inhomogeneous magnetic field. This displacement effect of the O_2-molecules causes further forces (F_1 and F_2), which produce an additional torque M. Since these forces are directly proportional to the oxygen concentration, this also applies to the torques. The equation for the torque M_{O_2} acting on the torsion filament, can be calculated for a gas mixture with the susceptibility χ_M as follows (Hengstenberg et al. 1980):

$$M_{O_2} = (\chi_M - \chi_K) \cdot V \cdot \mu_0 \cdot H \cdot L \cdot \frac{dH}{ds} \tag{3.73}$$

The occurring torques are in the order of 10^{-10} Nm (Kronmüller and Zehner 1980). Without further measures, the dumbbell would therefore move and make way for the O_2molecules that are pressing after it. This process is detected by optical scanning via a small mirror in the middle of the dumbbell and leads to an increase in the dumbbell current I_H via control electronics. This increase is proportional to the oxygen content c_{O_2} and serves as a measure of the concentration.

The optical scanning is done with a differential photodiode (Fig. 3.75). If the dumbbell moves due to an increasing or decreasing oxygen concentration, the radiation components of the two receiving diodes also change and the loop current I_H must be readjusted accordingly. This control process is completed when both radiation components are again identical (compensation procedure). The advantage of this method is the absolutely linear relationship between the O_2 concentration and the loop current I_H. An additional linearization of the signals is therefore not necessary. The different manufacturers offer measuring ranges from 0–1 vol% O_2 to 0–100 vol% O_2.

However, the flow rate is also limited to 0.3 L/min for these setups, in order not to disturb the dumbbell position by turbulent flows in the chamber. Bypass solutions from the manufacturers are available for larger flows. Typical temperature application ranges are from 0 to 45 °C, with integrated thermostatization of the measuring cell at 50° C. The time response is comparatively fast and is in the range of 2 seconds.

Magnetopneumatic Method

This method is based on a pressure increase Δp that occurs when the gas flows in a homogeneous magnetic field in the presence of oxygen. First investigations on this were

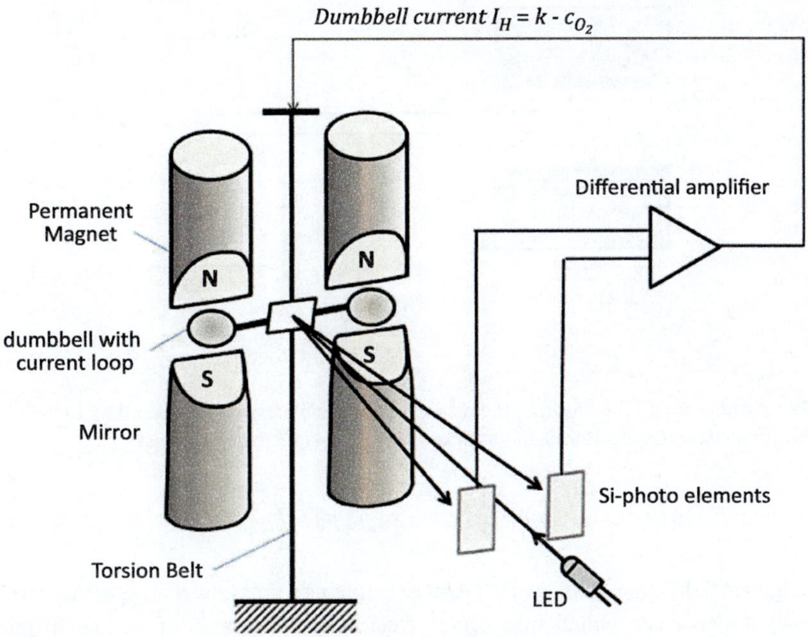

Fig. 3.75 Basic structure of the magnetomechanical oxygen measurement method. The dumbbell position is scanned by a focused LED optic and a differential photodiode

already carried out before 1950 by *K. F. Luft*[17] (Luft 1951). The method was later taken up and modified by Schlumberger (France). However, it took another 17 years before *Luft was* able to present a measuring device (Oxor) suitable for series production for use in mining (Luft and Mohrmann 1967). This method was also taken over by a company and produced in series (Oxygor of the company Maihak-Hamburg).

Figure 3.76 shows an experimental setup in which a gas pipe is arranged in the homogeneous part of a magnetic field. In this gap there is also 100% oxygen. If a small N_2 gas flow is now directed into this area, the pressure increases slightly by the amount Δp in the gas pipe. The oxygen molecules, which are in the magnetic field at the outlet of the gas pipe, press against the incoming nitrogen molecules due to their magnetic properties and the pressure Δp increases. The increase in pressure depends on the difference in the susceptibilities of the two gases and the magnetic field H and can be calculated as follows (Marx and Wiegleb 1995)

[17]Karl Friedrich Luft (1909–1999) German physicist and pioneer of physical gas measurement technology

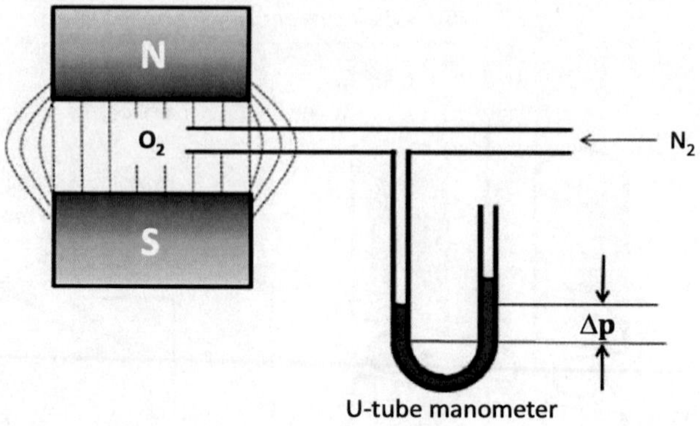

Fig. 3.76 Arrangement for measuring the magnetopneumatic pressure Δp according to the *Quincke rising-height method* (Georg Hermann Quincke (1834–1924) German physicist)

$$\Delta p = \frac{1}{2}\left(\chi_M - \chi_{N_2}\right) \cdot \mu_0 \cdot H^2 \tag{3.74}$$

For a magnetic field strength of $8 \cdot 10^5$/ A/m one obtains a pressure change of $\Delta p \approx 0.7$ Pa. This is therefore a very small measuring effect. The conversion of this small pressure change was probably one of the reasons why this method was relatively late in becoming ready for series production.

In its first setup (1959), *K.F.Luft* had already used a sensitive membrane capacitor to detect the change in pressure and then replaced it with a flow sensor in the second setup (1967). Today's oxygen sensors based on the magnetopneumatic method use both possibilities. In 1968 Hummel presented an oxygen meter which used a membrane capacitor as a transducer and a modulated electromagnet for excitation. This process was subsequently mass-produced by the company Hartmann & Braun (today ABB).

Probably the best-known device working on this basis today is the Oxymat from Siemens. Figure 3.77 shows the basic structure.

The reference gas (nitrogen N_2) is divided into two paths which generate a constant volume flow $V_1 = V_2$ via capillaries (1.). The total volume ($V_1 + V_2$) of the auxiliary gas is max. 5 mL/min. This gas flow also continues into the measuring chamber when no oxygen is present. It applies then:

$$V_1 = V_2 = V_3 = V_4 \tag{3.75}$$

ΔVIn this case, a possible cross flow is zero and the flow sensor (2.) does not register a volume flow. If there is now oxygen in the measuring chamber, it preferentially collects around the largest magnetic field. The electromagnet is located directly at the outlet of the volume flow *V4* in the chamber. This arrangement causes an increase in pressure Δp in the inlet area and the volume flow V_4 is reduced by the amount ΔV. It then applies:

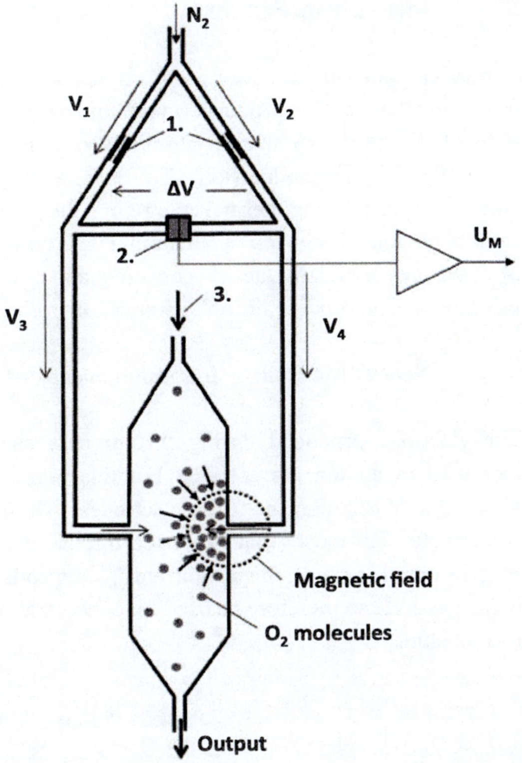

Fig. 3.77 Basic design of the Oxymat according to the magnetopneumatic measuring method with 1. capillaries, 2. microflow sensor, 3. gas inlet into the measuring cell

$$\Delta V = V_4 - V_4^* \tag{3.76}$$

The electromagnet is operated at a frequency of 8–25 Hz, which leads to a pulsation of the volume flow$\Delta V = 10^{-2}\,\text{mm}^3$ /volume % O_2 (Aidam 1979). Depending on which modulation frequency is set, response times (*t90*) from 100 ms (25 Hz) to 1 s (8 Hz) can be achieved. In addition to the fast display option, this design also offers another advantage. Since the sample gas only comes in contact with the measuring chamber, the chamber can be made of almost any material (stainless steel, tantalum, etc.). This method is particularly suitable in the field of chemical process engineering (process gas analysis), if e.g. oxygen contents in corrosive gases must be measured. The smallest available measuring range is specified by the manufacturer as 0.5 Vol-% O_2, with a detection limit of 500 ppm O_2.

3.4 Ionisation Process

Certain gaseous substances or groups of substances can be detected by prior ionization by measuring the reaction products (ions). Ionization is the release of electrons from the outer shell of the atoms. A neutral atom/molecule thus produces electrons (negative) and a positively charged residue (ion). To ionize a gas, energy must be supplied. This can be done e.g. by supplying thermal energy in a flame or by exposing the gas to radiation in the form of photons. Radioactive radiation (e.g. α emitter) can also be used for this purpose but is nowadays avoided due to concerns about the uncontrolled spread of radioactive substances. The process of ionization can therefore be described as follows:

$$\text{Neutral Molecule} + \text{Ionization energy} \rightleftharpoons \text{Positive ion} + \text{Electron} \quad (3.77)$$

The electrons produced during this process can be captured by two electrodes and measured as an ion current J_{Ion}. For this purpose a suction voltage of approximately ≈100 -300 V is connected to the electrodes. Within a certain voltage range the ion current is saturated. This means that the current J remains constant as the suction voltage increases. In this saturation range, the ion current J_{Ion} depends only on the number of ion pairs present in the gas and can therefore be used for concentration measurement. Figure 3.78 shows this relationship.

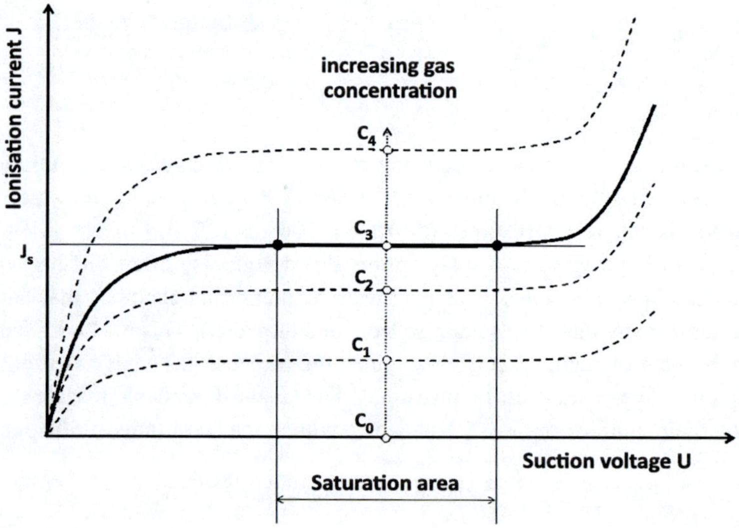

Fig. 3.78 Dependence of the ionization current J_{Ion} on the suction voltage U and the gas concentration c

Flame Ionization Detector FID

The flame ionization detector, or FID for short, was originally developed as a detector for use in gas chromatography. In the meantime, however, it has found a wide range of applications in exhaust gas analysis and environmental measurement technology. Today, the FID is the standard method for analyzing small hydrocarbon concentrations. The ionization is carried out in a small hydrogen flame, which provides combustion temperatures from 2500 to 3500 K. At these temperatures, hydrocarbons which usually consist of long C-chains, are partially ionized. The degree of ionization can be determined in principle from the *Eggert-Saha equation*, which indicates the proportion of molecules in a gas that are ionized at a certain temperature (Meschede 2010). The equation is:

$$\frac{n_i}{n_0} = \frac{(2\pi mkT)^{3/4}}{n_0^{1/2} h^{3/2}} \mathrm{e}^{-E_i/2kT} = 4.91 \cdot 10^{10} \frac{T^{3/4}}{n_0^{1/2}} \cdot \mathrm{e}^{-E_i/2kT} \tag{3.78}$$

Where n_i and n_0 are the particle number densities of the ionized and neutral molecules, m is the molecular weight, k is the Boltzmann constant, T is the absolute temperature and E_i is the ionization energy.

Ionization occurs when the particles collide with other atoms and molecules of the gas. Particles with high velocity have a higher probability to ionize others. Furthermore, the proportion of ionized particles depends on their ionization energy. Although the temperature T in the flame is quite high, it is not enough to ionize a large number n_i of molecules, because the ionization energy E_i is too high for these substances. Even if the exact conversion is not yet known in detail, it is assumed that carbon is formed via various intermediate stages during the simultaneous combustion of the hydrocarbons. The ionization energy of carbon is 3.9 eV. This low energy explains the high degree of ionization of hydrocarbons (HC) in the FID (Birkle 1979).

Processes in the Flame

A mixture of hydrogen and sample gas (containing the hydrocarbons to be measured) flows through the nozzle bore (A) into the measuring chamber and ignites the flame at the nozzle outlet (Fig. 3.79). This area contains the inner flame zone (B), also known as the pyrolysis zone. The temperatures in this zone are up to 1670 K (Staab et al. 1981). Due to the extreme heating, pyrolytic fission processes take place in this zone:

$$C_2H_6 + H^* \rightarrow C_2H_5^* + H_2 \tag{3.79}$$

$$C_2H_5'' + H^* \rightarrow 2\,CH_3^* \tag{3.80}$$

Fig. 3.79 Classification of the FID flame into different reaction zones (B→ C →D)

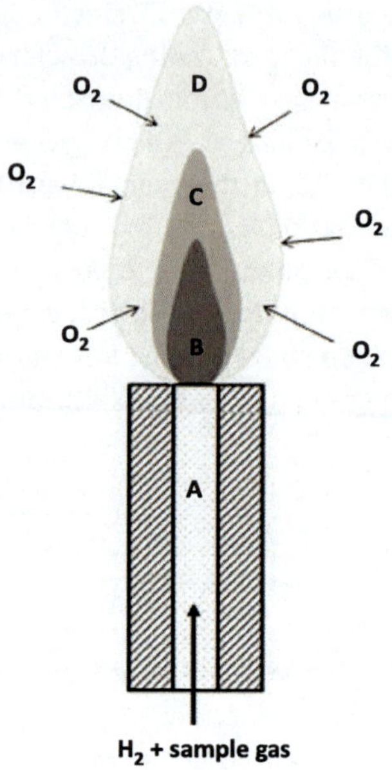

These cleavage processes produce radicals such as CH_2^*, CH^*, C^* which then enter the outer flame zone (C) of the oxidation zone. Although this zone is only a few 1/100 mm thick, the main reactions for ion formation take place in it. Oxygen O_2 diffuses in from the combustion air passing by from the outside and from the pyrolysis zone the hydrocarbon radicals enter the outer flame zone. The most important reaction that takes place there is:

$$CH^* + O^* \rightarrow CHO\ (\text{excited state}) \rightarrow CHO^+ + e^- \tag{3.81}$$

This process is called *chemical ionization.* The CHO^+ ion is the charge carrier responsible for the detector signal. The ion yield of this reaction is relatively low. Only about one in 500,000 hydrocarbon atoms forms an ion pair. However, since the basic ionization of the pure hydrogen flame is several orders of magnitude lower, the FID still has a relatively low detection limit, which goes as far as 50 ppb methane. Inorganic compounds such as CCl_4, CS_2, HCl etc., which cannot form CH^*radicals, are not detected by the FID or are detected with a much lower sensitivity.

The products of combustion from zone (C) are released into the outer atmosphere (D). Recombination processes can still occur on the way to the extraction electrodes and reduce the detector sensitivity.

Figure 3.80 shows the basic structure of a FID. The hydrogen flame of a few mm size burns in a small measuring chamber. The flame must be ignited before the measurement starts and sufficient combustion air must be available. Since the FID measuring method

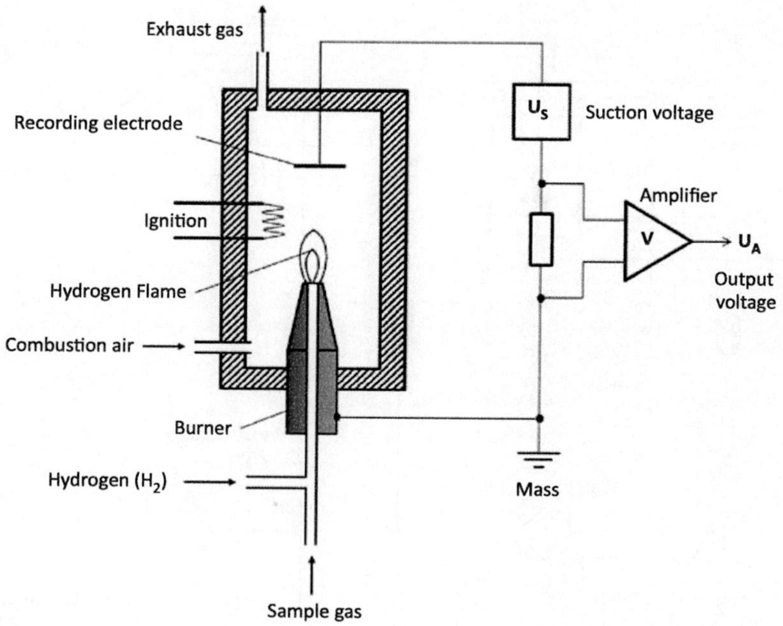

Fig. 3.80 Basic structure of a flame ionization detector (FID): The external components for controlling the gas flow are not shown

already reacts to extremely small traces of hydrocarbons, the combustion air must be conditioned. For this purpose, there are corresponding devices which carry out a catalytic combustion of the hydrocarbons in the combustion air and thus provide combustion air free of hydrocarbons. The volume flows for the combustion air, the hydrogen and the sample gas must be controlled very precisely, as these parameters are directly included in the measurement result.

Above the flame there is a collecting electrode connected to the suction voltage. The burner is connected to ground so that a closed circuit is established via the load resistance R. A voltage U_M then drops across the load resistor, which is proportional to the ion current J_{Ion}. The amplifier brings this small voltage to a level in the voltage range. The degree of amplification is v. The following relationship then applies:

$$U_M = J_{\text{Ion}} \cdot R \quad \text{and} \quad U_A = (J_{\text{Ion}} \cdot R) \cdot v \tag{3.82}$$

The ion current J_{Ion} generated by the hydrocarbons to be detected is in the order of 10^{-8} to 10^{-14} A. Even if no hydrocarbons are present in the sample gas, the hydrogen flame still generates a priming ionization current of 10^{-12} to 10^{-13} A. The reason for this is, among other things, minor impurities in the hydrogen or combustion air, which cannot always be

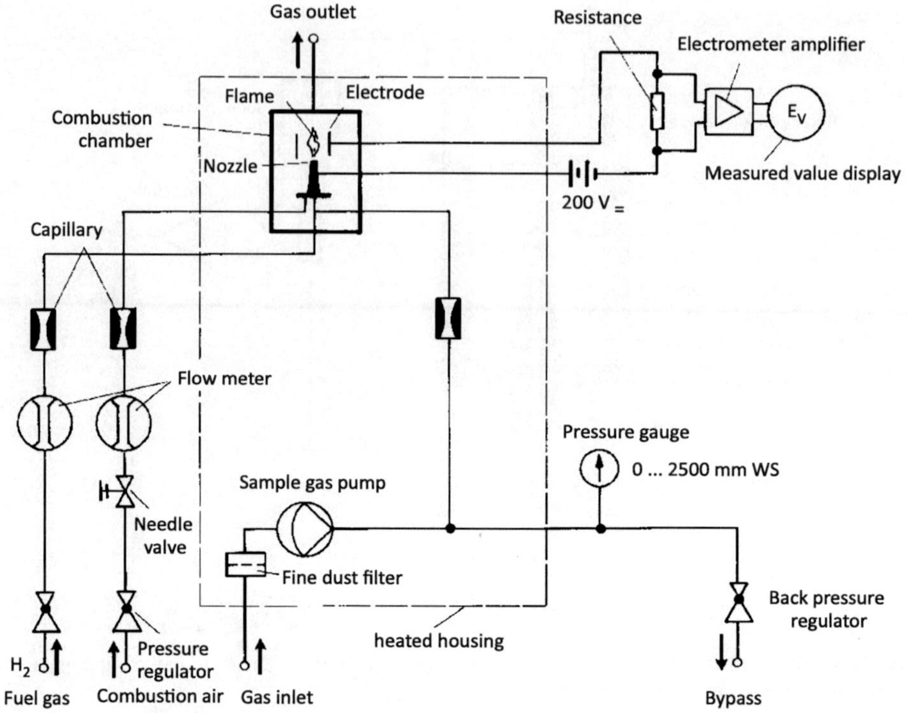

Fig. 3.81 Arrangement of the gas paths within an FID analyzer (Baumbach 1990)

completely excluded. Furthermore, contamination of the internal gas pipes (stainless steel pipes) and the sample preparation can be a cause for this priming stream. To avoid carry-over (*hang-up effect*), the gas lines are heated to temperatures of up to 200 °C. At these temperatures, higher hydrocarbons (long C-H chains) no longer condense in the gas preparation and are completely fed to the FID for analysis. Unexpected condensations in the gas treatment would lead to the fact that the zero point would only be reached after hours or even days due to constant evaporation of the condensed hydrocarbons.

As already mentioned, the individual gas flows must be regulated very precisely. For this purpose, capillaries (critical nozzles) are used, which are provided with a constant inlet pressure on the primary side via a pressure regulator (Fig. 3.81). This creates a constant volume flow for the fuel gas and the combustion air. The sample gas is sucked in via a heated sample gas pump and pumped past the capillary with a high-volume flow in a bypass. In the outlet there is a back pressure regulator, which then also ensures a constant flow through the capillary. There are also devices where the sample gas pump with the bypass is located behind the capillary and the sample gas is sucked in via the combustion chamber. This arrangement creates a negative pressure in the measuring chamber, which

Table 3.4 Structure factor for different alkanes. The relative indication is related to heptane =7

Substance	C number	Relative display	Deviation %
Methane	1	1	0
Ethane	2	2	0
Propane	3	3	0
Butane	4	3.8	−5
Pentane	5	5	0
Hexane	6	6.1	+1.7
Heptane	7	7	0
Octane	8	7.8	-2.5
Nonane	9	8.8	-2.2

has the advantage that the sample gas does not encounter the pump. However, this operation requires a high effort regarding control and is rarely used (Baumbach 1990).

Sensitivity of the FID

The sensitivity of the FID is determined by the ionization current J_{Ion}, which is given by a certain HC concentration c of in the sample gas. It can be calculated with m_{HC} = mass of the HC, t= time, q=substance-dependent proportionality factor, c=concentration of the HC, $\dot{V}$ = volume flow as follows (Oster 1963):

$$J_{\text{Ion}} = q \cdot \frac{m_{\text{HC}}}{t} = q \cdot c \cdot \dot{V} \tag{3.83}$$

For a constant volume flow $\dot{V}$, which must be specified internally, the following expression for the ion flow J_{Ion} is obtained in the presence of various hydrocarbons with different C-numbers, where A_n is the individual sensitivity factor for the respective gas:

$$J_{\text{Ion}} = A_1 \cdot c_1 + A_2 \cdot c_2 + \ldots + A_n \cdot c_n \tag{3.84}$$

For typical hydrocarbons such as alkanes, alkenes, etc., a sensitivity of the individual components is obtained which depends directly on the number of C atoms in the corresponding molecule (Table 3.4). The sensitivity factor A is identical for all components:

$$J_{\text{Ion}} = A \cdot c_1 + 2 \cdot A \cdot c_2 + 3 \cdot A \cdot c_3 + \ldots + n \cdot A \cdot c_n \tag{3.85}$$

Figure 3.82 shows the characteristic curves for the different alkanes (C_nH_m). This group of substances thus behaves linear, while the slope of the characteristic curve is directly dependent on the number of C atoms. However, this linearity is not given for all hydrocarbons. Figure 3.83 shows the relative indications (Heptane =7) for different groups of substances. The first group includes alkanes and aromatic hydrocarbons. With a few

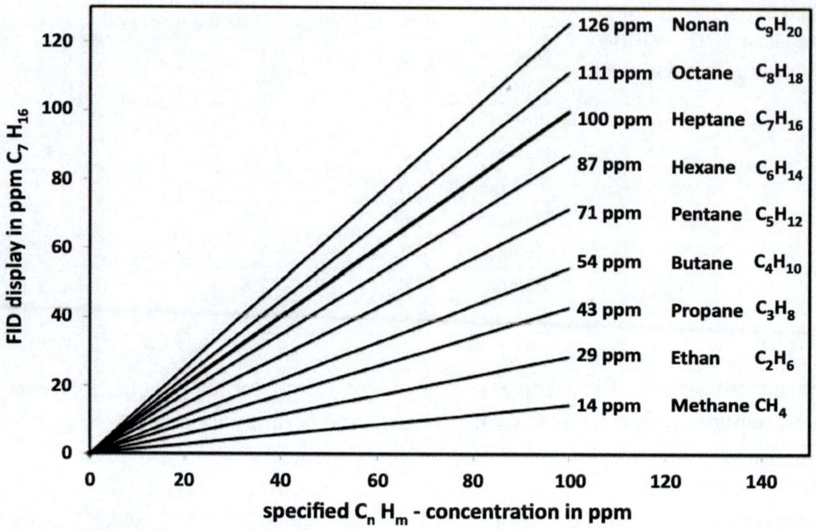

Fig. 3.82 Sensitivity curves for the alkanes (C_1 to C_9)

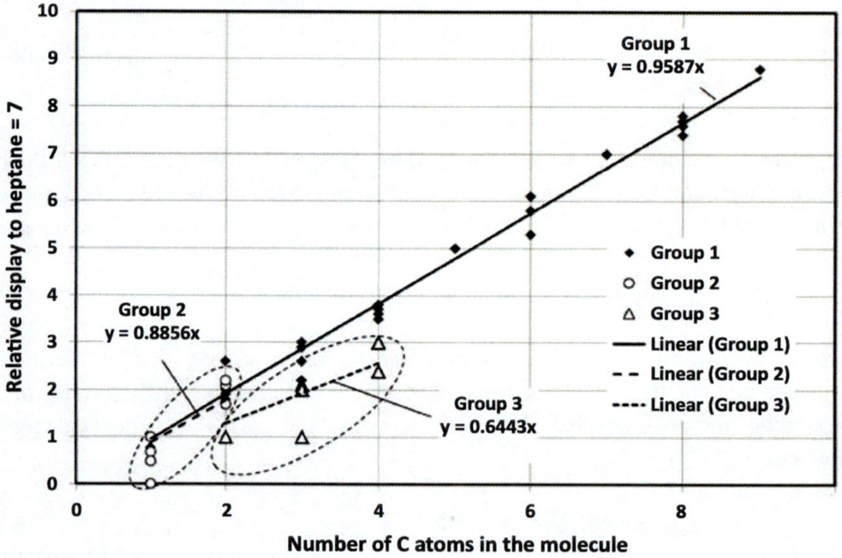

Fig. 3.83 Relative display as a function of the number of C atoms in the corresponding gas molecules. Group 1= Alkanes, alkenes, aromatics. Group 2= alcohols, Group 3 =acids, esters, ketones. (Profos and Pfeiffer 1992)

exceptions, these substances have a linear relationship ($y \approx 0.96$). The group of alcohols is very close to the trend line of group 1, but the relationship between the C count and the relative reading is lower ($y \approx 0.89$). Group 3 (acids, esters, ketones) shows very clear deviations. Here the curve is even flatter ($y \approx 0.64$). The deviations from the theory are the greatest here. The FID method has now become a standard method in various areas of application (motor vehicle engine test benches). In these applications, the ratios of the different hydrocarbons are mostly known or do not change significantly during the measurement.

Photoionization Detector (PID)

Another method uses electromagnetic radiation, in the ultraviolet spectral range, to ionize hydrocarbons. The radiant energy E of the photons depends on the frequency f or the wavelength λ. According to *Planck*[18] *it* results from the following relationship:

$$E = h \cdot f = \frac{hc}{\lambda} \tag{3.86}$$

The constant h is the so-called *Planck's quantum of action* ($h = 6.6260755 \cdot 10^{-34}$ J · s). The energy of a photon (light quantum) can therefore be calculated using the light wavelength λ.

Example

For the blue spectral range $\lambda \approx 450$ nm a photon therefore has the energy:

$$E = \frac{\left(6.6260755 \cdot 10^{-34}\ \mathrm{J \cdot s}\right) \cdot \left(3 \cdot 10^{8}\ \mathrm{m/s}\right)}{4.5 \cdot 10^{-7}\ \mathrm{m}} = 4.4 \cdot 10^{-19}\,\mathrm{J} \tag{3.87}$$

The ionization energy is generally given in electron volts (eV) and can be converted as follows:

$$1\ \text{Electron volt (eV)} = 1.60217733 \cdot 10^{-19}\ \mathrm{J} \tag{3.88}$$

$$1\ \mathrm{J}\ \text{(Joule)} = 6.2415064 \cdot 10^{18}\ \mathrm{eV} \tag{3.89}$$

A photon at $\lambda \approx 450$ nm therefore has an energy (in eV) of:

[18] Max Planck (1858–1947) German physicist and founder of quantum physics.

$$\frac{(4.4 \cdot 10^{-19}\ \mathrm{J})}{\left(1.6 \cdot \frac{10^{-19}\ \mathrm{J}}{\mathrm{eV}}\right)} = 2.7\ \mathrm{eV}$$

Equations 3.86 and 3.88 can be used to establish a simple conversion formula with which the light wavelength λ (in nm) can be converted into energy, with the unit eV:

$$\frac{1240}{\lambda[\mathrm{nm}]} = E[\mathrm{eV}] \tag{3.90}$$

Example
Light with a wavelength of 200 nm therefore has an energy of $E = 1240/200 = 6.2$ eV.
The reaction of a substance R in a PID can be specified as follows:

$$\mathrm{R} + h \cdot f \rightleftharpoons \mathrm{R}^+ + \mathrm{e}^- \tag{3.91}$$

To ionize a gaseous substance (R) by means of light radiation, however, higher energies in the range of 8–12 eV are required. According to Eq. 3.90, this corresponds to a wavelength range from 100 to 160 nm, which is also referred to as the vacuum UV range (VUV), since in this spectral range the air components (e.g. oxygen) already exhibit strong absorption bands.

Gas discharge lamps with different inert gas fillings are used as radiation sources (Fig. 3.84). This makes it possible to detect different groups of substances with different ionization potentials (Table 3.5).

The most important component of a PID is therefore the gas discharge lamp, which provides the measuring effect and partial selectivity. In practice, two different lamp types are used. The first type is the so-called hollow cathode lamp (HCL), in which the maximum emission of the discharge comes from a small zone of a hollow cathode. This emission is almost point-like, so that the radiation can be focused very well with a convergent lens. In this way, high luminous intensities can be achieved at the point where they are needed. Electrodeless discharge lamps (EDL[19]) have proven to be the best choice for hand-held measuring instruments, as they can be designed small and compact. The manufacturers specify lifetimes from 5000 to 10.000 h for both lamp types.

Flow Operation
Flow operation is mostly preferred in stationary measuring instruments in order to limit the time response to a few seconds ($t_{90} < 10$ s). Figure 3.85 shows such a structure. The gas

[19] **E**lectro**d**eless **D**ischarge **L**amp.

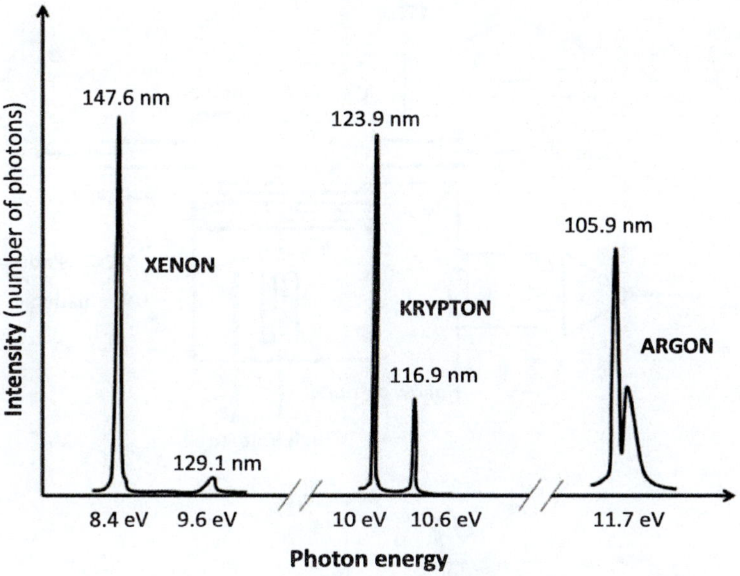

Fig. 3.84 Spectral distribution of the emission lines of the three main UV sources

Table 3.5 Spectral ranges for the detection of different substance groups

Spectral range	Substance group
Upto 9.5 eV	Benzene, aromatic hydrocarbon, amines
9.5–10.6 eV	Ammonia, ethanol, acetone
10.6–11.7 eV	Acetylene, formaldehyde, methanol

discharge lamp (in this case a HVAC) is located directly on the underside of the PID measuring chamber and thus forms a seal at this point. Two electrodes are arranged in the chamber. One of the electrodes is irradiated by VUV radiation ($h \cdot f$) and thus directly enters the PID measuring chamber, where ionization of the HC takes place. The ions are forced by the suction voltage to a current which drops above the load resistance R_L and generates a measuring voltage. The current is in the range of 1 nA/100 ppm isobutene. The measuring voltage is amplified ($\rightarrow U_{\mathrm{Meas}}$) and serves as proof of the HC concentration in the measuring cell. The complete gas measuring device is shown in Fig. 3.86.

Diffusion Process

Hand-held meters usually do not have an integrated gas pump, therefore the gas exchange takes place by diffusion. To keep the time response as low as possible ($t_{90} < 20$ s), the measuring chamber also has a correspondingly small volume. An EDL is used as a radiation source in this setup (see Fig. 3.87). Here too, the exit window is part of the measuring chamber. On the opposite side is a gas-permeable membrane. Below the

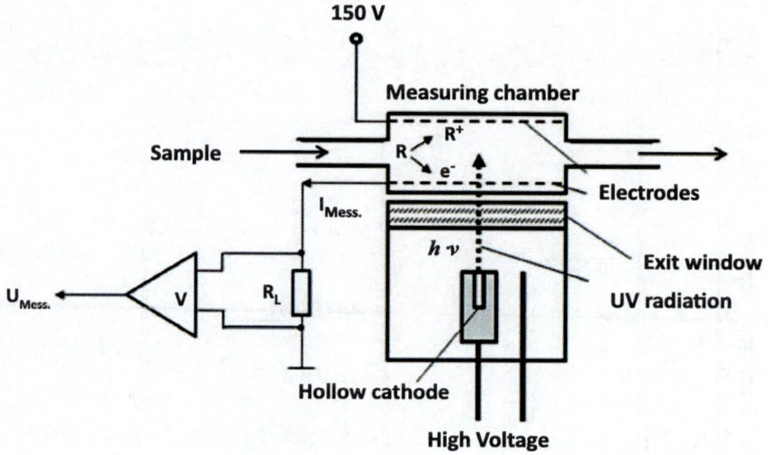

Fig. 3.85 Basic structure of a PID in flow-through operation with a HVAC as excitation source. Below is the design of a commercial PID unit. (Source: ACI GmbH-Berlin)

Fig. 3.86 PID table-top measuring device (lab pid) with integrated gas pump, for drawing ambient air into the device. (Source: ACI GmbH Berlin)

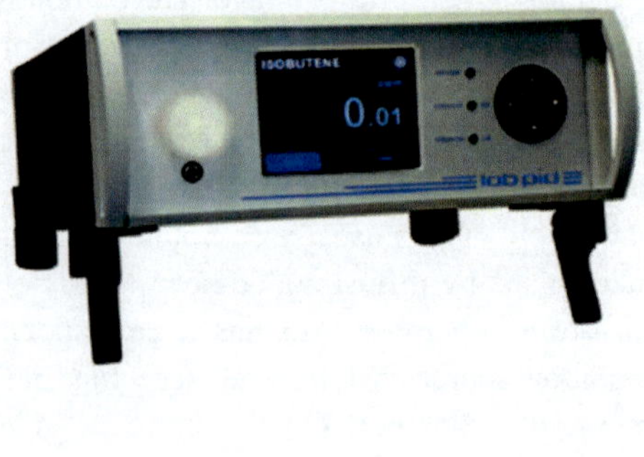

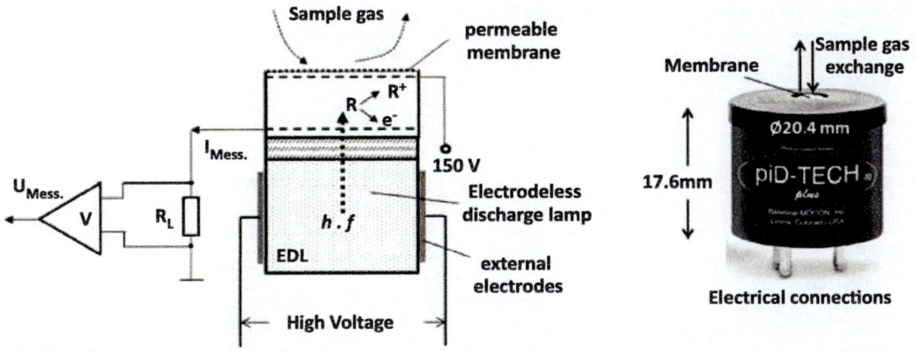

Fig. 3.87 Principle structure of a PID according to the diffusion process and real design (piD-Tech) of a commercial PID gas sensor. (Source: Baseline MOCON, Inc. USA)

membrane is the second electrode to which the suction voltage is connected. The signal processing is identical to flow operation.

Measuring Characteristics of the PID

PID gas sensors are usually calibrated with isobutene (C_4H_8), as this gas has a medium sensitivity and is stored as a test gas in compressed gas cylinders and is stable over a long period of time. If the PID is now flowed with a gas other than isobutene, the measurement result must be multiplied by appropriate correction factors. Benzene, for example, has a correction factor of 0.5. A display of 100 ppm isobutene then means a benzene content of 50 ppm. The PID is therefore more sensitive to benzene than to isobutene. The situation is different with ammonia, which has a correction factor of 10. In this case the ammonia concentration would be 1000 ppm, because the PID reacts much less sensitive to this gas. Before use, it must therefore be clear which gas is to be measured in order to then apply the appropriate correction factor. The manufacturers therefore offer extensive lists in which the correction factors for the different excitation wavelengths are listed. In Table 3.6 some correction factors are listed.

The characteristic curve (see Fig. 3.88) of the PID is linear in the lower range up to approximately 200 ppm. Above this concentration, the curve flattens out and changes to saturation from 2000 ppm. However, since the PID is mainly used for the detection of small concentrations, this is not a problem. The smallest available measuring ranges are at 2 ppm with a resolution of 0.5 ppb (piD-TECH® eVx, Baseline MOCON, Inc. USA). Figure 3.89 shows some PID sensors that are designed for different measuring tasks.

The accuracy of the PID method also depends on the accompanying gases and especially on the humidity. The manufacturers therefore recommend zeroing the matrix before each measurement without performing the component to be measured to eliminate these accompanying effects as far as possible. The influence of humidity can also be

Table 3.6 PID correction factors. (Source: RAE Systems Inc. USA)

Component	9.8 eV	10.6 eV	11.7 eV	IP eV
Acetone	1.2	1.1	1.4	9.71
Benzene	0.55	0.5	0.6	9.25
Chlorobenzene	0.44	0.4	0.39	9.06
Decane	4	1.4	0.4	9.65
Dimethyl disulfide	0.2	0.2	0.2	7.4
Ethanol		12	8	10.47
Ethylbenzene	0.52	0.5	0.5	8.77
Formaldehyde			0.6	10.87
n-hexane	300	4.3	0.5	10.13
Isobutanol	19	3.8	1.5	10.02
Isobutene	1	1	1	9.24
Isopropanol	500	6	2.7	10.12
Naphthalene	0.45	0.4	0.4	8.13
Perchloroethane	0.69	0.57	0.31	9.32
m-xylene	0.5	0.4	0.4	8.56

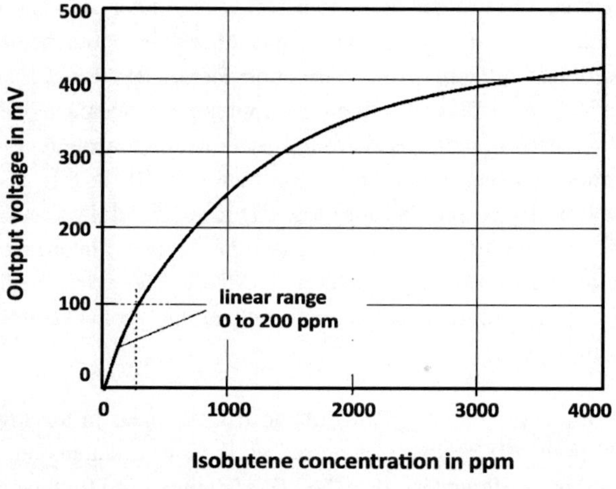

Fig. 3.88 Characteristic curve of a PID for the standard gas isobutene

compensated by using a humidity sensor if the gas humidity is measured simultaneously (ACI-GmbH-Berlin).

However, other effects also occur, which can be explained by the absorption of VUV radiation by accompanying components. The sensitivity in dry air, for example, decreases by up to 30% compared to the sensitivity in dry nitrogen. These influencing effects are

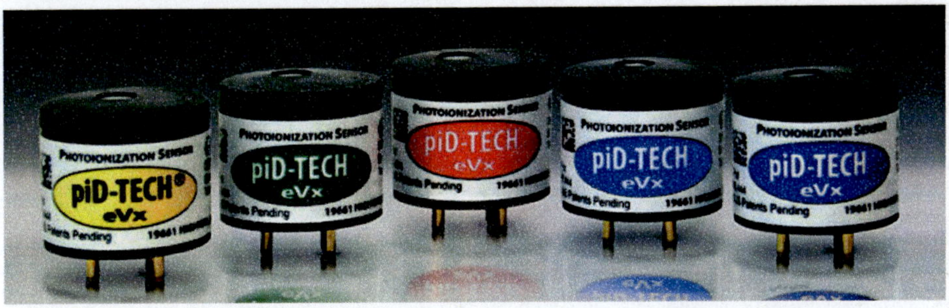

Fig. 3.89 Different PID sensors with an excitation of 10.6 eV or 10 eV for measuring ranges from 2 to 10.000 ppm. (Source: Baseline MOCON, Inc. USA)

called *quenching*. In ambient air, water vapor, carbon dioxide, carbon monoxide and methane are the main sources of quenching effects. The main application of PIDs is the monitoring of air quality of so-called VOC's.[20]

References

Aidam, E.: Ein sehr schneller magnetischer O_2-Messer nach dem Wechseldruckverfahren mit einem neuartigen Detektor-Element. DECHEMA Monographien. **67**, 231–243 (1979)

Baumbach, G.: Luftreinhaltung. Springer, Berlin (1990)

Bender, S.: Untersuchungen zum Betriebsverhalten temperaturregelbarer OFW-Gassensoren. Dissertation, RWTH-Aachen (2004)

Birkle, M.: Messtechnik für den Immissionsschutz. Oldenbourg, München (1979)

Brunck, O.: Clemens Winklers Lehrbuch der technischen Gasanalyse. 4. Aufl. Verlag von Arthur Felix, Leipzig (1919)

D'Ans, J., Lax, E.: Taschenbuch für Chemiker und Physiker. Bd. 1 (Physikalische-chemische Daten). Lechner, M.D. (Hrsg.). 4. Aufl. Springer, Berlin (1992)

Ebbinghaus, E.: Sauerstoffmessung auf magnetischer Grundlage mit stark unterdrücktem Nullpunkt. Zeitschrift für angewandte Physik. **V**, 294–297 (1953)

Göldner, H.-D., Horn, B., Liedtke, T., Marx, W.-R., Schaefer, W.: Verfahren zur Bestimmung der Gaskonzentration in einem Gasgemisch und Sensor zur Messung der Wärmeleitfähigkeit. Deutsche Patentschrift DE. **3711511**, C1 (1987)

Greiser, P., Heller, U.: Unterschiedliche CO_2 Messverfahren für die Überwachung von Feuerungsanlagen nach der 1. Bundesimmissionsschutzverordnung. Testotherm KG, Lenzkirch (1981)

Grienauer, H.: FTTCA Fourier-Transform-Thermal-Conductivity Analysis. ISA-Paper, Analysis Division, Anaheim (2006)

Grienauer, H.: FTTCA-Gasanalyse mit thermisch modulierten Wärmeleitfähigkeits-Sensoren mit Fourier-Analyse des Messsignals. 16. GMA/ITG-Fachtagung Sensoren und Messsysteme 2012. DOI https://doi.org/10.5162/sensoren2012/1.2.2

[20] Volatile Organic Compound =vaporizable organic components.

Grunewald, A.-U.: Verfahren und Einrichtung zur Bestimmung der Gaskonzentration in einem Gasgemisch. Offenlegungsschrift DE 19949327 A1 (1999)

Haran, E.N.: Gasdichtemessung mittels Akustischer Methode. Tech. Mess. **50**, 43–48 (1983)

Heise, M.M., Kirchner, H.-H., Richter, W.: Gasanalytik mit Gasdichtedetektoren. PTB-Mitteilungen. **91**(4/81), 266–272 (1981)

Hengstenberg, J., Sturm, B., Winkler, O.: Messen, Steuern, Regeln in der Chemischen Technik. Bd. II, Betriebsmesstechnik II (Physikalische Analytik). Springer, Berlin (1980)

Herwig, H., Moschallski, A.: Wärmeübertragung. Vieweg, Wiesbaden (2006)

Hildebrandt, I.: Entwicklung eines CO_2-Sensors für Inkubatoren nach dem Wärmeleitverfahren. Bachelor-Thesis, Fachhochschule Dortmund (2013)

Jochum, J.: Gasdichtemessung im pulsierenden Strömungsfeld. VDI-Fortschrittsberichte. Reihe 8, Nr. 122 (1986)

Karbach, A.: Messtechnik in der Versorgungstechnik. Springer, Berlin (1997)

Kronmüller, H., Zehner, B.: Prinzipien der Prozessmesstechnik. Bd. 2 Analysengeräte. Schnäcker, Karlsruhe (1980)

Krupp, H.: Theorie der thermomagnetischen Sauerstoffmessung. Dissertation, Technische Hochschule Karlsruhe (1953)

Kurtz, O., Barthelmess, J., Rüther, R., Wünsche, M., Donner, C.: Quartz crystal microbalance used to characterize electrochemical metal deposition. Journal f. Electrochemistry and Plating Technology J.E.P.T. **5**, 1–17 (2010)

Landolt-Börnstein: Zahlenwerte und Funktionen, II. Bd. Eigenschaften der Materie in ihren Aggregatzuständen, 5. Teil: Transportphänomene, Kinetik, homogene Gasgleichgewichte. Springer, Berlin (1968)

Lehrer: Ann. Phys. **81**, 229 (1926)

Lehrer, E., Ebbinghaus, E.: Ein Apparat zur Sauerstoffmessung in Gasgemischen auf magnetischer Grundlage. Zeitschrift für angewandte Physik. **II**, 20–24 (1950)

Luft, K.F.: Über eine neue Methode der magnetischen Sauerstoffanalyse. Zeitschrift für angewandte Physik. **III**(8), 300–303 (1951)

Luft, K.F., Mohrmann, D.: Neues Gerät zur paramagnetischen Sauerstoff-Messung. Chemie-Ingenieur-Technik. **30**(9/10), 575–578 (1967)

Marx, W.-R., Wiegleb, G.: Industrielle Gasanalyse Teil B (Paramagnetische Verfahren). Tech. Mess. **62**, 31–37 (1995)

Mashayekhi, P.: Eine massensensitive, elektronische Nase zur Erkennung, Unterscheidung und Qualitätskontrolle von Safran und Trüffel. Dissertation, Universität Bonn (2005)

Mayer, H.: Akustischer Gasanalysator. Deutsche Patentschrift DE 3046081 (1980)

Meschede, D.: Gerthsen Physik. 24. Aufl. Springer (2010)

Nerheim, A.G.: Anal. Chem. **35**, 1640 (1963)

Noble, F.W., Abel, K., Cook, P.W.: Anal. Chem. **36**, 1421 (1964)

Oster, H.: Siemens Zeitschrift. **37**, 481 (1963)

Pauling, L., Wood, R.E., Sturdivant, J.H.: J. Amer. Chem. Soc. **68**, 795 (1946)

Poling, B.E., Prausnitz, J.M., O'Connell, J.P.: The properties of gases and liquids. McGrow-Hill. (2001)

Profos, P., Pfeiffer, T.: Handbuch der industriellen Messtechnik. 5. Aufl. Oldenbourg (1992)

Richly, W.: Meß- und Analyseverfahren. Vogel Buchverlag, Würzburg (1993)

Staab, J., Baronick, J., Kroneisen, A.: Improving the Method of Hydrocarbon Analysis. SAE Technical Paper Series No. 810427 (1981)

Tipler, P.A., Mosca, G.: Physik für Wissenschaftler und Ingenieure. 7. Aufl. Springer-Spektrum (2014)

Vonderschmidt, S.: Paramagnetischer Sauerstoffsensor in Mikrosystemtechnik. Dissertation, TU Hamburg-Harburg (2012)

Wernekinck, U.: Gasmessung und Gasabrechnung. 4. Aufl. Vulkan, Essen (2009)

Wiegleb, G.: Gasanalyse nach dem Wärmeleitverfahren. Praxis der Naturwissenschaften – Chemie, 272–276 (1980)

Wiegleb, G.: Anordnung zur Messung der Wärmeleitfähigkeit von Gasen. Deutsche Patentanmeldung DE3502440.2 (1985a)

Wiegleb, G.: Schnellanzeigender Wärmeleitfähigkeitssensor zur Messung von Gaskonzentrationen, SENSOR'85-Konferenzunterlagen, Kap. 8.3.1–8.3.16 (1985b)

Wiegleb, G.: Gasanalyse nach dem Wärmeleitverfahren mit dem HYDROS. Versuchs- und Forschungsingenieur. **3**, 57–60 (1987a)

Wiegleb, G.: Aufbau und Wirkungsweise eines Wärmeleitfähigkeitsanalysators. GIT Fachzeitschrift für das Labor. **8**, 694–700 (1987b)

Zipser, L., Labude, K.: Akustische Gasanalyse Teil 3 Anwendungen. Tech. Mess. **58**(11), 427–432 (1991a)

Zipser, L., Labude, K.: Akustische Gasanalyse Teil 1 Grundlagen. Tech. Mess. **58**(12), 463–470 (1991b)

Related References

Bock, H.: Über die Wärmeleitfähigkeit von Gasgemischen. Ann. Phys. **6**(8) (1950)

Hauptmann, P.: Sensoren Prinzipien und Anwendungen. Hanser, München (1991)

Liu, Y., Chen, Y.-p., Song, H., Zhang, G.: Research on characteristics of hydrogen gas sensor based on palladium and yttrium alloy ultrathin film. The 14th International Meeting on Chemical Sensors. IMC 2012. doi: https://doi.org/10.5162/IMC2012/P2.8.8

Martin, P., James, A.T.: Biochem. J. **63**, 138 (1956)

Moseley, P.T., Tofield, B.C.: Solid State Gas Sensors. Adam Hilger, Bristol (1987)

Schawacht, T.: Bericht zur betrieblichen Praxis (paramagnetische O_2-Gassensorik). Fachhochschule Dortmund (2015)

Schmick, H.: Phys. Z. **29**, 633 (1928)

Schmitte, F.-J., Wiegleb, G.: Conductivity behaviour of thick film tin-dioxide gas sensors. Sensors Actuators B. **4**, 473–477 (1991)

Schunck, G.: Fühler zur Messung der Wärmeleitfähigkeit in Gasen. Deutsche Patentanmeldung 2952137 (1979)

Twisselmann, L.: Schnelle paramagnetische Sauerstoffbestimmung mit erzwungenem Gasaustausch an den Verdrängungskörpern. Dissertation, Technische Universität Berlin (1975)

van Herwaarden, S.: Physical principles of thermal sensors. Sensors and Materials. **8**(6), 373–387 (1996)

VDI-Wärmeatlas: Berechnungsblätter für den Wärmeübergang. 8. Aufl. Springer, Berlin (1997)

Wiegleb, G.: Funktion und Aufbau eines Wärmeleitfähigkeitssensors. Messen-Prüfen-Automatisieren **4**, 203–205 (1. Teil), **5**, 254–256 (2. Teil) (1986)

Wiegleb, G.: Industrielle Gassensorik. Expert Verlag, Renningen (2001)

Wiegleb, G., Böttcher, A.: Messungen von Chlorgasimmissionen mit dem BINOS IM 01-System. GIT Fachzeitschrift für das Labor. **5**, 497–501 (1988)

Wiegleb, G., Heitbaum, J.: Eigenschaften von Zinndioxid-Gassensoren in Dickschicht-Technologie. Dechema-Monographien Bd. 124 (1991a)

Wiegleb, G., Heitbaum, J.: Halbleitergassensor auf der Basis von Indiumzinnoxid. Exp. Tech. Phys. **39**(3), III (1991b)
Wiegleb, G., Heitbaum, J.: Eigenschaften polykristalliner Zinndioxidschichten. Verhandlungen der DPG-Frühjahrstagung in Münster DS. **24**(4), 985 (1991c)
Wiegleb, G., Heitbaum, J.: Semiconductor gas sensor for detecting NO and CO traces in ambient air of road traffic. Sensors an Actuators B. **17**, 93–99 (1994)

Physical-Chemical Gas Sensors

4

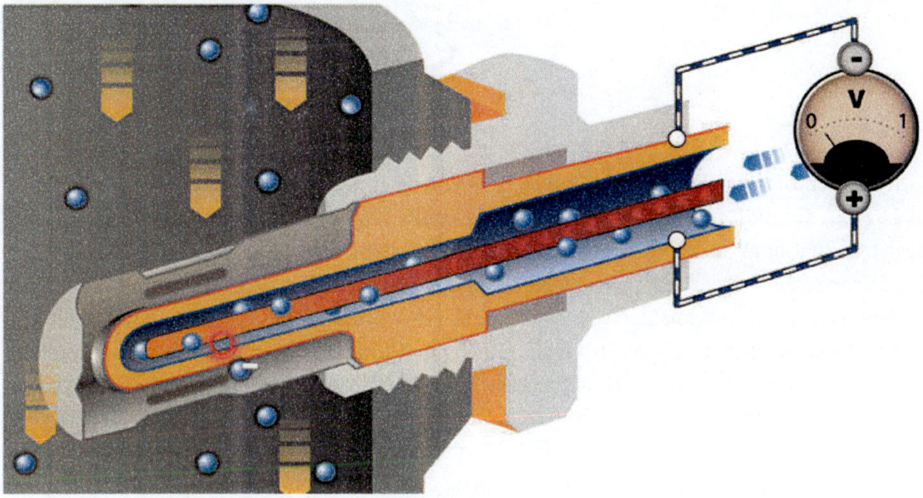

Lambda sensor from NGK Spark Plug Co. Ltd

4.1 Heat Tone Sensors (Pellistors)

Combustible gases such as natural gas, liquid gas, hydrogen or carbon monoxide can be determined by the so-called heat tone method or calorimetric method. This type of sensor is characterized by a high selectivity towards combustible gases.

G. Wiegleb, *Gas Measurement Technology in Theory and Practice*,
https://doi.org/10.1007/978-3-658-37232-3_4

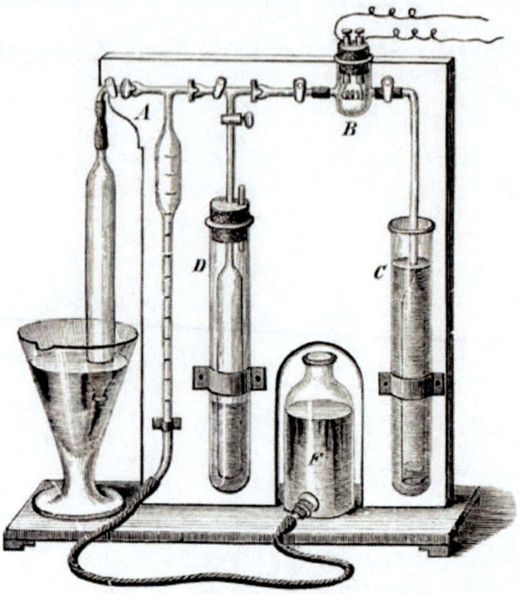

Fig. 94.

Fig. 4.1 Historical representation of an apparatus for determining the methane concentration of mine gas by catalytic combustion on an electrically heated platinum wire. The apparatus is called a Carburometer, a further development of the Grisoumeter (Brunck 1919)

This method was already known over 100 years ago for detecting methane in mine gas. In 1877, *J. Coquillion* published his research results on the combustion of methane gas on glowing platinum. Figure 4.1 shows an apparatus from this period, known as a *Carburometer*. The gas sample (e.g. mine gas from a mine) is placed in a glass flask (left in the picture) and is connected to the apparatus via a tap. By applying an electrical voltage, the platinum wire in combustion vessel B is heated to red heat and the methane from the sample is completely burnt off. The released carbon dioxide is then bound by potassium hydroxide solution in the absorption vessel D. Gas exchange in the apparatus is achieved by raising and lowering the level bottle F (→ pumping action). Due to the volume loss, the level in the measuring tube rises and can then be converted to the desired gas concentration. Vessel C serves as a hydraulic seal for the apparatus.

The first heat tone sensors also consisted of a thin, taut platinum wire that was heated to a temperature of about 1000 °C. At this temperature, the gases burn with the existing atmospheric oxygen and thus additionally heat the platinum wire. This temperature increase ΔTcan then be converted into a voltage signal with a measuring bridge, as with a thermal conductivity sensor. Unfortunately, due to the extremely high temperature, these sensors were not particularly stable in the long term. In the 1960s, this process was further developed by adding a catalytically acting additional layer to the platinum wire. As a result, the operating temperature ($T \approx 500^{\circ}$C) could be significantly reduced and the sensor lasted longer.

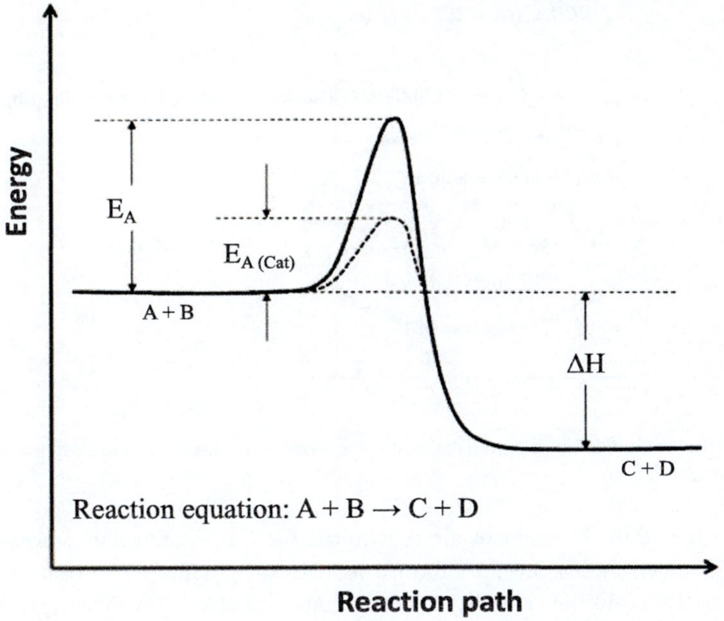

Fig. 4.2 Reaction paths for an exothermic chemical reaction with and without catalyst. The amount of heat ΔH released by the reaction leads to additional heating (heat tone effect)

To start a chemical reaction, a certain energy E_A is needed. In the pellistor, this activation energy is determined by the surface temperature. Figure 4.2 shows the reaction path. If a catalyst is now applied to the surface of the pellistor, the activation energy is reduced and the chemical reaction begins at lower temperatures. This effect is known as *heterogeneous catalysis* and is also of great importance for semiconductor gas sensors. Platinum, palladium or rhodium is used as catalyst. Combinations of these materials are also possible. Depending on the catalyst or catalyst mixture and base material, the activation energy for certain substances can be reduced in a targeted manner and thus have a certain influence on selectivity.

Figure 4.3 shows a cross-section of such a pellistor.[1] It consists of a small, spherical pill made of a highly porous alumina ceramic (γ-Al_2O_3) in which a thin platinum wire ($\approx$50 μ m) is embedded. When this platinum wire is heated by an electric current, a chemical reaction occurs both on the surface of the ceramic and inside the pellistor. The entire *surface of* the pill, which is only 1 mm in size, has a calculated area of 7850 m^2 (Jessel 2001). Due to the catalytically activated surface of the pill, the combustible gases react already at temperatures below 500 °C.

[1]Comes from English, and stands for pellet-resistor = Pellistor.

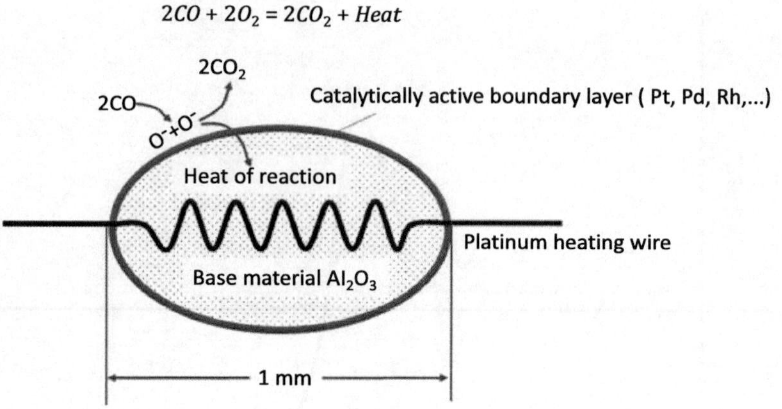

Fig. 4.3 Cross section through a pellistor and the chemical reactions in the catalytically activated boundary layer

The oxygen from the ambient air is required for this combustion process. Carbon dioxide and water vapour are then formed as reaction products. The heat Q_R released corresponds to the calorific value H_s of the substances involved. The quantity of heat Q_R is then proportional to the mass Δm of the combustible substances in the gas phase.

$$Q_R = H_s \cdot \Delta m \tag{4.1}$$

If this heat quantity Q_R would be transferred completely to the pellistor (mass m_P, heat capacity c_P), the following theoretical temperature increase ΔT would result:

$$\Delta T = \frac{Q_R}{c_P \cdot m_P} \tag{4.2}$$

For better handling, the pellistor is contacted with a pressure glass feed-through, which ensures not only the transmission of the electrical signals but also the mechanical support.

For practical reasons, the pellistor shown in Fig. 4.4 is always installed in a housing to protect the sensitive structure from external influences. The gas flow Δm reaching the pellistor is therefore determined by diffusion processes. Figure 4.5 shows such a structure suitable for practical applications. The pellistor is located in a closed chamber which is only accessible via a sintered metal frit (diffusion barrier). This means that only a certain amount of gas (n/t) reacts per unit of time, which then determines the heating line of the pellistor. According to *Fick's law*, the temperature increase $\Delta T$of the pellistor is obtained with the diffusion constant D (Jessel 2001):

$$\Delta T \approx D \cdot \frac{A}{l} \cdot c \cdot Q_R \tag{4.3}$$

The sensor signal S generated by the pellistor is proportional to the temperature increase ΔT . Therefore, the sensitivity E of the gas measurement results from:

$$E = \frac{S}{c} \approx D \frac{A}{l} Q_R \tag{4.4}$$

Since the diffusion constant D passes through:

$$D = \frac{\text{const.}}{\sqrt{M}} \tag{4.5}$$

can be represented by Eq. 4.4 as follows:

$$E \approx \frac{k}{\sqrt{M}} \tag{4.6}$$

The sensitivity of a pellistor is therefore inversely proportional to the root of the molar mass M. As a rule of thumb, Eq. 4.6 shows that the sensitivity E of a pellistor increases when the molecular weight decreases and vice versa. This relationship applies at least to hydrocarbons with a molar mass M of16 g /mol (methane) to142 g /mol (decane). Inorganic substances such as hydrogen H_2, carbon monoxide CO and carbon disulfide CS_2 behave differently.

Since the response time ($t_{90\%}$-time) also depends on the molecular weight M, a further rule of thumb formula can be defined:

$$t_{90\%} \approx k\sqrt{M} \tag{4.7}$$

With this equation the response time can be estimated, which is valid for molecules of different molecular mass M. If, for example, the response time of methane is determined experimentally for a certain pellistor sensor, the response time for another gas can be calculated approximately.

Example
The response time of methane ($M = 16$g/mol) is 14 seconds and was determined experimentally. What is the response time for *n-octane* ($M = 114$g/mol) if the same sensor design is to be used?

The constant k can be calculated by transforming Eq. 4.7:

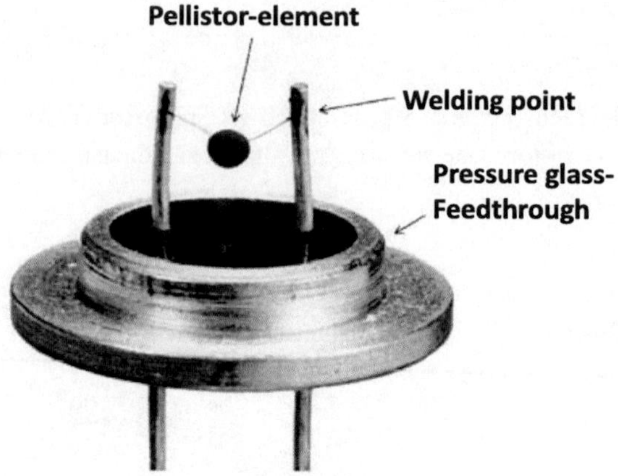

Fig. 4.4 Pellistor element mounted on a pressure glass feedthrough

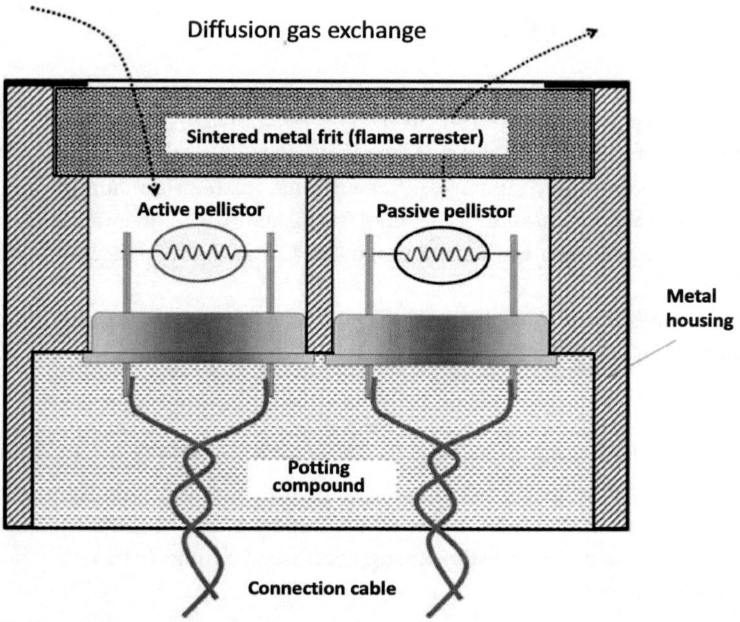

Fig. 4.5 Cross-section through an Ex sensor head with integrated pellistor element. The flame arrester is mechanically secured to hold the frit in position in case of an explosion in the chamber. The electrical connection side is protected by a special encapsulation

$$k = \frac{t_{90}}{\sqrt{M}} = \frac{14}{\sqrt{16}} = 3,5 \frac{\mathrm{s}}{\sqrt{\mathrm{g/mol}}} \tag{4.8}$$

With this value the new response time of *n-octane* is now calculated:

$$t_{90} = 3.5 \frac{\mathrm{s}}{\sqrt{\mathrm{g/mol}}} \sqrt{114\mathrm{g/mol}} = 38 \text{ seconds} \tag{4.9}$$

Sensor Construction

Pellistors are integrated in so-called measuring heads, which consist of a metal housing. The gas exchange here takes place almost exclusively by diffusion through a sintered metal frit. This frit has to fulfil two tasks. On the one hand, the pellistor is to be protected against mechanical influences, dust, moisture, etc., on the other hand this frit also acts as a flame barrier (see Chap. 18).

Since only flammable and therefore explosive gas mixtures are detected, the pellistor can trigger an explosion in the room to be monitored due to its high operating temperature. By closing the measuring chamber with the frit, a flame, as a result of an explosion in the measuring chamber, is cooled down and thus extinguished. This means that no flame from the measuring chamber penetrates into the hazardous area to be monitored. To prevent the sintered metal frit from being ejected by the explosion pressure, it must be mechanically fixed so that this is prevented under all circumstances. Measuring heads that are used for this application must prove their suitability by means of an appropriate test (Ex certificate).

Since the pellistor element also records the ambient temperature, a reference measurement is absolutely necessary. For this purpose, a second, almost identical pellistor is integrated into the measuring chamber.

However, this reference pellistor does not react to flammable gases because the catalytically active layer has been deliberately *poisoned.* Catalysts are known to be poisoned by lead-containing compounds, or poisoned and thus permanently deactivated. Pellistors are usually soaked in hot potassium hydroxide solution (KOH) and passivated in this way. This reference pellistor then only detects the changes in room temperature, while the gas measurement is virtually switched off. This provides a comparative measurement that also takes into account changes in the thermal conductivity of the gas mixture.

The sensor signals are evaluated by a Wheatstone measuring bridge (Fig. 4.6), with a downstream DC amplifier. The measuring and reference pellistors are located in the same bridge branch and are passed through with the same current intensity. The second bridge branch consists of fixed resistors with which the bridge is balanced in air. Since the sensor signals are comparatively large, the amplification factor is correspondingly small ($f < 50$).

Figure 4.7 shows an example of an Ex sensor head. The pellistor is located outside in the round stainless steel block with an upstream sintered metal frit. The evaluation electronics is integrated in the die-cast housing. The signal from the transmitter goes via a 4–20 mA

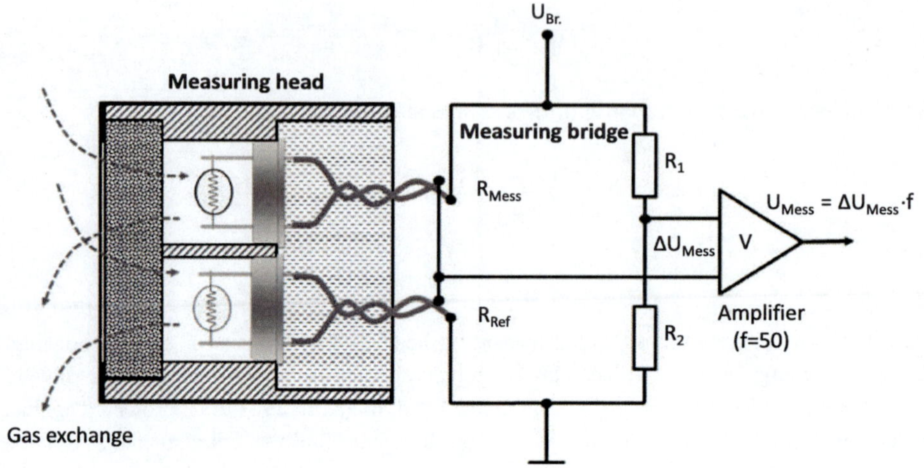

Fig. 4.6 Measuring head in connection with the evaluation electronics

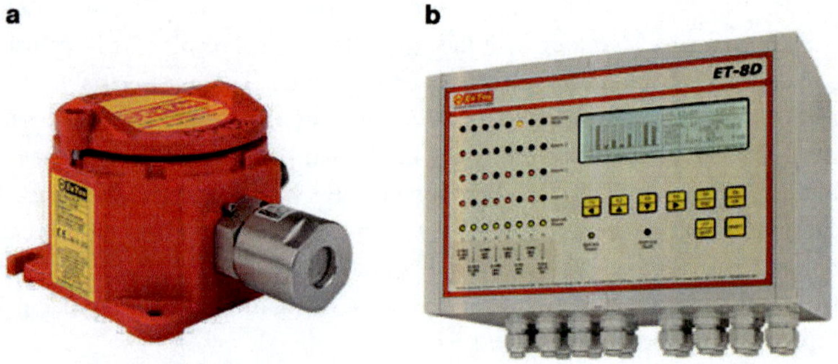

Fig. 4.7 Gas warning centre (**b**) for the evaluation of up to 8 transmitters (**a**) for monitoring explosive atmospheres (Ex zones). (Source: ExTox-Gasmess-Systeme GmbH Unna)

output to the gas warning centre (ET-8D), which can evaluate several measuring heads simultaneously. The gas warning center also supplies the measuring heads with the necessary electrical power ($P < 2$ W).

The sensitivity of the pellistor to different gases depends mainly on the temperature and the catalytic coating. The catalytic coating reduces the activation energy so that the combustion process can take place at lower temperatures. Since the activation energy is also different for the different gases, different sensitivity curves are obtained. Figure 4.8 shows this relationship using 6 gases. For most hydrocarbons, the optimum temperature for this pellistor is about 320 °C. Due to the higher activation energy, methane only reacts from this temperature and reaches the maximum at500° C. By selecting the pellistor temperature,

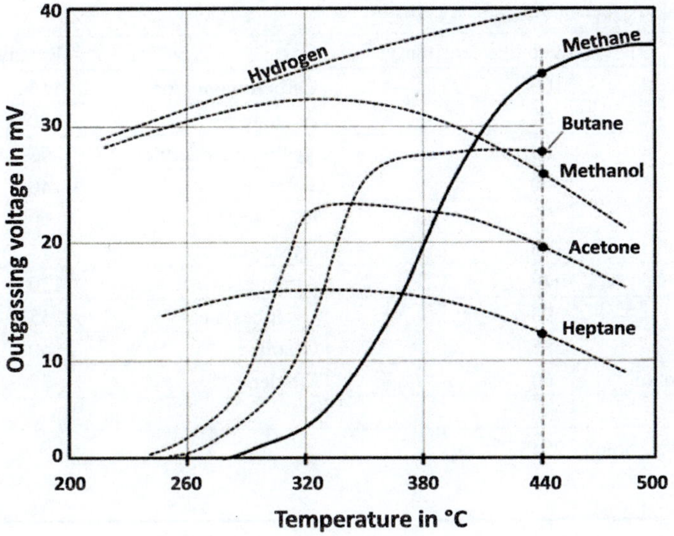

Fig. 4.8 Sensitivity of a Pd/ThO_2 pellistor for different gases at a concentration of 20% UEG (Göpel et al. 1991)

a certain selectivity can therefore be achieved. Remarkable is the high sensitivity of hydrogen, which is not exceeded by any of the mentioned gases.

The manufacturers issue corresponding sensitivity tables for the respective pellistor type, which indicate a relative sensitivity to methane or propane. However, these tables only give an indication of how the sensitivity to a change from methane to, for example, methanol behaves. For accurate measurements the sensors must be calibrated to the respective gas. Table 4.1 shows the changes in sensitivity for the pellistor of type 4P-50 CiTipel (Part No. PM 123-00) from City Technology UK.

The sensitivity of methane at different concentrations is shown in Fig. 4.9. Up to a concentration of approximately 10 vol% methane in air, the sensitivity curve rises steeply to reach a maximum. After that the curve drops slowly and continuously. At a concentration value of 100 vol.% methane the sensitivity is then zero. This behavior, typical for a pellistor, can be explained by the lack of oxygen. If the methane concentration rises, the atmospheric oxygen content decreases. However, since this oxygen is required for catalytic combustion in the pellistor, the sensitivity decreases. With 100% methane by volume, the oxygen content is of course zero, so that no more combustion can take place.

This behaviour leads to ambiguity in the measurement display. In practical instrument versions, additional measurement information is therefore required. A thermal conductivity sensor is used for this purpose. The sensitivity of the thermal conductivity sensor is independent of the oxygen concentration, so that ambiguity is easily detected.

Table 4.1 Relative sensitivities for a pellistor

Gas	Relative sensitivity	Gas	Relative sensitivity
Methane	100	Carbon monoxide	115
Propane	65	Acetone	70
n-Butane	65	Methyl ethyl ketone	55
n-Pentane	60	Toluene	40
n-Hexane	50	Ethyl acetate	60
n-Heptane	45	Hydrogen	115
n-octane	40	Ammonia	130
Methanol	95	Cyclohexane	55
Ethanol	85	Gasoline	60
Isopropyl alcohol	60	Ethylene	85
Acetylene	80		

Source: City Technology 2013

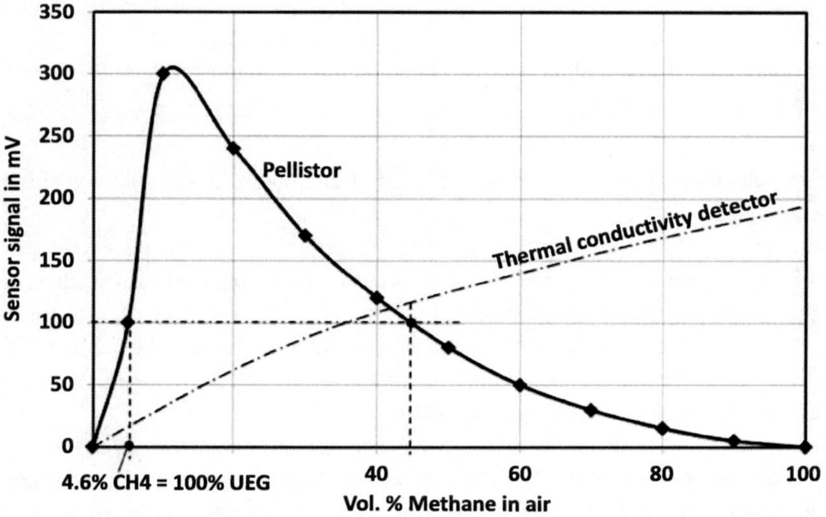

Fig. 4.9 Sensitivity curve of a heat tone measurement (pellistor) compared to a thermal conductivity detector used to detect the ambiguity of the pellistor characteristic above 10 vol% methane in air

Micropellistor

For some years now, miniaturized versions of the pellistor have also been used, which no longer use a *pill* in the conventional sense. The structures are built in microtechnology and are located on a silicon substrate. Figure 4.10 shows such a micropellistor. The resistors are integrated on a thin membrane that is suitable for the required operating temperature of

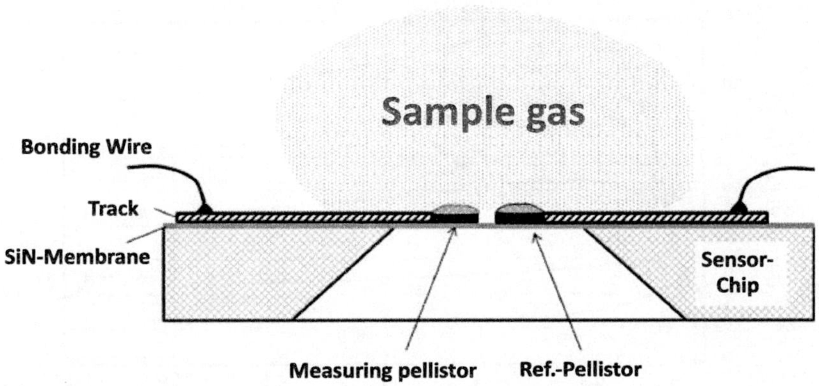

Fig. 4.10 Micropellistor in microtechnology

$T \approx 500^\circ$ C. A layer consisting of Al_2O_3 and the necessary catalysts is applied above the resistors. Also, in this case the reference resistor is covered with a poisoned layer. The signal evaluation is analogous to the known pellistors. An advantage of these miniaturized structures is the lower power consumption. This feature is particularly important for portable, battery-powered handheld measuring devices.

The power consumption can be reduced even further by pulsed operation, as the warm-up time for these low-mass superstructures is very short. The micropellistor from SGX Sensortech Ltd. England can be switched on for 0.2 seconds and then switched off for 4.8 seconds. This corresponds to a *duty cycle* of 4%. The average power consumption in this operating mode is just under 5 mW. With a battery capacity of 1100 mAh, operation for 23 days would theoretically be possible. The switch-on behavior of the micropellistor at different gas concentrations, from 0 to 4 vol% methane, is shown in Fig. 4.11. After switching on, the sensor voltage rises steeply and reaches a maximum after a. 200 ms After that the curve slowly drops. In pulsed mode, therefore, the maximum sensitivity is measured, while DC mode reflects the steady state and has a significantly lower sensitivity (−40%). Comparative data between a conventional pellistor and a micropellistor are listed in Table 4.2.

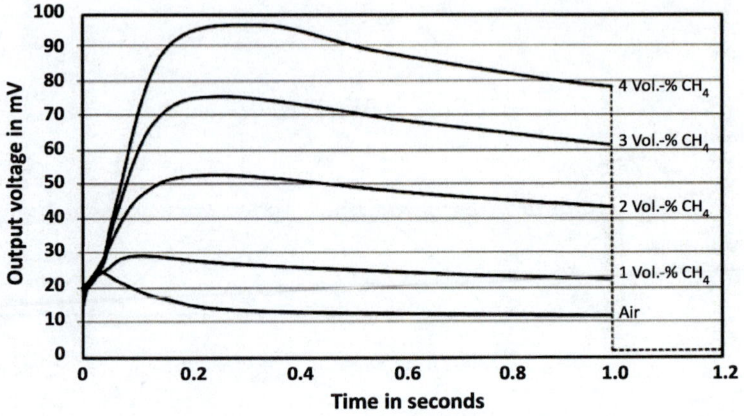

Fig. 4.11 Switch-on behaviour of the micropellistor Type: MiCs-MP-7217 SGX (Sensortech Ltd. England)

Table 4.2 Typical data for pellistors

Parameters	Pellistor	Micropellistor
Power consumption	300 mW (DC)	100 mW (DC) 5 mW in pulse mode
Sensitivity	35 mV/vol.-% methane	20 mV/vol% methane
Time constant	>20 s	<10 s
Drift behaviour	Identical behaviour	
Construction volume	No significant differences	

4.2 Electrochemical Cells

Electrochemical gas sensors use the release of electrons, which can be generated by certain chemical reactions, for gas analytical measurement. These electrochemical reactions always involve the transfer of charge carriers from one electrode to another phase (electrolyte). The electrolyte can be solid, gel-like or liquid. During this process chemical changes take place at the electrode as well as the charge transport through the electrolyte. Both the electrode reaction and the charge transport can be chemically influenced. This is the basis of the measuring process. One of the basic rules of electrochemical sensors is that at least a second electrode is needed to form an electrochemical cell (Wiegleb et al. 2001). The chemical reactions that take place in an electrochemical sensor can be very complex (Breuer et al. 1980). Therefore, only the essential electrochemical processes and the metrological properties are described below.

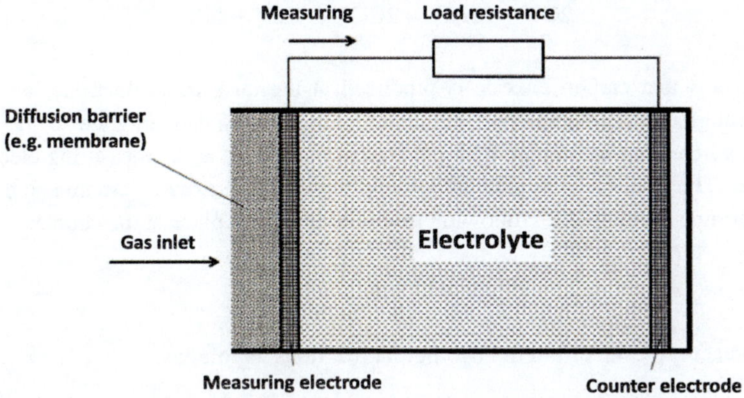

Fig. 4.12 Basic design of an electrochemical gas sensor using the amperometric method

Fig. 4.13 Partial view of the platinum electrode as a porous layer on the gas-permeable PTFE membrane. The first partial reaction therefore already takes place in the thin Pt layer

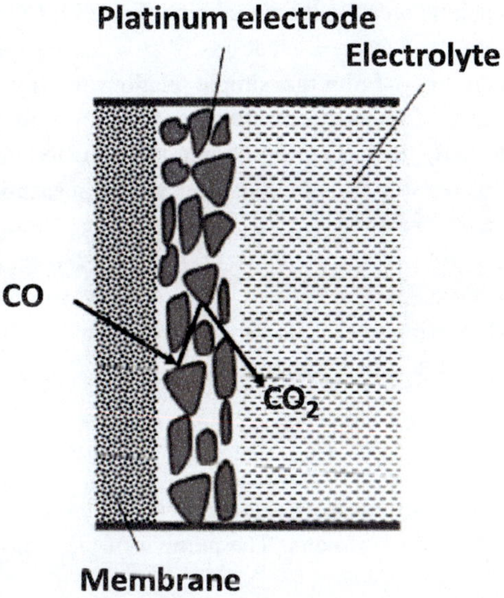

Liquid Electrolyte Sensors

Figure 4.12 shows the principle structure of an electrochemical gas sensor with a liquid electrolyte. As the electrolyte material is liquid, it must be in a container which is closed on all sides. Only from one side can the gas enter the interior of the gas sensor via a permeable membrane. This membrane also prevents the electrolyte liquid from leaking. On the back of the membrane is the measuring electrode, in the form of a thin conductive platinum or gold layer. This electrode layer is also permeable to the gas. The first chemical reaction already takes place at the measuring electrode (Fig. 4.13). For a CO sensor this would be the following reaction:

$$2CO + 2H_2O \rightarrow 2CO_2 + 4H^+ + 4e^- \quad (4.10)$$

This means that carbon dioxide is produced at the measuring electrode, which then passes through the membrane to the outside. The H^+ions then migrate to the counter electrode, where they recombine with the electrons released at the measuring electrode to form water. The electrons reach the counter-electrode via the external circuit and thus form the measuring current I_M. The following reaction then takes place at the counter electrode:

$$O_2 + 4H^+ + 4e^- \rightarrow 2H_2O \quad (4.11)$$

The total reaction results from the two partial reactions as follows:

$$2CO + O_2 \rightarrow CO_2 \quad (4.12)$$

The oxygen required for the overall reaction comes from the ambient air and also enters the reaction area via the membrane. For each CO molecule that passes through the membrane into the sensor, 2 electrons ($\rightarrow z = 2$) are created, which are detected in the measuring current I_M. From this simple relationship, it is immediately apparent that the characteristic curve of an electrochemical gas sensor with a liquid electrolyte exhibits linear behavior. Usually the measuring current is conducted via a load resistor and the voltage drop across this resistor R_L is then measured. The measuring voltage U_M *is* then obtained:

$$U_M = I_M \cdot R_L \quad (4.13)$$

According to *Faraday's first law,* the measuring current I_M can then be calculated from the previous considerations:

$$I_M = z \cdot F \cdot Q \quad (4.14)$$

The factor F is the *Faraday*[2] *constant* with $F = N_L \cdot e = 96486.7$ as/mol, and the$Q = dN/dt$ conversion in mol/s. The number of electrons released per gas molecule is called z. This results in I_M:

$$I_M = z \cdot F \cdot \frac{dN}{dt} \quad (4.15)$$

If the transport of the particles is mainly given by diffusion, then the following expression is obtained with *Fick's first law* for I_M:

---

[2]Michael Faraday (1791-1867) English natural scientist and important experimental physicist

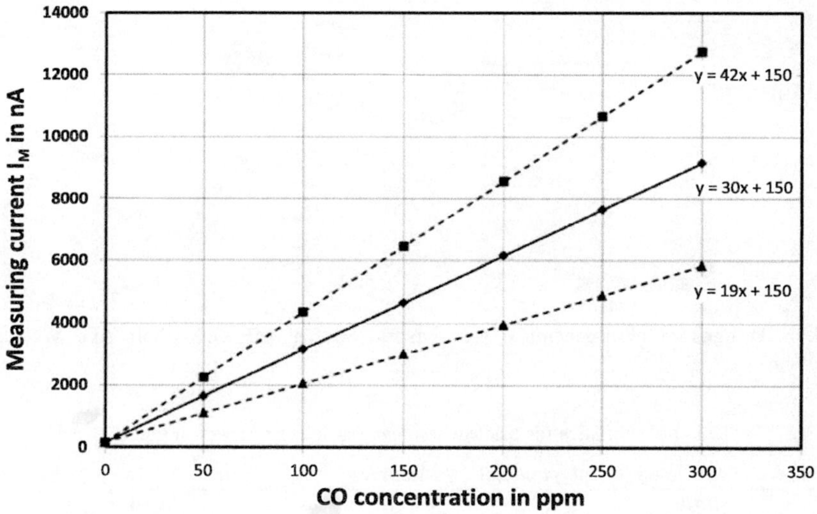

Fig. 4.14 Characteristic curve of an electrochemical CO gas sensor (Type CO 2E 300 Sensoric GmbH). The *solid line* shows the typical characteristic curve, while the *dashed lines* indicate the maximum permissible tolerance range

$$I_M = z \cdot F \cdot D \cdot A \cdot \frac{\Delta c}{\Delta x} \tag{4.16}$$

The measuring current I_M is therefore proportional to the entry area A into the gas sensor and proportional to the temperature-dependent diffusion coefficient $D(T)$.$\Delta c$is the difference in particle density (concentration) over the diffusion layer Δx . Since the particle density at the measuring electrode is zero, the measuring current also increases proportionally to c. Figure 4.14 shows a typical characteristic curve of an electrochemical gas sensor. Due to unavoidable leakage currents, the measuring current at $c = 0$ is different from zero. This value is called *zero-current*. In the example given (CO sensor), this zero-current is max. 150 nA. The sensitivity, S, for this gas sensor is$S = 30$nA/ppm ± 12nA/ppm. The tolerance range is comparatively large, but is typical for this type of gas sensor, and must therefore be taken into account in the gas measuring device.

Electrochemical gas sensors are particularly suitable for use in hand-held measuring devices, since no additional electrical energy is required to operate the sensors. They are also very small (Fig. 4.15), have a low weight and already provide a linear output signal. However, the large tolerance of the characteristic curve is a disadvantage. Some manufacturers therefore calibrate the sensors in the factory and write the calibration data into a memory, which is then read out in the gas measuring instrument. The sensors can thus be exchanged without any problems. This interchangeability is also absolutely necessary, as the life of these gas sensors is limited due to the chemical reaction in the electrolyte. Depending on the type of sensor and manufacturer, the service life is specified

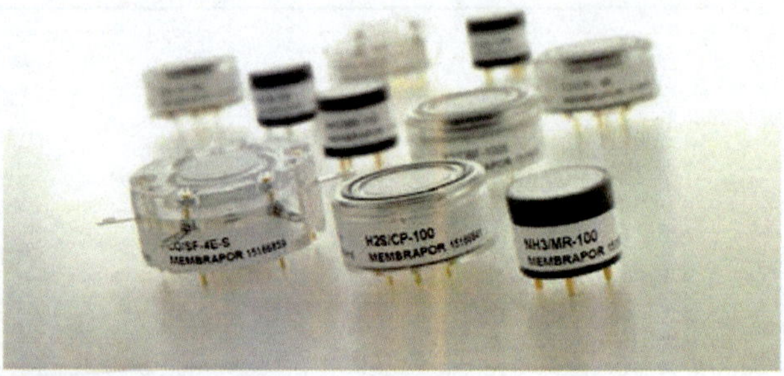

Fig. 4.15 Designs of electrochemical gas sensors (Source: MEMBRAPOR AG, Wallisellen, Switzerland)

Table 4.3 Typical characteristics for electrochemical gas sensors based on liquid electrolytes

Substance	MB in ppm	t_{90} *time* [s]	Lifetime	S[nA/ppm]	I_0 [nA]	Electrolyte
H_2S	0–30	<30	3 years	60	±10	Acid
CO	0–300	<35	3 years	30	<150	Acid
H_2	0–2000	<70	2 years	7,5	±100	Acid
Cl_2	0–5	<70	2 years	300	<15	Aqueous
AsH_3/ PH_3	0–1	<60	1–2 years	1400	±20	Organic
SiH_4	0–50	<60	1–2 years	130	±25	Organic
HCl	0–30	<30 (t_{50})	2 years	140	±100	Acid
HF	0–10	<40 (t_{50})	1 year	600	±100	Aqueous
F_2	0–1	<30 (t_{50})	1–2 years	1000	±20	Organic
HCN	0–30	<30	1–2 years	30	±5	Organic
$COCl_2$	0–1	<60 (t_{50})	1 year	650	±10	Organic
NH_3	0–100	<20 (t_{50})	1 year	90	±150	Organic
NO	0–100	<20	2 years	45	±30	Acid
N_2H_4	0–1	<30 (t_{50})	1 year	1200	±15	Organic
O_3	0–1	<60	1–2 years	≈2000	±20	Organic
NO_2	0–50	<30	2 years	200	±20	Organic

Source: sensoric GmbH Bonn

between 1 and 3 years (Table 4.3). Even during storage, ageing processes can affect the service life, so that a maximum storage time (several weeks) is specified. In order to extend the lifetime, the permeability of the membrane can be reduced, thus limiting the number of reactions per time. However, this advantage has a negative effect on the time response. The sensors then react more slowly. An acceptable value for the temporal response behaviour (t_{90} *time*) is 30–60 seconds.

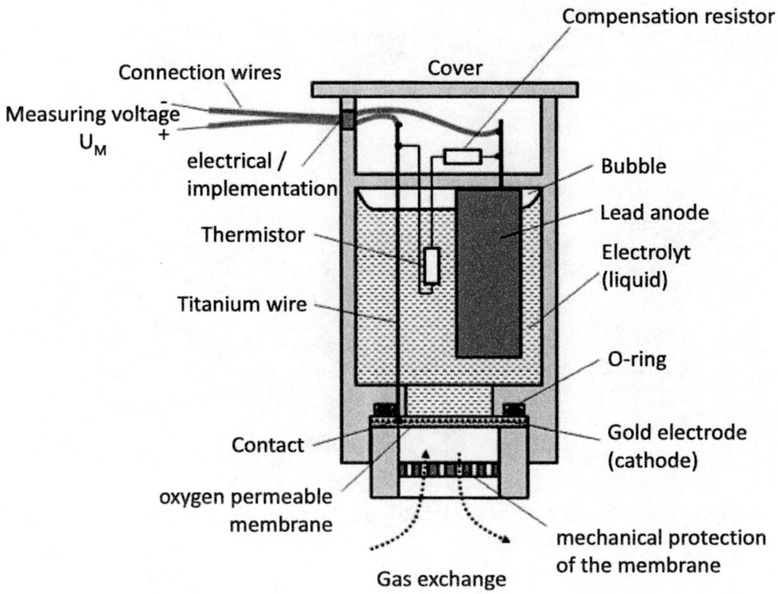

Fig. 4.16 Cross-section through an electrochemical oxygen sensor with a liquid electrolyte (e.g. acetic acid)

Electrochemical Oxygen Sensor

A special feature among the electrochemical gas sensors is the oxygen sensor. It is basically based on the same physical and chemical laws as described above. In contrast to the previous sensors, however, it is designed for high concentrations of up to 100% by volume. In most technical applications (e.g. combustion processes) an oxygen concentration in the range of 0 to 21 vol.% is expected. The maximum value corresponds to the oxygen content in the ambient air. Part of the oxygen is converted by the combustion process and thus this proportion is reduced accordingly. Conclusions about the combustion process can therefore be drawn from the oxygen content. The situation is similar in medical applications when it comes to breathing gas control or the control of a ventilator. In these cases, too, the oxygen concentration is in the volume percent range.

Figures 4.16 and 4.17 show such an oxygen sensor. A liquid electrolyte is contained in the sealed sensor housing (plastic). In this case it is diluted acetic acid. A solid lead anode is placed in the electrolyte, while the cathode is a thin gold layer on the gas-permeable membrane. The sensitive membrane is also provided with additional mechanical protection (diaphragm), through which gas can diffuse without any problems.

The following reactions take place at the electrodes:

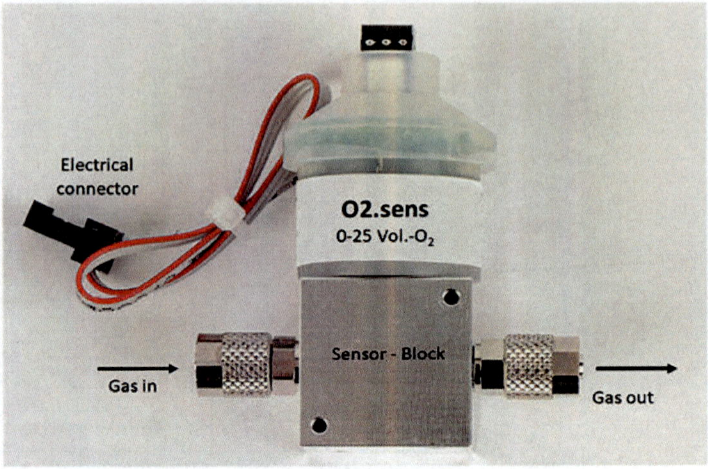

Fig. 4.17 Oxygen sensor based on liquid electrolytes. (Source: Wi.Tec-Sensorik GmbH, Wesel Germany) https://www.witec-sensorik.de

$$\text{Cathode-reaction}: \quad O_2 + 4H^+ + 4e^- \rightarrow 2H_2O \tag{4.17}$$

$$\text{Anode-reaction}: \quad 2Pb + 2H_2O \rightarrow 2PbO + 4H^+ + 4e^- \tag{4.18}$$

$$\text{Complete Reaktion}: \quad O_2 + 2Pb \rightarrow 2PbO \tag{4.19}$$

Per oxygen molecule, 4 electrons ($\rightarrow z = 4$) are created. The overall reaction is also important. It shows us that the oxygen in the sensor reacts with the lead of the anode to form lead oxide. The oxygen molecules that contribute to the measuring effect "consume" the anode continuously during the measuring time. The service life of this sensor variant thus depends directly on the O_2 concentration to be measured. The lifetime of this sensor is therefore given in $\%\text{-}O_2 \times$ hours. Typical values are $10^6\%\ h$. A sensor that works permanently in atmospheric oxygen ($\approx$21vol.%) therefore has a service life of approximately 5 years. If the average gas concentrations are lower, the sensor lives correspondingly longer. The operating temperature has an additional influence on the lifetime. If the operating temperature is increased from the specified 20 °C to 40 °C, the lifetime is halved by 50% ($\approx$2.5 years). At60 ° C you can then only expect 18 months.

The characteristic curve is also linear in this case and also has a wide bandwidth (±20%) with regard to sensitivity. The manufacturers usually offer several versions of the O_2 sensor. Figure 4.18 shows the characteristic curves for two different versions from the manufacturer Figaro. Type KE-25 has an average sensitivity of 0.7 mV/%-O_2, while type KE-50 has a sensitivity of 2.7 mV/%-O_2. The different voltages, $U_{M1}$and also U_{M2},result from the different load resistances R_L for both versions. . The measuring current I_M (per %-O_2) can be calculated for both versions as follows:

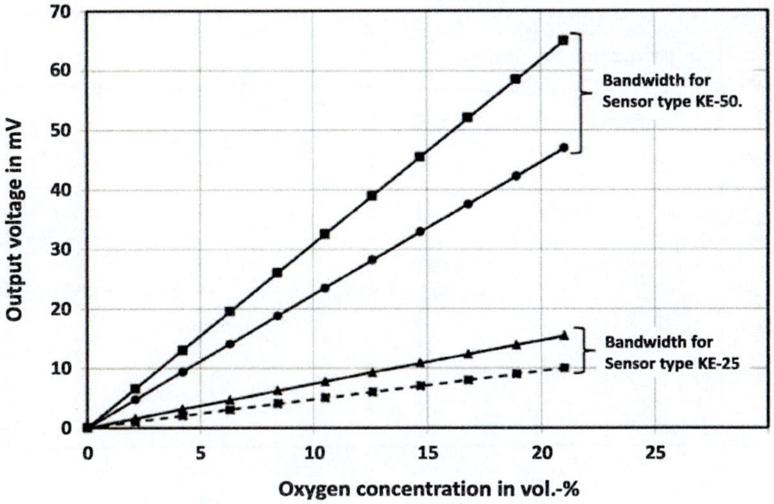

Fig. 4.18 Characteristic curves of the different sensor types

$$\text{Typ KE-25}: \qquad I_{M1} = \frac{U_{M1}}{R_{L1}} = \frac{0.6\text{mV}}{82\text{k}\Omega} = 7.23\text{nA}$$

$$\text{Typ KE-50}: \qquad I_{M2} = \frac{U_{M2}}{R_{L2}} = \frac{2.7\text{mV}}{520\text{k}\Omega} = 5.19\text{nA}$$

Although type KE-50 has a higher measuring voltage, the measuring current is lower. According to the manufacturer, both versions also differ in the thickness of the PTFE diaphragm. In addition to the lower measuring current, the version with the thinner diaphragm would also have to have a different time response. Figure 4.19 shows the gas jump from 100% nitrogen to ≈21vol% oxygen (air) for both versions. The version with the thin diaphragm thus has, as expected, a faster response behaviour, as diffusion is faster than with a thick diaphragm. However, the faster response is coupled with a shorter lifetime (–50%), since more oxygen molecules per time reach the sensor, and thus the lead anode consumes more quickly.

Pressure Dependence

According to Eq. 4.16, the measuring current is linearly related to the particle density in front of the membrane. If the particle density, for example, is increased by a pressure change, the measuring current increases linearly. The following relationship then applies:

$$I_M(p) = I_M(1013\text{hPa}) \cdot \frac{p}{1013\text{hPa}} \tag{4.20}$$

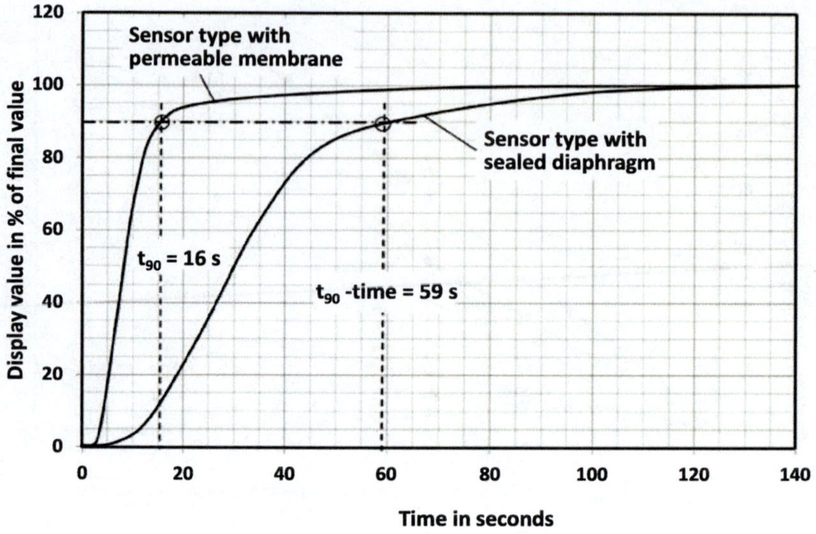

Fig. 4.19 Temporal response behaviour for two sensor versions with different, thick membranes

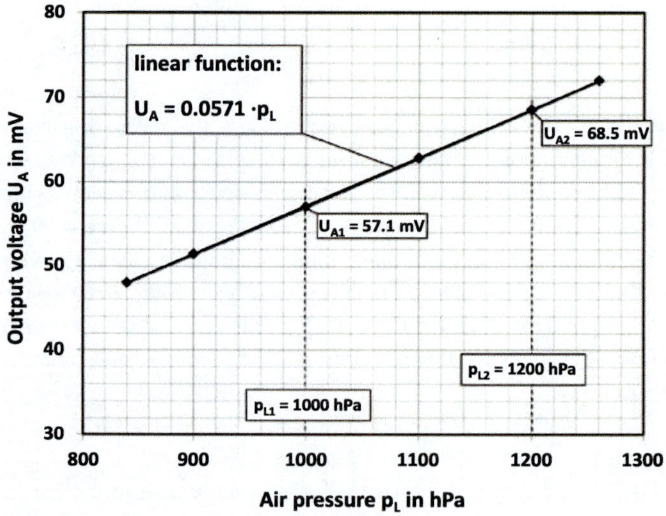

Fig. 4.20 Dependence of the measuring voltage on the air pressure for the KE-50 sensor at an oxygen concentration of ≈21vol.%

Figure 4.20 shows the course of the measuring voltage as a function of the air pressure p_L. Typically, a maximum pressure range from $p_L = 800$ up to 1200 hPa is specified.

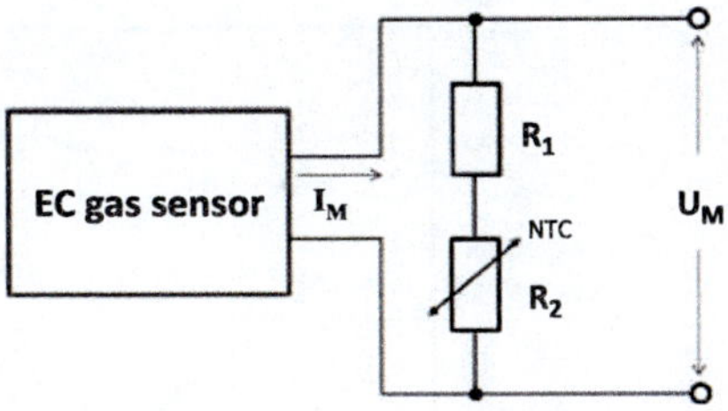

Fig. 4.21 External circuit of an electrochemical gas sensor for temperature compensation with a thermistor (NTC)

Temperature Compensation

Since the diffusion rate of the oxygen molecules through the membrane increases with temperature, the measuring current changes accordingly. For accurate measurements this temperature error must be compensated. The simplest possibility is to use a combination of fixed resistors and temperature-dependent resistors as load resistors. In the simplest case, a series connection of a fixed resistor R_1 and a thermistor R_2 can be used (see Fig. 4.21). The resistance value of the thermistor has an NTC[3] behavior and decreases with the high temperature, so that the voltage drop U_M also decreases at constant measuring current $I_{M.}$ *If* the measuring current increases due to an increase in temperature, the voltage drop remains constant and the measuring signal is temperature-compensated. However, the ratio between R_1 and R_2 must be selected so that this compensation is complete. If so,$R_1 \ll R_2$ the total load resistance is almost independent of the temperature. If so,$R_2 \ll R_1$ the penetration is much greater. For the sensor type KE-25 a combination of a fixed resistor $R_1 = 82\text{k}\ \Omega$and a thermistor R_2 with a resistance of 1 kΩ at 25 °C and B^4=3469 K is suggested. The temperature behavior of a thermistor can be described as follows (Karbach 1997):

$$R_2(T) = R_{25} \cdot e^{B \cdot \left(\frac{1}{T} - \frac{1}{298\,\text{K}}\right)} \tag{4.21}$$

The measuring voltage U_M *is* then obtained:

$$U_M = I_M(T) \cdot \left[R_1 + R_{25} \cdot e^{B \cdot \left(\frac{1}{T} - \frac{1}{298\,\text{K}}\right)}\right] \tag{4.22}$$

Figure 4.22 shows the measuring current as a function of temperature. $I_M(T)$increases with temperature, while the resistance R_{ges} decreases with temperature. This opposite behaviour can be adjusted by R_1 so that the measuring current then becomes independent of temperature. The result of this adjustment is shown in Fig. 4.23. An almost complete

[3]Negative Temperature Coefficent
[4]Material constant of the thermistor.

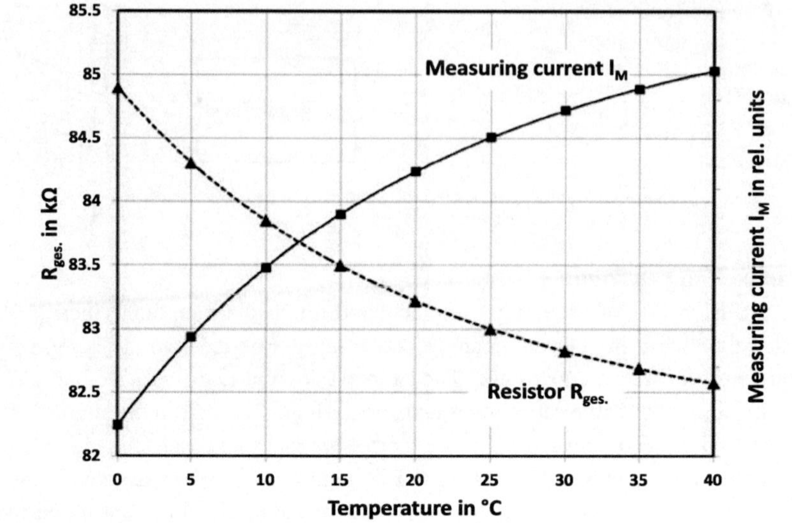

Fig. 4.22 Temperature behaviour of the total resistance R_{ges} and the measuring current I_M

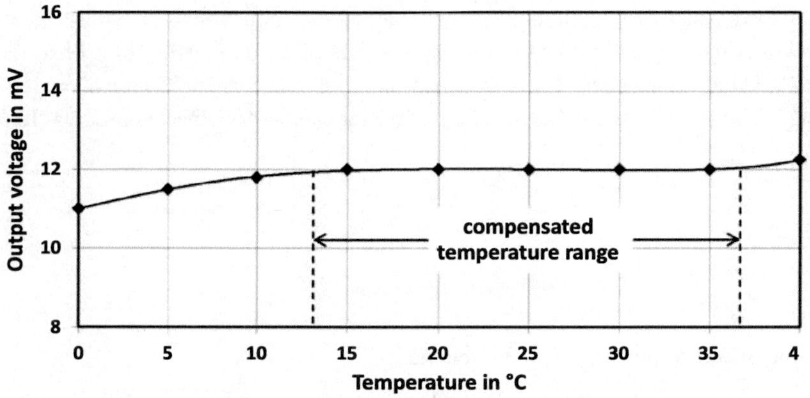

Fig. 4.23 Compensated output voltage U_M in the temperature range from 0 to 40 °C

compensation is achieved in the temperature range of 15–35 °C. At higher or even lower temperatures the error then increases. Since the thermistor is located in the electrolyte liquid, it naturally takes a certain time for the temperatures to adapt. For fast temperature changes, the thermistor can also be placed outside the sensor to reduce the reaction time.

Drift Behaviour

The irreversible change of the maximum measuring voltage is called drift. For this purpose, the sensor is flowed with a constant gas concentration (e.g. ≈21vol.-% O_2) over a long

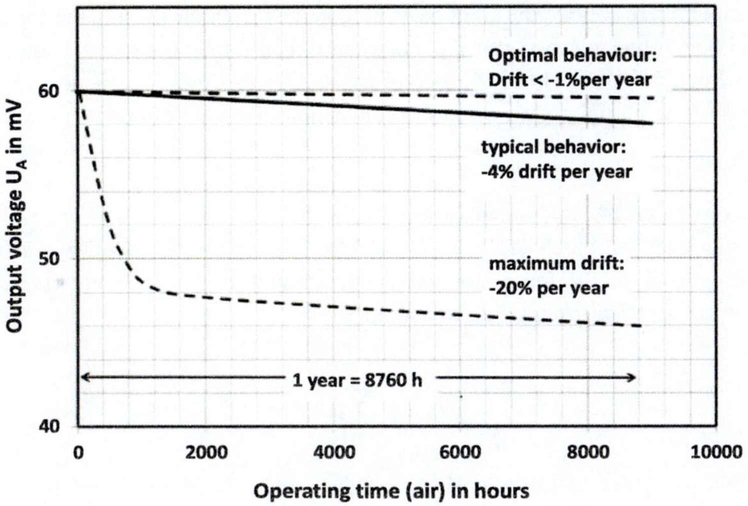

Fig. 4.24 Drift behaviour of electrochemical gas sensors in air with an oxygen content of vol.%≈21 O_2

period of time (several months). The signal changes are individually different for each sensor. Within a type series, considerable deviations can occur which are not predictable. Figure 4.24 shows three different drift behaviors over a period of 12 months. Both the optimum and the typical drift behavior has a linear behavior with time. A typical drift of 4%/year means that the measurement voltage is reduced from 60 mV to 57,6 mV. An oxygen meter that has such a sensor integrated would therefore only display 20.16% O_2 by volume instead of the specified 21% O_2 by volume. If a higher accuracy is required, the measuring device must be readjusted every few months. More critical, however, is the maximum drift of the sensor. It can be clearly seen that the greatest loss of sensitivity occurs in the first few weeks. With this knowledge, it would make sense to pre-age the sensor before delivery to the customer. After this aging process, the further drift is of a similar size to a typical sensor.

Solid Electrolyte Sensors

Gas sensors, based on solid electrolytes, are mainly used to measure oxygen at high temperatures. The largest quantities are used in the automotive industry, in the form of the so-called lambda sensor. Almost every vehicle today is equipped with at least one of these oxygen sensors. The base material is usually zirconium dioxide (ZrO_2), which, however, only develops its ion-conducting capabilities at high temperatures ($T > 300\,^\circ C$). By additionally doping the base material with yttrium oxide (Y_2O_3), the oxygen ion conductivity of the solid electrolyte can be significantly improved (Hengstenberg et al. 1980). Typical operating temperatures lie between 600 °C and 1000 °C.

In principle, there are two different basic arrangements which are suitable for operating these solid electrolyte sensors, namely the potentiometric method and the amperometric method (Ullmann 1993). However, before these two methods are dealt with, a brief explanation of the partial pressure and the conversion to usual volume concentration will follow.

Partial Pressure and Gas Concentration

The physical laws on which the solid electrolyte sensors are based always assume the partial pressures p_i of the respective gas concentrations c_i. Both quantities are closely related, but not identical. According to Dalton's law

$$p_{\text{ges}} = \sum_{i=1}^{k} p_i \tag{4.23}$$

From this regularity, the relation between the number of particles or the amount of substance n_i can be derived:

$$\frac{n_i}{n_{\text{ges}}} = \frac{p_i}{p_{\text{ges}}} \tag{4.24}$$

For an ideal gas, the ratios of substance quantity to concentration are equal to the ratio of concentrations. The following relationship therefore also applies, based on the gas concentration c_i in vol.%:

$$\frac{c_i}{c_{\text{ges}}} = \frac{p_i}{p_{\text{ges}}} \tag{4.25}$$

The conversion from partial pressure to concentration and vice versa is realized by changing Eq. 4.25:

$$c_i = \frac{p_i}{p_{\text{ges}}} \cdot c_{\text{ges}} \tag{4.26}$$

Example

An oxygen partial pressure of $p_i = 245$ hPa is contained in a gas mixture with an absolute pressure of $p_{\text{ges}} = 996$ hPa. The gas concentration c_i is then calculated as follows:

$$c_i = \frac{245\text{hPa}}{996\text{hPa}} \cdot 100\text{Vol.-\%} = 24,6\text{Vol.-\%O}_2 \tag{4.27}$$

The conversion from concentration to partial pressure is then obtained:

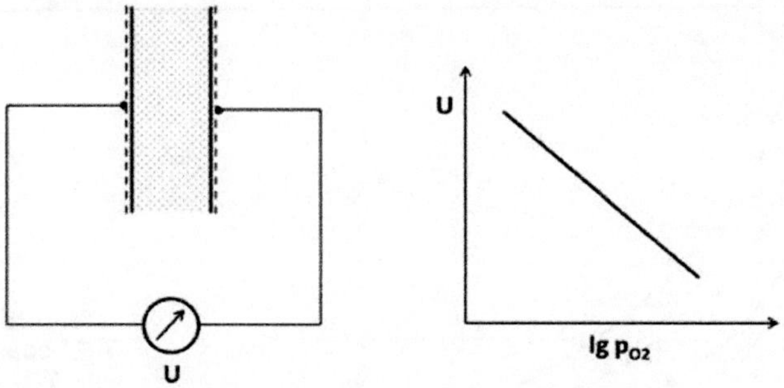

Fig. 4.25 Basic structure of a solid electrolyte sensor using the potentiometric method. The characteristic curve $U(p_{O_2})$ decreases logarithmically with the partial pressure of oxygen $p_{O_2} = p_1$

$$p_i = \frac{c_i}{c_{ges}} \cdot p_{ges} \tag{4.28}$$

Example
For an oxygen concentration of $c_i = 16.8$ vol%, the partial pressure should be calculated at a total pressure of $p_{ges} = 1125$ hPa. Equation 4.25 then results in:

$$p_i = \frac{16.8\text{Vol.-\%}}{100\text{Vol.-\%}} \cdot 1125\text{hPa} = 189\text{hPa} \tag{4.29}$$

Potentiometric Method
In this method, a currentless measurement of the electrical potential difference U between a measuring and reference electrode is carried out. The basic structure is shown in Fig. 4.25. The ZrO_2 base material is provided on both sides with a gas-permeable platinum layer which acts as an electrode.

If there are different partial pressures (p_I and p_2) of oxygen on both sides of the material, a voltage U_N can be measured at the electrodes. This voltage U_N (EMF[5]) is described by the *Nernst*[6] *equation:*

$$U_N = U_0 + \frac{R \cdot T}{4F} \cdot \ln\left(\frac{p_2}{p_1}\right) \quad \text{bzw.} \quad U_N = U_0 + c \cdot \ln\left(\frac{p_2}{p_1}\right) \tag{4.30}$$

[5]EMF Electromotive force.

[6]Walther Hermann Nernst (1864-1941) German physicist and chemist.

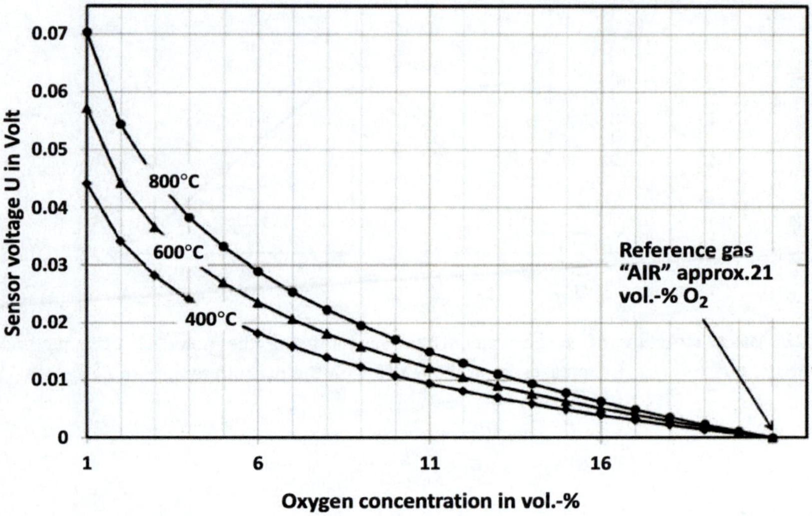

Fig. 4.26 Sensor voltage calculated according to the Nernst equation (Eq. 4.30) for different temperatures (given here in ° C) at $p_{ges} = 1013$ hPa. The measuring range 1–21 vol.% O_2 is suitable for controlling combustion processes

The offset voltage U_0 is very small and in special cases zero. In most applications it can therefore be neglected. Figure 4.26 shows the sensor voltage U_N for different temperatures and O_2concentrations. With the *Faraday constant* F = 96.487 C /mol, the factor *c* can be calculated for a typical operating temperature of $T = 750^\circ$ C

$$c = \frac{R}{4F} \cdot T = \frac{8.314\text{J/mol} \cdot \text{K}}{4 \cdot 96.487\text{C/mol}} \cdot (273 + 750)\text{K} = \frac{0.022\text{J}}{\text{C}} = 22\text{mV} \tag{4.31}$$

This value is of course only valid at the temperature of 750 °C. Equation 4.30 can then also be represented in a simplified form:

$$U_N = 22\text{mV} \cdot \ln\left(\frac{p_2}{p_1}\right) \tag{4.32}$$

Amperometric Method

By applying an external voltage *U* to the electrodes of the solid electrolyte, an electric current *I* flows through the solid electrolyte. If this is a pure ion conductor, *Faraday's laws* apply. The current flow is always associated with material transport of the corresponding ions. In an oxygen probe, O^{2-}-ions are transported. The current flow therefore simultaneously pumps oxygen through the electrolyte, so that one can also speak of a *gas pump* (Schaumburg 1992).

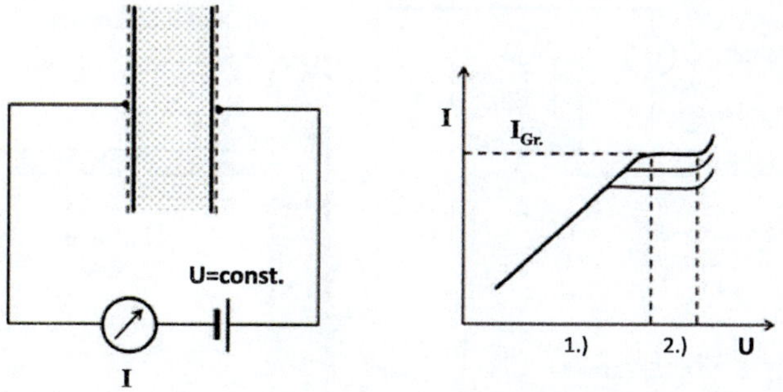

Fig. 4.27 Basic structure of a solid electrolyte sensor using the amperometric method

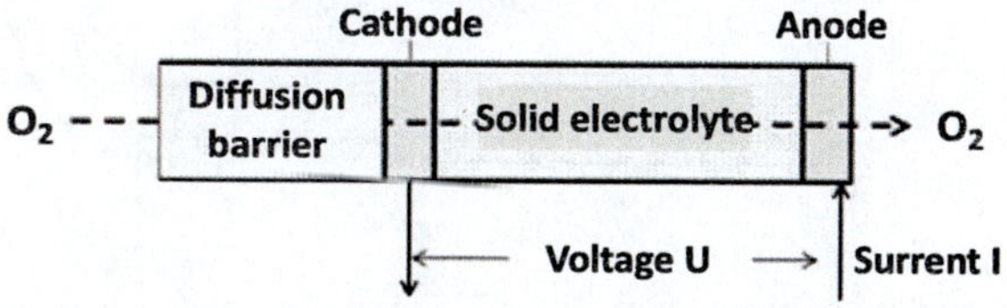

Fig. 4.28 Model to explain the limit current through the diffusion barrier at the oxygen inlet according to Dietz (1982)

Figure 4.27 shows the basic design of the sensor and its electrical behaviour (voltage dependence). The ion current I is linearly linked to the voltage U up to a limit value (I_{Gr}) (1). However, if the current continues to increase, the oxygen concentration c at the cathode is reduced and the current is limited (2). According to Dietz (1982), this process can be explained with the model in Fig. 4.28. The oxygen content at the cathode is limited by the transport through the porous diffusion barrier. *Fick's law* then applies to diffusion. The limit current can then be calculated as follows:

$$I_{Gr} = 4 \cdot F \cdot D \cdot \frac{Q}{L} \cdot c_{O_2} \tag{4.33}$$

D is the diffusion coefficient and Q is the sum of the cross-sectional areas of the pores. L is the effective length of the diffusion layer (Moseley et al. 1991).

If the voltage U is increased further, oxygen can still not reach the cathode, and the current goes into the limit. As can be seen from Eq. 4.33, this limiting current I_{Gr} is determined by the three variables D, Q, L and c_{O_2}.

Sensor Versions

The classic design of solid electrolyte gas sensors is shown in Fig. 4.29. Meanwhile there are also many sensor versions which are built up in a planar structure. However, the functions are identical. The sensor element consists of a ZrO_2 ceramic tube closed on

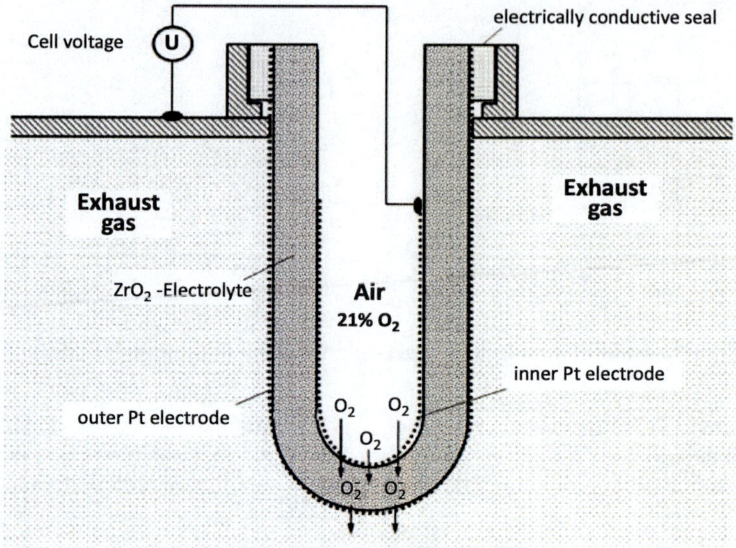

Fig. 4.29 Cross section through a solid electrolyte gas sensor

one side. The electrodes are applied inside and outside in the form of a gas-permeable platinum layer. The two electrodes are contacted at one point each, and led to the outside for further metrological processing of the cell voltage. The inner part of the sensor is in contact with the ambient air and thus has a constant oxygen content of ≈21 vol.%. The outer part of the sensor is completely enveloped by the sample gas (exhaust gas). Both areas must be separated from each other in a gas-tight manner to prevent gas exchange and thus falsification of the measured value. The entire part of the ZrO_2-ceramic must of course be brought to a constant operating temperature $T_B \approx 700^{\circ}$ C. In addition, the sensor must also be heated. In the simplest case, this temperature increase can also be supported by the hot sample gas, if it is the exhaust gas from combustion processes. In the automotive sector, the lambda sensor is operated under these conditions. The exhaust gas considerably reduces the heating power so that the operating temperature is reached much faster. For industrial O_2 measurements, the sensor can either be heated and constantly controlled by a heating element inside the probe (Hammer 2014), or the sensor can be located in a temperature-controlled furnace (see Fig. 4.36).

These O_2-gas sensors are usually operated directly in the exhaust gas flow. This mode of operation (in-situ) has the advantage that the sample gas does not have to be exhausted from the flue or stack by a pump. Since the sensor is heated to a high operating temperature ($T_B \approx 600$ up to $800\ ^{\circ}$ C), problems with droplet formation due to condensation of water vapour also occur. The mechanically sensitive ZrO_2 sensor element should therefore be protected against droplet impact by a metal sheet with openings for gas exchange. Some

Fig. 4.30 Lambda sensor for use in vehicles. (Source: NGK Spark Plug Co, Ltd.)

manufacturers also embed the gas sensor in a sintered metal sleeve that serves the same purpose.

Lambda Sensor

The lambda sensor (λ sensor) is mainly used in automotive engineering to detect the respective residual oxygen content in the combustion exhaust gas. From this, the combustion air ratio λ can be determined and adjusted. These sensors have been produced in millions of units since 1976. Lambda control is also used for catalytic exhaust gas purification in order to minimize the emission of pollutants such as nitrogen oxides, hydrocarbons and soot. Further applications of the lambda sensor can be found in the control of pellet heating systems and oil and gas boilers.

Figure 4.30 shows a lambda sensor with a corresponding protective plate. Industrial O_2-gas sensors also use such mechanical protection.

The characteristic curve of the lambda sensor is set in such a way that, at a λ value of ≈1, it generates a large signal change from approximately 200 mV to approximately 800 mV. The transition in this range is almost abrupt. The characteristic curve is shown in Fig. 4.31. This figure also shows the concentration curves of the typical exhaust gas components such as carbon monoxide (CO), nitrogen oxides ($NO + NO_2 = NO_x$) and unburnt hydrocarbons (HC). At a λ value of around 1 the emissions are lowest, so that one tries to adjust the fuel-air ratio within the λ control range.

Industrial Application

In industrial applications, one naturally wants to determine the exact oxygen content in volume %. For this purpose, the non-linear characteristic curve must be converted into a gas concentration c_{O_2} using appropriate algorithms. The typical, non-linear characteristic curve has already been shown in Fig. 4.26. If the concentration c_{O_2} or the O_2 partial pressure *p1* is plotted logarithmically, the *Nernst equation* produces a linear curve (see

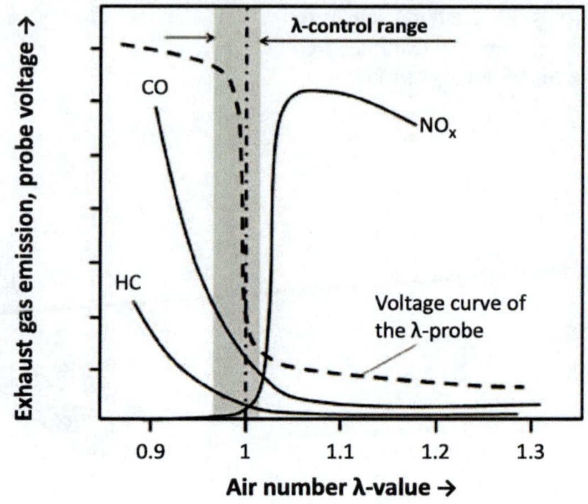

Fig. 4.31 Control range of the lambda sensor. The exhaust gas values refer to the catalytic aftertreatment and not the engine emission. (According to Kabaker 1988)

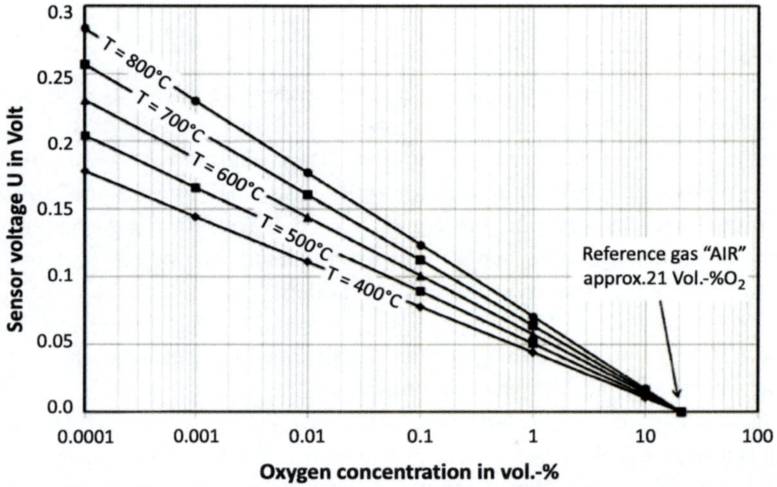

Fig. 4.32 Characteristic curve in semi-logarithmic representation. The concentration specification applies to a total pressure of 1013 hPa

Fig. 4.32). The characteristic curve intersects the *x-axis* at ≈21vol.% O_2. The slope of the characteristic curve shows the sensitivity of the process and is always the same per decade. For a change in concentration of 1 vol.% O_2 → 10 vol.% O_2, the voltage change ΔU is therefore the same as for a change of 0.01 vol.% O_2→ 0.1 vol.% O_2 or even 0.1 ppm O_2 → 1 ppm O_2. This method can therefore also be used to reliably measure traces of oxygen in high purity gases. The slope is only dependent on the probe temperature T_S. Figure 4.33 shows the dependence of the sensitivity per decade on the probe temperature.

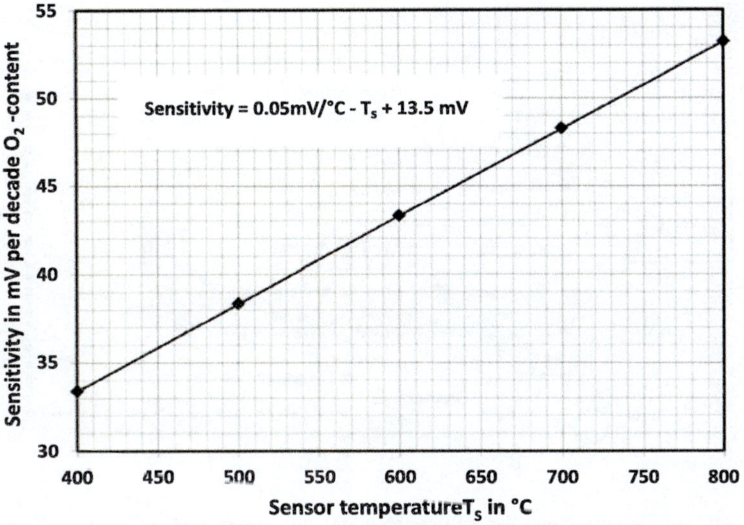

Fig. 4.33 Dependence of the sensitivity (mV/decade O_2) on the probe temperature T_S

The sensitivity changes, in the usual temperature range up to $T_S = 600$to 800 °C, from 43 mV to 53 mV, resulting in a sensitivity change of 0.05 mV/°C. This allows an estimation of the system deviations of the ΔT probe temperature.

Example

A ZrO_2 sensor shall measure the concentration range of 1–10 vol% O_2 in an exhaust gas (=measuring range). At a probe temperature of $T_S = 750$° C, 51 mV results for the sensor voltage U. A system deviation of ±1% of the target temperature of 750 °C would result in $\Delta T = 15$°C. This deviation would result in a change in the sensor voltage of $\Delta U = 0.75$ mV or 51 mV ±0.375 mV. The measurement error, based on the oxygen measurement, would then be 1.5%. To reduce this error, the probe temperature T_S should therefore be controlled to ±1° C.

However, for rough use in industrial combustion processes, the sensor design must still be modified as shown in Fig. 4.29. The outer part of the ZrO_2 sensor is provided with a protective cover for this purpose. Basically there are two possibilities for this. One possibility is to integrate the sensor assembly in a casing made of a sintered metal or sintered ceramic sleeve. Both materials are gas permeable and keep soot and dust away from the sensor structure. However, if there is a high dust load, the pores can become clogged and the gas exchange is interrupted. Another possibility is to place a protective plate (shield) around the sensor structure to protect it against mechanical influences (Fig. 4.34). In this case, the gas exchange must take place through small openings (mm range) in the plate. Dust and soot are not prevented by this. In some cases, depending

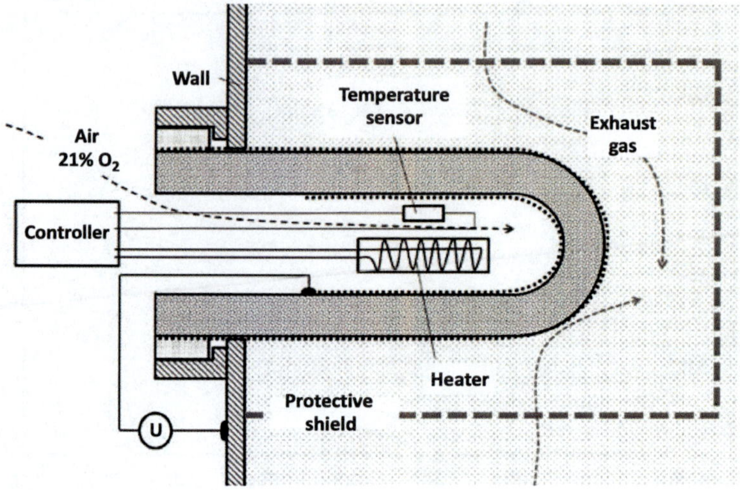

Fig. 4.34 ZrO_2 probe with integrated heating element and temperature sensor. The shock-sensitive structure is surrounded by a protective shield (e.g. perforated plate)

on the application, the manufacturers also select sensible combinations of both variants. Furthermore, a heating element and a temperature sensor (e.g. Pt 100 or thermocouple) are integrated inside the ZrO_2 sensor. By this arrangement, a uniform heating of the structure and a small control deviation ΔT_R of the probe temperature T_S is achieved. As only the internal temperature is controlled in this configuration, a slightly lower temperature is set on the outside, so that an additional temperature difference $\Delta T_{I \rightarrow A}$ occurs. External heating is also possible.

The entire superstructure is usually installed directly in an exhaust duct or chimney. Figure 4.35 shows a typical arrangement of such an installation. The sensor assembly is built up in the form of a lance with the ZrO_2 sensor, with the corresponding protective devices at its tip. Depending on the chimney diameter, these lances can have a length of several meters. A flange mounting on the outside of the chimney is provided for safe attachment of the structure. For maintenance work, the sensor structure can be accessed from this side by dismantling it. The transfer of electrical signals and heating energy is also located on this outside area. The electrical transfer then takes place at an evaluation and control electronics unit, in which the conversion of the sensor signals into the gas concentration takes place. The temperature controller is also located in this unit. For further evaluation of the measuring information, these electronic units have analog and/or digital interfaces which are connected to a control room or a process computer.

For applications in the laboratory or for high-precision analyses, outside a flue gas application, the ZrO_2 sensors are installed in a measuring device. Figure 4.36 shows such an installation.

In this case the ZrO_2 sensor is located in a thermostatted chamber which is purged with a reference gas (e.g. purified air). The volume flow is very low in this case and therefore does

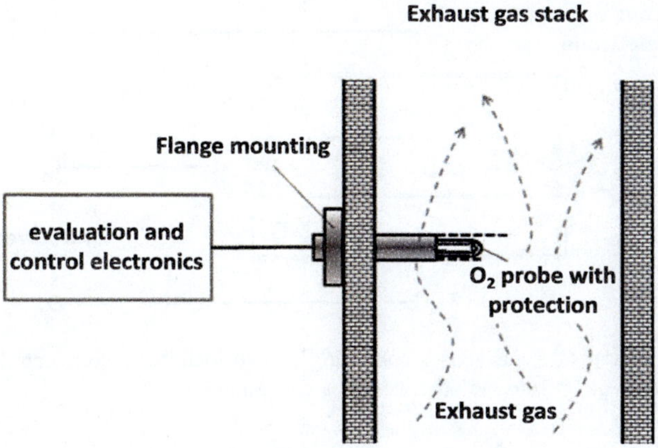

Fig. 4.35 Installation of an oxygen measurement based on the ZrO_2 probe in an industrial application for monitoring the O_2 concentration in an exhaust gas

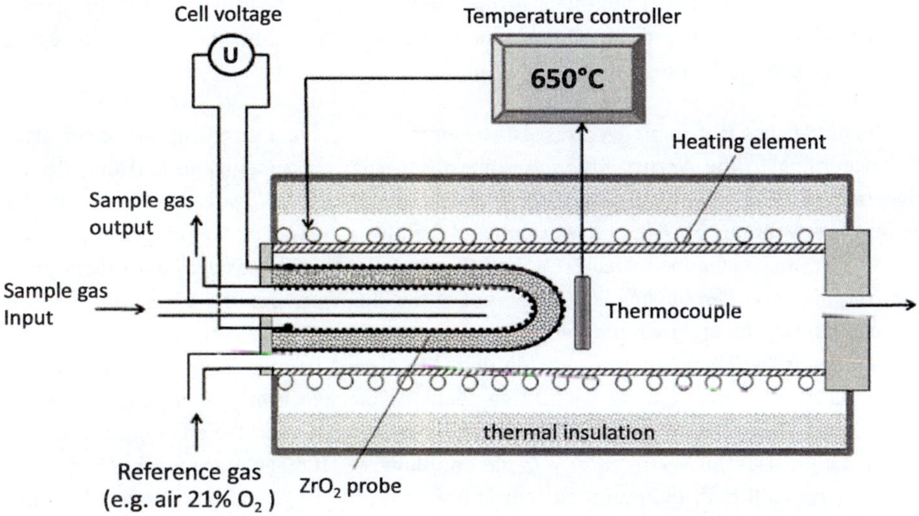

Fig. 4.36 Structure of an oxygen measurement with supply of the measuring and reference gas. (After Hengstenberg et al. 1980)

not disturb the temperature control. The sample gas to be analysed is fed into the ZrO_2 sensor via a pump. The physical measurement is otherwise identical with the method described above. Due to the good thermal insulation of the heated chamber, a high stability of the probe temperature and thus a high measuring accuracy is achieved. Furthermore, test gases can be used at regular intervals in this setup to calibrate the measuring device.

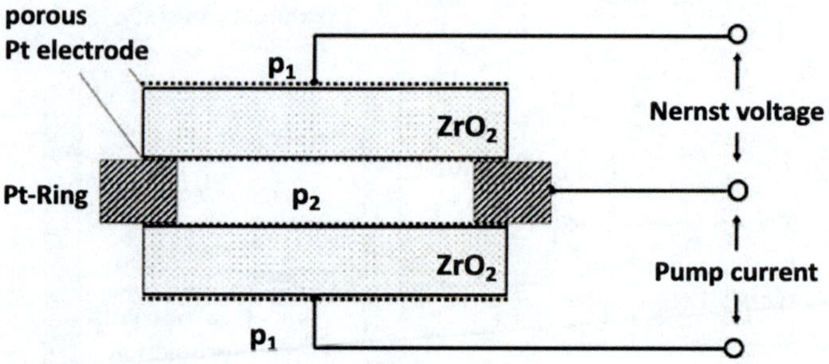

Fig. 4.37 Combination of a gas sensor consisting of two identical solid electrolytes, where one measures the Nernst voltage and the other acts as a gas pump

Combination Sensor

A combination of the potentiometric and amperometric methods was described by Haaland as early as 1977. This structure is shown in Fig. 4.37. It consists of two ZrO_2 discs connected by a thin platinum ring. Both disks are equipped with gas-permeable platinum electrodes. A small chamber volume is formed between the ZrO_2 discs, in which the partial pressure p_2 *of* oxygen is present. Outside the disks, i.e. in the ambient air or in the sample gas, the oxygen partial pressure p_1 is present.

The initial state is $p_1 = p_2$. A stress is measured between the Pt ring and the upper ZrO_2 disc according to the *Nernst equation* when the partial pressure inside and outside the chamber are different ($p_1 \neq p_2$). In the initial state, the *Nernst voltage* is therefore zero. If a voltage is now applied between the Pt ring and the lower ZrO_2 disk, the oxygen is pumped out of the chamber via the lower ZrO_2 disk (evacuated). This causes the *Nernst voltage to* rise continuously. The oxygen is pumped out of the chamber, creating a kind of reference that can be used for the measurement. If the polarity of the pump current is reversed, the oxygen enters the chamber from outside until the *Nernst voltage* is $p_1 = p_2$ zero again. Figure 4.38 shows the pumping and filling cycle. At the start time $p_1 = p_2$and the *Nernst voltage* is zero. At this point, the pump current is switched on and the chamber is pumped empty. This causes the *Nernst voltage to* rise continuously. If this voltage reaches a value of approximately 80 mV, the pump current is reversed and the chamber is now filled with oxygen again. The filling process is complete again when the *Nernst voltage* reaches zero (Dierksheide 2015).

Since the sensitivity per decade of oxygen concentration is always the same, the chamber is never empty (=oxygen-free). The value of 80 mV means at a sensor temperature of 700 °C that the oxygen concentration is only <1vol.%. The cycle time t_p is in the ms range depending on the concentration (Franx 1985). The cycle time is also a measure of the oxygen concentration c_{O_2}or partial pressure p_1 (Fig. 4.39). It applies:

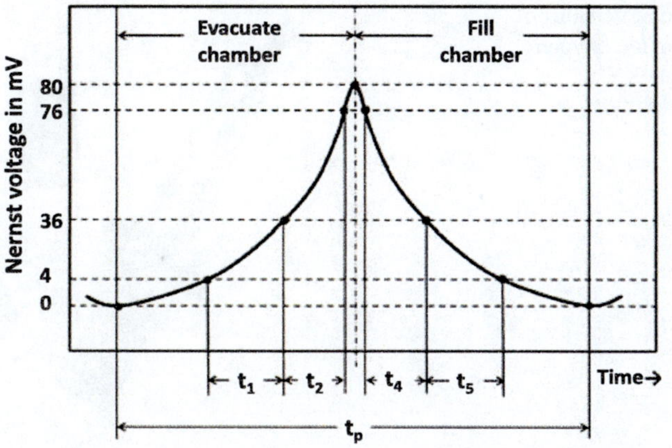

Fig. 4.38 Change in Nernst voltage during evacuation and filling of the chamber

Fig. 4.39 Course of the cycle times t_p and t_d as a function of the oxygen partial pressure p_I. The offset in the zero point is considerably greater at t_p than at t_d

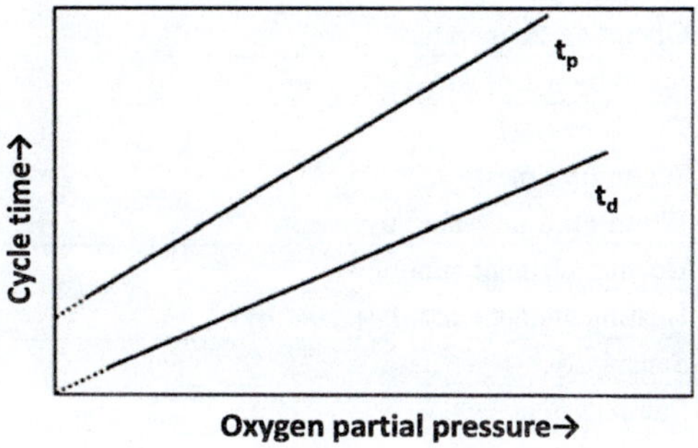

$$c_{O_2} \sim t_p \tag{4.34}$$

Alternatively, the change in slope can also be used as a measure for determining the oxygen concentration. Franx (1985) was able to show that the oxygen sensitivity to temperature changes in the range from 650 to 850 °C remains constant if the cycle time t_d is used for calculation:

$$t_d = [t_1 - t_2] + [t_5 - t_4] \tag{4.35}$$

The following relationship then applies, with α=sensitivity factor:

$$c_{O_2} = \alpha \cdot t_d \tag{4.36}$$

Fig. 4.40 Different versions of the zirconium dioxide oxygen sensors of the type XYA. (Source: First Sensor AG Puchheim)

Table 4.4 Properties of the combination sensor

Property	Specification
Measuring range	2 to 2000 hPa
Operating temperature	700 ° C
Stand by temperature	500° C
Rise time (t_{90})	4 seconds
Warm-up time	100 seconds
Warm-up time (stand by)	20° C
Heating element supply	4.35V ± 0.1V × 1.85 A (8 watts)
Heating element supply (stand by)	2 V × 0.85 A (1.7 watts)
Sensitivity	1 hPa/ms
Pump current	40μ A

Typical values for α are 1 hPa/ms. Due to manufacturing tolerances there is a permissible deviation of ±15%. Calibration of the sensors in the measuring device is therefore absolutely necessary.

There are several manufacturers who offer O_2-sensor elements based on the described technology (Fig. 4.40). The typical measuring characteristics and specifications are shown in Table 4.4.

Non-Nernst Sensors

In contrast to the *Nernst sensor,* the concentration *c of* the adsorbed substances is no longer in equilibrium with the surrounding gas atmosphere at the electrodes. A simple model to describe the processes at the interfaces was developed by Hammer (2001).

An example of this type of solid electrolyte sensor is the *CarboSen,*[7] which is suitable for detecting combustible gases such as hydrogen (H_2) and carbon monoxide (CO) in

[7] Company Lamtec GmbH Walldorf.

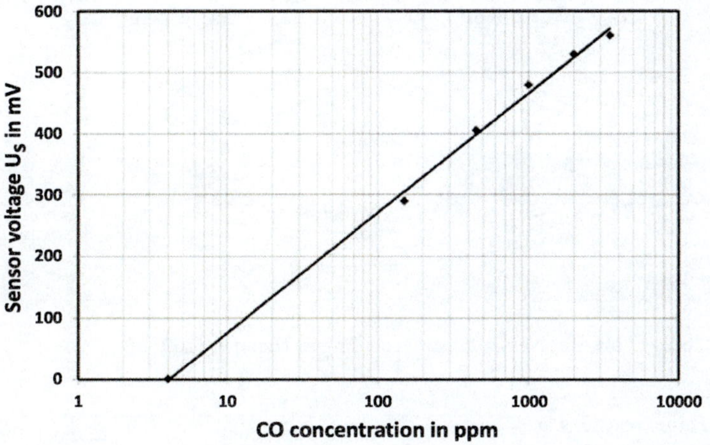

Fig. 4.41 Characteristic curve of CO measurement in the optimal concentration range from 0 to 1000 ppm in logarithmic representation. (From the technical data CarboSen ST from Lamtec GmbH Walldorf)

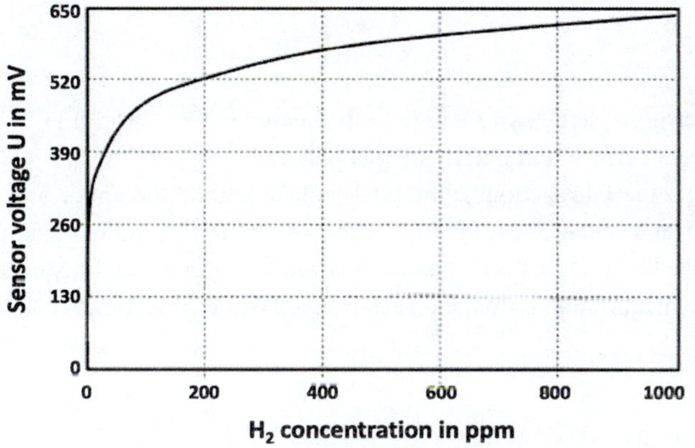

Fig. 4.42 Characteristic curve of the H_2 measurement in the optimum concentration range from 0 to 1000 ppm in linear representation. (From the HydroSen data sheet of Lamtec GmbH Walldorf 2010)

combustion processes (Figs. 4.41 and 4.42). It consists of a planar electrode structure, which is located above an electrolyte ceramic (ZrO_2). This functional ceramic is in turn applied to an Al_2O_3 carrier ceramic with a printed heater using the thick-film technique. Figure 4.43 shows the cross-section through this structure. The carrier material is not shown. At the surface, carbon monoxide CO from the gas phase binds to the surface and is adsorbed ($\rightarrow CO_{ad}$). This happens immediately with the oxygen in the air ($\rightarrow O_{2ad}$). The adsorbed oxygen can reach the anode by electron transfer at the cathode as an ion ($\rightarrow O^{2-}$)

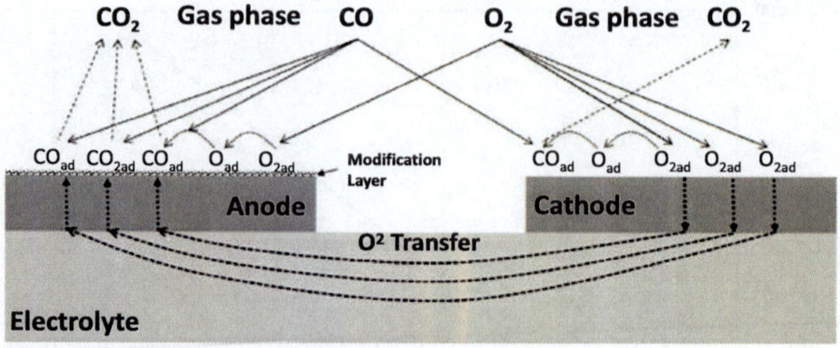

Fig. 4.43 Model of the non-Nernst sensor according to Hammer (2001)

Table 4.5 Partial reactions at the anode and cathode. The processes are back and forth reactions, with the preferred direction indicated by the arrows (according to Hammer 2001)

No.	Anode Reactions	Cathode Reactions
1	$1/2\ O_{2ad} \rightarrow O_{ad}$	$1/2\ O_{2ad} \rightarrow O_{ad}$
2	$CO \rightarrow CO_{ad}$	$CO \rightarrow CO_{ad}$
3	$CO_{ad} + 1/2O_2 \rightarrow CO_2$	$CO_{ad} + 1/2O_2 \rightarrow CO_2$
4	$CO_2 \leftarrow CO_{ad} + O_{ad}$	$CO_{ad} + O_{ad} \rightarrow CO2$
5	$O_{ad} + 2e^- \leftarrow O^{2-}$	$O_{ad} + 2e^- \rightarrow O^{2-}$
6	$O^{2-}_{(A)} \leftarrow O^{2-}_{(K)}$	

through the electrolyte, and react there with the adsorbed CO_{ad} to form CO_2, which is then desorbed from the surface and enters the gas phase.

The individual partial reactions, which take place both at the anode and at the cathode, are summarized in Table 4.5. Provided that the adsorption of the reaction product CO_2 at the electrodes can be neglected and that the desorbed CO_2 in the vicinity of the surface does not disturb the concentration ratios, the sensor voltage at the electrodes can be calculated as follows:

$$U_S = \frac{R \cdot T_S}{2 \cdot F} \cdot \ln\left[\frac{c_{CO_{ad,K}}}{c_{CO_{ad,A}}}\right] \tag{4.37}$$

Since both electrodes are located in the exhaust gas to be measured, no measurable voltage U_S should be detectable. The absorption behavior of the CO can be influenced by an additional coating (modification layer), which leads to a change in catalytic activity. As a result of this measure, more CO molecules are bound to the anode than to the cathode and a measurable sensor voltage U_S is generated. Figure 4.44 shows the different production stages of a CarboSen.

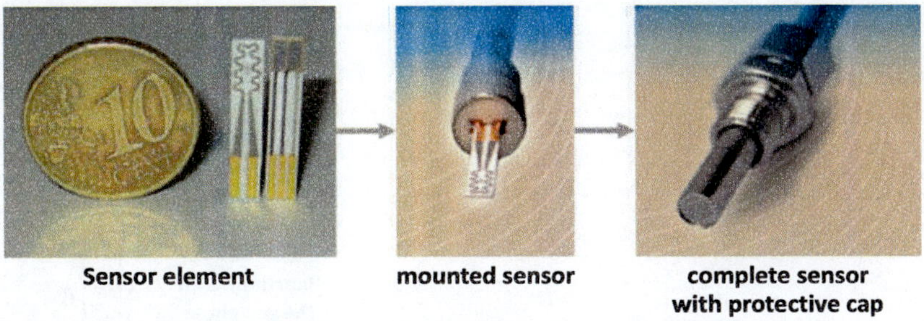

Fig. 4.44 Sensor design in thick-film technology. To protect against external influences, the sensor chip is surrounded by a sintered metal sleeve. (Source: Lamtec GmbH Walldorf)

4.3 Semiconductor Gas Sensors

Gas sensors based on metal oxide semiconductor materials have been used in gas measurement technology since 1968. The measuring method was first presented by *Taguchi*[8] and scientifically investigated by *Heiland.*[9] *The* first applications were in the field of leakage monitoring (explosion protection) of gas supply (natural gas +liquid gas) in buildings.

The basic principle of semiconductor gas sensors is comparatively simple. One uses the change in resistance ΔR_S which is caused by a component A with the concentration c in a gas mixture (A + B). It is important that a constant oxygen concentration is present in the gas mixture, and that the gas sensor has also reached a stable operating temperature of approximately 300 to 600 °C. The resistance value $R_S(c)$is then described by the following relationship:

$$R_S(c) = \alpha \cdot c^\beta \tag{4.38}$$

The constants α and β are determined empirically and result from the kinetic processes on the sensor surface. The β value lies between 0.5 and 0.9. The conversion of the resistance change ΔR_Sinto a voltage signal U_S is realized in the simplest case by a voltage divider (Fig. 4.45). The sensor voltage is then calculated:

$$U_S = U_V \left[\frac{R_S(c)}{R_V + R_S(c)} \right] \tag{4.39}$$

The voltage divider is usually calibrated in dry ambient air. It then applies:

[8]Naoyoshi Taguchi (1936-) Japanese inventor and entrepreneur.

[9]Gerhard Heiland (1917-2005) German physicist.

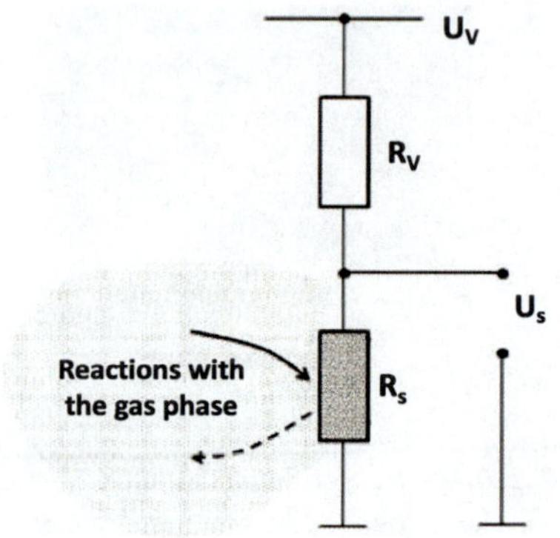

Fig. 4.45 Arrangement of the gas sensor with resistor R_S in a voltage divider. The voltage U_S is proportional to the gas concentration c in the surrounding gas phase

$$R_S(\mathrm{air}) = R_S \tag{4.40}$$

With this specification, the sensor voltage corresponds to half the supply voltage:

$$U_S = 0.5 \cdot U_V \tag{4.41}$$

The individual physical and chemical processes that lead to this change in resistance, on the other hand, are very complex and still the subject of research projects.

Mainly tin dioxide (SnO_2) and zinc oxide (ZnO) are used as base materials. Other possible substances are tungsten oxide (WO_3), iron oxide (Fe_2O_3), gallium oxide (Ga_2O_3) and indium oxide (In_2O_3). In addition to the pure substances, combinations such as indium tin oxide (IZO) are also used in order to be able to specifically change the sensor properties (Heitbaum and Wiegleb 1991).

The materials mentioned are mostly present in crystalline form. The lattice structure of the crystals is naturally interrupted at the surface (interface) to the gas phase. This disturbance of the crystal structure leads to surface states which may well lie in the forbidden zone of the semiconductor. The *Fermi energy* is therefore not identical with the level inside the crystal (Fig. 4.46). A charge exchange therefore takes place and a local charge distribution is formed on the surface. Oxygen from the ambient air intensifies this effect and a region with a strongly reduced electron distribution $N_e(z)$ is obtained, which is also called the *depletion zone* (Häusler 2004).

Another important parameter in the description of semiconductor gas sensors is temperature. Semiconductor materials have a negative temperature coefficient compared to electrical conductors (e.g. metals). If the semiconductor material is heated, additional charge carriers from the valence band enter the conduction band. This electron transfer causes a change in the basic resistance $R(T)$. The temperature behaviour can be described by the following equation:

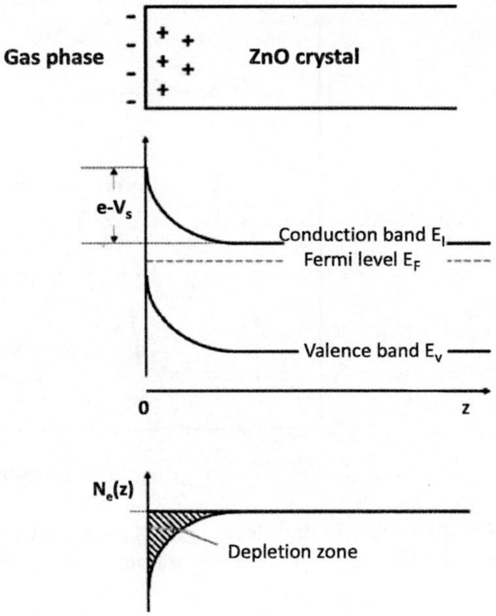

Fig. 4.46 Band model to explain the charge carrier distribution at the interface between the metal oxide crystal and the surrounding gas phase

$$R(T) = R_0 \cdot e^{[e \cdot V_S / k \cdot T]} \tag{4.42}$$

R_O is the basic resistance at room temperature, $e \cdot V_S$ the potential barrier, k the Boltzmann constant ($k = 1.38066 \cdot 10^{-23}$ J K^{-1}), and T the temperature in Kelvin.

Furthermore, the addition and diffusion processes in the layer structure are dependent on the temperature. Especially physisorption and chemisorption play an important role in explaining the mode of action of semiconductor gas sensors. Physisorption is a loose bond caused by electrostatic forces, dipole forces and *van der Waals forces*. The binding energy is below 25 kJ/mol or 0.5 eV per particle. Chemisorption is a strong ionic bond caused by an electron transfer between the adsorbate and the semiconductor surface. The binding energy in this case is 60 kJ/mol and, in contrast to physisorption, is usually not reversible.

Surface Reactions

At the SnO_2 surface, these bond forms can be present simultaneously and are mainly determined by the temperature T and the oxygen partial pressure p_{O_2} . At higher temperatures, chemisorption will mainly occur, whereas at lower temperatures physisorption (P_s) will predominate (Fig. 4.47). Chemisorption (Cs) is important for the detection of a gas reaction, since it is always associated with electron transfer. It therefore applies:

$$O_{2_{Ps}} + e^- \rightleftharpoons O^-_{e_{Cs}} \quad \textit{and} \quad O^-_{e_{Cs}} + e^- \rightleftharpoons 2O^-_{e_{Cs}} \tag{4.43}$$

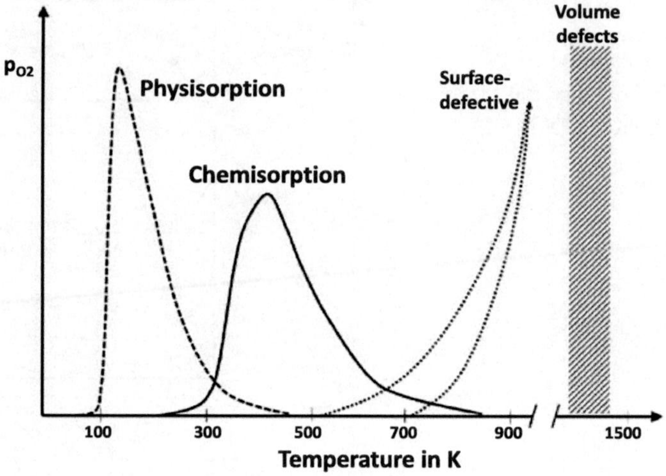

Fig. 4.47 Temperature dependence of surfaces and volume reactions relevant for semiconductor gas sensors: The data were determined from a thermal desorption spectrum. (Schaumburg 1992)

Thus, electrons e^- from the conduction band enter the chemisorbed bond by this electron transfer. As a reaction to this electron transfer, the resistance R_S increases. When these chemisorbed oxygen atoms O^- react with a pollutant molecule (e.g. CO), the bound electrons are released again and the resistance R_S decreases accordingly.

$$CO + O^-_{eC_s} \rightharpoonup CO_2 + e^- \tag{4.44}$$

The CO_2 molecule then desorbs from the surface and the free space can now be used by another molecule. In the stationary state, a resistance value R_S is therefore formed, which is characteristic of the number *n of* reactions described in Eqs. 4.43 and 4.44. This equilibrium is thus proportional to the concentration of the substances to be detected.

Sensor Setups

The first sensors (TGS types) built by *Taguchi,* consisted of a thin ceramic tube onto which a thin layer of SnO_2 was applied. Electrodes were attached to both ends of the tube, which were embedded in this SnO_2 layer, and thus had direct contact with the sensor layer. A heating coil was inserted into the tube, which brought the sensor assembly to the appropriate operating temperature. The heating coil was supplied with a constant heating voltage. The required operating temperature was then set under constant ambient conditions. If, for example, the ambient temperature changes, the sensor temperature also changes, and thus also the sensor properties.

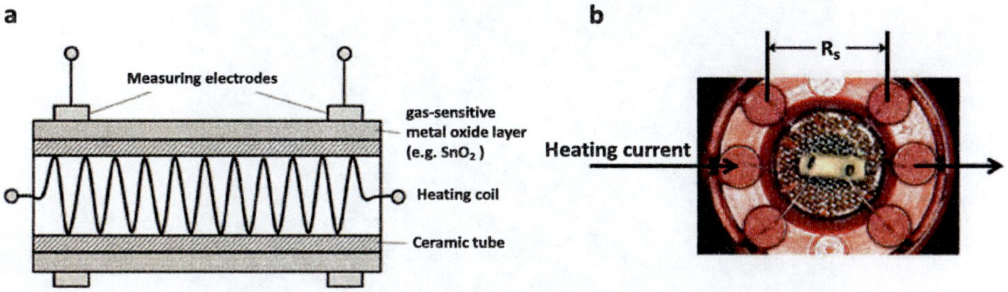

Fig. 4.48 Cross-section through the Taguchi Gas Sensor (TGS) (**a**) and mounting the sensor on a feedthrough in the housing (**b**)

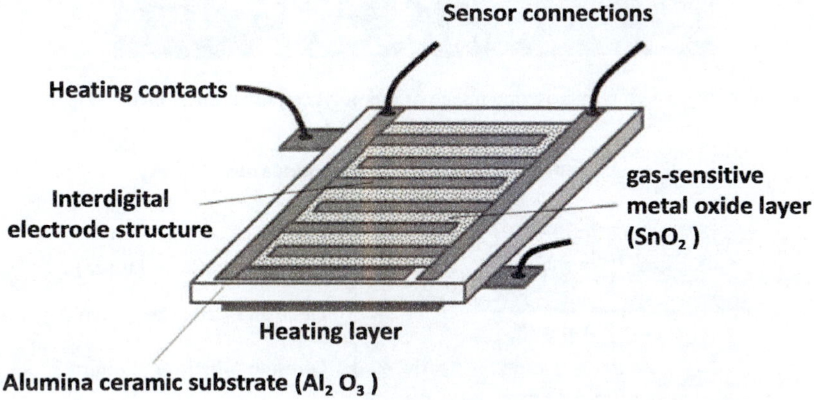

Fig. 4.49 Planar SnO_2 gas sensor with an electrode and heater structure in thin film technology. The gas-sensitive SnO_2-layer was realized in thick film technology (Wiegleb 1990)

Figure 4.48 shows this structure as a cross sectional view. All of the above steps were performed manually. Due to this type of production technology, the reproducibility and quality were not sufficient for today's requirements. Furthermore, the manufacturing costs were very high.

In the 80s of the last century, attempts were therefore made to construct the sensor element completely in thin-film technology. It turned out, however, that the SnO_2thin-film structures did not have the same service life (operating hours) as the manually produced TGS types with the comparatively thick layer structures in theμ m-range.

Another solution approach was to combine a miniaturized electrode and heater structure (thin-film technology) with a SnO_2 thick-film structure (Wiegleb 1990). This enabled longer lifetimes to be achieved, and at the same time made it possible to make use of the low-cost manufacturing processes of the layer technologies. The structure is shown in Fig. 4.49, and consisted of a planar Al_2 O_3-ceramic substrate (250 μm thickness). The thick-film structure was then printed on a 25 μm wide interdigital electrode structure made of gold and fired at

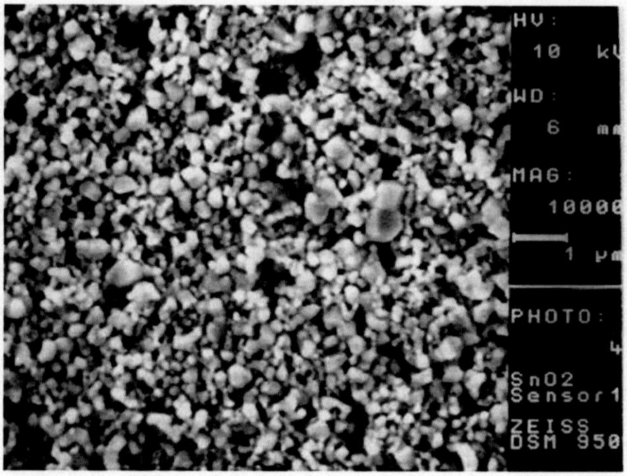

Fig. 4.50 Scanning electron microscope image of the polycrystalline SnO_2 layer (Wiegleb 1990)

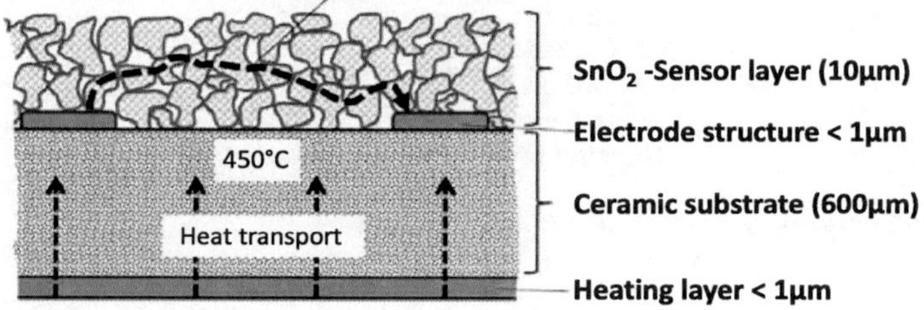

Fig. 4.51 Cross-section through a planar SnO_2 gas sensor with approximate values for the individual layer thicknesses (no scale representation)

high temperatures (thick-film process). Both, the electrode structure and the heater structure on the back of the planar structure, were applied by sputtering (thin film process).

The great advantage of this production technique is that the thin- and thick-film process suitable for mass production can be used. The thick-film technology then provides the same metrological properties as for the TGS superstructures. The thick-film paste, which is applied to the electrode structure with a squeegee using a screen printing process, consists of finely ground SnO_2 powder (polycrystalline). The baking process with a binder causes the layer to sinter together and is extremely durable. The diameter of the individual SnO_2 grains is$\gg$1µ m. Today, this manufacturing technology is used by almost all sensor manufacturers.

Figure 4.50 shows the surface of a SnO_2 thick film as an REM image. It can be clearly seen that the individual SnO_2 crystals touch each other, and thus form a contact surface.

Figure 4.51 shows a cross-sectional view through the entire sensor structure. The heater structure provides an even heating of the sensor structure from the bottom side. If the heater

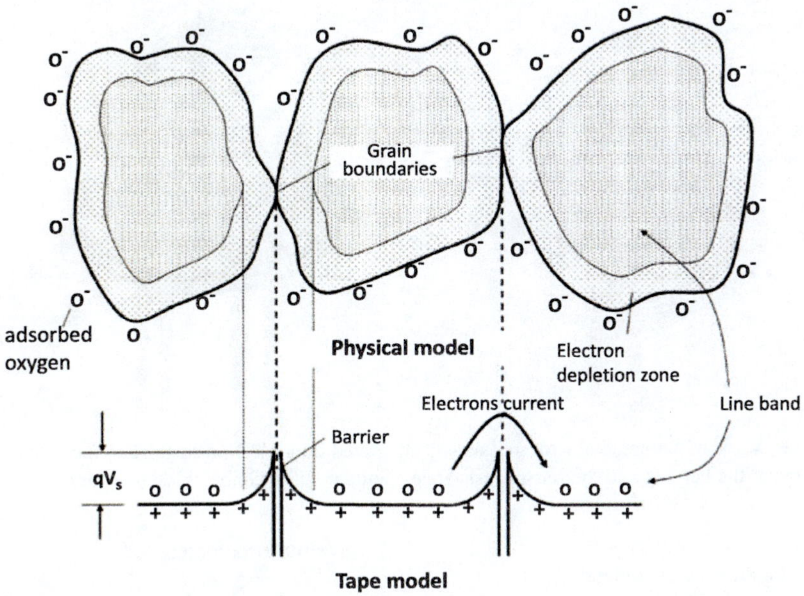

Fig. 4.52 Polycrystalline SnO_2 with the depletion layers and the transitions at the grain boundaries. In the *lower part of* the picture the potential barriers are shown, which result at the grain boundaries

is structured from platinum, it can be controlled to a constant temperature T_S via its resistance and the known temperature coefficient ($\alpha = 3.85 \cdot 10^{-3}\text{K}^{-1}$). This is a further advantage over the original TGS superstructures.

The individual SnO_2 crystals are located between the electrodes and touch each other at the grain boundaries. The current flow thus takes place via a branched network within the thick-film structure. As this layer is porous, the gas can also diffuse into the lower layers and thus contribute to the measuring effect. Figure 4.52 shows three of these crystals with the corresponding grain boundaries. Current flows through the three grains across the potential barriers. Accordingly, the grains have a depletion zone which does not extend over the entire grain, but only has a defined layer thickness. At the surface of each grain there are chemisorbed O^-, which are very reactive and react immediately with an existing partner. The transitions at the grain boundaries behave like *Schottky contacts,*[10] while an ohmic current conduction is present inside the grain.

With the help of thin-film and thick-film technology, semiconductor gas sensors can be miniaturized very well and produced cost-effectively in large quantities. The manufacturing costs per sensor cannot be undercut by any other measuring method. Figure 4.53 shows such a gas sensor. In this design, the sensor structure and the heating

[10] Walter Hans Schottky (1886-1976) German physicist and electrical engineer.

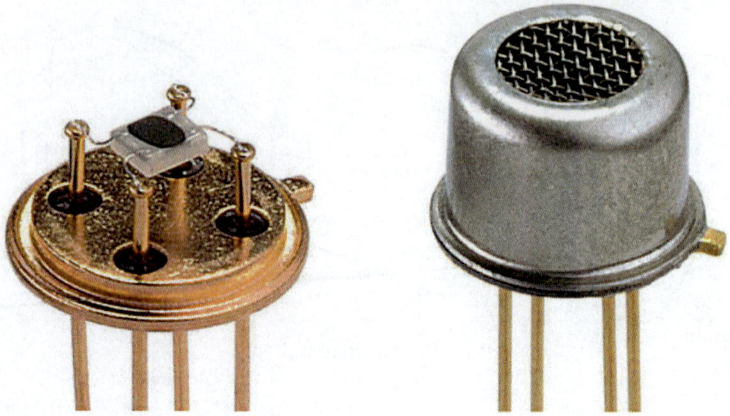

Fig. 4.53 View of a semiconductor gas sensor integrated into a TO package. The four connecting wires are for the heating and the sensor resistance. (Source: UST-GmbH, Geschwenda)

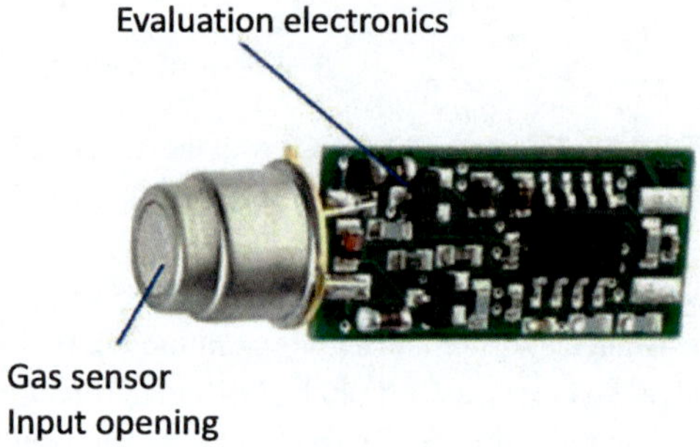

Fig. 4.54 Semiconductor gas sensor with an additional glass bead filtering behind the input opening and the connected evaluation electronics in SMD technology. (Source: UST-GmbH, Geschwenda)

element are located on one side of the chip. From a manufacturing point of view, this is a further cost advantage. The four connecting wires are contacted with the pressure glass feedthrough via a spot-welded joint. The mass of the sensor chip is so small that the gas sensor is held by the electrical wires. Of course, this construction must still be protected against external mechanical influences. For this purpose, a metal cap is placed over the structure and also firmly connected to the pressure glass feedthrough. At the top of the cap, there is an opening through which the gas can enter the interior of the assembly. A fine-meshed wire mesh and suitable membranes prevent the penetration of disturbing particles, dust, etc. A further protection possibility is to place a layer of$\approx 100\ \mu$ m thick glass beads in front of the sensor element to use the adsorption surface of the beads. Figure 4.54 shows a sensor element with such a glass bead filling in the upper part of the sensor (paragraph at the gas inlet). It is also possible to connect several elements into one unit (Fig. 4.55). An extremely miniaturized version is shown in Fig. 4.56.

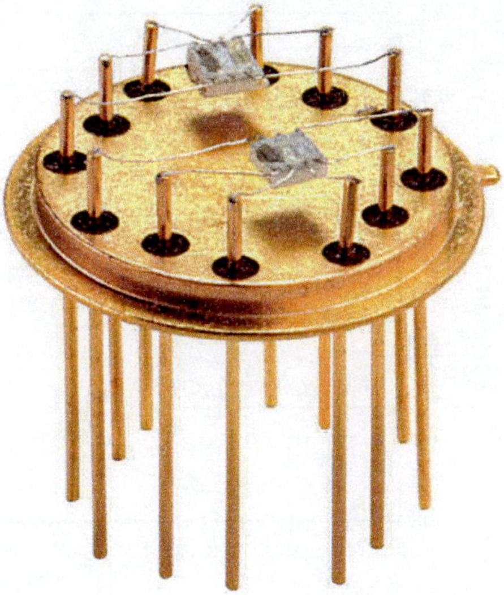

Fig. 4.55 Sensor setup with 2 different semiconductor gas sensors for simultaneous measurement of different components in a gas mixture. (Source: UST-GmbH, Geschwenda)

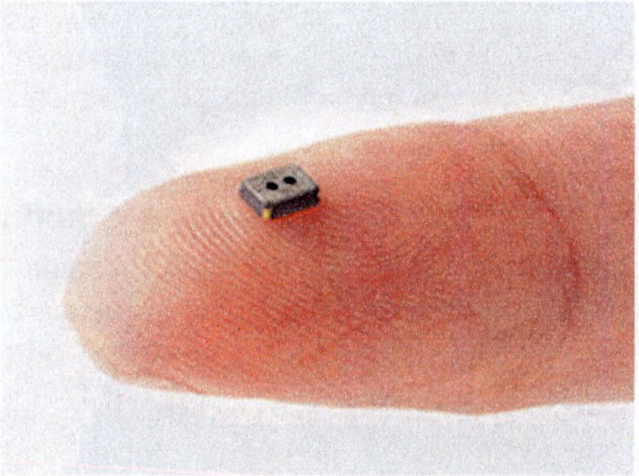

Fig. 4.56 Extremely miniaturised semiconductor gas sensor for installation in a smartphone. (Source: Figaro Engineering Inc. Japan)

Sensor Properties

After the physical properties of semiconductors and the chemical processes on the surface have been described, the resulting gas-sensitive properties will now be discussed.

If a SnO_2 sensor is put into operation at room temperature ($T \approx 20^{\circ}$C), it must first heat up to its operating temperature (e.g. $T_S = 450^{\circ}$ C). At this time there is not yet a sufficient number of charge carriers in the conduction band and the resistor R_S is high-impedance ($R \approx \infty$). In the heating phase, electrons are now introduced into the conduction band by thermal excitation, as can be expected for a semiconductor according to Eq. 4.42. Due to these increased temperatures, the reactions take place simultaneously, which lead to a binding of charge carriers as a result of chemisorption. As a result of this opposite effect, electrons are withdrawn from the conduction band. These two competing processes are only completed after a few minutes and only then does a stable sensor resistance develop.

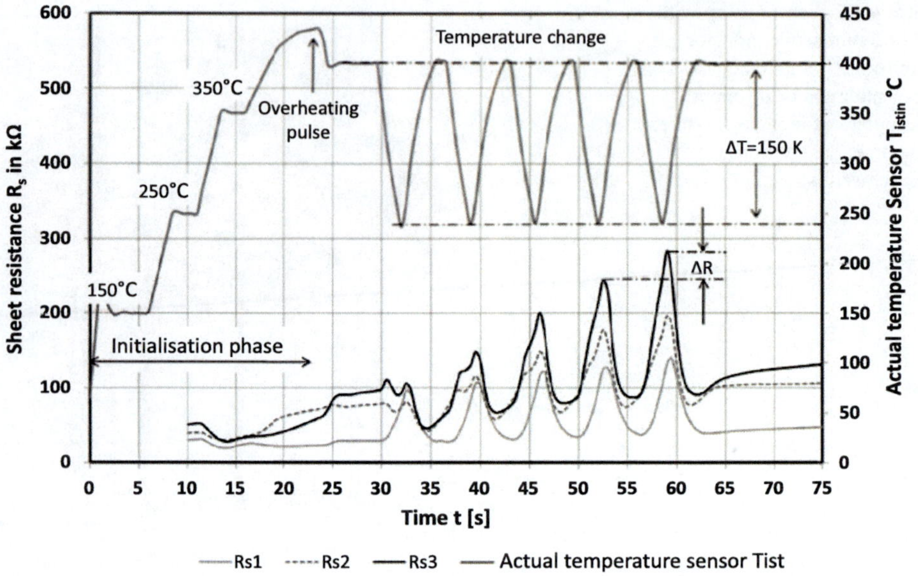

Fig. 4.57 Heating up phase and temperature cycles on the reversibly loaded *UST Triple Sensor®*, which consists of three different sensor structures (R_{s1},$R_{s2}R_{s3}$, . . .)

Some manufacturers guarantee the properties described in the data sheet only after a running-in phase of several hours or even days.

This process can be actively influenced, and thus also shortened within the scope of an intelligent heating phase. During an initialization phase, the sensor passes through a defined temperature ramp (150 °C → 250°C → 350°C) with an overheating pulse ($T \approx 435^{\circ}$C) as the start cycle. Subsequent rapid temperature changes ($\Delta T \approx 150$ K) are then used to evaluate whether the sensor has stabilized. Adsorbates are detected by the temperature changes and simultaneously desorbed again. During a few, less temperature changes ($n \geq 3$) the resistance R_S of the sensor is measured at the same temperature T, and compared with the previous value for an increase ($\Delta R > 0$). This thermal excitation of the semiconductor leads to an optimal chemisorption and thus to a faster binding of charge carriers. With an unloaded sensor, 3 temperature change pulses are usually sufficient to stabilize the sensor. The time sequence of these pulses can vary between 5 and 15 seconds, depending on the sensor design.

A further advantage is the real-time recognition whether the sensor has to pass through further temperature cycles for self-cleaning or not. This is the case if, for example,. extreme physisorption and possibly also chemisorption occurs on the surface and within the sensitive layer. Figure 4.57 shows the response behavior of the three different elements of a *Triple Sensor®*.[11] If the resistance difference ΔR continues to increase after the

[11] Registered trademark of UST-GmbH, Geschwenda

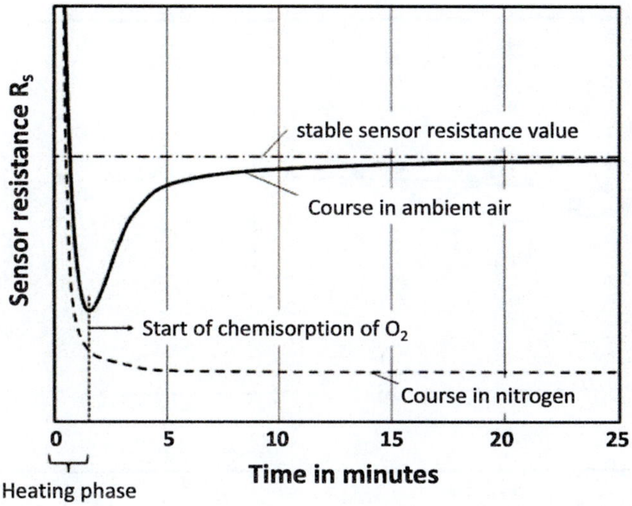

Fig. 4.58 Stabilization phase of a SnO_2 gas sensor after switching on (cold start) in ambient air and nitrogen

temperature changes, additional cycles must be run. Criteria stored in the software, which decide on further temperature cycles according to the application of the sensor, release the sensor for the measuring task according to certain criteria. These include the comparison with the stored sheet resistance R_0 in the basic state of the sensor. At the latest, however, as soon as the charge carriers are in equilibrium, the sensor is ready for operation (Kiesewetter and Müller 2015).

The temperature changes during the heating phase also detect whether non-desorbing substances have changed the sensor. Intelligent software can compensate for this change to a certain extent.

If this *cold start* is carried out in 100% vol. nitrogen (N_2) instead of in ambient air with an oxygen content of ≈21% vol. O_2, there is no chemisorption of oxygen from the environment and the sensor resistance R_S stabilizes at a low value (Fig. 4.58). In this state, the gas sensor is no longer sensitive to other gases because the chemisorbed oxygen is missing as a reaction partner. Only after a stabilization phase in ambient air can the sensor be used for measuring gases. However, the oxygen partial pressure p_{O_2} must remain constant even during the measurement. As a rule, one can assume that a constant oxygen content is always present during these measurements (room air analysis).

If one observes the basic resistance over a very long period of time (days, months, years), one will also notice significant changes. Figure 4.59 shows the result of such a measurement. Although the basic resistance is relatively stable in the logarithmic representation, the absolute values vary considerably ($\Delta R = 6\text{k } \Omega$). In relation to the basic resistance ($R_0 \approx 20\text{k } \Omega$), these are ±15%. For continuous measurements, electronic methods are therefore used to minimize this drift behavior.

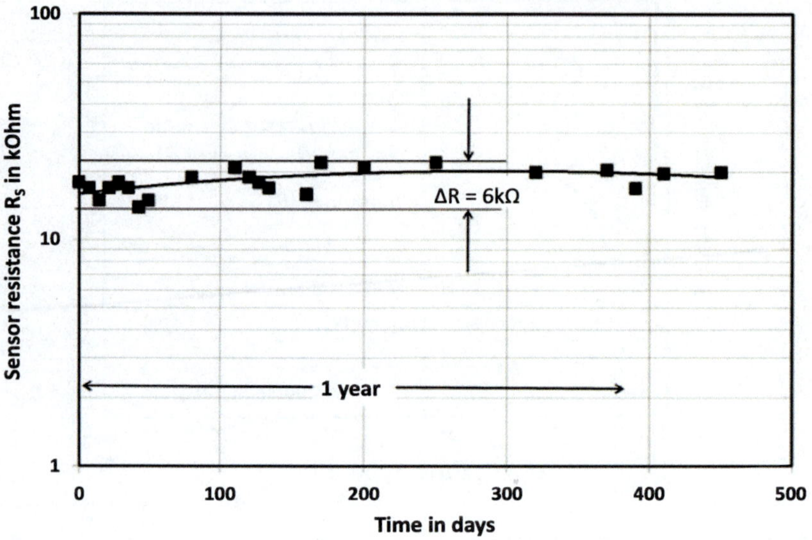

Fig. 4.59 Drift behaviour of a typical semiconductor gas sensor in ambient air over 15 months

This error must of course be considered in relation to the measuring effect. The question that now arises is, what influence does this drift have on the concentration measurement? For this purpose, the sensitivity curve for a specific gas is selected. In this specific case, the propane characteristic curve was considered. Figure 4.60 shows the propane characteristic curve in linear representation (both axes). The resistance curve $R_S(c)$ is described as a trend line with the following equation

$$R_S(c) = 110.42 \cdot c^{-0.66} \tag{4.45}$$

A change in the basic resistance R_S of -15% would be caused by a change in concentration $c_1 = 300$ ppm from to $c_2 = 400$ ppm according to Eq. 4.45. The error in the gas measurement would therefore only be $\Delta c = 100$ ppm. Related to a measuring range of e.g. MB $=$ 10.000 ppm this would then be only -1%. With this rough calculation, it was shown that the absolute error is very large, but in relation to the gas concentration c is an acceptable value.

Figure 4.61 shows that the gas sensors do not react selectively to a certain gas, but have different sensitivities for the individual gases. This is an advantage for air quality analyses or for detecting substances that are not normally present in the ambient air. If, on the other hand, you want to detect a certain species in the presence of other, similar gases, this method fails. The manufacturers of semiconductor gas sensors usually offer a whole range of different products. The individual sensors differ by different doping, which promises a better sensitivity for certain substances or groups of substances.

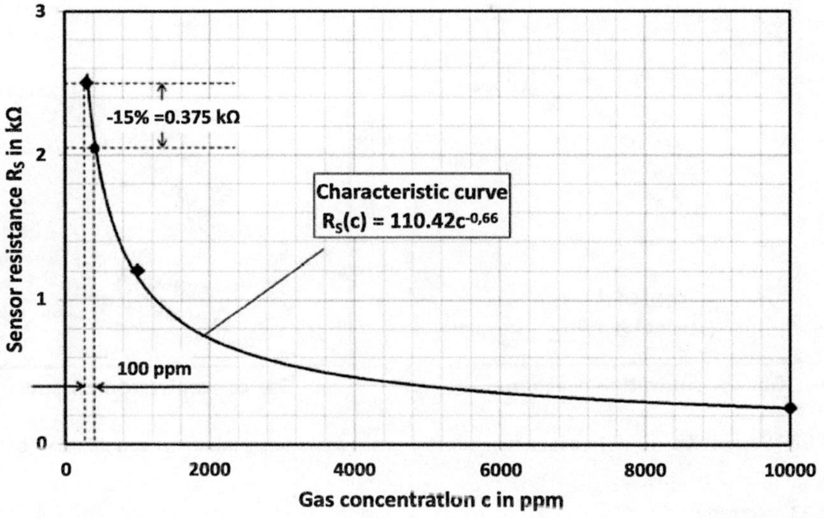

Fig. 4.60 Sensitivity curve (characteristic curve) in a more linear representation for the TGS 2610 sensor with propane gas. The measured values (data points) can be described by a curve according to Eq. 4.38 $[R_S(c) = \alpha \cdot c^\beta]$

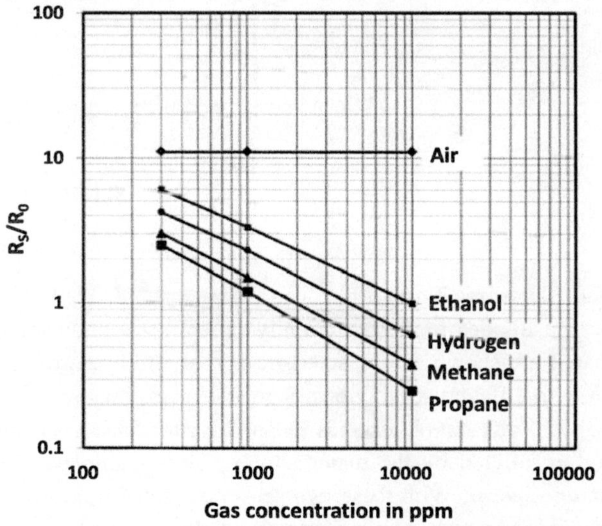

Fig. 4.61 Change of the resistance ratio (characteristic curve) for different gases depending on the concentration for a semiconductor gas sensor (Figaro-TGS 2610) for the detection of LPG (Liquid Petrol= Gas)

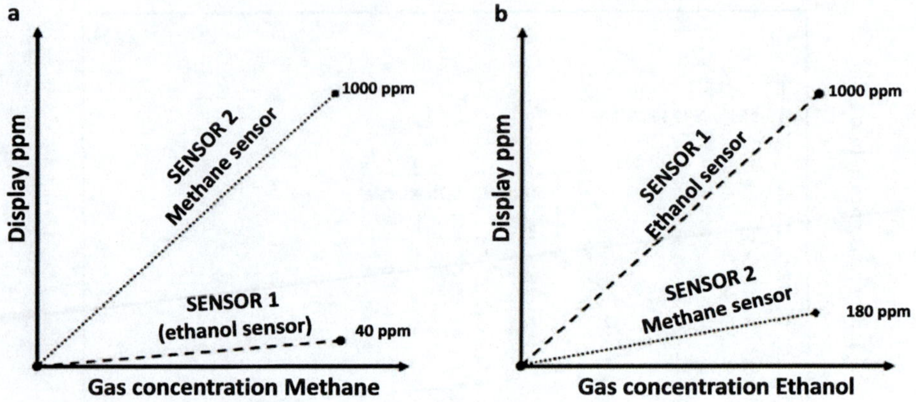

Fig. 4.62 Sensitivities of different semiconductor gas sensors to methane (**a**) and ethanol (**b**)

Fig. 4.63 Vectorial representation of the sensitivities for two gas sensors

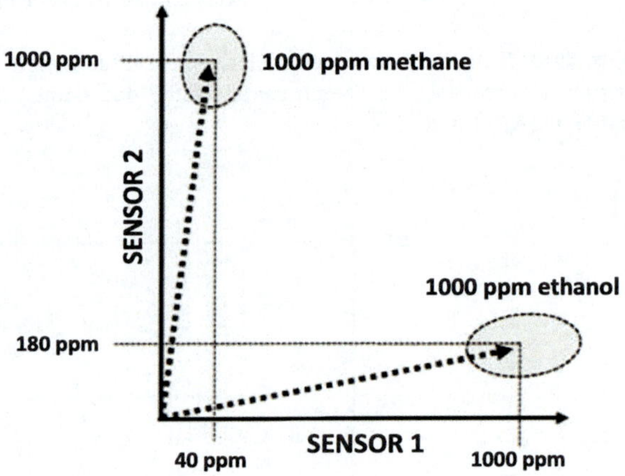

Neural Networks (Electronic Nose)

In Fig. 4.62, the sensitivities of two differently doped semiconductor gas sensors for methane and ethanol are shown as characteristic curves. Both sensors respond to these substances, albeit with different sensitivities. Sensor 1 reacts much more sensitively than sensor 2 when gassed with ethanol, whereas sensor 2 for methane has a higher sensitivity. Sensor 1 is therefore offered by the manufacturer as an *ethanol sensor*, while sensor 2 becomes a *methane sensor.* With these two sensors, a clear allocation to methane and ethanol measurement can be realized by entering the individual measurement results into a linear system of equations.

The sensitivities for a 2-component measurement can be shown very clearly in a vector diagram (Fig. 4.63). If the measurement results of sensor 1 are placed on the *x-axis,* and the results of sensor 2 on the *y-axis*, the vectors for the methane and ethanol measurement go in

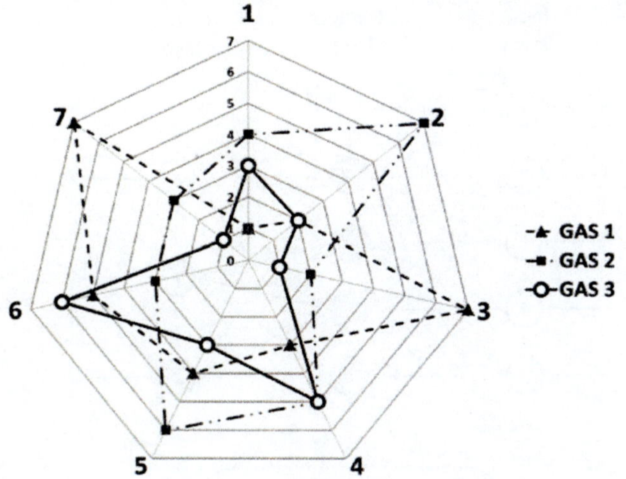

Fig. 4.64 Network representation of the sensitivities of 7 gas sensors for 3 different gases or gas mixtures

different directions. A selectivity of the measurement is therefore given. With three gases and three sensors, the third component is placed on the *z-axis* and one can also visualize this measurement. However, gas mixtures containing considerably more components cannot be visualized with this method.

If $n > 3$ components are now required, you still have the option of selecting a network display. Figure 4.64 shows the sensitivities of 7 gas sensors (sensor array), which react to the three given gases (1–3) with different sensitivities. The contours of these spanned nets are characteristic for the respective gas and show a certain pattern. This pattern can be used for the detection of gases and gas mixtures. It is important in this context to offer the sensor array the different gas mixtures in a learning phase, so that the given pattern is recognized in a later measurement.

Typical applications for this type of pattern recognition are, for example, in odour measurement (olfactometry). Since odour detection is usually carried out with the human nose, the use of gas sensor arrays is also referred to as *electronic noses*. The mathematical evaluation of these electronic noses is very complex. To simplify the process, it can be divided into three steps (Barschdorff and Becker 1990). In the input layer the sensor signals are recorded, which represent a multitude of *n individual information* (Fig. 4.65). In a *hidden layer,* these individual pieces of information are then compared with each other and subjected to weighting. This weighting and summation leads to a transfer function, which is part of the learning process described. Since this process is similar to the learning process in the human brain, it is also called a *neural network.*

In the literature, there are many examples of possible applications of neural networks, which allow a more reliable statement than with the human nose. The human nose has about 350 scent receptors, which can be compared with individual gas sensors. With these 350 scent receptors, about 5000 different odours can be detected. Figure 4.66 shows the

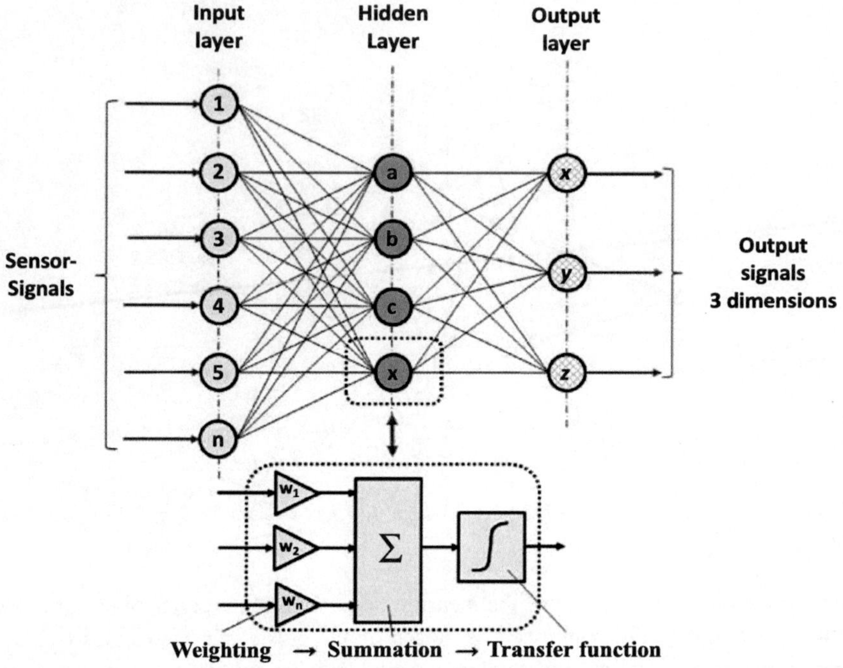

Fig. 4.65 Topology of a three-layer neural network with the computing element of the layer underneath (according to Barschdorff)

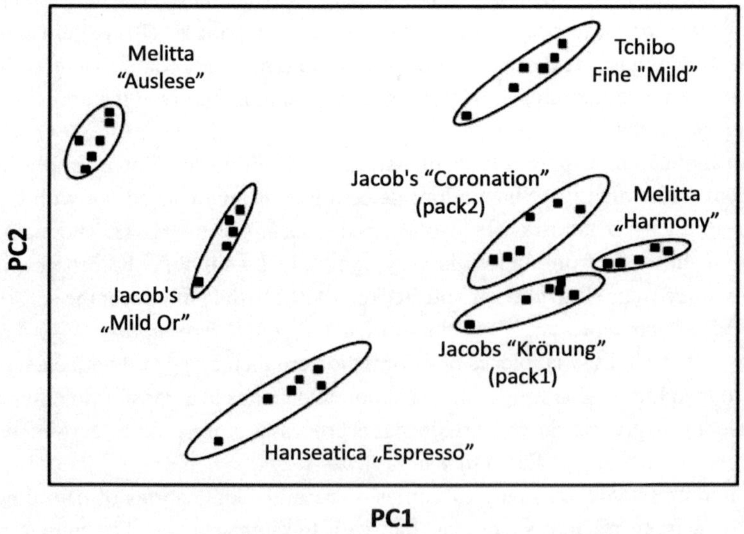

Fig. 4.66 Different pattern production of different types of coffee, to differentiate between varieties or to ensure production quality

response behaviour of an electronic nose for different types of coffee. The output signals of the neural network are laid on two axes (PC1 and PC2). Clear differences can be seen between the different types of coffee. With this method, for example, a quality inspection can be realized in the production of food.

4.4 Deposition in Thin Layers

Certain gases have the property of attaching themselves to solid surfaces by physisorption and chemisorption. This deposition (adsorption) can also lead to a solution of the gases in a thin boundary layer of the solid in a further step. These processes at the surface and/or in the boundary layer can change the electrical properties of this layer. These changes affect in particular the electrical resistance *R* and the capacitance *C*. But also other physical properties, such as the thermo force, the Hall voltage and the optical/magnetic properties, can be influenced by these surface reactions (Wedler 1970). In contrast to the semiconductor gas sensors already described, these processes also take place at room temperature.

Although these processes are not yet fully understood in detail, these effects can be used for some gas sensor applications. Figure 4.67 shows such a gas sensor. It consists of a substrate (glass, ceramic, silicon) on which electrodes are applied. Between the electrodes is the gas-sensitive layer, which is usually applied in very thin layers. In certain cases, the substrate is also heated in order to stabilize the temperature conditions or, in a regeneration phase, to force the desorption of substances from the layer.

A special feature in this context are certain humidity sensors that operate on the same principle. The water vapor present in the ambient air is deposited in a thin polymer layer, and thus changes the conductivity or capacity of this layer. This process is described in more detail in Chap. 10 and various sensor designs are also presented.

Hydrogen Sensor

Due to its chemical properties and small molecular size, hydrogen can penetrate solids and have a lasting effect on the properties of these solids. Nickel (Ni) and palladium (Pd) in

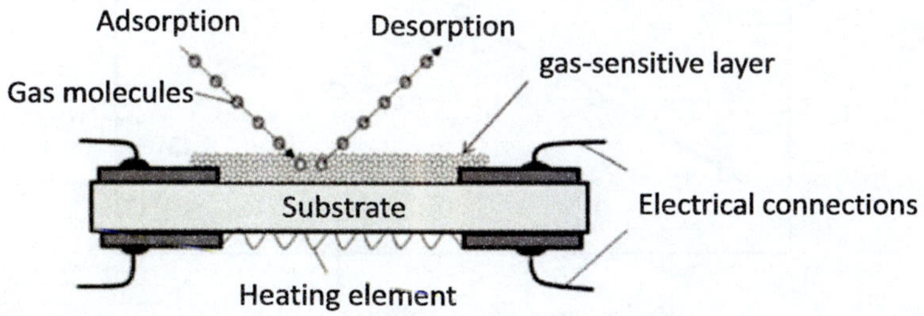

Fig. 4.67 Cross-section through a sensor setup for measuring the electrical properties of a gas-sensitive layer

particular are very well suited for use in H_2 gas sensors, and are now also used in gas analysis equipment. The advantage of these gas sensors is their low cross-sensitivity to other gases, as this property is only present in hydrogen. As these are adsorption and desorption processes, the time response of these gas sensors is very slow and is in the range of minutes. Therefore, quick indicating gas measurements are not possible with this technology (Figs. 4.68 and 4.69).

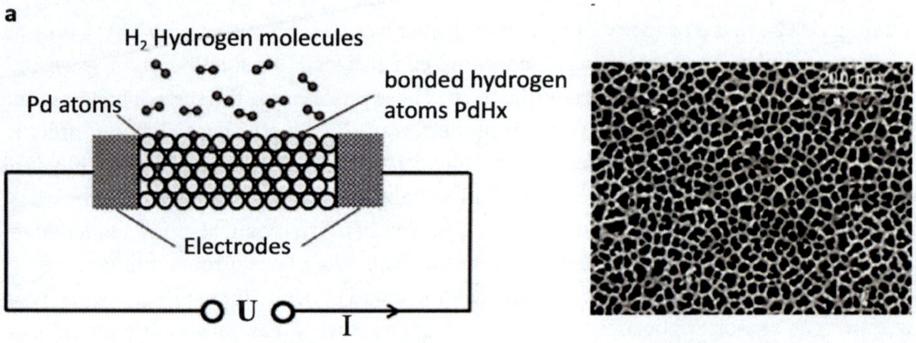

Fig. 4.68 Nanowires can absorb the H_2molecules faster than thin layers because the ratio between the mass and the surface is much more favourable. In **b** a network of 7 nm thick Pd-wires is shown. (Zeng et al. 2010)

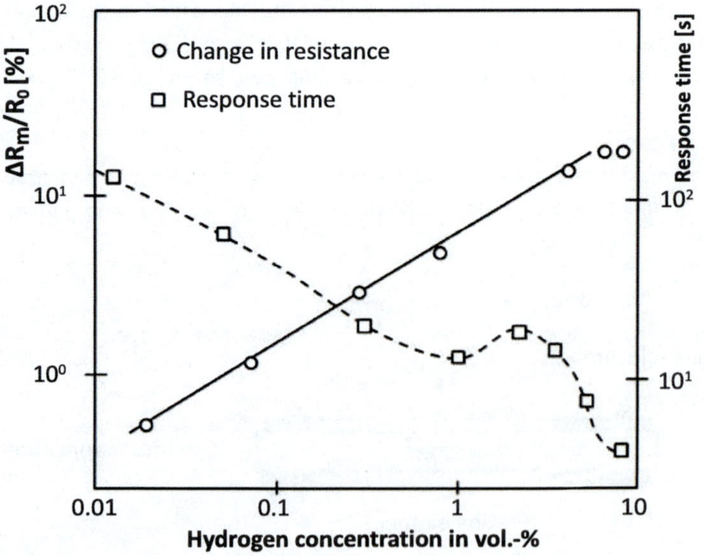

Fig. 4.69 Sensitivity and time response of an H_2-gas sensor based on nanowires with an average thickness of d = 4 nm. (Zeng et al. 2010)

In order to make the response behaviour faster, the layer thicknesses are always built up less. In the meantime, nanowires (d = 4 nm) made of palladium have been used to make the penetration layers even smaller (Zeng et al. 2010).

The time response of a structure with 4 nm wires is ≈10 seconds for hydrogen concentration in the volume % range. The sensitivity, as a percentage change in resistance, is about 20% for a gas concentration of 10 vol.% H_2.

Phthalocyanine Gas Sensor

The use of phthalocyanine as a sensor layer in gas sensor technology has been intensively investigated since the 1980s (Moseley et al. 1991). In particular, the comparatively high sensitivity of this substance to nitrogen oxides (NO and NO_2) drew attention to this substance. A distinction is made between pure phthalocyanine (H_2Pc) and metal-bound Pc. The structure of the Pc is very complex and has a molecular weight of 576 g/mol. The substance decomposes already at a temperature$T > 250^\circ$ C. Figure 4.70 shows the structural formula of Pc with bound copper (Cu) (→ CuPc). The integration of metals into the Pc structure can have a significant influence on the sensor properties. An important substance in this context is lead (→ PbPc).

The sensitivities of the different Pc variants are shown in Fig. 4.71. The lead-embedded Pc version (PbPc) shows a significantly higher sensitivity than the metal-free Pc version (H_2Pc).

The temporal response behaviour is also a problem with these gas sensors and can only be influenced by increasing the sensor temperature. The rise and fall time, for a gassing with 50 ppb NO and a sensor temperature of 170 °C, can be seen in Fig. 4.72. One can clearly see the very slow behaviour of this type of sensor. Figure 4.73 shows the dependence of the rise time and fall time for a temperature range from 120 to170° C. Higher sensor temperatures thus favour the temporal response behaviour very clearly. Unfortunately, a further increase in the temperature of $T > 250^\circ$ C can lead to damage to the Pc substance.

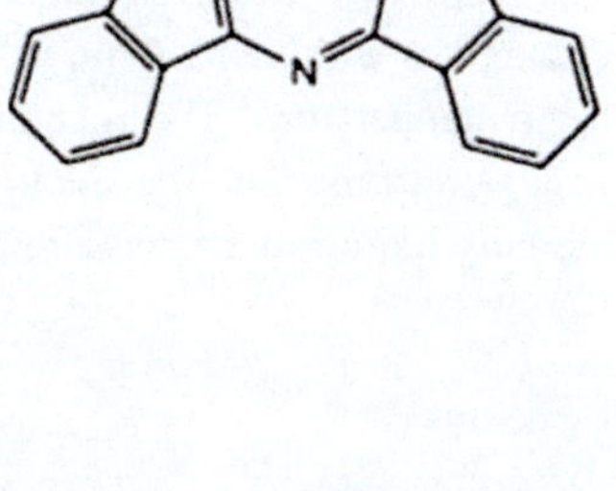

Fig. 4.70 Structural formula of copper phthalocyanine (CuPc)

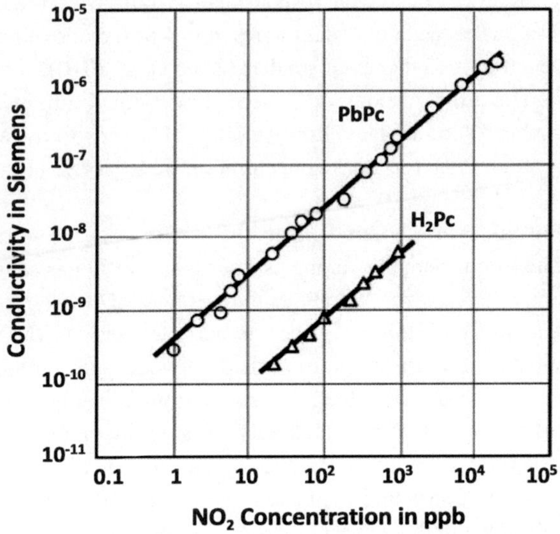

Fig. 4.71 Dependence of the conductivity ($1/R$) on the NO_2 gas concentration at a sensor temperature of 150 °C (Moseley et al. 1991)

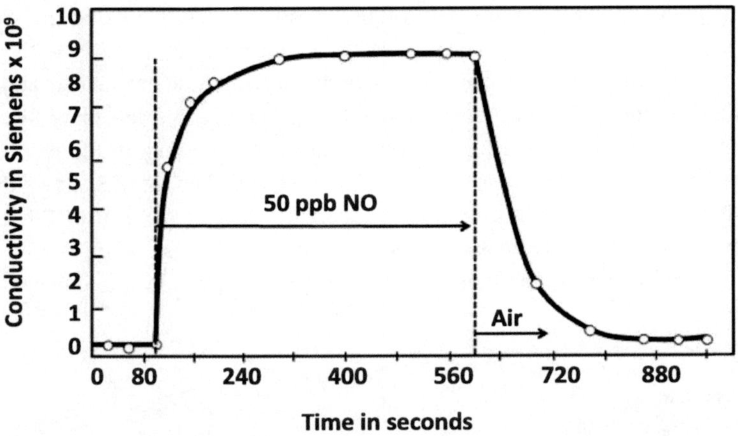

Fig. 4.72 Rise and fall time for a PbPc sensor at 170 °C for a gas jump from air at 50 ppb NO_2

Mercury Sensor

Mercury (Hg) dissolves very well in other metals and spontaneously forms an alloy (amalgam) with them. Gold in particular, reacts with mercury from the gas phase even at room temperature. The mercury vapour reaches the gold surface where it can initiate amalgam formation. The solubility of mercury in gold at 18 °C is already 0.18%. Although mercury has a vapour pressure (Fig. 4.74) of only 0.238 Pa at room temperature, this value

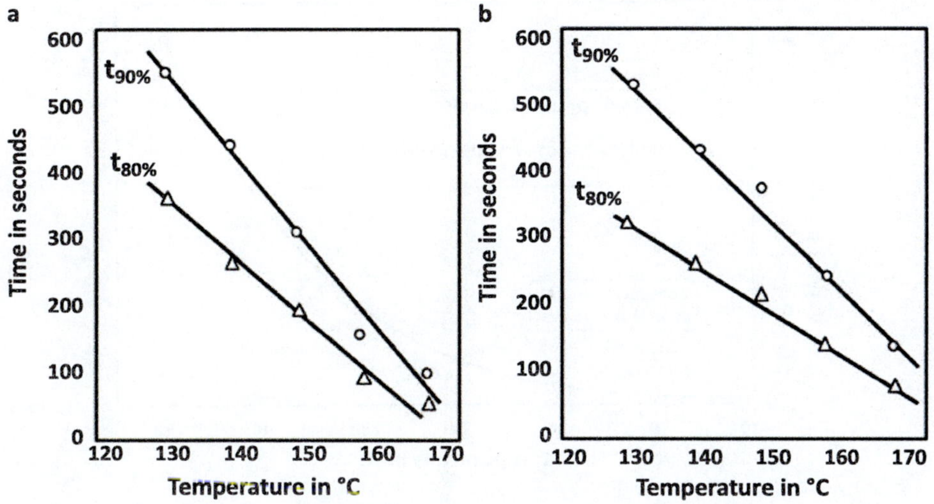

Fig. 4.73 Change in rise time (**a**) and fall time (**b**) for a gas jump from 50 ppm NO_2 at different temperatures. The times refer to the 80% or 90% value

already corresponds to a concentration of 2.4 ppm (≈20mg/m^3). The MAC[12] value for mercury in Germany was set at 0.1mg/m^3 (≈0.011 ppm). Air saturated with mercury vapour at room temperature is therefore equivalent to 25 times the MAK value.

The basic design of a miniaturized mercury gas sensor is shown in Fig. 4.75. The active sensor layer consists of gold and is applied in a thin film process. The layer thickness can vary between 10 nm and 100 nm (Schambach 2002). The resistance of the gold layer is measured as a value for the embedding of mercury in a boundary layer on the surface of the gold layer. The concentration of dissolved mercury is highest at the outermost edge, and decreases continuously with increasing layer thickness of the gold layer. Thus, a boundary layer can be defined in which the electrical resistance changes significantly, while no change can be detected in the lower part of the layer. Both *resistors* are arranged quasi parallel to each other, so that a parallel connection of the resistors is measured at the outer electrodes. Since this change in resistance is also very small at low mercury concentrations, it is useful to carry out the evaluation in a Wheatstone measuring bridge. A total of 4 of these resistors are used for this purpose, all of which are located on a micromechanical sensor chip (see Fig. 4.76). Two of these sensor resistors are provided with a passivation layer so that the mercury cannot reach the gold layer. These sensor resistors are required for the reference measurement. With this measure, temperature changes of the environment are

[12]Maximum allowable concentration.

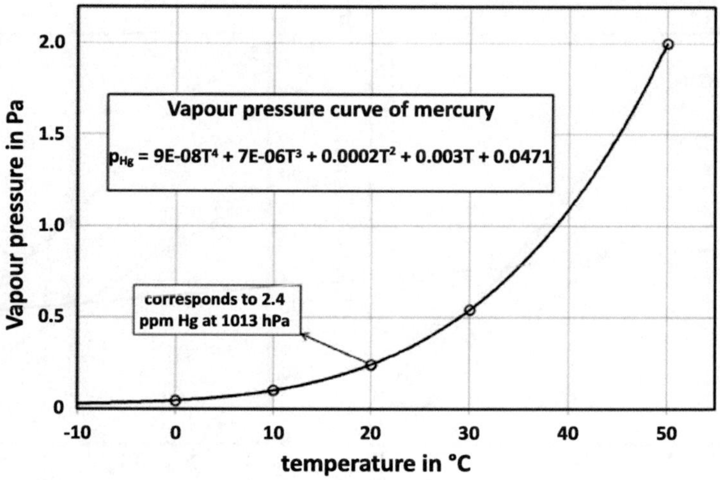

Fig. 4.74 Vapour pressure curve of mercury p_{Hg} between $T = -10^{\circ}$ C and 50° C. With the polynomial the intermediate values can be calculated

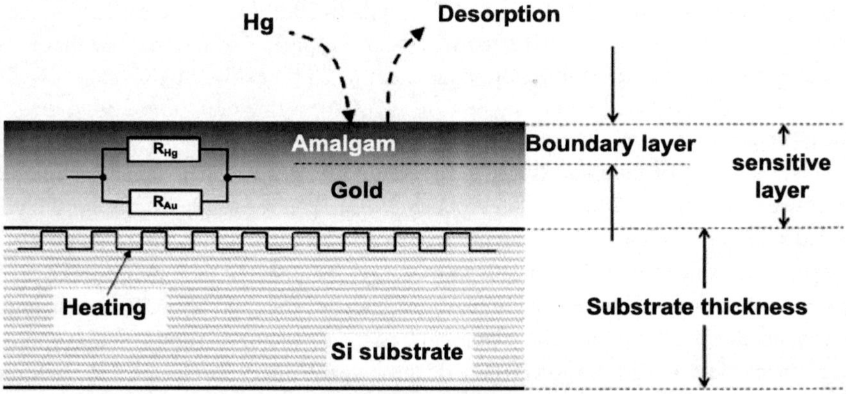

Fig. 4.75 Cross-sectional view through the active layer of the sensor structure

compensated for as far as possible. The gold resistors each have a value of $\approx$2.5k Ω, which changes by max. 75 Ω ($\approx$3%) during mercury gassing.

The outputs of the sensor resistors shown in Fig. 4.76 are connected in a measuring bridge, and the bridge voltage is evaluated with ΔU_A an instrumentation amplifier (Fig. 4.77).

The response behaviour shown in Fig. 4.78 shows a continuous increase, as the Hg concentration is permanently present during this time. If the Hg concentration had fallen to

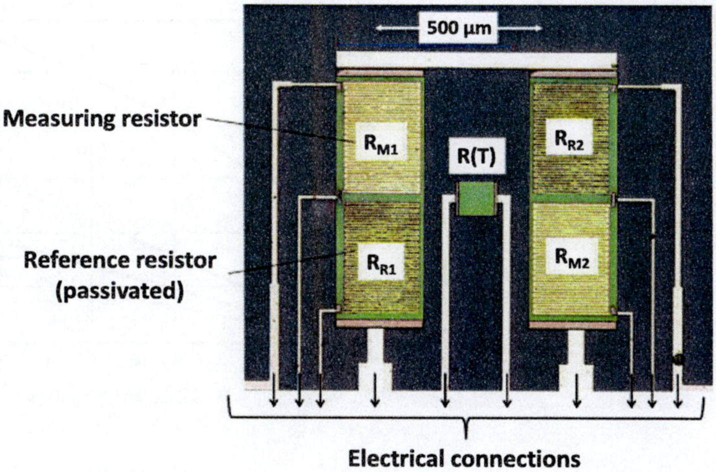

Fig. 4.76 Arrangement of the resistors on the microchip. The dimension of the active sensor structure, on a thin SiN-membrane, is ≈1 × 1 mm

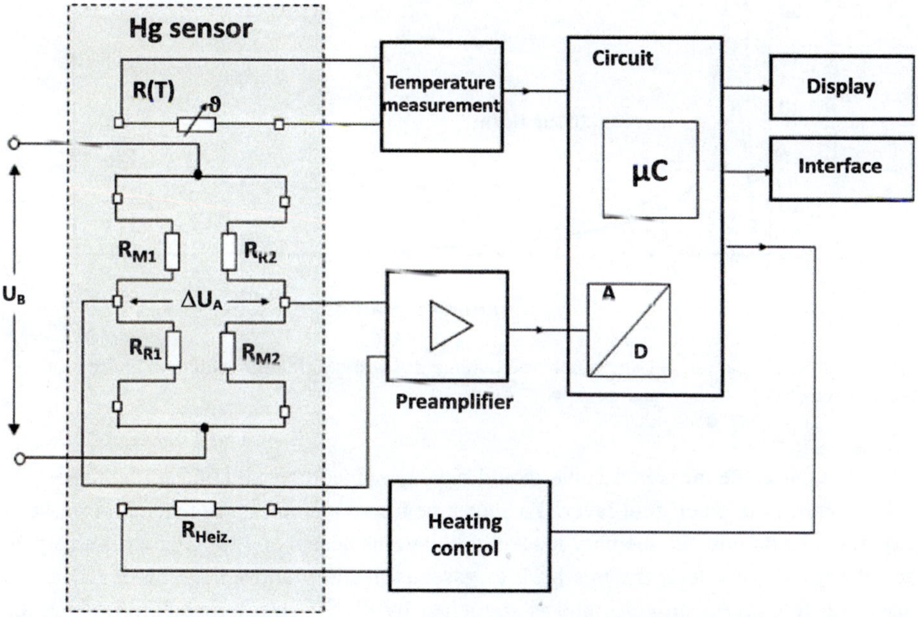

Fig. 4.77 Connection of the measuring and reference resistors in the measuring bridge. The output voltage ΔU_A is converted into the range from 0 to 300 mV with a suitable amplifier. The downstream evaluation electronics is equipped with a microcontroller and can thus take over further functions (temperature compensation linearization, heating control, ...) (Schambach 2002)

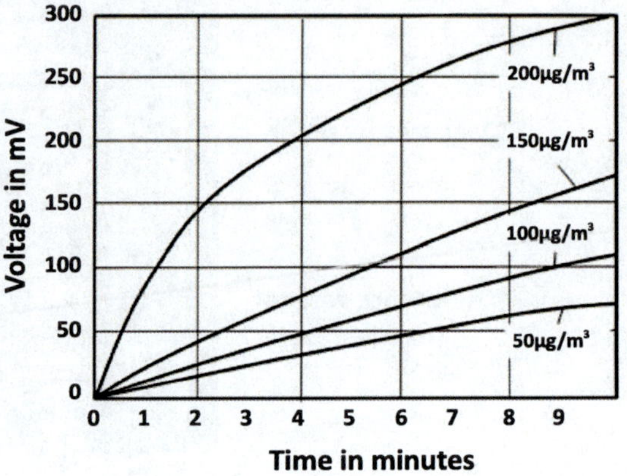

Fig. 4.78 Response of the mercury sensor to different Hg concentrations, as a function of time

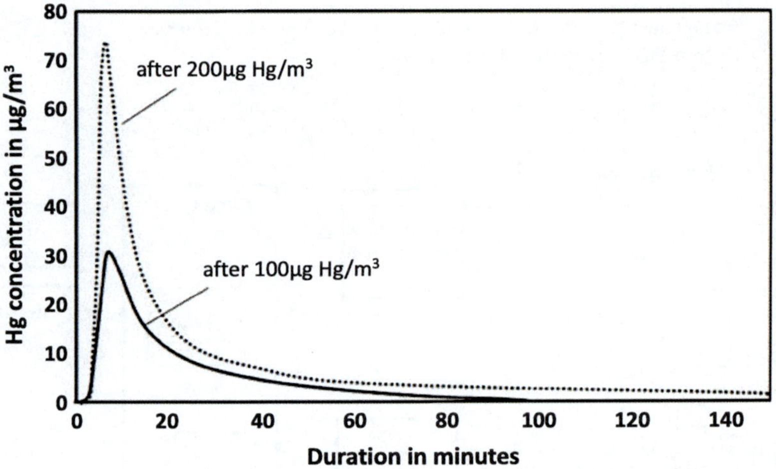

Fig. 4.79 Regeneration phases after measurements with different mercury concentrations. (Schambach 2002)

zero at a time *x*, the measured value would have remained constant, since the mercury is still present in the outer gold layer. To start a new measurement a regeneration phase is necessary. To do this, the mercury loaded gold layer is heated at 150 °C and the amalgam then decays at these temperatures back to gaseous mercury and gold. This regeneration phase can last several minutes and is shortened by higher chip temperatures, since the transformation of the amalgam depends very much on the temperature. Figure 4.79 shows this bake-out phase for different Hg loads.

4.5 Colour Reactions

Colour reactions are based on a chemical transformation that leads to a change in colour. These substances are sometimes also called indicators. The best known indicator is litmus paper. If there is a discoloration (red/blue), the presence of an acid or lye is indicated. Combined test strips, with different colour combinations, can even indicate the pH value of a liquid in gradations of pH 0, 1, 2, ..., 14. For the gas phase, these indicators exist in paper form, as a solid (granulate) or as a liquid.

Test Tubes

The oldest known method of detecting gases by defined colour reactions was described by *Lamb* and *Hoover* (Fig. 4.80). As early as 1919, the two applied for a patent, which is still widely used in this form today. In the German-speaking world, these test tubes are called *Dräger-Tubes*, as this company was the first in Germany to introduce the method. Many people have had unwanted contact with this technique in the past, as the police also used these test *tubes* to detect breath alcohol.

The test tube is filled with a substance that reacts chemically with the gas to be detected. This reaction either causes a direct colour change or is initiated by another substance. The substances are contained in a sealed glass tube, which is only opened shortly before the measurement, at the points provided for this purpose (predetermined breaking point) (Fig. 4.81). The gas to be detected is then sucked through the test tube by means of a hand pump and the corresponding chemical reactions take place (Fig. 4.82). In the front area, which has the first contact with the gas, the process is started and the gas reacts completely with the reagent. So it does not advance any further in the tube. The point at which the color change stops is therefore a measure of the gas concentration c_i in the sample. A scale is attached to the outside of the tube from which the gas concentration can be read. Since the transition is not sharp but flowing, the value between the marks must be estimated. An exact, quantitative gas analysis is thus not possible, but the displayed value at least indicates the existence of this substance and the magnitude of the concentration present. This method is still widely used for initial measurements in hazardous situations, or for checking *MAC values*. The great advantage is also that this method can be used to measure a large number of exotic gas components, for which otherwise expensive special measuring instruments are required. The *Dräger-Tube manual* lists more than 200 substances, some of which can be measured down to the ppb range (Fig. 4.83).

The *measuring range of* the test tubes is indicated on the glass bulb and always refers to a defined volume V. This volume is given by a certain number of strokes n with the gas detection pump provided for this purpose. The number of strokes n, can be between n = 1 and n = 30 depending on the substance. In some cases, the measuring range can also be halved by doubling the number of strokes. The scales for both measuring ranges can then be seen on the glass flasks. Some of the tubes are also provided with a pre-layer to filter out

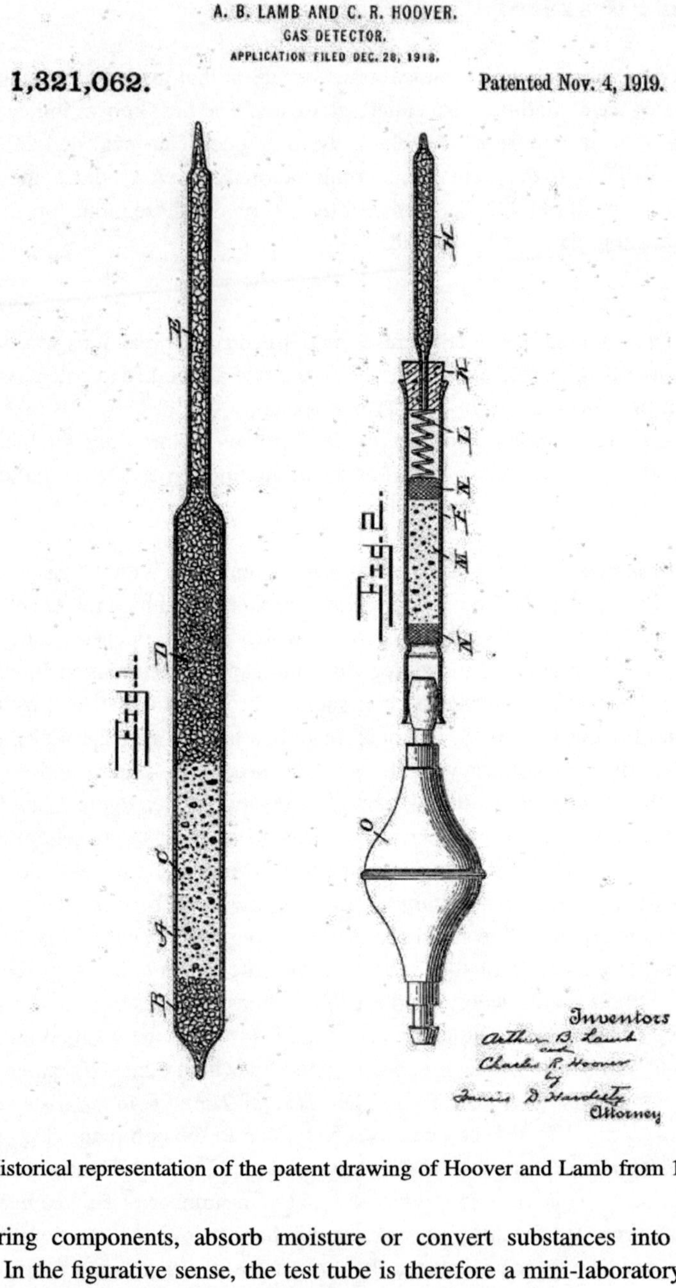

Fig. 4.80 Historical representation of the patent drawing of Hoover and Lamb from 1919

any interfering components, absorb moisture or convert substances into measurable substances. In the figurative sense, the test tube is therefore a mini-laboratory in which a wet chemical analysis is carried out independently (→ *Lab in Tube*).

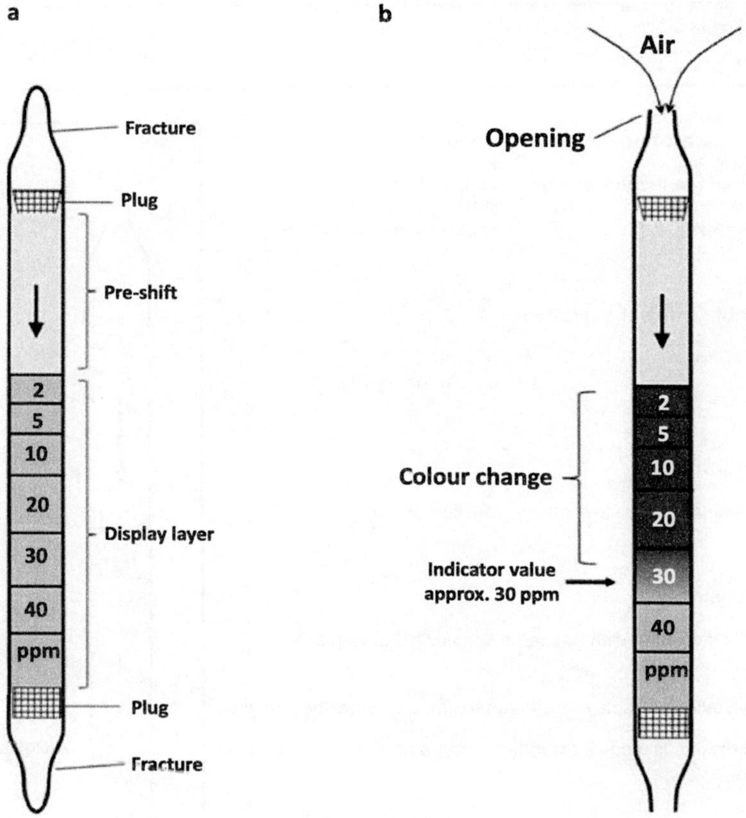

Fig. 4.81 Test tubes in closed (**a**) and open (**b**) form. When the measurement is complete, estimate the concentration value between the markers

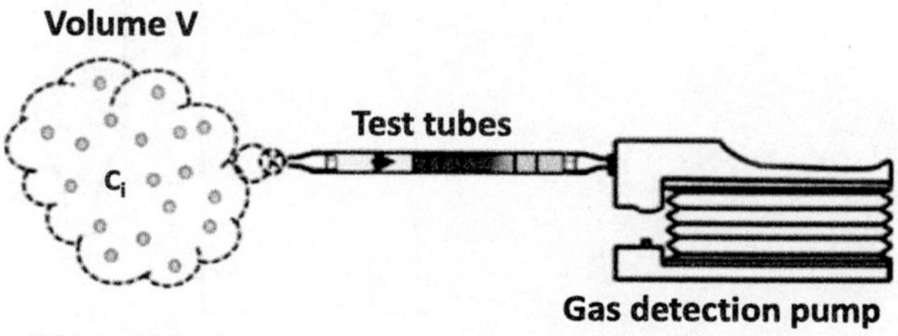

Fig. 4.82 Test tube in conjunction with the gas detector pump

Dräger-Tube

67 28411

Chlorine 0.3/b

Standard measuring range	:	0.3 to 5ppm
Stroke rate n	:	20
Duration of the measurement	:	approx. 8 min
Standard deviation	:	± 10 … 15 %
Color change	:	white —> yellow-orange

Permissible ambient conditions

Temperature	:	0 … 40°C
Humid	:	5 … 12 mg H_2O / L

Reaction principle

Cl_2 + o-Tolidine —> yellow-orange reaction product

Cross-sensitivity

Bromine is displayed with the same sensitivity but with a paler color.

Chlorine dioxide is displayed with approximately twice the sensitivity.

Nitrogen dioxide is also displayed, but with a paler color and lower sensitivity.

Measuring range extension

Measuring range 0.6 to 10 ppm at n = 10 strokes, multiply read scale value by 2.

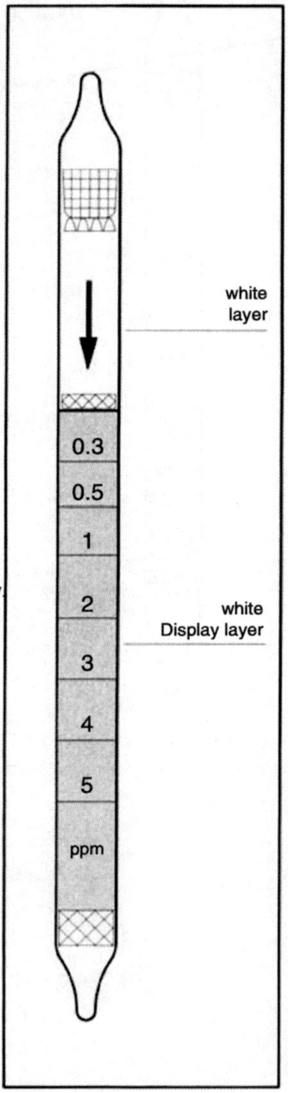

Fig. 4.83 Page from the Dräger-Test Tube Manual

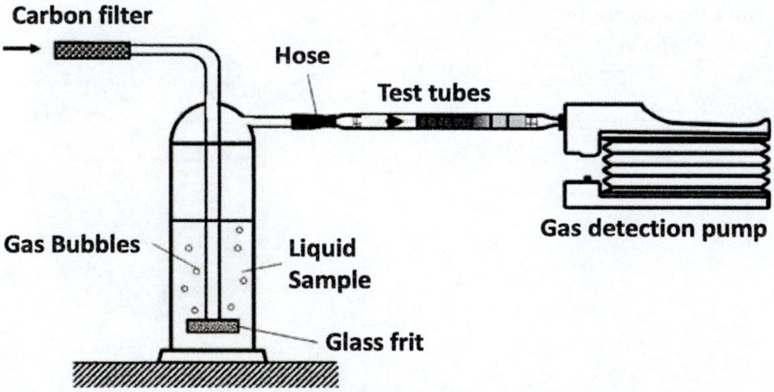

Fig. 4.84 Experimental set-up of the Dräger air extraction method for the analysis of VOCs in water samples. The carbon filter serves to filter out any VOC components from the ambient air, as these would otherwise falsify the gas measurement result

Dräger Air Extraction Process

This method can be used to detect vaporizable, organic components (VOC) in water. To do this, a water sample is filled into a gas washing bottle (see Fig. 4.84). The filling level is marked, as this certainly has an influence on the subsequent measurement. Typically, 250 ml of the water sample is filled into the bottle. The test tube then sucks in the ambient air through the gas washing bottle and small gas bubbles are formed in the bottle which bubble upwards. In these gas bubbles, the vaporizable components (VOC) of the water sample change into the gas phase. The process (stripping) can be described by *Henry's law*. The VOC concentration ($Y_{[mg/L]}$) in the water sample can be determined with the following empirical equation:

$$Y_{[mg/L]} = A \cdot B \cdot (X_{[ppm]} + C) \tag{4.46}$$

A is the calibration constant of the gas washing bottle ($A = 1.2$),$B + C$ are system constants from the measurement specification, and $X_{[ppm]}$ is the gas measured value of the corresponding test tube.

Example

The concentration of perchloroethylene in the contact water of a dry-cleaning system is to be determined. The system constants from the measurement specification are $B = 0.07$and $C = -0.1$. The gas measurement of the tube delivers a value of $Y = 0, 42$ ppm. With this information, the concentration in the liquid phase can now be calculated:

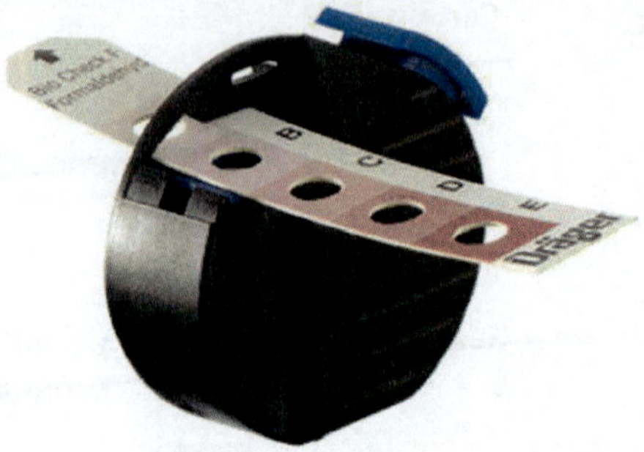

Fig. 4.85 Bio Check device for detecting formaldehyde in the ambient air. (Source: Dräger AG Lübeck)

$$Y_{[\mathrm{mg/L}]} = 1.2 \cdot 0.07 \cdot (0.42 - 0.1) = 0.03\mathrm{mg/L} \tag{4.47}$$

With a standard deviation from the measurement specification of 30%, the result is obtained:

$$Y_{[\mathrm{mg/L}]} = 0.03\mathrm{mg/L} \pm 0.01\mathrm{mg/L}$$

Indicator Badge (Bio-Check F)

This process is based on a biochemical principle, and also leads to a recognizable color change. The indicator responds only to formaldehyde. The handy device (Fig. 4.85) contains a cartridge which is opened with the push button on the side, and wets a white indicator layer by shaking it. The formaldehyde from the ambient air or room air, diffuses to the indicator layer and causes a discoloration from white to pink to red. The measuring time is 2 hours. The discolouration is compared with a colour scale which indicates 5 different exposure levels from A to E.

Colorimetric Method

This process is a special feature compared to the methods of colour reactions described so far. The measuring process can be automated by a clever device design, so that a continuous measurement is possible. Similar to litmus paper, there is a paper carrier that has been impregnated with an indicator substance. After drying, the indicator is embedded in the paper and is able to react with a certain gas. For the analysis of hydrogen sulfide (H_2S) this method has proven to be a standard procedure over many decades, as it reacts very selectively to this substance and can detect even very small concentrations in the ppb range. Lead acetate or silver nitrate is used as a colour indicator in this case. For other substances to be detected, specific indicators must be used.

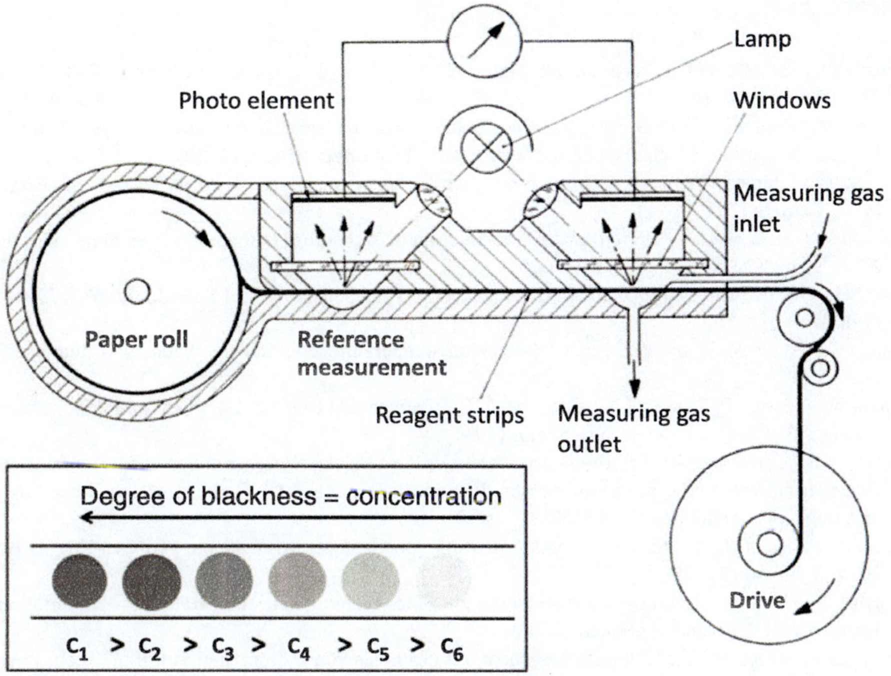

Fig. 4.86 Basic design of a colorimetric hydrogen sulfide gas analyzer

The chemical reaction that takes place in the presence of H_2S can be described as follows:

$$Pb[CH_3COO]_2 + H_2S \rightarrow PbS \downarrow + 2CH_3COOH \tag{4.48}$$

The reaction with the H_2S to be detected thus releases lead sulphide (PbS), which as a solid leaves a black colouring on the white paper. With this approach, a quasi-continuous measuring procedure can be realized. Figure 4.86 shows such a setup. The indicator paper is wound up on a roll, and is pulled through the measurement setup by a second roll containing a stepper motor (drive). The speed at which the indicator paper is transported is 1 mm/minute. In the measuring set-up there are two scanning devices with which the degree of density is measured via diffuse reflection. The reference measurement only records the basic reflection of the white indicator paper (reference measurement). The gas to be analysed is pumped through the measuring chamber. This chamber is designed in such a way that the measuring gas is pressed through the indicator paper, and thus there is sufficient contact time for a colour reaction. The degree of blackening depends on the hydrogen sulfide concentration c_{H_2S} and the exposure time t. Typical response times are 9 min (t_{90} *time*).

References

Barschdorff, D., Becker, D.: Neuronale Netze als Signal- und Musterklassifikation. Tech. Mess. **57**(11), 437–444 (1990)

Breuer, W., Deprez, J., Kitzelmann, D.: Spezielle elektrochemische Spurenmessverfahren. Messen, Steuern, Regeln in der chemischen Industrie, Bd. II. Springer, Berlin (1980)

Brunck, O.: Clemens Winklers Lehrbuch der technischen Gasanalyse. 4. Aufl. Verlag von Arthur Felix, Leipzig (1919)

Dierksheide, H.: Operating Principle and Construction of Zirconium Dioxide Oxygen Sensors of the XYA Series (2015)

Dietz, H.: Gas diffusion controlled solid electrolyte oxygen sensors. Solid State Ionics. **6**, 175–183 (1982)

Franx, C.: A dynamic oxygen sensor with zero temperature coefficients. Sensors Actuators. **7**, 263–279 (1985)

Göpel, W., Hesse, J., Zemel, J.N.: Sensors. A Comprehensive Survey. Chemical and Biochemical Sensors. Part 1, Bd. 2. VCH, Weinheim (1991)

Hammer, F.: Entwicklung eines miniaturisierten Festkörperelektrolytsensors aus der Raumfahrt zur Optimierung von Verbrennungsprozessen. Fortschritt-Berichte VDI. Reihe 8. Mess-, Steuerungs- und Regelungstechnik Nr. 935 (2001)

Hammer, F.: Sensorische Verbrennungsoptimierung von Gasfeuerungsanlagen. gwf-Gas-Erdgas Juli 2014, 376–380 (2014)

Häusler, J.: Charakterisierung von Gassensoren zur Überwachung belasteter Raumluft. Dissertation, Justus-Liebig-Universität Gießen (2004)

Heitbaum, J., Wiegleb, G.: Halbleitergassensor auf der Basis von Indiumzinnoxid. Exp. Tech. Phys. **39**(3), III (1991)

Hengstenberg, J., Sturm, B., Winkler, O.: Messen, Steuern, Regeln in der Chemischen Technik Bd. II, Betriebsmesstechnik II (Physikalische Analytik). Springer, Berlin (1980)

Jessel, W.: Gase-Dämpfe-Gasmesstechnik. Dräger Saftey AG, Lübeck (2001)

Kabaker, G.: Kfz-Lambda-Sonde. Sensor-Magazin. **44** (1988)

Karbach, A.: Messtechnik in der Versorgungstechnik. Springer, Berlin (1997)

Kiesewetter, O., Müller, J.: private Mitteilung (2015)

Moseley, P.T., Norris, J., Williams, D.E.: Techniques and Mechanisms in Gas Sensing. Adam Hilger, Bristol (1991)

Schambach, K.: Entwurf, Herstellung und Charakterisierung eines mikromechanischen Quecksilbersensors. Dissertation, TU-Dortmund, FH-Dortmund (2002)

Schaumburg, H.: Sensoren. Teubner, Stuttgart (1992)

Ullmann, H.: Keramische Gassensoren. Akademie, Berlin (1993)

Wedler, G.: Adsorption – Eine Einführung in die Physiosorption und Chemisorption. Verlag Chemie (1970)

Wiegleb, G.: Entwicklung eines Zinndioxid-Gassensors in Dickschicht-Technologie. Dissertation, Universität Witten-Herdecke (1990)

Wiegleb, G., Weber, M., Kitzelmann, D.: Industrielle Gassensorik, Kap. 2 Elektrochemische Sensoren. expert verlag, Renningen (2001)

Zeng, Q.X., et al.: Hydrogen gas sensing with networks of ultra small palladium nanowires formed on filtration membranes. Nano Lett. (2010). https://doi.org/10.1021/nl103682s

Related References

Coquillion, J.: Comptes Rendus. T. **84**, 458 (1877)

Haaland, D.M.: Internal reference solid electrolyte oxygen sensor. Anal. Chem. **49**(12), 1813–1817 (1977)

Separation Process

5

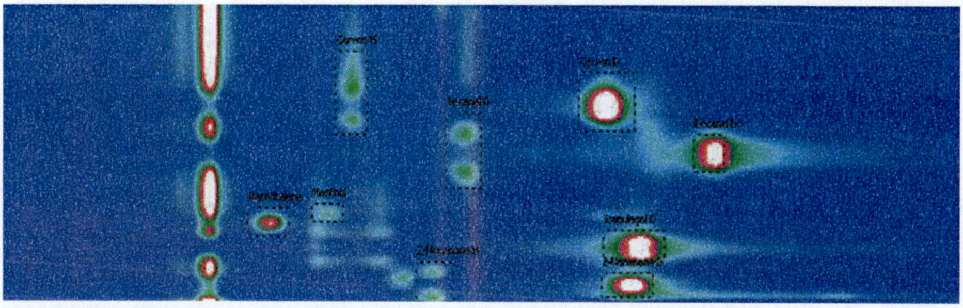

GC/IMS evaluation. Source: G.A.S. GmbH Dortmund

Given the large number of inorganic and organic compounds, it is understandable that a special, selective measuring method cannot be available for every component sought. For practical reasons, certain applications have therefore been decided on so-called sum parameters. A typical example is the measurement of the sum of all hydrocarbons (THC[1]) with an FID[2] in combustion processes (exhaust gas analysis). In addition, there are also important applications in chemical process engineering, where an exact analysis of certain substances in a complex matrix is required. This is required, for example, in the quality control of production processes to determine the purity of a substance (e.-g. pharmaceutical production process). Another important application is the composition

[1] Total Hydro Carbon.

[2] Flame ionization detector.

G. Wiegleb, *Gas Measurement Technology in Theory and Practice*,
https://doi.org/10.1007/978-3-658-37232-3_5

of natural gas deposits and the resulting calorific value determination in gas distribution and gas application (see Sects. 15.1 and 15.4).

These measurements are realized by separating the different substances from the matrix in a separation process. The separation takes place either in a temporal or spatial sequence. Non-selective detectors can therefore also be used as detection receivers, since selectivity is achieved through time or space.

5.1 Gas Chromatography (GC)

The most important and also best known separation method is the so-called *gas chromatography (GC)*. The separation process is caused by adsorption and desorption of the gases from the mixture (mobile phase) on defined surfaces (stationary phase), as this material property differs significantly for the individual gases. Such a gas chromatograph consists of a separation column in which the separation is carried out, and a detector, which has almost the same sensitivity for all gases. In order to be able to bring about a time offset of the individual substances, the entire set-up is flowed through with an inert carrier gas (helium, argon, nitrogen, air, ...). The detector signal indicates the zero point of this flow, which is also called the baseline. Then a defined gas sample is suddenly added to the carrier gas flow via a valve. In the downstream separation column, the adsorption and desorption processes now take place, which lead to a temporal separation of the substances. The components thus arrive at the detector in a defined sequence and are then detected by the detector. The time offset between feeding and detection is called retention time (t_R) and is characteristic for each substance. The entire time sequence is represented by the chromatogram and can be used for both qualitative and quantitative gas analysis. However, a constant carrier gas flow (eluent) and a constant temperature (→ isothermal chromatogram) are prerequisites for this clear temporal allocation. The separation column and the detector are therefore located in a thermostatic housing, which is designed for temperatures up to 400 °C. Figure 5.1 shows the basic design of a gas chromatograph.

Separation Process

In the separation column there is a mobile phase and a stationary phase. The carrier gas flow forms the mobile phase in which the gas or gas mixture can move freely. If the gas now comes into contact with the stationary phase, part of the substance in this phase can be bound by physical adsorption or solution (see also Sect. 2.4). The stationary phase consists of a thin tube (glass or stainless steel) which is coated on the inside with a thin liquid film (*adsorbate*) (→*capillary column*). Alternatively, the separation column can be filled with a substance (powder/granulate) (→ *packed column*) which also has a coating on the surface of the particles. The separation process can be described by *Henry's law,* and *Raoult's law* derived from it (Schomburg 1987). Figure 5.2 shows the mobile phase and the stationary phase for the equilibrium state and the dynamic state. The solubility can be described in equilibrium by the ratio of the gas concentration in the mobile phase ($c_{m,\ i}$) and the gas

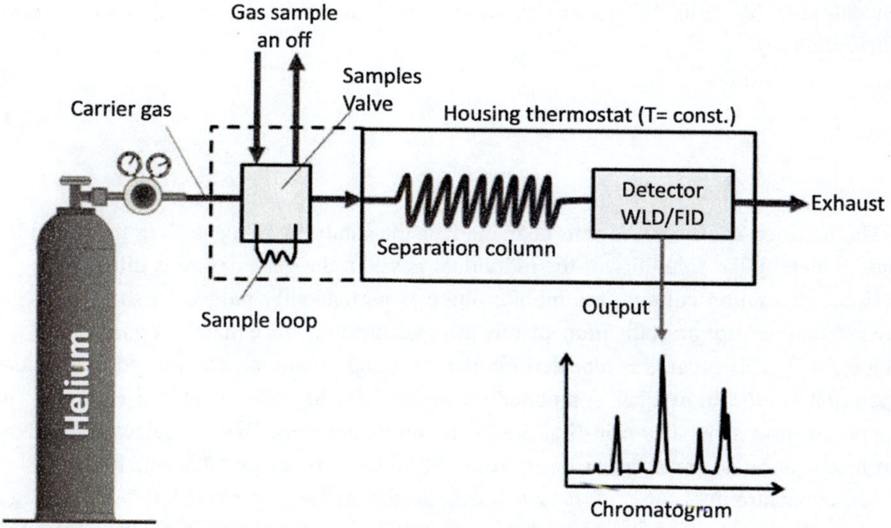

Fig. 5.1 Basic design of a gas chromatograph with the essential components (According to Clevett 1986)

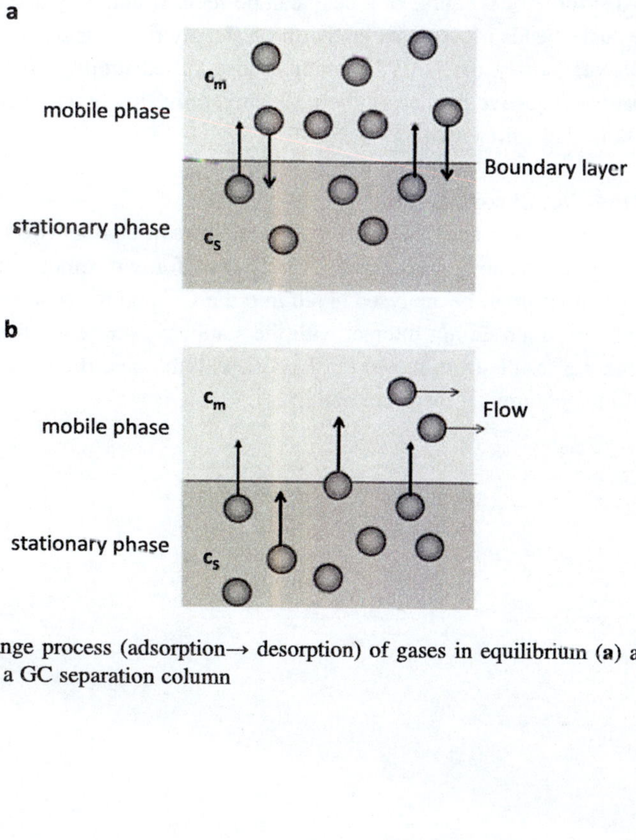

Fig. 5.2 Exchange process (adsorption→ desorption) of gases in equilibrium (**a**) and in dynamic operation (**b**) of a GC separation column

concentration ($c_{s,\ i}$) in the stationary phase. Equation 5.1 is also known as the *Nernst distribution law:*

$$K_i = \frac{c_{s,i}}{c_{m,i}} \tag{5.1}$$

The partition coefficient K_i thus characterizes the solubility of a gas (i) in the stationary phase, whereby the solubility of the individual gases in the gas mixture is different.

In the separation column, the mobile phase is permanently purged by the carrier gas flow, so that desorption starts immediately after adsorption, since there is a pure carrier gas flow $c_m \approx 0$. This process is repeated frequently in the separation column, so that a clean separation of the individual components can be seen at the end of the column. The separation process is shown in Fig. 5.3 in the repeating steps (b +c). Calculations show that a separation is already carried out from 100 of these processes (Clevett 1986).

The concentration profile of the separated substances has approximately the shape of a *Gaussian bell curve* and is called *peak*. However, the peak shape can deviate from this ideal Gaussian distribution if the relationship between $_{cm}$ and c_s becomes non-linear (→ isothermal curvature). This leads to a deformation which has a certain characteristic depending on the concentration range and material properties (polarity). The three possible variants are shown in Fig. 5.4.

In a, the ideal symmetrical course of a peak can be seen, which only occurs for a linear isotherm. If the curve bends off (convex isotherm), a steeply rising peak shape is obtained, which only falls off slowly (b). This behavior is also called *tailing.*[3] If the isothermal curvature is positive (concave isotherm), the peak rises only slowly and then falls steeply (c). This process is also called *leading.*[4]

Evaluation of the Gas Chromatogram

The total temporal peak sequence that results from a mixture (*n-components*) is called a gas chromatogram. Figure 5.5 shows such a gas chromatogram for two components (A + B). At time t = 0, the gas mixture to be analysed is fed into the GC and the time measurement is started. Components which do not interact with the stationary phase are first to reach the detector. The time t_m, results from the length L of the column, and the flow velocity u_T of the carrier gas in the column. It therefore applies:

$$t_m = \frac{L}{u_T} \tag{5.2}$$

[3]Follow English

[4]English lead

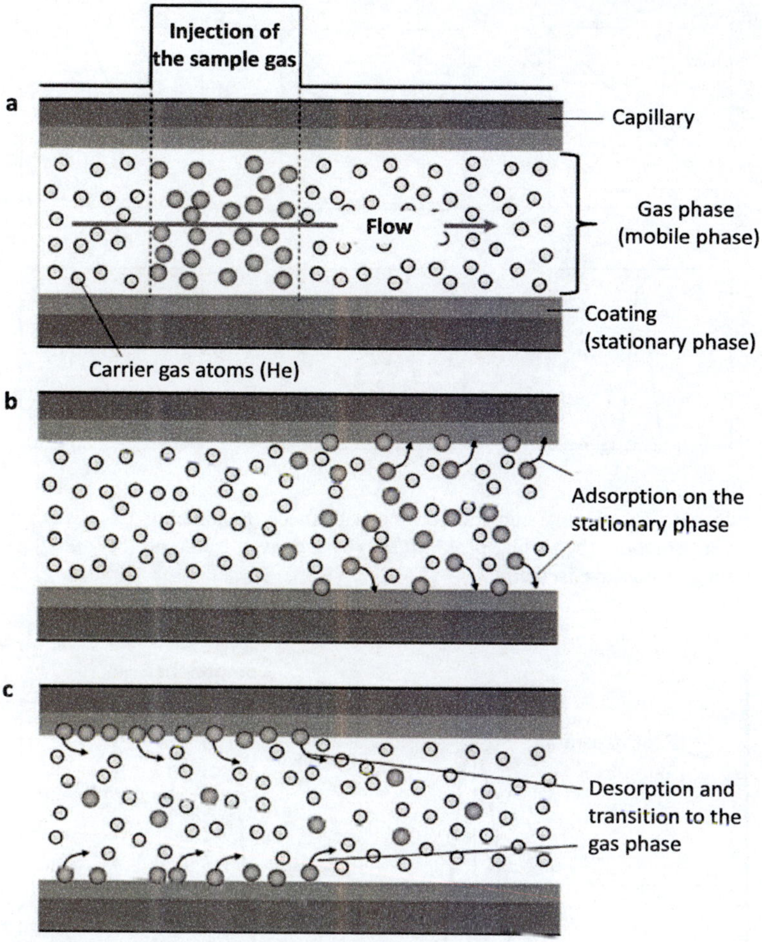

Fig. 5.3 Visualization of the separation process in a capillary column. The gas sample is suddenly added to the mobile phase and transported in the carrier gas flow (**a**). Afterwards the first adsorption in the stationary phase takes place immediately (**b**). If the concentration c_S *drops* due to the transport process, desorption takes place (**c**). The process **a** and **b** is repeated many times during the further transport in the column and then leads to a separation of the different gas components

The time t_m is also called dead time or breakthrough time of the mobile phase (Kronmüller and Zehner 1980). Component A reacts with the stationary phase on its way through the column. These reactions (adsorption and desorption) cause a time delay t_{sA}. The retention time $t_{RA}$for component A is therefore

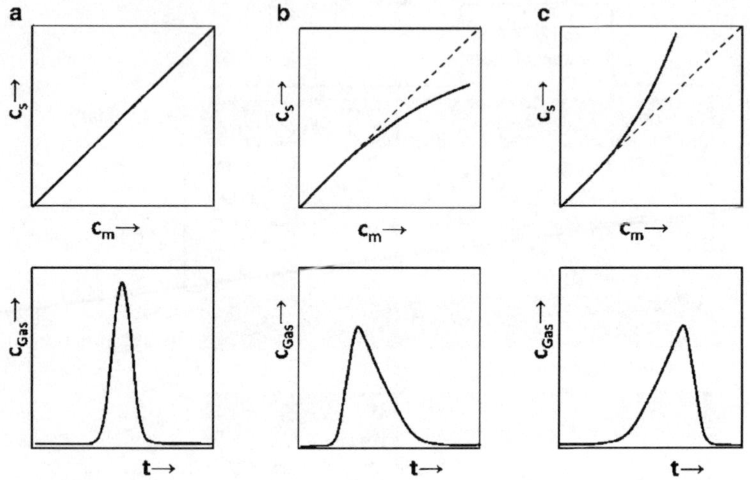

Fig. 5.4 Influence of isothermal curvature on the peak shape (Schomburg 1987). (**a**) symmetrical peak by a linear isotherm. (**b**) trailing peak (*tailing*) by a convex isotherm. (**c**) *leading* peak shape (*Leading*) through a concave isotherm

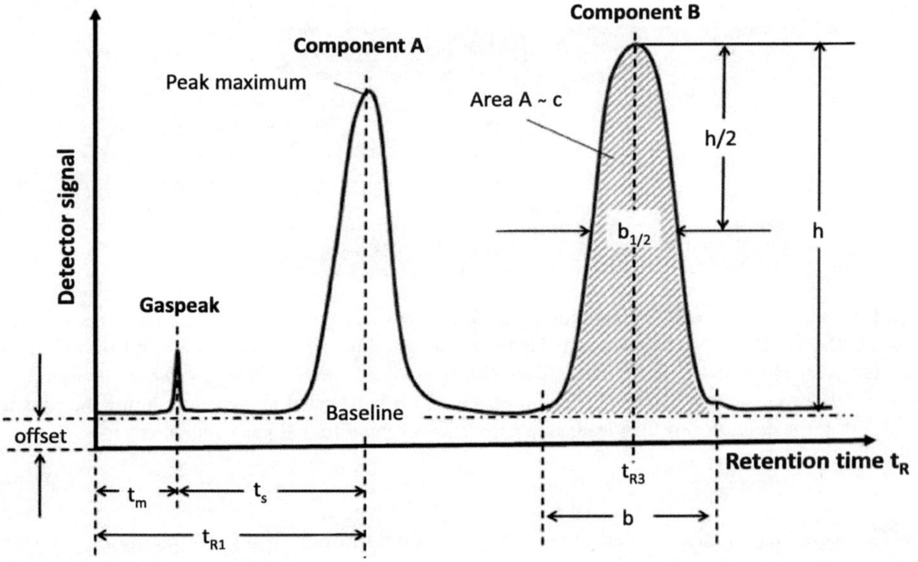

Fig. 5.5 Gas chromatogram with the different time values and the characterization of the peak form

$$t_{\mathrm{RA}} = t_m + t_{s\mathrm{A}} \tag{5.3}$$

A measure for the retention is the proportion of the gas quantity n ($n = c \cdot V$) in the mobile phase, and can be described by the delay factor Φ_i:

$$\Phi_i = \frac{c_{m,i} \cdot V_{m,}i}{c_{m,i} \cdot V_{m,i} + c_{s,i} \cdot V_{s,i}} = \frac{1}{1 + K_i \cdot \frac{V_{s,i}}{V_{m,i}}} \tag{5.4}$$

The reciprocal value $\beta = V_{m,\ i}/V_{s,\ i}$ is called the *phase ratio*. The distribution state is a dynamic equilibrium in which the gas is in constant exchange between the two phases.Φ_i, therefore, also indicates the proportion of time in the mobile phase. However, since gas transport only takes place in the mobile phase, the average migration velocity u_i is delayed by the separation column by a factor Φ_i relative to the carrier gas velocity u_T. It therefore applies:

$$u_i = u_T \cdot \Phi_i \tag{5.5}$$

With Eq. 5.2 one then obtains:

$$u_T = \frac{L}{t_m} \quad \text{and} \quad u_i = \frac{L}{t_{R,i}} \tag{5.6}$$

Set the Eqs. 5.4 and 5.6 in Eq. 5.5, then the following relationship is obtained:

$$\frac{L}{t_{R,i}} = \frac{L}{t_m} \cdot \frac{1}{1 + K_i/\beta} \quad \text{or} \quad t_{R,i} = t_m(1 + K_i/\beta) \tag{5.7}$$

The effectiveness (efficiency) of a chromatographic separation column is described by the so-called separation step number N. For a symmetrical peak (Gaussian shape), N can be calculated from the half-width $b_{1/2}$ or base width b (Down and Lehr 2004). It is valid:

$$N = 5,54 \cdot \left(\frac{t_R}{b_{1/2}}\right)^2 = 16 \cdot \left(\frac{t_R}{b}\right)^2 \tag{5.8}$$

Example

The retention time of a substance is 15 minutes, with a half-value width of 24 seconds. This results in the following number of separation stages:

$$N = 5,54 \cdot \left(\frac{15 \cdot 60}{24}\right)^2 = 7790 \tag{5.9}$$

The separation stage number N can also be represented by the length L of the separation column and the height of the separation stage (HETP[5]):

$$N = \frac{L}{\text{HETP}} \tag{5.10}$$

The HETP value is explained by the length of a column section within which the equilibrium between the mean concentration $c_{m,\ i}$of a component (i) in the mobile phase and its mean concentration $c_{s,\ i}$ in the stationary phase has been established.

The height of the separation stage is further dependent on the carrier gas velocity $u_{T,}$ and can be described by the *van Deemter equation*

$$\text{HETP} = A + \frac{B}{u_T} + C \cdot u_T \tag{5.11}$$

From Eq. 5.11, the optimum flow velocity u_{opt} can be determined at which the separation efficiency (1/HETP) is highest. Figure 5.6 shows this relationship graphically. Size A describes the turbulent diffusion *(eddy diffusion)* that occurs in packed columns on particles of the carrier material. A is therefore also dependent on the particle size d ($A \approx d$). In a capillary column, A = 0. The second term (B/u), stands for the diffusion in longitudinal direction of the column. This factor is therefore dependent on the temperature of the mobile phase. The product $C \cdot u$ characterizes the mass transfer between the stationary and the mobile phase. The constant C is therefore dependent on the layer thickness and porosity of the stationary phase. Furthermore, C is influenced by the column radius, column length and the diffusion coefficient of the stationary phase.

The *resolution R* of two peaks in a gas chromatogram is determined by the peak width b and the difference in retention times Δt. The narrower the peaks are, the easier it is to distinguish between them. Both, the peak width b and the retention times t_{Ri} are influenced by the column length L and the column temperature $T_{S.}$ The resolution can therefore be optimized with these parameters. A clear separation is obtained when the distance (Δt= retention difference) corresponds to the mean value of the two peak widths. The following relationship then applies for the resolution R:

[5]Height Equivalent to a Theoretical Plate.

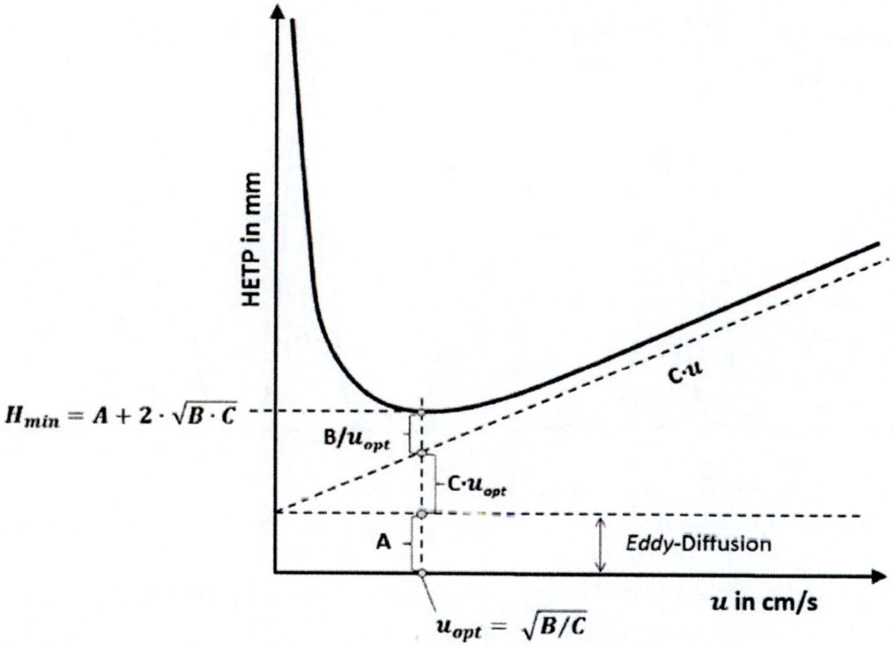

Fig. 5.6 Graphical representation of the *van Deemter equation* (after Clevett 1986)

$$R = \frac{(t_{RA} - t_{RB})}{0,5 \cdot (b_A + b_B)} \tag{5.12}$$

For a perfect separation of the two components (*A*, *B*), R = 1. Figure 5.7 shows the peak separation for two different situations. In the first case, the retention difference Δt_1 is not sufficient to achieve a complete separation of the two peaks. Although qualitative analysis is possible with this resolution (R < 1), the quantitative calculation can only be performed by appropriate approximations (dashed lines). In the second case, the retention difference Δt_2 corresponds exactly to the mean peak width, so that R = 1.

The quantitative analysis is carried out by calculating the area (integration) below the peak curve (see Fig. 5.5). The detector signal *D*(*t*) is integrated over the peak width *b*. The amount of substance (gas concentration *c*) is then calculated with a proportionality factor f_i:

$$c = f_i \cdot \int_0^b D(t) \cdot dt \tag{5.13}$$

The proportionality factor f_i must be determined experimentally for each individual component (*i*) by means of calibration measurements with known test gas concentrations

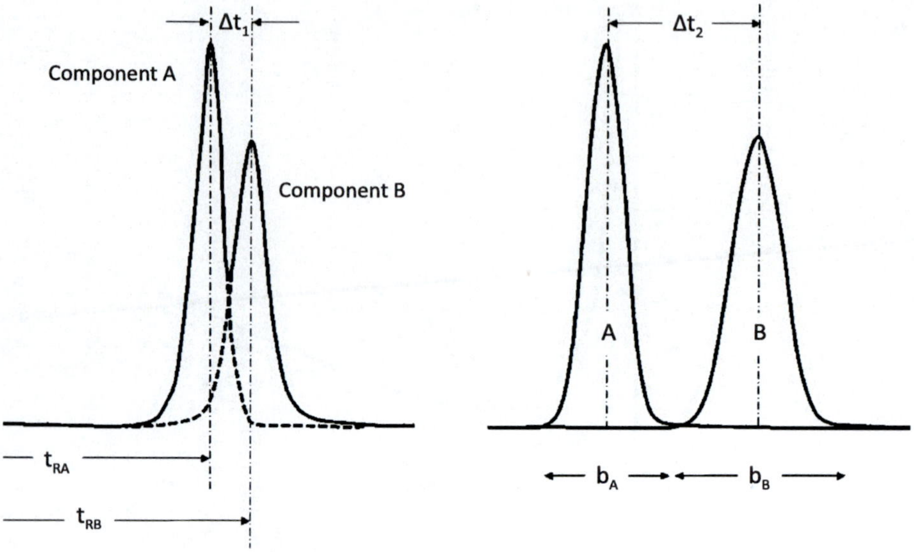

Fig. 5.7 Display of the resolution R of two adjacent peaks within a gas chromatogram

(c). This factor f_i also depends on the type of detector used (TCD, FID, PID, ...). Figure 5.8 shows various concentration peaks ($c_1 \ldots c_5$) from a chromatogram, which are converted into a concentration value c_n according to the area calculation according to Eq. 5.13. There is then a linear relationship between the given concentration and the calculated concentration. Figure 5.9 shows a complete gas chromatogram of natural gas (H-gas). The correlation between the retention time and the molar mass of the individual alkanes in the gas mixture is clearly visible.

Column Types

A basic distinction is made between packed columns and capillary columns (Fig. 5.10). Packed columns have an inner diameter in the mm range and are filled with a carrier material (e.g. diatomaceous earth). The material for the stationary phase is located on this carrier material. Today, packed columns are only used in a few cases because the separation properties of capillary columns are better. This column type has an inner diameter in the µm-range, and consists of a thin-walled stainless steel tube or a glass tube. Often high purity quartz tubes are also used because this material is much more temperature resistant. As the column length can be up to 100 m, the capillary columns are wound up.

The inner coating of the capillary can be made with different substances that form a liquid film a few µ meters thick. Depending on the application, long-chain hydrocarbon compounds such as *Apolan-87* (→ 87 C-atoms), *Carbowax* (e.g. Polyethylene glycol) or *siloxanes* (e.g. Polydimethylsiloxane) are used. The most important thing is that these

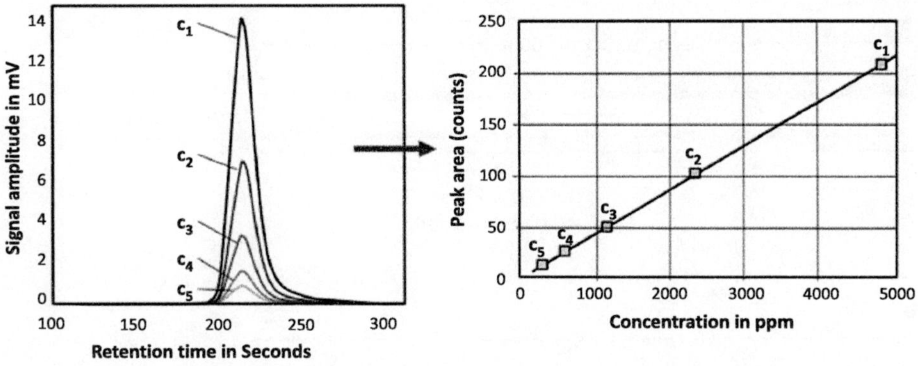

Fig. 5.8 Relationship between the peak height or peak area and the concentration c_n of the component in a gas mixture

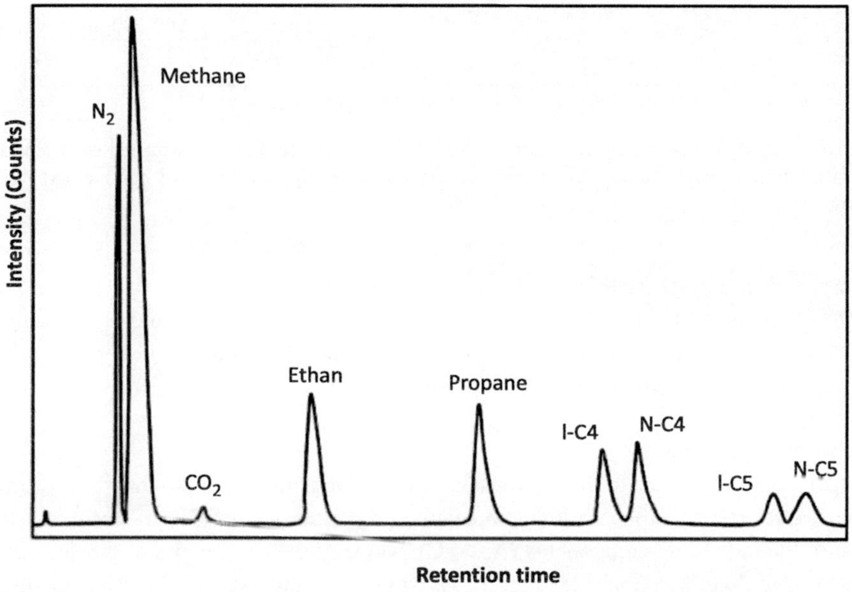

Fig. 5.9 Gas chromatogram of a specific natural gas quality (H-gas). The retention time is lowest for methane and increases with increasing molar mass of the individual alkanes (C_nH_{2n+2})

substances have a low volatility and do not evaporate due to the temperature in the oven. Nevertheless, during operation, especially at higher temperatures, the columns lose part of the stationary phase by *bleeding*. Furthermore, the absorption properties of the coating are very important for the substances to be separated. For this evaluation, the polarity of the coating is of interest. Polar components such as alcohols and water dissolve better in polar

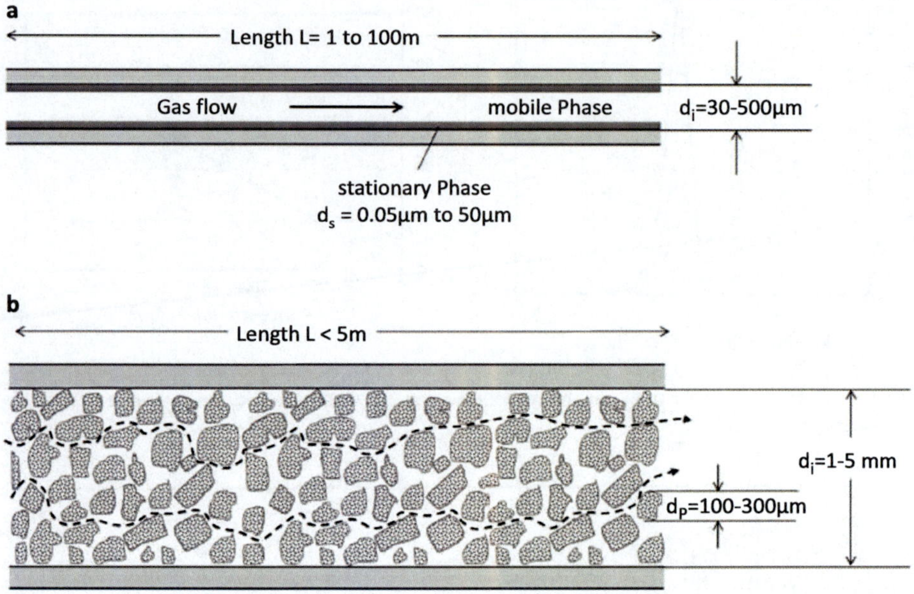

Fig. 5.10 Cross-sectional representation of the two possible column types with the corresponding size specifications (approximate values that can vary considerably). (**a**) Capillary column, (**b**) Packed column

coatings, while non-polar substances such as hexane or benzene dissolve in non-polar coatings (Down and Lehr 2004). Gas chromatographs, in which both groups of substances have to be analysed, therefore have several different separation columns integrated and can be switched on via a valve control, as required.

Temperature Influence

Normally, the separation column is adjusted to a constant temperature so that the separation properties are reproducible. At low temperatures, the isothermal GC provides good separation of the individual components (A, ..., C), whereby the peak width of the substances (C) that later elute, increases (see Fig. 5.11). At higher temperatures, this leads to significantly smaller peak widths in isothermal GC, but the peaks are now very close to each other and can no longer be separated ($R < 1$). Both requirements can be met with temperature programming. In this case, the temperature starts at a low value (like a) and increases to the higher value (like b) over time. This improves the separation and still gives a smaller peak width for all components.

Sample Injection

In the laboratory, the sample is injected, for example, by a syringe. The sample volume is usually only $\approx 10\ \mu L$ and is dosed by hand. In process measuring technology, the dosing is

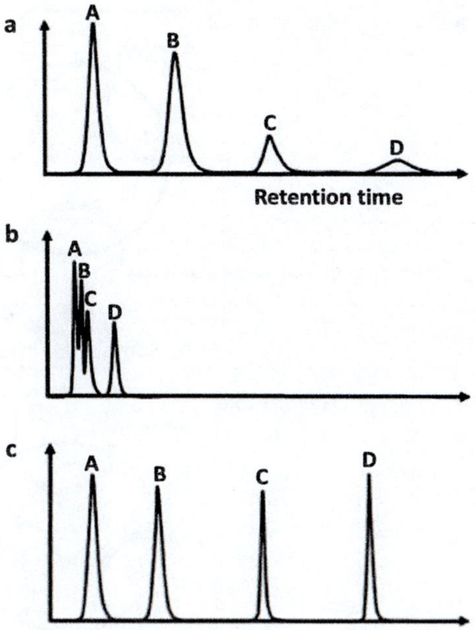

Fig. 5.11 Influence of column temperature on the gas chromatogram and combination with a temperature program (ramp). (**a**) Isothermal GC at low temperature, (**b**) Isothermal GC at high temperature, (**c**) GC with a temperature program

done automatically with special sample valves. In Fig. 5.12, such a valve is shown. It works according to the rotation principle and has 6 inputs, which are arranged 60° offset to each other. During sample loading (a), the sample volume is filled with the gas mixture to be analyzed. During this time period the separation column is exclusively flowed with the carrier gas. The sample volume is abruptly switched into the carrier gas flow by a 60° rotation, and transported into the separation column with the carrier gas (b). The valve remains in this position during the entire retention time and is only moved back to position (a) at the end. This procedure is repeated cyclically, so that a new updated gas analysis is available with each pass. With a typical cycle time of 1 minute, ≈0.5 millions of switching operations must be performed per year. This places very high demands on the valve technology.

GC Detectors

The detection of the individual peaks in a GC is done with appropriate detectors, which are selected according to the application. Figure 5.13 shows the time response for a fast detector, which allows the ideal mapping of the *Gaussian function* of a GC peak. In the real case, however, the peak is more or less deformed by the temporal detector response. The faster the detector responds, the less the peak deformation. Another quality feature is selectivity. The TCD is known as a broadband detector, and thus reacts equally to almost all components (1, 2, 3, 4). Selective detectors, on the other hand, can only detect certain species (2 + 3) and blank out the other peaks (1 +2) in the gas chromatogram which are not

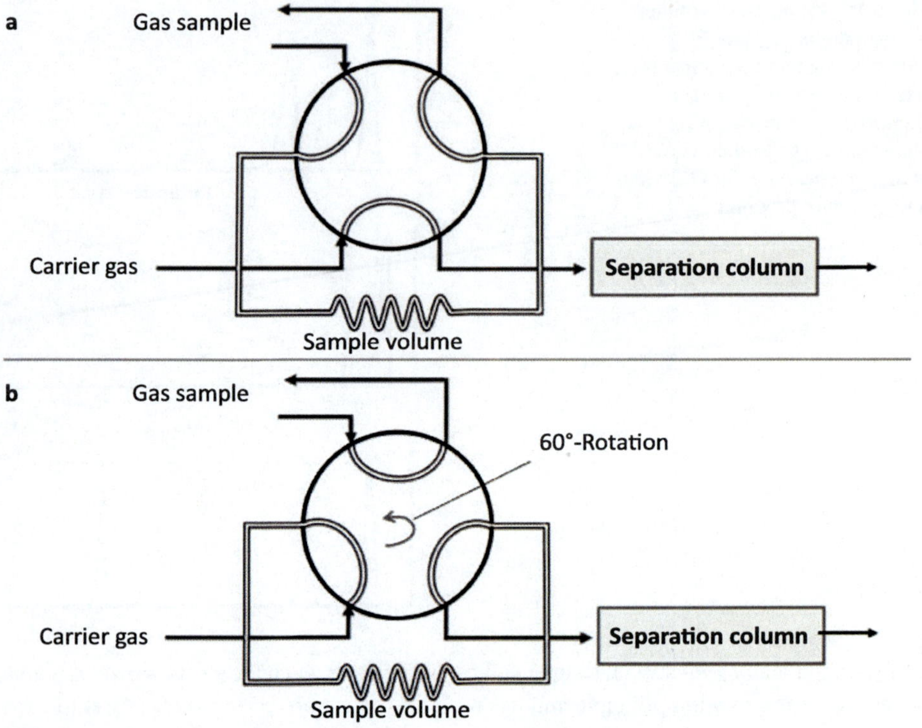

Fig. 5.12 Rotary valve with 6 connections for the sequential injection of a gas sample into a gas chromatograph (according to Clevett 1986). (**a**) Sample loading, (**b**) Injection

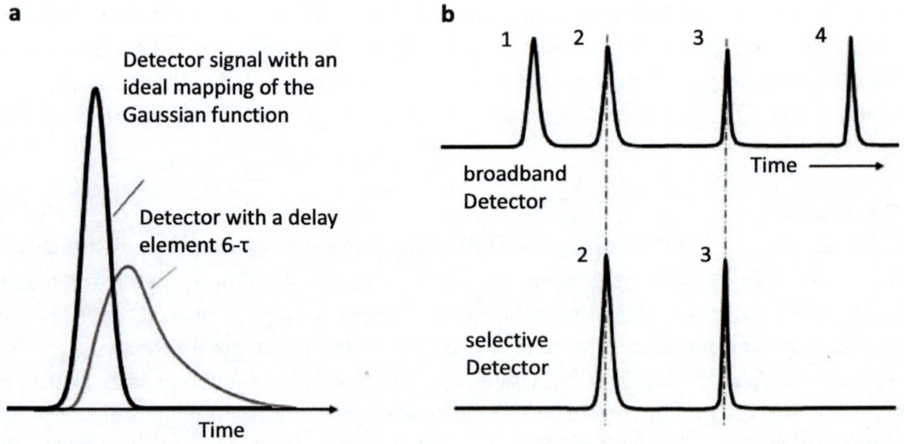

Fig. 5.13 Temporal response (**a**) and gas-specific response (**b**) of different detector types (broadband and selective) for use in gas chromatography (after Jentzsch and Otto 1970)

detected. The advantage of selective detectors is their higher sensitivity, which leads to better detection sensitivity.

TCD

Up to the 1960s, only thermal conductivity detectors (TCD) were used in GC technology (see Sect. 3.1). A great advantage of these detectors is their universal applicability, since only the difference between the carrier gas helium and the measuring component is detected. Since helium has an extremely high thermal conductivity, which is only exceeded by hydrogen, there is always a more or less large signal to detect the gaseous substances. However, a fast response time is important for use in GC. Since the TCD has a high flow dependence, the measuring cells are therefore specially dimensioned for GC applications in order to achieve the fastest possible response behaviour (t_{90} *time*) with a simultaneous low flow dependence (Jentzsch and Otto 1970).

MSD

The use of mass sensitive detectors (gas density balance) for gas chromatography was first described by Martin in 1956, and transferred to a commercial instrument by Nehrheim in 1963 (see Sect. 3.2). The method requires a reference gas (e.g. helium, nitrogen or sulphur hexafluoride). The detection limit is $1,4 \cdot 10^{-2}$ ppm for difluoro dichloromethane in helium (Jentzsch and Otto 1970). However, the areas of application are limited, and are usually only used for comparative measurements (calibration). The instrument is still manufactured today in almost unchanged form by *Gow Mac Inc.* in the USA.

FID

Flame ionisation detectors (FID) have been used since the 1960s for the sensitive detection of hydrocarbons (see Sect. 3.4). The advantages of this detection method are its selectivity for ionizable substances and its high resolution (low detection limit). However, since the burner has to be operated with hydrogen as fuel gas, the equipment required is naturally greater than for TCD.

ECD[6]

This detector type is characterized by a high sensitivity for compounds with electronegative groups (e.g. halogen compounds). A small amount of radioactive ^{63}Ni is present in the measuring cell in the form of a coating (Fig. 5.14). This radioactive layer emits β radiation. The carrier gas (argon with ≈10vol. % methane) is ionized by this radiation and a charge cloud of slow electrons is formed, which are absorbed by a collecting electrode. This generates a constant zero current I_{Mess} (nA range), which is processed in a downstream amplifier. If there are now other substances from the separation column in the carrier gas,

[6]Electron Capture Detector.

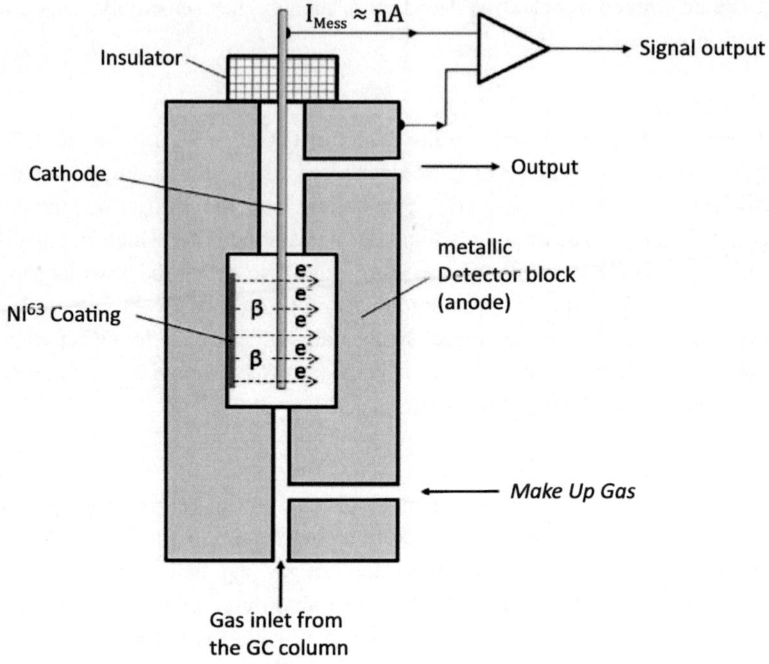

Fig. 5.14 Schematic structure of an ECD

the electrons are captured by these molecules and the zero current is reduced. This current reduction is then proportional to the gas concentration in the respective GC peak.

NPD[7]

This type of detector has similarities to an FID and is used for selective measurement of nitrogen and phosphorus compounds. Here, too, an oxyhydrogen gas flame[8] is used for ionization (Fig. 5.15). In addition, an alkaline bead (rubidium or cesium silicate) is located above the flame, which heats this bead to red heat. However, the temperature is lower than in the FID, so that the hydrocarbons are not ionized. If there are P-containing substances in the ECD, electrons are released via complex processes, which are then captured by the collecting electrode. The amplified ion current is then proportional to the concentration of the P compounds in the GC peak.

In a second operating mode, the hydrogen supply and the air supply are reduced so that the flame is extinguished. The hydrogen now burns, without flame, at the still hot alkali

[7]Nitrogen Phosphorus Detector.

[8]Oxyhydrogen gas = hydrogen/oxygen mixture.

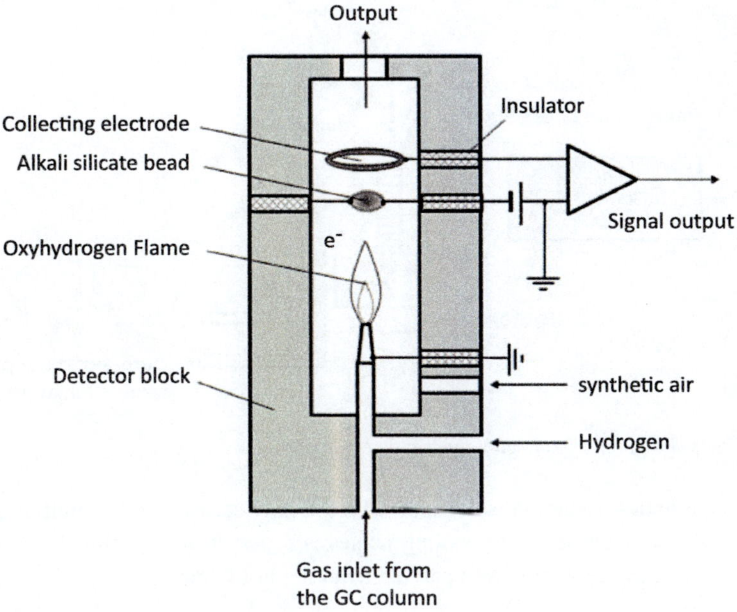

Fig. 5.15 Schematic structure of an NPD

silicate bead and a plasma is formed around the bead at 600–800 °C. The processes in the plasma around the bead also lead to an ion current in this case. In this phase, all N and P compounds are measured simultaneously as a sum.

PID

The photoionization detector ionizes the different hydrocarbons (HC) with the help of a short-wave UV radiation source ($\lambda < 200$ nm). The wavelength λ determines the ionization potential for the different groups of substances (see Sect. 3.4). By selecting the radiation source, a selective measurement between different groups of substances can be made. Aromatic hydrocarbons in particular can be detected very well with the PID (Driscoll et al. 1978).

FPD

Flame photometric measurements are mainly used in gas chromatography for the detection of sulphur (S) and phosphorus (P), in a wide variety of compounds. These compounds are burned and atomized in a hot hydrogen flame. The sulfur (S) then emits radiation at a wavelength of $\lambda_S = 394$ nm, while the phosphorus (P) has a strong radiation emission at $\lambda_P = 526$ nm. With appropriate inference filters (IF), the selective wavelength λ_S can then be filtered out, while the interfering wavelength ranges $\{\lambda_x - \lambda_y\}$ are blocked. This measure achieves a high selectivity to these two components. Figure 5.16 shows the basic structure

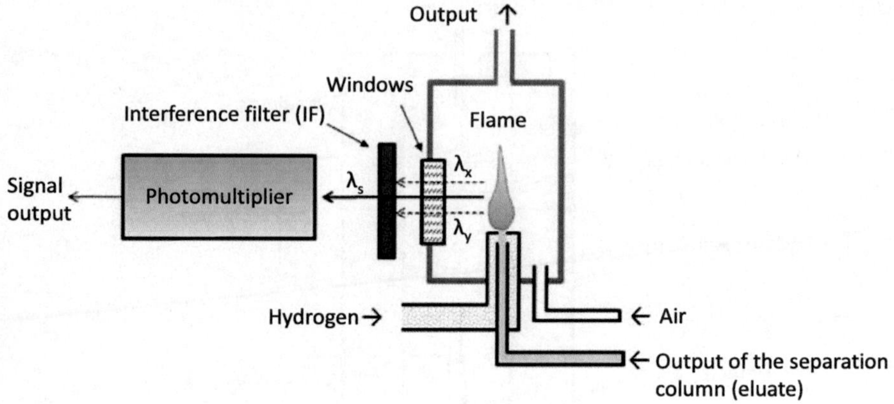

Fig. 5.16 Basic design of a flame photometric detector (FPD)

of an FPD. The optical radiation is detected with a highly sensitive photomultiplier (PM), which is usually also cooled. The cooling achieves a very low detection limit. For this purpose, the PM must be thermally separated from the hot flame.

The selection of the *suitable* detector type is of great importance for a gas analysis with GC technology. In Table 5.1, the individual detector methods are compared once again in order to have an optimal possibility of comparison.

Process Gas Chromatograph (PGC)

Process engineering processes or applications in the gas industry require automatic gas analyses of complex gas mixtures. Gas chromatographs can solve these tasks, especially when it comes to hydrocarbon compounds that are very difficult or impossible to measure with other methods. For this reason, GC's, which can perform automated analyses, have been used in process measurement technology for many years. For this purpose, the dosing devices (injection), the gas supply and the detection detectors must be controlled electronically. The gas chromatogram is recorded cyclically, i.e. a sample is injected via a valve and the peak sequence with the corresponding evaluations (→ calculate peak area) is recorded. A new analysis cycle is then started, and in the meantime the gas analysis is completed, stored and output. For a safe and long-term stable measuring operation, some monitoring functions are still necessary, which refer to the gas supply (→ flow control), the temperature, and the plausibility of the detector signals. Under these conditions, automated gas analyses can be performed with a GC.

The installation of these process gas chromatographs (PGC) is usually carried out in a control room, a special room in which the measuring and evaluation equipment is located. However, the control room should be placed as close as possible to the process to minimize time delays in sampling (→ gas transport via hose lines). These control rooms are also

Table 5.1 Comparison of the different detector types

Procedure	Application	Detection limit	Linearity
FID	**Selective**		
Flame ionization detector	Only ionizing components are measured in an H_2-flame	5 pg C/s	10^7
TCD	**Universal**		
Thermal conductivity detector	Components with different thermal conductivities	400 pg/mL	10^6
ECD	**Selective**		
Electron capture detector	E.g. molecules with heteroatoms	0.1 pg C/s	10^4
NPD	**Selective**		
Nitrogen phosphorus detector	For organic compounds containing N and P	0.1–0.4 pg	10^4
MSD	**Universal**		
Mass sensitive detector	Hardly any restrictions	10 pg-10 ng	10^5
PID	**Selective**		
Photo ionization detector	Selection by ionization potential (e.g. aromatic hydrocarbon)	2 pg benzene	10^7
FPD	**Selective**		
Flame photometer detector	Selection of emission lines by IF (e.g. sulfur compounds)	20 pg S/s 1 pg P/s	10^3 (S) 10^4 (P)

usually air-conditioned to ensure that the ambient conditions for the sensitive PGCs are constant.

Figure 5.17 shows such a PGC. It consists of two individual housings. The evaluation electronics are located in the upper part. The PGC can be operated on site via an indicating and adjustment unit without requiring an additional PC or laptop. Only in case of service, the technician may need to use a PC to troubleshoot the problem. The electronics communicate via a *BUS* with a control room, or via the Internet with any other location. It is also possible to operate several PGS's with one control unit. In the lower part, there is a heatable chamber in which the gas-carrying components such as separation column, valves, detectors etc. are located. The mechanical installation is made possible by wall rails, which also allow an optimal height adjustment for better operation.

Another type of installation is the direct installation on a gas pipeline. This type of installation is preferred in the natural gas industry, since not all points of the gas network have appropriate control rooms. A further advantage is the short gas technical connection, which leads to short dead times. Figure 5.18 shows such an arrangement.

A miniaturized PGC is required for this type of mounting. All modules required for operation are located in a flameproof enclosure (Ex-d), suitable for use in hazardous areas

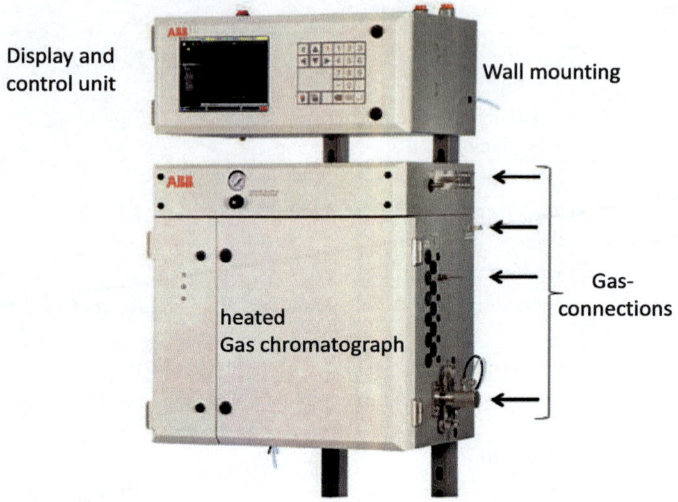

Fig. 5.17 Typical setup of a process gas chromatograph. (Source: ABB- Automation GmbH, Frankfurt am Main)

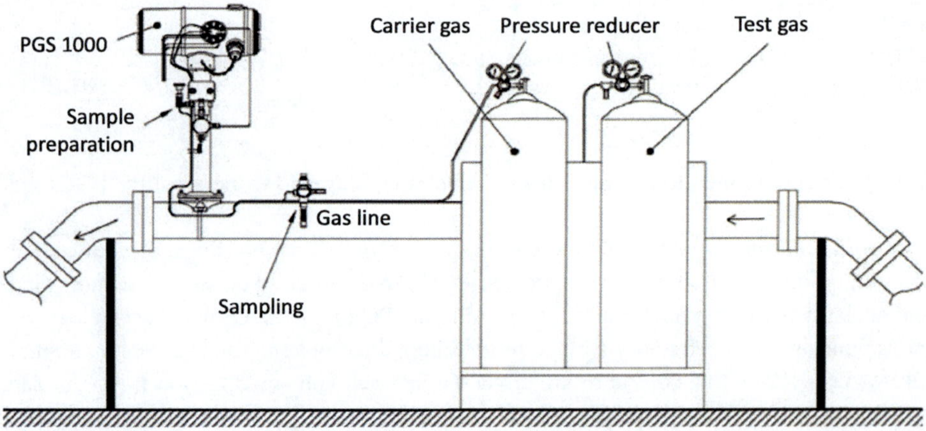

Fig. 5.18 Mounting a process gas chromatograph (*PGC 1000*) on a gas pipe with the associated gas supply (carrier gas and test gas) and the corresponding sampling. (Source: ABB-Automation GmbH, Frankfurt am Main)

(Ex-zone) (Fig. 5.19). The operation is preferably carried out in *REMOTE mode*[9] via a bus line from a central unit.

[9]Telemetry

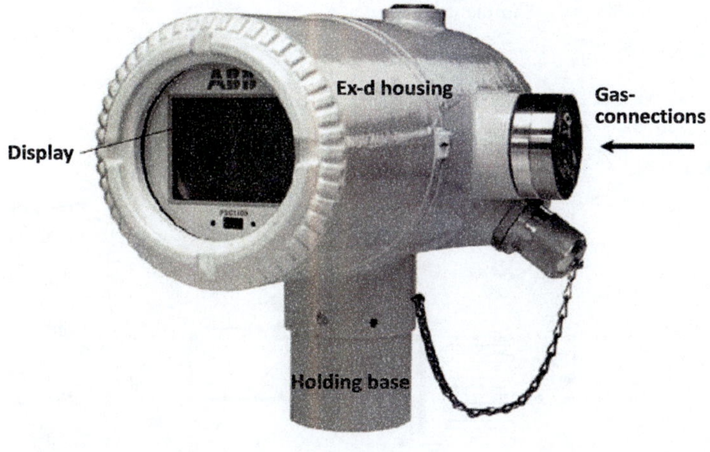

Fig. 5.19 Compact PGC of the *Totalflow* product series, for field use in the gas industry. (Source: ABB- Automation GmbH, Frankfurt am Main)

Micro-GC

The desire for an extremely small gas chromatograph that is easy to produce has been around for many years. Already in 1975, attempts were made to integrate a complete GC on a silicon wafer (Angel et al. 1983). Figure 5.20 shows such a micro GC as a *lab on chip*. The individual components (separation column, valve, TCD, pump, ...) are structured on the silicon wafer by etching processes. The electrical elements (heater, wires, ...) were deposited by coating processes (lithography, vapor deposition). However, the technical problems were too great at that time, so that this idea of the *lab on chip* could only be implemented very slowly.

Today, there are micro-GC's that are based on this basic idea and thus enable extremely small sizes. The individual components can certainly be realized with these microtechnical manufacturing processes. However, since the technological steps are very different for the individual components, separate production is advisable. Figure 5.21 shows such a micro-GC. It is located on a circuit board which also contains the control electronics for the individual components (see Sect. 15.1). In this example, two different separation columns are integrated in a planar structure, each connected to a separate TCD (detector). The detectors are also manufactured using microtechnology (see also Sect. 3.1) (Fig. 5.22).

Sensor-GC

The use of compressed and high-purity carrier gases is technically very complex and also expensive. Furthermore, the empty gas cylinders have to be regularly replaced by full cylinders, which means additional time and costs. For simple gas measurements there are therefore adapted GC's which use air as carrier gas. In this case, simple semiconductor gas

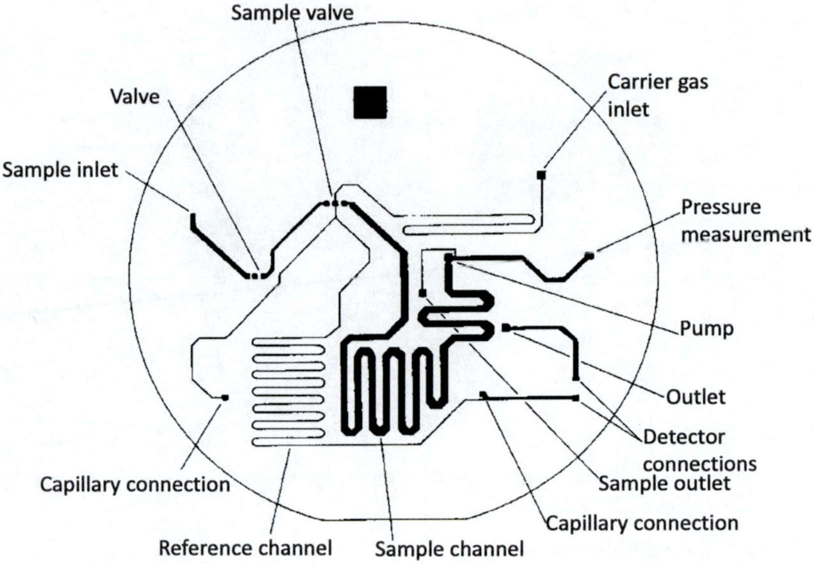

Fig. 5.20 Micromechanical GC(Lab on Chip) on a 3 inch silicon wafer. (Heuberger 1989)

Fig. 5.21 Complete GC on a circuit board with conventional components. (Source: Elster GmbH, Dortmund)

Fig. 5.22 Planar separation column, as part of a micro-GC. (Source: Elster GmbH, Dortmund)

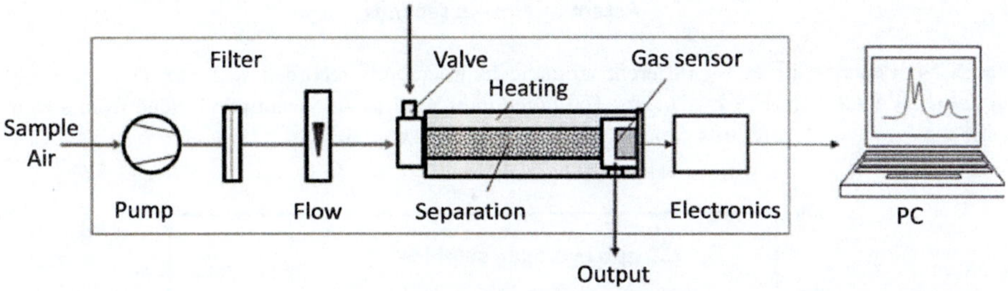

Fig. 5.23 Basic construction of a gas chromatograph with air as carrier gas and a semiconductor gas sensor for the detection of the peak sequence of a chromatogram (according to the data sheet E-ODSA-P2 110,914 of FIS Inc. Japan)

sensors, which do not require a carrier gas, and use purified air, are suitable as detection detectors. Figure 5.23 shows the principle structure of such a *sensor GC* (*SGC*). The air is sucked in and filtered by a pump. The downstream solenoid valve doses a defined quantity of gas to the carrier gas and is then fed into the separation column. The separation column is a packed column with different filling materials. At the outlet there is a broadband semiconductor gas sensor which is highly sensitive to the different hydrocarbons of the GC peaks.

Since this design is a very compact separation column, the separation performance is not particularly high. Figure 5.24 shows the retention times for different aromatic hydrocarbons. Between toluene and xylene, a clear separation of the peaks is possible, whereas there is a significant overlap between ethylbenzene and *m-xylene*. The measuring ranges are in the ppm range, with detection limits in the ppb range. Trace measurements are therefore certainly possible with this method. For the detection of hydrogen sulfide (H_2S) or methyl mercaptan, the manufacturer[10] specifies a range from 5 ppb to 1 ppm. The

[10] FIS Inc. Kitazono Japan

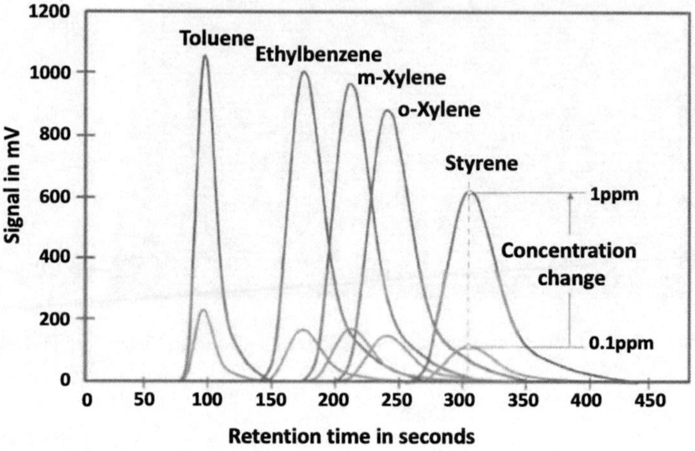

Fig. 5.24 Reaction times for different aromatic hydrocarbons recorded with the *VOC Analyzer Model: SGVA-P2* from FIS Inc. Japan. The *black lines* refer to a concentration value from 1 ppm while the *grey curves* refer to a concentration value from 100 ppb

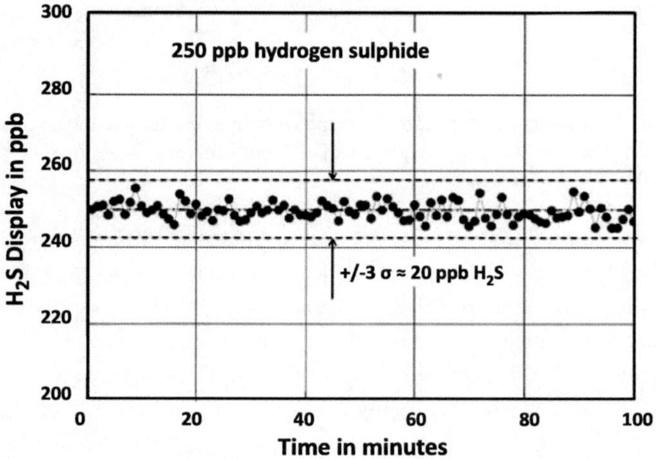

Fig. 5.25 Stability of the H_2S measurement over a period of 100 minutes (≈1 1/2 h) with the SGC (*ODSA-P2*)

stability of the gas analysis with the SGC over a measuring time of $t \approx 1\ 1/2$ h is shown in Fig. 5.25. A drift cannot be detected during this period, and the measured values have an absolute value of ≈250 ppb H_2S, with a standard deviation of $\pm 3\sigma < 20$ ppb.

5.2 Ion Mobility Spectrometry (IMS)

The detectors in GC technology often work on the principle of ionization (FID, PID, ECD), and are an important detection method for certain gaseous substances. With these detectors, however, only the *number* of ions or electrons is measured via the measuring current I_M. Which substances, or groups of substances have led to this ion formation can only be determined indirectly via the ionization potential. On the other hand, precise knowledge of the mobility of the ions in an electric field leads to further information that enables more precise substance data to be obtained. As early as 1905, Langevin[11] established a mathematical and chemical model description for the movement of ions in electric fields, which still forms the basis for the description of ion mobility spectrometry(IMS) today.

The IMS method is based on the fact that ions are generated under normal pressure (≈1 bar) and drift against the flow direction of a gas. This drift velocity v_D depends on the strength of the electric field E and the mass m_i of the ions. The different substances (=different masses m) therefore reach the detector one after the other, where they can be detected in the form of an electric current I_O. The movement of ions of different masses in an electric field is shown in Fig. 5.26.

The ratio of ion velocity v_D to electric field strength E is called ion mobility K. It applies:

$$\vec{v}_D = K \cdot \vec{E} \tag{5.14}$$

Under normal conditions (p_0 = 1013.25 hPa, T_0 = 273.15 K) the reduced ion mobility constant K_0 is obtained:

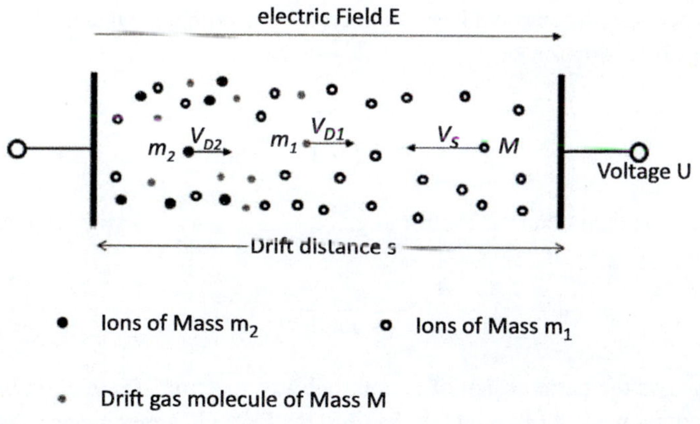

Fig. 5.26 Motion processes (drift velocity v_D) within an electric field E for ions of different masses (m_1 + m_2) and a countercurrent (drift gas with mass M)

[11] Paul Langevin (1872–1946) French physicist.

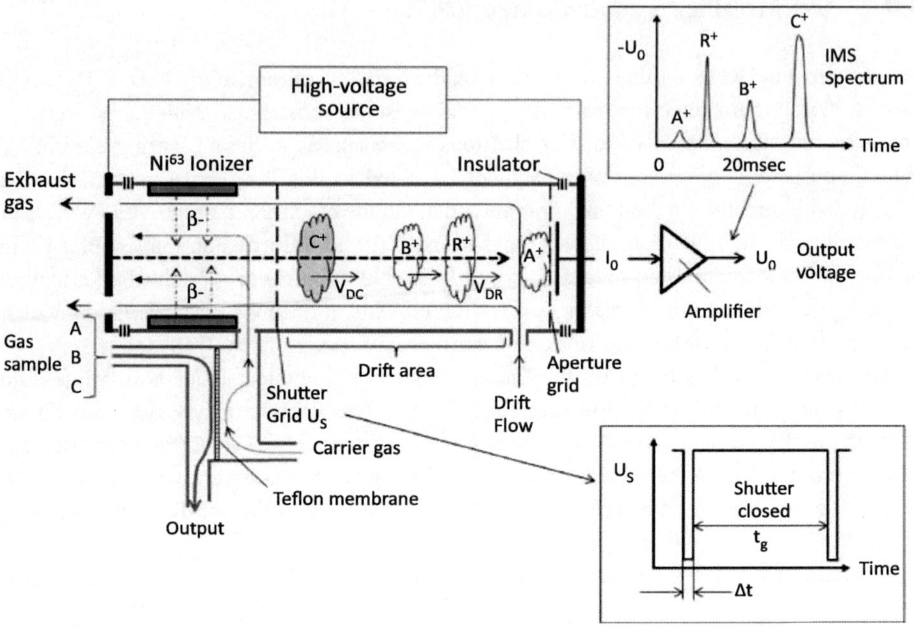

Fig. 5.27 Basic structure of an IMS. (According to Bacon et al. 1991)

$$K_0 = K\left[\frac{p_1 \cdot T_0}{p_0 \cdot T_1}\right] \tag{5.15}$$

K_0 is a substance-specific size and is given in $\text{cm}^2/V \cdot s$. With $v = s/t$ and $E = U/s$, you get the following expression for K:

$$K = \frac{s^2}{t_D \cdot U} \tag{5.16}$$

The drift time t_D is then obtained by forming:

$$t_D = \frac{s^2}{K \cdot U} \tag{5.17}$$

Figure 5.27 shows the basic design of an ion mobility spectrometer IMS. In this case, the gas sample (components A, B and C) is fed into the IMS via a membrane (Teflon). In the setup there is a radioactive ^{63}Ni source, which emits β radiation. This radioactive radiation ionizes the components A, B, and C. Alternatively, tritium with an activity of 50 MBq[12] is

[12]Bq =Becquerel, named after the French physicist Antoine Henri Becquerel (1852–1908) 1 Bq characterizes one radioactive decay per second.

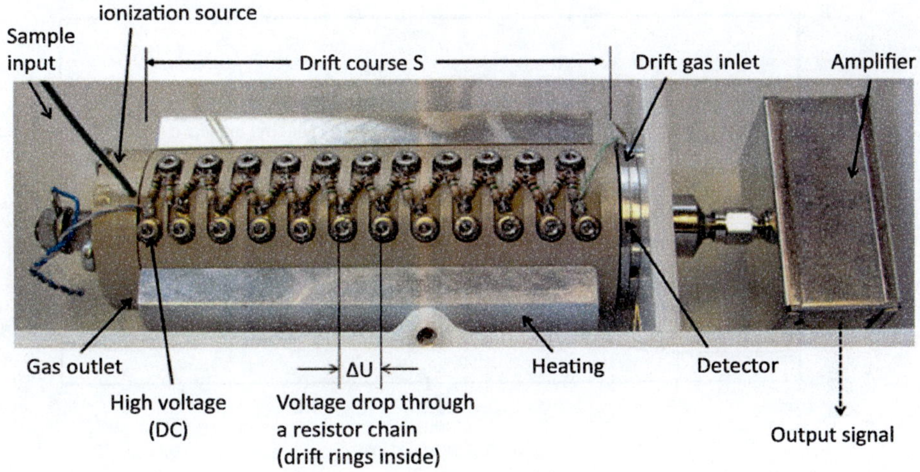

Fig. 5.28 Detailed view of an industrially manufactured IMS spectrometer. (Source: G.A.S. Gesellschaft für analytische Sensorsysteme mbH Dortmund)

used as β-radiator. The use of optical radiation in the UV range is also possible, and is also used in various measuring instruments. The ionized particles are set in motion in the external electric field E, which is generated by a high voltage source. However, a *shutter grating*, which is provided with a counter voltage U_S, prevents the further transport of the ions. Only when the voltage is switched off for a short time$\Delta t \approx 10\ \ \mu\text{m} - 300\ \ \mu\text{s}$ does the ions enter the drift space as a packet. There they move in the direction of the electric field E, against the flow of a drift gas (helium, argon or similar.). The heavy ions fall back in this drift distance of length s, while the lighter ions reach their destination (detector) faster. At the receiving electrode (detector), the electric current I_0 (≈ pA to nA) is measured and converted by an amplifier into a voltage U_0 (volt range), which is then displayed on the *y-axis* as intensity. The time sequence $y(t)$ of the pulses then characterizes the IMS spectrum and is used for quantitative and qualitative gas analysis. The entire IMS spectrum is available after $t \approx 50 - 100\,\text{ms}$.

Figure 5.28 shows the compact design of an IMS spectrometer. It is only about 25 cm in size and can therefore be integrated into small, handy measuring devices.

Figure 5.29 shows an IMS spectrum of *ethyl acrylate* and *methyl acrylate*. These substances are added to the natural gas in order to create an odour sensation (odorization) of the otherwise odourless natural gas. The spectrum was recorded with a *μIMS-ODOR*.[13] The length of the drift path was s = 4 cm, with an electric field strength of $E = 340\ V/cm$. The time sequence allows a separation of both substances, so that a selective analysis for

[13]The device was replaced by a new development *GC-IMS-ODOR*. Manufacturer: G.A.S. GmbH Dortmund.

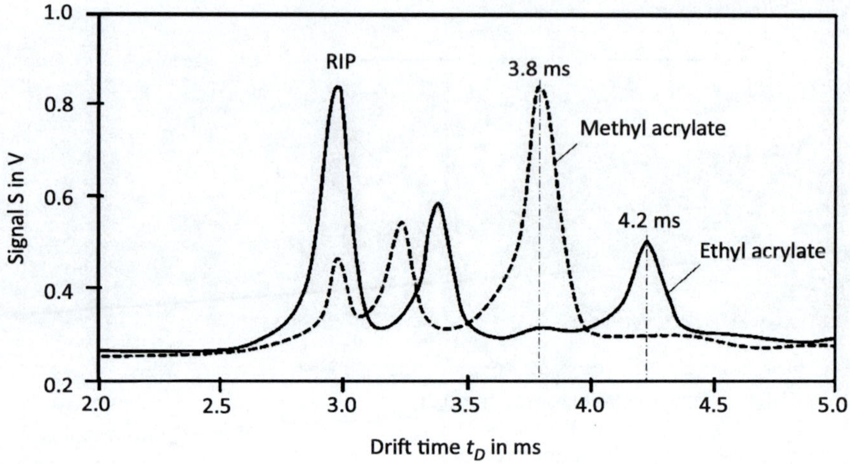

Fig. 5.29 IMS spectrum of the odorants ethyl acrylate and methyl acrylate. (Source: Sielemann et al. 2004 data)

both concentration values can be performed. After $t_D = 3, 8$ ms, for example methyl acrylate can be measured without cross sensitivities, and after $t_D = 4.2$ ms ethyl acrylate can be measured without cross sensitivities. The large peak at $t_D = 3$ ms results from the ionization of the air. The peak is therefore also called *reaction ion peak* and is abbreviated as *RIP*.

The quantitative analysis is performed in a similar way to gas chromatography. For this purpose, the respective IMS peak is integrated according to Eq. 5.13 and multiplied by a specific factor f_{IMS}. In this case too, this factor must be determined empirically.

A more detailed analysis (Revercomb and Mason 1975) of the ion mobility in the electric field leads to the following expression for K:

$$K = \frac{3 \cdot e}{16 \cdot N} \cdot \left[\frac{m + M}{m \cdot M}\right]^{1/2} \cdot \left[\frac{2 \cdot \pi}{k \cdot T}\right]^{1/2} \cdot \frac{1}{\sigma} \tag{5.18}$$

With the elementary charge $e = 1.602 \cdot 10^{-19}$ C, the density N of the drift gas, the mass m of the ions, the mass M *of the* drift gas molecule, the Boltzmann constant $k = 1.381 \cdot 10^{-23}$ J $\cdot$ K^{-1} and the impact cross-section σ. From Eq. 5.18 it can be deduced that the drift gas (M) has a considerable influence on K and thus also influences the drift time t_D. At constant pressure p and constant temperature T, Eq. 5.18 can be simplified as follows, if $q = Z \cdot e$ is used for the number of charges:

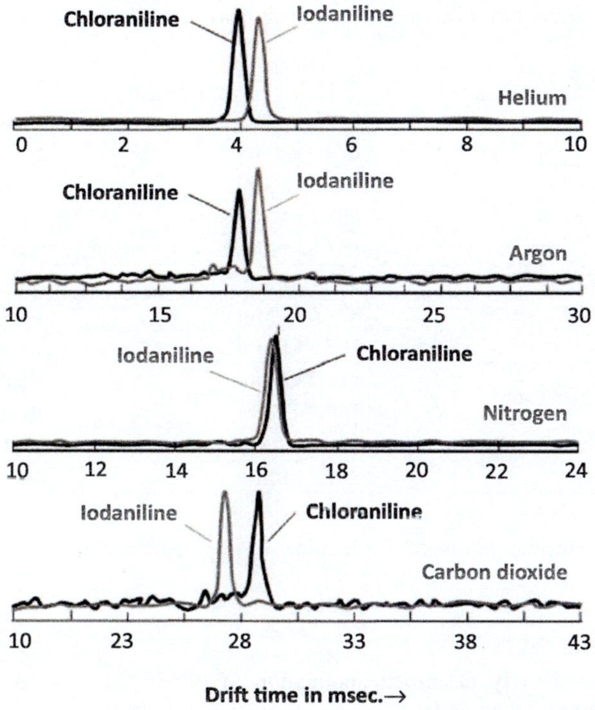

Fig. 5.30 Influence of the drift gas (Ar, He, CO_2) on the drift times t_D of the two specified substances. (Data source: Kanu and Hill 2004)

$$K \approx \frac{q}{\sigma} \cdot \sqrt{\frac{m+M}{m \cdot M}} \tag{5.19}$$

In fact, with different drift gases, a significant change in drift time can also be detected. Figure 5.30 shows an experimental result. The molar mass of chloraniline is 127.57 g/mol and for iodaniline 219.57 g/mol. The considerably heavier iodoaniline, with the light drift gas helium (4 g/mol), has a longer drift time than the chloroaniline. With a heavy drift gas, such as CO_2 (44 g / mol), this is reversed. In the middle range (nitrogen) the two drift times are the same. By using different drift gases, it is thus possible, in certain borderline cases, to achieve selectivity for certain applications (Asbury and Hill 2000).

Which substances or groups of substances can best be detected with the IMS is determined by the proton affinity of the substances relative to water vapour (Eiceman and Karpas 2005). Table 5.2 shows the different values for proton affinity. The table thus shows that there is no sensitivity to the alkanes (e.g. methane). The sensitivity series starts with the aromatic hydrocarbons and ends with the amines, which have an extremely high

Table 5.2 Proton affinity of different groups of substances. (According to Eiceman and Karpas 2005)

Substance group	Proton affinity	Example
Aromatic amines	930.0 kJ/Mol	Pyridine
Amines	899.0 kJ/Mol	Methylamine
Phosphorus compounds	890.6 kJ/Mol	Trimethyl phosphate
Sulphoxides	884.4 kJ/Mol	Dimethyl sulfoxides
	853.6 kJ / Mol	Ammonia
Ketones	832.7 kJ/Mol	2-pentanone
Esters	821.6 kJ/Mol	Methyl acetates
Alkenes	805.2 kJ/Mol	1-witches
Alcohols	789.2 kJ/Mol	Butanol
Aromates	750.4 kJ/Mol	Benzene
	691.0 kJ/Mol	**Water**
Alkanes	543.5 kJ/Mol	Methane

sensitivity. The possible fields of application of the IMS are then derived from this sensitivity series.

Applications of the IMS

In addition to the already described application of odorization control of natural gas (Staudinger 1997), the IMS is used in a wide range of applications, especially in trace gas analysis. In particular, the detection of explosives and drugs has been continuously improved in recent years. In addition to the detection of different drugs by analysing the vaporisable components in the air, the admixture of other substances also plays an important role.

Example

The main active ingredient of *ecstasy pills* is *3,4-methylenedioxymethylamphetamine* or MDMA for short. MDE and MBDB are similar substances that can be chemically derived from it. Figure 5.31 shows the IMS analysis of different *pills*. The different admixtures are clearly visible. The detection takes place already in the trace range, so that it is sufficient to detect small amounts in the ambient air. These devices are therefore important for police operations (→drug investigation) and to support doctors in emergency medicine.

In the military sector and also in the defence against terrorist attacks, the early detection of explosives and lethal gases plays a major role. The clear detection of these substances in the ambient air can help to thwart a possible attack and save human lives. Figure 5.32 shows the IMS spectrum of trinitrotoluene[14] (TNT explosive). A and B are different TNT

[14] 2-methyl-1,3,5-trinitrobenzene.

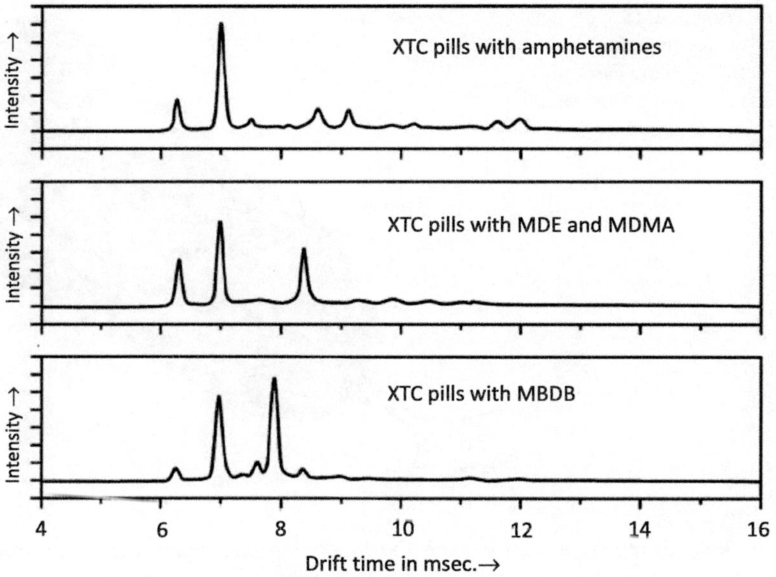

Fig. 5.31 Designer drugs (*Ecstasy XTC*) with different admixtures of amphetamines, MDE, MBDB. (According to data sheet of STEP GMBH, Pockau)

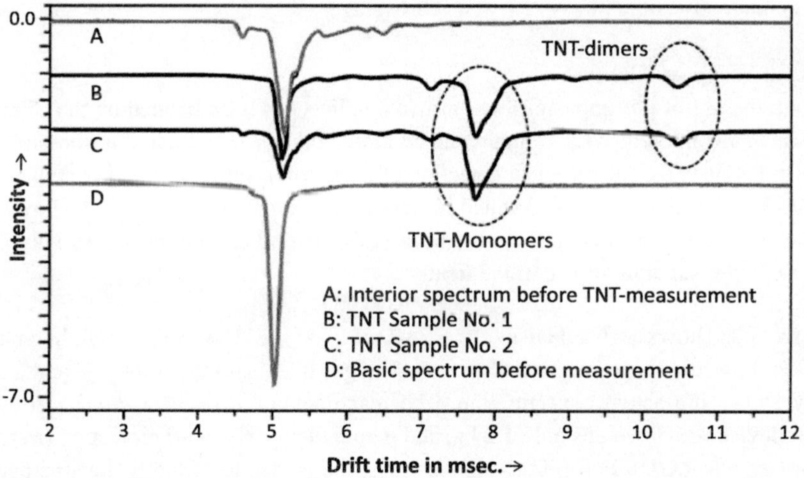

Fig. 5.32 IMS spectrum of TNT in ambient air with different intensities (B and C) (according to data sheet of STEP GMBH, Pockau)

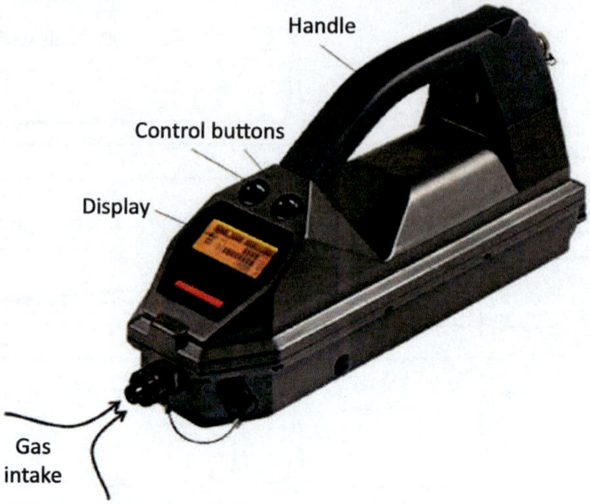

Fig. 5.33 Portable IMS measuring device (type Rais M 100) for detecting toxic gases in the ambient air. (Source: Bruker Daltonik GmbH, Leipzig)

samples, which differ significantly from the indoor spectrum and the basic spectrum. However, for anti-terrorist operations it is not absolutely necessary to carry out an exact quantitative analysis. As a rule, it is sufficient to clearly prove the presence of these substances in order to be able to react to them in a targeted manner. IMS measuring instruments are also used to monitor air conditioning systems in security-relevant facilities (e.g. Reichstag building) in order to detect the introduction of toxic gases into the air conditioning and ventilation system at an early stage.

Chemical Weapons Control

Although the use of poisonous gases in armed conflicts has been banned by the Chemical Weapons Convention (CWC), military and civilian facilities are prepared for possible use. This also includes the metrological detection of these toxic substances in the ambient air. The IMS in particular is very well suited for this application, since this method allows very fast and selective measurements. Furthermore, this method can be integrated into handy devices that are suitable for use in the field.

Figure 5.33 shows such a device of the *RAID M 100 type.* The weight including battery is 3.5 kg. It has the shape and size of an iron. The gas inlet is on the front side and a small display shows the actual concentration. The manufacturer *Bruker Daltonik* also offers another device which weighs only 1.2 kg, and is suitable for *personal* protection (*Personal Detector type μ-RAID*). For military applications, the device technology must be particularly robust. Figure 5.34 shows the application of the gas detector RAID M-100 in field use.

Fig. 5.34 Using the *RAID M-100* during a military exercise. The soldiers have appropriate protective clothing (gas mask) and are checking the surrounding air for dangerous war gases. (Source: Bruker Daltonik GmbH, Leipzig)

IMS-GC Coupling

Another possible improvement of the IMS technology is to perform a pre-separation of different substances from a complex matrix. By this measure, the selectivity of the IMS analysis can be significantly improved. The pre-separation can be carried out with a typical GC separation column, which has already been described in Sect. 5.1. Figure 5.35 shows the structure of such a coupling. The pre-separation is carried out in a GC and the eluted components are transferred in sequence to the downstream IMS. The retention time Δt_{GC} of the GC is at least 30 seconds and in special cases (→ higher separation performance of the GC) it can also take 3–10 minutes. The recording of an IMS spectrum, however, only takes $\Delta t_{IMS} < 100$ ms, so that several hundreds or even thousands of IMS spectra are available during one GC run. This information can be used to generate 3-axis displays that show the separation efficiency of the GC-IMS coupling. Figure 5.36 shows a gas analysis according to this procedure for a homologous series of ketones, with the sum formula $C_nH_{2n}O$.

On the *x-axis* is the drift time of the IMS measurement, whereas the intensity of the measured peaks is displayed on the *y-axis*. The retention time of the GC is then displayed on the *z-axis*. This 3D representation already shows quite clearly the possibilities of this evaluation method. However, the 3D information can also be reduced to a 2D representation (surface representation), which is also shown in Fig. 5.36 (background). With this representation method the information about the intensity is lost, but in a *false color representation* the intensity can be indicated by the degree of brightness or coloration. In

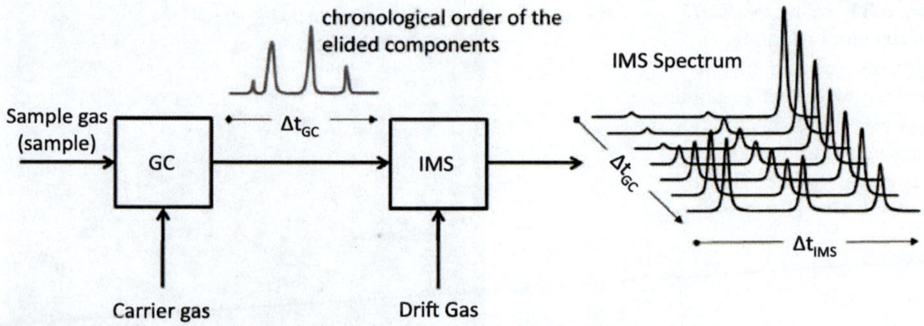

Fig. 5.35 Coupling a gas chromatograph (GC) with an ion mobility spectrometer (IMS). The carrier gas (GC) and the drift gas (IMS) can be identical (e.g. nitrogen or synthetic air)

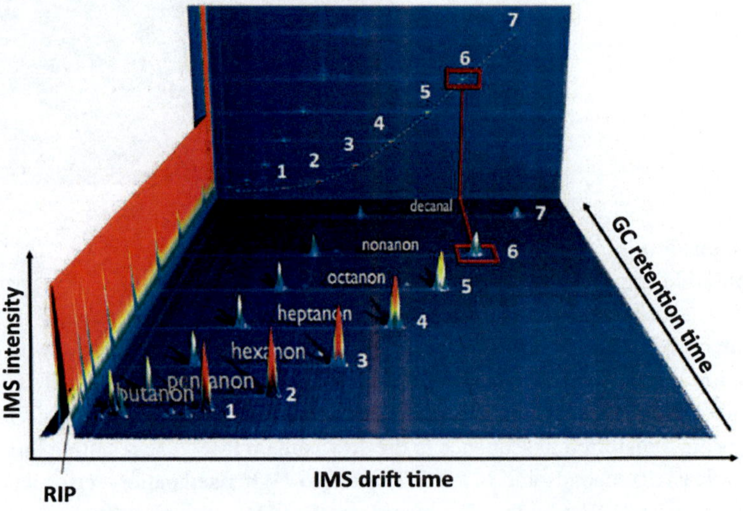

Fig. 5.36 GC-IMS spectrum of a homologous series $C_nH_{2n}O$ starting with first butanone (C_4H_8O) to seventh decanal. (Source: G.A.S. Gesellschaft für analytische Sensorsysteme mbH Dortmund)

Fig. 5.37, the analysis of a complex gas mixture is to be seen in a surface representation. One can clearly see the separation of the different substances and also the intensity is very well represented by the brightness (Wortelmann 2015; Table 5.3).

Another interesting measurement that can be performed with a GC-IMS coupling is the detection of flavouring substances in the food industry (Wortelmann 2013).

Example

During beer production, different aromatic substances are created which give the respective beer type a characteristic taste. An important substance in this context is *diacetyl*. This

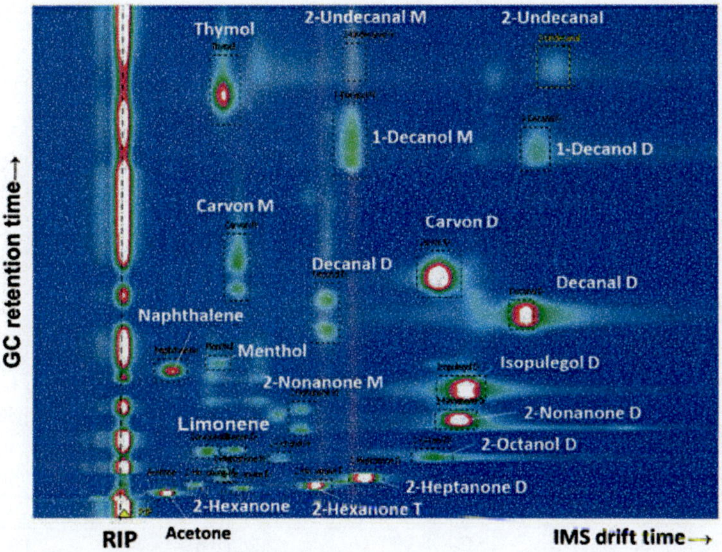

Fig. 5.37 Area representation of a GC-IMS evaluation of a complex gas mixture. (Source: G.A.S. Gesellschaft für analytische Sensorsysteme mbH Dortmund)

substance belongs to the ketone group and is the simplest representative of the diketone substance class. It has a distinct taste and smell of butter/butter caramel and is also a component of the natural butter aroma. In the brewing process, brewer's yeasts produce diacetyl which they further reduce to 2,3-butanediol. Top-fermented beers can contain up to 130 mg/L butanediol. If diacetyl is not broken down during the brewing process, the beer takes on an unpleasant off-taste. The diacetyl content must therefore be adjusted to 40 µg/L. The GC-IMS method is also suitable for this process control. Figure 5.38 shows the *headspace analysis*[15] of a beer sample. The diacetyl peak is clearly visible next to another aromatic substance (pantanedione).

Odorization

The process of adding odorous substances to natural gas is called *odorization* and takes place at very low gas concentrations. The mixing ratio must therefore be controlled in order to make an additional dosage if necessary. Especially for this application there are different measuring methods on the market. A GC-IMS coupling is also suitable for this application.

[15] Headspace analysis/vapor space analysis is a method for determining the concentration of volatile substances in a solution in the vapor space above the solution.

Table 5.3 Substances that can be detected with an IMS analysis device .(Data sheet of STEP Sensortechnik und Elektronik Pockau GmbH)

Substance	Formula	CAS[a]No.	DL[b] in ppb
Acetone	C_3H_6O	67-64-1	2
Ammonia	NH_3	7664-41-7	10
Benzene	C_6H_6	71-43-2	1
Butanol	$C_4H_{10}O$	71-36-3	1
Butyl acetate	$C_6H_{12}O_2$	123-86-4	3
Cyclohexane	C_6H_{12}	110-82-7	50
Dichloromethane	CH_2C_{12}	75-09-2	10
Ethyl acetate	$C_4H_8O_2$	141-78-6	2
Ethylbenzene	C_8H_{10}	100-41-4	1
Methanol	CH_4O	67-56-1	50
Methyl bromide	CH_3Br	75-28-5	50
Methyl tert-butyl ether	$C_5H_{12}O$	1634-04-4	50
Hexane	C_6H_{14}	110-54-3	2
Sulphur hexafluoride	SF_6	2551-62-4	5
Hydrogen sulphide	H_2S	7783-06-4	20
Sulphur dioxide	NO_2	10,102-44-0	10
Toluene	C_7H_8	108-88-3	5
Toluene diisocyanate	$C_9H_6N_2O_2$	584-84-9	10
Trimethylbenzene	C_9H_{12}	95-63-6	1
o-xylene	C_8H_{10}	95-47-6	1
m-xylene	C_8H_{10}	108-38-3	1
p-xylene	C_8H_{10}	106-42-3	1

[a]The CAS Number (also known as CAS Registration *Number* and CAS *Registry Number, CAS = Chemical Abstracts Service*) is an international designation standard for chemical substances
[b]Limit of detection

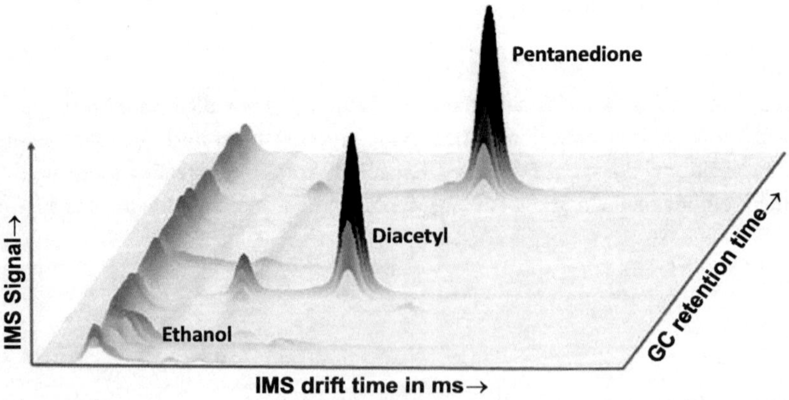

Fig. 5.38 Detection of flavouring substances in beer by means of *headspace analysis*. The measurement was performed with a special GC-IMS device of the *FlavourSpec®* type. (Source: G.A.S. Gesellschaft für analytische Sensorsysteme mbH Dortmund)

Example

The *GC-IMS-ODOR*[16] (see Figs. 5.39 and 5.40) is a fully automatic measuring instrument that has been specially developed for the precise quantification and result transfer of the concentration of methyl and ethyl acrylate (together *Gasodor®-S-Free®*) and *THT* in natural gas. It is based on a coupling between a GC and an IMS. The GC has a capillary separation column up to 30 m in which the pre-separation takes place. The IMS uses a

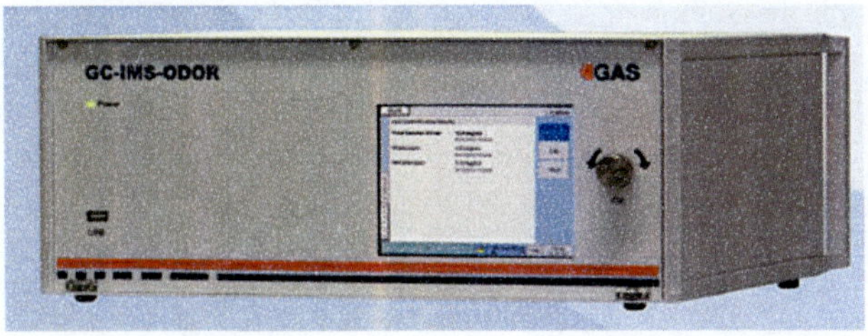

Fig. 5.39 Dimensions 449 × 435 × 177 mm (B × T × H). (Source: G.A.S. Gesellschaft für analytische Sensorsysteme mbH Dortmund)

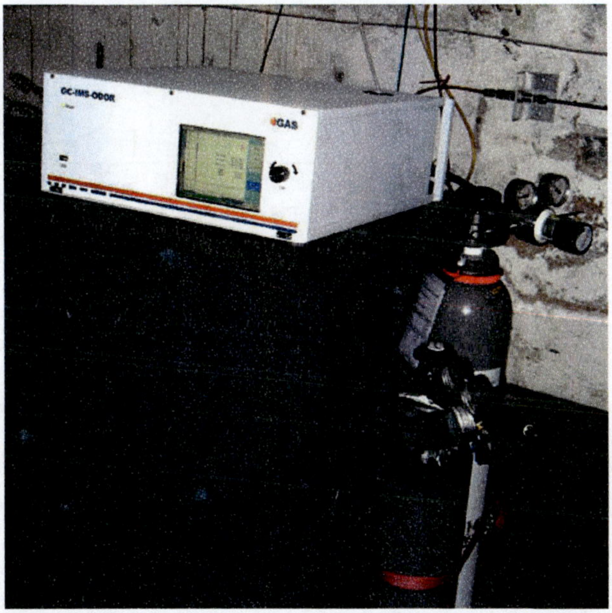

Fig. 5.40 *GC-IMS-ODOR* in practical use with the corresponding gas supply (Source: G.A.S. Gesellschaft für analytische Sensorsysteme mbH Dortmund)

[16]Manufacturer: G.A.S. Gesellschaft für analytische Sensorsysteme mbH Dortmund.

tritium ionisation source whose radiation (300 MBq) is below the limit subject to approval according to the *EURATOM Directive 96/26*. The system is operated with N_2 or synthetic air as purge and carrier gas, so that no noble gases are required. Measuring intervals can be selected individually and according to requirements. The manufacturer specifies the measuring range for *Gasodor® S-Free®* and for *THT* as 0–25 mg/m^3 (resolution 0.1 mg/m^3). The system can perform a self-check before a measurement. An endpoint calibration of the system using test gas can be performed manually or automatically at user-defined intervals to ensure maximum measurement accuracy. The *touch screen display* is used for manual operation of the instrument and for displaying results on site (see Fig. 5.38). Figure 5.41 shows the measurement results (spectra) for the two different odorant gases.

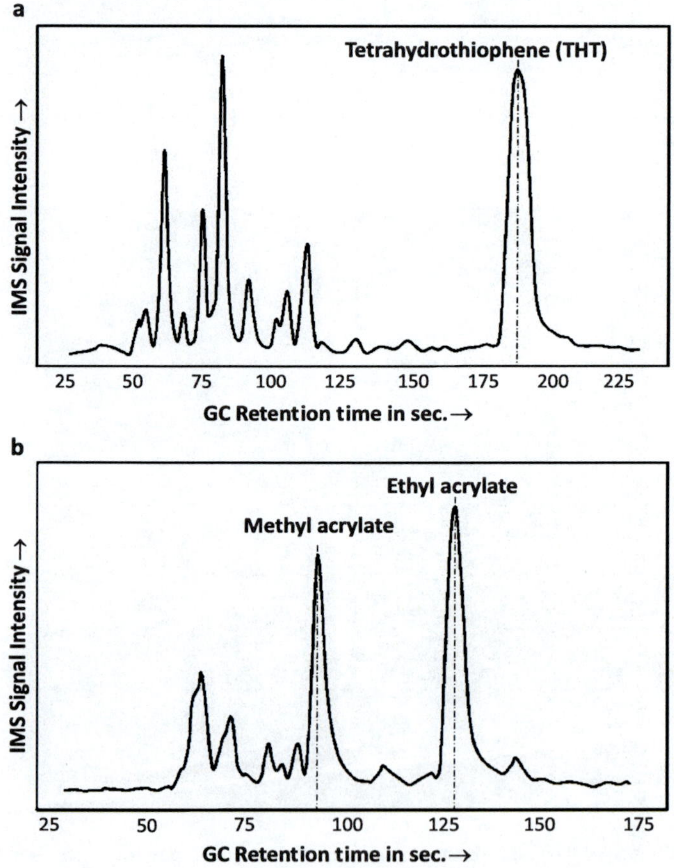

Fig. 5.41 (**a**) IMS signal (= total ion flow) for 9 mg/m^3 THT in H-gas and (**b**) 2.9 mg/m^3 methyl acrylate and 4.8 mg/m^3 ethyl acrylate in H-gas (=7.7 mg/m^3 *Gasodor®-S-Free®*). (Source: G.A.S. Gesellschaft für analytische Sensorsysteme mbH Dortmund)

5.3 Mass Spectrometry

The first mass spectrometer was built in 1919 by Aston[17]18 for scientific purposes to detect isotopes and isotope ratios. Mass spectrometers play an important role in basic physical research, for the determination of the atomic properties of matter. In a mass spectrometer, a beam of gaseous ions is generated which is separated according to mass m and charge q. From this, a mass spectrum is obtained which provides information about the molar mass of the substances and the relative quantities (concentrations).

Gaseous substances can be ionised (*EI*[18]) in a vacuum by electrical discharges (e.g. electron beams). These ions q are accelerated in an electric field E, and thus gain speed v. This relationship results from the law of conservation of energy ($\sum E = \text{const.}$) and can be described as follows:

$$\frac{1}{2} \cdot m \cdot v^2 = q \cdot U \tag{5.20}$$

From this, the velocity v of the ions can now be calculated:

$$v = \sqrt{\frac{2 \cdot q \cdot U}{m}} \tag{5.21}$$

Figure 5.42 shows the schematic structure of a mass spectrometer. After an acceleration zone (electric field E), the positively charged ions (+q) enter a homogeneous magnetic field B *and* are deflected by the Lorentz[19] force ($F = q \cdot v \cdot B$). The ions move on a circular path (r). The Lorentz force then corresponds to the centripetal force. It is therefore valid:

$$\frac{m \cdot v^2}{r} = q \cdot v \cdot B \tag{5.22}$$

Radius r of circled track then results by transformation of Eq. 5.22 and one gets

$$r = \frac{m \cdot v}{q \cdot B} \tag{5.23}$$

If Eq. 5.23 is transformed to v and both terms are squared, you get:

---

[17] Francis William Aston (1877–1945) English chemist and physicist.

[18] Electrical ionization.

[19] Hendrik Antoon Lorentz (1853–1928) Dutch mathematician and physicist.

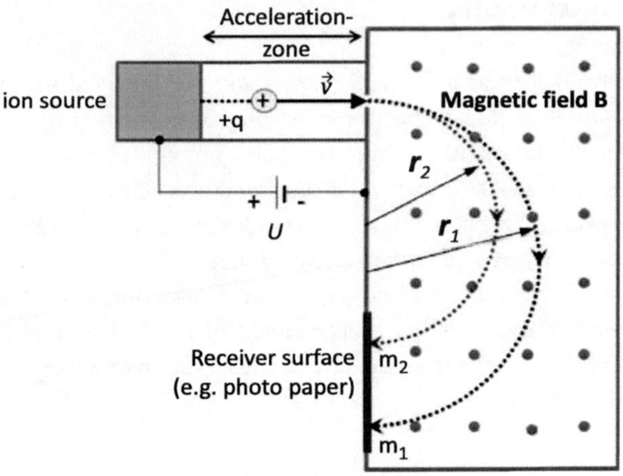

Fig. 5.42 Operating principle of a mass spectrometer. Positively charged ions are accelerated in an electric field and then enter a homogeneous magnetic field. The radii r_n of the circular paths depend on the mass m_n of the ions

$$v^2 = \frac{r^2 \cdot q^2 \cdot B^2}{m^2} \tag{5.24}$$

By inserting in Eq. 5.21 and further forming the result is then

$$\frac{m}{q} = \frac{B^2 \cdot r^2}{2 \cdot U} \quad or \quad r = \sqrt{\frac{2 \cdot m \cdot U}{q \cdot B^2}} \tag{5.25}$$

Equation 5.25 shows that the radius r depends on the ratio m/q. Therefore, the different ions meet at different places ($r_1, r_2, \ldots, r_n$) on the receiver surface. The first mass spectrometers had a photo plate as receiver surface, which was blackened by the impinging ions. Today, secondary electron multipliers (SEM) are used as receivers, which generate an electrical output signal proportional to the number of incident ions. With these devices, quantitative analysis is therefore also possible.

Time of Flight Mass Spectrometer (TOF Time of Flight)
The simplest variant of a mass spectrometer is to determine the time of flight t of the individual ions in a flight tube. Figure 5.43 shows such a TOF-MS (*Time of fly mass spectrometer*). The ions are brought to the velocity v_0 in the acceleration path and then fly with constant v_0 to the detector. The time of flight t depends on the mass m *of* the ions, based on the previous observations. The different ions thus arrive at the detector with a time delay and can then be assigned to a mass m with respect to the flight time. According to Eq. 5.21 one obtains with $v_0 = s/t$ the following expression:

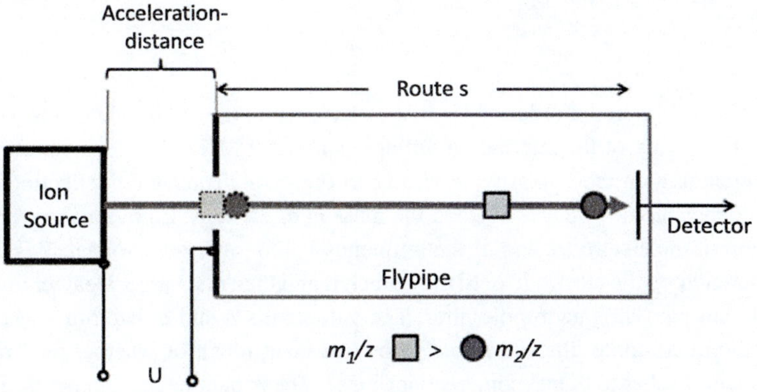

Fig. 5.43 Basic structure of a time-of-flight mass spectrometer (TOF-MS)

$$\frac{m}{q} = \frac{2Ut^2}{s^2} \tag{5.26}$$

Transformed after the time *t*, one then obtains the flight time as a function of the ion mass *m*:

$$t = s \cdot \sqrt{\frac{m}{q} \cdot \frac{1}{2U}} \tag{5.27}$$

Quadrupole Mass Spectrometer (QMS)

The quadrupole mass spectrometer consists of four metal rods arranged in a square (see Fig. 5.44). The distance between them is $2r_0$. The metal rods are electrically connected to each other along the *y-axis*. The same applies to the metal rods on the *x-axis*. A high-frequency voltage *V(t)* is now applied to the pair of electrodes, superimposed by a DC voltage *U*. The emerging ions that enter this field in the *z-axis are* subject to the following equations of motion (*Mathieu equations*):

$$\ddot{x} + (a + 2q \cdot \cos\xi) \cdot x = 0 \tag{5.28}$$

$$\ddot{y} - (a + 2q \cdot \cos\xi) \cdot y = 0 \tag{5.29}$$

With the *Mathieu parameters*:

$$\xi = \frac{\omega t}{2}, \qquad a = \frac{8eU}{mr_0^2\omega^2}, \qquad q = \frac{4eV}{mr_0^2\omega^2}$$

The DC voltage U is in the order of 1000 V (1 kV), while the amplitude can be from V to 3 kV. The frequency of the alternating voltage is about 2 MHz. The ions are attracted or repelled depending on their polarity, as shown in Fig. 5.44. The lateral deflection depends on the frequency ω, the voltages $V/U$and the mass m *of* the ions. Light ions move faster in the direction of the electrodes and, if the frequency is too low or the voltage V is too high, they are attracted by the electrode until they touch it and then discharge. Heavier ions on the other hand can pass the quadrupole filter. The value pairs a and q determine whether the ions pass the quadrupole filter on stable paths (resonant ions), or whether they reach the electrodes and discharge there (non-resonant ions). These parameters can be used to limit the mass range of the ions that can pass the rod system. Figure 5.45 shows the possible

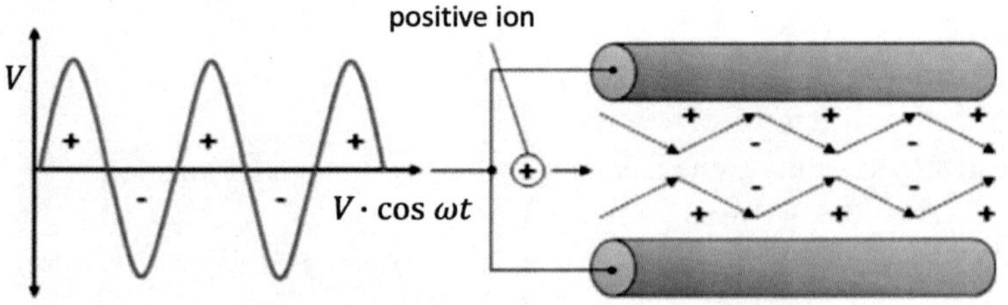

Fig. 5.44 Deflection of the ion in a quadrupole by an alternating electric field $V(t)$

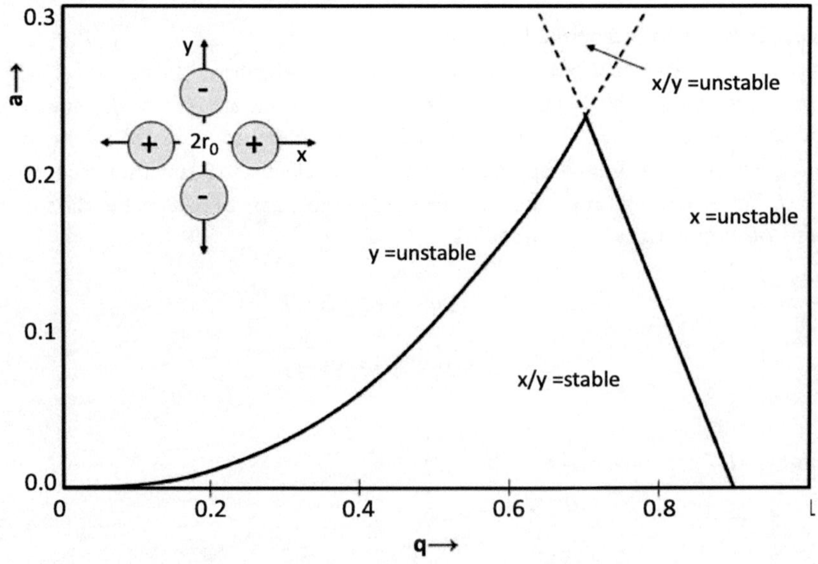

Fig. 5.45 Diagram of a quadrupole filter (*x-y representation*) and the associated stability diagram

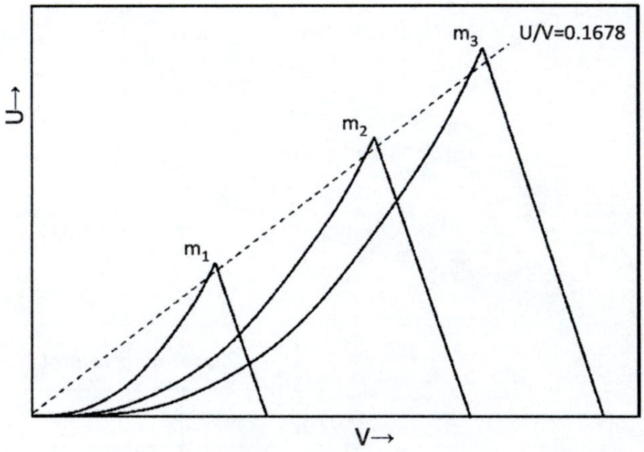

Fig. 5.46 Stability regions for different ions of mass $m_1 < m_2 < m_3$ as a function of the voltages U and V

ranges resulting from a and q. The ratio $a/q = 2U/V$ forms the working line, where within the stable area (*x/y-stable*) the ions can pass through the QMS. For different masses, this results in a stability diagram as shown in Fig. 5.46. The width of this mass range Δm is set by the ratio of U/V. For an optimum ratio of $U/V = 0,\ 1678$, a mass spectrum with completely separated mass peaks is obtained. The spectrum is scanned by a proportional increase of the voltages U and V. The mass scale is then proportional to the voltage ($m \approx V$).

The basic design of a quadrupole mass spectrometer is shown in Fig. 5.47. In the simplest case, the required gas ions are produced by bombarding electrons which are created in a directly heated hot cathode and accelerated in an electric field. Figure 5.48 shows such an ionization chamber in which an ion current I *is* created by electron bombardment. The inlet of the gases to be analyzed is via special inlet valves. The pressure in this impact space is about 10^{-4} mbar. The energy of the colliding electrons has a great influence on the ion formation, and is adjusted by the acceleration voltage U_B. The ionization of the gases in the collision space starts, depending on the type of gas, at 10–20 eV and runs through a maximum at approximately 100 eV. Reference data are usually recorded at 70 eV.

The length of the stick electrode is approximately 20 cm. After the resonant ions have passed through, they reach the detector where they generate a signal current which is processed in a downstream amplifier. The entire system is in a vacuum chamber at a pressure of $p < 10^{-5}$ mbar (high vacuum).

Commercial QMS have a typical mass range from 1 to 2000 u with a resolution of 1 u. Figure 5.49 shows a mass spectrum of a complex hydrocarbon mixture. The distances between the lines result from a mass difference of $\Delta m = 1$. In addition to the ions of the individual gases, however, the spectrum also contains fragments of these substances, which

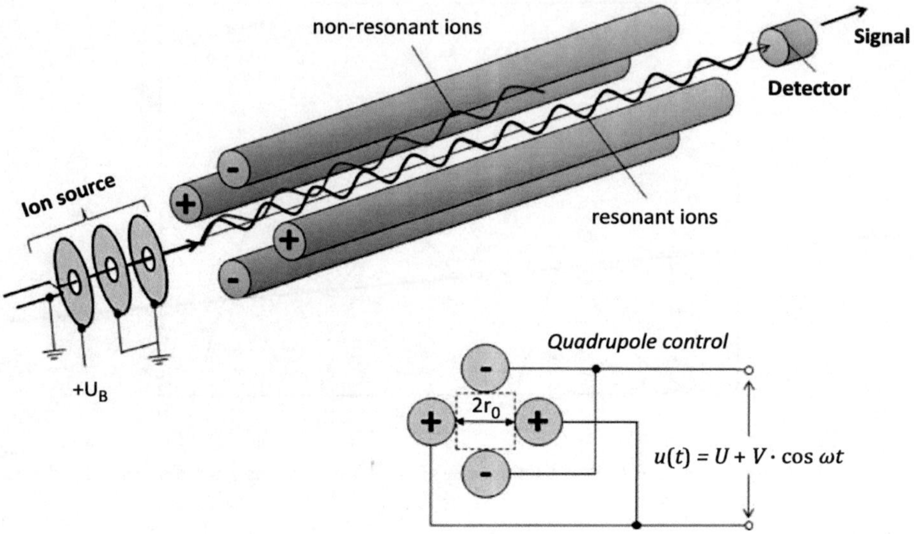

Fig. 5.47 Structure of a quadrupole mass spectrometer with the control $u(t)$. Passage of the resonant ions and *filtering of* the non-resonant ions

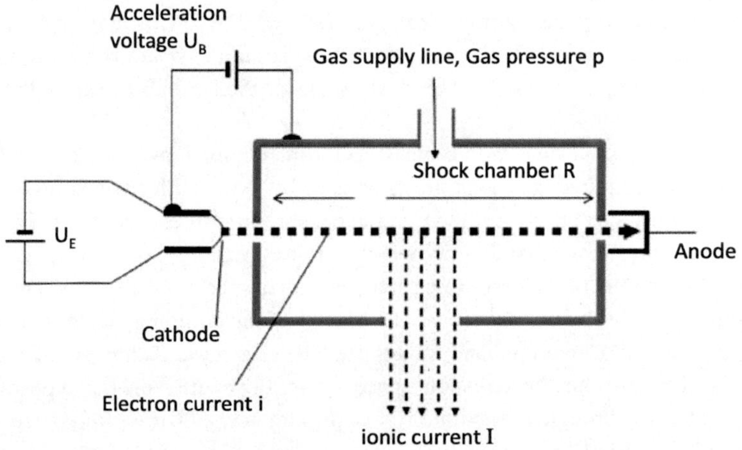

Fig. 5.48 Schematic diagram of the ionization chamber of an electron impact ionization source

must be interpreted accordingly. Especially at high ionization energies, an increased fragmentation of the gaseous components takes place.

In order to reduce this fragmentation, the possibility of chemical ionization (CI) is used, by adding certain gases (reactant gas), which also lead to useful results at ionization energy below 20 eV. Figure 5.50 shows the mass spectrum of SO_2 for different types of excitation.

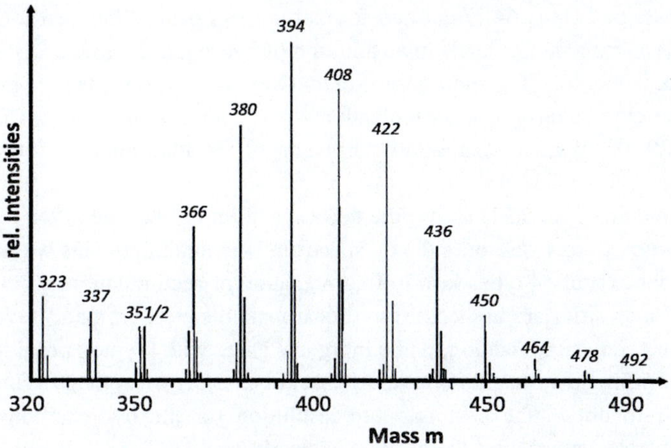

Fig. 5.49 Mass spectrum of a complex hydrocarbon mixture from the molecular ion range. (According to Budzikiewicz and Schäfer 2005)

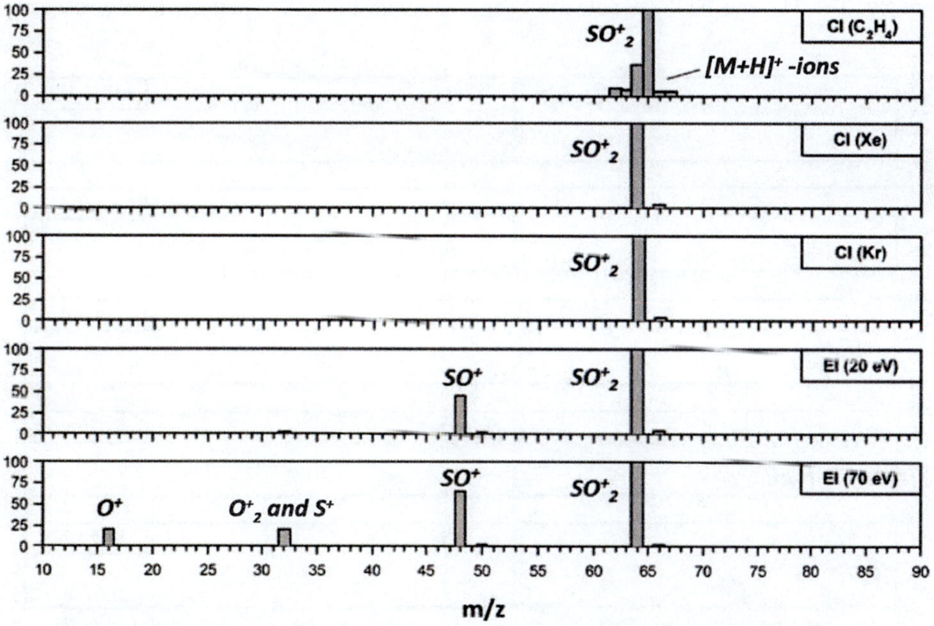

Fig. 5.50 CI and EI mass spectra of sulfur dioxide (SO_2). O^{+2} and S^+ cannot be differentiated from each other due to the same atomic mass. (Source: Frischmuth 2001)

The molar mass of SO_2 is 64 g/mol and forms the main peak. Chemical ionization with ethene (C_2H_4) as reactant gas leads to an addition of hydrogen (H) to the SO^+_2 ion, so that the main peak is then 65. The inert gases xenon (Xe) and krypton (Kr), do not falsify the spectrum. The pure electron impact ionization already leads to a SO^+-ion formation as a fragment at 20 eV. With an excitation energy of 70 eV additional O^+, O^+_2 and S^+ are formed.

A similar picture (Fig. 5.51) shows the mass spectrum of benzene (C_6H_6) with a molar mass of 78 g/mol. In this case only the CI spectrum with xenon remains without fragmentation. All other excitation types lead to fragmentation or accumulation.

The signal intensities are highest for hydrocarbons (alkanes, aromatics) with ethene as reactant gas of chemical ionization, while inorganic gases with krypton provide the highest values. Figure 5.52 shows an automotive exhaust gas analysis with ethene as reactant gas of a CI in order to obtain the best possible resolution for the hydrocarbons. A change (reduction) in these values after the catalyst is clearly visible.

A comparison measurement between a mass spectrometer (MS) and a Fourier transform infrared spectrometer (FTIR) is shown in Fig. 5.53. Here, both devices were selected for NO, and the change over time during a *test run* on a roller test bench was recorded. Depending on the speed *v*, the NO content increases significantly or decreases synchronously. The MS and FTIR signals are congruent, with the MS signal obviously reacting

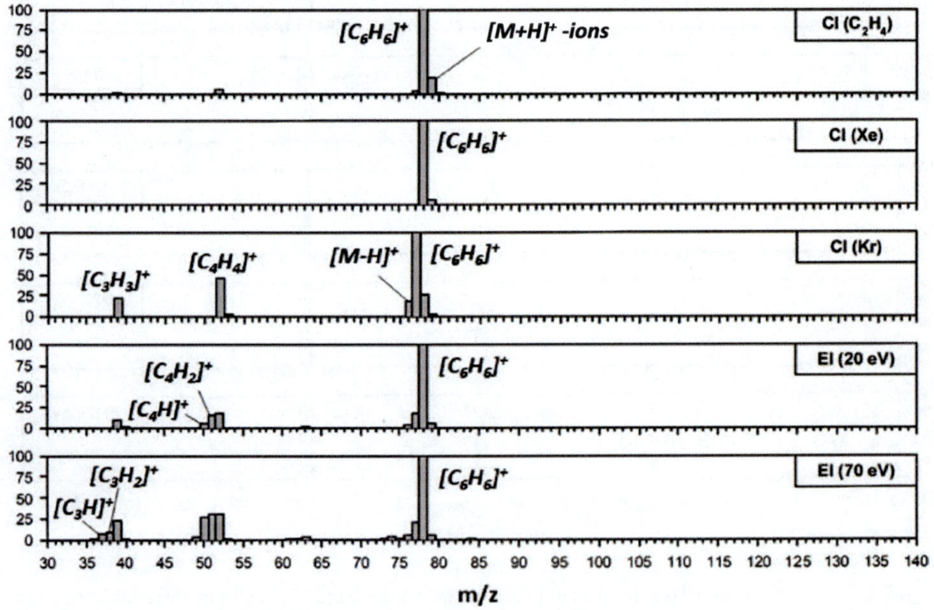

Fig. 5.51 Mass spectrum of benzene (C_6H_6). (Source: Frischmuth 2001)

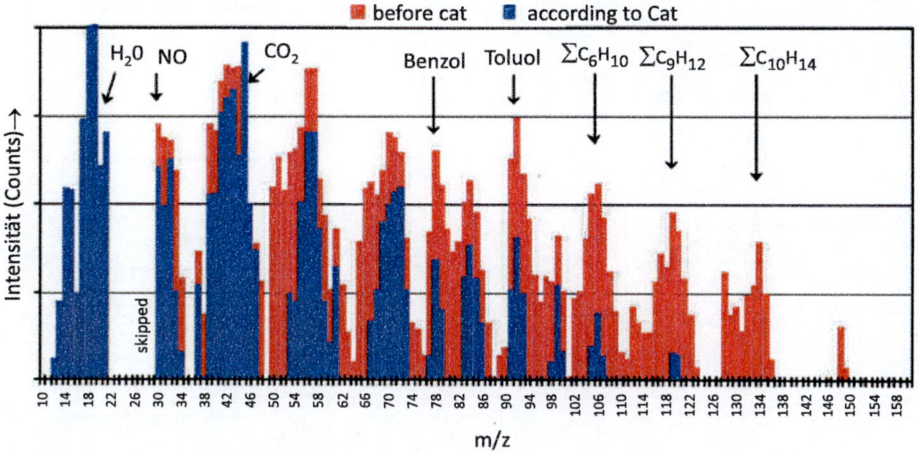

Fig. 5.52 CI mass spectrum of the exhaust gas of a petrol engine at idling speed. (Source: Frischmuth 2001)

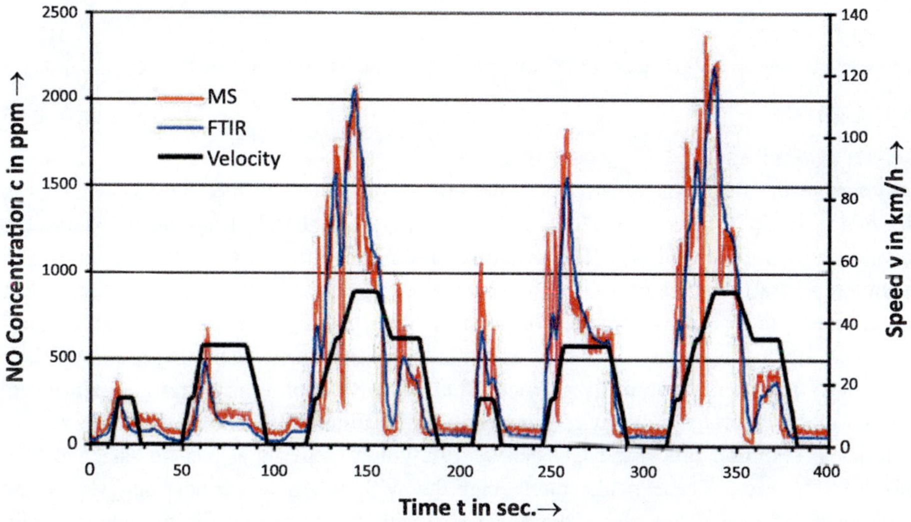

Fig. 5.53 NO measurement as a comparison between an MS and an FTIR unit in the exhaust gas of an Otto engine upstream of the catalytic converter. (Source: Frischmuth 2001)

somewhat faster to changes in concentration than the FTIR signal. The MS was operated with krypton as the reactant gas in order to obtain the highest possible sensitivity to the inorganic gases.

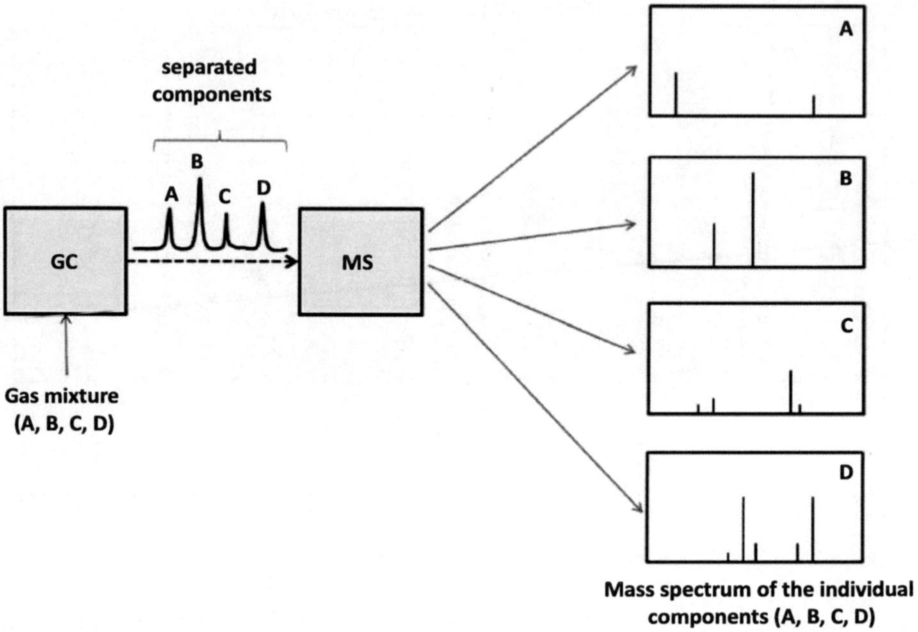

Fig. 5.54 Coupling the GC with a MS for better separation of different substances and substance groups

GC-MC Coupling

To improve the detection sensitivity, coupling the mass spectrometer (MS) with a gas chromatograph (GC) is useful. The GC-IMS-coupling has already been discussed and the advantages have been shown. The situation is similar with the GC-MS coupling. The upstream separation stage of a GC allows the filtering out of interfering components that can influence an MS measurement. This measure significantly increases selectivity. However, this method is hardly ever used in process measurement technology, since it is a particularly maintenance-intensive device. The GC-MS coupling is therefore mainly carried out in the laboratory for environmental analyses. Figure 5.54 shows the combination. Within the retention times, the GC delivers individual peaks that can be assigned to a specific substance. These peaks then enter the MS, where a further analysis of the constituents is carried out. Since the MS has a high detection sensitivity, substances in the trace range can also be detected. A disadvantage of the MS is the use of vacuum technology, which is an additional obstacle to process applications.

References

Angel, J., Terry, S., Barth, P.: Silicon micromechanical devices. Sci. Am. **248**(4), 44–55 (1983)

Asbury, G.R., Hill, H.H.: Using different drift gases to change separation factors (alpha) in ion mobility spectrometry. Anal. Chem. **72**, 580 (2000)

Bacon, A.T., Getz, R., Reategui, J.: Ion-mobility spectrometry tackles tough process monitoring. Chemical Engineering Progress, Juni 1991

Budzikiewicz, H., Schäfer, M.: Massenspektrometrie – Eine Einführung. 5. Aufl. Wiley-VCH, Weinheim (2005)

Clevett, K.J.: Process Analyzer Technology. Wiley (1986)

Down, R.D., Lehr, J.H.: Environmental Instrumentation and Analysis Handbook. Wiley (2004)

Driscoll, J.N., Ford, J., Jaramillo, L., Becker, J.H., Hewitt, G., Marshall, K.J., Onishuk, F.: The photoionization detector in gas chromatography. Am. Lab. **10**, 137–147 (1978)

Eiceman, G., Karpas, Z.: Ion Mobility Spectrometry. CRC Press, Boca Raton (2005)

Frischmuth, M.: Massenspektrometrie mit chemischer Ionisierung (CI-MS) zur dynamischen Erfassung limitierter und nichtlimitierter Komponenten im Automobilabgas. Dissertation, Universität Lüneburg (2001)

Heuberger, A.: Mikromechanik. Springer (1989)

Jentzsch, D., Otto, E.: Detektoren in der Gas-Chromatographie. Akademische Verlagsgemeinschaft, Frankfurt am Main (1970)

Kanu, A.B., Hill, H.H.: Ion mobility spectrometry: recent developements and noval applications. LPI April/Mai 2004

Kronmüller, H., Zehner, B.: Prinzipen der Prozessmesstechnik Teil 2. Schnäcker, Karlsruhe (1980)

Revercomb, H.E., Mason, E.A.: Theory of plasma chromatography/Gasous electrophoresis – a review. Anal. Chem. **47**, 970 (1975)

Schomburg, G.: Gaschromatographie (Grundlagen, Praxis, Kapillartechnik). 2. Aufl. VCH (1987)

Sielemann, S., Teepe, M., Vautz, W., Baumbach, J.I.: Bestimmung des Odorierungsgehaltes in Erdgas mittels Ionenbeweglichkeitsspektrometrie. GWF Gas+Erdgas. **145**(11), 643–646 (2004)

Staudinger, W.: Odorierung. Vulkan, Essen (1997)

Wortelmann, T.: FlavourSpec®, Trace Detection of Volatile Organic Compounds in the Headspace of Liquids and Solids. Analytics Seminar. Firmenseminar (2013)

Wortelmann, T.: private Mitteilung (2015)

6 Basics of Radiation Absorption

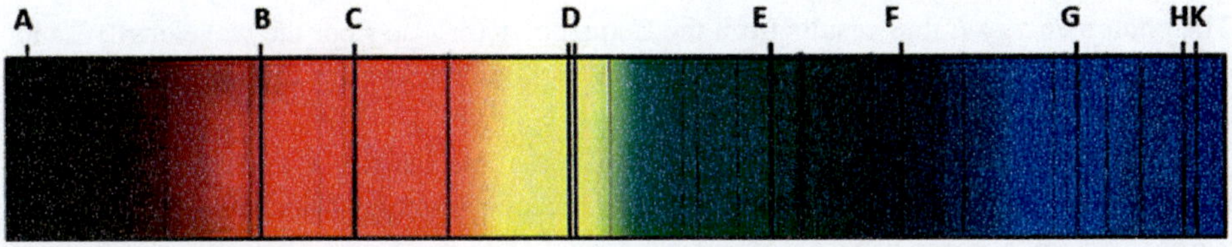

Fraunhofer Lines

With the invention of the spectroscope by *Fraunhofer*[1] in 1814, it became possible to study the radiation distribution of the solar spectrum more precisely. He discovered that the spectrum had no radiation components at certain points in the otherwise continuous course of the spectrum. These *Fraunhofer lines,* named after him, are created by absorption of optical radiation in the gas envelope of the sun. The spectral position of the absorption lines allows conclusions to be drawn about the composition (→ gas analysis) of the solar atmosphere. *Fraunhofer* marked the most prominent lines with the letters A to K, while the other weaker lines were assigned numbers. The strong lines A and B, however, are not caused by absorption in the solar atmosphere, but by atmospheric oxygen (O_2) in the Earth's atmosphere. The absorption line $\lambda = 760.83$ nm is also used for technical gas analysis (laser spectroscopy) of oxygen.

[1] Joseph Fraunhofer, since 1824 Knight of Fraunhofer (1787–1826) German optician and physicist.

G. Wiegleb, *Gas Measurement Technology in Theory and Practice*,
https://doi.org/10.1007/978-3-658-37232-3_6

Only later investigations by *Kirchhoff*[2] and *Bunsen*[3] made it possible to assign the Frauenhofer lines to the individual elements. The theoretical foundations for this were only laid with the introduction of *Planck*'s quantum mechanics.

6.1 Physical Principles of Radiation Transitions

The interaction of electromagnetic radiation with gaseous substances takes place through absorption, emission and scattering of the radiation. According to the laws of quantum mechanics, the energy transfer that is exchanged during these interactions can only take place in precisely defined energy packets, the so-called *quantal* (Fig. 6.1). According to *Planck*, the smallest energy packet (photon) results from:

$$E = h \cdot \nu = \frac{h \cdot c}{\lambda} \tag{6.1}$$

The photon energy E thus results from the frequency ν (or also f) or the wavelength λ and can also be expressed with the speed of light $c \approx 300.000 \ \mathrm{kms}^{-1}$. The constant h is the *Planck constant* ($h = 6.626 \cdot 10^{-34} \mathrm{J} \cdot \mathrm{s}$).

Gas molecules can absorb the radiation by different mechanisms, which are determined by the photon energy (→ wavelength) and the molecular structure. There are three different types of energy absorption in a molecule:

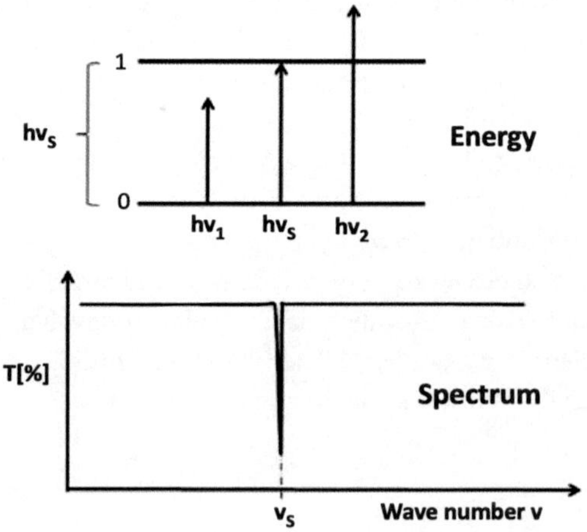

Fig. 6.1 Scheme of absorption of radiation based on energy $E = h\nu$ and wavelength (wave number) as a spectrum (according to Schrader 1980)

[2] Gustav Robert Kirchhoff (1824–1887) German physicist.

[3] Robert Wilhelm Eberhard Bunsen (1811–1899) German chemist.

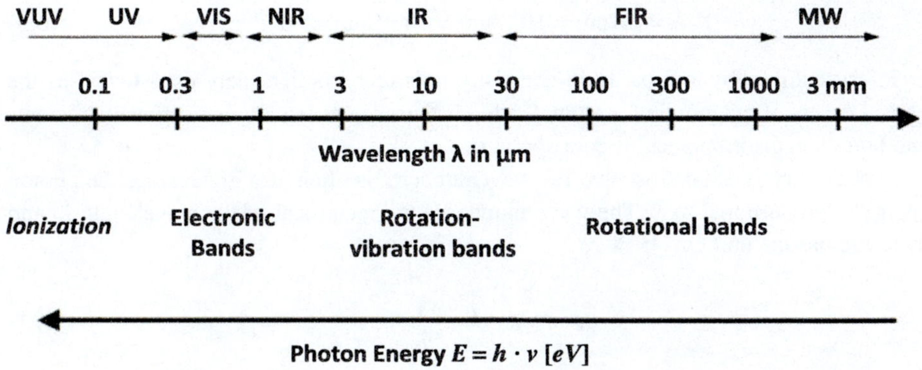

Fig. 6.2 Distribution of the different energy shares that can be transferred to a gas molecule when absorbing radiation

- The smallest energy level (E_r) is required for rotation excitation of the molecules.
- The medium energy level (E_v) is required for the excitation of vibrations.
- The highest level (E_e) allows the valence electrons to be raised to the next higher shells (electron level).

The transfer of energy to a gas molecule by radiation, can also consist of the sum of the three energy levels:

$$\sum E = E_r + E_v + E_e \tag{6.2}$$

With Eq. 6.1, the wavelength ranges of the individual energy components can be calculated and are shown in Fig. 6.2.

Units

The information for the spectral position can be expressed in different forms. The known form is the specification as wavelength λ. This form has proven to be useful for the physical approach and is therefore the preferred form for the following illustrations. In the visible and UV range, the wavelength is given in nm (nanometers $=10^{-9}$ m). In the spectral range above 1000 nm, i.e. from the near infrared (NIR), the wavelength is given in μm (micrometers $=10^{-6}$ m). Above 1000 μm, the transition between the far infrared (FIR) and the microwaves (MW), the wavelength is given in mm (millimeters $=10^{-3}$ m). In older textbooks, you can also find the *Angström unit*[4] (Å) for the wavelength, which was common at that time:

[4] Anders Jonas Ångstrom (1814–1874) Swedish physicist.

$$1\ \text{Å} = 0.1\text{nm} = 10^{-4}\mu\text{m} = 10^{-7}\text{mm} = 10^{-10}\text{m} \tag{6.3}$$

In crystallography and partly in chemistry, this unit is still widely used today, as the typical orders of magnitude for atomic radii, distances between atoms in crystal structures and bond lengths in molecules correspond to ≈ 1 Å.

In chemistry and spectroscopy, the wavenumber $\widetilde{\nu}$ is often also given since the photon energy is proportional to $\widetilde{\nu}$. The wave number is the reciprocal of the wavelength λ, and therefore has the unit cm^{-1}:

$$\widetilde{\nu} = \frac{\nu}{c} = \frac{1}{\lambda} \tag{6.4}$$

The conversion from cm^{-1} to μm can be done as follows:

$$\lambda(\text{in }\mu\text{m}) = \frac{10.000}{\widetilde{\nu}(\text{in cm}^{-1})} \tag{6.5}$$

Furthermore, there is the possibility to specify the spectral position by the unit electronvolt (eV). The electronvolt is a unit of energy which results from $E = h\nu$, and can be converted as follows:

$$1\ \text{electronvolt (eV)} = 1.60217733 \cdot 10^{-19}\text{J} \tag{6.6}$$

This unit is mainly used in vacuum UV, i.e. wavelength range $\lambda < 200\lambda = 200$ nm. According to this, a photon has an energy (in eV) of:

$$\frac{\left[6.626 \cdot 10^{-34}\text{Js} \cdot 3 \cdot 10^{8}\text{m/s}\right]/2 \cdot 10^{-7}\text{m}}{\left(1.6 \cdot \frac{10^{-19}\text{J}}{\text{eV}}\right)} = 6.22\text{eV} \tag{6.7}$$

A simplified conversion formula with which the light wavelength λ (given in nm) can be converted to energy (given in eV) is:

$$\frac{1240}{\lambda[\text{nm}]} = E[\text{eV}] \tag{6.8}$$

Energy Levels

Figure 6.3 shows the gradations of the different energy levels. In the basic level E_0 the molecules are in absolute resting position. Due to thermal excitations ($T \ll -273°\text{C}$), single gas molecules then reach the first rotation and oscillation levels, based on the *Boltzmann distribution.* If the gas molecule is now exposed to a radiation that provides

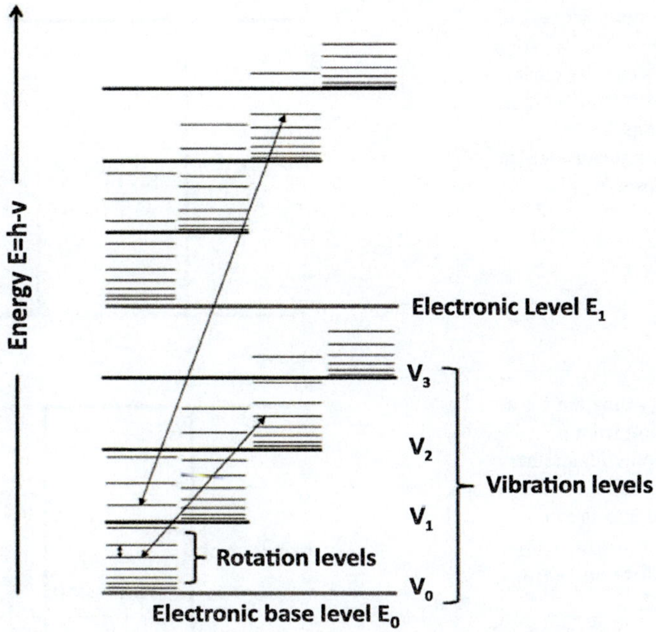

Fig. 6.3 Representation of the three different energy levels for a gas molecule in an energy diagram. The division is made into two electron states E_0 and E_1, as well as the oscillation and rotation levels. The arrows show examples of possible transitions between the different levels (Herzberg 1973)

exactly the amount of energy ΔE required for a higher level, this amount of energy is absorbed by the molecule. According to Eq. 6.1, this is linked to a wavelength λ and the corresponding absorption lines can be seen in the spectrum. These absorption lines are comparable to the Fraunhofer lines in the solar spectrum.

According to the physical law of energy conservation, this energy must either remain in the molecule or be released again in another form of energy. Typically, however, the molecules remain in this excited state for only a short period of $t \approx 10^{-8}$ seconds before returning to the ground state. The position of the ground state is determined by the thermodynamic state (temperature and pressure) of the gas. The dwell time t is also related to the pressure p of the temperature T, since energy transfer can already take place through collisions between molecules. If this energy is completely transferred into translational energy of the molecules, the temperature of the gas rises. In a connected gas volume, the pressure would therefore increase. The radiant energy would thus be converted into thermal energy.

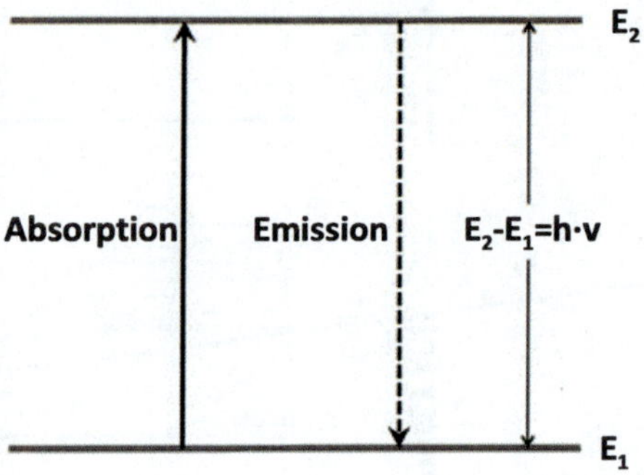

Fig. 6.4 Elementary process of interaction between the radiation and the gas molecule by raising it from a low level E_1 to a higher level E_2. The emission is delayed, by spontaneous or also stimulated emission

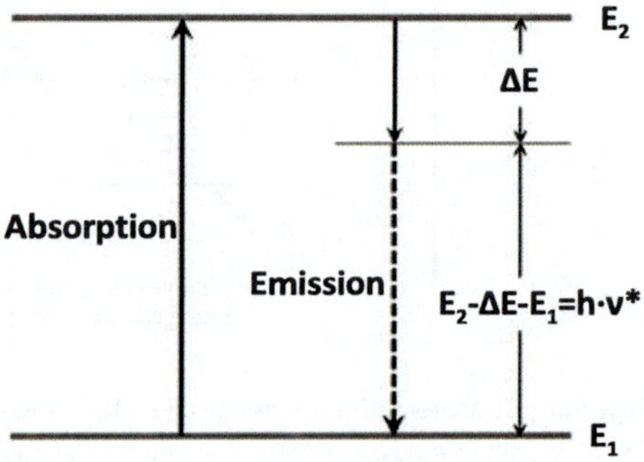

Fig. 6.5 Energy diagram for a radiation emission from a previous collision with another molecule. The amount of energy ΔE is converted into thermal energy, so that a smaller amount is available for the emission

Fluorescence

Another possibility is that the absorbed radiation energy is also released again by radiation. If this release (emission) takes place without further intermediate steps, the emitted wavelength is identical to the irradiated wavelength (Fig. 6.4). In this case one also speaks of resonance fluorescence.

In this state, the excited gas molecules can also interact with other molecules and transfer part of the energy ΔE to them without radiation. The *remaining energy* is then released by emission and the molecule returns to its ground state (Fig. 6.5). This type of fluorescence is subject to the rules according to *Stokes,*[5] according to which the emission wavelength of the fluorescence must always be longer than that of the excitation line.

$$\lambda_{\text{Absorption}} > \lambda_{\text{Emission}} \tag{6.9}$$

Figure 6.6 shows such a fluorescence spectrum with the corresponding excitation spectrum. The difference between the two maxima is called Stokes shift. This fluorescence process

[5] Sir George Gabriel Stokes (1819–1903) Irish mathematician and physicist.

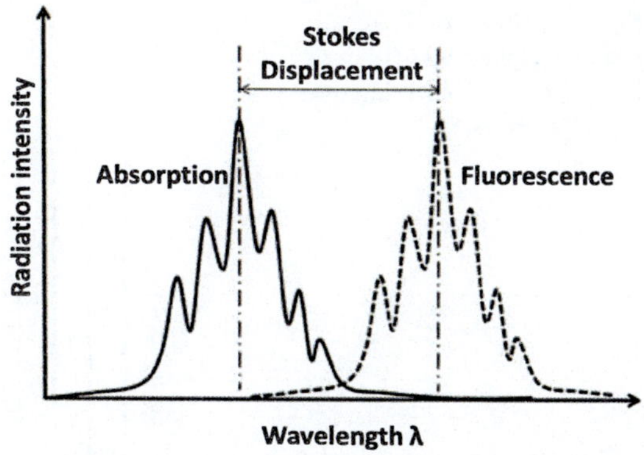

Fig. 6.6 Shift of the emission wavelengths (fluorescence) compared to the excitation spectrum (absorption) according to Stokes

can be considerably disturbed by certain substances if the accompanying substances cause increased transitions without radiation. This is caused in particular by collisions at the molecular level. The fluorescence radiation is greatly reduced by this process. In this context, one therefore also speaks of a fluorescence *quenching* (Bock 1980).

Another form of light is *phosphorescence*, which in contrast to fluorescence is much slower and can decay over minutes or even hours. Chemical reactions can also cause excitation of gas molecules, which manifests itself in the emission of radiation. This process is known as *chemiluminescence* and is used very successfully for the detection of low concentrations of nitrogen oxide (NO).

Smekal-Raman Scattering

Furthermore, there is the possibility that a molecule interacts with radiation, in which absorption occurs only *experimentally* (Meschede 2010). For this purpose, the molecule is transformed into a state in which an arbitrary energy level E_W is assumed. Since this *virtual level* E_W is not allowed quantum mechanically, the photon is immediately emitted. This does not have to result in a return to the basic level, but can also take place at one of the neighboring rotation levels. If the excitation energy starts from the ground state (Fig. 6.7a), the emission of the radiation to the rotational levels is lower, and thus corresponds to the rules according to *Stokes*. If the excitation energy starts already at a rotation level (Fig. 6.7b) that was occupied by thermal excitation, emission lines are obtained that have a higher energy. These shorter wavelength radiation components are called *anti-Stokes* lines (Fig. 6.8). This type of light scattering is called the *Smekal-Raman effect*. Although this effect of *Raman*[6] was not proven until 1928, *Smekal*[7] was able to predict the effect as early as 1923, based on theoretical investigations.

[6] Sir Chandrasekhara Venkata Raman (1888–1970) Indian physicist.

[7] Adolph Gustav Stephan Smekal (1895–1959) Austrian physicist.

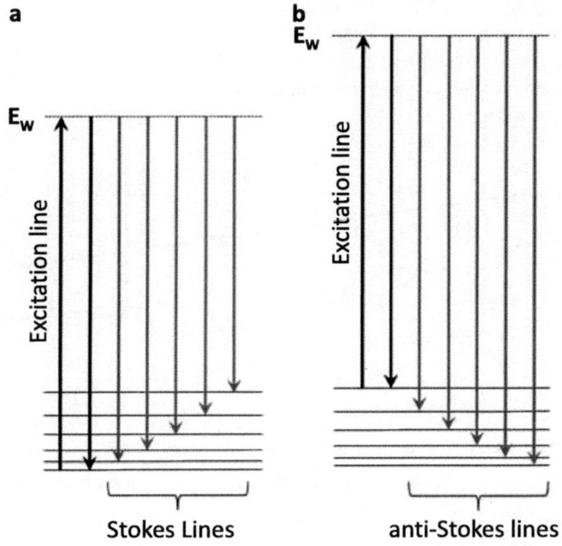

Fig. 6.7 Origin of Stokes (**a**) and anti-Stokes lines (**b**) during Smekal-Raman scattering over a virtual level ($_{EW}$)

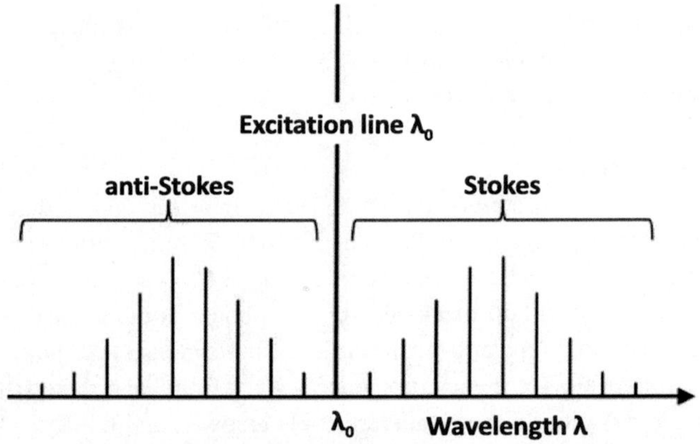

Fig. 6.8 Spectral distribution of Stokes and anti-Stokes lines in Smekal-Raman scattering. The excitation line λ_0 is the main component with the highest intensity in the Raman spectrum and is caused by Rayleigh scattering

However, most of the emitted radiation occurs without any wavelength shift $\Delta\lambda$. The excitation light is therefore only deflected by *Rayleigh scattering.* Since these scattering effects are caused by spontaneous emission, the radiation is emitted in any direction. This means that the original orientation of the excitation light is lost. A powerful laser is usually used as the excitation source, whereby the excitation wavelength λ_0 is of secondary importance. The *Smekal-Ramam effect* does not occur via a resonance condition, so that excitation can also take place in the visible spectral range in order to observe the rotation lines of the gases.

Infrared Spectra

The most important spectral range used for gas analysis purposes lies between 1 μm and 12 μm. In this range, almost all gases have a more or less distinct absorption band, which is characterized by rotational oscillation transitions. In principle, only gas molecules with a permanent dipole moment can absorb IR radiation. These are, for example, substances like CO, HCl, NH_3, etc., while noble gases (e.g. helium) and molecules with the same atoms like H_2, N_2, are excluded from this technique. An exception is oxygen, which has unpaired electrons and therefore also has paramagnetic properties. Oxygen has rotation bands in the microwave range and a weak absorption band at $\lambda \approx 760$ nm.

Rotations

First, the lines of rotation for a diatomic molecule are to be described. In this case, the molecule is regarded as a rigid rotator, in which the rotational energy can be calculated according to classical mechanics as follows:

$$E_{\text{rot}} = \frac{I \cdot \omega^2}{2} \tag{6.10}$$

In Eq. 6.10, I is the mass moment of inertia of the molecule and ω is the angular frequency of the rotational movement. With the angular momentum $L = I \cdot \omega$ one then obtains:

$$E_{\text{rot}} = \frac{L^2}{2 \cdot I} \tag{6.11}$$

From the solution of the *Schrödinger equation*[8] for the rotation, we obtain the quantization of the angular momentum L. The eigenvalues of the square of the angular momentum are:

$$L^2 = J \cdot (J+1) \cdot \hbar^2 \tag{6.12}$$

The magnitude ħ results from the uncertainty relation of *Heisenberg*[9] ($\hbar = \frac{h}{2}\pi$). J is the rotational quantum number. This results in the energy levels of the rotation:

[8]Erwin Schrödinger (1887–1961) Austrian physicist.

[9]Werner Heisenberg (1901–1976) German physicist.

$$E_{rot} = \frac{J \cdot (J+1)\hbar^2}{2 \cdot I} = J \cdot (J+1) \cdot B \qquad J = 0, 1, 2, 3 \ldots \tag{6.13}$$

B is the rotational constant or characteristic rotational energy of the molecule in question (Atkins 1990):

$$hcB = \frac{\hbar^2}{2I} \rightarrow B = \hbar/4\pi cI \tag{6.14}$$

The transitions between the individual adjacent rotation levels apply only to the values $\Delta J = \pm 1$. The energy $\Delta E = h\nu$, between two adjacent rotation levels, is given by:

$$h\nu = \frac{(J+1) \cdot (J+2)\hbar^2}{2 \cdot I} - \frac{J \cdot (J+1)\hbar^2}{2 \cdot I} = \frac{2(J+1)\hbar^2}{2 \cdot I} \tag{6.15}$$

If one now introduces the wave number $\widetilde{\nu} = \nu/c$ instead of the frequency ν, and replaces $\hbar = \frac{h}{2}\pi$, one obtains:

$$h \cdot c \cdot \widetilde{\nu} = \frac{J \cdot (J+1) \cdot h^2}{8 \cdot \pi^2 \cdot I} \tag{6.16}$$

$$\widetilde{\nu} = 2 \cdot (J+1) \cdot \left[\frac{h}{8 \cdot \pi^2 \cdot I}\right] \tag{6.17}$$

$$\widetilde{\nu} = 2 \cdot (J+1) \cdot B \tag{6.18}$$

The distance $\Delta\widetilde{\nu}$ between two adjacent rotation lines is therefore 2*B*. From an infrared spectrum, the constant 2*B* can thus be *read* and the moment of inertia *I* of a molecule can be calculated.

$$I = h/8\pi^2 B \tag{6.19}$$

The moment of inertia of a rigid rotator can also be described by the laws of classical mechanics. In the simplest case, it consists of two atoms arranged at a distance *r0* from each other.

For different atoms (→ different atomic masses m_1 and m_2) the axis of rotation passes through the center of gravity *S*, which is located at a distance r_1 or r_2 from the different masses (Fig. 6.9). Related to the axis of rotation, the moment of inertia results:

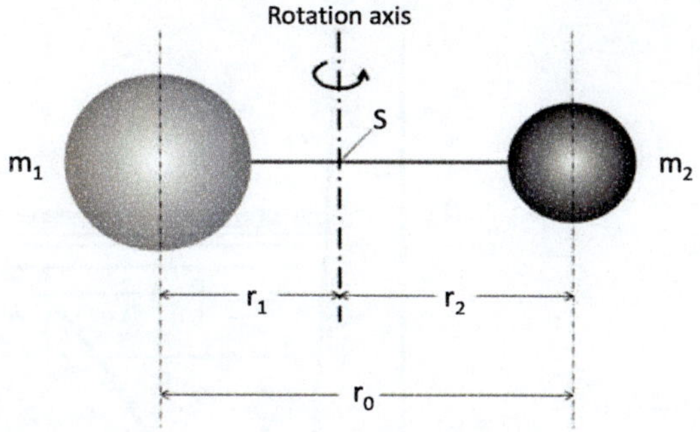

Fig. 6.9 Dumbbell model of a diatomic molecule, with the distance r_0 to each other

$$I = m_1 r_1^2 + m_2 r_2^2 \tag{6.20}$$

Furthermore, the following applies to the torques:

$$m_1 r_1 = m_2 r_2 \qquad \text{and} \qquad r_0 = r_1 + r_2 \tag{6.21}$$

From this, one then receives:

$$I = \left[\frac{m_1 \cdot m_2}{m_1 + m_2}\right] \cdot r_0^2 \tag{6.22}$$

The expression in brackets in Eq. 6.22 is called reduced mass μ:

$$\mu = \frac{m_1 \cdot m_2}{m_1 + m_2} \tag{6.23}$$

With the reduced mass μ one gets the expression for a molecule with two different atoms:

$$I = \mu \cdot r_0^2 \tag{6.24}$$

Vibrations

The vibration energy E_V can also be calculated from the *Schrödinger equation* and is

$$E_V = [v + 1/2] \cdot h\nu \qquad \textit{with} \qquad v = 0, 1, 2, 3, \ldots \tag{6.25}$$

Here v is the vibration quantum number, and ν the frequency of the vibration. The criterion is $\Delta v = \pm 1$. The natural frequency ν of an oscillator results from the restoring force k_F, which corresponds to the binding forces at the molecular level, and the mass, which is also in this case the reduced mass μ:

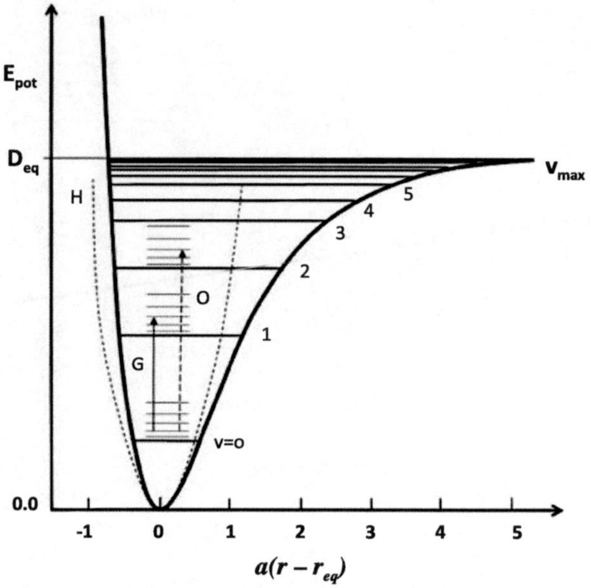

Fig. 6.10 Morse potential for the anharmonic oscillator. The harmonic oscillator (parabola) is shown in *dashed lines* for comparison. The transition *G* is the fundamental oscillation, while *O* shows the harmonic oscillation *(overtone)*

$$\nu = \frac{1}{2\pi}\sqrt{k_F/\mu} \tag{6.26}$$

Since a molecule is only approximately a harmonic oscillator, the energy distances are not equal and become narrower as the quantum number increases. This creates additional lines, the so-called *hot bands*. These *hot bands are* mainly created when the transition from a thermally occupied oscillation state (→ *Boltzmann distribution*) occurs. Furthermore, higher transitions ($\Delta v = \pm 2, \pm 3, \ldots$) can be excited, which then lead to harmonics. Since the transition probabilities are much lower than for the fundamental oscillation, the intensities are also significantly lower. Harmonics are often found in the NIR range.

The anharmonic oscillator can be described by the *Morse potential*[10] for the potential energy (Fig. 6.10). It shows the actual course and explains the possibility of dissociation[11] above D_{eq}. The Morse potential can be described by the following equation:

[10]Philip McCord Morse (1903–1985) American physicist.

[11]Dissociation (from lat. Dissociare separate) describes processes in chemistry, in which the division of a chemical compound into two or more molecules, atoms or ions takes place.

$$E_{pot} = h \cdot c \cdot D_{eq} \cdot \left[1 - \exp\left[-a \cdot \left(r - r_{eq}\right)\right]\right]^2 \tag{6.27}$$

The depth of the potential well is given by D_{eq} and the parameter a can be described by the following relation:

$$a = \omega \cdot \left[\frac{\mu}{2hcD_{eq}}\right]^{1/2} \tag{6.28}$$

To get to the possible oscillation levels one has to solve the Schrödinger equation for the Morse potential. One then receives:

$$E_v = [v + 1/2]h\nu - [v + 1/2]^2 \cdot x_{eq} \qquad \text{mit} \qquad x_{eq} = \frac{a^2\hbar}{2\mu\omega} \tag{6.29}$$

The second term essentially describes the anharmonicity of the oscillation. It gains in importance with increasing quantum number v, since it is subtracted from the first term (harmonic oscillator). The parameter x_{eq} is therefore also called the anharmonicity constant. According to this approach, the number of oscillation levels is finite (v_{max}) and above this limit, dissociation of the molecule occurs.

The transitions between the rotation levels and the vibration levels also depend on the occupation probability n_n. According to the *Boltzmann distribution,* the number n_n of molecules can be calculated which are located at the different energy levels E_n:

$$n_n = n_0 \exp[-E_n/kT] \tag{6.30}$$

At room temperature ($T \approx 300$ K), most molecules are in the ground state (v = 0) of the vibrational energy, and therefore have the energy $E_0 = 0.5 \cdot h\nu$. The next oscillation level (v = 1) already has an energy level of 0.1 to 0.3 eV (≈4 – 12 μm), which can only be reached by thermal excitation at $T >$ 1000K temperatures. The rotation levels are much smaller, so that even at room temperature there is enough thermal energy ($k \cdot T$) to occupy these levels.

Example

It shall be calculated how many CO molecules manage to pass from the basic level to the first oscillation level due to thermal excitation. The spectral position of the CO absorption band is 4.67μm.

According to Eq. 6.8, the following value results for the energy gap E_N from the wavelength $\lambda_{CO} \approx 5$ μm = 5000 nm:

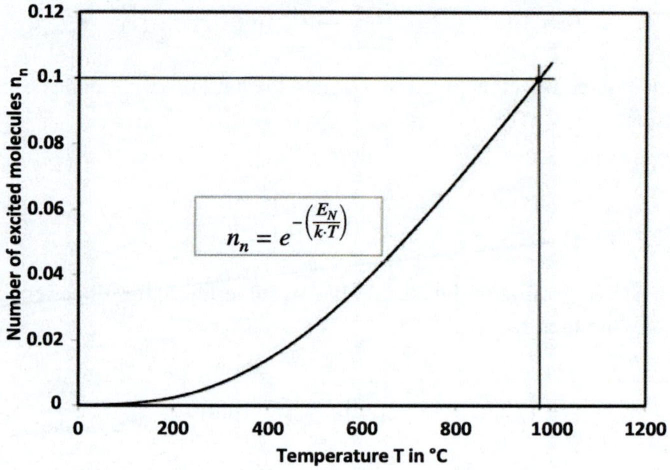

Fig. 6.11 Course of the number n_n (100 % = 1) of molecules with sufficient thermal energy to reach the first level of vibration

$$\frac{1240}{5000} = 0.248\text{eV. For } E_N/k \text{ you then get :}$$

$$\frac{0.248\ \text{eV} \cdot 1.6 \cdot 10^{-19}\text{J/eV}}{1.3807 \cdot 10^{-23}\text{J/K}} = 2873/\text{K}$$

This value $n_n(T)$ is now used for the calculation. The result is shown graphically in Fig. 6.11. At a temperature of 1000 °C approximately 10% of the molecules are in the excited state. At room temperature, only 65 ppm are in the excited state.

Rotational-Vibration Transitions

Figure 6.12 shows the rotation levels and the transition from the base level to the first vibration level. These transitions play a decisive role in IR gas measurement technology, since excitation into the higher vibration level can only occur by absorption of radiation. In the ground state v = 0 the molecule has the initial energy:

$$E_J = \frac{1}{2} \cdot h \cdot \nu + J \cdot (J + 1) \cdot B \tag{6.31}$$

From this level it can change to the next higher oscillation state (v = 1) and the rotational state $J + 1$. The following final energy is reached:

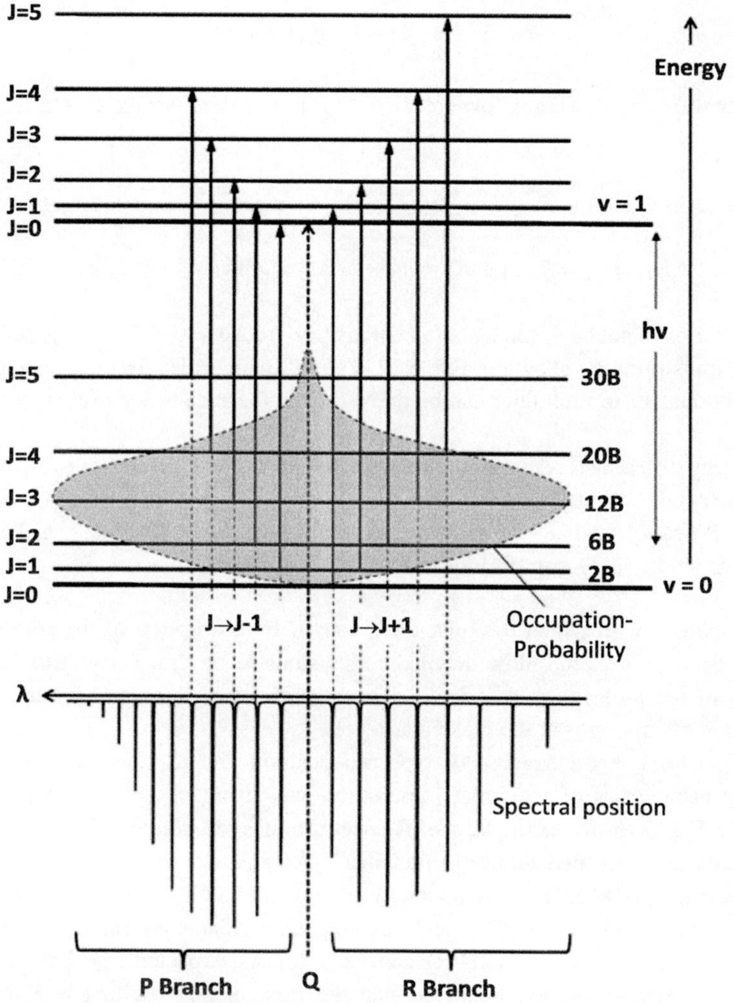

Fig. 6.12 Illustration of the possible rotational oscillation states for a diatomic gas molecule (e.g. CO). The occupation probability (*shaded area*) of the rotational levels results from thermal excitation, according to the *Boltzmann distribution*

$$E_{J+1} = \frac{3}{2} \cdot h \cdot \nu + (J + 1) \cdot (J + 2) \cdot B \tag{6.32}$$

Another possibility is that the molecule changes to the vibrational state v = 1 and the rotational state $J - 1$. The final energy is then:

$$E_{J-1} = \frac{3}{2} \cdot h \cdot \nu + J \cdot (J-1) \cdot B \tag{6.33}$$

The energy differences ΔE are now calculated for both cases. For the first case the result is then:

$$\Delta E_{J \to J+1} = E_{J+1} - E_J = h\nu + 2(J+1)B \quad \text{with} \quad J = 0, 1, 2, \ldots \tag{6.34}$$

$$\Delta E_{J \to J-1} = E_{J-1} - E_J = h\nu - 2JB \quad \text{with} \quad J = 1, 2, \ldots \tag{6.35}$$

In this case, J must not be 0, since only one transition from J = 0 to J = 1 is permitted. The individual transitions are shown in Fig. 6.12. The following rules for the spectral position of the individual absorption lines can be derived from these considerations:

- The energy differences result for the **R-branch** with $J \to J+1$ from($h\nu + 2B$), ($h\nu + 4B$), ($h\nu + 6B$) etc.
- For the **P branch** with $J \to J - 1$ you get ($h\nu - 2B$), ($h\nu - 4B$), ($h\nu - 6B$) etc.
- The distance in the **Q gap** (forbidden transition) is $4B$.

Real gas spectra, with which the transitions caused by absorption of the corresponding radiation can be observed, allow detailed conclusions to be drawn about the molecular properties of the molecules. This method is therefore very important for the scientific investigation of structures at the molecular level.

In gas analysis, knowledge of the spectral position and the factors influencing the absorption behavior is of particular importance, which are described in the following Chap. 7. In Fig. 6.13, for example, the IR spectrum of hydrogen chloride HCl is shown. This spectrum was recorded with c = 1 ppm in a 1 meter cuvette ($c \cdot L = 1$ ppmmeter). The pressure was p = 1013 hPa at a temperature of $T = 25°C$. The spectral resolution of the spectrometer was 0.06cm^{-1}. In this spectrum, both, the fundamental oscillation at 3.47 μ m and the harmonic at 1.77 μ m (NIR) can be seen. As expected, the harmonic has a significantly lower absorption behavior than the fundamental oscillation. Furthermore, the fine structure within the band caused by the rotational transitions can be recognized. To be able to see this fine structure better, the scaling from 2600 to 3100 cm^{-1} was set in Fig. 6.14.

According to the theory, the distances should be exactly 2B and be equidistant for all transitions. The *x-axis* was represented in wave numbers that are proportional to the energy. However, the distances become smaller with increasing wave number (higher energy). The reason for this is the changing core distance r at higher energy and the associated centrifugal forces acting on the rotating and vibrating atoms. The energy differences therefore become smaller. The following then applies to the energy levels

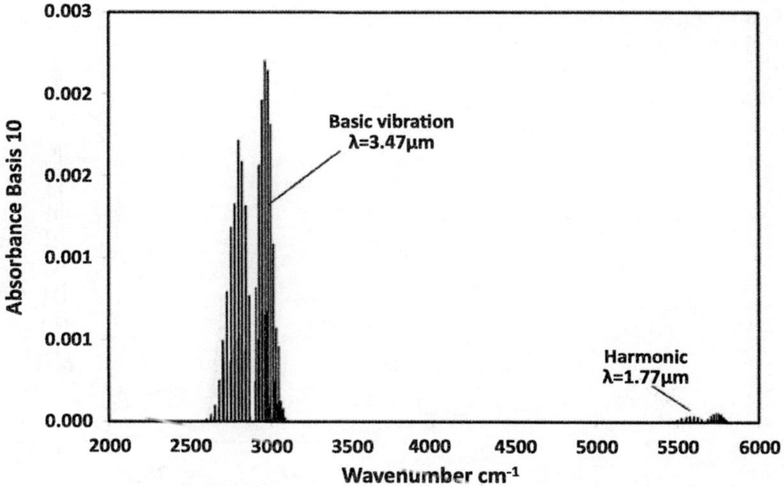

Fig. 6.13 Infrared spectrum of hydrogen chloride (HCl)

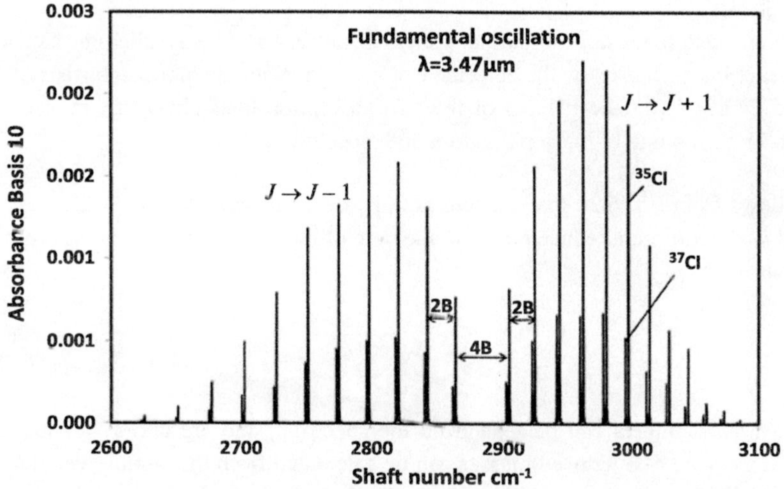

Fig. 6.14 Infrared spectrum of the fundamental oscillation of HCl between 3.2 μm and 3.8 μm

$$E(J) = BJ(+1) - DJ^2(J+1)^2 \tag{6.36}$$

The constant D describes the influence of centrifugal force and is therefore called the *centrifugal strain constant*. For most molecules $D < 10^{-4}$.

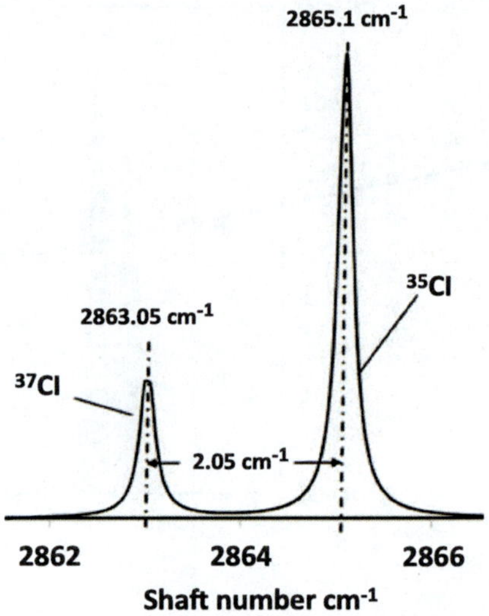

Fig. 6.15 Rotation lines of HCl caused by different chlorine isotopes (^{35}Cl and ^{37}Cl). (Data from the IR spectrum Fig. 6.14)

Isotope Effect

Furthermore, next to the largest lines, at a small distance, a second, smaller line can be seen. This smaller line is caused by the chlorine isotope ^{37}Cl. Natural chlorine consists of 75.8% ^{35}Cl and 24.8% ^{37}Cl. The heights of the two absorption lines show this ratio (≈1 : 3). Figure 6.15 shows one of these pairs, in a high spectral resolution.

It follows from Eq. 6.26 that for two almost identical molecules with the same force constant k_F, but different reduced mass μ, the ratio of the wave numbers can be represented as follows:

$$\frac{\widetilde{\nu}_1}{\widetilde{\nu}_2} = \sqrt{\mu_2/\mu_1} \tag{6.37}$$

The two wave numbers can be read from the spectrum, and are and $\widetilde{\nu}_1 = 2865.1\mathrm{cm}^{-1}$ $\widetilde{\nu}_2 = 2863.5\mathrm{cm}^{-1}$. The reduced masses can be calculated from the atomic weights:

$$\mu_1 = \frac{1.00794 \cdot 34.9689}{1.00794 + 34.9689} = 0.9797 \quad \textit{and} \quad \mu_2 = \frac{1.00794 \cdot 36.9659}{1.00794 + 36.9659} = 0.9812 \tag{6.38}$$

Now, the ratios from the spectrum and the calculation are compared:

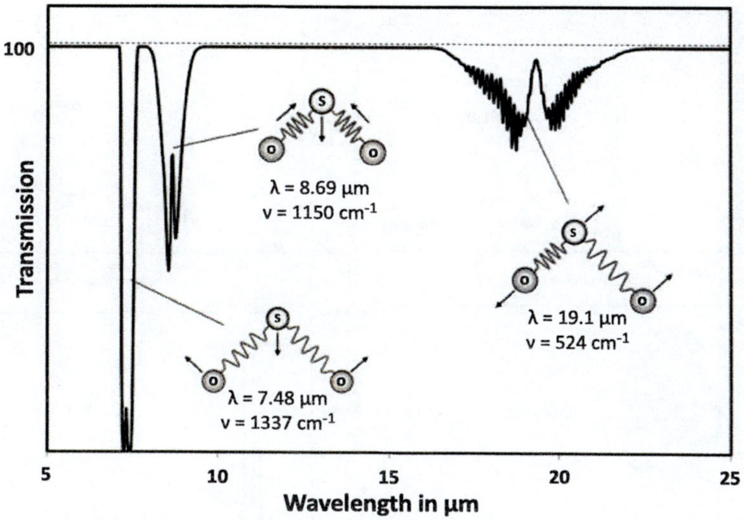

Fig. 6.16 Infrared absorption spectrum of sulfur dioxide (SO_2) with the three fundamental oscillations at 19.1 µm, 8.69 µm and 7.48 µm

$$\frac{\widetilde{\nu}_1}{\widetilde{\nu}_2} = \frac{2865.10}{2863.05} = 1.000716 \qquad and \qquad \sqrt{\frac{0.9812}{0.9797}} = 1.000765 \tag{6.39}$$

The difference between the two values is small, and is explained by measurement errors and deviation from the behavior of the harmonic oscillator. It is interesting in this context that the isotope line $\widetilde{\nu}_2$ with the reduced masses μ_1 and μ_2 from the atomic weights can nevertheless be determined quite accurately:

$$\frac{\widetilde{\nu}_1}{\sqrt{\mu_2/\mu_1}} = \widetilde{\nu}_2 \rightarrow \widetilde{\nu}_2 = \frac{2865.1}{1.000765} = 2862.9\text{cm}^{-1} \tag{6.40}$$

According to the experimental values, the distance $\Delta\widetilde{\nu}$ between the two lines is 2.05 cm^{-1}, while the calculation gives a value of 2.19 cm^{-1}. The deviation is therefore $<7\%$.

Multiatomic Gas Molecules

However, many gaseous substances consist of more than 2 atoms and thus show additional absorption lines, which cannot be expected in diatomic molecules. Due to the different possibilities *to oscillate* and *rotate,* a wide variety of absorption structures result, which differ significantly depending on the type of gas. Figure 6.16 shows the absorption spectrum of sulfur dioxide (SO_2). It consists of three bands, which can be explained by the different directions of oscillation.

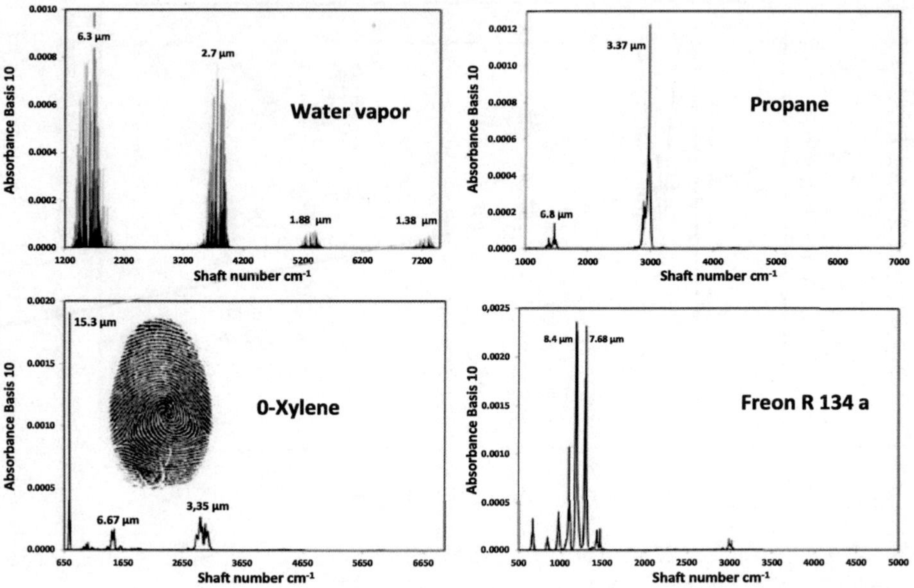

Fig. 6.17 Absorption spectra of different gases. The spectra were recorded under identical conditions (c = 1 pp., L = 1 m, p = 1013 hPa, T = 298 K) and are therefore comparable

Each direction of oscillation has a specific energy, which of course determines the spectral position. Within the band, the described rotation-oscillation transitions also take place. In addition to the different oscillation possibilities and the corresponding rotational oscillation bands, there is also the influence of the atomic masses within the molecular structure of the molecules. Furthermore, isomers, i.e. compounds with the same molecular formula but different structures, also have different absorption spectra. In Fig. 6.17, four examples of different gases are shown, by which the diversity of the absorption structures can be recognized. It is therefore appropriate to compare the multitude of these possibilities with a human fingerprint. Even if there are spectral overlaps between the different gases, they can be separated and selectively measured with correspondingly high-resolution spectroscopic methods (laser spectroscopy, FTIR).

6.2 Laws of Radiation Absorption

Absorption photometers, which operate in the infrared (IR) and ultraviolet (UV) spectral range, are of great importance in gas measurement technology. They use the selective absorption of gases at defined wavelengths to obtain the gas concentration from this

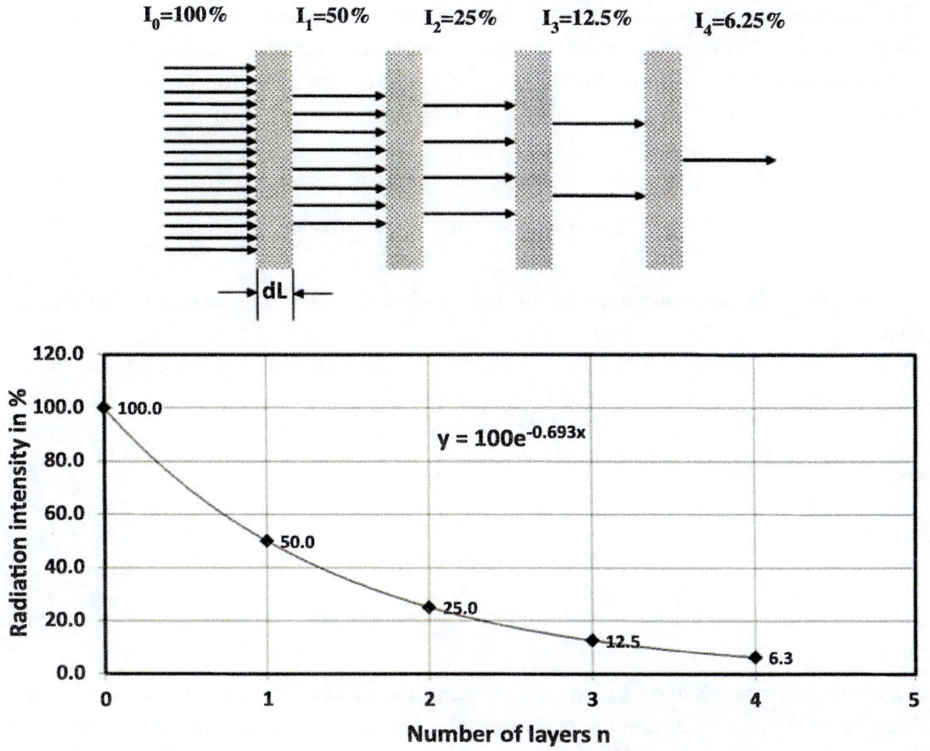

Fig. 6.18 Model for the derivation of the Lambert-Beer law (after Bock 1980)

information. Therefore, the regularities which are subject to this absorption effect also play an important role. The physical effects and influencing factors are therefore described in more detail below.

Lambert-Beer Law

When monochromatic radiation passes through a gaseous medium, a certain amount of the radiation is absorbed by the medium. A prerequisite for this absorption is that the spectral position of the radiation matches a transition in the energy scheme. To calculate the proportion of radiation that results from the absorption, the model shown in Fig. 6.18 can be used. The incident radiation $I_0 = 100\%$ passes through a thin layer dL in which a certain amount of radiation is absorbed. In this case, this proportion was fixed at 50%, so that after this layer, only 50% of the original radiation remains. The same process happens in the second layer, which reduces the original radiation to 25%. The same process happens in the third and fourth layer, so after these $n = 4$ layers the radiation intensity has been reduced to 6.25%. If you now enter the result in a diagram, you get a falling intensity curve, which can be approximated by an e-function. If one now assumes that the layer thickness

dL is infinitesimal, the decrease in intensity dI will be proportional to the layer thickness dL and the intensity I. Furthermore, the specific absorption behavior of the different gases has to be considered by a proportionality factor α. Therefore, the following relationship applies:

$$\mathrm{d}I = \mathrm{d}L \cdot \alpha \cdot I \quad or \quad \frac{\mathrm{d}I}{I} = \alpha \cdot \mathrm{d}L \tag{6.41}$$

In a next step, the integration along the light path L through the gaseous medium takes place:

$$\int_{I_0}^{I} \frac{\mathrm{d}I}{I} = -\int_{0}^{L} \alpha \cdot L \tag{6.42}$$

So it follows:

$$\ln \frac{I}{I_0} = -\alpha \cdot L \quad or \quad I = I_0 \cdot \exp(-\alpha \cdot L) \tag{6.43}$$

In gas measurement technology, the gas concentration c also plays a decisive role on the absorption behavior. Furthermore, pressure and temperature will influence the gas density, which changes the number of interacting molecules. Due to this density change, more or less radiation is absorbed. For this purpose, the model just described is extended.

To do this, we observe the gas mixture to be investigated (concentration c) at a pressure p and temperature T in a cuvette (see Fig. 6.19). The entire gas layer has a length L and the incoming radiation has an intensity I_0. At the end of the cuvette the radiation intensity $I(c)$ has decreased to the value due to absorption. If there is no absorbing gas concentration in the cuvette (c = 0), both intensities are equal $I_0 = I(c)$.

The ratio of outgoing radiation $I(c)$ to incoming radiation I_0 is called transmission T or transmittance D:

$$T = \frac{I(c)}{I_0} \tag{6.44}$$

The proportion of radiation that is absorbed by the gas is called absorption A:

$$A = \frac{I_0 - I(c)}{I_0} = 1 - \frac{I(c)}{I_0} \tag{6.45}$$

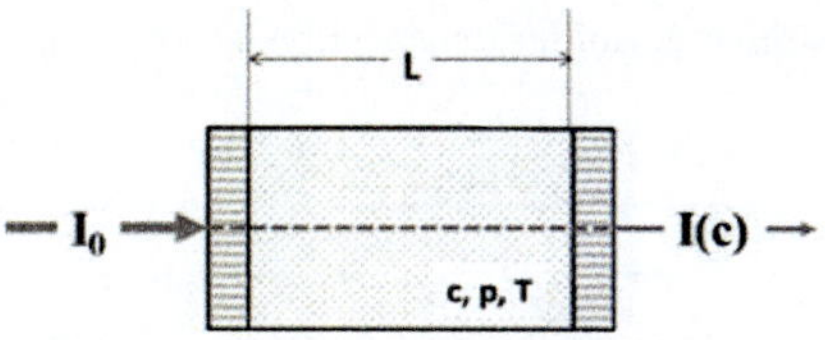

Fig. 6.19 Radiation absorption in a cuvette under defined boundary conditions. The cuvette is connected with windows that are transparent for the spectral range under investigation

Both radiation components ($A + T$), add up to 1 if no additional light scattering (S) occurs in the cuvette. Furthermore, reflection losses (R) at the window surfaces can influence this calculation. It is then valid, depending on the point of view:

$$A+T+S+R=1 \qquad \text{or} \qquad A+T=1 \tag{6.46}$$

The absorption A in the cuvette increases with increasing cuvette length L, since the number of interactions with the gas molecules increases along this path. The same happens naturally at a higher pressure p in the cuvette, which leads to a higher particle density. According to the laws of gases, the temperature T has an opposite effect. With increasing temperature T, the gas density decreases. Since radiation absorption can occur at different wavelengths λ, the spectral course of the absorption coefficient α (λ) must be known. All influencing factors are now summarized in a law known as the *Lambert*[12]*-Beer law*[13] (Bock 1980):

$$I(c)=I_0\exp(-c\cdot\alpha(\lambda)\cdot L) \qquad \text{or} \qquad I(c)=I_0\exp\left(-\alpha(\lambda)\cdot\frac{p_1T_0}{p_0T_1}\cdot c\cdot L\right) \tag{6.47}$$

The absorption coefficient is indicated in the gas measuring technique with cm $^{-1}$, if the dimensionless value 1 is used for the gas concentration c ($1 = 100$ vol.% or for 10^6 ppm). Equation 6.47 can also be represented with the decimal logarithm, in which case the absorption coefficient $\alpha(\lambda)$ changes$\rightarrow\alpha^*(\lambda)$:

$$I(c)=I_0\cdot 10^{-\alpha^*(\lambda)\cdot c\cdot L} \tag{6.48}$$

An important quantity that can be derived from the *Lambert-Beer law* is the extinction:

[12] Johann Heinrich Lambert (1728–1777) German mathematician and physicist.

[13] August Beer (1825–1863) German mathematician, chemist and physicist.

Table 6.1 Illustration of the three quantities transmission, absorption, extinction/absorbance (Bock 1980)

Transmission [%]	Absorption [%]	Extinction/Absorbance
0.0	100	
0.1	99.9	3.000
1.0	99	2.000
5.0	95	1.301
10	90	1.000
20	80	0.699
30	70	0.523
40	60	0.398
50	50	0.301
60	40	0.22
70	30	0.155
80	20	0.097
90	10	0.046
100	0	0.000

$$\text{Extinction } E_{10} = -\log_{10}(1/T) = -I_0/I(c) \tag{6.49}$$

or

$$\text{Extinction } E_e = -\ln(1/T) = -I_0/I(c) \tag{6.50}$$

In international linguistic usage, the *absorbance* is called *Absorbance A* and has also become increasingly widespread in the German-speaking world. However, *absorbance* must not be confused with absorption (Table 6.1).

The conversion between the *absorbance* values (base 10) to the natural logarithm (Napierian) is done by multiplication with ln(10) = 2.302585… ≈ 2.303.

$$2.303 \cdot \log_{10}\left[\frac{I_0}{I(c)}\right] = \alpha(\lambda) \cdot c \cdot L \tag{6.51}$$

According to Eq. 6.47, the radiation intensity decreases exponentially with the concentration c, if the other parameters are kept constant. Figure 6.20 shows the course based on a gas measurement from 0 to 10,000 ppm (=1 vol.%). As absorption coefficient a value of $\alpha = 30\ \text{cm}^{-1}$ was chosen, and a cell length of $L = 10$ cm was used. The greatest sensitivity (signal change/concentration) is at the beginning, which runs in an almost linear range up to c = 1000 ppm. After that, a curvature sets in which can be linearized up to a concentration range of 4000 ppm, using appropriate algorithms. Above this range, the measurement gets more and more saturated and a linearization becomes more and more

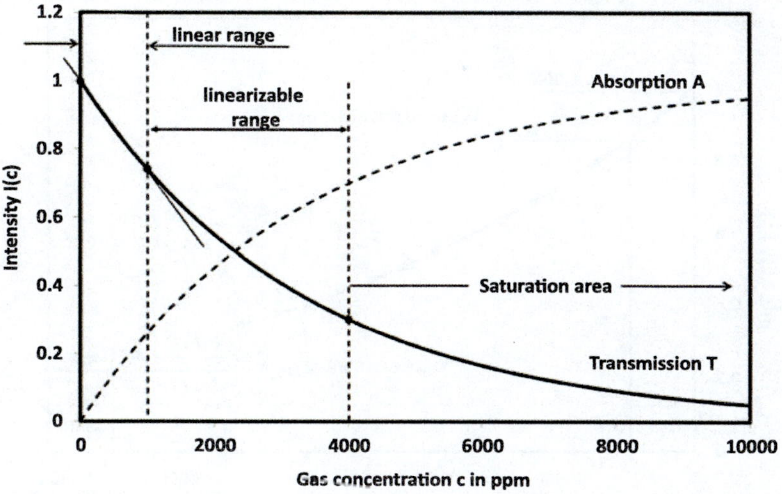

Fig. 6.20 Course of transmission *T* and absorption *A* as a function of the gas concentration *c*

difficult, and is connected with correspondingly large deviations. For a gas analyzer that shows this behavior, the cell length would have to be reduced to 4 cm to remain in the linearizable range. For different cell lengths the product of $c \cdot L$. In this specific case, the linearizability is given by 4000 ppm (=0.4%) for a cell length of 10 cm. The product is therefore *40%-mm*. With a cell length of 4 cm and a maximum concentration of 1 vol.%, the same product is obtained as the result.

Detection Limit

An important parameter in the evaluation of gas measuring methods is the smallest gas concentration that can still be detected. This point will be repeatedly referred to in the following chapters. A detailed description is given in particular in Sect. 12.4. On the basis of the course of the curve, according to Lambert-Beer, an estimation should be carried out at this point. Figure 6.21 shows the curve of a typical gas analyzer (absorption photometer). The measuring points were represented by an exponential trend line with which all points of the characteristic curve can be calculated. At the zero point (c = 0), the radiation intensity is 1.000 and takes on a value of 0.861 at a concentration *c* of 500 ppm. The difference for this concentration change around 500 ppm is therefore 0.139. Related to the concentration, this is then 0.139/500 ppm = 0.000278 /ppm. Modern detector systems are capable of reliably detecting signal changes from $\Delta I = 10^{-5}$. If this property is transferred to the concrete example, this means a concentration change Δc of only 0.036 ppm = 36 ppb. This value corresponds to the physical limit for the detection of a component with the concrete structure. In the endpoint, the sensitivity is much lower. The detection limit at this point of the characteristic curve is then 0.14 ppm = 140 ppb. In practical cases, however,

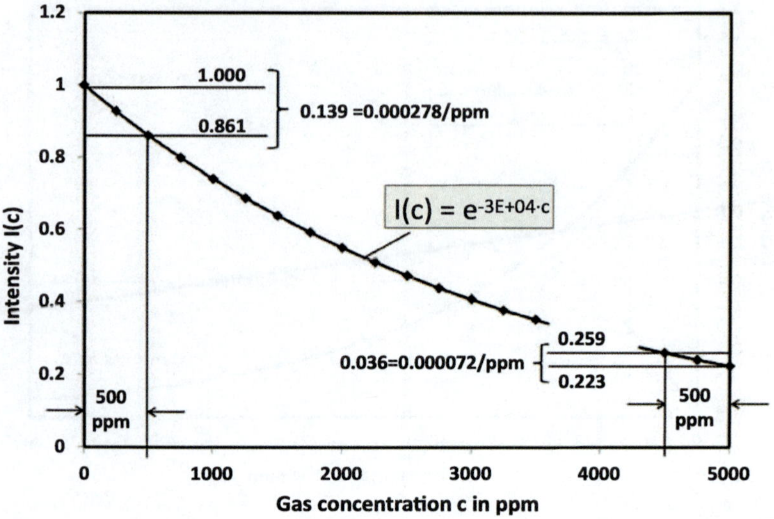

Fig. 6.21 Determining the detection limit from the characteristic curve of the gas analyzer

the detection limit is obtained from the real measurement data and subsequent statistical evaluation. As a rule, these values are one order of magnitude higher, since external interference is the limiting factor.

Line Shape and Influencing Variables

The contour of an absorption or emission line is influenced by various physical processes and laws, which also lead to corresponding interference effects in gas measurement. The transition between two energy levels leads to a spectral line that is defined very precisely. The characteristic of such a spectral line is defined by the spectral position ν_0 of the absorption maximum, the line width $\Delta\nu$ at 50% of the maximum absorption and the maximum distribution function (absorption, transmission, cross section, etc.). Figure 6.22 shows such a real absorption line of CO_2, with the corresponding definition quantities.

In principle, this line should be arbitrarily sharp, since the reference $E = h \cdot \nu$ applies to the photon. There are, however, several influencing factors that lead to a wider absorption line. The first limitation is the *Heisenberg uncertainty relation*:

$$\Delta E \cdot \Delta t \geq \hbar \tag{6.52}$$

Since the lifetime τ of an excited state, at a higher energy level, is limited in time, one obtains a fuzziness of energy:

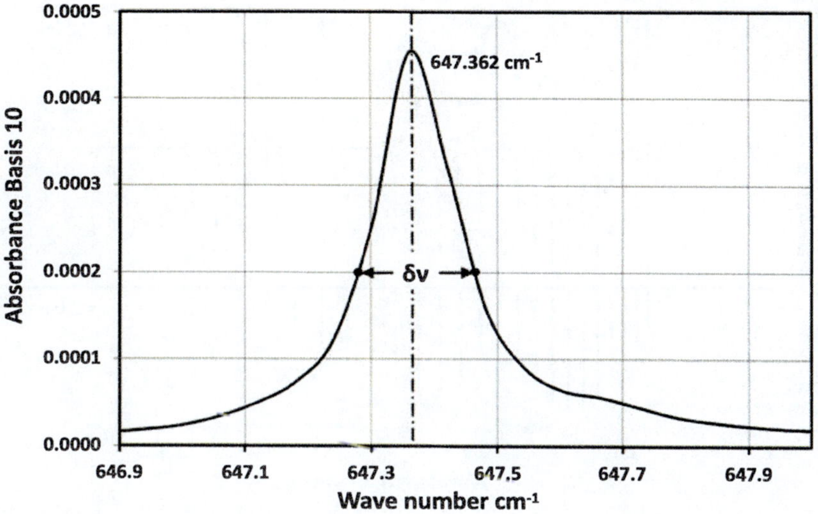

Fig. 6.22 Representation and description of an absorption line. The half-width $\delta\nu = \gamma/2\pi$

$$\Delta E \geq \hbar/\tau \qquad \text{or also} \qquad \delta\nu = \left[5.31 \ \text{cm}^{-1}\right]/[\tau/P \ s] \tag{6.53}$$

The typical lifetime of an electronically excited state is $\approx 10^{-8}$ s. This results in a natural line width of $5 \cdot 10^{-4}\text{cm}^{-1}$. According to the laws of classical physics, this process can be compared to a damped harmonic oscillator. The spectral power density then results from (Soltwisch 2010):

$$P_\omega(\omega) = P_0 \cdot \frac{\gamma/2\pi}{(\omega_0 - \omega)^2 + (\gamma/2)^2} \tag{6.54}$$

This function describes the Lorenz profile of a spectral line, with the half-width $\gamma = \delta\omega$ or $\gamma = \delta\nu \cdot 2\pi$. This minimum spectral width is also called the natural line width.

Pressure Broadening

The lifetime τ, at an excited level, is additionally shortened by impacts with other gas molecules. Each collision therefore influences the retention at the higher level and thus also the line width $\delta\nu$ of the transition. These collisions can occur with a similar molecule (CO → CO), as well as with a foreign impact partner (CO → N_2). In the case of foreign collision partners, one also speaks of a *widening of the collision*, which also has an effect on the characteristic curve behaviour of optical gas measuring methods (→ carrier gas dependency). Figure 6.23 shows the absorption spectrum of hydrogen chloride (HCl) at

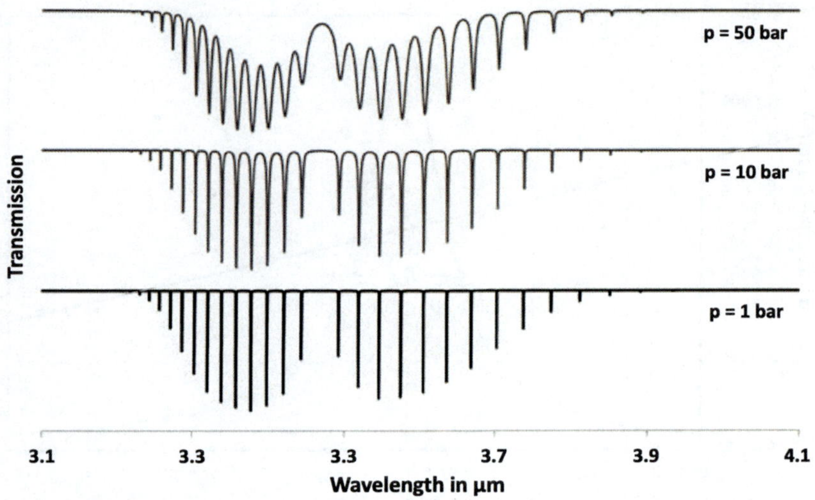

Fig. 6.23 Change in the band structure of HCl at different pressure values (1–50 bar) in the cuvette. The data was calculated using the MolExplorer program, which is based on the HITRAN library (PAS-Tech GmbH Hamburg)

different pressure conditions. The widening of the absorption lines due to an increase in pressure is clearly visible.

This total line widening can be calculated from p_S= partial pressure of the carrier gas, p= absolute pressure, γ_E= coefficient of intrinsic pressure widening, and γ_S= coefficient of shock widening by a foreign gas (carrier gas) as follows:

$$\delta\nu(p) = 2\gamma_E \cdot (p - p_S) + 2\gamma_S \cdot p_S \tag{6.55}$$

If there are no foreign impact partners, Eq. 6.55 is reduced to the following expression, which only describes the impacts with each other and thus reflects the pressure broadening.

$$\delta\nu(p) = 2\gamma_E \cdot p \tag{6.56}$$

The probability of impact is naturally related to the pressure p and is lowest in the vacuum range (e.g. vacuum). To be able to measure the *pure* absorption line, one must therefore work in the vacuum range at $p < 1$ hPa. Figure 6.24 shows individual, high-resolution absorption lines at four different absolute pressures above 1 bar.

Carrier Gas Dependency

According to Eq. 6.55, the half-value width $\delta\nu$ changes if there is a foreign gas in the gas mixture. In gas analysis, this has serious effects on the characteristic curve. Typically, the

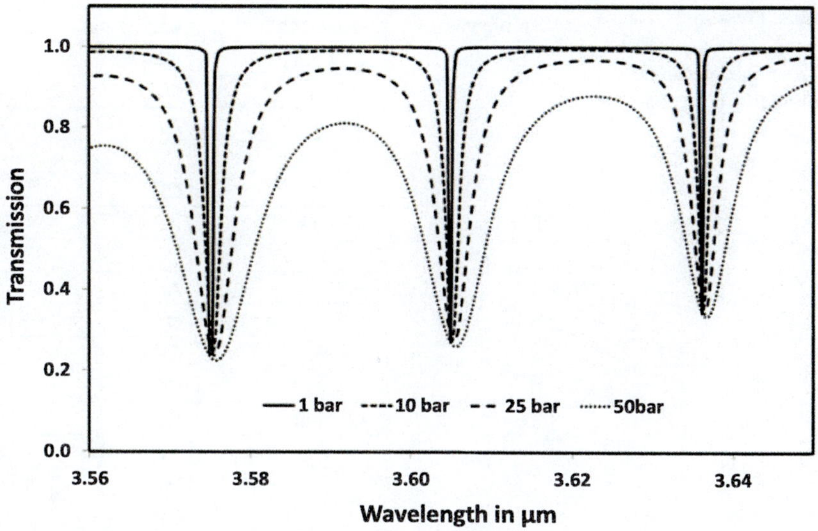

Fig. 6.24 Change in the line structure of HCl within the absorption band at different pressure conditions

characteristic curve for a certain gas is recorded with nitrogen (N_2) as zero gas. If the accompanying components change in a real gas mixture, this can lead to a false reading. This error is called *carrier gas dependency*. In most applications, however, this influence is neglected, since the error is <1% of span. Depending on the application, this error must then be determined experimentally and corrected if necessary.

In the literature (Burch et al. 1962), there are references to this influencing variable, which are called *foreign broadening coefficient (FBC)*. Table 6.2 lists the coefficients for different gases and gas mixtures.

Pressure Shift

Furthermore, the central position ν_0 of the absorption line shifts according to the following equation:

$$\nu_0^* = \nu_0 + \delta \cdot p \tag{6.57}$$

This shift is very small and only becomes relevant in high pressure conditions. The order of magnitude for δ is ≈0.01cm $^{-1}$/bar. Figure 6.25 shows the widening in a pressure range of 1–50 bar. The coefficient in this case is 0.013 cm $^{-1}$ /bar.

Table 6.2 Pressure broadening coefficient (FBC) for different gas mixtures. (According to Burch et al. 1962)

Sample gas	Wavelength	Carrier gas	FBC
N_2O	4.49 μm	He	0.73
		O_2	0.83
		Ar	0.78
		H_2	1.23
		CH_4	1.08
N_2O	7.78 μm	He	0.70
		H_2	1.21
		Ar	0.83
		O_2	0.72
		CO_2	1.17
		CO	0.97
CO	4.67 μm	He	0.64
		Ar	0.78
		H_2	0.85
		CH_4	1.12
CO_2	4.26 μm	He	0.59
		O_2	0.81
		H_2	1.17
		Ar	0.78
CH_4	3.32 μm	He	0.56
		Ar	0.82
		CO_2	1.25

Doppler Broadaning

Molecules move freely in space according to the laws of the kinetic theory of gases. If a measurement is made by passing a light beam through the gas volume, it hits molecules with different velocities v_M and directions of flight. The velocities result from *Maxwell's velocity distribution* and are given by the temperature *T*. This situation is shown in Fig. 6.26. In extreme cases the radiation hits the molecules 1 + 2 which move in the direction of the radiation (1) and against the radiation (2). If the molecule moves towards the radiation, the radiation received becomes shorter-waved according to the wavelength shift:

$$\lambda_1 = [1 - v_M/c] \cdot \lambda \tag{6.58}$$

If the molecule moves away from the radiation source, a longer-wave signal is obtained:

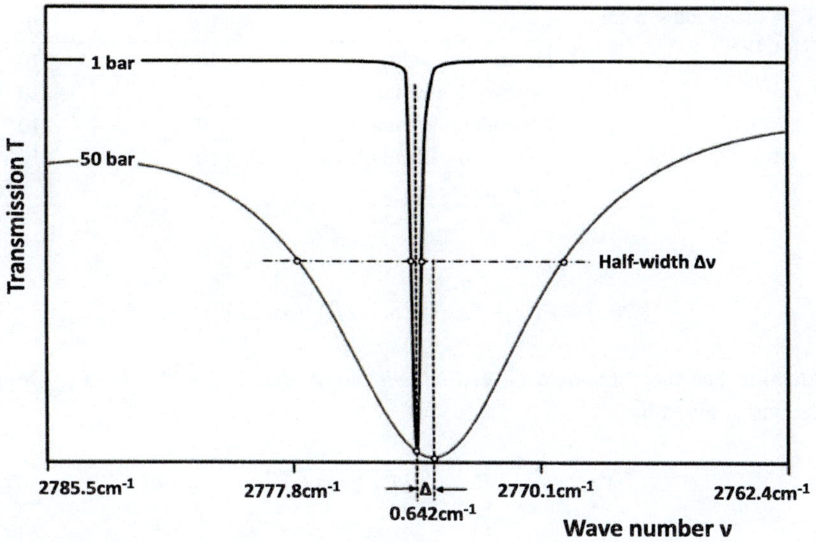

Fig. 6.25 Widening and shifting of an HCl absorption line in a pressure area of 1 → 50 bar

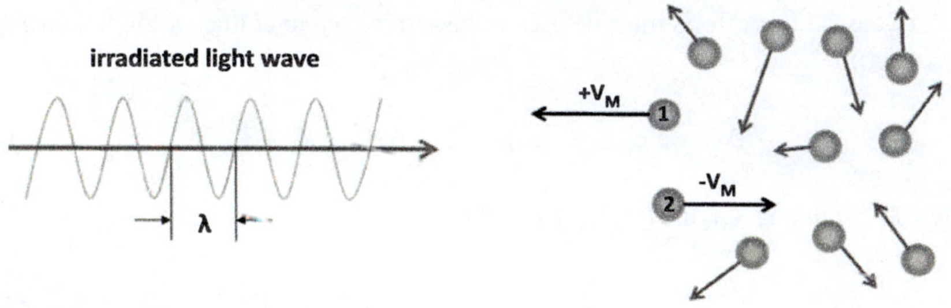

Fig. 6.26 Model to explain the *Doppler effect* on the spectral line width

$$\lambda_2 = [1 + v_M/c] \cdot \lambda \tag{6.59}$$

Within the gas, the molecules naturally move in all directions equally, statistically speaking, whereby the most probable speed v_W is used as the velocity.

$$v_W = \sqrt{2kT/m} \tag{6.60}$$

Based on this consideration, the absorbed radiant power ($P\omega$) for a Doppler-broaded spectral line can be calculated (Soltwisch 2010):

Table 6.3 Line widths in cm^{-1} (Schrader 1980)

Effect	UV	IR
Natural line width	10^{-3}	10^{-8}
Doppler line width	10^{-2}	10^{-3}
Pressure broadening 1 mbar	10^{-5}	10^{-5}
Pressure broadening 1 bar	10^{-2}	10^{-2}
Laser line width	10^{-3}–10^{-6}	…10^{-10}

$$P(\omega) = P(\omega_0) \cdot \exp\left[c(\omega - \omega_0)/(\omega_0 \cdot v_W)\right]^2 \tag{6.61}$$

This function has the form of a *Gaussian bell curve*. The half-width for the Doppler-broadad line is given by:

$$\delta\nu = \frac{\nu}{c} \cdot \sqrt{(8kT \cdot \ln 2)/m} \tag{6.62}$$

The Doppler width $\delta\nu$ thus increases with the wave number and is proportional to $T^{1/2}$.

The line width $\Delta\lambda$ or $\Delta\nu$ of an absorption or emission level therefore depends on different physical factors. Since these influencing factors can also occur simultaneously, the sum of the various effects must be used to determine the total line width. It then applies (Atkins 1990):

$$\Delta E = \Delta E_{\text{push}} + \Delta E_{\text{Doppler}} + \Delta E_{\text{naturally}} + \ldots \tag{6.63}$$

For the corresponding service life τ_{total} you then get:

$$\frac{1}{\tau_{\text{total}}} = \frac{1}{\tau_{\text{push}}} + \frac{1}{\tau_{\text{Doppler}}} + \frac{1}{\tau_{\text{naturally}}} + \ldots \tag{6.64}$$

Typical line widths are listed in Table 6.3.

Temperature Influence

The temperature T of the gas molecules has an influence on the velocity distribution, and thus also on the line width due to the Doppler effect. With increasing temperature T, there is also the possibility for single molecules to reach a higher energy level of rotation according to the *Boltzmann distribution.* From these higher rotation levels, additional rotational-vibration transitions are now possible, according to the selection criteria. At higher gas temperatures, therefore, additional rotation lines would have to be visible in the IR spectrum. Figure 6.27 shows the IR spectrum of hydrogen chloride (HCl). Between 200 K and 400 K, 4 additional lines are visible. Between 400 K and 800 K there are

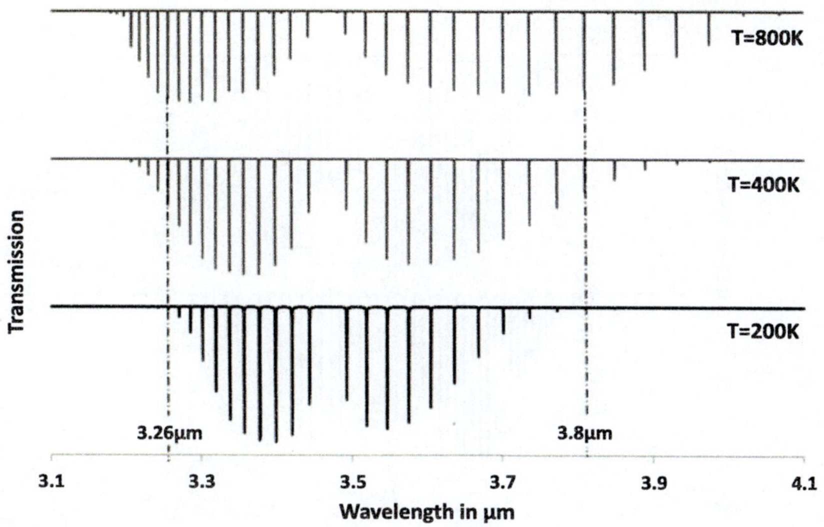

Fig. 6.27 Change in the band structure of HCl at different temperatures

only 2 additional lines, but the line thickness increases significantly due to the increasing occupation probability. Between 200 K and 600 K there are about 6 new lines in the chosen example, i.e. additional ≈1 lines per 100 K. This influencing effect must therefore be considered for gas analytical measurements at large temperature changes (T > 50 K) (Wiegleb 1984; Gatzmanga 2005).

In gas measurement technology, one works mainly in a temperature range from 5 to 50 °C. The occupation probabilities change in this temperature range from $\Delta T = 45$ K, even if no additional energy levels are excited. Due to these different occupations of the rotation levels, the intensities of the individual lines change. Figure 6.28 shows the carbon monoxide spectrum for both temperatures. The spectra initially look very similar. No new rotational-oscillation transitions can be seen between the 5 °C spectrum and the 50 °C spectrum. However, when the ratio of the two spectra is plotted, these shifts become quite clear. The lower rotation levels (region A.), are *depopulated* by excitation of the molecules in them and the lines weaken. The higher rotation levels (area B. + C) benefit from this process, which causes these lines to become stronger. This effect can be observed in IR photometers, which leads to a temperature error in the sensitivity. Naturally, this temperature error is superimposed by the influence of the gas density change.

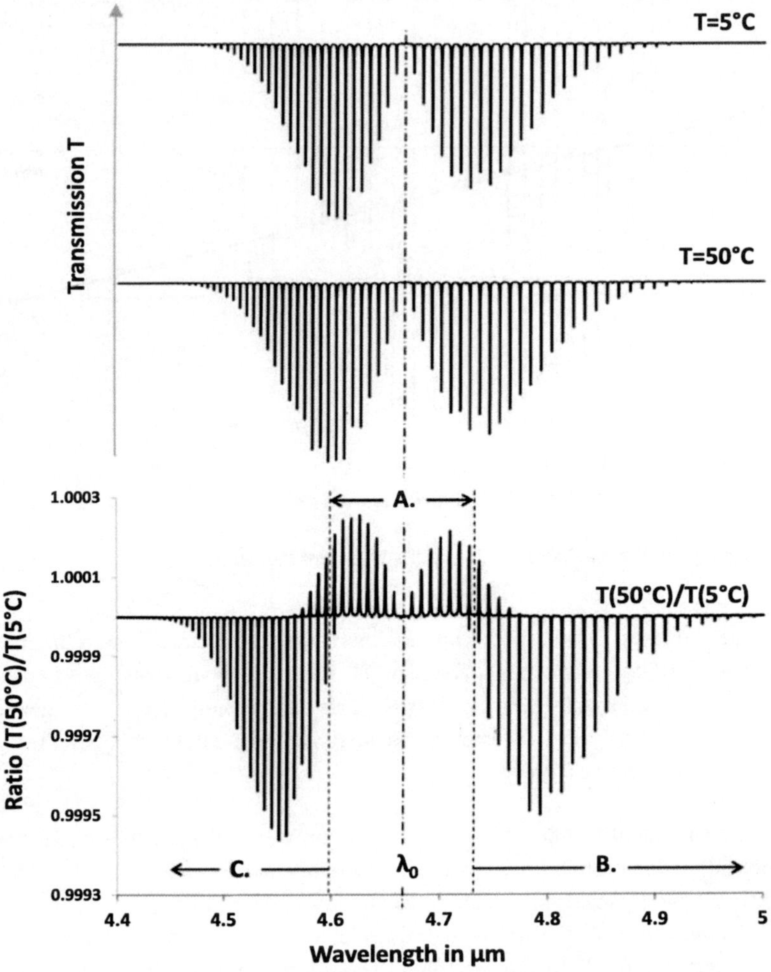

Fig. 6.28 Comparison of the CO spectra (transmission) at 5 °C and 50 °C and as quotient of both transmission values

References

Atkins, P.W.: Physikalische Chemie. VCH, Weinheim (1990)

Bock, R.: Methoden der Analytischen Chemie. Bd. 2: Nachweis und Bestimmungsmethoden (Teil 1). Verlag Chemie, Weinheim (1980)

Burch, D.E., Singleton, E.B., Williams, D.: Absorption line broadening in the infrared. Appl. Opt. **1**(3), 359–363 (1962)

Gatzmanga, H.: Entwicklung einer Hochtemperatur-online-Analytik bis 300 C auf der Basis von Infrarotanalysatoren. Sachlicher Abschlussbericht zum AiF-Vorhaben 13706 BR (2005)

Herzberg, G.: Einführung in die Molekülspektroskopie. Steinkopf, Darmstadt (1973)
Meschede, D.: Gerthsen Physik. 24. Aufl. Springer (2010)
Schrader, B.: Infrarot- und Ramanspektrometrie. Ullmanns Enzyklopädie der technischen Chemie. 4. Aufl., Bd. 5. Verlag Chemie, Weinheim (1980)
Soltwisch, H.: Physik III (Atom-und Quantenphysik) Skript WS 2010/11 VL25/5 (2010)
Wiegleb, G.: Gasanalysensysteme für erhöhte Prozesstemperaturen. Tech. Mess. **51**, 385–393 (1984)

Related References

Ansyco GmbH: Technische Information 201305-02 V3. 29.11.2015 http://www.ansyco.de Zugegriffen: November 2015
Clarke, M.G.: Industrial Air Pollution Monitoring. Chapman & Hall (1998)
Penner, S.: Quantitative Molecular Spectroscopy and Gas Emissivities. Addison-Wesley Company, Reading (1959)
Wiegleb, G.: Forschungsbereich Gasmesstechnik im Institut für Mikrosensorik (IfM) an der Fachhochschule Dortmund. GASWÄRME International. **56**(5/2007), 373–376 (2007)
Winkler, L.W.: Bestimmung von gelöstem Sauerstoff in Wasser. Deutsche Chemische Gesellschaft (Berlin). **21**, 2843 (1888)
Zöchbauer, M., Wiegleb, G.: New optical methods for applications in process analysis and medical instruments. Exp. Tech. Phys. **40**(2), 241–257 (1994)

7 IR Absorption Photometer

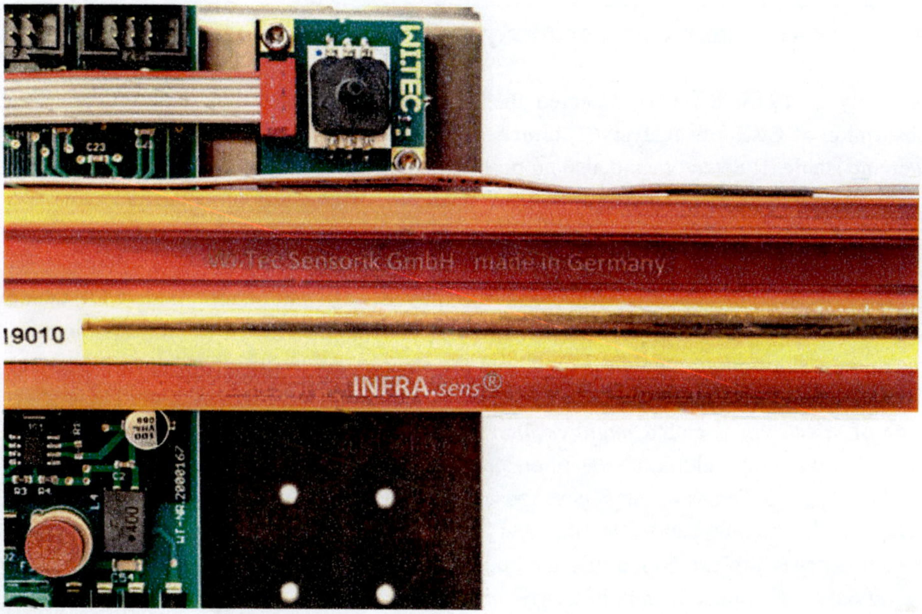

INFRAsens, www.witec-sensorik.de

The infrared spectral range was discovered accidentally by *Herschel*[1] *in* 1800, when he wanted to measure the temperature of the different colors of sunlight. To do this, he let the

[1]Friedrich Wilhelm Herschel (1738–1822) German-British astronomer.

G. Wiegleb, *Gas Measurement Technology in Theory and Practice*,
https://doi.org/10.1007/978-3-658-37232-3_7

sunlight pass through a prism and placed a thermometer in each color range. He noticed that beyond the red end of the visible spectrum, the thermometer showed the highest temperature, although no visible radiation was visible there. From the observed temperature rise, he concluded that the solar spectrum continues below (→ *infra*) the red.

This infrared range, along with the visible and ultraviolet spectral range, forms the basis of the optical methods of gas measurement technology. As early as 1865, *Tyndall*[2] reported on the absorption of infrared radiation by various gases. He investigated in particular carbon dioxide, nitrogen oxide, and alcohol vapor. In 1881, *Röntgen*[3] discovered that the absorption of infrared radiation in a closed measuring cell filled with ammonia leads to a significant increase in pressure (Röntgen 1881). Many years later, this opto-pneumatic effect provided the basis for the first practical IR gas analyzers according to K.F. Luft. Parallel to this, Bell[4] observed the same effect, using sound propagation for detection. Bell mainly investigated solids and liquids. This effect is therefore also known as the *Bell effect*, while for gaseous substances the *Röntgen Tyndall effect* is used (Hill and Powell 1968).

7.1 Opto-Pneumatic Gas Analysers

As early as 1926, Schmick patented the first IR gas analyzer that used the selective absorption of gases for analysis (Schmick 1926). The problem at that time was that there were no stable IR detectors and also no possibilities of spectral selectivation. A thermopile or bolometer was used to detect thermal radiation, which fell onto the detector in a broadband manner without any further filtering. Interference filters, as we know and use them today, were still unknown at that time.

The selectivation was realized by a gas-filled cuvette (e.g. carbon monoxide-CO), which was placed in one of the two beam paths (Fig. 7.1). This provided a reference, since this radiation is not further attenuated by CO in the sample gas. A cuvette with nitrogen was inserted in the measuring path, so that the total absorption by CO was measured here. This type of selectivity is called *negative filtering*. To reduce the cross-sensitivities to other gases (see Fig. 7.2), additional gas filters containing the interfering gases were installed in the beam path to filter out the superimposed spectral range. Since no saturation could be achieved due to the band structure and with cell lengths of a few centimeters, this possibility was limited. Since this method was also very susceptible to interference, it found only few applications in industrial practice.

[2] John Tyndall (1820–1893) Irish physicist.

[3] Wilhelm Conrad Röntgen (1845–1923) German physicist.

[4] Alexander Graham Bell (1847–1922) British and later American speech therapist, inventor and major entrepreneur.

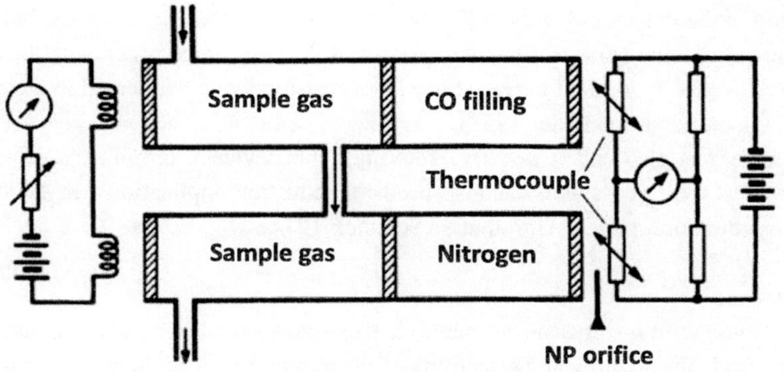

Fig. 7.1 Basic design of the first photometric gas analyzer according to Schmick

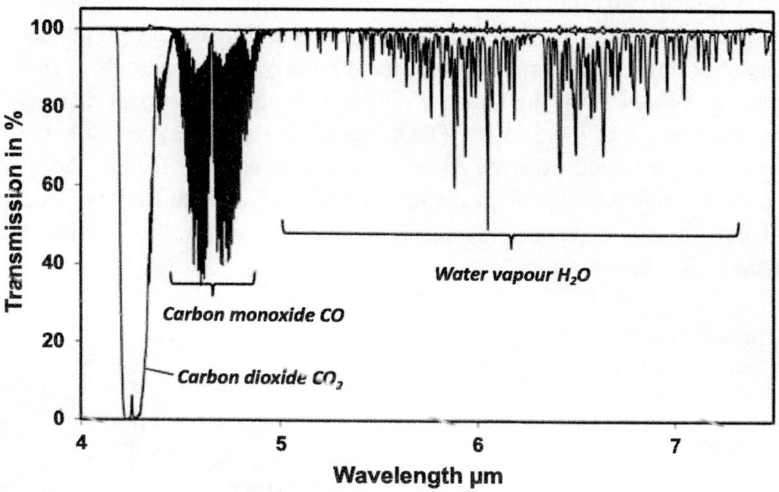

Fig. 7.2 Typical gaseous components ($CO/CO_2/H_2O$) from a combustion process (exhaust gas analysis) with overlapping areas leading to cross-sensitivity

Inframeter

A further development of this method was described by Naumann in 1953 (Fig. 7.3). In this case the thermocouples were in a closed chamber (f) which was filled with the type of gas to be measured. The temperature ΔT of the platinum wires ($i + h$) was increased by absorption of the IR radiation absorbed by the gas molecules. Then the heat transfer to the thin platinum resistors (h, i, k, l) took place.

During pre-absorption, by the gas to be measured in the measuring cuvette (c), the temperature increase decreased. This change was proportional to the gas concentration in the measuring cuvette (c). For a gas concentration of 2000 ppm CO in the sample gas, the temperature in the detector chamber (*f*) changed by ≈0.2°C (Naumann and Schulz 1960). The reference chamber (*g*) was filled with nitrogen (N_2) and compensated with the resistors (*k* + *l*) for changes in ambient temperature. All 4 resistors were connected in a measuring bridge (Naumann and Schneider 1953).

This process is also called negative filtering. The device was built by the Siemens company for several years and was also used for industrial applications. In 1967, it was replaced by the more modern Ultramat 1 (Schunck 1974).

URAS Method

The first photometric gas measuring methods used negative filtering and a constant light irradiation. Both, the stability and selectivity of the measurement could not be achieved to a sufficient degree with these setups. Only with the completely new approach, the selective gas detector of *K.F. Luft*, this problem was solved in 1938 (Lehrer and Luft 1938). The method consisted of two beam paths, a measurement path and a reference path, which were separated from each other (Fig. 7.4). Both beam paths were irradiated with an incandescent lamp (nickel-chromium wire, ≈7 Watt electrical power). A chopper wheel chopped the beam paths in phase so that the radiation fell into the two measuring chambers of the detector at a frequency of $f = 6.25$ Hz. The hermetically sealed measuring chamber of the gas detector contained the gas type to be measured (e.g. CO). Due to the modulated radiation, the pressure p then periodically increased equally in both halves of the chamber, so that without selective pre-absorption in the analysis chamber, the pressure difference between the measuring chambers was $\Delta p = 0$.

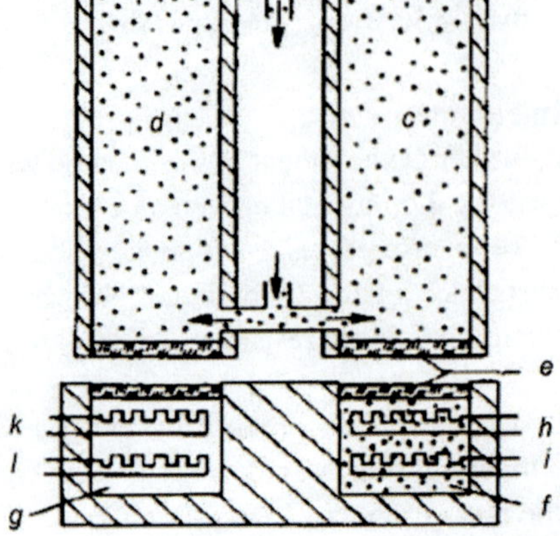

Fig. 7.3 Inframeter of the company Siemens & Halske Karlsruhe. The radiation source *a* consisted of a heated ceramic plate, which radiated to both sides and led through mirror *b* into the measuring and reference chamber (*d* + *c*). The sample gas was equally in both chambers

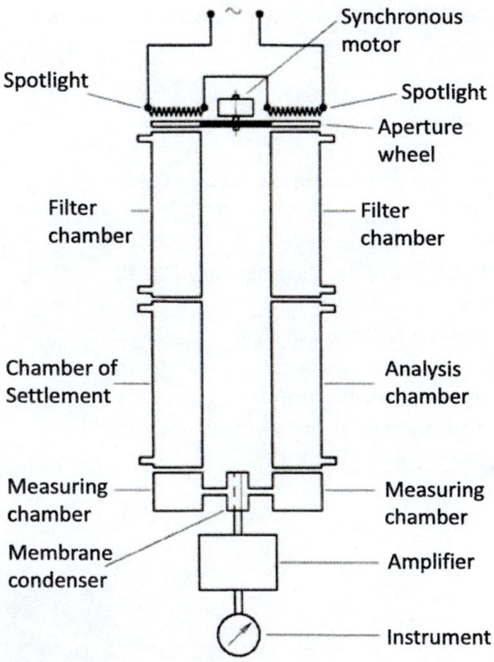

Fig. 7.4 Basic structure of the first *BASF-URAS* according to Lehrer und Luft

If a part of the radiation in the analysis chamber was absorbed by the unknown sample gas, a little less radiation energy (ΔE) fell into the left measuring chamber compared to the right measuring chamber. This resulted in a small, but measurable pressure difference of $\Delta p \approx 0.1$ µbar. This small pressure difference was measured by a membrane capacitor, which separated both measuring chambers from each other. In 1943, *Luft* published his first results and experiences with the *ultra-red absorption recorder (Ultrarot-Absorptionsschreiber, URAS)*[5] in a detailed paper (Luft 1943). He already established a theory of radiation absorption and was thus able to calculate the corresponding physical quantities. In Table 7.1, the results from this period are summarized.

With this setup, which was built at BASF in Oppau and was therefore also called *BASF-URAS*, air could already achieve a CO measurement range of 0–500 ppm, with a detection limit of 10 ppm. Figure 7.5 shows the device. The analysis chamber consisted of a glass tube and had a length of 30 cm. The glass tube was completely gold-plated inside to ensure good transmission of IR radiation from the source to the detector. An approach that is still used in today's IR gas analyzers. He also varied the pressure and gas concentration in the measuring chamber to achieve the optimum measuring effect. He found that the optimal measuring effect is around 10 vol.% for a gas mixture and also varies for different gases.

[5]Until the 1960s, the term *ultrared was* used instead of *infrared* in the German-speaking world.

Table 7.1 Measurement results for the first URAS by Luft (1943)

	10 ppm CO in the analysis chamber	50 ppm CO in the analysis chamber	Measuring chamber covered
Measuring voltage in mV at the capacitor	0.1	5.5	100
Measuring voltage in mV at the amplifier	0.8	40	800
Capacity change in pF	$3.2 \cdot 10^{-4}$	$1.6 \cdot 10^{-2}$	$3.2 \cdot 10^{-1}$
Deflection of the diaphragm in µm	$8 \cdot 10^{-4}$	$4 \cdot 10^{-2}$	$8 \cdot 10^{-1}$
Pressure difference in µbar	0.016	0.8	16
Energy difference of the measuring chambers in J	$3 \cdot 10^{-7}$	$1,5 \cdot 10^{-5}$	$3 \cdot 10^{-4}$
Temperature difference of the measuring chambers in °C	$1 \cdot 10^{-5}$	$5 \cdot 10^{-4}$	$1 \cdot 10^{-2}$

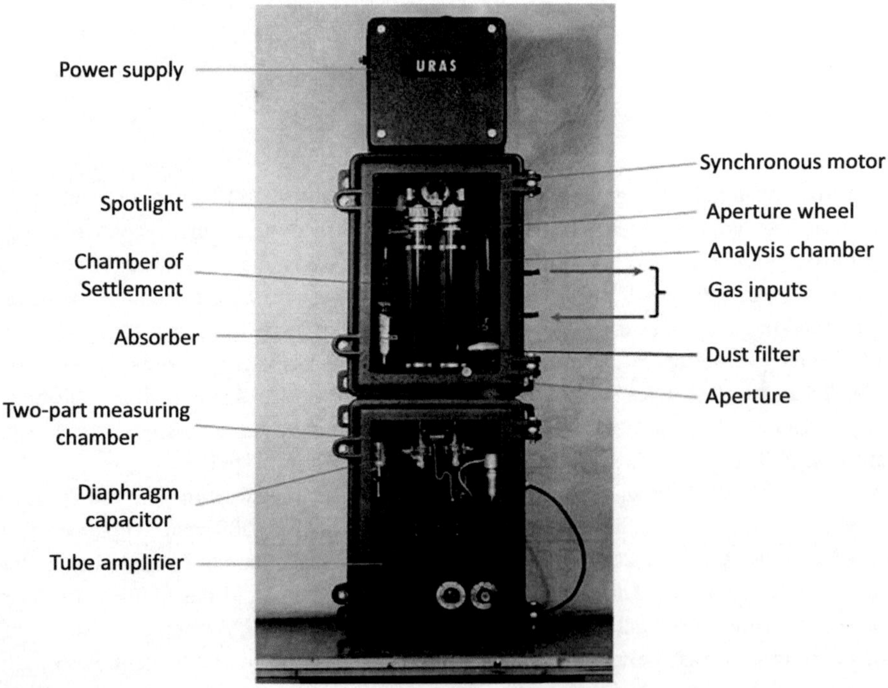

Fig. 7.5 Structure of the first URAS at BASF in Oppau 1938 The device consisted of three stable cast iron housings which were connected to each other. The entire structure had a height of over 1 meter

For gases with strong absorption, the maximum is naturally at lower gas concentrations. Luft explained this with the absorption of the radiation in the first few millimeters behind the entrance window. The amount of heat released is then quickly transferred to the window at a high gas concentration and thus does not contribute to an increase in the measuring effect (pressure increase). This also explains the maximum. Furthermore, he determined the selectivity of the measurement by introducing the concept of cross-sensitivity and quantifying it for comparison purposes.

$$\text{Cross-sensitivity} \quad Q = \frac{A_2 \cdot c_1}{A_1 \cdot c_2} \tag{7.1}$$

In Eq. 7.1, A_1 is the slope at concentration c_1, and A_2 is the slope at concentration c_2. The cross-sensitivities determined were between 10^{-3} and 10^{-5}, so that one could certainly speak of a high selectivity of this method, which was not achieved by any other method at that time. By the end of the war in 1945, several hundred of these gas detectors had been produced, most of which were used in the plants of I.G. Farbenindustrie (Luft et al. 1993).

In the 1950s, the method was taken up by the company Hartmann & Braun (Frankfurt am Main), and further developed into a gas analyzer ready for series production. The basic design was completely adopted and replaced by components suitable for series production. Figure 7.6 shows the first series-produced instrument, which was delivered from 1955, for CO, CO_2, and CH_4. The essential difference was the now very compact design. Instead of

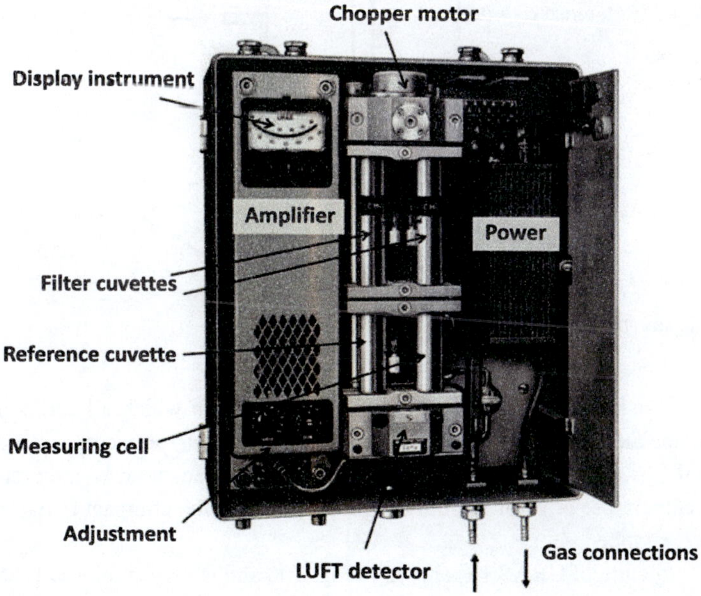

Fig. 7.6 Structure of the first device series of type URAS 1

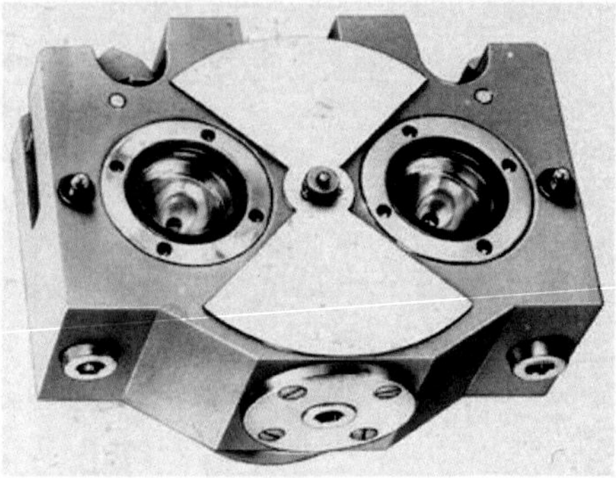

Fig. 7.7 Radiator arrangement for URAS 1 with integrated chopper wheel and motor drive

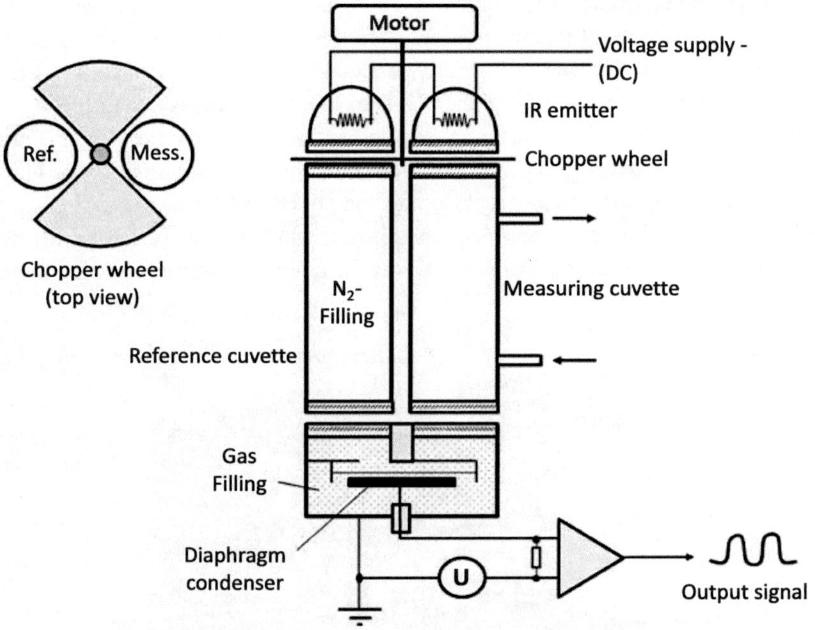

Fig. 7.8 Basic structure of URAS 1

the three cast housings, a sheet metal housing was chosen in which all components were integrated. In the meantime, the tube amplifiers were also much more compact and required less electrical power. Furthermore, a circulating air thermostat was integrated, which limited the temperature error of the unit. Figure 7.7 shows the compact radiator unit with the associated chopper.

The basic structure of URAS 1 is shown in Fig. 7.8, and is very similar to BASF URAS. However, by using components suitable for series production, the URAS 1 was much more

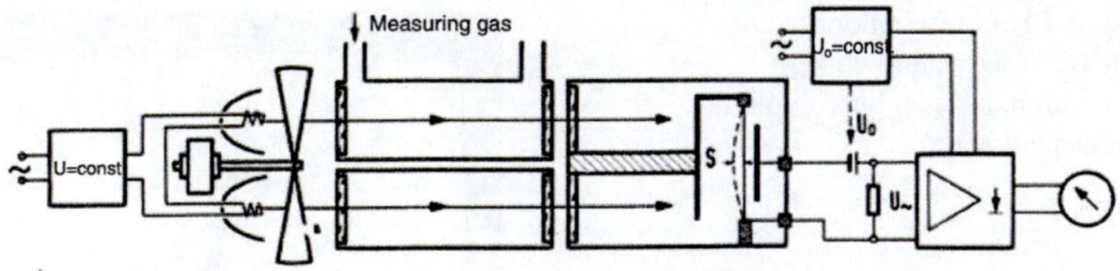

Type of measurement signal	Electrical power	Radiation		Temperat. Pressure	Distance capacity	AC voltage (controlled)	Direct current
Performance level	2 x 7 W	2 x 4 W modulated: 2x1W effective		4.10^{-5} W	2.10^{-12} W	3.10^{-14} W	5.10^{-4} W
or amount of the measuring signal at Fullscale 100 ppM CO			Absorb. Radiation-Power: 5.10^{-5} W (cuvette l:250 mm)	T_{eff} = 17.10^{-4} grd. p_{eff} = 2.6 µm WS	S_{eff} = $6.5.10^{-3}$ µm C_{eff} = 3.10^{-3} pF	U~ 2mV	J_x = 300 µA
Disturbance variable	Change in voltage Change in helix	Speed change	Air pressure change Pollution	Microphony (vibration)		Change from Uo	Changing the characteristic curve of the amplifier
			Temperature change				
Countermeasure	Voltage constant holder	Synchron-ous motor	Barometer aperture — Thermostat	Vibrating metal and dimensioning of the receiver		Voltage constant holder	negative feedback

Fig. 7.9 Distribution and decrease of the power level from the emitter (electrical power) to the detector (AC voltage) by 14 orders of magnitude. The temperature change of $1.7 \cdot 10^{-4°}$ C per 100 ppm CO thus leads to a response value (noise limit) of $\approx 10^{-6°}$C. (Engelhardt 1963)

reliable and therefore quickly found its way into a wide range of applications. The first measuring tasks were in the purity testing of operating gases in the chemical industry, indoor air monitoring for harmful and explosive gases and plant physiology. The analysis of flue gases (CO, CO_2) was also carried out with the URAS 1 and was to become one of the main applications in the following years. A special version of the URAS 1 was operated with a chopper frequency of 25 Hz, so that very short response times for a gas jump from under 150 ms became possible. This requirement came from medical technology for lung function testing and anesthetic gas monitoring.

The further development of the URAS at Hartmann & Braun in these years was mainly carried out by W. Schaefer (Schaefer 1961), who also refined the theory established by K.F. Luft (Schaefer 1965). Figure 7.9 shows some practical results of these theoretical considerations. A fundamental problem that arises with all photometric gas measuring instruments is the proportion of radiation absorbed by the measuring gas, and the energetic conversion in the detector. These effects are very small, and are therefore considerably disturbed by external influences. The first gas analyzers, therefore, had special problems with these influences, as at that time there were no electronic compensation possibilities. For example, mechanical orifices (barometer orifice) were pushed into the beam path for pressure compensation. These diaphragms were connected to the movable part of a barometer box and partially shadowed the beam path. If the air pressure changed, the

Fig. 7.10 Linearization (in vol. % CO) of the output voltage *U* via an individual imprint on the display scale

orifice shifted and released an additional radiation flow, which, when balanced accordingly, compensated the influence of the air pressure. A further problem was to assign the non-linear output signals to a concentration. Until the introduction of transistor technology, this was achieved by individually printing the scale with the corresponding characteristic curve (Fig. 7.10).

UNOR Method

At the beginning of 1957, K.F. Luft moved to Bergbau-Forschung GmbH in Essen-Kray, where he established the physics department for gas measurement technology. There he started the further development of the URAS method. For this purpose, he did not set up the two measuring and reference chambers next to each other as in the URAS, but placed them one behind the other. The radiation therefore had to pass through the measuring chamber before it reached the reference chamber. This measure therefore provided an optical filtering of the radiation before it reached the reference chamber. This had far-reaching consequences for the metrological properties. Figure 7.11 shows the principle structure of this new type of measuring arrangement (Luft 1958).

The absorption spectrum (Fig. 7.11a), after passing through the measuring cuvette (20 cm), shows only slight attenuation at a CO concentration of 100 ppm, while the transmission values behind the front chamber (VK) are already close to zero. The concentrations in both detector chambers are identical and are in the volume % range (≈5 – 20 vol.%). After the 2nd gas layer in the detector, which is significantly longer than the 1st layer, the transmission values are then already in saturation ($T = 0\%$).

Figure 7.12 shows the detailed transmission curves for an absorption line of CO at 2176.28 cm^{-1}. The spectroscopic data are based on real values from the HITRAN[6]

[6]**HI**gh **R**esolution **TRAN**smission molecular absorption database.

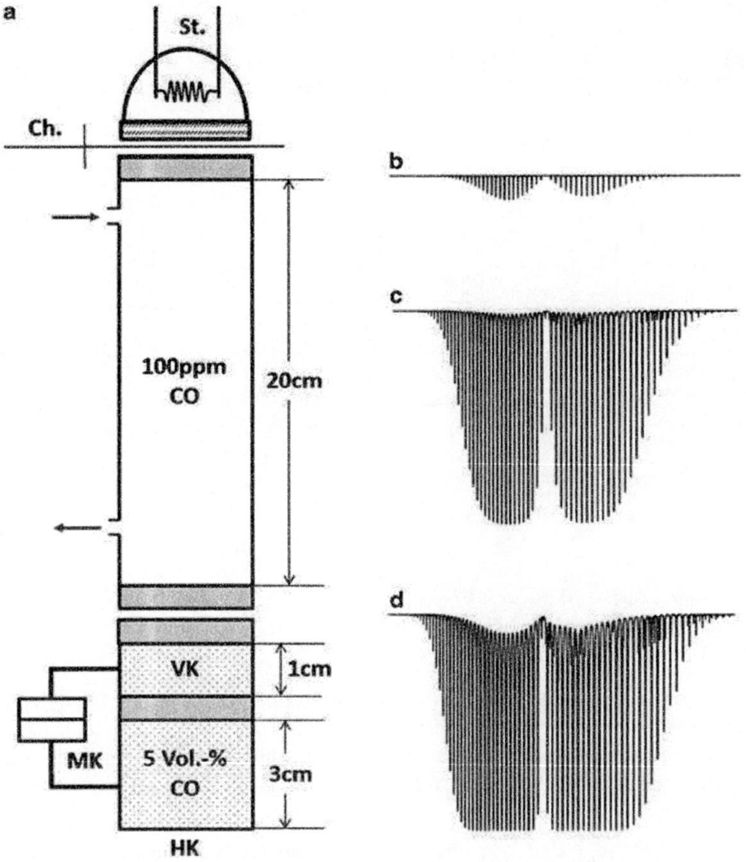

Fig. 7.11 Infrared gas analyzer with successive gas layers according to K.F. Luft. M = Membrane condenser, VK= front chamber, HK = rear chamber, Ch = Chopper. (**a**) Spectrum behind the measuring cuvette, (**b**) Spectrum after the first gas layer (VK), and (**c**) Spectrum after the second gas layer (HK)

database. The individual positions were each shown for a CO concentration in the measuring cuvette from 0 ppm (=without pre-absorption) and 100 ppm (with pre-absorption).

In the first case, the emitter intensity (a) passes through the measuring cuvette to the detector without attenuation. In the 1st layer (e) of the detector, a large proportion of radiation from the spectral environment of λ_0 is absorbed. In the 2nd layer (g) further energy components are absorbed, but these are further away from the central wavelength λ_0. In the central region λ_0, most of the radiation has already been absorbed due to the higher absorption coefficient. The differences of the radiation components $D(\lambda)$ then

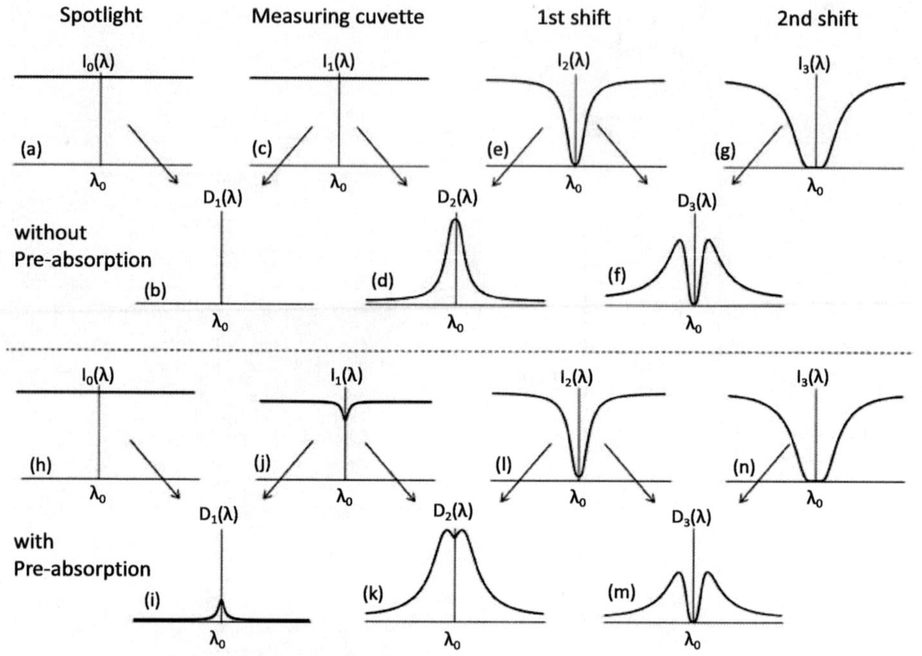

Fig. 7.12 Intensity distributions ($I\lambda$) for a CO absorption line at 4.595 μm, before and after the different gas layers (cuvette, detector), and the differences ($D\lambda$) between the layer passages. The graph is shown for the two cases that arise with and without sample gas (with and without pre-absorption) in the cuvette. (According to Kessler 1967)

indicate the spectral range that was absorbed in the respective layer. In the 2nd layer, the radiation absorption $D3(\lambda)$ is highest in the side flanks and zero λ_0 in the center.

If a selective pre-absorption by the gas in the measuring cuvette now takes place, the conditions change accordingly. The difference $D_1(\lambda)$ indicates a small change (i). The difference $D_2(\lambda)$ also changes in this case, because the portion with the highest absorption has already been removed (k) in the central area λ_0. An insert (*dip*) can be clearly seen that is proportional to the pre-absorption in the measuring cuvette. The difference $D_3(\lambda)$ between the 1st and 2nd layer in the detector, however, does not change in the central area, since saturation has already been reached there.

Since the intensity difference $D_3(\lambda)$ does not change due to the gas in the measuring cuvette, this value can be used as a reference (Fig. 7.13). The pressure increases in the two layers result from the respective integrals:

$$p_k = \int D_2(\lambda)\mathrm{d}\lambda \quad \text{and} \quad p_m = \int D_3(\lambda)\mathrm{d}\lambda \tag{7.2}$$

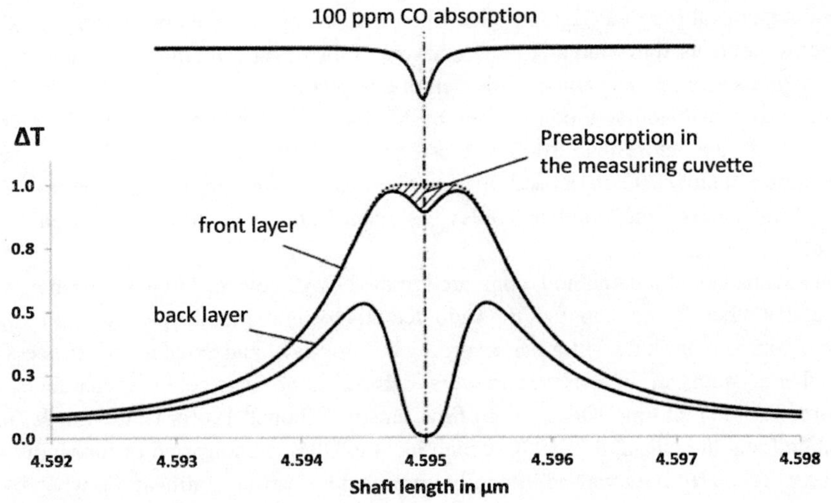

Fig. 7.13 Display of the transmission values T and the differences ΔT

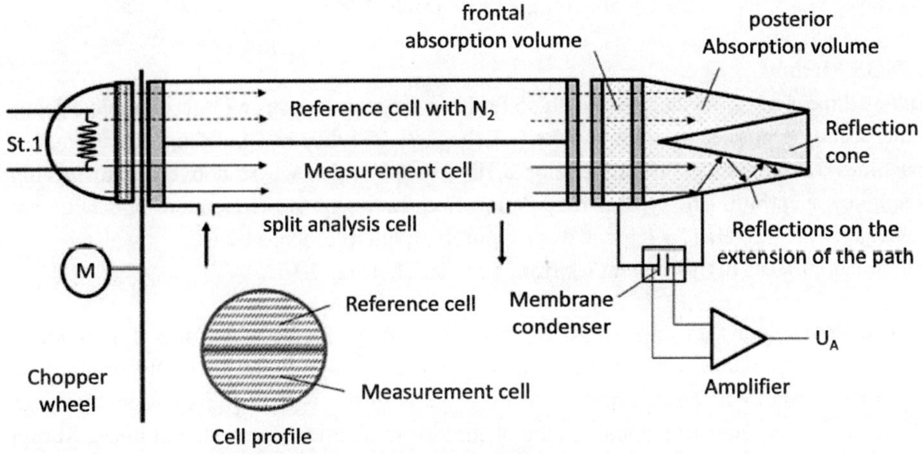

Fig. 7.14 Basic structure of UNOR 1 according to K.F. Luft

In balanced condition, the pressure difference $\Delta p = p_k - p_m$. To reach this state, the integral of the 1st layer and the 2nd layer must be identical. Since the largest part is already absorbed in the 1st layer, the 2nd layer must be longer than the first. Therefore, a reflection cone was installed in the setup (Fig. 7.14), which causes a path extension by forced wall reflections. This setup also has a split cuvette, which is divided into a measurement and reference side. This provides an additional possibility for comparison with a reference beam that passes through a reference side filled with nitrogen (→ no pre-absorption). The chopper wheel is used to irradiate the two halves of the cuvette alternately, so that several comparison measurements can be made within one second.

This body was manufactured in series by Maihak AG Hamburg from 1963 onwards. Only 4 years later, more than 1000 of these gas measuring instruments were in use, of which approximately 700 were used in West German coal mining. By opening up further areas of application, the number of devices manufactured could be increased to ten thousand by 1977.

The advantages of this method, compared to the URAS devices known until then, were so great that other device manufacturers adopted this method after the patents expired. The Siemens company built the Ultramat according to this model and extended the process to a 4-chamber system, in which the pressure compensation between the chambers was measured for the first time with a micro flow sensor (Schunck 1966). Hartmann & Braun started building the URAS 3 in 1980, which like the Ultramat consisted of four individual chambers. The selectivity was additionally improved by partial stratification, whereby the membrane condenser was still used as the detection element. Between the 1st layer and the 2nd layer, partially permeable windows were inserted, so that the incident radiation was weakened in the rear chamber. Compared to a two-layer receiver, the effectiveness of the rear layer was thus reduced (Fabinski and Ascherfeld 1980).

BINOS Method

Since all measuring devices described so far have more or less strong microphonic effects, which become noticeable with shocks (→ shock sensitivity) and building vibrations (→ machines), *Schunck* decided to develop a IR-detector which works above this disturbing frequency spectrum. The typical modulation frequencies of the Uras, Unor and Ultramat devices are at 5–20 Hz. To achieve this immunity to interference, a transducer was required which can process high frequencies from $f > 100$ Hz (Fig. 7.15).

The *BINOS* consists of a single-layer detector (Fig. 7.16), which has a very small chamber volume, and a reference chamber, which was arranged centrally around the measuring chamber. The purpose of this arrangement is to achieve insensitivity to vibrations, since the mechanical center of gravity is identical for both volumes. Abrupt movements of the instrument therefore do not lead to undesired gas movement within the closed detector chambers (Fig. 7.17). At the compensation point, between the two chambers, there is a microflow sensor (Fig. 7.15) which can detect extremely fast changes due to the low thermal mass (Schunck 1974). Figure 7.18 shows the characteristics of this detector arrangement. The sensitivity of the microflow sensor increases linearly with frequency at a constant volume flow. From approximately 1 kHz the curve then bends slightly. The frequency behavior of the gas movement between the measuring and reference chamber shows an opposite behavior. From the zero point it rises steeply at low frequencies and then falls slowly with a $\approx 1/x$ -function. The sensitivity curve for the IR detector results from both properties. Since the noise signal remains almost constant, the signal-to-noise ratio between 100 Hz and 1 kHz is also constant.

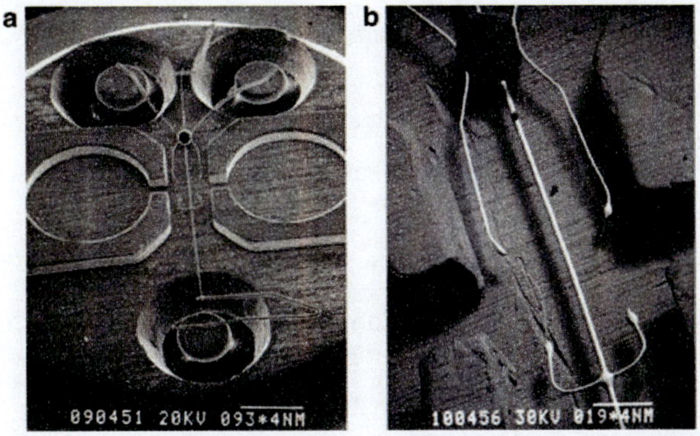

Fig. 7.15 Design (**a**) of the microflow sensor according to Schunck for use in an IR detector. Detailed view (**b**) of the differential anemometer

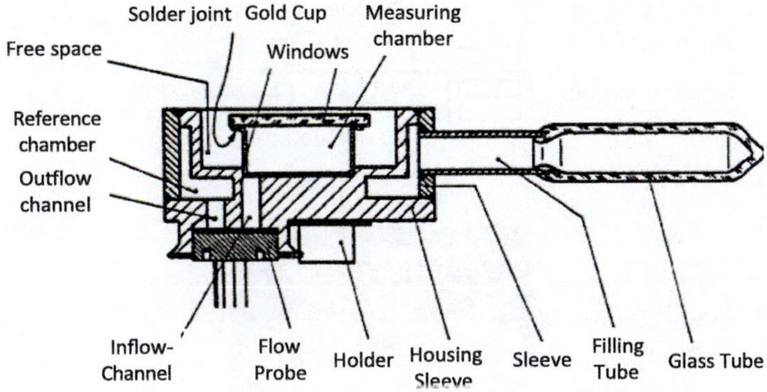

Fig. 7.16 Cross-section through the construction of the adhesive-free gas detector, which can be degassed up to 450 °C (Schunck and Randow 1977)

The flow sensor was realized in the form of a differential anemometer. Figure 7.15 shows an SEM image of the micro flow sensor arrangement (Fig. 7.15a), and the differential anemometer (Fig. 7.15b). The flow is caused by the pressure increase Δp and comes out of the measuring chamber through the right circular orifice. The flow is transferred at one point (orifice plate opening) into a flow channel (Fig. 7.16), and then passes through the left orifice plate into the reference chamber. The micro flow sensor is located between the two orifice openings. Since the *flow* is only a fast pulsation with a small volume, the gas molecules move virtually on the spot, around a point. The differential anemometer has a heating wire ($\approx$20 μm) in the middle, which also carries the whole assembly at the same time. The holding wires for the measuring resistors have a diameter of $\approx$10 μm. The measuring resistors consist of thin platinum wires in the μm-range. The whole assembly is assembled manually under a microscope.

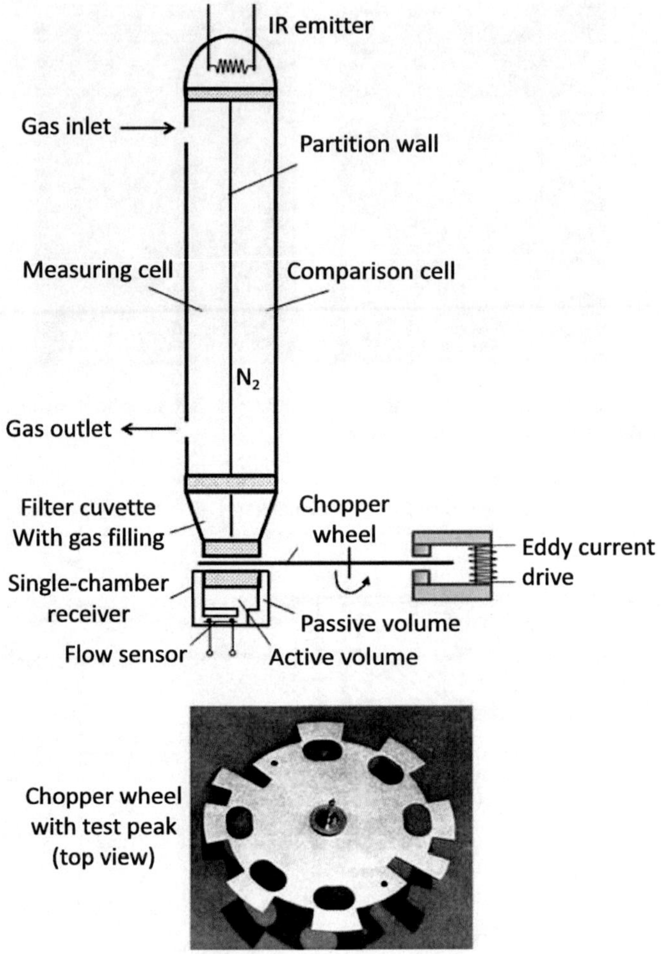

Fig. 7.17 Basic structure of the BINOS method according to Schunck

Another special feature of this detector is that it is completely free of glue. The entire construction is manufactured exclusively by welding, brazing and glass soldering. This allows the entire detector assembly to be cleaned (outgassed) at high temperatures before filling with the gas. Especially water vapor adheres to the surfaces and spreads in the measuring chamber after the filling process. By baking out at temperatures $>200^{\circ}C$, with simultaneous purging with pure nitrogen, the detector is freed from water vapor and therefore no more H_2O cross-sensitivities can be detected. A further advantage of these adhesive-free connections is their good long-term stability. Adhesive connections are never completely tight, so that the trapped gases can diffuse through the adhesive layer over a long period of time ($\approx 5 - 10$ years). This leads to a loss of the gas filling and a change in

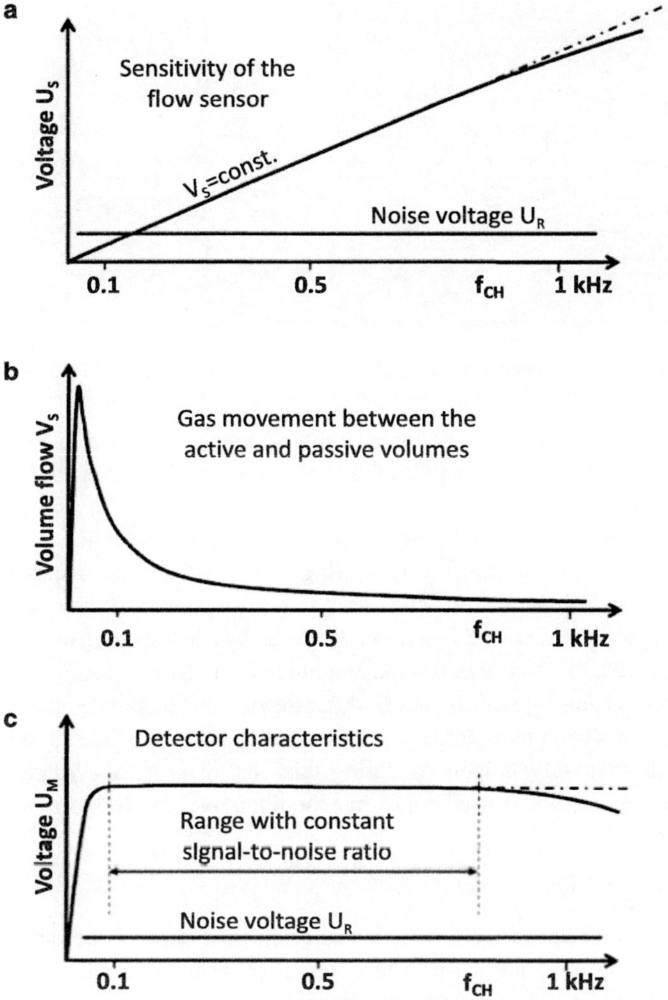

Fig. 7.18 Properties of the microflow sensor and the resulting detector characteristics

the receiver characteristics over time (→ sensitivity drift). Bonded receivers can run dry and must be replaced after a few years. The BINOS detector does not have this problem. Figure 7.19 shows a size comparison, which shows the small size of the BINOS detector.

The chopper drive in the BINOS is provided by an eddy current motor. Since the chopper disc consists of a thin aluminum sheet, an eddy current is generated in the disc when an external magnetic excitation is applied (Fig. 7.17). This eddy current generates a magnetic field that is opposite to the excitation field. A second excitation coil is located at a different position to the first excitation coil, and is also driven at a different time so that the

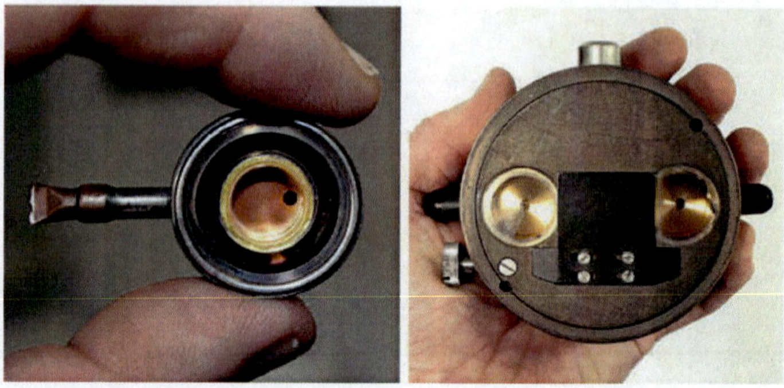

Fig. 7.19 Size comparison between the BINOS 1 detector and the URAS 2 detector

eddy current produces an attractive force, and the chopper disc then rotates at a constant speed of 1600 U/min. The principle of this motor drive was taken over from electric electricity meters.

The chopper disc releases the beam path for the measuring and reference channel with a time delay and thus modulates the measuring effect. After three measuring cycles, a so-called test peak is generated which is used to readjust the sensitivity. This test peak is produced by a 1/4 opening of the beam path followed by a 3/4 opening of the beam path. In sum (*peak to peak*), this test peak has the amplitude $0.5 \cdot (I_M + I_R)$. Aging of the radiation source leads to a change of both I_M and I_R. After the run, the amplitude is readjusted so that the conditions are always constant (Fig. 7.20). Since I_M *is* also included in the peak height, the linearity is favorably influenced during this control process (Wiegleb 1983). The BINOS measuring method is used today, in a modified form, by Emerson in the *X-Stream* product series.

URAS 10 System

The further development of the URAS 4 was abandoned because the device design was very complex and only paid off for a few special applications. Furthermore, this device could only detect one gas component at a time. Therefore, a completely new instrument concept was developed, which was much easier to manufacture and with which a multi-component analysis was now possible. The developments in the field of microelectronics supported this approach in a special way. The use of powerful microprocessors now made it possible to perform a wide range of computing and control tasks that could not be performed before. The IR detectors of this type of device consist of very compact units which can be connected in series. Two gas measurements are thus possible in one beam path. In a second beam path, which is modulated with the same chopper, two further measurements are possible (Fig. 7.21). In total, up to four components can be measured simultaneously with one analyzer of the URAS 10 series. Both beam paths can have different cell lengths to cover different concentration ranges. The measuring ranges within one beam path are determined by the common cell length.

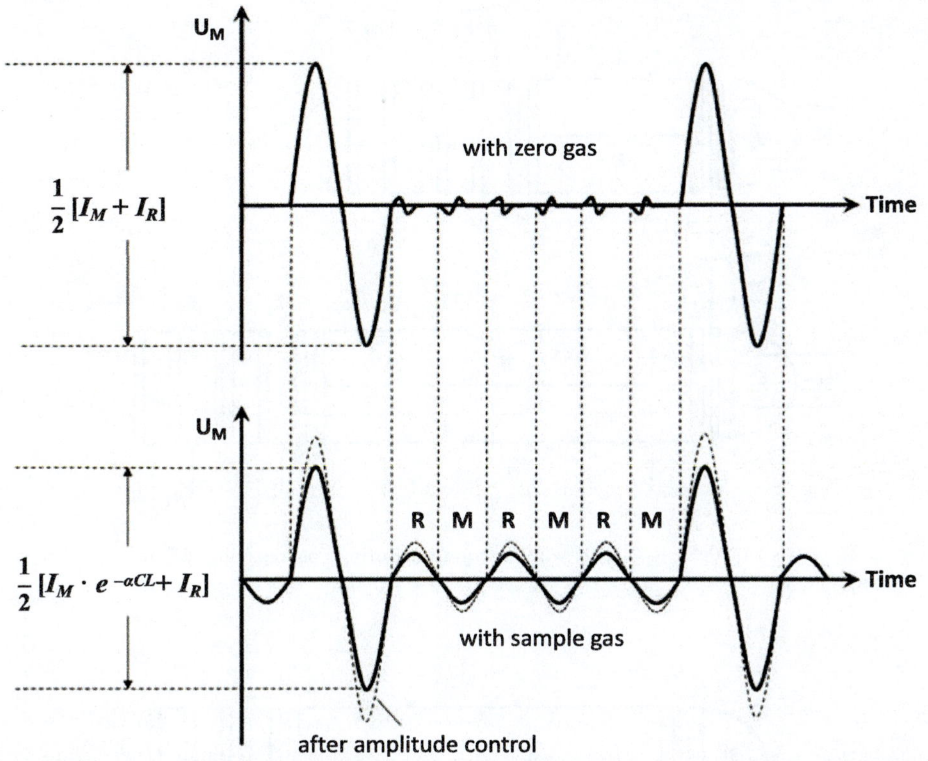

Fig. 7.20 Signal curves in the BINOS procedure after half a chopper wheel rotation. The test peak height is reduced during gassing according to the Lambert-Beer law, so that a linearization effect occurs after control at a constant peak height

A special feature of the URAS 10E is the possibility of an automatic calibration in the zero point and in the end point (Fabinski 1985). The zero-point adjustment is realized by flushing with ambient air, which may have to be cleaned from CO_2 and humidity by a filter unit. After a defined stabilization phase, the current value is taken and stored. This value is defined as zero.

The endpoint calibration is usually performed with test gas from a pressurized gas cylinder, which must flow through the cuvette after zero adjustment. This type of calibration is very complex and expensive. One usually needs trained personnel on site, or an automatic valve control system to feed the test gas into the analyzers. The gas cylinders must be checked regularly and replaced if necessary. The delivery times for test gases are 4 – 6 weeks. The URAS 10 has an optional gas-filled calibration cuvette in the beam path (Fig. 7.22). With this option it is now possible to carry out internal calibrations within the

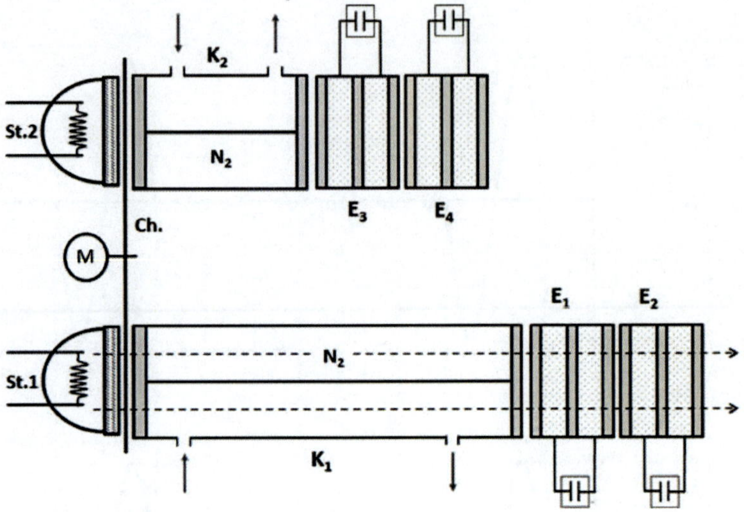

Fig. 7.21 Design of the URAS 10E with four gas measuring channels and different cell lengths (Ascherfeld et al. 1990)

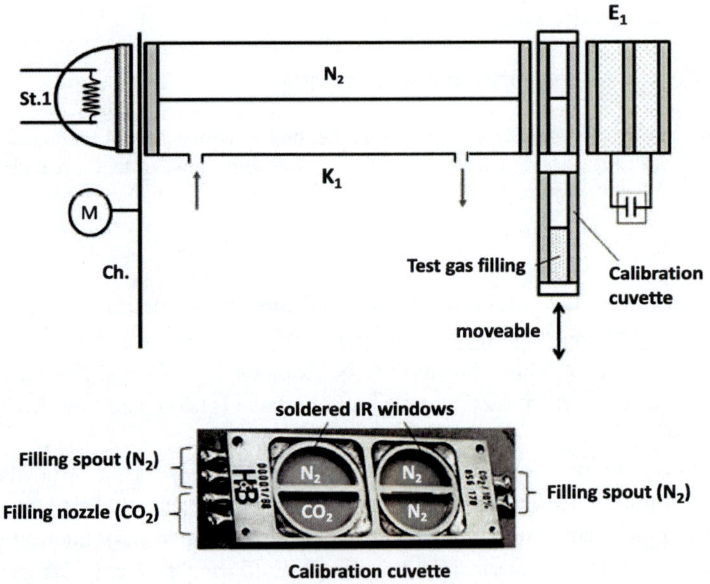

Fig. 7.22 Structure of the URAS 10E with integrated calibration cuvette (after Fabinski 1985). Top view of the calibration cuvette (CO_2 filling) with the sector related gas fillings. The segments of the calibration cuvette are arranged in the above sectional view rotated by 90° to better represent the beam path

instrument, which contributes to a considerable reduction in costs and an increase in reliability (Wiegleb 1992). Test gas from gas cylinders is no longer required. Under normal operating conditions the arrangement of the cuvette is such that no absorption takes place. In this position, only the part of the cuvette filled with nitrogen (N_2) is in the beam path. However, if the end point is to be checked, the cuvette is pushed by an electric motor into a position in which the test gas filling is congruent with the measuring side of the cuvette. The concentration of the test gas filling is selected so that the product $c \cdot L$ is in agreement with the end point concentration.

Example
If a concentration of 200 ppm CO is to be measured in a 200 mm cuvette, the result is a product $c \cdot L$ of 0.02 vol.% CO 200mm = 4% mm. Thus, if a path length of L of 5 mm (in the calibration cell) is measured, a concentration of 0.8 vol.% CO would have to be filled into the calibration cell to obtain the same product.

Sample: 0.8 vol.% CO 5mm = 4% mm

The long term stability (>10 years) of the gas filling is extremely important to be able to perform such a calibration. Also, in this case, glued joints are not allowed. The entire structure is therefore sealed by a soft solder connection, which is also applied to the window surfaces (CaF_2 or BaF_2). To be able to carry out this soldering connection, the window material must first be metallized at the specified soldering points. The entire assembly is then heated in a furnace, above the melting point of the solder, and the joint is closed. The second pair of panes is joined to the flask in the same way, with the solder now having a lower melting point so that the first joint is not broken by the second heating. The cuvette is then heated cyclically and evacuated before the defined calibration gas can be filled in. The metallic sealing points of the filling nozzles are additionally soldered. The manufacturing process allows a stability of the gas concentration over many years.

The URAS is today produced by ABB Automation GmbH (Frankfurt am Main) under the product name URAS 14 (Fig. 7.23).

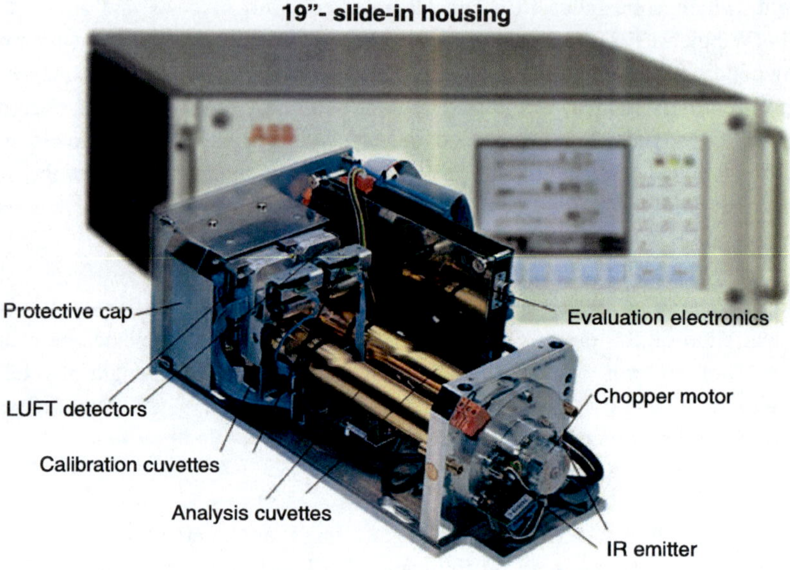

Fig. 7.23 URAS 14 module (foreground) with two measuring channels for simultaneous gas analysis of up to 4 measuring components. In the background a 19″-rack housing can be seen, in which the URAS 14 module is integrated. (Source: ABB Automation GmbH Frankfurt am Main)

7.2 Filter Photometer

The IR measurement methods described so far are all based on the selective absorption by the type of gas to be measured in the *gas detector*. This selectivity is therefore given by the absorption lines of the different gases without the need for further spectral limitation. Although this method was practically born out of necessity, as there were no usable optical filters available at that time, this method has considerable advantages with regard to the measurement properties. Particularly noteworthy are the cross-sensitivities and the detection limits. However, these advantages are bought by a complex detector technology and a complex mechanical arrangement with movable components (chopper wheel).

With the filter photometers you go a different way. They use miniaturized radiation detectors, which are mass produced and therefore very inexpensive. Typical detector principles suitable for this type of gas measurement technology are *thermopiles* and *pyroelectric detectors*. Thermopiles are produced in millions of pieces for clinical thermometers/ear thermometers, and pyroelectric detectors can be found in every motion detector. Both detector principles are broadband, i.e. they cannot perform selective measurements. To achieve a spectral limitation, narrow-band interference filters are therefore connected upstream of these detectors, which provide the necessary selectivity.

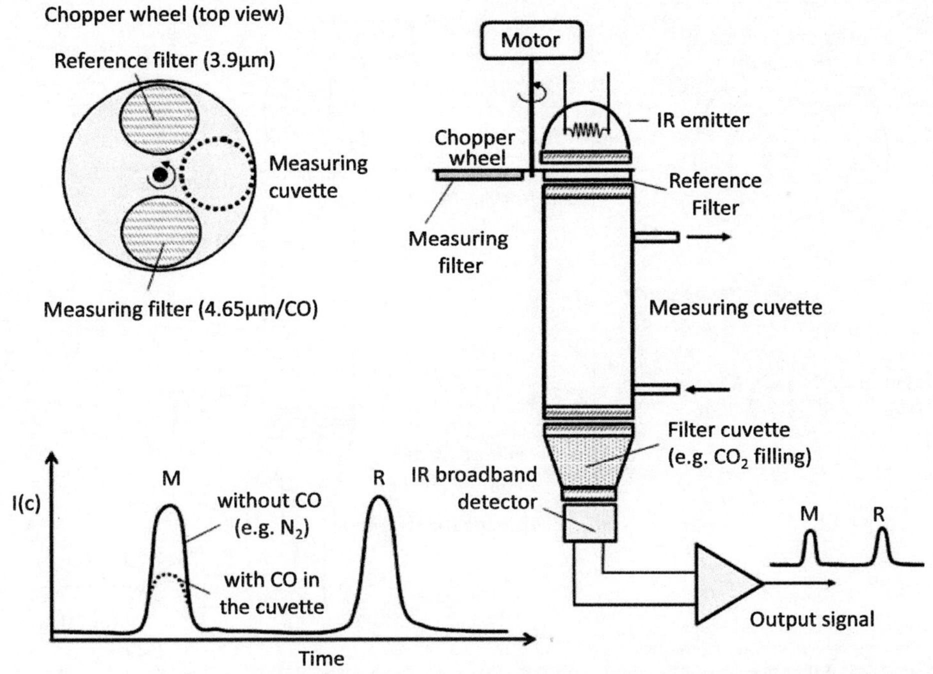

Fig. 7.24 Structure of a filter correlation method for carbon monoxide CO and a reference wavelength. In addition, the cross-sensitivity to CO_2 can be reduced by using a filter cuvette (gas filtering)

Filter Correlation

The first filter processes came onto the market as early as the 1960s and used interference filters that were mounted on a rotating chopper disc. Figure 7.24 shows such a method for two interference filters. Whenever an interference filter appears in front of the beam path, the measurement signal rises for a short time (measurement peak) and then disappears again. The dark area that arises between the individual peaks is used to calibrate the electrical zero point. Usually at least 2 interference filters are required for a measurement and reference peak. The measurement peak changes as a function of the gas concentration c in the cuvette and is thus used to determine the concentration. Both the peak height and the peak integral can be used as a basis for calculation. Since the measuring wavelength λ_M and the reference wavelength λ_R lie in different spectral ranges, this is also referred to as a wavelength comparison or *filter correlation method.*

The two signals, $M =$ *measuring* signal and $R=$ reference signal are calculated in the evaluation electronics as follows:

$$\text{Modulation} = \frac{R - M}{R} \tag{7.3}$$

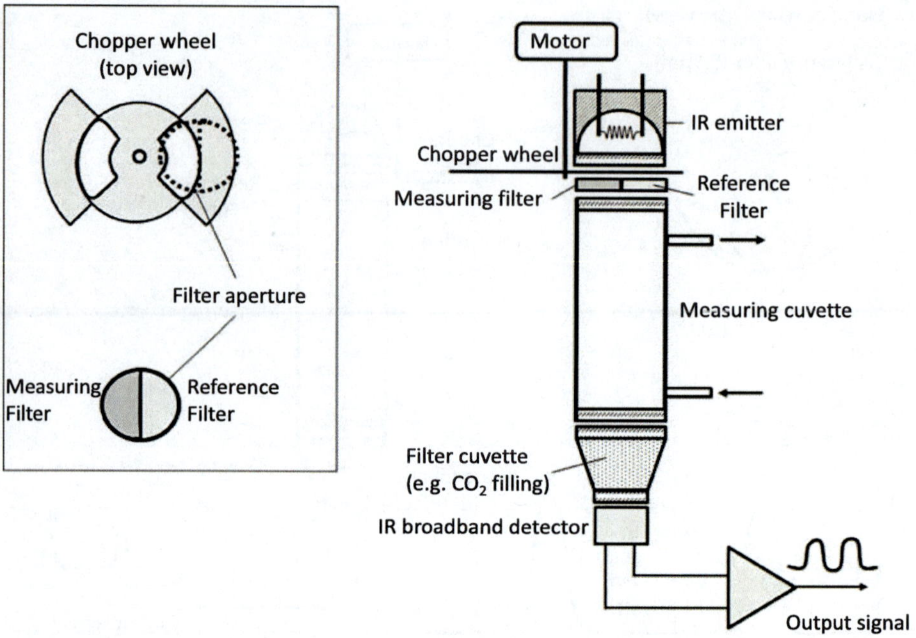

Fig. 7.25 Filter correlation method (BINOS 100) with fixed filters (Wiegleb 1985)

This type of calculation has the advantage that a change in transmission in the cuvette due to contamination or aging effects/drift of the radiation source is compensated. Contamination or a radiation source drift is usually independent of the wavelength, so that this influence affects both channels (*M* and *R*) equally. This is a major advantage over methods that compare a cuvette filled with nitrogen with a cuvette filled with sample gas (→ URAS principle). If there is contamination in the measuring cuvette, this has a direct effect on the measurement result. Methods that work according to the *substance comparison* therefore require good sample preparation.

With this method, multi-component devices can also be constructed by placing additional interference filters on the chopper disc.

A variant of this procedure is shown in Fig. 7.25 (Lehnert et al. 1987). In this setup, the interference filters are placed in front of the cuvette, side by side and are irradiated alternately by the chopper disk. The advantage of this arrangement is that the heavy interference filters do not have to be placed on a disc, which requires a corresponding mechanical stability of the chopper drive. Furthermore, the filter surfaces are continuously irradiated. By this measure the measuring time is fully utilized, which has a positive effect on the signal-to-noise ratio. In this case, a phase-sensitive amplifier is used in the signal processing, which ensures the necessary interference suppression of other frequencies (Wiegleb 1988a, b).

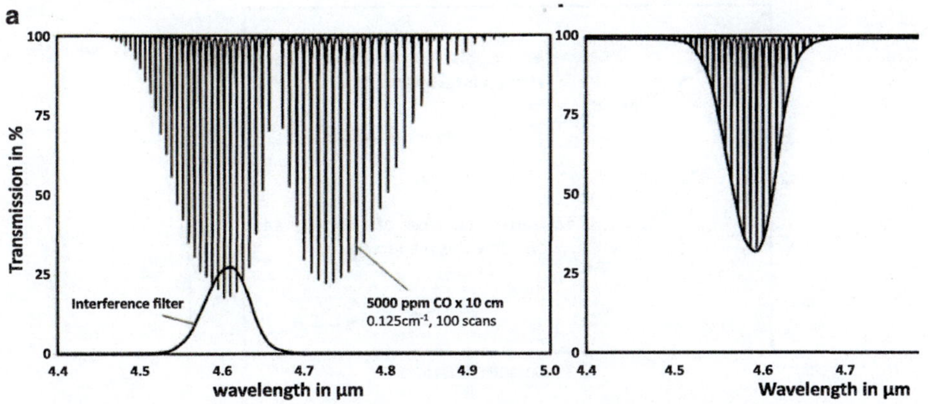

Fig. 7.26 Absorption band of 5000 ppm CO compared to the passband curve of the interference filter (**a**). The passband curve, in combination with the interference filter, is shown in (**b**)

Ballast Radiation

Figure 7.26 shows the spectrum of carbon monoxide (CO) recorded with a 10 cm cuvette. The absorption lines go partially below 25% of transmission, which means that almost 80% of the energy is absorbed within the line width. To achieve spectral narrowing, an interference filter is connected upstream, which has a transmission curve at the maximum of the left band (4.6 µm). The central wavelength λ_{max} of the interference filter is selected so that as many strong lines as possible within an absorption band can contribute to the measuring effect. The spectral passband is thus determined by the interference filter and in this specific case results in an envelope below which the individual CO lines can be seen. However, absorption by the CO only takes place in the individual lines. The radiation between the lines also falls on the detector and thus generates a signal but is not changed by the gas. This part is therefore also called *ballast radiation* I_B, as it does not contribute to the measuring effect. The *Lambert-Beer law* must therefore be adapted as follows:

$$I(c) = I_B + I \cdot e^{-\alpha(\lambda)\cdot c\cdot L} \tag{7.4}$$

Furthermore, the following relationship applies to the irradiated intensity I_0:

$$I_0 = I_B + I \quad \text{and} \quad I_B = I_0 \cdot \beta \quad \text{or} \quad I = I_0(1-\beta) \tag{7.5}$$

The β *value* indicates the proportion of radiation that cannot be absorbed by the gas, since there are no relevant absorption lines in this spectral range. If both expressions in Eq. 7.4. are used, the following expression for the modified *Lambert-Beer law* is obtained after transformation:

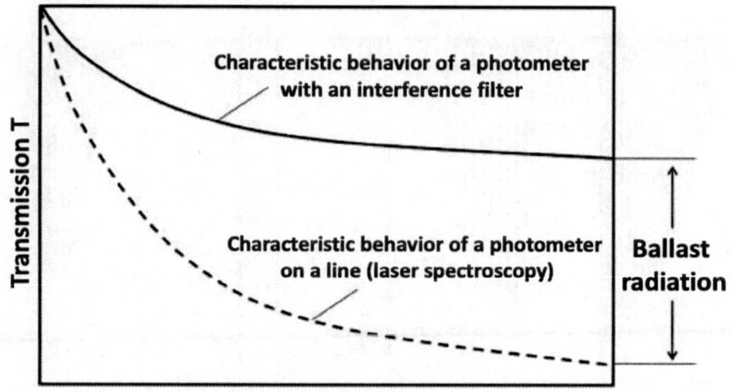

Fig. 7.27 Derivation and definition of the ballast radiation in a filter photometer. The optimum absorption can only be achieved by using a narrow-band laser whose emission wavelength is congruent with the absorption line (ideal case)

$$I^*(c) = I_0\left[\beta + (1-\beta)\cdot e^{-\alpha(\lambda)\cdot c\cdot L}\right] \tag{7.6}$$

To calculate the proportion of *ballast radiation,* the spectral data of a high-resolution IR spectrum must be available. From these data, the β value can be determined by integration. It can also be estimated experimentally (Fig. 7.27). In a filter photometer, the CO concentration of $c = 5000$ ppm shown in the example leads to an intensity change of $\approx$10% ($T = 90\%$), while the absorption lines in the passband of the interference filter are on average at $T \approx 25\%$. It then applies:

$$I^*(c) = I_0\left[\beta + (1-\beta)\cdot e^{-\alpha(\lambda)\cdot c\cdot L}\right] = 0.9 \quad \text{and} \quad I(c) = I_0 \cdot e^{-\alpha(\lambda)\cdot c\cdot L} = 0.25 \tag{7.7}$$

With $I_0 = 1$ you get the following expressions:

$$0.9 = \left[\beta + (1-\beta)\cdot e^{-\alpha(\lambda)\cdot c\cdot L}\right] \quad \text{and} \quad 0.25 = e^{-\alpha(\lambda)\cdot c\cdot L} \tag{7.8}$$

If you insert the second term into the first, you get:

$$0.9 = [\beta + (1-\beta)\cdot 0.25] \tag{7.9}$$

If one dissolves this expression after β, a value of 0.87 results. By this example, one recognizes that the ballast radiation lies with $\approx$87% and only 23% of the radiation can contribute to the measuring effect. This is one reason why CO measurement in the IR range with filter photometers is so problematic. In the CO_2-spectrum there are much more absorption lines, which also absorb much more. The proportion of ballast radiation is therefore much lower.

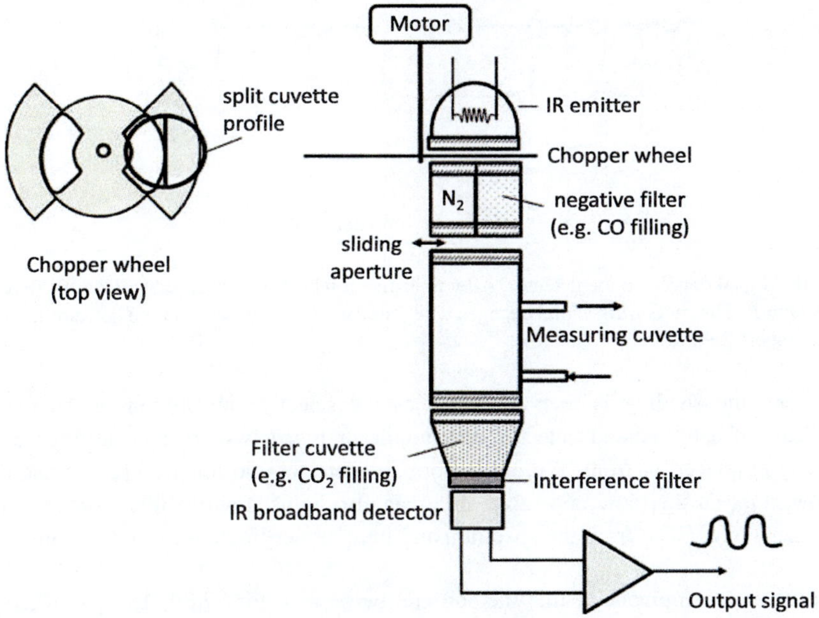

Fig. 7.28 Design according to the principle of negative filtering (Wiegleb 1988a, b)

Gas Filter Correlation (Negative Filtering)

Selective gas filters are an alternative to the use of interference filters. Figure 7.28 shows such a setup. A divided filter cuvette is filled with nitrogen on one side and with the gas to be detected on the other side (e.g. NO or CO). Both sides are alternately irradiated by an IR source and the chopper. There is a high concentration of CO in the filter cuvette so that all lines are almost completely absorbed, and only the *ballast radiation* passes through. On the left side all radiation passes through. Selective absorption by the CO now takes place in the measuring cuvette. The radiation that passes through the gas-filled filter cuvette is already saturated around the absorption lines so that no further radiation is absorbed. This path serves as a reference value (R). On the left side the absorption by the CO takes place ($M \rightarrow M^*$). In the detector, the difference between the two signals is then measured. So the *ballast radiation* is subtracted from the signal on the left side and does not appear any further. This method is therefore also called *negative filtering*. Alternatively, the CO gas filter can also be mounted on a chopper wheel and used as shown in Fig. 7.24. This type of construction is then referred to as *gas filter correlation*. The absorption by the CO sample gas is also very low, but the cross-sensitivities are considerably smaller due to this measuring method. Interfering gases that cause cross-sensitivity are absorbed on both the left and right side of the photometer setup, so that this effect is almost completely compensated. As there is already a noticeable absorption in the filter cuvette, the signal is smaller overall than on the left side (N_2 filling). The design is therefore adjusted with a mechanical diaphragm that projects into the beam path on the left side ($\rightarrow R \approx M$). After

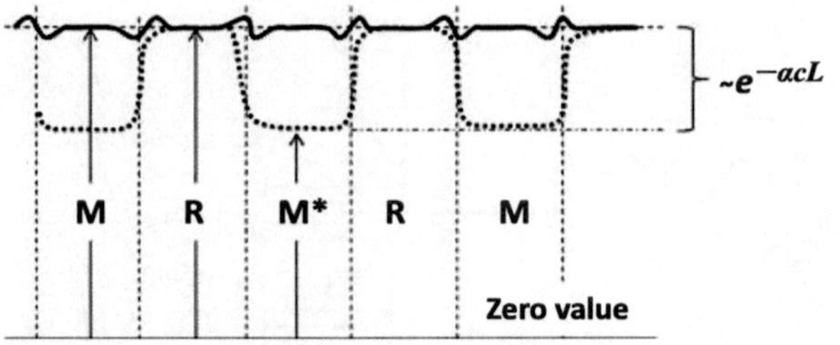

Fig. 7.29 Signal curves on the detector. After adjustment with the mechanical orifice, the following applies $M = R$. The measuring signal changes when the sample gas flows over the detector according to the Lambert-Beer law

adjustment, the shading is then the same as the selective absorption in the gas cell (Fig. 7.29). If a broadband detector (thermopile or pyro) is used, an interference filter must also be placed in front of the detector. Gas detectors do not require this additional filter measure. Furthermore, a positive filter can also be used which filters out interfering components (e.g. CO_2) from the spectrum and thus further limits the cross-sensitivity.

A further development of this method can be seen in Fig. 7.30. The gas filters and interference filters are located on a filter wheel which is driven by a stepper motor. Depending on the manufacturer, up to 14 different filter elements can be placed on this wheel, so that a multi-component analysis can be performed. The radiation of the IR source is focused on the input aperture via imaging optics (deflection mirror). The filter wheel and the chopper wheel are in the beam path. The position 1, 2, . . ., n of the individual filters is approached sequentially, whereby the filter wheel remains in this position for a certain period of time (Δt). In this phase, the IR radiation is modulated (*chopped*) by the chopper wheel and, after passing through a long-distance cell, reaches the IR detector. The modulated signal is then further processed in the evaluation electronics and stored as measured value S_n. After a complete rotation of the filter wheel, all necessary measured values (S_1 to S_n) are available, and the calculation of the gas composition can now be carried out. In this calculation phase, the remaining cross-sensitivities are also determined and offset against each other. The selectivity of the entire gas analysis is thus considerably increased. By using a long-distance cell, the absorption is very high, so that even small gas concentrations can still be measured reliably. This type of multi-gas analyzer (e.g. MCA 10 HWIR)[7] is used in the exhaust gas analysis of large combustion plants (power stations) according to the 13th BImSchV,[8] and in waste incineration plants according to the 17th BImSchV.

With the MCA 10 HWIR, up to 12 IR components can be detected simultaneously. The measuring ranges are adapted to the respective measuring task. Typical measuring ranges (manufacturer specifications) are summarized in Table 7.2.

---

[7]Dr. Födisch Umweltmesstechnik AG, Markranstädt.

[8]BimSchV = Federal Immission Control Ordinance (*B*undes-*Im*missions-*Sch*utz *V*erordnung).

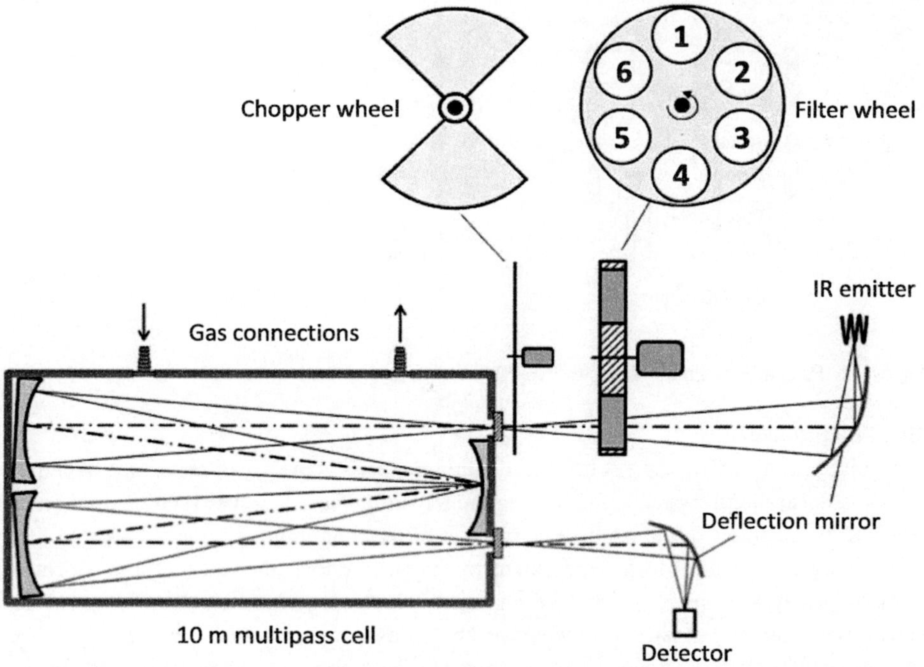

Fig. 7.30 Multigas analyzer (MCA 10 HWIR) with a heated long-path cuvette. (According to documents of Dr. Födisch Umweltmesstechnik AG)

Table 7.2 Typical measuring ranges of the MCA 10 HWIR analyzer. Measuring range 1 is certified according to DIN EN 15267-3 (MCERTS Performance Standards)

Component	Measuring range 1	Measuring range 2	Measuring range 3
Carbon monoxide CO	0. . .75 mg/m^3	0. . .300 mg/m^3	0. . .5000 mg/m^3
Carbon dioxide CO_2	0. . .25 vol.-%	0. . .50 vol.-%	–
Nitrogen monoxide NO	0. . .200 mg/m^3	0. . .400 mg/m^3	0. . .3000 mg/m^3
Nitrogen dioxide NO_2	0. . .50 mg/m^3	0. . .500 mg/m^3	–
Nitrous oxide N_2O	0. . .50 mg/m^3	0. . .3000 mg/m^3	–
Ammonia NH_3	0. . .10 mg/m^3	0. . .50 mg/m^3	0. . .500 mg/m^3
Sulphur dioxide SO_2	0. . .75 mg/m^3	0. . .300 mg/m^3	0. . .2500 mg/m^3
Hydrochloric acid HCl	0. . .15 mg/m^3	0. . .90 mg/m^3	0. . .5000 mg/m^3
Hydrofluoric acid HF	–	0. . .20 mg/m^3	–
Water vapour H_2O	0. . .40 vol.-%	–	–
Methane CH_4	0. . .50 mg/m^3	0. . .500 mg/m^3	–
Hydrocarbons Cges	0. . .15 mg/m^3	0. . .30 mg/m^3	–
Oxygen O_2	0. . .25 vol.-%	–	–

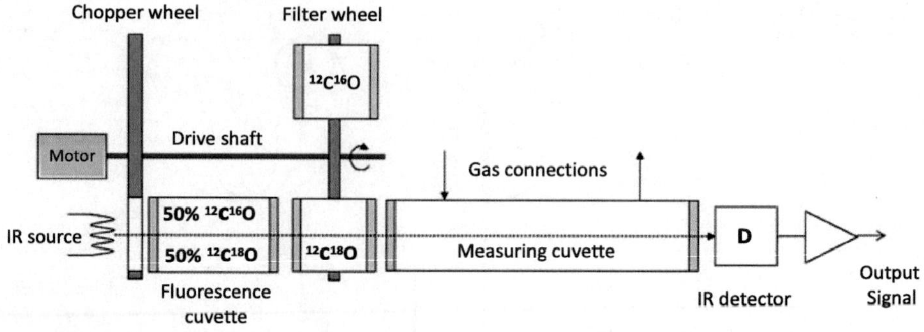

Fig. 7.31 Principle diagram of the Andros 7000 CO analyser (Link et al. 1971)

IR Fluorescence

A special feature among the gas filter correlation methods is the use of resonance radiation, which is released during relaxation during the transition from a higher rotational vibration level. If, for example, a gas volume is exposed to IR-radiation, the molecules are excited and release their energy to a large extent by impacts among themselves in the form of thermal energy (→ heating). However, a part of this energy is also released by fluorescence, whereby this radiation is selective and is used as a *radiation source* for a gas measurement. Figure 7.31 shows such a structure. The fluorescence cuvette is irradiated by a broadband IR source, whereby the radiation is intermittent by a chopper wheel. The cuvette is filled with CO, which consists of two different oxygen isotopes (^{16}O and ^{18}O). Both CO gases emit different lines due to the isotope effect. A second filter wheel, which is located behind the fluorescence cuvette, has two rotating gas cells filled with pure isotopes $^{12}C^{16}O$ and $^{12}C^{18}O$ respectively. The two wheels are driven by a common shaft so that the phase relationship is constant. The phase relation between the chopper wheel and the filter wheel is 90°. The fluorescence cuvette is first irradiated and the gas is excited to emission. In the emission phase, the irradiation is interrupted and one of the two gas filters is located in front of the beam path. This process is repeated alternately.

Since the natural CO gas consists of 99.8% $^{12}C^{16}O$, only this radiation is absorbed in the measuring cuvette. The $^{12}C^{18}O$ radiation is not absorbed and therefore serves as a reference. With this method, a very selective CO measurement can be realized. Since the fluorescence radiation is very low, cooled (−30°C) semiconductor detectors are used (Schaefer 1974). However, due to technical difficulties, the method has not yet been able to establish itself for use in monitoring exhaust emissions.

NDIR Gas Sensors

Since the 1990s, various instrument manufacturers have been developing simple NDIR photometers based on interference filter correlation (Pohl and Wiegleb 1996). In comparison to the described IR photometers, modulated radiation sources are used in these setups.

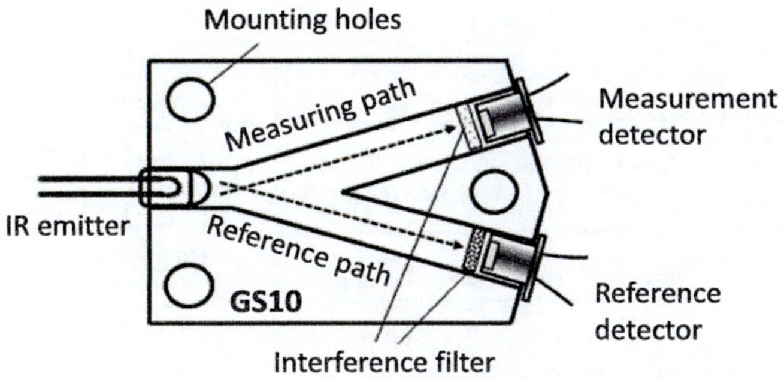

Fig. 7.32 NDIR gas sensor (type GS 10) for measuring 5 vol.% CO_2. The optical path length is $L \approx 20$ mm. With another interference filter (3.4 μm), a methane concentration of 100 vol.% can be measured with the same setup (Wiegleb 1996). (Manufacturer Sensors Europe GmbH Erkrath)

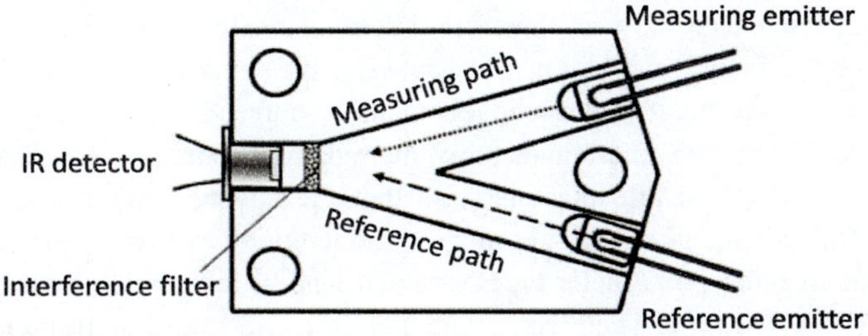

Fig. 7.33 Setup after the irradiation method with a reference source that is only switched on for calibration purposes. (Schmale 2000)

The modulation is usually donc by switching the radiation source on and off, with a frequency of $f = 1 - 25$ Hz. A mechanical chopper is therefore not necessary, so that very small and robust gas sensors can be realized. However, at least two radiation detectors are required to make use of this advantage. One radiation detector has, as entrance window, an interference filter with a passband curve at the maximum (λ_M) of gas absorption, while a second detector serves as reference. The wavelength of the reference measurement (λ_R) is ≈4 μm, since there are no absorption bands of the gases. Thermopile detectors and pyroelectric detectors are preferably used as radiation detectors. Since both detector types are mass produced, the manufacturing costs are correspondingly low. The sensor manufacturers usually also use so-called double detectors, which lead to a further cost reduction. Figure 7.32 shows an NDIR gas sensor equipped with two separate IR detectors. Alternatively, this setup can also be equipped with two radiation sources and one detector (Fig. 7.33). Only one radiation source is switched on during measurement operation. The reference measurement, with the second radiation source, is carried out at large time intervals (1× per day). Thus, the aging of the measuring source can be compensated.

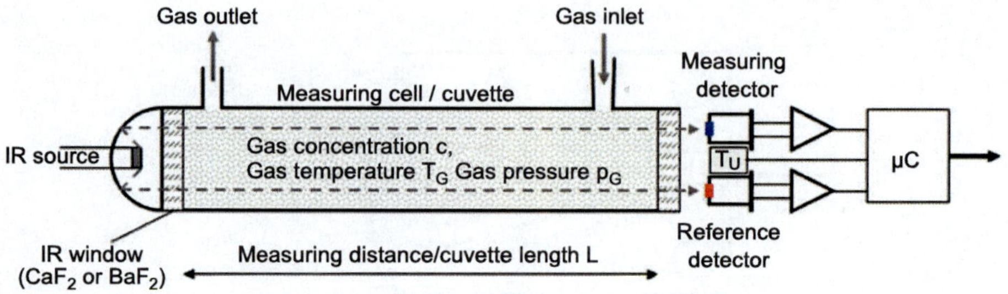

Fig. 7.34 Dual-channel filter photometer with modulated radiation source and a double detector

In the simplest case, a miniature incandescent lamp can be used as the radiation source, but its usable radiation range only extends to 4.5 µm. A CO measurement at 4.65 µm is therefore only possible with major restrictions with regard to the signal-to-noise ratio. For the wavelength range $\lambda > 4.3$ µm, therefore, micromechanical surface emitters are used which, according to Planck's radiation law, can cover the entire required spectral range up to 12 µm. A further advantage of these novel radiation sources is the very good modulation capability up to 25 Hz. The small diameter of the radiation source of $d \approx 8$ mm therefore also allows very small cuvette diameters, so that the cuvette volume can be reduced accordingly. This advantage comes into play in the temporal response behavior (flushing time). As with all other photometer types, the cell lengths are physically defined by the *Lambert-Beer law.* The typical cell lengths are therefore also from $L \approx 1$ mm for high CO_2 concentrations (100 vol.%), to $L \approx 250$ mm for very small SF_6 measurements below 50 ppm. To compensate for the influence of temperature, these IR detectors usually have an integrated temperature sensor that directly measures the chip temperature (Fig. 7.34). Pyroelectric detectors usually already have an integrated first amplifier stage that uses an FET[9] for voltage measurement. Some detector manufacturers also integrate operational amplifiers, which then generate the required signals via *current* measurement (*current mode*).

Figure 7.35 shows the spectral position of the CO_2 absorption band, and the interference component CH_4. The CO_2 measurement is not disturbed by an overlap with the methane band, whereas the reference measurement includes a small portion of the CH_4 band. In the presence of methane, the reference measurement is therefore influenced.

Modulation

The two signals, for the measuring channel (*M*) and the reference channel (*R*), are calculated by forming a quotient. In comparison to the filter photometers described above, the two signals (*M* and *R*) are always present simultaneously, so that even short-term common mode effects ($\gg$1 second) are compensated. The modulation (Mod.) reflects

[9] FET = Field effect transistor.

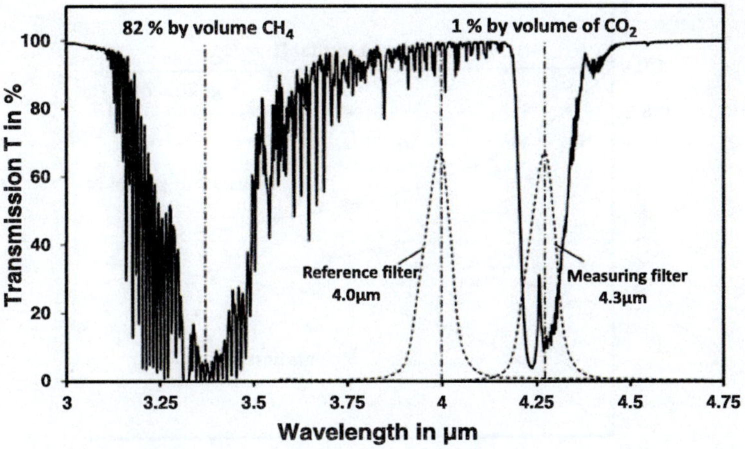

Fig. 7.35 IR spectrum of L-gas with a CO_2 content of ≈1 vol.% CO_2 and ≈82 vol.% CH_4, with the corresponding transmission curves of the interference filters (measurement and reference radiation) for a CO_2 measurement (L-gas is natural gas with a low calorific value (L = low))

the radiation portion that is absorbed by the gas. The modulation is therefore comparable with the absorption *A* (Eq. 6.45).

$$\text{Mod.} = \frac{(R - M)}{R} = 1 - \frac{M}{R} \tag{7.10}$$

A significant advantage of this calculation method is the compensation of the emitter drift due to aging effects, as well as the compensation of contamination in the cuvette. This effect is shown in Fig. 7.36.

At the beginning of the measurement, the measuring and reference signal is ≈1 Volt. The modulation then results from:

$$\text{Mod.} = 1 - \frac{1\,\text{V} \cdot \text{e}^{-\alpha cL}}{1\,\text{V}} = 1 - \text{e}^{-\alpha cL} \tag{7.11}$$

If the radiation intensity changes, due to aging effects or contamination, this change affects both channels equally. In the concrete example, 20% of the radiation is to be lost due to these influences and the sensor voltage drops to 0.8 Volt as a result. It therefore applies:

$$\text{Mod.} = 1 - \frac{\cancel{0.8\text{V}} \cdot \text{e}^{-\alpha cL}}{\cancel{0.8\ \text{V}}} = 1 - \text{e}^{-\alpha cL} \tag{7.12}$$

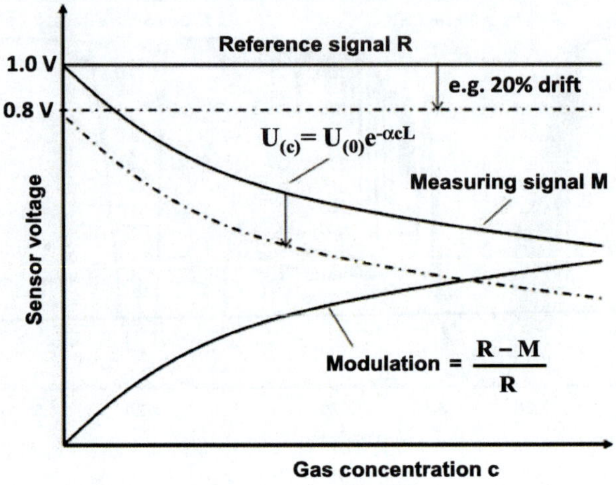

Fig. 7.36 Course of the measurement and reference signals in case of contamination or ageing of the IR radiation source

A uniform change of the radiation intensities is therefore completely compensated by this evaluation. In practice, however, there are deviations from this ideal case, so that measurable changes in modulation can be observed over a longer period of time.

The modulation calculation also leads to a compensation of external influences, such as, for example, rapid temperature changes due to air currents, vibrations etc., which are not always predictable. Figure 7.37 shows the signal curves for an NDIR gas sensor that was operated for approximately 20 hours at room temperature and zero gas flow (N_2). The static room temperature varied only by $\Delta T = \pm 1^{\circ}C$. Mostly problematic are fluctuations of the room temperature caused by drafts, opening doors or windows. Ideally, the raw signals change equally in both channels, so that the modulation calculation delivers results almost without errors. Changes in the static temperature (e.g. $10 - 50^{\circ}C$) must also be electronically compensated by measuring the detector temperature (T_U).

NDIR gas sensors of the described design can be built very small and compact. However, the cell length L is determined by physical laws that cannot be *miniaturized*. The only possibility to realize a larger cell length L with small overall dimensions is a so-called folded optical beam path. For this purpose, mirrors are used which deflect the radiation and thus create a longer optical path ($n \cdot L$). However, multi-reflection cells (Armerding et al. 1991) are very complex and expensive and must be adjusted optically with great precision. A simple optical arrangement is shown in Fig. 7.38. The IR source and the IR detector are located on one side next to each other. The opposite mirror has the radius of curvature r, where the distance between the mirror and the detector or radiator corresponds to the radius r. The center of this circle is located exactly between the two optical components. According to the laws of imaging optics, the radiation of the IR source

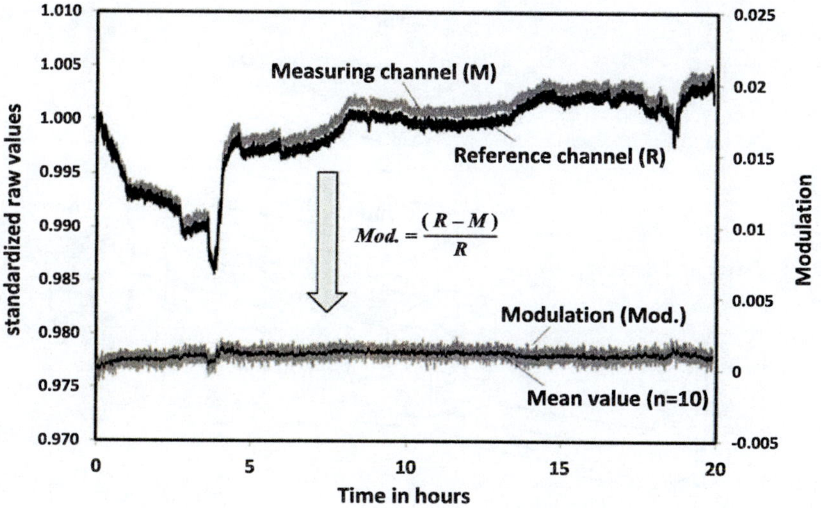

Fig. 7.37 Examination of a NDIR gas sensor over a period of 20 hours (day-night cycle) in the laboratory

Fig. 7.38 Structure of an optically folded IR gas sensor

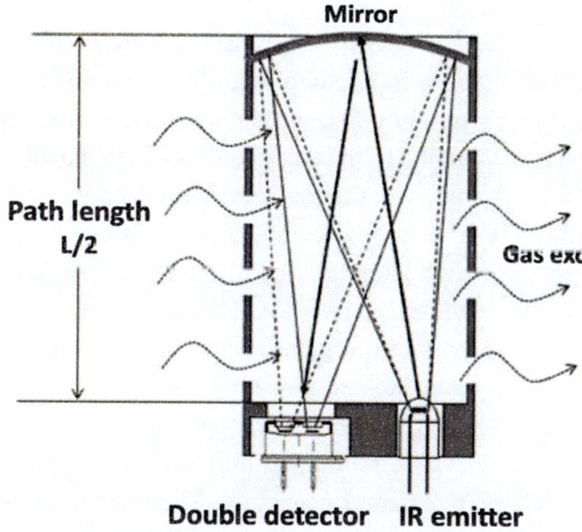

is directly imaged onto the entrance window of the IR detector. This design eliminates the need for complex adjustment. The optical path length is therefore twice as large as the mechanical construction. The gas exchange can take place both from the side (ventilation slots) and from above.

The structure shown in Fig. 7.39 is a very compact version of an optically folded system. The entire structure was realized in plastics technology (injection molding) and can therefore be produced extremely cost-effectively. However, the choice of plastics must

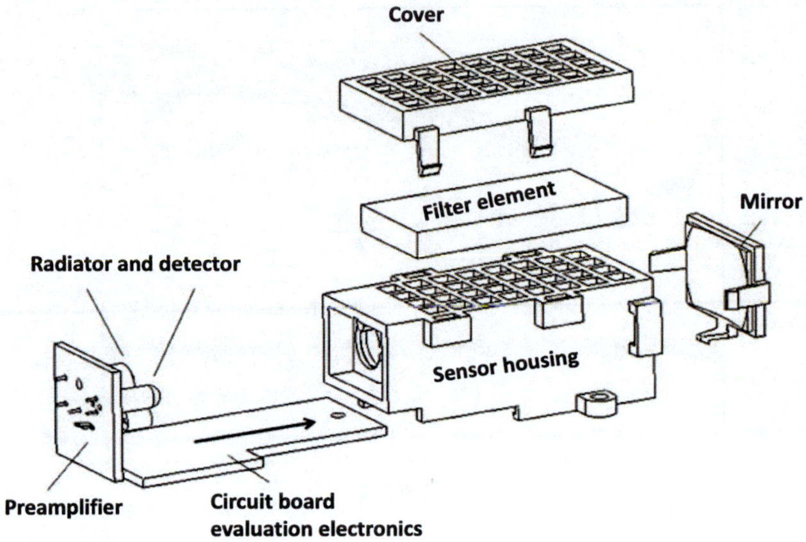

Fig. 7.39 Modular sensor unit with folded optical system and integrated evaluation electronics (Wiegleb et al. 2005)

be such that the thermal properties do not lead to deformation of the sensor housing. Even the smallest changes in geometry can lead to a distortion of the optical beam path, which can cause corresponding measurement errors. In the development phase, great importance was therefore attached to the composition of the plastic mixture. The mirror, too, is realized in plastics technology and metallized with gold in a further production step (sputtering or electroplating). The optical components (emitter, detector) are located on a circuit board (preamplifier) which is directly connected to a base board. A microcontroller is located on this base board, which controls the radiation source and performs further signal processing. The entire optoelectronic structure can be prefabricated and tested. It is then pushed into the sensor housing and fixed in place. The gas exchange takes place from above, through a filter element (foam), which is to keep the optical beam path dust-free. The filter element is held by a cover which is clipped into the sensor housing. The filter element can thus be replaced relatively easily if necessary. The main application for this sensor design is in the ambient air monitoring of natural gas and carbon dioxide (Wiegleb 2006).

The structure shown in Fig. 7.40 was further reduced within the scope of an R&D project (μ-GAS[10]) by making the measuring chamber very flat and only having a height of 4 mm. With a total length of ≈50 mm the usable cell length L was almost 100 mm. To improve the transmission of the radiation via internal reflections, the entire structure was coated with a highly reflective gold layer. Measurements of air quality (CO_2 content) in a

[10]BMBF project at the University of Applied Sciences Dortmund.

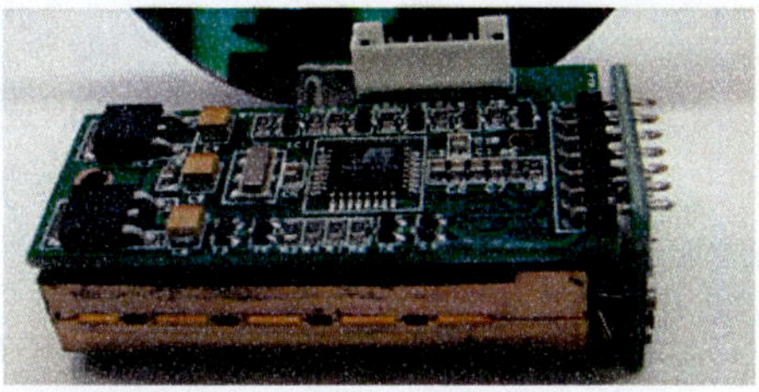

Fig. 7.40 μ-GAS sensor with integrated μC evaluation electronics. The gas exchange takes place via the *laterally* visible *ventilation slots* by diffusion (Wiegleb 2009)

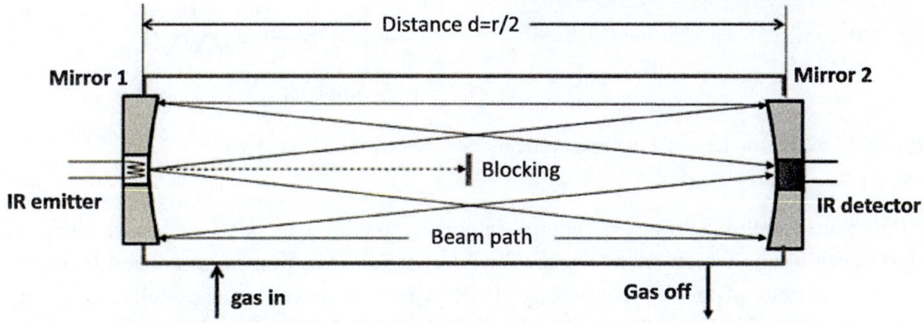

Fig. 7.41 Optical beam path for a setup with triple passage through the measuring cell

vehicle were used to demonstrate the performance of this setup (Köhne and Wiegleb 2008). The detection limit for these measurements was <20 ppm CO_2.

Another variant of the folded optics is shown in Fig. 7.41. The two mirrors have a defined radius of curvature r and are located at a distance $d = r/2$ from each other. The radiation of the IR source reaches mirror 2 with a certain aperture angle γ, which reflects the radiation in a parallel beam path onto mirror 1. From there, the radiation is deflected a second time in the direction of mirror 2, and the radiation is now concentrated on the entrance window of the detector. The radiation thus travels three times the distance d, so that the optical path L through the gas is three times the value of d ($L = 3d$). Also, this construction does not need to be adjusted in a complicated way. By moving the radiation source, the optimum point is determined when the signal quantity passes through a maximum during adjustment at the detector.

7.3 Photoacoustic Gas Analysers (PAS[11])

The photoacoustic effect in gases was discovered and scientifically investigated independently by Röntgen and Tyndall as early as 1881. However, it was only with the development of powerful and miniaturized microphones that this effect could be made useful for

[11] Photo Acoustic Spectroscopy.

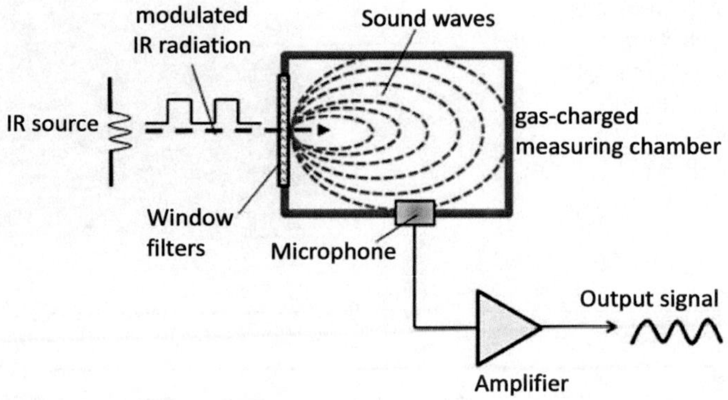

Fig. 7.42 Basic structure of a photoacoustic measuring cell for gas analysis

gas measurement technology. Another important prerequisite was the availability of interference filters, which are necessary for the spectral limitation of broadband IR radiation. In the field of laser spectroscopy, PAS is also used very successfully, since the necessary spectral bandwidth can already be specified with a laser, and the modulation of the radiation is much easier. Furthermore, lasers have a very high spectral intensity, which is necessary for a low detection limit.

Figure 7.42 shows a simple structure of a photoacoustic measuring cell that works with a broadband IR source (thermal radiation).

The IR radiation is usually modulated by a mechanical chopper wheel and tuned to the respective gas with a narrow-band interference filter. The gas mixture to be analyzed is in the measuring cell. Since selective absorption causes an increase in pressure in the measuring cell, which is then to be detected by a microphone, the measuring cell must be closed during the measuring process. As a rule, electrically controlled solenoid valves are used for this purpose. The pressure increase (sound amplitude) is proportional to the gas concentration c (Fig. 7.43). To exclude disturbing wall effects (*bark effect*) caused by broadband radiation absorption, the interior of the cell is gold-plated. After the measuring process, the solenoid valves are opened so that new sample gas can flow in. This process is repeated cyclically. In this case, it is therefore a discontinuous measurement, as the cycle times are in the range of seconds. Simplified versions use a diffusion barrier through which the sample gas must pass. This barrier also serves as a closure of the measuring cell for the photoacoustic pressure increase. Although the pressure is also reduced slightly via this diffusion barrier, the measuring effect is still detectable. This type of photoacoustic measurement is only used for simple measurement tasks, such as indoor air monitoring. This is because the diffusion barrier also allows sound from outside to enter the measuring cell, so that this ambient sound can disturb the actual gas measurement. For industrial measurements, with high demands on measurement accuracy, the measuring cell must therefore be closed during each measurement process.

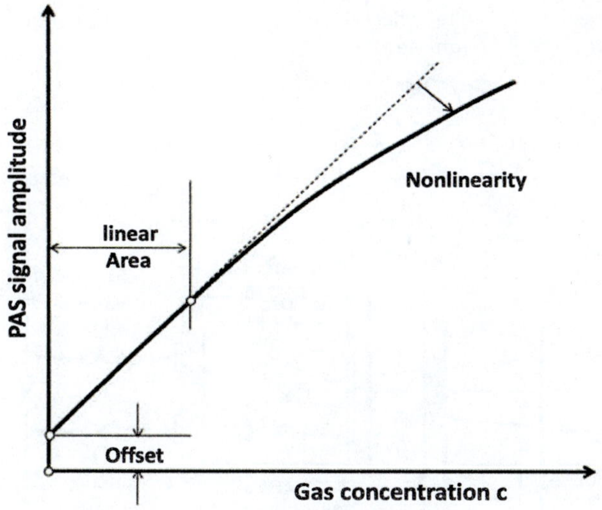

Fig. 7.43 PAS characteristic curve with an offset at zero point

The characteristic curve of a photoacoustic system starts at zero point. Without absorption by the gas to be detected ($c = 0$) in the cell, the PAS signal is theoretically also zero. Due to low absorption at the wall, or elsewhere in the cell, a small portion of the incident radiation is absorbed and provides a constant offset (Fig. 7.43). This offset is kept very low by the manufacturer through constructive measures (polishing, gold plating, . . .). If the gas concentration in the cell now increases, a proportional PAS signal is generated, which results from the *Lambert-Beer law*. For low concentrations the signal is linear, while it is buckled for high concentrations. In this case, the PAS signal must be electronically linearized. This curve is exactly the opposite of the photometer setup. In a photometer, the raw signal is given by the transmission T, i.e. you have a large signal that changes only slightly with small gas concentrations. With the PAS method it is the other way round. The signal is zero at zero point and increases with the gas concentration c. It is therefore proportional to the absorption A in the measuring cell. For the detection of small gas concentrations in the ambient air, this method therefore has advantages. The detection limits are specified by the manufacturer and, depending on the type of gas, are usually $\gg$1 ppm. For sulfur hexafluoride SF_6, this method even achieves a detection limit of 4 ppb, at $\lambda = 10.6\,\mu\text{m}$.

INNOVA System

Figure 7.44 shows an industrial gas analyzer of the INNOVA type. Two microphones are installed in the measuring chamber to compensate for structure-borne and ambient noise from outside (Staab 1994). The radiation of the IR source is focused on the entrance window of the measuring chamber via a large-surface, gold-plated reflector, so that a high

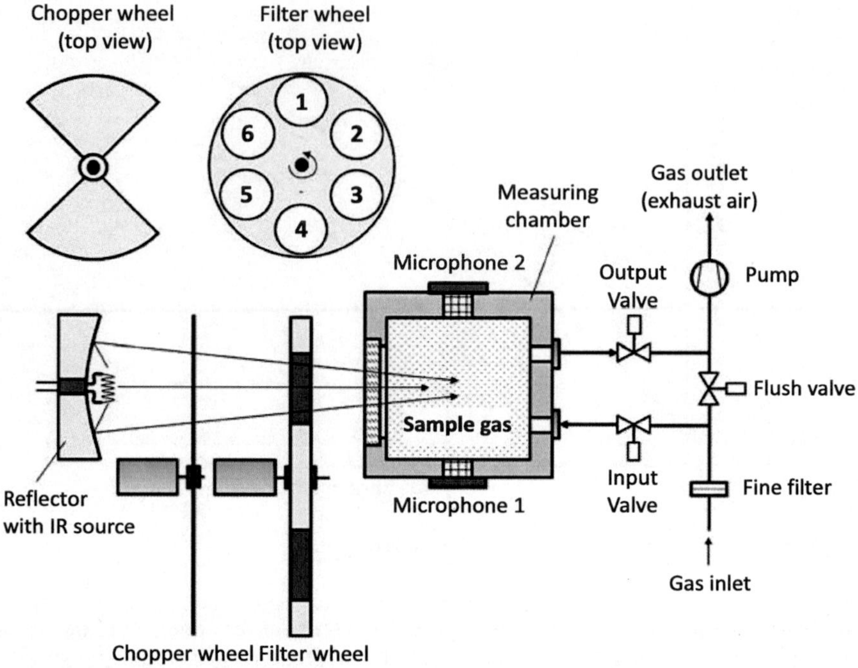

Fig. 7.44 Physical structure of the INNOVA 1412i. (According to the manufacturer's specifications of LumaSense Technology Inc. USA)

radiation component is available for the measuring effect. The broadband radiation is in turn limited by an interference filter. In this case, however, the interference filter is located on a filter wheel (turret), which can accommodate up to 6 interference filters. The individual IF filters are matched to different gases, so that with this setup up to 6 gases can be measured simultaneously. A total of 27 different filters are offered by the manufacturer[12] for a spectral range from 3.4 to 14.1 µm. With these filters, the entire available IR spectrum can be covered almost completely.

The measuring cycle consists of two phases. First, an interference filter (1) for measuring component *A* is in front of the measuring chamber. The chopper wheel is used to modulate the radiation, thus generating a sound frequency f_0 in the measuring chamber, which is picked up by the microphones (1 + 2). The concentration c_A of gas *A* is then calculated from the sound amplitude. Now, the second filter is placed in front of the measuring chamber and the gas *B* is measured. The gases *C*, *D*, *E* and *F* are measured one after the other in the same way. At the end of this measuring phase the contents of the

[12]LumaSense Technology Inc. USA.

measuring chamber are flushed out and new sample gas is introduced. A dwell time of $\Delta t = 2$ s results in a cycle time of 12 seconds, plus flushing time. For most applications this time interval is sufficient, so that one can definitely speak of a (quasi) continuous process.

7.4 IR Components for Photometer

The construction of an IR photometer always consists of three components:

- Radiation source
- Cuvette
- Detector

Furthermore, an additional interference filter is usually required for spectral narrowing. The interference filter is usually integrated into the detector as an entrance window, and is thus a component of the detector unit. In the following, the properties and special features of these components are described, since the measurement properties of these components have a significant influence on.

IR Radiation Sources

The generation of infrared radiation is preferably done by thermal radiation sources. These IR radiators are brought to a temperature that is significantly above room temperature, so that infrared radiation is emitted in accordance with *Planck's law of radiation*. The spectral and spatial distribution of the radiation is of great importance for use in gas measurement technology and can be calculated with appropriate models (Figs. 7.45 and 7.46). For this purpose, a thin plate with area A is considered, which has been heated to a temperature T (Fig. 7.47). This plate emits diffuse radiation according to *Lambert's law*. Since the radiation components in directions that are inclined by the angle φ correspond to the cosine,

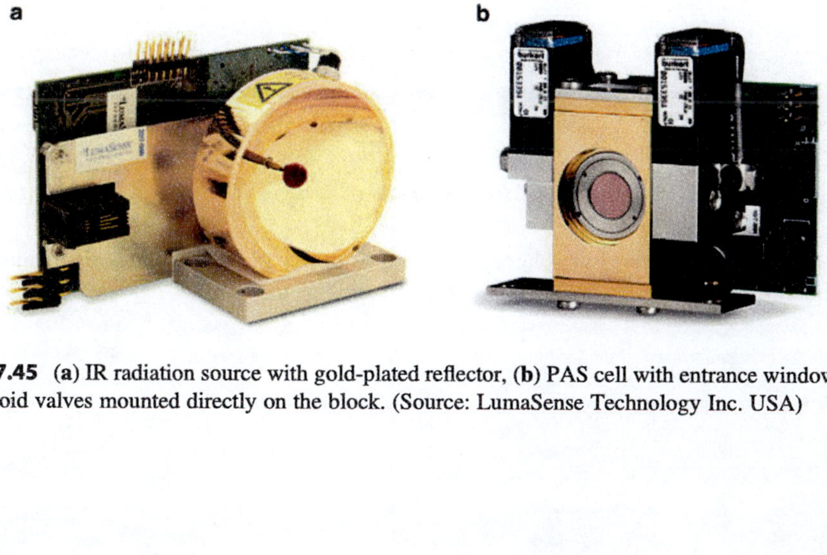

Fig. 7.45 (a) IR radiation source with gold-plated reflector, (b) PAS cell with entrance window and solenoid valves mounted directly on the block. (Source: LumaSense Technology Inc. USA)

Fig. 7.46 Filter wheel (revolver) with 6 interference filters installed. (Source: LumaSense Technology Inc. USA)

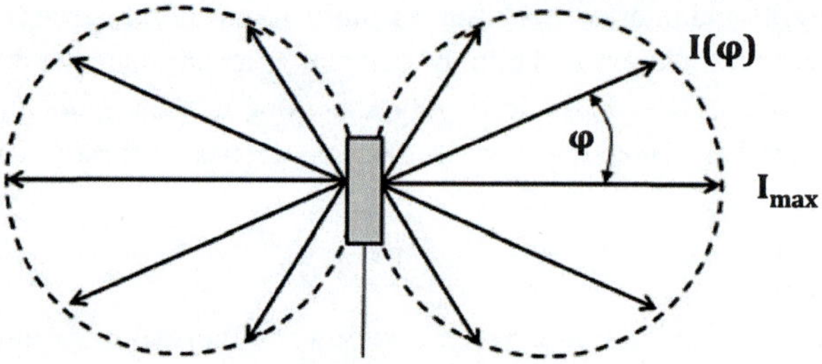

Fig. 7.47 Distribution of radiation emission from an area *A* according to Lambert

the radiation distribution ($I\varphi$) must form a circle. This type of area radiation is therefore also called *Lambertian radiator* with the following distribution curve:

$$I(\varphi) = A \cdot L \cdot \cos\varphi \tag{7.13}$$

The radiant power P_E emitted from this area *A* can be calculated by the *Stefan-Boltzmann law* as follows:

$$P_E = \varepsilon \cdot \sigma \cdot A \cdot T^4 \tag{7.14}$$

The radiant power is therefore related to the fourth power of the absolute temperature *T*. The factor σ is the Stefan[13]-Boltzmann constant, and has the value:

[13] Josef Stefan (1835–1893) Austrian mathematician and physicist.

$$\sigma = 5.6703 \cdot 10^{-8}\,\mathrm{W} \cdot \mathrm{m}^{-2}\mathrm{K}^{-4} \tag{7.15}$$

The quantity ε is the emissivity and describes how well the surface can emit the radiation. The value is between 0 and 1 depending on the surface condition. Shiny metal surfaces (e.g. mirrors) have a good reflectance R ($\approx$0.9) and a bad emissivity ($\varepsilon \approx 0.04$). In contrast, a surface blackened with soot has an emissivity of $\varepsilon = 0.95$. The degree of reflection is correspondingly low. The following relationship therefore applies:

$$\varepsilon + R = 1 \tag{7.16}$$

It can also be shown that the emissivity ε and the absorption coefficient A are identical (Hahn 2007).

Black Radiator

The spectral distribution ($P\lambda$) of the emitted radiation is given for an ideal *black body* and results from *Planck's radiation law*:

$$\mathrm{d}P_\lambda = \frac{2\pi hc^2}{\lambda^5} \cdot \frac{A}{\exp\left(\frac{hc}{k\lambda T}\right) - 1} \mathrm{d}\lambda \tag{7.17}$$

Figure 7.48 shows the course of the spectral radiation density as a function of the wavelength λ for an ideal black body ($\varepsilon = 1$). For surface temperatures <700° C, the

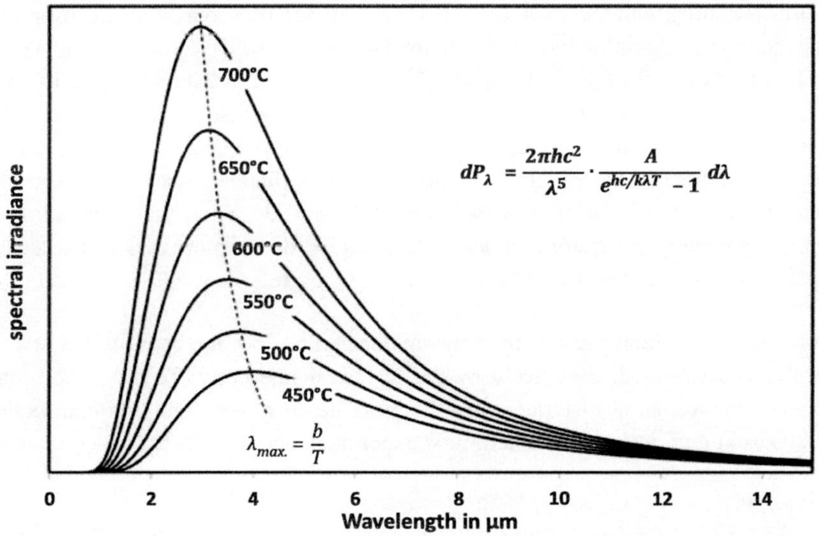

Fig. 7.48 Spectral radiation density of an ideal black body according to Planck

radiation distribution only begins at a wavelength $\lambda \approx \mu$ m and is therefore not yet visible to the human eye. Furthermore, the distribution function has a maximum λ_{max}, whereby the maximum shifts with temperature T towards the short-wave spectral range. This shift can be calculated by the first derivative of Planck's radiation formula (Eq. 7.17), and the so-called shift law according to Wien[14] is then obtained:

$$\lambda_{max} = \frac{2897.8 \ \mu\text{m} \cdot \text{K}}{T} \tag{7.18}$$

The value is ≈2897.8 µm · K also called the *Wien- constant* ($b = 2.897756 \cdot 10^{-3}$ m · K). The distribution of the emitted radiation, above and below λ_{max}, can be calculated from Eq. 7.17 The radiation fraction up to the maximum is 25% while the fraction above the maximum is 75%. This distribution is independent of the temperature T (Budzier and Gerlach 2010).

Example
If we consider the sun as an ideal *Planck radiator*, the surface temperature T of the sun can be calculated from the maximum of the spectral distribution of sunlight, which is at 500 nm = 0.5 µm:

$$T = 2897.8 \ \mu\text{m}\,\text{K}/0.5 \ \mu\text{m} \approx 5800\,\text{K} \approx 6000^{\circ}\text{C} \tag{7.19}$$

Cavity Radiators
An ideal black radiator can only be approximated by a so-called cavity radiator, since there is no surface coating with a ε-value of 1. This type of radiation source is used for physical investigations and calibration purposes. Figure 7.49 shows such a test radiator. It is usually heated to a temperature of $T = 500$ K (≈227°C) and emits a radiation distribution that is very close to that of an ideal black body radiator. For this purpose a cavity is blackened on the inside ($\varepsilon \approx 0.95$). In the cavity there are emission, absorption and reflection of radiation, but due to the geometrical arrangement in the cavity, the radiation passes through the cavity and emits an almost ideal blackbody radiation to the outside. These radiation sources are used in the laboratory for calibration and determination of the properties of IR detectors. The specifications of these IR detectors therefore always refer to a *500 K emitter*.

In practice, IR emitters are mostly grey emitters, with a ε-value between 0.8 and 0.95. This value is determined exclusively by the surface of the body. To obtain the highest possible ε-value, different materials must therefore be examined. The optimum radiation conditions can therefore only be determined experimentally.

[14] Wilhelm Wien (1864–1928) German physicist.

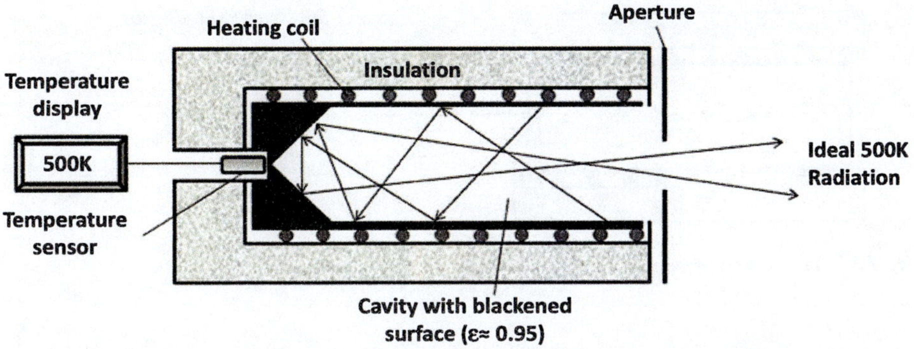

Fig. 7.49 Schematic diagram of an infrared radiator (cavity radiator) that comes very close to an ideal black body radiator for calibration purposes

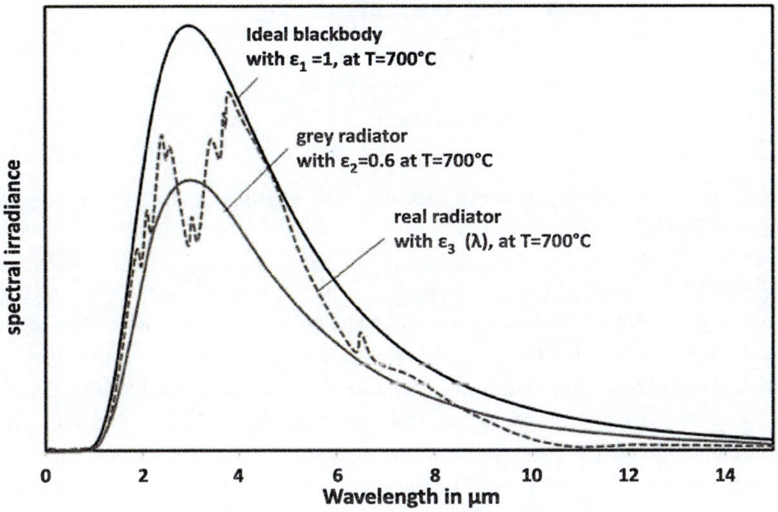

Fig. 7.50 Spectral distribution of the different emitter types, at the same surface temperature ($T = 700^{\circ}$C)

Grey Emitter

A closer look at real emitters also shows that the ε-value is not a real constant, but can also change with the wavelength λ. In this case, one also speaks of selective emitters, which only emit radiation in certain wavelength ranges (Fig. 7.50). This results in intensity distributions that lie below the ideal curve and show gaps at certain points in the spectrum. Open or porous structures are advantageous here, as they allow an approach to a cavity radiator. Since interactions (absorption/reflection) also take place in this porous layer, this leads to an increase in the ε-value.

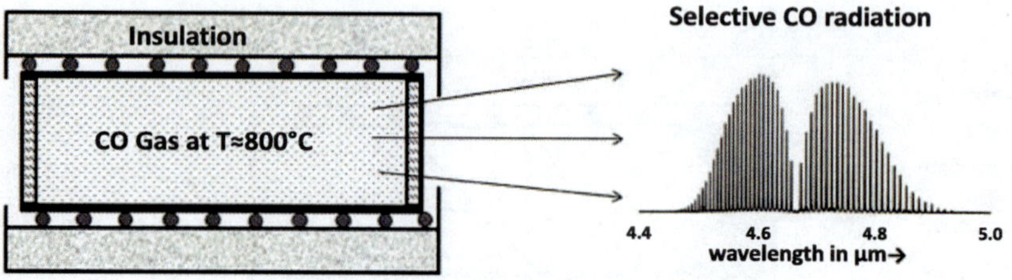

Fig. 7.51 Generation of selective gas radiation by heating a volume of gas

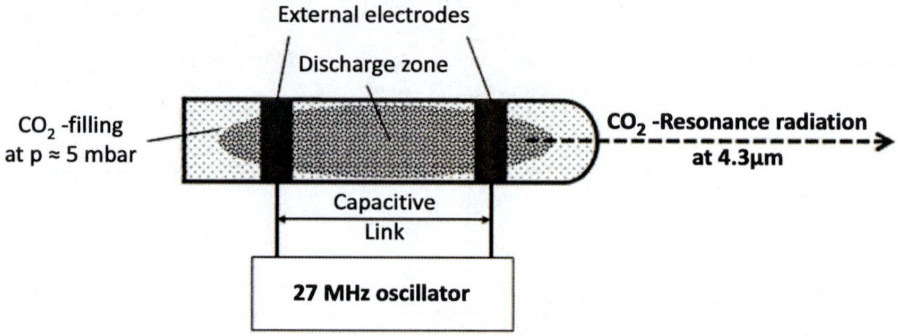

Fig. 7.52 Setup for generating selective radiation (→ resonance radiation) for the IR range (Arnoudse et al. 1992)

Gas Radiation

A special feature among IR emitters is the radiation emission of gas molecules. Especially hot gases belong to this category. If a volume of gas is heated, the molecules reach higher rotational vibration levels due to thermal excitation because of the Boltzmann distribution. When they relax, part of this energy is released again by radiation and the individual transitions can then be seen in the emission spectrum. The spectral position of these emission lines is thus congruent with the absorption lines of the gases (Fig. 7.51). In this context, one also speaks of selective IR radiation. In principle, this radiation can therefore also be used for gas measurement. Due to the high temperatures and the associated technical requirements, this possibility has hardly been used for gas analysis up to now. Only in the field of *remote* measurement is this method occasionally used.

Another method of generating selective gas radiation is to excite and ionize the trapped gas by means of an electrical discharge (Arnoudse et al. 1992). To do this, a glass tube is filled with the gas to be excited at a pressure p of a few mbar and hermetically sealed (Fig. 7.52). All materials that are transparent in the IR range are suitable as exit windows. (e.g. MgF_2). The electrical excitation is performed with external electrodes (capacitive coupling) to prevent a chemical reaction between the electrode material and the gas.

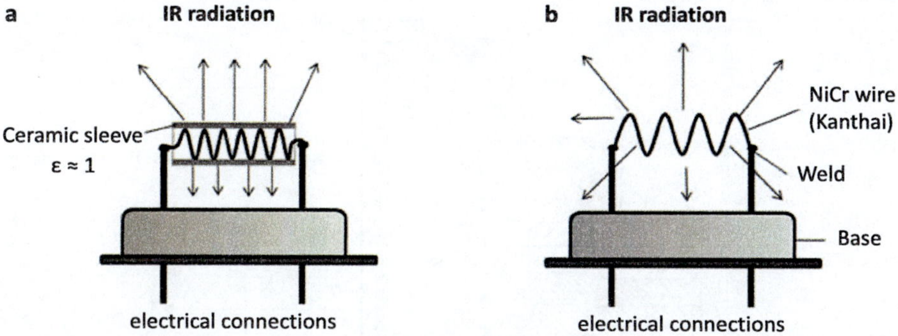

Fig. 7.53 IR radiation source with an NiCr or Kanthal wire (**b**) and an additional ceramic sleeve (**a**) to improve the radiation distribution and increase the emission

Usually high-frequency oscillators in the MHz range are used for this purpose. Ionization in the gas phase excites not only electronic levels but also rotational oscillation levels which, after relaxation to the ground state, emit selective radiation. For CO_2, this radiation is at 4.3 μm and 2.7 μm. This selective IR radiation can then be used to operate a corresponding absorption photometer. This method is only of limited use for CO, since CO_2 reacts to CO during ionization and carbon can be released. The application possibilities of this method are therefore limited.

Glow Emitter

For practical applications, the glow emitter has established itself in gas measurement technology. It usually consists of a high-alloy NiCr[15]wire or Kanthal,[16] which is heated by an electric current. After initial heating, this wire forms an oxide layer that protects the underlying metal and increases the emission coefficient ε. A bare metal surface has a ε value of ≈0.15, while the oxide layer has a ε value of ≈0.7. Figure 7.53 shows the structure of such a radiator. The wire is wound into a coil and welded to the connecting legs of a pressure glass feedthrough (socket). This type of connection guarantees a high temperature resistance. Since each individual winding radiates individually, an inhomogeneous radiation distribution is obtained, which can lead to a drift in the gas detector if the position of the wire's changes. Some manufacturers therefore embed the wire winding in a ceramic sleeve, which also has a better ε-value (>0.9). This measure results in a homogeneous and stable radiation distribution compared to simple wire windings.

[15]Nickel-chrome wire.

[16]Trade name for FeCrAl alloys.

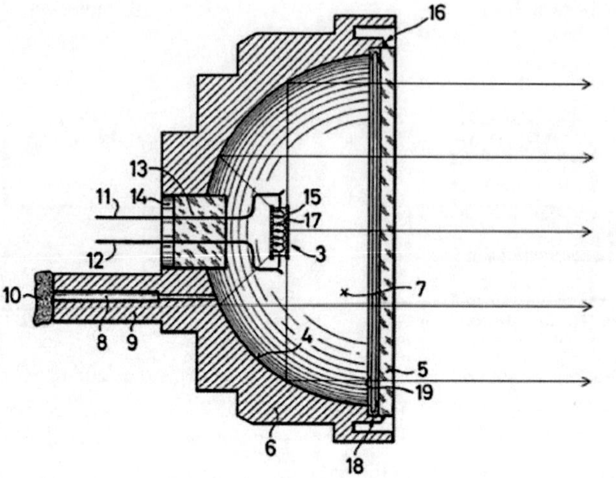

Fig. 7.54 Infrared radiation source for low pressure operation. (Modlinski and Wiegleb 1988)

Low Pressure Operation

The heat transfer from the filament takes place via the usable IR radiation and additionally also via the heat conduction of the surrounding air. This heat conduction leads to energy losses, which the IR radiation then lacks. Furthermore, convection in the surrounding gas phase of the wire winding causes further cooling of the wire. This convection can further lead to a position dependence of the output signal. Figure 7.54 shows an IR radiator that works in negative pressure mode and thus prevents disturbing convection. Furthermore, the parasitic heat conduction is greatly reduced, so that the radiation yield is more favorable overall. The CaF_2 window (5) is connected to the reflector housing (6) by a glass solder connection (16) without adhesive and the wire winding (17) is surrounded by a ceramic sleeve (15). The radiating unit (3) is located at the focal point ($r/2$) of the spherical reflector (4) with radius r, so that the radiation emerges almost parallel (small opening angle θ) from the window. The pressure glass feedthrough (14), with the connecting wires (11, 12), is also connected to the housing with a temperature-resistant and gas-tight glass solder connection (13). The internal volume is evacuated via a gas connection (9) and then sealed with a metallic connection (10).

Incandescent lamps can only operate in DC mode due to their large thermal mass. This means that they require an additional mechanical chopper to modulate the radiation. However, the advantage of these radiators is the high electrical output of over 10 watts with which they can be operated. The IR radiation is therefore also correspondingly high, which is particularly important for the analysis of low gas concentrations (→ signal-to-noise ratio).

Miniature Incandescent Lamps

For simple applications where high measurement accuracy is not required, miniature incandescent lamps are the ideal IR radiation source. Like incandescent lamps, these emitters work with a wire coil, which in this case, however, is made of tungsten. Tungsten has a very high melting point (3380 °C) and can therefore also be operated at very high temperatures (≈1000°C). However, the metallic surface has a disadvantageous effect on the emission characteristics, as the ε-value is comparatively small. This disadvantage is partly compensated by the very high operating temperatures of the filaments. In a spectral range up to 4.5 μm, there is therefore sufficient radiant energy available. The wire coil is melted into a glass bulb, which also has a vacuum filling. The usable spectral emission range is only limited by the glass material of the bulb. This spectral range is shown in Fig. 7.55. Gases such as methane, ethane, propane, butane and carbon dioxide can be measured quite well with it, while carbon monoxide (CO) can only be measured to a limited extent. Nitric oxide (NO), sulfur dioxide (SO_2) or sulfur hexafluoride (SF_6) cannot be measured with these radiation sources. However, these radiation sources have proven to be useful for monitoring CO_2 in the room air (ventilation control), especially since the production costs are comparatively low.

Due to the absorption of radiation in the glass layer, the glass bulb heats up and in turn emits IR radiation (secondary radiation), which is superimposed on the primary IR radiation of the wire coil. This further restricts the modulation capability of a miniature incandescent lamp. The maximum operating frequency is 1 – 5 Hz.

Miniature incandescent lamps must be pre-aged before initial operation to ensure stable measurements. During the first days of aging, the intensity changes by up to 5% from the

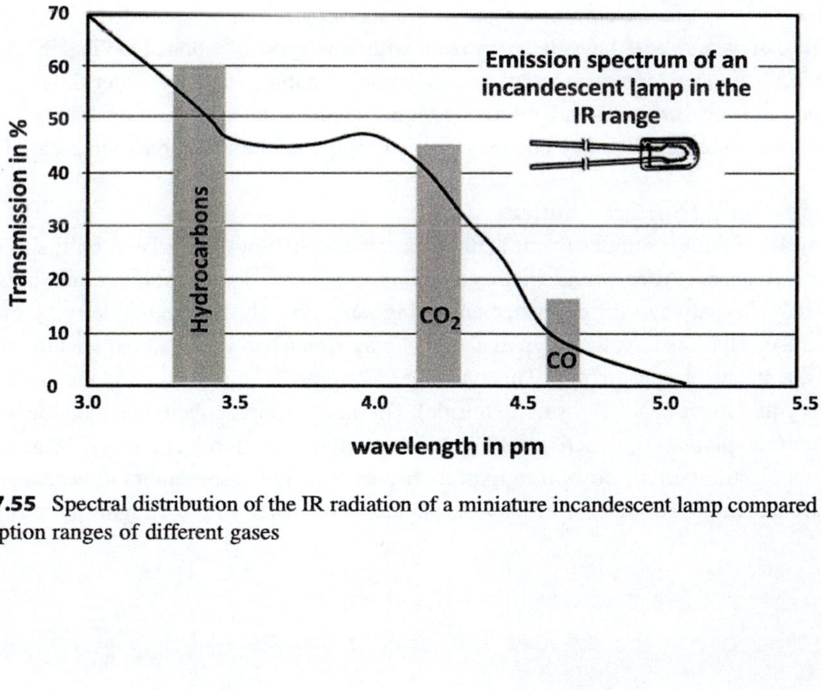

Fig. 7.55 Spectral distribution of the IR radiation of a miniature incandescent lamp compared to the absorption ranges of different gases

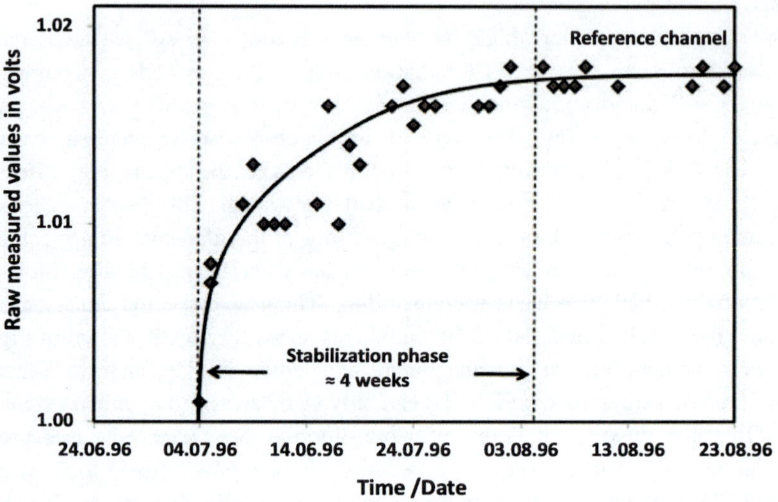

Fig. 7.56 Running-in phase of a gas sensor after the first switch-on. This running-in phase is carried out at the manufacturer's premises, so that only pre-aged gas sensors are delivered (Type GS10, Sensors Europe GmbH Erkrath)

basic intensity. After 4 weeks at the latest, this running-in phase is completed and the radiation sources can be installed (Fig. 7.56). The service life of these miniature incandescent lamps is then years >10 if the maximum permissible operating voltage is reduced by 10 – 20%. AC operation also has a life-prolonging effect. A modulation with 1 Hz and a *duty cycle* of 50% leads to a doubling of the service life. If AC operation continues to be linked to a change in the current direction, the service life can be extended even further. Lifetimes of years >20 have been proven with this type of control (Wiegleb 2015). Figure 7.57 shows a long-term measurement of the radiation intensity. After the running-in phase, the measurement and reference signal changes by a maximum of 5%. After 10 years, the radiation intensity changes very strongly and the radiation source then fails.

Micromechanical Surface Emitters

Thermal IR radiation sources manufactured in microtechnology have been on the market for several years. After initial quality problems, these radiation sources are now an interesting alternative to the described annealing radiators. The design is relatively simple (Fig. 7.58). The basis is a microchip made of silicon, which has a trapezoidal opening in the center due to anisotropic etching. This opening (2.2 mm × 2.2 mm $\rightarrow A \approx 5\ \mathrm{mm}^2$) is closed by a very thin membrane (e.g. silicon nitride). On this membrane there is a metallic heater structure (e.g. platinum), which is applied by sputtering (thin-film technology) (Fig. 7.59). When an electric current flows through this heater structure, the membrane heats up and emits IR radiation. To increase efficiency, an additional layer with the highest possible ε-

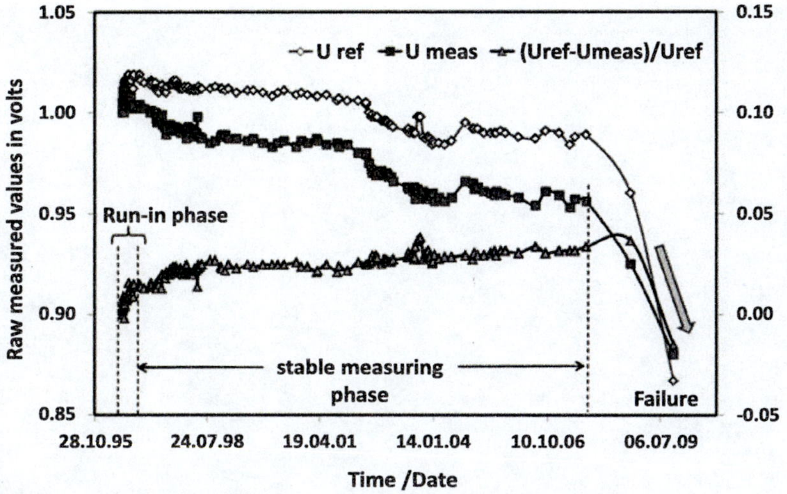

Fig. 7.57 Measuring the measurement and reference signals of an NDIR gas sensor (GS10) over a period of 14 years. With the calculated values of U_{ref} and U_{meas}, one obtains a compensated, temporal signal curve (Source: University of Applied Sciences Dortmund)

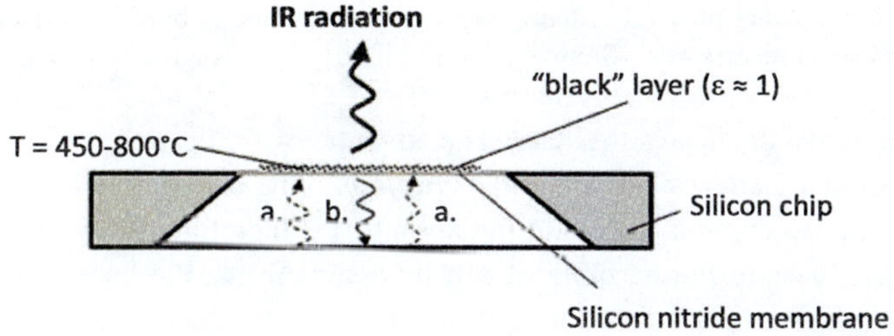

Fig. 7.58 Cross-section of an IR radiator structure built up using microtechnology. The main part of the IR radiation is emitted from the upper side of the membrane and from there reaches the following photometer

Fig. 7.59 Wafer with several microchips (IR radiation source) before separation. (Source: Axteris-Microsystem, Kägiswill/ Switzerland)

Fig. 7.60 Surface structure of an IR radiator with a porous coating to bring the ε-value as close as possible to the ideal value of $\varepsilon \approx 1$. (Source: Axteris-Microsystem, Kägiswill/Switzerland)

value is placed above the heater structure. The advantage of this arrangement is that it is a radiating surface that behaves like a *Lambert radiator*. The back of the membrane has no additional coating, so that the backward radiation (b.) is therefore also very low. Another advantage of this design. Furthermore, it can be assumed that the backward radiation is reflected at the pressure glass feed-through (a.) and thus returned to the membrane. The efficiency of this emitter arrangement is therefore particularly favorable. Typical membrane temperatures are at $T \approx 450^{\circ}$C, so that in the visible range the surface does not radiate and remains black. Only at higher temperatures ($T > 800^{\circ}$C) a dark red discoloration can be detected.

Important for an efficient conversion of the electrical power ($P = U \cdot I$) into the desired optical radiation power ($P\lambda$) is the surface structure of the emitter. Since the emission coefficient of the membrane surface ε does not have optimum properties, these emitters are provided with an additional layer. Figure 7.60 shows the surface of such an emitter. The structure is very porous and a three-dimensional structure is clearly visible. The ramifications of this structure have a very advantageous influence on the emission coefficient, which then comes close to the optimum value of 1 (Fig. 7.61).

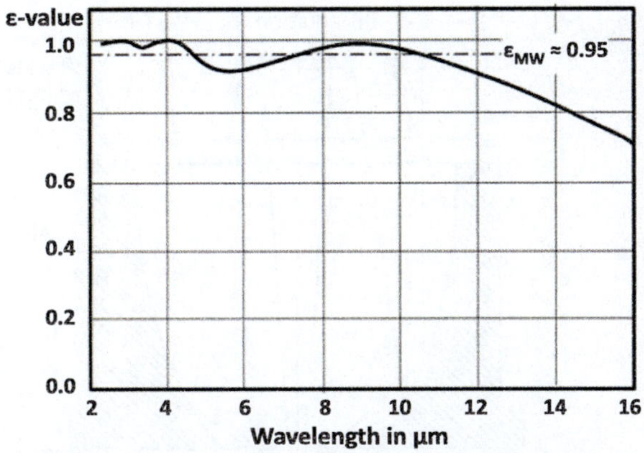

Fig. 7.61 Emission value (ε) of the *black* layer on the heater structure of the IR radiator. The average value of $\varepsilon_{MW} \approx 0.95$ refers to the usable spectral range from 2 to 12 µm

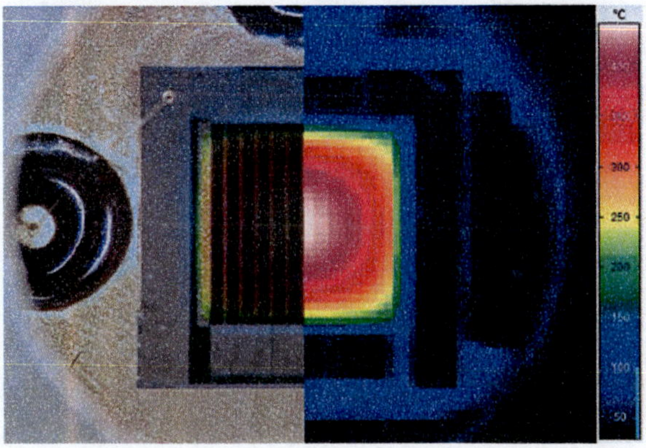

Fig. 7.62 Inhomogeneous temperature distribution of a surface radiator. (Source: Axteris-Microsystem, Kägiswill/ Switzerland)

The temperature distribution on the membrane shows a gradient that decreases from the center to the edge (Fig. 7.62). This gradient is caused by physics and has no further influence on the gas measurement.

To further optimize the radiation yield, the IR radiation is deflected with a parabolic light funnel (Fig. 7.63). This measure provides a preferred direction (smaller opening angle) for the light rays (Fig. 7.64). In this way, the lateral beams are also additionally coupled into the optical system. The reflector is made of aluminum and, if necessary, is also gold-plated in order to transmit the IR radiation with lower losses. In addition, the assembly can also be closed with an IR window (CaF_2, BaF_2) to keep disturbing environmental influences and dust out of the system. However, a completely hermetic encapsulation can be problematic if gas exchange with the ambient air is no longer possible. Outgassing can lead to an accumulation of pollutants in the emitter structure, which can then destroy the layer structure. Some manufacturers therefore install a vent in the pressure glass feedthrough to prevent gas-tight encapsulation. Unfortunately, the natural CO_2 content of the ambient air then also enters the radiator structure and leads to a disturbing

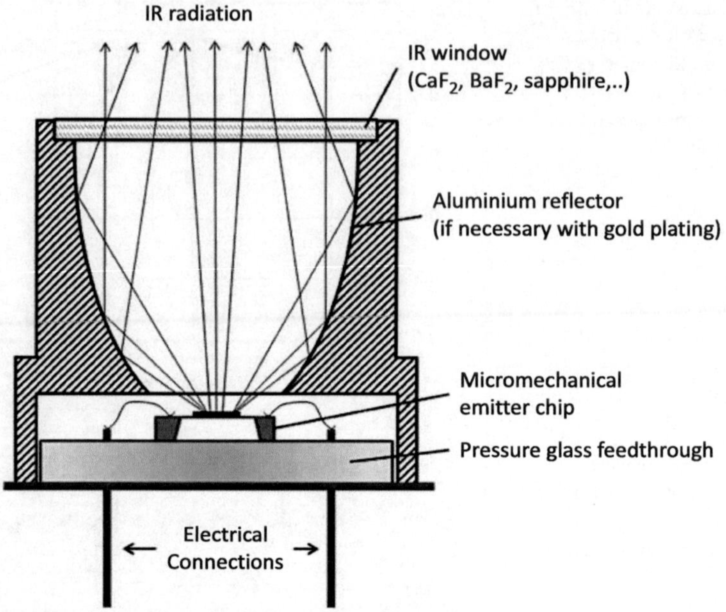

Fig. 7.63 Completely mounted IR radiator with a parabolic light funnel for bundling the lateral radiation components

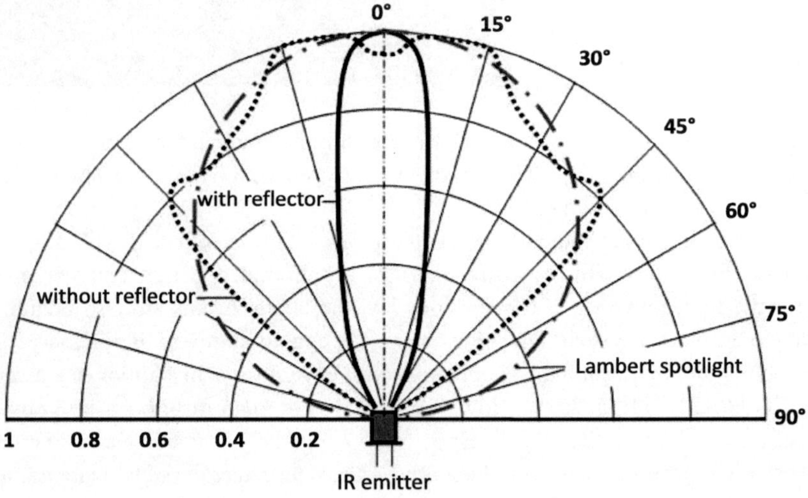

Fig. 7.64 Spatial distribution of area radiation with and without reflector compared to a Lambert radiator. The values are each normalized to 1 for the maximum

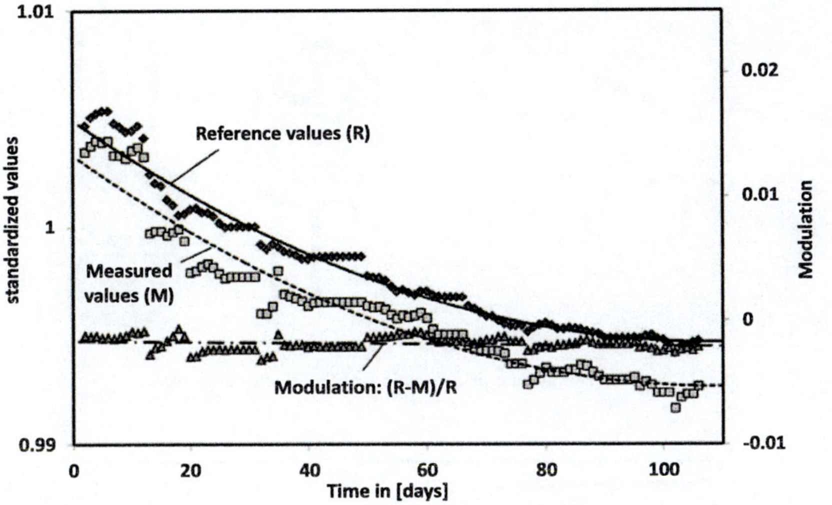

Fig. 7.65 Running-in behavior of a micromechanical IR radiator in the first 100 days after switching on (burn-in phase)

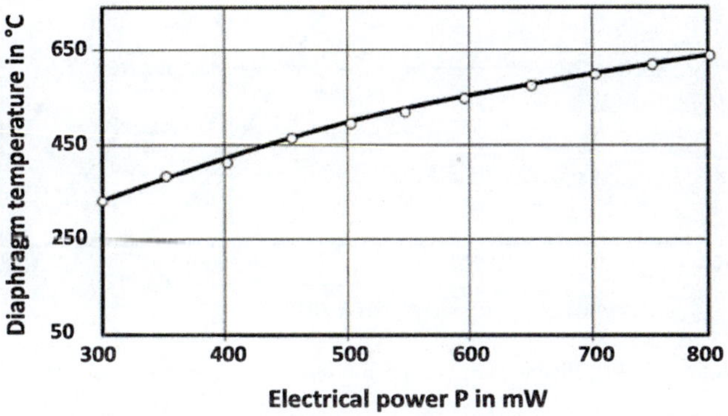

Fig. 7.66 IR radiator setup (TO-39) with ventilation. (Source: Axteris-Microsystem, Kägiswill/ Switzerland)

pre-absorption. Other manufacturers use other materials that do not emit so much gas, so that these structures may then also be encapsulated. These radiation sources must also be burnt in over a longer period in order to achieve stable measurement results (Fig. 7.65).

The surface temperature T of the layer is determined by the supplied electrical power P. The power is usually below 1 Watt, which leads to temperatures of $T < 800^{\circ}\text{C}$ (Fig. 7.66). The functional relationship $T(P)$ is shown in Fig. 7.67 for a specific radiator (type EMIR-S200).

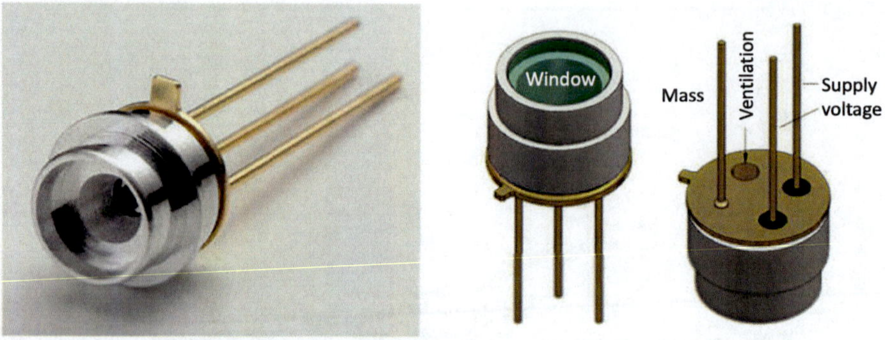

Fig. 7.67 Temperature of the Membrane as a function of the electrical power

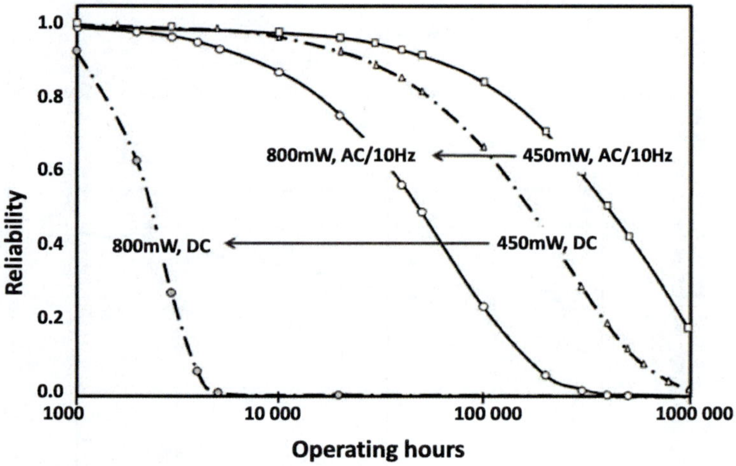

Fig. 7.68 Reliability for $R(t)$ different operating modes of micromechanical radiation sources

The lifetime of these radiation sources depends mainly on the operating mode. The higher the operating temperature T, the greater the probability of failure. The operating temperature T is given by the electrical power P, which is specified by the manufacturers. Figure 7.68 shows the reliability $R(T)$, as a probability of survival as a function of time t, according to the *Weibull distribution*.[17] The reliability $R(t)$ can be calculated as follows:

[17] Ernst Weibull (1887–1979) Swedish engineer and mathematician.

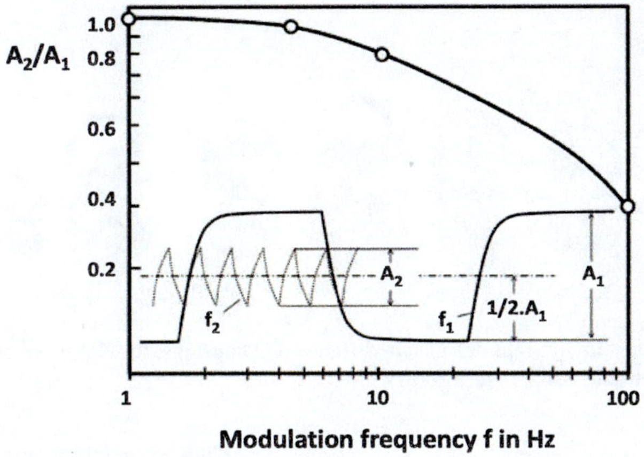

Fig. 7.69 Dependence of the modulation capability on the switching frequency f

$$R(t) = e^{-\left(\frac{t}{\tau}\right)^{\beta}} \tag{7.20}$$

The failure slope β essentially determines the *Weibull distribution* and must be determined experimentally. The characteristic lifetime τ corresponds to the MTTF,[18] in which 63% of all components have failed.

Electrical Modulation Capability

A major advantage of micromechanical radiation sources is the low mass of the membrane, so that heating can be completed within a few milliseconds. The modulation of the radiation is usually achieved by switching the electrical power supply on and off. The increase and decrease of the radiation follow this specification and leads to a time delay. Figure 7.69 shows the frequency response of the radiation sources from 1 to 100 Hz as a ratio of the amplitudes A_2/A_1. Furthermore, two examples of modulated radiation at two different frequencies are shown in this figure. The radiation pattern, at the frequency $f_1 \approx 3$ Hz, almost completely reproduces the on and off cycle of the control voltage. During this time, the radiation emission reaches the 100% value and then completely returns to zero. The resulting amplitude is A_1 and represents the maximum possible value. At a higher frequency f_2, the radiation emission is no longer completely switched on and off, as the switching times are too short compared to the inertia of the system. The amplitude then only reaches the value A_2. The ratio of both values is shown in the diagram on the *y-axis* and is also called the modulation depth. The curve shows that the degree of modulation drops significantly at $f = 10$ Hz, and only reaches a value of 40% at $f = 100$ Hz.

[18]Mean time to failure.

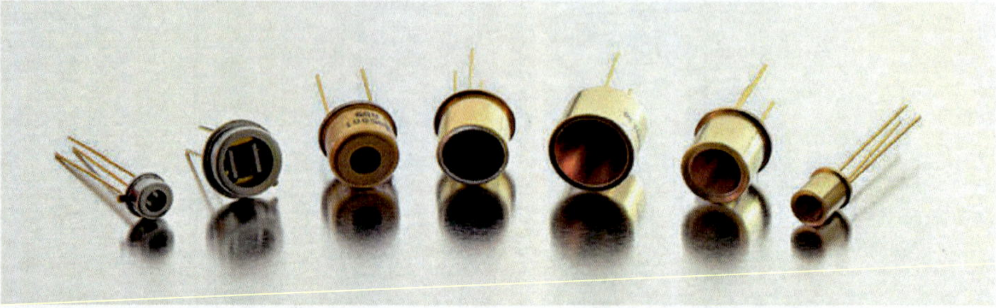

Fig. 7.70 IR emitters in different designs, in different TO housings, for NDIR gas analysis. (Source: micro-Hybrid Electronic GmbH Hermsdorf)

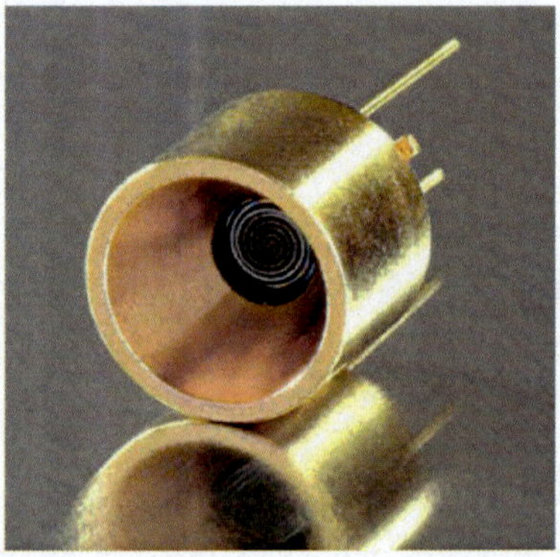

Fig. 7.71 IR radiation source with a spiral of Kanthal and a gold-plated reflector. (Source: micro-Hybrid Electronic GmbH Hermsdorf)

As the thermal detectors (e.g. thermopile) usually react more slowly ($\tau \approx 30$ ms), the radiation sources are not always the limiting factor for a sensor design.

In the meantime, there are different designs of these IR emitters, which are offered by various manufacturers (Fig. 7.70). These designs are of great importance for applications in gas sensor technology, since they influence the coupling of IR radiation into the optical setup. In most cases, these IR emitters are integrated in a so-called TO housing (Fig. 7.71). However, there are also SMD[19] versions that can be used for miniaturized gas sensors (Fig. 7.72).

IR Light-Emitting Diodes

Since a few years, there are also light emitting diodes for the IR range between 2 µm to 5 µm. In principle, these LEDs are suitable for use in gas sensor technology. The big advantage of these radiation sources is the perfect modulation, up to very high frequencies

[19] SMD = Surface mounted devices (surface mounting).

Fig. 7.72 Micro emitter as SMD chip. (Source: micro-Hybrid Electronic GmbH Hermsdorf)

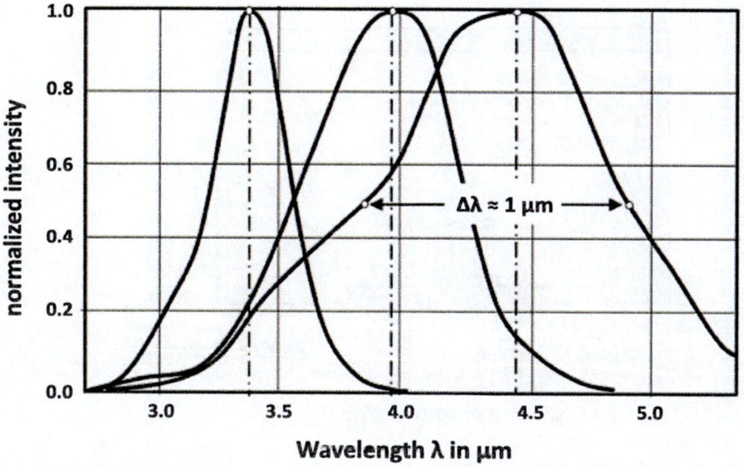

Fig. 7.73 Spectral distribution of three different IR LEDs (Manufacturer's data of LED Microsensor NT St. Petersburg, Russia)

(≈100 MHz). Furthermore, the power consumption is very low with $P_{el} < 1$ mW, so that battery-powered devices can also be realized.

However, the spectral distribution is not limited to a narrow range as in a diode laser but extends over a spectral range of up to 1 μm (Fig. 7.73). Typical absorption bands of gases are, however, much narrower, so that a large part of the radiation remains unused in gas measurement (ballast radiation). Furthermore, the radiation of the LED is not directed but behaves like a *Lambert lamp* (Fig. 7.74).

According to the manufacturer, the service life is approximately 80,000 hours. For better utilization of the radiation power, the chip is also installed in a reflector housing. Figure 7.75 shows such a structure. Since the emission characteristic of the IR LED changes very strongly with temperature, the LED chip is regulated to a constant

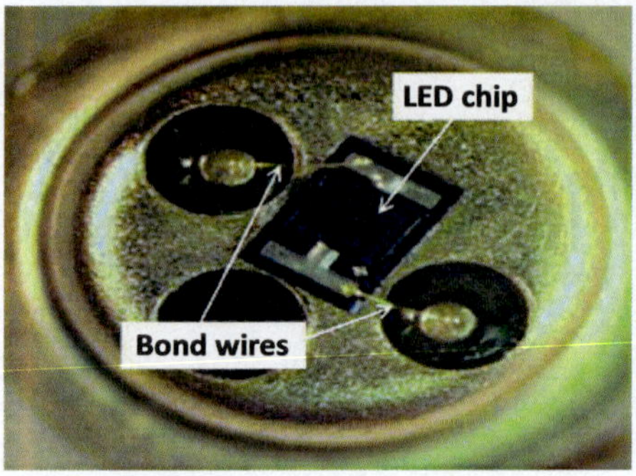

Fig. 7.74 LED chip for generating IR radiation. Mounted on a TO socket (LED Microsensor NT St. Petersburg, Russia)

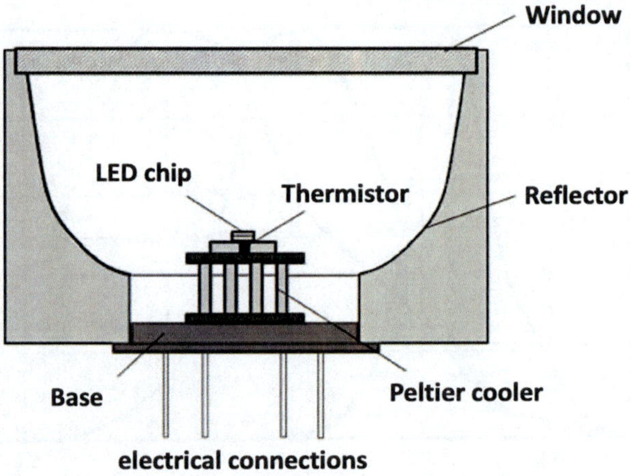

Fig. 7.75 Mounting an IR LED in a reflector housing for better utilization of the radiation

temperature by a Peltier element. This measure, however, means that the advantage of low power consumption is lost again.

Figure 7.76 shows the radiation characteristics of an IR LED at $\lambda \approx 4\,\mu\text{m}$, for two different temperatures. One can clearly see the decrease of the radiation intensity with temperature, as well as a shift of the emission maximum to longer wavelengths. Furthermore, the CO_2 absorption band can be seen in the emission curve, since the spectral distribution was obviously measured in ambient air ($\approx$500 ppm CO_2). A large part of the radiation emission therefore takes place as ballast radiation, which does not contribute to a measuring effect.

Furthermore, the radiation maximum λ_0 also shifts with the diode current (Fig. 7.77). The shift λ is about 100 nm for a current change from 50 mA→220 mA. This shift is probably also due to a temperature effect since the temperature influence is very dominant.

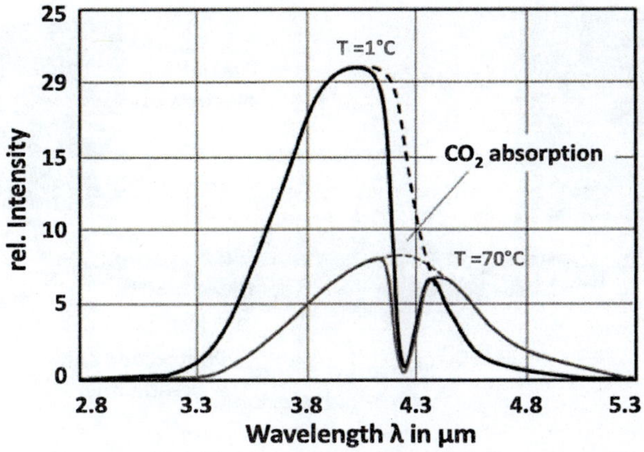

Fig. 7.76 Influence of temperature on the radiation characteristics of an IR LED

Fig. 7.77 Influence of the diode current on the radiation characteristics of an IR LED

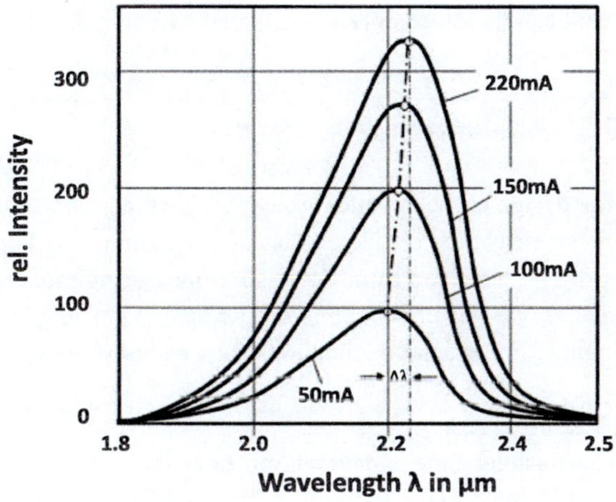

The manufacturer[20] of the IR-LED offers an evaluation kit to test the usability of the IR-LED. The structure is shown in Fig. 7.78. It consists of a cuvette with 3-fold reflection to increase the measurement path.

The gas exchange takes place by diffusion. The IR-LED is controlled by an external board which allows different operating modes. The evaluation electronics essentially consists of a synchronous detector. For a gas concentration of 500 ppm CO_2, the setup provides a signal change of $\Delta I \approx 1\%$. In contrast, a comparative measurement with a conventional IR filter method provides a $\Delta I \approx 13\%$ signal change (Döring 2013). This

[20]LED Microsensor NT St. Petersburg, Russia.

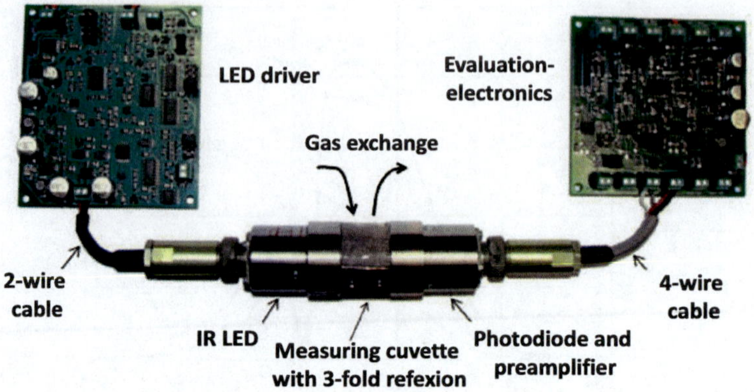

Fig. 7.78 CDS-3 test setup for CO_2 measurements with an IR LED

large difference can be attributed to the high proportion of ballast radiation. An improvement can therefore only be achieved by using an additional interference filter.

7.5 Cuvettes

The design of the cuvette through which the gas to be measured flows has a great influence on the measuring characteristics of the photometer. Typically, the cuvette consists of a tube with two windows through which a measuring radiation can pass. The gas is fed into the cuvette via gas connections on the side. Typical gas flows are in the range of ≈1 liters/minute. For special applications (fast response times), ≈4 liters/minute are also used.

Flushing Time

The flushing time τ depends on both the cuvette volume V_K, and the volume flow V_D through the cuvette. In a cuvette that is filled with a gas of concentration c_0 and purged with zero gas (e.g. N_2), the concentration $c(t)$ behaves according to the following exponential law

$$c(t) = c_0 \exp - V_D \cdot t / V_K \tag{7.21}$$

So after the time τ the concentration $c(t)$ has dropped to the e^{-1} -fold. For a simplified approach, it can be assumed that the flushing time corresponds to 3 times the volume of the cuvette or is $t_{Aus} = 3 \cdot \tau$. After this time, the cuvette has been rinsed or filled to 95% (Fig. 7.79). With the volume flow V_D and the cuvette volume V_K, *the* following relationship is obtained for the flushing time of a cuvette:

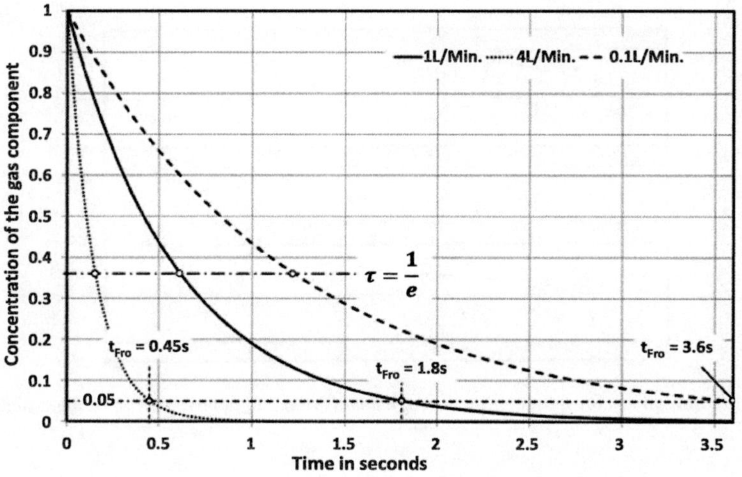

Fig. 7.79 Flushing behavior of a cell with a volume of 10 cm³, for different volume flows V_D of 0.5 L/min, 5 L/min and 4 L/min

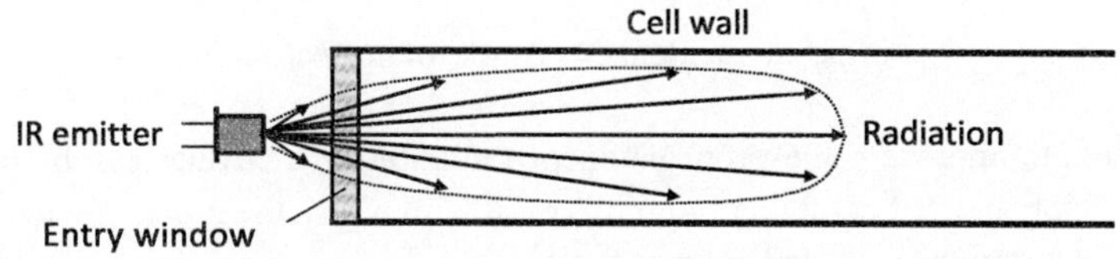

Fig. 7.80 Divergent coupling of IR radiation into a cuvette

$$3 \cdot \tau = t_{\text{Aus}} \approx \frac{3 \cdot V_K}{V_D} \tag{7.22}$$

Example

A cuvette has an inner diameter of 8 mm and a length of 20 cm. The volume flow rate is 1 L/min. The purging time is then calculated:

$$t_{\text{Aus}} \approx \frac{3 \cdot \pi \cdot 0.8\,\text{cm}^2 \cdot 20\,\text{cm}}{4 \cdot 1000\,\text{cm}^3 \cdot 60^{-1}\,\text{s}^{-1}} \approx 1.8\,\text{s}$$

Wall Reflection

Figure 7.80 shows the coupling of the radiation of an IR source into a cuvette. Due to the radiation distribution, which is determined by the imaging optics of the radiator, the individual beams enter the opening (entrance window) of the cuvette at different

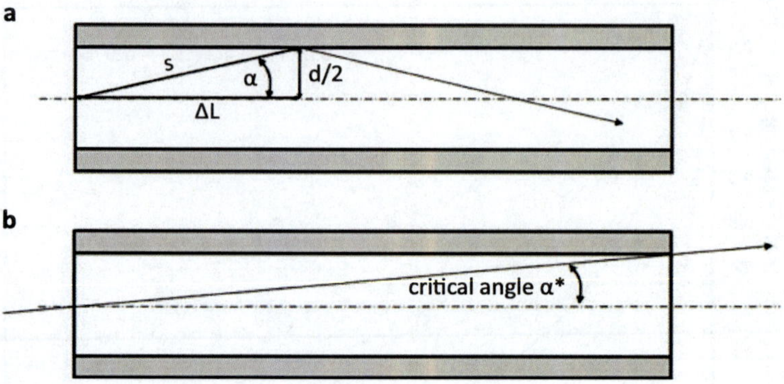

Fig. 7.81 Arrangement for light conduction in a cuvette (**a**) and determination of the critical angle α^* (**b**)

angles α. This inevitably leads to wall reflection. According to the rules of geometric optics, the principle applies to a reflection at the interface:

$$\text{Angle of incidence} = \text{Angle of reflection}$$

With this simple approach, the propagation of radiation in a cuvette can be calculated. Meanwhile, sophisticated computer programs exist with which the propagation of complex rays can be determined in detail by *ray tracing*. At this point only the simple physical relationships will be described.

If one assumes that a cuvette is internally mirrored for better radiation transmission, the entire incident radiation reflects at the interface (wall). In practice, this state is not reached and must therefore be corrected with the corresponding reflection coefficients (<1). Figure 7.81 shows the arrangement of a cuvette for light conduction. The radiation enters the system via a beam angle α, assuming that the radiation source is positioned centrally.

The section s can then be calculated by a geometrical approach as follows:

$$s = \frac{d}{2 \cdot \sin \alpha} \tag{7.23}$$

The distance s is the path that a light beam travels through the gas, passing through a part of the cuvette (ΔL). The following ΔL then applies to the section:

$$\Delta L = \frac{d}{2 \cdot \tan \alpha} \tag{7.24}$$

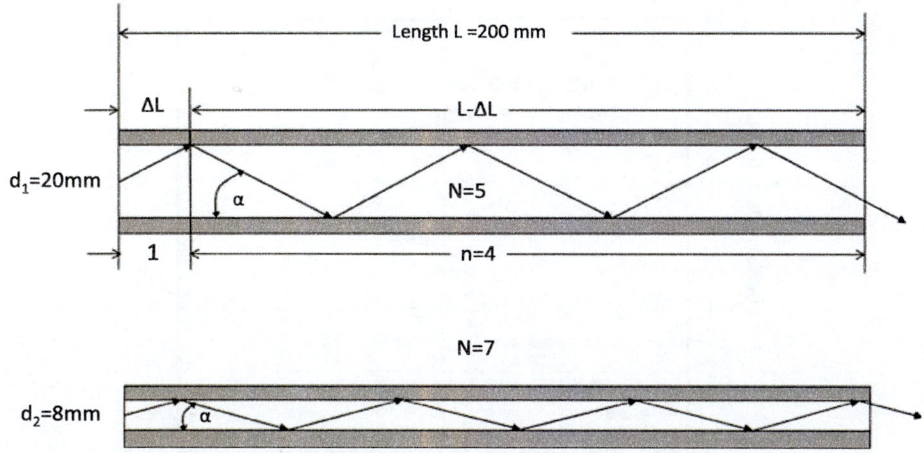

Fig. 7.82 Ray path through a cuvette reflecting from inside to calculate the wall reflections and the total optical path ($L^*\alpha$) through the gas layer

The critical angle α^* indicates the opening range of the radiation which is not transmitted by a wall reflection. This applies to all angles smaller than α^*. The limiting angle is then calculated:

$$\alpha^* = \tan^{-1}(d/2L) \tag{7.25}$$

Figure 7.82 shows the transmission of a single beam through the entire cuvette length L. This representation is shown for two different inner diameters (d_1 and d_2). It can be seen immediately that the number N of wall reflections depends not only on the angle of incidence α but also on the diameter d. The smaller d is; the larger N becomes. If one now assumes that one wall reflection occurs in each interval $2 \cdot \Delta L$, one obtains a total of n reflections in the path $L - \Delta L$. On the distance L there are then $N = n + 1$. It applies therefore:

$$N = \frac{(L - \Delta L)}{2 \cdot \Delta L} + 1 = \frac{L}{2 \cdot \Delta L} + \frac{1}{2} \tag{7.26}$$

If we now use Eq. 7.24 for ΔL, we obtain an expression for the number N of wall reflections (Fig. 7.83) in a cuvette, with the length L, the diameter d and the angle of incidence α:

$$N = \frac{L}{d} \cdot \tan\alpha + \frac{1}{2} \tag{7.27}$$

The total length L^*, that a light beam under the angle of incidence α travels through the gas layer inside the cuvette is calculated:

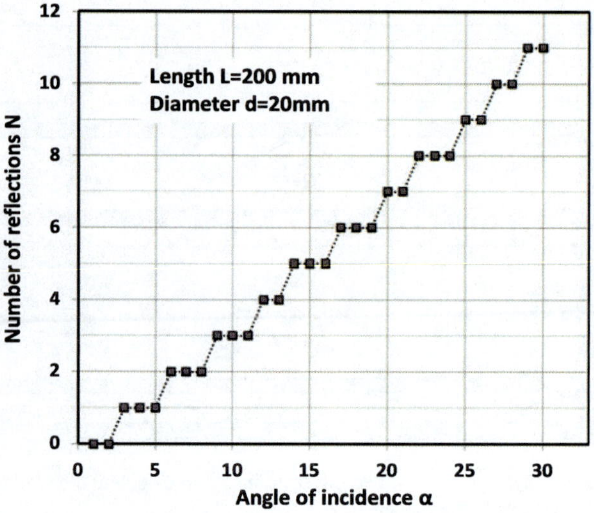

Fig. 7.83 Number of wall reflections N at different angles of incidence α for a cell length $L = 200$ mm and an inner diameter of 20 mm

$$L^* = s \cdot \left[\left(\frac{L - \Delta L}{\Delta L}\right) + 1\right] = s \cdot \left(\frac{L}{\Delta L}\right) \tag{7.28}$$

If you now use Eq. 7.53 for s, and Eq. 7.54 for ΔL, you obtain the entire optical path $L^*(\alpha)$ of a light beam through the cuvette:

$$L^*(\alpha) = \frac{L}{\cos \alpha} \tag{7.29}$$

The gain then results from $L^* - L$. Figure 7.84 shows the dependence of the total length L^* and the gain as a function of the angle of incidence.

Example

One cell has a total length of $L = 250$ mm and an inner diameter of $d = 8$ mm. The IR emitter with reflector has an aperture angle of 40°, while the version without imaging optics (Lambert emitter) has an aperture angle of 90°. The number of wall reflections can be calculated for the emitter with reflector as follows:

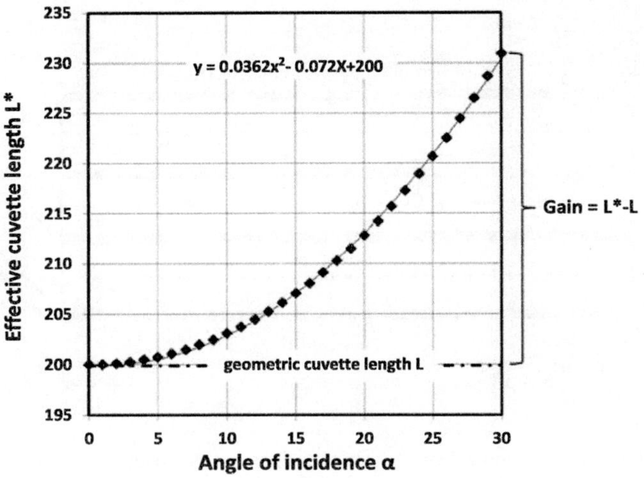

Fig. 7.84 Effective cell length L^*, with different angles of radiation α

$$\text{with Reflector} \quad N_1 = \frac{250\text{mm}}{8\text{mm}} \cdot \tan 20^\circ + \frac{1}{2} = 11.88 \rightarrow 11 \text{ Reflections}$$

$$\text{without Reflector} \quad N_2 = \frac{250\text{mm}}{8\text{mm}} \cdot \tan 45^\circ + \frac{1}{2} = 31.75 \rightarrow 31 \text{ Reflections}$$

The optical path through the cuvette can be calculated for both cases as follows:

$$\text{with Reflector} \qquad L_1^* = \frac{250\,\text{mm}}{\cos 20^\circ} = 266\,\text{mm},$$

$$\text{without Reflector} \qquad L_1^* = \frac{250\,\text{mm}}{\cos 45^\circ} = 354\,\text{mm}$$

The linear increase of the cell length L is limited by the technical limitations of the instrument, since the analyzers are usually integrated into 19″ housings. In these housings, the cell length is limited to $L < 300$ mm. By deflecting the radiation with two mirrors, the cell length can be more than doubled. In the example from Fig. 7.85, the total length is $2L + d + L'$. However, this gain in distance is offset by the unavoidable losses at the mirror surfaces, so that only an overall improvement of ≈50% can be expected. In practice, this considerable additional expenditure is therefore only rarely used.

Process Cuvettes

For applications in chemical process engineering, the cuvette must be sufficiently resistant to the components present in the sample gas. Especially corrosive components (HCl, CL_2, SO_2, HF, . . .) can damage the cell and thus change the transmission behavior permanently. This inevitably leads to a drift of the measuring signals, which must be prevented. With

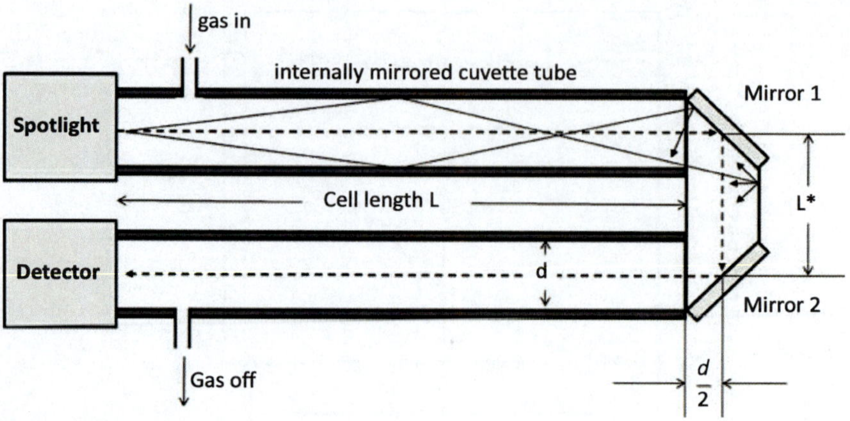

Fig. 7.85 Analysis cuvette with a 180° deflection to increase the distance *L*

appropriate design features, this goal can be achieved to a large extent. The choice of materials in particular plays a decisive role. In this case, stainless steel is used as the cell material, which has a corresponding corrosion resistance. Higher quality materials are Hastelloy[21] Monel, Inconel, Tantalum, Titanium, etc.

Pressure-resistant Cuvette

For certain applications, the analyzer must be connected directly to the processes to be investigated. Pressure reduction is not possible in such cases, so that the measuring cuvette must be designed for the respective gas pressure (Heimel et al. 1983). Figure 7.86 shows such a design. The cuvette consists of a solid stainless-steel block in which the gas connections are connected to the supply tubes (stainless steel) via cutting ring fittings. The IR windows must be of an appropriate thickness (d) to withstand the internal pressure. The window thickness d results from the process pressure p and the area A or the diameter $\varnothing(d \approx \alpha \cdot p \cdot A)$. The coefficient α depends on the material and must be known before the determination. For safety reasons, the window thickness is determined with double safety. The windows in the cuvette are sealed by O-rings.

In the case of extremely toxic gases (e.g. phosgene), an additional sheathing is added (double-jacketed cell), which prevents the escape of these gases into the environment in the event of a gas leak via a flushing device (Wiegleb 1984).

[21] Trade name for alloys of nickel, chromium, molybdenum.

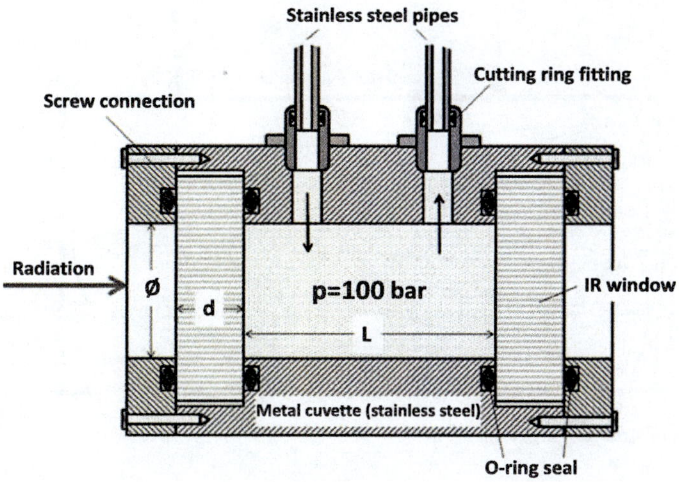

Fig. 7.86 Measuring cuvette in a pressure-resistant version

Multi-Pass Cuvettes

According to the *Lambert-Beerschen law,* the absorption A is proportional to the product of $c \cdot L$. In order to obtain a usable absorption A even at very low gas concentrations c, the optical path L through the sample gas must therefore be increased. A so-called multipass cell is suitable for this purpose, in which the light beams are reflected back and forth several times to create a correspondingly long path L. Constructions which allow a 2- and 3-fold reflection, with simple optical possibilities, have already been presented. However, with these setups, only distances of meters <1 can be realized. For a path extension of >10 meters, other paths must therefore be taken. For gas analysis, two different cuvette setups have so far proved to be practicable: the *White cell* and the *Herriott cell.* Both cell types were named after their respective inventors.

The White cell (White 1942) consists of three mirrors with the focal length f. The radiation enters the cell through a window and is focused by the mirror SP2 on mirror 2 (Fig. 7.87). The beam is then reflected towards mirror 3 and then focused on the exit window. In both arrangements, the beam passes through the measuring cell at least 4 times. Since the mirrors SP2 and SP3 can be rotated, further reflections are also possible, which then allow *n times the* passage.

Heriott cells are mostly used for laser spectroscopy (Fig. 7.88). These multipass cells allow an even more compact design because the laser radiation is already bundled.

It is important for both cuvette setups that the reflection ($R \approx 1$) of the radiation is as high as possible to keep the losses low. Furthermore, the reflection properties of the mirrors must remain stable over a long period of time to prevent signal drift. The basic material used for the reflectors is usually aluminum, which is given a mirror-like surface finish by

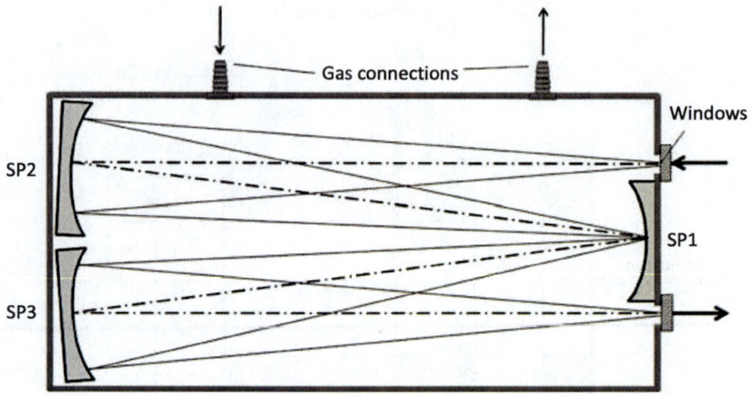

Fig. 7.87 Basic structure of a white cell

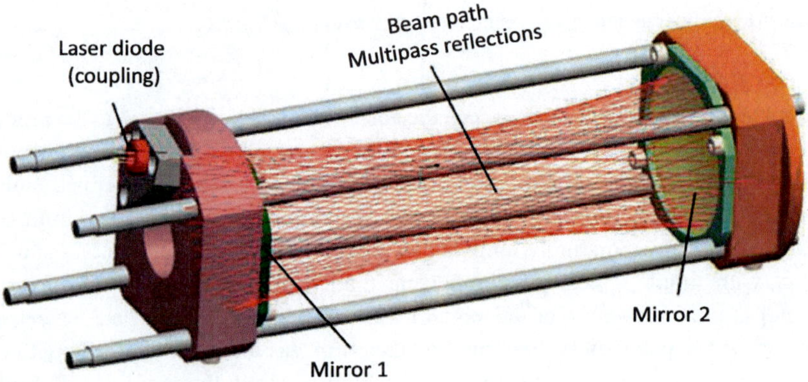

Fig. 7.88 Structure of a Heriott gas measuring cell for laser radiation. (Source: Fraunhofer Institute for Physical Measurement Techniques Freiburg)

high-speed diamond cutters. Since pure aluminum is very sensitive, a suitable protective layer must be provided. Figure 7.89 shows a cross-section of the layer structure. In total, two metallic protective layers are applied, which are covered by a dense gold layer. Finally, a highly reflective gold layer is applied, which then provides the required degree of reflection. The protective layers are applied by *sputtering*[22] in a vacuum chamber.

Concentrator Optics

The transfer of radiation from the measuring cuvette to the detector plays an important role in the design of IR gas sensors. For an optimal signal-to-noise ratio, all available radiation

[22] *Sputtering* is an adhesive thin film method, as an alternative to evaporation.

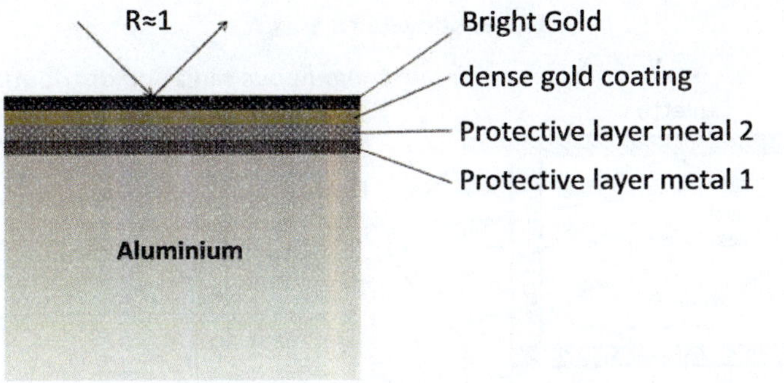

Fig. 7.89 Layer structure of the mirrors

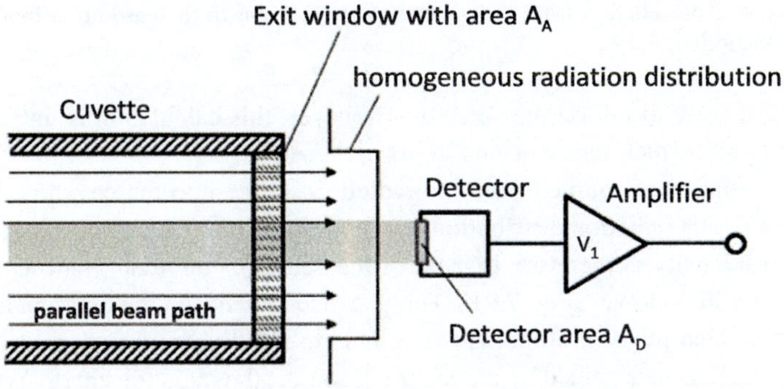

Fig. 7.90 Transition of the radiation from the cuvette to the detector with a homogeneous radiation distribution (parallel radiation)

should reach the detector. Normally, however, the area (aperture) of the cuvette is much larger than that of the detector. The receiving area A_D of typical IR detectors is 1 × 1 mm or 2 × 2 mm, while the aperture at the exit window of the cuvette has a diameter between $d_K = 4$ mm and 25 mm (≈1 inch). Assuming that the radiation distribution in the cuvette is uniform, only a very small part of the IR radiation from the cuvette will hit the detector (Fig. 7.90). If the inside diameter of the cuvette is, for example, $d_K = 8$ mm, the ratio φ between the useful radiation ϕ_K in the cuvette and the radiation received ϕ_D by the detector ($d_D = 2$ mm) would be as follows:

$$\varphi = \frac{\Phi_K}{\Phi_D} = \frac{\pi \cdot d_K^2}{4 \cdot d_D^2} = \frac{201}{16} \approx 12.6 \tag{7.30}$$

According to this calculation, only ≈8% of the usable radiation is received by the detector. The signal-to-noise ratio could therefore be improved by a factor of 12.6 if all available

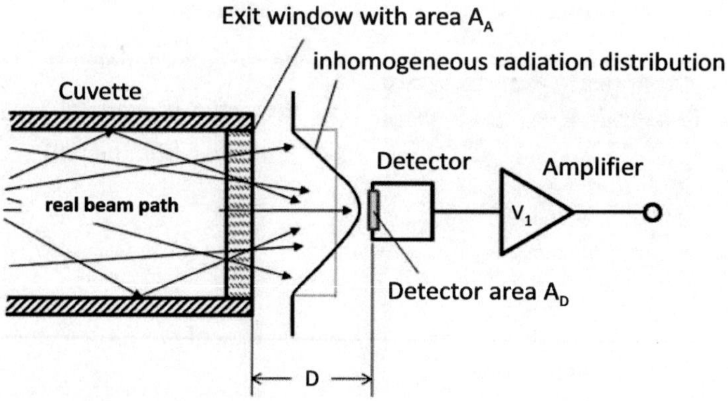

Fig. 7.91 Transfer of radiation from the cuvette to the detector in the case of an inhomogeneous radiation distribution

useful radiation were to fall on the detector. However, this calculation is only valid for a parallel beam path, which rarely occurs in practical setups.

Since the radiation is coupled from the radiation source into the cuvette with a given aperture angle θ, this radiation distribution is of course also found at the end of the cuvette. The radiation intensity is therefore highest in the center of the exit window, so that the possible factor will be lower (Fig. 7.91). The distance D between the exit window and the detector surface also plays an important role. Due to the divergence of the real radiation distribution, which is determined by the radiation source, the radiation energy is distributed over an increasingly large area as D increases. The radiation component falling on the detector surface A_D at distance D is then proportional to $1/D^2$.

A common method is to use a converging lens to concentrate the radiation on the detector (Fig. 7.92). In the IR range, however, lenses made of CaF_2 or sapphire must be used, although these are associated with quite high costs. It must also be considered that the measuring radiation passes through the distance f (focal length) and that the water vapor and CO_2 content of the ambient air influence the measuring radiation on its way (*pre-absorption*). With UV photometers, these disadvantages do not play a role, which is why convergent lenses made of quartz glass are preferred in these setups. As a first approximation, the gain factor G in radiation can be calculated by the ratio of the two areas A_D and A_K:

$$G = \frac{A_K}{A_D} \tag{7.31}$$

The downstream electronic amplifier can therefore, at the same output voltage, reduce the gain by a factor G if a lens is used. In this context, one also speaks of an *optical gain*.[23]

[23] Optical amplification.

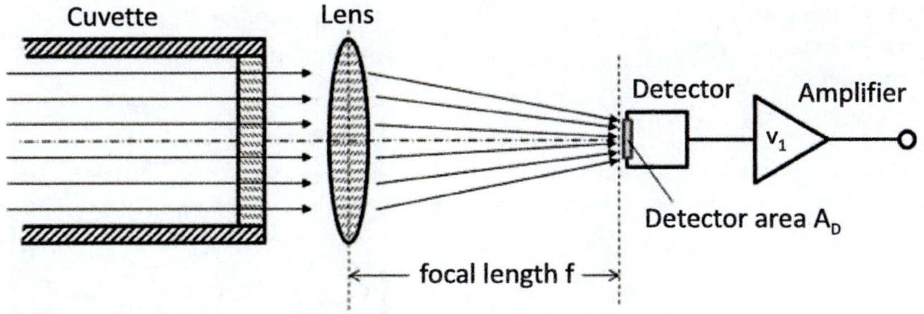

Fig. 7.92 Concentration of radiation with the aid of a converging lens (*Optical Gain*)

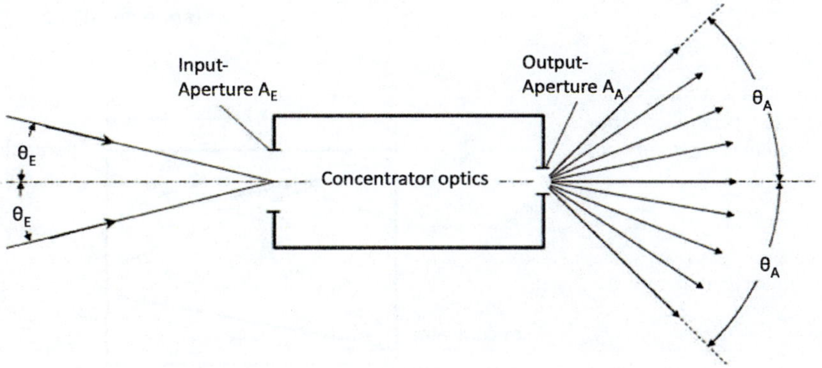

Fig. 7.93 Change in divergence due to the use of a concentrator optic

In general, the divergence is increased if optics are used to concentrate the radiation on a smaller aperture (Fig. 7.93). The divergence becomes greater the higher the concentration factor or gain factor G is (Welford and Winston 1989). It therefore applies:

$$G = \frac{A_E}{A_A} \approx \frac{\theta_A}{\theta_E} \tag{7.32}$$

In IR photometers, concentrators are often used which guide the radiation to the detector surfaces by wall reflections. In the simplest case, conical arrangements are used for this purpose, which, with internally polished and, for example, gold-plated surfaces, ensure optimum wall reflection (Fig. 7.94). The area ratio is usually 1 : 4 up to 1 : 10. In addition, most of these optics are equipped with a gas filling to prevent the influence of pre-absorption by air components and to perform positive gas filtering to reduce possible cross-sensitivities. For a CO measurement, for example, 100 vol.% CO_2 is filled in to reduce the cross-sensitivity to CO_2.

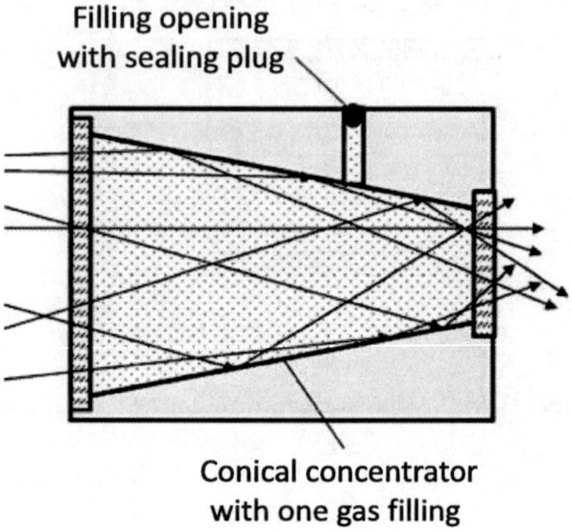

Fig. 7.94 Concentrator optics with integrated gas filtering, to reduce cross-sensitivities

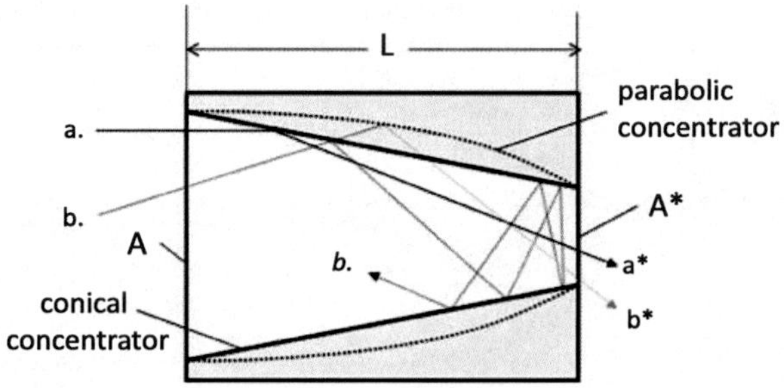

Fig. 7.95 Conical concentrator compared to a parabolic shape for two different angles of incidence

Conical concentrators can be produced relatively easily and are therefore often used for this purpose. However, the ratio between the input area A and the output area A^* is very limited, as any concentration on a small area is not possible. Figure 7.95 shows two radiation patterns (a. + b.) in a cone. The beam a. can pass through the cone without any problems, while the beam b., which is guided at a steep angle in the cone, leaves the cone again on the input side due to multiple reflections. This beam cannot therefore pass the cone. However, if the conical wall surface is replaced by a parabolic shape, this beam can also pass the concentrator. A parabolic shape therefore leads to an improvement of the transmission properties.

A parabolic concentrator gets its shape from two identical parabolas (1 + 2), whose focus is located at the edge of the exit aperture A^* (Winston 1970). The axes of the two parabolas are offset by the angle $2 \cdot \theta$ to each other. The irradiation angle θ is thus specified and all beams irradiated at this angle can pass the concentrator at the edge (focus). For irradiation angles smaller than θ, the beams are distributed over the A^* output surface. Rays that have a larger angle of incidence cannot pass the concentrator and are reflected back.

This type of compound *parabolic concentrator* surfaces is known as a *compound parabolic concentrator (CPC)* or *Winston Cone. In* addition to the planar beam paths

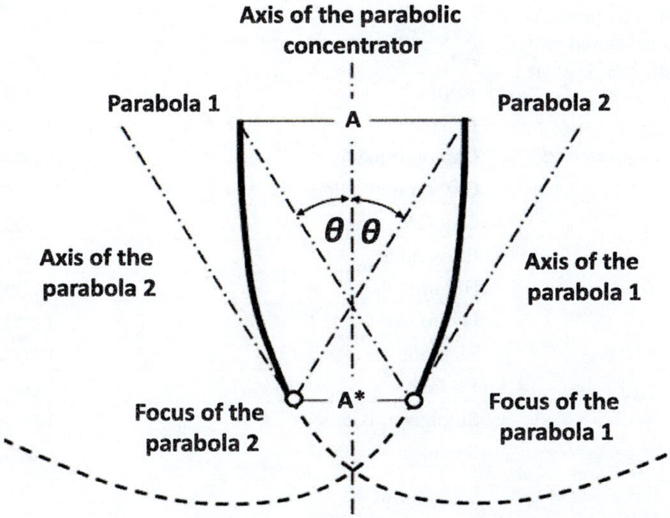

Fig. 7.96 Design of a composite parabolic concentrator according to Winston (1970)

Fig. 7.97 Tangential radiation into a CPC leading to a spiral path

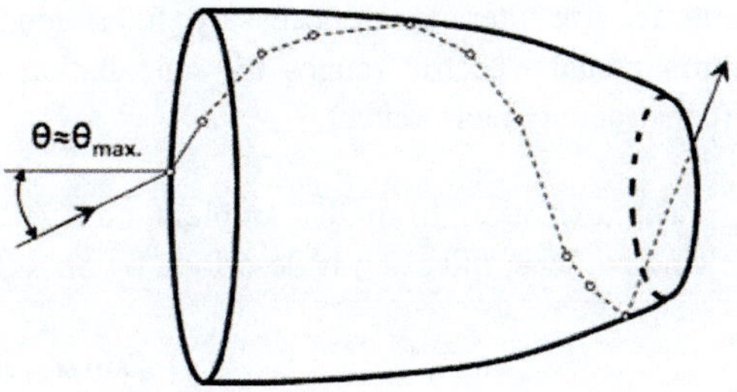

Fig. 7.98 Winston Cone for scientific applications, with a 4.2 K bolometer as IR detector. (Source: Infrared-Laboratories Inc., Tucson USA)

shown in Fig. 7.96, there are also other beam paths, of which a typical path can be seen in Fig. 7.97. If a light beam is irradiated at the maximum permissible angle θ_{max}, tangential to the surface of the CPC, it passes through a spiral path. In this path, the light beam is often reflected before it leaves the CPC at the exit aperture A^*. The distance travelled is then correspondingly longer than with a simple reflection. Figure 7.98 shows a commercial setup of a CPC.

Table 7.3 Refractive indices for some gases measured at the light wavelength $\lambda \approx 589$ nm (sodium D-line), at $T = 0\,^\circ\mathrm{C}$ and $p = 1013$ hPa (*Handbook of Chemistry and Physics 1979*)

Gas type	Refractive index n
Air	1.000293
Argon	1.000281
Benzene	1.001800
Carbon dioxide	1.000448
Carbon monoxide	1.000340
Chlorine	1.000773
Ethyl ether	1.001512
Helium	1.000036
Hydrogen	1.000132
Methane	1.000444
Pentane	1.001711
Sulphur dioxide	1.000686
Steam	1.000249

Interference Filter

In most cases, spectral narrowing is achieved using interference filters. Only with gas-filled detectors according to *K.F. Luft,* and selective radiation sources (e.g. laser), the use of interference filters is not necessary. Interference filters exploit the wave nature of light to limit certain spectral ranges by cancellation (destructive interference) or amplification (constructive interference).

The transition from the ambient air (gas phase), to the optically denser medium (window, lens, filter . . .) is described by *Snellius' law of refraction*:[24]

$$n_1 \cdot \sin \alpha = n_2 \cdot \sin \beta \tag{7.33}$$

The refractive index n is related to the speed of light c_0 in vacuum, and the speed of light in the medium c as follows:

$$c_0 = c \cdot n \tag{7.34}$$

The refractive index is between 1 and 2.5, depending on the medium, and gases have a very low refractive index, so that in general it can be assumed $n_1 \approx 1$ (Table 7.3).

If the light beam passes through the optical material, the rules of geometrical optics result in a parallel shift according to the following equation:

[24] Willebrord van Roijen Snellius (1580–1626) Dutch astronomer and mathematician.

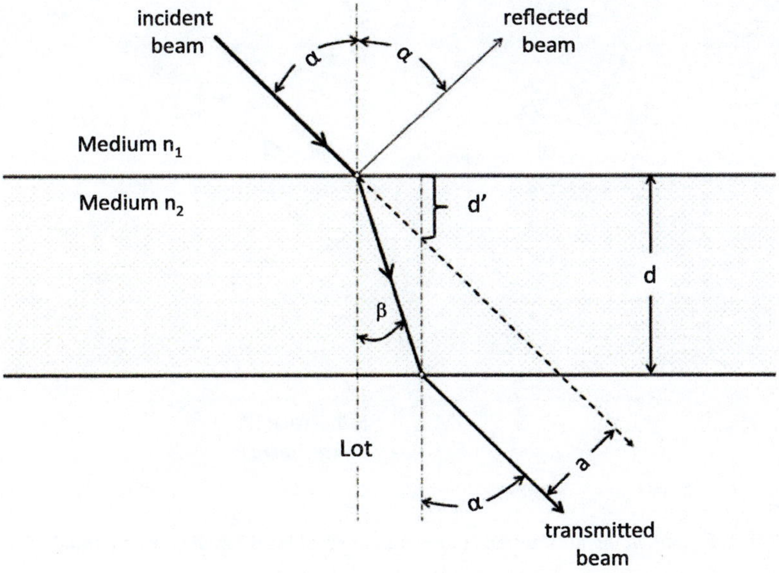

Fig. 7.99 Optical path of a light beam through a plane-parallel plate of thickness d

$$a = \frac{d \cdot \sin(\alpha - \beta)}{\cos \beta} \tag{7.35}$$

Figure 7.99 shows the geometric conditions that arise when a light beam passes through a so-called plane-parallel plate (medium n_2).

Light rays are basically reflected at each interface with different refractive indices (n_1, n_2). The reflectance R can be calculated using the *Fresnel equation.*[25] *The following* applies to a perpendicular incidence ($\alpha = 0$)

$$R = \left(\frac{n_1 - n_2}{n_1 + n_2}\right)^2 \tag{7.36}$$

According to Eq. 7.36, one obtains for a transition from air $n_1 \approx 1$ to CaF_2, with $n_2 = 1.432$ a reflectance of $R \approx 3.2\%$. Figure 7.100 shows the different reflections on a simple layer structure. If the light beams are reflected on thin layers ($d\lambda$), the individual beams (1. + 2.) can interfere with each other.

The geometrical path difference s between the two beams is

[25] Augustin Jean Fresnel (1788–1827) French physicist and engineer.

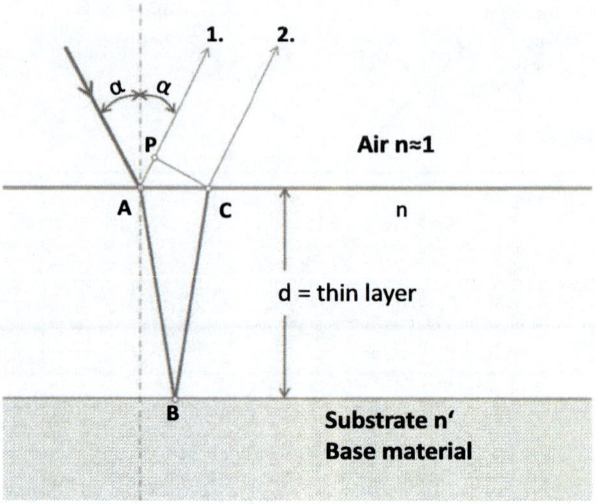

Fig. 7.100 Reflection at the air interface ($n = 1$) to the thin layer n and from the thin layer n to the substrate n'

$$s = \overline{AB} + \overline{BC} - \overline{AP} \tag{7.37}$$

The optical path $\delta = s \cdot n$ can now be calculated from the geometrical path difference s. At the interface B there is still a phase shift of $\lambda/2$. It is therefore valid:

$$\delta = 2d\sqrt{n^2 - \sin^2\beta} - \lambda/2 \tag{7.38}$$

For the vertical incidence you then get:

$$\delta = 2dn - \lambda/2 \tag{7.39}$$

A prerequisite for constructive interference (amplification) is a path difference of $\delta = k\lambda$. If you enter this value in Eq. 7.39 and then resolve it to λ, you obtain the wavelengths λ for which amplification applies:

$$\lambda = \frac{4dn}{2k + 1} \tag{7.40}$$

An erasure takes place at $\delta = (k + 1/2)\lambda$. For the corresponding wavelengths λ then applies:

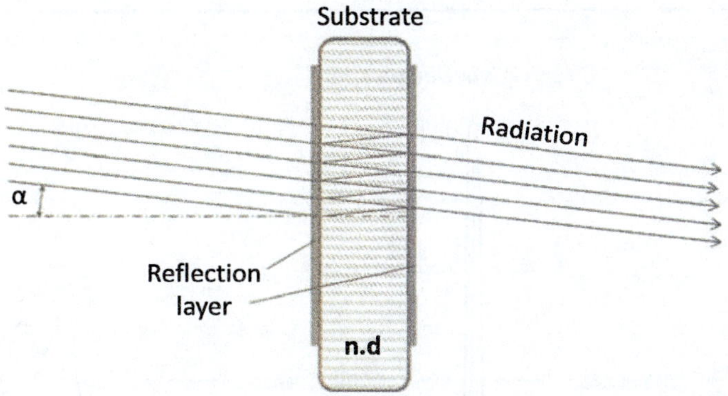

Fig. 7.101 Simple structure of a Fabry-Perot etalon with a transmission curve according to Eq. 7.41

$$\lambda = \frac{2dn}{k} \tag{7.41}$$

An extinction, according to Eq. 7.41 can be used for an anti-reflection layer, e.g., if the optical layer thickness $dn = \lambda/2$ is selected for the desired spectral range.

The intensity I_D of the transmission and the intensity I_R of the reflection can be described by the following equations, which are also called *Airy function*[26] (Demtröder 1993):

$$I_R = I_0 \frac{F \cdot \sin^2(\delta/2)}{1 + F \cdot \sin^2(\delta/2)} \quad \text{and} \quad I_R = I_0 \frac{1}{1 + F \cdot \sin^2(\delta/2)} \tag{7.42}$$

The simplest usable arrangement constructed according to these principles is called Fabry[27]-Perot[28]-etalon[29] (Fig. 7.101). It consists of a transparent pane on which a reflective layer with a reflectance of R has been applied on both sides. The transmittance ID of an etalon is periodic, with the spectral position of the maximum being given by $\lambda_n = \delta/k_n$. With these setups, it is therefore not possible to limit the transmission to only one wavelength λ_0, and this must be achieved by additional measures. Interference filters have a similar structure as the Fabry-Perot etalon. The spectral limitation to only one central wavelength λ_0 is realized by additional coatings. This layer structure determines the essential transmission properties of the filter.

A typical passband curve of a narrowband interference filter is shown in Fig. 7.102. The central wavelength λ_0 in this particular case is about 9 μm. Important for applications in gas

[26] Sir George Biddell Airy (1801–1892) English mathematician and astronomer.

[27] Maurice Paul Auguste Charles Fabry (1867–1945) French physicist.

[28] Jean-Baptiste Alfred Pérot (1863–1925) French physicist.

[29] French étalon = Sample piece.

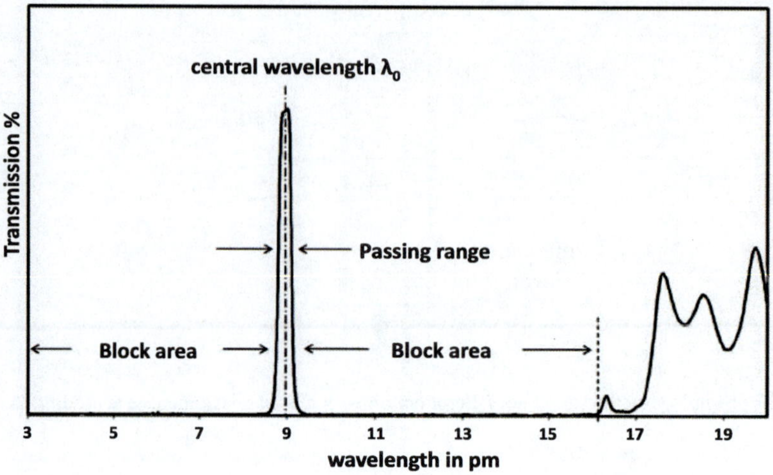

Fig. 7.102 Definitions of the different spectral ranges for an interference filter

measurement technology, is the so-called blocking range in which no unwanted radiation passes through the interference filter. This spectral range must therefore be precisely defined to suppress unwanted cross sensitivities. The spectral range above 16 μm could therefore lead to cross-sensitivities, as some gases absorb IR radiation in this range. However, when CaF_2 window materials are used, this range is already blocked from 12 μm onwards, as the transmission of CaF_2 there approaches zero. The blocking range must therefore be matched to the transmission behavior of the other optical components.

The central wavelength λ_0 (CWL[30]) can be first approximated by the intersection of the passband curve with the course of the half transmission $1/2T_{max}$:

$$\lambda_0 = \frac{\lambda_1 + \lambda_2}{2} \tag{7.43}$$

However, Eq. 7.43 only applies to very narrow-band interference filters. Since the symmetry of the passband curve only applies to a reciprocal wavelength, a modified equation must be used for broadband filters (bandpass) (Morelli 2006) (Fig. 7.103):

$$\lambda_0 = \frac{2\lambda_1\lambda_2}{\lambda_1 + \lambda_2} \tag{7.44}$$

The *slope* in the passband of the interference filter is the 80% value and the 5% value of the passband curve:

[30]Central Wavelength.

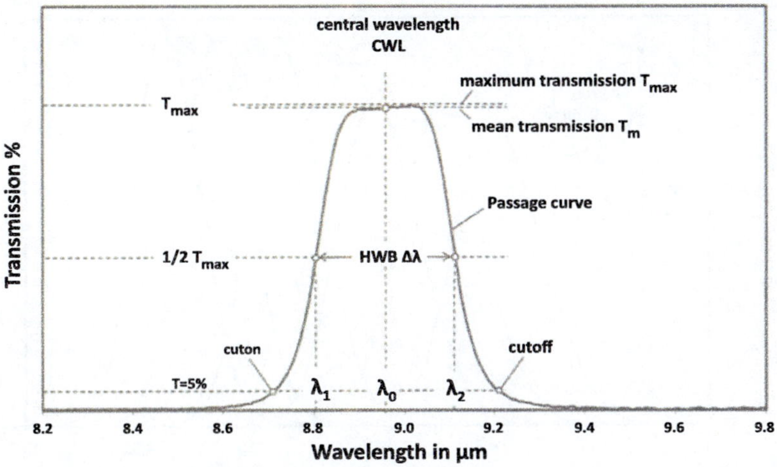

Fig. 7.103 Definition of the transmission characteristic of interference filters. (According to Morelli 2006)

$$\%\text{-Slope} = \frac{\lambda_{80\%} - \lambda_{5\%}}{\lambda_{5\%}} \cdot 100 \tag{7.45}$$

The greatest influence on the passband is the angle dependence of the radiation. The path difference δ changes very strongly with the angle of incidence α, so that the central wavelength λ_0 also shifts accordingly. However, this shift is non-linear and leads to considerable problems in gas measurement technology, since the bandwidth of the pass-band curve, for example, is of the same order of magnitude as the bandwidth of an absorption band. Instrument manufacturers, therefore occasionally tilt the filters in order to achieve an adjustment which then leads to a better selectivity (Fabinski and Deptolla 1980).

In addition to a shift in the transmission curve (Fig. 7.104), a significant decrease in maximum transmission can also be observed when the angle of incidence α changes. However, the area below the transmission curve remains almost constant (Fig. 7.103).

The angular dependence λ_α of the central wavelength λ_0 (Fig. 7.105) can be calculated as follows:

$$\frac{\lambda_\alpha}{\lambda_0} = \sqrt{1 - \frac{\sin^2\alpha}{n^2}} \tag{7.46}$$

The refractive index n of the distance layer is between 1.7 and 3.5, which can be assumed for most interference filters $n \approx 2.5$. However, if the interference filter is not illuminated with a constant angle of incidence α, but with an irradiation beam θ, the changes are even more serious (Fig. 7.106). In practice, one therefore tries to limit the opening of the

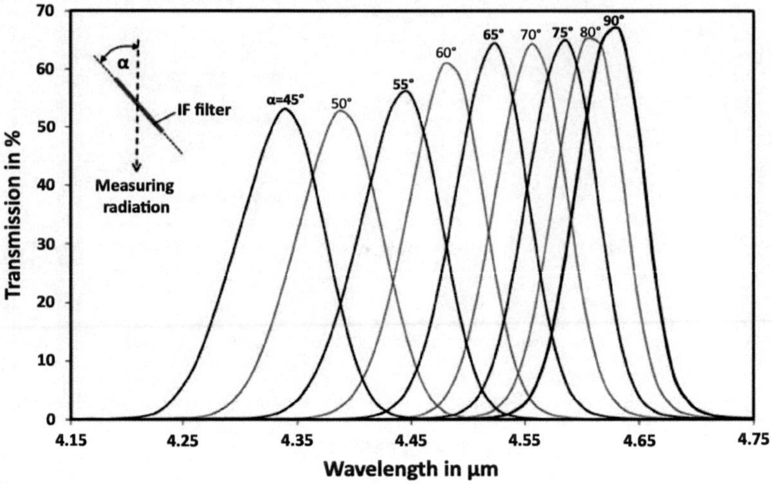

Fig. 7.104 Passband curve of an interference filter with a central wavelength of 4.63 µm (CO measurement) and different angles of incidence

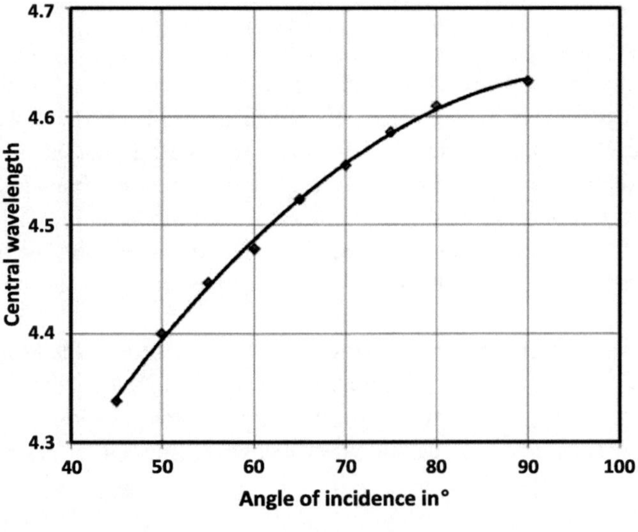

Fig. 7.105 Non-linear change of the central wavelength with a variable angle of incidence α

irradiation angle θ as much as possible. However, since the divergence is determined by the radiation source, these measures are kept within limits. Solutions that work with parallel beam paths are possible in principle, but this path is rarely taken. The effort (cost, space, adjustment, …) regarding the imaging optics (lenses and mirrors) is very high.

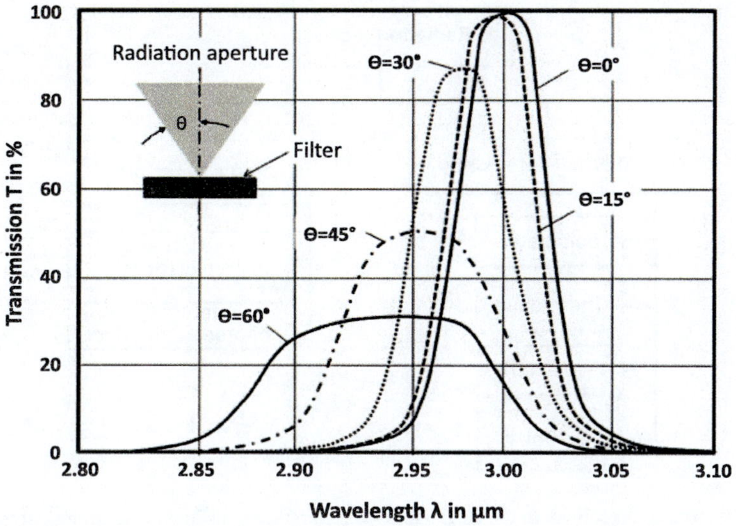

Fig. 7.106 Transmission behavior of the transmission curve for different light beams with the divergence θ. (According to Morelli 2006)

Temperature changes of the interference filters also lead to a shift of the central wavelength. However, this shift is very small. The influence can be explained by a change in the optical path $s = d \cdot n$. As the temperature T increases, the geometric distance d of the applied layers expands. Furthermore, the refractive index n changes with temperature. Both influencing factors therefore lead to a shift of the transmission curve according to the following equation:

$$\lambda(T) = \lambda_0(1 + \Delta T \cdot \gamma) \tag{7.47}$$

However, the coefficient γ is very small and lies between $0.3 \cdot 10^{-4}$/K and $1 \cdot 10^{-4}$ /K depending on the filter design. With increasing temperature, therefore, the wavelength shifts $\lambda(T)$ to longer wavelengths.

7.6 Detectors

In the infrared spectral range, radiation receivers can be divided into two groups: *thermal detectors* and *quantum detectors*. Figure 7.107 shows the different types of detectors that are basically suitable for gas measurement technology.

The classification shown above results from the mechanism of action of the two different detector types. While the quantum detectors enable a direct conversion of the

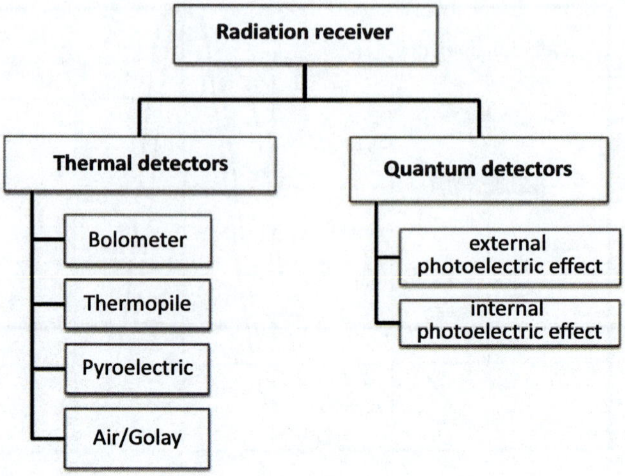

Fig. 7.107 Radiation receivers in the IR range used for applications in gas measurement technology

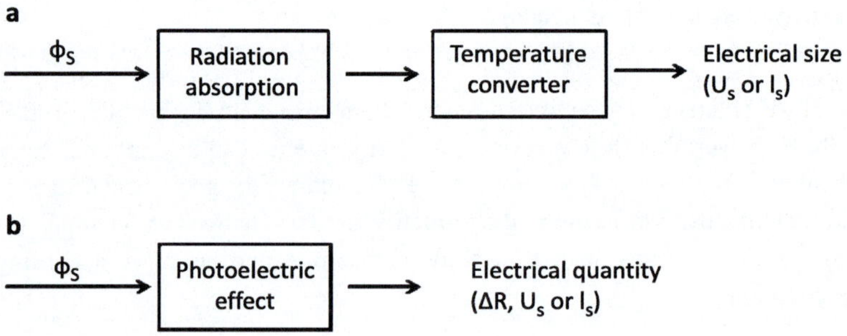

Fig. 7.108 Signal paths or mode of operation of thermal detectors (**a**) and quantum detectors (**b**)

radiation energy into an electrical signal, the thermal detectors only do this by means of a temperature measurement. Figure 7.108 shows the signal paths of the two detector types.

The thermal detectors have found the most widespread use in gas measurement technology. This type of detector first converts the incident IR radiation into heat, and then measures the resulting temperature increase ΔT with different methods. This temperature increase is a measure for the radiation power Φ_S. The temperature measurement is usually measured in the form of a voltage change ΔU_S caused by a change in radiation power $\Delta\Phi_S$. The ratio of the two quantities is called voltage sensitivity/sensitivity or, in international terms, the sensitivity or *responsivity* R_V:

$$R_V = \frac{\Delta U_S}{\Delta \Phi_S} \tag{7.48}$$

The sensitivity is stated in volts/watt (V/W) in the data sheets of the manufacturers, whereby the test conditions for this information usually refer to a 500 K blackbody radiator and the ambient temperature is 25 °C. Furthermore, the frequency f must be specified to obtain a comparable value. However, the frequency range varies between 0 Hz (=DC) and a few kHz, making comparisons between the individual manufacturers' specifications difficult. The bandwidth Δf of the measurement is given in 1 Hz.

A practicable comparative value for different detector types is the so-called specific detectivity D^* (i.e. *D star*). This D^* results from the ratio of the sensitivity and the effective value of the noise figure, related to the root of the detector area A_S.

$$D^* = \frac{\sqrt{A_S}}{\widetilde{U}_R} \cdot R_V \tag{7.49}$$

The unit for the detectivity D^* is given as cm · Hz$^{1/2}$/ W. Another important comparison value for detectors is the noise voltage $\widetilde{U}_R$, which is already considered in the specification D^*. Depending on the detector design, this noise voltage is composed of different sources. One of the main causes is resistance noise (*Johnson noise*), which is caused by the statistical movement of charge carriers in a conductive material. Since this movement increases with temperature T, the noise quantity is also very strongly dependent on the temperature and results from

$$\widetilde{U}_R = \sqrt{4 \cdot k \cdot T \cdot R \cdot \Delta f} \tag{7.50}$$

Example

An IR detector has an internal resistance of $R = 65\ \mathrm{k\Omega}$. The noise voltage is then calculated at $T = 25\,^\circ\mathrm{C}$:

$$\widetilde{U}_R = \sqrt{4 \cdot 1.38 \cdot 10^{-23}\,\frac{\mathrm{J}}{\mathrm{K}} \cdot 298\,\mathrm{K} \cdot 6.510^4\,\Omega \cdot 1\,\mathrm{Hz}} = 33\,\mathrm{nV/Hz^{1/2}} \tag{7.51}$$

If the noise voltage is related to the sensitivity, the noise *equivalent* power NEP is obtained:

$$\mathrm{NEP} = \frac{\widetilde{U}_R}{R_V} = \frac{\Delta \Phi_S}{\Delta U_S} \cdot \widetilde{U}_R \tag{7.52}$$

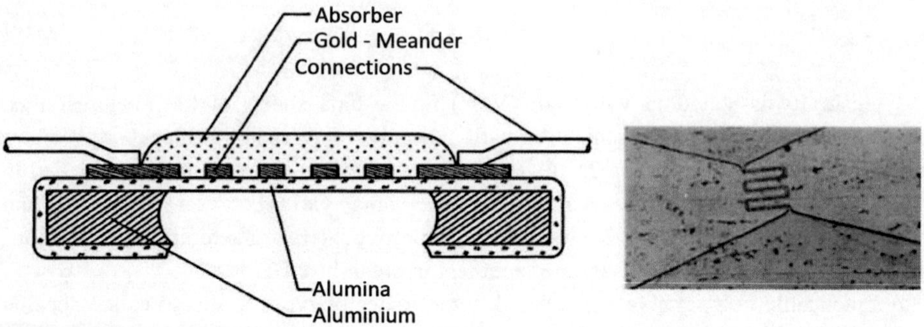

Fig. 7.109 Cross-section through the sensor structure (*left*) and photographic view (*right*) of the 50 μm × 50 μm large resistance meander. (Source: BMBF-FB W79-33, 1979)

The NEP value, therefore, indicates a radiation power that generates a signal in the detector that has the same value as the noise signal. The signal to noise ratio[31] (S/N) is then S/N = 1.

Bolometer

The first IR detectors consisted of thermometers that were directly irradiated with IR radiation and reacted to this radiation by increasing the temperature. Herschel was the first to prove the existence of IR radiation in this way. Resistance thermometers were then used to generate electrical signals. In 1878, the bolometer was invented by Langley.[32] The first bolometers consisted of blackened platinum wires that were directly exposed to IR radiation. The evaluation was done with a measuring bridge and a similar reference resistor which was not illuminated. However, this type of radiation measurement technique was limited to slow signals because the thermal mass of the resistance wires did not allow dynamic measurements. With the introduction of thin-film technology, this limitation changed. With this technology it was possible to produce very thin resistors with an extremely low mass. The first bolometer of this kind was developed at the Battelle Institute Frankfurt for aerospace. Figure 7.109 shows the structure of this bolometer arrangement. The resistor was applied to a thin aluminum oxide membrane ($d \approx 100$ nm) and covered with a blackened layer (absorber) to increase the absorption of radiation.

As resistance material a gold layer of 25 nm thickness was applied, which had a resistance value of about 120Ω. With this construction, a sensitivity of 1000 V/W at an NEP value of $8.5 \cdot 10^{-9}$ W/Hz$^{1/2}$ was already achieved. Today's bolometers mainly use silicon and vanadium oxide as resistor material, because the sensitivity and the noise behavior are much better.

---

[31] *Signal to noise ratio.*

[32] Samuel Pierpont Langley (1834–1906) American astrophysicist and flight pioneer.

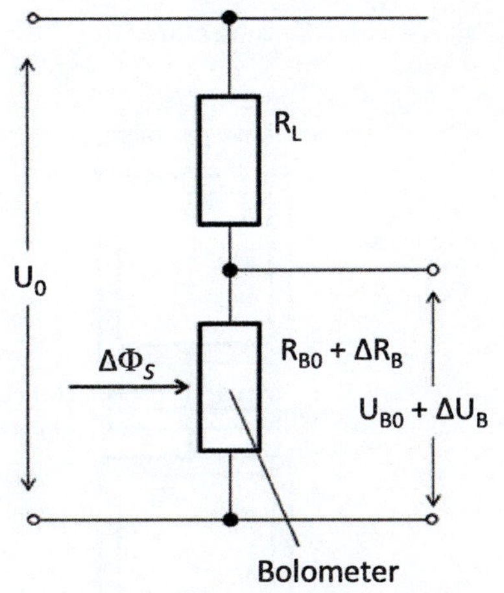

Fig. 7.110 Basic circuit (voltage divider) for a bolometer

The evaluation of the bolometer signals is carried out in the simplest case with a voltage divider as shown in Fig. 7.110. The change in resistance ΔR_B, caused by the irradiation $\Delta\Phi_S$ provides a voltage change of ΔU_B.

Bolometers are rarely used in photometric gas measurement technology, although they are in principle suitable for this measurement task. Commercially available bolometers are mostly used as radiation detectors for temperature measurement. Another application of bolometers is the use in night vision devices and IR cameras (Budzier and Gerlach 2010).

Luft and Golay Detectors

Both detector types are closely related and were developed almost simultaneously but independently of each other. Both detector types are shown in Fig. 7.111. Since these detector types use a gas as an absorption medium or transmission medium for a pressure increase, these types of IR detectors are also called pneumatic detectors.

The Luft detector was developed especially for gas analysis and consists of a closed chamber in which a gas mixture is enclosed (Luft 1943). The gas absorbs the IR radiation only in the spectral range in which the characteristic absorption bands are located. The energy absorption leads to a temperature increase in the gas phase and thus to a pressure increase, which can be measured with a membrane capacitor between the measuring chamber and a reference chamber. The electrical measurement signal is therefore present in the form of a change in capacitance. The Luft detector is therefore a selective detector which can be adjusted to a specific gas by means of the appropriate filling. Details of the applications in gas measurement technology can be found in Sect. 7.1.

The *Golay*[33] *detector* is a broadband detector (Fig. 7.112). The IR radiation is first absorbed by a blackened layer and only transfers this thermal energy to a surrounding gas

[33]Marcel Jules Edouard Golay (1902–1989) Swiss engineer and physicist.

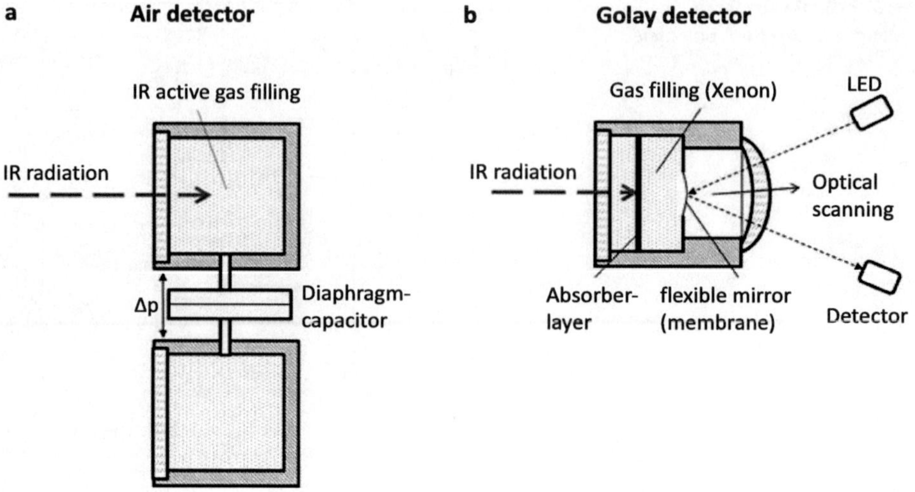

Fig. 7.111 On the two pneumatic IR detectors for *Luft* (**a**) and *Golay* (**b**)

Fig. 7.112 Spectral sensitivity of the two pneumatic IR detectors

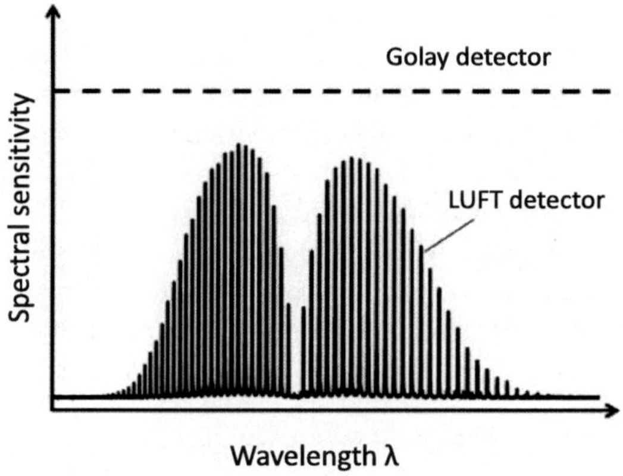

in a second step. Xenon is used as a filling gas to keep heat conduction losses low. In this case too, there is an increase in pressure in the chamber, which in a further step leads to a deflection of the flexible membrane. In this setup, the deflection is detected by optical scanning. The Golay detector is mainly used for metrological purposes where high accuracy is required.

Thermopile Detectors

The term thermopile is a synonym for the mode of operation. *Thermo* stands for the thermocouple and *pile* for a stack. A thermopile is therefore a stack of thermocouples.

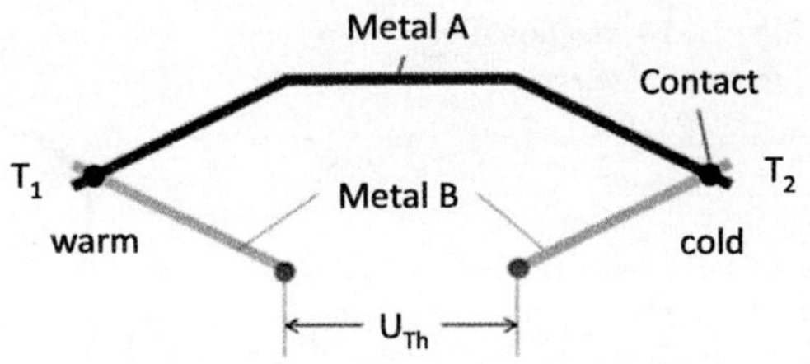

Fig. 7.113 Temperature measurement with a thermocouple

Table 7.4 Seebeck coefficients from the thermoelectric voltage series

Material	α in μ V/K
Bismuth	−72
Constantan	−35
Nickel	−15
Platinum	0
Carbon	3
Aluminium	3.5
Rhodium	6
Copper	6.5
Gold	6.5
Silver	6.5
Iron	19
NiCr	25

Thermopile detectors use the so-called Seebeck[34] effect to detect a temperature change ΔT_S due to radiation absorption. The Seebeck effect is based on contact points of two different conductors ($A + B$). Figure 7.113 shows an arrangement consisting of two contact points so that a closed circuit is formed. If these contact points are now exposed to the different temperatures $T_1 > T_2$, a thermoelectric voltage U_{Th} is obtained which can be calculated as follows:

$$U_{\mathrm{Th}} = (S_A - S_B) \cdot (T_1 - T_2) \tag{7.53}$$

S_A and S_B are the Seebeck coefficients resulting from the thermoelectric voltage series. Some values are given in Table 7.4.

Thus, for a pairing of iron and constantan you get a value of 54 μV/K. Since the temperature changes ΔT_S due to IR radiation are only in the mK range, these voltage values (≈50 nV) are much too low. Thermopile detectors, therefore, use a series connection of up to 200 such thermocouples to generate correspondingly useful voltage values. With 200 elements, a coefficient of 10.8 mV/K is obtained. For 1 mK, these are then 10.8 μV.

[34] Thomas Johann Seebeck (1770–1831) German physicist.

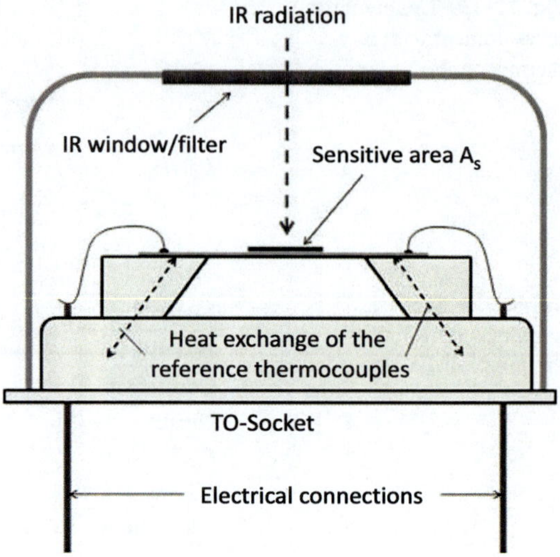

Fig. 7.114 Sectional view of a thermopile detector

Figure 7.114 shows a cross-section of such a thermopile detector. It consists of a silicon chip on which the thermocouples have been applied using microstructure technology. The entire assembly is placed on a TO socket and is sealed gas-tight with a cap. In the upper part of the cap is the entrance window, which can also be designed as an interference filter to allow a defined radiation to pass through. In the center of the microstructure the thermocouples are located on a very thin membrane. This type of arrangement results in a low mass m, which makes the time response correspondingly fast. For a better use of the IR radiation, this central part is blackened with a corresponding layer ($\varepsilon \approx 1$). In the center the temperature T_1 is measured.

The reference elements are applied directly to the silicon chip and therefore have a very good thermal contact with the environment. Heat is transferred from the chip via the metallic TO socket to the environment, so that the reference temperature T_2 corresponds approximately to the ambient temperature $T_{U.}$ For the exact measurement of the reference temperature, there is another resistive temperature sensor on the microchip which takes over this task. Figure 7.115 shows such a microchip. The receiving area varies depending on the detector version and manufacturer between 0.2 mm × 0.2 mm and 3 mm × 3 mm.

Thermopile detectors were mainly developed for use in ear thermometers and are therefore produced in high volumes. The production technology is very sophisticated and therefore also cost-effective. For the use in gas analysis these detectors are equipped with the appropriate interference filters. A further requirement for the use in gas measurement technology are several receiving channels to be able to perform a reference measurement. Figure 7.116 shows a selection of different thermopile detectors, which have been adapted for the different applications in gas measurement technology.

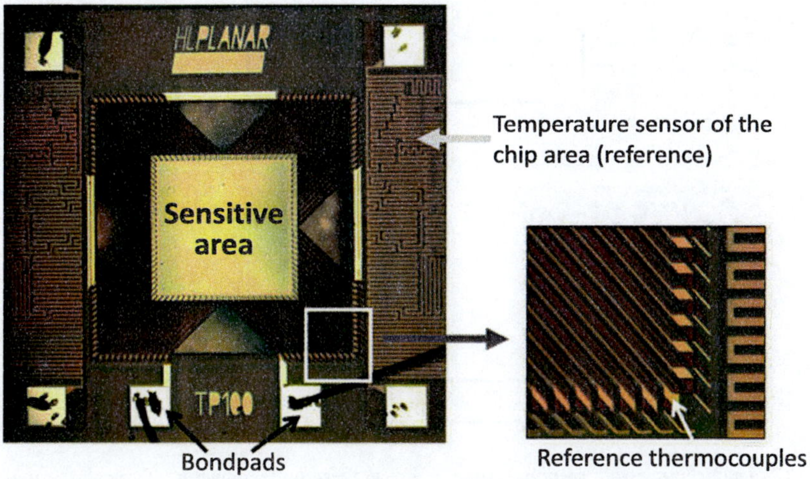

Fig. 7.115 View of the microstructure of a thermopile detector, with the arrangement of the measuring and reference elements. (Source: Dortmund University of Applied Sciences)

Fig. 7.116 Thermopile detectors for different applications in gas measurement technology as single-channel, two channel and four channel detectors. (Source: micro Hybrid Electronic GmbH Hermsdorf)

An essential quality feature of thermopile detectors is the time response. The use of membranes as thin as possible results in a very low mass for the radiation-sensitive surface A_S. This has a positive effect on the time response. Furthermore, these detector types can be filled with different gases to reduce the thermal conductivity within the construction. This measure also has a positive effect on the time response behavior. The measurement of the temporal response behavior is realized with a radiation source, which is abruptly shaded by a mechanical shutter and then released again. Thus, the radiation flux Φ_S is directed to the detector almost infinitely fast. This device is then used to measure the step response of the detector. Figure 7.117 shows such a measurement result.

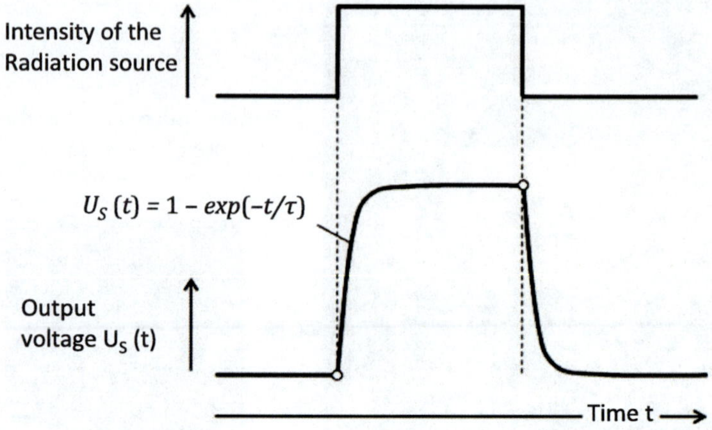

Fig. 7.117 Determining the *response* time τ for a thermopile detector. Typical thermal time constants for thermopile detectors can be found at $\tau \approx 10$ ms

Pyroelectric Detectors

The basis for this detector type is the so-called *pyroelectric effect*. Certain materials change the polar dipole moment of the molecules when they undergo a temperature change ΔT. This effect produces surface charges that can be tapped with external electrodes. The charge exchange between the electrodes is usually conducted via a high-impedance resistor R_L. The measured value is the voltage drop U_S which results from the ohmic law:

$$U_S = I_P \cdot R_L \tag{7.54}$$

Figure 7.118 shows the schematic structure of a pyrolelectric detector. The preferred materials are single crystals consisting of lithium tantalate ($LiTaO_3$), lithium niobate ($LiNbO_3$), or triglycine sulfate (TGS). If the variable radiation flux $\Phi_S(t)$ hits the crystal with the sensitive surface A_S, the radiation is almost completely absorbed. The degree of absorption of the surface is made as large as possible by a coating, similar to the thermopile detectors. The absorption of the radiation energy leads to a delayed temperature change $\Delta T(t)$, which in turn causes the charge change $\Delta Q(t)$. The voltage according to Eq. 7.54 is therefore also time-dependent $\rightarrow U_S(t)$.

After the charge exchange, $Q = 0$ and so is the voltage U_S. Metrologically, this means that there is no signal change at a constant radiation flow ($\Phi_S =$ const.). Only when the radiation is $\Phi_S = 0$, the charge Q is shifted in the other direction and a negative signal is generated until this charge exchange is also completed. Pyroelectric detectors can therefore only be operated with alternating light. Rapid temperature changes (gradients), which can affect the detector from outside, cause a similar effect. Pyroelectric detectors must therefore

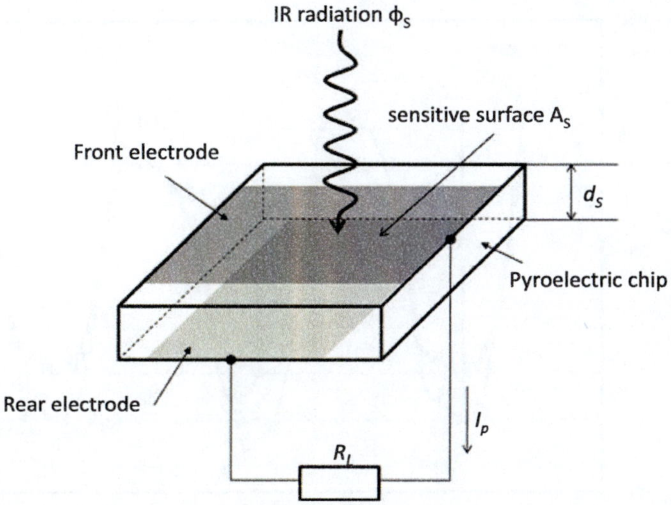

Fig. 7.118 Principle function of a pyroelectric detector

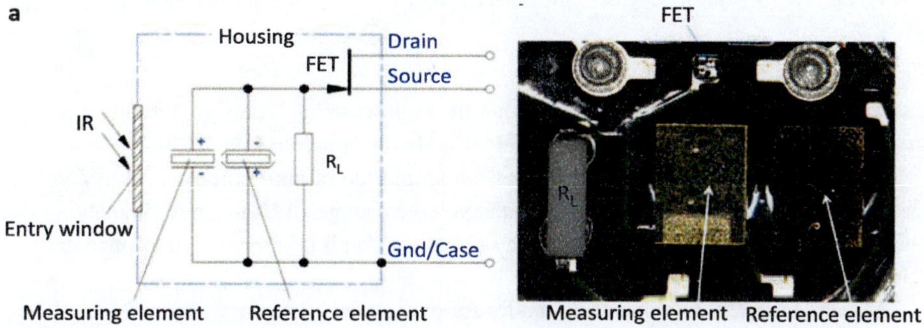

Fig. 7.119 Wiring of a temperature compensated pyrodetector with integrated FET amplifier (**a**) and view of the opened detector with the same components (**b**)

be protected against these influences by thermal insulation from the environment. Alternatively, the manufacturers also offer compensated versions. These use a second crystal, which is not exposed to radiation, for temperature compensation. To achieve compensation, however, the crystal must be inserted with the opposite polarity to the measuring element. Both elements are therefore connected antiparallel. Figure 7.119 shows a typical constellation of a temperature compensated pyrodetector. In this case, amplification is provided by an FET transistor (voltage mode), which is also integrated in the detector housing.

The use of an integrated FET amplifier results in an offset voltage in the voltage range at the output (source). The actual measurement signal is superimposed on this offset voltage

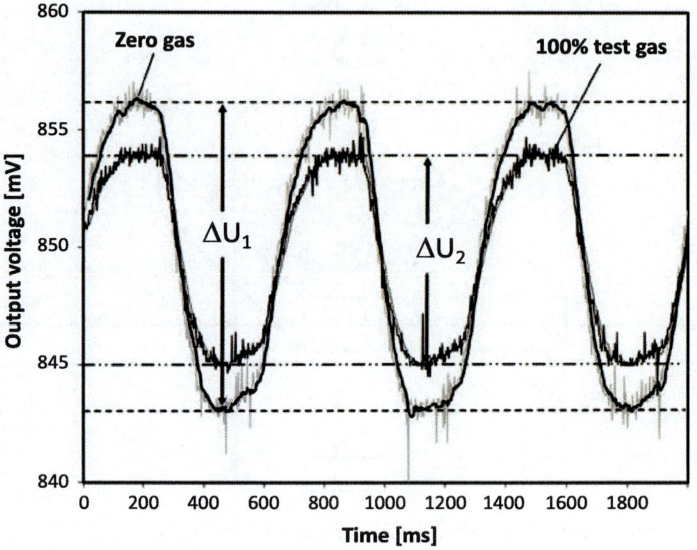

Fig. 7.120 Signal path of a pyroelectric detector with an FET amplifier integrated in an IR gas sensor

and must be decoupled by a capacitor for further processing. The signal curve, a sensor arrangement modulated with approximately 1.6 Hz, is shown in Fig. 7.120.

The actual measurement signal ΔU_1 has an amplitude of approximately 15 mV with a zero gas flow. If a sample gas is supplied, the voltage changes ΔU_2 to approximately 8 mV. The difference of 7 mV was caused by the sample gas and is converted into a concentration value in the subsequent electronics.

There are basically two operating modes for pyroelectric detectors. Figure 7.121 shows the internal circuitry for both operating modes. In current mode, the equalizing current I_P, which is caused by the exchange of charge, is amplified by an operational amplifier. The output voltage results from:

$$\Delta U_S = I_P \cdot R_f \tag{7.55}$$

During voltage operation the voltage drops directly above resistor R_G and is output in the case shown with an operational amplifier as voltage follower. Both operating modes have advantages and disadvantages. In particular, the frequency behavior is different in both operating modes (Fig. 7.122). With current mode, higher modulation frequencies can generally be achieved than with voltage mode.

Multi-channel detectors are usually placed directly in the beam path, at the exit of the measuring cuvette. The divergence of the radiation source then determines the intensity distribution at this point, which usually has a maximum in the middle (Fig. 7.91). However,

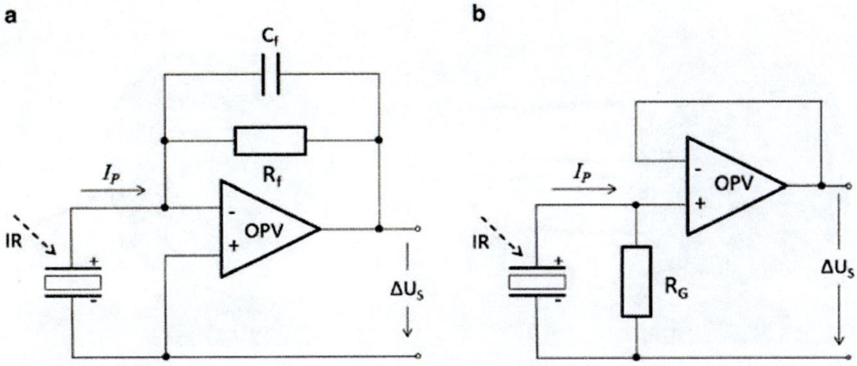

Fig. 7.121 Internal wiring for current (**a**) or voltage (**b**) operation of the pyrodetectors

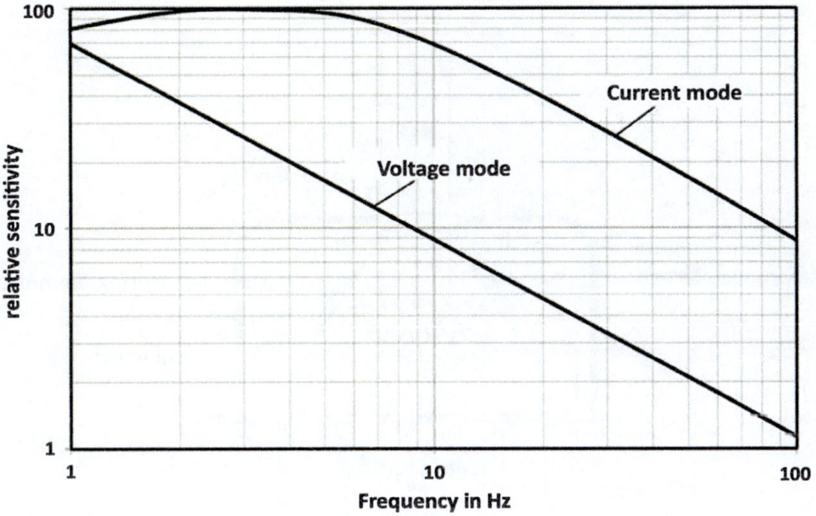

Fig. 7.122 Frequency response of pyrodetectors in voltage and current mode

multiple detectors do not have a detector element at this point (Fig. 7.123). The entry windows are in the flank of the radiation distribution, so that lower radiation intensities can be expected for the individual elements. This problem can be solved by an internal beam splitter in the detector setup. The radiation enters the detector from the center and is directed to the individual detector elements by reflectors. This results in a uniform illumination of the detectors, which also compensates for a possible inhomogeneity of the radiation source. Figure 7.124 shows such a setup. The entrance window is exactly in the middle and has an aperture of 4 mm. A cuvette with a small inner diameter (e.g. 1–4 mm) can be placed directly in front of this narrow aperture.

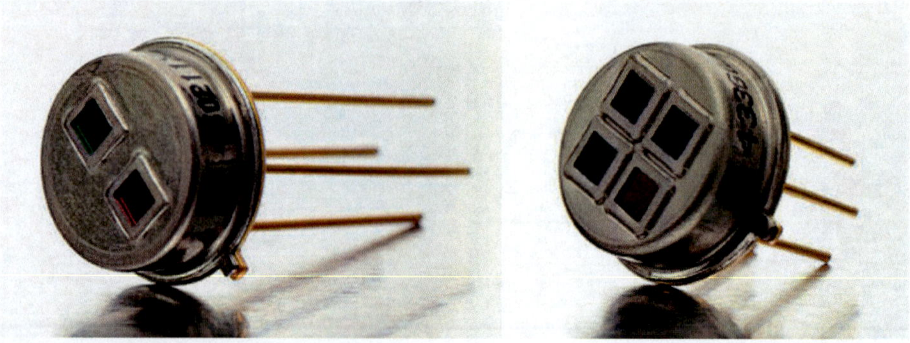

Fig. 7.123 Pyroelectric detectors in two-channel and four-channel versions. (Source: micro-Hybrid Electronic GmbH Hermsdorf)

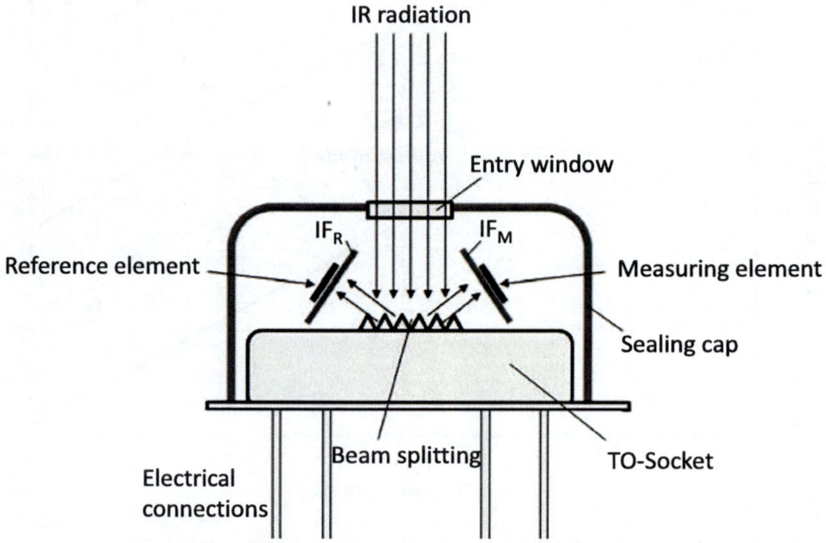

Fig. 7.124 Cross-section through a pyroelectric detector (type LIM-001, Infratec GmbH Dresden) with integrated beam splitting for better utilization of the radiation distribution and compensation of inhomogeneous beams

Comparison Thermopile/Pyro

Both types of detectors work according to the thermal method and theoretically have similar properties regarding detectivity (D^*). However, lithium tantalate ($LiTaO_3$) detectors available on the market achieve a higher detectivity than thermopile detectors. For practical applications, however, further properties must be considered. Thermopiles, for example, are better suited for high ambient temperatures because they do not use temperature-sensitive crystals (→ *Curie temperature*). Pyroelectric detectors can also be used for very high modulation frequencies, up to the kHz range, with appropriate external circuits. In

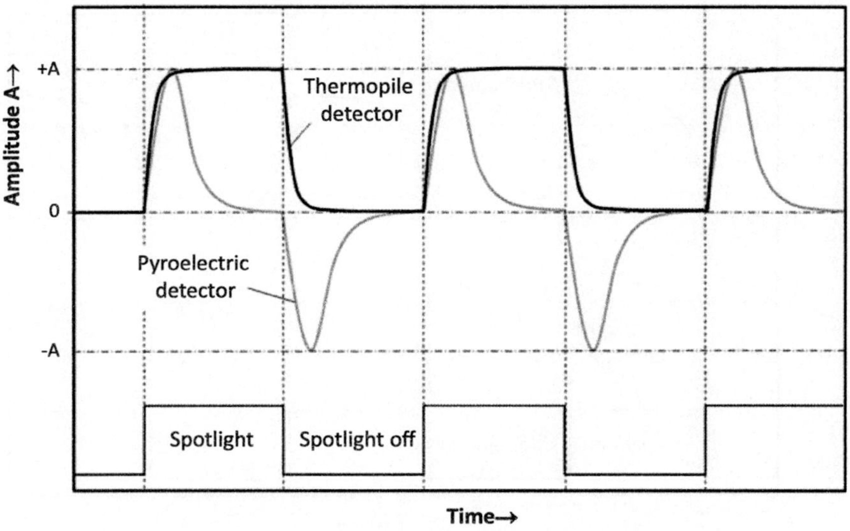

Fig. 7.125 Temporal response of a thermopile detector, compared to a pyroelectric detector, at a low modulation frequency of $f < 1$ Hz

general, the time response of both detector types is completely different. Figure 7.125 shows the signal curve for a small modulation frequency $f < 1$ Hz. The thermopile detector starts from the zero level and after a certain time reaches the maximum value, which it maintains until the radiation source is switched off. The switch-off behavior is symmetrical to the switch-on behavior and decays with the same time constant τ the pyroelectric detector also rises steeply after the radiation source is switched on and then, after passing through a maximum, drops back down to the zero value. After switching off the radiation source, a new change of charge takes place, which now runs in the opposite direction and thus indicates a negative course.

If the modulation frequency is increased (e.g. $f \approx 5$ Hz) the pyroelectric detector shows a different behavior (Fig. 7.126). Since the time for the complete charge exchange is no longer sufficient, the signal no longer reaches the zero value, but is drawn into the negative range immediately after the radiation source is switched off. The thermopile detector follows the change in radiation with a delay since the time for complete thermal equilibrium is no longer sufficient here either. Decisive for this behavior is the thermal mass of the sensor element. Typical time constants for thermopile detectors are given with $\tau \approx 10$ up to 100 ms.

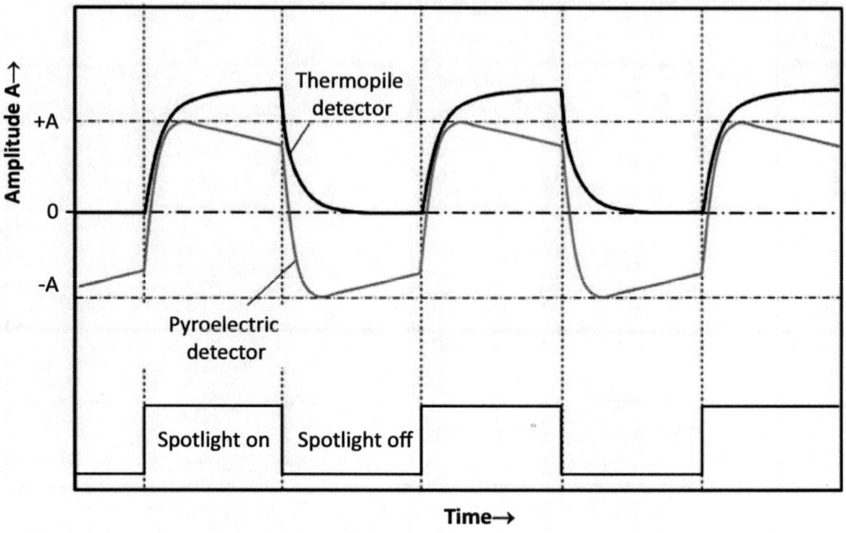

Fig. 7.126 Time response of a thermopile detector, compared to a pyroelectric detector, at a modulation frequency of $f \approx 5$ Hz

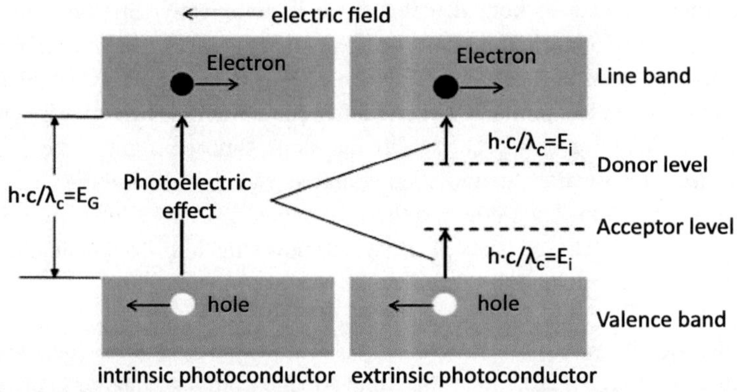

Fig. 7.127 Conduction processes in a semiconductor by photoexcitation. Most IR quantum detectors can be described as intrinsic photoconductors

Quantum Detectors

In gas measurement technology, mainly infrared detectors are used, which work according to the internal photo effect. Basically, there are two types of design which generate either a photoconductor or a photovoltage.

The photoconductivity is caused by the absorption of light quanta of energy hc/λ. Figure 7.127 shows the energy levels in a semiconductor material. They consist of the

Table 7.5 Energy gap E_G and wavelength limit λ_c for selected intrinsic photoconductors (Keyes 1980)

Material	T (K)	E_C (eV)	λ_c (μm)
CdS	295	2.4	0.52
CdSe	295	1.8	0.69
CdTe	295	1.5	0.83
GaP	295	2.24	0.56
GaAs	295	1.35	0.92
Si	295	1.12	1.2
Ge	295	0.67	1.8
PbS	295	0.42	2.9
PbSe	195	0.23	5.4
InAs	195	0.39	3.2
InSb	77	0.23	5.4
$Pb_{0.2}Sn_{0.8}Te$	77	0.1	12
$Hg_{0.8}Cd_{0.2}Te$	77	0.1	12

valence band and the conduction band, which are energetically separated by a gap E_G (Table 7.5). If this energy is high enough ($E > E_G$), electrons are lifted from the valence band into the conduction band of the semiconductor material, leaving holes in the valence band. If both charge carriers are mobile in the electric field, this is called *intrinsic photoconductivity*. If the electron-hole pairing takes place at an impurity, then only one type of charge carrier is mobile in the electric field and this is called *extrinsic photoconductivity*.

IR detectors that generate a photovoltage consist of an *n-type* and a *p-type* material that are in contact with each other on one surface. A concentration gradient of electrons from the *n-layer* to the *p-layer* is formed, which causes an electric field. If IR radiation falls into the *pn boundary layer*, electron-hole pairs are generated. These are separated due to the electric field. The electrons go to the *n-zone* and the defect electrons to the *p-zone*, provided that there is no external electric field. This charge separation causes a photovoltage to occur. This type of detector is called a photoelement. If a voltage is applied from the outside in the reverse direction, a depletion zone is formed between the *n*- and *p-layer*. Only a very small current flows, the resistance is high. If radiation now hits this depletion zone, electrons and defect electrons are formed which generate a current. This current is proportional to the incident radiation intensity. These detectors are called photodiodes. Photodiodes or photoelements can also be operated as photoconductors. Important materials for photoelements or photodiodes are

- Mercury cadmium telluride HgCdTe
- Silicon Si
- Indium antimonide InSb.

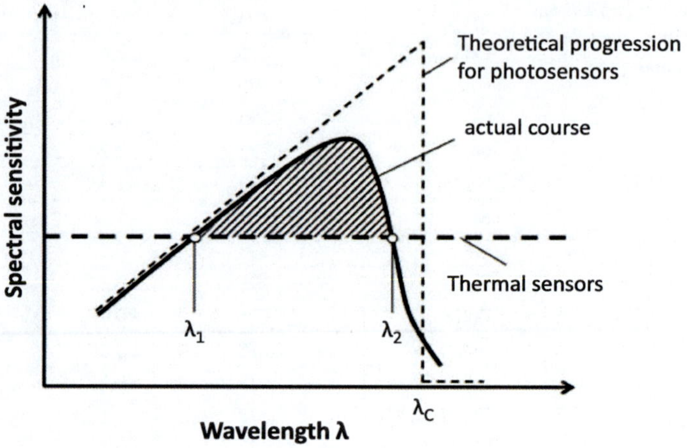

Fig. 7.128 Spectral sensitivity of a photoconductor, compared to a thermal detector (e.g. thermopile)

For both, the photoelement and the photodiode, the energy of the incident light must be greater than the energy gap between the valence band and conduction band (Table 7.5). The size of the energy gap is therefore decisive for the long-wave limit of the detector's sensitivity (Fig. 7.128). The advantages of photodetectors are their simple design, their high sensitivity (about 100 times more sensitive than a pyroelectric detector) and their speed. The disadvantage of these detectors is the limited spectral range. To achieve a high D^* (specific detectivity) in the mid-infrared, thermal noise is reduced by cooling (Salzer et al. 2015).

The advantages of photoconductors compared to thermal detectors are very limited. Figure 7.128 shows the spectral sensitivity of a photoconductor compared to a thermal detector. In the spectral range between λ_1 and λ_2, the photoconductor has a better sensitivity, while outside this zone the thermal detector is better. By active cooling of the photoconductor, this spectral range is significantly increased. Figure 7.129 shows this influencing factor for the common materials (lead sulfide PbS, lead selenide PbSe). The spectral range $\lambda > 3\,\mu m$ which is interesting for gas measurement technology is only covered by the PbSe detector. With active cooling, this detector type can be used up to 5 µm and then has considerable advantages over a thermal detector.

However, active cooling requires considerable technical effort. Liquid air is not suitable as a cooling medium and is only possible in laboratories. In gas measuring instruments, on the other hand, *Peltier elements* are used which are very small and compact. To achieve a high temperature difference ΔT between the detector temperature T_D and the ambient temperature T_U, multi-stage Peltier coolers are often used. Figure 7.130 shows a detector setup with a 3-stage Peltier cooler. The individual coolers are connected in series, so that temperature differences of $\Delta T = 50$ K can be achieved with this arrangement. A

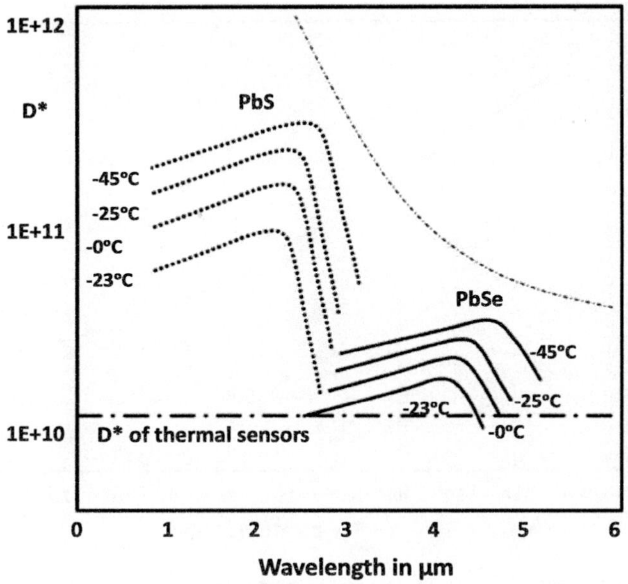

Fig. 7.129 Influence of temperature on the spectral sensitivity of photoconductors (PbS and PbSe)

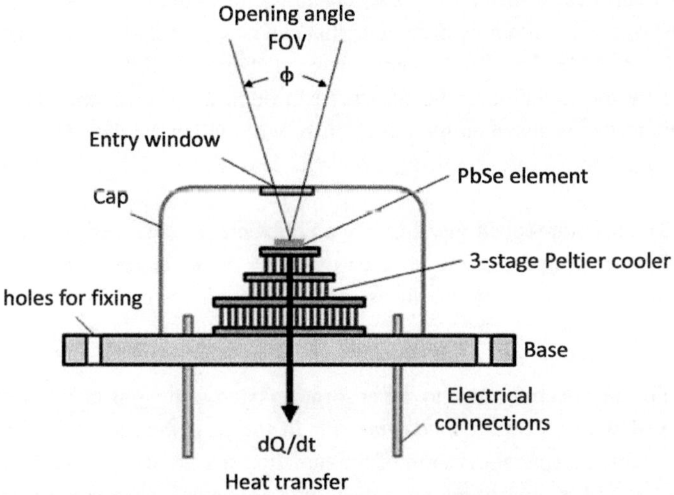

Fig. 7.130 Cross-section through a PbSe detector with a 3-stage Peltier cooling

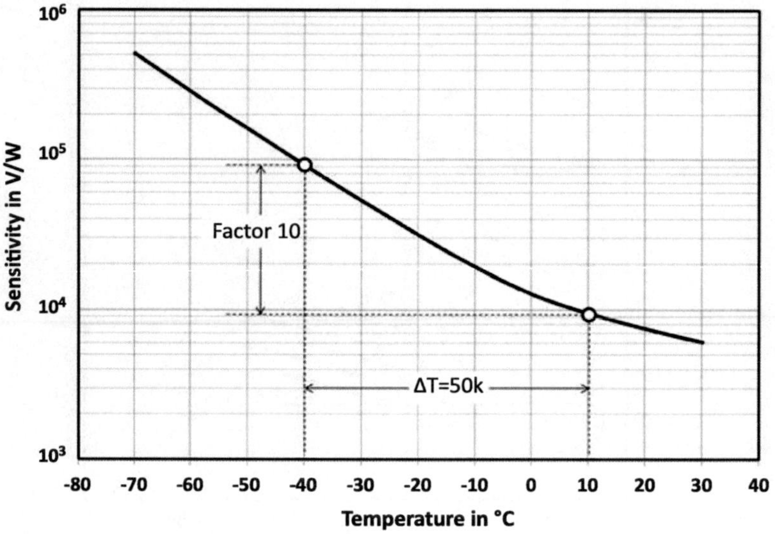

Fig. 7.131 Sensitivity curve of a PbSe detector with temperature

prerequisite, however, is that the power loss d*Q*/d*t* of the d*Q*/d*t* cooling elements is dissipated by good heat transfer from the rear to the environment. For this purpose, the sockets consist of a solid, metallic base plate that can be connected to an external heat sink via screws. The actual IR detector element is located at the end of the last cooling stage and is illuminated by the radiation to be measured through a window opening. Next to the detector element there is also a temperature sensor with which the detector temperature T_D is measured. This temperature information can also be used to control the detector temperature.

Figure 7.131 shows the sensitivity curve of a PbSe detector. The sensitivity changes by a factor ≈ 10 when the temperature is lowered by $\Delta T = 50$ K. Here, it becomes clear what advantages active cooling has on the measuring characteristics of a PbSe detector.

Interferometer

The limitation of the spectral range to the absorption band of the gas to be investigated can also be achieved with so-called *spectrometers.* In the past, prisms were used in the IR range, which enabled a spatial division of the radiation due to dispersion.[35] For differentiation purposes, the photometric methods described above are therefore still referred to as *non-dispersive* today. Later, diffraction gratings were used as dispersive elements, which were subsequently replaced by interferometers. In the IR range, this method has become

[35] Change of the refractive index n with the wavelength λ.

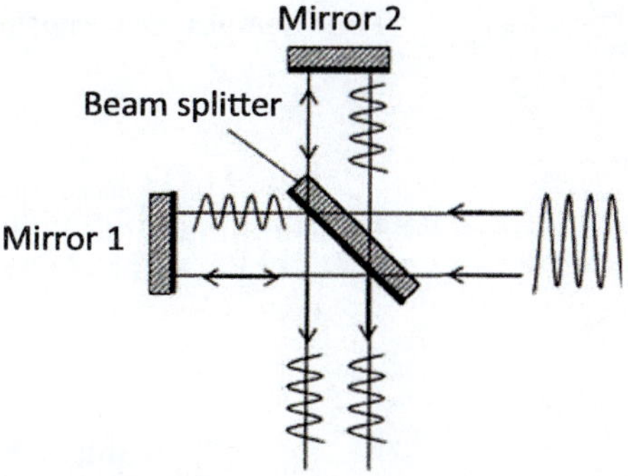

Fig. 7.132 Classical Michelson interferometer with two mirrors and a beam splitter. One of the two mirrors is movable and thus generates an alternating interference pattern in the direction of observation

generally accepted, especially in the laboratory. Two methods are used in gas measurement technology, which are described below.

FTIR Interferometer
In an interferometer, light waves are superimposed, which are determined by the amplitude A, wavelength λ and phase position φ. The simplest and best-known interferometer was developed by Michelson[36] and is still used today as the basis for many instrument setups. Figure 7.132 shows the basic principle of the Michelson interferometer.

A light wave (monochromatic radiation $\Delta\lambda \approx 0$) reaches a beam splitter, which splits the light wave into two separate paths. One light wave is reflected in the direction of the mirror 2, while the other light wave passes through the beam splitter and hits the mirror 1. Both light waves are reflected by the respective mirror back to the beam splitter and can now interfere with each other at the interface of the beam splitter. If both light waves have travelled the same path s, they are also in phase $\varphi = 0$ at the interface and add up. If one of the two mirrors is shifted $\Delta s = \lambda/4$, the two waves are extinguished at the interface, because the crests of one wave meet the troughs of the other wave. The phase shift in this case is $\varphi = \pi$. If one of the two mirrors is moved $s(t)$ slowly, the interference pattern changes permanently from light (constructive interference) to dark (destructive interference). The amplitude response can be described by a sine function. If a second wavelength λ_1 is coupled in addition to the monochromatic radiation λ_2, a superposition of these waves occurs. The result at the output of the radiation part is a beat. In Fig. 7.133, this course of the interferogram ($n = 2$) is shown in the time domain.

If further wavelengths ($n = 3, 4, \ldots, \infty$) are now added, increasingly complex interferograms are obtained. From the interferogram, however, the spectrum can be calculated with the help of the Fourier transformation. For a continuous analysis, the conversion must be carried out very quickly. The method is therefore also called *fast*

[36] Albert Abraham Michelson (1852–1931) American physicist.

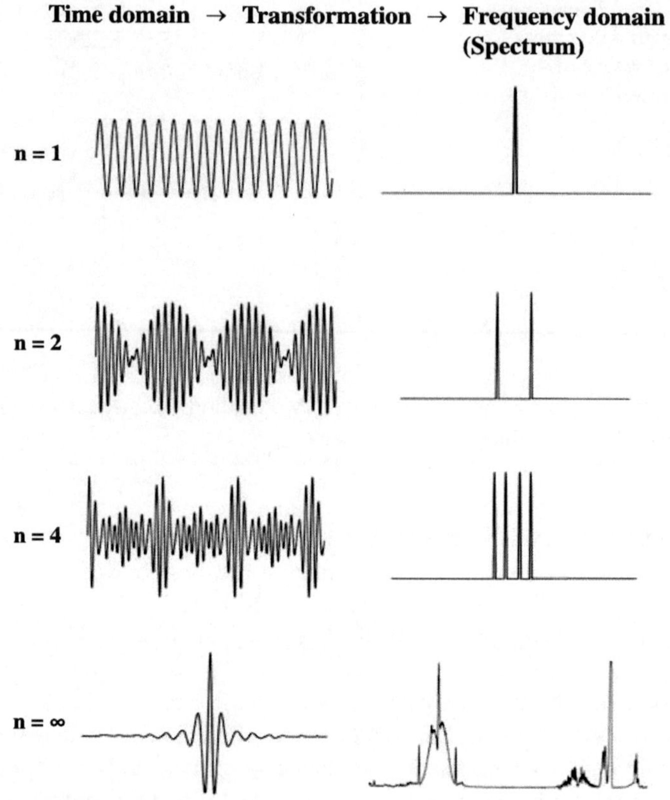

Fig. 7.133 Superposition of waves in an interferometer in the time domain (interferogram) and frequency domain (spectrum)

Fourier transformation FFT. This mathematical method requires a great deal of computing power, so that FTIR spectroscopy has only been able to establish itself as a practicable gas measurement method in recent decades.

Figure 7.134 shows a complete FTIR spectrometer. The central part is the Michelson interferometer already described. In this setup, the broadband IR radiation is generated by a powerful incandescent radiator and coupled into the interferometer via a mirror optic. One of the two mirrors is located on a movable mechanism and can be moved in a defined way. This movement is usually carried out at a constant speed and is repeated continuously. A complete pass is called a *scan*. The exact position $s(t)$ of the mirror is determined by means of an adjustment laser. Since the radiation of the adjustment laser also passes through the interferometer, a sinusoidal change in intensity is obtained at the output, the maxima of which correspond exactly $\Delta s = \lambda$. The IR radiation now enters a measuring cell in which

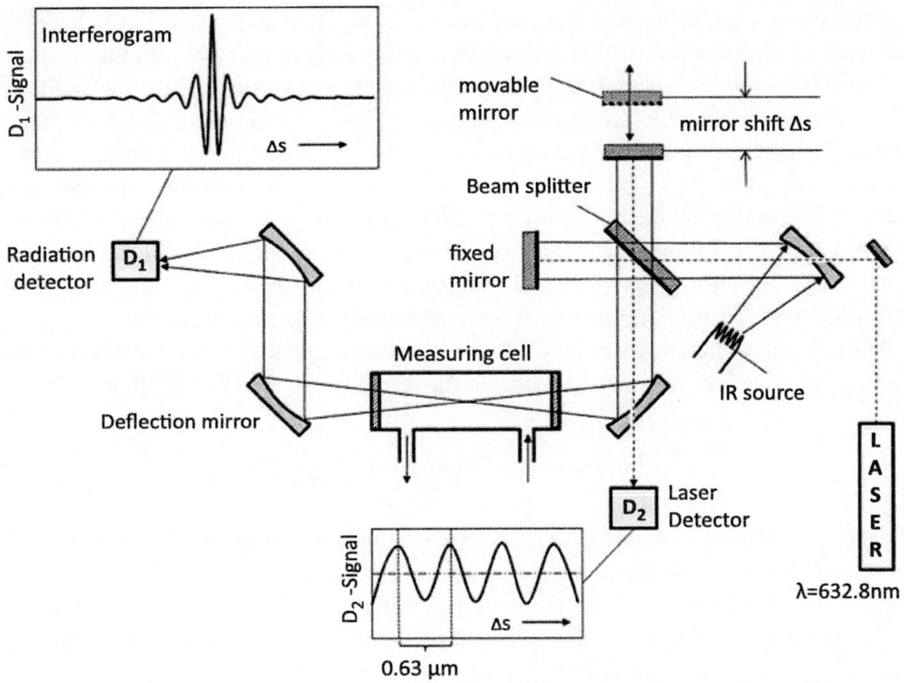

Fig. 7.134 Schematic diagram of an FTIR spectrometer, based on a Michelson interferometer

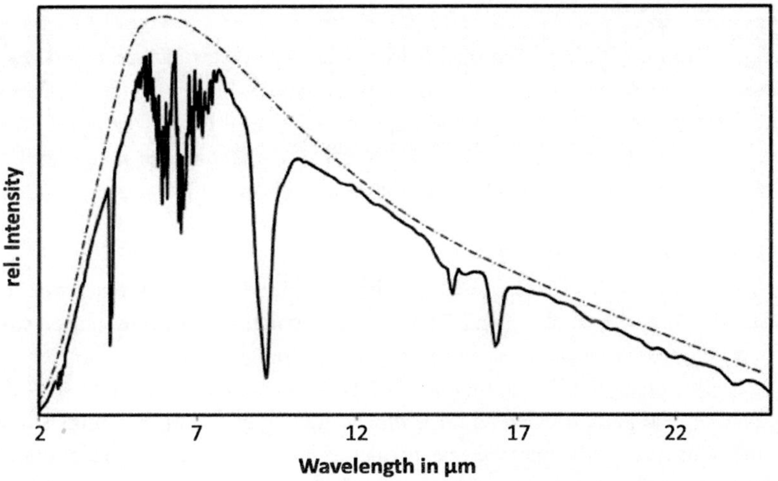

Fig. 7.135 Background spectrum of an FTIR spectrometer. The *dashed dotted line* shows the emission spectrum of a thermal radiator. The absorption bands are caused by carbon dioxide and water vapor contained in the ambient air

the selective absorption of radiation by the gases to be detected takes place. The radiation is then focused on a detector, which delivers an interferogram as measurement information.

The FTIR process is therefore in principle a radiation process. Since the radiation distribution of the IR source corresponds to Planck's law, one can expect a spectrum that reflects this course. Figure 7.135 shows such a spectrum. The dotted line corresponds to Planck's law, while the spectrum below it reflects the real course of the radiation. As a result of absorption by water vapor and carbon dioxide from the ambient air in the interferometer setup, some areas are already provided with absorption lines. This spectrum is stored as a *background spectrum* and compared with the following spectra. In this case the background spectrum forms the reference and must be checked regularly.

The spectral resolution $\Delta\nu$ of an FTIR spectrometer is determined by the mirror stroke Δs. The larger the mirror stroke, the smaller the resolution. It therefore applies:

$$\text{Spectral resolution}: \quad \Delta\nu = \frac{1}{\Delta s} \tag{7.56}$$

For a mirror stroke of 1 cm you get a spectral resolution of one wave number $= 1\ \text{cm}^{-1}$. If the mirror stroke on 10 cm is increased, the resolution improves to 0.1 cm^{-1}. In gas analysis, resolutions of 0.5^{-1} cm are required, which corresponds to a mirror stroke of $\Delta s = 2$ cm. The signal to noise ratio (*S/N*) can be improved by the number *n* of scans. The following relationship applies here:

$$N^* = \frac{N}{\sqrt{n}} \tag{7.57}$$

A noise figure *N* of e.g. 0.33 in one scan is improved by a factor of 3.16 by 10 scans, and then amounts to only $N^* = 0.104$. So the signal/noise ratio was increased by the factor 3.16. However, if 100 scans are performed, the improvement of 10 is to be expected. Since the time duration then also increases by a factor of 100, the number *n* required for the respective application must always be considered in each individual case.

Fabry-Perot Interferometer (FPI)

The spectral position of interference filters λ_0 is determined by the internal layer structure and is only variable to a small extent. The central wavelength position can be shifted to smaller wavelengths by mechanical tilting. The tuning range $\Delta\lambda$ is approximately 200 nm for an angle adjustment of 35°. Another possibility is to vary the temperature *T* of the filter to change the optical path $d \cdot n$. With this influencing variable, a tuning range suitable for gas measurement can also be realized (Zöchbauer 1993). Also in this case the tuning range is very small and only suitable for certain gases (Wiegleb and Zöchbauer 1994).

Typically, a Fabry-Perot interferometer is tuned with a mechanical change, the distance *d* of two plane-parallel plates (mirrors). Conventional FPI are difficult to adjust and also

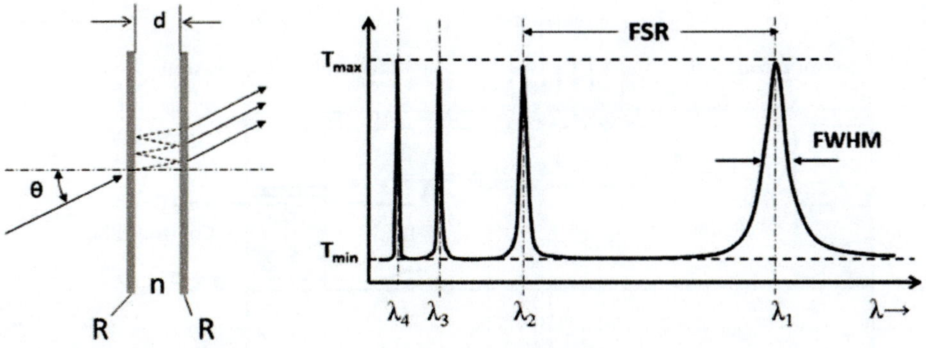

Fig. 7.136 Principle function of the Fabry-Perot interferometer

very cost intensive. The use of microsystem technology has made it possible to develop a tunable FPI that combines many advantages (Neumann et al. 2003). Figure 7.136 shows the basic structure of such an FPI. The two semi-transparent mirrors with reflection coefficient R are located at a distance d from each other. If a light beam is transmitted through the first semitransparent mirror and reflected by the second mirror, it passes through the gap twice and can then constructively interfere with the other beams. It is for a maximum transmission $T_{\max}$ for $\theta = 0$ the following condition:

$$\lambda_m = \frac{2d}{m} \tag{7.58}$$

The variable m is the ordinal number (1, 2, 3, . . .). The distance between $m = 1$ and $m = 2$ is called the *free spectral range* (*FSR*). The half-value width *FWHM* depends mainly on the reflectance R of the partially transparent mirrors. The ratio of FWHM and FSR is called Finesse $\widetilde{F}$:

$$\widetilde{F} = \frac{\text{FSR}}{\text{FWHM}} \tag{7.59}$$

The realized version of the FPI is shown schematically in Fig. 7.137. Both reflectors are located within a microstructure, with one of the mirrors being firmly attached to the structure. The second mirror is mechanically movable via a suspension and can move in the direction of the first mirror. This movement is caused by electrostatic forces which are generated by applying a variable control voltage. The IR radiation reaches this FPI structure via a broadband filter. The filter is required to suppress the higher orders, since only the 1st order is to be used within the FSR (Fig. 7.138). The radiation can then, as already described, interfere between the two reflectors, and thus transmit a characteristic wavelength. If the control voltage is now increased, the second mirror moves towards the

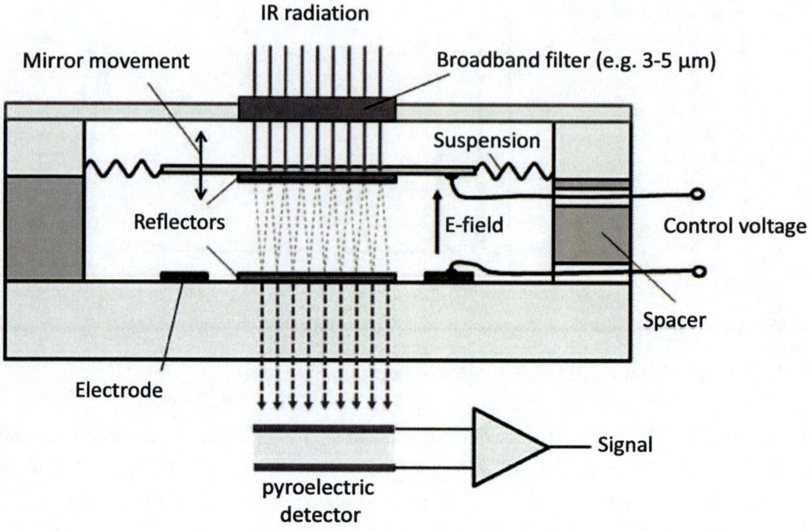

Fig. 7.137 Schematic structure (cross-sectional view) of the Fabry-Perot interferometer according to Neumann

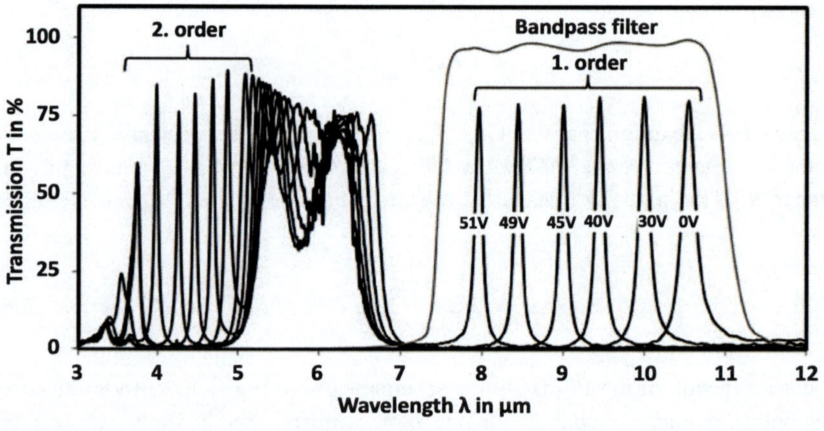

Fig. 7.138 1st order passband between 8 μm and 11 μm and in the 2nd order between 4 μm and 5 μm (Ebermann 2014)

first mirror and the transmission curve is shifted (Fig. 7.139). At the output of the FPI there is a pyroelectric detector which then converts the radiation into an electrical signal.

The microstructure of the moving mirror is an important part of the FPI. This part is shown in Fig. 7.140. In the left picture you can see the complete structure. In the center, you can see the square mirror surface. The suspension is realized by a free etching, the detail of

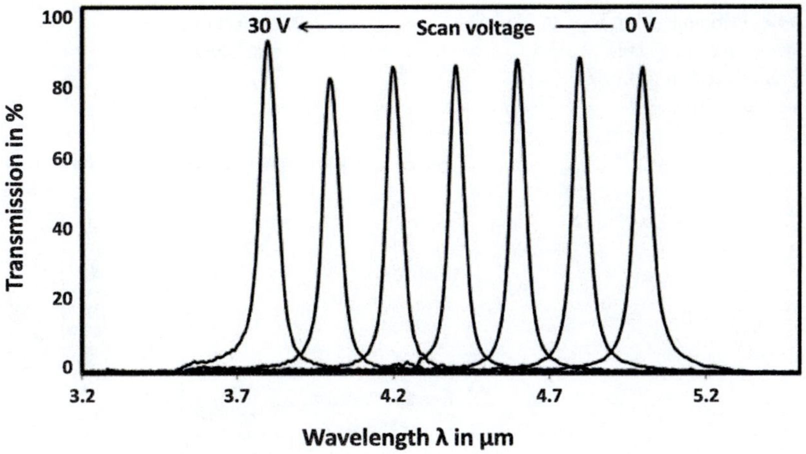

Fig. 7.139 Scan area of a tunable FPI

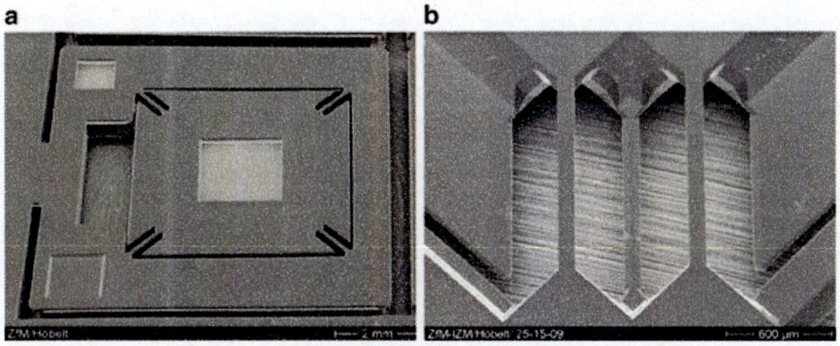

Fig. 7.140 (**a**) Microstructure of the FPI, (**b**) Detailed view of the suspension/suspension. (Source: Infratec GmbH Dresden)

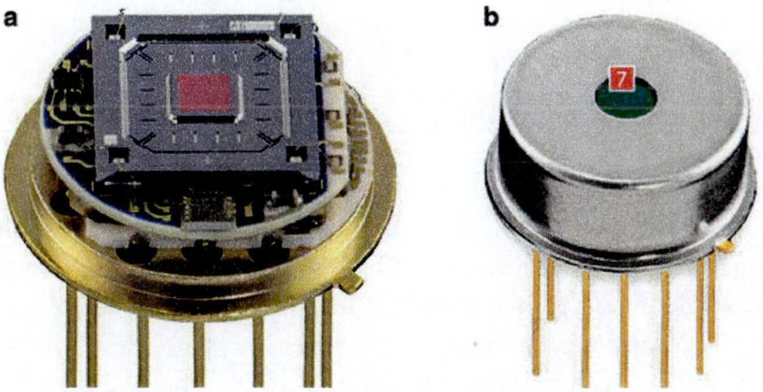

Fig. 7.141 Completely assembled assembly without (**a**) and with (**b**) sealing cap. (Source: Infratec GmbH Dresden)

Fig. 7.142 Principle setup for simultaneous gas analysis in two spectral ranges resulting from the 1st and 2nd order FPI (Ebermann 2014)

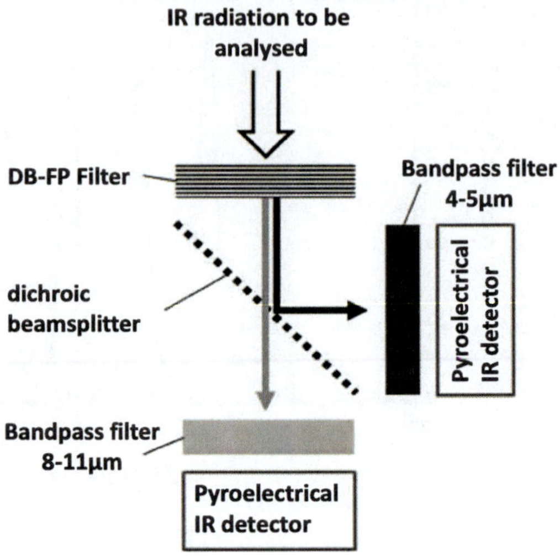

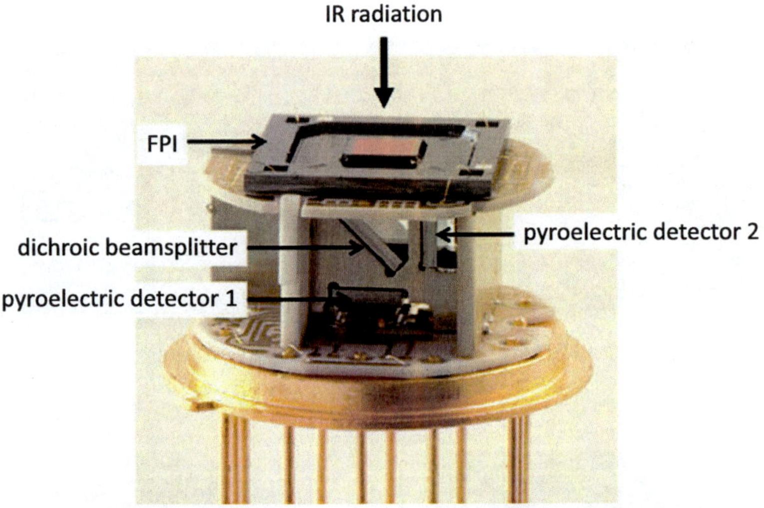

Fig. 7.143 Realized setup of a double FPI for gas analysis in two spectral ranges. (Source: Infratec GmbH, Dresden)

which can be seen in the right part of the picture. In each corner three thin connecting bars are visible, which are arranged in a vertical offset. By this measure torsional movements are prevented. The width of the webs was chosen so that the electrostatic forces are sufficient to ensure a given displacement path.

The entire assembly is then integrated into a TO8 housing and hermetically sealed (Fig. 7.141). The diameter is only ∅15 mm and the height 6.5 mm. The entrance window has an aperture of 3 mm. Since the movable mirror also reacts to vibrations and acceleration, the assembly can also be supplied with an integrated ASIC to reduce these influencing factors.

The tuning range of the FPI is 1.2 μm or 2.5 μm depending on the version. For measurements that are to be performed in the lower wavelength range at 3–5 μm as well as at 8–10 μm, two independently operating FPIs are required. The upper spectral range can be realized with the 1st order, within the FSR, while the 2nd order is then at ≈4.5 μm. This measurement problem can also be solved with a combined setup (Ebermann 2014).

The setup with which both spectral ranges can be covered is shown in Figs. 7.142 and 7.143. The incident radiation is first sent through the FPI. Both spectral ranges (1st and 2nd order) are thus let through. Then the radiation passes through a dichroic beam splitter and reaches the two pyroelectric detectors, which are additionally preceded by a band-pass filter to limit the corresponding order.

References

Armerding, W., Walter, J., Comes, F.J.: A White cell type multiple reflection system for tropospheric research. Fresenius J. Anal. Chem. **340**, 661–664 (1991)

Arnoudse, P.A., et al.: Instrumentation for the breath-by-breath determination of oxygen and carbon dioxide based on nondispersive absorption measurements. Anal. Chem. **64**, 200–224 (1992)

Ascherfeld, M., Fabinski, W., Vogt, R.: Erweiterte Möglichkeiten und Anwendungen mit dem NDIR-Fotometer Uras 10E. Technisches Messen. **57**, 11–17 (1990)

Budzier, H., Gerlach, G.: Thermische Infrarotsensoren. Wiley VCH, Weinheim (2010)

Demtröder, W.: Laserspektroskopie. Springer (1993)

Döring, S.: Vergleichende Messungen zwischen thermischen IR-Strahlern und IR-Leuchtdioden für schnellanzeigende CO_2-Gasmessungen. Bachelor-Thesis, Fachhochschule Dortmund (2013)

Ebermann, M.: Mikromechanische Mehrband-Fabry-Perot-Interferometer für die Anwendung in der Infrarot-Spektroskopie. Dresdner Beiträge zur Sensorik, Bd. 54. TUDpress, Dresden (2014)

Engelhardt, H.: Neuere Methoden zur automatischen Gasanalyse. Neue Technik. **5**(4), 182 198 (1963)

Fabinski, W.: Verfahren zum Nachjustieren von Infrarot-Gasanalysatoren. Deutsches Patent DE3522949 vom 27.6.1985 (1985)

Fabinski, W., Ascherfeld, M.: Funktion und Aufbau des NDIR-Prozessfotometers Uras 3. Technisches Messen **47**, 257–264 (V7232-3) (1980)

Hahn U.: Physik für Ingenieure. Kap. 3, 345. Oldenburg (2007)

Heimel, H., Randow, A., Röß, R., Wiegleb, G.: Küvette für Gasanalysengeräte. Deutsche Patentschrift DE 3328171 (1983)

Hill, D.W., Powell, T.: Non-Dispersive Gas Analysis. Adam Hilger, London (1968)

Kessler, G.: Über die Absorption ultraroter Strahlung in hintereinanderliegenden Gas-Schichten und ihre Bedeutung für photometrische Analyseverfahren. Dissertation, Technische Hochschule, Clausthal (1967)

Keyes, J.R.: Optical and Infrared Detectors. Topics in Applied Physics. Springer (1980)

Köhne, S., Wiegleb, G.: Gassensorik in der Fahrzeugtechnik. Design & Elektronik 07/2008, 62–63 (2008)

Lehnert, M., Modlinksi, U., Wiegleb, G.: Mikroprozessorgesteuertes Gasanalysengerät in Kompaktbauweise. Technisches Messen. **54**, 423–429 (1987)

Lehrer, E., Luft, K.F.: Verfahren zur Bestimmung von Bestandteilen in Stoffgemischen mittels Strahlenabsorption. DRP 730478 (1938)

Link, W.T., McClatchie, E.A., Watson, D.A.: A Fluorescent Source NDIR Carbon Monoxide Analyzer. Joint Conference on Sensing of Environmental Pollutants. Palo Alto USA 8.–10. Nov. 1971 (AIAA Paper 71-1047) (1971)

Luft, K.F.: Über eine neue Methode der registrierenden Gasanalyse mit Hilfe der Absorption ultraroter Strahlen ohne spektrale Zerlegung. Zeitschrift für technische Physik. **24**, 97–104 (1943)

Luft, K.F.: Über nichtdispersive Ultrarotanalysengeräte. Fresenius Zeitschrift für analytische Chemie. **164**, 100–108 (1958)

Luft, K.F., Schäfer, W., Wiegleb, G.: 50 Jahre NDIR-Gasanalyse. Technisches Messen. **60**, 363–371 (1993)

Modlinski, U., Wiegleb, G.: Infrarot-Strahlungsquelle, insbesondere für ein mehrkanaliges Gasanalysengerät. Deutsches Patent DE 3809160.7 vom 8.3.1988 (1988)

Morelli, D.-W.: Interference Filter Handbook. JDS Uniphase Corporation USA (2006)

Naumann, K., Schneider, E.: Siemens Zeitschrift **27**, 8/18 (1953)

Naumann, A., Schulz, G.: Das Inframeter, ein neues Betriebsgerät zur Infrarot-Gasanalyse ohne spektrale Zerlegung. Chem. Ing. Techn. **32**(10), 669–764 (1960)

Neumann, N., Heinze, M., Stegbauer, H.J., Hiller, K., Kurth, S.: Ein mikromechanisches, durchstimmbares Fabry-Perot-Filter für die nichtdispersive Gasanalytik im Spektralbereich 3…5 µm. 6. Dresdner Sensor Symposium (2003)

Pohl, R., Wiegleb, G.: Miniaturization of infra-red gas sensors. Proceedings of „Sixth Meeting on Chemical Sensors" Gaithersburg USA July (1996)

Röntgen, W.C.: On tones produced by the intermittent irradiation of a gas. Philosophical Magazine Series. **5**(**11**), 308–311 (1881)

Salzer, R., Thiele, S., Suemmchen, L., Bezugla, E.: IR-Gerätetechnik (innerer Fotoeffekt). www.chemgapedia.de (2015)

Schaefer, W.: Gasanalyse mit dem URAS bei kompliziert zusammengesetzten Messgasen. Chem.-Ing.-Technik. **33**(6), 426–430 (1961)

Schaefer, W.: Bestimmung der Schwingungsrelaxationszeit in CO/N_2-Gasgemischen aus der Analyse des Frequenzganges eines Ultrarot-Gasanalysators. Zeitschrift für angewandte Physik. **19**(1), 55–60 (1965)

Schaefer, W.: Die Anwendung von spektroskopischen Methoden und Lasern zur Bestimmung von gasförmigen Schadstoffen in der Luft. PTB-Mitteilungen. **2**, 84–92 (1974)

Schmale, U.: Entwicklung eines langzeitstabilen Niedrigenergie IR-Gassensors. Dissertation, RWTH, Aachen (2000)

Schmick, H.: Einrichtung zur Bestimmung eines Bestandteiles in einem Gemisch, insbesondere von Gasen mit Hilfe der Absorption der vorzugsweise ultraroten Gesamtstrahlung. DRP 465 899 (5.9.1926) Siemens & Halske (1926)

Schunck, G.: Einrichtung zur Messung schwacher Gasströmungen, vorzugsweise in einem Gasanalysegerät. Deutsches Patent 1573098 vom 30.11.1966 (1966)

Schunck, G.: Schnelle Meßfühler für kleine Gasströme. Dissertation Technische Hochschule, Karlsruhe (1974)

Schunck, G., Randow, A.: Pneumatischer Infrarot-Strahlungsdetektor mit einer vakuumdichten Kammer und einem strahlungsdurchlässigem Fenster. Deutsches Patent DE 27 20 636 C2 vom 7.5.1977 (1977)

Staab, J.: Industrielle Gasanalyse. Oldenbourg (1994)

Welford, W.T., Winston, R.: High Collection Nonimaging Optics. Academic Press Inc. (1989)

White, J.U.: Long paths of large aperture. J. Opt. Soc. Am. **32**, 285 (1942)

Wiegleb, G.: Lichtquelle für nichtdispersive Gasanalysengeräte als Fluoreszenzlampe. Deutsche Patentanmeldung DE 33 16 771 A1 vom 7.5.1983 (1983)

Wiegleb, G.: Gasanalysensysteme für erhöhte Prozesstemperaturen. Technisches Messen. **51**, 385–393 (1984)

Wiegleb, G.: Einsatz von LED-Strahlungsquellen in Analysengeräten. Laser und Optoelektronik. **3**, 308–310 (1985)

Wiegleb, G.: NO_x-Messung mit dem BINOS-System, Messen-Prüfen-Automatisieren 5/1988, 259–261 (1988a)

Wiegleb, G.: Infrarot Gasanalysator BINOS 100 mit Mikroprozessorsteuerung, Messen-Prüfen-Automatisieren 10/1988, 430–434 (1988b)

Wiegleb, G.: Mehrkomponenten-Gasanalysengerät mit µP-Steuerung und interner Kalibrier- und Korrekturmöglichkeiten. VDI-Berichte. **960**, 625–634 (1992)

Wiegleb, G.: Sensoreinrichtung zur Erfassung der Gaskonzentration mit mehreren selektiven Strahlungsempfängern. Deutsche Patentanmeldung DE 196 04 16, 6.2.1996 (1996)

Wiegleb, G.: Ein neuartiges Gaswarnsystem zur Leckageüberwachung in Industrie- und Gewerbeanlagen. GASWÄRME International. **55**(4/2006), 1–4 (2006)

Wiegleb G.: Abschlussbericht: MICROGAS FKZ: 1746X04 (2009)

Wiegleb, G.: Lebensdauer von IR-Strahlungsquellen. Interner Bericht der Fachhochschule Dortmund (2015)

Wiegleb, G., Zöchbauer, M.: New optical methods for applications in process analysis and medical instruments. Exp. Tech. Phys. **40**, 241–257 (1994)

Wiegleb, G., Knaup, G., Plotz, F.: Sensoreinheit zur Erfassung eines Fluid, insbesondere zur Erfassung von Erdgas, Kohlenwasserstoffen, Kohlendioxid oder dgl. in Umgebungsluft. Internationales Patent WO 2005/121752 A1 (2005)

Winston, R.: Light collection within the framework of geometrical optics. J. Opt. Soc. Am. **60**, 245–247 (1970)

Zöchbauer, M.: Entwicklung eines thermisch abstimmbaren Fabry-Perot-Korrelationsspektrometers für die Gasanalyse. Dissertation, Universität Frankfurt am Main (1993)

Related References

Engelhardt, H.: Wege und Grenzen neuer Verfahren der physikalischen Gasanalyse. VDI-Zeitschrift. **103**(28), 1393–1404 (1961a)

Engelhardt, H.: Wege und Grenzen neuer Verfahren der physikalischen Gasanalyse. VDI-Zeitschrift. **103**(29), 1435–1440 (1961b)

Erbel, M., Wiegleb, G., Gehrke, A., Korbmacher, S.: Miniaturisierter Infrarot-Gassensor mit integrierter Mikromembranpumpe zur Detektion von Kohlendioxid in der Umgebungsluft. Technisches Messen. **64**(4), 147–151 (1997)

Fabinski, W., Deptolla, U.: Verfahren zur Erhöhung der Selektivität von NDIR-Fotometern. Technisches Messen **47**, 113–119 (V 723-38) (1980)

Fabinski, W., Grunewald, A., Norman, W., Hielscher, B., Wiegleb, G., Wolff, C.: A new analyzer for measuring TOC and TN content in sewages, Journal of Process Analytical Chemistry, 121–126 (1994)

Golay, M.: A Pneumatic Infra-Red Detector. The Review of Scientific Instruments. **18**(5), 357–362 (1947)

Golay, M., Zahl, H.: Pneumatic Heat Detector. The Review of Scientific Instruments. **17**(11), 511–515 (1946)

Haisch, M., Fabinski, W., Wiegleb, G., Fuß, W., Hering, P.: Messen der $^{13}CO_2$-Konzentration in Atemluft mit einem nichtdispersiven IR-Spektrometer. Verhandlungen der DPG-Frühjahrstagung in Berlin MO. **12**(5), 359 (1993)

Herriott, D.R., Schulte, J.H.: Folded optical delay lines. Applied Optics. **4**, 883 (1965)

Jamieson, J.A., et al.: Infrared Physics and Engineering. McGraw-Hill, New York (1963)

Johnson, J.B.: Thermal Agitation of Electricity in Conductors. Phys. Rev. **32**, 97–109 (1928)

Korbmacher, S., Wiegleb, G.: Miniaturisierte Infrarot-Gassensoren zur Erfassung von explosiven und toxischen Gasgemischen. Kongressband OPTO, Leipzig. **6**(115–120), 294–305 (2003)

Luft, K.F.: Der UNOR, ein neues Gasanalysengerät für den Bergbau. Glückauf Fachzeitschrift für Rohstoff, Bergbau und Energie. **98**, 493–495 (1962)

Mook, K., Wiegleb, G.: Rauchgasanalyse mit dem BINOS-System. Technisches Messen. **50**, 423–428 (1983)

Nyquist, H.: Thermal Agitation of Electric Charge in Conductors. Physical Review. **32**, 110–113 (1928)

Riemer, W.: Nichtdispersive IR-Gasanalyse in der Fotosynthese. Labor-Praxis. **4**(4), 40–43 (1980)

Schaefer, W.: Ultrarot-Analysatoren und andere Betriebsphotometer. VDI-Berichte. **97**, 15–23 (1966)

Schaefer, W.: Fotometrische Analysenmessgeräte. VDE- Bezirksverein Frankfurt. Betriebsanalysenmesstechnik zur Qualitätssicherung und Umweltschutz (1980)

Schunck, G.: Nichtdispersive Infrarot-Gasanalysatoren für Industrieprozesse und Umweltschutz. DECHEMA-Monographien Bd. 80, Teil 2 (1976)

Schunck, G., Randow, A.: Verfahren zur Messung des optischen Absorptionsvermögens von Proben und Vorrichtung zur Durchführung des Verfahrens. Deutsches Patent DE 26 14 181 vom 2.4.1976 (1976)

Wiegleb, G.: IR-Gasanalyse mit Mikroprozessorsteuerung, Labor Praxis 10/1987, 118–1122 (1987)

Wiegleb, G.: Miniaturisierung physikalischer Gassensoren, 6. Dresdner Sensor-Symposium 9.12.2003 (2003)

Wiegleb, G.: µ-GAS – Innovative CO_2-Sensorik in der Fahrzeugtechnik. Forschungsbericht 2008 der Fachhochschule, Dortmund (2008)

Wiegleb, G.: Gasanalysenvorrichtung (Filter-Korrelationsmessung) Deutsches Patent DE 3544015 vom 13.12.1985 (1985)

Wiegleb, G.: A new calibration technique for industrial gas analyzers, Process Contr. Qual. **3**, 273–281 (1992)

Wiegleb, G.: Gasmonitor zur Überwachung von Gasleckagen in Wohngebäuden. Energie wasser-praxis 11/2006, 8–11 (2006)

Wiegleb, G., Randow, A., Röß, R.: SO_2-Messung mit dem neuen UV-BINOS. Technisches Messen. **50**, 143–150 (1983)

Zehner, B.: Signal und Rauschen beim Differential-Hitzdrahtanemometer. Dissertation Technische Hochschule, Karlsruhe (1979)

UV Absorption Photometer

8

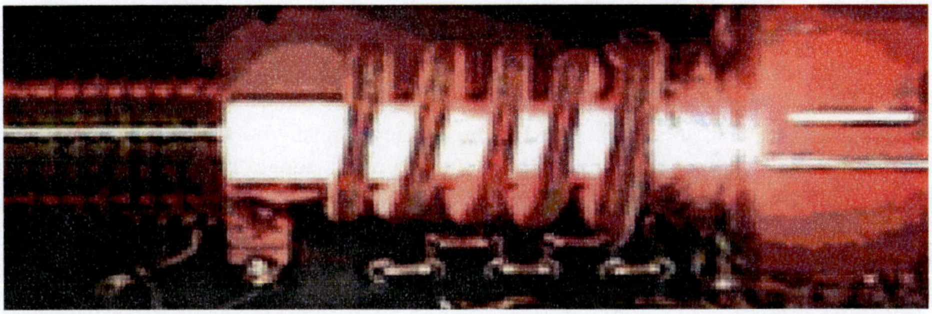

Inductively coupled gas discharge

8.1 Fundamentals

The ultraviolet spectral range is also suitable for gas analytical measurements and is therefore also used intensively for certain applications. Compared to the IR range, however, there are some special features to be considered. Physically, the absorption spectra differ due to additional electronic transitions, since the photon energy in this spectral range is greater than in the IR range. Also, the absorption coefficients in the UV range are usually higher, which leads to a better resolution (detection limit). Furthermore, there are no disturbing influences of air components such as water vapor and carbon dioxide.

Besides these advantages, there are also some disadvantages. The spectral range usable for gas measurement technology only goes from $\lambda \approx 200$ nm to $\lambda \approx 400$ nm. In this comparatively narrow range, overlaps between the individual absorption bands of the gases

G. Wiegleb, *Gas Measurement Technology in Theory and Practice*,
https://doi.org/10.1007/978-3-658-37232-3_8

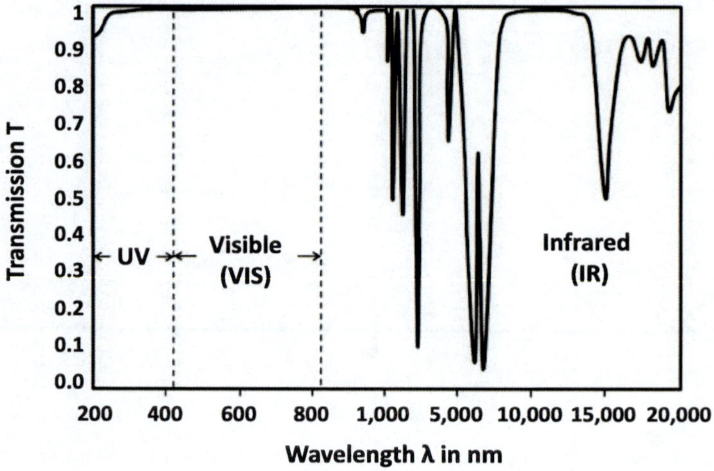

Fig. 8.1 Transmission of radiation through a 10 m thick layer of air, without additional air pollution (Down and Lehr 2005)

cannot be avoided. Cross sensitivities can therefore lead to limitations. Figure 8.1 shows the UV range in comparison with the IR range. The wavelength axis (λ) is plotted as a logarithmic representation and thus illustrates the narrow UV range. The absorption lines in the IR range are mainly caused by H_2O and CO_2, which are always present in the air.

Figure 8.2 shows the UV absorption spectra of sulfur dioxide, nitrogen monoxide and nitrogen dioxide in a 10 cm cuvette and a concentration of $c \approx 1000$ ppm in each case. The NO_2 measurement at $\lambda_3 = 390$ nm does not affect the other accompanying gases, whereas the SO_2 measurement at $\lambda_2 = 285$ nm is disturbed by NO_2. This cross-sensitivity can be compensated by a simultaneous measurement of both gases. In the range of the NO measurement at $\lambda_1 = 226.5$ nm, all gases overlap (see detailed illustration in Fig. 8.3), so that the NO measurement can only be detected reliably and free of cross-sensitivities with special detection methods such as molecular resonance absorption.

The physical laws of radiation absorption are identical to the relationships already described in Sects. 6.1 and 6.2. The basic construction of a UV photometer does not differ significantly from an IR photometer and consists of similar components. However, the optical elements, such as radiation sources, detectors and filters are much more complex and therefore much more expensive than IR components. Furthermore, there are limits to the miniaturization of the UV setups by the radiation sources, which are only slowly softened with the introduction of UV light-emitting diodes.

Due to the advantages described above, however, UV measurement technology has become established as the standard method for certain gases. This applies in particular to the following gases:

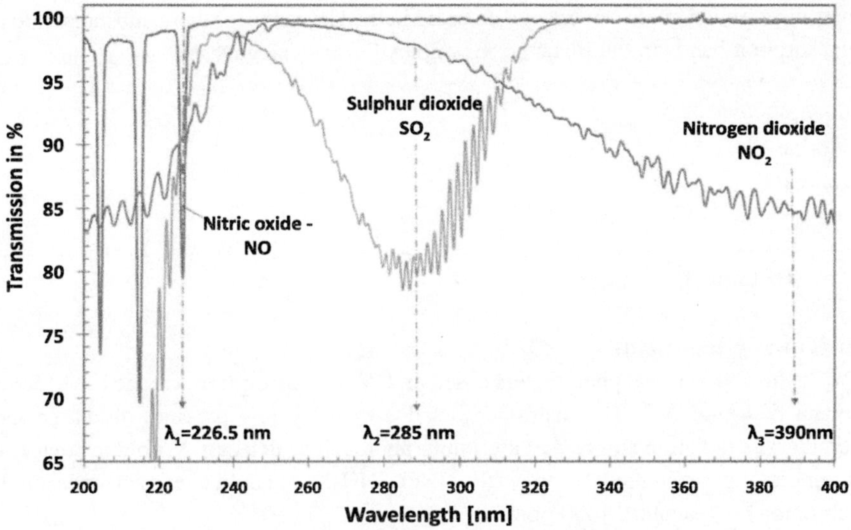

Fig. 8.2 UV spectrum of important gases occurring in emission measurement

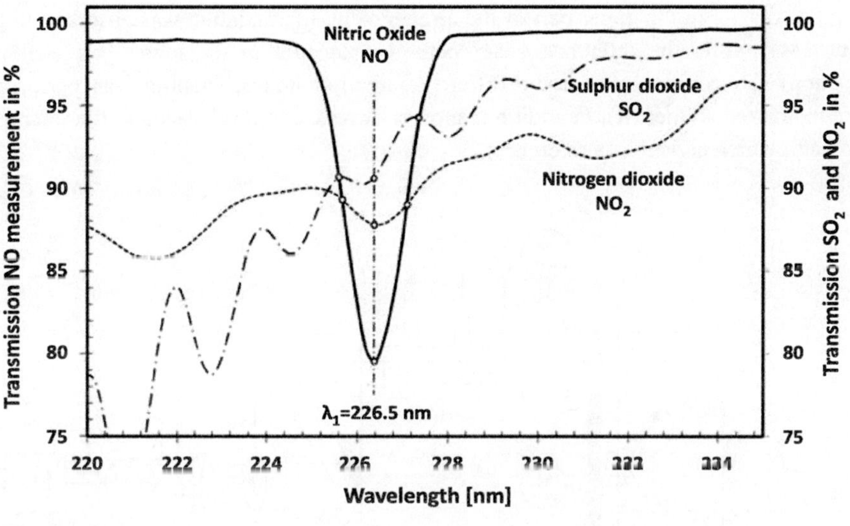

Fig. 8.3 Detailed view of the superposition of the NO absorption band by sulfur dioxide SO_2 and nitrogen dioxide NO_2

- Sulfur dioxide SO_2
- Ammonia NH_3
- Nitrogen monoxide NO
- Nitrogen dioxide NO_2
- Ozone O_3.

Furthermore, certain gases can only be detected in the UV range, as these substances do not have absorption bands in the IR range. Examples for this category are:

- Mercury Hg
- Chlorine Cl_2
- Bromine Br.

8.2 UV Gas Analyzers

LIMAS (first Generation)

One of the first operational photometers based on UV-VIS absorption was the LIMAS from Hartmann & Braun AG. The instrument used a mercury low pressure discharge lamp (Hg emitter) as radiation source and a vacuum photocell as detector. With this device, the most important gases in the UV range (SO_2, NO_2, NH_3, ...) could already be measured in concentration ranges under 1000 ppm.

Figure 8.4 shows the optical path of the LIMAS. The radiation of the Hg emitter was bundled via a lens onto a beam splitter. The beam splitting or combining was carried out with reflectors S_2. In the upper part of the arrangement, the radiation was absorbed in the cuvette (*M*), while the reference beam path was located in the lower part (without absorption). To compensate for the different radiation fluxes, shading was performed with a motorized shutter (*R*). In addition, another cuvette could be placed in the reference beam path, which served as a reference.

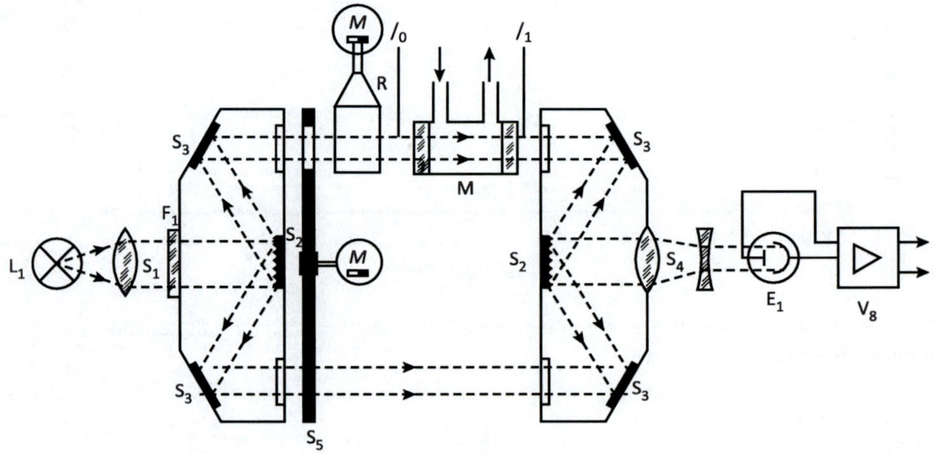

Fig. 8.4 Optical beam path of LIMAS 1 with an Hg emitter (L_1), measuring cuvette (*M*), beam splitter/combiner (S_2), detector (E_1) and amplifier (V_8)

However, due to the instability of the device and the limited lifetime of the radiation source, it found only a limited application range compared to the IR photometers (e.g. URAS).

RADAS

Meinel achieved his breakthrough in the field of UV gas measurement technology in 1972, with a completely new process for resonance absorption of nitrogen monoxide (NO). He used a discharge lamp filled with air (N_2 and O_2) in which the excited oxygen and nitrogen molecules react to form NO. The following reactions are observed in the discharge zone:

$$N_2 + O_2 \xrightarrow{\text{Excitation}} 2\,NO^* \tag{8.1}$$

$$NO^* \xrightarrow{\text{Radiation emission}} NO + h\nu \tag{8.2}$$

$$2\,NO \xrightarrow{\text{Recombination}} N_2 + O_2 \tag{8.3}$$

The excited NO^*, after the transition to the ground state, in turn emits NO specific resonance radiation at $\lambda_R = 204.4$ nm, 214.6 nm and 226.5 nm. The radiation emission of the $\gamma(0/0)$ band at 226.5 nm proved to be most suitable for NO absorption measurement. The structure of this emission band consists of many transitions, not all of which contribute to radiation absorption. Only the radiation from the transitions between the ground levels $(0, 0)$ can be absorbed in the measuring cuvette. Other transitions, which are assigned to a higher thermal temperature due to excitation (occupation distribution) in the discharge plasma, cannot therefore be absorbed at room temperature. In this context, one also speaks of so-called *hot* and *cold* bands. The occupation distribution of the rotational levels of the NO corresponds to a temperature of $T = 1540$ K (Meinel 1975).

Figure 8.5 shows the emission spectrum of NO at $\lambda \approx 226.5$ nm as a fine structure. The hot and cold lines are interlocked like a comb and can only be separated with high-

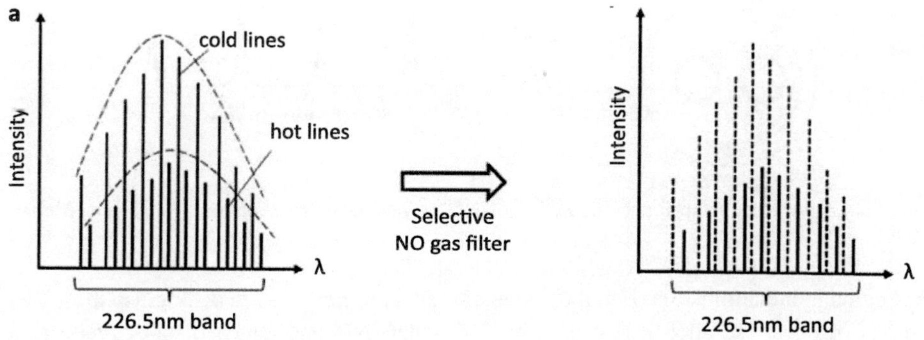

Fig. 8.5 Superposition of the *hot* and *cold* lines in the emission spectrum (**a**) and absorption of the cold lines by the selective gas filter (NO filling)

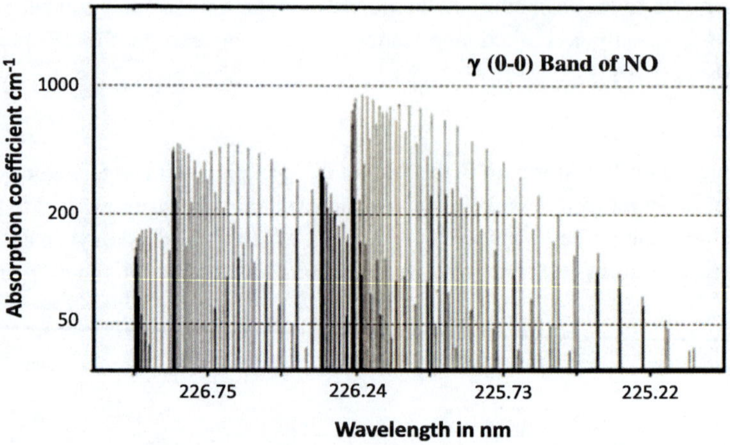

Fig. 8.6 Absorption spectrum $\gamma(0/0)$ of the band of NO at 226.5 nm, with a large number of rotation lines. The line widths $\Delta\lambda$ lie between 0.0005 and 0.002 nm (Meinel 1975)

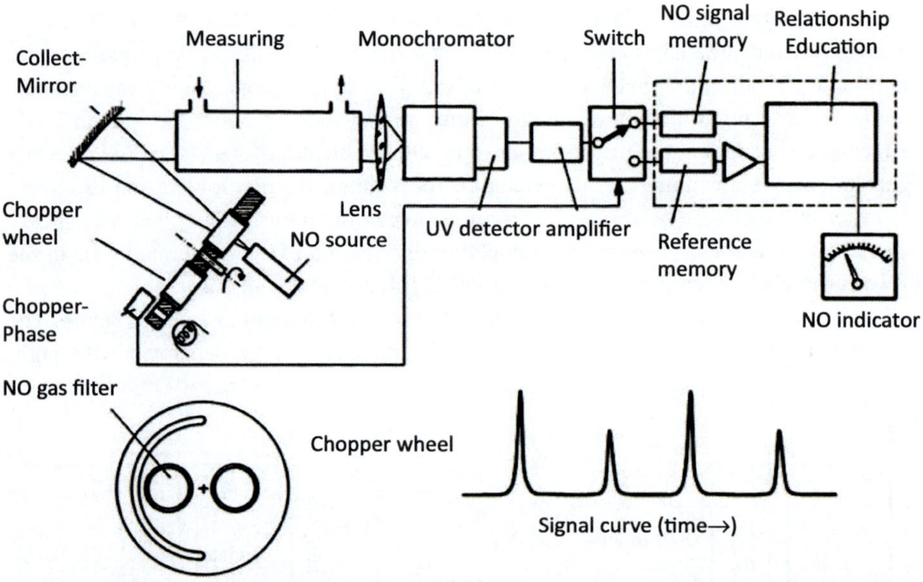

Fig. 8.7 Meinel test setup for the detection of NO by resonance absorption at 226.5 nm (Meinel 1975)

resolution monochromators (Fig. 8.6). Meinel used a different way to distinguish these two groups of lines. He filled a cuvette with 100 vol.% NO and attached this cuvette to a chopper wheel, which was moved alternately into the beam path (see Fig. 8.7).

As early as 1978, Hartmann & Braun AG presented the first commercial analytical instrument (RADAS 1) on this physical basis (Zöchbauer et al. 1978). The selective NO radiation source consisted of a hollow cathode lamp (HCL) filled with an oxygen/nitrogen mixture at a pressure of $p \approx 5$ mbar (operating voltage 300 V, current 1–5 mA). Since the oxygen encounters the cathode material (metal) in the discharge zone, the oxygen filling is used up during the operating time, so that the lifetime is greatly reduced. The HCL therefore had a reservoir of approximately 1 L to achieve a lifetime of at least 1 year. Furthermore, the lamp was only switched on for the short moment of the measurement, when the filter wheel was in the corresponding position, to further increase the service life (Fig. 8.8). The radiation was split by a partially transparent mirror (beam splitter) in front of the measuring cuvette (MC) and reached a reference detector (E_1). Depending on the filter wheel position, there were two signals that were measured one after the other. The measuring detector (E_2), at the end of the cuvette, showed a similar signal pattern, whereby the radiation of one peak (I_{M2}) could be changed by the NO gas to be measured (Fig. 8.9). So-called *photomultipliers,* which have an internal physical amplification, were used as detectors.

The signals of the two detectors were offset against each other by a double quotient formation. Since a total of 4 different signals are processed (see Eq. 8.4), this is also referred to as a *four-channel method.* This method has several advantages, since theoretically the drift of the detectors, ageing of the radiation source and contamination of the cuvette are compensated.

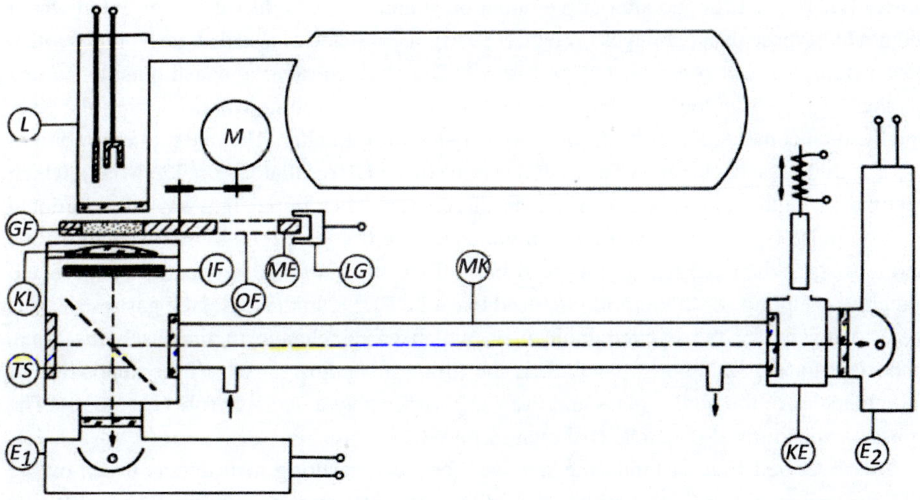

Fig. 8.8 Structure of the RADAS 1. GF = gas filter, BS = beam splitter, L = radiation source (HKL), IF = interference filter, LG = light barrier, ME = modulation unit, KL = collimator lens, KE = calibration unit, OF = opening (Zöchbauer 1979)

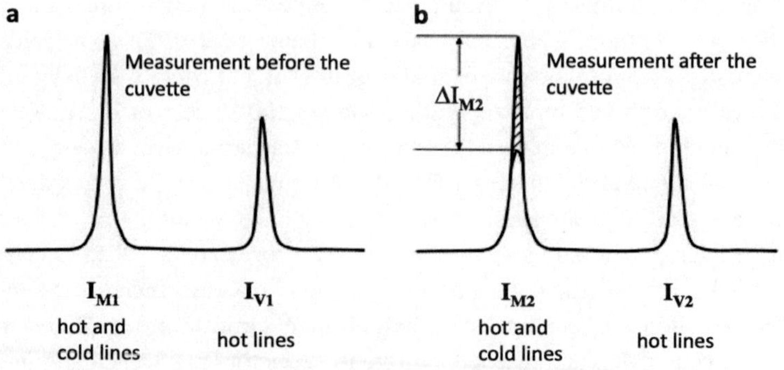

Fig. 8.9 Signal characteristics of the two detectors E_1 (**a**) and E_2 (**b**)

$$A = \frac{I_{V2}/I_{V1} - I_{M2}/I_{M1}}{I_{V2}/I_{V1}} \tag{8.4}$$

For a regular check, a calibration unit (CU) was additionally integrated, which pushes a cuvette with a defined NO concentration into the beam path, and thus checks the calculation and recalibrates it if necessary. The method has proven itself particularly in emission measurement technology, for the detection of low NO concentrations (measuring range <100 ppm) in large power plants and in the automotive industry (exhaust gas analysis). The weak point of this measuring technique was the service life and size of the radiation source (HCL). In 1986, an alternative radiation source was developed and patented which did not have these disadvantages (Wiegleb 1986). It consisted of a sealed gas volume with a considerably smaller reservoir ($V_1 \approx 30\,\text{cm}^3$). The electrodes were placed outside, so that no chemical reaction took place inside the UV lamp. The lifetime was well over one year, and the total construction volume was also considerably smaller. The external excitation of this electrodeless discharge lamp (EDL[1]) is done by an HF oscillator ($f \approx 125$ MHz), which generates a high field strength in the discharge zone. For initial ignition, an additional ignition pulse (spark) is applied, which generates the first electrons in the discharge zone and thus starts the external excitation. A basic distinction is made between capacitive and inductive coupling. Both methods are used to excite EDLs. In this case, the gas reservoir is used to regenerate the excited molecules, which can exchange in the discharge zone. Important for the lifetime of the EDL is the filling procedure, in which the deposition of O_2 molecules in the quartz glass and the water vapor play a decisive role (Fig. 8.10). The emission spectrum of the EDL is shown in Fig. 8.11.

The described EDL is today the standard for gas measuring instruments based on the NO resonance absorption method and allows the integration into small portable gas

[1] EDL = **E**lectrodeless **D**ischarge **L**amp.

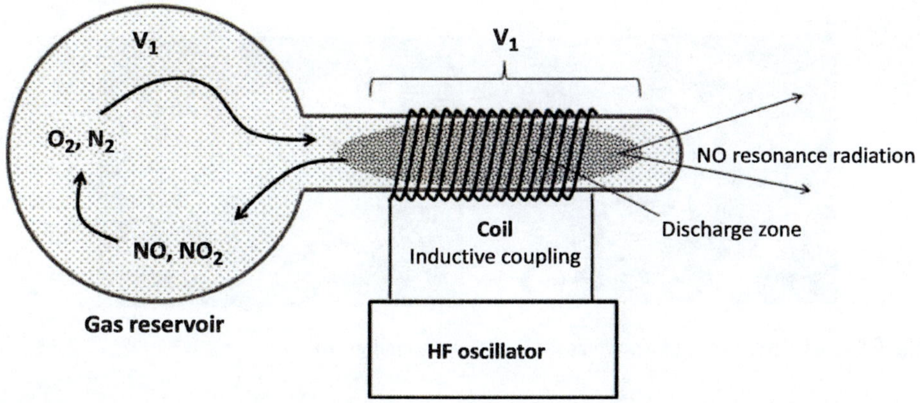

Fig. 8.10 Electrodeless discharge lamp (EDL) for generating NO resonance radiation with inductive coupling of electrical energy. (Wiegleb 1986)

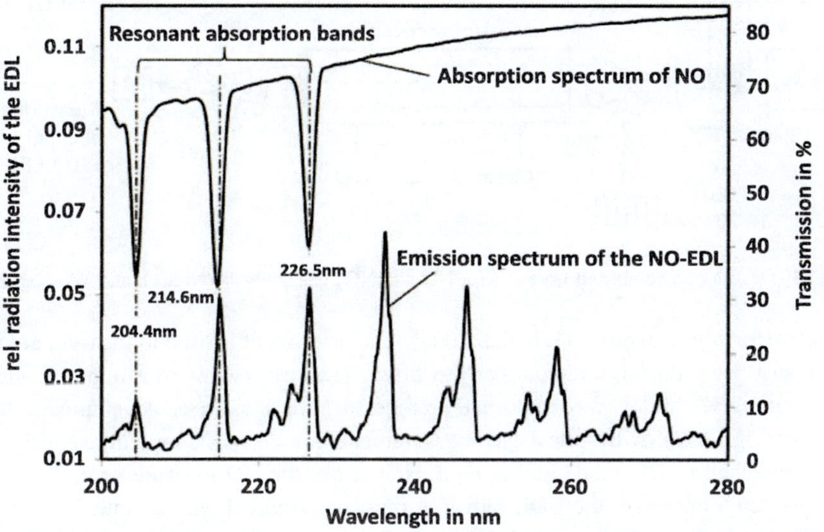

Fig. 8.11 Emission spectrum of the EDL in coincidence with the three NO absorption bands at 204.4 nm, 214.6 nm, 226.5 nm

analyzers. The complete structure of the EDL unit is shown in Fig. 8.12. The outer dimensions are approx. 9 cm × 8 cm × 4 cm.

UVRAS NO*x* Gas Analyzer (ULTRA.sens)

On this basis, a NO_x analyzer was developed for use in portable emission measuring instruments, which for the first time also made it possible to measure NO and NO_2

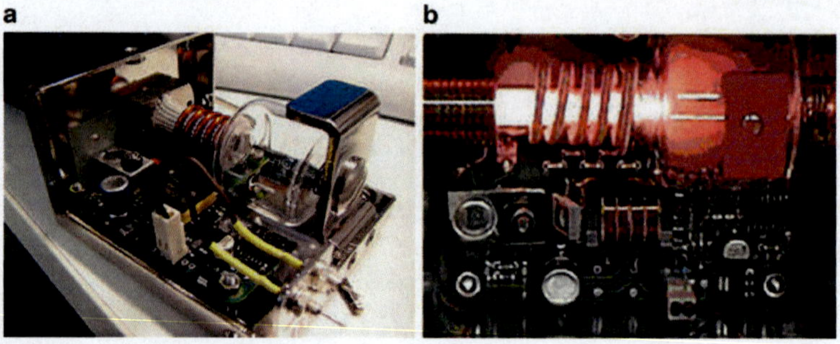

Fig. 8.12 EDL with control electronics (**a**). EDL in operation (**b**)

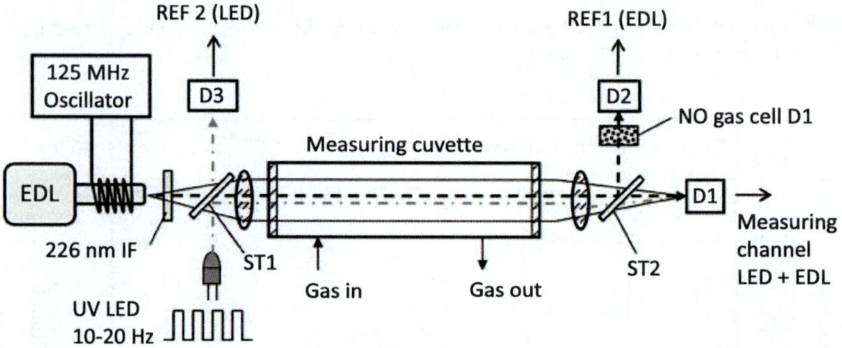

Fig. 8.13 Design of a combined UVRAS for NO and NO_2 gas analysis (Nietsch and Wiegleb 2003)

concentration simultaneously (Wiegleb 2004). The instrument is used to measure automobile exhaust gases during a regular car test drive. The size, power consumption, and the insensitivity to vibrations were important prerequisites for this device development. In the setup (see Fig. 8.13), two optical gas measurements are performed simultaneously. The EDL is switched in DC mode and is used exclusively for NO measurement. To narrow down the 226.5 nm band, the EDL radiation must be provided with an interference filter (IF) so that only this radiation enters the measuring cuvette. The radiation, at the exit of the cuvette, is split into two paths via a beam splitter (ST2) and directed to two separate photodiodes (detector D_1 and D_2). In front of detector D_2, a gas cell with a 100 vol.% NO filling is placed, which serves to absorb the selective radiation (cold lines). By this measure, a reference for the EDL radiation at 226.5 nm is obtained.

In addition to the EDL radiation, the radiation of a UV LED ($\lambda \approx 390$ nm) is coupled in via the beam splitter BS1, which is intended for NO_2 measurement. However, to distinguish between EDL and LED radiation, the LED must be modulated with a frequency of $f = 10 - 20$ Hz. The resulting signal curve at the measuring detector is shown in Fig. 8.14.

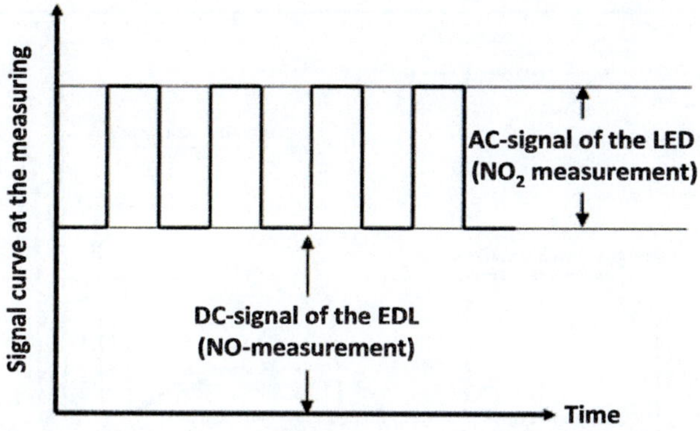

Fig. 8.14 Signal curve on the measuring detector for EDL (DC) and LED (AC) radiation

In this case, the gas concentration c is also calculated via modulation (*Mod*). The measured and reference values are determined before each measurement with zero gas and stored for normalization ($D_1^*(\mathrm{DC}), D_1^*(\mathrm{AC}), D_2^*, D_3^*$). The modulation of the NO and NO_2 measurement is done by the following equations:

$$\mathrm{Mod}_{\mathrm{NO}} = \frac{D_2/D_2^* - D_1(\mathrm{DC})/D_1^*(\mathrm{DC})}{D_2/D_2^*} \tag{8.5}$$

$$\mathrm{Mod}_{\mathrm{NO_2}} = \frac{D_3/D_3^* - D_1(\mathrm{AC})/D_1^*(\mathrm{AC})}{D_3/D_3^*} \tag{8.6}$$

The detection limits for NO and NO_2 are below 1 ppm, with a response time of $t_{90}<2$–3 seconds. The method is therefore an alternative to the costly CLD analyzers, which can also only be used for stationary measurements.

Figure 8.15 shows a comparison measurement between a CLD and a UVRAS device. First, both devices were exposed to zero gas (air) and then to a test gas (≈400 ppm NO). The NO_2 channel showed a NO_2 concentration at this transition, since the NO_2 test gas reacted with the atmospheric oxygen to form NO. After the measuring cuvette was purged, the NO_2 value dropped to zero. The zero point was then calibrated with nitrogen. The exhaust gas measurement, in which only NO was present, showed a clear correlation between the two measuring methods. Above 600 ppm NO, however, the CLD analyzer went into the limit, while the UVRAS analyzer continued to provide plausible measured values. Figure 8.16 shows a compact UVRAS module (ULTRA.sens®), for measuring up to 3 gas components (NO, NO_2, SO_2).

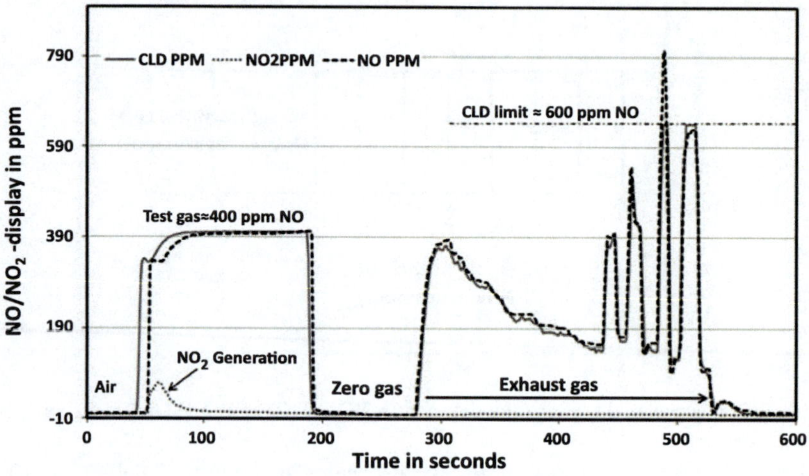

Fig. 8.15 Comparison measurement between a chemiluminescence analyzer (CLD) and a UVRAS NO_x analyzer (*dashed* curve)

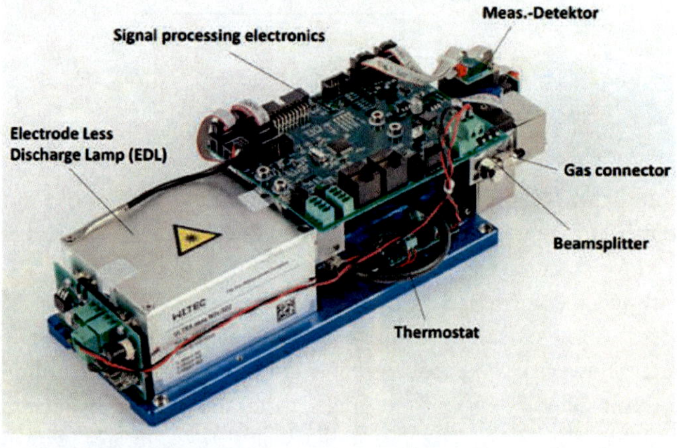

Fig. 8.16 ULTRA.sens using the UVRAS method for simultaneous gas analysis of NO, NO_2 and SO_2 in exhaust gases (automotive and CEMS). This method also allows the measurement of H_2S in biogas in the ppm range. (Source: https://www.witec-sensorik.de/en/, Wesel Germany)

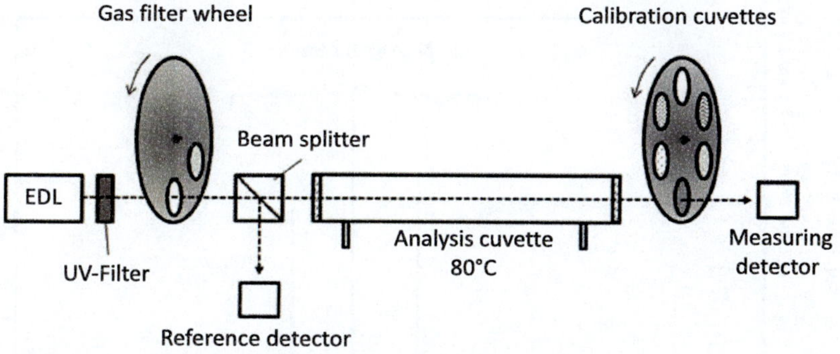

Fig. 8.17 Basic structure of the LIMAS 11 UV

A further development of the RADAS technology, based on the EDL described above, also led to a competitive measurement technology[2] for stationary exhaust gas measurements in the automotive industry (Baronick et al. 2001; Leippe et al. 2004).

LIMAS 11 UV (2nd Generation)
The new generation of devices, as successor to the RADAS process, is marketed today by ABB Automation GmbH (Frankfurt am Main) as *LIMAS 11UV.* The design has remained like the RADAS method (Fig. 8.17). The HCLwas replaced by an EDL and instead of the photomultipliers, Si photodiodes are now used as radiation receivers. The filter wheels can be equipped with different combinations, so that several gases can be measured simultaneously. The EDL also provides emission lines of corresponding strength in other spectral ranges, which can be used for gas analytical measurements (Fig. 8.18).

Application Example
For the control of SCR[3] catalysts, it is necessary to measure the ammonia (NH_3) content in addition to the NO_x measurement. Since the EDL also has emission lines in the spectral range of the NH_3 absorption bands, the method is therefore also suitable for this analysis. Figure 8.19 shows the absorption spectrum of 500 ppm NH_3 in a 10 cm cuvette. There is only a coincidence at 204.5 nm. This band is therefore very suitable for measuring the NH_3 concentration. In a 10 cm cuvette, the radiation is already >70% absorbed by the NH_3 at 204.5 nm. Detection limits for NH_3 from 1 ppm are therefore unproblematic.

The structure of the LIMAS 11 (UV module+slide-in housing) is shown in Fig. 8.20. The compact design is very well suited for integration into more complex measuring systems.

[2]DUV-RAS from ABB Analytical Frankfurt am Main.
[3]SCR =Selective Catalytic Reduction.

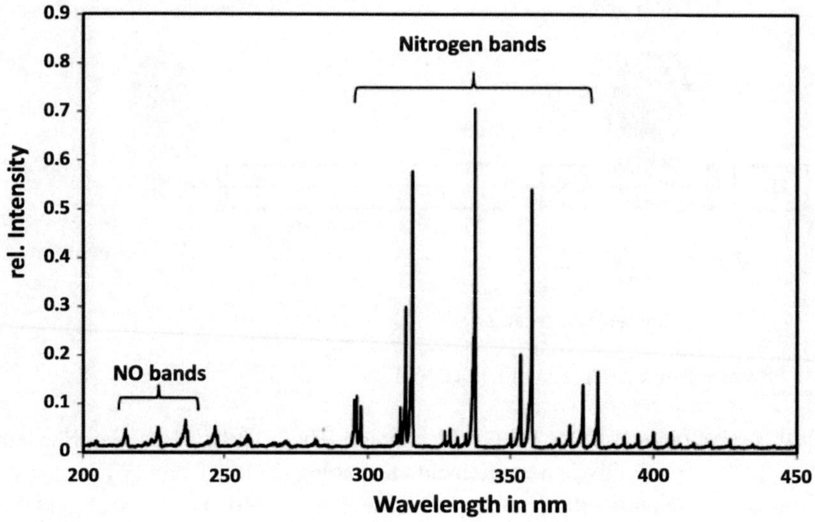

Fig. 8.18 Total emission spectrum of NO-EDL

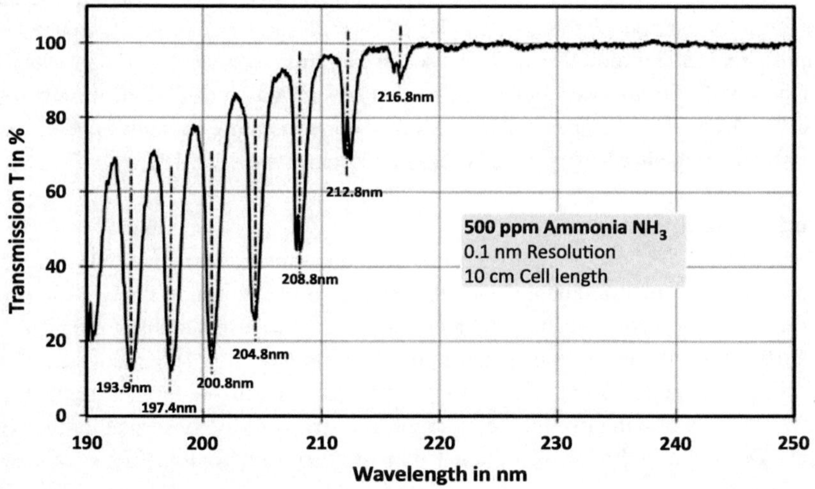

Fig. 8.19 UV absorption spectrum of ammonia

DEFOR (NDUV/UVRAS)

In the meantime, the method has also been adopted by the company Sick-Maihak. The physical structure is shown in Fig. 8.21. A second filter wheel has been integrated for extended application possibilities, with which additional gas components can be detected.

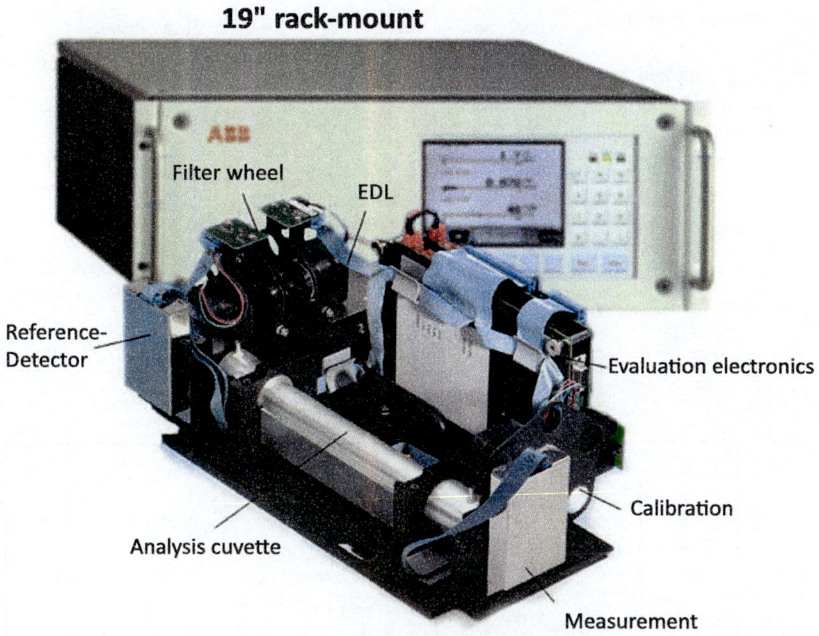

Fig. 8.20 LIMAS 11 UV module (*foreground*) and *in the background* as 19″ -rack-mounted housing. (Source: ABB Automation GmbH, Frankfurt am Main)

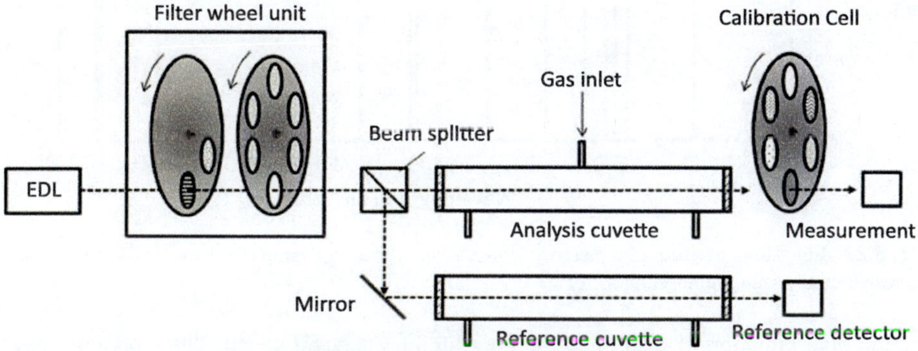

Fig. 8.21 UV photometer of the company Sick-Maihak Hamburg

In this case, the reference measurement is only carried out after passing through a separate reference cuvette. The possible measuring ranges of the DEFOR are listed in Table 8.1.

As an alternative to using an EDL, mercury low pressure discharge lamps can also be used as radiation sources for UV photometers. Advantageous with these radiation sources

Table 8.1 Possible measuring ranges of the DEFOR. Measuring ranges marked with (*) refer to a daily calibration and temperature changes of max. 2 °C (air-conditioned rooms)

Gas type	Measuring ranges
Cl_2	0 ... 125 ppm / 0 ... 100 vol.%
COS	0 ... 250 ppm / 0 ... 100 vol.%
CS_2	0 ... 50 ppm / 0 ... 30 vol.%
H_2S	0 ... 25 ppm / 0 ... 100 vol.%
NH_3	0 ... 50 ppm / 0 ... 100 vol.%
NO	0 ... 10 ppm / 0 ... 100 vol.%
NO_2	0 ... 50 ppm / 0 ... 100 vol.%
NO_2	0 ... 10 ppm / 0 ... 100 vol.%
SO_2	0 ... 25 ppm / 0 ... 100 vol.%
SO_2	0 ... 10 ppm / 0 ... 100 vol.%

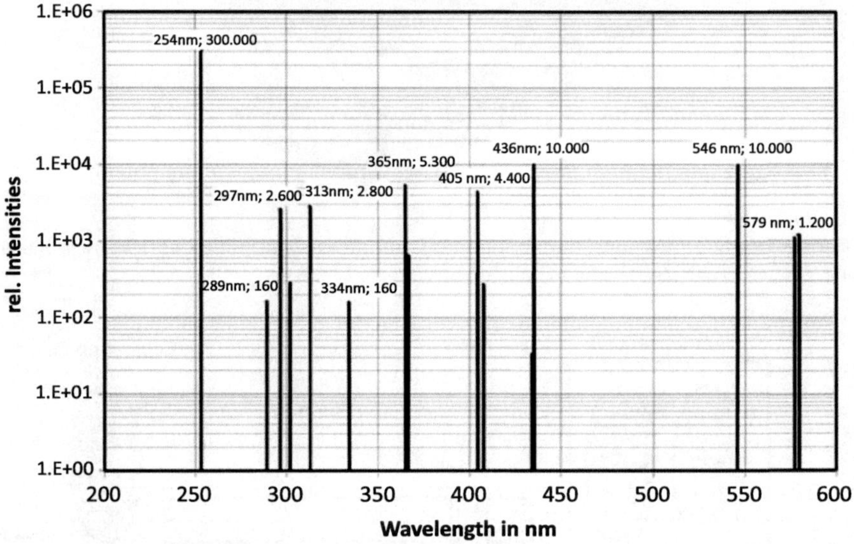

Fig. 8.22 Emission spectrum of a mercury low-pressure discharge lamp (PenRay®) in a logarithmic representation of the radiation intensity

are the high radiation intensity and the long lifetime (several years). The emission lines, however, lie in the spectral range $\lambda > 254$ nm, with the 253.7 nm line (exactly $\lambda = 253.6521$ nm) dominating the spectrum in terms of intensity (Fig. 8.22) (Table 8.2).

The 253.7 nm line is particularly suitable for mercury detection, as it is also a resonance absorption. In this case it is called atomic absorption, as mercury (Hg) is present in atomic form and does not form a molecule.

Table 8.2 Relative intensities of the emission lines, relative to the 436 nm line (=10, 000) of a PenRay Hg lamp

Wavelength in nm	Rel. intensity
253.6521	300,000
289.3601	160
296.7283	2600
302.1504	280
312.5674	2800
313.1555	1900
313.1844	2800
334.1484	160
365.0158	5300
365.4842	970
366.2887	110
366.3284	650
404.6565	4400
407.7837	270
434.7506	34
435.8335	**10,000**
546.0750	10,000
576.9610	1100
579.0670	1200

Source: Sansonetti et al. (1996)

Ozone Measurement

An important gas measurement, which is carried out at $\lambda = 253.7$ nm, concerns ozone (O_3). The determination of the ozone concentration in the ambient air is of great interest, as this gas is used as an indicator for the evaluation of the so-called *summer smog*. Ozone has a very broad absorption spectrum (continuum) in the UV range between 200 and 300 nm with a maximum at $\lambda \approx 256$ nm (Fig. 8.23). This band is also called the *Hartley band*. The absorption coefficient is very high and is $\alpha = 308.32\text{cm}^{-1}$, for the Hg line at $\lambda = 253.7$ nm (Griggs 1968). This wavelength is therefore very well suited for ozone measurement in ambient air, because due to the high absorption coefficient even very small concentrations $c \gg 1$ ppm can be measured. Furthermore, the defined absorption coefficient[4] also allows an absolute calibration of ozone measuring instruments, since according to the *Lambert-Beer law*, at a known pressure p and temperature T, the intensity ratios $I(c)/I_0$ can be used to infer the concentration c (Becker et al. 1975; Schurath and Wendler 1975).

Ozone also has weaker absorption bands in other spectral ranges. The *Huggins band* ranges from 300 to 360 nm, with an absorption coefficient of $\alpha = 0.15\text{cm}^{-1}$ at $\lambda = 334.15$

[4]According to the *National Bureau of Standards* (USA), the absorption coefficient of O_3 is given as $k = 133.9 \pm 1.9\text{atm}^{-1}\text{cm}^{-1}$, for base 10.

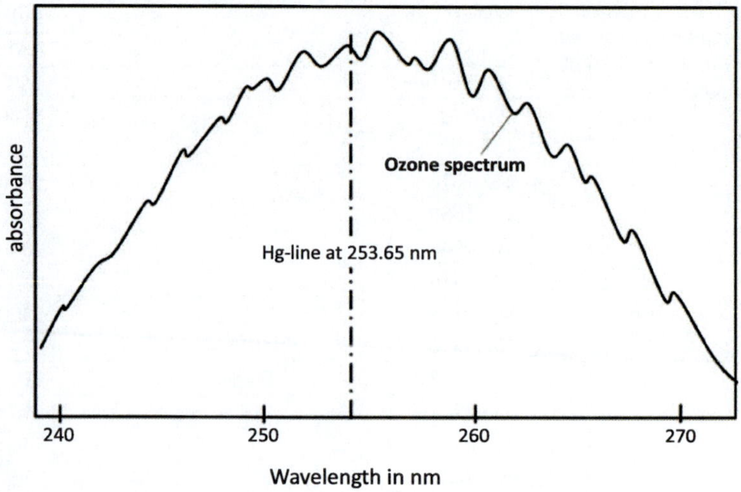

Fig. 8.23 Absorption spectrum of ozone in the spectral range between 240 nm and 270 nm with a maximum at $\lambda_{max} \approx 256$ nm

nm. The weakest ozone band (*Chappuis band*) extends into the visible spectral range of 440–850 nm, with an absorption coefficient of $\alpha = 0.127 \text{cm}^{-1}$ at $\lambda \approx 577$ nm (Griggs 1968). In this spectral range, ozone, for example, can be measured in high concentration ranges (volume %) using an LED as radiation source (Wiegleb 1985). High concentration ranges of ozone play an important role in monitoring ozone generators.

Fluorescence Emitters

Since the Hg radiation mainly occurs at 254.7 nm, this radiation is limited to a few gas measurements. By using phosphors, however, this radiation can be transformed into the longer wavelength spectral range, so that other gases can then also be measured.

Figure 8.24 shows such a radiation source, which was developed especially for gas measurement technology. The Hg radiation is excited by internal electrodes (3, 4) and emits it in all directions. The lamp is additionally coated with a fluorescent layer (5) which is excited by the 253.7 nm radiation and emits a corresponding fluorescence spectrum (secondary radiation). In the direction of observation, an opening (6) is left free which is not covered by the fluorescence layer. This measure makes the thickness *d of* the fluorescence layer uncritical, since only the emission of the rear layer is detected. Different fluorescent substances can thus be used to cover different spectral ranges for gas analysis purposes (Fig. 8.25). The radiation is coupled out with a lens (8), which ensures optimum utilization of the radiation energy.

The temperature dependency of UV radiation is of particular importance for the usability of the instruments, since in practical application the measuring instruments are

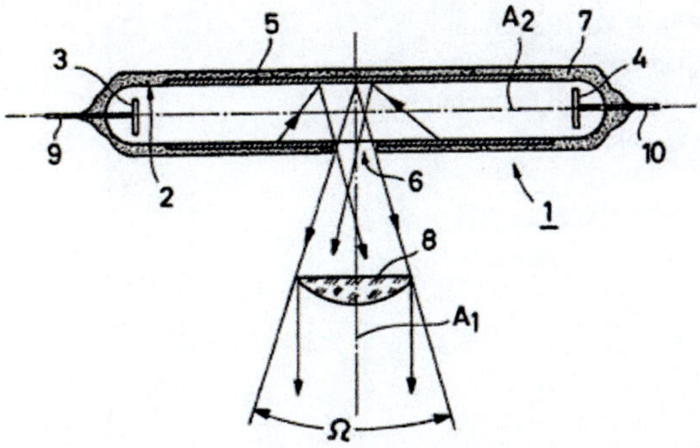

Fig. 8.24 Mercury low-pressure lamps with an additional fluorescent layer (5) for radiation emission at $\lambda \approx 285$ nm. (Source: Patent specification DE 33 16 771, Wiegleb 1984)

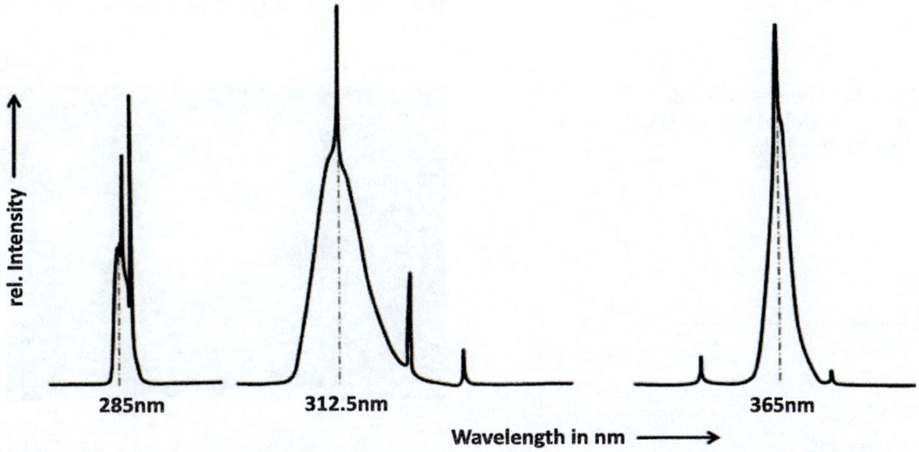

Fig. 8.25 Fluorescence radiation with different luminescent materials for different gases. 285 nm → Sulphur dioxide SO_2; 312.5 nm → Chlorine Cl_2; 365 nm → Nitrogen monoxide NO_2

subject to natural temperature changes (e.g. 0 to 45 °C). Furthermore, these measuring instruments are also mainly temperature controlled, so that the temperature is permanently set at 60 °C.

The intensity reaches a maximum at $T \approx 60$°C. Typically, the photometer assembly is therefore controlled at this temperature. For higher temperatures, there is a plateau between 100 °C and 150 °C where the radiation remains constant and only decreases further at temperatures $T > 60$°C (Wiegleb 1984) (Fig. 8.26). This change is mainly related to the Hg vapor pressure in the discharge zone, which naturally increases with rising temperature. Fluorescent lamps are also offered by various manufacturers for analytical instruments (Fig. 8.27) which are very compact and have a high radiation yield. The lifetime is stated by the manufacturers as approximately 8000 h, which refers to a drop to the 50% value of the initial intensity. In practical cases the life span is usually much longer.

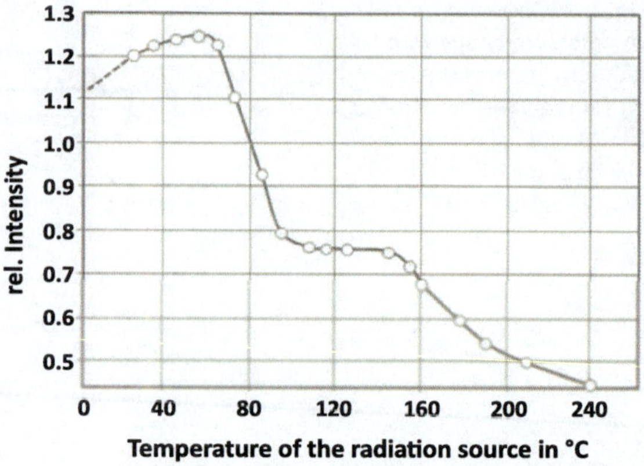

Fig. 8.26 Radiation distribution of the fluorescent lamp at high temperatures. (Wiegleb 1984)

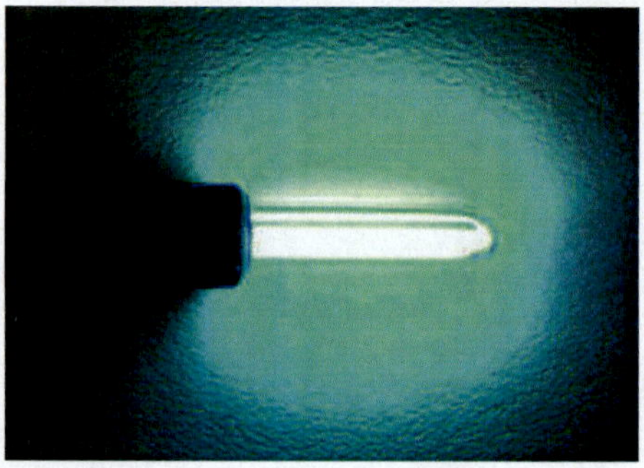

Fig. 8.27 Fluorescent lamp from Ultra-Violet Products USA of the PenRay type®

UV-BINOS

Based on this radiation source (Fig. 8.24), a gas analyzer (UV-BINOS) was constructed, with which sulfur dioxide (SO_2) could be measured in small concentrations (measuring range 100 ppm SO_2) for the first time (Wiegleb 1983). The measurement of small SO_2 concentrations became mandatory with the introduction of the Ordinance on Large Combustion Plants (13th BImSchV) in 1983, as the emission limits were continuously lowered (Baumbach 1990). The detection limit for this method was, with a cuvette length of 20 cm, at ≈0.1 ppm SO_2. The existing cross-sensitivity to NO_2 was compensated for with a second measuring channel in which the NO_2 concentration was measured simultaneously. A further advantage of UV measurement technology is the lack of cross-sensitivity to water vapor, which occurs mainly with IR measurement of SO_2.

The radiation from the fluorescent lamp was bundled in the instrument with a lens and passed through an interference filter into the analysis cuvette (Fig. 8.28). The cuvette is divided into two halves, one of which is filled with nitrogen and thus serves as a reference. The two halves are then illuminated (modulated) one after the other by a mechanical chopper wheel and converted into a voltage signal by the UV detector. The special features

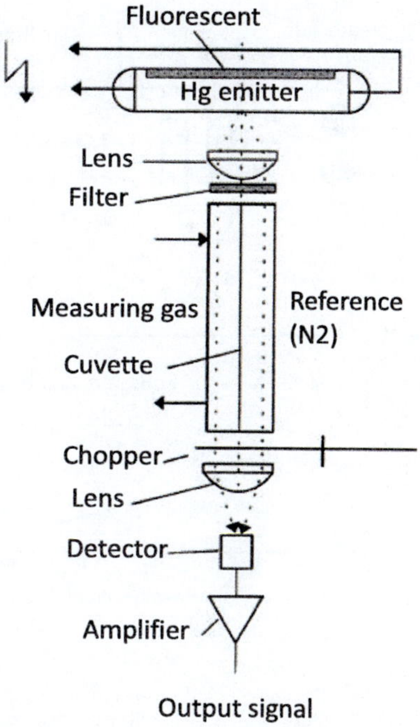

Fig. 8.28 UV-BINOS for the detection of SO_2 (Wiegleb 1983)

of radiation modulation have already been described in detail in Chap. 7 (BINOS method). The product is marketed today by Emerson-Rosemount under the label *X-Stream.*

8.3 New Measurement Methods

Diode Array Spectrometer

In the UV range, gas measuring instruments are also used, which use a spectral decomposition of UV radiation with dispersive elements (diffraction grating). These devices are usually based on miniaturized spectrometers that have an integrated diode array as a detector. Figure 8.29 shows such a setup. A broadband deuterium lamp[5] is used as the radiation source, which, depending on the manufacturer and design, has a spectral distribution from 200 nm to 400 nm. Therefore, the entire UV range can be covered with one radiation source. Alternatively, xenon flash lamps are also used as radiation sources.

The radiation of the D_2 lamp is bundled with a lens and then directed into a cuvette. There the absorption of the corresponding wavelength ranges takes place. With a second lens, the radiation is then focused on the entrance slit of a spectrometer. The holographic grating

[5]Deuterium (D_2) is an isotope of hydrogen.

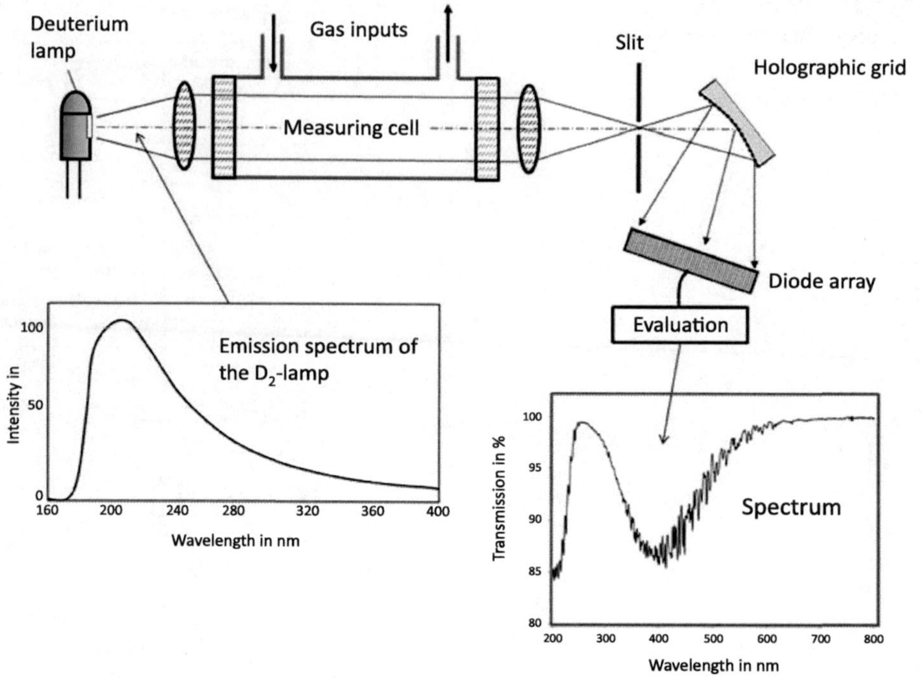

Fig. 8.29 Basic structure of a diode array spectrometer

spatially splits the entire spectrum of the radiation source and displays it on the diode array. The diode array consists of at least 256 individual receiving diodes, each of which receives a different wavelength λ of the bandwidth. For each detector element, these are $\lambda = 200\text{nm}/256 \approx 0.8$ nm. With this spectral resolution, most gas measurements in the UV range can be performed. Overlaps with other gas spectra must be compensated by a corresponding calculation (chemometrics).

An advantage of this spectroscopic technique is the flexible adaptation to the respective measuring task since the spectral selection is possible via software. Furthermore, this measurement technique is often used in combination with fiber optics. The optical fiber is constructed as a bundle to be able to transmit as much energy as possible. The radiation is usually coupled in via appropriate lenses to keep the losses as low as possible. An additional advantage of fiber optics is the separation between the gas measurement and the electronic units (radiation source, spectrometer, evaluation electronics, . . .).

Especially in process engineering, many areas where gas is used are designated as Ex-zones.[6] All electronic equipment must be able to demonstrate appropriate protective measures, otherwise operation in this zone is not permitted. If these devices are installed

[6]Hazardous areas where explosive gases may occur.

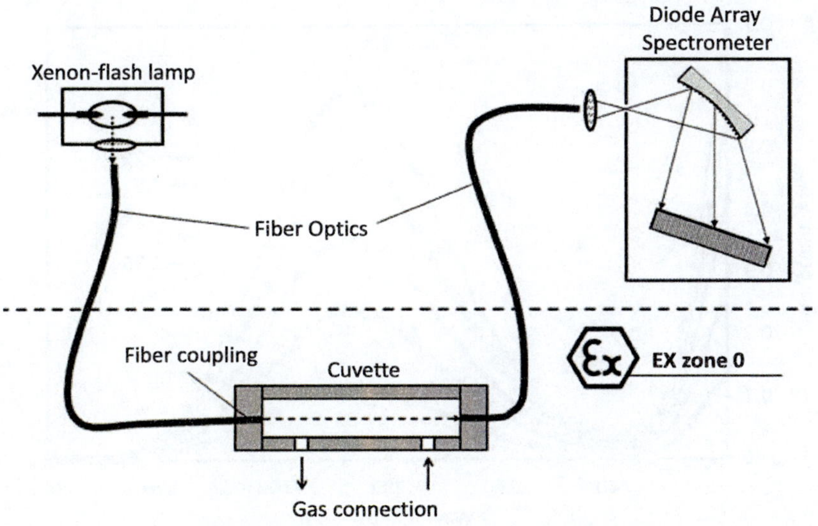

Fig. 8.30 Installation of a gas measurement based on a diode array spectrometer with fiber-optic transmission of the measuring radiation

outside the Ex-zone, these costly protective measures are completely unnecessary. Figure 8.30 shows such an installation for an Ex-application. The measurement is carried out in the cuvette and the radiation is transmitted through the fiber optics, which does not pose any danger in the Ex-zone.

Furthermore, the externally arranged cuvette can also be heated to high temperatures (e.g. 300 °C) or designed for high pressures (e.g. 200 bar). A typical application of UV spectroscopy is the simultaneous measurement of sulfur dioxide (SO_2) and hydrogen sulfide (H_2S) in natural gas processing plants. Since both components have a large overlap in the UV range, the use of UV spectroscopy is useful. Figure 8.31 shows the absorption spectrum of SO_2 with different concentrations. Each data point represents a specific wavelength that can be assigned to an element of the diode array. In this case, the spectrometer had a diode array consisting of 1024 elements (Barshad 2012). The SO_2 measurement can be recorded between 280 nm and 290 nm with almost no cross-sensitivity. The H_2S concentration is measured between 220 nm and 230 nm and must be corrected accordingly due to the overlap with SO_2.

LED Photometer ULTRA.sens

The first LED-based UV gas analyzers were introduced to the market in the 1980s, for measuring chlorine gas (Cl_2) and nitrogen dioxide (NO_2) (Wiegleb 1985). Further gas measurements were not possible at that time because no commercial LEDs were available at 400 nm. Due to further developments in the field of light-emitting diode technology, UV-LEDs are now also on the market, which emit a sufficient radiation intensity under

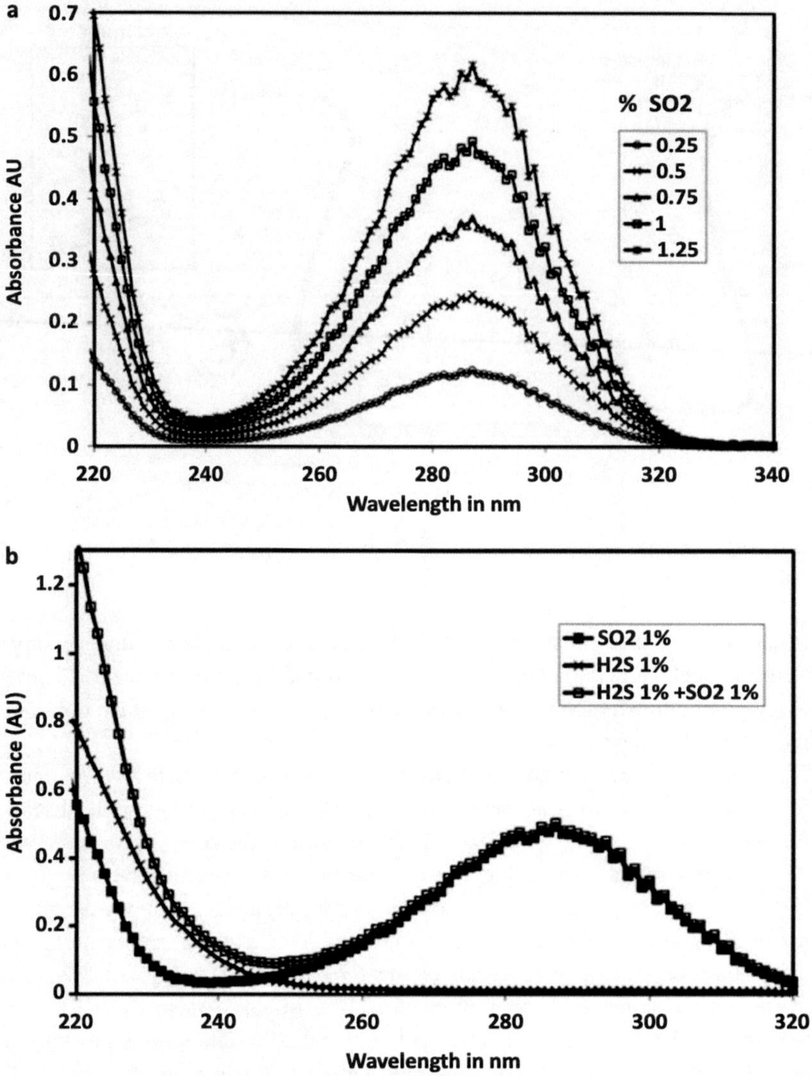

Fig. 8.31 SO_2 spectrum with different concentrations in the volume % range (**a**) compared to a superposition of the SO_2 spectrum with the H_2S spectrum (**b**)

300 nm. With these UV-LED's, very compact structures can be realized, which can also be designed to measure several components. Figure 8.32 shows such a measurement setup, which was developed for SO_2-analysis (Sridhavan 2012). The LED radiation source is modulated with a frequency $f = 100$ Hz and has an emission maximum of $\lambda = 290$ nm. DC operation is also possible. Via a beam splitter a part of the LED radiation reaches a

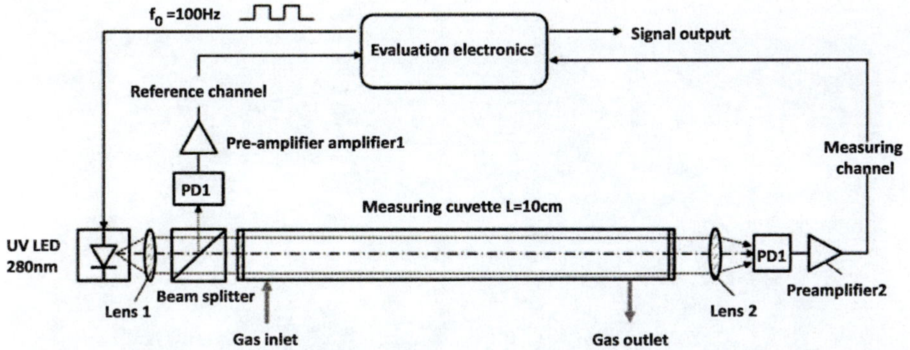

Fig. 8.32 Optical path of the UV LED photometer *ULTRA.sens* https://www.witec-sensorik.de/en/products/ultra-sens/

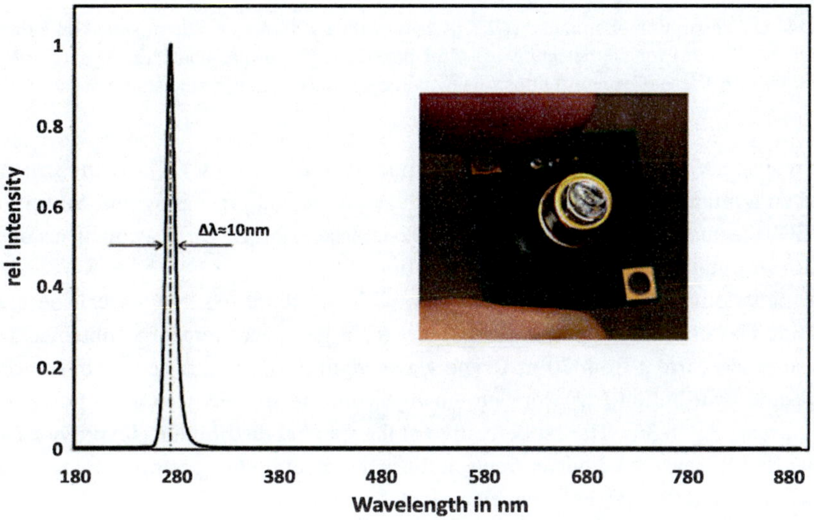

Fig. 8.33 Emission spectrum of a UV LED. (Source: Wi.Tec-Sensorik GmbH Wesel Germany)

reference detector (PD1). The remaining radiation is directed into the measuring cuvette and then focused on detector PD2.

A major advantage of UV LED technology is the narrow band emission of UV radiation. Figure 8.33 shows the emission spectrum of this UV LED. The bandwidth is only $\lambda \approx 10$ nm. Additional optical filtering is therefore completely unnecessary. The electrical power P, which is required for the safe operation of the UV LED, is $P_{\text{elek}} \approx 100$ mW. Due to the small size and low power consumption, very compact designs

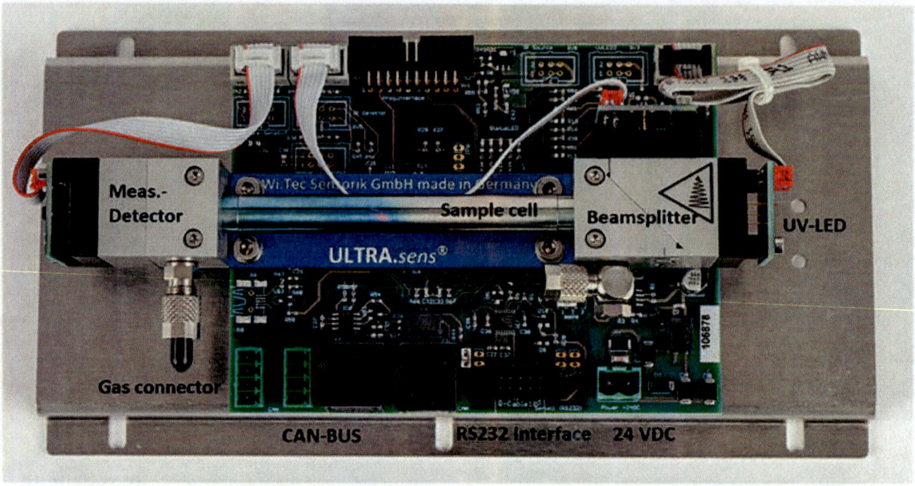

Fig. 8.34 UV absorption photometer (*ULTRA.sens*) with a 100 mm cuvette mounted on a Eurocard (100 mm × 160 mm) for the detection of sulfur dioxide (SO_2) in gas analyzers. (Source: Wi.Tec-Sensorik GmbH, Wesel Germany) https://www.witec-sensorik.de/en/products/ultra-sens/

can be realized. Figure 8.34 shows one such module (*ULTRA.sens*[7]). The entire structure is located on a printed circuit board on which the entire signal processing and control of the UV-LED is located. A CAN-interface, RS 232-interface and USB are available as outputs. A 4–20 mA output is also available as an option.

The radiation intensity increases linearly with the diode current, as can be seen in Fig. 8.35. The optimum signal-to-noise ratio for a gas measurement is obtained with a specified diode current from 20 mA. The wavelength is λ_0 not affected by this variation. The spectral distribution I (λ) does not change with the supplied electrical power or the diode current (Fig. 8.36). The 1st derivative of the spectral distribution ($I\lambda$) provides a zero crossing for all measured diode currents at the same point in the spectrum. In this case, the zero crossing is at the specified wavelength of $\lambda_0 \approx 275$ nm.

An important quality feature is the noise signal at zero point. The detection limit (NG) was determined by 3 times the standard deviation at zero point and was at NG $\approx$0.5 ppm SO_2. With a moving average over 30 data points, this value was then reduced to 0.33 ppm SO_2 (Fig. 8.37). Another limitation of LED technology is the service life of the UV LED, which is specified by the manufacturer as $\approx$1000 h. An endurance test showed that the service life is indeed at 4000 h. The service life of the UV LED in the *ULTRA.sens*

[7] www.witec-sensorik.de

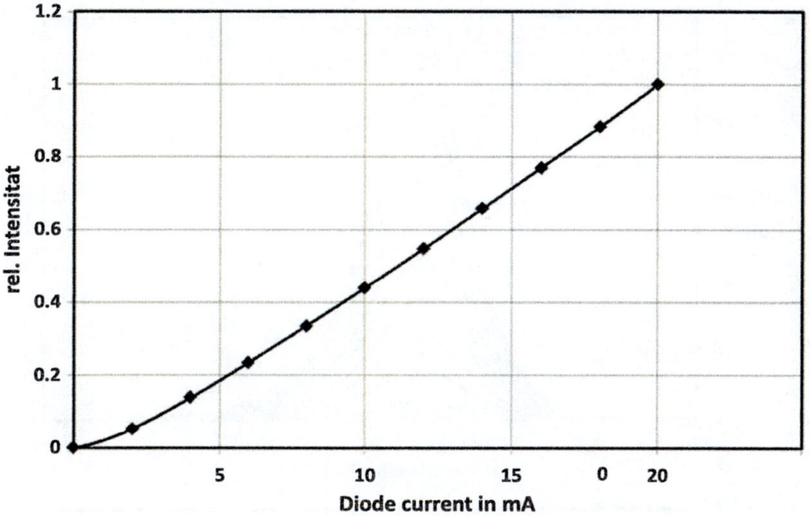

Fig. 8.35 Intensity change depending on the diode current of the UV LED

module is specified as >20, 000 h, which is achieved by intelligent pulsing of the UV LED (Fig. 8.38).

Optical Hydrogen Sensor
The physical principle of an optical hydrogen sensor was already described by Butler in 1994. A corresponding sensor arrangement was patented in 2007 (Wienecke et al. 2007). The transmission through a very thin palladium layer ($d \approx 20$ nm) changes considerably when molecular hydrogen (H_2) penetrates this palladium layer. The hydrogen then accumulates in this layer and partly undergoes chemical reactions, forming palladium hydride (PdH_x). This intercalation causes a change in the electronic structure around the Fermi edge, which leads to a change in the optical transmission properties. Figure 8.39 shows the principle structure of such an H_2 sensor. Figure 8.40 shows the spectral changes caused by hydrogen intercalation.

The commercially available H_2 sensor (type MOHS 2)[8] is designed for a gas concentration of 0.1 to 4 vol.% hydrogen. Due to the time-dependent adsorption processes, the time constant is approximately 60 seconds. The manufacturer specifies a lifetime of 2 years.

[8]Materion GmbH Wismar.

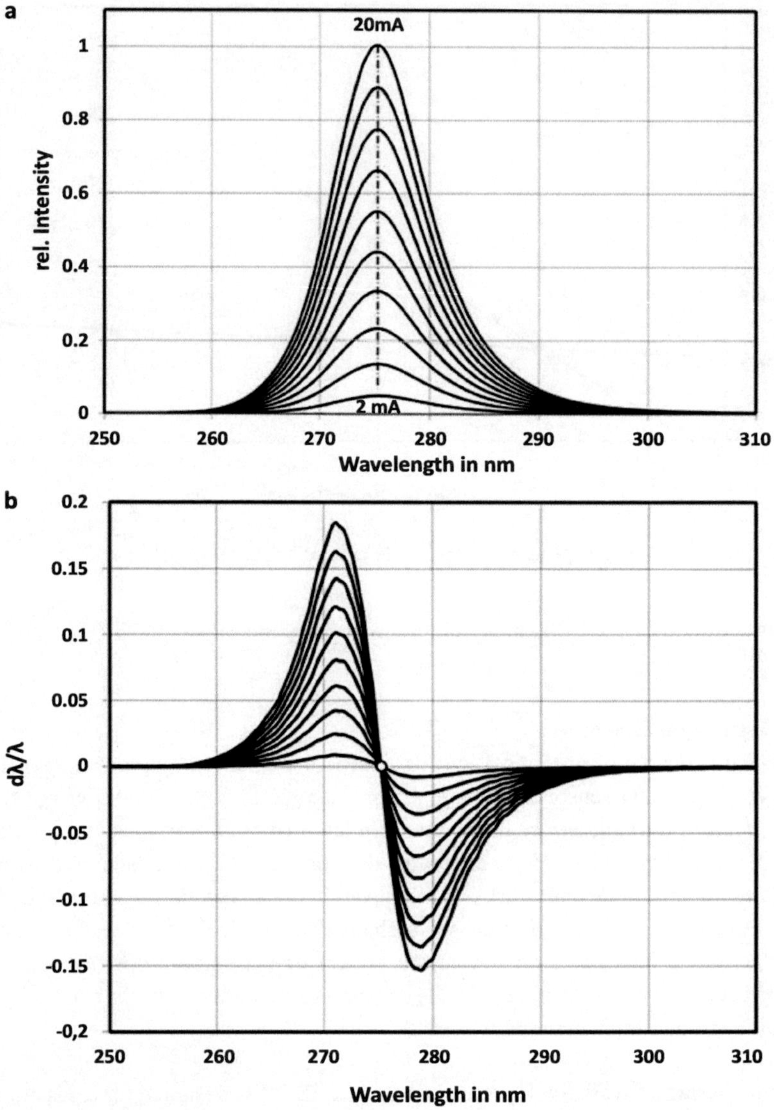

Fig. 8.36 Spectral distribution of LED radiation at different diode currents (**a**). 1. derivation of the intensity distribution $I(\lambda)$ for different diode currents (**b**)

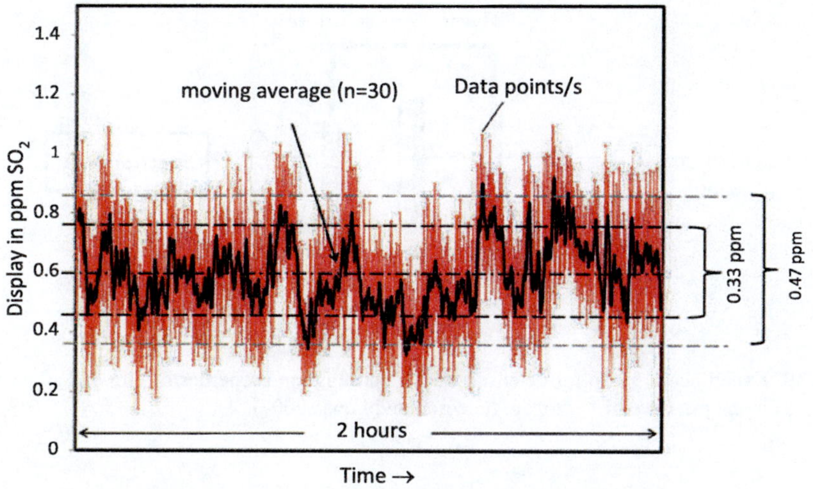

Fig. 8.37 Specification of the zero-point noise as 3-fold standard deviation in ppm SO_2

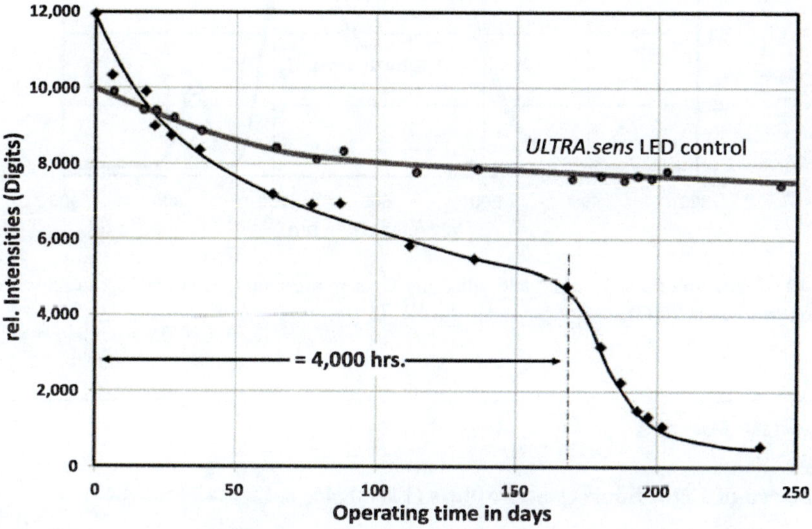

Fig. 8.38 Lifetime test of the UV LED at 40 °C and specified control compared to *ULTRA.sens* LED control

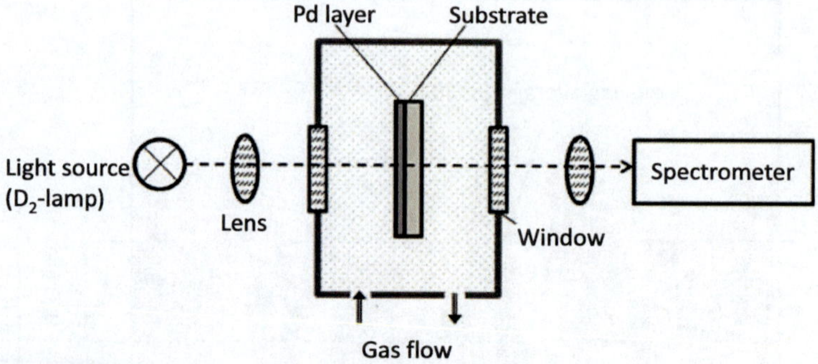

Fig. 8.39 Experimental setup for determining the transmission properties of thin palladium layers (8–13 nm) in the presence of hydrogen. (Gleeson and Lewis 2007)

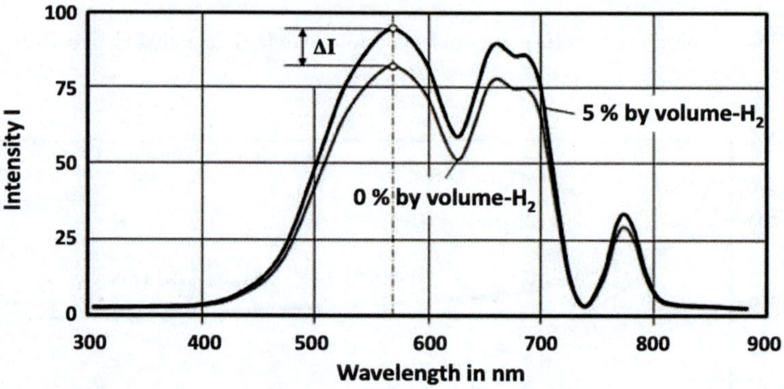

Fig. 8.40 Spectrum of a 8–13 nm thick palladium layer in nitrogen (=0 vol.% H_2) and 5 vol.% H_2. (Gleeson and Lewis 2007)

8.4 UV Spectra

UV Spectra of Light Sources and Filters (Figs. 8.41, 8.42, 8.43 and 8.44)

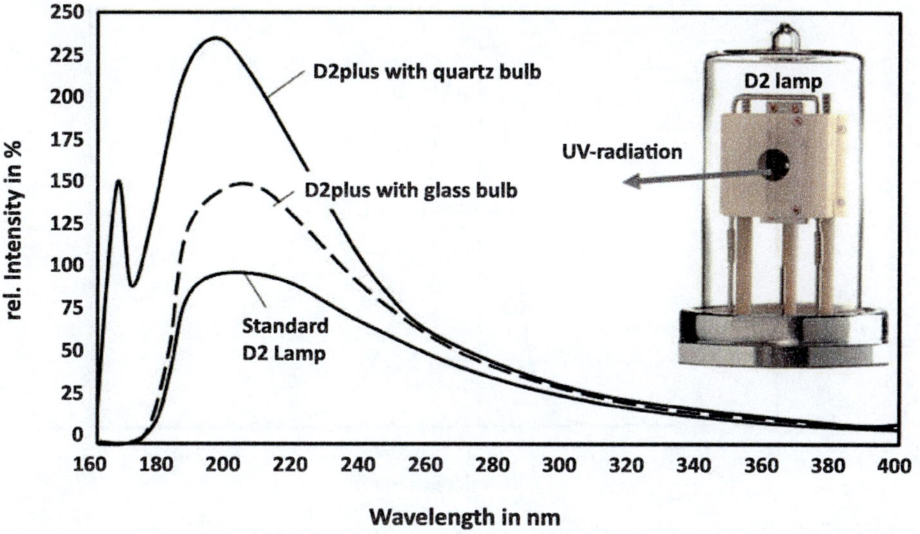

Fig. 8.41 Emission spectrum of different deuterium lamps. (According to Heraeus Nobelight, Hanau)

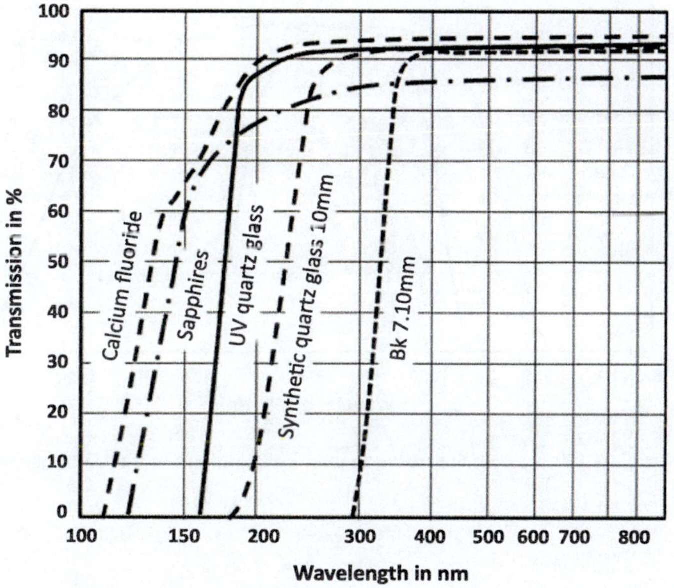

Fig. 8.42 Transmittance of different glass types in the UV range

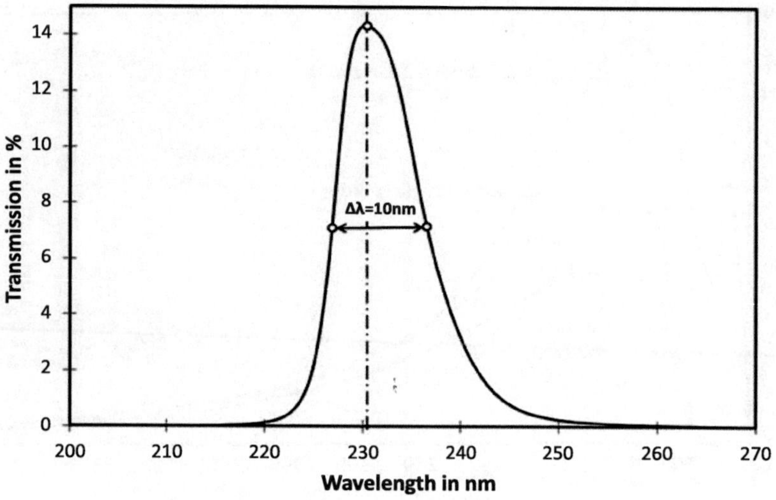

Fig. 8.43 Typical transmission curve of an interference filter in the UV range

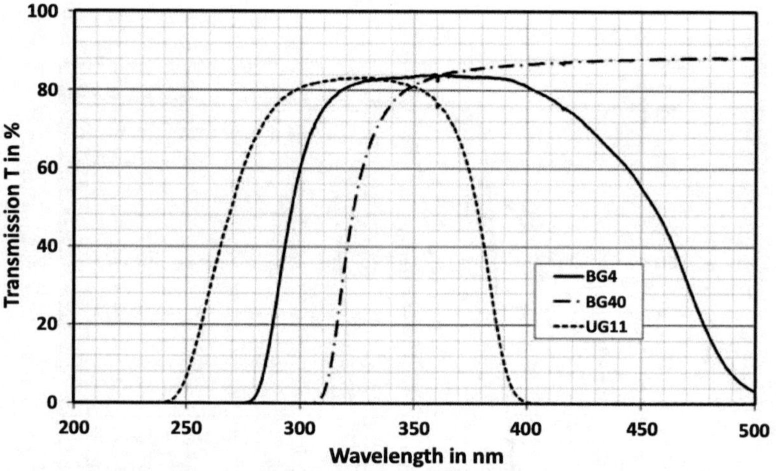

Fig. 8.44 Color glass filters as an alternative to interference filters in the UV range. (According to Schott)

UV Gas Spectra (Figs. 8.45, 8.46, 8.47, 8.48, 8.49 and 8.50)

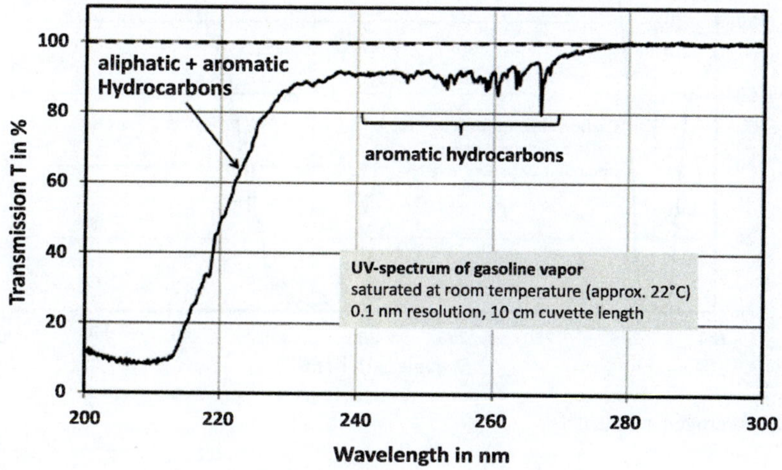

Fig. 8.45 Petrol vapor

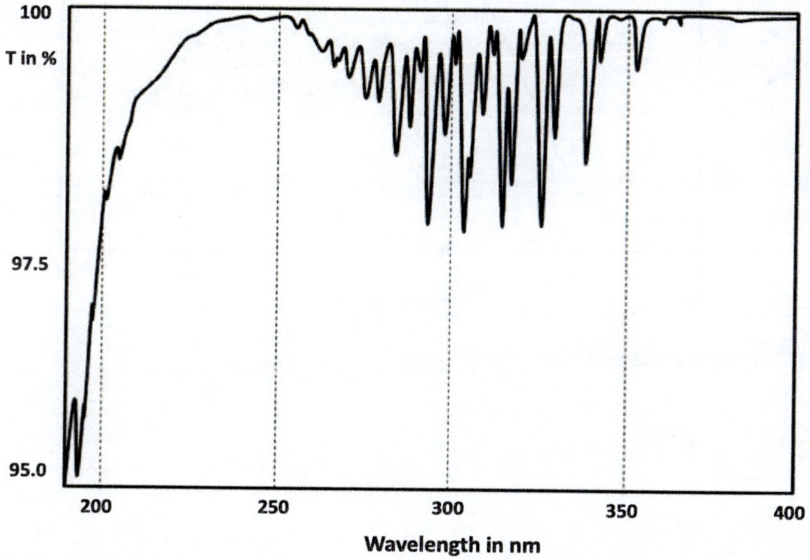

Fig. 8.46 Formaldehyde

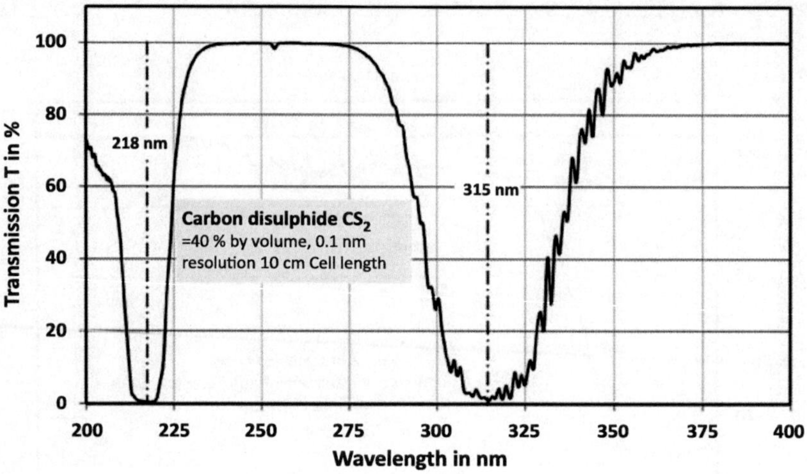

Fig. 8.47 Hydrogen sulfide CS_2

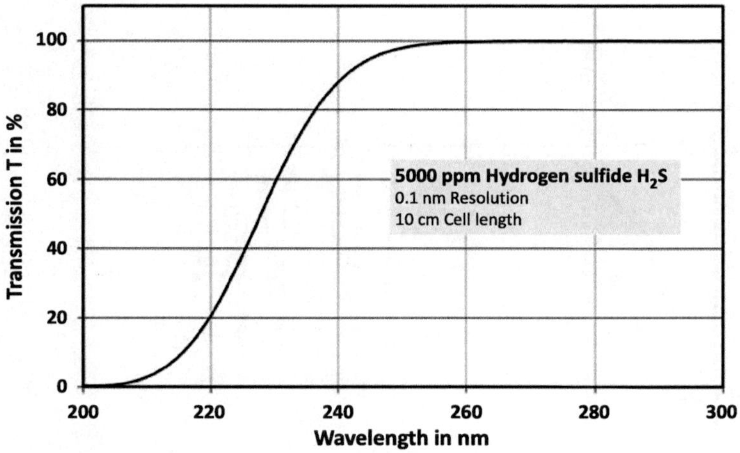

Fig. 8.48 Hydrogen sulfide H_2S

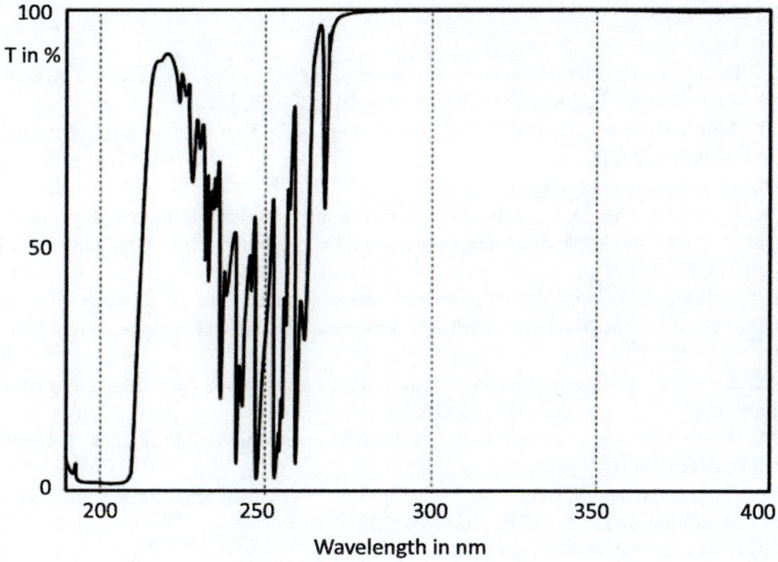

Fig. 8.49 Benzene vapor C_6H_6 saturated at19 °C

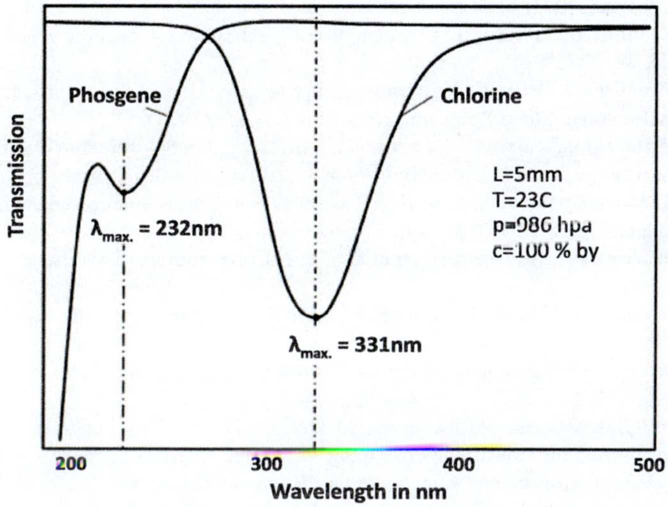

Fig. 8.50 Chlorine gas and phosgene

References

Baronick, J.D., et al.: Evaluation of an UV Analyzer for NO_x Vehicle Emission Measurement. Society of Automotive Engineers Inc. Paper 2001-01-0213 (2001)

Barshad, Y.: Sulfur Recovery Analyzers; UV Diode Fiber Optics Technology. Applied Analytics Inc. USA White Paper (2012)

Baumbach, G.: Luftreinhaltung. Springer (1990)

Becker, K.-H., Hendrichs, A., Schurath, U.: Ein transportables Gerät zur Kalibrierung von Ozonanalysatoren durch Messung der optischen Absorption. Staub-Reinhaltung der Luft. **35**, 326–329 (1975)

Butler, M.A.: Micro mirror optical fiber hydrogen sensor. Sens. Actuat. B. **22**, 155–163 (1994)

Down, R.D., Lehr, J.H.: Environmental Instrumentation and Analysis Handbook. Wiley-Interscience (2005)

Gleeson, K., Lewis, E.: Response changes of thin film palladium based optical fiber hydrogen sensor over time. J. Phys. Confer. Ser. **76**, 1–6 (2007)

Griggs, M.: Absorption coefficient of ozone in the ultraviolet and visible regions. J. Chem. Phys. **49**(2), 857–860 (1968)

Leippe, G., et al.: Neues System zur Messung von Stickstoffkomponenten im Einsatz von SCR-Katalysatoren. MTZ. **5**(2004), 392–399 (2004)

Meinel, H.: Deutsches Patent 2246365 (1972)

Meinel, H.: Detection of nitric oxide by resonance absorption technique. Zeitschrift für Naturforschung. **30a**, 323–328 (1975)

Nietsch, I., Wiegleb, G.: NO_x Gassensor auf der Basis der UV-Resonanzabsorption, 6. Dresdner Sensor-Symposium (2003)

Sansonetti, C.J., Salit, M.L., Reader, J.: Wavelengths of spectral lines in mercury pencil lamps. Appl. Optics. **35**(1), 74–77 (1996)

Schurath, U., Wendler, W.: Über die Verwendbarkeit von ozonhaltigem Sauerstoff zur Kalibrierung von Ozonanalysatoren. Staub-Reinhaltung der Luft. **35**, 329 (1975)

Sridhavan, S.: Entwicklung eines UV-Gassensors zur Messung von Schwefeldioxid (SO_2) Bachelor-Thesis an der Fachhochschule Dortmund/Wiegleb (2012)

Wiegleb, G.: Lichtquelle für nichtdispersive Gasanalysengeräte als Fluoreszenzlampe. Deutsche Patentanmeldung DE 33 16 771 A1 vom 7.5.1983 (1983)

Wiegleb, G.: Gasanalysensysteme für erhöhte Prozesstemperaturen. Technisches Messen. **51**, 385–393 (1984)

Wiegleb, G.: Einsatz von LED-Strahlungsquellen in Analysengeräten. Laser und Optoelektronik. **3**, 308–310 (1985)

Wiegleb, G.: Lampe für die Erzeugung von Gas-Resonanzstrahlung. Deutsches Patent DE 3617110 vom 21.5.1986 (1986)

Wiegleb, G.: UV-Gassensor zum Nachweis von Stickoxiden TRAFO Forschungsreport 2004 der AiF (Arbeitsgemeinschaft für industrielle Forschung – Otto von Guericke) (2004)

Wienecke, M., et al.: Hydrogen Sensor. Internationale Patentanmeldung WO 2007/121935 A1 (2007)

Zöchbauer, M.: Ein neues Betriebsmessgerät zum fotometrischen Nachweis von Stickoxid durch Resonanzabsorption. TIZ-Fachberichte. **10**, 638–642 (1979)

Zöchbauer, M., Fabinski, W., Staab, J.: Ein Betriebsfotometer nach dem Resonanzabsorptionsverfahren zur Messung von Stickoxid. Technisches Messen atm. **45**, 11–15 (1978)

Related References

Komhyr, W.D.: Operation-Handbook. Ozone Observation with a Dobson Spectrophotometer WMO/ TD-1469 (1980)

Nietsch, I., Wiegleb, G.: A Novel NO_x Gas Sensor Based on Resonance Absorption Technique in the UV-Range. SENSOR 2005 conference proceedings II B7.4 (2005)

Silva, S.F., Coelho, L., Frazao, O., Santos, J.L., Malcata, F.X.: A review of palladium based fiber optic sensors for molecular hydrogen detection. IEEE Sens. J. **12**(1), 93–102 (2012)

Wiegleb, G.: Gasanalysenvorrichtung (Filter-Korrelationsmessung) Deutsches Patent DE 3544015 vom 13.12.1985 (1985)

Wiegleb, G., Randow, A., Rößß, R.: SO_2-Messung mit dem neuen UV-BINOS. Technisches Messen. **50**, 143–150 (1983)

9 Radiation Emission and Laser Technology

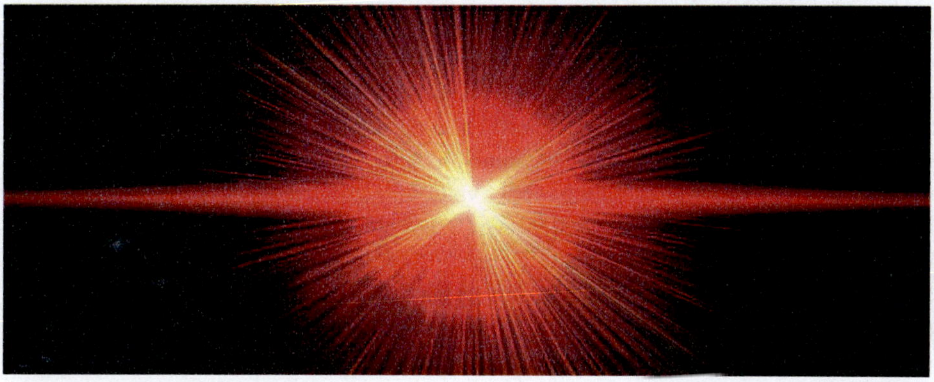

Scattering of laser light. (Source: fotolia)

9.1 Radiation Emission

Measurement methods based on the emission of radiation are also used very successfully in gas measurement technology. Compared to absorption photometers, or spectrometers, these methods work from the zero point. In absorption measurement, on the other hand, there is a high zero signal (I_{max}) at the zero point, which only changes by the amount ΔI caused by the gas concentration c to be measured (Fig. 9.1). For very small gas concentrations (ppm range) this change is also correspondingly small and must be subtracted from the high basic signal I_{max}. Since the basic signal changes due to external influences (e.g. temperature) or due to aging of the components (e.g. emitter), this change is

G. Wiegleb, *Gas Measurement Technology in Theory and Practice*,
https://doi.org/10.1007/978-3-658-37232-3_9

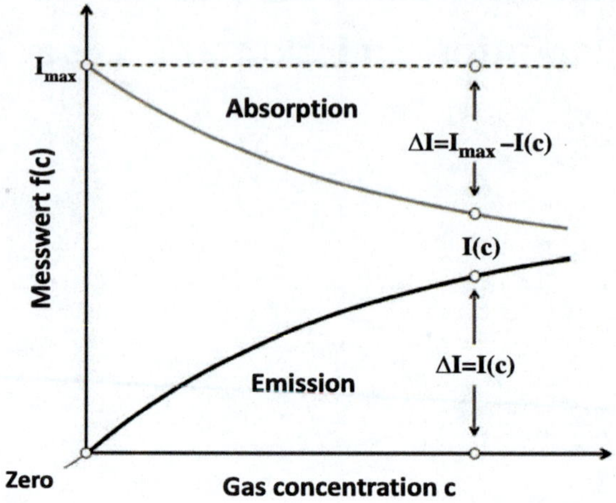

Fig. 9.1 Comparison of the methods according to radiation absorption and emission

also included in the signal evaluation. The external influences therefore lead to a limitation of these procedures for very small gas concentrations (< ppm range). Furthermore, the noise levels at high signals are also considerably higher due to the existing noise sources and thus represent an additional limitation. The amplification of the signals is also limited since the signals must not enter the electronic limitation. The maximum level of the signal voltage is usually determined by the supply voltage.

If, on the other hand, the gas measurement is considered in the emission, the measured values in the zero point are also zero. Depending on the method, however, small offset voltages may also be present at the zero point, but these are suppressed by suitable measures. The degree of signal amplification can therefore be selected much higher for this type of measurement than for absorption photometers. In general, the measuring methods that work based on radiation emission are much more sensitive, so that even smaller gas concentrations in the ppb or even ppt range can be measured.

Chemiluminescence Detector (CLD)

This measuring method uses the fluorescence radiation that is produced because of a chemical reaction. The chemiluminescence is thus a specific luminous phenomenon that is produced by the released chemical binding energy. It is proportional to the number of molecules in the measuring chamber and can therefore be used for quantitative gas analysis. To obtain a reproducible measurement signal, it is necessary to keep the external conditions, such as the pressure *p* in the measuring chamber, the temperature *T* and the volume flow (particles per time) constant. Mainly nitrogen monoxide (NO) is measured in the trace range with this method. In rare cases it is also possible to measure ozone (O_3) (van Heusden 1974; Baumbach 1990). A reaction partner is required for the reaction of the sample gas, which must be provided by a defined ozone concentration for NO measurement. In the measuring chamber several reactions take place simultaneously, which can be described as follows:

$$NO + O_3 \rightarrow NO_2^* + O_2 \tag{9.1}$$

$$NO + O_3 \rightarrow NO_2 + O_2 \tag{9.2}$$

$$NO_2^* \rightarrow NO_2 + h\nu \tag{9.3}$$

$$NO_2^* + M \rightarrow NO_2 + M \tag{9.4}$$

Altogether 4 different reaction equations can be represented, of which only one leads to the emission of radiation ($h\nu$). Furthermore, these reactions take place at different speeds. A constant proportion of approximately 10% leads to excited NO_2 molecules according to Eq. 9.1. 90% of the NO molecules produce NO_2 according to Eq. 9.2, but this is not excited. The transition of the excited NO_2 molecules is then associated with the emission of radiation (→ chemiluminescence). Furthermore, the excited NO_2 molecules can also collide with other molecules in the gas mixture. These collision partners M absorb the energy of the excited state and lead to a non-radiative transition of the NO_2 molecule to the ground state. The non-radiative energy release is called *quenching* and increases strongly with pressure (→ mean free path), since the probability of collision increases due to the higher gas density. CLD devices are therefore operated at an absolute pressure of $p = 5 - 25$ mbar.

The fluorescence radiation starts at $\lambda \approx 600$ nm and extends to $\lambda \approx 3.3$ μm (see Fig. 9.2). With an upstream bandpass filter, however, only a part of the emission spectrum is used. For practical reasons, a range from 600 to 900 nm has proven to be optimal for CLD measurements. The complete structure of a CLD measuring instrument is shown in Fig. 9.3. The ozone required for the measurement is generated by a so-called *silent discharge*, in which atmospheric oxygen (O_2) passes through a high electric field and in this way partially reacts to form O_3. Figure 9.4 shows a cross-section of such a structure. A high voltage in the kV range is connected between the inner and outer electrode. In the resulting electric field (E), an energy transfer takes place in which the oxygen molecules from the supplied air are converted into ozone. The ozone content is then approximately 2 vol.% and is therefore present in excess. Important for a constant ozone concentration is an air intake free of water vapor, as both ozone generation and chemiluminescence radiation are influenced by water vapor. The vacuum pump sucks both gas flows (ozone and sample gas), each via a capillary, through the measuring cell, where the reactions described above take place.

The chemiluminescence radiation is bundled with a collecting lens and focused on a detector. Photomultipliers or large-area photodiodes are usually used as detectors. The output voltage U_M, after the amplifier, increases linearly with the NO concentration and

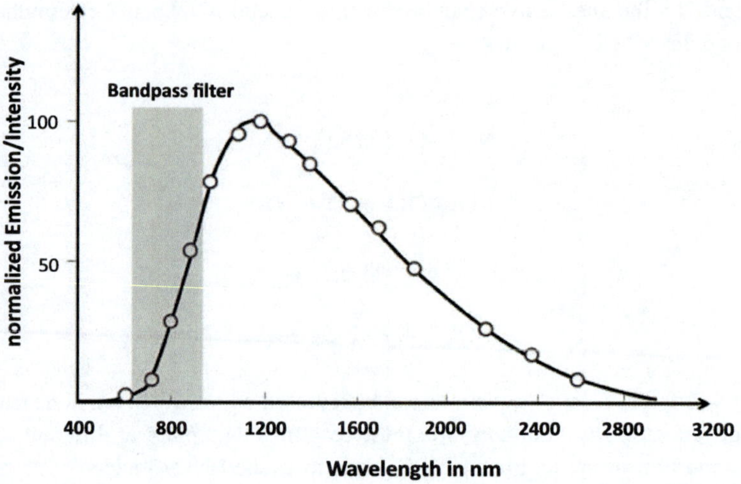

Fig. 9.2 Emission spectrum of chemiluminescence radiation during a reaction of nitrogen monoxide (NO) and ozone (O_3 in negative pressure operation (Clough and Trush 1967)

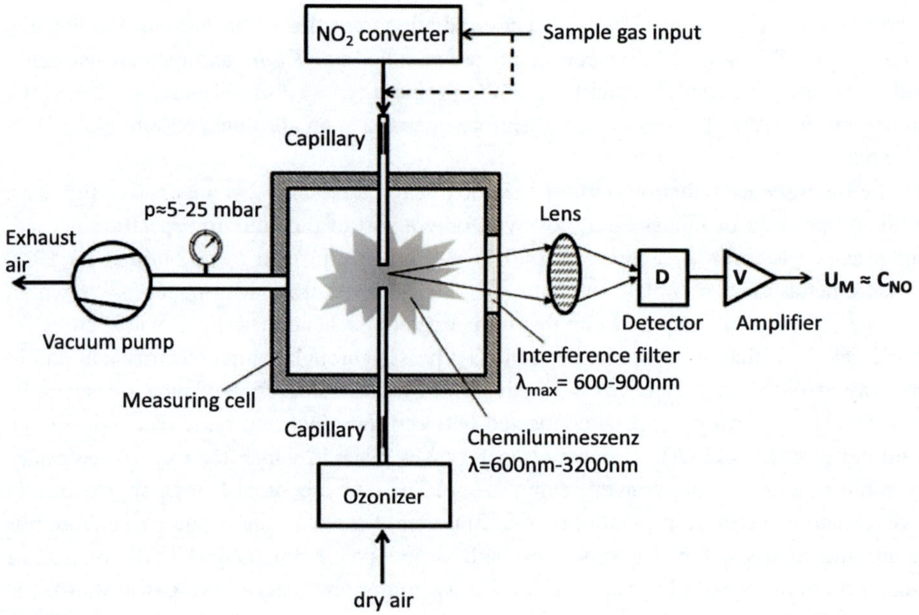

Fig. 9.3 Basic design of a CLD measuring instrument for the detection of nitrogen oxides (NO*x*)

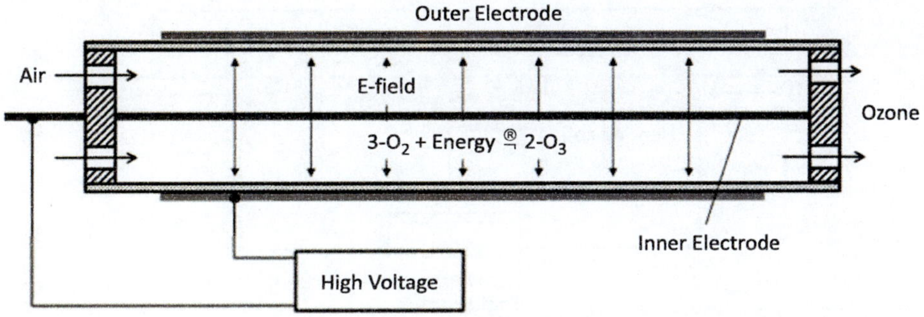

Fig. 9.4 Cross-section through an ozone generator using the silent discharge method

usually does not need to be linearized. Only for devices that operate under atmospheric pressure and measure high NO concentrations, additional linearization is required.

NO_2 measurement is not possible with this method, so the NO_2 must first be converted into NO by catalytic conversion:

$$NO_2 \xrightarrow{\text{Catalysis}} NO + \frac{1}{2} O_2 \tag{9.5}$$

Suitable catalysts are metal or metal oxide surfaces which are heated to temperatures $T_{Cat} > 300°C$. The degree of conversion is specified by the manufacturers of these devices as >96% for a new catalyst. Since the degree of conversion changes over the course of operation, it must be checked regularly. Figure 9.5 shows such a catalyst. A known material for this application is molybdenum (e.g. granules), which is placed in a stainless-steel tube. Around the stainless steel pipe there is a heating coil which provides the necessary temperature T_{Cat} by means of an electric current (→ power P). This temperature can then be measured by a temperature sensor (e.g. Pt 100) and controlled to a constant value.

An alternative to thermal catalysis is photolysis. Nitrogen dioxide is broken down into nitrogen monoxide and an oxygen atom (radical) under the influence of UV light at a wavelength of $\lambda < 410$ nm. This highly reactive oxygen can then form ozone with an oxygen molecule, which in turn reacts with the nitrogen monoxide to form NO_2. The following reactions are therefore possible (Schwarzenbach and Hüglin 2011):

$$NO_2 + h\nu \rightarrow NO + O^* \tag{9.6}$$

$$O^* + O_2 \rightarrow O_3 \tag{9.7}$$

$$NO + O_3 \rightarrow NO_2 + O_2 \tag{9.8}$$

In summary, the equations shown above result in the following chemical equilibrium:

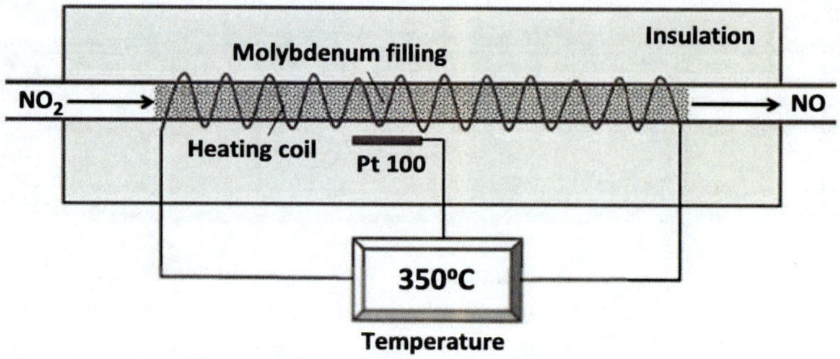

Fig. 9.5 Design of a catalytic converter for the conversion of NO_2 to NO

$$NO_2 + O_2 + h\nu \rightleftarrows NO + O_3 \tag{9.9}$$

Based on this conversion method, comparative measurements were carried out over a period of 6 years. Figure 9.6 shows the results between a conventional molybdenum converter (Mo) and a photolysis converter (*BLC – blue light converter*). The BLC type always showed slightly lower values than the molybdenum converter (Mo) in this series of measurements.

CLD instruments are mainly used in environmental monitoring networks to measure atmospheric NO and NO_2 concentrations. Typical measuring ranges for this application are 50 ppb to 25,000 ppb (25 ppm). The detection limits are 0.5 ppb (ECO Physics, model CLD 66).

Another important application of CLD devices is in medical technology. The measurement of exhaled nitrogen oxide is a simple, precise and direct method for determining respiratory tract inflammation. The detection limit for this application must be <0.1 ppb and a response time t_{90} from <100 ms.

UV Fluorescence

In UV fluorescence, the gas sample to be analyzed is exposed to excitation radiation that is specifically absorbed by the gas. Since this type of absorption is an excitation to a higher electronic level, the transition to the basic level is affected by emission of radiation. The excited molecules remain in this state for a certain period of time ($t \approx 10^{-8}$ up to 10^{-4} seconds) before changing to the ground state. The emission of radiation is called fluorescence and is shifted to longer wavelengths due to Stokes' rules. Figure 9.7 shows the excitation radiation for an SO_2 measurement and the associated fluorescence radiation. The maximum of the fluorescence radiation is $\lambda_{max} \approx 265$ nm.

Excitation and emission thus take place in two different processes, which can be represented by the following formal relationship (Jahnke 2000):

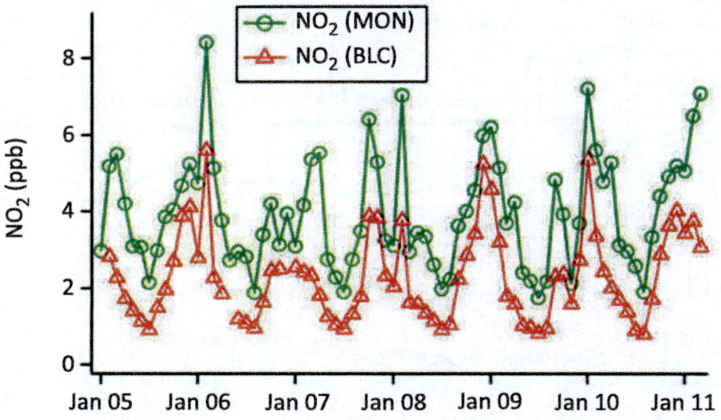

Fig. 9.6 Comparison measurement over a period of 6 years between a conventional molybdenum converter (Mo) and a photolysis converter (BLC) in an urban environment (Schwarzenbach and Hüglin 2011)

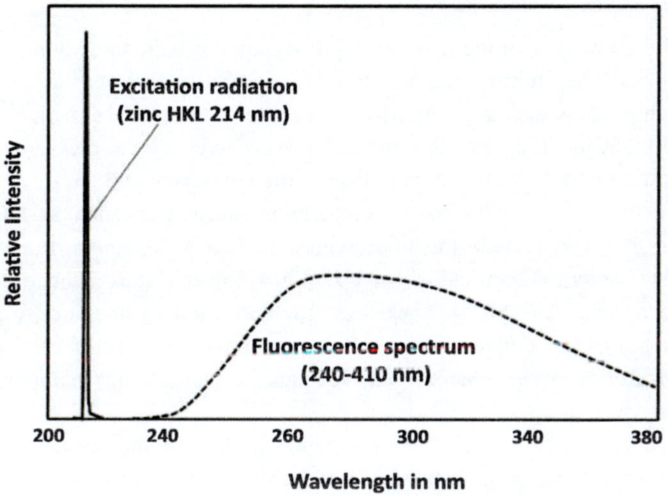

Fig. 9.7 Excitation and fluorescence spectrum for gas measurement of SO_2

$$SO_2 + h\nu_1 \xrightarrow{\text{Excitation}} SO_2^* \xrightarrow{\text{Fluorescence}} SO_2 + h\nu_2 \tag{9.10}$$

UV fluorescence has established itself as the standard method for low SO_2 concentrations (Immission measurement technology). Figure 9.8 shows the instrument design. The excitation is performed by a hollow cathode lamp (HCL) with zinc as base material. Zinc has a strong emission band at $\lambda \approx 214$ nm. A lens is used to focus this radiation into a measuring

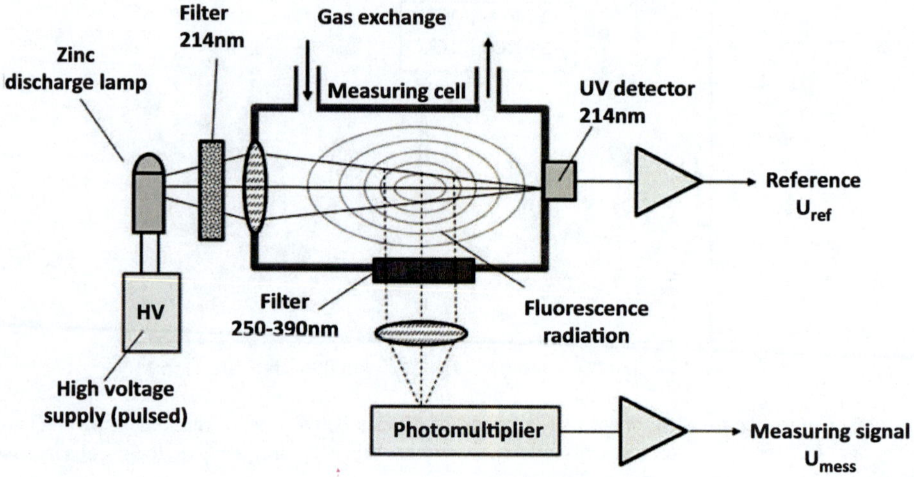

Fig. 9.8 Basic design of a fluorescence analyzer for measuring SO_2

chamber, and the intensity of the 214 nm line is monitored with an additional photodiode (reference signal). The fluorescence radiation is measured with a sensitive photodetector (e.g. Photomultiplier), which is positioned at an angle of to the 90° direction of the beam. The radiation intensity (U_{mess}) of the fluorescence received by the measuring detector is directly proportional to the SO_2 concentration in the measuring cell.

In this process, too, the typical accompanying substances such as CO_2, O_2, N_2, hydrocarbons, etc., can impede the fluorescence radiation (*quenching*). The maximum error for an SO_2 determination can show up to 30% lower values (Jahnke 2000). These errors can be reduced by suitable measures, such as vacuum operation (vacuum) and upstream gas filters (*scrubbers, cutters*). It is also possible to calibrate the measuring instruments with test gases that already contain the interfering components in the corresponding concentrations.

The method is mainly used for environmental protection in measuring stations and measuring vehicles (Immission measurements). Further areas of application are workplace and room air measurement. The response times are therefore also comparatively large and lie for the t_{95} *time* between 40 s and 120 s. Up to a concentration of 1000 ppb the method is absolutely linear and deviates beyond that from the linear behaviour. The measuring ranges are therefore only offered up to a maximum value of 10,000 ppb (10 ppm) (ansyco,[1] SO_2-analyzer AF21M). A detection limit of <1 ppb SO_2 is achieved by averaging over 120 seconds.

[1] **ansyco** analytical systems and components GmbH, Karlsruhe.

Fluorescence Quenching O_2-Sensor

Fluorescence quenching, which is caused by accompanying substances of the fluorescent material, has been known for many years. In the described processes (CLD and UV-fluorescence), this effect occurs – as an unwanted interference – and is suppressed by suitable measures. However, this *quenching effect* can also be used for a gas measurement if the fluorescent material is a solid which is brought into contact with the gas to be analyzed.

In this case, the fluorescent substance (*fluorophore*) is also raised to a higher electronic level by external light (→ excitation) and then releases the energy either by emission of radiation (fluorescence) or by shocks with accompanying substances (*quencher*).

As early as 1919, Stern[2] and Volmer[3] published a scientific description of this phenomenon. In this work, a mathematical relationship between the fluorescence intensity I_0 of the fluorescent substance in the absence of the quencher, and the fluorescence intensity I in the presence of the quencher and its concentration $[Q]$ could be derived. The lifetimes (decay time τ) behave like the intensities:

$$\frac{I_0}{I} = \frac{\tau_0}{\tau} = 1 + k_{SV} \cdot [Q] \tag{9.11}$$

The fluorescence quenching of certain dyes by oxygen was first observed by Kautsky[4] (1939). In the 1990s, the first oxygen sensors were presented that operate according to this principle (Colvin et al. 1996). Figure 9.9 shows one such gas sensor. In this example, the fluorescence layer is irradiated with a blue LED that is matched to the absorption spectrum of the fluorophore. The detector behind it receives the fluorescence radiation and passes this information $U(t)$ on to the evaluation electronics via an amplifier. The LED is modulated with a frequency of $f = 10^4$ up to 10^5 Hz, since the typical lifetime τ is in the order of a few microseconds. The temporal course of the fluorescence radiation follows an e-function, with the decay time τ:

$$U(t) = U_0 \cdot e^{-t/\tau} \tag{9.12}$$

Figure 9.10 shows this temporal decay behavior of the fluorescence radiation. The excitation source is switched off at a time t_0 and the afterglow is then observed. The decay constant τ can be determined from this curve and is a measure of the O concentration present.

[2]Otto Stern (1888–1969) German physicist.

[3]Max Volmer (1885–1965) German chemist.

[4]Hans Kautsky (1891–1966) Austrian chemist.

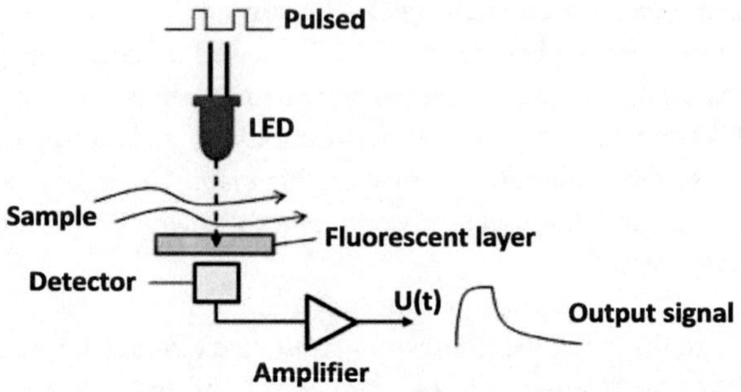

Fig. 9.9 Design of an oxygen sensor according to the principle of fluorescence quenching

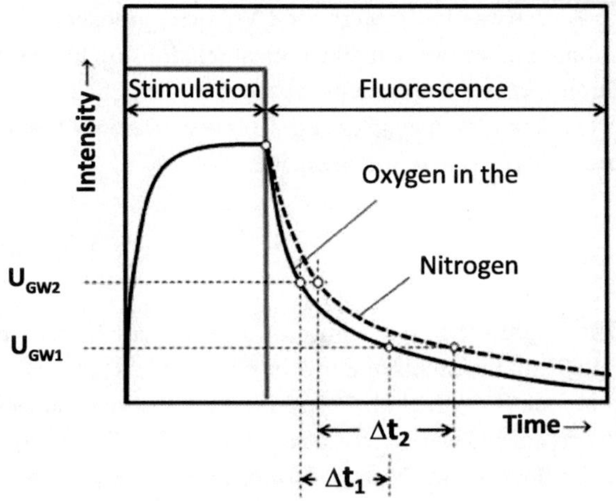

Fig. 9.10 Time course of fluorescence radiation at different O_2 concentrations and determination of the decay time τ

The decay time τ can then be calculated from the time course $U(t)$, if the output voltages are determined by the limit values U_{GW1} and U_{GW2}:

$$\tau = \frac{\Delta t}{\ln\left[U_{\mathrm{GW2}}/U_{\mathrm{GW1}}\right]} \tag{9.13}$$

Figure 9.11 shows the measured time course of fluorescence radiation for different O_2 concentrations from 0 vol.% to 21 vol.%, which lead to decay times between $\tau \approx 2\ \mu s$ and $\tau \approx 5\ \mu s$.

As an alternative to the pulsed control of the LED described above, it can also be operated with a sinusoidal control (frequency f). The fluorescence radiation (emission) follows the excitation process with a time delay (Fig. 9.12), whereby this phase shift ($\Delta\phi$) can be calculated as follows:

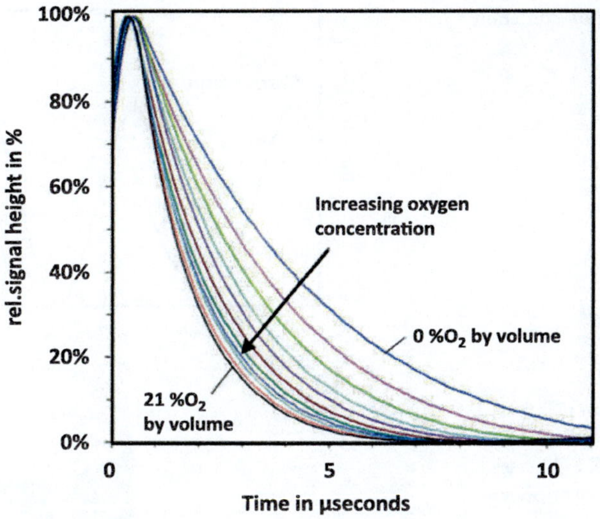

Fig. 9.11 Time course of fluorescence radiation at different O_2 concentrations. (According to OxySense, White Paper OxySentry System)

$$\tan \phi = 2\pi f \tau \quad (9.14)\leftarrow$$

These O_2-gas sensors can be designed very small and compact. Sensors available on the market have a diameter of $\varnothing = 20$ mm and a height (without contacts) of $h = 13$ mm (Fig. 9.13). The gas enters the sensor structure via a PTFE membrane, which results in a time constant t_{90} of ≈30 seconds. The measuring range is specified as 0–25 vol.% O_2 or as partial pressure 0...300 mbar. The lifetime of such an O_2-gas sensor is limited because the fluorophore decomposes over time and then no longer has the original properties.

Compared to electrochemical O_2-gas sensors, the optical O_2-gas sensor is characterized by its insensitivity to rapid pressure fluctuations (Fig. 9.14).

Another application of this technique is the non-contact measurement of oxygen concentration in the packaging industry. The oxygen content has a significant influence on the shelf life of food and must therefore be measured. A non-destructive method consists of incorporating a fluorescent dye into the packaging, which can be detected from the outside by optical scanning. Figure 9.15 shows such a setup (White et al. 2003). The dye (*Oxydot*) is either in the liquid (e.g. beer) or above it in the gas phase (head space). While in the liquid phase the dissolved oxygen can be measured, in the gas phase it is the outgassed oxygen content. The fluorimeter contains the excitation LED ($\lambda \approx 450$ nm) and the detection of the fluorescence radiation ($\lambda \approx 600$ nm), which is transmitted to the measuring instrument by fiber optics. The measuring device contains a sensitive photodetector (e.g. photomultiplier). The evaluation, according to the methods already described, is

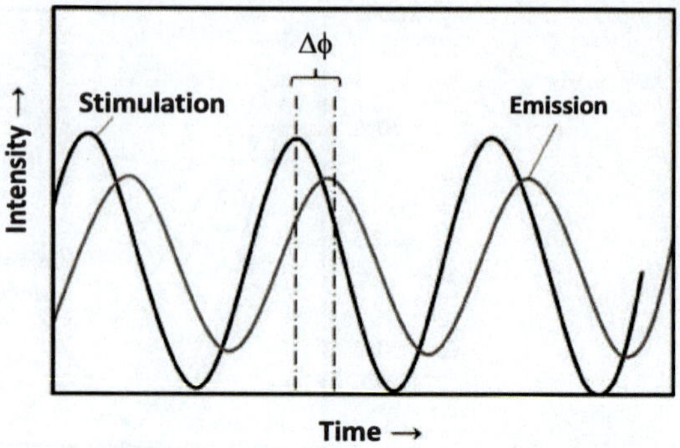

Fig. 9.12 Timing of excitation and emission

Fig. 9.13 O_2-gas sensor based on the principle of fluorescence quenching (Type XYO Series Optical Oxygen Sensor Source: First Sensor)

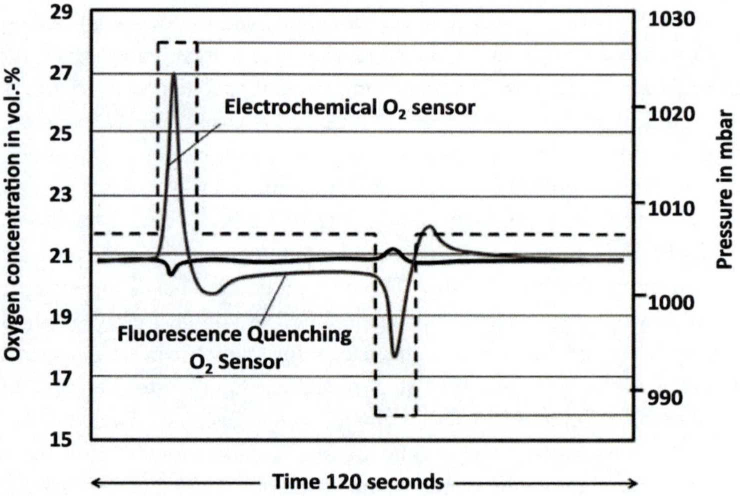

Fig. 9.14 Comparison between an electrochemical oxygen sensor and a fluorescence quenching oxygen sensor

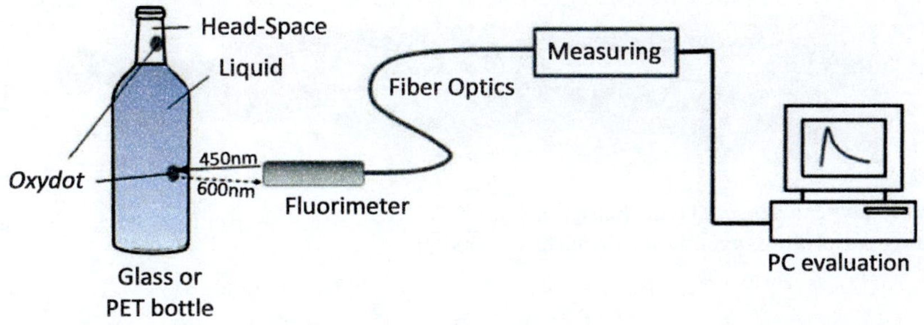

Fig. 9.15 Use of optical oxygen measurement in the food industry (White et al. 2003)

then carried out in a PC. The fluorimeter must be placed at max. 1 cm in front of the *Oxydot* to avoid interferences.

In the meantime, the fluorescence quenching method is also used for highly accurate analyses of dissolved oxygen in the deep sea. CONTROS[5] offers a system that can determine the content of dissolved oxygen up to a sea depth of 6000 m (Fig. 9.16). The response time is only 3 seconds for the $t_{63\%}$ -time. The accuracy is specified with $\pm 1\%$.

Gas Detection with IR Cameras

Gaseous pollutants can only be detected in the UV and IR range due to the absorption characteristics described. The only exceptions are nitrogen oxide (NO_2) and bromine (Br_2), which can be detected as brown gas clouds in the visible spectral range. This means that most gases are not visible to the human eye and are therefore a particular source of danger. Particularly in processes (e.g. biogas plants), in which large quantities of gas are converted by non-specialist personnel (farmers), leaks occur time and again with, for example fatal accidents. Gas warning devices are useful protective measures in biogas plants, with which these pollutants (methane, carbon dioxide, hydrogen sulfide, . . .) can be detected. In most cases, however, gas detectors are only designed for spot measurements, i.e. they only measure at one point in the plant. If gas escapes at another point, this leakage cannot be detected due to the point measurement or only with a long-time delay. For inspection purposes, therefore, a new type of procedure is used to make undesired gas leaks visible.

Infrared cameras have been used for many years to measure the surface temperature of buildings (energy saving) or of components in electrical engineering (hot-spot detection). Figure 9.17 shows two examples of such a thermographic representation of the temperature distribution, as a false-color display. Thermal imaging cameras are mostly used in a spectral range from $\lambda \approx 8\,\mu m$ to 13 μm, to be able to record the usual temperatures from 0 to 50 °C. In a spectral range of 3–5 μm the temperature ranges from 500–1000 K are optimally covered. Since the camera chip is sensitive for the entire IR range, it can also be

[5]Kongsberg Maritime Contros GmbH, Kiel.

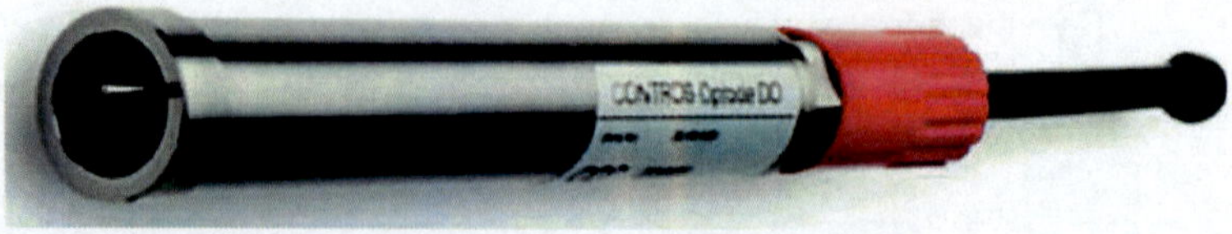

Fig. 9.16 Oxygen sensor (HydroFlashTM O_2, CONTROS) based on fluorescence quenching, for measuring dissolved oxygen in the deep sea to 6000 m

a b

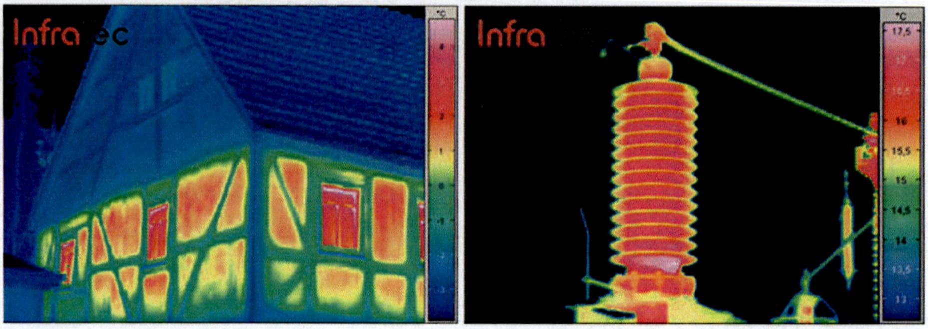

Fig. 9.17 Thermographic images of a half-timbered house (**a**) and a high-voltage insulator (**b**). (Source: InfraTec GmbH Dresden)

used for the detection of IR-active gases. Therefore, an interference filter must be placed in front of the camera lens for selectivity, which is used to tune to a specific gas (Fig. 9.18).

The physical principles for the visualization of gas leakages are derived from the *Lambert-Beer law*. This has already been discussed in detail in Sect. 6.2. To be able to prove an absorption of IR radiation by a gas cloud, a corresponding background radiation must be available. In this case, the temperature of the background (HG) must be greater than the temperature of the gas cloud (GW):

$$T_{HG} > T_{GW} \tag{9.15}$$

If this condition is met, the background emits IR radiation, which is detected by the gas camera. If there is a gas cloud between the gas camera and the background, part of the radiation is absorbed and appears as a dark spot in the thermographic image. The part of the radiation, which is absorbed by the gas cloud can be calculated as follows:

$$\Delta I = I_0 \cdot \left(1 - e^{-\alpha \cdot c \cdot L}\right) \tag{9.16}$$

The absorption coefficient α is determined by the type of gas. The quantity L describes the spatial extension of the gas cloud, and c is the concentration of the component to be

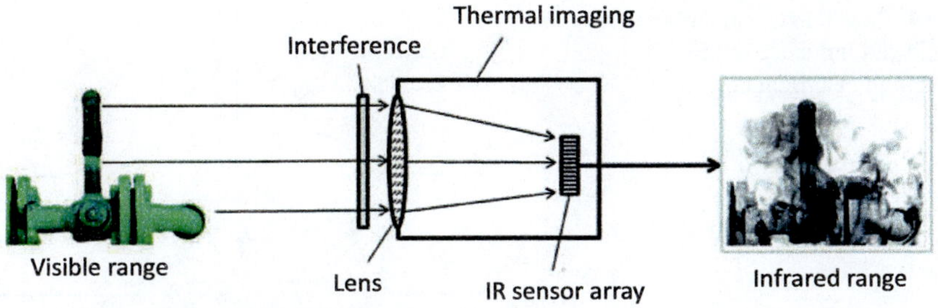

Fig. 9.18 Basic arrangement for locating gas leaks in process control. (Source: ICOdata GmbH, Neu-Isenburg)

detected in the gas cloud. Since the concentration distribution within the gas cloud is inhomogeneous, only the integral quantity can be recorded with the gas camera:

$$c \cdot L = \int_0^L c(L) \cdot \mathrm{d}L \tag{9.17}$$

Figure 9.19 shows the passage of IR radiation through a gas cloud with the concentration profile $c(L)$. The spatial expansion is reduced to the product $c \cdot L$. However, this method is quite suitable for locating leaks, since only the leak must be found. In this context, however, it is important to know which leak can still be reliably detected with the described method. The leakage rate in g/h is therefore an important parameter for comparative measurements. The manufacturers specify a value of <1 g/h over a distance of $d \approx 100$ m.

Figure 9.20 shows an arrangement in which a picture is taken at a known gas concentration (e.g. 100 vol.% CO_2) against a background that is also defined ($T = 500$ K). With this arrangement the sensitivity of the gas camera can be determined.

In Fig. 9.21, the human respiration process has been visualized with a gas camera, which was recorded in front of a 50 °C background. The CO_2 concentration (c) in the breathing air is $c < 5$ vol.% and can still be reliably detected with the IR image displayed (Vollmer and Möllmann 2010).

Another possibility is to locate a hot gas source due to the radiation emission. The background radiation is then less than the gas radiation. In this case:

$$T_{\mathrm{HG}} < T_{\mathrm{GW}} \tag{9.18}$$

This situation occurs, for example, when exhaust gases or gas flames are to be measured. In Fig. 9.22 such an emission process is shown.

The reliable detection of a gas cloud must also be possible over a large distance ($d \approx 100$ m). Figure 9.23 shows such an application for checking leaks in a natural gas pipeline. Natural gas consists of over 90% methane (CH_4) with a molecular weight of 16 g/

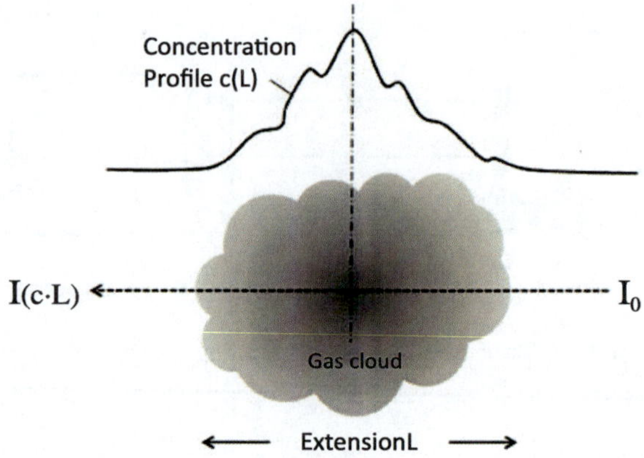

Fig. 9.19 Spatial extension of a gas cloud and influence on the concentration determination

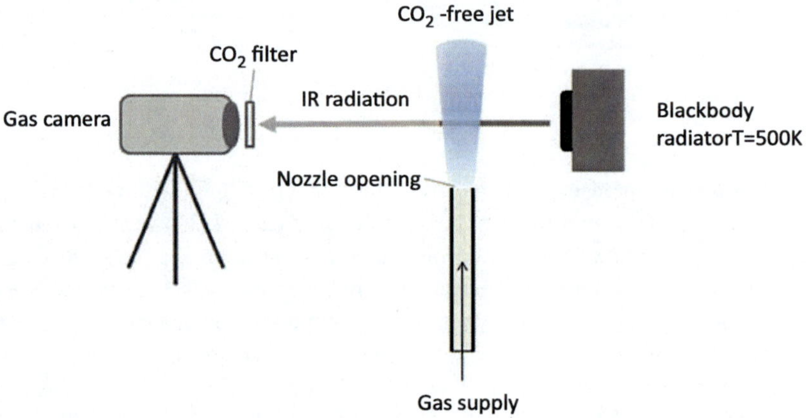

Fig. 9.20 Arrangement for determining the sensitivity of the gas camera to a specific gas (e.g. CO_2)

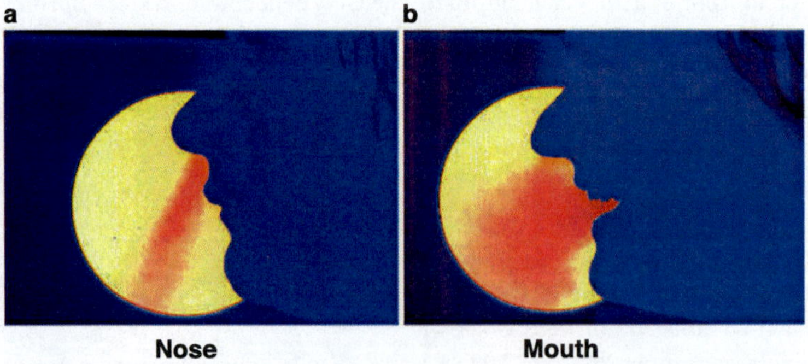

Fig. 9.21 Breathing process through the nose (**a**) and mouth (**b**) against a 50°C background. (Source: Vollmer and Möllmann 2010, Wiley-VCH Verlag GmbH & Co. KGaA.)

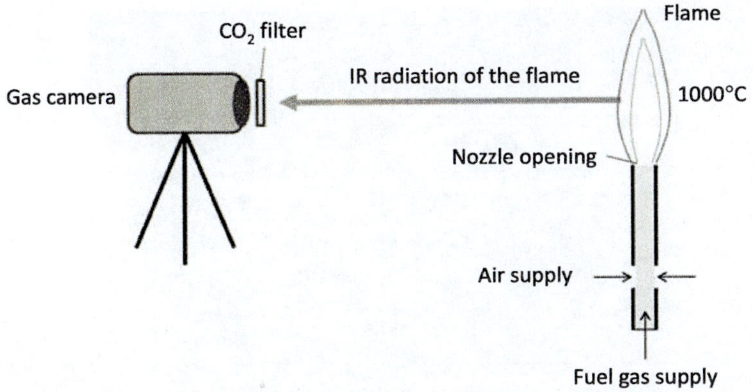

Fig. 9.22 Detection of a radiation emission due to increased gas temperatures

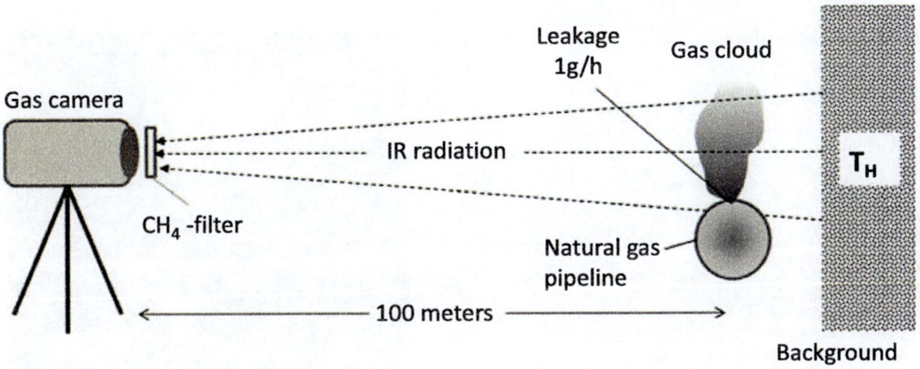

Fig. 9.23 Setup for detecting methane leakage from a natural gas pipeline over $d = 100$ m

mol. Under normal conditions, one mole of a gas occupies a volume of 22.4 L. A leakage rate of 1 g/h therefore means the escape of 1.4 L of natural gas per hour or $\approx$23 cm^3/minute. In the laboratory, CO_2 leakage values of 1 mL/minute in front of a defined background radiation of $T_H = 35^{\circ}$C could already be detected (Vollmer and Möllmann 2010).

The design of a gas camera is almost identical to that of a thermal imaging camera (Fig. 9.24). The lens can be used for the entire IR range from 3 to 12 μm. Adapted filters for the different gases are placed between the lens and the IR sensor array. Hydrocarbons (VOC[6]) are detected in the spectral range from 3.2 to 3.5 μm, while sulfur hexafluoride (SF_6) is detected at 10.3 to 10.7 μm. The manufacturers specify over 30 gases that can be

[6]Volatile Organic Carbon.

Fig. 9.24 View of the *EYE-C-GAS* camera from *OPGAL*. (Source: ECOdata GmbH, Neu-Isenburg)

Fig. 9.25 Detection of a gas leak in a leaking biogas plant. (Source: ECOdata GmbH, Neu-Isenburg)

detected with this technique. Cooled InSb-arrays (3.2–3.5 μm) or QWIP[7] arrays (10.3–10.7 μm) with a resolution of 320 × 240 (76,800 pixels) are used as IR detectors.

The main application in Germany is the monitoring of 8000 biogas plants, most of which are leaking and release the climate-damaging methane in high concentrations into the environment. Figure 9.25 shows such a leak, which can only be localized in a short time by using a gas camera. A measurement with a *sniffer*[8] is also possible but compared to a gas camera it is very time consuming and faulty. In the gas and petrochemical industry, very high demands are placed on dense production plants and pipeline systems. In these plants, the devices used must also meet high safety standards, which are not required in biogas plants. In particular, suitability for explosion hazards is mandatory in petrochemical plants.

[7] Quantum Well Infrared Photodetector.

[8] Portable gas detector sniffer.

Therefore, there are now also gas cameras (*EYE-C-GAS*) which have an Ex-approval according to the ATEX guidelines.

9.2 Laser Spectrometer

Laser spectroscopy has been used in gas measurement technology for many years because the advantages of narrow band ($\Delta\nu \approx 10^{-4}\,\mathrm{cm}^{-1}$) and high intensity in a single emission line promise high selectivity and low detection limits. Since the invention of the optical laser by Maiman[9] in 1960, one of the first applications of laser spectroscopy for the detection of gases has been investigated (Görlich 1969).

Gas Laser

First attempts were made to use the IR lines of gas lasers (e.g. HeNe) for an absorption measurement. However, since these lines could not be matched to the respective absorption lines of the gases, the spectral overlaps were rather random and could only be used for metrological purposes to a limited extent. With the introduction of tunable CO and CO_2 lasers this limitation could be overcome. Both laser types can *laser* at different wavelengths, so that a tuning element (holographic grating) could be used to move the laser to a defined line. With this method, the first suitable gas measuring instruments based on laser technology were built between 1970 and 1980 (Sigrist 1994). Figure 9.26 shows such a laser spectrometer based on a tunable gas laser.

By rotating the grating, at the end of the laser resonator, the wavelength λ_L at which the laser amplification takes place is determined. The grating is operated in a so-called *Littrow*[10] arrangement, in which the light is reflected ($\alpha = \beta$). The laser gas (CO, CO_2, ...) is enclosed in a tube which is closed with windows at the so-called *Brewster*[11] angle θ_B. For the transition to the gas phase $n_1 \approx 1$.

$$\theta_B = \tan^{-1} n_2/n_1 \tag{9.19}$$

Due to this measure the reflections at the window surfaces are zero and the light has a defined polarization. At the output of the gas laser, the radiation is modulated by a mechanical chopper and passed via a beam splitter (ST_1) to a spectrum analyzer. In this analyzer the current wavelength λ_L of the laser is determined. Furthermore, the beam splitter directs the main part of the laser radiation into the gas measuring system. A mirror (Sp) is used to redirect the beam by 90° and another beam splitter (ST_2) is used to decouple

[9] Theodore Harold Maiman (1927–2007) US physicist.

[10] Otto von Littrow (1843–1864) Austrian physicist and astronomer.

[11] Sir David Brewster (1781–1868) Scottish physicist.

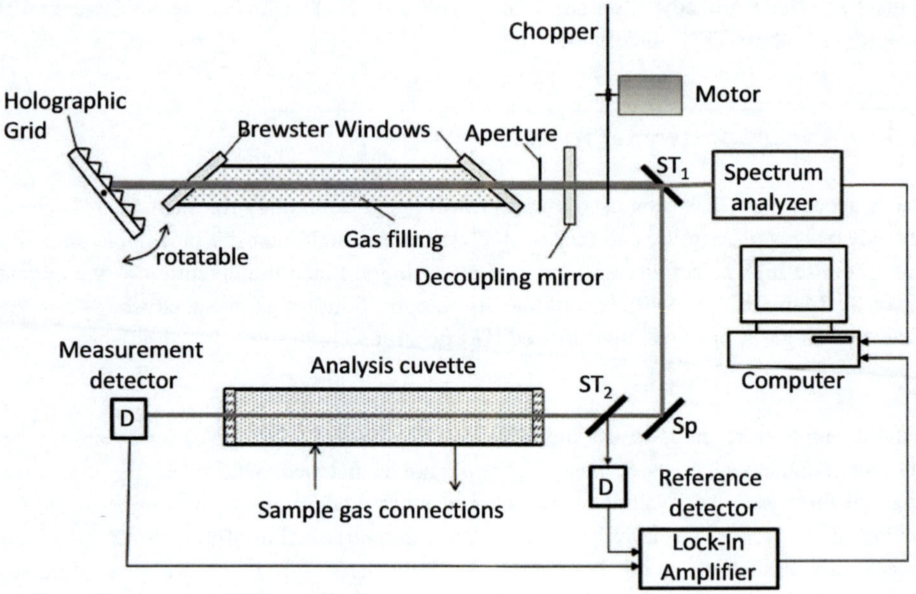

Fig. 9.26 Laser spectrometer with a tunable gas laser

the reference radiation, which is detected by an IR detector. In the analysis cuvette, the radiation is then absorbed by the gases to be detected. This proportion is detected by the measuring detector. Both detector signals are connected to a lock-in amplifier, which is in turn connected to the computer at the output. Further signal processing is then done in the computer. With these setups, highly selective gas measurements in the ppm range could already be carried out. However, this type of gas measurement technology was mainly used for scientific purposes. The application range of different gas lasers is summarized in Table 9.1.

Lead Salt Laser

With the introduction of so-called lead salt lasers ($PbS_{1-x}Se_x$), which were available as semiconductor materials, tunable semiconductor lasers could be used for gas analysis for the first time (Preier et al. 1974; Hesse and Preier 1975). The usable spectral range was between 2 and 12 μm, so that all important gas components could be detected with this technology.

Lead salt lasers consist of different mixtures. Mainly tin (Sn), selenium (Se), tellurium (Te), germanium (Ge) and cadmium (Cd) are added. With the mixtures, different emission ranges of the lasers can be achieved, so that almost the entire IR region can be covered with different laser types. Figure 9.27 shows the emission ranges of the different mixtures and the absorption ranges of the most important gases.

Table 9.1 Example of the application areas of different gas lasers (Sigrist 1994)

Laser gas	Wavelength range	Sample gas
HeNe	3.39 μm	CH_4, Butane,
CO	4.8–8.4 μm	CO, NO, ...
CO_2	9.2–10.8 μm	NH_3, SF_6, C_2H_4, SO_2, ...
CO_2 Isotopes	9–12 μm	22 Gases
HF	2.7–2.9 μm	HF, NO, CO_2
DF	3.6–4 μm	CH_4

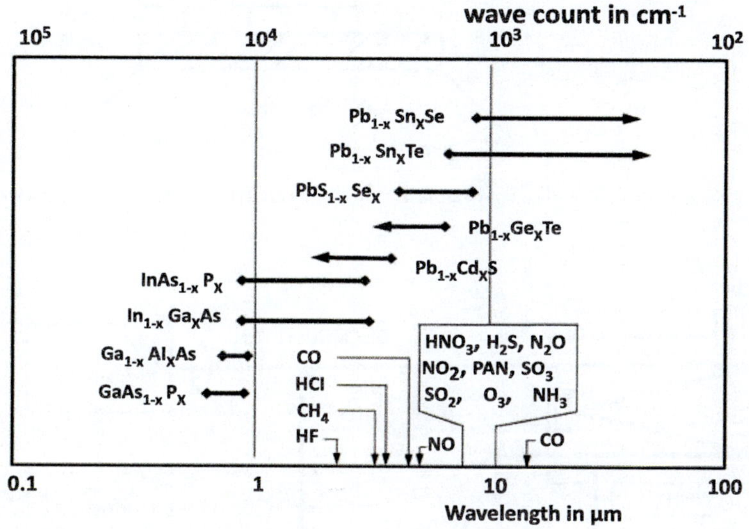

Fig. 9.27 Absorption lines of important gases and the emission ranges of different semiconductor lasers. (Hinkley 1972)

The laser structure consists of a layer structure, which is mainly used for electrical and thermal contacting (Fig. 9.28). The laser structure is in very good thermal contact with a copper plate (*heat sink*), which is regulated by an external cooler to the required chip temperature between 70–100 K. The laser radiation is emitted to the side and has to be passed on by an appropriate optical system. Although the actual lead salt laser was only a few millimeters in size, it required complex and bulky peripherals for safe operation. In particular, mode selection (→ monochromator) and cryogenic cooling ($T < 100$ K) required a lot of space. Figure 9.29 schematically shows a complete spectrometer setup. Especially the safe operation of the radiation source requires much adjustment effort. These boundary conditions could only be met in the laboratory, so that this technique could not be used for industrial gas measurement either. A summary of this work can be found in Eng et al. (1980). The detection limits for this instrument technology were in the ppb range and in some cases even significantly lower (Table 9.2).

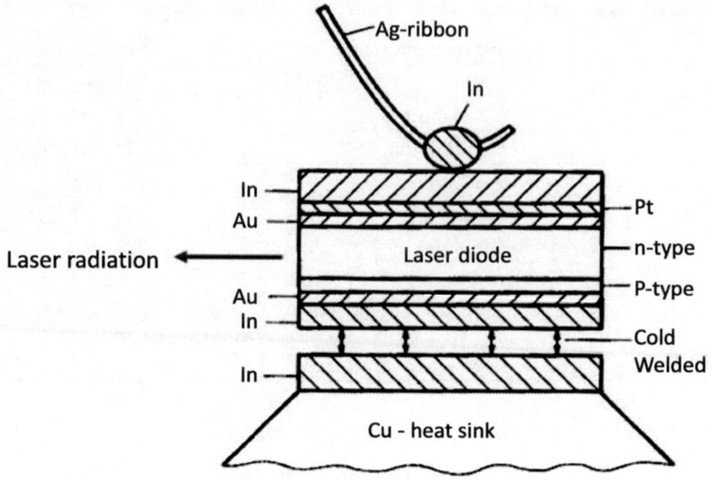

Fig. 9.28 Cross-section through a lead salt laser for IR gas analysis (Hesse and Preier 1975)

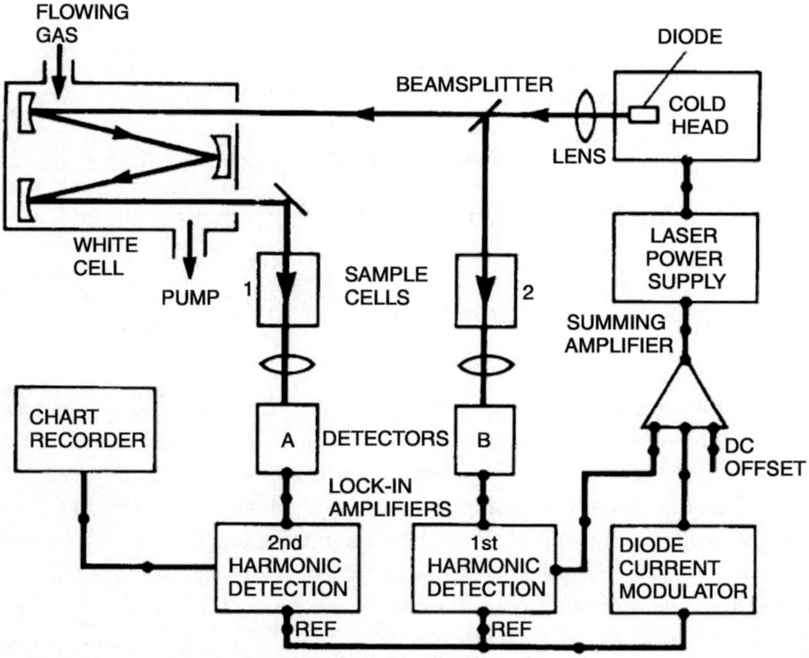

Fig. 9.29 The white cell had an optical length of 200 m and was operated in negative pressure at $p \approx 25$ mbar. *Sample Cell 2* was filled with a high concentration of SO_2 and served to stabilize the laser wavelength. *Sample Cell 1* was used for calibration ticks. With the setup, SO_2 concentration could be measured from 100 ppt. (Source: Eng et al. 1980)

Table 9.2 Minimum sensitivity (detection limits) for a diode laser system based on lead salt lasers

Gas	Wave number in ν_0 cm^{-1}	Sensitivity (ppb)
SO_2	1140	3
O_3	1050	0.5
N_2O	1150	2
CO_2	1075	300
H_2O	1135	50
NH_3	1050	0.05
PAN	1150	−0.3
CH_4	1300	0.03
SO_2	1370	0.3
NO_2	1600	0.02
NO	1880	0.03
CO	2120	0.01
CO_2	2350	0.001

Source: Butler et al. (1978)

Coherent Anti-stokes Raman Spectroscopy (CARS)

With the introduction of tunable high-power lasers, a new type of Raman scattering became possible, which is called nonlinear Raman spectroscopy. If two different laser frequencies are irradiated into a sample through a lens, an enormous field strength $\boldsymbol{E}$ is formed in the focus. Due to the nonlinear influence of this field strength $\boldsymbol{E}$, the emission of the anti-stokes light waves can also take place in the form of laser radiation. A prerequisite for this effect is that the frequencies of the two incident waves differ by exactly the Raman-active oscillation frequency (Demtröder 1993):

$$\omega_{\text{vib}} = \omega_2 - \omega_1 \tag{9.20}$$

The two laser beams therefore only need to be adjusted to each other in the difference. The fundamental wavelength λ_0 does not play a role in this approach since Raman scattering takes place over virtual levels that do not correspond to any real energy level. Figure 9.30 shows the transitions that lead to the emission of anti-Stokes radiation. Furthermore, according to the conservation of momentum the following condition must be fulfilled:

$$2 \cdot \vec{k_2} = \vec{k_1} + \vec{k_a} \tag{9.21}$$

This type of Raman scattering is called *CARS*.[12] The angle α_2 at which the anti-Stokes radiation is emitted thus results from this wave vector relationship (Eq. 9.21). The respective wave vector $\vec{k}$ results from:

[12] Coherent Anti-Stokes Raman Scattering.

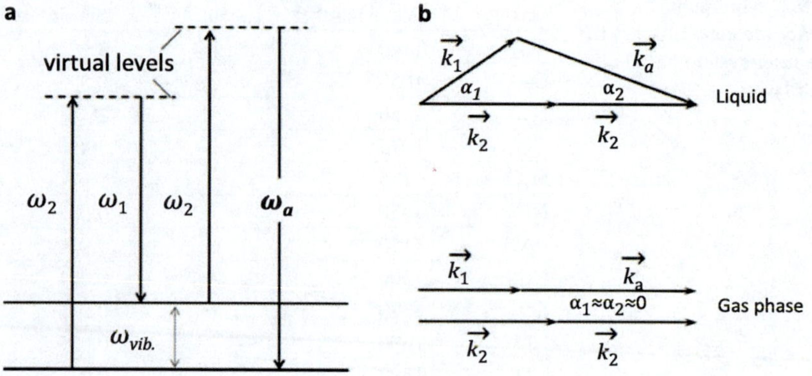

Fig. 9.30 Energy (**a**) and momentum conservation (**b**) in Coherent anti-Stokes Raman Scattering for liquids and gases

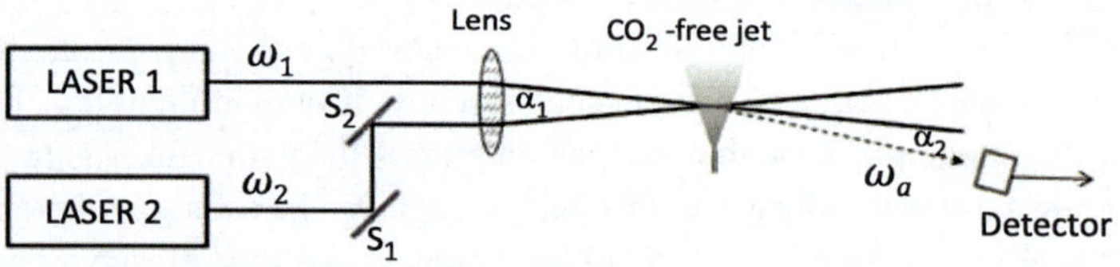

Fig. 9.31 CARS experiment for measuring the spatially resolved concentration distributions in a CO_2 free jet (Wiegleb 1980)

$$\vec{k} = \frac{\omega \cdot n}{c} \tag{9.22}$$

Equation 9.22 can therefore also be described as follows:

$$2 \cdot \omega_2 \cdot n_2 = \omega_1 \cdot n_1 + \omega_a \cdot n_a \tag{9.23}$$

In media, which only have a low dispersion (e.g. gases), the following applies $n_2 \approx n_1 \approx n_a$. Therefore, the angle $\alpha_1 \approx \alpha_2 \approx 0$.

Figure 9.31 shows such a CARS structure. The two lasers (1 + 2) were set up as dye lasers, which were pumped by a frequency-doubled neodymium YAG[13] laser with a wavelength at $\lambda \approx 532$ nm. Dye laser 1 was operated with Rhodamin 6G and had a fixed wavelength at $\lambda \approx 560$ nm, while the second laser was operated with Rhodamin B and could be tuned with a grating in a *Littrow arrangement.* The radiation of laser 1 was flipped

[13]**Y**ttrium **A**luminium **G**arnet.

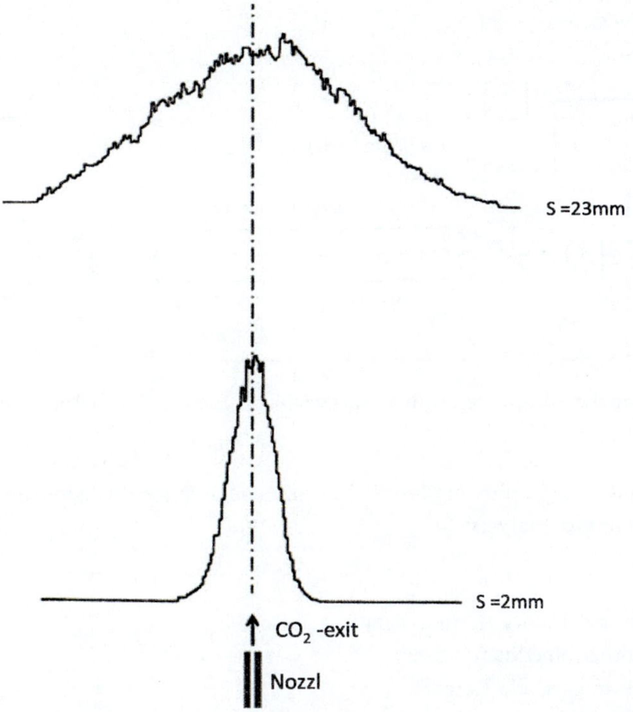

Fig. 9.32 Spatially resolved CO_2 measurement in a free jet. Pure CO_2 escaped from the nozzle (Ø = 1 mm) and mixed with the surrounding air so that the CO_2 concentration decreased with increasing distance (*s*) from the nozzle (Wiegleb 1980)

with two mirrors (S_1 and S_2) so that both laser beams could be irradiated at the angle α_1. The sample was then in focus. For gaseous materials $\alpha \approx 0$, a dichroic mirror had to be used for S_2. The anti-Stokes wavelength λ_a was then recorded and evaluated with a photodiode. In the example given, a CO_2 free beam was chosen which could be moved relative to the focus by a stepper motor. This arrangement made it possible for the first time to perform a spatially resolved gas measurement of the CO_2 distribution. The result of this measurement is shown in Fig. 9.32 (Wiegleb 1980).

This method also allows gas analytical investigations in flames and combustion processes in turbines and engines. The advantage is the spatially resolved information within the flame. Further applications of the CARS method can be found today in medical technology and biology (CARS microscopy).

Laser Diode Spectrometer

The further development of laser technology, especially for use in communications technology, also led to better laser sources in gas measurement technology. The disadvantages of lead salt lasers could thus be eliminated, as these new laser types also work at room

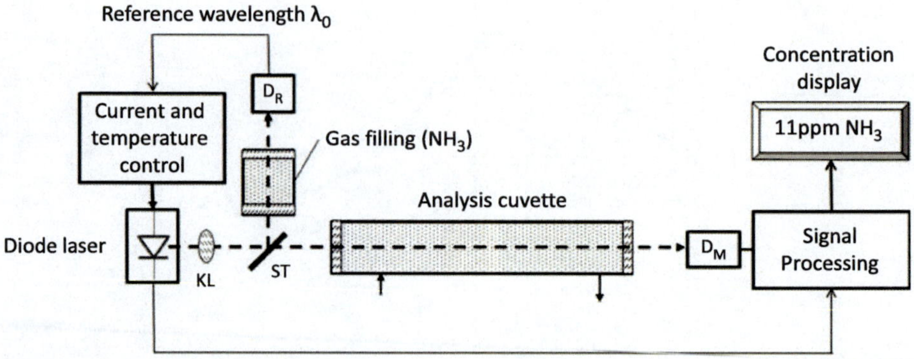

Fig. 9.33 Absorption photometer with a tunable diode laser and wavelength stabilization for ammonia (NH_3)

temperature. Today, the following laser types are available for this application, which are suitable for use in gas analysis:

- VCSEL (**V**ertical **C**avity **S**urface **E**mitting **L**aser)
- DFB (**D**istributed **F**eedback **L**aser)
- QCL (**Q**uantum **C**ascade **L**aser).

Figure 9.33 shows a laser photometer based on these diode lasers. The radiation of the laser source must first be bundled with suitable optics, since the diode laser has a high divergence due to the small cavity. In the simplest case, a lens (KL) with a short focal length is used. The emission wavelength of the laser can be adjusted within certain ranges by means of the temperature and the diode current. This tunability is a great advantage, as it enables the spectral separation (*scan*) of individual absorption lines of the gases. Since such a scan measures the intensity on the line as well as directly next to the line, an additional reference measurement is not necessary since both are recorded with each scan anyway. This modulation technique achieves a high selectivity and a high stability of the gas measurement. To be able to find a specific absorption line (λ_0) in the spectrum of the gas to be measured, some instrument manufacturers install a gas-filled cell through which a reference beam is passed. The emission wavelength λ_0 of the laser can thus be reliably adjusted at any time, even if no sample gas is present in the analysis cuvette.

However, tunability is only possible within very narrow limits. The spectral range in which the laser is to operate must be defined in advance and individually tuned to each gas. The manufacturers of these lasers can only meet these requirements to a limited extent in terms of production technology, and usually only find the appropriate laser wavelength by selection. This type of production is very complex and expensive, since not all manufactured lasers are suitable for use in gas measurement technology. Identical diode

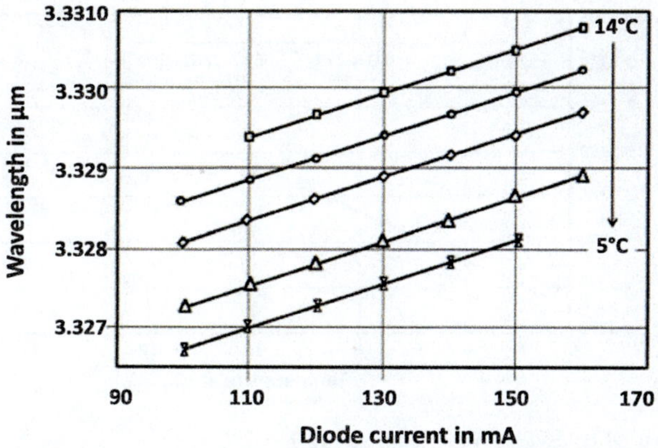

Fig. 9.34 Tuning a DFB laser with the diode current at different temperatures

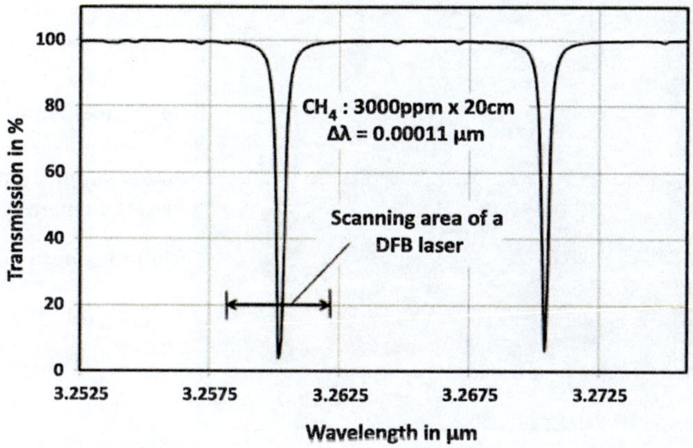

Fig. 9.35 Scan range of a DFB laser, compared to a typical line width and the distance between two rotation lines, in the IR spectrum of methane CH_4

lasers used in communication technology do not have to meet these high spectral requirements and are therefore much cheaper. Figure 9.34 shows the tunability of a DFB laser with temperature and diode current (Rhine et al. 2012). The range in which the diode current is tuned lies between 100 mA and 160 mA. The emission wavelength λ_0 then shifts only by $\Delta\lambda \approx 1 - 2$ nm. In this case, a temperature change of 10 K leads to a change of the wavelength by $\Delta\lambda \approx 2 - 3$ nm. The entire scanning range is therefore at $\Delta\lambda \approx 4$ nm.

However, the tunability of $\Delta\lambda \approx 4$ nm is usually sufficient to scan an absorption line. Figure 9.35 shows a part of the CH_4 absorption spectrum at $\lambda \approx 3.4$ µm. The CH_4 line width ($\Delta\lambda$) is at 1 nm and the distance to the next CH_4 line is approximately 10 nm. For the scanning range to be sufficient to tune the laser to a line, the fundamental wavelength λ_0 of

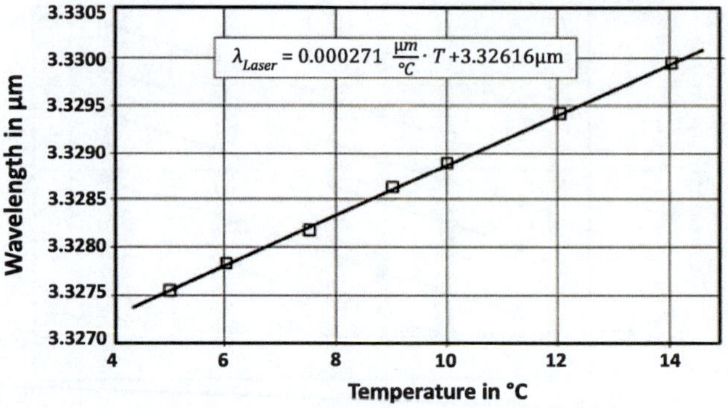

Fig. 9.36 Tuning a DFB laser with the temperature T

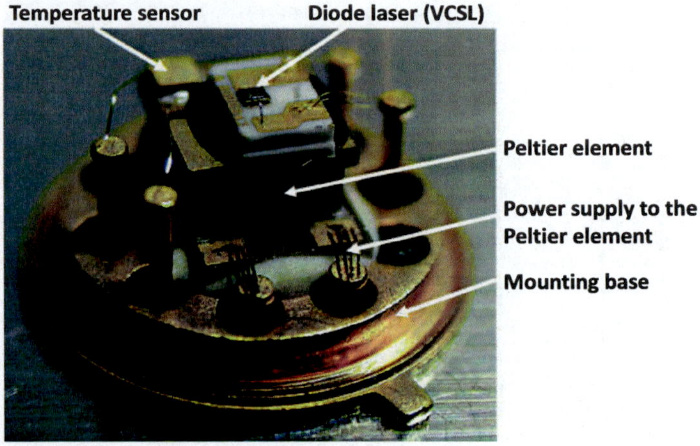

Fig. 9.37 Construction of a diode laser (VCSEL) with a Peltier element for temperature control. (Source: Vertilas GmbH Garching)

the laser must match the absorption line exactly (→ selection). A *presetting* can also be done with the temperature. The tuning of the laser with the temperature is shown in Fig. 9.36. The temperature coefficient is 0.000271 µm/°C or 0.271 nm/°C. A tuning around $\Delta\lambda = 4$ nm is therefore achieved with a temperature change of $\Delta T \approx 15^{\circ}$C.

The spectral tuning of the diode laser can therefore be carried out both with the temperature T and with the diode current $I_{D.}$ The temperature T is usually specified with a Peltier element, with which temperature differences from $\Delta T \approx 40$ K are possible. Figure 9.37 shows such a setup. The miniature Peltier element is arranged on a base (pressure glass feed-through), through which the heat is dissipated to the outside. The laser chip is located on a ceramic substrate with good thermal conductivity to ensure heat transfer to the Peltier element. The ceramic substrate is fixed to the Peltier element with a

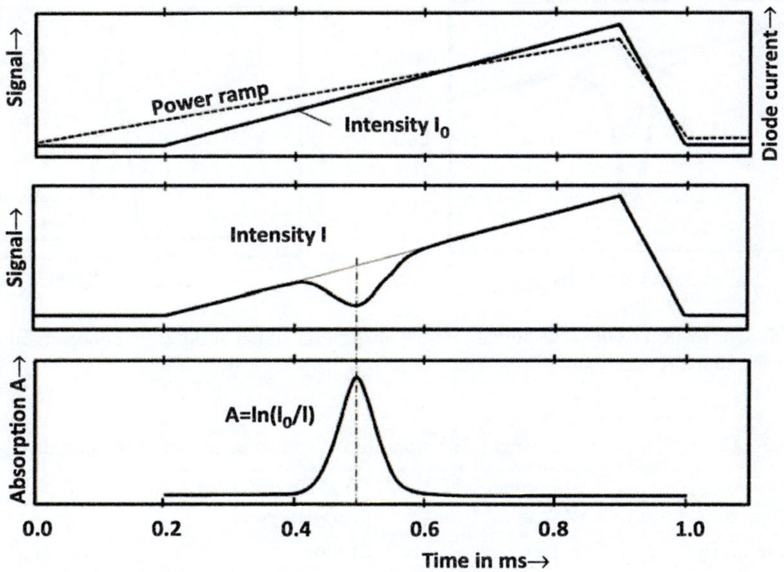

Fig. 9.38 Driving the laser diode with a current ramp (**a**) to scan the wavelength λ in the area $\Delta\lambda \approx 2 - 4$ nm. The selective absorption (**b**) then takes place within this scanning range. By conversion of the intensity signals the absorption *A* (**c**) is obtained

heat conductive adhesive. Furthermore, a temperature sensor is placed on the Peltier element, with which the temperature *T* is set or controlled. This temperature is also used to determine the start wavelength λ_0 for the tuning range $\Delta\lambda$. The fine tuning is usually done with a current ramp. Figure 9.38 shows such a scanning process. The current ramp leads to a continuous increase of the radiation intensity, which is detected by the measuring detector. Figure 9.38a shows the detector signal without the gas to be measured in the measuring cuvette. If sample gas is now introduced, partial absorption occurs during the scanning process (Fig. 9.38b). The signal change I_0 due to the current ramp must be offset against this signal *I*. A value is then obtained which corresponds to the physical absorption *A and* increases proportionally with the concentration c_x (Fig. 9.38c). In Fig. 9.39, the evaluations for different gas concentrations c_0, c_1, ..., c_4 during the scanning process (detector signal) and the absorption calculated from it are shown.

DFB and VCSEL lasers operate in the near infrared range, so that only the weak harmonics of the absorption bands can be detected (Table 9.3). Figure 9.40 shows the entire IR spectrum of ammonia between 1.4 μm and 14 μm. The strongest lines are at 9.4 μm and cannot be detected with these laser types. Although the usable absorption lines at 1.5 μm are extremely weak (Fig. 9.41), NH_3 concentrations below 1 ppm can still be reliably detected with a laser spectrometer. Quantum cascade lasers (QCL) operate in a spectral range $\lambda > 3.5$ μm, so that with these laser setups, even the strongly absorbing bands

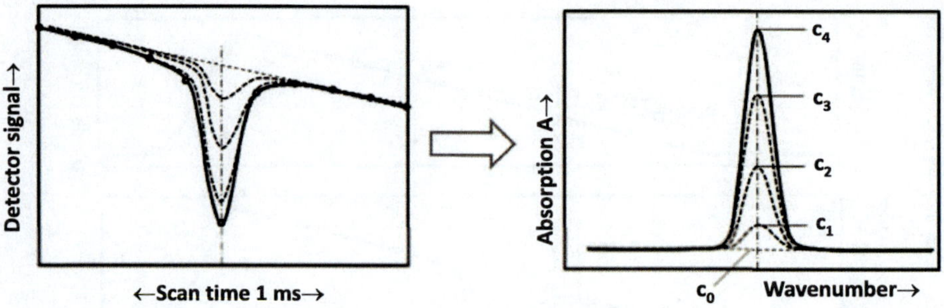

Fig. 9.39 Conversion of the scan signal into the absorption signal at different concentrations (c_1 – c_4). c_0 is the signal for zero gas

Table 9.3 Absorption lines and typical detection limits that can be achieved with diode lasers

Gas type	Formula	Wavelength (nm)	Detection limit
Steam	H_2O	1390/1802	60 ppb
Carbon monoxide	CO	1960	3 ppm
Carbon dioxide	CO_2	1570	30 ppm
		2330	500 ppb
Nitric oxide	NO	1800	60 ppm
		2650	1 ppm
Nitrogen dioxide	NO_2	680	340 ppb
Nitrous oxide	N_2O	2260	1 ppm
Methane	CH_4	1650	600 ppb
Acetylene	C_2H_2	1520	80 ppb
Hydrofluoric acid	HF	1310	10 ppb
Hydrochloric acid	HCl	1790	150 ppb
Hydrogen sulfide	H_2S	1570	290 ppb
Ammonia	NH_3	1500	800 ppb
Formaldehyde	HCHO	1930	50 ppm
Phosphine	PH_3	2150	1 ppm
Oxygen	O_2	760	80 ppm

Source: Vertilas GmbH Garching

in the middle and far IR range can be detected. This type of laser is therefore preferred for highly sensitive gas measurements. Unfortunately, the costs for the QCL are still very high, so that this technology can only be used in special cases. Figure 9.42 shows such a measuring device for plasma research (Iseni et al. 2014).

Oxygen Measurement

For some years now, the weak absorption band of oxygen at 760 nm has also been usable for gas measurement with laser diodes (Wiegleb and Zöchbauer 1994). This O_2 absorption

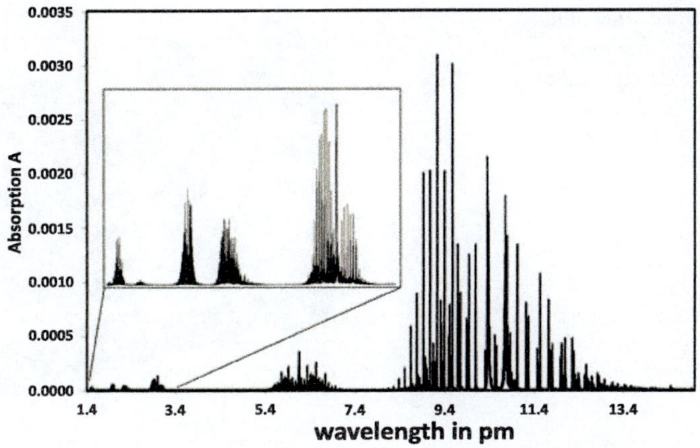

Fig. 9.40 NH_3 spectrum in the entire IR spectral range. In the magnified image, the weak NH_3 harmonic at 1.5 μm, which can be used for a gas measurement with VCSEL or DFB lasers, is shown on the *far left*

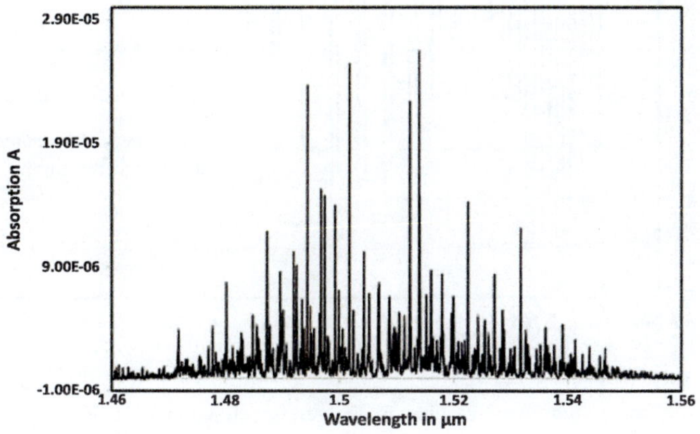

Fig. 9.41 Section of the NH_3 harmonic in the range around 1.5 μm, from Fig. 9.40

band is shown in Fig. 9.43. The strongest absorption lines thus lead to an absorption of only ≈6% for pure oxygen in a 100 cm long measuring distance. Since the typical measuring ranges for oxygen in combustion processes are in the volume % range anyway, this method can certainly be used for gas analysis. Meanwhile several manufacturers offer O_2-measuring instruments on this basis. The advantage of this method is that a direct O_2 measurement in a combustion process or in the exhaust gas stack is possible.

ABB Automation GmbH offers a modern in-situ device (LS4000), with which measuring distances (e.g. in the stack) from 0.5 to 20 m, at a process pressure of 20 bar and a

Fig. 9.42 Detailed view of the *Q-MACS Process fiber optic* with a Quantum Cascade Laser QCL. (Source: Neoplas control GmbH Greifswald)

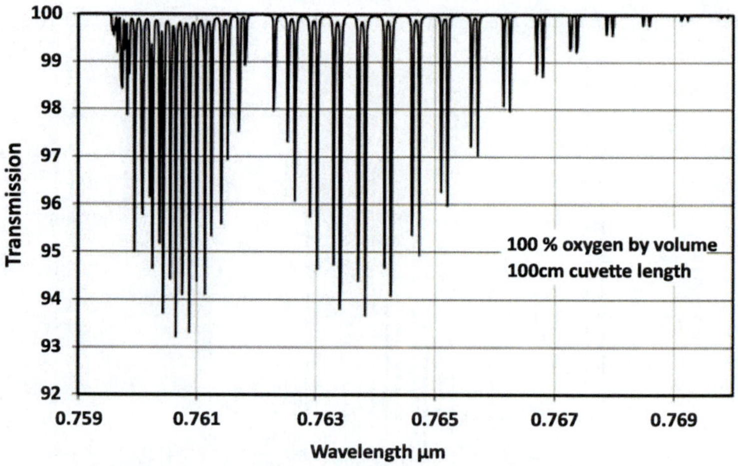

Fig. 9.43 Absorption spectrum of oxygen in the red spectral range at 761 nm. (Data source: HITRAN database)

temperature of up to 1500 °C are possible. The oxygen measuring ranges are indicated with 0...1 vol.% O_2 (smallest measuring range) to 0...100 vol.% O_2 (largest measuring range). The detection limit ($4 \cdot \sigma$) is 100 ppm O_2, with a response time of $t_{90} < 5$ seconds.

Photoacoustic Laser Measurement

Diode lasers are also very well suited for use in photoacoustic spectroscopy (PAS). As in photometric gas measurement, the laser is tuned to the absorption line and modulated with a fixed frequency *f*. If the gas to be detected is now inside the PAS cell, the absorbed radiation energy is converted into heat by inelastic impacts of the colliding molecules (→ temperature increase). Since the PAS cell is sealed off during the measurement, this

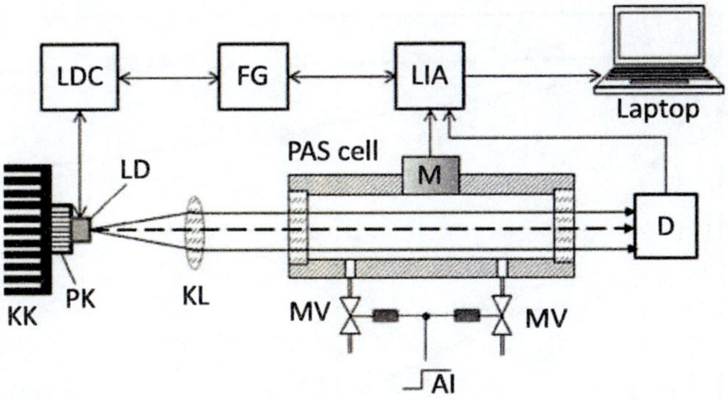

Fig. 9.44 Schematic structure of a PAS gas detection system

temperature increase leads to a periodic pressure increase in the order of magnitude of $\Delta p \approx 10^{-2}$ up to 10^{-1} Pa. The pressure increase is linearly related to the gas concentration in the PAS cell over a wide range. In contrast to photometric gas measurement, the PAS signal is generated from the zero point. If there is no gas in the measuring cell, the microphone does not receive any sound pulses. The zero point is therefore very stable with this method.

Figure 9.44 shows such a structure (Rhine et al. 2012). The diode laser (LD) is brought to the required temperature with a Peltier cooler and scanned with the diode current across the absorption line. This task is performed by the Laser Diode Controller (LDC). The scanning frequency (≈2 kHz) is specified by a function generator (FG) and is then identical to the frequency of the sonic pulses in the PAS cell. The measurement signal is received by a microphone (M) and is forwarded to a lock-in amplifier (LIA). The LIA receives the required reference frequency from the FG. In addition, the transmitted radiation of the laser diode is received by a detector and serves to form a reference signal. Since the PAS effect naturally depends on the irradiated intensity, this must be detected by the detector and offset against the PAS signal. This measure is then also used to stabilize the sensitivity.

Cavity Ring-Down Spectroscopy CRDS

The propagation of light occurs at a speed $c \approx 300,000\,\mathrm{km\,s^{-1}}$. This speed is so great for human perception that it is estimated to be infinitely fast. Since there were no measuring methods, the speed of light was considered infinitely fast until the 17th century. It was not until 1849 that an optical experiment was carried out by Fizeau,[14] in which he was able to measure the difference in transit time Δt over a distance $s \approx 2 \times 8.63$ km using a gear method (Tipler and Mosca 2014). In this case, the difference Δt was ≈57 μs and could be

[14] Armand Hippolyte Louis Fizeau (1819–1896) French physicist.

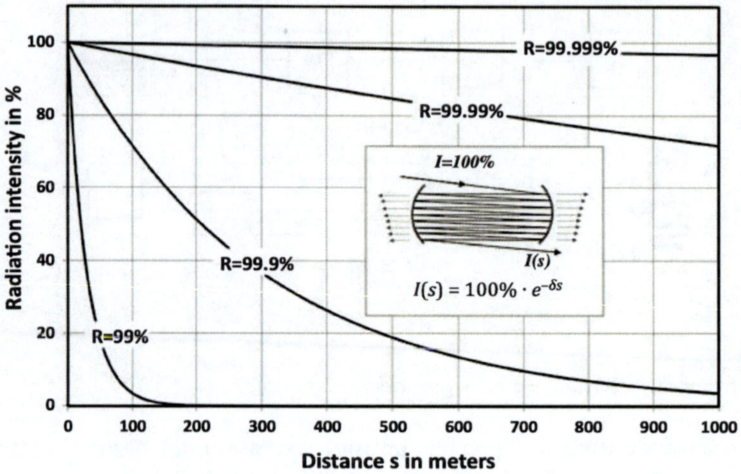

Fig. 9.45 Radiation intensities in a long path cuvette with different mirrors

reliably measured with this purely mechanical method. With today's electronic methods, this time measurement is completely unproblematic.

But, it was not until the 1990s that O'Keefe and Deacon (1988) and Lehmann and Romanini (1996) succeeded in using this method of transit time measurement for gas measurement technology (Busch and Busch 1999). In the CRDS method, a light beam is coupled into a cuvette and passes through this measuring section by multiple reflections. The radiation absorption by the gas takes place along this path and the light beam weakens increasingly until it has completely disappeared. The decay over time can be described by an e-function and is then a measure of the gas concentration in the cuvette. To get a good temporal resolution of this decay behavior, very long distances of several kilometers are required. To realize this boundary condition in a measuring cuvette from <1 m, the beams must be reflected several thousand times. This requires highly reflective mirrors that must have the lowest possible reflection losses. Figure 9.45 shows the change in radiation intensity I as a function of the distance $s = n \cdot L$ travelled. The distance between the two mirrors in this example is $L = 30$ cm. For each reflection, the same percentage of the radiation is transmitted by the mirrors (=Transmission T). However, the prerequisite is that no absorption or scattering occurs in the mirror. It is then valid:

$$I(s) = I \cdot R^n = I \cdot R^{s/L} \tag{9.24}$$

With a reflectance of $R = 99\%$, the injected radiation I dropped below 5% after only 100 meters, while a reflectance of $R = 99.999\%$ after this distance emitted less than

$\Delta I \approx 0.5\%$ of the radiation. Therefore, a reflectance of $R > 99.99\%$ is a prerequisite for a spectroscopic application.

If there is a radiation detector at the output of the long-travel cuvette, it receives the radiation intensity I_0 after the first pass:

$$I_0 = I_{\text{Laser}} \cdot T^2 \tag{9.25}$$

The coupled radiation intensity is the unattenuated (100%) laser intensity I_{Laser}. According to Eq. 9.24, the following intensity is then obtained for one pass ($=2 \cdot L$):

$$I(n) = I_0 \cdot R^{2n} = I_0 \cdot \mathrm{e}^{\ln R \cdot 2n} \tag{9.26}$$

The time t, which is needed for a certain number n, is obtained from $v = s/t$ with $v = c$ and $s = 2nL$ the following expression:

$$t = n \cdot \frac{2L}{c} \quad \text{or} \quad n = \frac{c \cdot t}{2L} \tag{9.27}$$

If you now use this expression in Eq. 9.26, you get the time change $I(t)$ which is recorded by the detector:

$$I(t) = I_0 \cdot \mathrm{e}^{\frac{c \cdot t \cdot \ln R}{L}} \tag{9.28}$$

For a high reflectance of $R > 0.9999$, $\ln(R)$ can also be replaced by $(R - 1)$:

$$I(t) = I_0 \cdot \mathrm{e}^{-\frac{c \cdot t \cdot (1-R)}{L}} = I_0 \cdot \mathrm{e}^{-t/\tau_0} \tag{9.29}$$

The magnitude τ_0 is a measure of the *ring down time* in which the intensity has dropped to 1/e:

$$\tau_0 = \frac{L}{c \cdot (1-R)} = \frac{s}{c} \tag{9.30}$$

The expression $s = L/(1 - R)$ is the effective path length through the absorption cell. Table 9.4 shows some examples for different degrees of reflection (R) and the resulting decay times τ_0 as well as the corresponding path lengths s.

Figure 9.46 shows the optical setup of a CRD spectrometer (Berden and Engeln 2009). The laser is operated using the pulse method, whereby the pulse duration t_P must be shorter than the dwell time in the cavity t_c to avoid superimposition effects. The absorption cell forms a resonator (cavity) in which the laser beam is reflected *n-fold.* At the output of the cell, the time response $I(t)$ is measured.

Table 9.4 Optical path length s of the resonator and ring-down times τ_0, as a function of the reflectance R of the mirrors for a resonator length $L = 30$ cm (Vogelsang 2011)

Degree of reflection R in %	Ring-down time τ_0 in µs	Optical path length s in m
99	0.1	30
99.9	1	300
99.99	10	3000
99.995	20	6000
99.999	100	30,000

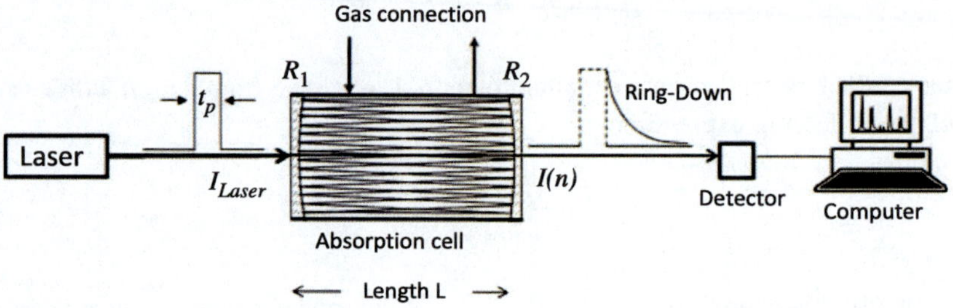

Fig. 9.46 Design of a CRD spectrometer

If the absorption cell also contains a gas with the concentration C,[15] which absorbs the laser radiation, the intensity changes according to the *Lambert-Beer law*, with the corresponding absorption coefficient α of the gas. After the first pass, Eq. 9.25 then changes as follows:

$$I_0 = I_{\text{Laser}} \cdot T^2 \cdot \mathrm{e}^{-\alpha \cdot C \cdot L} \tag{9.31}$$

or after *n passes*:

$$I(n) = I_0 \cdot R^{2n} \cdot \mathrm{e}^{-2n\alpha CL} = I_0 \cdot \mathrm{e}^{2n(\ln R - \alpha CL)} \tag{9.32}$$

The time response $I(t)$ is then, analogous to Eq. 9.29:

$$I(t) = I_0 \cdot \mathrm{e}^{-ct(1 - R + \alpha CL)/L} \tag{9.33}$$

The decay time τ must now also be supplemented by the term $\alpha \cdot L$, which leads to a faster decrease of the radiation intensity in the absorption cell:

[15] In order to avoid confusion with the speed of light c, the concentration is shown as a capital C at this point.

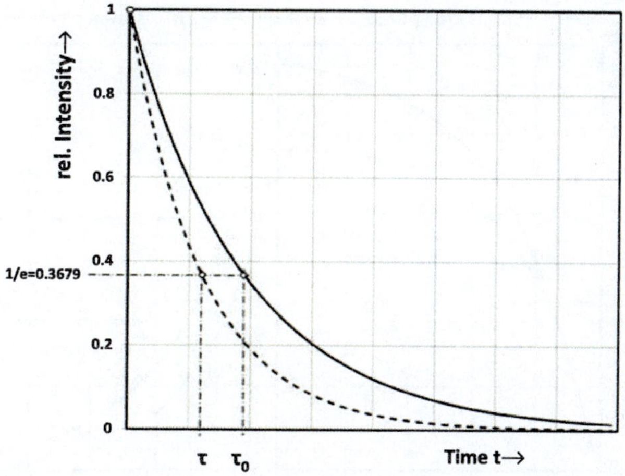

Fig. 9.47 Decay curves in the absorption cell without gas and with gas (*dashed curve*) leading to different τ values

$$\tau = \frac{L}{c(1 - R + \alpha CL)} \qquad (9.34)\leftarrow$$

Figure 9.47 shows both cases, with and without additional absorption by the gas to be measured. The concentration C can then be determined by transforming the above equations:

$$\alpha \cdot C = \frac{1}{c} \cdot \left[\frac{1}{\tau} - \frac{1}{\tau_0}\right] \quad \text{or} \quad C = \frac{1}{\alpha \cdot c} \cdot \left[\frac{1}{\tau} - \frac{1}{\tau_0}\right] \qquad (9.35)\leftarrow$$

The evaluation of the phase shift or the time difference $(\tau_0 - \tau)$ at a certain intensity decrease (e.g.1/e), is also called CAPS (*cavity-attenuated-phase-shift-spectroscopy*).

The calculation of the change in intensity $\Delta I = I_0 - I$, which is a quasi-stationary state at the end of the pulse duration, is called CEAS (*cavity-enhanced-absorption-spectroscopy*) and is the simplest variant of evaluation (Fig. 9.48). In principle, the method can also be operated using the DC method (O'Keefe et al. 1999) or with incoherent radiation sources (gas discharge lamp, IR emitters, . . .) (Schiwon 2004). The concentration is calculated with the intensities I_0 (without gas) and I with gas (van Helden et al. 2013):

$$\mathrm{CEAS}_{\mathrm{Signal}} = \frac{I_0 - I}{I} = \alpha \cdot C \cdot \frac{L}{(1 - R)} \qquad (9.36)\leftarrow$$

respectively;

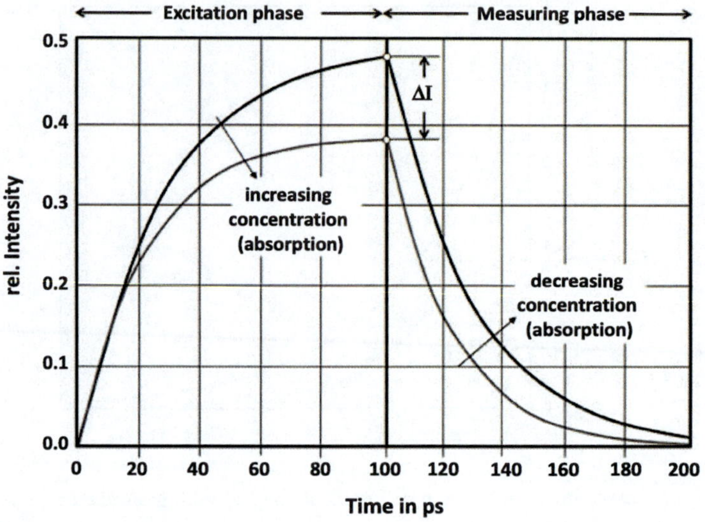

Fig. 9.48 Intensity I in the absorption cell (cavity) for a rectangular irradiation. The *upper curve* shows the intensity curve without gas (empty) compared to selective absorption (*grey curve*) by the gas present. (Source: ansyco 2013)

$$C=\frac{1}{\alpha c}\left[\frac{k}{\sqrt{I}}-\frac{k}{\sqrt{I_0}}\right] \qquad (9.37)\leftarrow$$

With this type of laser spectrometer, extremely low gas concentrations in the sub-ppm range can be reliably detected. Figure 9.49 shows an industrial measuring instrument (*LASAR*) based on CEAS technology. The CEAS device is in a 19″ plug-in unit. A vacuum control ($p \approx 50$ mbar) is required to operate the device to reduce band widening caused by the influence of pressure. The pump unit is in another 19″ rack, below the CEAS device.

Since the entire measuring system was designed for the exhaust gas analysis of combustion engines, the exhaust gas must be extracted via an external sample preparation. The setup has a two-stage filtration with a filter rate of 99.999% for particle sizes up to 0.1 μm. This effort is also necessary, as contamination would cause lasting damage to the highly reflective mirrors. To prevent condensation of water vapor, the filters are heated to a temperature of $T > 100\,^\circ\mathrm{C}$. The exhaust gas is then sucked in through a critical nozzle, which in conjunction with the pump ensures a constant volume flow of ≈0.35 L/min. The pressure in the absorption cell is constantly controlled at $p \approx 50$ mbar. This measure (negative pressure operation) additionally prevents water vapor or other components (e.g. long chain HC) from condensing out on the way from the sample preparation to the CAES instrument. Heated lines are therefore not necessary. The manufacturer offers four different modules, each of which can measure 3 gases simultaneously (Table 9.5).

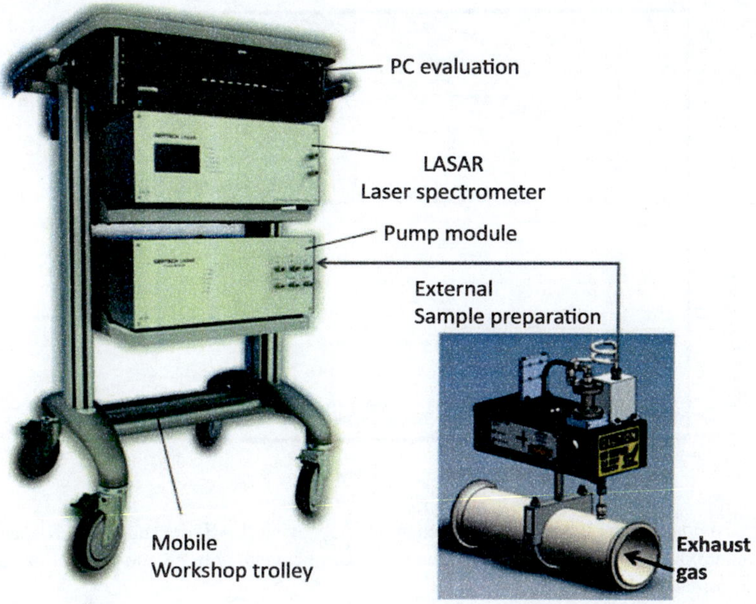

Fig. 9.49 SEMTECH-LASAR measuring system for use in a workshop with external sample preparation for the exhaust gas to be analyzed. (Source: Sensors Inc. Michigan USA)

Table 9.5 Measuring range and combinations of gas measurements

Module	Measuring ranges	
NH_3	NH_3	1000 ppm
	CO_2	30 vol.%
	H_2O	30 vol.%
N_2O	N_2O	50 ppm
	NH_3	100 ppm
	CH_4	200 ppm
NO	NO	800 ppm
	CO_2	20 vol.%
	H_2O	20 vol.%
NO_2	NO_2	800 ppm
	CO_2	20 vol.%
	CH_4	10 vol.%

Source: Sensors Inc. Michigan USA

NH_3 measurement is particularly important for the SCR[16] process, in which ammoniacal agents (*urea/AdBlue*) are added to the exhaust gas to reduce the nitrogen oxide

[16]Selective Catalytic Reduction.

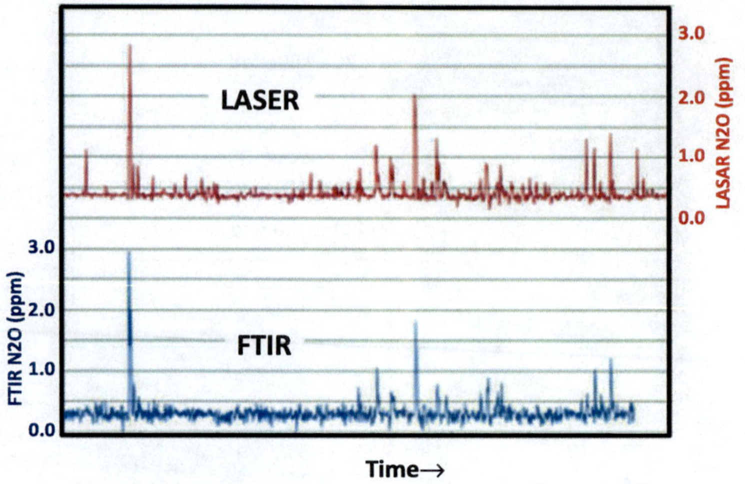

Fig. 9.50 Comparison measurement between an OEAS device (LASAR) and an FTIR spectrometer, using a conventional white multipass cell. Due to the large cuvette volume, the purge time and the gas flow rate is correspondingly high

(NO) content. Excess active ingredient leads to undesirable NH_3 emissions, which must be reliably detected even in traces (≈ppm). The high level water vapor content in the exhaust gas is usually problematic and can interfere with the NH_3 measurement. The LASAR device has therefore solved this problem.

The method was also investigated in comparison to other measuring methods (e.g. FTIR spectroscopy). Better results could be achieved with these comparative measurements than with established methods. Figure 9.50 shows a measurement on a real exhaust system (N_2O measurement). In particular, the detection limit (→ zero-point noise) is better with the CEAS method than with the FTIR measurement. A further technical advantage of the instrument is the low sample gas volume of only 350 ml/min. for the LASAR instrument, compared to 15 L/min. for the FTIR method.

The characteristic curve of the OEAS method (LASAR) is very linear and has only small deviations, which are smaller than 1% (see Fig. 9.51).

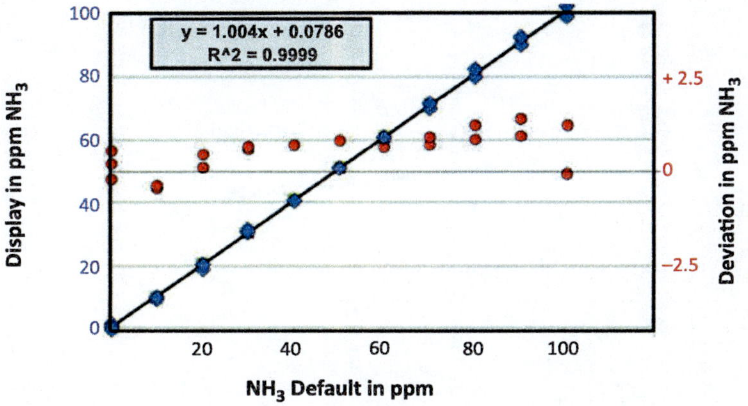

Fig. 9.51 Linearity check of the LASAR device for a concentration range from 0 . . . 100 ppm NH_3. The mixture was made with a gas divider

9.3 *Remote* Measurement Procedure

With the help of optical measuring methods, gas analytical investigations can be carried out over long distances, which are not possible with other methods. As a rule, the gas to be analyzed must be fed to the analyzer via a pump and hose line, or the mobile analyser is placed in the area where the gas atmosphere to be measured is located. Optical methods therefore have the advantage that the light beams can transport the measurement information over long distances at the speed of light. In the meantime, therefore, many methods have been established which make use of these advantages. In particular, the use of laser technology has made it possible to realize novel measuring methods which today play an important role in the field of environmental measurement technology and safety engineering. In particular, the following effects are used:

- Absorption
- Scattering
- Fluorescence
- Running times.

Laser beams can transmit the light power over long distances and are therefore particularly suitable for telemetry applications. They have a diameter ($\varnothing = 2 \cdot R_0$) of a few millimeters, which increases significantly at greater distances due to the beam divergence. The smallest radius R_0 in the beam waist (Fig. 9.52) results from the resonator size L of the laser setup (Naumann and Schröder 1992).

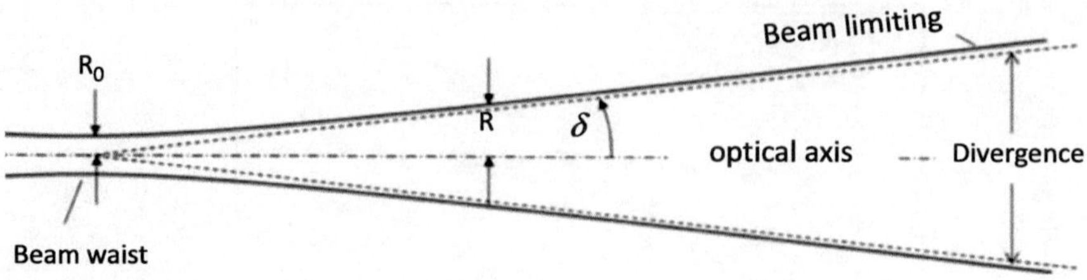

Fig. 9.52 Limitation of a laser beam during its propagation in *z-direction*

$$R_0 = \sqrt{\frac{L\lambda}{2\pi}} \tag{9.38}$$

The further radiation path is hyperbolically limited and can be described with the following equation:

$$R = R_0 \cdot \sqrt{1 + \left(\frac{z \cdot \lambda}{\pi \cdot R_0^2}\right)^2} \tag{9.39}$$

The fanning is described by the divergence δ of the radiation, which also results from the emission wavelength λ

$$\delta = \frac{\lambda}{\pi \cdot R_0} \tag{9.40}$$

For a HeNe laser with an emission wavelength of $\lambda = 633$ nm, the result is a radius R_0 of 0.224 mm. This results in a divergence of $\delta \approx 1$ mrad. Gas lasers with a large resonator therefore have a lower divergence than laser setups with small resonators. Semiconductor lasers with resonator lengths in the mm range are therefore highly divergent and must be adapted with external optical components (lens systems). But even a divergence of $\delta = 1$ mrad ($\approx 0.057^\circ$) leads to a light spot of 2 m diameter after only one kilometer of travel.

For this reason, telescopes are connected upstream of the lasers, which cause the laser beam to expand. The magnification Γ results from the two focal lengths of the lenses:

$$\Gamma = \frac{f_1}{f_2} \tag{9.41}$$

The laser beam is expanded with this magnification, so that the following relationship results for the radius R':

$$R' = R\Gamma = R \cdot \frac{f_1}{f_2} \qquad (9.42)\leftarrow$$

However, this widening reduces the divergence by the same factor. It then applies:

$$\delta' = \frac{\delta}{\Gamma} = \delta \cdot \frac{f_2}{f_1} \qquad (9.43)\leftarrow$$

Example
A laser with a diameter of 1 mm and a divergence of 1 mrad leads to a light spot to 1 km of about 2 m. However, if the laser steel is widened by a factor of 10 ($2R = 10$ mm), a light spot of ≈20 cm is obtained.

Fence Line Monitoring
The simplest interaction is the absorption of radiation on its way through a gas cloud (Fig. 9.53). Figure 9.54 shows such a process. The laser emits the radiation intensity I_0, which is reduced to the value I_1 by selective absorption in the gas cloud. The *Lambert-Beer law* applies to this case, whereby the calculation of the concentration c depends on various factors. First, the expansion L of the gas cloud and the concentration distribution must be known. Within the gas cloud, the gas concentration cannot be assumed to be homogeneous. Under these conditions only the product of $c \cdot L$ can be determined. Gas measuring instruments that work according to this method therefore also indicate the measuring ranges and detection limits in ppm · m. For a measured value of 10 ppm · m, there can be a concentration c of 100 ppm over 10 cm, as well as a concentration c of 1 ppm over 10 meters.

In addition to the linear setup (Fig. 9.54a), in which the transmitter (laser) and receiver (detector) are spatially separated, it is also possible to place both units in one location (Fig. 9.54b). This construction has the advantage that the radiation passes the gas cloud twice, thus increasing the absorption effect. Furthermore, all active components are placed in one place, which simplifies the installation. For short distances, the setup a is often used,

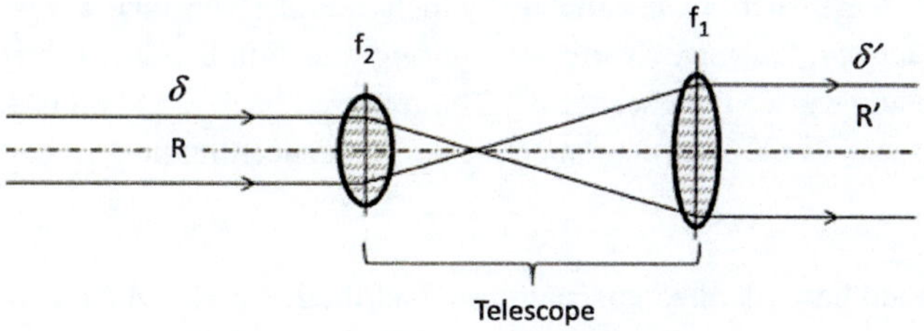

Fig. 9.53 Telescope for reducing the beam divergence of lasers

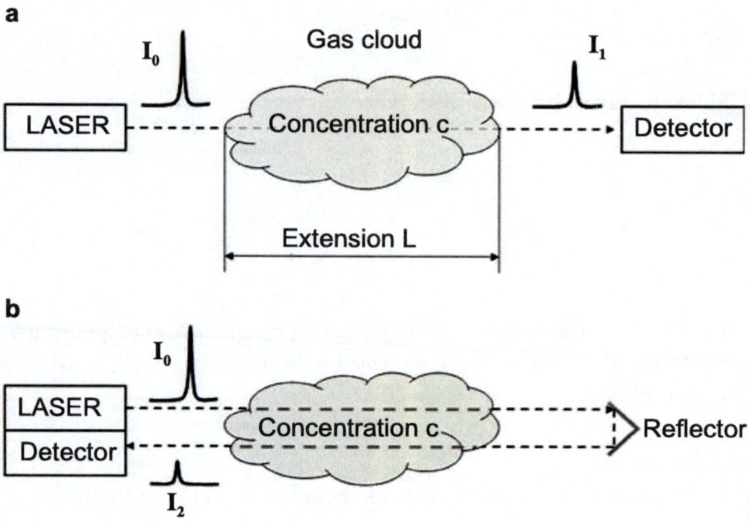

Fig. 9.54 Fence line measurement with two different setups

Fig. 9.55 Fence-line arrangement (OPSIS (OPSIS AB, Furulund, Sweden)-DOAS (Differential Optical Absorption Spectroscopy)-Systems) for laser measurement of exhaust gases from shipping in Essen-Kettwig. (Source: FH-Düsseldorf, K. Weber)

while setup b is used for large distances. In this case a reflector must be installed on the opposite side and optically aligned.

Figure 9.55 shows such an installation, which was used to take a gas measurement across Lake Baldeney. One can clearly see a strong radiation that is reflected on the other side of the lake and causes light scattering. The reflected back radiation is then somewhat weakened and receives the information about the gas concentration.

Example

An SO_2- gas cloud has a defined concentration distribution $c(R)$, which is to be measured with a laser setup (Fig. 9.56). The absorption coefficient of SO_2 at the laser wavelength is $\alpha = 23.5\,\text{cm}^{-1}$. The question arises how the laser radiation I_0 changes through this gas cloud.

According to the Lambert-Beer law: $I(c) = I_0 \cdot e^{-\alpha \cdot c(R) \cdot L}$

By integration, $c(R) \cdot L$ can be calculated. For this purpose, the integration range must first be determined by calculating the zero points:

$$-3.8R^2 + 35.34 = 0$$

$$3.8R^2 = 35.34$$

$$R = \pm\sqrt{\frac{35.34}{3.8}} = \pm 3.05\text{m}$$

$$c \cdot L = 2 \cdot \int_0^{3.05} (-3.8R^2 + 35.34)\,\text{d}R$$

$$c \cdot L = 2 \cdot (3.8 \cdot 3.05^2 + 35.34) = 143.7\,\text{ppm} \cdot \text{m}$$

The product $c \cdot L$ is therefore (=143.7 ppm · m = 14370 ppm · cm) and thus corresponds to an average concentration of 143.7 ppm · m/6.1 m = 23.6 ppm. The change in intensity $I_0 = 1$ can now be calculated as follows:

$I(c) = I_0 e^{-23.5\,\text{cm}^{-1} 14{,}370\,\text{ppm}\cdot\text{cm} 10^{-6}} = 0.713$. The laser radiation that passes through the gas cloud is thus attenuated by 28.7%.

Another possibility of fence line monitoring is to monitor a particularly endangered area on all four sides. The laser radiation must be deflected at the corners with mirrors. If an exhaust cloud passes through this *optical fence*, radiation is absorbed and registered by the detector. Depending on the wind direction, the gas cloud emerges at different points (Fig. 9.57).

The advantage of the fence line measurement is a large area monitoring of a plant or a terrain with a laser beam. In the practical implementation, however, there are many things to consider. Especially the weather influences (rain, snow, fog, . . .) lead to an impairment of the measurement, because these influences can only be compensated to a certain degree. Heavy snowfall or dense fog leads to a complete scattering of the laser radiation, so that no usable information reaches the detector. Furthermore, it must be ensured that the optical interface is not contaminated or shaded by insects or animals (birds).

Direct Flue Gas Measurement in the Stack (Cross Stack)

In emission measurement technology in power plants, waste incineration plants or other power engineering facilities, a direct gas measurement in the stack or flue gas duct has become established, which is called *in-situ measurement.* The light radiation passes through the stack (*cross stack*) and thus records the integral value of the gas concentration c under the real pressure and temperature conditions (p_i, T_i). Figure 9.58 shows two possible arrangements which are frequently used in practice. The main difference is that in the first case (A.) the transmitter and receiver units are placed on one side, while in the second case (B.) a separation is made. In case A., a reflector must be placed on the opposite

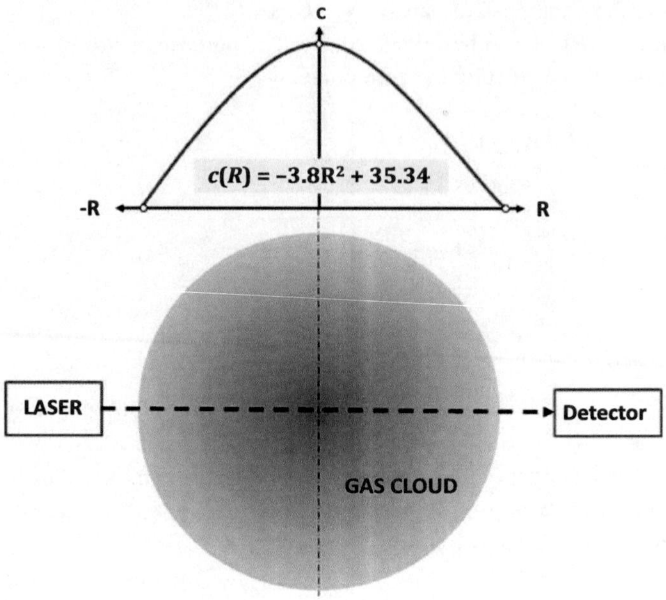

Fig. 9.56 Gas cloud with a known concentration distribution $c(R)$

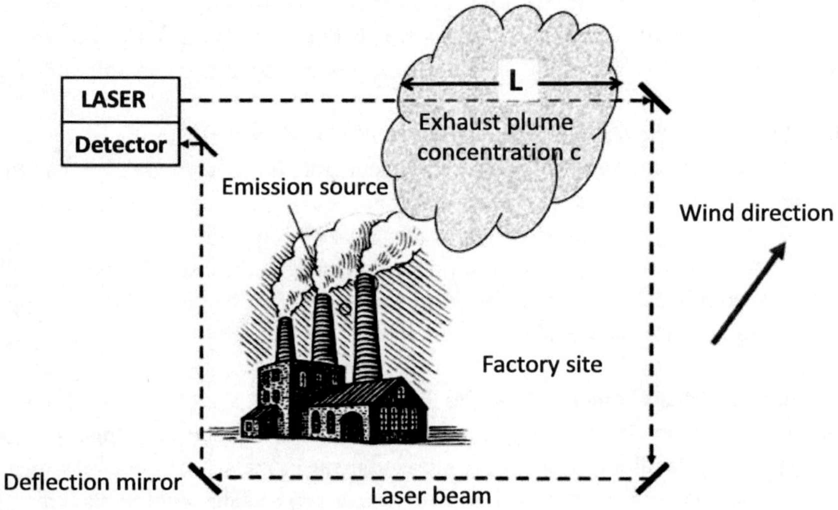

Fig. 9.57 *Fence line monitoring* around a factory site to detect an exhaust cloud as a result of an unforeseen incident or accident

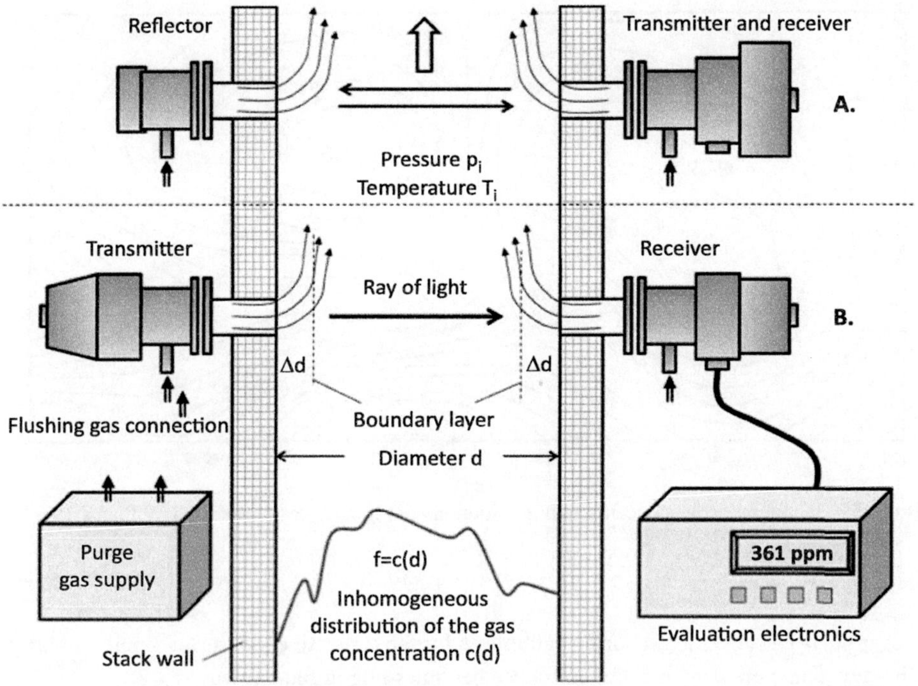

Fig. 9.58 Different arrangements of the transmitter and receiver units for an optical emission measurement in a stack

side to redirect the radiation back towards the receiver. The advantage of this arrangement is the double passage through the stack with an increased measuring effect. With this arrangement, however, the reflector must be adjusted very precisely and must also be stable over a long period of time. In large power plants, with considerable building vibrations, this cannot always be guaranteed. If, on the other hand, the sensor and receiver are separated, the radiation from the transmitter can be aligned in such a way that the receiver is *over-radiated.* This makes the light spot much larger than the receiver aperture. Inaccurate adjustments or movements/vibrations do not have as strong an effect in this arrangement as in the first case.

To protect the optical interfaces (windows) from dust and other deposits, they are permanently surrounded by an air curtain. For this purpose, ambient air is sucked in via a pump and cleaned by a dust filter. The purge air thus treated is then conducted between the windows and the exhaust gas, thereby displacing the unwanted particles. In this case, the air cushion spreads into the chimney and forms a thin boundary layer Δd that is not identical to the exhaust gas. The effective measuring distance is then reduced to d^*.

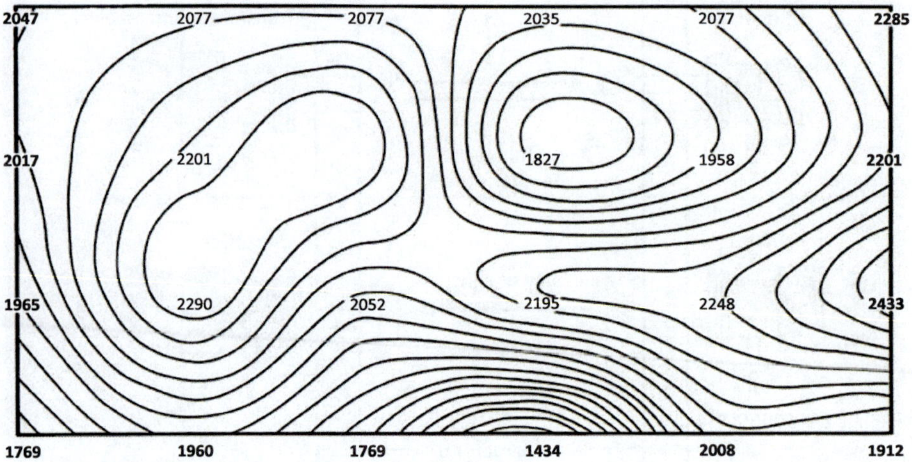

Fig. 9.59 Inhomogeneous concentration distribution $c(d)$ in a stack (Brooks and Williams 1976)

$$d^* = d - 2 \cdot \Delta d \tag{9.44}$$

This influence is considered during calibration for the respective measuring point. For large chimney diameters $d \ll \Delta d$, this influence becomes negligible anyway.

A great advantage of the in-situ measurement technique is the integral measurement of the gas concentration c over the entire stack diameter d. Due to streak formation, the concentration distribution $f = c(d)$ is inhomogeneous and depends on the structural conditions. Figure 9.59 shows such a distribution in a flue gas duct (Brooks and Williams 1976). The points of equal concentration related to a line, which leads to a picture that reminds of an isobaric map. The concentration values vary within a range of $\pm 25\%$.

Tunnel Monitoring

Tunnel installations shall be monitored for CO and NO concentration to control ventilation or to check the functioning of the ventilation system. Optical methods are also suitable here, as measurements can be taken directly in the ambient air without any time delay (Fig. 9.60). Furthermore, there are no disturbing influences from rain and snowfall which could impair the optical method. The CO or NO/NO_2 concentrations in a tunnel result from the exhaust gases of vehicles entering the tunnel. The higher the vehicle density, the more exhaust gases are produced per unit time. The performance of the ventilation system must therefore be permanently adapted to the current situation. The concentration values increase towards the middle of the tunnel, while low concentration values are measured at the ends. Depending on the tunnel length, several gas measuring devices are installed. The distance between the transmitter and receiver unit is specified as 10–20 m, depending on the manufacturer. The distance between the individual measuring units is between

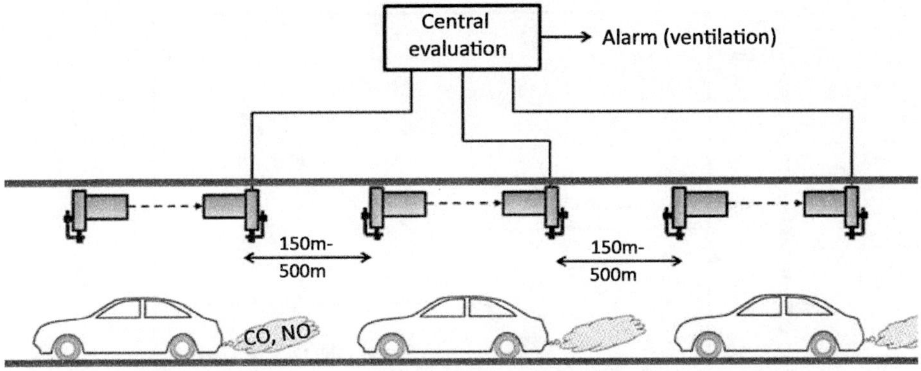

Fig. 9.60 Tunnel monitoring with optical gas measuring devices

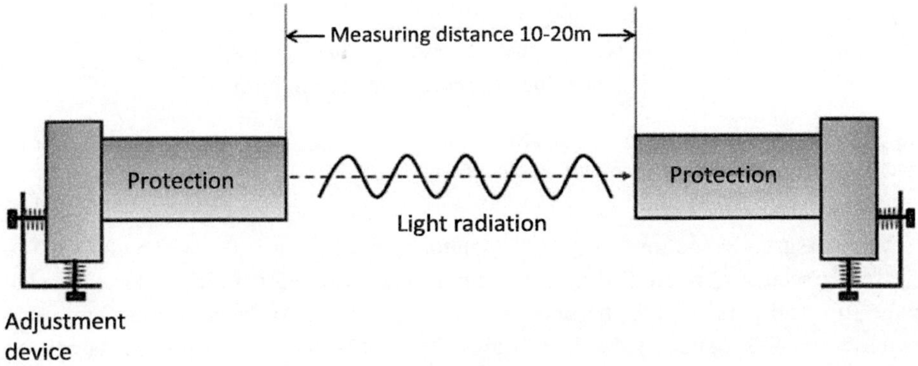

Fig. 9.61 Arrangement of a measuring unit for optical monitoring of the gas concentration in a tunnel

150 and 500 m, depending on the installation site (country-specific). In a tunnel with a length of 10 km approximately 20–60 measuring systems are then installed.

The transmitter and receiver units are equipped with a protective tube to protect these superstructures from the influence of dust, which can also occur in a tunnel system. Figure 9.61 shows such an arrangement. The adjustment of the two measuring heads is particularly important so that the optical transmission from the transmitter unit to the receiver is as loss-free as possible. This adjustment must be carried out before the first commissioning. Then the measuring units are calibrated. The information from the individual measuring stations is forwarded to a central unit for evaluation. There, the gas concentrations are analyzed and, if necessary, an alarm or fan control is triggered if the gas concentrations are exceeded.

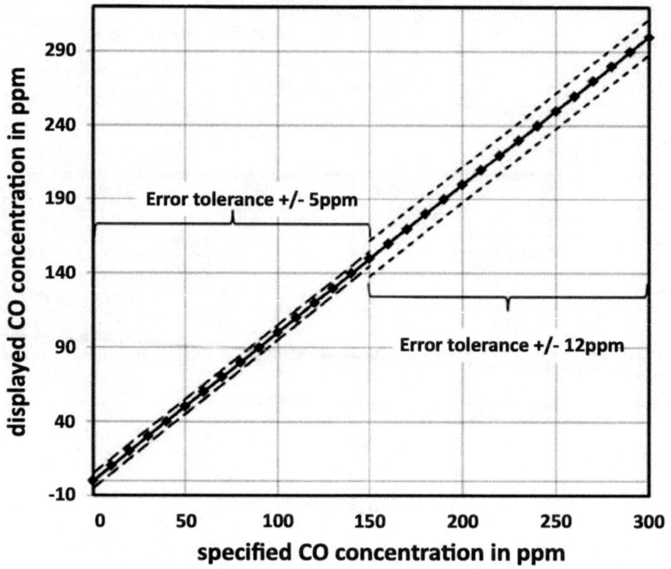

Fig. 9.62 Representation of the characteristic curve and the associated measurement errors of an optical gas measurement system for tunnel monitoring

The measuring range for this type of monitoring equipment is 0. . .300 ppm CO. The measuring accuracy is specified by the manufacturers with ±5 ppm in the lower display range (0. . .150 ppm). For high concentrations (0. . .300 ppm) the measuring error also increases to ±12 ppm. Figure 9.62 shows the calibration curve with the associated measurement errors.

LIDAR

In the field of environmental metrology, various methods have been established over the last decades which allow remote sensing of pollutants or environmentally relevant substances over a large distance. The LIDAR method will be explained here as a representative of these measurement methods.

The term LIDAR refers to all optical measuring methods which, in addition to determining the gas concentration, also allow the localization of the gas distribution. The spatial distribution of the gas concentration plays an important role in many areas. The best-known distribution of this type is the ozone layer, which has a pronounced structure at altitudes up to 60 km, calculated from the ground. This structure is altered by many external influences and is still the subject of scientific studies today. The word *LIDAR* is derived from the English description ***Light Detection And Ranging.***

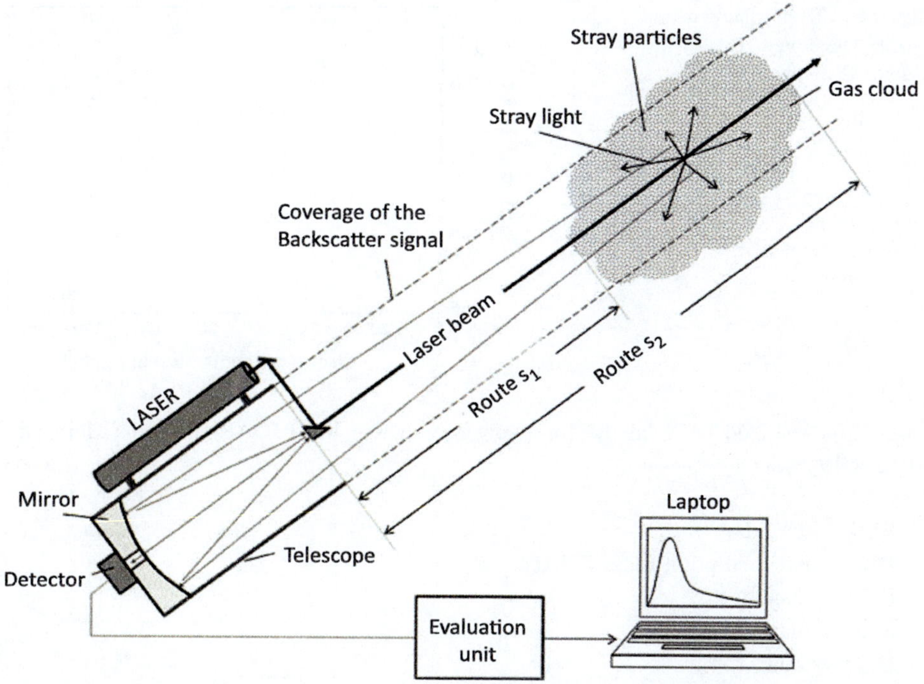

Fig. 9.63 Principle representation of a LIDAR instrument with atmospheric backscattering (according to Klein and Werner 1993)

Figure 9.63 shows the principle arrangement of a LIDAR device. The laser is usually used as a pulsed version, which emits short light pulses (Δt_P) containing high energy. This ensures that this laser beam can penetrate the atmosphere over many kilometers. On its way through the atmosphere, it is scattered by particles/aerosols (Mie scattering) and molecules (Rayleigh scattering). However, the scattered radiation is not directed and spreads in all directions. The part that is emitted in the direction of the LIDAR device enters the telescope and is focused in bundled form on a detector. The temporal difference (Δt_S) between the different locations where light scattering occurs can be determined from the differences in propagation time. It then applies:

$$\Delta t_{S1} = \frac{2s_1}{c} \quad \text{and} \quad \Delta t_{S2} = \frac{2s_2}{c} \tag{9.45}$$

In a stratospheric ozone measurement, the laser beam travels a distance up to 50 km. The delay time resulting from Eq. 9.45 is then 0.3 ms. On the way back to the telescope, the scattered radiation passes the same distance again and can cause a further absorption (+ scattering). The evaluation of the different runs with the different delay times is quite complex and is realized with appropriate computer programs. Depending on the structure, a

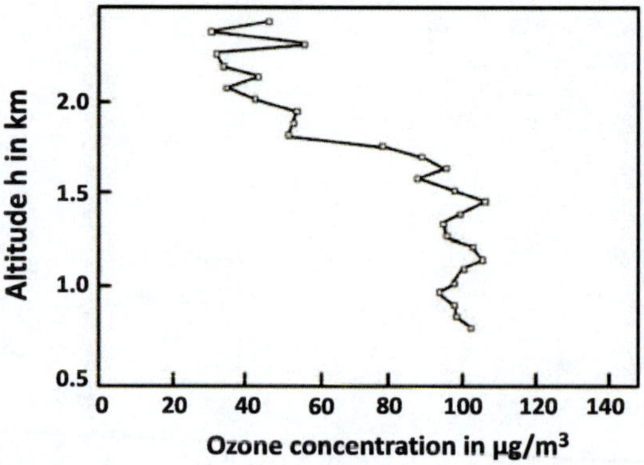

Fig. 9.64 Tropospheric ozone profile, taken on 4.10.1991 in Itzehoe (Bösenberg)

distinction is made between the following variants of the LIDAR method (Klein and Werner 1993):

- Backscatter lidar
- Differential absorption Lidar (DIAL)
- Fluorescence lidar
- Raman lidar
- Doppler lidar.

Stratospheric Ozone Layer

The maximum concentration of the ozone layer is located at an altitude of about 30 km. Ozone is formed there permanently by solar radiation and can be described by the following reaction equation:

$$3 \cdot O_2 + h\nu \leftrightarrow 2 \cdot O_3 \tag{9.46}$$

However, the ozone is not stable and under normal conditions decays back to oxygen within a few minutes. However, in the stratosphere there are other conditions (low temperature, low pressure) under which the ozone is stable for a longer period of time. Depending on the altitude, the life span there can last several days. In the night hours, the ozone concentration in this layer therefore hardly decreases (Fig. 9.64). However, due to the seasonal change in solar radiation, the concentration in this layer changes considerably. In the spring months, the ozone concentration is higher by a factor of ≈2 than in the autumn months (Fig. 9.65). First measurements of the ozone layer were carried out with a weather balloon, but only up to an altitude of $h = 20 - 30$ km. The pronounced profile of the ozone layer (see Fig. 9.66) reaches up to an altitude of 100 km and can therefore not be measured with a weather balloon.

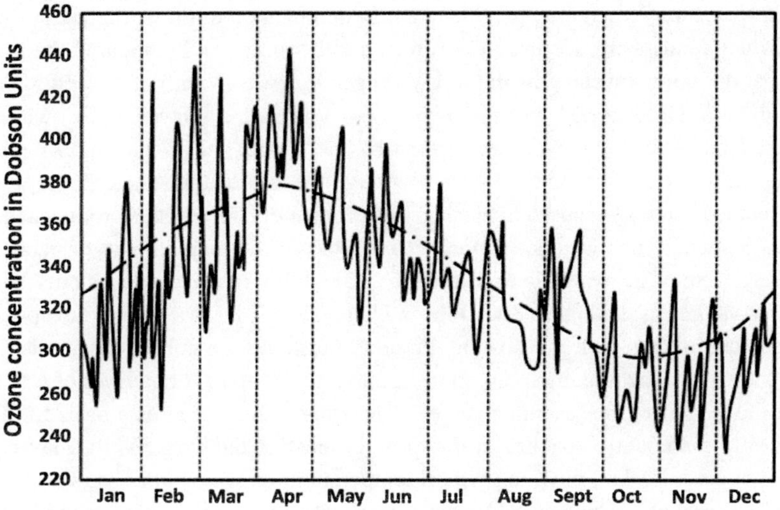

Fig. 9.65 Seasonal variations of the ozone layer measured with a Dobson photometer (measured in Belgium)

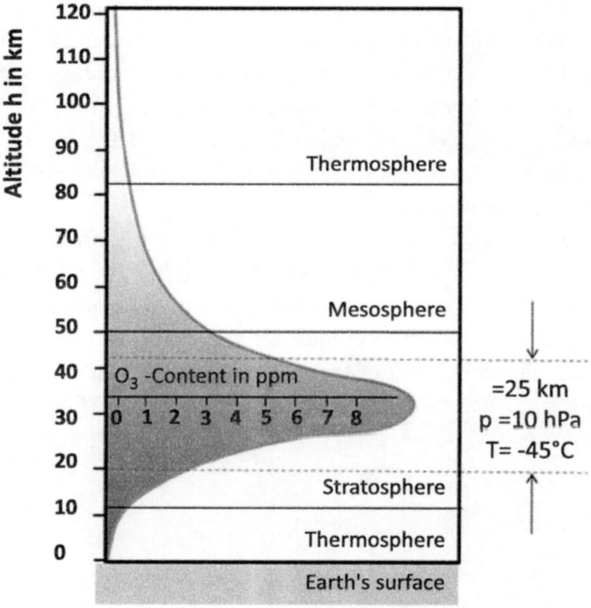

Fig. 9.66 Vertical distribution (altitude profile) of the ozone layer

In 1924, *Dobson*[17] therefore developed a different method of measuring ozone concentration in the atmosphere. He used the sun as a light source and compared the radiation received at different wavelengths in the UV range. At a wavelength of $\lambda_1 = 305.5$ nm the solar radiation is absorbed by the ozone (→ measuring wavelength), whereas at $\lambda_2 = 325.4$ nm there is no more absorption by the ozone. This wavelength is used for reference purposes. As the sunlight passes through the entire atmosphere (layer thickness L), this method can only be used to measure the product of $c \cdot L$ For this product a new unit has been defined, which is called *Dobson Unit (DU).* If the total amount of ozone in the atmosphere were to be available as 100 (0 °C, 1013 hPa), the layer would only be 3 mm thick. This value was defined as 300 Dobson Units (DU). 1 DU therefore corresponds to a layer thickness of 0.01 mm pure ozone. Figure 9.66 shows the concentration distribution (profile) of ozone. The maximum O_3 concentration ($c \approx 9$ ppm) is in a layer of $h \approx 30$ km. This profile is subject to seasonal changes. This value can be used as a basis for a rough estimate of the amount of ozone. In this profile, most of the ozone is in a layer with a thickness of $\Delta h \approx 25$ km, in which typical temperatures of $T \approx -45$°C and a pressure of $p \approx 10$ hPa prevail. If one would now assume the O_3 concentration as constant, for simplification, the following value for the Dobson unit results:

$$c \cdot L \cdot \frac{T_0.T}{p \cdot p_0} = 25 \cdot 10^6\,\text{mm} \cdot 9 \cdot 10^{-6} \cdot \frac{273\,\text{K} \cdot 10\,\text{hPa}}{228\,\text{K} \cdot 1013\,\text{hPa}} = 2.72\,\text{mm} = 272\,\text{DU} \qquad (9.47)\leftarrow$$

This result corresponds to the measured values, which lie between 150 DU and 500 DU.

The arrangement for measuring the ozone concentration according to Dobson is shown schematically in Fig. 9.67. But now, however, other measuring methods (LIDAR, satellites) have also been developed, with which above all the height profiles can be recorded. It is interesting to note that these measuring instruments are still compared and calibrated with the Dobson photometer.

Investigations into the decrease in ozone concentration (→ *ozone hole*) in Antarctica were carried out using laser techniques, which were later supplemented by satellite measurements (Fig. 9.68). The measuring instruments in the satellites use not only solar radiation but also radiation from stars to measure the ozone layer. Since the satellite is in orbit, it can measure different areas of the ozone layer and then combine them into one image. These images are then visualized in a false-color display (Fig. 9.69).

[17] Gordon Miller Bourne Dobson (1889–1976) British physicist and meteorologist.

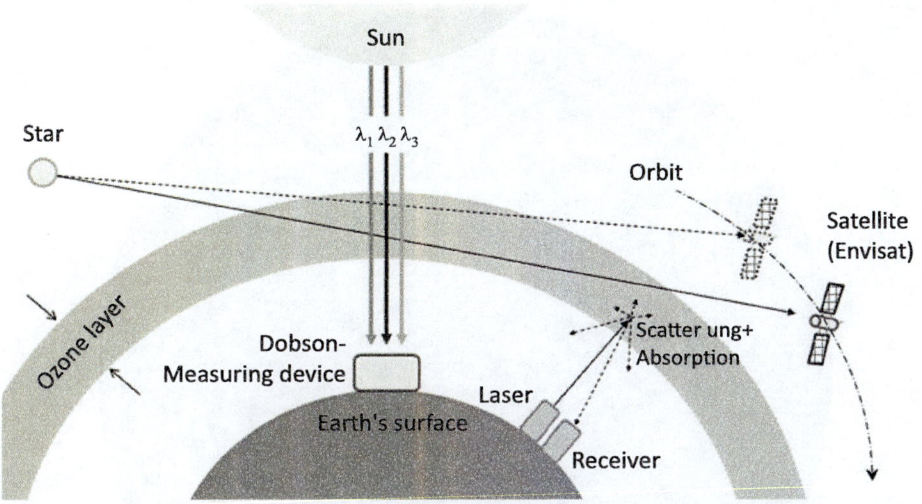

Fig. 9.67 Different ways of measuring the ozone layer, with optical remote measurement methods

Fig. 9.68 LIDAR mission. (Source: Starfire Optical Range on Kirtland US Air Force Base, New Mexico)

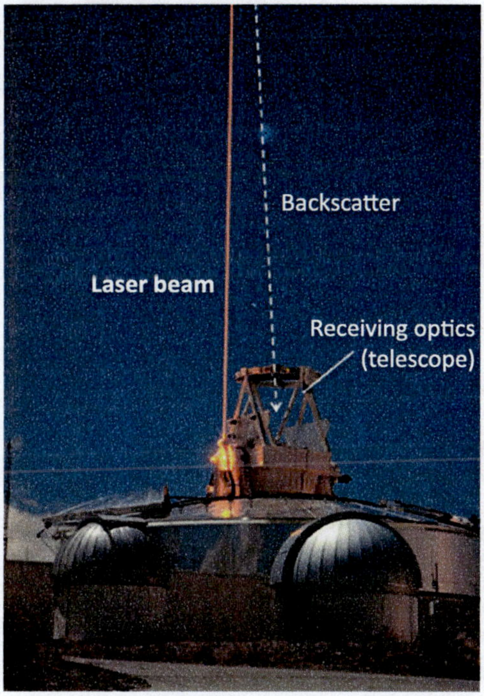

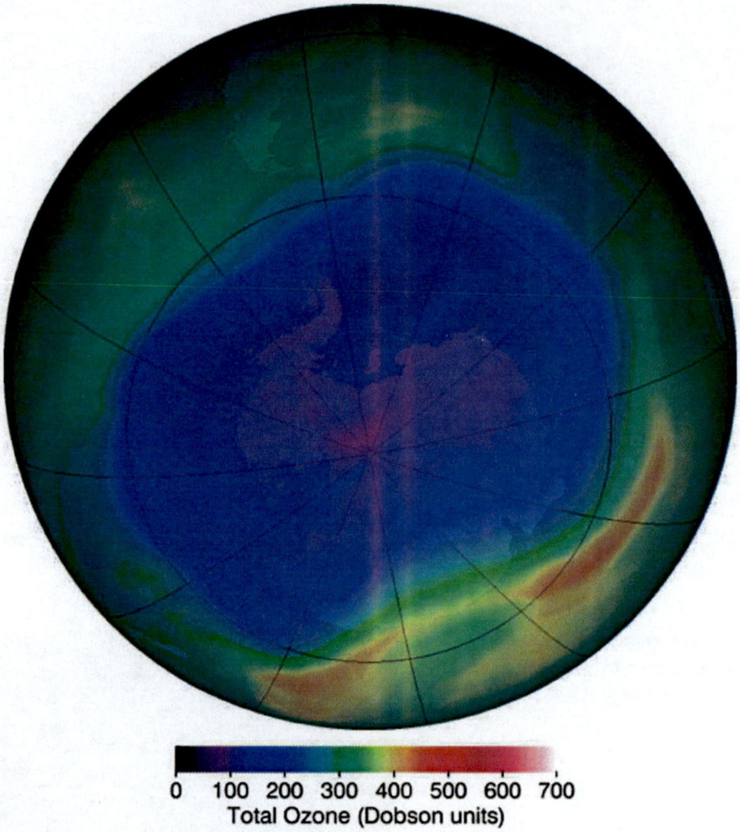

Fig. 9.69 Illustration of the largest expansion of the ozone hole above the South Pole to date on 24 September 2006, with an expansion of 29.6 million km^2. (Source: NASA-USA)

References

Baumbach, G.: Luftreinhaltung. Springer (1990)

Berden, G., Engeln, R.: Cavity Ring-Down Spectroscopy Techniques and Applications. Wiley (2009)

Brooks, E.F., Williams, R.L.: Flow and Gas Sampling Manual. EPA 600/2-76-203 (1976)

Busch, K.W., Busch, M.A.: Cavity-Ringdown Spectroscopy, An Ultratrace-Absorption Measurement Technique. ACS Symposium Series 720 (1999)

Butler, J.F., Nill, K.W., Mantz, A.W., Eng, R.S.: Applications of Tunable-Diode-Laser IR Spectroscopy to Chemical Analysis. Am. Chem. Soc., AVS Symposium Series. **85**, 12–23 (1978)

Clough, P.N., Trush, B.A.: Mechanism of chemiluminescent reaction between nitric oxide and ozone. Faraday Soc. **63**, 915–925 (1967)

Colvin, A.E., et al.: A novel solid-state oxygen sensor. John Hopkins Tech. Digest. **17**(4), 377–385 (1996)

Demtröder, W.: Laserspektroskopie – Grundlagen und Techniken. Springer (1993)

Eng, R.S., Butler, J.F., Linden, K.J.: Tunable diode laser spectroscopy: an invited review. Optical Eng. **19**(6), 945–960 (1980)

Görlich, P.: Laser in der Spektrometrie. Experimentelle Technik der Physik. **17**(2), 105–125 (1969)

Hesse, J., Preier, H.: Lead Salt Diodes. Festkörperprobleme XV, 229–251 (1975)

Hinkley, E.D.: Tunable infrared lasers and their applications to air pollution measurements. Opto-Electron. **4**, 69 (1972)

Iseni, S., Reuter, S., Weltmann, K.-D.: NO_2 dynamics of an Ar/Air plasma jet investigated by in situ quantum cascade laser spectroscopy at atmospheric pressure. J. Phys. D: Appl. Phys. **47**, 1–7 (2014)

Jahnke, J.A.: Continuous Emission Monitoring. Wiley (2000)

Kautsky, H.: Quenching of luminescence by oxygen. Trans. Faraday Soc. **35**, 216–219 (1939)

Klein, V., Werner, C.: Fernmessung von Luftverunreinigungen mit Lasern und anderen spektroskopischen Verfahren. Springer, Berlin (1993)

Lehmann, K., Romanini, D.: The superposition principle and cavity ring-down spectroscopy. J. Chem. Phys. **105**(23), 10263–10277 (1996)

Naumann, H., Schröder, G.: Bauelemente der Optik. Hanser, München (1992)

O'Keefe, A., Deacon, D.A.G.: Review of Scientific Instruments **59**, 2544 (1988)

O'Keefe, A., Scherer, J.J., Paul, J.B.: CW integrated cavity output spectroscopy. Chem. Phys. Lett. **307**, 343–349 (1999)

Preier, H., Herkert, R., Pfeiffer, H.: Growth of Pb_{1-X} Se_X single Crystals by sublimation. J. Cryst. Grow. **22**, 153–158 (1974)

Rhein S., Bruhns H., Wolff M., Nähle L., Fischer, M., Koeth J.: Photoacoustic Methane Detection using a noval DFB-type Diode Laser at 3.3 µm. The 14th International Meeting on Chemical Sensors IMCS 2012. https://doi.org/10.5162/IMC2012/2.5.6 (2012)

Schiwon, R.: Cavity Enhanced Absorption Spectroscopy von dünnen Filmen mit Ferninfrarotstrahlung. Dissertation. Ruhr-Universität, Bochum (2004)

Schwarzenbach, B., Hüglin, C.: Selektive Messung von Stickstoffdioxid mit Photolysekonverter. Messbericht der EMPA Dübendorf Schweiz, http://www.empa.ch/nabel (2011)

Sigrist, M.W.: Air Monitoring by Spectroscopic Techniques. Wiley (1994)

Tipler, P.A., Mosca, G.: Physik für Wissenschaftler und Ingenieure. 7. Aufl. Springer Spektrum, Heidelberg (2014)

van Helden, J.H., Lang, N., Macherius, U., Zimmermann, H., Ropcke, J.: Sensitive trace gas detection with cavity enhanced absorption spectroscopy using a continuous wave external-cavity quantum cascade laser. Appl. Phys. Lett. **103**, 131114 (2013)

van Heusden, S.: Ozon, NO und NO_2-Detektion auf der Grundlage der Chemilumineszenz. Philips technische Rundschau. **33**(12), 375–383 (1974)

Vogelsang, M.: Cavity-Ring-Down-Spcktroskopie als nachweisstarke Detektionsmethode in analytischen Fließsystemen. Dissertation, Universität Ulm (2011).

Vollmer, M., Möllmann, K.-P.: Infrared Thermal Imaging. Wiley-VCH, Weinheim (2010)

White, R.I., Boulton, C.A., Mungy, A.: A Novel: Non-invasive method of measuring in-pack oxygen concentration and its application in the study of staling of fruit-flavoured alcoholic beverages. FCI-Environmental, 1–7 (2003)

Wiegleb, G.: Aktive Ramanstreuung in organischen Flüssigkeiten und Gasen. Diplomarbeit, Universität Essen (1980)

Wiegleb, G., Zöchbauer, M.: New optical methods for applications in process analysis and medical instruments. Exp. Techn. Phys. **40**(2), 241–257 (1994)

Related References

Engeln, R., Berden, G., Peeters, R., Meijer, G.: Cavity enhanced absorption and cavity enhanced magnetic rotation spectroscopy. Rev. Sci. Instrum. **69**, 3763–3769 (1998)

Peeters, R., Berden, G., Apituley, A., Meijer, G.: Open-path trace gas detection of ammonia based on cavity-enhanced absorption spectroscopy. Appl. Phys. B. **71**, 231–236 (2000)

Vollmer, M., Möllmann, K.-P.: CO_2-Detektion mit IR Kameras: Grundlagen, Experimente und Anwendungen. Technisches Messen **79**, 65–72 (2010)

10 Humidity Measurement in Gases

The term *moisture* is generally understood to mean the presence of water in liquid and gaseous form. If this water content is related to the ambient air, it is referred to as humidity. Gas humidity is the proportion of water for any gas or gas mixture. The water content in solids is described via material moisture and is not dealt with further in this chapter.

Moisture measurement is an important part of gas measurement technology, since in almost all applications a more or less large water vapor content is present. In combustion processes for example, water is formed as a reaction product from the HC compounds, whereas in medical technology (breathing gas analysis) the water content originates from the lungs. Since the water content is very closely linked to the gas temperature, it can change over a wide range and thus have a massive influence on the concentration ratios in a

G. Wiegleb, *Gas Measurement Technology in Theory and Practice*,
https://doi.org/10.1007/978-3-658-37232-3_10

gas mixture. Furthermore, the resulting condensation of the water plays an important role in the gas treatment of the sample gas (see Sect. 14.1).

10.1 Fundamentals

The concentration of gaseous water (H_2O) in a mixture is given in gas measurement technology in the usual concentration units of volume %, ppm or g/m³. In addition to these specifications, the water vapor concentration c_{H_2O} can also be described by the term *moisture f*. A distinction is then additionally made between absolute and relative humidity. The concentration is derived from the vapor pressure (see Sect. 2.1). Above a layer of water, depending on the water temperature T, a characteristic saturation vapor pressure p_S is formed which, according to *Dalton*'s law, adds up to the pressure p_t of the dry gas. For the total pressure p the following then applies:

$$p = p_S + p_t \tag{10.1}$$

The vapor pressure p_S, which is set at a certain temperature T, is also called *saturation vapor pressure*. This term is used to express that it is the maximum possible vapor pressure that can occur at a certain temperature. The temperature dependence of p_S is described by the Magnus formula:

$$p_S(T) = a \cdot \exp\left[\frac{\beta \cdot T}{\alpha + T}\right] \tag{10.2}$$

The coefficients a, α and β are different for water and ice, and are summarized in Table 10.1. Since this is an empirical formula, there are different values for these coefficients in the literature, but they differ only slightly from each other (Wernecke 2003). In general, the values of Sonntag (1990) are used for this calculation, which are also found in the VDI/VDE guideline 3514.

The accuracy of the Magnus formula is ±0.6% for a temperature range $T < 60\,^\circ$C. Above 60 °C the deviation from the current table values (Cerbe 2008) increases to 2.3%. In this case, above 50 °C the coefficients for the temperature range of 50–180 °C must be used.

Figure 10.1 shows the table values (Table 10.2) and the Magnus formula (Eq. 10.2). The results are almost congruent in the graphical representation. The difference formation Δp then shows the deviations (second axis labeling) between both values.

Below 0 °C the water usually freezes. Under certain conditions, however, the water can cool down even further (supercooled water) without solidifying. The vapor pressure above this supercooled water surface is then greater than that of an ice surface of the same temperature. Figure 10.2 shows this relationship for a temperature range from −16 to 5 °C (Hengstenberg et al. 1980).

Table 10.1 Coefficients of the Magnus formula (VDI/VDE 3514)

Condition	*T-range* (°C)	*a* (hPa)	*β*	*α* (°C)
Above water	Up to 50	6.112	17.62	243.12
Above water	50 to 180	5.95	16.819	227.3163
Over ice	−80 to 0.01	6.112	22.46	272.62

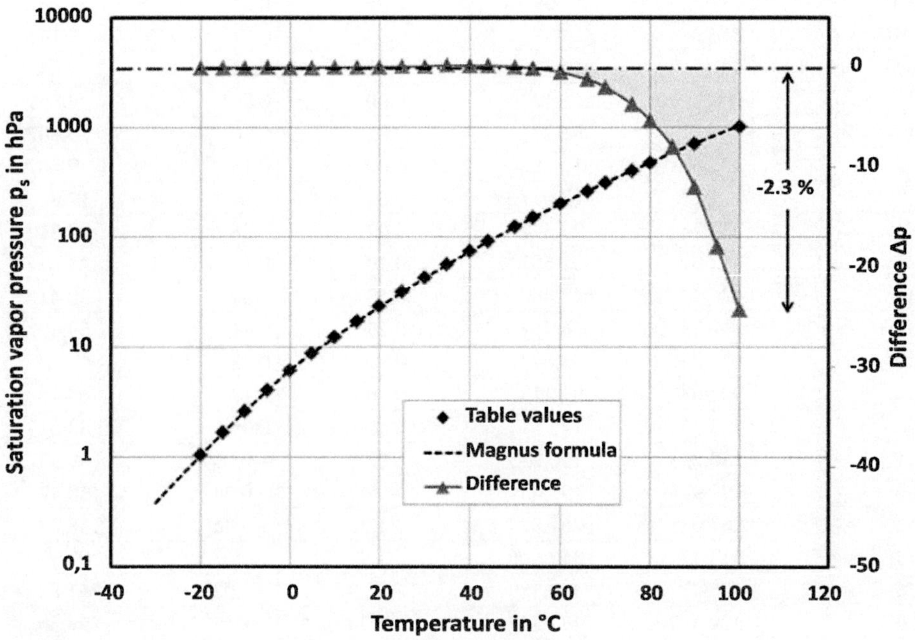

Fig. 10.1 Saturation vapor pressure as a function of temperature T

The maximum humidity that can be present at a defined temperature is called saturation humidity f_S, and is calculated from the maximum water vapor mass m_W and the volume V_f of humid air:

$$f_S = \frac{m_W}{V_f} \tag{10.3}$$

The commonly used unit of saturation humidity is g/m^3. In most cases, however, the gas contains less water vapor and the saturation quantity are not reached. The absolute humidity f then results from the current water vapor mass m_A, and the volume V_f and is also given in g/m^3:

Table 10.2 Saturation vapor pressure p_S and moisture content f_S for a temperature range from −30 to 100 °C. The literature values refer to Cerbe (2008)

Temperature	p_S Magnus	p_S Literature	p_S Delta	% rel. Error	f_S in g/m^3
−30	0.380	–	–	–	0.33
−25	0.633	–	–	–	0.55
−20	1.033	1.039	0.01	0.62	0.884
−15	1.653	1.659	0.01	0.37	1.387
−10	2.599	2.599	0.00	0.01	2.141
−5	4.017	4.031	0.01	0,34	3.247
0	6.112	6.112	0.00	0.00	4.851
5	8.717	8.726	0.01	0.10	6.802
10	12.260	12.282	0.02	0.18	9.407
15	17.017	17.057	0.04	0.24	12.84
20	23.326	23.392	0.07	0.28	17.31
25	31.601	31.697	0.10	0.30	23.07
30	42.337	42.467	0.13	0.31	30.41
35	56.128	56.361	0.23	0.41	39.47
40	73.675	73.844	0.17	0.23	51.24
44	90.963	91.118	0.16	0.17	62.45
50	123.452	123.513	0.06	0.05	83.31
54	150,294	150.215	−0.08	−0.05	99.93
60	199.933	199.458	−0.47	−0.24	130.42
66	263.035	261.827	−1.21	−0.46	168.34
70	313.977	312.006	−1.97	−0.63	198.42
76	406.077	402.389	−3.69	−0.92	251.83
80	479.489	474.147	−5.34	−1.13	293.66
90	713.870	701.824	−12.05	−1.72	423.88
100	1038.449	1014.18	−24.27	−2.39	598.14

$$f = \frac{m_A}{V_f} \tag{10.4}$$

The relative humidity φ results from the quotient of the absolute humidity f and the saturation humidity f_S multiplied by 100% since the relative humidity is given in %.

$$\varphi = \frac{f}{f_S} \cdot 100\% \tag{10.5}$$

Alternatively, the vapor pressure can also be used to calculate the relative humidity φ. Analogous to Eq. 10.5, the following expression is then obtained with the current vapor pressure p_d and the saturation vapor pressure p_S:

Fig. 10.2 Vapor pressure curve for ice and supercooled water (Hengstenberg et al. 1980)

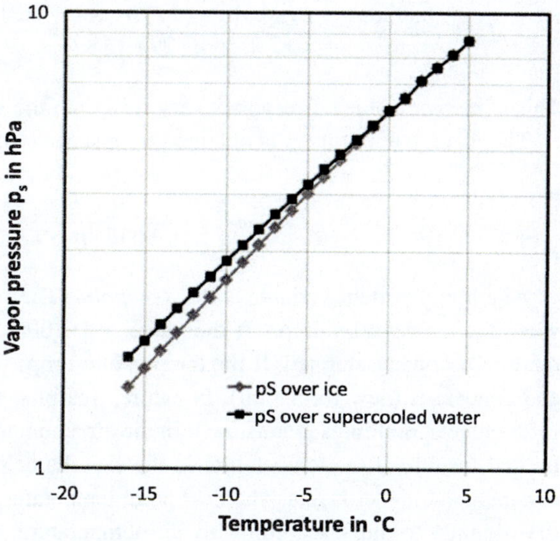

$$\varphi = \frac{p_d}{p_S} \cdot 100\% \tag{10.6}$$

Furthermore, the gas humidity *f* can also be converted using the equation of state for ideal gases:

$$p_d \cdot V_W = n \cdot R \cdot T = m_W \cdot R_W \cdot T \cdot M_W^{-1} \tag{10.7}$$

Equation 10.7 can be used to calculate the theoretical water vapor density $\varrho_{H_2O} = m_W / V_W$ in the standard state ($T_0 = 273.15\,\text{K} - 0\,^\circ\text{C}$, $p_0 = 1013.25$ hPa). However, the value of $\varrho_{H_2O} = 0.8038\,\text{kg/m}^3$ is only one calculation parameter since the water vapor pressure p_d at $0\,^\circ\text{C}$ can be a maximum of 6.112 hPa. With this value the saturation humidity f_S is now calculated (Cerbe 2008):

$$f_S = 0.8038\,\text{kg/m}^3 \cdot \frac{p_S \cdot T_0}{p_0 \cdot T_S} \tag{10.8}$$

Example

The saturation humidity f_S should be calculated for a temperature $T_S = 20\,^\circ\text{C}$. In Eq. 10.1, the real numerical values are used:

$$f_S = \frac{0.8038\,\text{kg}}{\text{m}^3} \cdot \frac{23.326\,\text{hPa} \cdot 273.15\,\text{K}}{1013.25\,\text{hPa} \cdot 293.15\,\text{K}} = \frac{0.017242\,\text{kg}}{\text{m}^3} = 17.24\,\text{g/m}^3 \tag{10.9}$$

This numerical value corresponds very well with the value 17.32 g/m^3 from Table 10.2.

The relative humidity φ is obtained by inserting Eq. 10.8 into Eq. 10.5:

$$\varphi = f \cdot \frac{p_0 \cdot T}{0.8038 \cdot p_S \cdot T_0} \cdot 100\% \tag{10.10}$$

Another important variable is the *dew point*. The relative humidity φ of a gas increase when the temperature drops. If the value $\varphi = 100\%$ is reached, the humidity in the gas mixture becomes saturated. If the temperature drops further, the excess water precipitates and condenses (dew formation). In nature, this process takes place on the surface of the earth (streets, buildings, plants, …). In the air, condensation nuclei (e.g. dust particles) are needed for this. If the temperature in the air falls below the dew point, it is called *fog* or *clouds* in the higher layers. The dew point temperature T_{TP} can be calculated by changing the Magnus formula according to the temperature, if the vapor pressure p_d is known (Wernecke 2003):

$$T_{\text{TP}} = 243.12 \cdot \left[17.62 \cdot \left(\ln \frac{p_d}{6.112}\right)^{-1} - 1\right]^{-1} \tag{10.11}$$

Similarly, the *frost point temperature* T_{FT} can be calculated, at which the air above an ice surface is completely saturated with water vapor. This value is of particular interest in meteorology.

$$T_{\text{FP}} = 272.62 \cdot \left[22.46 \cdot \left(\ln \frac{p_d}{6.112}\right)^{-1} - 1\right]^{-1} \tag{10.12}$$

Example

The dew point temperature T_{TP} is to be determined\for a gas mixture with a relative humidity of $\varphi = 12\%$ at 1 bar and 25 °C. The saturation vapour pressure p_S can be calculated using the Magnus formula or taken from the table. At a temperature of 25 °C, $p_S = 31.697$ hPa. With Eq. 10.6 one obtains:

$$p_d = \varphi \cdot p_S = 0.12 \cdot 31.697\,\text{hPa} = 3.804\,\text{hPa}$$

Since this value does not change during cooling, we can estimate the dew point from Table 10.2. It lies between -5°C and -10°C. By interpolation we obtain a value of $T_{TP} \approx -5.8^{\circ}$C. The more precise calculation is done with Eq. 10.11. It therefore applies:

$$T_{TP} = \frac{243.12^{\circ}\mathrm{C}}{\frac{17.62}{\ln(3.804/6.112)} - 1} = -6.4^{\circ}\mathrm{C}$$

10.2 Measurement Procedure

The measurement of air humidity or gas humidity can be carried out with very different methods. Figure 10.3 shows the most important measuring methods with the corresponding measuring ranges. The information on the measuring ranges refers to common measuring instruments available on the market. In particular, the concentration range, which is given by the dew point temperature or the water vapor concentration, plays an important role in the selection of the respective method in addition to the accuracy.

Hair Hygrometer

The probably best known and oldest moisture measuring method is the so-called hair hygrometer. With this method, the change in length ΔL of a hair is measured by binding water vapour in the hair. The change in length is approximately 2.5% for a change in relative humidity φ from 0 to 100%. Both, human and animal hair (horse) are used as *hair*, as are artificial polymer fibers (nylon, cellulose). The hairs are bound to form a *hair harp*, in which the hairs/fibers lie next to each other and are mechanically connected at the ends. This hair harp is tensioned by a spring. If the length of the hair/fibers changes, the change in length ΔL is converted into a change in angle by a mechanism and displayed. A simple arrangement for this can be seen in Fig. 10.4.

This type of humidity measurement is usually found in simple weather stations or air conditioning systems. With the help of this technology, simple humidity switches can also be constructed, with which, for example fans in damp rooms can be switched. The accuracy is <5% relative humidity. The response time is very high due to the diffusion and adsorption processes and is 10–12 minutes for the 90% value. Natural hair can be used for the temperature range from -35 to 50 °C, while artificial fibers are used between 0 °C and 80°C.

Psychrometer

Another relatively old method is the so-called *psychrometer*. It is based on the transfer of energy that occurs when water vapor evaporates in a gas stream. Figure 10.5 shows such an arrangement. The air first flows past thermometer 1, which measures the ambient

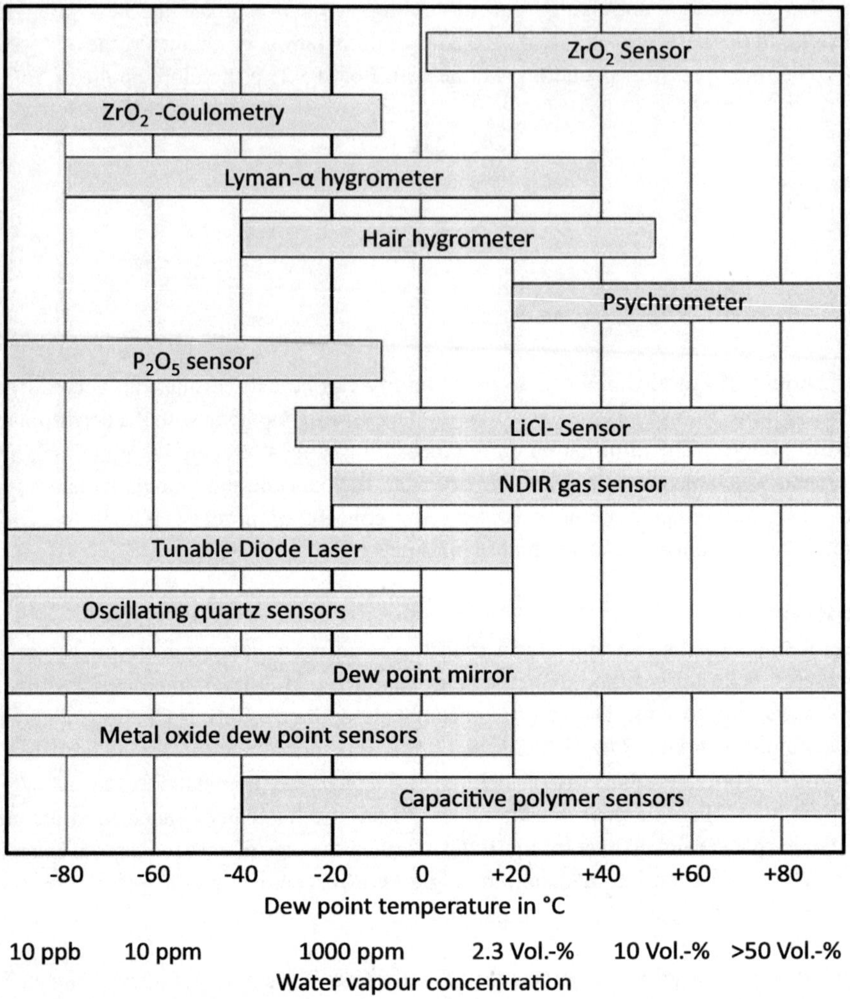

Fig. 10.3 Overview of the most important moisture measuring methods and their measuring ranges. Dew points from −100 to −30°C are called *trace moisture*. From −30 to +30°C we have the *climate humidity* and from +30 to +100°C the *high humidity* (Tränkler and Obermeier 1998)

temperature T_1. Thermometer 2 has a damp sock at the measuring point, which is permanently moistened by a water tank. The air flow causes the water to evaporate and thus reduces the temperature T_2 by a certain amount ΔT. This temperature difference is the greater, the drier the gas to be measured is in the flow. In extreme cases, when the gas is already completely saturated (→ 100% relative humidity), no more water can be absorbed and the temperature difference is zero. The maximum temperature difference is then present when the air is completely dry (→ 0% relative humidity). The vapor pressure p_d

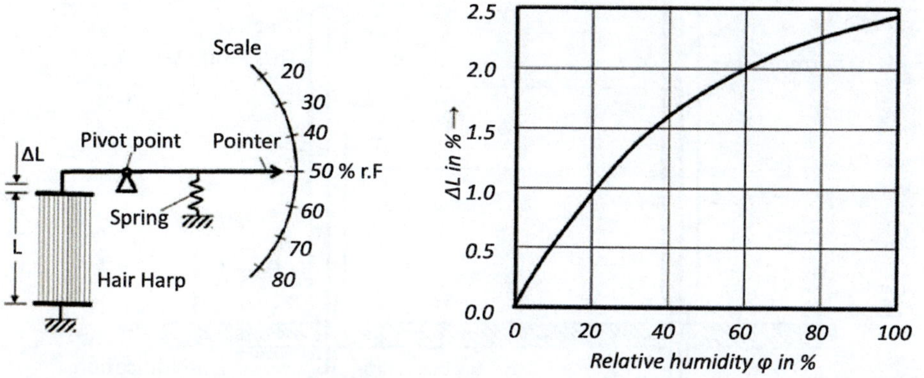

Fig. 10.4 Structure of a hair hygrometer and the expected change in length ΔL due to water retention in the hair/fibre. (After Karbach et al. 1997)

can be calculated with the psychrometric ΔT difference using an empirical formula by *"Sprung"*:

$$p_d = p_S(T_F) - p \cdot A \cdot \Delta T \tag{10.13}$$

A is the device-dependent psychrometer constant and p *is* the current air pressure. Depending on the device type, the constant lies between $A = 5.8 \cdot 10^{-4}\,\mathrm{K}^{-1}$ and $A = 5.8 \cdot 10^{-4}\,\mathrm{K}^{-1}$. For the *Assmann aspiration psychrometer*, an improved psychrometer equation was given by Sonntag (1988)

$$p_d = p_S(T_F) - p \cdot 0.646 \cdot 10^{-3} \cdot \Delta T \cdot \left[1 + 0.944 \cdot 10^{-3} \cdot T_F\right] \tag{10.14}$$

The further conversion of the vapor pressure p_d into the relative humidity φ is carried out with Eq. 10.6.

In meteorology and environmental measurement technology, *Assmann's aspiration psychrometer* is a standard procedure with which air humidity φ can be measured with an accuracy of $\Delta\phi \approx 1\%$. Figure 10.6 shows the basic structure. It consists of two thermometers for the dry and wet temperature. The air flow is generated by a fan, which is driven by a clockwork. The measurement takes about 5 minutes until stable temperature values are obtained. The air humidity φ is then calculated using Eq. 10.6 or determined with the aid of tables or diagrams (see Fig. 10.9). However, this manual measurement is no longer desired today, as automatic recording is not possible with this method. Modern designs have an electrically driven fan and two Pt 100 measuring resistors for precise temperature determination. Humidification takes place continuously from a reservoir, so

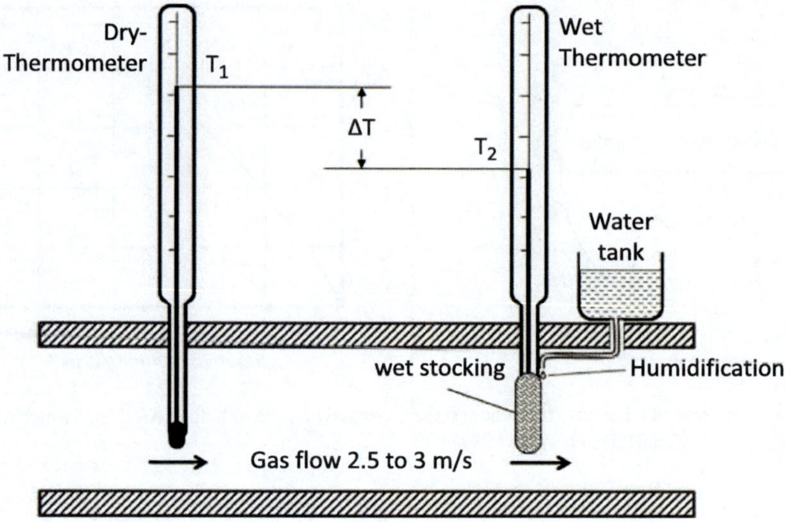

Fig. 10.5 Basic structure of a psychrometer with continuous humidification

that digital data recording is also possible with these electrical aspiration psychrometers (Wernecke 2003).

A simpler version is the *slinging psychrometer* or *vortex hygrometer*. This handy instrument is mainly used in the nautical field. This instrument is shown in Fig. 10.7. The two thermometers are again located on a plate. At the temperature measuring points, the plate has openings (apertures) through which the air can flow. The stocking must be moistened before each measurement. Afterwards the psychrometer is set in a rotating motion with the hand crank so that the air can flow through the openings and cause the desired measuring effect. After a few minutes the spinning process is stopped and the temperatures are read. There are tables for each instrument, from which the humidity can then be seen. The accuracy of the measurement is ±5% relative humidity (Fig. 10.8).

In practice, mainly tables and charts are used to determine the relative humidity φ. According to Eq. 10.6, the respective saturation vapor pressure p_S for the temperature T_F is determined anyway using tables or the Magnus formula. With p_d, the relative humidity φ must then also be calculated. These rather complicated calculations are not necessary when using tables or psychrometer tables, which have been specially created for this procedure. However, the accuracy of the readings is not very good, so with exact values only the calculation method is possible. Figure 10.9 shows the dependence of relative humidity φ on the psychrometric difference for three different ambient temperatures. The psychrometric table in Fig. 10.9 allows a direct reading of the relative humidity φ. For this purpose, the drying temperature and ΔT must be known. In the example given, $\Delta T = 10^\circ$C and the drying temperature is 30°C. The point of intersection is thus on the curve for 40% relative humidity. However, if the drying temperature is 35°C and $\Delta T = 15.1^\circ$C, the value for

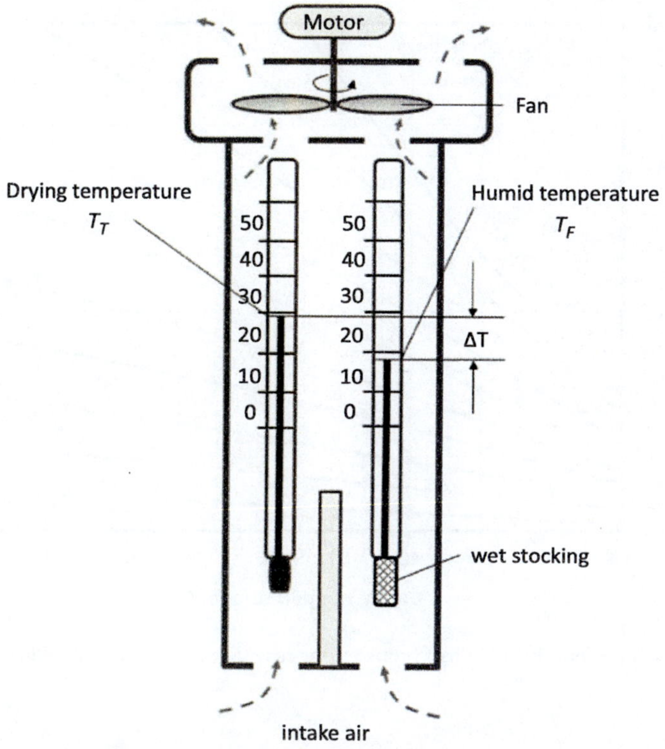

Fig. 10.6 Technical design of the Assmann aspiration psychrometer

relative humidity is between the curves for 20% and 30%. Since the intersection point is closer to the 20% curve, it is estimated to be ≈22%. An exact calculation with the Magnus formula gives a saturation vapor pressure $p_s = 23,\ 182$ hPa at $T_F = 19.9^\circ$C. This value results in a vapor pressure of $p_d = 13,\ 372$ hPa using the psychrometer Eq. 10.14. The saturation vapor pressure $p_s = 56,\ 128$ hPa can be read directly from Table 10.2. Now one must still calculate the relative humidity φ with Eq. 10.6:

$$\varphi = \frac{p_d}{p_S} \cdot 100\% = \frac{13,372\,\text{hPa}}{56,128\,\text{hPa}} \cdot 100\% = 23.8\% \tag{10.15}$$

The estimate was therefore off by $\Delta\phi = 1.8\%$ relative humidity.

Dew Point Hygrometer

The direct determination of the dew point or the dew point temperature T_{TP}, enables a very accurate measurement of the gas humidity. Figure 10.10 shows this method. A mirror is cooled to a defined temperature from the rear side via a Peltier element. During the cooling

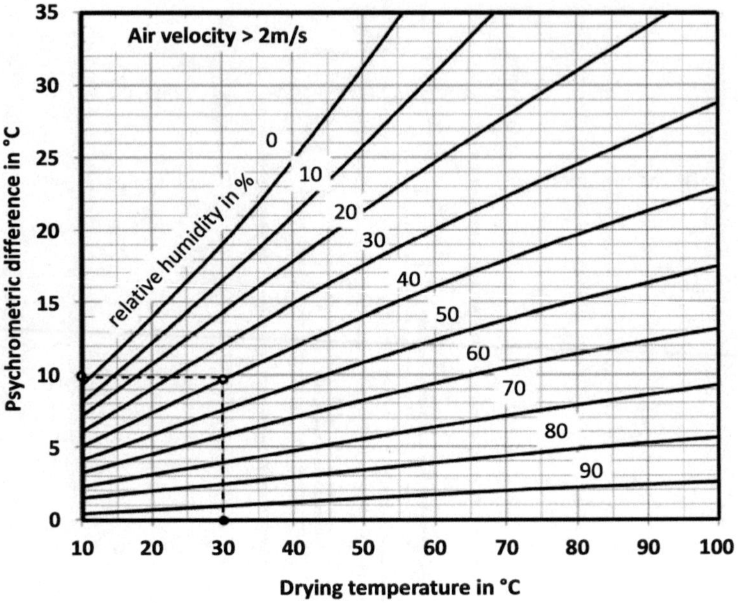

Fig. 10.7 View of a centrifugation psychrometer for easy determination of humidity

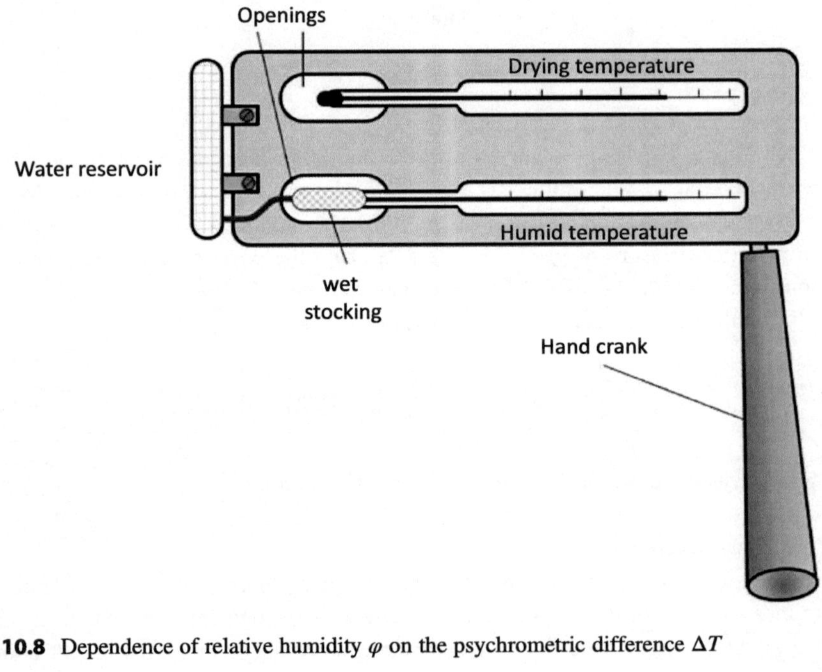

Fig. 10.8 Dependence of relative humidity φ on the psychrometric difference ΔT

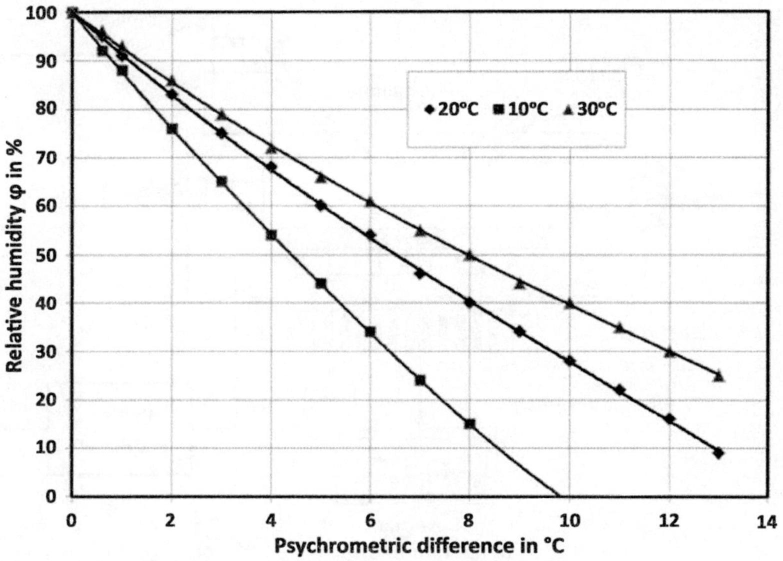

Fig. 10.9 Psychrometric table for determining the relative humidity of air, based on the psychrometric difference ΔT and\the air temperature (→ drying temperature)

process, the gas mixture to be measured flows into the measuring chamber. However, the volume flow must be very small in order not to disturb the thermal equilibrium at the mirror surface. If the mirror is now cooled down further and further, condensation will form on the mirror at a certain temperature. The temperature that the mirror assumes at this point in time corresponds exactly to the dew point temperature T_{TP} of the gas mixture. The condensation process is observed with a reflex light barrier. If dew occurs, the reflection of the radiation is abruptly stopped and the signal from the photodiode is greatly reduced. The cooling process is now stopped by reducing the power supply to the Peltier element. This causes the water vapor to evaporate from the mirror surface and the light path is released again. The cooling process is now restarted until another condensation occurs. Thus, a temperature control around the dew point temperature T_{TP} is performed. The temperature measured in this way can be displayed directly. If the ambient temperature T_U is measured at the same time, the saturation humidity can be determined from it. This information is then used to calculate and display the relative humidity φ. The accuracy for this measuring method is $<0.2^{\circ}$C for the indication of the dew point temperature T_{TP}.

The mirror is usually made of gold-plated copper or polished stainless steel to achieve good corrosion resistance. Furthermore, the gas must be cleaned of dust before entering the chamber, as this can seriously interfere with the measurement. Nevertheless, during operation, the mirror surface becomes soiled, which can lead to incorrect measurements.

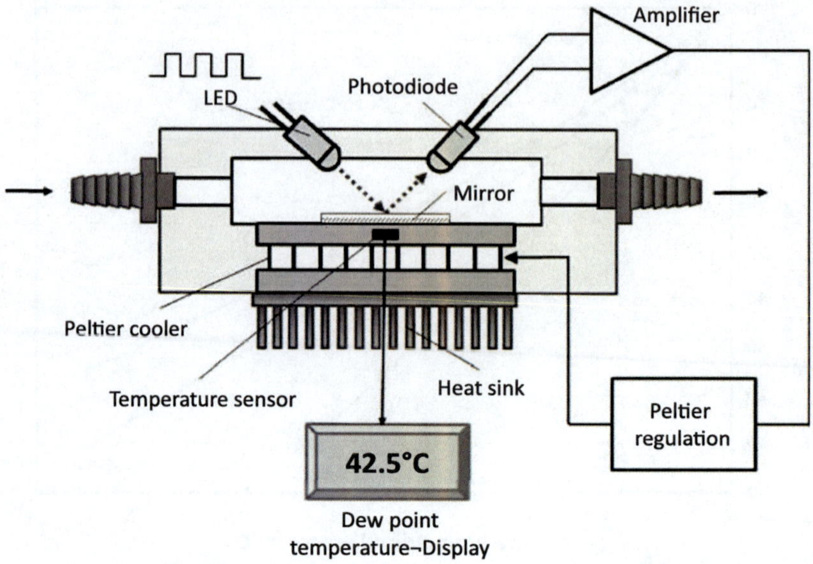

Fig. 10.10 Basic structure of a dew point hygrometer with optical scanning. The mirror surface is approximately 10 mm^2

The mirror must therefore be cleaned occasionally. Some manufacturers have therefore integrated appropriate compensation measures. One possibility is to subject the mirror to a defined temperature cycle to concentrate finely distributed dirt in a few places. In the *PACER process*,[1] a cooling process is first carried out, which leads to a condensation of water vapor on the mirror surface. These substances partially dissolve in the resulting *puddle*. In a second stage, the mirror is then heated to a high temperature so that the water evaporates again. As a result of this process, the impurities collect in a few places on the mirror surface. In total, the amount of dirt remains constant, but the clean area percentage increases. Due to this measure, an exact dew point measurement is possible. At the end of the heating process, a calibration takes place in which the mirror properties (reflectance) are stored (Clevett 1986). The manufacturer states that the cleaning cycle can be reduced by at least a factor of 10. The cleaning process takes about 180 seconds. During this time, the last measured value is stored (frozen) and output. Current values are only displayed after the cleaning phase (Fig. 10.11).

Figure 10.12 shows a process measuring instrument (*OPTIDEW*) based on a dew point hygrometer. In this arrangement, the dew point humidity sensor (probe) can be connected externally via a cable up to 250 m. This device also has a compensation option that reduces the effects of contamination. However, this is a software solution that registers and then

[1] Programmable Automatic Contaminant Error Reduction.

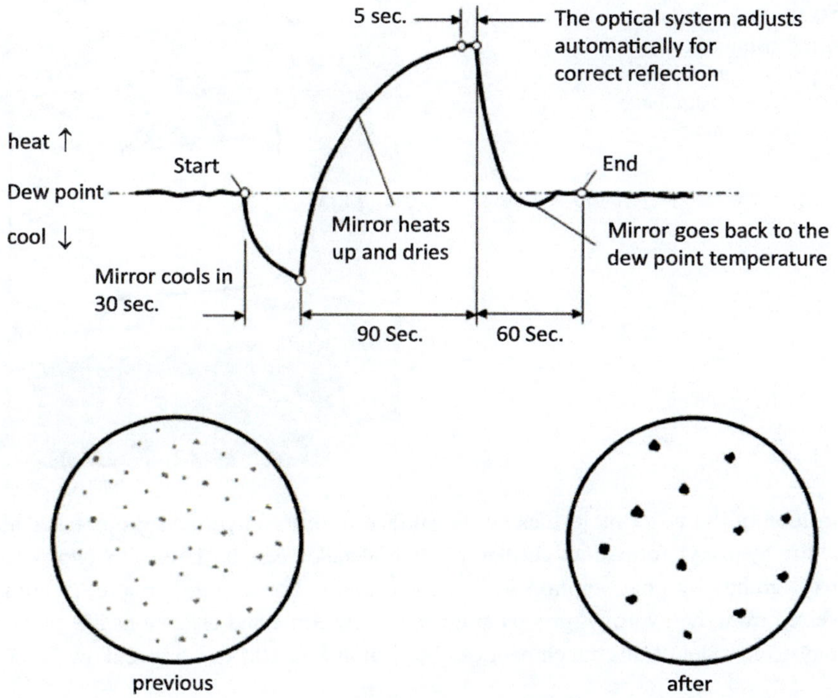

Fig. 10.11 Temperature cycle and mirror surface before and after the *PACER process*. (Clevett 1986)

compensates for slow drift. This *DCC system*[2] behaves like a *self-learning system* that adapts its settings to the operating conditions, and thus matches the real needs for correction. As an option, the manufacturer also offers a high-temperature version for use up to 130 °C.

Metal Oxide Humidity Sensors

Metal oxides such as Al_2O_3, are used as sensitive layers in humidity sensors due to their interaction with water, whereby the water attaches itself to the surfaces of the metal oxides. The layer structure of an Al_2O_3 humidity sensor is shown in Fig. 10.13, as a cross-sectional view. The sensor is constructed like a capacitor. The base electrode is made of pure aluminum, on which a porous Al_2O_3 layer is present. A water vapor permeable gold layer is applied as counter electrode. To achieve high sensitivity, the sensitive layer must have the largest possible surface area. The water vapor passes through the gold electrode into the pores of the sensitive layer and deposits there. The connection to the metal oxide surface is achieved by chemisorption and physisorption. During chemisorption,

[2]Dynamic Contamination Correction.

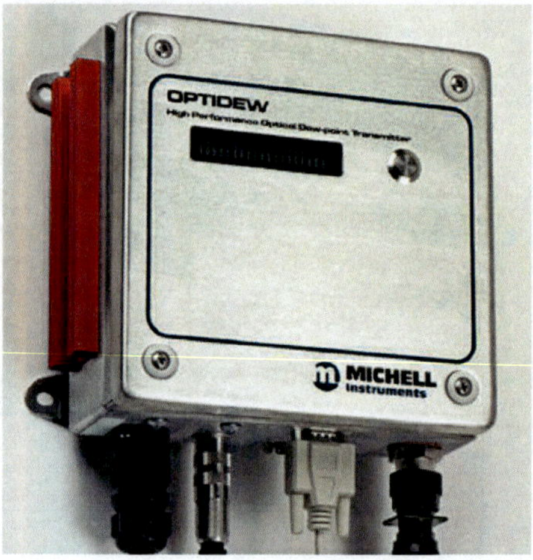

Fig. 10.12 Process moisture measuring instrument with a housing protection class IP 66. (Source: Michell Instruments GmbH, Friedrichsdorf)

dissociation of the water molecules on the surface to form a hydroxyl group takes place. After this hydroxyl formation, further water molecules can be bound to two adjacent hydroxyl groups by physisorption in hydrogen double bonds. Furthermore, protons are transferred from the hydroxyl groups to the water molecule and H_3O^+ ions are generated. The physical model of this attachment can be seen in Fig. 10.14 (Korotcenkow 2013).

The number of water molecules bound by adsorption is in equilibrium with the surrounding humidity and influences the relative *primitiveness* as well as the electrical conductivity of the sensitive layer. The metal oxide sensor can be easily evaluated by impedance measurement, whereby there is a good correlation to absolute humidity. Al_2O_3 sensors can be used at room temperature for dew point ranges from below -110 to +70°C. The permissible operating temperatures are between −40°C and +70°C. At temperatures higher than 70°C, the sensitive layer of an Al_2O_3 sensor is irreversibly damaged. The accuracy of the dew point measurement is ±3°C for low dew points (−110 to −65°C), and ±2°C in the higher range (> − 65°C). Due to aging effects in the sensitive layer, the sensor must be recalibrated regularly to maintain the accuracy range (Gevatter 1998).

Polymer Sensors

The design of these sensors is similar to the Al_2O_3 humidity sensors. Here, too, the sensitive polymer layer is located between two electrodes that work like a plate capacitor. The capacity C of the sensor changes almost linearly with the relative humidity φ:

$$C(\varphi) = C_0 \cdot (1 + \delta \cdot \varphi) \tag{10.16}$$

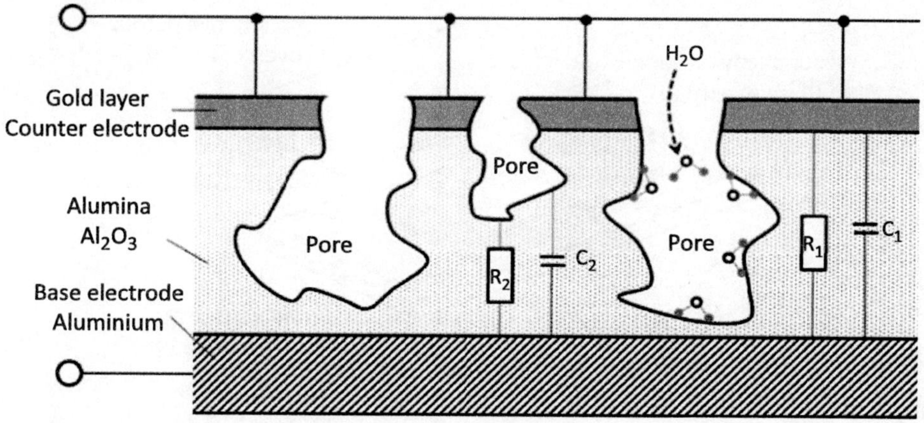

Fig. 10.13 Cross-section through an Al_2O_3 humidity sensor. The resistors and capacitors characterize the electrical equivalent circuit of the sensor layer

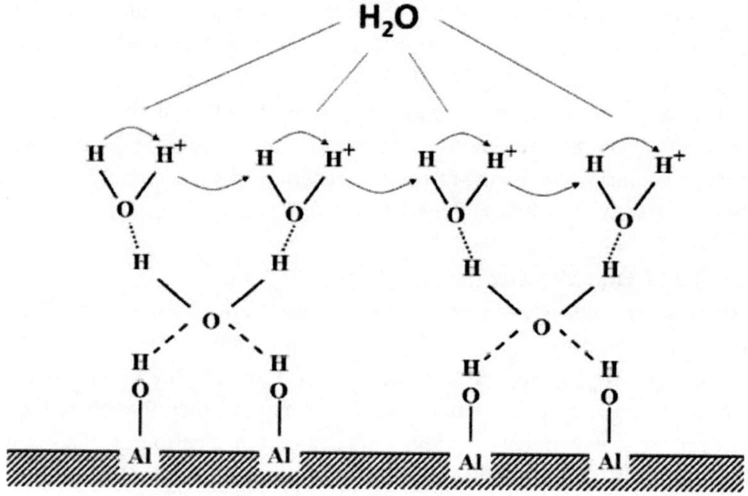

Fig. 10.14 Model of the structure of the adsorbed water layers on an Al_2O_3 surface

The proportionality factor has a value of $\delta \approx 0.002/\%$. With a basic capacity of $C_0 = 200$ pf at 100% relative humidity, the capacity at $C(\varphi) = 240$ pf would increase. Per percent of relative humidity, there would then only be a 0.4 pf change.

In the literature, PMMA (polymethyl methacrylate), polyamides, or CAB (cellulose acetate butyrate) are given as polymer material. The most suitable material of the individual sensor manufacturers is usually found out by extensive tests and usually not disclosed. The sensitive layer has a thickness of 0.1 to 1 μm (Fig. 10.15).

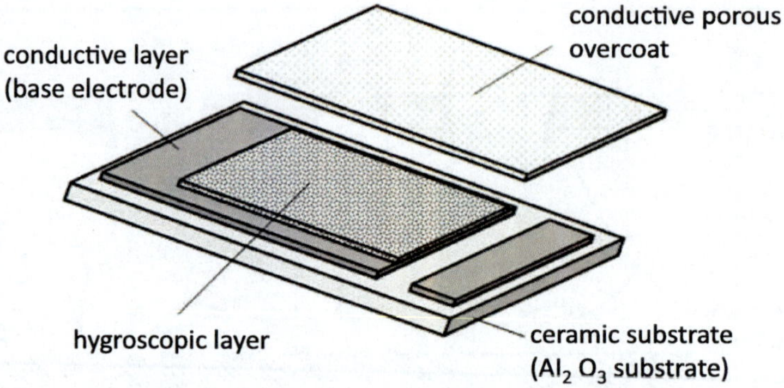

Fig. 10.15 Layered structure of a polymer humidity sensor

In addition to capacitive polymer-based humidity sensors, there are also resistive versions that cause a change in resistance (ΔR). The layer structure is similar. However, the materials used in this case are different ceramic materials containing cobalt, germanium, zinc, titanium dioxide or nickel oxide. Furthermore, mixtures of these materials are also common (Wernecke 2003). Figure 10.16 shows such a sensor. These sensor types are mainly used in applications where a good long-term stability of the sensitive layer is required. Examples are the automotive industry or air conditioning technology. The measuring range is determined by the respective materials and is then only within a narrow concentration range (e.g. 10–30% r. F. or 40–90% r. F.).

Miniaturization of Humidity Sensors

Humidity sensors are used today in many areas of measurement technology. Especially in the medium humidity range (air conditioning technology) these sensors find many applications. Due to the simple design of the sensor structure, the humidity sensors can also be easily miniaturized. Some sensor manufacturers have therefore started to combine the miniaturized sensor element directly with evaluation electronics. The control and evaluation of the sensor signals takes place directly in one module. Figure 10.17 shows such a combination. It is a detail enlargement (Fig. 10.17b) of an arrangement which is accommodated in a TO housing (Fig. 10.17a) with an outer diameter of $d \approx 9$ mm.

A further integration step is shown in Fig. 10.18. In this setup, the sensor element and the microelectronics were mounted on a circuit board, which can be placed as an SMD chip directly on another circuit board. The manufacturer recommends to thermally isolate the device from other components of the PCB in order not to influence the local temperature conditions. For the calculation of the relative humidity, the ambient temperature T_U is required, which is also part of this sensor design.

The humidity sensor can be controlled or communicated directly via an external microcontroller. For this purpose, the sensor has a serial interface (2-wire). The data

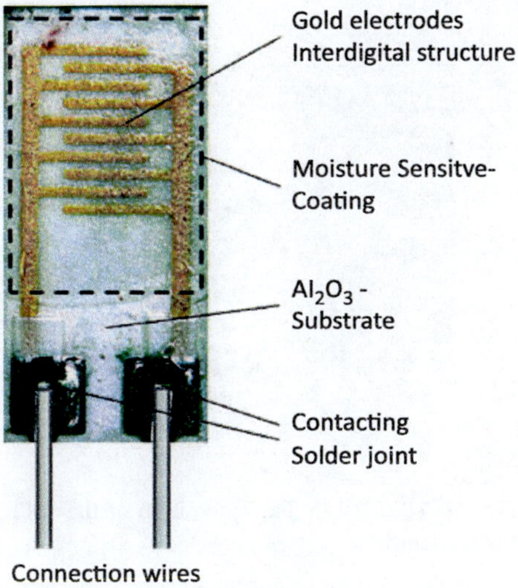

Fig. 10.16 Resistive humidity sensor with an interdigital sensor structure (gold electrodes)

Fig. 10.17 Combination of a miniaturized humidity sensor with an evaluation circuit/microelectronics. (Source: InnovativeSensor Technology IST AG, Ebnat-Kappel, Switzerland)

transmission is carried out with a resolution of 14 bit (internal A/D converter). Figure 10.19 shows the external wiring of the humidity sensor.

Figure 10.20 shows a selection of extremely small microsensors that have been specially developed for mass applications in humidity measurement technology. In addition to the actual humidity sensor, these sensors also include a temperature sensor for compensation purposes and conversion to other humidity units. In addition to the sensor technology, the

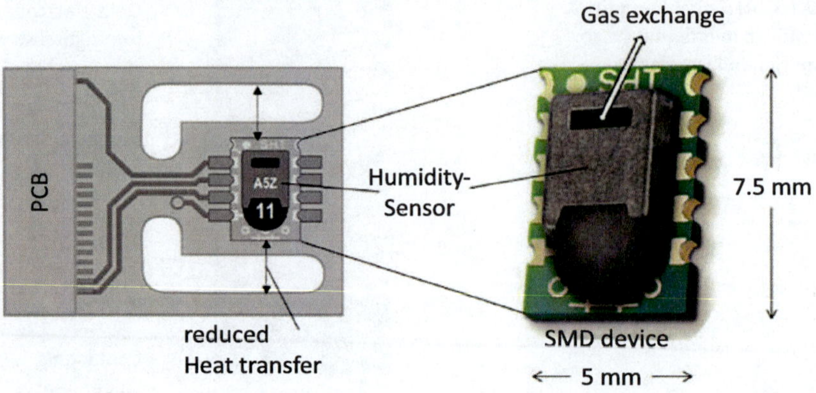

Fig. 10.18 Design and integration of the SHT humidity sensor. (Source: SENSIRION AG Staefa Switzerland)

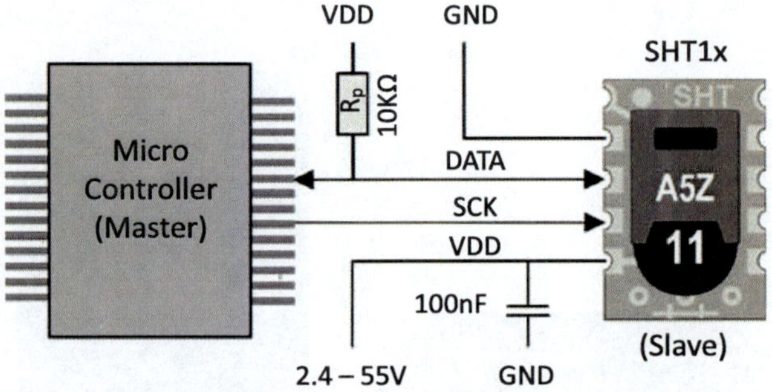

Fig. 10.19 External wiring and data exchange with a micro-controller acting as master. (Source: SENSIRION AG Staefa Switzerland)

SMD[3] components also integrate the complete signal processing, which provides compensated and linearized output signals. The manufacturer specifies an accuracy of humidity measurement of ±2% relative humidity. The temperature measurement has an accuracy of ±0.3°C.

The response time for an abrupt change in humidity is shown in Fig. 10.21. The humidity sensor was abruptly exposed to high humidity ($\varphi = 90\%$) via a solenoid valve, and the output signal was recorded. In this example the $t_{90\%}$ time is ≈13 seconds. The manufacturer specifies 8 seconds as the $t_{63\%}$ -time.

[3]Surface Mounted Devices = surface mounting.

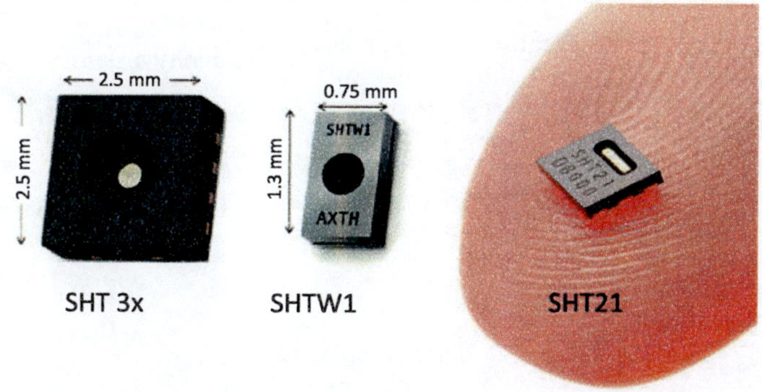

Fig. 10.20 Miniaturized humidity sensors with integrated evaluation electronics in SMD designs. (Source: SENSIRION AG Staefa Switzerland)

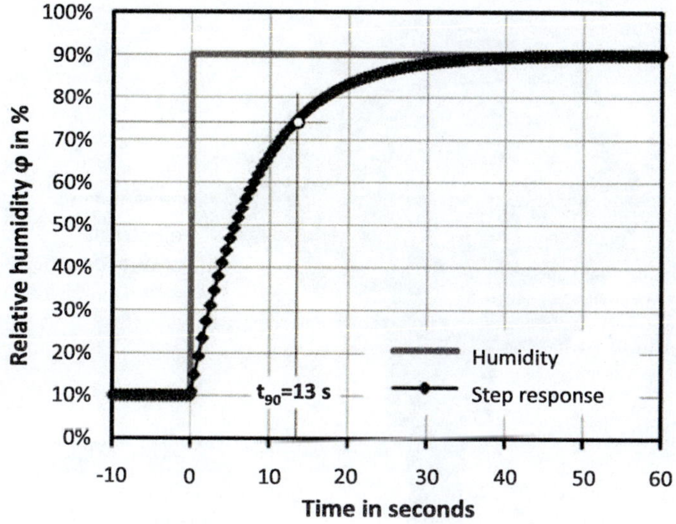

Fig. 10.21 Temporal response of a humidity sensor of the SHT series for a rapid change in humidity values. (Source: Testing Guide for SHTxx Relative Humidity & Temperature Sensor Series. SENSIRION AG Staefa Switzerland)

Industrial Forms of Execution

For industrial applications, the designs shown so far are usually not directly suitable. The sensitive sensor elements are integrated into corresponding measuring heads for this application area, which allow safe and reliable handling. Figure 10.22 shows the individual stages of this integration. The humidity sensor is integrated into a measuring head, which may also contain an additional temperature sensor (Figs. 10.23 and 10.24). The measuring head provides appropriate mechanical protection and ensures the safe transmission of the

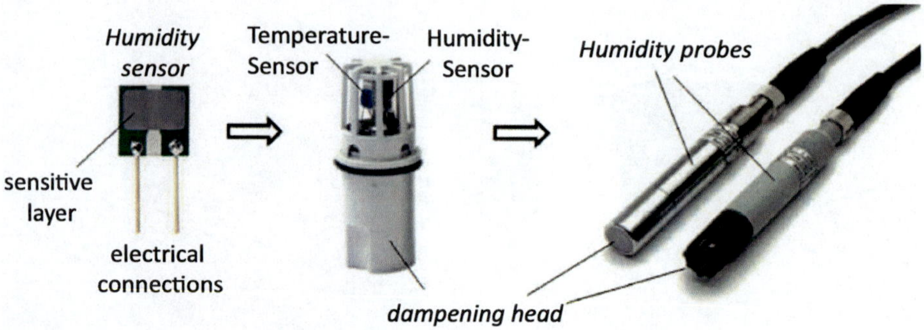

Fig. 10.22 Integration steps from the sensor element via a humidity measuring head to a humidity probe. This example is a handheld probe for mobile use in air conditioning technology. (Source: Michell Instruments GmbH, Friedrichsdorf)

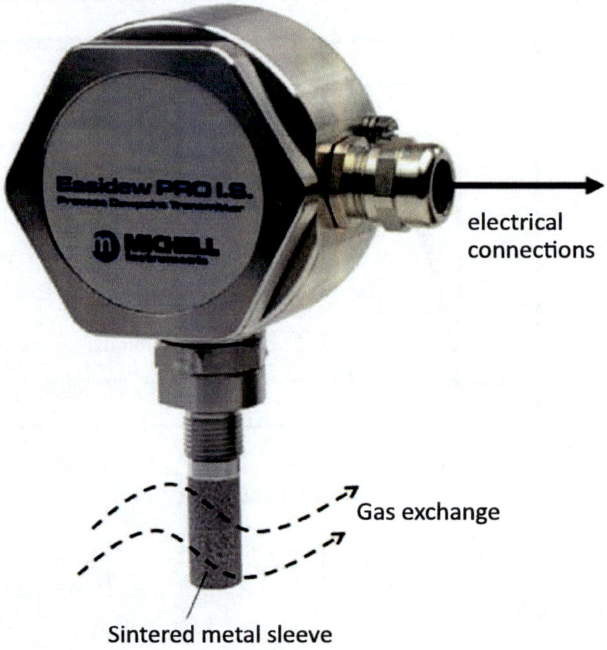

Fig. 10.23 Stationary moisture measurement for industrial applications. (Source: Michell Instruments GmbH, Friedrichsdorf)

measured data to the device electronics. The measuring head is then installed in a humidity probe. The humidity probe shown here is a hand-held probe which is used in climate and environmental measurement technology for mobile investigations. The measuring head is exchangeable and can be quickly replaced in case of service.

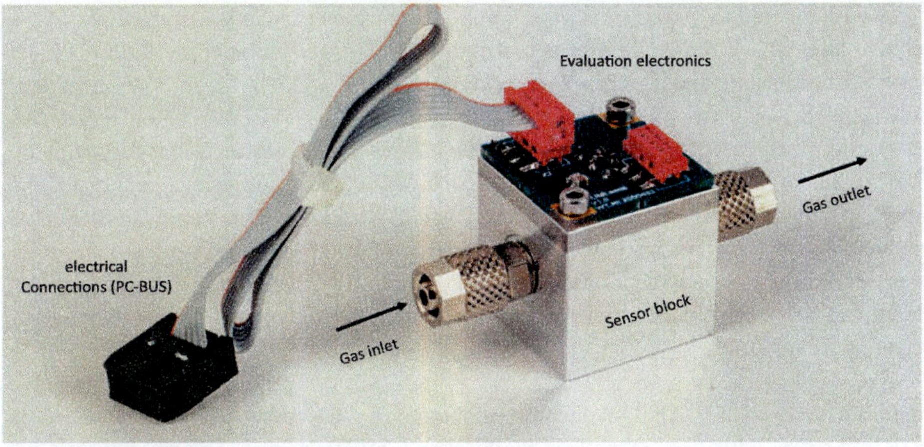

Fig. 10.24 Miniaturized humidity sensor (HUMI.sens®) for industrial OEM applications, with integrated I2C interface and flow operation (Source: https://www.witec-sensorik.de/en/technologies/humidity-sensor/, Wesel Germany)

For stationary measurements in the industrial sector, the humidity sensor is usually additionally protected with a gas-permeable sintered metal frit. In this application, the entire structure is in a sealed metal housing (e.g. IP 65). These versions are also available for the Ex-area.

Lithium Chloride Dew Point Sensor

Lithium chloride (LiCl) is a hygroscopic salt that absorbs moisture from the ambient air. With this property, a dew point sensor can be constructed which directly measures T_{TP} the dew point temperature. Figure 10.25 shows such a structure. It consists of a thin tube with a layer of tissue impregnated with LiCl. Around this tissue layer there are two platinum wires (electrodes 1 and 2), which are wound helically without touching each other.

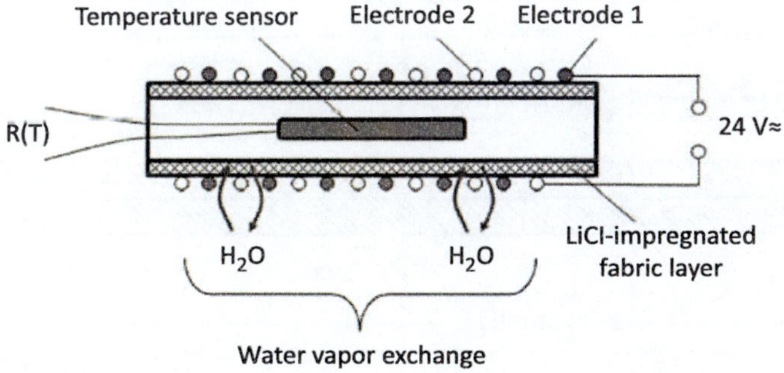

Fig. 10.25 Cross-section through a LiCl dew-point sensor

The LiCl solution lowers the water vapor partial pressure on the surface to ≈12% relative humidity. If there is a higher moisture content in the ambient air, the LiCl layer absorbs this moisture and an electrolyte solution is formed with the water. If a voltage U is now applied to the two electrodes, a current I flow. To avoid polarization effects, an alternating voltage is used for this purpose. Due to the electrical loss conduction $P = U \cdot I$, heat is now released in the LiCl layer, which leads to evaporation of the water. This reduces the electrical conductivity and the heating current begins to drop. At a certain point, at which neither absorption nor release of water vapor takes place, the system is in equilibrium (→ conversion temperature). This transformation temperature indicates the partial pressure of water vapor in the sample gas. The integrated temperature sensor records this value.

The LiCl humidity sensor displays - due to its principle - only relative humidity'$s\varphi > 12\%$. The achievable accuracy for the dew point temperature is $\approx 1^{\circ}C$. The method is very robust and works almost hysteresis-free. However, if the LiCl sensor is not used for a longer period, it must be protected from the humid outside air in a package. Furthermore, the LiCl coating must be renewed occasionally.

Coulometric Humidity Sensor

This sensor type has a similar structure to the LiCl dew point sensor. However, phosphorus pentoxide (P_2O_5) is used as the hygroscopic material in this case. This process was first described by *Keidel as* early as 1956 (Keidel 1959). In the literature, reference is often made to the *Keidel measuring cell.* Figure 10.26 shows a cross-sectional representation of this sensor type. The P_2O_5-coating is applied on the inside of this construction and the two electrodes are also placed there. The gas to be measured is passed through the thin tube (capillary). The length L is up to 1 m. The water contained in the sample gas is completely absorbed by the phosphorus pentoxide and metaphosphoric acid is formed first:

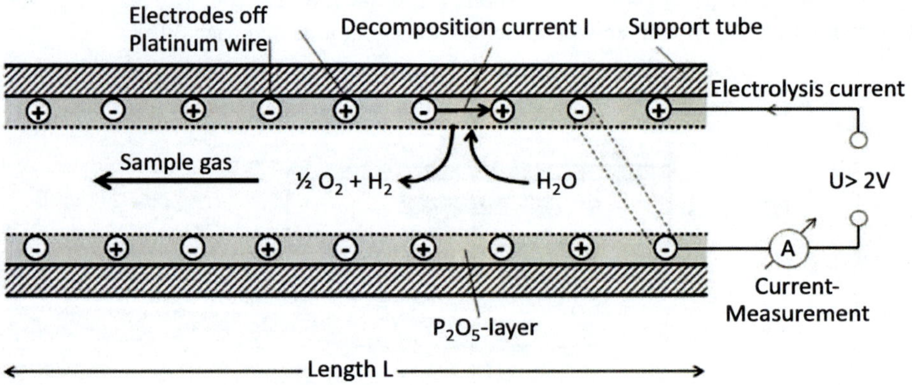

Fig. 10.26 Cross-section through a coulometric humidity sensor

$$P_2O_5 + H_2O \rightarrow 2HPO_3 \tag{10.17}$$

If a voltage U *is* applied to the electrodes, which must be greater than 2 V (decomposition voltage), the metaphosphoric acid is converted into hydrogen and oxygen and phosphorus pentoxide is released again:

$$2HPO_3 \rightarrow H_2 + \frac{1}{2}O_2 + P_2O_5 \tag{10.18}$$

In total, therefore, only the water vapor is decomposed by electrolysis. According to Faraday's law, a charge of 96485.34 is required for the conversion of 1 mol monovalent substances ($z = 1$). For water, 18.0153 g/mol. This results in the following relationship for the measuring current I_M, which is set at a certain absolute humidity f and a defined volume flow $\dot{V}$:

$$\begin{array}{ccc} \text{Value time } z & \text{Absolute Humidity} & \text{Volume} - \text{flow} \\ & \searrow f \downarrow & \downarrow \dot{V} \end{array}$$

$$I_M = \underbrace{9.64853365 \cdot 10^4 \frac{As}{mol} \cdot \frac{2}{18.0153 \frac{g}{mol}}}_{\text{Faraday's Law}} \cdot \underbrace{\left[\frac{g}{m^3}\right] \cdot \left[\frac{m^3}{s}\right]}_{H_2O - \text{Mass in g/s}} \tag{10.19}$$

Simplified, the equation can be summarized as follows:

$$I_M = 1.071149 \cdot 10^4 \cdot \frac{As}{g} \cdot f \cdot \dot{V} \tag{10.20}$$

With this method, water vapor concentrations in the trace range between 0.1 ppm 2000 ppm can be measured with an accuracy of < ± 5 ppm. The volume flow must be constantly controlled during the measurement, as it is directly included in the measurement result. Typical flow rates lie between 10 L/h and 100 L/h. The response time is about 30 seconds for the 90% value.

Example

A sample gas contains 1 ppm water vapor at a temperature of 25 °C. The volume flow is 100 mL/min. The measuring flow I_M shall be calculated.

The absolute humidity f can be determined from the concentration (1 ppm = 10^{-6}). 18 g H_2O occupies a volume of 22.4 L under normal conditions. 1 ppm is 10^{-3} L/m^3.

$$f = \frac{10^{-3}\,\mathrm{L}\cdot\mathrm{m}^{-3}\cdot 18\,\mathrm{g/mol}}{22.4\,\mathrm{L/mol}} = 0.000804\,\mathrm{g}\cdot\mathrm{m}^{-3}$$

The volume flow can be converted as follows:

$$\dot{V} = \frac{0.1\,\mathrm{L}}{60\,\mathrm{s}\cdot 1000\,\mathrm{L/m^3}} = 1.667\cdot 10^{-6}\,\mathrm{m^3/s}$$

The measuring current can now be determined as follows:

$$I_M = 1.071149\cdot 10^4\cdot\frac{\mathrm{As}}{\mathrm{g}}\cdot\frac{0.000804g}{\mathrm{m}^3}\cdot\frac{1.667\cdot 10^{-6}m^3}{\mathrm{s}} = 14.34\mu\mathrm{A}$$

This value refers to standard conditions. The conversion to real conditions is done via the temperature ratio:

$$14.34\mu\mathrm{A}\cdot\frac{273}{298} = 13.14\ \mu\mathrm{A}$$

HUMItrace Sensors

Figure 10.27 shows a P_2O_5 sensor in a planar structure. The phosphorus pentoxide coating in this example is on an interdigital structure. The structure, which is also to be used in space travel (Mars mission), was specially developed for small water vapor concentrations (T_{TP} – 80°C). For the frost point range from −70 to −40°C, the sensor shows a linear output signal (40 to 1000 μA), with the display between 30 L/h and 30 L/h being almost constant. A flow dependence could be limited with this sensor design (Lorek et al. 2010).

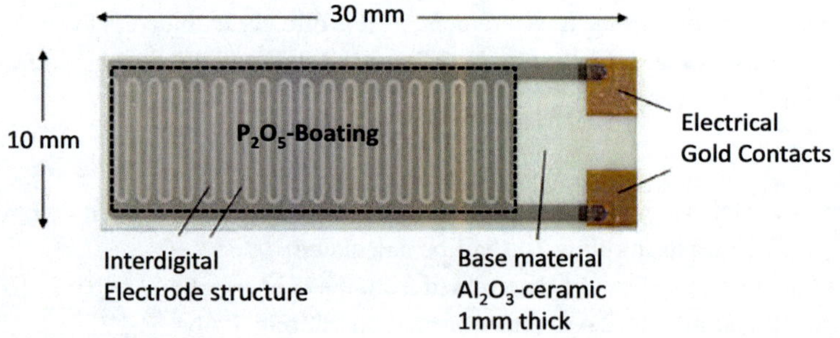

Fig. 10.27 Planar sensor structure of a P_2O_5 humidity sensor (HUMITRACE) for the trace range from −70 to 45 °C. (Source: Dr. Wernecke Feuchtemesstechnik GmbH, Potsdam)

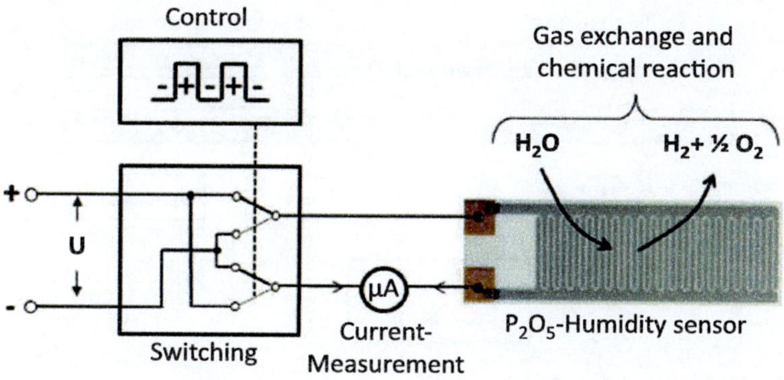

Fig. 10.28 Wiring of the P_2O_5 humidity sensor with alternating polarity to suppress electromigration at the electrodes. (Source: Koncz et al. 2010 and Dr. Wernecke Feuchtemesstechnik GmbH, Potsdam)

This system is controlled by a special power supply which includes a polarity reversal of the supply. This measure suppresses electromigration at the electrodes (Koncz et al. 2010). Figure 10.28 shows the control and the evaluation.

Coulometric ZrO_2 Solid Electrolyte Cell
Another new development is water vapor measurement with a coulometric measuring cell based on a ZrO_2solid electrolyte (Lawrenz 2014). In most cases, solid electrolytes are used to measure the oxygen content (see Sect. 4.2). To do this, the partial pressure difference between the sample gas and a reference gas is used to determine the oxygen concentration using the Nernst equation. If, on the other hand, a voltage U_E *is* applied externally, electrolysis of the substances present in the ZrO_2 probe will occur, depending on the voltage value. For water, the decomposition voltage is $U_E > 1.2$ Volt, at an ambient temperature of $T = 750^{\circ}$C. Since oxygen decomposes at a voltage of 0.4 Volt, both components are always measured simultaneously at $U_E > 1.2$ Volt. The electrolysis current (= measuring current) I_M results from Faraday's law, as with the Keidel measuring cell:

$$I_M = F \cdot z \cdot \left[\frac{\dot{m}(H_2O)}{M(H_2O)} + \frac{\dot{m}(O_2)}{M(O_2)} \right] \tag{10.21}$$

In order to be able to draw conclusions about the water vapor concentration, a second ZrO_2 measuring cell is therefore required in which the oxygen content is determined (see figure). The water vapor content is then determined from both quantities. This method can be used to measure both oxygen and moisture content in a measuring range from 0.1 ppm to 2000 ppm. Humidity measurements in air are not possible, however, since both measurements are carried out in different ZrO_2 cells, and thus high deviations are observed at high oxygen concentrations. Measurements in inert gases (N_2, Ar, ...), on the other

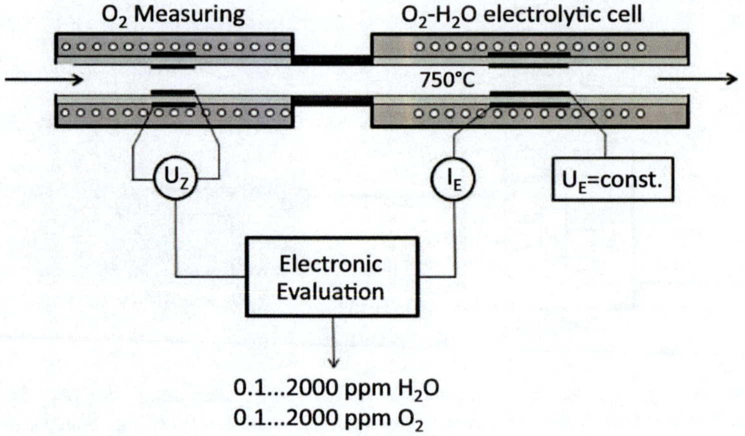

Fig. 10.29 Basic structure of a potentiometric oxygen measurement and a coulometric water vapor/ O_2measurement. (According to Lawrenz 2014)

hand, have clear advantages over the P_2O_5 method. Especially the response times and the accuracy are much better. Furthermore, this cell cannot dry out.

ZrO_2 Humidity Measurement

The oxygen measurement method, based on solid electrolytes, has already been described in Sect. 4.2. This measuring method can also be used to indirectly determine the water vapor concentration in a waste gas (Fig. 10.29). The structure for this is shown in Fig. 10.30. It consists of a typical zirconium dioxide probe, which is usually used for oxygen measurement in exhaust gases. The oxygen concentration is calculated using the Nernst equation:

$$U_N = \frac{R \cdot T}{4F} \cdot \ln\left(\frac{p_2}{p_1}\right) \tag{10.22}$$

Simplified, this equation can also be described at 1000 hPa via the O_2-concentrations with constant $C = RT/4F$:

$$U_N = C \cdot \ln\left(\frac{c_2}{c_1}\right) \tag{10.23}$$

For a probe temperature of $T = 750^{\circ}\mathrm{C}$, $C = 22$ mV. The oxygen partial pressure p_1 describes the O_2-concentration of the exhaust gas, while p_2 indicates the O_2-concentration of the reference gas. As a rule, ambient air with an oxygen content of $c \approx 20.8$ vol.% is used as reference gas.

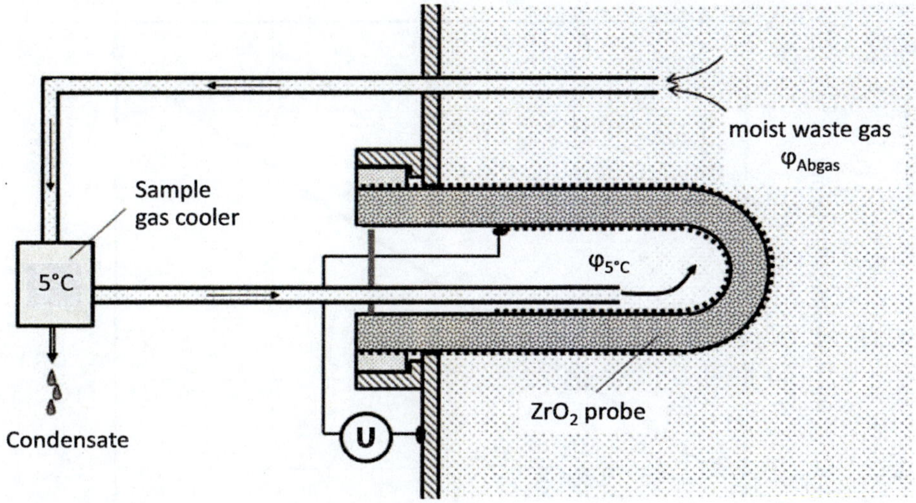

Fig. 10.30 Schematic representation of an indirect humidity measurement using the increase in O_2 concentration due to condensation of the water vapor

In the setup shown in Fig. 10.30, however, the exhaust gas is removed from the process and passed through a sample gas cooler, which condenses almost all of the water content. For a cooler temperature of $T_K = 5^\circ$C, the remaining H_2O fraction corresponds to a water vapor partial pressure of $p_d = 8.7$ hPa. Because the water vapor component has been removed from the sample gas, the other gas concentrations increase by a factorf_{H_2O} (see Sect. 14.1, Eq. 14.17). The oxygen content c_2 downstream of the sample gas cooler is then calculated:

$$c_2(O_2) = c_1(O_2) \cdot f_{H_2O} \tag{10.24}$$

If you insert Eq. 10.24 into Eq. 10.23, you get the following expression:

$$U_N = C \cdot \ln\left(\frac{c_1 \cdot f_{H_2O}}{c_1}\right) = C \cdot \ln\left(f_{H_2O}\right) \tag{10.25}$$

respectively;

$$U_N = C \cdot \ln\left[\frac{p_0 - p_d(5^\circ C)}{p_0 - p_d}\right] \tag{10.26}$$

By dissolving Eq. 10.26 to p_d, one obtains the desired water vapor partial pressure in the exhaust gas:

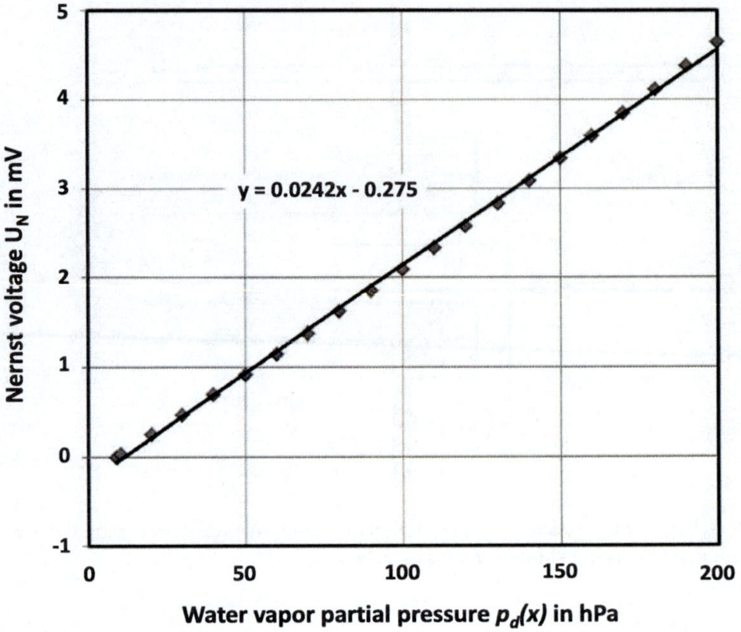

Fig. 10.31 Nernst voltage U_N as a function of the water vapor partial pressure $pT_d(x)$ in the exhaust gas

$$p_d = p_0 - \frac{p_0 - p_d(5\,^{\circ}\mathrm{C})}{\exp[U_N/C]} \tag{10.27}$$

Figure 10.31 shows the characteristic curve for a vapor pressure range up to 200 hPa, which corresponds to a saturation temperature of 60 °C The concentration in this case would be $c_1 \approx 20$ vol.% H_2O. The characteristic curve is almost linear over the entire range.

However, this method is not suitable for low moisture contents, as the possible sources of error are too large. The main problem are the relatively low cell voltages (<1 mV), which are additionally superimposed by thermal asymmetry voltages. This method can only be used for very high steam concentrations, such as those found in the steam generation of power plants or in drying processes in large bakeries (Lawrenz 2015).

Oscillating Quartz QCM[4]

Oscillating crystals can also be used for moisture measurement in the trace range. For this purpose, the quartz is coated with a moisture-sensitive polymer. The water vapor then accumulates in this layer and leads to a change in the resonance frequency f_0. This

[4]Quartz Crystal Microbalance.

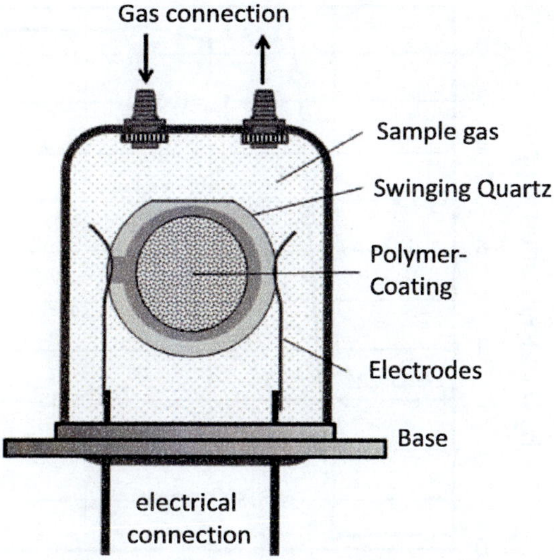

Fig. 10.32 Structure of a QCM sensor for measuring moisture in the trace range

frequency change is caused by the increased mass Δm due to the water vapor deposition. Although this change is very small, water vapor concentrations down to the ppb range can be measured with this method. For this purpose, however, the temperature of the device must be controlled very precisely since the fundamental frequency of the oscillator is also influenced by the temperature. Figure 10.32 shows the structure of a QCM sensor. The oscillating quartz is installed in a closed chamber through which the sample gas flows. The entire assembly is in a thermostat which is controlled to a constant temperature and shields the sensitive sensor from changes in the environment. There is also a second QCM sensor in the thermostat, which does not react to water vapor, however, and thus represents a reference. Naturally, the process also reacts very sensitively to soiling, as particles can deposit on the polymer layer and thus simulate a non-existent water vapor concentration. The sensor is therefore regularly exposed to dry gas and test gas in the moisture analyzers to compensate for possible drift causes. This adjustment cycle is carried out automatically via appropriate valve controls.

Typical measurement ranges for this method are 0.5 to 700 ppm with an accuracy of ±5%, or 0.1 to 2000 ppm with an accuracy of ±10%. The response times are in the minute range. Figure 10.33 shows the time response for a water vapor concentration of 10 ppm.

Condensation Sensors

Protection against condensation is very important for many applications, as water in liquid form can cause more damage than in gaseous form. For this purpose, dew sensors are used which generate a corresponding signal shortly before condensation. Condensation sensors work like humidity sensors but have a high sensitivity near the dew point. Figure 10.34

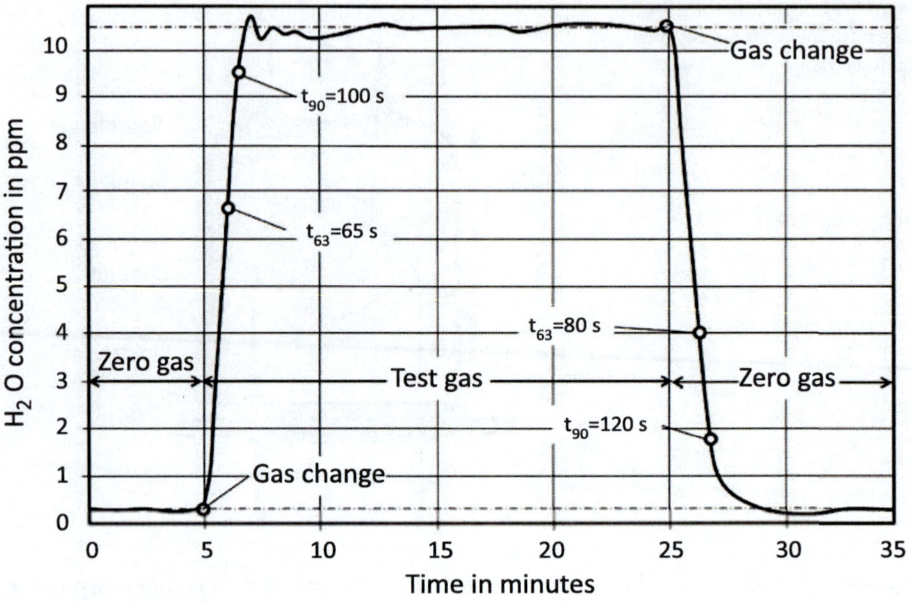

Fig. 10.33 Temporal response of a QCM sensor in the lower ppm range. (According to data from Michell Instruments GmbH, Friedrichsdorf)

shows the operating principle. The sensitive sensor layer consists of a polymer material in which graphite particles are embedded. An electric current can flow between the electrodes through these graphite particles, which is linked to an electrical resistance R_S of the sensor layer. R_S is typically in the range of ≈1 kOhm, with a relative humidity of <50%. As the air humidity rises, the polymer layer absorbs more and more water molecules, which causes the layer to swell. This increases the distance between the conductive graphite particles and the resistance increases. In the case of strong swelling, which occurs near the dew point, the bonds between the graphite particles are completely broken, so that the flow of current is no longer possible. The resistance then increases by several orders of magnitude ($R_S \approx M\Omega$). This sharp increase in sensor resistance is used to detect the threat of condensation (Fig. 10.35).

Figure 10.36 shows the resistance curve at R_S (φ), a relative humidity φ of 30 to 100%. With a linear representation of the sensor resistance, significant changes in resistance can only be observed from 90% relative humidity. The sensor acts like a switch that is closed above 90% relative humidity. The logarithmic curve (second *y-axis*) shows the slow increase between 30% and 85%, and the steep slope from 90% relative humidity. The response time (t_{90}) is less than 1 minute for the maximum humidity change.

The possible applications for this sensor are manifold:

Fig. 10.34 Principle function of a dew sensor

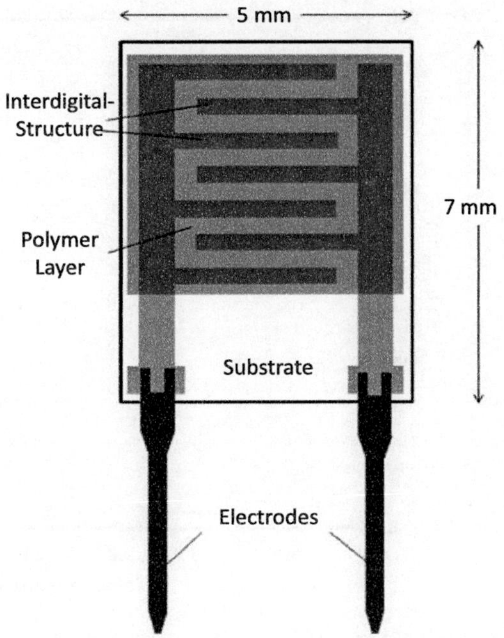

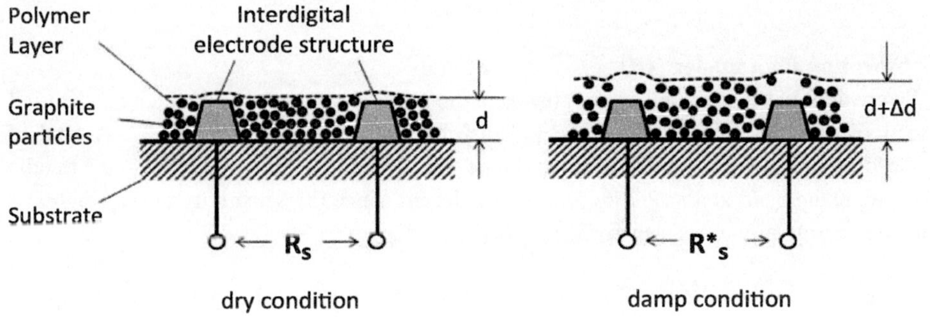

Fig. 10.35 General view of a dew sensor (Type SHS-A4L Hygrosens Instruments GmbH Löffingen)

- Air conditioning and ventilation technology
- Monitoring of sanitary rooms
- Chilled ceiling switch
- Moisture protection in electrical devices and installations
- Detection of condensate on external walls
- Leakage detection.

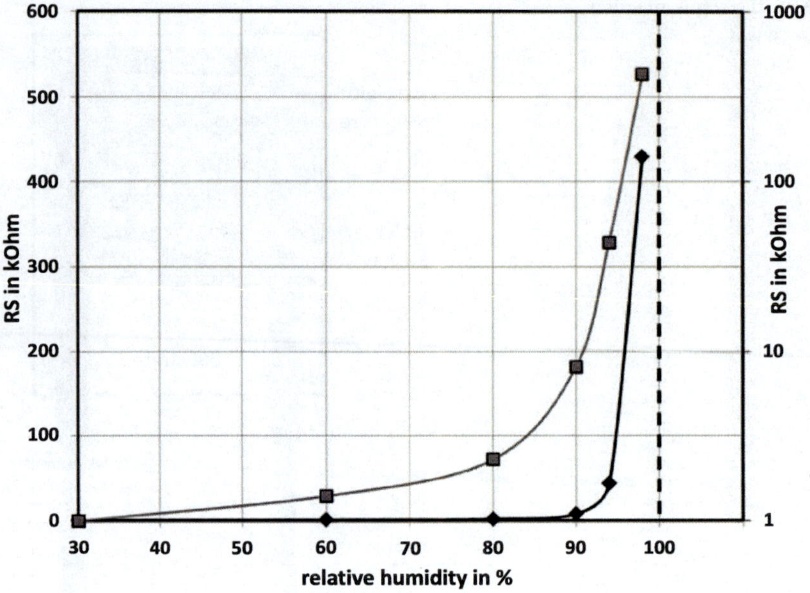

Fig. 10.36 Characteristic curve behavior of a dew sensor (Type SHS-A4L Hygrosens Instruments GmbH Löffingen) in linear and logarithmic representation

Absorption Photometer (IR)
Water vapor has distinct absorption bands in the infrared spectral range which can be used for moisture measurement. The spectral distribution is shown in Fig. 10.37. A total of 4 absorption bands are available which can be used. For an H_2O measurement in the middle IR range, the band at $\lambda \approx 2.7\mu m$ is usually selected, since the sensitivity is very high and there is hardly any cross sensitivity to other gases.

The measurement in the IR range is carried out using a 2-beam method (see Chap. 7). The broadband IR radiation is generated by a thermal radiator (e.g. miniature incandescent lamp or micro radiator), and enters a gold-plated absorption cuvette (e.g. 10 cm length). There is the selective absorption by the water vapor takes place. The double detector has two receiving channels, for the reference radiation (e.g. 3.95 μm) and the measuring radiation (e.g. 2.7 μm). The concentration c_{H_2O} is calculated from the signal change in the measuring channel. The dew point for this measurement must always be below the cuvette temperature to prevent condensation. Depending on the cuvette length (L) used and the selected measuring wavelength (λ), this method can be used to measure water vapor from the volume % range to the ppm region. Figure 10.38 shows such an IR photometer (*INFRA.sens*®), which allows water vapor measurements.

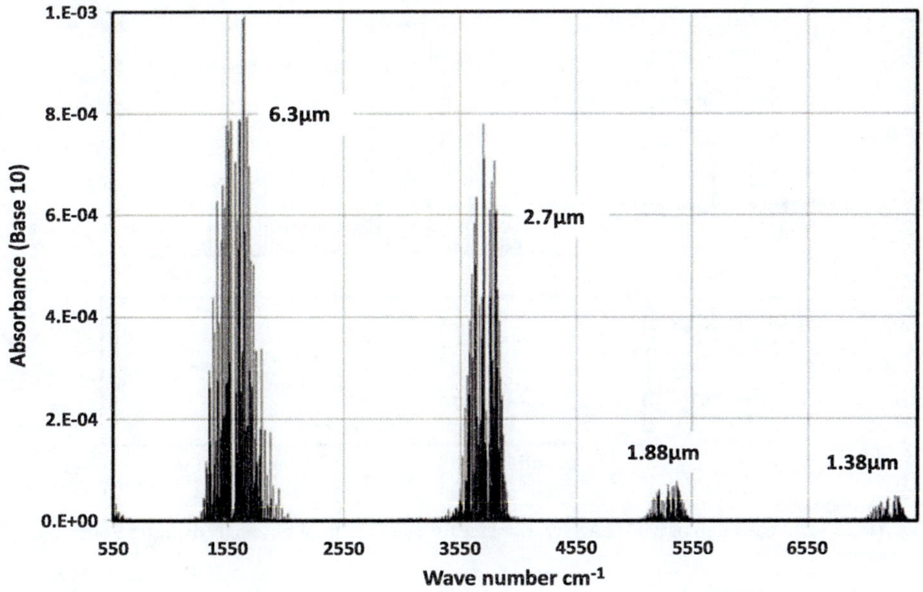

Fig. 10.37 IR absorption spectrum of water vapor

Fig. 10.38 *INFRA.sens*® infrared measuring system. (Source: https://www.witec-sensorik.de/en/applications/humidity-measurement/ Wesel Germany)

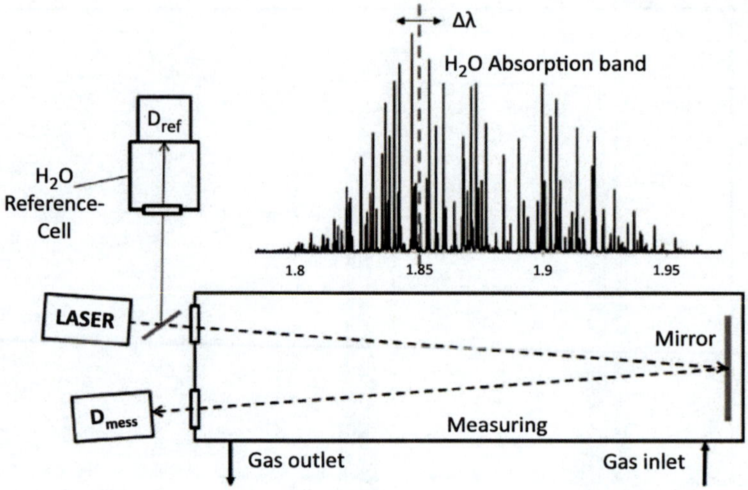

Fig. 10.39 Optical setup of a TDLAS hygrometer for use in natural gas analysis (after Potter 2011)

TDLAS[5]-hygrometer
The use of diode lasers is now also widespread in trace moisture measurement. Especially the moisture measurement of natural gas is very successfully performed with this method (Potter 2011). The advantage of diode lasers is the narrow band of emission wavelengths, which allows selective measurement, even in the presence of other interfering components. Figure 10.39 shows such a *TDLAS setup*. The diode laser was selected for an emission wavelength of $\lambda = 1.854$ μm. The tunability ($\Delta\lambda$) is sufficient to detect the maximum absorption line in the immediate vicinity of the laser line. The laser radiation passes through the measuring cell twice and is then detected by the measuring detector. In front of the measuring cuvette, there is a beam splitter that directs a small portion of the laser radiation into a reference cell. This reference cell is used to stabilize the wavelength and thus also the measuring accuracy of the arrangement. A typical measuring range for natural gas monitoring is 1000 ppm H_2O. The resolution/detection limit is specified by the manufacturer at 1 ppm. The time response is in the range of <2seconds.

Figure 10.40 shows an industrial measuring device based on laser spectroscopy (TDLAS). The entire assembly is located in a flameproof enclosure for applications in the Ex zone.

Lyman-α-hygrometer
In the VUV range, the absorption of the radiation takes place by the H atoms of the water molecules. However, this type of measurement is not selective. However, if the

[5]**Tuneable Diode** Laser Absorption Spectroscopy.

Fig. 10.40 View of a TDLAS hygrometer for use in the natural gas industry in an explosion-proof housing. (Source: Michell Instruments GmbH, Friedrichsdorf)

measurement is carried out in pure air, possible error influences can be compensated. Figure 10.41 shows such a photometric humidity sensor. The radiation is generated by a hydrogen lamp (e.g. EDL), which emits so-called Lyman-α lines at $\lambda = 121.6$ nm. Alternatively, krypton lamps were used for this purpose (Buck 1973). This radiation is absorbed by the H atoms of the water molecules, according to the Lambert-Beer law. Only magnesium-fluoride panes (MgF_2) are used as window materials, which allow sufficient transmission in this spectral range. However, the transmission properties are not stable, so that an unwanted drift can occur. Furthermore, there are cross-sensitivities to oxygen and ozone, which represent an additional limitation of the measuring properties. Nevertheless, this type of humidity measurement technology is used in the field of meteorology. When measuring humidity in the stratosphere, rapid changes of the moisture content occur, for example, when these measurements are made outside an aircraft flying through different dew point areas. The response time of the Lyman α instrument is in the millisecond range ($\approx$100 Hz). Figure 10.42 shows a comparison measurement between a Lyman-α hygrometer and a dew point hygrometer. The fast response of the Lyman α hygrometer can be clearly seen. Approximately 5 seconds before the dew point hygrometer, the corresponding humidity values are displayed.

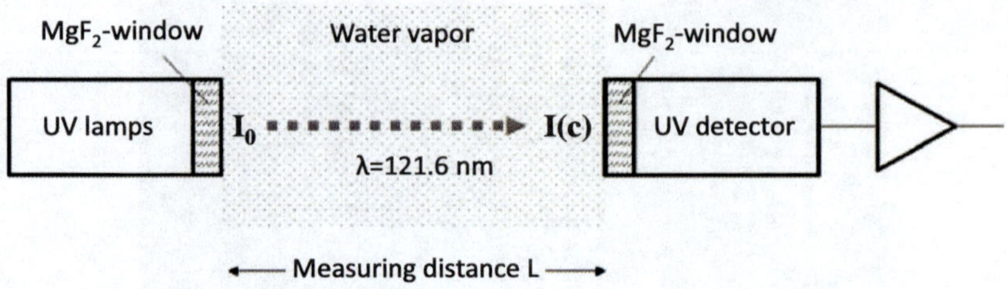

Fig. 10.41 Basic structure of a Lyman α humidity sensor

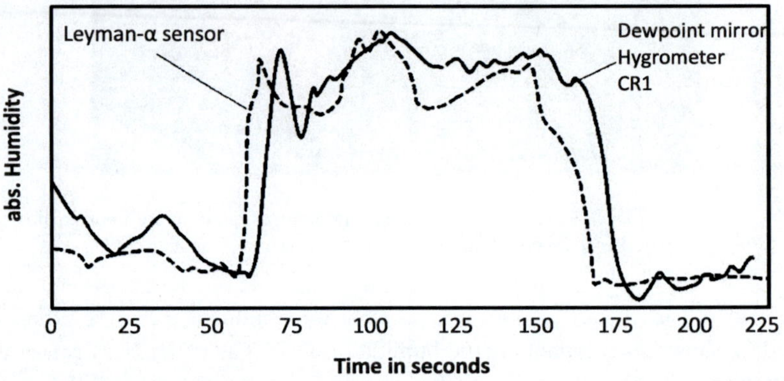

Fig. 10.42 Comparison measurement between a Lyman α hygrometer and a special nitrogen-cooled chilled mirror hygrometer (Tränkler and Obermeier 1998)

10.3 Calibration

The calibration of humidity measuring instruments is carried out via a defined humidity, which can be produced with different methods. The methods differ mainly in the area of application, i.e. for which humidity range the devices/sensors are to be calibrated. In this chapter, only the most important procedures, which belong to the standard procedures, are explained.

Calibration Salt

The simplest method is to use the moisture content above a saturated salt solution. Hygroscopic salts lower the vapor pressure in the gas phase above to the defined value. This humidity value is almost independent of the ambient temperature T_U. Figure 10.43 shows such a structure. In a beaker there is a saturated salt solution, which consists of 80–90% undissolved salt. Above this layer the solution is completely saturated. For an

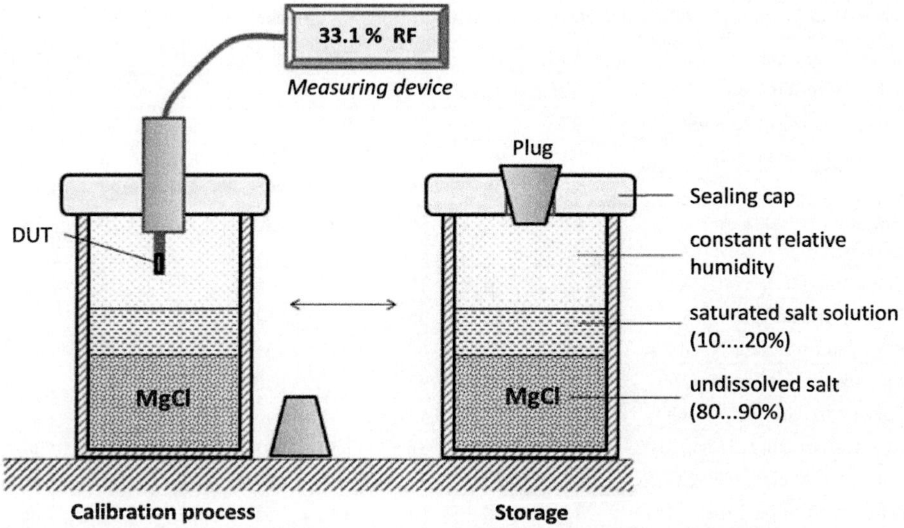

Fig. 10.43 Testing humidity sensors with humidity standards (saturated salt solutions)

exact measurement, it is important to keep a certain ratio between the surface $A_{lös}$ and the height h_P of the sample space (Wernecke 2003). It applies:

$$\frac{\sqrt{A_{lös}}}{h_P} > 1 \tag{10.28}$$

If the beaker is closed, a defined relative humidity is formed in this space, which depends only on the property of the salt. For magnesium chloride ($MgCl_2$), this humidity is 32%. Table 10.3 lists further salts that produce a different saturation humidity. With these substances, e.g. a hygrometer, within the measuring range, allows a check and if necessary, an adjustment to be carried out. For this purpose, the test sample is held in the beaker, whereby, if possible, no leaks from outside should disturb the measurement. After a few minutes (note the response time of the test specimen) a stable measured value is obtained, which is compared with the corresponding humidity standard. For accurate measurements, the test samples should be stored at a constant humidity of 30–40% 24 h before measurement.

Saturation Method

Probably the most important and most accurate method works via the saturation vapor pressure p_s. If an inert gas (e.g. nitrogen) is pumped through a water reservoir and split into many small bubbles with a sinter disk, these bubbles saturate on their way through the water. At the outlet, the gas flowing through has a relative humidity of 100%. The

Table 10.3 Humidity standards for saturated salt solutions (Wernecke 2003)

Saline solution	15°C	20°C	25°C	30°C	35°C	40°C	45°C	50°C
	Relative humidity (%)							
Potassium sulphate K_2SO_4	97	97	97	96	96	96	96	96
Potassium nitrate KNO_3	94	93	92	91	89	88	85	82
Potassium chloride KCL	87	86	85	85	84	82	82	81
Ammonium sulphate $(NH_4)_2SO_4$	81	81	80	80	80	79	79	78
Sodium chloride NaCl	76	76	75	75	75	75	75	75
Sodium nitrite $NaNO_2$	–	65	65	63	62	62	59	59
Ammonium nitrate NH_4NO_3	69	65	62	59	55	53	48	42
Magnesium nitrate $Mg(NO_3)_2$	56	55	53	52	50	49	46	–
Potassium carbonate K_2CO_3	44	44	43	43	43	42	–	–
Magnesium chloride $MgCl_2$	34	33	33	33	32	32	31	30
Potassium acetate CH_3CO_2K	21	22	22	22	21	20	–	–
Lithium chloride LiCl	13	12	12	12	12	11	11	11

corresponding vapor pressure p_d results from the water temperature T_W. To be able to carry out a calibration for the entire characteristic curve, this maximum value must be diluted with dry gas so that a lower humidity is also obtained. Figure 10.44 shows such a setup. The dry carrier gas is divided into two gas streams in front of the bead vessel. The first gas flow $\dot{V}_1$ passes directly through a flow controller (MFC) into the mixing vessel, while the second gas flow $\dot{V}_2$ is passed over the bead vessel and saturates with water vapor. The beaded vessel is in a cryostat which controls the water temperature (dew point) T_{TP}. However, this temperature must be below the ambient temperature T_U, *to* prevent condensation in the continuing hoses. To control the humidity values, some manufacturers also offer an additional dew point measurement, which physically determines the dew point exactly and thus serves as a reference measurement. Both units are flowed through in parallel to obtain the best possibility of comparison. The concentration c_{H_2O} at the outlet of the mixing vessel is then calculated:

$$c_{H_2O} = \frac{p_s}{p_0} \cdot \frac{\dot{V}_2}{\dot{V}_1 + \dot{V}_2} \cdot 100\% \tag{10.29}$$

Example

At a water temperature of 20 °C, a saturation vapour pressure of 23.4 hPa is established at the outlet of the bead vessel. The mixing ratio is 10 : 1. At an air pressure of 995 hPa, the following moisture content results at the outlet of the mixing vessel:

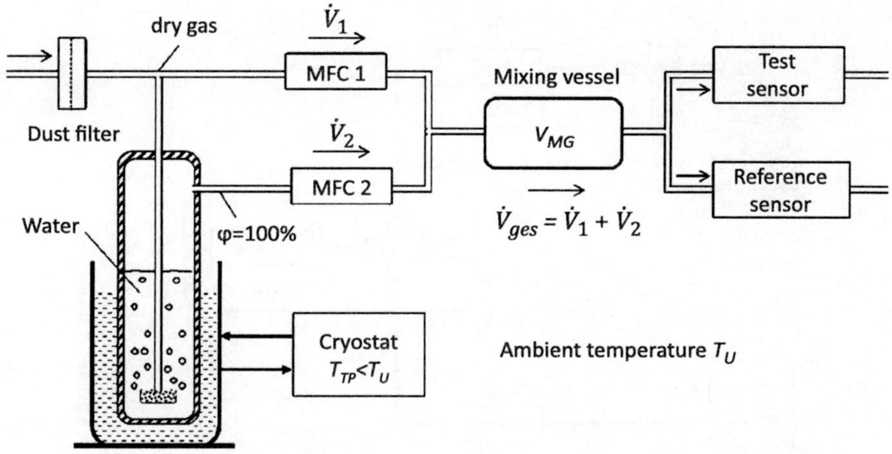

Fig. 10.44 Setup for calibrating humidity measuring instruments using the dew point method

$$\frac{23.4\,\text{hPa}}{995\,\text{hPa}}\cdot\frac{\frac{0.1\,\text{L}}{\text{min}}}{\frac{1.1\,\text{L}}{\text{min}}}\cdot 100\% = 0.314\,\text{Vol.0\%}H_2O$$

CEM Process[6]

The saturation method is mainly used at dew point temperatures below the ambient temperature. If the dew points are above the ambient temperature, the gas flows from the bead vessel to the mixer and, if necessary, also to the measuring instrument must be completely heated to prevent condensation. However, this option is costly and not very practical.

For these applications, a defined amount of water $\dot{m}_{H_2O}$ is completely evaporated and then immediately mixed in a defined gas flow $\dot{V}_{DL}$. After evaporation, the water takes up a volume $\dot{V}_{H_2O}$, which is calculated from the molar mass for H_2O of 18,015 g/mol, and the molar volume for ideal gases of 22,414 L/mol:

$$\dot{V}_{H_2O} = \dot{m}_{H_2O}\cdot\frac{22,424\,\text{L/mol}}{18,015\,\text{g/mol}}\cdot\frac{T_{VD}}{273.15\,\text{K}} \tag{10.30}$$

[6]Controlled Evaporator Mixer.

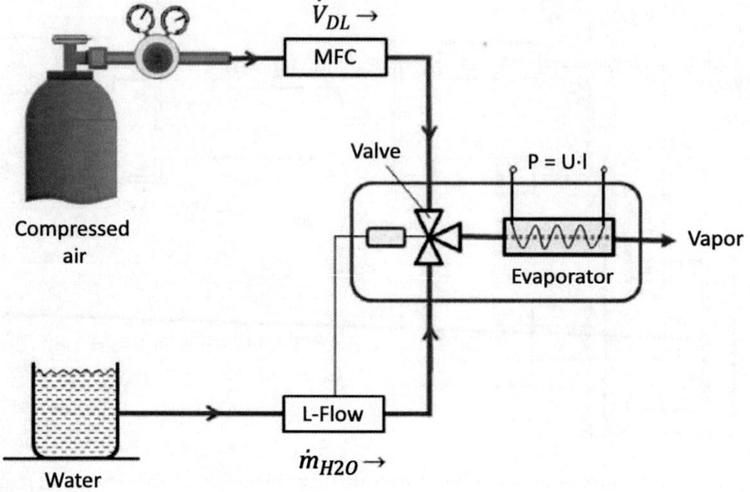

Fig. 10.45 Arrangement for generating a defined water vapor concentration by controlling the mass flow (water) and volume flow of the carrier gas. (According to documents from Bronkhorst-Nord GmbH, Kamen)

The evaporator temperature T_{VD} changes the molar volume due to thermal expansion and must therefore be taken into account. The mixture has a water vapor concentration c_{H_2O} which is indicated in volume %:

$$c_{H_2O} = \frac{\dot{V}_{H_2O}}{\dot{V}_{DL} + \dot{V}_{H_2O}} \cdot 100\,\text{Vol.0\%} \tag{10.31}$$

The concentration in g/m^3 is then calculated as follows:

$$c_{H_2O} = \frac{\dot{m}_{H_2O} \cdot 1000\,\text{L/m}^3}{\dot{V}_{DL} + \dot{V}_{H_2O}} \tag{10.32}$$

Figure 10.45 shows this unit. The gas flow is set to a constant flow rate by a controller (MFC1). The dry carrier gas is supplied from a compressed gas cylinder. The water is dosed directly from a storage tank via a mass *flow* meter (*L-Flow*). The solenoid valve always allows only a part of the water to enter the vaporizer zone and thus controls the concentration. The vaporizer consists of an electrically heated tube which ensures a fast transition to the gas phase. The outlet of the vaporizer consists of a thin stainless-steel tube with an outer diameter of $d \approx 3$ mm. This output can be connected directly to the measuring instrument to be checked, or it can be used for other purposes.

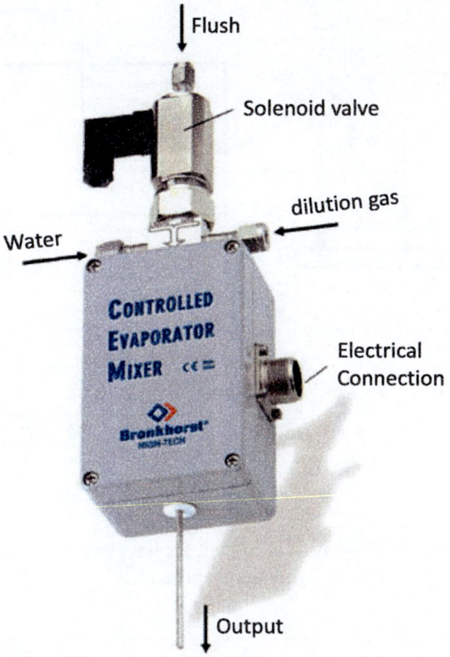

Fig. 10.46 View of the CEM unit for the controlled generation (evaporation) of water vapor concentrations. (Source: Bronkhorst-Nord GmbH, Kamen)

HovaCAL® Process

This method works according to a similar principle as the CEM method just described. Here too, the water is fed with a carrier gas stream into an evaporator, which ensures a fast and efficient conversion into the gas phase (→ principle of dynamic evaporation). The gas has a temperature of up to180 °C behind the evaporator unit. The *HovaCAL® process* also offers the possibility to mix up to 6 different gases and vapours. A more detailed description of these possibilities is given in Chap. 12 (Fig. 10.46).

The simplest version is shown in Fig. 10.47. In this example, the water to be evaporated is in a beaker standing on a scale. A peristaltic pump is used to pump a certain amount of water into the evaporator. The exact determination of the amount of water ($\dot{m}_{H_2O}$) that enters the evaporator per time is obtained by evaluating the electronic balance. For this purpose, the output of the balance is connected via a digital interface to a computer, in which a special evaluation program (ViewCAL211) is used to perform the corresponding calculations according to Eq. 10.31 or Eq. 10.32.

In a further development, the balance was replaced by a flow sensor (MFS), so that the current mass flow ($\dot{m}_{H_2O}$) is recorded directly via the MFS. According to the manufacturer, the accuracy is comparable to that of an electronic scale. The general view of the instrument (*HovaCal digital 21-MF*) is shown in Fig. 10.48.

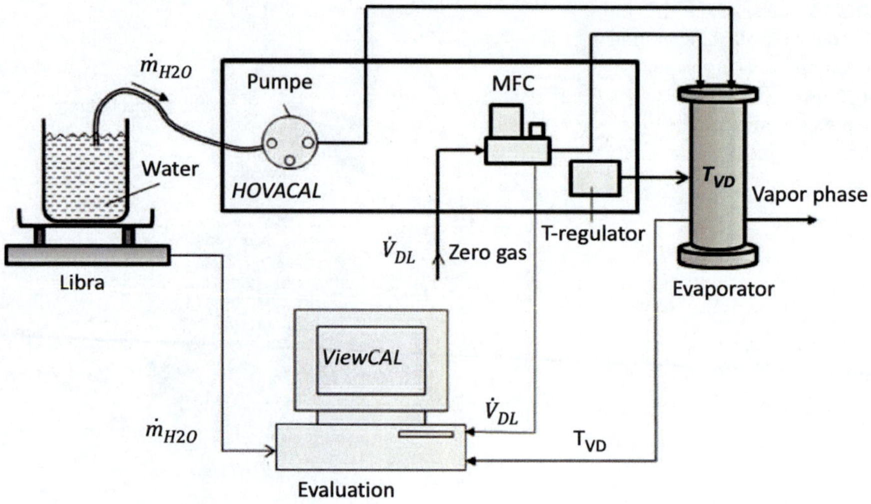

Fig. 10.47 HovaCAL® unit for the generation of defined water vapour concentrations according to the principle of dynamic evaporation

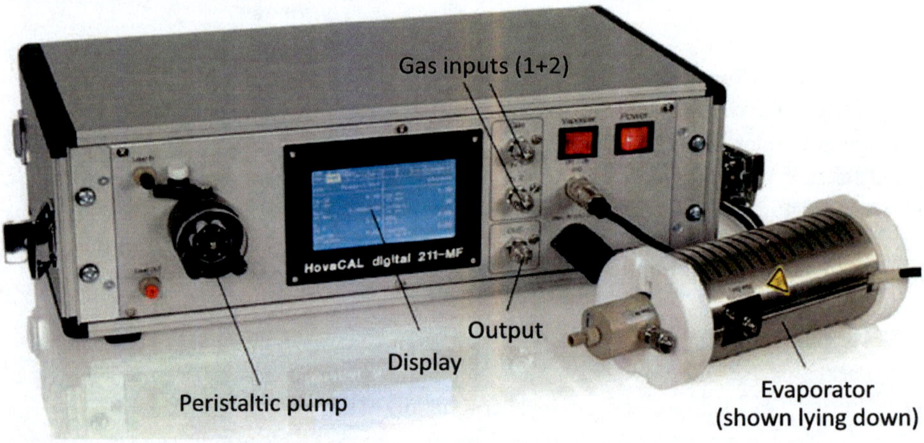

Fig. 10.48 View of the *HovaCal digital 21-MF* with an integrated MFS for the water supply. (Source: IAS GmbH, Oberursel)

Trace Moisture

Another possibility, the defined humidification of a dry carrier gas, is based on permeation. Water molecules can migrate through certain plastics (permeation, see Sect. 2.4), whereby this plastic layer represents a barrier that must be overcome. The permeation rate is therefore usually very low. This method is therefore also suitable for producing very low water vapor concentrations.

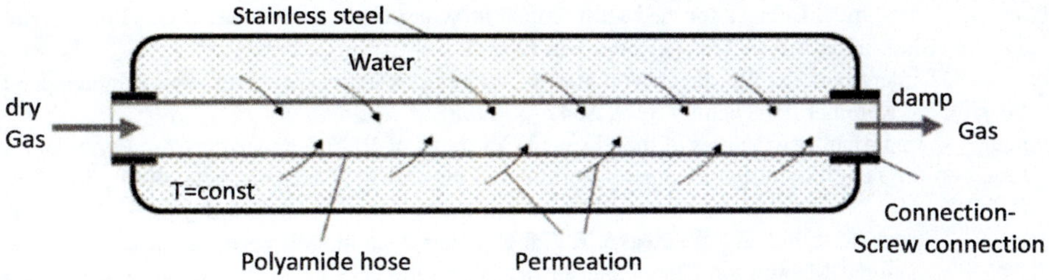

Fig. 10.49 Calibration of trace moisture by permeation of water through a polyamide tube

Figure 10.49 shows such an arrangement. It essentially consists of a polyamide plastic hose which is surrounded by water. The water temperature is regulated to a certain value to keep the permeation rate constant. For polyamide, a water vapour permeation coefficient of $2.44 \pm 0.12\,\text{g} \cdot \text{mm} \cdot \text{m}^{-2} \cdot \text{d}^{-1}$ at $T = 30^{\circ}\text{C} \pm 0.1^{\circ}\text{C}$ was determined (Wernecke 2003). With this method, H_2O concentrations in the lower ppm range can be generated, which are relevant for trace moisture sensing. For a dew point of -80 to -30°C, concentration values from $\approx$0.5 ppm to 380 ppm H_2O must be generated to calibrate such a trace moisture analyzer. Usually, dew point mirror hygrometers are also used as a reference (Tiebe et al. 2012).

References

Buck, A.L.: Developement of an improved Lyman-alpha-Hygrometer. Atmos. Technol. **2**, 213–240 (1973)

Cerbe, G.: Grundlagen der Gastechnik. Hanser, München (2008)

Clevett, K.-J.: Process Analyzer Technology. Wiley Interscience, New York (1986)

Gevatter, H.-J. (Hrsg.): Handbuch der Meß- und Automatisierungstechnik. Springer, Berlin (1998)

Hengstenberg, J., Sturm, B., Winkler, O.: Messen, Steuern, Regeln in der chemischen Technik. Bd. II Betriebsmesstechnik II. Springer, Berlin (1980)

Karbach, A., et al.: Messtechnik in der Versorgungstechnik. Springer, Berlin (1997)

Keidel, F.A.: Determination of Water by Direct Amperometric Measurement. Anal. Chem. **13**, 2034–2048 (1959)

Koncz, A., Lorek, A., Wernecke, R., Schwanke, V.: Coulometrischer Feuchtesensor mit Wechselspannungsquelle. Europäische Patentanmeldung vom 22.12.2010 (2010)

Korotcenkow, G.: Handbook of Gas Sensor Materials. Springer, Berlin (2013)

Lawrenz, U.: Präzise Spurenfeuchtemessung in der industriellen Fertigung. messtec drives Automation. **5**, 50–52 (2014)

Lawrenz, U.: private Mitteilung (2015)

Lorek, A., Koncz, A., Wernecke, R.: Developement of a gas flow independent coulometric trace humidity sensor for aerospace and industry. In: Proceeding the First European Conference on Moisture Measurement Aquametry in Weimar 5.–7.10 2010. S. 289–296 (2010)

Potter, D.R.: Analytical Devices for the Measurement of Water vapor in the Natural Gas Process and Transmission Industry. GAS2011, Rotterdam (2011)
Sonntag, D.: Internationaler Vergleich von Assmannschen Aspirationspsychrometer. Tagungsband Feuchte 88, Kammer der Technik Suhl, S. 47–57 (1988)
Sonntag, D.: Important new Values of the Physical Constants of 1986, Vapour Pressure Formulation based on the ITS-90, and Psychrometer Formulae. Zeitschrift für Meteorologie. **40**(5), 340–344 (1990)
Tiebe, C., Hübert, T., Lorel, A., Wernecke, R.: New planar trace humidity sensor. IMCS2012 The 14th International Meeting on Chemical Sensors in Nürnberg 294–297. https://doi.org/10.5162/IMCS2012/3.4.2 (2012)
Tränkler, H.-R., Obermeier, E.: Sensortechnik – Handbuch für Praxis und Wissenschaft. Springer, Berlin (1998)
VDI/VDE-Richtlinie 3514 Blatt 1 Gasfeuchtemessung, Kenngrößen und Formelzeichen (2006)
Wernecke, R.: Industrielle Feuchtemessung Grundlagen, Messmethoden, technische Anwendungen. Wiley-VCH, Weinheim (2003)

Related References

Campbell, G.S., Tanner, B.D.: A Krypton Hygrometer for measurement of atmospheric water vapor concentrations. Moisture and Humidity. ISA Research Triangle Park. S. 609–614 (1985)
Datenblatt 11.12.01 Feuchtenormale RCI GmbH, Offenbach (2001)

11 Flow Measurement Technology

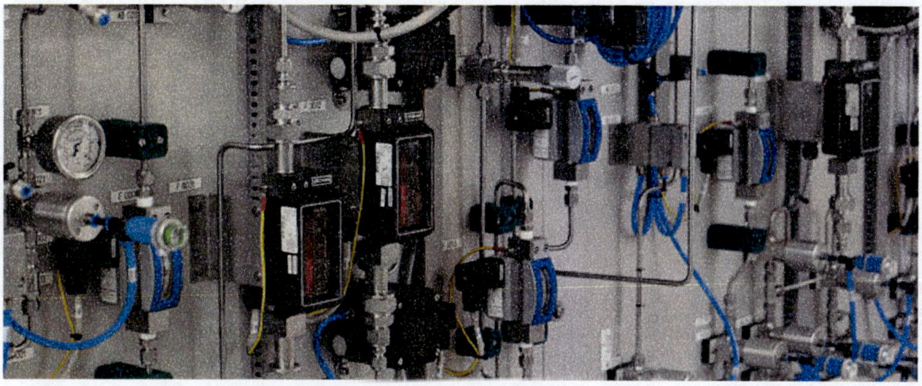

Control of gas flows. Source: KROHNE Messtechnik GmbH Duisburg

In many technical application areas, the flow rate plays an important role in addition to the gas concentration *c*. Especially when it comes to calculating the quantity of a substance in a gas mixture, both variables are always required. For this reason, various measuring methods and devices can be found in these applications which measure the gas flow with a corresponding accuracy. For these tasks a multitude of different measuring methods are available.

Basically, in flow measurement technology, a distinction is made between the volume flow $\dot{V}$, which is specified in example L/min or m^3/h, and the mass flow $\dot{m}$, which typically has kg/h or g/s as a unit. The conversion between both units is done via the density ϱ. A further distinction is made in the form of gas transport. Gas can be defined as flow in hoses or pipes or can move freely in space (air flows). In industrial applications, guided flows are

G. Wiegleb, *Gas Measurement Technology in Theory and Practice*,
https://doi.org/10.1007/978-3-658-37232-3_11

mostly used to transport materials over a certain distance. The best known example of this is the transport of natural gas over many hundreds of kilometres, from the point of withdrawal to the end consumer. Free air flows are often measured in air conditioning and ventilation technology and in meteorology.

In this chapter the most important flow measurement methods for gases are presented and described, with the emphasis on pipe flows.

11.1 Introduction

Flow is the transport of a fluid (gas or liquid) at a constant velocity v. Figure 11.1 shows this process, in which a gas volume $\mathrm{d}V = a \cdot b \cdot c$ moves freely over a distance Δs. In Sect. 2.7 the physical principles of gas flows have already been described, so that we can concentrate in this section on the resulting metrological quantities.

If the movement shown in Fig. 11.1 takes place in free space, it is called a flow or flow velocity. If this process takes place within a spatial boundary (e.g. pipe), a flow is present. In this case, the flow velocity is sometimes also referred to as flow velocity. In fluid mechanics, the velocity is usually represented by w, whereas in flow measurement technology the velocity is indicated by v.

The flow rate passing through a hose or pipe of diameter d is calculated from the ratio of the quantity (volume or mass) and the time unit Δt:

$$\text{Volume flow} = \frac{\text{Volume}}{\text{Time}} \quad \text{or} \quad \text{Mass flow} = \frac{\text{Mass}}{\text{Time}} \tag{11.1}$$

The diameter d results in an area $A = \pi d^2/4$ for a round cross-section. The following therefore applies to the volume flow:

$$\dot{V} = \frac{\pi d^2}{4} \cdot v \tag{11.2}$$

and for the mass flow:

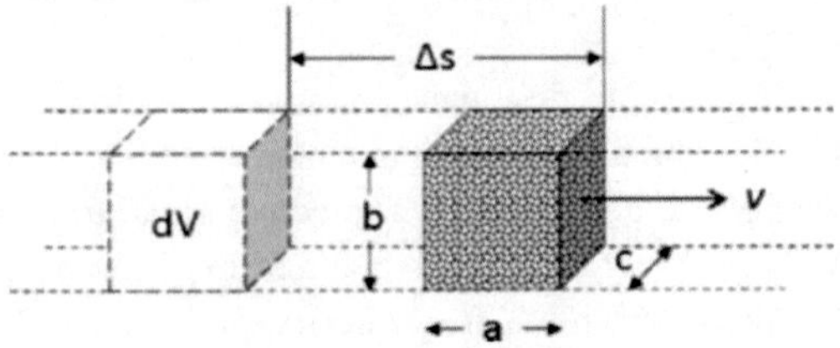

Fig. 11.1 Representation of the transmission (flow) of a gas volume $\mathrm{d}V = a \cdot b \cdot c$ in space with the velocity v

$$\dot{m} = \varrho_{Gas} \cdot \frac{\pi d^2}{4} \cdot v \tag{11.3}$$

11.2 Measurement Procedure

The flow $(\dot{m}; \dot{V})$ in pipelines can be measured either directly by volumetric or gravimetric methods or indirectly by the flow velocity v. The individual methods differ considerably in terms of handling and accuracy, sometimes considerably. In this subchapter, therefore, the methods commonly used in gas measurement technology for small and medium flow rates are presented and described.

Volumetric Methods

The oldest measurement methods for gas flow were based on a volumetric method. Figure 11.2 shows such a setup. To do this, take a glass tube closed at one end, with a known base area A, fill it with water and immerse it in reverse in a water bath. If a gas now flows over the tube from below, the gas accumulates in the upper part and thus displaces the water previously present. From the height h and the base area A the collected volume $V_{Gas} = A \cdot h$ is then calculated. For exact measurements the temperature and pressure of the gas must also be known. The volume flow $\dot{V} = V_{Gas}/\Delta t$ can then be determined by means of a time measurement (Δt). With this simple experiment it is indeed possible to measure a volume flow based on the SI units (length and time). However, this method is not suitable for practical measurements outside the laboratory.

As early as 1816, *Clegg*[1] constructed the first gas meter, which enabled continuous gas measurement. In a gas meter, the written filling process is cyclically repeated and additionally converted into a rotary movement. This rotary motion can then be used to connect a counter that registers the total volume flow.

Figure 11.3 shows one such historical *gas meter*. The gas was introduced into the instrument from the side. A displacement body, which was under water, was filled with the gas and generated a buoyancy force, which was converted into a rotational movement. This rotary motion could be displayed and registered by a mechanical counter.

Drum Gas Meter

Modern gas meters have a very similar design. Figure 11.4 shows a cross-sectional view of such a drum meter. The gas enters the centre of the device and fills chambers 1–4 one after the other. The chambers are mechanically connected to each other (→ drum) and rotate around a horizontal axis. A slight overpressure produces a force F on the chamber, which

[1] Samuel Clegg (1781-1861) British chemist and engineer.

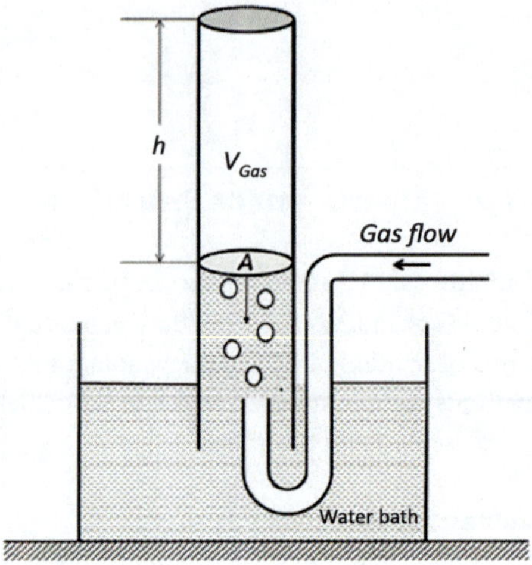

Fig. 11.2 Simple experiment for recording a gas volume

Fig. 11.3 Historical representation of a volumetric gas measuring device (Brunck 1919)

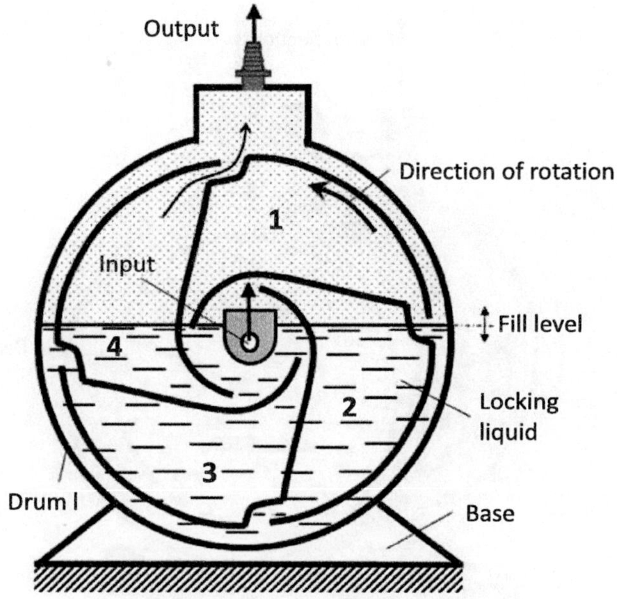

Fig. 11.4 Cross-section through a drum gas meter (Strohrmann 1987)

leads to a torque M. The torque then moves the drum counterclockwise. The filling process of chamber 1 is completed in the illustration and the gas enters the space above the barrier liquid, which is connected to the outlet. Due to the curved shape of the chambers, chamber 2 is closed above the liquid level and this chamber is now filled with the gas until chamber 2 has also arrived in the position of chamber 1 as shown and is vented. The gas in chamber 4 is completely displaced in this cycle and transported to the outlet. This process is repeated cyclically, so that a rotational movement is created, with which the gas quantity is detected. The rotational movement is also connected to a counter that measures the number of revolutions and thus records the gas flow. Modern gas meters of this type also have an additional increment disc (pulse generator) which can be scanned optically and thus offers a high resolution of the current gas flow.

Figure 11.5 shows a gas meter of this type with which, depending on the type, a flow range of 1–60 L/h or 100–18.000 L/h can be recorded. The accuracy is stated as 0.2%, which applies to a flow range of 80% of the maximum range. For the entire measuring range (0 ... 100%) an accuracy of 0.5% is specified. However, this high accuracy is only possible if the barrier fluid level is exact and the device is absolutely horizontal. Therefore the unit must be adjusted accordingly before operation. These devices are mostly used in the laboratory as *experimental gas meters.* Clean tap water is usually sufficient as sealing

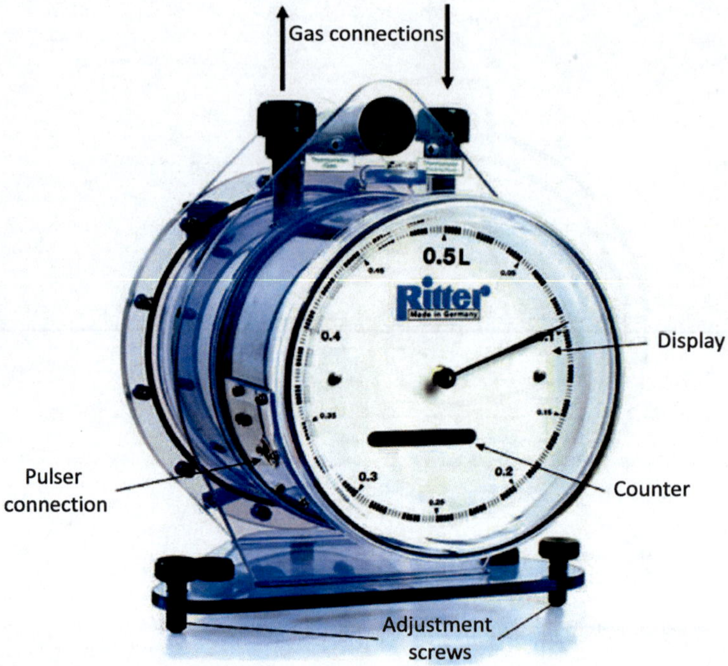

Fig. 11.5 TG 50 series drum gas meter. (Source: Dr.-Ing. RITTER Apparatebau GmbH & Co. KG, Bochum Germany)

liquid. Alternatively, low-viscosity oil can also be used. Furthermore, the pressure loss is very low and is only 0.1–0.4 hPa. The measuring behaviour is also shown in Fig. 11.6 on the basis of the error influences.

Drum gas meters are among the most accurate volume measurement methods. PTB[2] also uses this method as calibration standard (Strohrmann 1987).

Rotary Piston Gas Meter

This gas meter consists of two pistons whose outer shape corresponds to a *Cassini curve*. Both pistons are rotatably mounted in the middle. Since the two pistons are only partially in contact with the gas in their different positions, the pressure of the gas flow is applied to the pistons by a force. Figure 11.7 shows the different positions (1 → 4). In the first position, the upper piston has the gas pressure only on the left side, so that the force effect creates a torque which drives the piston. In the case of the lower piston, the pressure forces cancel each other out, so that it is force-free. The lower piston is then moved by the upper piston in this position via a synchronous gear. In position 2, both pistons have the same contact

[2] Physikalisch Technische Bundesanstalt in Braunschweig.

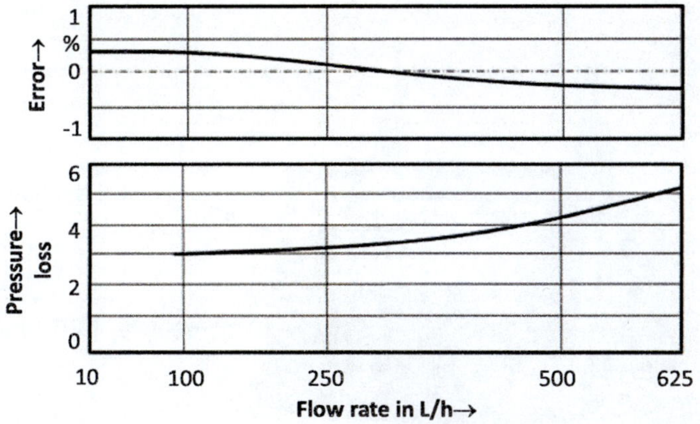

Fig. 11.6 Error curve and pressure losses of an experimental gas meter

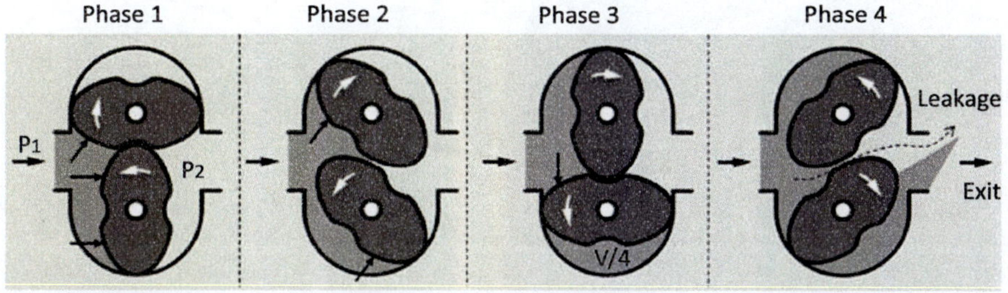

Fig. 11.7 Individual rotational phases and mode of operation of a rotary piston meter (Source: ABB/Frenzel 2011, Hogrefe 1994)

surface and therefore continue to move in opposite directions. Position 3 is identical to position 1, except that the conditions for both pistons are reversed. In position 4 the gas volume is then released. Only 1/4 of the total volume is transmitted per passage.

Both pistons have only a small distance of $\approx 0.05 - 0.3$ mm to each other. Under the influence of the pressure difference Δp, a small gap current flows through the gap, which is not measured. However, this measuring error is corrected by an appropriate adjustment. Rotary piston meters are very sensitive to contamination and must be cleaned occasionally. Nevertheless, this method has established itself as a reliable procedure in process engineering.

MilliGascounter

The related *MilliGascounter* method (Fig. 11.8) is based on a similar method to the drum gas meter. This method was developed especially for the detection of very small gas

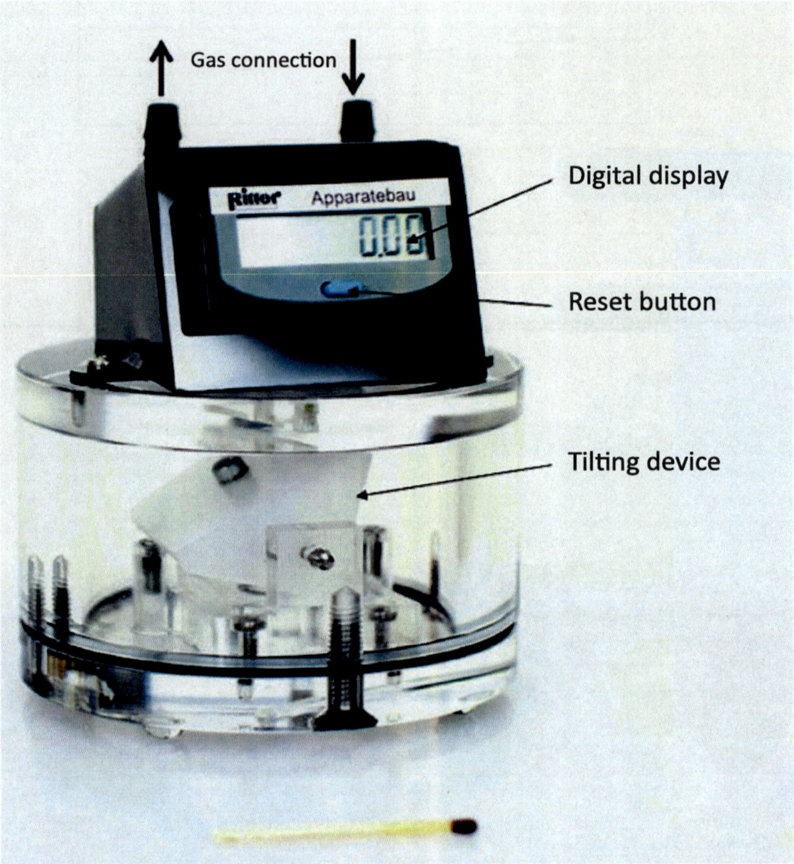

Fig. 11.8 View of the *MilliGascounter®* Type MGC-1 PMMA in comparison to the size of a match. (Source: Dr.-Ing. RITTER Apparatebau GmbH & Co. KG, Bochum Germany)

volumes (Scherer 1990). Figure 11.9 shows the operating principle. The gas flows in a chamber filled with barrier liquid. In this chamber a tilting container is arranged, in which the gas volumeΔV ($\approx$3 mL) collects in the upper part. The displacement of the sealing liquid creates a buoyancy force so that the container tilts to the left. The previously accumulated gas volume is thereby released and can escape via the outlet. This tilting process causes the gas supply to be directed into the left chamber, which now fills with gas. Also in this position, the displacement of the barrier liquid creates a buoyancy force and the container now tilts to the right. The process is repeated in a certain time interval Δt, depending on the volume flow. The number n of tipping movements is registered and multiplied by the volume ΔV ejected per tipping operation. With this calculation, you get the total gas volume $V_{ges} = n \cdot \Delta V$ or also by conversion, realted to the time Δt, the volume

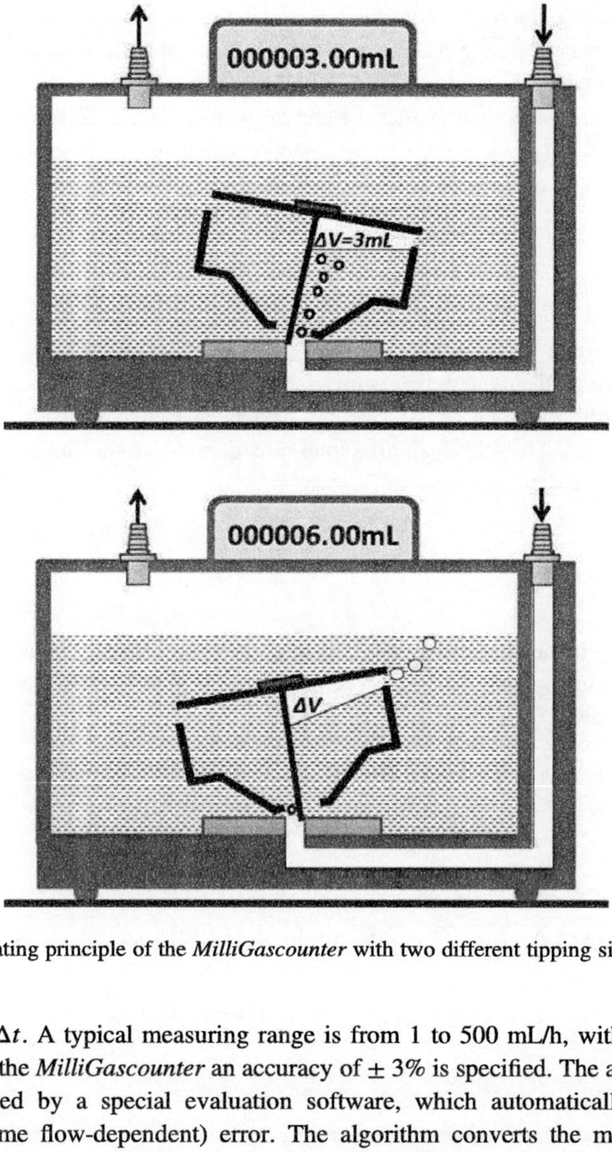

Fig. 11.9 Operating principle of the *MilliGascounter* with two different tipping situations

flow $\dot{V} = \Delta V/\Delta t$. A typical measuring range is from 1 to 500 mL/h, with calibration at 500 mL/h. For the *MilliGascounter* an accuracy of ± 3% is specified. The accuracy can be further improved by a special evaluation software, which automatically corrects the dynamic (volume flow-dependent) error. The algorithm converts the measured values into the true volume at the respective current volume flow, based on the calibration curve. The remaining residual error is then less than ±1%, over the entire measuring range.

Since the measurements involve very small gas volumes, dead space volumes ($_{VT}$) play a major role. Dead space volume is the volume of the entire gas pipe system between the gas source and the *MilliGascounter*. If the measurement of the gas volume is started at ambient pressure (pressure in dead space volume current= atmospheric pressure p_A), a volume flow initially only causes an increase in the gas pressure in the dead space volume. Only after the minimum input pressure $p_T \approx 5$ mbar is exceeded does the *MilliGascounter* start measuring the gas volume. Due to the design of the *MilliGascounter*, this overpressure $p_T \approx 5$ mbar remains in the dead space volume after the end of the test. This dead volume V_{Missing} must therefore be added to the measurement result and is calculated as follows:

$$V_{\text{Missing}} = V_T \cdot \left[\frac{p_A + p_T}{p_A} - 1\right] \tag{11.4}$$

Differential Pressure Method with Probes

In a flowing medium, the dynamic pressure p_{dyn} is an important variable which can be used to measure the flow velocity v. Pressure probes have been used to measure velocity in gases for over 200 years. The physical principles are derived from Eq. 2.169–2.171. The following applies to dynamic pressure

$$p_{\text{dyn}} = p_G - p_S \tag{11.5}$$

In this case p_G is the total pressure in a pipe and p_S is the static pressure. The dynamic pressure is generated by the flow and can be calculated with the gas density ϱ_g and the flow velocity v as follows

$$p_{\text{dyn}} = \varrho_g \cdot \frac{v^2}{2} \tag{11.6}$$

Example

A flow velocity v of 1.0 m/s generates the following dynamic pressure with air as medium:

$$p_{\text{dyn}} = 1.293 \frac{\text{kg}}{\text{m}^3} \cdot 0.5 \cdot \frac{1\ \text{m}^2}{\text{s}^2} = \frac{0.6465\ \text{kg}}{\text{ms}^2} \approx 0.65\ \text{Pa}$$

The expected pressure changes are therefore very small and must be recorded with high-resolution manometers or pressure sensors. The simplest arrangement consists of a U-tube (see Fig. 11.10). A pressure difference of 9.8 Pa leads to a height difference of h = 1 mm, if water is used as barrier fluid (=1 mmWS).

Even smaller pressure differences are recorded with so-called inclined-tube manometers. Figure 11.11 shows such a setup. The pressure difference $\Delta p = p_1 - p_2$ is also in this case determined by the altitude h and the density of the barrier fluid ϱ_{Fl}. It therefore applies:

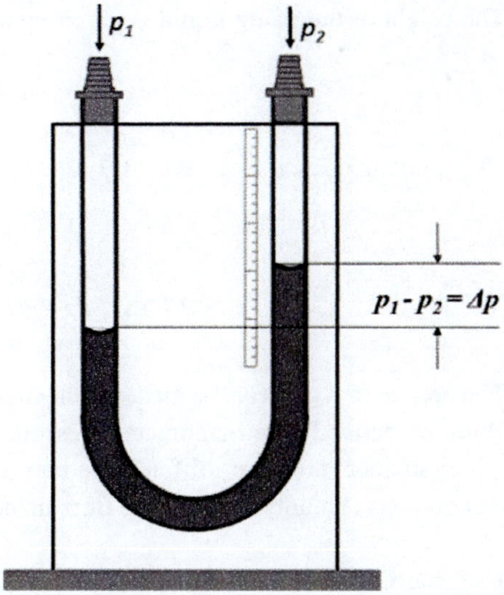

Fig. 11.10 U-tube manometer for recording small pressure differences ($\Delta p > 10$ Pa)

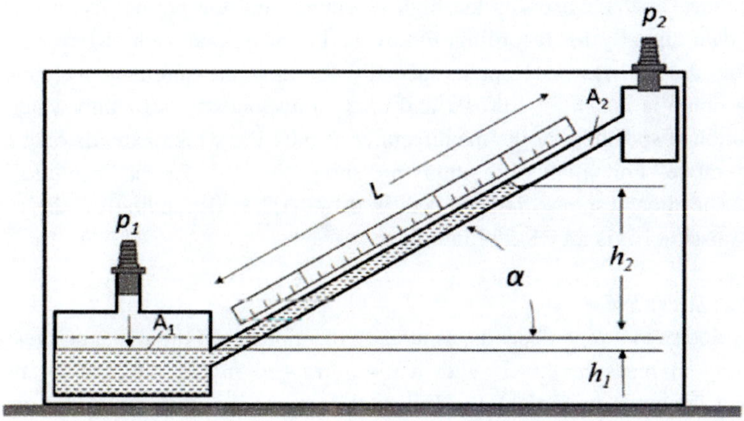

Fig. 11.11 Inclined-tube manometer for recording small pressure differences. (according to Fiedler 1992)

$$p_1 - p_2 = \varrho_{Fl} \cdot g \cdot (h_1 + h_2) \tag{11.7}$$

Also applies to the displaced volume from the storage tank (left):

$$A_1 \cdot h_1 = A_2 \cdot L \quad \text{and} \quad h_1 = \frac{A_2}{A_1} \cdot L \tag{11.8}$$

The height of the rising liquid is given by the length *L*:

$$h_2 = L \cdot \sin\alpha \tag{11.9}$$

Using both expressions in Eq. 11.7, the equation for determining the differential pressure Δp is obtained:

$$\Delta p = p_1 - p_2 = \varrho_{\mathrm{Fl}} \cdot g \cdot \left[\frac{A_2}{A_1} + \sin\alpha\right] \cdot l \tag{11.10}$$

The area ratio A_2/A_1 can be made much smaller than $\sin\alpha$, so that it can then be neglected. With an inclined tube manometer pressure differences can be recorded from $\Delta p \approx 2$ Pa. Even smaller pressure differences can only be measured with elaborate precision manometers (Prandtl manometer, Betz manometer) (Fiedler 1992).

Differential Pressure Sensors

Electronic differential pressure sensors are now also able to solve these measuring tasks. A great advantage of the sensors is the high resolution and the possibility to transmit the measured data digitally for recording. Figure 11.12 shows one such differential pressure sensor. This design was developed especially for flow measurement technology. The measuring range is $\Delta p = \pm 100$ Pa and thus corresponds to $\approx$10 mm water column. The resolution is specified by the manufacturer as 0.04 Pa, which extends over the entire measuring range. For small differential pressures $\Delta p < 1$ Pa the resolution is even 0.001 Pa. The time response is specified as 0.142–1.3 s, depending on the resolution. Data transmission is via an RS-232 interface.

Pitot and Prandtl Tube

The flow velocity in a pipe depends, as already mentioned, on the dynamic pressure. The simplest way to measure this is with a pipe arranged in the direction of flow. This arrangement is shown in Fig. 11.13. The total pressure is measured via the opening in the direction of flow (*Pitot tube*[3]), while the static pressure is measured on the tube wall. The connected U-tube manometer then forms the pressure difference Δp, so that the value can be used directly to calculate the flow velocity *v*:

$$v = \sqrt{\frac{2 \cdot \Delta p}{\varrho_g}} \tag{11.11}$$

[3]H. Pitot (1695-1771) French physicist.

Fig. 11.12 Differential pressure sensor with digital interface (Source: SENSIRION AG Staefa Switzerland)

A further improvement of this differential pressure measurement was introduced by *Prandtl.*[4] In the dynamic pressure measuring tube named after him, both p_G and p_S are measured. The design is shown in Fig. 11.14. This type of probe is most common today and is also used, for example, to measure the speed of airplanes. In general the Prandtl tube is mainly used for air flows (e.g. in wind tunnels). However, it is hardly used in pipelines.

Aichelen Point

In a pipeline, however, the flow velocity is not constant when viewed over the diameter. The velocity profile varies depending on the flow (laminar or turbulent). In round pipes, with a sufficiently long inlet section, a rotationally symmetrical flow profile is always formed. The maximum velocity v_{max} for all flow profiles is in the pipe centre (r = 0). By experimental investigations it was found out that the average velocity $\overline{v}$ is present at one point of the flow profile. This location is called *Aichelen point* (*rm*) after the discoverer. With this information (→ single-point measurement) the total flow in a pipe can be calculated (Aichelen 1947). For laminar flows with a Reynold number $Re < 2300$

[4]L. Prandtl (1875-1953) German physicist.

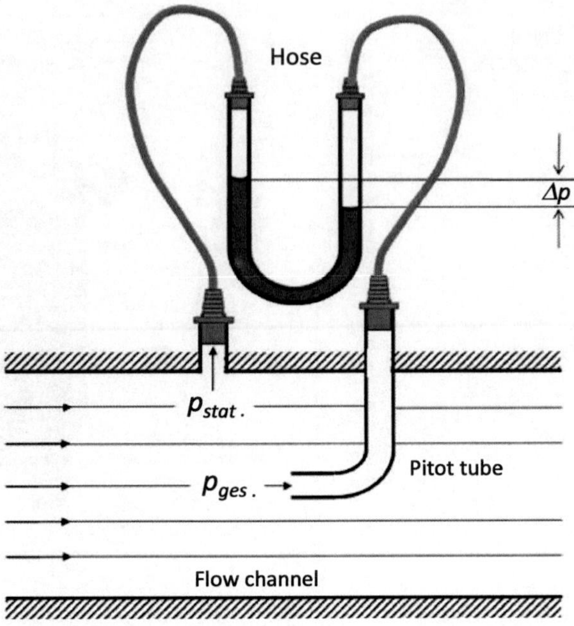

Fig. 11.13 Arrangement for measuring the differential pressure with a Pitot tube in a pipeline to determine the flow velocity v

$$r_m = 0.71 \cdot R \tag{11.12}$$

and for turbulent flows with $Re > 2300$:

$$r_m = 0.77 \cdot R \tag{11.13}$$

However, this measuring method cannot be used for disturbed flow profiles, which occur very frequently.

Annubar Method

In these cases, a multi-point measurement is the remedy. A known method (*Ellisson-Annubar*) uses a total of 4 measuring points at the points $r_1 = 0.866 \cdot R$ and $r_2 = 0.5 \cdot R$. The four measuring points are tapped via the pressure p_1, which represents the average value over these 4 points. The counter pressure p_2 is taken from the back of the probe in the middle of the pipe (see Fig. 11.15).

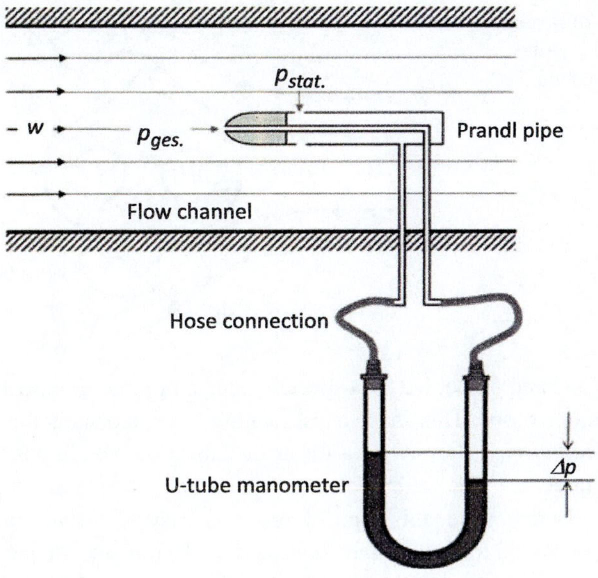

Fig. 11.14 Arrangement for measuring the static and total pressure with a dynamic pressure measuring tube according to *Prandtl*. If both outputs are connected to a U-tube pressure gauge, the differential pressure Δp is displayed immediately

Fig. 11.15 Dynamic pressure probe (*Annubar*) for determining the average velocity over the pipe cross-section at four measuring points

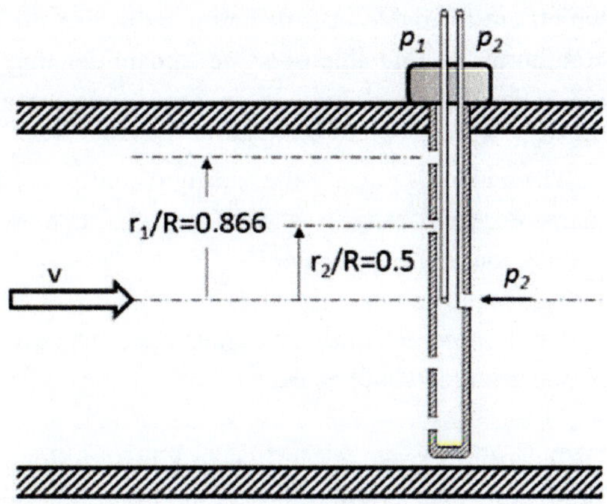

With this arrangement, the calculation of the flow velocity v and thus the volume flow $\dot{V}$ can only be performed with an additional calibration.

Deltaflow and TORBAR

A further development of this *Annubar probe principle* is the *deltaflow* concept from Systec Controls and the *TORBAR-Pitot* tube from ABB (Frenzel et al. 2011). Both concepts

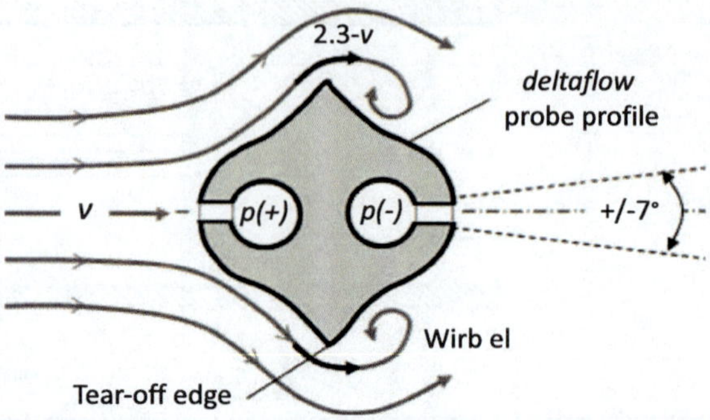

Fig. 11.16 Mode of operation and alignment of the probe profile (*deltaflow*) within the pipe flow

do not consist of a round probe, but of a specially shaped profile. Figure 11.16 shows the shape of the *deltaflow* probe. This shape causes a high flow velocity at the break-off edge, which is 2.3 times the pipe flow v. As a result of these flow conditions, the differential pressure Δp also increases. The calibration constant of the probe is not affected by this and remains constant. Furthermore, this type of probe is easy to install, since an inclined position $\pm 7°$ does not lead to any changes in accuracy. A complex adjustment is therefore not necessary.

The probe has several measuring points, which are located on both sides of the probe tube. These measuring points are specified by drilling a hole in the probe profile. Figure 11.17 shows an installation in a pipe. On both sides there are 6 measuring points each, which are available at 3 different radii. For differential pressure measurement (Δp) any commercially available pressure transmitter that meets the accuracy requirements can be used. For the installation in ex-zones, one can therefore also use appropriately certified pressure measuring instruments.

The advantage of this measuring technique is that it is possible to measure in small pipe diameters (mm range) as well as in large pipes (up to 15 m). Figure 11.18 shows an installation in a biogas plant.

The pressure range goes up to 350 bar and temperatures which are only limited by the material properties of the materials used. The manufacturer specifies a maximum operating temperature of 1240° C here.

Differential Pressure Method at Constrictions

In this process, the pipe flow is modified in a defined manner by built-in obstacles in the form of constrictions such as orifices and nozzles. In particular, the flow velocity v can be influenced. The resulting pressure change Δp is called *differential pressure*. This differential pressure is then used to calculate the volume flow $\dot{V}$ and the mass flow $\dot{m}$. Due to its high accuracy and simple handling, this method has become established for measuring pipe flows. There are various national and international standards in which this method has been

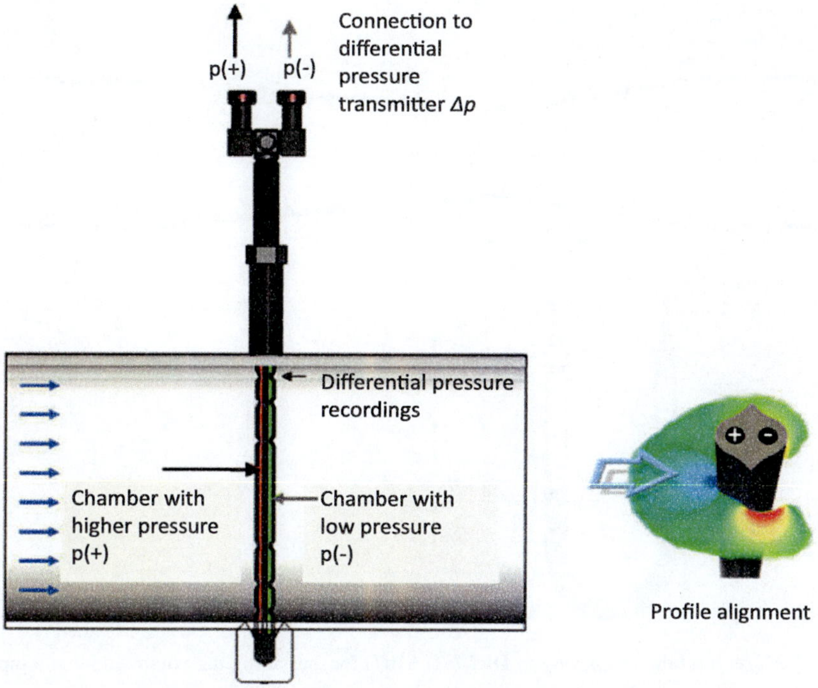

Fig. 11.17 Arrangement of the *deltaflow probe* in a pipeline (source: Systec Controls GmbH, Puchheim)

Fig. 11.18 Real installation (biogas plant) of a *deltaflow* flow measurement on a pipeline (source: Systec Controls GmbH, Puchheim)

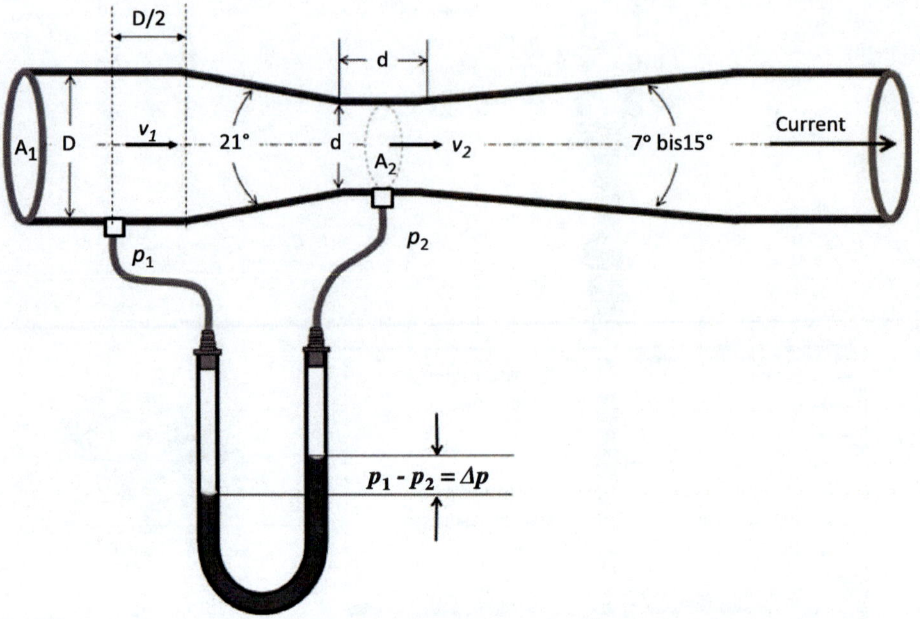

Fig. 11.19 Venturi tube (according to DIN ISO 5167) for measuring the volume flow in a pipeline

defined and described. Especially DIN 1952, VDI/VDE 2040 and DIN EN ISO 5167 are recommended.

Figure 11.19 shows a *venturi tube* with which such a flow measurement can be performed. In this design, the flow is guided into the constriction (d) and out again almost without loss due to the small opening angles. According to the continuity equation, the following relationship results:

$$A_1 \cdot v_1 = A_2 \cdot v_2 \quad \text{or} \quad \frac{\pi D^2}{4} \cdot v_1 = \frac{\pi d^2}{4} \cdot v_2 \tag{11.14}$$

At the constriction (d) the flow velocity v_2 thus increases in accordance with the opening ratio m:

$$m = d^2/D^2 = A_2/A_1 \tag{11.15}$$

$$v_1 = m \cdot v_2 \tag{11.16}$$

The pressure ratios (p_1, p_2) in the pipe (D) and at the constriction (d) can be calculated from the *Bernoulli equation* for a horizontal position (h = 0) as follows

$$p_1 + \frac{\varrho \cdot v_1^2}{2} = p_2 + \frac{\varrho \cdot v_2^2}{2} \tag{11.17}$$

With $v_2 = v_1/m$ and inserting into the *Bernoulli equation*, the flow velocity in the pipe system (D) is obtained by resolution according to v_1:

$$v_1 = \frac{m}{\sqrt{1-m^2}} \cdot \sqrt{\frac{2 \cdot \Delta p}{\varrho}} \tag{11.18}$$

The volume flow is then calculated by multiplying by the pipe area A_1:

$$\dot{V} = \frac{\pi}{4} \cdot d^2 \cdot \frac{1}{\sqrt{1-m^2}} \cdot \sqrt{\frac{2 \cdot \Delta p}{\varrho}} \tag{11.19}$$

Friction losses always occur at the pipe walls, which must be taken into account when calculating the volume flow. For this purpose, a flow coefficient C is introduced, which must be determined experimentally ($C_{\text{Aperture}} \approx 0.6$; $C_{\text{Nozzle}} \approx 0.75$; $C_{\text{Venturi}} \approx 0.98$). Furthermore, there is a change in density of the gas at the constriction, which has to be corrected via the expansion coefficient ε. The following therefore applies to the volume flow:

$$\dot{V} = \frac{C}{\sqrt{1-m^2}} \cdot \varepsilon \cdot \frac{\pi}{4} \cdot d^2 \cdot \sqrt{2 \cdot \Delta p/\varrho} \tag{11.20}$$

and applies to mass flow:

$$\dot{m} = \frac{C}{\sqrt{1-m^2}} \cdot \varepsilon \cdot \frac{\pi}{4} \cdot d^2 \cdot \sqrt{2 \cdot \Delta p \cdot \varrho} \tag{11.21}$$

For apertures (see Fig. 11.20), the relationships listed can be formally transferred. Since these orifices represent an abrupt obstacle, this leads to high losses. The *C-value* is therefore correspondingly low. Furthermore, a significant pressure loss $p_0 - p_4$ can be observed behind the orifice plate, which corresponds to an energy loss of the pipe flow. At high volume flows, this loss is considerable. More favourable in these applications are the methods of measuring probes, such as *deltaflow* and *TORBAR*, which hardly cause any pressure reduction in the pipe.

Running-in Section

The positioning of the measuring point is particularly important for all differential pressure methods. Especially behind pipe bends, valves and constrictions, the flow profile is considerably disturbed. However, if the flow distribution is not stable, the differential

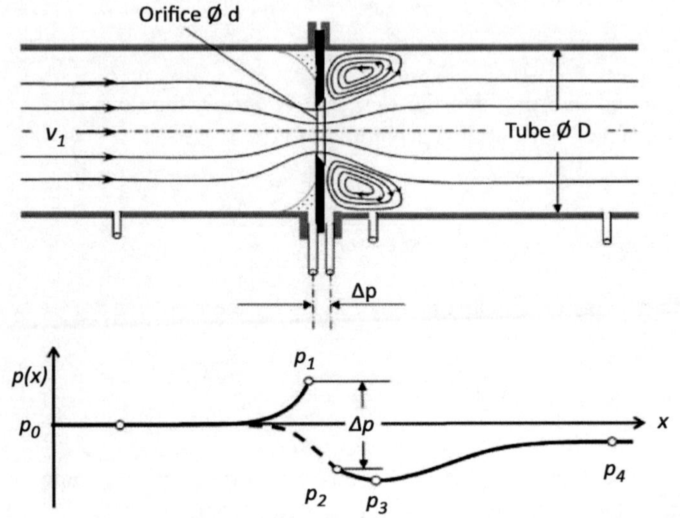

Fig. 11.20 Flow conditions at a standard orifice plate. The pressure Δp must be measured directly before and after the orifice plate to obtain the highest value. Due to the losses caused by the orifice plate, the pressure ($\rightarrow p_4$) behind the orifice plate is reduced

pressure may not have the required stability. Therefore, a certain inlet distance A is required for accurate measurements. In principle, therefore, should amount to $A > 10 \cdot$ Diameter (D). For the outlet section then applies to $A > 4 \cdot D$. Figure 11.21 shows the installation situation after a pipe bend. The inlet conditions also depend on which components (throttle, valve, controller, elbow, etc.) are present before the flow measurement. If the conditions on site are known exactly, the inlet sections can also be reduced and the expected measuring error can be reduced by a correction factor. For this purpose, the boundary conditions are entered into a computer program,[5] which then determines the corresponding correction factor from a large number of already known installation situations and individual calibration data. With this calculation method, the run-in distance can be reduced by the factor ≈5.

Variable Area Flowmeter

A very simple and robust method is the volume flow measurement with so-called floats. This method is often used for control purposes, as the position of the float intuitively indicates the flow rate. The higher the float rises in the pipe, the higher the flow rate. Figure 11.22a shows such a flow meter in cross-section. In this example, the float consists of a ball and can move freely vertically in a conical tube. Other common float shapes are shown in Fig. 11.22b. Commercially available variants of variable area flow meters are shown in Fig. 11.23.

[5] *ImproveIT* from Systec Controls GmbH Puchheim.

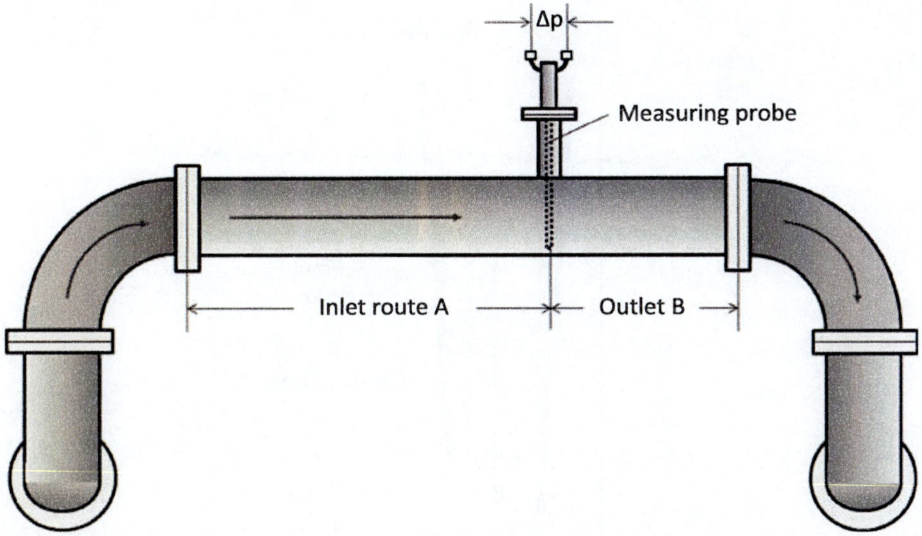

Fig. 11.21 Arrangement of a differential pressure measurement after an obstacle

In a state of suspension, the weight force F_G of the ball is equal to the flow resistance F_W. The buoyancy force F_A is neglected in gas measurement technology, whereas it must be taken into account for liquid flows. It therefore applies:

$$mg = c_W \cdot A \cdot \frac{\varrho_G}{2} \cdot v^2 \tag{11.22}$$

If you use the area formula for $AA = \pi \cdot D^2/4$, you get

$$mg = c_W \cdot \frac{\pi}{4} \cdot D_S^2 \cdot \frac{\varrho_G}{2} \cdot v^2 \tag{11.23}$$

According to Fig. 11.22a, the volume flow $\dot{V}$ with the flow velocity v results from the following relationship:

$$\dot{V} = \frac{\pi}{4} \cdot \left[D_K^2 - D_S^2\right] \cdot v \tag{11.24}$$

If you convert Eq. 11.24 to v and use this expression in Eq. 11.23, you get the following expression after further conversion:

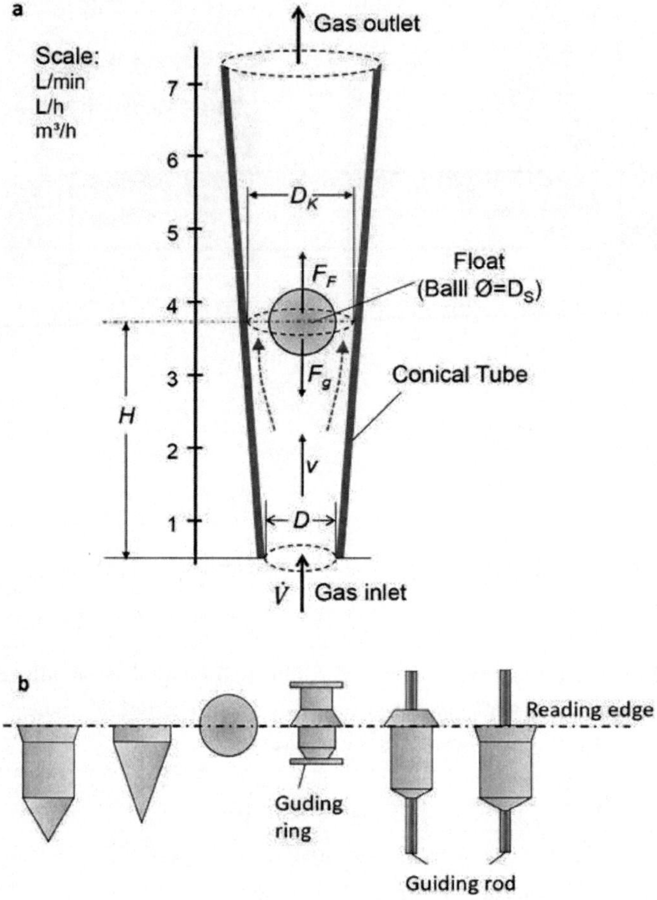

Fig. 11.22 Design (sectional view) of a variable area flowmeter (**a**) and selection of possible designs of float (**b**)

$$\dot{V} = \sqrt{\frac{\pi}{2 \cdot c_W}} \cdot \left[\frac{D_K^2}{D_S^2} - 1\right] \cdot \frac{D_S}{\varrho_G} \cdot \sqrt{m \cdot g \cdot \varrho_G} \tag{11.25}$$

By introducing the flow coefficient a the expression for the volume flow $\dot{V}$ and the mass flow$\dot{m}$ is obtained:

$$\dot{V} = a \cdot \frac{D_S}{\varrho_G} \cdot \sqrt{m \cdot g \cdot \varrho_G} = a \cdot D_S \cdot \sqrt{m \cdot g / \varrho_G} \tag{11.26}$$

Fig. 11.23 Variable area flowmeters of the DK 46, DK 47, DK 48 and DK 800 series for measuring gases with glass cones of different lengths for graduated accuracy requirements All devices are equipped with a fine adjustment valve for precise adjustment of a flow rate value (Source: KROHNE Messtechnik GmbH, Duisburg)

$$\dot{m} = a \cdot D_S \cdot \sqrt{m \cdot g \cdot \varrho_G} \tag{11.27}$$

The height H of the float is contained in the flow coefficient a, since D_K increases with the height $H$and thus allows a higher volume flow $\dot{V}$ to pass through. For a applies:

$$a = \sqrt{\frac{\pi}{2 \cdot c_W}} \cdot \left[\frac{D_K^2}{D_S^2} - 1\right] \tag{11.28}$$

The characteristic curve of the variable area flowmeter is determined by the taper of the tube. Since the inside diameter D_K increases linearly with height, the area A_K changes quadratically with increasing height H. The following relationship therefore applies to the height H:

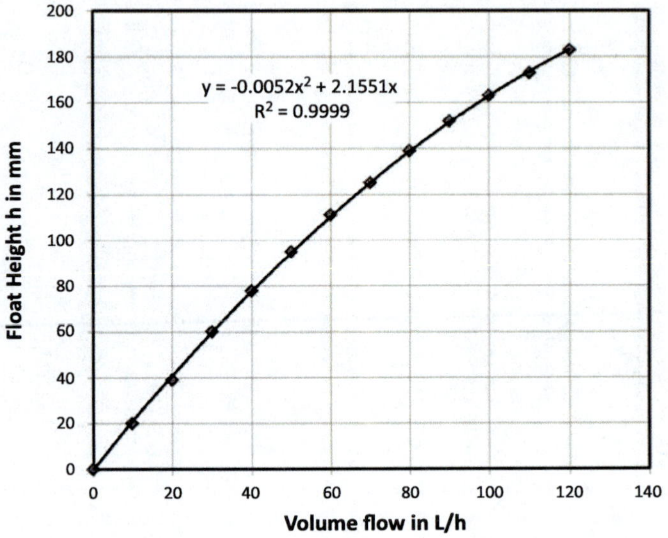

Fig. 11.24 Typical characteristic curve of a variable area flowmeter

$$H \sim \sqrt{\dot{V}} \tag{11.29}$$

The characteristic curve of a variable area flowmeter is therefore non-linear and corresponds approximately to a root function. Figure 11.24 shows the typical characteristic curve. The characteristic curve for a specific gas (usually air) is usually printed on the pipe. The accuracy of the measurement is, depending on the accuracy class, 1% of the measured value up to 6% of the measured value. Figure 11.25 shows the resulting measurement errors for the different flow rates and accuracy classes. In the lower third of the measuring range there are sometimes considerably higher deviations. For control purposes only, an accuracy class of 6 is usually sufficient. The height *H* is read off at the reading edge, which is different for the individual float shapes (see Fig. 11.22b). For the spherical shape frequently used in gas measurement technology, the reading edge is mostley the central point of the ball.

Due to turbulence within the riser pipe, the float is not always at rest, making accurate readings difficult. This wobbling movement is improved with a guided float. In this case, the float glides almost frictionless on a vertical rod from bottom to top and can no longer break out laterally. Alternatively, the float can also be fitted with guide rings which have appropriate ventilation openings through which the gas flow is guided. In some cases, ribs are also inserted in the riser pipe to hold the float in position (Hogrefe 1994).

Variable area flowmeters are always calibrated by the manufacturer for a specific gas. The scale applied to the tube usually refers to air or nitrogen in standard condition. If other gases are to be measured with the same flowmeter, the scale is not correct and must

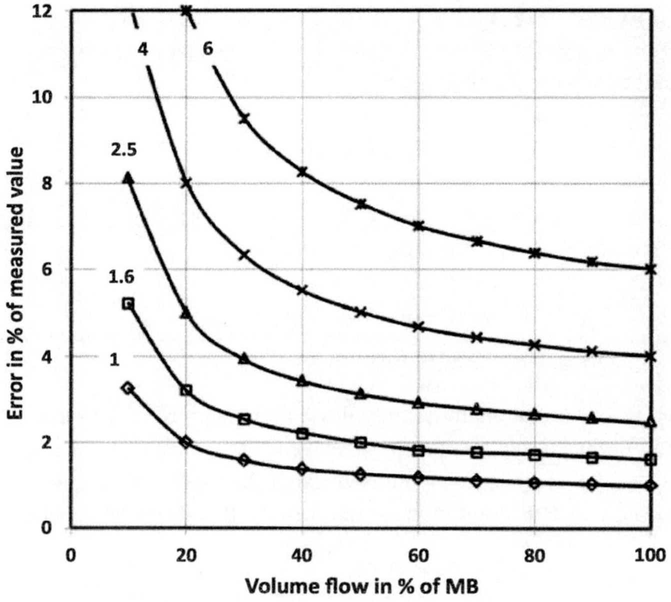

Fig. 11.25 Accuracy classes (1 = 1 %) of variable area flowmeters

therefore be converted. According to Eq. 11.6, the volume flow is directly related to the gas density:

$$\dot{V} \sim \sqrt{\frac{1}{\varrho_G}} \tag{11.30}$$

If such a flowmeter calibrated with air is now used for hydrogen (H_2), the displayed scale values must be multiplied by a correction factor K_V to obtain the true flow rate. The following correction value therefore applies in this case:

$$K_V = \sqrt{\frac{\varrho_{\mathrm{Air}}}{\varrho_{\mathrm{H_2}}}} = \sqrt{\frac{1.293}{0.089}} = 3.812 \tag{11.31}$$

If the pressure p and the temperature T change, the following relationships apply:

$$K_T = \sqrt{\frac{T_1}{T_2}} \quad \text{and} \quad K_p = \sqrt{\frac{p_2}{p_1}} \tag{11.32}$$

Ultrasonic Flowmeter (Fig. 11.27)
Sound propagates in a gas at the speed of sound c. The speed of sound depends on the type of gas and the temperature T and is given by

$$c = \sqrt{\kappa \cdot R_i \cdot T} \tag{11.33}$$

Within dry air, sound speed at $T = 0\quad ^\circ\mathrm{C}$ has value of 331.6 m/s. If a sound wave now spreads out within a gas flow v, the speeds overlap:

$$\begin{array}{ll} \textit{in current direction} & c + v \\ \textit{against the current} & c - v \end{array}$$

Since speed is a vectorial quantity, only those parts that run in the same directions will be effective. In Fig. 11.26 such a course is displayed. The sound signal is transmitted from a transmitter (A) to a receiver (B) diagonally through the pipe, covering a distance $L = D/\sin \alpha$. The time t_1, which a sound impulse requires for this distance, is calculated:

$$t_1 = \frac{l}{c + v \cos \alpha} \tag{11.34}$$

If the sound impulse is now sent from B to A, the time t_2 increases, since the propagation occurs against the flow velocity v. It now applies:

$$t_2 = \frac{l}{c - v \cos \alpha} \tag{11.35}$$

For the difference in transit time one obtains $\Delta t = t_2 - t_1$ from Eq. 11.33 and Eq. 11.35. After some transformations the following expression is used:

$$\Delta t = t_2 - t_1 = \frac{2 \cdot L \cdot v \cdot \cos \alpha}{c^2 - v^2 \cdot \cos^2 \alpha} \approx \frac{2 \cdot L \cdot v \cdot \cos \alpha}{c^2} \tag{11.36}$$

Solving the equation to v gives the mean flow velocity $\overline{v}$ in the pipe:

$$\overline{v} = \frac{c^2}{2 \cdot L \cdot \cos \alpha} \cdot \Delta t \tag{11.37}$$

To compensate for the temperature dependence of the sound velocity c, the following approximation for the sound velocity is used:

$$c^2 \approx \frac{L}{t_1} \cdot \frac{L}{t_2} = \frac{L^2}{t_1 \cdot t_2} \tag{11.38}$$

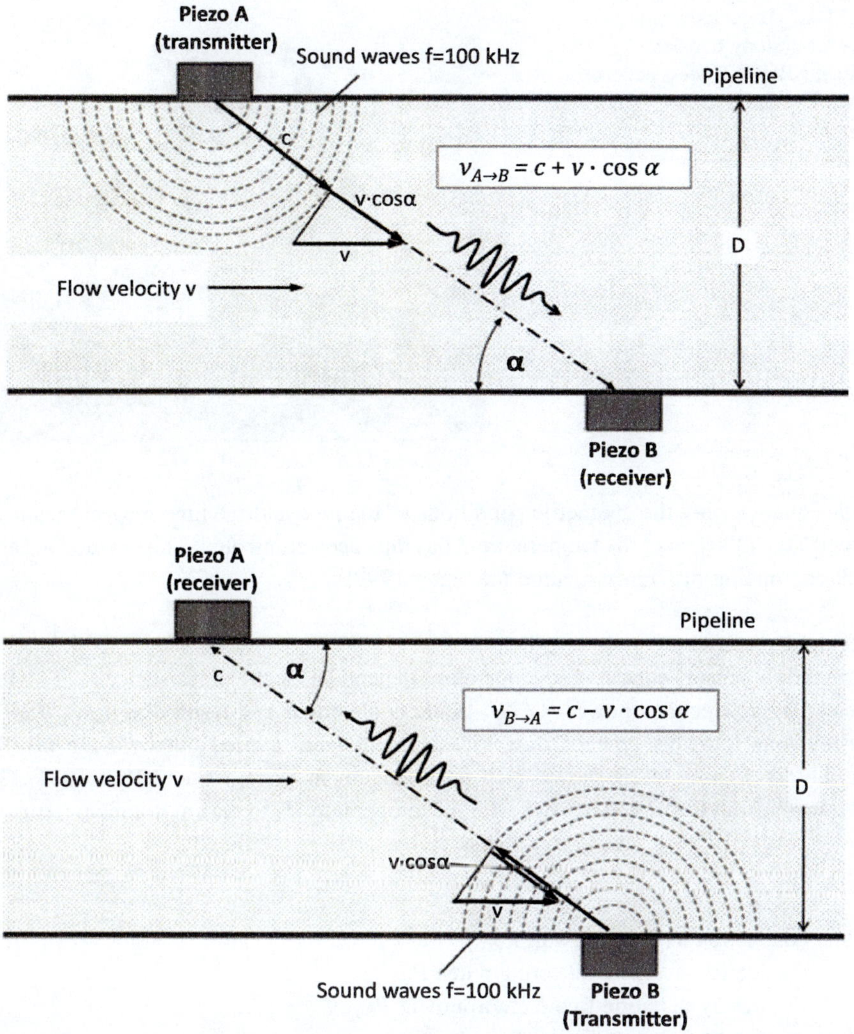

Fig. 11.26 Principle mode of operation of an ultrasonic flowmeter. (according to Hoffmann 2000)

For the flow velocity $\overline{v}$, the following relationship now results from the use of c_2 in Eq. 11.37:

$$v = \frac{L}{2 \cdot \cos \alpha} \cdot \frac{\Delta t}{t_1 \cdot t_2} \tag{11.39}$$

Fig. 11.27 12-jet ultrasonic gas meter for custody transfer. (Source: KROHNE Messtechnik GmbH, Duisburg)

In this equation only the geometric conditions of the pipe and the time measurements are present. The influence of the temperature T has thus been eliminated. This method is called the direct running difference method (Gevatter 1999).

Thermal Processes

Thermal flow sensors consist of a temperature-dependent electrical resistance $R_0(T)$ which is heated by an electrical power $P = U \cdot I$ and cools down *as* a result of a flow. The first investigations into this type of flow measurement were carried out by *King in* 1914 (Fig. 11.27). Figure 11.28 shows such an arrangement (Wiegleb and Röß 1984). The heat transfer from the heated resistor with the temperature T_1 to the environment (temperature T_U) takes place via several mechanisms:

- Heat conduction to the pipe wall P_{WL1}
- Heat conduction via the connecting wires P_{WL2}
- Heat transfer by radiation to the environment P_{Str}
- Heat transfer by free convection P_{K1}
- Heat transfer by forced convection P_{K2}.

Only the last part of the heat transfer caused by forced convection depends on the flow velocity v and is therefore used to determine the flow rate. The following applies to the temperature-dependent resistance value $R(T)$

$$R(T) = R_0(1 + \alpha \cdot (T_1 - T_U)) \tag{11.40}$$

The supplied electrical power P is converted into the heat flow $\dot{Q}$:

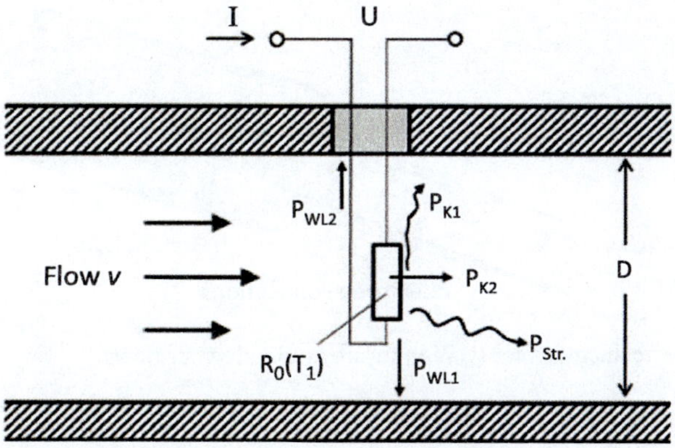

Fig. 11.28 Heat transfer mechanisms of a heated electrical resistance $R_0(T)$ in a flow channel with the diameter D

$$P = U \cdot I = \frac{U^2}{R(T)} = I^2 \cdot R(T) = \dot{Q} \tag{11.41}$$

The dissipated heat flow $\dot{Q}$ is divided into the five individual components described above. Since it can generally be assumed that the proportions remain almost constant, they can be taken into account by means of corresponding constants (K_1, K_2). In certain cases, for example, when the thermal conductivity of the gas mixture changes significantly, this influence must be considered separately. For the general case then applies:

$$\dot{Q} = (T_1 - T_U) \cdot (K_1 + K_2 \cdot v^n) \tag{11.42}$$

If the sensor resistance $R(T)$ in a measuring bridge is controlled to a constant temperature, *King's law* applies:

$$I^2 \cdot R(T) = (T_1 - T_U) \cdot (K_1 + K_2 \cdot v^n) \tag{11.43}$$

The exponent n is theoretically $n = 0.5$. Due to the described accompanying effects of heat transfer, it changes depending on the sensor design and is then $0.25 < n < 0.5$.

Hot-Wire Anemometer (HWA)

Air flows can be measured with so-called hot-wire anemometers (HWA), which consist of a thin heating wire and behave according to Eq. 11.43. Figure 11.29 shows such a setup. In these arrangements, the wire is very short (mm range) and very thin (μm range) in order to achieve the lowest possible mass and short response times. Also in this case the measuring resistor is connected in a Wheatstone bridge. The circuit design is shown in Fig. 11.30. By feeding back the output voltage to the measuring bridge (feedback), a constant temperature

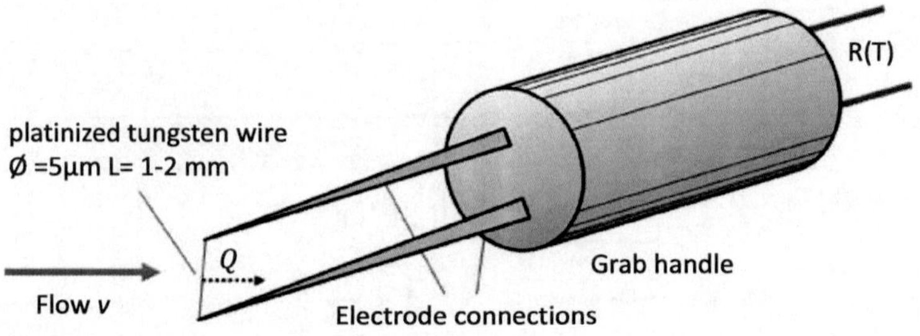

Fig. 11.29 Hot-wire anemometer (HWA) for measuring low air flows

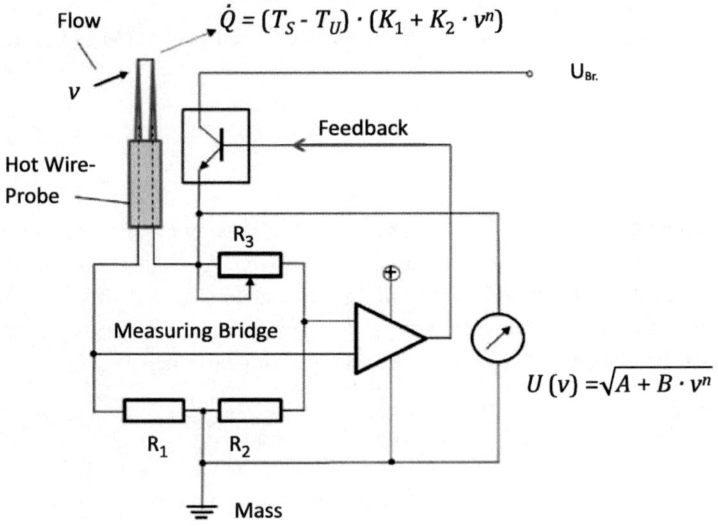

Fig. 11.30 Measuring circuit for evaluating a hot-wire anemometer (HWA)

T_S of the measuring resistor is obtained. The cooling by the current v is then compensated U_{Br} by increasing the bridge voltage. The bridge voltage $U(v)$ then has the following course:

$$U^2(v) = A + B \cdot v^n \tag{11.44}$$

$$U(v) = \sqrt{A + B \cdot v^n} \tag{11.45}$$

The signal curve $U(v)$ from Eq. 11.45 is shown in Fig. 11.31 for different coefficients (A, B). The coefficients must be determined experimentally by experiments separately for each type of probe. In the case that the temperature and thus the density is constant, the following applies to A

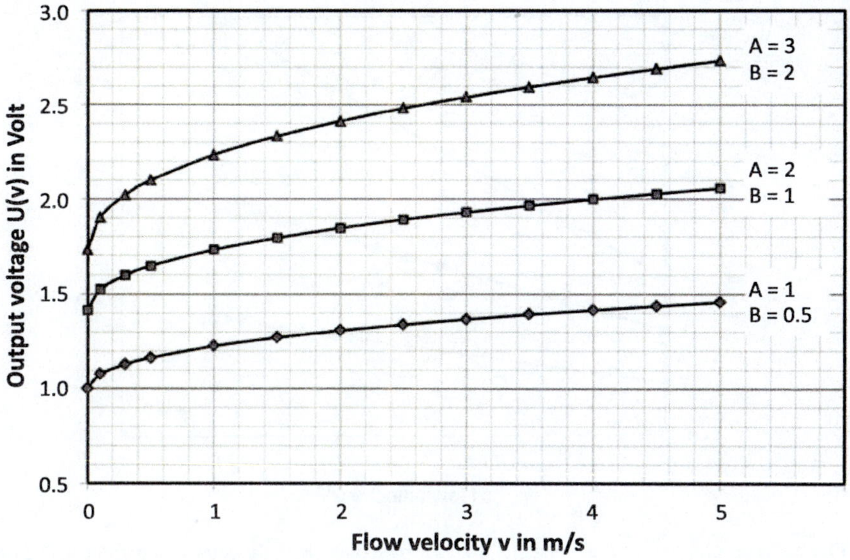

Fig. 11.31 Signal curve $U(v)$ for different coefficients (A, B) for a flow range from 0 to 5 m/s

$$U^2_{v=0} = U^2_0 = A \tag{11.46}$$

If the calibration curve is set as a function

$$U^2 = f(\sqrt{v}) \tag{11.47}$$

a straight line is obtained which intersects the *y-axis* at point A and where B represents the slope (Fig. 11.32). This procedure is used for the experimental determination of the coefficients (Hagemeier 2011).

Thermal Mass Flow Sensor

The heat flow $\dot{Q}$ which is transferred by the resistor $R(T)$ to the gas flowing past leads to an increase in the temperature (ΔT) of the gas. The following relationship therefore applies between the heat flow and the mass flow $\dot{m}$:

$$\dot{Q} = \dot{m} \cdot c_p \cdot \Delta T \tag{11.48}$$

Therein c_p is the specific heat capacity of the gas at a constant pressure. With the supplied electrical power one then obtains

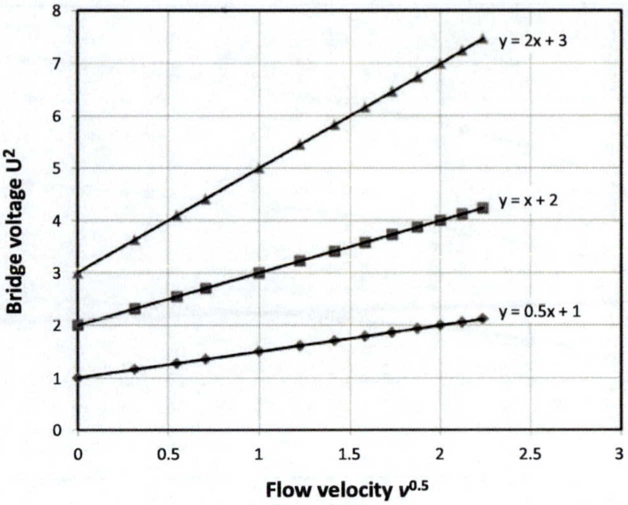

Fig. 11.32 Determination of the constants *A* and *B* by experimentally determined values. The point of intersection with the *y-axis* gives the *A-value* and the slope of the respective straight line gives the *B-value*

$$\dot{m} = \frac{I^2 \cdot R(T)}{c_p \cdot \Delta T} \tag{11.49}$$

Thermal flow meters therefore always indicate the mass flow $\dot{m}$.

Both Eq. 11.43 and Eq. 11.48 contain information on temperature differences which have a direct effect on the measurement result. For this reason, technical designs will always require a reference measurement to compensate for the effects of the ambient temperature T_U. Figure 11.33 shows such a sensor design, which was developed for use in corrosive gas mixtures (Wiegleb 1984b). It consists of two Pt temperature sensors, of which only one is exposed to the gas flow (T_1). The second temperature sensor is located in a blind hole and measures both changes in the ambient temperature T_U and the heat transfer due to the heat conduction of the gas mixture. The two sensor elements are connected to a Wheatstone bridge and provide an output voltage ΔU_{Br} which is proportional to $\sqrt{\dot{m}}$.

Figure 11.34 shows a measurement at two different temperatures (300 K and 410 K). However, the volume flow $\dot{V}$ is plotted on the *x-axis*. The conversion is done via gas density, which depends on the temperature. It is therefore valid for the temperature ratio:

$$\frac{410\ \mathrm{K}}{300\ \mathrm{K}} \approx \frac{79\ \mathrm{mV}}{58\ \mathrm{mV}} \approx 1.36 \tag{11.50}$$

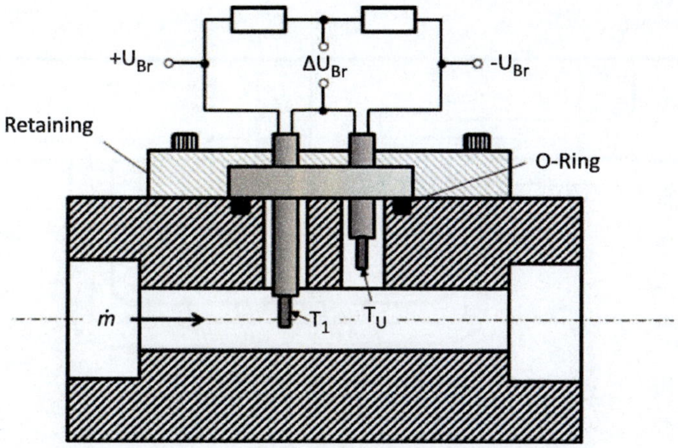

Fig. 11.33 Thermal mass flow meter for corrosive gases (Wiegleb 1984b)

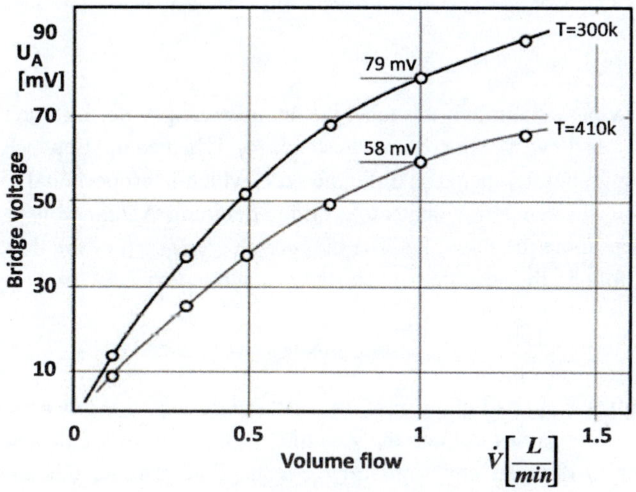

Fig. 11.34 Output signal of a thermal mass flow meter

Mass Flow Meter

A frequently used variant of the thermal mass flow method, very for precise measurements, is shown in Fig. 11.35. Two measuring resistors are located on the outside of a thin-walled metal tube (capillary tube). Usually these resistors are wound around this tube as a coil in the form of a platinum or nickel wire. Both resistors are connected in a Wheatstone measuring bridge and heated by the bridge current. This heating creates a symmetrical temperature profile without gas flow and the measuring bridge is balanced ($\rightarrow R_{S1} = R_{S2}$). If

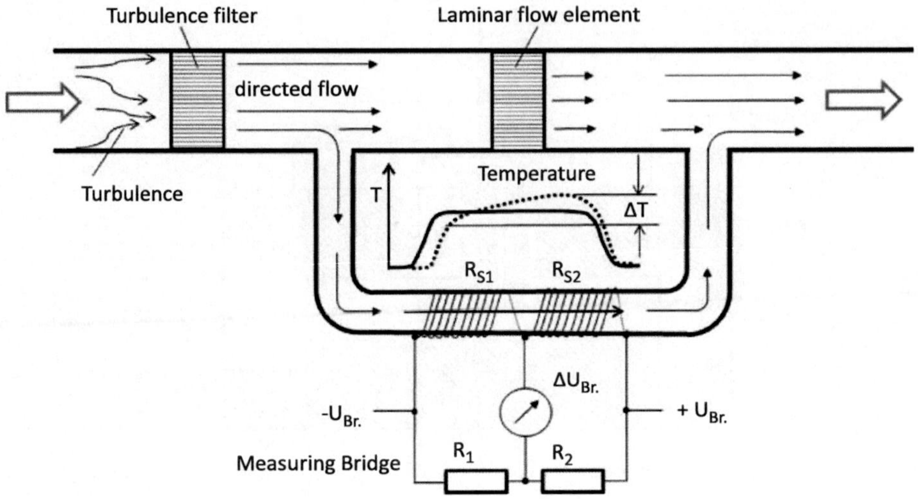

Fig. 11.35 Illustration of the operating principle of a thermal mass flow meter in a bypass arrangement

a gas flow is now passed through this tube, the temperature profile shifts so that the sensor resistance R_{S1} cools down and the sensor resistance R_{S2} is heated up. Between both resistors there is a corresponding temperature difference ΔT, which is proportional to the gas flow. This temperature difference now leads to a bridge detuning ΔU_{Br} and is used for further evaluation to determine the flow rate. The output voltage U_{Signal} of the flow meter results from the mass flow $\dot{m}$, the specific heat capacity c_p and a device factor K:

$$U_{\mathrm{Signal}} = K \cdot c_p \cdot \dot{m} \tag{11.51}$$

However, only a small part of the gas flow is passed through this thin capillary tube. The main part of the flow passes through the structure. In the inlet of the flow meter there is a turbulence filter to dampen undesired flow effects. Furthermore, this filter also keeps disturbing particles out of the system. The measuring system is located in a bypass which can be adjusted by the laminar flow element. The ratio between the main flow and the bypass is thus kept constant.

Figure 11.36 shows a typical sensor setup used for measuring small and medium gas flows. The lower metal block (stainless steel) contains the gas connections and the flow elements. The upper part of the housing houses the sensor system with the associated evaluation electronics. Due to the compact design of these flowmeters, several units can also be interconnected to create more complex gas mixtures. For this purpose the manufacturers offer base plates on which the flow meters are screwed. The gas connections are then connected to the base plate via O-rings and completely seal the gas supply. The gas

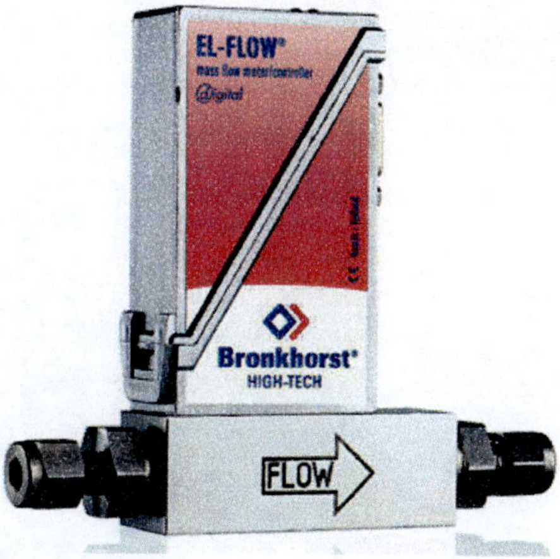

Fig. 11.36 Mass flow meter for the precise measurement of gas flows. (Source: Bronkhorst Deutschland Nord GmbH, Kamen)

is fed through corresponding holes in the base plate. Figure 11.37 shows such a setup, with 6 flow meters and downstream controllers (MFC).

For many applications in gas measurement technology, however, the volume flow is required as an output variable. The conversion of the mass flow into a volume flow is done via the gas density. It is therefore valid:

$$U_{\text{Signal}} = K \cdot c_p \cdot \dot{m} = K \cdot c_p \cdot \varrho \cdot \dot{V} \tag{11.52}$$

In most cases, these mass flow meters are calibrated for a specific gas (e.g. Nitrogen). If a different gas is now used, a corresponding conversion must be made when a volume flow rate is specified. For this purpose it is necessary to know the gas density and the specific heat capacity c_{p1} of the gas which is now to be measured. The conversion factor C then results from the following relationship:

$$C = \frac{c_{p0} \cdot \varrho_0}{c_{p1} \cdot \varrho_1} \tag{11.53}$$

Example

A thermal mass flow sensor has a measuring range of 2 L/min and was calibrated with nitrogen. Argon is now to be measured. The correction factor is accordingly:

$$C = \frac{1.25 \cdot 1.038}{1.784 \cdot 0.523} = 1.391$$

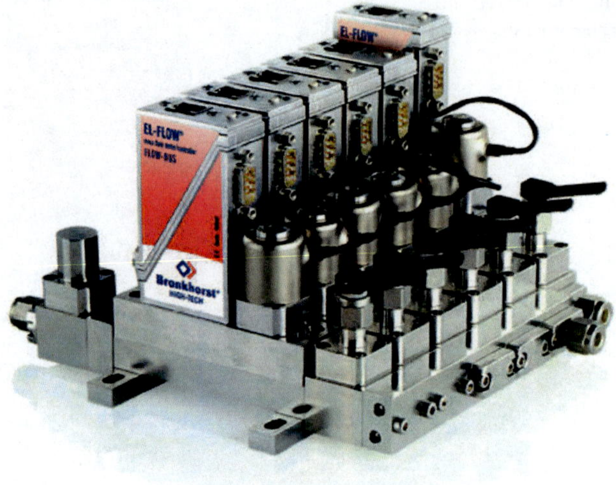

Fig. 11.37 Arrangement of 6 mass flow controllers (MFC) for the production of multi-gas mixtures. (Source: Bronkhorst Deutschland Nord GmbH, Kamen)

The measuring range for argon in this setup would therefore be 2.8 L/min. All display values must therefore be multiplied by the factor $C = 1.391$.

For gas mixtures consisting of *n-components,* the mixing rule can be applied. For this purpose, the conversion factors *Cn* of the individual components and the respective volume fraction *Vn* must be known. For the conversion factor, the following relationship then results C_{mix}:

$$\frac{1}{C_{\text{mix}}} = \frac{V_1}{C_1} + \frac{V_2}{C_2} + \cdots + \frac{V_n}{C_n} \tag{11.54}$$

Example

A gas mixture consists of the following components

10% oxygen by volume	$C_{O_2} = 0.98$
20% argon by volume	$C_{Ar} = 1.39$
30% methane by volume	$C_{CH_4} = 0.76$
40% helium by volume	$C_{He} = 1.41$

The conversion factor C_{mix} is then calculated:

$$\frac{1}{C_{mix}}=\frac{0.1}{0.98}+\frac{0.2}{1.39}+\frac{0.3}{0.76}+\frac{0.4}{1.41}=0.924$$

It follows that for $C_{mix} = 1.082$.

The conversion factors C can be read from the manufacturers' tables or calculated using Eq. 11.53. Some manufacturers store the characteristic curves for up to 25 different gases in the electronics of the flow meters and can therefore react directly to the different types of gas. The *Bronkhorst* company also offers an Internet version[6] for these conversions.

Micromechanical Flow Sensors

In addition to the possibility described above of winding the temperature sensors in the form of thin wires on a capillary tube, microsensors can also be used for this purpose. The advantage of this technique is that the resistors can be structured almost arbitrarily and are highly reproducible due to the manufacturing process. Since the structures and thus also the masses are very small, very fast measurements can also be realized. There are only limitations with regard to corrosion resistance. Since the microstructures are in direct contact with the sample gas, no corrosive gases can be measured, as these would then damage the sensor structure. Figure 11.38 shows a planar structure consisting of a low-resistance heating resistor R_H, which has a resistance value of 45 Ω at 0 °C. The maximum heating voltage is 3 volts, resulting in a heating power of $P_H < 200$ mW. For this setup, an overtemperature to the environment of $\Delta T \approx 30$ K is recommended. The ambient temperature T_U is measured with a separate temperature sensor R_T. Both resistor structures consist of a thin platinum layer, which is structured using appropriate processes. The front part, on which the heating element is located, is exposed to the flow v, so that this resistor cools down accordingly. The mathematical description and characteristic behavior of this sensor is identical to that of a hot-wire anemometer (HWA).

Although this design is very simple and robust, the power consumption of almost 0.2 watts is too high for certain applications (battery operation). Further miniaturization of this technology leads to a reduction in power consumption, since the thermal mass of these microstructures is lower by a factor of 10–100. Figure 11.39 shows one such microstructure (type MFS02[7]). In this case, the resistor structure is located on a thin membrane, so that the thermal mass is extremely small. A heating element is arranged between the two measuring resistors (R_1 and R_2), which heats up to a defined temperature by means of an electrical power P. The required heating power in this case is only$P \approx 10$ - 50 mW. If the flow is $v = 0$, both measuring resistors have the same resistance value $R_1 = R_2$. If a flow v now

[6] http://www.fluidat.com

[7] Manufacturer: Innovative Sensor Technology IST AG, Ebnat-Kappel, Switzerland.

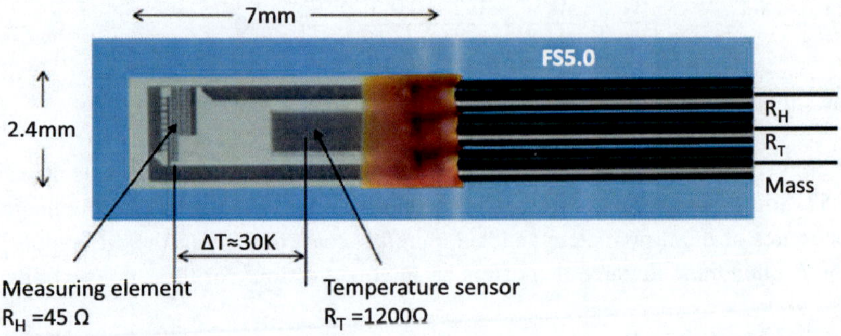

Fig. 11.38 Structure of a planar flow sensor (type FS5.0) with a measuring and reference resistor for compensation of the ambient temperature on a 0.2 mm thick ceramic substrate. (Source: Innovative Sensor Technology IST AG Ebnat-Kappel Switzerland)

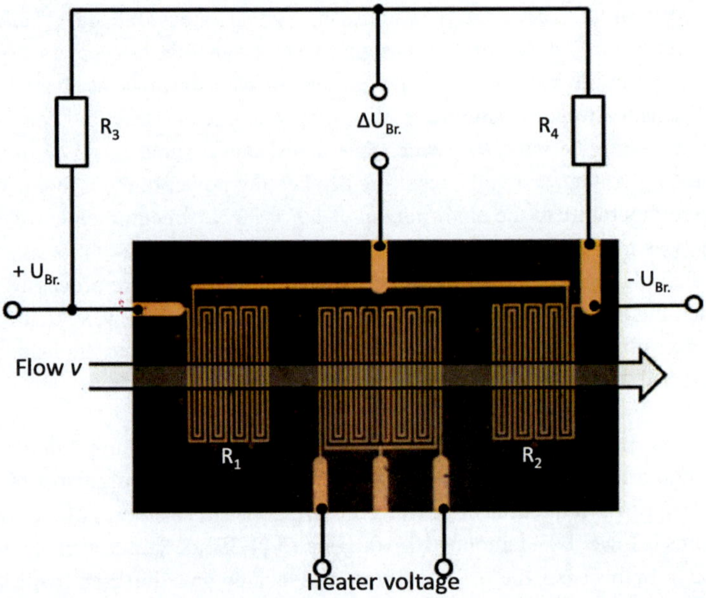

Fig. 11.39 Microstructure used to measure a gas flow The resistors R_1 and R_2 are connected to R_3 and R_4 to form a Wheatstone measuring bridge. If the flow is in the opposite direction (180°), the polarity of ΔU_{Br}

takes place above the microstructure, a certain amount of heat is transferred from the heating element to the flowing gas, whereby the temperature of the resistor R_2 increases noticeably. It then applies $R_2 > R_1$. In the measuring bridge, the resistance difference ΔU_{Br} is converted into a voltage change ΔR. Due to the low mass of the structure, the time constant $t_{63} < 10$ ms. With increasing heating power the sensitivity is also increased.

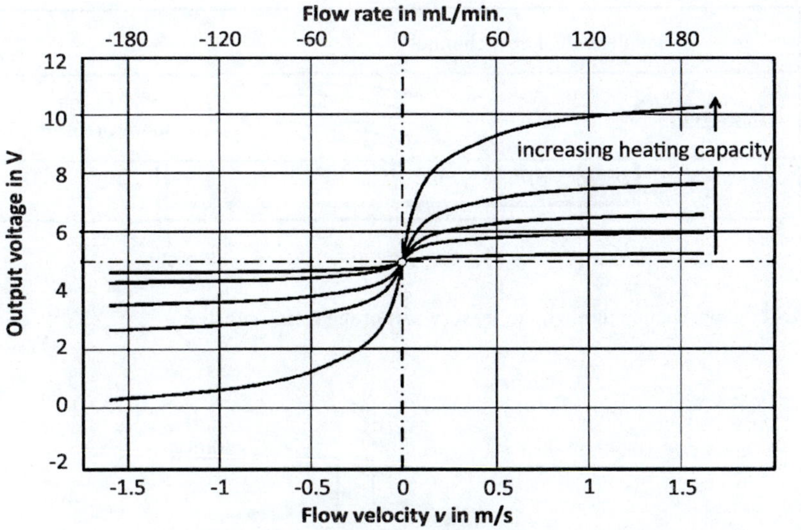

Fig. 11.40 Sensitivity curves for heating outputs from 6.6 to 41.3 mW. The zero point is electronically set at 5 volts in order to be able to record a flow in the opposite direction. (according to data from IST AG Ebnat-Kappel Switzerland)

Figure 11.40 shows the amplified output signal for different heating powers between 6.6 mW and 41.3 mW. If the flow is directed in the opposite direction, the signal reverses, as R_1 heats up and R_2 cools down $R_2 > R_1$. With this type of differential anemometry, directional detection can also be realized.

CMOSens®

A similar sensor element is also used in the setups described below. The microchip on which the heaters and temperature sensors are located is part of a microsystem. The signal amplification as well as the AD conversion and signal processing is already carried out in this microsystem, which the manufacturer refers to as a *CMOSens® chip*. The sensor part is shown in Fig. 11.41 in a cross-sectional view. The microchip is located in a flow channel with a through-hole of 0.9 × 1.5 mm. The flow passes through this channel and leads to the known shift in the temperature profile. The resulting resistance changes ($\Delta R = T_2 - T_1$) are converted into corresponding volume flows in the evaluation electronics. An advantage of this arrangement is the fast response time of $t_{98} < 100$ ms. The manufacturer specifies only oxygen, nitrogen, air, methane and natural gas as permissible measuring gases. Corrosive gases are generally excluded from the measurement. A very compact sensor design can be seen in Fig. 11.42, which is designed for a flow rate of ±200 L/min. Since this sensor can be flowed in both directions, the measuring method also allows bidirectional measurement.

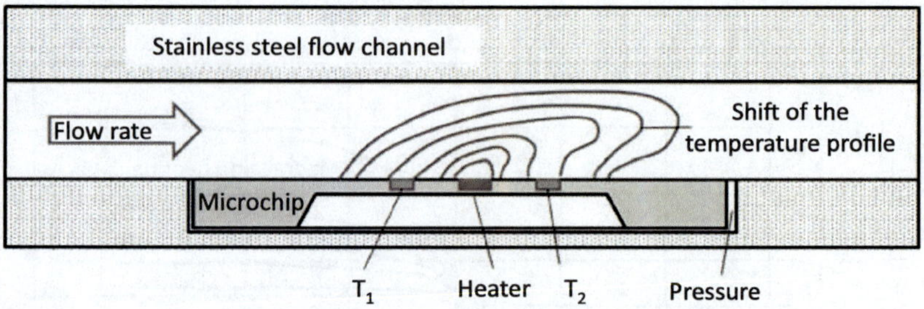

Fig. 11.41 Cross section through the sensory part of the *CMOSens® chip*

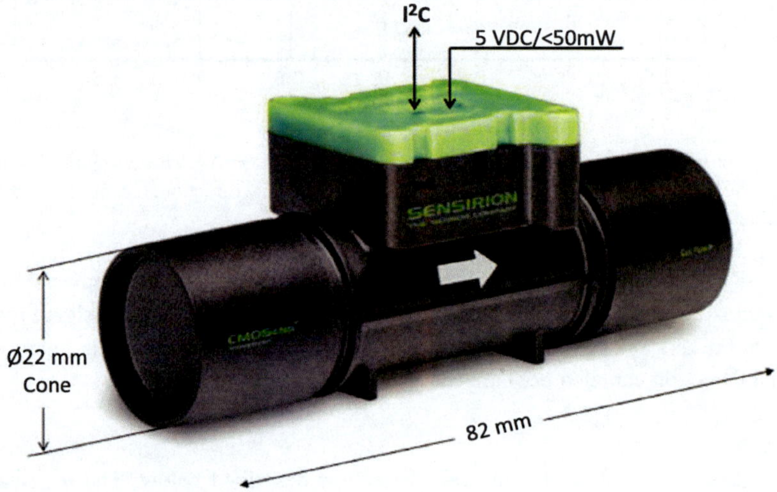

Fig. 11.42 Flow sensor SFM300 for a bidirectional flow range of ±200 L/min. (Source: SENSIRION AG, Staefa, Switzerland)

A *mass flow controller (MFC)* is also available in this technology for use in gas control. Due to the fast response time of the microsensor, the MFC can also approach a new flow value (*setpoint*) within <100 ms. Conventional MFCs that are not equipped with a microsensor require 0.2–1 s for this control process. The entire design is also very compact and is shown in Fig. 11.43.

New Developments

A new development in the field of differential anemometry is the use of thermocouples as temperature sensors. Figure 11.44 shows such a sensor design. The entire microstructure is located on a very thin membrane, which occupies an area of1.4 m × 0.344 mm A resistor (260–300 Ω), which functions as a heating element, is arranged in the center. Depending on

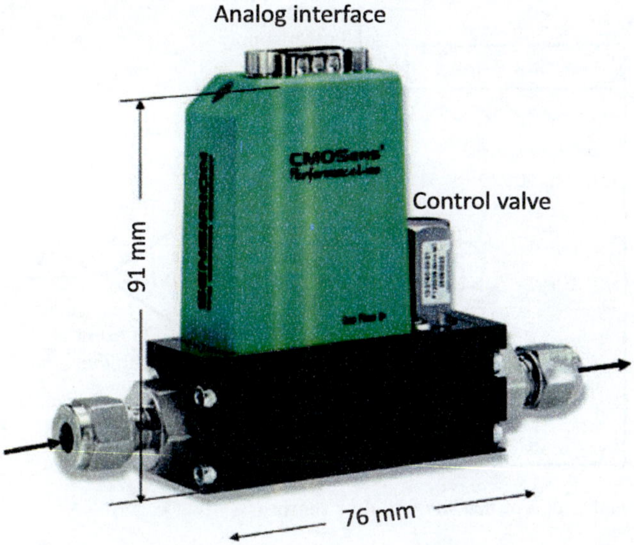

Fig. 11.43 Mass flow controller (MFC) based on flow measurement with a microsensor. (Source: SENSIRION AG, Staefa, Switzerland)

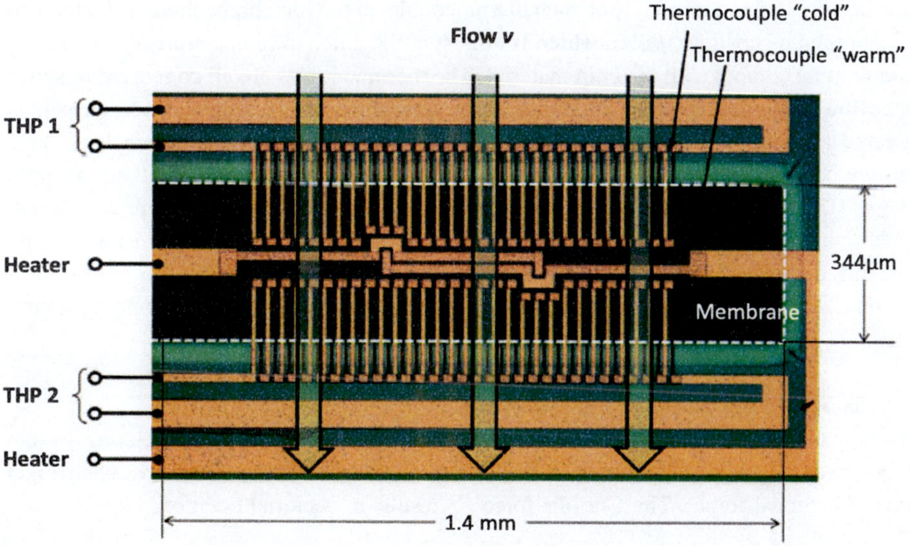

Fig. 11.44 Microstructure of a flow sensor (type SiF0920) with 30 thermocouples each for detection of the flow velocity v. (Manufacturer: SiFlo Technologies LLC)

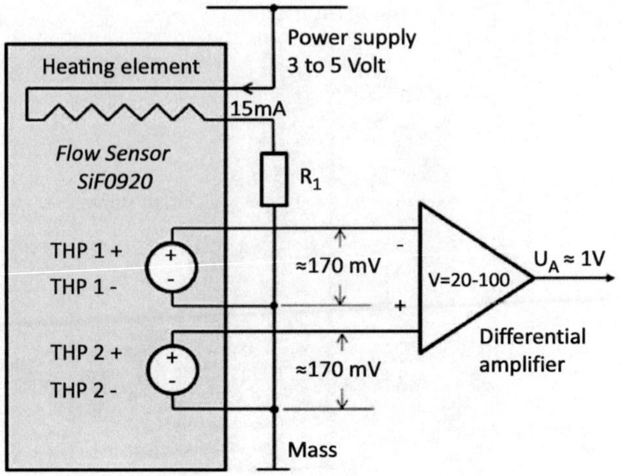

Fig. 11.45 External circuit of the SiF0920 flow sensor

the voltage applied (3–5 V), the heating power lies between 30 mW and 90 mW. Directly next to the heating element there are two rows of thermocouples on the membrane. Due to this immediate proximity to the heating element, these thermocouples also have almost the same heating temperature T_H. For each thermocouple on the membrane there is a reference element on the substrate (*bulk*), which is almost at the ambient temperature T_U due to the good thermal contact with the chip material. The thermocouples are all connected in series so that the output voltage $T_U = T_H$ is *zero* at the same temperature. If the heating element is operated at maximum power, a temperature is set on the membrane, which leads to a measurement voltage of ≈170 mV and can be measured on the respective thermocouple rows (THP1 and THP2). With a differential amplifier the voltage values are subtracted, resulting in a voltage at the amplifier output $U_A \approx 0$ (see Fig. 11.45). If a flow v now occurs over this microstructure, the flow profile shifts from THP1 to THP2. This results in a differential voltage at the preamplifier input, which now leads to a corresponding output voltage $U_A(v)$.

Coriolis Mass Flow Meter

This type of sensor measures the force F_C acting on a body m moving in a rotating system (angular velocity ω) at the velocity v. This force acts in addition to the known centrifugal force (centrifugal force). The Coriolis force F_C results in vectorial notation:

$$\overrightarrow{F_C} = 2 \cdot m \cdot \left(\vec{v} \times \vec{\omega}\right) \tag{11.55}$$

and for the amount:

$$F_C = 2 \cdot m \cdot v \cdot \omega \cdot \sin\varphi \tag{11.56}$$

In a right-angled system, the result is $\sin\varphi = 1$:

$$F_C = 2 \cdot m \cdot v \cdot \omega \tag{11.57}$$

Figure 11.46 shows the operating principle of such a sensor design. The gas flows through the sensor block and is redirected in a loop at a distance L from the block. If this loop is now set into oscillation by an external excitation, the loop moves upwards at a certain point in time at the angular velocity ω. The force F_C is then exerted on the mass element Δm of the mass flow $\dot{m}$ and points downwards. In the return of the loop the flow velocity v is reversed, so that due to the vector product ($\omega \times v$) the force F_C also points in another direction (downwards). These forces create a torque M on the loop, which leads to a bending around the angle Θ. The torque results from:

$$M = L^2 \cdot \omega \cdot \dot{m} \tag{11.58}$$

The angle of rotation Θ can be calculated from the mechanical spring constant K of the loop and the torque M

$$\Theta = \frac{M}{K} \tag{11.59}$$

The angle of rotation Θ is thus proportional to the mass flow $\dot{m}$ and can be measured with appropriate sensors. Figure 11.46 shows the resulting torsional movements which occur at different times t of the period of oscillation T.

In real flowmeters based on the Coriolis effect, very different loop shapes can be found, which serve to increase the measuring effect or to suppress disturbing effects (Tränkler and Obermeier 1998). Figure 11.47 shows such an arrangement. The structure consists of a double loop. The excitation is made by an electromagnet, which is driven at a defined frequency. Each loop has a separate control. The movement due to the Coriolis effect is scanned at the common connection point of the loops with a corresponding sensor. For adaptation to different measuring ranges, the mass flow can also take place in a bypass. Figure 11.48 shows the same structure, with some additional technical details. A standard design is shown in Fig. 11.49.

11.3 Flow Calibration

Pipe Flows

Flowmeters must always be individually adjusted and calibrated, as the device characteristics depend on many manufacturing tolerances. Therefore it is necessary to

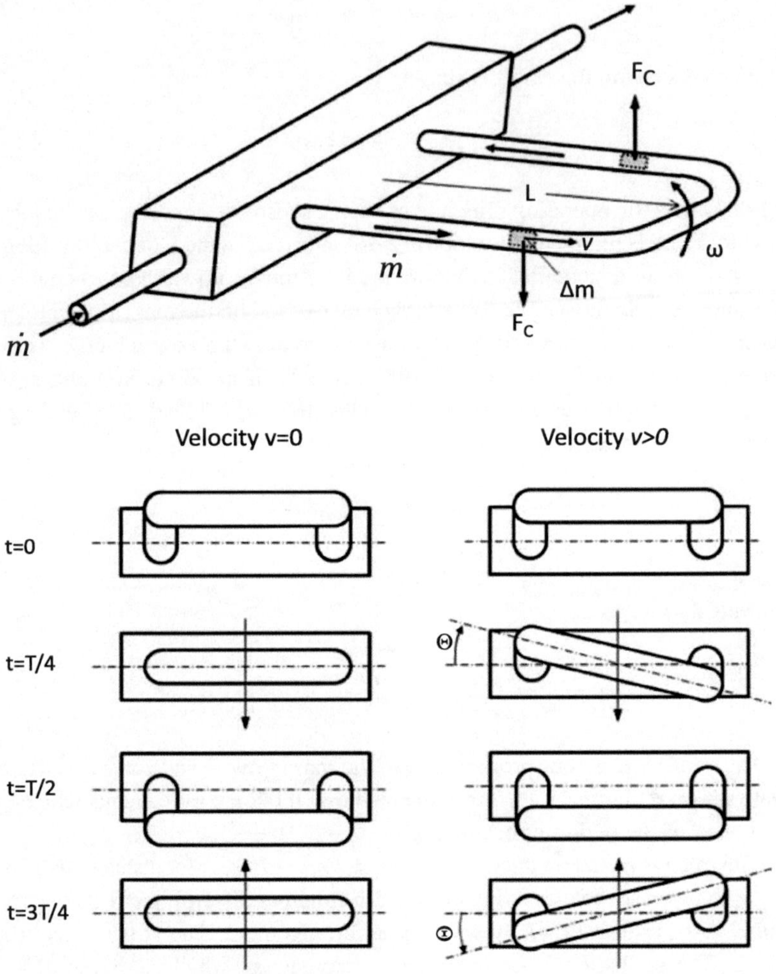

Fig. 11.46 Force effects of a flowing gas in an oscillating loop. Shown below are the torsional movements due to the torques M caused by the Coriolis force F_C

produce a *standard flow rate* as accurate as possible. In the field of calibration, different methods are used by PTB,[8] which are only of limited use in industrial practice.

For the area of the flow probes, flow channels are used which have a corresponding nominal diameter (NW = DN = D) in order to obtain the real flow conditions at the later measurement location as far as possible. In the simplest case, this calibration device

[8] Physikalisch Technische Bundesanstalt in Braunschweig Germany.

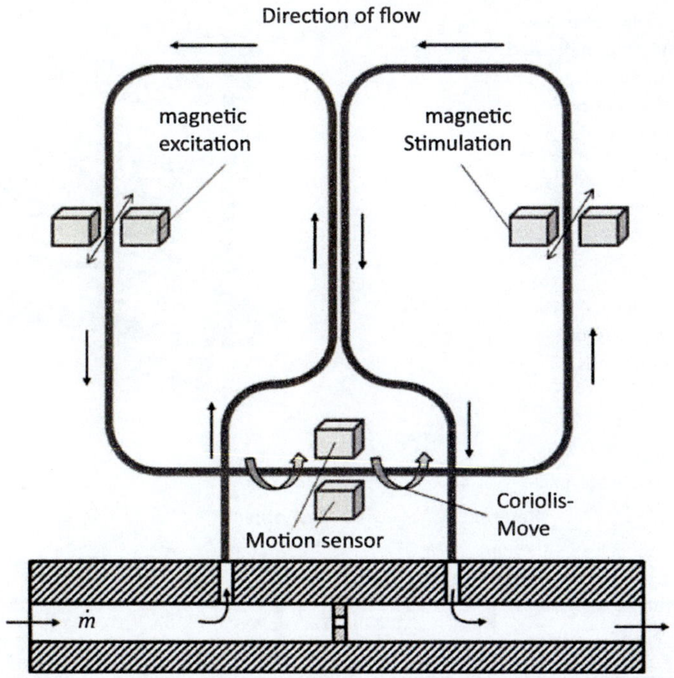

Fig. 11.47 Operating principle of a Coriolis flowmeter with a double loop and common scanning

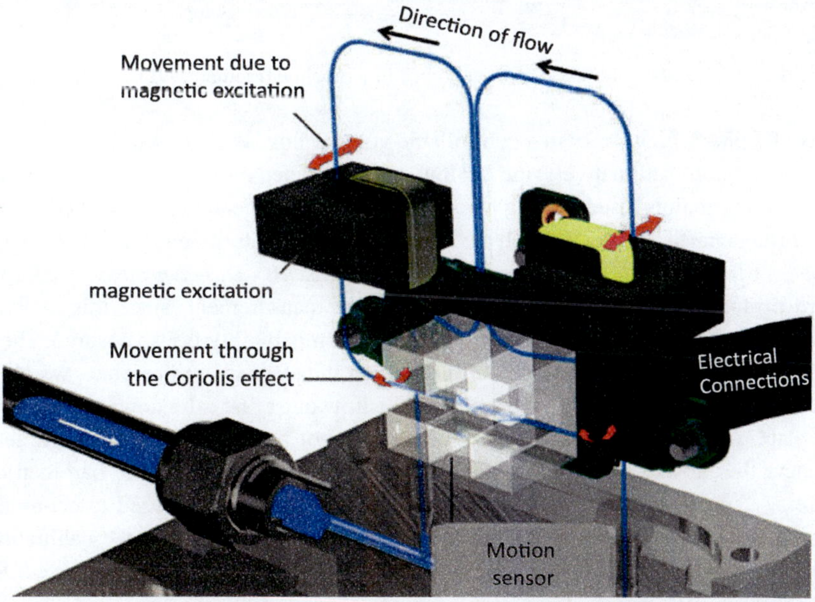

Fig. 11.48 Operating principle of the *Cori-Flow* flowmeter with a double loop. (Source: Bronkhorst Deutschland Nord GmbH, Kamen)

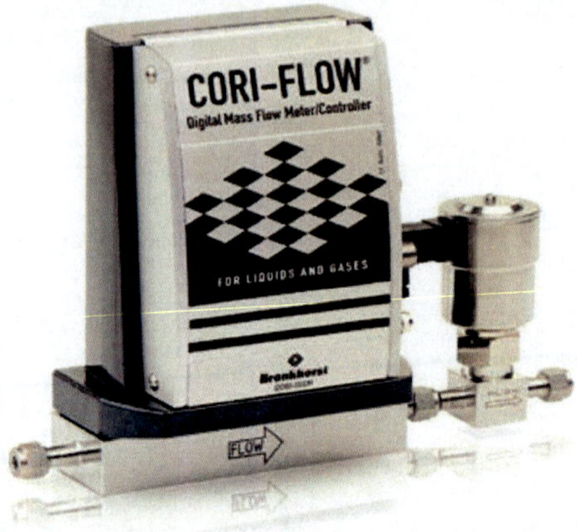

Fig. 11.49 Coriolis mass flow meter *Cori-Flow* with integrated valve for controlling the gas flow. (Source: Bronkhorst Deutschland Nord GmbH, Kamen)

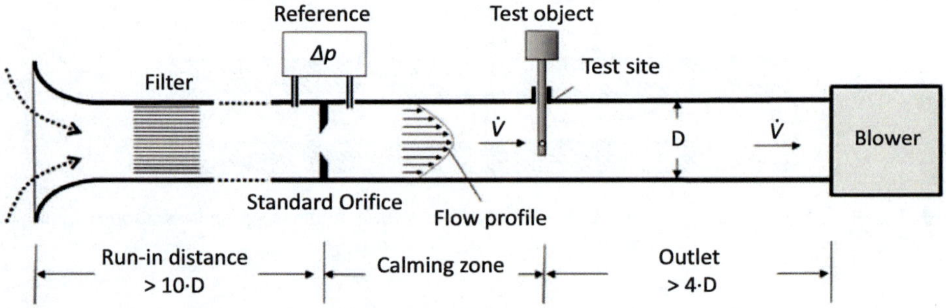

Fig. 11.50 Setup for the calibration of flow probes in a pipeline (Fiedler 1992)

consists of a pipe which sucks in a controllable volume flow $\dot{V}$ via a blower. Figure 11.50 shows such a setup. Alternatively, the air flow can be pumped into a circuit (closed system). This method is mainly used for high flow rates. To calm the flow, the sucked-in air is guided through baffles (filters), which are designed to prevent turbulence. In addition, there must be a sufficient inlet distance, which must be at least $10 \cdot D$. A Δp-measurement via a standard orifice plate is usually used as a reference measurement, since this method is clearly described by the relevant standards and thus comparability is also ensured. The test specimen is located at an appropriate distance from the standard orifice plate, which also represents a disturbance of the flow profile and the flow must first calm down again after the orifice plate. After the test location, an outlet distance of $\approx 4 \cdot D$ is required before the air then enters the blower. In addition, the ambient temperature T_U and the barometric air pressure p_{baro} must of course also be recorded. This method can be used to calibrate all measuring instruments that are suitable for use in pipelines. However, the calibration is very complex and costly if very large pipes with a diameter of 1–2 m are involved. Such

Fig. 11.51 Calibration stand for the volumetric calibration of flow meters of different measuring principles with ambient air. Flow range $\dot{V} = 5000$ L/h to 2,500,000 L/h, nominal size range DN25 to DN300. (Source: KROHNE Messtechnik GmbH, Duisburg Germany)

test stands are therefore rare. Figure 11.51 shows such a test rig, which can be used to calibrate flowmeters for different pipe diameters.

Soap Bubble Counter

For the area of small quantities, other methods have been established that can be used in any laboratory. Figure 11.52 shows the so-called *soap bubble counter*. It consists of a glass tube with a scale on the outside. The total volume V_0 of the glass tube is first determined by filling it with a liquid (water or mercury) and then weighing the mass m. The volume flow is measured by measuring the time Δt taken for a soap bubble to pass through the total volume V_0 or a corresponding fraction thereof. At the beginning of the measurement, therefore, the mirror of the soap solution is pressed by pressing the bellows until the gas inlet is opened, so that a bubble is formed which is transported upwards by the inflowing gas. When the bubble passes the first mark on the scale, a stopwatch measures the time Δt until the bubble has reached the last mark and the total volume V_0 has been passed. It then applies:

$$\dot{V} = \frac{V_0}{\Delta t} \tag{11.60}$$

The conversion to normal conditions for the volume flow is realized by the general gas equation:

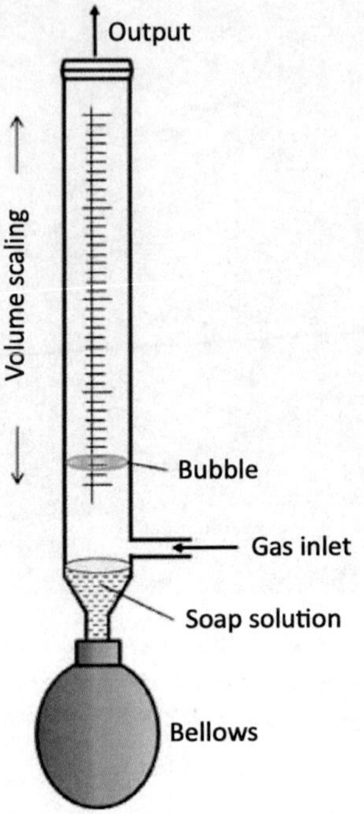

Fig. 11.52 Soap bubble counter for determining the volume flow in the laboratory

$$\dot{V}_0 = \dot{V} \cdot \frac{p_{\text{baro}} \cdot 273.15\ \text{K}}{1013\ \text{hPa} \cdot T_U} \tag{11.61}$$

A further correction calculation must be made if the gas to be measured is a dry gas. As the water evaporates from the bubble and enters the gas phase, this proportion must be taken into account. The correction is realized by the following equation (Down and Lehr 2005):

$$\dot{V}_c = \dot{V}_{\text{Measurement}} \cdot \frac{p_{\text{baro}} - p_{\text{Steam}}}{p_{\text{baro}}} \tag{11.62}$$

Sliding Piston with Hg Seal

Another possibility is to use a sliding piston instead of the soap bubble. The advantage of this arrangement (see Fig. 11.53) is that the piston position can be determined by a lateral optical scanning (e.g. transmitted or reflected light barrier) or by distance or speed measurement. It is important that the movement within the glass tube is as friction-free as possible and that no gas can escape at the sealing points. The so-called *Hg seal* has proven to be the optimal solution to this problem. The piston is mostly made of Teflon (PFTE) and has an annular groove which can be filled with mercury through cross holes. For this purpose the piston is inserted into the tube and the filling process can be carried out.

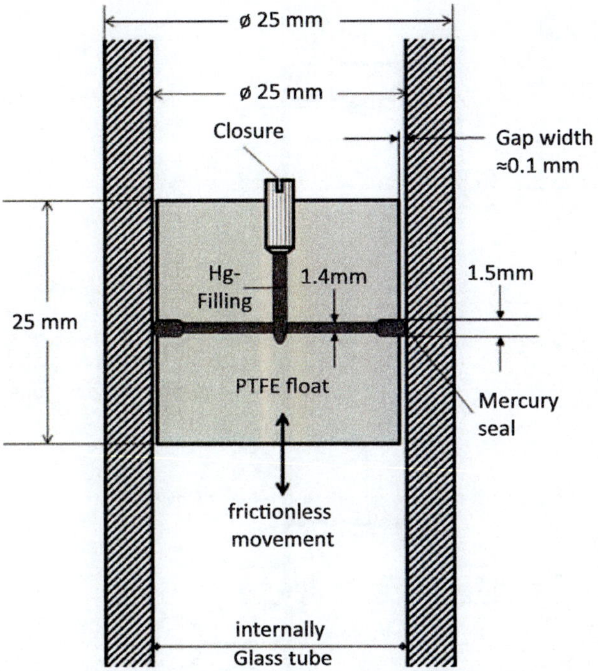

Fig. 11.53 Construction and arrangement of the sliding body with a Hg seal (Wiegleb 1984a)

Then the filling opening is closed with a screw to prevent mercury from escaping. The gap between the piston and the inner wall is only 1/10 mm and is sealed with the escaping mercury (Wiegleb 1984a). Due to the high surface tension, the mercury remains in position (annular groove) and does not leak. The coefficient of friction between the mercury and the glass wall is very low so that the frictional force F_R can be neglected. Since there is no force F_N between the two, only the gravitational force $m \cdot g$ of the piston acts on the gas volume below. The pressure p_G of this gas volume is measured and is included in the calculation of the volume flow. Furthermore, there is a temperature sensor directly at the gas inlet, which is also used for compensation or conversion.

In order to achieve the highest possible accuracy, calibrated glass tubes are used, whose inner diameter D is exactly defined. In the simplest case, the height measurement h is recorded using an ultrasonic method. Laser interferometers have also been used for exact measurements (Choi et al. 2009).

Figure 11.54 shows the complete construction of the flow calibration device. The test sample is placed in front of the calibration device and switched on via a solenoid valve. The use of light barriers is common for this type of height measurement. The manufacturers of these calibration devices adjust the position of the light barriers at the factory to be able to

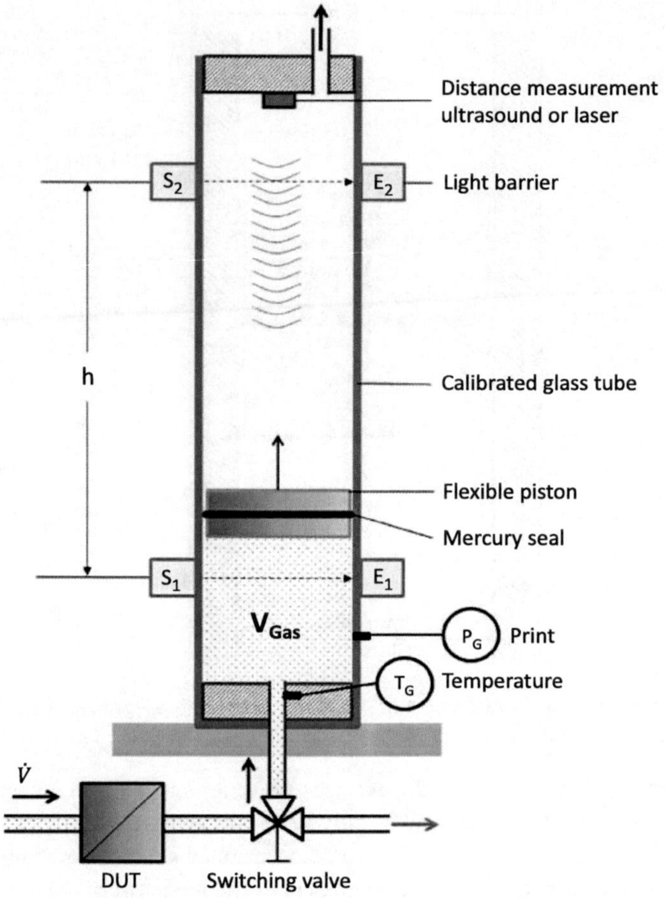

Fig. 11.54 Overall setup of a volume flow measurement with a sliding piston. The test sample is switched before the calibration device

specify a defined gas volume V_O. The setup in Fig. 11.55 has a total of 4 different height positions (h_1 to h_4) with which four volumes can be specified to cover different measuring ranges. The first light barrier (h_O) serves as the start position for the simultaneous time measurement ($\Delta t \approx 4 - 8$ s).

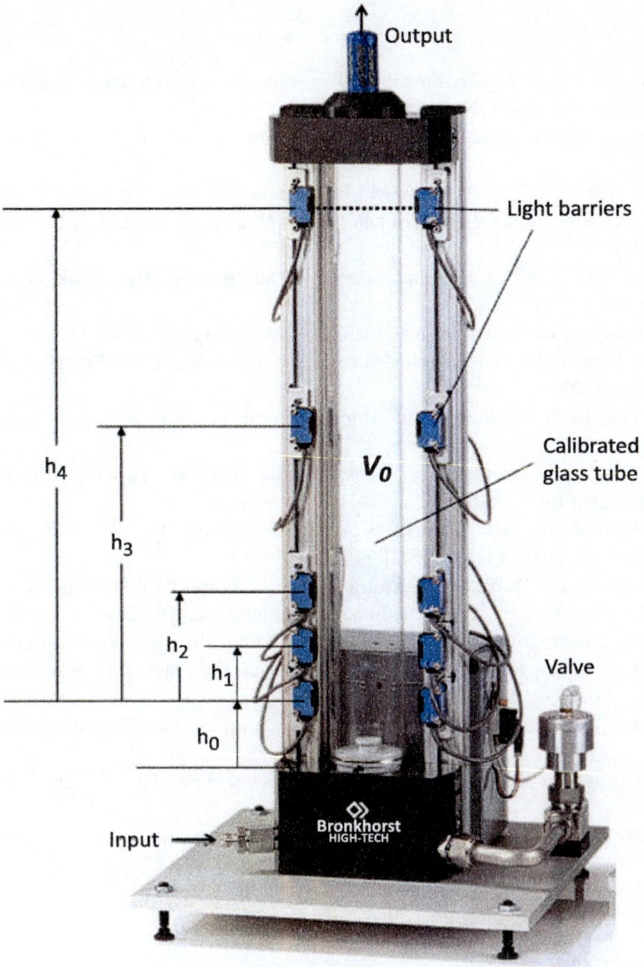

Fig. 11.55 Commercial calibration stand on the basis of a sliding piston (*Piston Prover*). With different light barriers different measuring ranges can be realized. (Source: Bronkhorst Deutschland Nord GmbH, Kamen)

References

Aichelen, W.: Der geometrische Ort für die mittlere Geschwindigkeit bei turbulenter Strömung in rauhen und glatten Rohren. Z. Naturforsch. **2a**, 108–110 (1947)

Brunck, O.: Clemens Winkler's Lehrbuch der technischen Gasanalyse. 4. Aufl. Verlag von Arthur Felix, Leipzig (1919)

Choi, H.M., Park, K.A., Oh, Y.K., Choi, Y.M.: Improvement and uncertainty evaluation of mercury sealed piston prover using laser interferometer. Flow Measurement und Instrumentation. **20**(4–5), 200–205 (2009)

Down, R.D., Lehr, J.H.: Environmental Instrumentation and Analysis Handbook. Wiley-Interscience (2005)

Fiedler, O.: Strömungs- und Durchflussmesstechnik. Oldenbourg, München (1992)

Frenzel, F., et al.: Industrielle Durchfluss-Messtechnik – Grundlagen und Praxis. ABB Automation Products GmbH. (2011)

Gevatter, H.-J.: Handbuch der Mess- und Automatisierungstechnik. Springer, Berlin, Heidelberg (1999)

Hagemeier, T.: Praktikumsanleitung zur Messung der örtlichen Geschwindigkeit. Universität Magdeburg, Fakultät für Verfahrens- und Systemtechnik (2011)

Hoffmann, F.: Grundlagen der Ultraschall-Durchflussmessung für den industriellen Einsatz. KROHNE Messtechnik GmbH, Duisburg (2000)

Hogrefe, W.: Handbuch der Durchflussmessung. Fischer & Porter GmbH, Göttingen (1994)

Scherer, P.: Gas-Meßvorrichtung. Deutsche Patentanmeldung DE 4006508 C2 vom 2.3.1990 (1990)

Strohrmann, G.: Messtechnik im Chemiebetrieb. 4. Aufl. Oldenbourg, München (1987)

Tränkler, H.-R., Obermeier, E.: Sensortechnik – Handbuch für Praxis und Wissenschaft. Springer (1998)

Wiegleb, G.: Mittel und Methoden zur Messung des Gasdurchflusses. Maschinenmarkt. **90**, 104–106 (1984a)

Wiegleb, G.: Thermischer Massendurchflussmesser für korrosive Gase. Tech. Mess. **51**, 43–45 (1984b)

Wiegleb, G., Röß, R.: Durchflusssensor für extreme Umweltbedingungen. VDI-Berichte Nr. **509**, 43–45 (1984)

Related References

Bonfig, K.W.: Technische Durchflussmessung. 2. Aufl. Vulkan, Essen (1987)

Bonfig, K.W.: Sensoren und Sensorsysteme. 5. Aufl. expert (1991)

Böswirth, L.: Technische Strömungslehre. Vieweg, Wiesbaden (2004)

Hardy, J.E., et al.: Flow Measurement Methods and Applications. Wiley-Interscience (1999)

Homa, T., Berg, M., Mell, H.-H.: Handbuch für Schwebekörper-Durchflussmesser. 1. Aufl. (03/VA-FLOW-DE 10.2005) ABB Automation Products GmbH (2005)

Lemme, H.: Sensoren in der Praxis. Franzis, München (1993)

Profos, P., Pfeiffer, T.: Handbuch der industriellen Messtechnik. Oldenbourg, München (1994)

VDI/VDE Richtlinie 3513 Schwebekörper-Durchflussmesser Berechnungsgrundlagen. Beuth, Berlin (1971)

VDI/VDE 2040 Berechnungsgrundlagen für die Durchflussmessung mit Blenden, Düsen und Venturirohren. Blatt 1–5. Beuth, Berlin (1987–1996)

Calibration and Test Methods

12

Source: Messer Group, Krefeld

The calibration and testing of gas measuring instruments (analysers) is an important area in gas measurement technology and is described in detail in this chapter. Compared to other physical measurement quantities such as pressure, temperature, force, length, etc., there are many influencing factors in gas measurement technology that must be taken into account in order to achieve correspondingly accurate measurement results. The conditions under which the measurement results are obtained therefore have a great influence on the quality of the results. Changes in air pressure, temperature influences, flow dependencies, draught and humidity are only examples of how diverse these external influences can be.

Another important area is the production of appropriate calibration gases, which are an essential prerequisite for the assessment of the measuring characteristics of gas measuring

G. Wiegleb, *Gas Measurement Technology in Theory and Practice*,
https://doi.org/10.1007/978-3-658-37232-3_12

instruments. Therefore, all common static and dynamic methods used in the laboratory and in the test field are described. Furthermore, the production of certified test gases, which are available in compressed gas containers (test gas cylinders), is also dealt with.

12.1 Introduction

The calibration and testing of gas measuring instruments has been extensively described in many different standards (e.g. DIN[1]) and VDI/VDE guidelines. In addition there are national standards (e.g. USA) and also international standards (EN-ISO). In addition to the general standards, there are also industry-specific standards, such as:

- Exhaust gas measurement technology in the automotive industry (EPA,[2] EEA[3])
- Flue gas analysis in power plants (BImSchV[4])
- Gas analysis technology in the chemical industry (NAMUR[5])
- Natural gas industry and natural gas application (DVGW[6])

in which the specific requirements for gas detection technology for the respective industry are described. However, the list is incomplete and only represents the most common areas of application. Further regulations result from the requirements of the professional association (e.g. MAK-Monitoring) or from everyday requirements such as the Dispensing System Ordinance (Schankanlagenverordnung), in which the CO_2 measurement in *beer cellars* is described.

In addition, there are also guidelines and recommendations of state institutions, such as the BAM,[7] PTB,[8] UBA[9] or NIST.[10] It is obvious that not all requirements resulting from the standards, directives, regulations etc., can be described comprehensively at this point. When presenting the individual methods and procedures, however, reference is always made to the current standards, in which a detailed description of the properties can be found. In these works, gas measuring instruments are always given different names depending on the application or industry. Typical terms are for example:

[1] German Institute for Standardization e. V.
[2] Environmental Protection Agency.
[3] European Environmental Agency.
[4] Federal Immission Control Ordinance.
[5] Standards working group for measurement and control technology in the chemical industry.
[6] German Association of the Gas and Water Industry
[7] Federal Institute for Materials Testing.
[8] Physikalisch-Technische Bundesanstalt.
[9] Environment Federal Office.
[10] National Institute of Standards and Technology.

- Gas sensor
- Gas warning device
- Gas analyzer
- Gas probe
- Exhaust gas meter
- Gas monitor,
- And so on

In this chapter only the term *analyzer* is used, which is described in DIN 51897 as follows:

> *"An analyser produces a clear correlation between the content of the substance under consideration (analyte, gas component) in the sample and a physical or chemical property of this substance and outputs it as an electrical measurement signal".*

The assessment of the measurement quality of analysers is usually done by using calibration gases. For this purpose, a defined gas mixture c is flowed through the analyzer and the measurement signal c^* is viewed, which is the response to this gas concentration. Usually 10 different gas concentrations (calibration points) are needed to derive a good correlation between c and c^*. These pairs of values (c_n/c_n^*) are used for an calculation by an *n-th* degree polynomial or a linear regression. The calibration function then describes the correlation between the given gas concentration (test gas) and the displayed concentration value of the analyser. A graphical representation of the calibration function is called a *calibration curve*. The inverse function is called *analysis function* and describes the conclusion of the displayed gas concentration to the specified concentration value of the test gas or sample gas (VDI 2449 part 2).

Figure 12.1 shows this relationship. The concentration of the specified gas mixture is plotted on the *x-axis*, which usually goes from 0 to 100% of the measuring range end value (Full Scale). The entire measuring span is also referred to as the measuring range (MR). *Suppressed measuring ranges* are also common in gas measurement technology, which for example are valid between 80–100 of the maximum gas concentration.

The ideal calibration function has a linear behaviour, ranging from zero point (ZP) to the end point (EP). In practice, however, real calibration functions deviate more or less from this ideal behavior. Depending on the location within the calibration curve, the errors are designated as:

- Zero point error (offset),
- Endpoint error (sensitivity),
- Linearity error (between ZP and EP).

These errors will be discussed in more detail in the presentation of test procedures (Sect. 12.4).

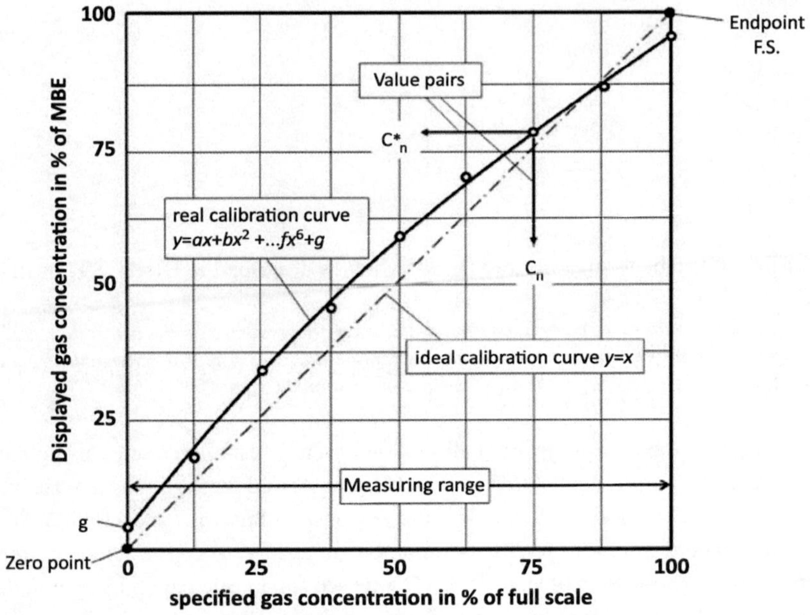

Fig. 12.1 Display of the calibration function in a calibration curve

12.2 Zero Gas and Test Gas

Checking the calibration function requires a precise knowledge of the composition of the calibration gases (test gases) in order to describe the measuring characteristics of the analyser. Nitrogen (N_2), which is supplied in compressed gas cylinders (up to 300 bar), is usually used as the zero gas. Companies and facilities with a high nitrogen demand have liquid tanks with a correspondingly high storage capacity. In addition to Nitrogen, other high purity gases such as Argon (Ar), Oxygen (O_2), Helium (He), etc., are also suitable. However, Nitrogen has become generally accepted for reasons of simple and safe handling. Furthermore, the costs are considerably lower compared to other zero gases.

Purity Statement

The purity of the Nitrogen depends on the requirements for measuring accuracy. The manufacturers offer different qualities for this purpose, which is defined by a short designation (e.g. 5.7). The first digit indicates the number of *nines* in the percentage for the pure gas content. The second digit indicates the first decimal place, which has a deviation of "9". Both digits are separated by a full stop. The term *five-point seven* (5.7) therefore indicates a concentration of 99,9997% nitrogen by volume. The remaining 0.0003% by volume (=3 ppm) is distributed among admixtures (minor constituents),

Table 12.1 Different nitrogen qualities with the specified minor constituents

Quality	Purity in %	Auxiliary components in ppm	
Technical nitrogen	>99.8	k. A.	
VERISEQ® (pharma)	>99.99	O_2	<3
		H_2O	<5
		CO	<5
		CO_2	<5
		Ar	<5000
		Odour	No
Nitrogen 5.0	>99.999	O_2	<3
		H_2O	<5
		CW	<0.2
Nitrogen 5.3	>99.9993	O_2	<2
		H_2O	<2
		CW	<0.1
Nitrogen 5.5	>99.9995	O_2	<1
		H_2O	<1
		CO	<0.5
		CO_2	<0.5
		CW	<0.1
		Hal. CW	<0.001
Nitrogen 6.0	>99.9999	O_2	<0.5
		H_2O	<0.5
		CO	<0.1
		CO_2	<0.1
		H_2	<0.5
		CW	<0.1
Nitrogen 7.0	>99.99999	H_2O	<0.05
		O_2	<0.03
		CO	<0.03
		CW	<0.03
		CO_2	<0.03
		H_2	<0.03
		Hal. CW	<0.001

Source: According to data from Linde AG, publication HiQ® Specialty Gases 2014

which are described in more detail in a certificate. Nitrogen is available in 7 quality levels. Table 12.1 lists the different qualities with the possible minor constituents.

Synthetic Air

In certain applications, however, Oxygen (O_2) is always required as a component of the zero gas. Examples of this are the testing of heat tone sensors in safety engineering or the calibration of respiratory gas measuring devices in medical technology. In the simplest case, purified compressed air is suitable here. Since this compressed air is generated with compressors whose pistons are lubricated with oil, the compressed air contains increased hydrocarbon components and possibly also odorous substances. Furthermore, the ambient air also contains Carbon Dioxide (400 to 2000 ppm) as well as other secondary components that could interfere with gas measurement. For analytical purposes these components must be removed with appropriate filter elements. In this case, it is easier to use synthetic air consisting of 20% Oxygen by volume and 80% Nitrogen by volume. Synthetic air is also available in different quality levels.

Quantity of Gas in the Compressed Gas Cylinder

A 50 L pressurized gas container filled with methane is under a pressure of 120 bar at an ambient temperature of 20°C. According to the ideal gas law, there should be a volume under normal pressure of $50\text{L} \cdot 120\text{bar}/1\text{bar} = 6000\text{L} = 6\text{m}^3$ in the cylinder. Taking into account the real gas factor (eq. 2.120) $Z = 0.8206$ of methane, however, the following gas volume results (Schreckenberg 1985):

$$V^* = V_0 \cdot \frac{p \cdot M}{\varrho \cdot R \cdot Z \cdot T} = 0.05\text{m}^3 \cdot \frac{120 \cdot 16.043}{0.671 \cdot 0.083144 \cdot 0.8206 \cdot 293.15} = 7.172\text{m}^3 \quad (12.1)$$

The real value is therefore approximately 20% higher than when calculated according to the ideal gas law. However, if the real gas factor is $Z > 1$, these gases (Carbon Monoxide, Nitrogen, Hydrogen, ...) have a correspondingly lower content.

Gas Supply Systems

In order to ensure that the quality of the gases from the storage tank also reaches the actual application (→ calibration), special precautions are necessary during installation. Not every fitting that has been proven in the technical field can be used for high purity special gases. The gas suppliers also offer the suitable gas withdrawal system for the different applications. The range of possible devices goes from simple cylinder pressure reducers to complete central gas supply systems (Fig. 12.2). For demanding applications these supply systems are completely made of stainless steel. For simpler applications brass fittings and copper pipes are sufficient.

Zero Gas Generators

An alternative possibility for the use of N_2 cylinder gases or liquid nitrogen are so-called nitrogen generators. These generators produce a high concentration of N_2 (95.5–99.9% by volume) from the ambient air via corresponding membranes, which can be used for many gas analytical purposes (zero gas). Oxygen generators also work according to the same procedure. Furthermore, zero gas can also be generated from the ambient air by removing the interfering components. This can be done for example, by certain filter elements

Fig. 12.2 Central gas supply in stainless steel design for applications with high purity gases. (Source: Messer Group GmbH, Krefeld)

(*scrubbers*) or by chemical reactions and catalytic combustion. Various equipment manufacturers offer this type of zero gas generator for use in gas measurement technology. The costs for the devices are mostly quite high and have to be compared to the costs for gas consumption. An advantage of the gas generators is their easy handling, which no longer requires the transport of heavy gas cylinders with the space required (floor space).

Safety Regulations

The handling of compressed gas containers is subject to a large number of regulations which must be observed during use. In addition to the statutory regulations, the professional associations also provide relevant instructions and specifications for handling gases. Depending on the industry, these regulations can vary and also describe other focal points. Furthermore, the gas industry also provides appropriate instructions for the safe handling of hazardous gases. The *Industrial Gases Association* alone has almost 100 safety instructions, technical information, leaflets and position papers, all of which can be accessed via the Internet.[11]

In gas measurement technology, the required gases are usually used in the laboratory or in an operating plant to check or adjust gas analyzers. Therefore, it is possible to concentrate on a few notes on safety engineering at this point.

The gases are usually supplied directly by the gas manufacturer or via a specialist freight forwarder. Storage during operation must take place in ventilated rooms, which in the simplest case consist of a lockable shelter. The compressed gas containers can be stored there. An iron chain is used to secure them against *falling over*. In the laboratory area 10 L cylinders and 50 L cylinders are the usual useful size. The transport from the warehouse to

[11] http://www.industriegaseverband.de, EIGA: European Industrial Gases Association.

the laboratory is to be carried out by trained personnel. While the 50 L cylinders are transported exclusively with a cylinder trolley, 10 L cylinders can also be carried into the laboratory *under the arm. It* is important that a cap is always screwed on during transport. In this state the bottle cannot become a danger in case of a fall (→no explosion). Only the weight of the bottle can cause corresponding damage. It is therefore recommended to wear safety shoes. The weight of an empty 10 L aluminium cylinder is 11,5 kg while a 10 L stainless steel cylinder for the high pressure range (200 bar filling pressure/300 bar test pressure) already weighs 32 kg. If the cylinder is then also filled with a *heavy gas* (e.g. sulphur hexafluoride) another ≈13 kg is added, so that the gas cylinders can then weigh up to 45 kg. A 50 L gas cylinder already has an empty weight of up to 93 kg. The transport must therefore be carried out with a cylinder trolley on which the cylinder can be pushed in a secured position. Figure 12.3 shows the transport of two 10 L gas cylinders. The step to the building was supplemented by a ramp to ensure a safe passage.

The gas cylinder valves are connected via corresponding side nozzle threads and connected to a pressure reducer. To avoid confusion, these are equipped with different valve connections, which are defined by DIN 477–1 and DIN 477–5. Test gases with an oxygen content of <21% always have a *left-hand thread* (M19 × 1.5 LH), while test gases with an oxygen content of >21% have a G3/4 connection.

Fig. 12.3 Safe transport of 10 L test gas cylinders with a gas cylinder trolley (Source: Fachhochschule-Dortmund)

Labelling of Gas Cylinders

The standard DIN EN 1089–3 represents a system of colour coding of gas cylinders which provides additional information about the properties of the gas content (toxic, flammable, oxidising, inert). These are already visible, even if the hazardous goods label is not yet legible due to excessive distance. The mandatory labelling of the gas content is carried out on the hazardous goods label (see Fig. 12.4).

The colour coding is only defined for the bottle shoulder. For medical gases the cylindrical part is white. Gases and gas mixtures are marked according to the general rule (see Fig. 12.5a). Common gases for industrial and medical applications and gas mixtures for inhalation are specially marked (see Fig. 12.5b and c). In the case of medical gas mixtures for inhalation containing less than 1.000 ppm NO (nitrogen monoxide) in nitrogen, a special colour *turquoise blue* is used on the bottle shoulder. The two main components of the gas mixtures can alternatively be marked in a ring on the shoulder (see Fig. 12.5d). The colour of the cylindrical cylinder shell is not specified in the standard (except for medical gases). In order to facilitate as uniform an allocation to the main areas of application as possible, the member companies of the Industriegasverband (IGV 2010) have agreed on the following colouring in addition to the medical gases:

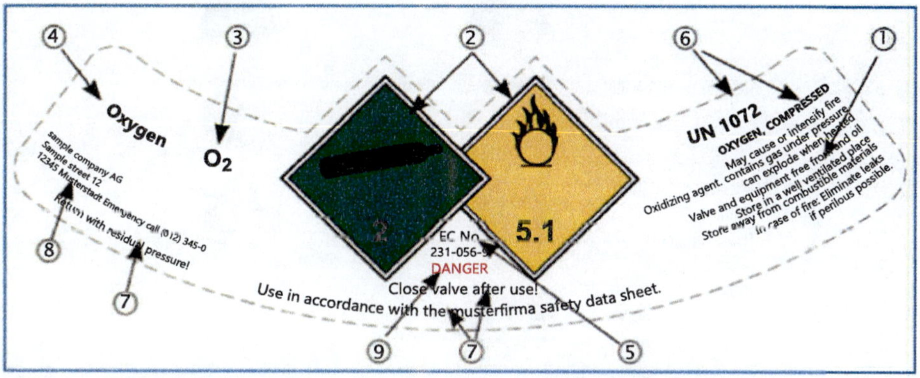

① Danger and safety information
② Hazard label according to ADR/RID
③ E.g. composition of the gas mixture or purity specification of the gas
④ Trade name of the gas manufacturer
⑤ EC number for single substances not applicable for gas mixtures.
⑥ UN number and name of the substance
⑦ Note from the gas manufacturer
⑧ the name, address and telephone number of the manufacturer
⑨ Signal word

Fig. 12.4 View of a hazardous goods sticker. The shape and design of the illustrated sticker may vary depending on the manufacturer. Text and symbols comply with the legal regulations. In the case of special gases, the information can be distributed over several stickers due to lack of space. (Source: Industriegaseverband e. V. Berlin)

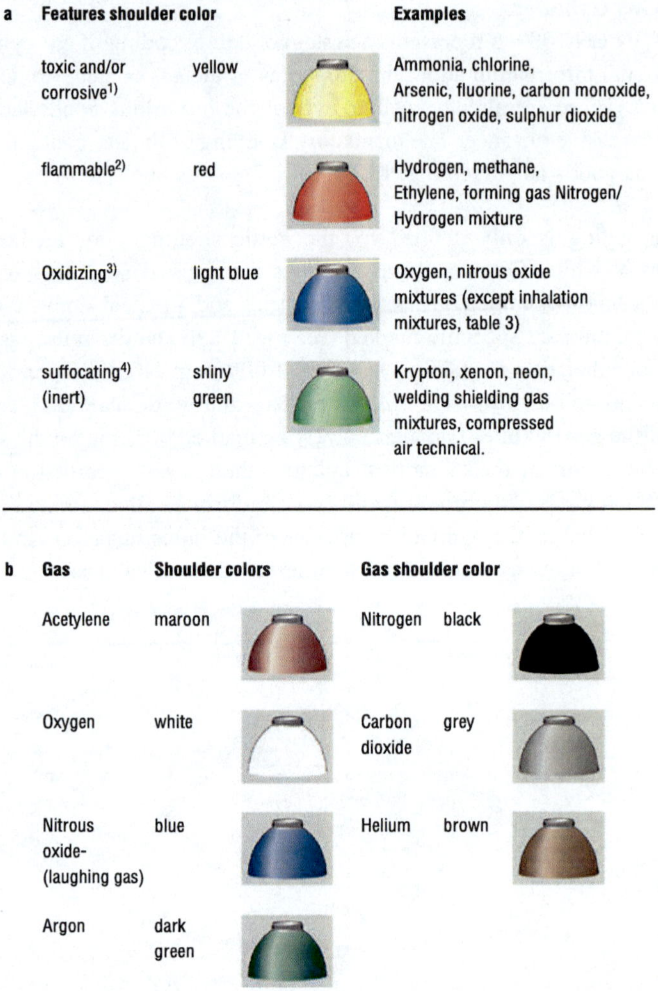

Fig. 12.5 (**a**) General labelling rule for gases and gas mixtures. (**b**) Special labelling for common gases (Source: Industriegaseverband e. V. Berlin). (**c**) Special marking for inhalation mixtures with a ring-shaped marking in the colours of the two components of the gas mixture. (**d**) Special marking for inert gas mixtures with a ring-shaped marking in the colours of the two components of the gas mixture. (Source: Industriegaseverband e. V. Berlin)

- Industrial gases: Grey or the same colour as the shoulder, but not white
- Special/specialty gases: Not specified.
- Medicine/inhalation gases: White.
- Breathing air cylinders used by fire brigades and rescue services have a yellow or red luminous warning colour.

c

Gas/gas mixture	Shoulder colors	
Synthetic air/ Compressed air for breathing purposes For oxygen concentrations between 20 – 23%	white/black	
Oxygen/helium mixture For all oxygen concentrations	white/brown	
Mixture oxygen/carbon dioxide For all oxygen concentrations	white/grey	
Oxygen/nitrous oxide mixture For all oxygen concentrations	white/blue	

d

Gas/gas mixture	Shoulder colors	
Carbon dioxide/nitrogen	grey/ black arz	
Carbon dioxide/oxygen	grey/white	
Argon/Oxygen	dark green/white	
Argon/Nitrogen	dark green/black	

Fig. 12.5 (continued)

The standard does not apply to bundle and trailer cylinders, fire extinguishers and gas cylinders for liquid gas.

Safety in the Laboratory

In the laboratory or in the plant to be monitored, the gas cylinders are also placed in a safe place. Since the gases are used in close proximity to the analyser, the gases are also placed there. This can be ensured in a central gas supply system, which is connected to the laboratory via an appropriate pipeline. In the simplest case, the gas cylinder is attached to the laboratory bench or an appropriate wall bracket. This fastening ensures that the gas cylinder cannot fall over during operation. There is a particular danger especially if a pressure reducer has been connected to the gas cylinder. If the cylinder falls over in this

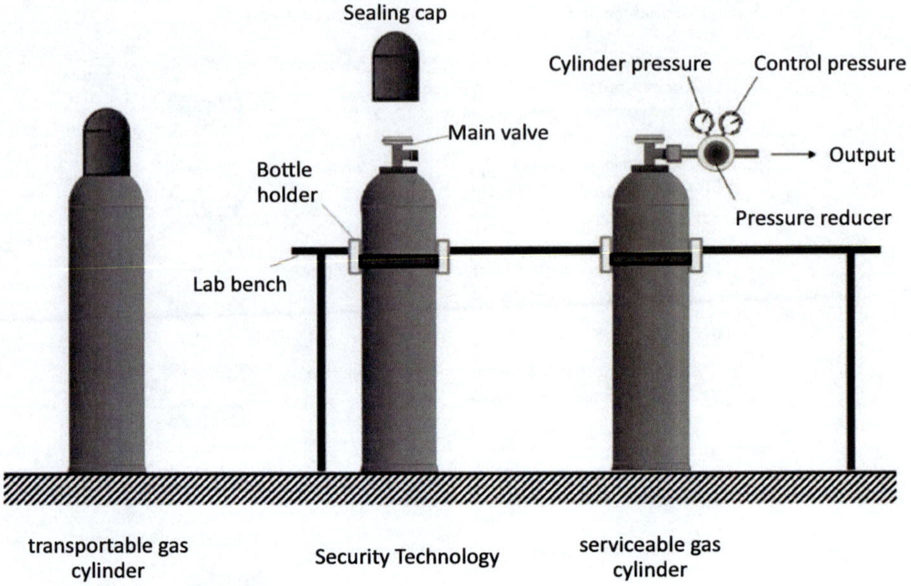

Fig. 12.6 Safe handling of compressed gas containers at the workplace (e.g. Laboratory)

state, the main valve can be torn off and the pressurised cylinder contents are released abruptly. In this case the gas cylinder will move like a *rocket* in the laboratory and may even break through walls. Therefore, it is very important to secure the cylinder during gas sampling. A gas cylinder must therefore never be transported with a connected pressure reducer. For transport, the cap must always be screwed on as a fuse. Figure 12.6 shows the fuse at a typical workplace. The corresponding cylinder holders can be obtained from specialist laboratory dealers. During the tests, make sure that there are always two people in the laboratory. In case the test setup or the gas pipes have a leak and the gases can spread uncontrolled in the room, good ventilation should always be provided. Furthermore, in certain cases (explosive and/or toxic gases) gas warning devices are recommended. In special cases this is also mandatory.

Test Gases Production and Accuracy

Test gases are gas mixtures that are specially produced for testing analysers. The requirements for the accuracy of the gas mixture and the purity of the starting materials are therefore very high. In principle, all gases can be mixed together in any concentration ratio. Exceptions are mixtures of combustible gases with oxygen, which can lead to an explosive gas mixture. Furthermore, vapours are subject to a maximum filling pressure to prevent condensation in the compressed gas cylinder. The concentration range extends from the lower ppb range to the volume % range. In most cases, these gas mixtures are based on a *basic gas*, such as nitrogen or synthetic air.

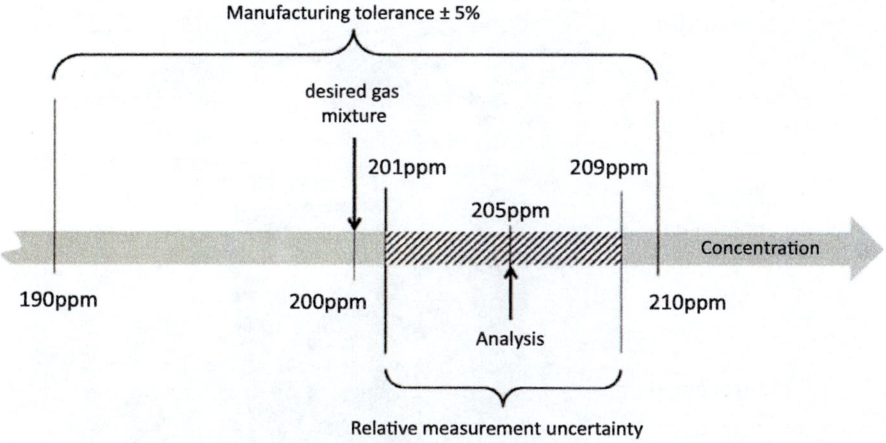

Fig. 12.7 Manufacturing tolerances and measurement uncertainties in the production of test gases, using the example of a 200 ppm gas mixture

The test gas concentration of the individual components can be selected at will, taking into account the above-mentioned restrictions. However, due to the production processes, the manufacturer can only maintain a tolerance of ±5%, relative to the desired concentration value. If the desired gas concentration is 200 ppm, the supplied test gas is then between 190 ppm and 210 ppm. The exact value of the gas mixture is determined with appropriate analytical methods. In the example shown in Fig. 12.7, the analysis value is within the manufacturing tolerance at 205 ppm. The measurement uncertainty of the analysis is indicated with ±2% of the analysis value. The supplied test gas thus has a concentration value between 201 ppm and 209 ppm. The stability of the test gases is stated by the manufacturer in months. Usually the stability is 12 or 24 months. After that the test gases are still usable, but the manufacturer no longer guarantees the data stated on the certificate of analysis. Gas mixtures with less reactive components such as CO_2 have a much longer stability, while Chlorine gas mixtures or Nitrogen oxides have a lower stability.

Test gases are usually produced by gravimetric methods, which are highly accurate. The individual components are filled into the compressed gas container (gas cylinder) via a balance. The weighing result then leads to the calculation of the concentration in the gas cylinder. Figure 12.8 shows such a filling process. The gas cylinder holding 50 L is placed on a balance. The empty weight of an 50 L bottle is from 67 kg (200 bar) to 93 kg (300 bar) and is stored as *tare.* To produce a test gas of 1000 ppm Carbon monoxide (CO) in Nitrogen (N_2), pure Carbon monoxide is required under normal conditions $10^{-3} \cdot 50\text{L} \cdot 200\text{bar}/1\text{bar} = 10\ \text{L}$. The molecular weight of CO is 28.0104 g/mol, so 10 L weigh CO under standard conditions (1013 hPa, 273,15 K)$28.0104\text{g/mol} \cdot 10\text{L}/22.414\text{L/mol} = 12.5\ \text{g}$. The Nitrogen content then weighs $28.0134\text{g/mol} \cdot 9990\text{L}/22.414\text{L/mol} = 12.4856\ \text{kg}$. The real gas factors Z are 1.0847 for Carbon monoxide at 200 bar

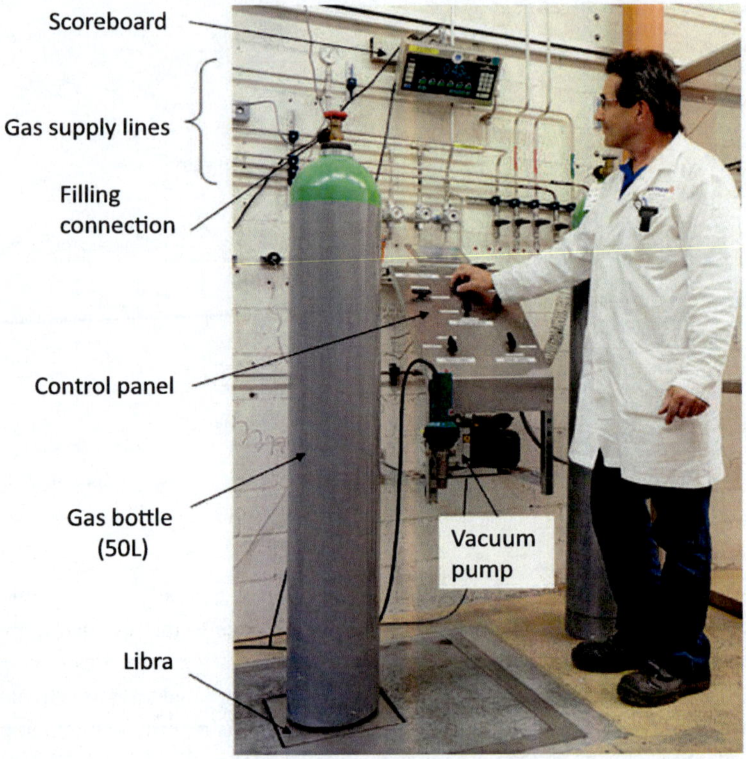

Fig. 12.8 Filling a test gas cylinder using the gravimetric method. (Source: Messer Group GmbH, Krefeld Germany)

and 1.0711 for nitrogen almost identical. For an exact calculation, however, these must be taken into account. This example shows what requirements must be met for the accuracy of the balance. Therefore, the balance is checked at regular intervals with calibrated weights and readjusted, if necessary. Figure 12.9 shows such a check.

After the filling process, the gas cylinder must be mixed for some time so that the components in the gas cylinder are evenly distributed (→*homogenised*). This is done on a device that regularly rotates the gas bottles in a horizontal position, thus forcing the mixing process. After completion of the mixing process, the exact determination of the components is then carried out with the appropriate analysis methods.[12] The analysis result is applied to the gas cylinder together with the hazardous goods label (see Fig. 12.10).

In addition, a certificate is issued on which all the data of the analysis are listed. The certificate is delivered with the respective gas cylinder from the manufacturer to the user.

[12]Gas chromatography, spectroscopy, CLD, mass spectrometry, absolute chemical methods, ...

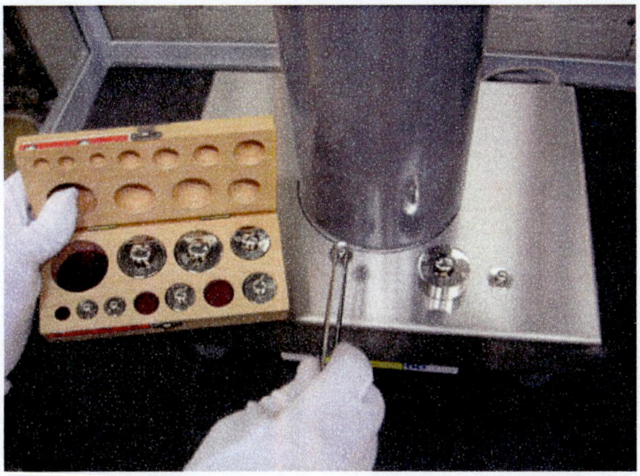

Fig. 12.9 Checking the balance used for mixing gases with calibrated weights. (Source: Messer Group GmbH, Krefeld Germany)

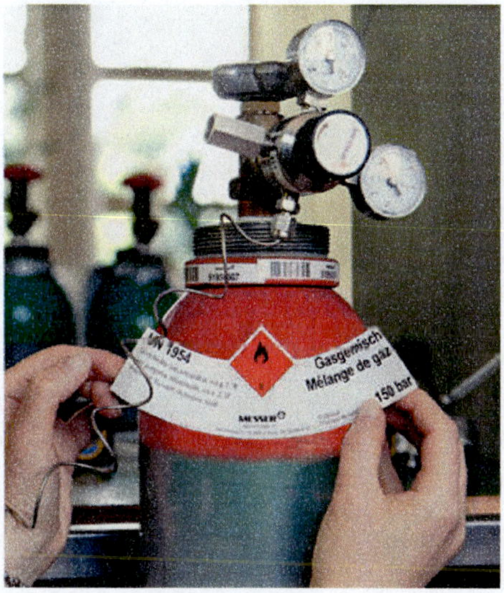

Fig. 12.10 Filled compressed gas container is provided with a corresponding hazardous goods sticker. (Source: Messer Group GmbH, Krefeld Germany)

The content of this certificate is described in detail in DIN EN ISO 6141 (*Contents of certificates for calibration gas mixtures*). Figure 12.11 shows an example.

In addition to the large and heavy compressed gas cylinders, more and more test gases are being filled in so-called pressure cans. The content of these pressurized cans (e.g. 1 L) is a few litres of useful gas (e.g. $1\ \mathrm{L} \cdot 10\ \mathrm{bar} \approx 10\ \mathrm{L}$). It is advantageous that they are easy to transport and, if necessary, several pressure cans can be used to test the complete calibration curve of an analyzer. In most cases, less than 1 L of gas is required for one measuring

Westfalen AG

Technische Gase / Westfalengas / Tankstellen

Westfalen AG
48136 Münster
USt.-ID-Nr. DE 126 117 135

UN 1956 VERDICHTETES GAS, N.A.G., (STICKSTOFF, SCHWEFELDIOXID), 2.2, (E)

Bitte Lagertemperatur beachten !

Produktmanagement Spezialgase
Ihr Ansprechpartner Britta Lange
Niederlassung Münster
Fon 0251/695-229
Fax 0251/695-706
b.lange@westfalen.com

Analysenzertifikat (certificate of analysis)
Fertigungsauftrag (order) **3960131**
Dieses Zertifikat stimmt überein mit den Empfehlungen der DIN ISO 6141 (This certificate is in accordance with the recommodations of DIN ISO 6141)

Seite 1 von 1

Kundenmaterial: (material no.)		Behälternummer: (cylinder no.)	27600502731439	Gemischnummer: (gasmixture no.)	G328063

Bestandteil (component)	**Sollwert (nominal value)**	**Istwert (analytical value)**	**Einheit [1]) (unit)**	**Analysentoleranz [2]) (analytical accuracy)**
Schwefeldioxid 3.0 [SO2]	6.000	6.067	ppm	± 2,00 % relativ
Stickstoff 5.0 [N2]	Rest	Rest		

[1]) Konzentrationsangaben sind, falls nicht anders vermerkt, als Stoffmengenanteile (= Molanteile) zu verstehen (Concentration data are, if differently does not note, to understand as mole fraction of the component (= parts per mole)).
[2]) Die Analysentoleranz gibt, falls nicht anders vermerkt, die Unsicherheit des Gehaltes einer Komponente als Produkt der Standardunsicherheit mit dem Erweiterungsfaktor an (The analytical accuracy gives, if differently does not note, the uncertainty of the content of a specified component as a product of the standard uncertainty and the coverage factor).

Fülldruck ideal bei 15° C: (ideal filling pressure at 15° C)	150,00 bar	Behältervolumen: (cylinder volume)	10,00 l	Fülldatum: (date of filling)	03.2015
Min. Verwendungsdruck: (min. pressure of utilization)	5,00 bar	Herstellgenauigkeit: (filling tolerance)	± 10,00 % relativ	Lagertemperatur: (storage temperature)	-10° C bis +40° C
Ventilanschluss: (valve outlet)	M 19 x 1,5 links			Haltbarkeit ab Fülldatum: (stability since date of filling)	12 MON
Produkt: (product)	A040101105	Prüfgas/Gasgem gem Spez T10 MFI Alu			

Datum: 13.03.2015 (date)

Prüfer/Ersteller: (tester/author) Denis Bergschneider SGZ Hörstel

Qualitätsmanagementsystem
DIN EN ISO 9001 Reg.-Nr. 1709
Technische Gase/Westfalengas
QM-Beleg: F-TG-49.4/01; QME 10

Aufsichtsrat: Dr. Wolf-Albrecht Prautzsch (Vors.)
Vorstand: Wolfgang Fritsch-Albert (Vors.), Reiner Ropohl, Dr. Carsten Wilken
Commerzbank AG Münster (BLZ 400 400 28) Konto-Nr. 3 954 492
Sparkasse Münsterland Ost (BLZ 400 501 50) Konto-Nr. 7 138 · Registergericht Münster: HRB 186

Fig. 12.11 Example of a certificate of analysis according to DIN EN ISO 6141

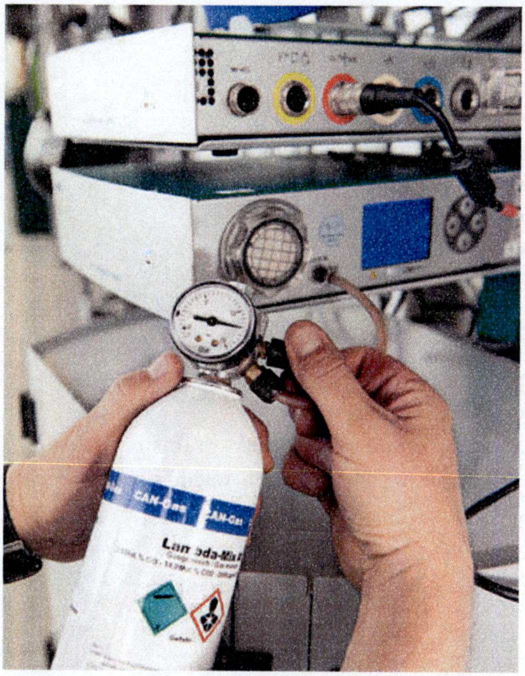

Fig. 12.12 Using a compressed gas can filled with test gas. (Source: Messer Group GmbH, Krefeld Germany)

point. With a flow rate of 1 L/min and a stable measurement value of the analyser after 30 seconds, approximately 20 tests could be performed with a pressurised gas cylinder from 10 L useful capacity. These pressure cans are particularly suitable for service purposes. Figure 12.12 shows the use of this option.

12.3 Gas Mixing Devices

The use of test gases from compressed gas containers (*gas cylinders*) is useful for the control of a defined gas concentration value, within the measuring range (MR) of an analyzer. In conjunction with a zero gas, the calibration curve can be recorded using two points. If the calibration value of the test gas is close to the end point (Span), the analyzer can also be adjusted. Especially from a practical point of view this kind of control during operation is recommended. If several calibration points are checked within the measuring range, a corresponding number of test gases is required. In these cases, the required calibration gas mixtures are prepared by dynamic or static methods.

The basic procedures for preparing calibration gas mixtures are divided into the following methods according to DIN EN ISO 6145:

- Volumetric pumps
- Continuous injection
- Capillaries
- Critical nozzles
- Thermal mass flow controllers (MFC)
- Diffusion
- Saturation
- Permeation
- Electrochemical production.

In this subchapter the procedures are described and also devices available on the market are presented. Figure 12.13 shows the basic structure of an arrangement for the production of calibration gas.

The procedures described in this chapter differ mainly in the way the defined partial volume flows $\dot{V}_n$ are generated. The concentration c_{ges} after the mixing chamber is determined by the mixing rule:

$$\dot{V}_1 \cdot c_1 + \dot{V}_2 \cdot c_2 = \left[\dot{V}_1 + \dot{V}_2\right] \cdot c_{\text{ges}} \tag{12.2}$$

$$c_{\text{ges}} = \frac{\dot{V}_1 \cdot c_1 + \dot{V}_1 \cdot c_2}{\dot{V}_1 + \dot{V}_2} \tag{12.3}$$

If a base gas (=zero gas) is used for *gas* 2, then $c_2 = 0$. Equation 12.3 is then simplified as follows:

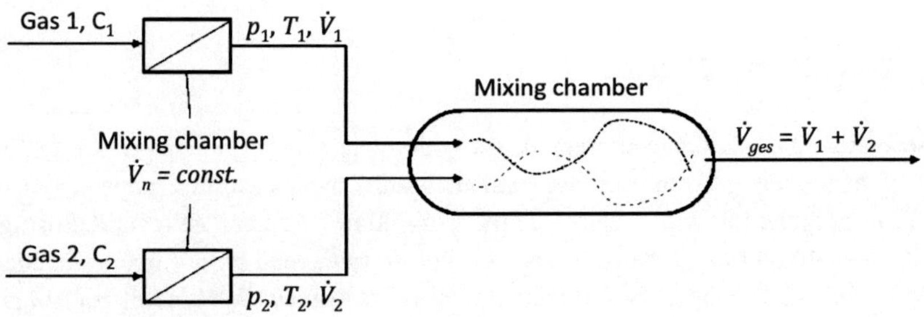

Fig. 12.13 Principle arrangement for the volumetric-dynamic production of calibration gas mixtures, which generate constant volume flows with appropriate devices

$$c_{ges} = c_1 \cdot \frac{\dot{V}_1}{\dot{V}_1 + \dot{V}_2} \tag{12.4}$$

As a rule, ideal gases can be assumed for the volume specifications. In particular, if test gas mixtures are used which have a concentration of <1 volume %, the real gas behaviour is not important. The real gas behaviour is taken into account by the real gas factor Z, which results from the *virial equation* (see Sect. 2.5). The following relationship therefore applies to the quantity of substance *n* and the pressure virial coefficient $B'(T)$

$$p \cdot V = n \cdot R \cdot T \cdot Z = n \cdot R \cdot T \cdot [1 + B'(T) \cdot p] \tag{12.5}$$

With the virial equation one then obtains the coefficient $B'(T)$:

$$B'(T) = B'_0 + (B'_{30} - B'_0) \cdot T(^\circ\mathrm{C})/30 \tag{12.6}$$

The values for B'_0 and B'_{30} can be found in DIN EN ISO 14912 for over 120 gases. Furthermore, calculated values of *Z* are also available there, which refer to a temperature of $T = 15°\mathrm{C}$. An extract is shown in Table 12.2.

The table shows that the real gas factor *Z* leads to a volume change of $\Delta V \gg 1\%$ for most gases. Only for substances like propane, butane, sulphur hexafluoride and sulphur dioxide the errors resulting from the real gas behaviour are >1%. For accurate calibration rings,

Table 12.2 Data on the real gas behaviour of important gases

Component	Formula	M	B'_0	B'_{30}	$Z(15\,^\circ\mathrm{C})$
Acetylene	C_2H_2	26.0373	−8,4	−5,8	0.99290
Air	–	28.9625	−0.58	−0.27	0.99958
Ammonia	NH_3	17.0306	−14.9	−9.7	0.98770
Argon	Ar	39.9480	−0.96	−0.61	0.99922
n-Butane	C_4H_{10}	58.1222	−42.2	−28.9	0.96445
Carbon dioxide	CO_2	44.0095	−6.69	−4.75	0.99428
Carbon monoxide	CO	28.0101	−0.66	0.31	0.99952
Ethane	C_2H_6	30.0690	−9.86	−7.16	0.99149
Helium	He	4.0026	0.53	0.47	1.00050
Hydrogen	H_2	2.0159	0.605	0.574	1.00059
Methane	CH_4	16.0425	−2.36	−1.63	0.99801
Nitrous oxide	N_2O	44.0129	−7.18	−5.08	0.99387
Propane	C_3H_8	44.096	−20.97	−14.79	0.98217
Sulphur hexafluoride	SF_6	146.0560	−15.2	−10.8	0.98700
Sulphur dioxide	SO_2	64.0650	−24	−17	0.97950

Source: DIN EN ISO 14912

with volumetric methods for the production of calibration gases, this real gas behaviour must therefore be taken into account.

Volumetric Piston Pumps

The use of piston pumps is particularly suitable for a defined production of calibration gas mixtures, as this is an absolute procedure (Schierjott 2012). Figure 12.14 shows the central part of these gas mixing devices. It consists of a piston in a cylinder which is filled with the test gas or the zero gas. The piston can be filled one after the other with zero gas and test gas or two identical pistons can be used. The piston is connected to an eccentric which is coupled to an electric motor via a gear. The speed n can be adjusted via the gear and indicates the number of filling processes per time (Δt). The cylinder is movably mounted, so that the gas supply and gas discharge takes place through valve openings, which are opened and closed by the cylinder. In the position shown (Fig. 12.14) the gas is forced out of the cylinder. The opening for gas supply is closed. The active content delivered per revolution can be determined from the geometrical quantities h and d:

$$V_H = h \cdot \pi \cdot d^2/4 \tag{12.7}$$

The geometric volume flow $\dot{V}_{\text{geo}}$ is then obtained by multiplying the stroke volume V_H by the stroke number n:

$$\dot{V}_{\text{geo}} = V_H \cdot n \tag{12.8}$$

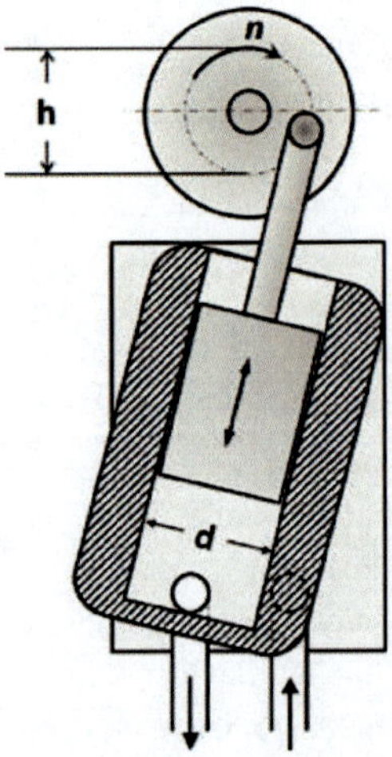

Fig. 12.14 Delivery piston of a gas mixing pump

Table 12.3 Example for volume calculation

	Piston pump 10	Piston pump 1
Cylinder type	D41 10759-4	D41 10759-5
Piston d in mm	14.7	4.65
Height h in mm	13	13
Volume in cm^3	22.0632	2.2077
Strokes n in min^{-1}	327	327
Flow rate in $\dot{V}_{geo}$ cm^3/min	721.466	7.2192

Data source: Wösthoff Messtechnik GmbH, Bochum

Table 12.3 shows the data for two different pistons (1 : 10). The calculated values already give an indication of how manufacturing tolerances may affect the mixing result. The influencing variables $h \pm \Delta h$ and $d \pm \Delta d$ are included in the error analysis as measurement uncertainty. The manufacturer therefore checks these deviations and determines the respective values with an accuracy of Δh, $\Delta d < 1/000$ mm. The final test of these devices is carried out by a symmetry test of the two piston pumps at a mixing ratio of 50 : 50 with nitrogen and oxygen. The concentration of 50 vol.% O_2 is measured with an appropriately accurate O_2-analyzer. If the inputs of the piston pumps are now swapped, a 50 vol.% O_2 display should also be obtained. On the basis of the deviations, the deviations for the real delivery volume $\dot{V}_{geo}$ can thus be concluded. An extended verification also of other, arbitrary settings (n) is carried out with soap film flow meters (Sommer 1988) or according to Eq. 12.8 with the linearity test (Schierjott 2015). Further error considerations can be found in DIN 51898–1.

In the calculations, the volume $\dot{V}_{geo}$ is equated with the volume $V(p, T)$ of the gas portion produced during the period considered. However, at high flow velocities, the volume delivered may deviate from the geometric volume $\dot{V}_{geo}$ due to dynamic effects. Therefore the complete approach is as follows:

$$\dot{V}(p,T) = F \cdot \dot{V}_{geo} = F \cdot V_H \cdot n \tag{12.9}$$

Here F is the delivery factor for the gas delivered by the piston pump at the stroke rate n. The deviation of the delivery factor is less than 10^{-4} with a suitable design, so that F~1. The exact value also depends on the nature of the inlet and outlet openings and the viscosity η of the gas.

Figure 12.15 shows the individual phases that belong to an entire cycle. In position 1 the piston rises and sucks in the gas. The exhaust valve is closed. In position 2, the piston has reached the top dead center and the cylinder reverses. The descending piston now pushes the gas through the outlet. In this position 3 the outlet valve is open. In position 4 the piston has now reached the bottom dead center and the cylinder is turned around again. This means that the system is filled and emptied per revolution. With this type of gas delivery,

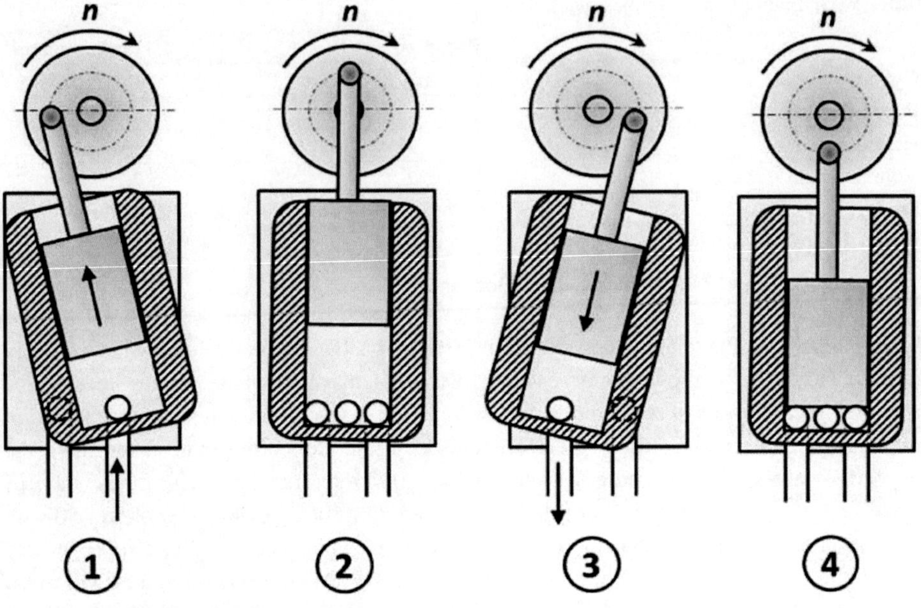

Fig. 12.15 Filling and emptying the cylinder during a period

the volume is ultimately only dependent on the geometric parameters and is therefore traceable to the SI unit meter [m].[13]

However, an essential prerequisite for accuracy is stable pressure and temperature conditions in the cylinder during operation. Figure 12.16 shows a complete system that works with two delivery pistons. In the simplest case, one piston (A) delivers the test gas, while the second piston (B) delivers the zero gas. The stroke rate n_A and n_B can be set separately for both pistons. The changeover valves are, as already described, part of the delivery piston. At the inlets to the two pistons, the pressure difference must be as Δp small as possible to limit the error. For $\Delta p < 20$ Pa the error is enclosed <0.2‰.

With the same piston volume $V_A = V_B$ and $F_A = F_B$ the mixing ratio is only changed by the stroke rate n_A and n_B. The concentration c_M at the outlet of the mixing vessel is then calculated:

$$c_M = c_A \cdot \frac{\dot{V}_A}{\dot{V}_A + \dot{V}_B} = c_A \cdot \frac{F_A \cdot V_A \cdot n_A}{F_A \cdot V_A \cdot n_A + F_B \cdot V_B \cdot n_B} = c_M \cdot \frac{n_A}{n_A + n_B} \tag{12.10}$$

[13]Traceability according to DIN 6143: *Property of a measurement result or the value of a standard to be related to suitable standards, generally international or national standards, through an uninterrupted chain of comparative measurements with stated measurement uncertainties.*

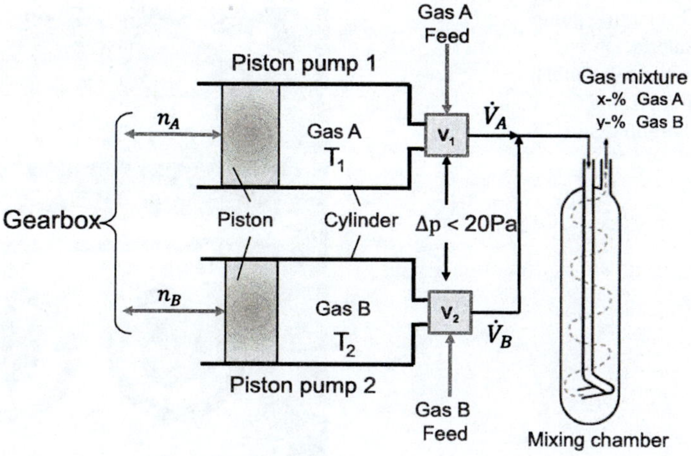

Fig. 12.16 View of a mixing device using the volumetric piston pump method

By varying the number of strokes n_A and n_B theoretically any gas concentration c_{Mx} between the zero gas and the test gas can be set. The systems available on the market work with a fixed gear reduction, so that only discrete steps are possible. With two delivery pistons these are, for example, 10%-steps. Concentrations of 0%, 10%, 20%, ..., 90% and 100% of the respective test gas concentration c_A can be set. For example, if a certified test gas with a concentration of $c_A = 6067$ ppm is used, the possible concentration steps are 0 ppm, 606.7 ppm, 1213.4 ppm, ..., 5460.3 ppm and 6067 ppm. If a volume corresponding to 1/10 of V_B is specified for cylinder A, concentration steps of 0%, 1%, 2%, ..., 9% and 10% are possible. For the calibration gas produced in this way, these would be 0 ppm, 60.67 ppm, 121.34 ppm, ..., 546.03 ppm and 606.7 ppm. The switching steps are set externally via corresponding control elements (see Fig. 12.17). During operation it must be ensured that the sum of both settings always results in 100%. The setting means40 % : 60% a 40% mixing of the test gas with the zero gas. However, if a setting of is selected, the40 % : 70% result is only a mixed gas concentration of 36.36%. By *"wrong settings"*, intermediate values can therefore also be achieved, which can be calculated using the following equation:

$$c_M = c_A \cdot \frac{\%\text{Value}(A) \cdot 100}{\%\text{Value}(A) + \%\text{Value}(B)} \tag{12.11}$$

The gear settings result in different filling times of the pistons, which produce corresponding concentration values after mixing. Figure 12.18 shows these times for the zero gas piston (ZG) and the test gas piston (TG). In this example, the zero gas piston has the speed n_1 and the test gas piston has the same speed. The resulting gas mixture is

Fig. 12.17 Gas mixing pump *DIGAMIX* (Manufacturer: Wösthoff Messtechnik GmbH, Bochum) with 2 pistons for generating calibration gases in 10% steps

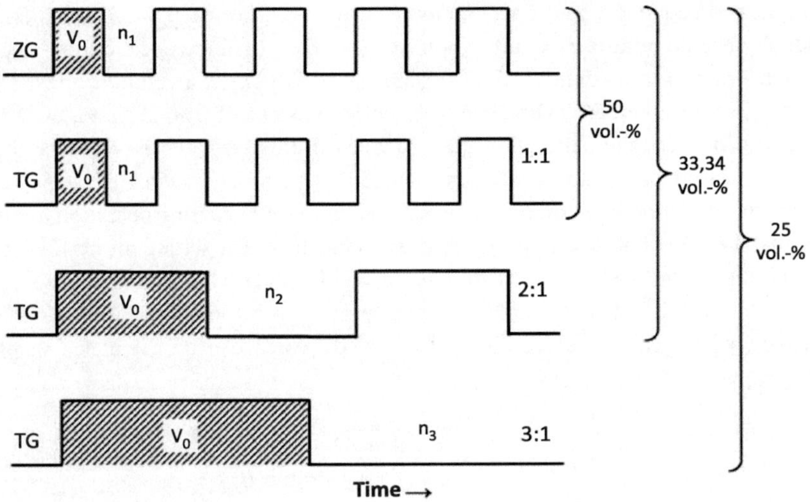

Fig. 12.18 Temporal phases for the zero gas piston (ZG) and the test gas piston (TG) resulting in different mixing ratios

therefore 50%. With different numbers of strokes n_x, one thus obtains the different gas concentrations c_x.

The time required to complete the mixing process is determined by the flow rate of the pump and the volume V_K of the mixing chamber. According to Eq. 12.47 with $V_K = 150$ mL and a delivery rate of 721 cm³/min after 60 seconds a concentration value is obtained which corresponds to 99.2% of the final value. In practice, therefore, it is usually waited 1–2 minutes until a stable concentration value has been reached.

For exact mixing, a pressure difference of $\Delta p < 20$ Pa is required at the pump inlets. This keeps 0, 2‰ the error in the production of gas mixtures to under. To meet this requirement, the differential pressure Δp can be checked with a U-tube manometer. The gases $A + B$ are led over a bead vessel, which allows the excess gas to flow off in case of an oversupply of gas. It therefore applies:

$$\Delta \dot{V}_A = \dot{V}_{zA} - \dot{V}_A \tag{12.12}$$

By readjusting the amount of gas supplied $\dot{V}_{zA}$, the proportion of $\Delta \dot{V}_A$ is kept as small as possible. If the concentration mixture is changed, for example, from 20 : 80 to 10 : 90 more gas A and less gas B is required. The volume flows $\dot{V}_{zA}$ and $\dot{V}_{zB}$ must then be readjusted manually. Furthermore it must be ensured that the pressure difference is $\Delta p \approx 0$. The manufacturer also offers a corresponding pressure regulator for this case. Figure 12.19

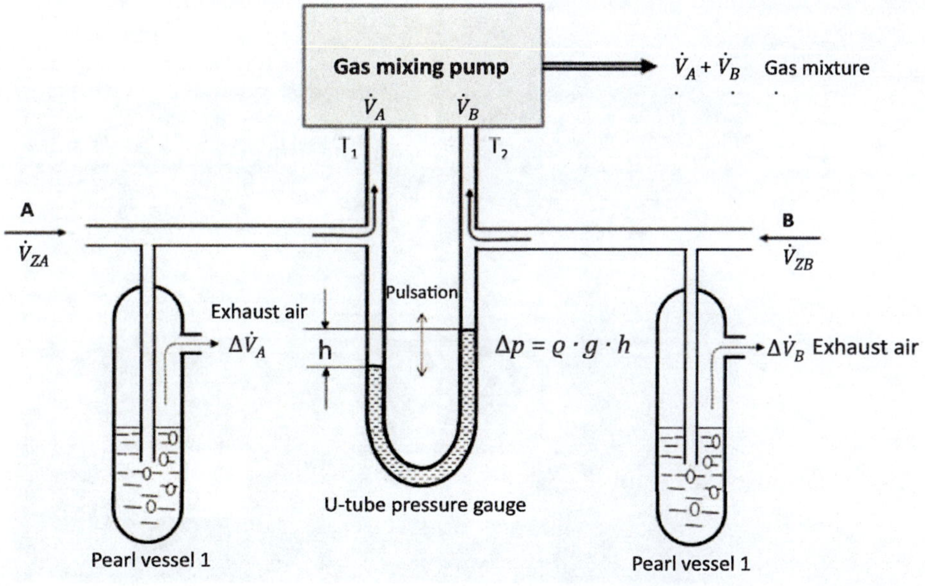

Fig. 12.19 Arrangement of the bead vessels and the U-tube manometer to check the flow and pressure conditions

shows the arrangement of the bead vessels and the U-tube manometer. The difference h signals a corresponding pressure difference Δp according to the hydrostatic pressure of the liquid column (h). Since the pistons always deliver the gas in portions only, a pulsation occurs at the inlet of the pump, which is detected by the U-tube manometer. In this case the height h changes by ± 1 mm.

The temperatures of gases A and B must also be identical for an exact gas mixture. In the gas mixing device, the piston pumps are thermally coupled, so that $T_1 \approx T_2$. The manufacturer recommends that the pumps *are warmed up* about 30 minutes before use, so that a stationary temperature distribution of the pump system is achieved. Temperature sensors (Pt100/Pt1000) can optionally be integrated to check and, if necessary, correct the temperature error. These sensors then record the individual temperatures of the two pumps and thus of gases A and B.

Figure 12.20 shows a gas mixing pump which has 4 pistons integrated and thus allows an even finer subdivision of the gas concentrations. For gas A and gas B there is one piston each for the range of 1–10% and one piston for the range of 10–100%. The mixing ratios can therefore be adjusted in 1% steps over the entire range from 1% ... 99%.

The 4 pistons are also used to mix 100% gases down to the ppm range. Figure 12.21 shows such an arrangement. With the first two pistons a maximum ratio of 100 : 1 can be achieved, resulting in a mixing concentration ($A + B$) of 0.99%. This gas mixture is in turn fed to the 1%-piston and mixed 100 : 1. The concentration c_A at the outlet of the second stage is then 98 ppm. In general, the following equation can be set up for any setting of the 4 pistons:

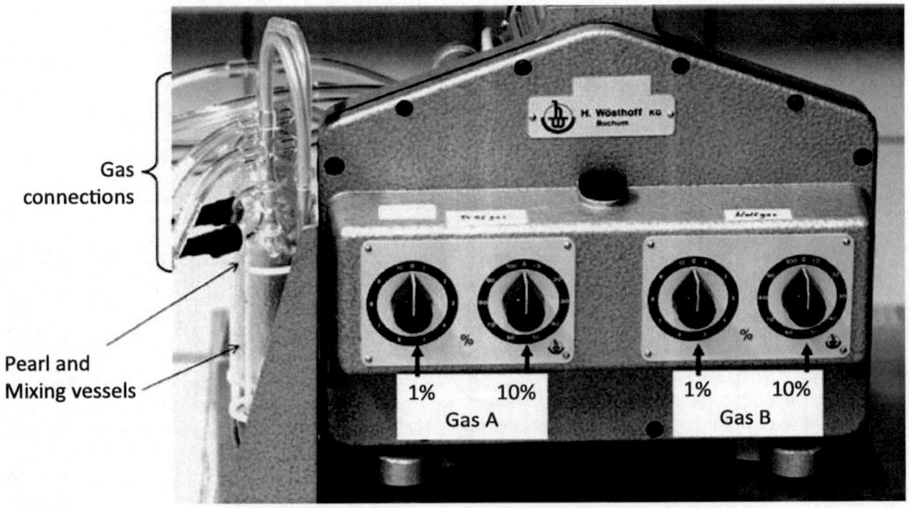

Fig. 12.20 Gas mixing pump *DIGAMIX* (Manufacturer: Wösthoff Messtechnik GmbH, Bochum) with 4 pistons for generating calibration gases in 1% steps

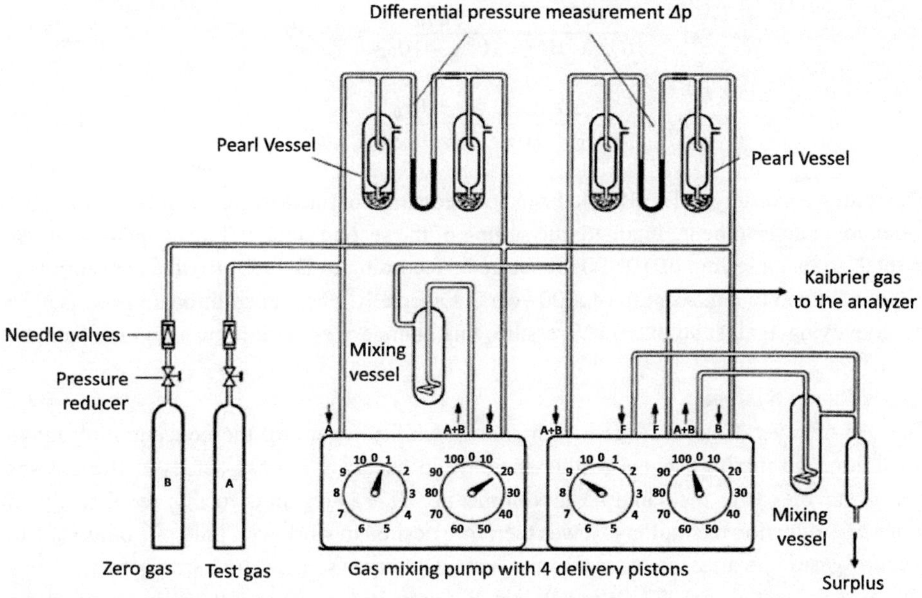

Fig. 12.21 Example of a multi-stage gas mixture with 4 pistons. The fifth piston (*F*) is only used for gas delivery, to deliver gas against a pressure in the analyzer or in the gas line (filter element or similar) if necessary. (Source: Wösthoff Messtechnik GmbH, Bochum)

$$c_A = \frac{\mathrm{I}(1\%)}{\mathrm{I}(1\%) + \mathrm{I}(100\%)} \cdot \frac{\mathrm{II}(1\%)}{\mathrm{II}(1\%) + \mathrm{II}(100\%)} \cdot 10^6 \text{ppm} \tag{12.13}$$

With I(1%) = adjustment of the 1% piston in gas mixing system I, I(10%) = adjustment of the 10% piston in gas mixing system I, II(1%) = adjustment of the 1% piston in gas mixing system II, II(10%) = adjustment of the 10% piston in gas mixing system II.

Example

With the settings in Fig. 12.21 the following gas concentration is obtained at the outlet of the second mixing stage.

$$c_A = \frac{1\%}{1\% + 20\%} \cdot \frac{9\%}{9\% + 100\%} \cdot 10^6 \text{ppm} = 3932 \text{ppm}$$

The highest gas concentration that can be set with this type of pump operation results from

$$c_{A1} = \frac{10\%}{10\% + 10\%} \cdot \frac{10\%}{10\% + 10\%} = 25.0\text{Vol.-}\% \quad (12.14a)$$

$$\text{or} \quad c_{A2} = \frac{10\%}{10\% + 10\%} \cdot \frac{10\%}{10\% + 90\%} = 5.00\text{Vol.-}\% \quad (12.14b)$$

The setting according to Eq. 12.14a leads to a reduction of the flow rate by a factor of 5, but if the flow rate is to be maintained, the setting of the second stage (II) must again be 100% in total, so that a setting of 10 : 90 is required. According to Eq. 12.14b, this concentration value is limited to a maximum of 5.00 vol.%. Especially when an additional piston is used for conveying against an external pressure, this boundary condition must be maintained.

Single Piston Systems

The use of several flasks leads to a higher flexibility regarding the concentration range (→multi-stage mixing), but a further increase in accuracy is only achieved if all flasks always have exactly the same basic volume V_H. *There* are limits to this requirement in terms of production technology. It was therefore desired to work with only one pump and to feed the gases one after the other into the same delivery system. A first arrangement of this kind was presented in 1987 (Wiegleb and Randow 1987). Figure 12.22 shows the basic structure of the system. The two gas flows *A* and *B* were fed pressure-free via bead vessels to a solenoid valve, which led the two gases one after the other into a mixing vessel with volume V_O. A pump sucked the gas from the mixing vessel via a capillary and delivered it to the analyzer. The resulting gas mixture was calculated from the opening times t_1 for gas *A* and the opening times t_2 for gas *B*.

The company Wösthoff GmbH in Bochum uses a similar process and developed it up to series production readiness. In 2001 a corresponding series device was presented. In this device (*DIGAMAT*) a piston pump is used, whose filling processes are synchronized with

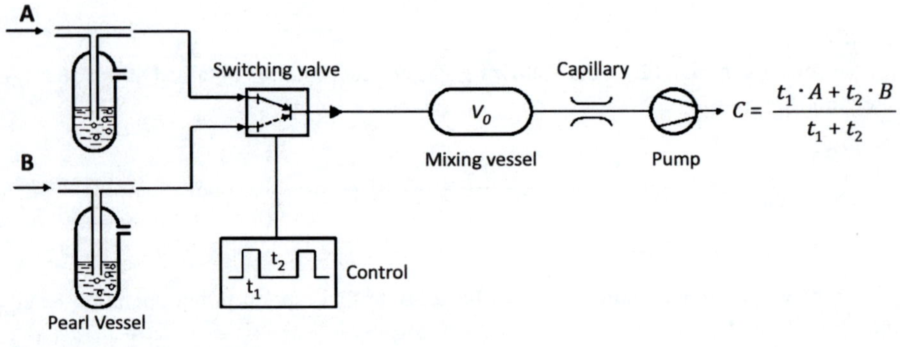

Fig. 12.22 Mixing device with a solenoid valve for switching between two gas flows and subsequent mixing in a separate vessel (Wiegleb 1987)

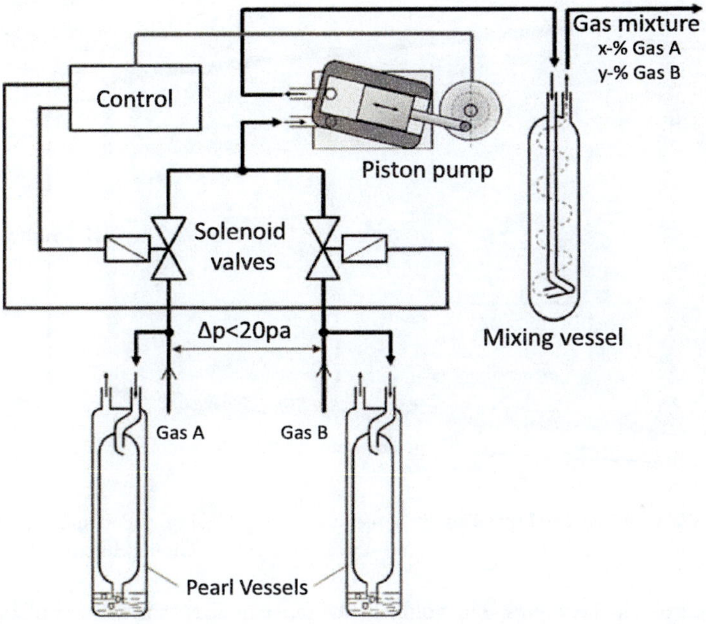

Fig. 12.23 Basic design of a gas mixing device with a piston and delayed mixing of the components in a mixing vessel

the control of the opening times t_1 and t_2. By this measure the accuracy of the calibration gas production could be improved.

Figure 12.23 shows the structure of the *DIGAMAT*. The inputs are also set to $\Delta p < 20$ Pa with appropriate bead vessels. Alternatively, an inlet pressure regulator can be connected upstream. The pumping process is only complete when the gas (*A*) is fed from the piston into the mixing vessel. This point in time is scanned at the eccentric with a Hall sensor and serves as the starting point for the change to the second gas (*B*). The first valve is therefore closed and the second valve opened, so that now gas *B* enters the piston. In this design, too, the number of pump strokes is decisive for the mixed concentration. To generate a gas concentration of 1 vol.% from 100% gases, 1 piston filling with gas *A* must be carried out and then 99 fillings with gas *B*. The two gases are then homogenized in one or more mixing vessels to provide a stable calibration gas concentration at the outlet. The mixing process is completed after about 3 minutes and depends on the flow rate. The procedure has not yet been conclusively described in a standard. Since 2014, however, a corresponding draft of DIN 51898–3 has been available in which this single-piston method is described. Figure 12.24 shows the current device design.

The DIGAMAT has up to 2 piston pumps which are connected to a solenoid valve unit. The two pistons have a volume ratio of 10 : 1. With the arrangement shown in Fig. 12.25, concentration ranges from 0.1 to 100.0 vol.% can be set in 0.1 vol.% steps by skilful control

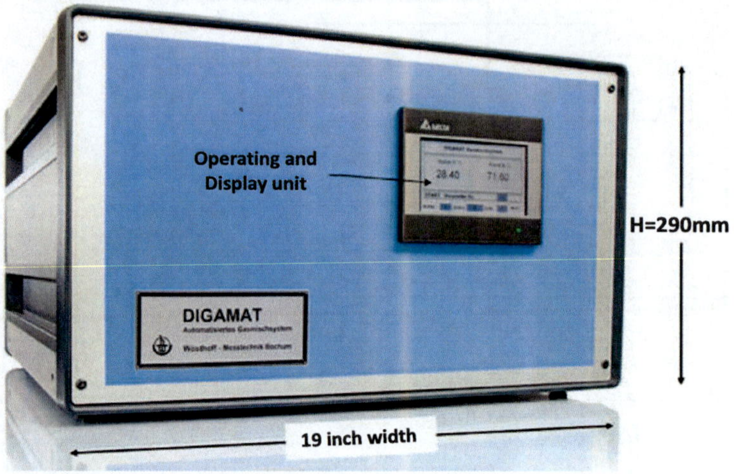

Fig. 12.24 View of the *DIGAMAT* with a menu-driven operating and display unit. The gas connections are located at the rear. (Source: Wösthoff Messtechnik GmbH, Bochum Germany)

of the two pumps via the valves. The volume flow rate can be set via the period T and varies between 12–70 L/h. For higher flow rates, larger piston pumps are available which can deliver 40–220 L/h. However, the setting time depends on the volume flow rate and is delayed at a smaller flow rate.

Figure 12.26 shows the inside view of the DIGMAT. You can see the multi-stage mixing device there to reduce the setting times.

Capillary Dosing Unit

A capillary is a tube with a very small inner diameter (radius $=r$) and an expansion (length $=L$), where $L \gg r$. The flow through a capillary can be described by *Hagen Poiseuille's law,* which has already been derived in Sect. 2.7. Due to the volume flow $\dot{V}$, a pressure drop Δp is created in the capillary that is proportional to $\dot{V}$. The volume flow with the dynamic viscosity is then calculated from the following expression:

$$\dot{V} = \frac{\pi \cdot r^4 \cdot \Delta p}{8 \cdot \eta \cdot L} \tag{12.15}$$

For a given capillary the Eq. 12.15 can be simplified and described by a linear relationship. The constant quantities are combined to a factor K_A, which can be determined experimentally. To do this, the volume flow (gas A) from the capillary is fed into a flowmeter of known accuracy. With the pressure difference Δp_A read off, a calibration curve can then be generated for several pairs of values, the slope of which corresponds exactly to the factor K_A. It is then valid:

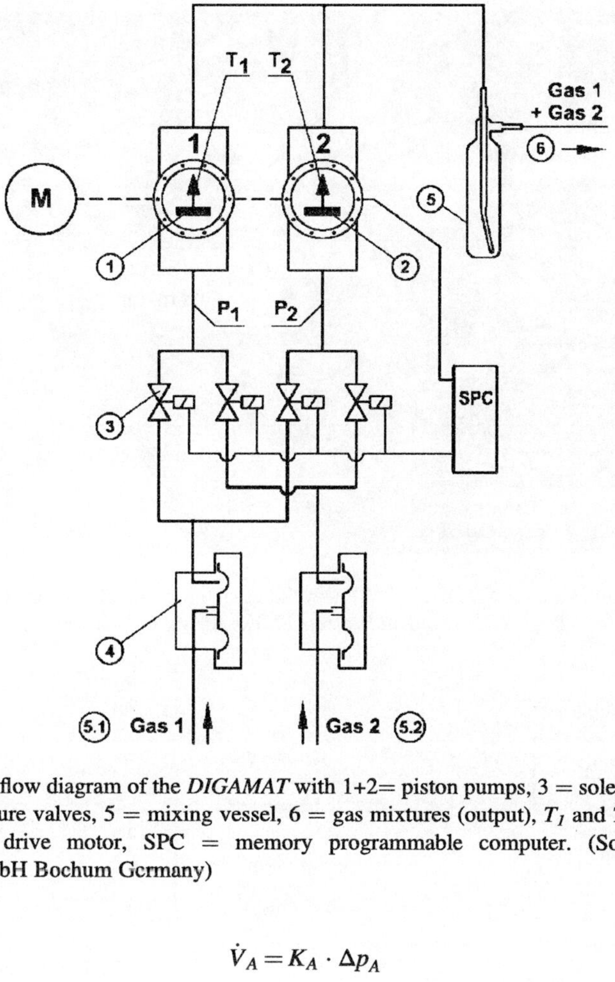

Fig. 12.25 Gas flow diagram of the *DIGAMAT* with 1+2= piston pumps, 3 = solenoid valves, 4 = differential pressure valves, 5 = mixing vessel, 6 = gas mixtures (output), T_1 and T_2 are the period durations, M= drive motor, SPC = memory programmable computer. (Source: Wösthoff Messtechnik GmbH Bochum Germany)

$$\dot{V}_A = K_A \cdot \Delta p_A \tag{12.16}$$

In practice, the capillaries are also thermostatted to reduce the temperature influence of viscosity. A deviation of the control temperature T_R from $\Delta T_R < \pm 1$ K should not be exceeded in order to keep the error of the volume setting $\Delta \dot{V} < \pm 2\%$. Figure 12.27 shows the design of such a gas mixing device, also known as a *gas blender* or *gas divider*.

In order to also obtain a larger dynamic range for generating different gas concentrations, the use of several capillaries is necessary. For this purpose, the device manufacturers have come up with completely different configurations. There are combinations with identical capillaries as well as combinations with different capillaries available. Furthermore, the total number of capillaries is important for the respective application (e.g. Number of support points). In Fig. 12.28 such a combination of 10 identical capillaries is shown. The input of test gas and zero gas into the respective capillaries is

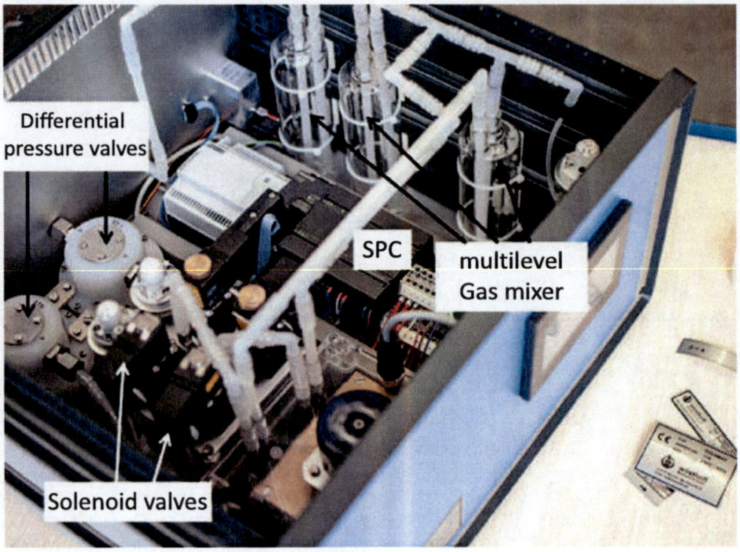

Fig. 12.26 Interior view of the DIGAMAT with the essential components for generating the gas mixtures according to Fig. 12.25. (Source: Wösthoff Messtechnik GmbH Bochum Germany)

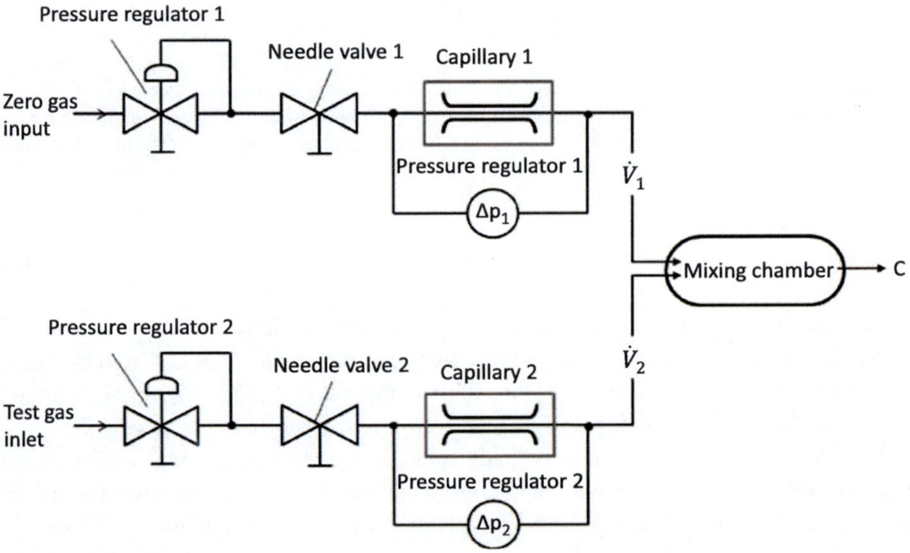

Fig. 12.27 Mixing system with two capillary tubes for the generation of different volume flows $\dot{V}_1$ and $\dot{V}_2$ which lead to a defined gas concentration c

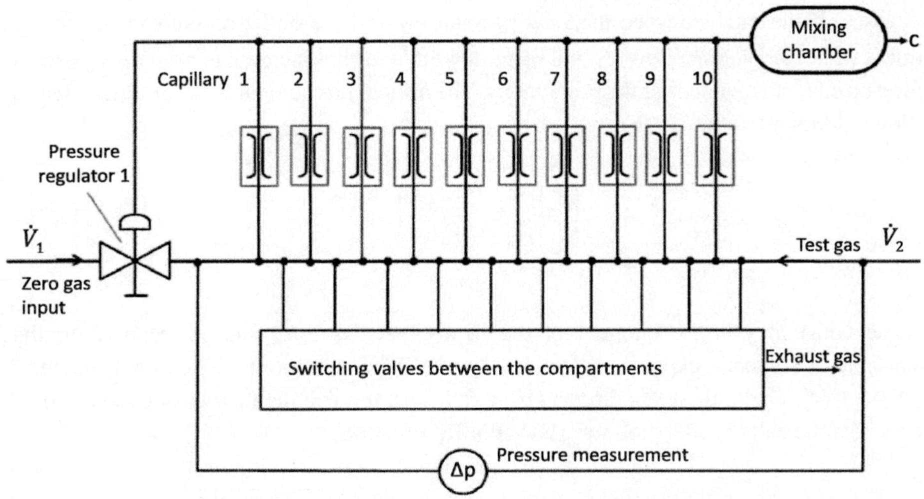

Fig. 12.28 Gas divider with $m = 10$ identical capillaries to generate a maximum of 11 calibration points to accommodate a calibration function (according to DIN EN ISO 6145-5)

done by solenoid valves, which connect a defined number (m) of capillaries in parallel. With this arrangement a calibration curve with 11 calibration points can be realized. If the zero gas flows through all capillaries, the zero point of the analyzer is measured. If the test gas flows through all capillaries, the end point of the analyzer is displayed. The flow rate through all capillaries results from the flow rate of each capillary:

$$\dot{V}_{\text{ges}} = \dot{V} \cdot m \tag{12.17}$$

In the case of a mixture, the volume flow rates result from the number n of capillaries through which gas A flows. It then applies approximately:

$$\dot{V}_A = n \cdot \dot{V} \quad \text{and} \quad \dot{V}_B = (m - n) \cdot \dot{V} \tag{12.18}$$

The concentration c_A of component A is then calculated:

$$c_A = \frac{\dot{V}_A}{\dot{V}_A + \dot{V}_B} = \frac{n \cdot \dot{V}}{n \cdot \dot{V} + (m - n) \cdot \dot{V}} = \frac{n}{m} \tag{12.19}$$

Critical Nozzles

A critical nozzle essentially consists of a constriction with diameter d, through which a gaseous medium flows. If the inlet pressure p_1 increases, the volume flow $\dot{V}_1$ of the gas also increases. If the ratio between the inlet pressure p_1 and the outlet pressure p_2 reaches a critical value, the volume flow $\dot{V}_1$ will increase with a further increase in pressure upstream of the nozzle, independent of the pressure p_2. The critical pressure ratio can be calculated as follows (Marsoner et al. 1976)

$$\left[\frac{p_2}{p_1}\right]_{\text{Crit}} = \left[\frac{2}{\gamma+1}\right]^{\frac{\gamma}{\gamma-1}} \tag{12.20}$$

The constant γ is the quotient of the heat capacities c_p/c_v and corresponds to the isentropic or adiabatic exponent κ for ideal gases. Depending on the type of gas, the ratio is 0.4 ... 0.55. Under these conditions ($p_2 < 0.5 \cdot p_1$), the volume flow through the critical nozzle can be calculated as follows (DIN EN ISO 6145-6)

$$\dot{V}_{1(0)} = \frac{p_1 \cdot 273.15\text{K}}{1013.25\text{hPa}} \cdot A \cdot \sqrt{\frac{R}{T_1 \cdot M}} \cdot \sqrt{\gamma \cdot \left[\frac{2}{\gamma+1}\right]^{\frac{\gamma+1}{\gamma-1}}} \tag{12.21}$$

With the temperature T_1 of the gas in front of the nozzle, the general gas constant $R = 8.31451\text{J/mol} \cdot \text{K}$, the molar mass M of the gas, in grams per mole and the cross-sectional area A of the nozzle. This equation can be simplified for constant temperatures (T_1). One then obtains:

$$\dot{V}_{1(0)} = p_1 \cdot D \cdot d^2 \cdot \sqrt{1/M} \tag{12.22}$$

Figure 12.29 shows the design of a gas mixing device with two critical nozzles. The pressure regulators (1 + 2) ensure that p_1 and p_3 are constant. Behind the two critical nozzles (1 + 2) one thus receives constant volume flows, which unite in the mixing chamber and a constant gas concentration c results. The concentration c then results from:

$$c = \frac{\dot{V}_2}{\dot{V}_1 + \dot{V}_2} \tag{12.23}$$

With the device shown in Fig. 12.29 only one gas concentration can be set at a time. Commercial devices usually use several nozzles, which are interconnected by a corresponding valve control (Schmidt 2014). On the inlet side, both gases are fed to the device at a pressure of 3 bar (±1 mbar). The gas flow can be varied by combination, so that the number Z of possible gas concentrations increases with the number N of nozzles. With 4 nozzles 16 gas concentrations can therefore be set. The nozzles are usually located in a metal block so that the temperatures of the nozzles are identical. In Fig. 12.30 such an

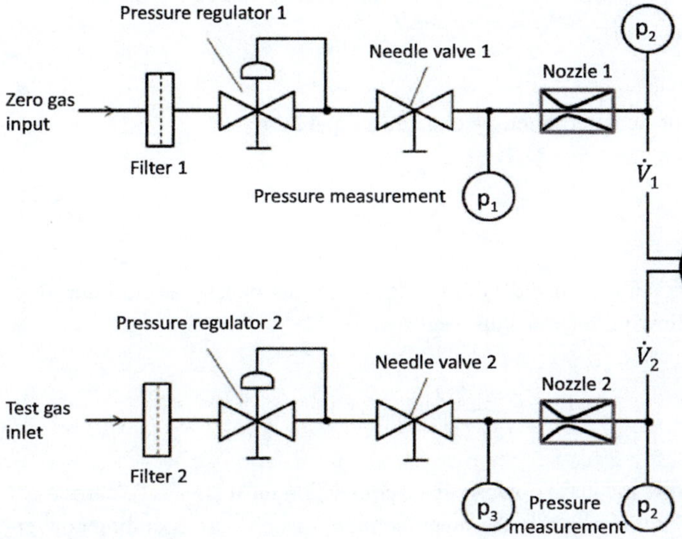

Fig. 12.29 Process for producing a binary gas mixture with critical nozzles. (According to DIN EN ISO 6145-6)

Fig. 12.30 Gas mixing device with 4 critical nozzles and corresponding external valve switching. (According to Schmidt 2014)

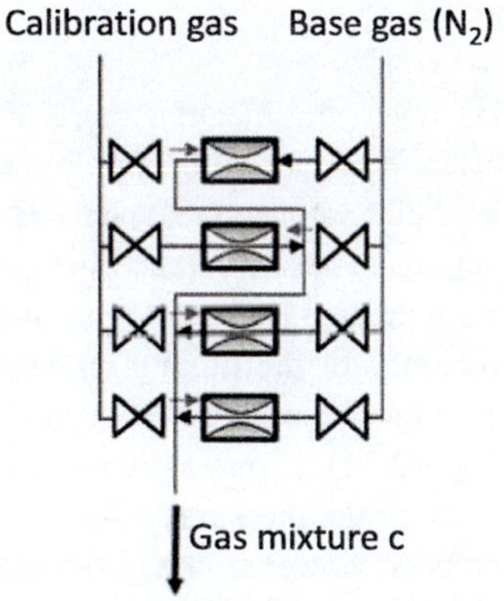

arrangement is shown. The degree of dilution can be done in steps from 1 : 1000 to 1 : 1, with an accuracy of ±0.2%.

Thermal Mass Flow Controllers (MFC)

A widely used method for the dynamic production of calibration gas mixtures uses thermal mass *flow controllers* (*MFC*) to generate the required volume flows $\dot{V}_1$ and $\dot{V}_2$. The functional principle and design of this *MFC* has already been described in Sect. 11.2. The heat flow that is transferred to the gas by the flow results from

$$\dot{Q} = \dot{m} \cdot c_p \cdot \Delta T \tag{12.24}$$

The mass flow can thus be determined by changing Eq. 12.24:

$$\dot{m} = \frac{\dot{Q}}{c_p \cdot \Delta T} \tag{12.25}$$

In order to be able to conclude on the volume flow, the gas density is introduced in Eq. 12.25, so that the following relationship results:

$$\dot{V} = \frac{\dot{Q}}{c_p \cdot \varrho \cdot \Delta T} \tag{12.26}$$

In principle, the MFC shows the mass flow, so that a conversion must always be carried out (Slemeyer et al. 1992). For this purpose, the manufacturers specify corresponding conversion factors C, with which the different gases can be adapted. This factor can also be calculated by the ratio of gas density and heat capacity:

$$C = \frac{c_{p0} \cdot \varrho_0}{c_{p1} \cdot \varrho_1} \tag{12.27}$$

However, the manufacturers of calibration equipment using this method determine the actual flow rate using appropriately accurate flow methods. Drum gas meters, gas measuring bell and piston test benches with mercury or graphite seals have proven to be suitable options. These reference methods are described in detail in DIN EN ISO 6145-1. The manufacturers of these gas mixing devices also use these possibilities to improve the accuracy of the mixing ratios. With standard MFC, the square addition of all possible sources of error leads to a mixing error that increases sharply at low volume flows (see Fig. 12.31). If one assumes that the calibration gas mixture should have an accuracy of <2% of the measured value, in the *worst case* only a range of 20–100% can be used. In practice, however, the errors are much smaller, so that with standard MFC an interval of 10–100% can be specified as a usable range. Some manufacturers select the MFC or check it with reference methods according to DIN EN ISO 6145-1. With this measure a dynamic range of 5–100% is possible (Schmäh 2015).

Figure 12.32 shows a gas mixing device with two MFCs with which a binary gas mixture can be produced. The concentration of the calibration gas mixture results from the two adjustable partial flows:

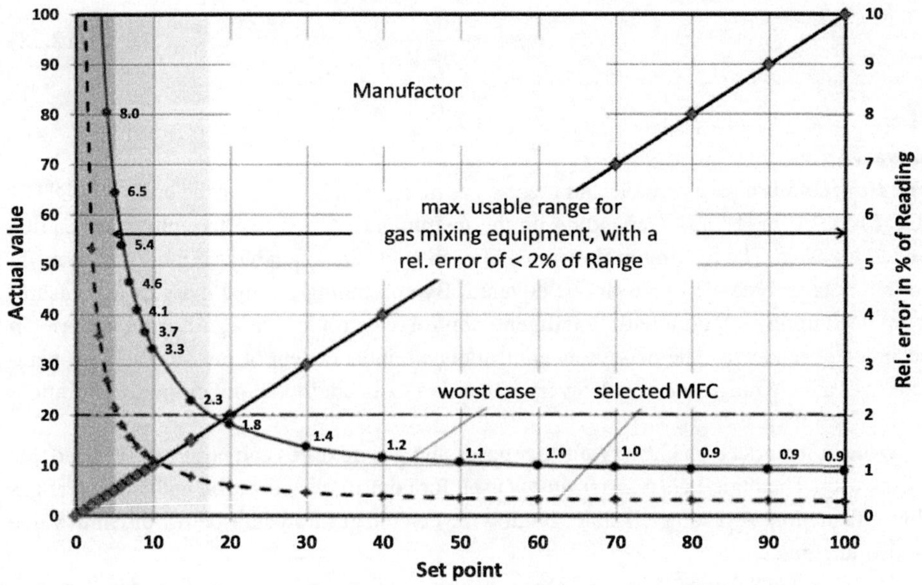

Fig. 12.31 Mixing device for producing binary gas mixtures with thermal mass flow controllers. (According to DIN EN ISO 6145-7)

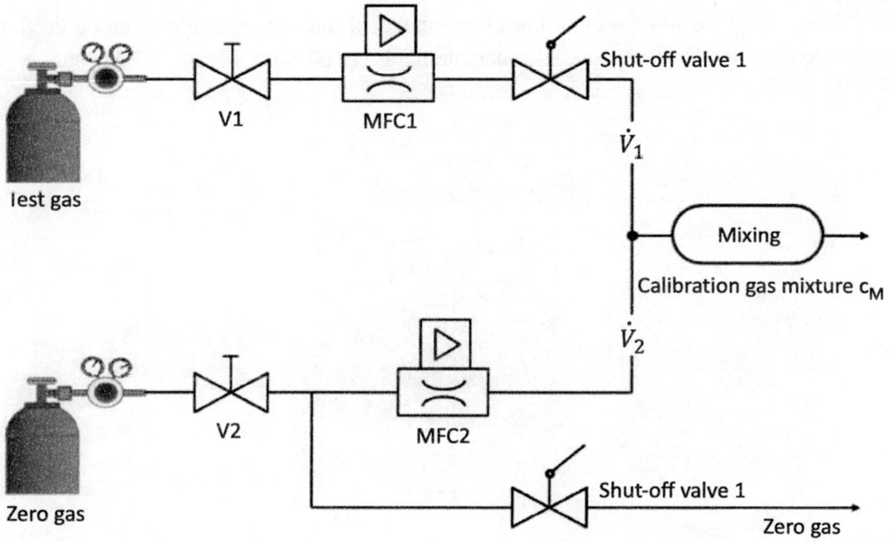

Fig. 12.32 The measurement errors of the individual MFCs refer to both the zero point and the end point or sensitivity. As a result, the flow rates $\dot{V}_n$ in the lower range (<20% of the setting range) show a strongly increasing error

$$c_M = \frac{\dot{V}_1}{\dot{V}_1 + \dot{V}_2} \quad (12.28)$$

HovaGAS

The *HovaGAS* is a very variable and precise gas mixer which works on the basis of DIN EN IOS 6145-7 (Fig. 12.33). Depending on the instrument version two to eight gases can be mixed together. The maximum flow rates of each MFC are available from ml/min to m^3/h, so that a wide dynamic range can be covered. By combining thermal mass flow measurement technology (MFC) with intelligent control electronics, a significant increase in accuracy is achieved. The mass flow controllers are independent of pressure and temperature over a wide range and are also very linear due to an additional polynomial calibration.

With an integrated gas library, further gases and gas mixtures can be stored, expandable by the user. The digital *Gasmaster unit* is used for control, visualization and automatic gas flow calculation. A parameterisable shutdown when the gas flow falls below the limit value is also integrated.

The *HovaGAS* device (Fig. 12.33) can be operated manually, via a user-friendly touch screen, or remotely with a PC or notebook via the comfortable *Lab-View* based software *ViewGAS*. Figure 12.34 shows the visual part of the user interface.

Permeation Tube

The permeation of gaseous or even liquid substances through a membrane can be used to generate calibration gas mixtures under defined conditions. Figure 12.35 shows a

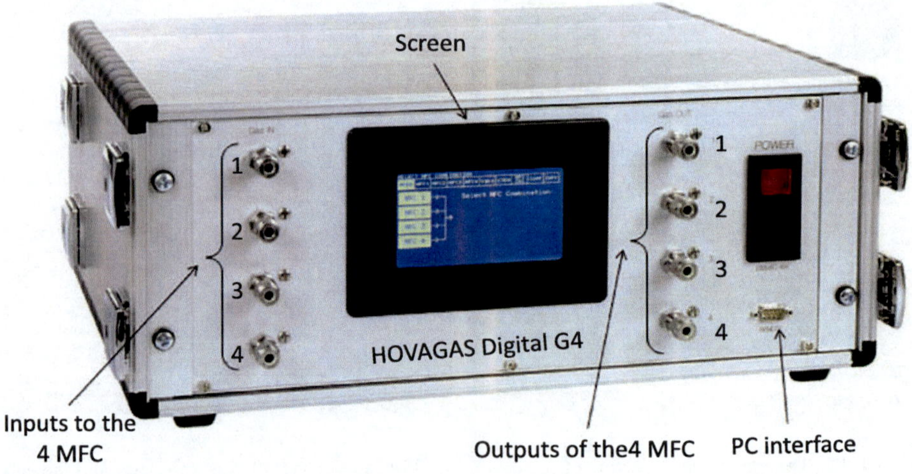

Fig. 12.33 View of a *HovaGAS device* with 4 MFC. (Source: IAS-GmbH, Oberursel Germany)

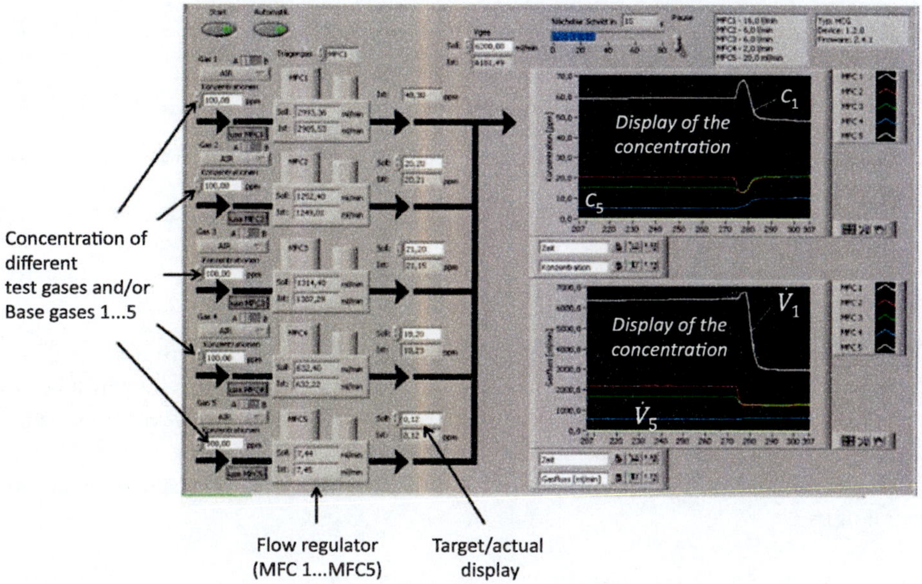

Fig. 12.34 User interface of the *ViewGas* software with setting options for the different gas flows, concentrations and gas types. (Source: IAS-GmbH, Oberursel Germany)

Fig. 12.35 Cross-section of a permeation tube used for the calibration of gas analysers in the trace range

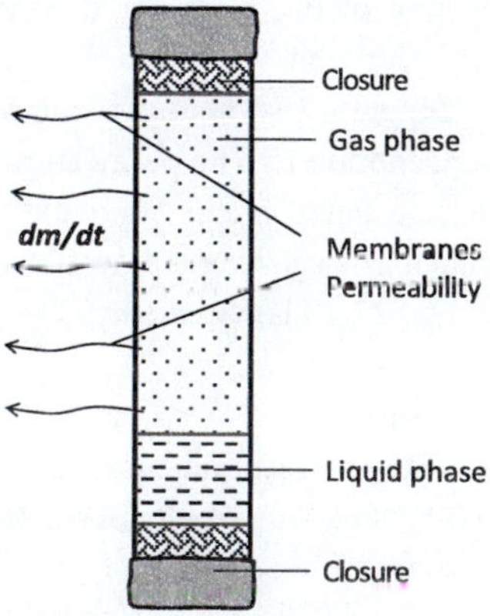

permeation tube which is used for this purpose. The calibration component is located in the tube and penetrates through a suitable membrane (e.g. Teflon or polyethylene) into a carrier gas stream $\dot{V}_V$. The permeation rate $\dot{m}_P$ of the calibration component and the carrier gas flow thus determine the mass concentration β in the gas mixture:

$$\beta = \dot{m}_P / \dot{V}_V \tag{12.29}$$

In this case, the conversion to a volume concentration c is performed with the reciprocal density $K = 24.46/M$ (VICI 2012):

$$c = \frac{\dot{m}_P \cdot K}{\dot{V}_V} \tag{12.30}$$

The permeation rate $\dot{m}_P$ is specified by the manufacturer of the tubes (→certificate) and has the unit ng/min. The permeation rate can also be measured directly by mounting the tube on a microbalance and weighing either continuously or periodically. However, this value changes very strongly with temperature, so that the measurement must be carried out in a thermostatted setup. The temperature dependence of the permeation rate can be calculated as follows:

$$\log \dot{m}_{P0} = \log \dot{m}_{P1} + \alpha[T_1 - T_0] \tag{12.31}$$

The index "*0*" refers to values at temperature $T = 0^{\circ}\mathrm{C}$ and the index "*1*" to the current temperature. The temperature coefficient α is between 0.030 and 0.034. A temperature change of 0.1 K in the thermostatted structure already leads to an error of 1% in the concentration.

Figure 12.36 shows a complete structure suitable for the production of very low gas concentrations. The permeation tube usually has a length L of 1–10 cm and is located in a thermostatted oven. With the temperature the permeation rate is brought to a constant value. The carrier gas flow absorbs the released gas quantity and a concentration according to Eq. 12.30 is obtained.

Example

An SO_2 permeation tube has a permeation rate of 1250 ng/min. The carrier gas flow is 0.6 L/min. This results in the following concentration $K = 0.382$:

$$c = \frac{1250 \cdot 0.382}{\frac{600\mathrm{cm}^3}{\mathrm{min}}} = 0.796\mathrm{ppm} = 798\mathrm{ppb}$$

If this concentration is now mixed with a further dilution stage, the value is reduced even further. A dilution of 10 : 1 then leads to a concentration of ≈80 ppb.

This example shows that the production of calibration gas mixtures using the permeation method is particularly suitable for very low concentrations. The contents of the tubes

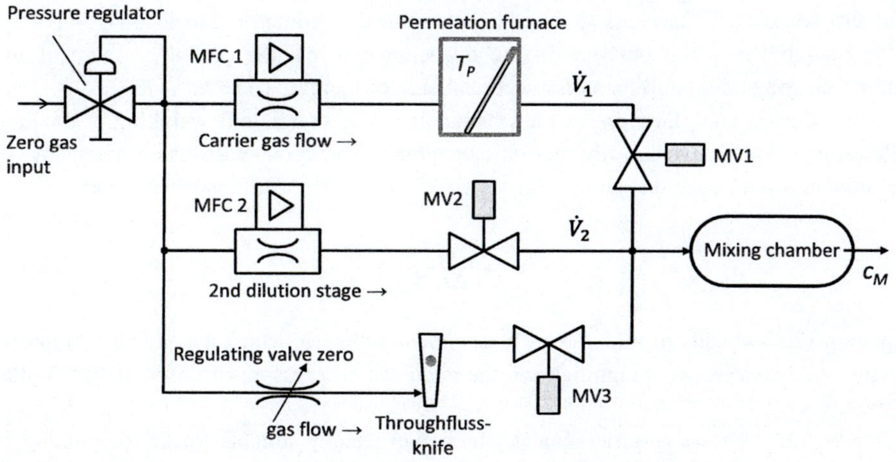

Fig. 12.36 Arrangement for generating gas concentrations in the ppm to ppt range

will be used up over time, so that either a new tube must be inserted or the old tube must be refilled. Both options are available on the market for over 250 components from different manufacturers.

Diffusion Process

This method uses the diffusion behaviour (see Sect. 2.4) of gases to produce a defined calibration gas concentration. Especially organic compounds, which are present in liquid form at room temperature, can be produced with this method as mole fractions of 10^{-9} to 10^{-3}. The vaporizable liquid is located in a reservoir which is connected to a mixing chamber via a diffusion tube of suitable length L and diameter d. The emerging diffusion mass flow $\dot{m}_D$ mixes with the base gas $\dot{V}_0$ and thus produces a defined calibration gas concentration. The diffusion mass flow essentially depends on the

- Diffusion coefficient D into the base gas
- the vapour pressure p_v at the vessel temperature T_B and
- the dimensions (d, L) of the diffusion tube.

The calculation of the diffusion volume flow $\dot{V}_D$ can therefore basically be calculated using the following equation (Nelson 1992)

$$\dot{V}_D = \frac{A}{L} \cdot D \cdot \ln\left[\frac{p}{p - p_V}\right] \tag{12.32}$$

Furthermore, Eq. 12.32 can also be used to determine the geometric dimension (A/L) of the diffusion tube. p is the pressure in the diffusion cell, pV the partial pressure of the calibration component and the cross-sectional area of the tube ($A = \pi d^2/4$).

The diffusion mass flow $\dot{m}_D$ can also be achieved by continuous weighing of the tube with a micro float balance, or by periodic weighing. The average diffusion mass flow of component A out of the diffusion cell over time is given in Δt grams per minute:

$$\dot{m}_D = \frac{\Delta m}{\Delta t} \cdot \overline{p}/p_0 \tag{12.33}$$

Equation 12.33 already provided for a pressure correction, in which a mean air pressure in the interval between two weighings and the reference air pressure (usually 1013,325 hPa) is used.

Figure 12.37 shows an arrangement which is especially suitable for the production of calibration gas mixtures by the diffusion method (DIN 6145-8). In the diffusion cell there is a sufficient amount of liquid which leads to a diffusion mass flow $\dot{m}_D$ at a defined

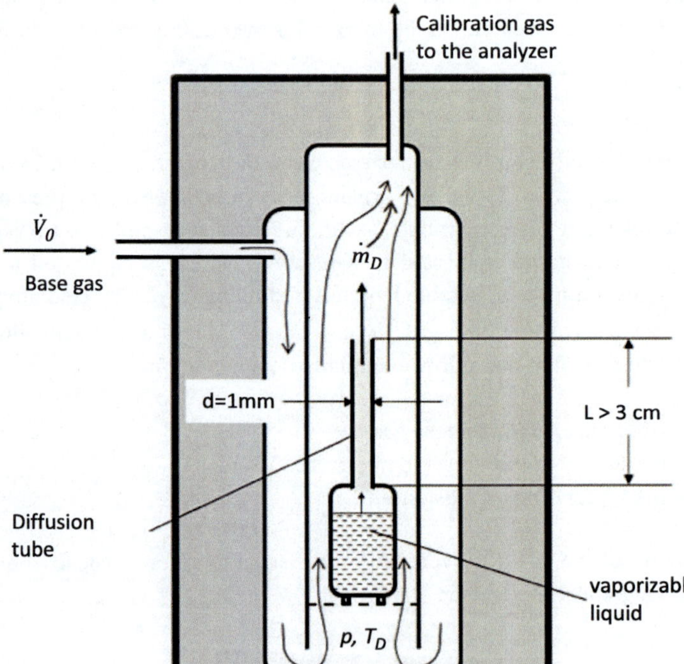

Fig. 12.37 Device for generating a calibration gas mixture using the diffusion method according to DIN EN ISO 6145-8

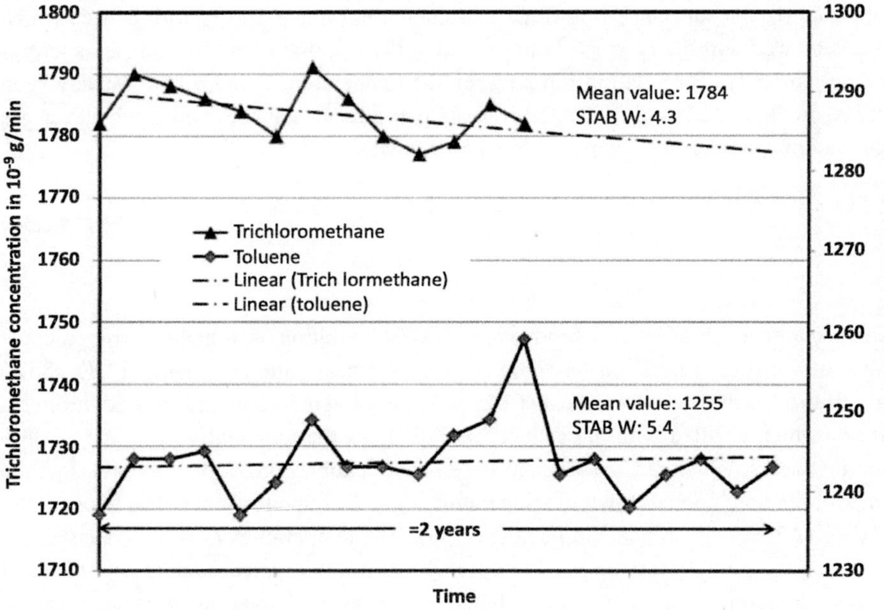

Fig. 12.38 Stability behaviour of diffusion cells over a longer period (1–2 years) with different substances. (Data source: DIN EN ISO 6145-8)

temperature ($T_D \pm 0.15$ K). The base gas, with the volume flow $\dot{V}_0$, is introduced laterally into the structure and thus enters the outer chamber of the mixing device. The base gas can adapt to the temperature T_D on its way to the diffusion cell. The volume flow is then redirected and now flows parallel to the diffusion mass flow to the outlet. The mixing process takes place on the last section. If necessary, this mixture can be additionally homogenized in a further vessel. Thus, the volume flow does not influence the diffusion flow which is susceptible to faults. This is an essential requirement for a stable and accurate calibration gas mixture at the outlet of the device.

The stability of these diffusion cells is very good, since the physical boundary conditions do not change during the operating time. Only when the substance has completely evaporated, failure occurs. Diffusion cells can therefore be used over many years for the stable calibration of certain gas measuring instruments. Figure 12.38 shows a long-term measurement (1–2 years) for toluene and trichloromethane.

Saturation Method

The vapour pressure p_D of a pure substance depends exclusively on the equilibrium temperature T of the substance. If a complementary gas (e.g. Nitrogen) is passed through a vaporizable liquid, the gas is completely saturated with the vaporizable substance after a sufficiently long time t. An essential requirement is that the contact surface between the gas and the vaporizable substance is as large as possible. For this purpose, a frit is used which produces a large number of small bubbles which then saturate with the substance correspondingly quickly. The relationship between the temperature T and p_D has already been described in Sects. 2.1 and 10.3. The vapour pressure p_D can be converted into a gas concentration c at a barometric pressure p_B as follows:

$$c = \frac{p_D}{p_B} \tag{12.34}$$

For accurate calibration gas production, 100% saturation is required. For practical reasons, a two-stage saturation has proven to be the best solution. Figure 12.39 shows such a setup according to DIN EN ISO 6145-9. The complementary gas is passed through a *saturator* which is filled with the evaporable liquid. At the temperature T_1, $\approx 95 - 98\%$ saturation occurs, which is not sufficient for an exact calibration. In the second stage, the gas mixture is now cooled down to temperature T_2 *in a* defined manner. The temperature difference ($\Delta T = T_1 - T_2$) should be at least 5 K. The temperature T_2 is then decisive for determining the partial pressure p_D. Figure 12.40 shows the typical temperature curve of the vapour pressure p_D of an evaporable liquid. The data for the various substances can be taken from relevant literature references (e.g. D'Ans and Lax 1992; Lide 2007; Landolt and Börnstein 1960).

The use of a circulation pump is necessary if a recirculation system for the calibration gas is used. This system then works continuously and eliminates any delays in the state of equilibrium. After the gas has passed through the flow path several times to set the equilibrium state, the calibration gas mixture can be taken from the sampling point and fed to the analyser.

HovaCAL[14]-Process

In this process, the liquid to be evaporated is pumped out of a container using a pump. The mass flow rate $\dot{m}_L$ is measured with a highly accurate thermal flow meter (MFM). Alternatively, the weighing method described in Sect. 10.3 can be used. According to the manufacturer (IAS-GmbH), the accuracies to be achieved are almost identical. With two mass flow controllers (MFC 1 and MFC 2) the base gas and, if necessary, additional test gases can be added (see Fig. 12.41). Up to 5 MFCs can be integrated in the maximum expansion stage. The evaporation of the liquid in the downstream evaporator then leads to a corresponding gas concentration at the outlet of the evaporator. The maximum evaporator temperature is $T_{VD} = 180°C$. If an undiluted liquid (100%) is completely evaporated, the following gas concentration c results at the outlet of the evaporator, with V_m molar volume and $M =$ molar mass

[14]Hot Vapor Calibration.

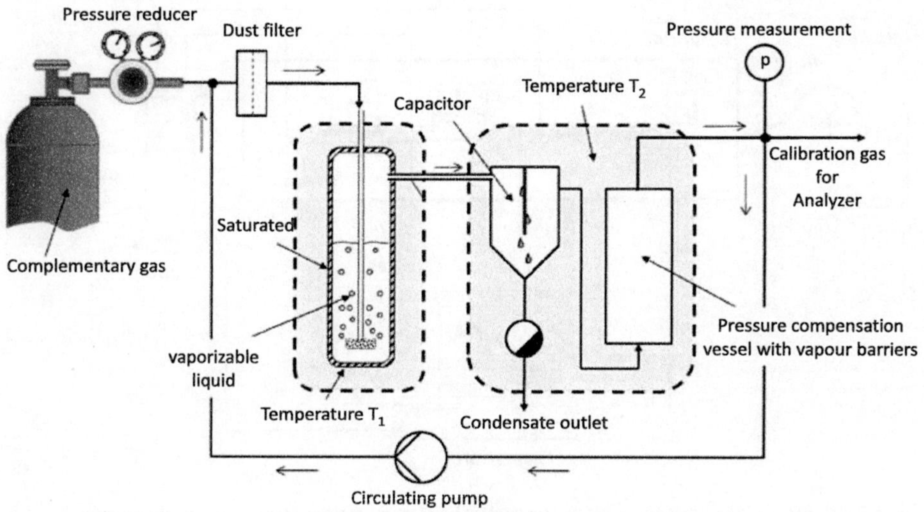

Fig. 12.39 Construction of a device for the production of calibration gas mixtures according to the saturation method. (Source: DIN EN ISO 6145-9)

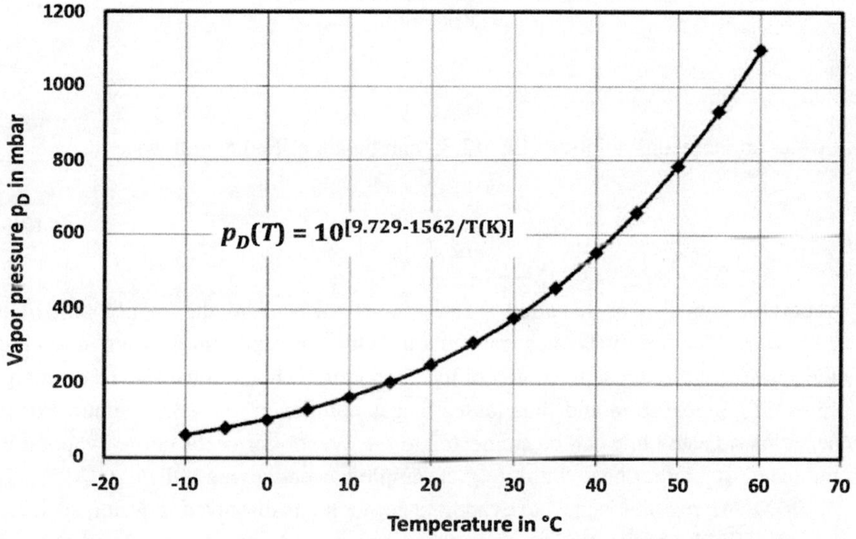

Fig. 12.40 Example of a vapor pressure curve according to which the respective gases can be calibrated

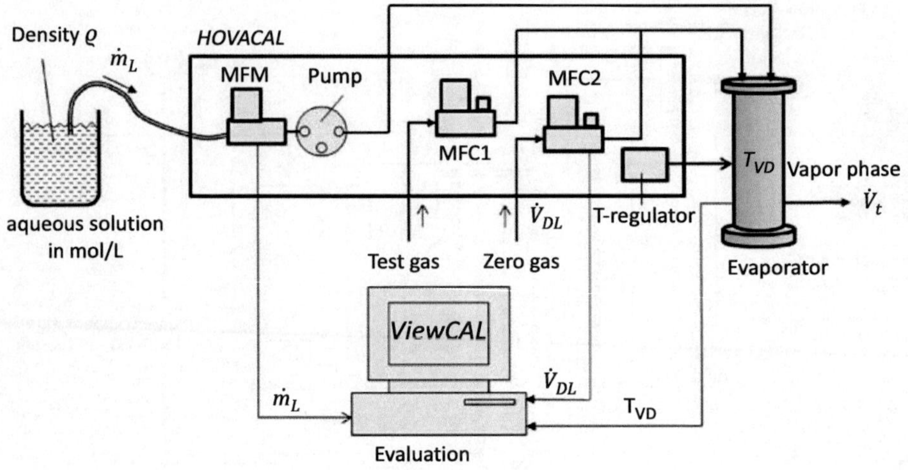

Fig. 12.41 *HovaCAL* arrangement for the controlled evaporation of an *n-normal* solution for the calibration of special gas types (NH_3, HCl, . . .)

$$c = \frac{\dot{m}_L}{\dot{V}_t} = \frac{\dot{m}_L}{\dot{V}_{\mathrm{DL}} + \dot{m}_L \cdot V_m / M} \tag{12.35}$$

For water and aqueous solutions Eq. 12.35 can be simplified as follows:

$$c_{\mathrm{H_2O}} = \frac{\dot{m}_L}{\dot{V}_{\mathrm{DL}} + \dot{m}_L \cdot 1.244} \tag{12.36}$$

The evaporator can also be installed a few metres away from the supply unit of the *HovaCAL unit.* The supply is then carried out with correspondingly long hose lines. Since the gas-steam mixture at the outlet of the evaporator is hot, it must also be discharged hot to prevent condensation and thus losses of gas component or water vapour. For this purpose, either a heated line can be connected to the evaporator or the entire evaporator is accommodated in a larger heated unit (e.g. a sample gas conveying unit).

If the *HovaCAL process* is used to evaporate components dissolved in water, such as B. HCl, NH_3 or HF, the concentration c_K of these components can be calculated using the following equation:

$$c_K = \frac{22.414\text{L/mol} \cdot a_K \cdot (\dot{m}_L/\rho_L)}{\dot{V}_t} \cdot 10^6\text{ppm} \qquad (12.37)$$

Where a_K is the concentration of the component in aqueous solution in mol/L, $\dot{m}_L$ the flow rate of liquid (aqueous solution) in g/min, ρ_L the density of the liquid in g/L and $\dot{V}_t$ the total gas flow in L/min. The total gas flow $\dot{V}_t$ is the sum of Eq. 10.30 and $\dot{V}_{DL}$. Usually the concentration c_K is related to standard conditions, so that the temperature correction in Eq. 10.30 is not necessary.

Example
With a 0.001 molar HCl solution and a liquid transport of 1 g/min, the following HCl concentration is obtained at the outlet of the evaporator with a base gas flow of 5 L/min.

$$c_K = \frac{22.414\frac{\text{L}}{\text{mol}} \cdot 0.001\frac{\text{mol}}{\text{L}} \cdot \frac{1\text{g/min}}{1\text{g/mL}}}{5000\frac{\text{mL}}{\text{min}} + \frac{22.414\text{mL/mol}}{18.015\text{g/mol}} \cdot 1\frac{\text{g}}{\text{min}}} \cdot 10^6\text{ppm} = 3.589\text{ppm}$$

The concentration in mg/m^3 is obtained with Eq. 2.104:

$$\frac{3.589\text{mL}}{\text{m}^3} \cdot \frac{36.46\text{g/mol}}{22.414\text{L/mol}} = 5.839\frac{\text{mg}}{\text{m}^3}$$

Calibration of Mercury Analyzers (Schmäh 2009)
The measurement of mercury vapours is mainly carried out in waste incineration plants and the chemical industry. In this application both elemental mercury (Hg^0) and ionic mercury ($HgCl_2$) are to be considered. They behave very differently with regard to transport in the gas phase. In contrast to Hg^0, $HgCl_2$ also adsorbs on surfaces at higher temperatures, so that lower results can result. Since only small quantities of substances (≈µg/h^3) are transported, adsorption effects on filter or pipe surfaces have a dramatic effect. Lower results in the two-digit percentage range are not uncommon.

Various methods are available for the generation of test gases containing Hg^0 and $HgCl^2$. In order to be able to test the entire system, however, test gas containing $HgCl_2$ must be added, otherwise the reduction stage of the measuring system cannot be checked. Due to the different adsorption behaviour of elemental and ionic mercury, test gas containing $HgCl_2$ is also required for testing the sampling system. It has been shown that the transport of $HgCl_2$ in the gas phase is strongly favoured by water vapour. Dry $HgCl_2$ test gas cannot be used even at temperatures from 180 to 200 °C, as $HgCl_2$ is adsorbed by all surfaces in contact with the gas and disappears from the gas phase. However, the presence of water shifts the adsorption-desorption equilibrium from the surface to the gas phase, as water and $HgCl_2$ molecules compete for the available surface space. Due to the high concentration ($\approx 1 - 10\%$ by volume) water adsorption is predominant compared to $HgCl_2$concentration (ppb). Therefore the $HgCl_2$test gas must contain water vapour, especially since this composition also corresponds to the real gas matrix (e.g. Flue gas).

The generation of water vapour containing $HgCl_2$ test gas is carried out by dosing and evaporation of aqueous solutions containing $HgCl_2$ under addition of a carrier gas with the *HovaCAL device*. The concentration of both mercury and water vapour in the gas phase can be defined from the material flows and the mercury content of the reagent solution by precise measurement and control of the liquid and gas flow. The mercury concentration c_{Hg} results from the mercury chloride solution a_{Hg} in mol/L, the dilution gas flow $\dot{V}_{DL}$ in mL/min and the liquid flow $\dot{m}_L$ in g/min:

$$c_{Hg} = \frac{22.414 \cdot a_{Hg} \cdot \dot{m}_L}{1244 \cdot \dot{m}_L + \dot{V}_{DL}} \cdot 8.954 \cdot 10^9 \mu g/m^3 \tag{12.38}$$

Comparative measurements show that the mercury concentrations produced show relative deviations of less than 1% in a range from 0 to 80 μg/m³. Figure 12.42 shows the result of a TÜV test. The relationship between the target value and the actual value is absolutely linear. The maximum deviation for this measurement was −0.98% from Span (TÜV 2007).

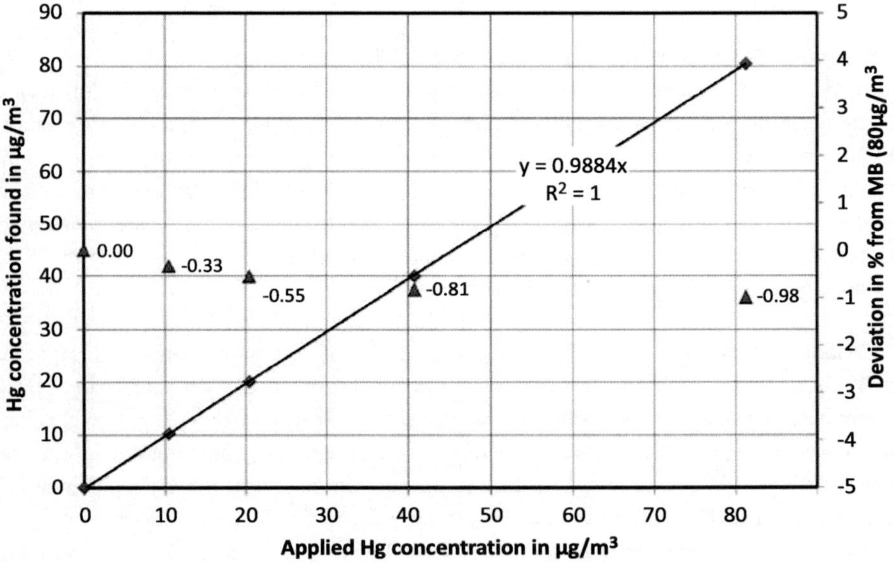

Fig. 12.42 Linear regression of the difference between the nominal value and the actual value for an Hg calibration function. (Data source: TÜV-Rheinland 2007)

Syringe Injection Method

This method is used for the continuous production of calibration gas mixtures, with two or more components from pure gases or other gas mixtures, by continuous injection of the calibration component into a base gas stream by means of a syringe. It is also possible to inject a vaporizable liquid into the base gas stream. The volume flows, from which the volume fractions are determined, can be calculated from the individual flow rates. With this method it is also possible to increase the volume fraction of the calibration component in the mixture in small steps. It is therefore a useful method for determining other characteristics of gas analysers, such as the lower detection limit and accuracy. The uncertainty of the volume fraction that can be achieved with a binary mixture is <5% and the range of application is 10^{-5} to 10^{-2}. Figure 12.43 shows the basic structure according to DIN EN ISO 6145-4.

Calibration component A, which is either in the gas phase or in the liquid phase, is delivered into a base gas stream from a syringe whose plunger is continuously driven by a speed-controlled motor. The concentration c_A of calibration component A in a mixture with the base gas B is given by

$$c_A = \frac{\dot{V}_A}{\dot{V}_A + \dot{V}_B} \tag{12.39}$$

If calibration component A is injected as a liquid, a corresponding conversion to the gas phase must be carried out. Since the density is calculated from $\varrho = M/V_m$, the volume flow of calibration component A can be calculated as follows:

$$\dot{V}_A = \frac{\dot{V}_{A(\mathrm{Fl})} \cdot \varrho_{A(\mathrm{Fl})}}{\varrho_{A(\mathrm{Gas})}} \tag{12.40}$$

With $\dot{V}_{A(\mathrm{Fl})}$ = volume flow of infected liquid A, $\varrho_{A(\mathrm{Fl})}$= density of injected liquid A at the mixing temperature, $\varrho_{A(\mathrm{Gas})}$= density of injected component A in the gas phase.

Dynamic Calibration Unit SYCOS[15]

For high calibration gas concentrations, there is a risk that the vaporized component will condense in the subsequent line. This danger already exists when the dew point temperature is $T_{\mathrm{TP}} \approx T_U$. Furthermore, at very high concentrations, which require a dew point of $T_{\mathrm{TP}} \gg T_U$, the injection block must be heated to guarantee complete evaporation. A heated line must therefore be connected to the outlet of the evaporator, which leads to the test

[15]Gasmet data sheet SYCOS K-DPG 5F 1107 D Rev. 2013.

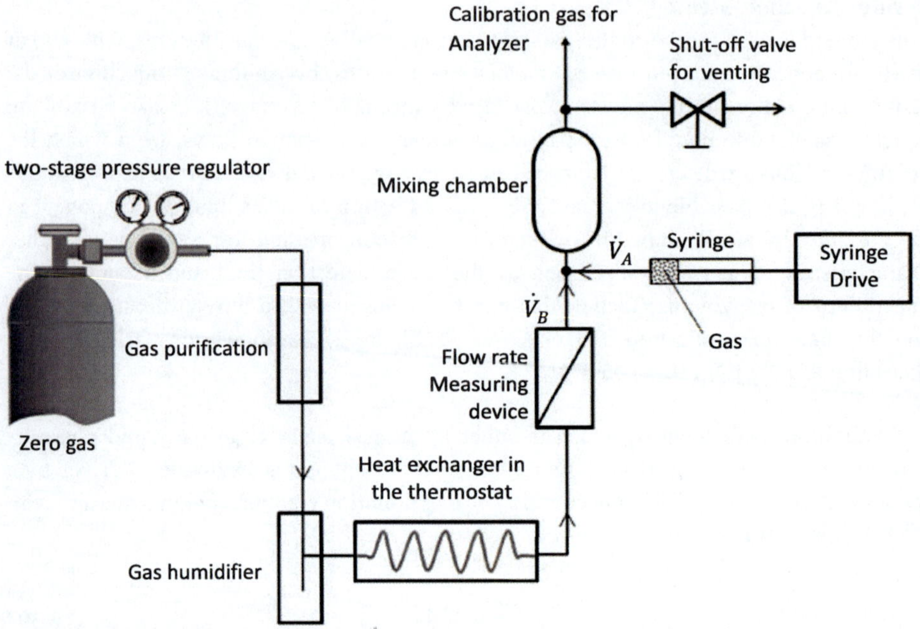

Fig. 12.43 Basic design of a device for continuous dosing of a calibration component A into a carrier gas flow B to generate a calibration gas mixture according to Eq. 12.32. (Source: According to DIN EN ISO 6145-4)

device (analyzer) (see Fig. 12.44). Figure 12.45 shows the construction of the SYCOS K-DPG/M13 (Table 12.4).[16]

This method enables the production of calibration gases in a wide concentration range, since both the zero gas flow and the mass flow of the calibration component can be varied. The process is essentially limited by the vapour pressure of the calibration component. For the production of test gases in the ppm range, dosing is carried out via the needle of the syringe through a septum. The septum in the septum holder is easily exchangeable.

A further field of application is the production of defined water concentrations greater than 40% by volume or the humidification of test gases which are present in the ppm range in pressure bottles. For water calibration or the preparation of test gases in the volume percent range, an injection nozzle connected to the syringe (*Luer Lock*) by a line1/16″ is used instead of an injection needle.

[16]Gasmet Technologies GmbH, Karlsruhe.

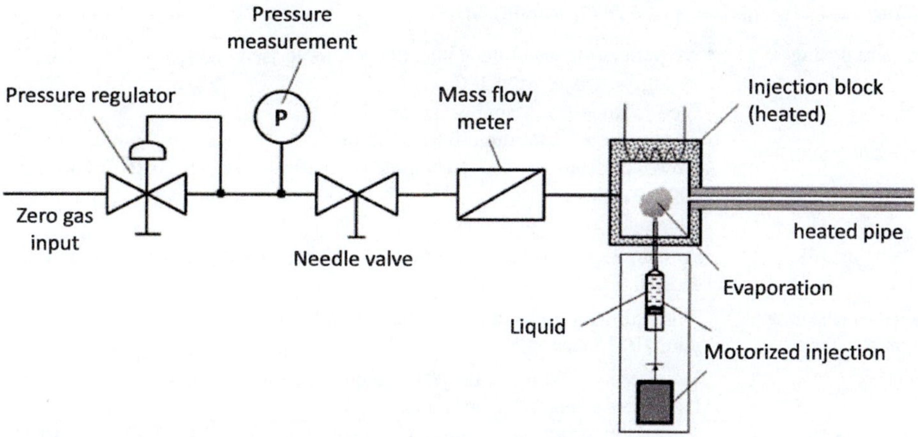

Fig. 12.44 Use of the method for evaporable liquids whose dew point after mixing is below the ambient temperature

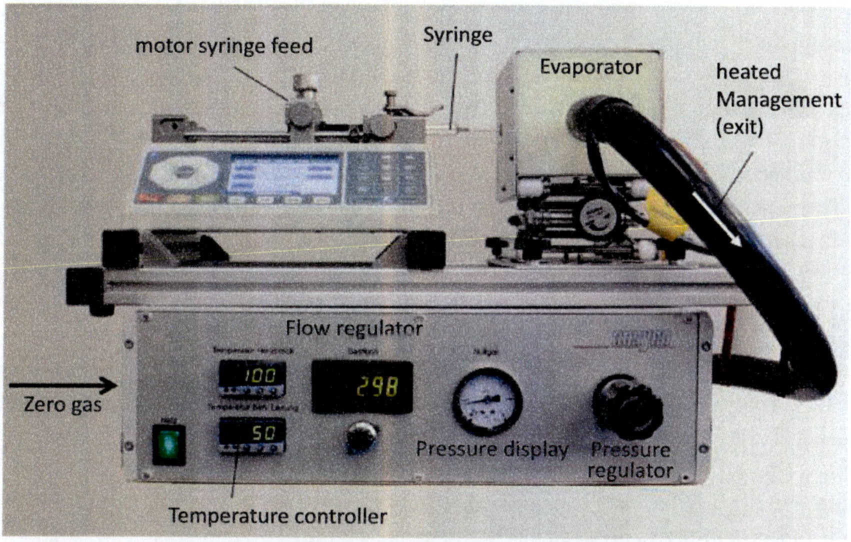

Fig. 12.45 Dynamic calibration unit *SYCOS K-DPG/M13* for the production of test gases from volatile solvents and gases. (SourceGasmet Technologies GmbH, Karlsruhe Germany)

Calibration Gas Generator HovaCAL-Digital SP

This process is also based on the injection of liquids and evaporation into a defined base gas flow. The *HovaCAL® digital SP* generates dry and humidified gases and gas mixtures as well as test gases from organic and inorganic components. The gases are dosed into the evaporator by mass flow controllers (MFC), the liquids (water or organic solvents) are

Table 12.4 Specified data of SYCOS K-DPG/M13

Liquid dosing	Programmable continuous injection via motorised syringe Syringe sizes: 10 µl to 100 ml Feed mechanism: Stepping motor Stepper motor resolution: 0.012 micron Flow rate: from 10 Picoliter/minute for a 10 µl syringe to 0.25 l/min for a 100 ml syringe Nominal thrust $F = 300$ N
Zero gas inlet	Pressure regulator max. Inlet pressure 35 bar, outlet pressure adjustable to approx. 2 bar
Flow measurement/ control	Thermal mass flow controller (MFC), calibrated to air (nitrogen), 0.5–5 l/ min, 1013 mbar, 0 °C
Injector	Stainless steel with exchangeable septa holder or injection nozzle, temperature controlled up to max. 180 °C
Heated pipe	2 m length, PTFE core 6 mm AD; Pt 100 element, regulated to max. 180 °C
Structure	Syringe pump and injection block on base plate control unit in 19″ 3 HE housing

Source: data sheet of Gasmet GmbH, Karlsruhe

dosed into the evaporator by a micro dosing system (Vautz and Schmäh 2009). The system consists of a supply unit with temperature and gas flow control, an evaporator unit and the syringe dosing unit. The system is operated either via a touch screen or remotely via the *viewCAL* software. Due to gas dosing via mass flow controllers and liquid dosing via a syringe pump, the gas-vapour mixture produced is independent of pressure and temperature. The push-pull syringe pump sucks in the liquid automatically and doses continuously. This means that while one syringe is dosing, the second is being filled. Continuous operation is possible via rotary slide valves with a low dead volume. The precision syringes can be exchanged by the user, which results in a large dosing range.

The device is particularly suitable for organic solvents in the medium concentration range (low volume % up to the 100 ppm range). Figure 12.46 shows the instrument version with the two syringes and the display and operating options.

Volumetric-Static Method

An obvious possibility to produce gas mixtures is to fill the two components ($A + B$) into a container and then to mix them until a homogeneous gas concentration is achieved (*cocktail principle*). Basically, DIN 6144 describes exactly this procedure for the production of test gases.[17] Since this is not a continuously available gas mixture, it is called a

[17] In DIN 6144 the term **test gas** is explicitly chosen, whereas in DIN 6145 only the term **calibration gas** is used. DIN 5187 has the following justification: *According to TRG 102 (March 1985 edition) of the DruckbehV, "TEST GAS" is a gas mixture that is to be used in measurement technology. Since most gas analytical equipment requires calibration, gases and gas mixtures for this application are called "calibration gases".*

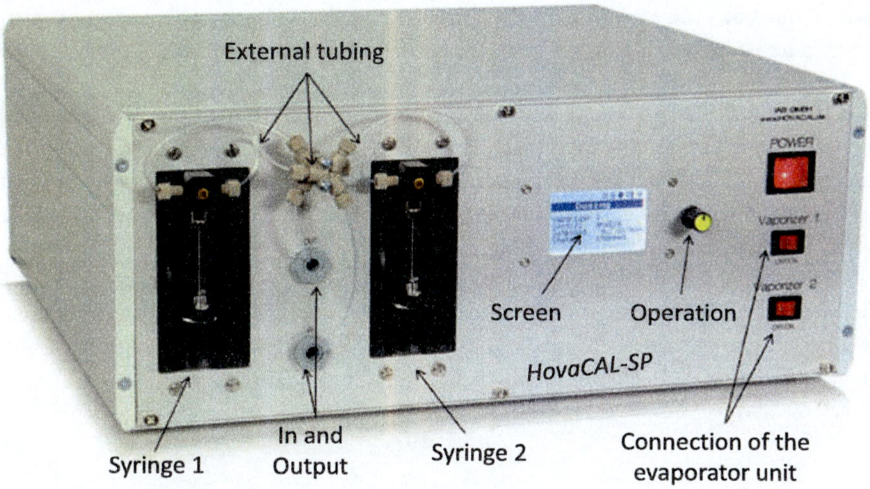

Fig. 12.46 Device after the procedure of syringe injection. The *HovaCAL-digital SP* works with 2 syringes which ensure a continuous calibration operation. (Source: IAS GmbH, Oberursel)

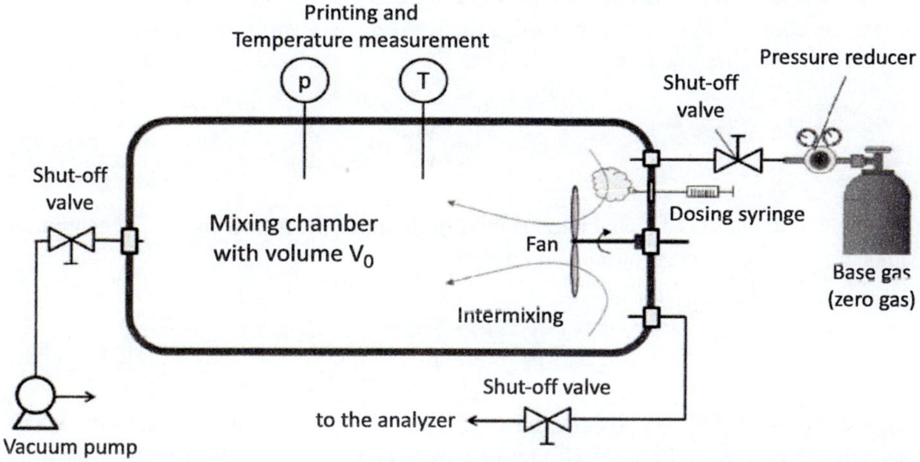

Fig. 12.47 Mixing apparatus for the production of test gases using the volumetric-static method according to DIN EN ISO 6144

volumetric-static method. Figure 12.47 shows the basic structure of such a mixing device. The vessel in which this gas mixture is produced is either made of glass or stainless steel to avoid undesired surface reactions with the gases. The volume is usually several litres. In DIN 6144, a real measurement is given as an example, which was carried out with a vessel volume of 111.5 L. In another example the volume is only 14 L. To use the gas from the vessel, it must be produced with an overpressure (≈2 bar). This overpressure ensures that the gas is transferred from the vessel to the analyser for testing and calibration purposes.

The container must be completely free of foreign matter before filling. For this purpose the chamber is evacuated (5 mbar) and filled with the base gas. Afterwards the gas from the chamber is checked with an appropriately sensitive analyzer and, if necessary, evacuated again and filled with base gas. The process is repeated until no more foreign substances can be detected and the conditioning is completed.

After conditioning, pure base gas at atmospheric pressure ≈1 bar is in the chamber. The exact pressure and temperature must be recorded for further calculations. The admixture is now injected into the chamber through the septum by a syringe filled with the gas or liquid. The gas mixture then has the following concentration after homogenisation:

$$c_x = \frac{p_1 \cdot V(x_G)}{p_2 \cdot V_G + p_1 \cdot V(x_G)} \tag{12.41}$$

With c_x= volume fraction of admixture x (≈ concentration), V_G= volume of the base gas in the mixing chamber, $V(x_G)$= volume of the gaseous admixture in the syringe, p_1= pressure of the gas mixture before the last injection of base gas, p_2= pressure of the gas mixture after the last injection of base gas. Since the volume $V(x_G)$ is significantly smaller than V_G, Eq. 12.41 can be simplified:

$$c_x = \frac{p_1 \cdot V(x_G)}{p_2 \cdot V_G} \tag{12.42}$$

If an evaporable liquid is injected into the mixing chamber, the following gas concentration results after complete evaporation and mixing:

$$c_x = \frac{p_1 \cdot V(x_l)}{p_2 \cdot V_G + p_1 \cdot V(x_l)} \tag{12.43}$$

$V(x_l)$ is the volume of the added admixture after complete evaporation of the substance. The conversion from the liquid phase (V_l) to the gaseous phase $V(x_l)$ is done by the following equation:

$$V(x_l) = \frac{V_l \cdot \varrho \cdot V_m}{M} \tag{12.44}$$

With V_l the volume of the admixture in liquid state, ϱ= density of the liquid, M= molar mass of the liquid. V_m= Molar volume of the liquid. Also in this case $V(x_l)$ is much smaller than V_G, so that Eq. 12.34 can also be simplified:

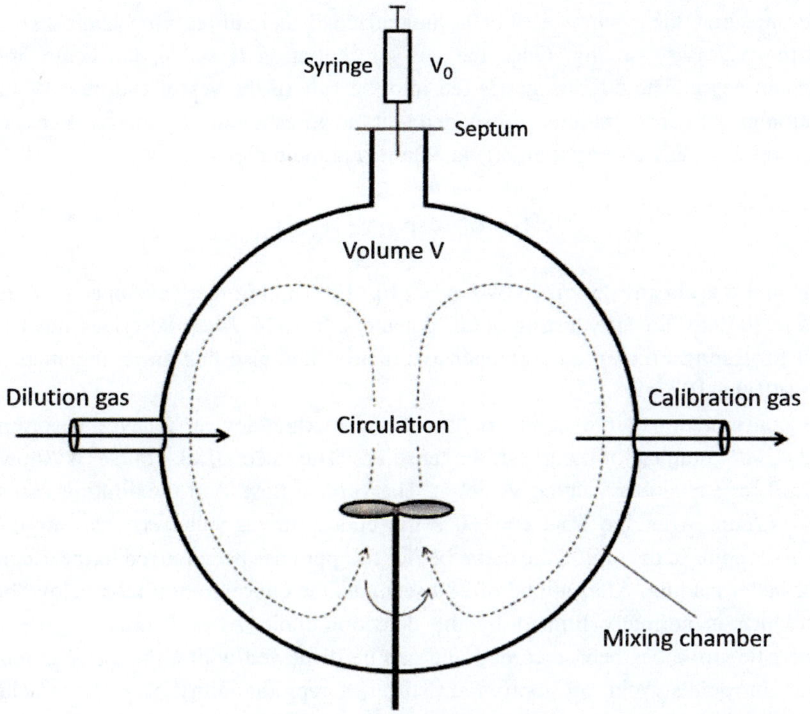

Fig. 12.48 Design of a mixing device based on the principle of exponential dilution (after McKinley 1974)

$$c_x = \frac{p_1 \cdot V(x_I)}{p_2 \cdot V_G} \tag{12.45}$$

Exponential Dilution

A variant of the static method consists in a dynamic dilution of a given gas concentration c_O by a defined flow $\dot{V}$ of a base gas. Figure 12.48 shows such an arrangement. The basic volume V of the chamber is filled with a test gas of concentration c_O. Alternatively, this test gas concentration can also be generated by injecting a defined quantity of gas via a syringe into the base volume V. In this case the following applies:

$$c_0 = \frac{V_0}{V + V_0} \approx \frac{V_0}{V} \tag{12.46}$$

After injection, the gas must still be homogenized. This requires a mechanical stirrer to ensure the necessary mixing. Once the gas concentration is stable, the actual mixing process can begin. The dilution gas is fed into the side of the vessel and mixes with the calibration gas of concentration c_0. Provided that the gases are ideally mixed, an exponentially decreasing gas concentration $c(t)$ is obtained at the outlet:

$$c(t) = c_0 \cdot \exp -t \cdot \left[\dot{V}/V\right] \tag{12.47}$$

This method was already described by *Lovelock in* 1960 and further developed by *Otte and Jentsch in* 1970 for linearity testing of GC detectors. In 1976, *Ritter* described this EDF[18]-method for testing FIDs in a comprehensive treatise and also had an arrangement for it patented (Ritter 1975).

The concentration curve at the outlet of the mixing device behaves according to Eq. 12.47 if complete mixing can be ensured. The theoretical course is shown in Fig. 12.49 for a chamber volume V = 1 L. The volume flow $\dot{V}$ of the dilution gases was set to 1 L/min. After the time t = 60 s the concentration value $c(t)$ has dropped to 1/e = 367.9 ppm. In the figure, the curve below 100 ppm has been moved to the secondary axis for better reading. After a total of 280 seconds, the concentration falls below the 1% value, which is normally limited by the detection limit. After 5 minutes, the entire characteristic curve has been recorded and can be displayed with almost any number of interpolation points. With an appropriate stirrer, a reproducibility of <1% is achieved (Ritter and Adams 1976). An essential prerequisite for using this method is a fast response of the analyzer and a small measuring chamber volume compared to the basic volume *V*.

Electrochemical Production

If an electric current *I* flows through an electrolyte, the electrolyte decomposes into its constituent parts according to *Faraday's law*. If, for example, aqueous hydrochloric acid (HCl) is decomposed by this electrolysis, chlorine gas (Cl_2) is deposited at the anode (+) and gaseous hydrogen (H_2) at the cathode (−). A prerequisite for this is that the applied voltage is higher than the decomposition voltage *Uz* (Table 12.5).

The following gases can be generated electrochemically:

- Oxygen (O_2)
- Hydrogen (H_2)
- Hydrogen cyanide (HCN)
- Hydrogen sulfide (H_2S)
- Chlorine (Cl_2)
- Bromine (Br_2)
- Chlorine dioxide (ClO_2)

[18] **E**xponential **D**ilution **F**lask.

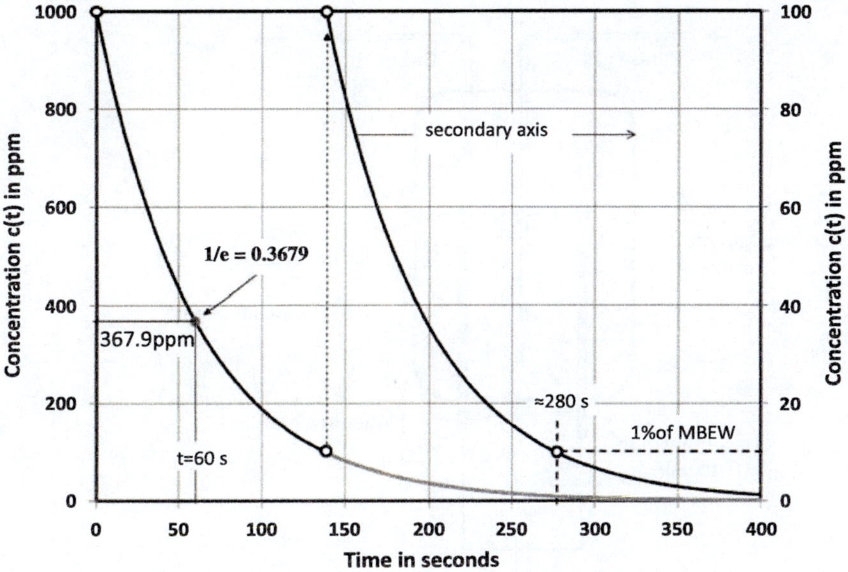

Fig. 12.49 Concentration curve at the$c(t)$ outlet of the dilution device

- Ammonia (NH_3)
- Nitrogen monoxide (NO)
- Nitrogen (N_2)
- Carbon dioxide (CO_2)
- Hydrogen phosphide (PH_3)
- Arsine (AsH_3)
- Ozone (O_3).

The advantages of this process are that stable gas mixtures, ranging from 0.1 to $250 \cdot 10^{-6}$ volume fractions, can be produced within a few minutes. The main areas of application are in immission measurement technology and environmental protection (Wiegleb and Böttcher 1988).

The mass m of the deposited substance is derived from Faraday's first and second law, according to the following equation (Kuchling 2014)

$$m = \frac{I \cdot t \cdot M}{z \cdot F} \tag{12.48}$$

This includes the Faraday constant $F = 96.458, 3365$ C/mol, $M=$ molar mass, $z=$ valence (number of electron charges carried by an ion) and t the time. For the resulting volume flow $\dot{V}_A$ of the calibration gas, the following relationship results:

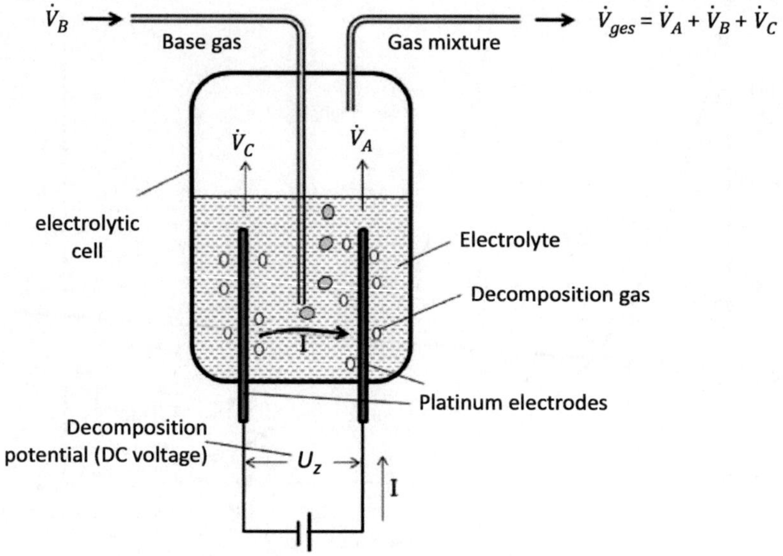

Fig. 12.50 Structure of an electrochemical cell for the production of calibration gas mixtures using the electrochemical method according to DIN EN ISO 6145-11 with common gas supply

$$\dot{V}_A = K \cdot \left[\frac{I \cdot V_m}{z \cdot F}\right] \cdot \left[\frac{T_Z}{T_0}\right] \tag{12.49}$$

K is a dimensionless factor that describes the efficiency of the cell. It is determined experimentally. *Vm* is the molar volume of the gas (=22.414 K/Mol), T_Z is the cell temperature and $T_0 = 273.15$ K.

Figure 12.50 shows an electrochemical cell for gas generation. Two platinum electrodes protrude into an aqueous electrolyte and are connected to a voltage source. A battery is used as the voltage source, which, in comparison to power supplies, has no disturbing residual ripples from the 50 Hz network. The current *I*, which flows through the cell, is a measure for the generation of the gas quantity and must therefore be recorded very accurately (<1%). Since the decomposition products can occur at both the anode and cathode, both components accumulate above the electrolyte surface and are discharged from the cell mixed with the base gas.

If separation of the two sides is desired, a gas-tight membrane must be inserted between the two electrodes. This version is also shown in Fig. 12.51. Thus, at the outlet there is a mixture of the base gas and the decomposition products. Since usually only one calibration gas component is required, a general separation is recommended. Gas which is produced on the left side (cathode) is then vented. The calibration gas concentration *c* at the outlet of the cell is given by

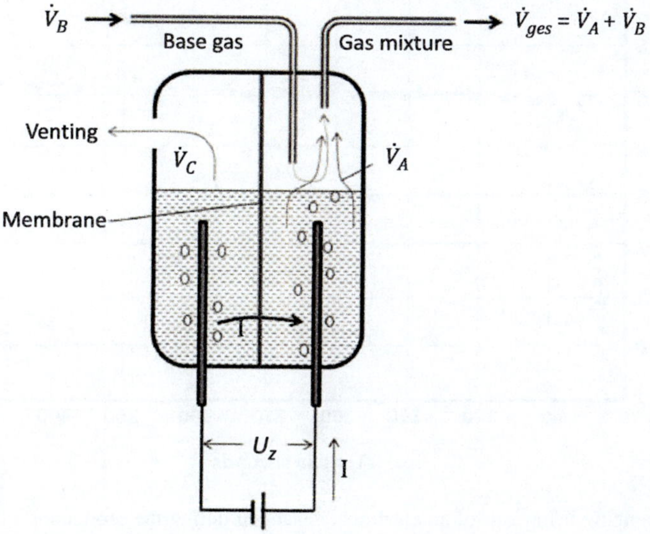

Fig. 12.51 Structure of an electrochemical cell for producing calibration gas mixtures using the electrochemical method according to DIN EN ISO 6145-11 with separate gas supply

$$c = \frac{\dot{V}_A}{\dot{V}_A + \dot{V}_B} \tag{12.50}$$

As the volume flow of the base gas (*B*) is much higher than the volume flow of the calibration gas (*A*), Eq. 12.50 can be simplified. It then applies:

$$c = \frac{\dot{V}_A}{\dot{V}_B} \tag{12.51}$$

By inserting Eq. 12.49 into Eq. 12.51 one obtains the following expression:

$$c = \left[\dot{V}_B\right]^{-1} \cdot K \cdot \left[\frac{I \cdot V_m}{z \cdot F}\right] \cdot \left[\frac{T_Z}{T_0}\right] \tag{12.52}$$

The constants can be summarized in the size δ, so that the following simple expression results for a constant temperature (room temperature):

$$c = \delta \cdot I / \dot{V}_B \tag{12.53}$$

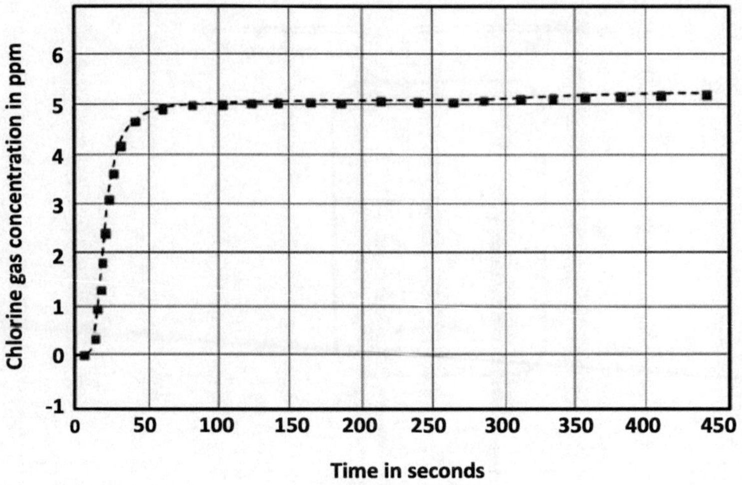

Fig. 12.52 Operating behaviour of an electrochemical cell during the production of a gas mixture with chlorine as calibration component. The cell has reached the desired content in less than 2 min $> 90\%$ and remains stable at the volume fraction level of $5 \cdot 10^{-6}$ according to 4 min. The overall measurement uncertainty is $<5\%$. (Source DIN 6145-11)

Example

A calibration gas mixture of bromine (Br) in nitrogen is to be produced with the electrochemical process. Bromine is diatomic, so $z = 2$. Zinc bromide is used as electrolyte as a 1 mol/L solution. The current I that is produced at a decomposition voltage of 1.81 volts is 0.1 mA. The cell temperature is 20 °C. The efficiency of the cell is also indicated as $K = 0.998$. The base gas flow is set to 5 L/h with an MFC. With Eq. 12.52 one obtains:

$$c = \frac{3.6\text{ml}}{5} \cdot 0.998 \cdot \left[\frac{0.1\text{mA} \cdot \frac{22.414\text{L}}{\text{mol}}}{2 \cdot 96.485, 34 \frac{\text{As}}{\text{mol}}}\right] \cdot \left[\frac{293.15}{273.15}\right] = 8.96\text{ppm}$$

The stability of the test gases produced with the electrochemical method is very good. Figure 12.52 shows such a calibration gas generation as a function of time. Already after 1 minute the chlorine gas content has stabilized to a value of $c \approx 5$ ppm and hardly changes in the following minutes. Since chlorine gas is very reactive, all parts in contact with the gas (hoses, valves, vessels, etc.) must first saturate before the value finally stabilizes. This phase can also last from several minutes to hours.

12.4 Test Procedure

In order to be able to establish comparability between different gas measuring instruments, the measuring characteristics must be known and, above all, comparable. For this purpose, appropriate standards and guidelines have been drawn up which are intended to serve precisely this purpose. There is no generally applicable standard or guideline for gas measuring instruments, gas analysers or gas sensors. In the VDI guideline 2449 part 2, for example, most of the terms relevant to *air pollution control* are defined. A transfer of this terminology to other areas, outside air pollution control, is quite common. VDI Guideline 4203 Part 2, however, defines the test procedures for measuring gaseous and particulate emissions. Other standards regulate the immission measurement of air pollutants or exhaust gas measurement technology in the automotive sector (e.g. OIML R99/ISO 3930). The terms commonly used in gas analysis are defined in DIN 51897. Furthermore, there are special specifications which refer to the different industries (e.-g. medical technology) and areas of application.

This chapter attempts to describe and define the essential and generally valid properties of gas measuring instruments (gas analysers and gas sensors). Emphasis is placed on how these properties can be measured.

Calibration Curve

The calibration curve is plotted using the procedures described in Sect. 12.3. It describes the function $f(c_i)$ between the specified gas concentration *ci* and the electronic display $A(c_i)$ provided by the gas detector. The relationship between the two quantities is usually a linear function:

$$A(c_i) = a_0 + bc_i \tag{12.54}$$

This requirement only applies to gas detectors with an integrated linearization function. The relationship between the specified gas concentration c_i and the displayed value $A(c_i)$ is shown in Fig. 12.53 as a calibration curve. Since most measuring methods provide non-linear output values, electronic linearization is always present in the products available on the market today. The only exceptions are simple gas detectors or OEM gas modules that are integrated into other gas detectors.

4–20 mA Interface

The constant value a_0 describes a deviation of the zero point (offset) which is present at $c_i = 0$. In certain applications, this offset is deliberately set in order to obtain a monitoring function. This possibility is usually used with current outputs (20 mA) to detect a line interruption. If the value *zero is* measured for $A(c_i)$, this can be caused by a concentration$c_i = 0$. Zero would also be displayed if the line was interrupted, from the gas meter to a control room. A distinction between the two possibilities is therefore not

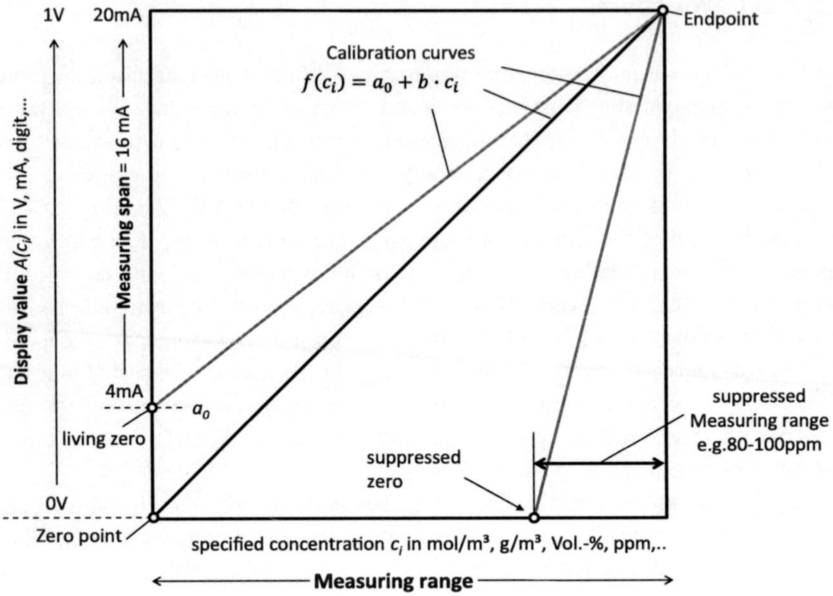

Fig. 12.53 Display of the different calibration curves as normal function, with raised zero point (*life-zero*) and suppressed measuring range

possible. If, on the other hand, the zero point is raised to 4 mA, a $c_i = 0$ value of 4 mA is displayed for a value of 4 mA, whereas 0 mA is displayed for a line break. This type of zero point display is also called *living zero* or *life zero*. The 4 mA can also be used to supply energy to low-power sensors or provide additional diagnostic information in the form of gradations (... 2 mA, 3 mA, ...). Digital data communication (HART[19]) can also be carried out via this interface.

Sensitivity

The sensitivity *E* of the gas measurement is described by the slope factor *b*, which is a constant quantity in a linear function. In general, sensitivity is described as the change in slope resulting from the first derivative of *f*(*c*). The following relationship therefore applies:

$$\text{Sensitivity } E = \frac{\Delta U_A}{\Delta c} \tag{12.55}$$

[19]**H**ighway **A**ddressable **R**emote **T**ransducer (HART).

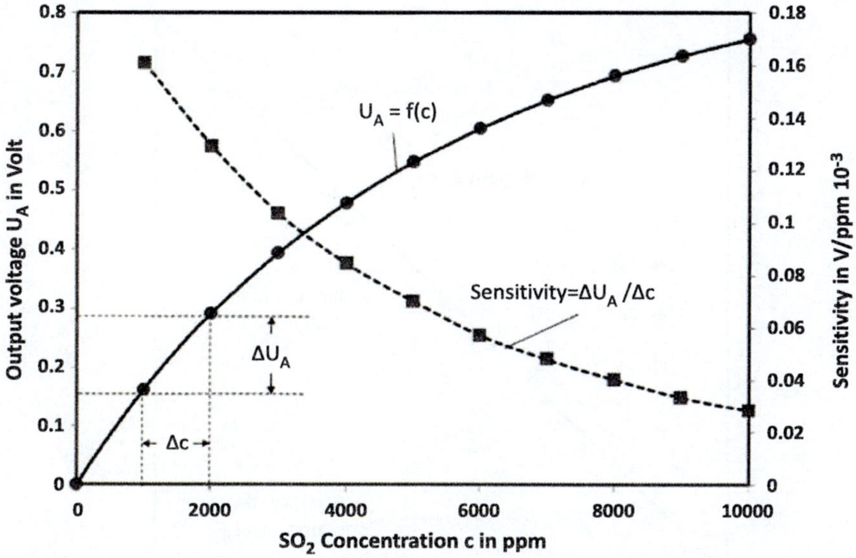

Fig. 12.54 Nonlinear calibration function $U_A = f(c)$ and the sensitivity curve $E(c) = \Delta U_A \Delta c$ determined from it

Figure 12.54 shows a non-linear calibration function $U_A = f(c)$ which is typical for a photometric gas measurement (see Chap. 6). As the gas concentration c increases, the sensitivity E decreases progressively and then runs towards zero (saturation).

Measurement Error

Important for a neutral assessment of the measurement quality of a device or a process are the expected measurement errors. The concentration value c_i recorded by the gas meter leads to an expected display value A_i according to Eq. 12.56. The uncertainty in the preparation of the calibration gas mixture results in a minimum error F_{min} which is linearly included in the measurement result via the slope of the calibration function (Fig. 12.55). Furthermore, the gas measuring instrument also has a total measuring error, which is composed of a large number of individual sources of error ($F_1, F_2, \ldots, F_n$). These sources of error will be discussed in more detail in the course of this chapter.

The total error of the gas measurement F_{ges} then results from the sum of the error squares (Hengstenberg et al. 1980):

$$F_{ges} = \sqrt{F_1^2 + F_2^2 + \cdots + F_n^2} \tag{12.56}$$

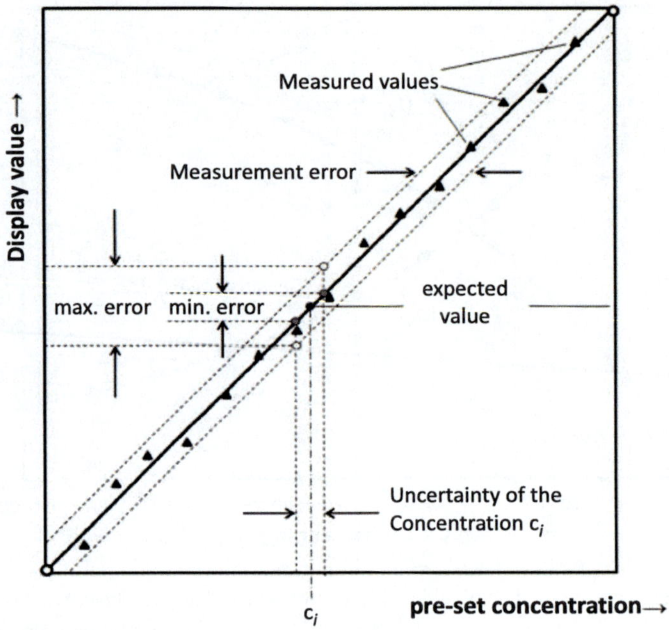

Fig. 12.55 Representation of the measurement errors from the uncertainty of the calibration gas concentration c_i and the total error of the gas analyzer F_{ges}

In gas measurement technology, it is common practice to specify the individual error influences separately and also to define them precisely.

Detection Limit and Noise Behaviour

The lower detection limit (LDL) is an important parameter for the assessment of gas measuring instruments. Since the sensor signals are always overlaid by statistical fluctuations (noise), there is a limit below which a signal change can no longer be separated from the noise. According to VDI 2449 part 2, the detection limit is therefore referred to as the smallest change caused by a gas concentration Δc which differs from zero with a statistical certainty of 95%. Figure 12.56 shows such an experiment which was used to determine this limit. The signal at zero point (→ zero gas) has a statistical noise margin of $\approx 4\sigma$. Within this bandwidth are ≈95% of all measured values from this interval. If a test gas mixture with the concentration Δc is now applied to the analyzer, the mean output signal changes by exactly the same amount as the noise bandwidth. In this picture you can see that the gas concentration Δc leads to a detectable change in concentration. This is also referred to as the signal-to-noise ratio (SRV[20]), which is the detection limit =1. The solid

[20] **S**ignal to **N**oise **R**atio = SNR.

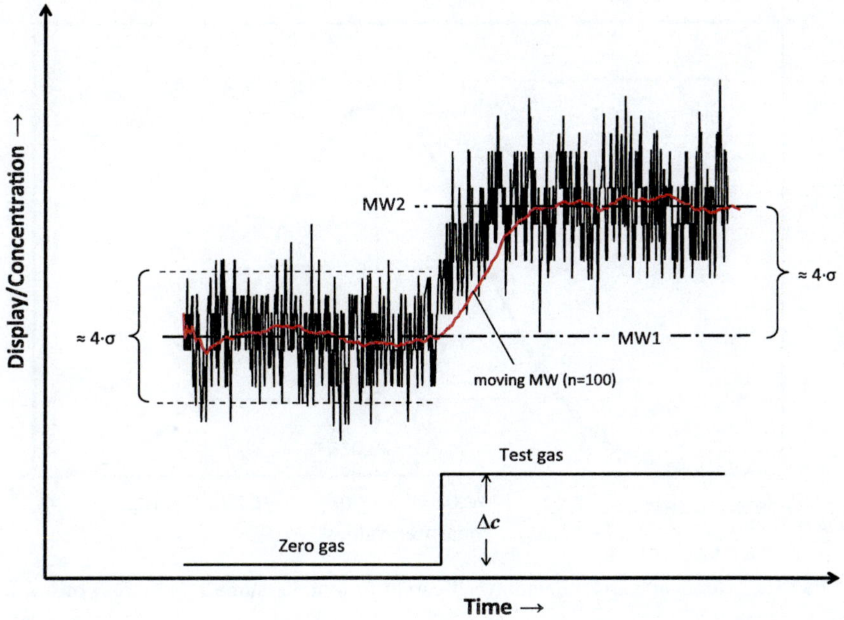

Fig. 12.56 Signal curve of an analyser at zero point with a gas jump Δc that corresponds exactly to the lower detection limit LDL=$\Delta c \approx 4 \cdot \sigma$

line in Fig. 12.56 shows a moving average of the measurement signals. Here, too, it can be seen immediately that averaging (lowpass filter, RC element) leads to an improvement in the *signal-to-noise ratio* and thus also reduces the detection limit. On the other hand, however, this also increases the response time.

This statistical certainty can be described with the normal distribution (*Gaussian distribution*). The starting point for the distribution curve is the mean value $\overline{x}$ of a numerical series of *n-values*:

$$\overline{x} = \frac{1}{n} \sum_{i=1}^{n} x_i \tag{12.57}$$

Another important variable, which is used for the dispersion σ of the values x_i around the *true value*, which is assumed to be the mean value $\overline{x}$. The scatter of the total measured values, as mean square deviation, is also called standard deviation (*STABW*) and can be calculated as follows:

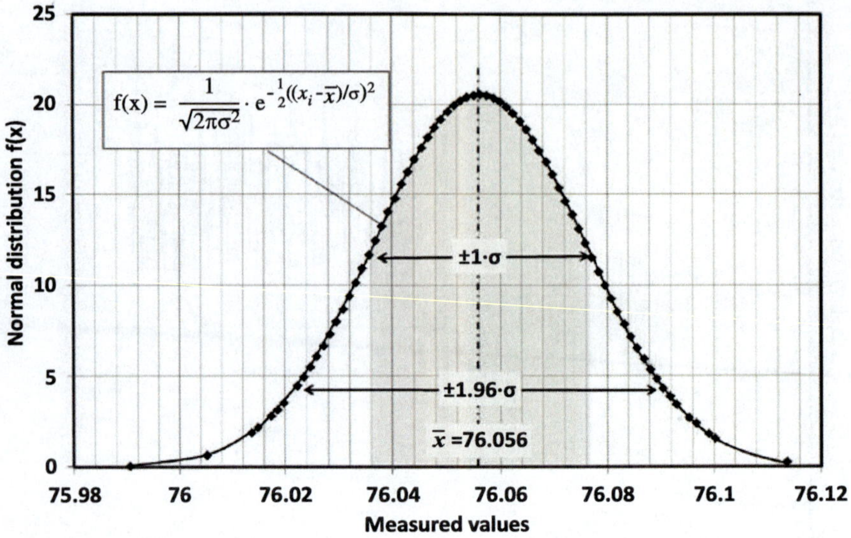

Fig. 12.57 Gaussian normal distribution (bell curve) for a finite number ($n = 300$) of measured values

$$\sigma = \sqrt{\frac{1}{n} \cdot \sum_{i=1}^{n} (x_i - \bar{x})^2} \tag{12.58}$$

With the standard deviation σ and the mean value $\bar{x}$ the normal distribution according to Gauss can now be calculated:

$$f(x) = \frac{1}{\sigma\sqrt{2\pi}} \cdot e^{-\frac{1}{2}\left[\frac{x_i - \bar{x}}{\sigma}\right]^2} \tag{12.59}$$

Figure 12.57 shows the normal distribution of a real series of measurements from 300 values recorded with a UV gas sensor (ULTRA.sens®). The simple standard deviation $\pm\sigma$ contains 68.3% of all measured values. This proportion is shown in the figure as a grey area. The statistical certainty of 95% requires a confidence limit V of:

$$V = \pm 1{,}96 \cdot \sigma \tag{12.60}$$

In general, the relationship between the confidence limit V and the normal distribution can be illustrated as follows:

Table 12.5 Decomposition voltages for 1-normal solutions that can be used as electrolyte

Substance	Decomposition potential in V
Perchloric acid ($HClO_4$)	1.65
Hydrochloric acid (HCl)	1.31
Nickel chloride ($NiCl_2$)	1.85
Zinc bromide ($ZnBr_2$)	1.8
Cobalt chloride ($CoCl_2$)	1.78
Lead nitrate [$Pb(NO_3)_2$].	1.52
Copper sulphate ($CuSO_4$)	1.49
Silver nitrate ($AgNO_3$)	0.7
Hydrobromic acid (HBr)	0.94
Hydroiodic acid (HI)	0.52
Sodium hydroxide (NaOH)	1.69
Potassium hydroxide (KOH)	1.6
Ammonium hydroxide (NH_4OH)	1.74

Source: DIN EN ISO 6145-11

$$V = \pm u \cdot \sigma \tag{12.61}$$

The factor u is called fractile and depends on the statistical certainty P (Tables 12.4 and 12.5). Table 12.6 provides some data on this. However, these data apply only to a large number of measured values ($n \rightarrow \infty$).

With a finite number of measured values, only one approximate solution is obtained. For a sample with a finite number of measurements n, the fractiles in Table 12.7 apply.

From this table (12.7) it can further be seen that at least 100 measured values must be available for the gas-analytical determination of the detection limit. The fractile is then assumed with u = 2, so that for the lower detection limit (LDL) applies:

$$\text{LDL} = V = \pm 2 \cdot \sigma \tag{12.62}$$

For the practical determination of the detection limit, all other disturbance factors (pressure, temperature) which may affect the analyser must be constant. Furthermore, there shall be no signal drift during the recording of the measured values which would significantly interfere with the determination of the detection limit. The time distribution of the measured values used for the Gaussian distribution from Fig. 12.57 is shown in Fig. 12.58. The individual points that lie within the detection limit are shown in black. Only 15 values (bright points) are outside the limit of $\pm 1.96 \cdot \sigma$. All measurement points are distributed around the mean value $\bar{x}$.

Occasionally the detection limit is also determined by a so-called *peak-to-peak measurement* (Lieck 1998). For this purpose, the upper (maximum) and lower (minimum) values are recorded during a specified measurement period. The distance between the two

Table 12.6 Fractiles of the normal distribution (Stetter 1992)

P	50%	68.3%	95%	99%	99.9%
U	0.674	1.00	1.960	2.576	3.201

Table 12.7 Fractiles for a finite number of measurements (Adunka 1998)

n	P = 90%	P = 95%	P = 99%
2	6.31	12.71	63.66
3	2.92	4.3	9.93
5	2.13	2.78	4.6
10	1.83	2.26	3.25
20	1.73	2.09	2.86
50	1.68	***2.01***	2.68
100	1.66	***1.98***	2.63
200	1.65	***1.97***	2.60
∞	1.65	***1.96***	2,.58

values then gives the detection limit (LDL). Since the maximum and minimum peaks recorded in each case can also be caused by individual events and external interference sources (EMC), they are not statistically relevant (→ *outliers*). After the determination of the detection limit shown in Fig. 12.59, the value of LDL = 0.12 ppm would therefore be 3 times higher than the statistical evaluation with LDL = 0.038 ppm.

An important way to improve the detection limit is to filter the signals. The simplest method uses a low-pass filter (RC element), which integrates the signals and thus averages them over a certain period of time. This time period results from the product $R \cdot C = \tau$. This filter function can be simulated in the software by a moving average value. Figure 12.60 shows the curves that result from different averaging over *n values*. The bright curve shows the raw values ($n = 1$). Averaging over 10 values already results in a significant reduction of the noise signal by the factor $f \approx 3$. Averaging over 100 values provides an improvement by a factor of $f \approx 10$. The improvement of the detection limit is linked via the function $f(n) \approx n^{1/2}$ to the number of measured values that are included in the averaging.

Usual averaging times in gas measurement technology range from 0.1 second to 10 seconds. Averaging in the minute range (50–100 seconds) is also used for ambient air quality measurements, as these are very small concentrations which are always limited by the detection limit. However, averaging cannot be carried out with any number of measured values, as this has an extreme effect on the response times and very slow changes (→*fluctuations*) in the range from 0.1 Hz to 0.01 Hz cannot be influenced sufficiently by averaging. These fluctuations then form the detection limit for the analyzer.

Setting Time

If an analyser is exposed to a change in gas concentration over time, the analyser should react to these changes as quickly as possible and display the current measured value $c(t)$.

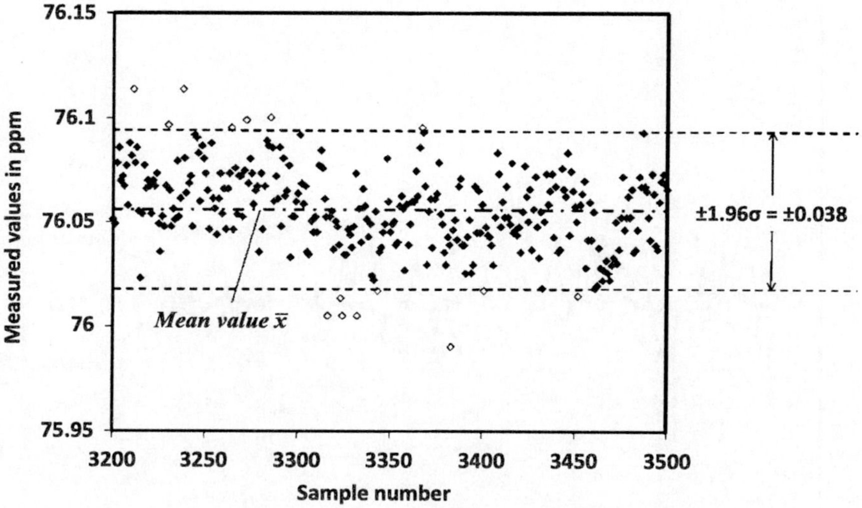

Fig. 12.58 Display of the confidence limit (detection limit) for 300 measured values. With a statistical certainty of 95%, 285 measured values are within the confidence limit and 15 measured values are outside it

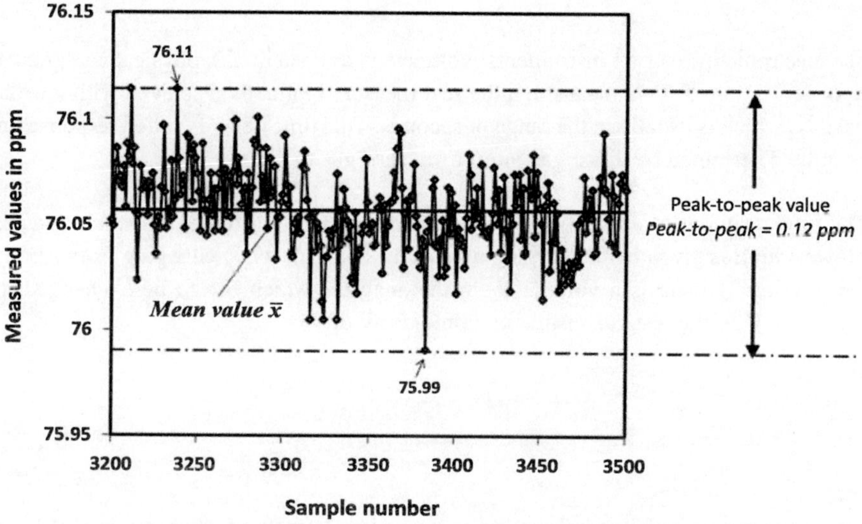

Fig. 12.59 Calculating the detection limit from the *peak-to-peak* method

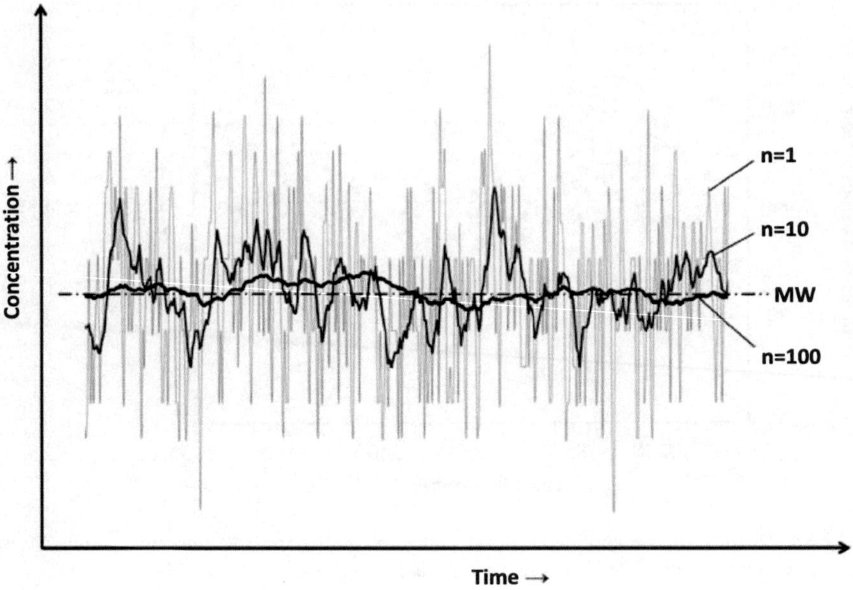

Fig. 12.60 Noise signal of a gas analyzer during averaging over *n measured values* ($n = 1,10,100$)

Unlike electronic measuring instruments (voltmeters, ammeters, ...), physical analysers are more or less damped. This means that the real measured value is displayed with a certain time delay, which is usually in the range of seconds. This time delay is called response time and can be determined by a fast gas jump from zero gas to a test gas.

The time response of an analyzer is determined by many different influencing factors. The lowest limit is given by the purging time of the analyser when changing from zero gas to the test gas. If there is a volume V_0 in the analyzer which has to be exchanged, the volume flow $\dot{V}$ of the test gas results in a time delay of:

$$c(t) = c_0 e^{-\dot{V} \cdot t / V_0} \tag{12.63}$$

Example

For an IR gas sensor with a sample cell ($d_i = 8\text{mm}$, $L = 10\text{cm}$) and a volume flow of 1 L the purge time is to be determined. To obtain a stable value, max. 1% residual gas should remain in the sample cell. For a 99% purging of the cuvette then applies:

$$t = -\frac{V_0}{\dot{V}} \cdot \ln 0.01 = \frac{\pi \cdot 0.8\text{cm}^2 \cdot 10\text{cm}}{4 \cdot 16.67\text{cm}^3/\text{s}} \cdot 4.6052 = 1.4\,\text{Seconds} \tag{12.64}$$

This means that 1.4 seconds are required for the rinsing time.

For a rough calculation the triple chamber volume V_0 can also be used and divided by the volume flow $\dot{V}$. For the example given, this would then be $t \approx 1$ seconds.

If the gas exchange takes place by a diffusion process, the exchange takes correspondingly longer. Besides this physical time constant, the evaluation electronics also contribute to the delay. Especially if the output signal is damped in order to reduce the detection limit, the time response increases.

Furthermore, the averaging of the signals, to improve the detection limit, has a negative influence on the time response. With increasing averaging, the detection limit improves, but the rise time increases steadily. The optimum averaging time must therefore be determined for the respective application. In the example in Fig. 12.61, the detection limit hardly changes at n > 10, while the rise time increases linearly. In this case averaging over 4–8 values would be recommended.

In addition to the analyzer, in certain cases the upstream elements such as filters, pumps, hoses also play a role and influence the temporal behavior of the entire system. Since the purging of these elements takes place upstream of the analyzer, the analyzer does not *notice* this process, which is part of the time delay. This part of the time delay is called dead time t_{Tot}.

The determination of the time response of gas analysers is specified in VDI Guideline 2449 Part 2. Figure 12.62 shows a signal curve from which the corresponding times can be read. At time t_0 the zero gas is abruptly switched over by a solenoid valve and replaced by test gas (jump function). In order to be able to detect an exact definition of the change over time, the 10% value of the stable 100% value is used as the starting point for determining the rise time. The time span between t_0 and $t_{10\%}$ is referred to as dead time. The rise to the 90% value serves as the end point of the rise time t_{An}. The setting time is therefore t_{Ein}:

$$t_{\text{Ein}} = t_{\text{Tot}} + t_{\text{An}} \tag{12.65}$$

Some manufacturers also specify the time constant τ for the temporal behaviour, which describes the 63% value and is defined as follows:

$$c(t) \approx c_o \cdot \left[1 - \text{e}^{-t/\tau}\right] \tag{12.66}$$

The fall time can be determined analogous to the rise time by determining the time interval between the 90% value and the 10% value. For linear calibration functions both times should be identical $t_{\text{An}} = t_{\text{Ab}}$. However, if the calibration function is non-linear, there may be significant differences between the two values. Since the sensitivity E decreases with

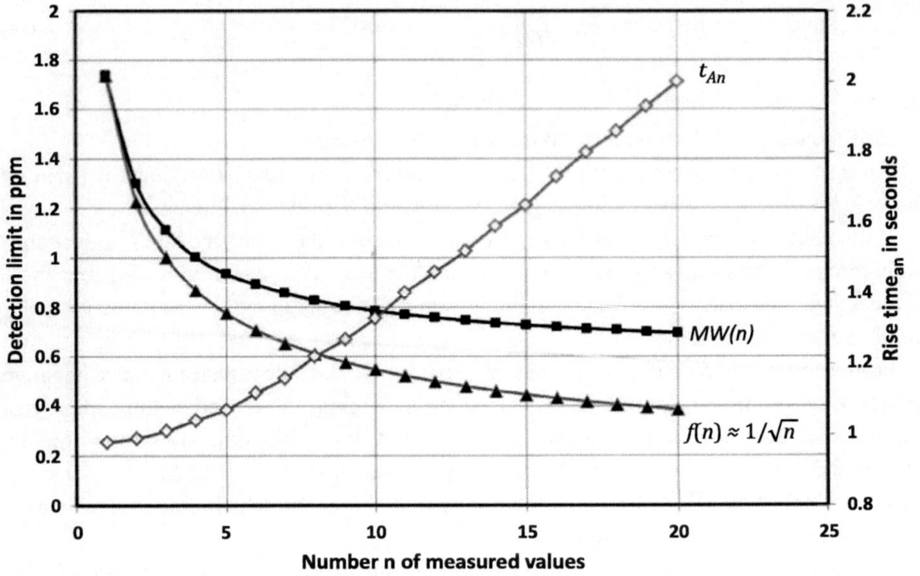

Fig. 12.61 Change of the lower detection limit (LDL) and the rise time t_{An} by averaging over *n-values*

increasing gas concentration, the end point (90% value) will reach a stable value faster than the opposite case. The change from test gas to zero gas then leads to a range of increasing sensitivity and the endpoint (10% value) is delayed. Therefore, then $t_{\mathrm{An}} < t_{\mathrm{Ab}}$.

Pressure Influence

Gas analyzers are usually operated at atmospheric pressure p_A. This means that there is a pressure in the measuring system from $p_A = 1013\mathrm{hPa} \pm 100$ hPa. As the number n of molecules per volume (→density) changes with a pressure change, this has a significant influence on the measurement result. According to the general gas equation:

$$p \cdot V = n \cdot R \cdot T \tag{12.67}$$

n is proportional to p. A linear pressure behaviour can therefore be expected. A pressure change of $\Delta p = 100$ hPa should therefore cause a signal change of 10%. Since additional pressure errors are caused by other physical effects, this pressure influence must be determined experimentally. Figures 12.63 and 12.64 show corresponding setups with which both the positive pressure (1013 hPa + 100 hPa) and the negative pressure range (1013 hPa–100 hPa) can be covered.

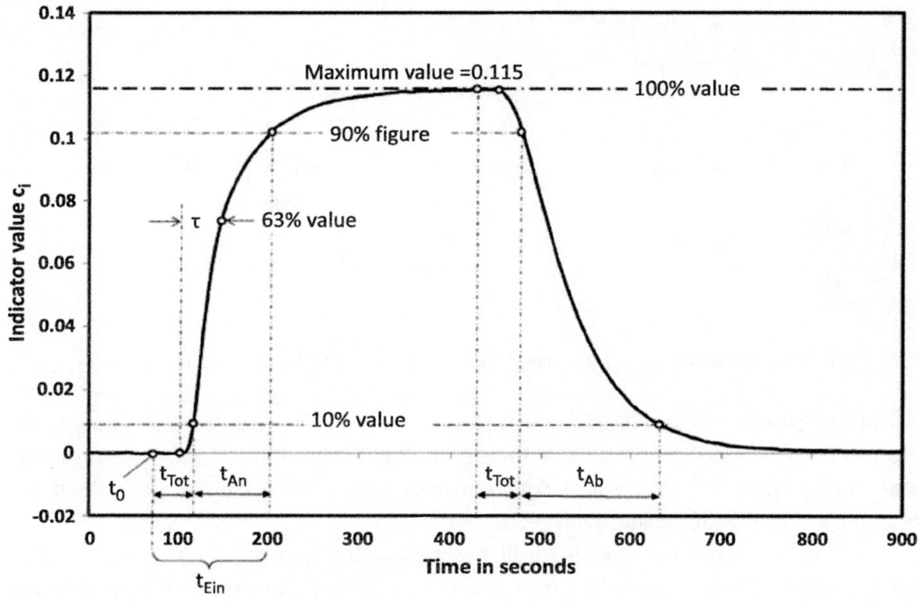

Fig. 12.62 Temporal behaviour of an analyser for a gas change from zero gas→ Test gas zero gas according to VDI Guideline 2449 Part 2

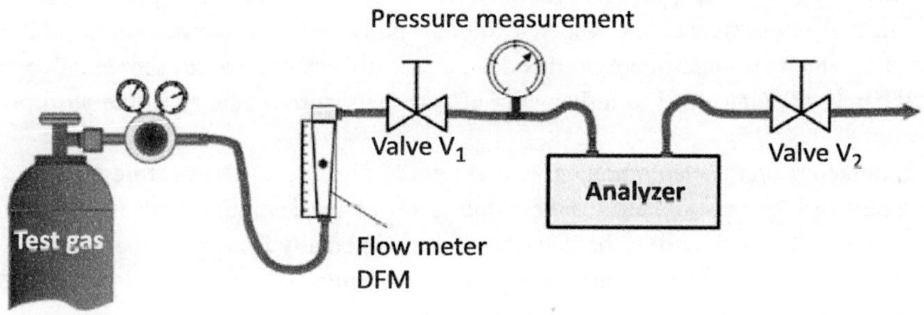

Fig. 12.63 Test arrangement for determining the pressure error in the overpressure range

For the overpressure range, the analyzer is equipped with a valve in the outlet to throttle the flow rate and increase the pressure p in the analyzer. The increase in pressure is measured with an appropriate measuring device or pressure sensor. Both the pressure and the flow rate can be adjusted with valves 1 and 2. To determine the pressure dependence, the pressure is slowly increased and the display of the analyser is noted at the corresponding pressures. The pressure characteristic curve can then be drawn up from this information.

In the vacuum range, a pump is required to suck in the gas and thus create a vacuum in the analyzer. The negative pressure is adjusted with valve 1. The test gas from the

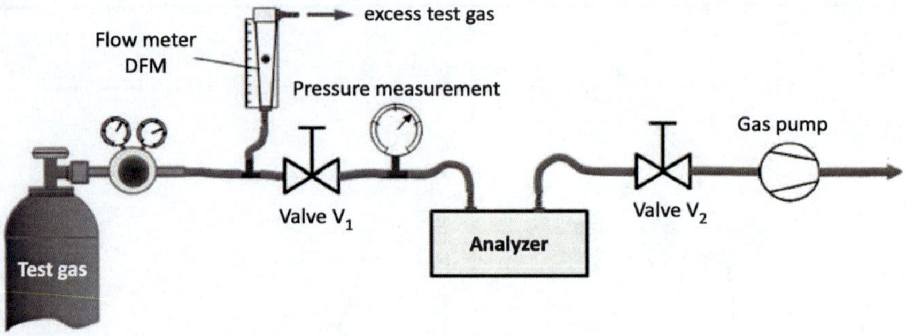

Fig. 12.64 Test arrangement for determining the pressure error in the vacuum range

pressurized gas cylinder is offered in excess for this purpose and checked with a flow meter (FM). This ensures that no *false air* is sucked into the analyzer and falsifies the measurement result. These values are also determined step by step and noted down with the measurement results of the analyzer.

Figure 12.65 shows the overall result for the positive pressure ($+\Delta p$) and negative pressure range ($-\Delta p$) of a CO_2 IR gas analyzer. As expected, the result is a linear function between the pressure p in the analyzer and the displayed output signal. According to the gas equation, the influence would have been directly proportional to the pressure p. However, the slope of the CO_2 in this setup is very small $a = 0.0018$ compared to the theory of $a = 0.001$. The slope is also different for the different gases and must always be determined experimentally in these cases. Compared to other gases, CO_2 has a pronounced real gas behaviour, which also contributes to this deviation. Furthermore, spectroscopic influences (shock band widening) lead to influencing effects that cause a change in the absorption properties.

The influence on the calibration curve is shown in Fig. 12.66. The overpressure causes an increase in the measurement results, while a lower concentration is displayed in the underpressure. The influence to be determined experimentally is typically between 1–2% of the measured value for a pressure change around 10 hPa. The zero point is not affected by this pressure influence and always shows zero.

Temperature Influence

The influence of temperature on the output signals in a gas measurement is much more serious than the pressure influence just described. According to the *first law of Gay-Lussac*, the display value $c(T)$ of a gas measuring instrument calibrated at temperature T_{cal} should behave as follows with a temperature change ΔT

$$c(T) = c_{cal} \cdot T_{cal} \cdot \frac{1}{T_{cal} \pm \Delta T}. \tag{12.68}$$

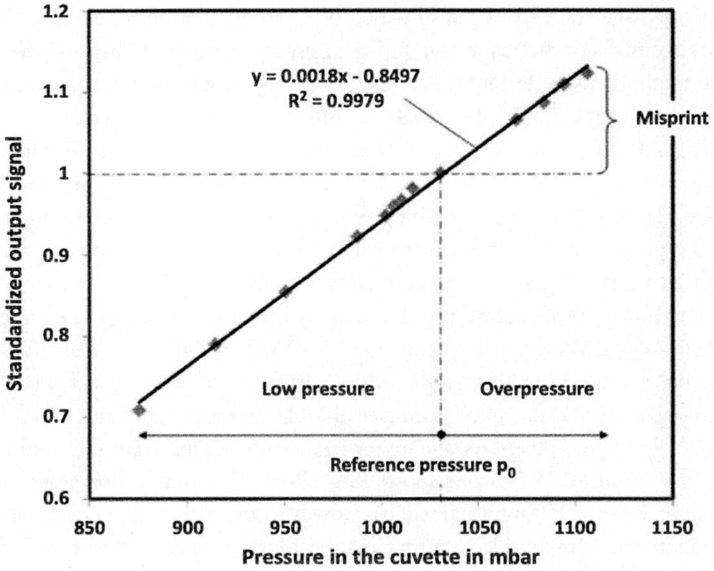

Fig. 12.65 Change in the output signal of a CO_2 gas analyzer based on IR absorption at different pressures in the sample gas cell

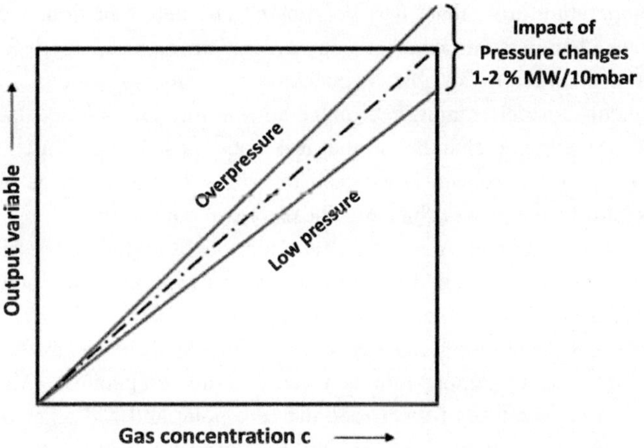

Fig. 12.66 Changing the calibration function at different pressures in the analyzer

The resulting change is only 1/273 per °C or 0.366%/K. In the zero point ($c = 0$) there should be no change according to Eq. 12.68. However, practical experience shows that the analyser behaves completely differently. The influence of temperature according to

Eq. 12.68 is only one source of error to which other sources of error can be added. These error sources result from the respective physical process on which the analyzer is based. Photometric methods use interference filters, for example, whose transmission behaviour changes with temperature. The temperature shift of the central wavelength λ_0 is only $\approx 1 \cdot 10^{-4}$ /K, but this influence can still affect the measurement result. Furthermore, the radiation sources and detector elements also have a temperature error, which can be considerable. The sensitivity of a thermopile detector can change, for example, by −0, 36%/K. At a typical ambient temperature of 5 to 45 °C this influence already results in an error of ≈15%, which affects both the zero point and the end point.

The temperature error therefore has the greatest influence on the measurement result. The instrument manufacturers use various methods to limit this error. Effective measures are, for example, *Thermostatting of* the entire physical structure to a temperature that should be 5–10 K above the maximum permissible ambient temperature T_U to ensure adequate control. At an ambient temperature of $T_U = 45$ °C, the analyzer would therefore have to be controlled at 55 °C. Thermostatting allows the total temperature error to be reduced by a factor ($f \approx 10$ empirical value). However, this factor depends strongly on the size of the analyzer. Miniaturized sensors can be thermostatted better than large, bulky devices. The increased analyzer temperature also has the advantage that possible condensation or adsorption is prevented or suppressed. The disadvantages, however, are high electrical power consumption and faster aging of the components.

Another frequently used option is electronic compensation of the temperature error. To do this, the temperature error must first be known. The determination is carried out in so-called climate chambers, in which the analyzers are exposed to different ambient temperatures T_U, *in* order to be able to conclude the temperature error. Figure 12.67 shows the test setup for determining the temperature error. The test specimen is located in the middle of the climatic chamber so that it is not exposed to a temperature gradient. During the test, the test specimen is exposed to a flow of gas which is brought to the chamber temperature via a heat exchanger. The electrical connections (supply voltage and signal outputs) are led to the outside. The gas is also brought to the outside via hoses at the outlet of the test specimen and fed to the exhaust air. Figure 12.68 shows an opened climatic chamber in which a test specimen is located.

For a simple check of the temperature error, the analyzer is examined at three temperatures. The starting temperature is usually room temperature. The analyser is therefore used at a temperature of 25 °C and the zero point and end point are adjusted to the required values. ZP_{25} is then set to *zero* and S_{25} corresponds to the regular measuring span (e.g. 1000 ppm). Afterwards the climate chamber is shut down to the lowest temperature value. Figure 12.69 shows an example that starts at 5 °C after the adjustment. The zero point has dropped to the value ZP_5 and thus allows a first error consideration. The temperature error is then $TPF_1 = NP_5/20$ K. The temperature error of the sensitivity results from the difference between the two readings and is $TPF_2 = S_{25} - S_5/20$ K. Similarly, the temperature error is determined at 45 °C for the zero point and the sensitivity.

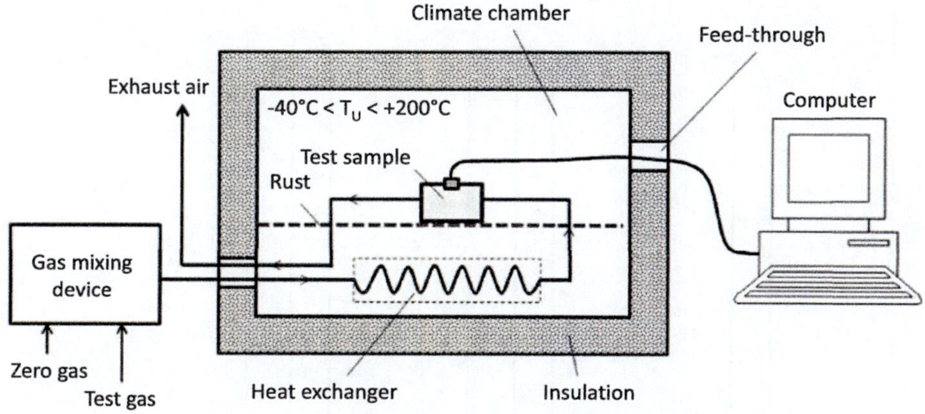

Fig. 12.67 Arrangement for determining the temperature error in a climate chamber

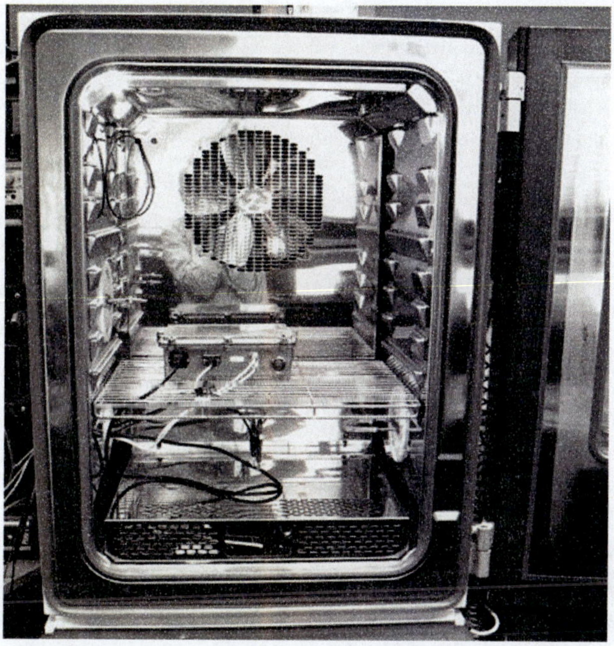

Fig. 12.68 View into an open climate chamber with the connected test sample

Figure 12.70 shows real values for such a climatic chamber measurement with two UV gas analyzers (ULTRA.sens®) for NO and NO_2. In this case, however, the errors were determined at a total of 10 temperatures. Between 0 °C and 40 °C the error for the sensitivity is a linear function. The slope represents the uncompensated temperature

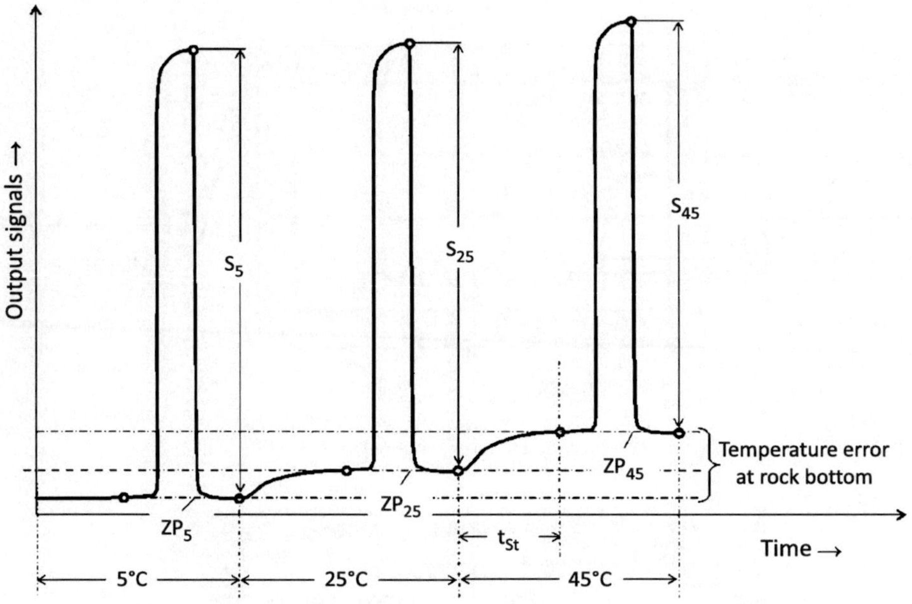

Fig. 12.69 Typical signal curve during a climatic test

error. After this measurement, the NO structure has an error of 1.2%/K while the NO_2 structure has an error of 0.3%/K. With this function the temperature error in a microcontroller can be almost completely compensated. The device manufacturers specify the residual error in %/10 K or also %/10 °C. Typical temperature error specifications are formulated as follows:

- <1% of span per 10 °C at zero point
- <2% of the measured values per 10 °C for the sensitivity.

Cross Sensitivities

The influence of other gases on the measurement result of an analyzer for a specific component is called *cross-talk or cross-interference*. The reasons for this measurement error can be of very different nature. Absolutely selective gas analyzers that only respond to one measurement component and do not take other components in the gas mixture into account are practically unavailable. The influence of these interfering gases must be quantified for comparison purposes. The formal calculation of the cross-sensitivity Q is performed according to:

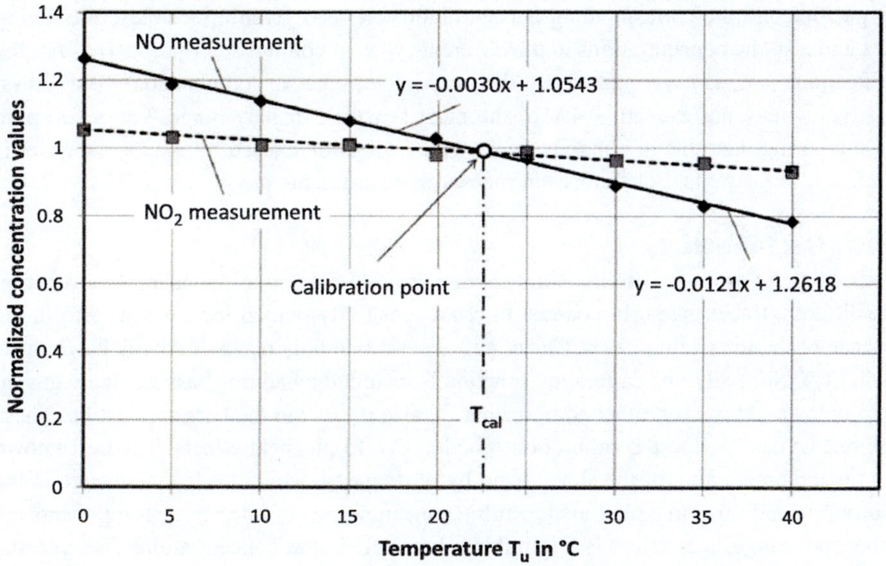

Fig. 12.70 Sensitivity change of a UV-NO/NO_2 gas analyzer (without compensation) in the temperature range between 0 °C and 40 °C. The calibration was performed at $T_{cal} = 22.5°C$

$$Q = \frac{A_2/c_2}{A_1/c_1} = \frac{A_2 \cdot c_1}{A_1 \cdot c_2} \tag{12.69}$$

The cross-sensitivity is thus the ratio between the two readings of the interfering gas A_2 and the sample gas A_1, each divided by its own gas concentration ($c_{1,\,2}$ Hengstenberg et al. 1980). Q is a dimensionless quantity that is only used for comparison purposes. The selectivity S is the reciprocal of the cross-sensitivity.

Example

A gas analyzer generates an output voltage of 1 volt at a concentration of $c_1 = 10$ vol.%. An interfering component with the concentration $c_2 = 100$ vol.%, however, produces an output signal from 125 mV. The cross-sensitivity is therefore:

$$Q = \frac{0.125\text{V} \cdot 10\text{vol.}0\%}{1\text{V} \cdot 100\text{vol.}0\%} = 1.25 \cdot 10^{-2} \tag{12.70}$$

Since the cross-sensitivity Q is not a constant quantity that is independent of the concentration values c_n, the influences of typical gas concentrations are usually always given. For emission measuring instruments, the typical gases that may occur in waste gas are specified and the corresponding concentrations are also given. Since these interfering gases and also the concentrations in the different types of combustion (gas/coal) differ, the certifications are always related to the respective type of combustion. The cross-sensitivities may not exceed ± 4% of the most sensitive display range. For a 100 ppm SO_2 measuring instrument, a maximum ±4 ppm indication caused by interfering gases is therefore still tolerable. For immission measuring instruments the limit is ±6%.

Carrier Gas Dependency

Another type of cross-sensitivity is the *carrier gas dependence* of the calibration function with different tracer gases. In contrast to cross-sensitivity, which increases at zero in the presence of the interfering gases, carrier gas dependence only occurs in the deflection (→ sensitivity). Normally, the calibration function is recorded by adding base gas (zero gas) to the sample gas. However, other components are also present in the later application. These components can change the calibration function due to physical effects. The best known effect is the broadening of the shock band by other gases, which leads to a change in the absorption conditions in a photometer. If the concentrations of interfering components are in the ppm range, this effect is negligible. However, if the concentrations rise into the double-digit volume-percent range, this effect can have a significant influence on the characteristic curve. A typical example is the simultaneous CO_2 and CH_4measurement in a biogas plant. If the calibration curves for the CO_2 and CH_4 sensors are each determined in nitrogen as zero gas, significant deviations are obtained for a real biogas mixture. The calibration should therefore also be carried out in the matrix that occurs in practice in order to be able to exclude these error influences. Figure 12.71 shows the behaviour for an admixture at different concentration ratios.

Flow Dependence

Typically, analysers are flowed with a gas flow rate of 0.25–1 L/min. The manufacturers always specify a flow range for the respective instrument for which the instrument was specified. Typical specifications are:

- normal 10 ... 90 L/h, min. 90 L/h[21]
- normal 20 ... 60 L/h, max. 90 L/h[22] .

If the analyser is flowed with this flow rate, all specified data apply. For particularly sensitive reacting measuring instruments, an error is also specified that occurs when the flow rate changes (*example: <0.5% of the measuring span for a flow rate change of ± 10 L/ h*).

There can be many reasons for this flow dependence:

[21] Thermal conductivity analyzer module Caldos 17 (ABB-Frankfurt).

[22] Oxygen analyzer module Magnos 17 (ABB-Frankfurt).

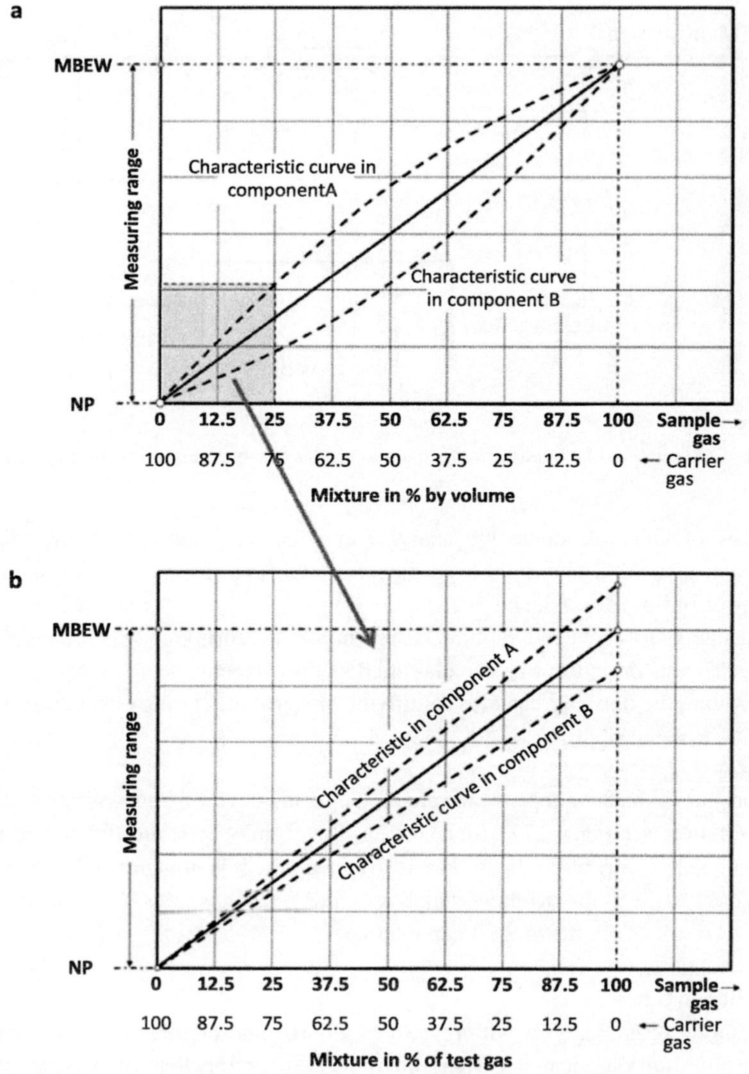

Fig. 12.71 Influence of the carrier gas dependence on the calibration curve. In a the complete concentration range from 0 ... 100 vol.% is shown. The largest deviation from the ideal characteristic curve is found within the calibration curve. For smaller admixtures (test gas mixture) the greatest deviation is found in the end point and thus changes the sensitivity multiplicatively

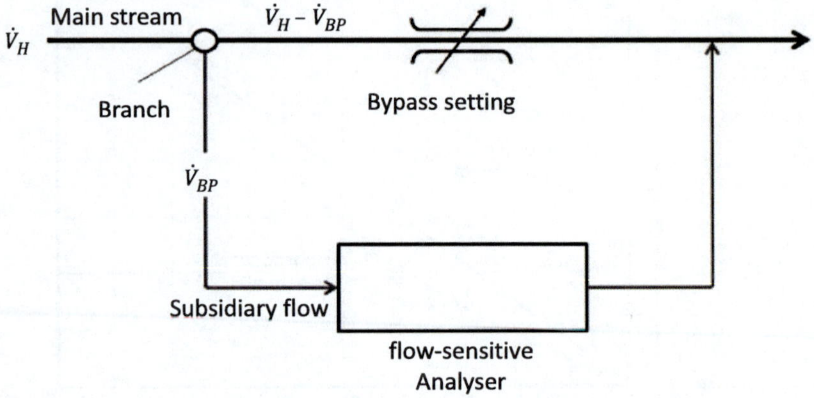

Fig. 12.72 Construction of a bypass arrangement to reduce the bypass gas flow through the analyser

- The flow of air cools down the analyzer and the temperature structure within the structure is disturbed. The change in temperature due to a change in flow is then defined as an error in flow dependence.
- The changing flow rate causes the pressure in the measuring cell and changes the gas density. The resulting error is also classified as flow dependence.
- Furthermore, the flow rate can also disturb the physical measuring effect and thus create a flow dependence.

If the errors due to the flow dependence are too large, the device manufacturers will suggest a bypass solution (see Fig. 12.72). Since the gas flow cannot be reduced at will to keep the flow effect small, a part of the main flow is diverted which is not harmful to the analyser. By this measure, the main gas flow, which provides a fast gas transfer, is maintained and the analyser receives the flow rate it can tolerate.

Position Dependence

The positioning of an analyzer with respect to a horizontal orientation is very important, especially for mobile applications. Therefore the possible errors that can occur if the base is not horizontal must be known. A simple test setup is shown in Fig. 12.73. For this purpose the analyzer is lifted on one side by a defined tilt angle $\alpha = 10°$. With an analyzer width of 480 mm (19″), a height of $h \approx 80$ mm would thus be required to reach a 10° -position. This measurement must of course be carried out for all axis directions.

In particular, analysers that operate according to a thermal process are subject to gravitational force, since free convection is caused by g. In these cases, a shift of the temperature distribution within the measuring cell can occur, which then becomes noticeable as a positional dependence. Usually this error is <1 % /10°. For stationary devices

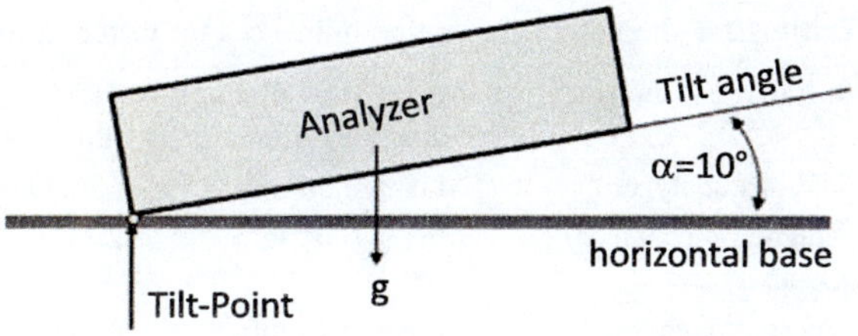

Fig. 12.73 Arrangement for checking the positional dependence of a gas analyser

(wall mounting), the position error is calibrated by a zero point adjustment so that it no longer appears.

Mechanical Loads (Shock and Vibration)

The mobile use of gas detectors inevitably leads to requirements regarding mechanical stability and insensitivity to shock and vibration. Stationary gas detection instruments can also be subjected to mechanical loads in certain installation locations (power stations, turbines, engine test benches, ...). The analyser must therefore be designed to meet these requirements and the measurement results must not be distorted (Table 12.8). In order to be able to check the mechanical properties, appropriate tests are carried out. The influences of vibrations are carried out on a so-called *shaker* (see Fig. 12.74). For this purpose the analyser is firmly attached to the shaker. The shaker can be adjusted in terms of vibration amplitude y_0 and frequency $f = \omega/2\pi$ so that different installation locations can be simulated. Furthermore, these values can also be run through automatically by a stored program in order to meet the corresponding requirements of standards and directives. The shaker is usually also rotatably mounted, so that the different installation situations can also be recorded with regard to the gravitational force. Simultaneously to the vibration test, the analyzer signals are recorded in order to determine the influence of the vibration test if necessary. Destruction usually occurs within the analyzer if it is a matter of systems capable of vibration which come into$y(t)$ resonance due to external excitation.

Mobile measuring instruments must above all withstand the *shock test* (drop test). In this test, a hand-held measuring instrument is dropped from a certain height (e.g. h = 1 m) onto a concrete floor to prove its robustness to this everyday requirement.

Leak Test

Gas measuring instruments are usually flown through by the gas to be measured. For this purpose, the internal measuring chambers of the device must be gas-tight to the environment. Leaks (leakages) can cause the gas to escape from the system and, in the case of toxic gases, lead to accidents (poisoning). The same applies to flammable (explosive) gases. Furthermore, if there is a negative pressure in the measuring system, air could enter the measuring chamber and thus lead to reduced results in the gas analysis. Both cases must be excluded by a leak test of the entire gas measuring section.

Table 12.8 Permissible vibration influences for different physical gas analyzers

Infrared gas analyzer	Max.± 0.04 mm at 5 ... 55 Hz, 0.5 g at 55 ... 150 Hz slight temporary influence on measured value near the radiator modulation frequency
UV gas analyzer	Max.± 0.04 mm/0.5 g at 5 ... 150 Hz
Magnetomechanical O_2-analyzer	Max.± 0.04 mm at 5 ... 20 Hz
Thermomagnetic O_2-analyzer	Max.± 0.04 mm at 5 ... 60 Hz
Thermalconductivity analyzer	Max.± 0.04 mm at 5 ... 30 Hz
Micromechanical WLD	Max.± 0.04 mm at 5 ... 55 Hz, 0.5 g at 55 ... 150 Hz
Laser analyzer	Max.± 0.6 mm around the optical axis, f < 500 Hz

Data source: ABB Automation GmbH, Frankfurt am Main

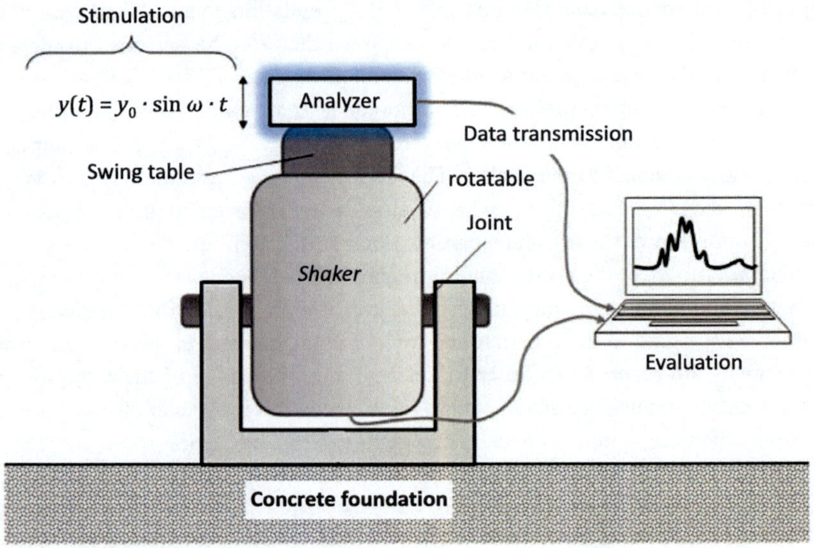

Fig. 12.74 Arrangement for measuring the influence of external vibrations on the analyser, using a vibration table/shaker

A very simple test method is to apply a small overpressure $p_{Ü}$ to the test specimen and observe the change in overpressure $\Delta p/\Delta t$ over time. The *leakage rate LR* can then be calculated as follows:

$$LR = V_0 \cdot \frac{\Delta p}{\Delta t} \tag{12.71}$$

V_O is the volume of the measuring chamber in litres. The leakage rate is therefore given in mbar · L · s^{-1}.

Example

A gas analyzer with a measuring chamber volume from $V_0 = 0.133$ L is provided with an overpressure from 211 hPa. After 5 minutes the pressure on 206 hPa has dropped. The leakage rate LR is therefore.

$$LR = 0,133L \cdot \frac{211 - 206\text{mbar}}{5 \cdot 60\text{s}} = 2,2 \cdot 10^{-3}\text{hPa} \cdot L \cdot s^{-1}$$

With this type of measurement, the temperature must be kept constant, since the pressure would fall to this value even if the temperature were to fall. In the example given, this would be $-1,4$ hPa/K. Typical manufacturer's analyser specifications indicate leakage rates that range between LR $< 1 \times 10^{-3}$ hPa l/s and LR $< 1 \times 10^{-6}$ hPa l/s.

Figure 12.75 shows a test arrangement with which this test can be performed. The gas outlet is closed while an overpressure $p_{\ddot{U}}$ is applied at the gas inlet. The overpressure is created with a bellows and fixed with a locking screw (Fig. 12.76). The build-up is now observed for about 5 minutes and the pressure drop after this time is noted. The leakage rate is then calculated according to Eq. 12.71.

Drift Behaviour

A *drift* is the irreversible change of the sensor signals. These changes lead to a shift of the zero point and/or the end point. The drift is always linked to time, as the signal change does not happen immediately. The information about the drift is represented as follows:

- Zero point drift $\leq$1% of span per week
- Sensitivity drift $\leq$1% of the measured value per week.

The causes of the signal changes can be very diverse and are also not predictable. The most important causes of drift behavior can be found in

- Ageing processes of the components used (e.g. Radiation sources in photometers)
- Poisoning effects through chemical reactions with the sample gas (pellistor, semiconductor gas sensors, ...)
- Contamination of the measuring chamber by dust and/or precipitation
- Consumption of the sensor reagents (liquid electrolytes, ...).

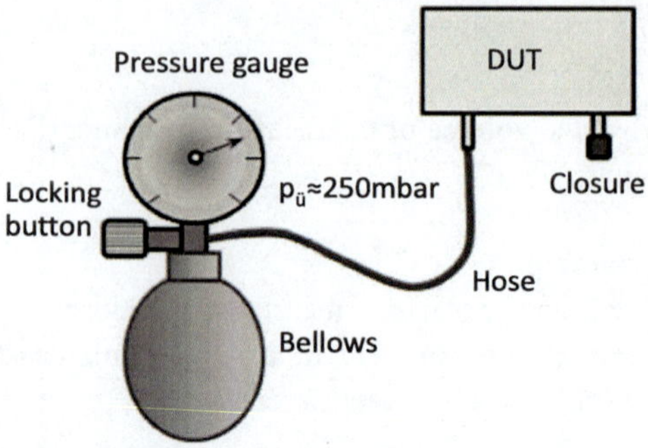

Fig. 12.75 Experimental arrangement for determining the leakage rate of gas analysers

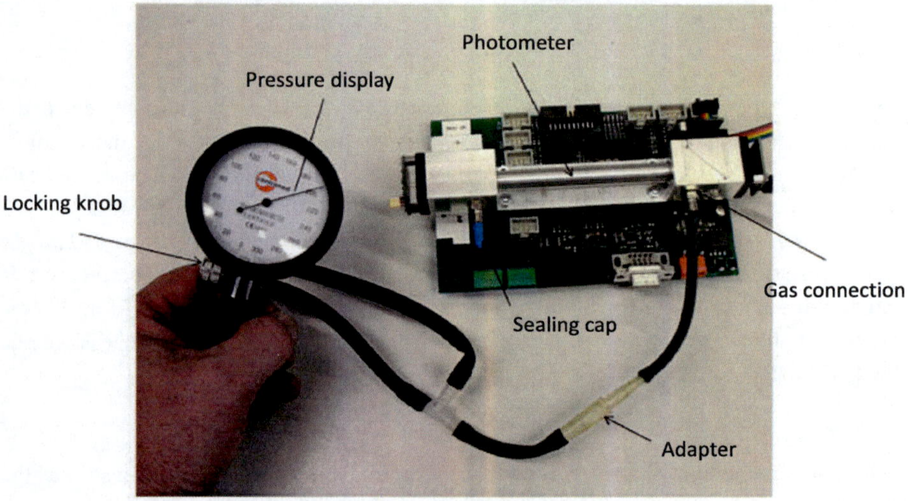

Fig. 12.76 View of the leak test with a real gas module (*ULTRA.sens*® Wi.Tec-Sensorik GmbH, Wesel). A *sphygmomanometer* can be used as pumping and measuring unit

Since these changes usually occur very slowly, the analyzers must be examined over a longer period of time in order to obtain a reliable statement about the drift behavior. Typical time periods are weeks to months. The signal course during a drift analysis is shown in Fig. 12.77.

The analyser is initially flowed with zero gas and test gas. Both the zero point and the end point are adjusted in the period t_{P1}. The measuring phase can be carried out both in the laboratory, under defined test conditions, and under real conditions on the process. However, during a laboratory test, not all influencing factors that have an effect on the drift behaviour are completely recorded. Especially the influence of accompanying gases from the process and dust can only be determined on the real process. Therefore, suitability tests for combustion plants always require testing in the laboratory and in the process in order to determine all influencing factors.

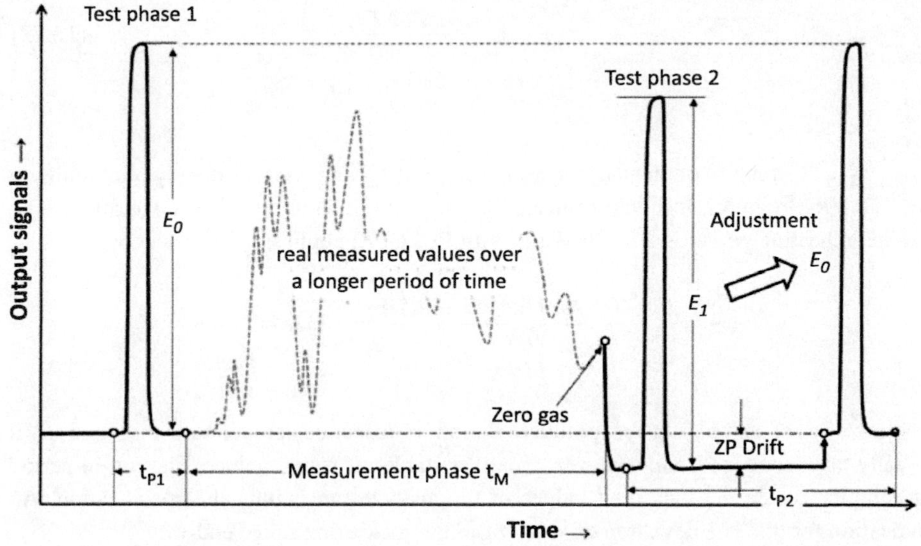

Fig. 12.77 Signal curves at the beginning (t_{P1}) and end of the test phase (t_{P2}). Afterwards, the adjustment to the specified zero and full scale value takes place. The signals during the measuring phase are of minor importance

Since the sensor signals are also influenced by other influences such as pressure and temperature, these influences must either be kept constant or be included in the drift determination by means of subsequent calculation.

Heating Time / Running-in Period

If an analyzer is switched on again after a longer pause (≈24 h), the measuring signals will not immediately adjust to a stable value. Depending on the physical structure of the analyzer, the individual components may have to heat up, for example, the IR radiation source of an IR photometer. The power loss of the radiation source also transfers thermal energy to other components, which must then also be thermally stabilized. Semiconductor gas sensors must also saturate with oxygen on the active sensor surface before a stable sensor signal is obtained. Temperature-sensitive analyzers are also thermostatted in certain cases. This means that they are heated by an active heater to a temperature that is approximately 10 K above the expected operating temperature. This temperature is controlled, whereby control deviations from $\Delta T \approx \pm 1$ K are common. Depending on the mass m_A of the analyzer and the maximum available power P, the warm-up phase can take minutes or even hours. This warm-up time t_{AWZ} can be calculated with a rough calculation:

$$t_{\text{AWZ}} = \frac{m_A \cdot c_A \cdot \Delta T}{P} \tag{12.72}$$

Example

The physical structure of an analyser weighs $m_A = 1$ kg and consists mainly of aluminium ($c_A = 0.896$kJ/kg · K). It should be heated from room temperature (22.5 °C) to 50 °C. The available heating power is P = 20 Watt. With Eq. 12.72 applies:

$$t_{\text{AWZ}} = \frac{1\text{kg} \cdot 0.896\text{kJ/kg} \cdot \text{K} \cdot 27.5\text{K}}{20\text{W}} = 1232\text{s} \approx 21 \text{ min}$$

Since larger setups always produce an inhomogeneous temperature distribution, it usually takes longer for the analyzer to deliver stable measured values. The run-in period is complete when the measured values of the analyzer are within the specification. An indication for this is a deviation of <1% from the measuring range end value.

Figure 12.78 shows a typical running-in phase of an analyzer. At switch-on time t_0 the analyzer is at ambient temperature (e.g. Room temperature). After a longer transport in winter this temperature can also be significantly below room temperature. After switching on, the temperature of the analyzer rises with an e-function and then stabilizes at the set control value (→ heating time). During this phase the measuring signals of the analyzer can behave in different ways. *Overshoots* often occur when two opposing temperature characteristics of the individual analyzer components compete with each other. A further period of time can also elapse before a stable measurement value is obtained from the analyzer. Important for the completion of the running-in period is the stability of the measurement signals.

Linearization and Linearity

Physical gas measuring instruments are, with a few exceptions, always non-linear. This means that when the gas concentration rises, the display value does not change in direct proportion to this rise. In many cases the calibration curve is therefore *curved* and loses sensitivity as the gas concentration increases. The curve is similar to an e-function, a logarithmic curve or a root function, which can usually be traced back to a physical law. The correction of this non-linear behavior is called *linearization*. The linearization can only be realized with appropriate electronic means, which are nowadays performed with microcontrollers. The first gas analysers (1920–1960) did not have these possibilities and the characteristic curve was printed on the analogue display scale of the gas measuring instrument (see Fig. 7.10). With the introduction of transistor amplifiers, the linearization functions were adapted by logarithmic amplifiers or diode characteristics (1960–1980). The first gas detectors with microprocessors appeared on the market at the beginning of the

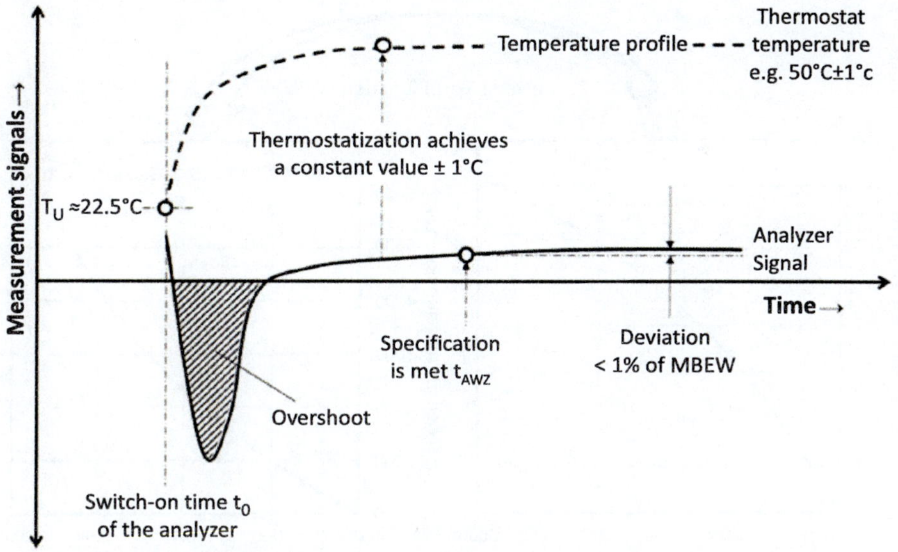

Fig. 12.78 Typical inlet behaviour of a gas analyser with a thermostat

1980s and had corresponding algorithms for linearization integrated (Lehnert et al. 1987; Wiegleb 1987).

For this type of linearization, the first step is to record the calibration curve with a representative number of interpolation points. In practice, 10 calibration points have proven to be sufficient and practical for this task. The interpolation points can either be selected at the same distance from the concentration values, or more data points can be recorded in the lower concentration range, as the sensitivity is higher there. Which type is the best for the respective analyzer has to be found out by experiments. In a second step an *inverse function* is formed from the determined data points by swapping the axes. A higher-order polynomial ($n = 2, 3, \ldots, 6$) is then placed through the existing data points. For this purpose, *Excel* or a similar program is used (e.g. *Mathcad*). The polynomial thus determined is the *linearization function* of the analyzer. Figure 12.79 shows this data for a typical gas analyzer.

The display value of the analyzer is converted to the gas concentration by this function. The result is a linear curve between the specified gas concentration and the display value of the analyzer. In Fig. 12.80 this curve is shown. The function $y(x) = 0.9999x$ with $R^2 = 1$ indicates already that it concerns here an extremely linear function. Nevertheless there are differences between the set value and the actual value, which are shown in Fig. 12.80 on the

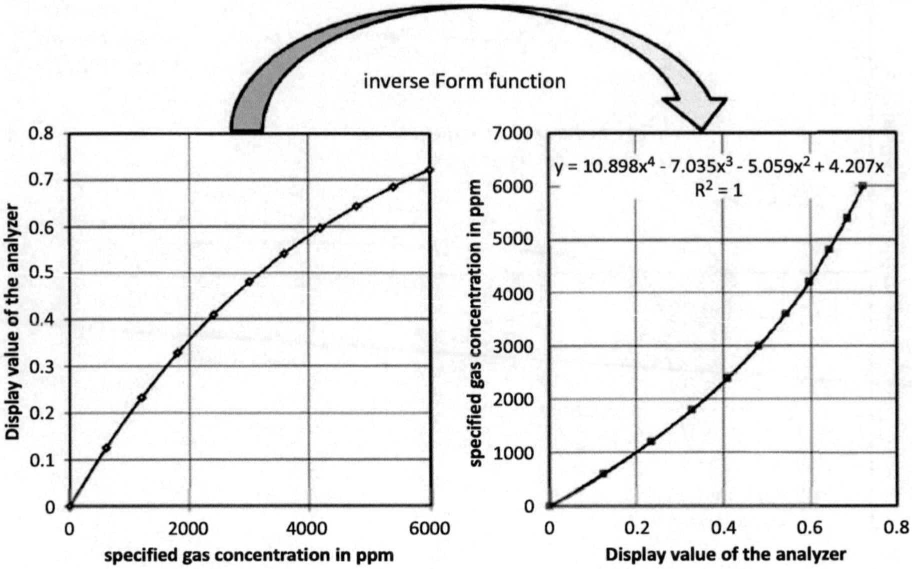

Fig. 12.79 Recording the calibration curve and converting it to the inverse function to determine the linearization function

secondary axis. The maximum deviations are ± 6 ppm. For a measuring range of 6000 ppm this linearity error is therefore ≈ ± 0.1% of the MBEW. The manufacturers of gas analysis instruments usually specify a value of ±1% of the MBEW, but this is usually not reached.

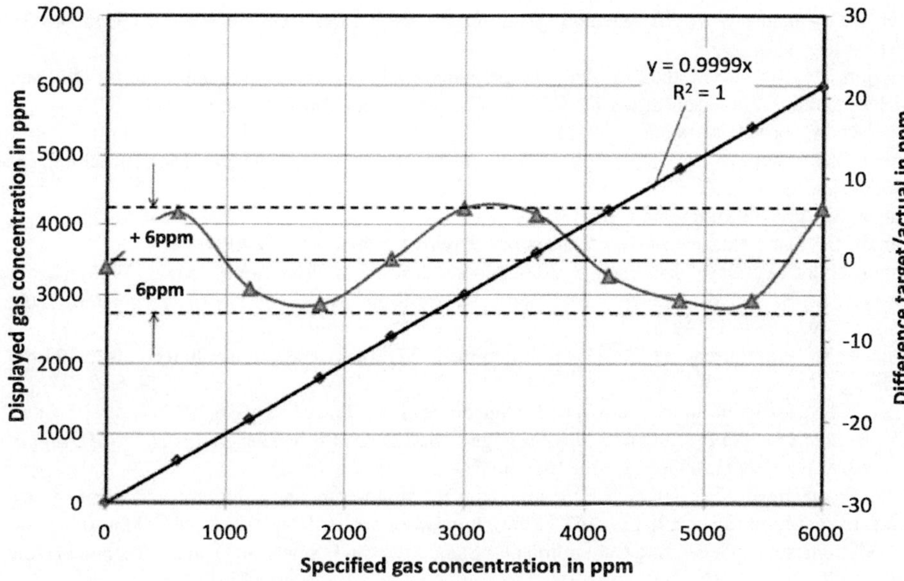

Fig. 12.80 Linearized calibration curve $y(x) = 0.9999x$ with the deviations between the target value (= preset gas concentration) and the actual value (= displayed gas concentration)

References

Adunka, F.: Meßunsicherheiten-Theorie und Praxis. Vulkan, Essen (1998)

D'Ans, J., Lax, E.: Taschenbuch für Chemiker und Physiker. 4. Aufl. Springer, Berlin (1992)

Hengstenberg, J., Sturm, B., Winkler, O.: Messen, Steuern, Regeln in der Chemischen Technik. 3. Aufl., Band III. Springer (1980)

Industriegaseverband IGV: Druckschrift Nr. 20D 30.0–04/10. Informationen zur Euro-Norm DIN EN 1089–3 Farbkennzeichnung von Gasflaschen in Deutschland (2010)

Kuchling, H.: Taschenbuch der Physik. 21. Aufl., Hanser Verlag 2014

Landolt, H., Börnstein, R.: Zahlenwerte und Funktionen aus Physik, Chemie, Astronomie, Geophysik und Technik, 6. Aufl. Springer, Berlin (1960)

Lehnert, M., Modlinksi, U., Wiegleb, G.: Mikroprozessorgesteuertes Gasanalysengerät in Kompaktbauweise. Tech. Mess. **54**, 423–429 (1987)

Lide, D.R. (Hrsg.): CRC handbook of chemistry and physics, 88. Aufl. CRC, Boca Raton, FL, 2007–2008 (2007)

Lieck, G.: Nachweisgrenze und Rauschen. Laborpraxis, Juni, 62–67 (1998)

Marsoner, H.J., Stolze, A., Klepp, G.: Untersuchung über die Eichgasherstellung mittels kritischer Düsen bei Umgebungsdrücken von 1 bar bis 4 bar. Tech. Mess. **43**(12), 365–400 (1976)

McKinley, J.J.: How to Calibrate Ambient Air Monitors. Instrumentation Technology, June, 45–49 (1974)

Nelson, G.O.: Gas Mixtures. Preparation and Control. Lewis Publishers, Chelsea (1992)

Ritter, J.J.: Exponential Dilution Flask. United States Patent 3,911,723 14. Okt. 1975

Ritter, J.J., Adams, N.K.: Exponential dilution as a calibration technique. Anal. Chem. **48**(3), 612–619 (1976)

Schierjott, P.: Absolute Method for dynamic preparation of calibration gases approved by DIN and ISO standards. Druckschrift der Fa. Wösthoff Messtechnik GmbH Bochum (2012)

Schierjott, W.: private Mitteilung (2015)

Schmäh, M.: Vortrag zur Überprüfung und Justierung vom Quecksilber-Messeinrichtungen. VDI-Wissensforum (2009)

Schmäh, M.: private Mitteilung (2015)

Schmidt, G.: Sonic Nozzles for Gas Calibrators. Envirotech IET, 14–15 (2014)

Schreckenberg, W.: Gase-Handbuch. 3. Aufl. Messer-Griesheim, Frankfurt am Main (1985)

Slemeyer, A., Balser, M., Kübler, P.: Rechnergestützte Kalibrierung von Gasanalysegeräten. Tech. Mess. **5**(92), 58–62 (1992)

Sommer, K.: Prüfeinrichtung für Gasmischpumpen. Metrologische Abhandlungen. **8**(2), 91–99 (1988)

Stetter, H.: Messtechnik in Maschinen und Anlagen. B.G. Teubner, Stuttgart (1992)

TÜV: Bericht über die Prüfung der Leistungseigenschaften des Prüfgasgenerators *HovaCAL digital 211 MF*, Nr. 936/21205306/A vom 15.3.2007

Vautz, W., Schmäh, M.: HovaCAL a generator for multi-component humid calibration gases. Int. J. Ion Mobil. Spec. **12**, 139–147 (2009). https://doi.org/10.1007/s12127-009-0030-0

VICI Metronics Inc.: Generating Calibration Gas Standards with Permeation Devices. Technical Note 1001 (2012)

Wiegleb, G.: IR-Gasanalyse mit Mikroprozessorsteuerung. Labor Praxis. **10**, 118–122 (1987)

Wiegleb, G., Böttcher, A.: Messungen von Chlorgas-Immissionen mit dem BINOS IM 01-System. GIT Fachz. Lab. **5**, 497–501 (1988)

Wiegleb, G., Randow, A.: Einrichtung für die Herstellung bestimmter Konzentrationen gasförmiger Stoffe sowie zum Mischen verschiedener gasförmiger Stoffe in einem vorgegebenen Verhältnis. Deutsche Patentanmeldung DE 3716289 A1 5.5.1987

Related References

Anders, B., Hässelbarth, W.: Analytische Überprüfung von Kalibriergasgemischen nach DIN EN ISO 6143, Beuth Verlag (2014)

Barbe, J., Grabhorn, H.: Zuverlässige Kalibriergase für gasanalytische Verfahren. Labor Praxis. **36**, 52–61 (2012)

Below, L. et al.: Messgeräte für Gase – Messgeräte für den Kohlenstoffdioxidanteil in Brenngasen. PTB-Prüfregeln. Bd. 24, Eigenverlag der PTB (1998)

Engelhardt, H.: Verfahren zur Herstellung von Prüfgasen für die Eichung von Gasanalysatoren – eine Übersicht. Tech. Mess. **2**, 43–52 (1977)

Faulhaber, R., Staab, J.: Analysengeräte mit Qualitätszertifikat. Tech. Mess. **2**, 61–66 (1977)

Hart, H., Lotze, W., Woschni, E.-G.: Meßgenauigkeit. 3. Aufl. Oldenbourg-Verlag, München (1997)

Hengstenberg, J., Sturm, B., Winkler, O.: Messen, Steuern, Regeln in der Chemischen Technik. 3. Aufl., Band III. Springer (1980)

Jentsch, D., Otte, E.: Detektoren in der Gaschromatographie. Akademische Verlagsgemeinschaft, Frankfurt am Main (1970)

Kettrup, A.: Apparatur zu dynamischen Herstellung von Prüfgasen. GIT Fachz. Lab. **6**, 556–558 (1982)

Kuchling, H.: Taschenbuch der Physik. 21. Aufl., Hanser Verlag 2014

Lange, N.A.: Handbook for chemistry. 15. Aufl. McGraw-Hill, New York (1999)
Lovelock, J.E.: Gas Chromatography. R.P.W. Scott (Hrsg.). Butterworth Inc., Washington D.C., 26 (1960)
Müller, H.: Konzeption eines automatischen Kalibrierprozesses für die Qualifizierung von IR-Gassensoren unter Berücksichtigung des Gasdichtefehlers. Diplomarbeit, FH-Heilbronn (2010)
Pichelmaier, J.: Kalibriertechnik für Gassensoren. Tech. Mess. **57**(11), 419–423 (1990)
Pichelmaier, J.: Befeuchtung von Prüfgas für die Kalibrierung von Gas- und Feuchtesensoren. Tech. Mess. **58**(12), 471–477 (1991)
Wiegleb, G.: A new calibration technique for industrial gas analyzers. Process. Control. Qual. **3**, 273–281 (1992)

Standards

DIN 51897 Gasanalyse-Begriffe (2000)
DIN 51898-3 Gasanalyse – Absolutes volumetrisches Verfahren zur dynamischen Herstellung von Kalibriergasen – Teil 3: Ein-Kolbenpumpensysteme (2014 Entwurf)
DIN EN ISO 14912 Gasanalyse – Umrechnung von Zusammensetzungsangaben für Gasgemische (2006)
DIN EN ISO 6141 Gasanalyse – Inhalte von Zertifikaten für Kalibriergasgemische (2015)
DIN EN ISO 6142 Gasanalyse – Herstellung von Prüfgasen – Wägeverfahren (2006)
DIN EN ISO 6142-1 Gasanalyse – Herstellung von Prüfgasen – Teil 1: Wägeverfahren für Prüfgase der Klasse I (2014 Entwurf)
DIN EN ISO 6143 Gasanalyse – Vergleichsverfahren zur Bestimmung und Überprüfung der Zusammensetzung von Kalibriergasgemischen (2006)
DIN EN ISO 6144 Gasanalyse-Herstellung von Prüfgasen-Volumetrisch-statisches Verfahren (2006)
DIN EN ISO 6145-1 Gasanalyse-Herstellung von Kalibriergasgemischen mit Hilfe von dynamisch-volumetrischen Verfahren. Teil 1: Kalibrierverfahren (2008)
DIN EN ISO 6145-10 Gasanalyse-Herstellung von Kalibriergasgemischen mit Hilfe von dynamisch-volumetrischen Verfahren. Teil 10: Permeationsverfahren (2008)
DIN EN ISO 6145-11 Gasanalyse-Herstellung von Kalibriergasgemischen mit Hilfe von dynamisch-volumetrischen Verfahren. Teil 11: Elektrochemische Herstellung (2008)
DIN EN ISO 6145-2 Gasanalyse-Herstellung von Kalibriergasgemischen mit Hilfe von dynamisch-volumetrischen Verfahren. Teil 2: Kalibrierverfahren (2008)
DIN EN ISO 6145-3 Gasanalyse-Herstellung von Kalibriergasgemischen mit Hilfe von dynamisch-volumetrischen Verfahren. Teil 3: Kalibrierverfahren (2008)
DIN EN ISO 6145-4 Gasanalyse-Herstellung von Kalibriergasgemischen mit Hilfe von dynamisch-volumetrischen Verfahren. Teil 4: Kontinuierliches Spritzen-Injektionsverfahren (2008)
DIN EN ISO 6145-5 Gasanalyse-Herstellung von Kalibriergasgemischen mit Hilfe von dynamisch-volumetrischen Verfahren. Teil 5: Kapillardosierer (2011)
DIN EN ISO 6145-6 Gasanalyse-Herstellung von Kalibriergasgemischen mit Hilfe von dynamisch-volumetrischen Verfahren. Teil 6: Kritische Düsen (2008)
DIN EN ISO 6145-7 Gasanalyse-Herstellung von Kalibriergasgemischen mit Hilfe von dynamisch-volumetrischen Verfahren. Teil 7: Thermische Durchflussregler (2011)
DIN EN ISO 6145-8 Gasanalyse-Herstellung von Kalibriergasgemischen mit Hilfe von dynamisch-volumetrischen Verfahren. Teil 8: Diffusionsverfahren (2008)
DIN EN ISO 6145-9 Gasanalyse-Herstellung von Kalibriergasgemischen mit Hilfe von dynamisch-volumetrischen Verfahren. Teil 9: Sättigungsverfahren (2011)
VDI-Richtlinie 2448 Blatt 2: Statistische Auswertung von stichprobenartigen Emissionsmessungen an geführten Quellen. Ermittlung der oberen Vertrauensgrenze (1997)

VDI-Richtlinie 2449 Blatt 1: Prüfkriterien für Meßverfahren. Ermittlung von Verfahrenskenngrößen für die Messung gasförmiger Schadstoffe (Immission) (1995)
VDI-Richtlinie 2449 Blatt 1: Prüfkriterien für Meßverfahren. Prüfkriterien von Messverfahren. Allgemeine Methode zur Ermittlung der Unsicherheit kalibrierfähiger Messverfahren (2001)

13 Dust Measurement Technology

Physically speaking, dust is a special case of the distribution of a solid in a gaseous medium. In common parlance, the term is used for particles suspended in the air (suspended dust including soot and smoke) or a separated pile of fine particles.

Dust can be of very different origins, ths, it can have natural or anthropogenic causes, be made of organic or inorganic materials. Its properties can also be very different. Differentiations can be made, for example, according to the particle size, whereby coarse dust with a particle diameter of $d_P > 10\,\mu m$ and fine dust with a particle diameter of

G. Wiegleb, *Gas Measurement Technology in Theory and Practice*,
https://doi.org/10.1007/978-3-658-37232-3_13

$d_P < 10\,\mu m$ are differentiated. Particle sizes $d_P < 2.5\,\mu m$ are respirable fine dust and $d_P < 0.1\,\mu m$ are called ultra-fine particles. However, other properties such as particle shapes, adhesion tendency, conductivity or electrostatic chargeability also show the diversity of the dust term.

From the point at which the dust is generated to its effect, that is, from emission to immission, it is also subject to various processes, for example, dilution or distribution, and can enter into chemical reactions with other air pollutants.

Dust can have very different effects on humans and the environment. It can simply be disturbing, but also explosive or hazardous to health. In this respect, the measurement of dust emissions is an important field in gas measurement technology.

13.1 Introduction

There can be various reasons for carrying out measurements of dust emissions. Accordingly, these can be discontinuous or continuous.

Discontinuous measurements for air pollutants including dust shall be used for the time-limited random sampling of the emission behaviour of an installation. However, these measurements shall allow conclusions to be drawn about the continuous behaviour of the installation and shall take account of possible variations over time of the air pollutants to be measured. The reasons for this can be of various kinds:

- Verification of compliance with the emission limit
- Self-monitoring, safety inspections
- Controls to determine the condition of the installation or in case of complaints
- Acceptance measurements (proof of guarantee)
- Measurements within the scope of a licensing procedure (e.g. for plant extensions, conversions, etc.)
- Preparation of the declaration of issue
- Functional test or calibration of continuous emission measurement systems
- Cause analysis of a certain emission behaviour (e.g. in case of operational disturbances)
- Prediction of the emission behaviour for certain operating conditions.

Officially ordered measurements are supported by §§26, 28 BImSchG (Federal Immission Control Act) among others. The measurement requirements are specified in individual ordinances to the BImSchG and TA Luft.

Continuous emission monitoring is ordered by the competent monitoring authority for installations requiring a permit and in special cases for installations not requiring a permit (§29 BImSchG). Concrete requirements are also contained in the various plant-related BImSchV and TA Luft.

To ensure the representativeness of the measurements, both discontinuous and continuous measurements must meet certain requirements. These are laid down in a number of

Table 13.1 Selected European standards and VDI guidelines for dust measurement

Name	Title	Date of issue (verification)
DIN EN 13284–1	Stationary source emissions - determination of mass concentrations of low level particulate matter - part 1: Manual gravimetric method	2002–04
DIN EN 13284–2	Stationary source emissions - determination of dust mass concentrations at low dust concentrations - part 2: Automatic measuring systems	2004–12
DIN EN 14181	Emissions from stationary sources - quality assurance for automatic measuring systems	2015–02
DIN EN 15259	Measurement of emissions from stationary sources - requirements for measurement lines and measuring stations and for the measurement task, measurement plan and measurement report	2008–01
VDI 2066 sheet 1	Measurement of particles - dust measurements in flowing gases - gravimetric determination of dust loading	2006–11 (2012–04)
VDI 2066 sheet 5	Measurement of particles - dust measurement in flowing gases; fractional dust measurement using the impaction method - Cascade impactor	1994–11 (2015–05)
VDI 2066 sheet 8	Measurement of particles - dust measurement in flowing gases - measurement of the soot number in combustion plants for fuel oil EL	1995–09 (2011–04)
VDI 2066 sheet 10	Measurement of particulate matter - measurement of dust in flowing gases - measurement of emissions of PM 10 and PM 2.5 from the sources listed above by the impaction method	2004–10 (2015–05)
VDI 3950	Emissions from stationary sources - quality assurance for automatic measuring and electronic evaluation equipment	2006–12 (2012–04)

European standards and VDI guidelines. If standards exist for certain situations, these are to be regarded as binding. A selection of corresponding standards and VDI guidelines for dust measurement is listed in Table 13.1.

13.2 Discontinuous Emission Monitoring for Dust

The discontinuous measuring methods serve on the one hand to gravimetrically assess the dust content per unit volume of an exhaust gas stream and on the other hand to determine the particle size distribution of the dust particles. They are often used as reference measuring methods, since calibration with external standards, for example, test gas, is not possible. In principle, the discontinuous manual methods have proven themselves in practice for decades. However, problems occur in particular when measuring low dust contents <0.5 mg/m^3 and in water-saturated exhaust gases.

13.3 Gravimetric Determination of Dust Content

The basis of the process is the isokinetic extraction of a dust-laden partial gas volume flow from a main volume flow, the separation of the dust contained therein at a filter and the volume-related determination of the dust mass. In manual and semi-automatic processes, this is determined by differential weighing of the dried filter before and after sampling.

Requirements for Measuring Sections and Measuring Stations

When carrying out measurements and for the quality of the measured values generated by them, a careful selection of the measurement location is of great importance. The following requirements must be observed:

- Position and shape of the measuring section in the exhaust gas duct
- Position of the measuring plane in the measuring section
- Number, position and condition of the measuring openings
- Condition of the measuring platform.

Since the distribution of exhaust gas velocity and dust concentration over the measuring cross-section may be inhomogeneous, a preliminary investigation at the selected measuring point is recommended with regard to the inlet and outlet sections and the presence of internals, bends, etc.

The measuring cross section should be arranged within a straight measuring section with a constant shape and size and trouble-free inlet and outlet. The selection criteria for the correct placement of the sampling point are defined according to VDI 2066 sheet 1 or DIN EN 13284-1. In addition, according to the latter regulation, the measuring cross-section must be selected in such a way that a device-dependent minimum volume flow is maintained. In the case of dynamic pressure sensors (standard device for measuring the exhaust gas velocity), this must correspond to a differential pressure of 5 Pa.

The maximum deviation between the axis of the flow and the axis of the exhaust duct shall not exceed 15°. The maximum ratio between the minimum and maximum exhaust gas velocity in the duct must not exceed a value of 1 : 3.

For a representative result in dust measurement, so-called network measurements according to VDI 2066 sheet 1 or DIN EN 13284-1 must be carried out. According to this, the measuring cross-sections are divided into partial areas with the same area sizes (rectangular cross-sections into similar partial areas, round cross-sections into circular rings) and the measuring points are determined on the centres of gravity of the rectangular areas (see Fig. 13.1) or intersections of the measuring axes with the centre of gravity lines of the circular rings (see Fig. 13.2).

According to VDI 2066 Sheet 1 and DIN EN 13284-1, the general and tangential methods exist for determining the positions for the measuring points in round exhaust ducts, which are to be regarded as equivalent (see Fig. 13.2). With both methods, the

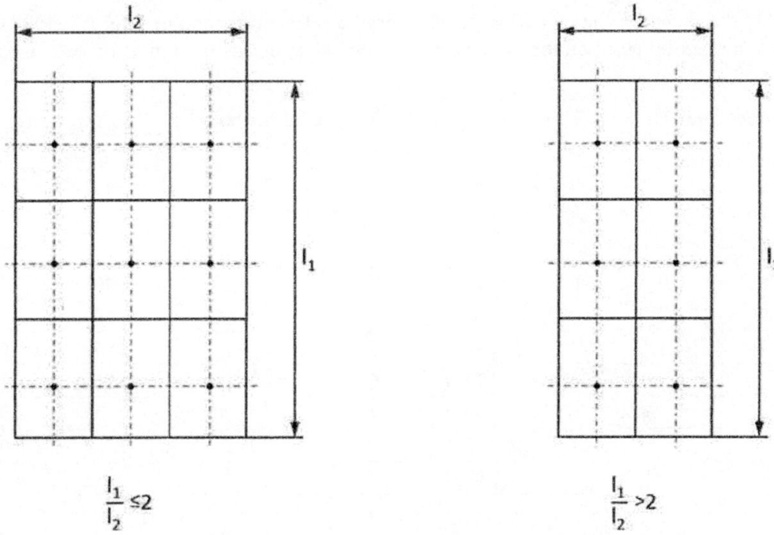

Fig. 13.1 Position of the measuring points in the rectangular duct cross-section according to VDI 2066 sheet 1 or DIN EN 13284-1

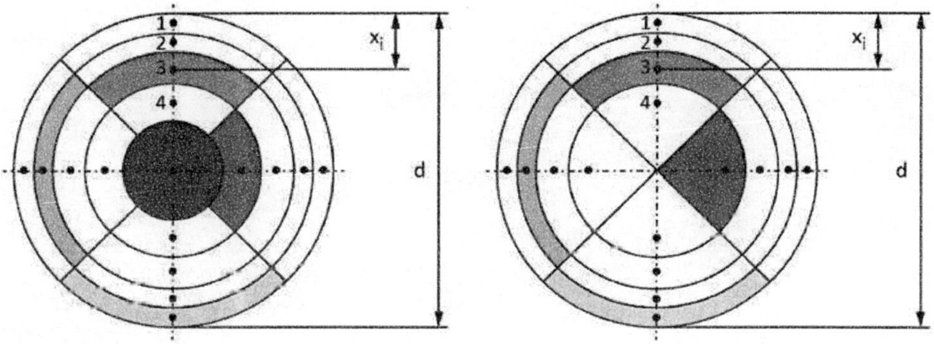

Fig. 13.2 Position of the measuring points in the round duct cross-section according to VDI 2066 sheet 1 or DIN EN 13284-1

measuring cross-section is divided into equally large partial areas, with the measuring points being located in the middle of each partial area. In the general method, an additional measuring point is located in the middle of the exhaust duct.

The calculation of the distance of the measuring points from the duct wall is done according to the following formula:

Table 13.2 Percentage values for K_i depending on the selected method of DIN EN 13284-1 (i= index of the measuring point on the diameter, n_d= number of measuring points on each measuring axis)

General procedure					Tangential method				
i	Ki				i	Ki			
	$n_d = 3$	$n_d = 5$	$n_d = 7$	$n_d = 9$		$n_d = 2$	$n_d = 4$	$n_d = 6$	$n_d = 8$
1	11.3	59	4.0	3.0	1	14.6	6.7	4.4	3.3
2	50.0	21.1	13.3	9.8	2	85.4	25.0	14.6	10.5
3	88.7	50.0	26.0	17.8	3		75.0	29.6	19.4
4		78.9	50.0	29.0	4		93.3	70.4	32.3
5		94.1	74.0	50.0	5			85.4	67.7
6			86.7	71.0	6			95.6	80.6
7			96.0	82.2	7				89.5
8				90.2	8				96.7
9				97.0					

Table 13.3 Minimum number of measuring points according to both regulations

	Minimum number of measuring points	
Measuring cross section [m^2]	Round exhaust ducts	Rectangular exhaust ducts
<0.1	1	1
0.1–1.0	4	4
1.1–2.0	8	9
>2.0	At least 12 and 4/m^2	At least 12 and 4/m^2

$$x_i = K_i \cdot D \tag{13.1}$$

x_i	Distance of point i from the duct wall
K_i	Percentage value, depending on the selected method
D	Channel diameter

The percentage value K_i can be taken from Table 13.2 depending on the selected method.

In general, it is sufficient to define two measuring axes perpendicular to each other. Should it be necessary to increase the number of measuring points (e.g. in case of unfavourable flow conditions), the corresponding calculation formula for x_i *can* be found in the guidelines. The determination of the measuring points in rectangular flue gas ducts is carried out analogously to VDI 2066 sheet 1 or DIN EN 13284-1. 20 measuring points are usually sufficient for large flue gas ducts. Depending on the measuring cross-section, a minimum number of measuring points must be observed. Table 13.3 provides a corresponding overview.

Performance of Manual Gravimetric Measurements
In the run-up to the measurements, and in the case of measurements ordered by the authorities, a measurement plan in accordance with the specifications of DIN EN 15259 must be submitted. In addition, it is necessary to subject the filter sleeve and the plane filter from the measuring apparatus to a pre-treatment before starting and then to weigh them out. According to VDI 2066 sheet 1 and DIN EN 13284-1, the parts are stored for this purpose at 180 °C for a period of 1 h in a drying cabinet and then cooled to room temperature in the desiccator for at least 8 h.

Since the difference in mass weighing before and after measurement is used to determine the dust content, the post-treatment should be of the same type. The standards require drying for 1 h at 160 °C and cooling for 8 h. Depending on the properties of the dust, especially its thermal stability, other temperatures can also be selected. The principle structure of sampling is shown in Fig. 13.3.

Extractive sampling for particulate detection must be carried out isokinetically. This means that the partial volume flow is extracted at approximately the same velocity as in the main exhaust gas flow in order to prevent segregation phenomena during sampling. A sufficient nozzle diameter must be used for this purpose. In addition, knowledge of the flow conditions in the measuring cross section is required. If the exhaust gas velocity at the measuring point is not known, it must be measured using a dynamic pressure probe or Prandtl pitot tube with a connected micromanometer. From its hook-shaped probe, which is directed against the exhaust gas flow, total pressure and static pressure are measured. The measurement is carried out via differential pressure manometer. The dynamic pressure is then calculated as follows:

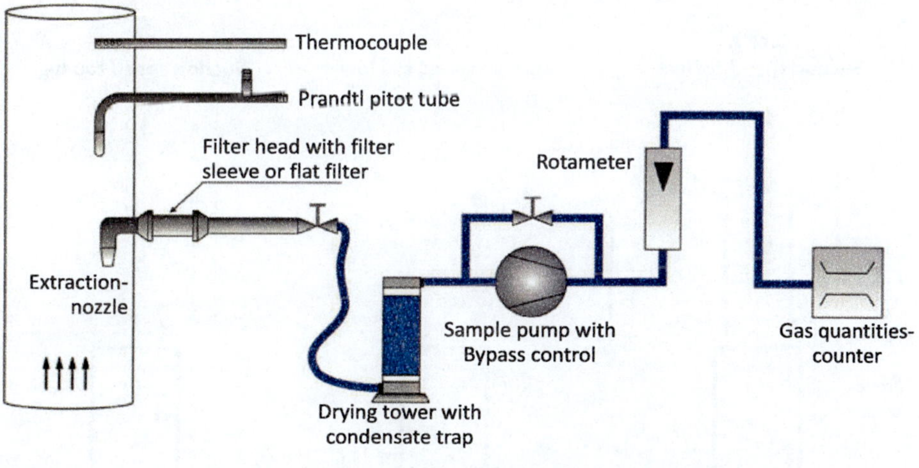

Fig. 13.3 Measuring system for gravimetric dust measurement with external filter

$$p_{\text{dyn}} = p_{\text{ges}} - p_{\text{stat}} \tag{13.2}$$

p_{dyn}	Dynamic pressure
p_{ges}	Total print
p_{stat}	Static pressure

This information can then be used to calculate the flow velocity.

$$v = k \cdot \sqrt{\frac{2 \cdot p_{\text{dyn}}}{\rho}} \tag{13.3}$$

v	Gas velocity [m/s]
k	Factor for consideration of the geometry of the pitot tube (Prandtl tube k = 1)
p_{dyn}	Dynamic pressure at the Prandtl tube [pa]
ρ	Gas density in operating condition [kg/m^3].

In the event that measurements are made in combustion exhaust gases instead of in air, the gas composition must also be determined to determine the exhaust gas density ρ Furthermore, the temperature, pressure and moisture content of the flue gas are important for determining the dust load in the flue gas. The influence of an extraction system that is not velocity balanced is shown in Fig. 13.4.

The gas flow in front of the probe opening is influenced by suction that is not adapted to the real exhaust gas velocity. If the extraction speed is selected too low, a

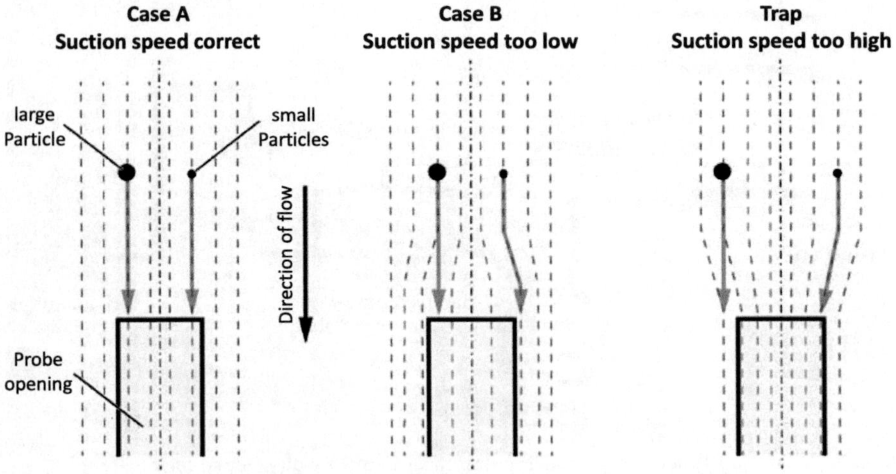

Fig. 13.4 Influence of suction errors on the sampling

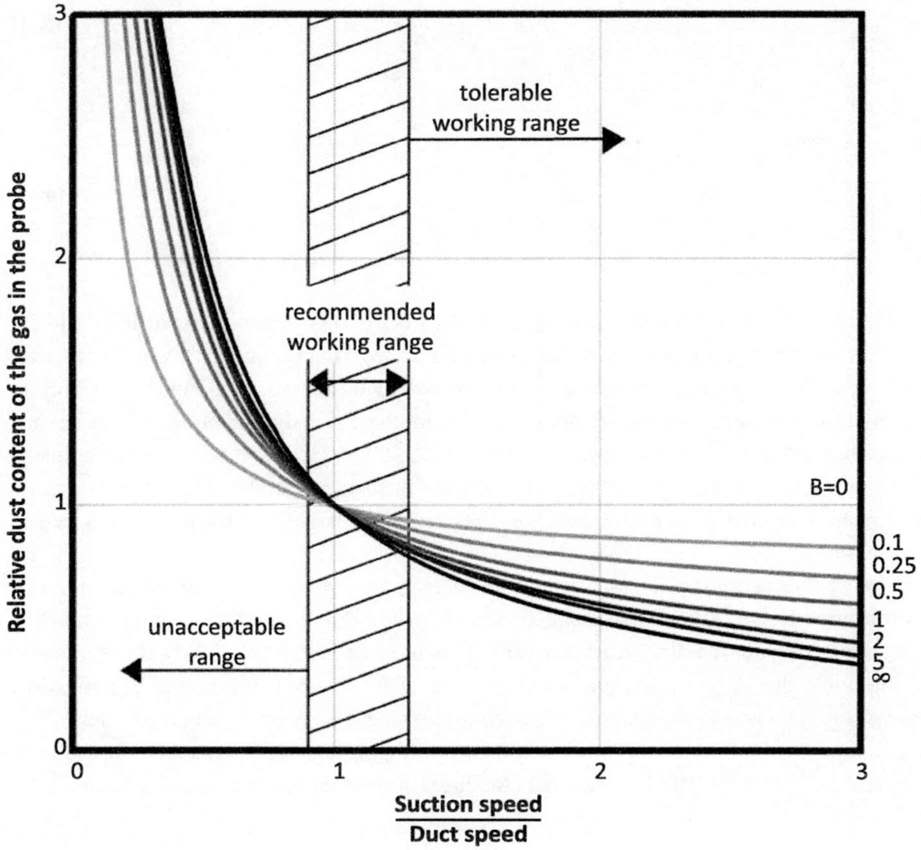

Fig. 13.5 Error at different speed ratios (recommended working range − 5 to +15%)

disproportionately large number of larger (heavier) particles are captured due to the mass inertia. If the extraction speed is too high, the effect is reversed, whereby the segregation effects are more significant if the extraction speed is too low than if the required extraction speed is exceeded. In applications where a correct setting of the extraction is not easily possible (e.g. for pulsating exhaust gas flows), the extraction speed is selected above the flow speed determined at the measuring point (max. 10%).

Figure 13.5 shows the relative dust content of the gas in the probe as a function of the ratio of extraction speed to duct speed.

This shows that the error rises steeply at lower extraction speeds, while it is comparatively low in the range of higher partial flow speeds. The dimensionless similarity parameter B used for evaluation is calculated as follows:

$$B = \frac{v_O \cdot v_S}{D_A \cdot g} \tag{13.4}$$

B	Similarity parameter
v_O	Channel velocity [m/s]
v_S	Sinking speed of the particles [m/s]
D_A	Probe diameter [m]
g	Acceleration of gravity [m/s^2]

Figure 13.5 also shows the limiting cases $B = 0$ for very small solid particles and for $B = \infty$ very large particles. In addition, the different working ranges are given. In practice, work is usually carried out in the slightly overisokinetic range. This applies not only to gravimetric dust content measurement, but also to continuously operating extractive dust measuring instruments. For network measurements, the extraction speed is to be adjusted by changing the exhaust gas volume flow at each measuring point in the measuring cross section in such a way that isokinetic or slightly overisokinetic extraction is ensured in each case.

During dust sampling, a measurement of the sampled volume is necessary at the same time in order to be able to make statements about the dust content per m^3 volume. It is important to ensure that the measuring arrangement has a high degree of tightness in order to minimize the error on the gas volume meter. By offsetting the partial volume flow against the sampling time, the partial gas volume extracted via filter sleeve and plane filter is obtained. With all manual procedures, the sampling time usually extends over cycles of 30 minutes. The calculation of the dust content is simplified according to the formula

$$c_{\mathrm{iNtr}} = \frac{m}{V_{\mathrm{iN}}} \tag{13.5}$$

c_{iNtr}	Dust concentration in standard condition dry [mg/m^3]
m	Balanced mass [mg]
V_{iN}	Dry partial volume extracted in standard condition [m^3]

The specification of the dust content refers to the dust concentration in the standard state dry. The standardization is carried out with the extracted partial volume flow. The difference between the balanced mass of the filter sleeve and the plane filter before and after the measurement is referred to as mass m. The common unit for this is mg. In addition to the mass of the dust at the filters, the mass of the dust in the sampling device after recovery of the deposits by rinsing and evaporation of the rinsing solution with subsequent weighing shall be determined and added.

The extracted partial volume in standard condition is calculated as follows:

$$V_{\text{i.N.}} = \frac{T_n}{T_g} \cdot \frac{p_k}{p_n} \cdot V \tag{13.6}$$

V	Suctioned partial volume [m3]
T_n	Standard temperature (273,15 K)
T_g	Gas meter temperature [K]
p_n	Standard pressure (1013 mbar)
p_k	Absolute pressure in the duct [mbar]

Sampling Options

Depending on the measuring task, different versions of the measuring filter can be selected for sampling. The guidelines VDI 2066 sheet 1 and DIN EN 13284-1 provide specific design regulations and recommendations for the filter material. Depending on the properties of the sample gas, the filter housing is located either in the exhaust duct (in-stack arrangement) or outside the exhaust duct (out-stack filtration). For saturated exhaust gases (water, SO_3, etc.) the out-stack apparatus must be used. The different designs are shown in Figs. 13.6 and 13.7.

Both directives contain notes on the filter materials to be used. According to these, for example, glass fibre filters should not be used because of possible reactions with SO_3. In addition to the already proven quartz fibre filters, PTFE filters are also used, but their application is limited to a temperature range of up to max. 230 °C. When evaluating the measurement, the dust deposits in front of the filter must be explicitly taken into account. They are washed out with water and acetone and determined as evaporation residue.

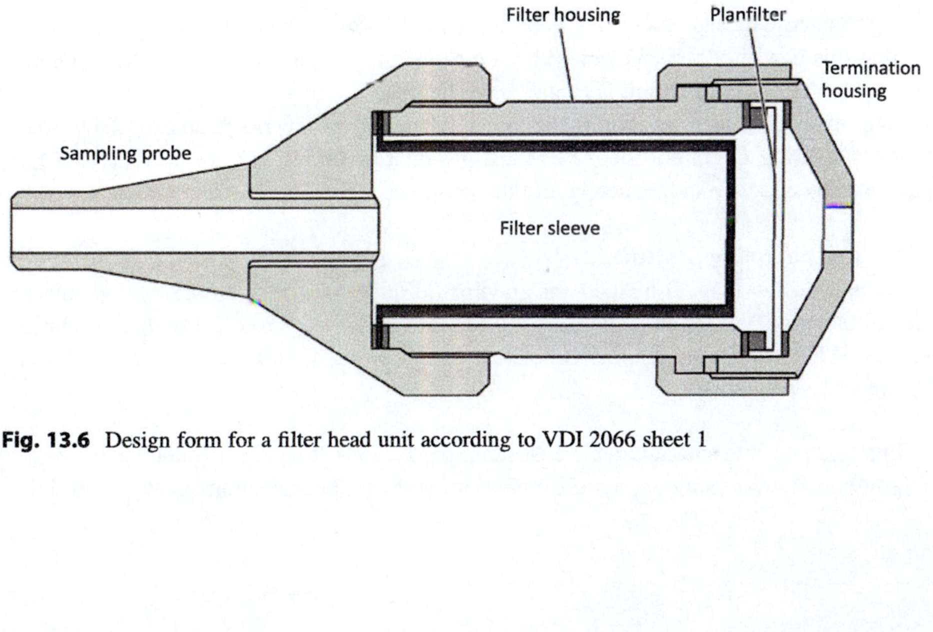

Fig. 13.6 Design form for a filter head unit according to VDI 2066 sheet 1

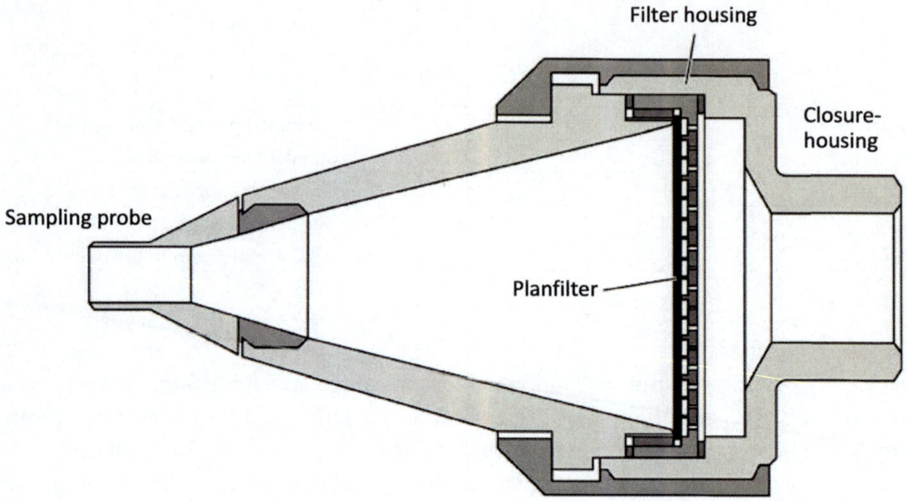

Fig. 13.7 Design form for a plane filter head unit according to VDI 2066 sheet 1

Special requirements for dust measurement are made in exhaust gases saturated with water vapour. The measuring arrangements listed so far are not applicable in this case because the filters would soak through moisture in a very short time and thus lose their separation efficiency. In addition, the formation of salts or the washing out of soluble dust components could falsify the measurement result. For this reason, a process has been developed in which dilution with hot, dry air immediately behind the extraction probe lowers the dew point and thus reduces the relative humidity in the extracted partial flow. This dilution method was discussed in 1993 as a preliminary draft of VDI 2066 sheet 9, but was later withdrawn.

If gravimetric measurements are to be carried out in exhaust gases saturated with water vapour, this can only be achieved with an out-stack arrangement in which the sampling tube upstream of the filter unit is heated accordingly.

If the exhaust gas temperature is slightly above the water dew point, an external heating of the measuring filters outside the exhaust gas duct as shown in Figs. 13.6 or 13.7 has proven to be effective in practice before the gravimetric measurements are started.

Automatic Sampling Systems

In addition to the manual method for gravimetric dust measurement, various automated processes now exist, for example from the SHC500 Gravimat product family from SICK AG or iTES from Paul Gothe GmbH, as well as the GMD 12 from Dr. Födisch Umweltmesstechnik AG.

Firstly, there are semi-automatic dust sampling systems (Fig. 13.8) which continuously measure the flow velocity or pressure conditions at the probe and automatically control the

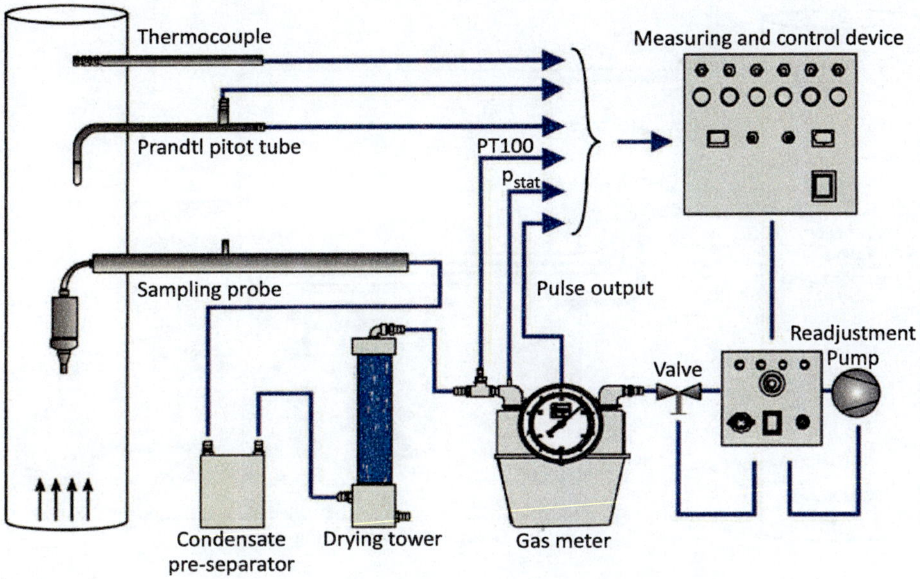

Fig. 13.8 Semi-automatic sampling for gravimetric dust measurement

extraction speed. However, the determination of the amount of dust collected at the filter is still done by weighing the sampler before and after the measurement.

On the other hand, there are fully automated systems (Fig. 13.9) that do not require any complex filter preparation and processing and provide real-time measured values. At the same time the exhaust gas humidity and velocity, optionally the oxygen and carbon dioxide content in the exhaust gas are measured. Due to the high acquisition costs, such devices are hardly used in the market.

Gravimetric Measurement with on-Site Evaluation by Hot Weighing

The problems with both manual and automatic gravimetric dust measurement are the time-consuming preparation and post-processing of the filters and the delayed evaluation. It is not possible to make a statement about the actual dust content at the measuring location during the measurements. As the evaluation of the samples must usually be carried out in accredited laboratories, several days or even weeks can pass between the measurement date and the submission of the measurement report.

In the meantime, the GMD 13[1] gravimetric measuring system (see Fig. 13.10) has been developed, which is combined with a precision hot scale to provide initial information about the dust concentration in the exhaust gas directly at the measuring point. The GMD

[1] Dr. Födisch Umweltmesstechnik AG, Markranstädt.

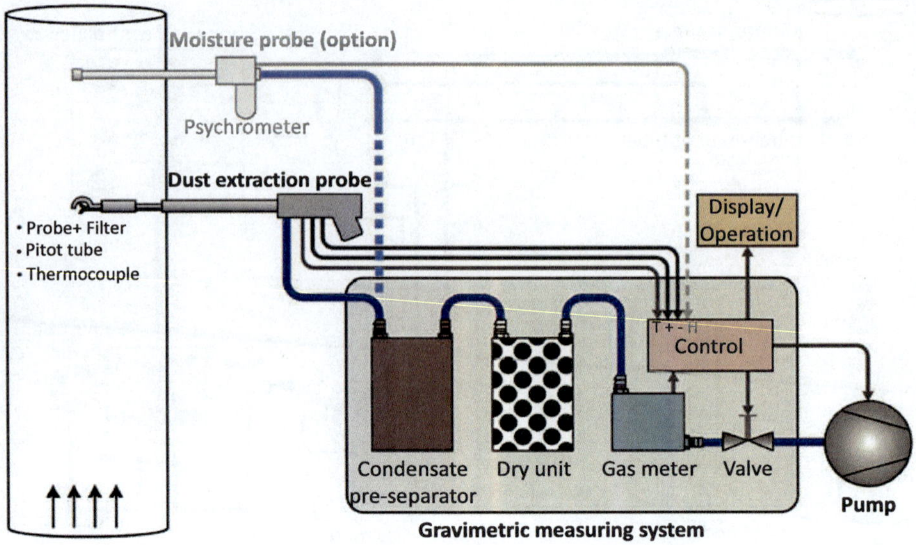

Fig. 13.9 Principle of automated gravimetric dust measurement (GMD 12 of Dr. Födisch Umweltmesstechnik AG)

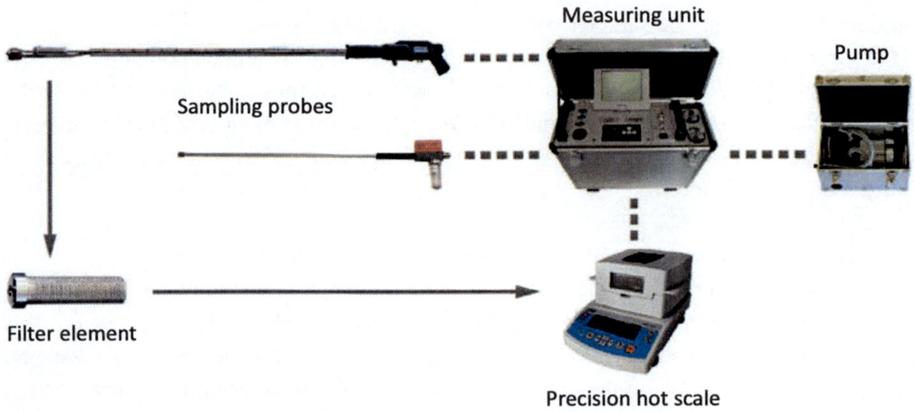

Fig. 13.10 Gravimetric measuring system GMD 13

13 is the world's first measuring system to feature integrated hot weighing with the possibility of on-site evaluation without additional laboratory equipment. The weighing is completely controlled and evaluated by the measuring unit. Simple and safe system design and project-oriented software ensure precise measurement results. The GMD 13 records the sample gas velocity, sample gas pressure and sample gas volume taken.

Furthermore, the sample gas humidity can be measured with a separate sampling probe. Thus, all relevant parameters for the determination of the dust content under standard conditions are recorded by the system and the sample gas extraction is automatically controlled isokinetically.

The essential further development of the GMD 13 consists of the patented method of hot weighing of the filter integrated in the measuring system. This eliminates the need for expensive conditioning before and after dust extraction and makes it possible to determine the measured dust concentration on site in real time. The filters used consist of pressed glass fibres and are protected by a robust filter holder. Weighing of the complete filter element is permitted according to DIN EN 13284-1. This takes place before and after the gravimetric measurement. The controller guides through the weighing process and takes over the measured values. The weighing is semi-automated with a weighing accuracy of less than 1.0 mg. By weighing the same filter element several times, the weighing accuracy can be improved to 0.4 mg with a reasonable amount of time. A stable average value is obtained after only 3–5 weighings, so that precise measurement results can be achieved.

13.4 Determination of the Particle Size Distribution

The determination of the particle size distribution is usually done by laboratory analysis of the particles obtained from gravimetric dust content measurements. This requires dispersion of the available particle collective (pile) in air or aqueous solution. However, the obtained dispersity state deviates from the original state in the gas flow by an unknown amount. Furthermore, minimum quantities are required for the particle size analyses, which are difficult to obtain, especially when the dust content in the exhaust gas is low.

To determine the particle size distribution of particles, a variety of methods can be applied which are based on different physical measuring principles. For example a, mechanical, electrical and optical separation methods are state of the art.

The following selected methods are used to determine the particle size distribution:

- Mechanically:
  - Screen classification
  - Density separation
  - Flow separation (measuring cyclone, cascade impactor)
  - Visual analysis (centrifugal force)
  - Sedimentation analysis.
- Electric:
  - Mobility measurement
  - Field disturbance measurement (Coulter counter).
- Optically:
  - Measurement of particles (microscopy, image analysis)
  - Particle counting

– Laser diffraction.

Flow separation methods are particularly suitable for the in-situ determination of the particle size distribution of dusts in their original disperse state. For this purpose a. cascade impactors are used. They are suitable for partial flow extraction in the sense of VDI guideline 2066 with simultaneous separation into fractions. In connection with a gravimetric evaluation the mass distribution of the aerodynamic particle diameter can be determined. In particular, VDI Guideline 2066 Sheet 5 contains detailed information on the performance and evaluation of cascade impactor measurements. According to VDI 2066 Bl. 10, the equivalent diameter (aerodynamic diameter) of

> *"Diameter of a sphere of the unit particle density 1 g/cm³, which, under the influence of external mechanical forces in equilibrium of forces, has the same migration velocity towards the gas as the particles under investigation".*

A dust particle of unknown shape and particle density has the same aerodynamic diameter as a spherical particle with the density 1 g/cm³ if the sinking speeds of the particles are identical.

$$v_S = \frac{\rho_P \cdot d_P^2 \cdot g}{18 \cdot \mu_G \cdot \chi} \tag{13.7}$$

v_S	Sinking speed [m/s]
ρ_P	Particle density [g/cm³]
μ_G	Dynamic viscosity of the gas [kg/ms]
d_P	Particle diameter [µm]
g	Acceleration due to gravity (9.81 m/s²)
χ	Dynamic form factor

In practice, the dynamic form factor is often assumed to have the value 1. Thus, the following relationship results for the sink rate:

$$v_S = \frac{\rho_0 \cdot d_{ae}^2 \cdot g}{18 \cdot \mu_G} \tag{13.8}$$

ρ_0	Unit particle density (1 g/cm³)
d_{ae}	Aerodynamic particle diameter [µm]

In the example shown (Fig. 13.11), the irregular particle can be assigned an aerodynamic particle diameter of 0.73 µm.

The mathematical connection can be illustrated as follows:

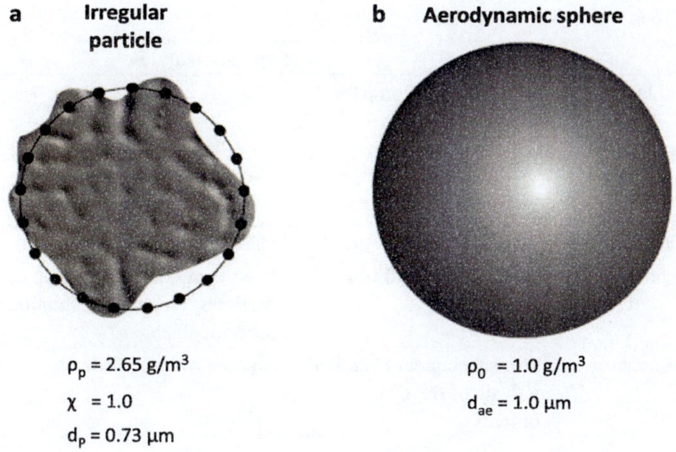

Fig. 13.11 Comparison of the particle diameters of an irregular particle (**a**) and an aerodynamic sphere (**b**)

$$d_P = \sqrt{\frac{\chi \cdot \rho_0}{\rho_P}} \cdot d_{\mathrm{ae}} \tag{13.9}$$

d_P	Particle diameter [µm]
d_{ae}	Aerodynamic particle diameter [µm]
ρ_P	Particle density [g/cm^3]
ρ_0	Unit particle density (1 g/cm^3)
χ	Dynamic form factor

Cascade impactors can be used for different process engineering measurement tasks as well as for emission and immission measurements. With their help, dust particle size analyses can be carried out, for example, which serve to determine the fractional separation efficiency of dust separators. This is necessary, for example, with regard to the measurement of emissions of fine dusts, as they do not sediment in the environment and adsorb gaseous pollutants. Depending on the separation technology used, in the fine dust range, some of the particles are deposited in the considerable deficits in the fractional separation efficiency (e.g. with cyclone separators).. Reduced separation efficiency in the particle size range from 0.3 to 1.0 µm is also known for so-called high-performance separators, such as bag filters or electrostatic precipitators. Depending on the technical requirements, different types of impactor can be selected (Table 13.4).

Dust measurement using a cascade impactor makes use of the different inertia of particles. Several combinations of nozzle and baffle plates (partly with a filter) are

Table 13.4 Types and use of impactors

Impactor type	Application	Special feature
Multi- or multiple nozzle impactor	In the range 0.5 to μ15 m	
Multi- or multi-nozzle impactor with pre-separator	With more than 15% mass fraction >10 m	
Low pressure impactor	For grain sizes <1 μm	Classification up to 0.05 μm
Electrical low pressure impactor	For grain sizes 30 nm-10 μm	Upstream corona charging of the particles, electrical insulation of the individual stages
High-volume impactor	With subsequent chemical analysis of the dust contents	Suction volume flow up to 1000 l/min

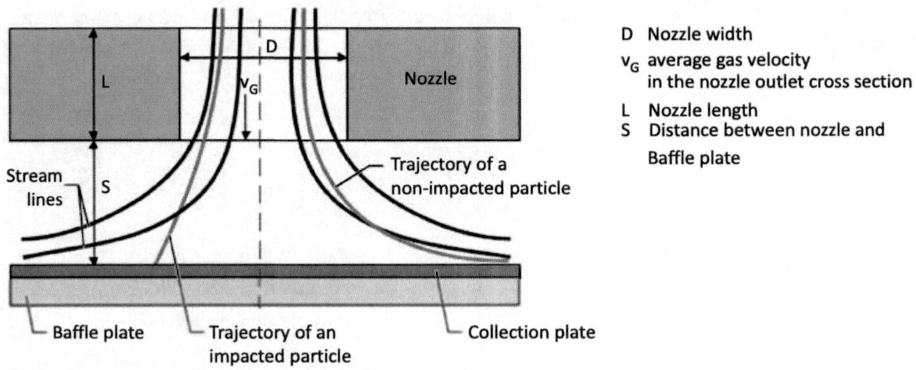

Fig. 13.12 Impactor principle

connected in series. The cascade impactor can be equipped with round or slotted nozzles (Fig. 13.12).

When the dust-laden gas stream hits the baffle plate, the larger (more inert) particles are separated, while the smaller ones follow the gas stream.

The nozzle diameter, which reduces from stage to stage, causes an increase in the gas velocity and thus the separation of the individual grain fractions. Before the gas flow exits, there is an absolute filter which absorbs the residual dust content of the gas. Figure 13.13 shows the basic design of a multi- or multi-nozzle impactor.

Similar to the gravimetric measurement, the cascade impactor must be pre-treated before the measurements are performed in order to obtain correspondingly error-free measured values. The absolute filter is dried, equilibrated (desiccator) and weighed. In addition, it must be ensured that all components are cleaned before the measurement. Then it is

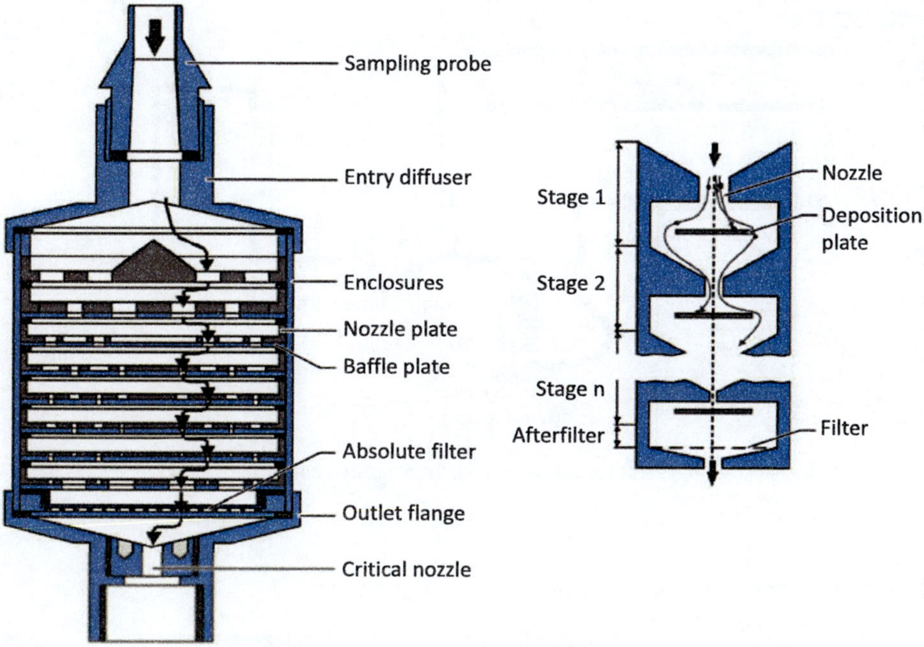

Fig. 13.13 Schematic structure of a multi- or multi-nozzle impactor

advisable to determine and record the tare weights of the associated nozzle and baffle plates of each stage.

The design of sampling with the impactor can essentially be compared to gravimetric dust measurement.

Isokinetic suction is performed in compliance with VDI 2066 sheet 1. However, the extraction speed must not be changed, otherwise the separation characteristics of the impactor will vary and the determined particle size distribution will be falsified. Therefore, measuring cross-sections with approximately the same velocity distribution should be selected.

Ideally, the impactor should be placed directly in the exhaust duct (in-stack sampling, see Fig. 13.14). The partial gas flow reaches the actual impactor via the sampling probe on a short path. However, especially with small measuring cross-sections, the impactor must be placed outside the exhaust pipe (Fig. 13.15).

The measurement shall take into account specific characteristics of the measuring point with regard to the expected dust concentration, the waste gas temperature (filters shall be pretreated according to the waste gas temperature), the waste gas humidity or dew point temperature and the flow distribution in the waste gas. If there are large differences in the velocity distribution, a network measurement can be carried out (possibly parallel measurement with second impactor). The duration of sampling results from the dust content in

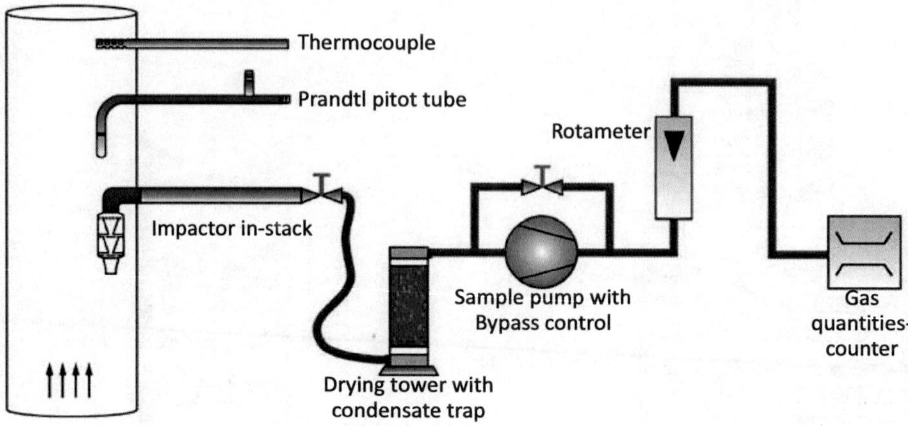

Fig. 13.14 In-stack arrangement for cascade impactor measurement

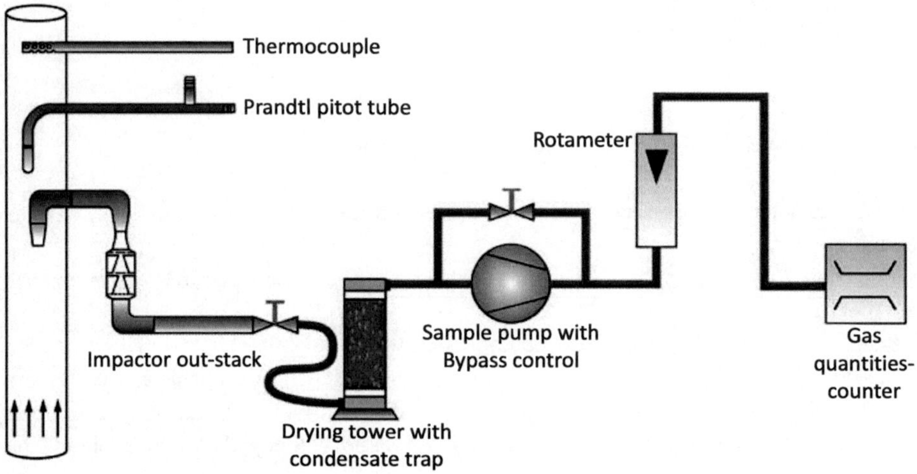

Fig. 13.15 Out-stack arrangement for cascade impactor measurement

the flue gas, the loadability of the impactor, the possible extraction speed, the detection limit of the weighing and the available time frame. Parallel to the impactor measurement, a total dust measurement should definitely be carried out.

After the measurement, the individual separation stages of the impactor are removed one after the other to determine the residues and the nozzle plate and flapper plate are balanced together. Then they must be cleaned with suitable cleaning agents (cloth, brushes). Afterwards the empty weights can be determined again. To evaluate the absolute filter, it is dried again and then weighed. The grain size is determined on the basis of the aerodynamic diameter already defined.

$$d_{ae50}(i) = K_1(i) \cdot K_2 \cdot \sqrt{\frac{1}{C \cdot (d_{ae50}(i))}} \tag{13.10}$$

d_{ae50}	Aerodynamic particle diameter [µm]
K_1, K_2	Constants
i	Impact level
C	Cunningham factor

d_{ae50} is the aerodynamic diameter of the grain size, 50% of which is deposited on an impactor stage. K_1 is a constant for the impact stage i of a given impactor, independent of the respective measurement conditions. In K_2 the measuring conditions are entered, i.e. this constant is the same for all interaction stages. The constants are calculated as follows:

$$K_1(i) = \sqrt{\frac{9 \cdot \pi \cdot \mathrm{St}_{50}(i) \cdot D_i^3 \cdot N_i}{4 \cdot \rho_0}} \tag{13.11}$$

$$K_2 = \sqrt{\frac{\mu_G}{V_m}} \tag{13.12}$$

St_{50}	Stokes number
D_i	Nozzle diameter of the impactor stage i [m]
N_i	Number of nozzles of the impactor stage
ρ_0	Unit particle density (1 g/cm^3)
μ_G	Dynamic viscosity of the gas (depending on T and gas composition) [kg/ms]
Vm	Volume flow through the impactor, in operating condition [m^3/ s]

In the calculation, wall losses, which always occur at the extraction probe, transition part, etc., in impactor measurements compared to total dust measurements, must be taken into account. A rough formula for this can be found in VDI 2066 sheet 5. Wall losses depend on the type of dust and particle size and increase with increasing particle size. In addition, the design of the cascade impactor influences the magnitude of the wall losses. In the design shown in Fig. 13.13, the deflection and rebound effects are minimized by special constructive design means (use of blind holes).

The QCM impactor is a special design of cascade impactor. In this device, a quartz microbalance is installed in each stage instead of a filter or a separating plate. This consists of 2 measuring crystals, whereby the dust particles are deposited on one, while the other serves as a reference. The measure for the dust load is the change in the natural frequency of both crystals. Within a few seconds, an automatic printer outputs the mass concentration for each size range and the relative distribution as a histogram. At the same time, it shows which stages need to be cleaned after overloading. The detectable particle size range extends from 0.05 to µ25 m.

Another device with direct display during the measurement is the electrical low-pressure impactor. The dust particles are specifically charged via a corona charge and separated on the individual separation stages. An integrated electrometer transmits the measured current to an evaluation unit. With this impactor, particle sizes from 0.03 to µ10 m can be detected.

13.5 Emission Measurements of Fine Dust (PM2.5/PM10)

The negative influence of fine and ultrafine particles on human respiration is undisputed and has been proven in various studies. As a result, uniform legal limit values have been introduced, at least throughout Europe. Figure 13.16 shows how the various particle sizes affect the human body.

To protect health and the environment, the European Council adopted Directive 1999/30/EG in April 1999, which, among other things, set immission limits for particulate PM10 in the air, which have been binding since 2005 (Table 13.5).

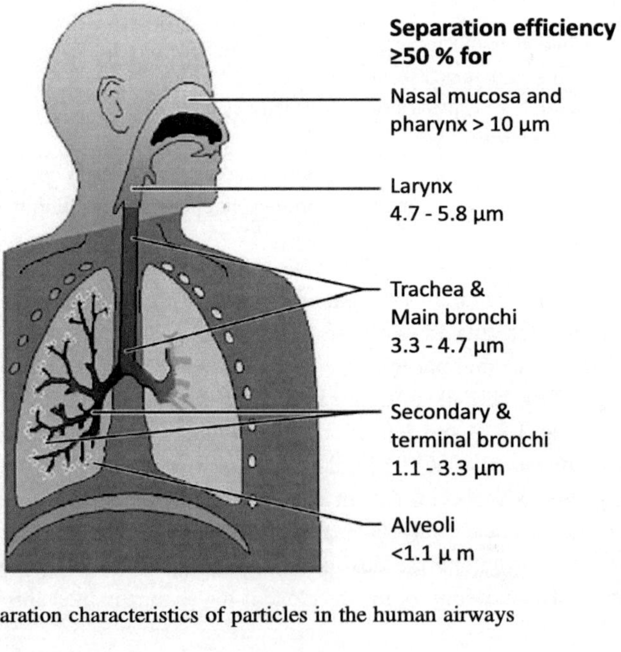

Fig. 13.16 Separation characteristics of particles in the human airways

Table 13.5 Immission limit values according to Directive 99/30/EG (margins of tolerance are not given in the table)

	Averaging period	Limit value	Time since the limit value was reached
Level 1			
24-hour limit value for the protection of human health	24 h	50 μg/m³ PM10 must not be exceeded more than 35 times per year	January 01, 2005
Annual limit	Calendar year	40 μg/m³ PM10	January 01, 2005

In 2007, the EU Parliament adopted Directive 2008/50/EG, which sets mandatory upper limits for very small particles (PM2.5):

- A target value for PM2.5 of 25 μg/m³ has been in force since 2010 (this value is to be aimed at, it is not binding).
- A limit value for PM2.5 of 25 μg/m³ has been in force since 2015.
- From 2020, this limit value for PM2.5 will be reduced to 20 μg/m³.

PM10 and PM2.5 particles are those that pass through a size-selective air inlet that has a collection efficiency of 50% for an aerodynamic diameter of 10 and 2.5 μm respectively. The dust fractions PM10 (thoracic fraction) and PM2.5 (alveolar fraction) make up on average approximately 80% and 50% of the mass of the total dust. Figure 13.17 shows how the individual particles are produced and which processes or elimination mechanisms they undergo.

In the atmosphere, fine dust particles serve as condensation nuclei which, in combination with increased air humidity, lead to accumulation effects. Above a certain mass of adsorption, the process tends to sedimentation. As a result, the concentration of fine dust in the atmosphere decreases with increasing air humidity and especially during rain.

However, analyses carried out so far have shown that it is difficult to comply with the required limit values. A more or less continuous monitoring of the most important source groups (road traffic and industry/trade with a share of 60–80% of PM10 pollution) will therefore be inevitable in the future.

The control of particulate matter emissions can be done discontinuously with cascade impactors or cyclones. VDI Guideline 2066 Sheet 10 contains detailed explanations on the measurement of PM10 and PM2.5 emissions at guided sources using the impaction method (reference method).

With the impactor shown in Fig. 13.18, a volume flow of 2–3 m³/h is extracted, depending on the operating state. Due to the relatively high volume flow, the collection

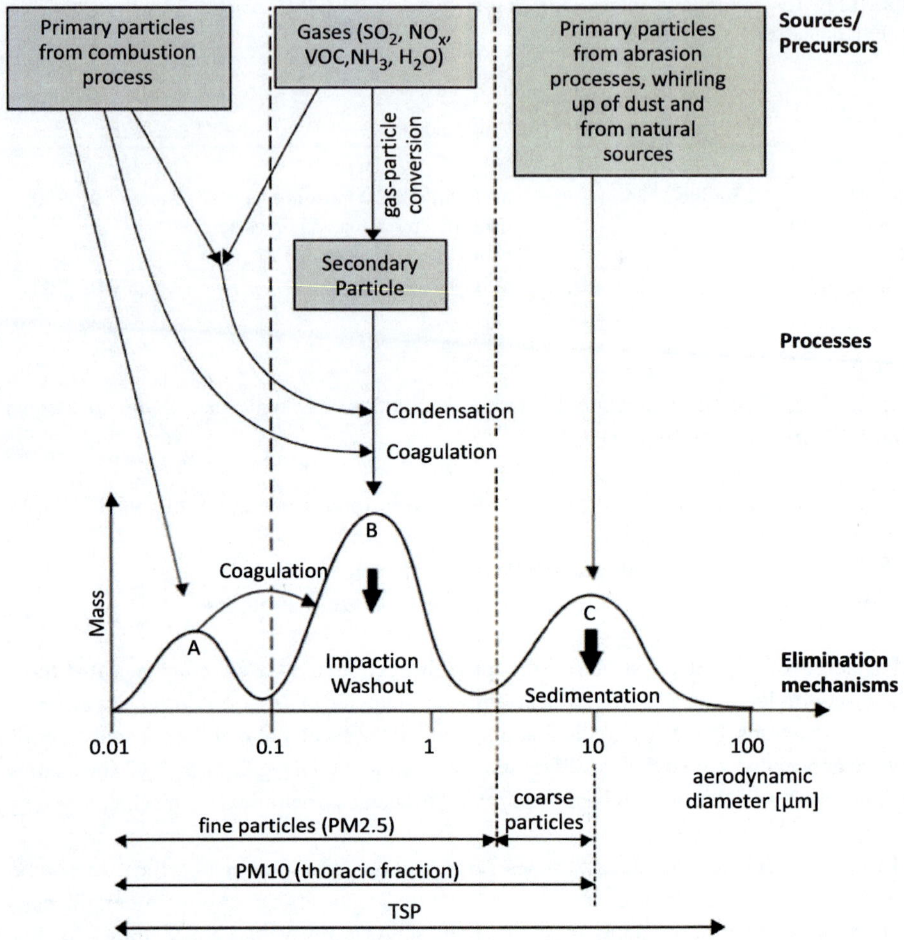

Fig. 13.17 Simplified representation of the size distribution of atmospheric aerosol near the source and the most important processes (A - ultrafine particles; B - accumulation mode; C - coarse particles)

times can be kept low (for dust concentrations of 30 min/m^3 (i. N.) approx. 30 min). Figure 13.19 shows a cascade impactor according to VDI 2066 sheet 10 for comparative measurements on a dust duct.

In order to obtain reliable information on the PM10 fractions in total dust emissions, it is essential to determine them in the emission range from anthropogenic sources. For this reason, an increased development of continuous measuring instruments for fine dust PM10 and PM2.5 has been taking place since 2000.

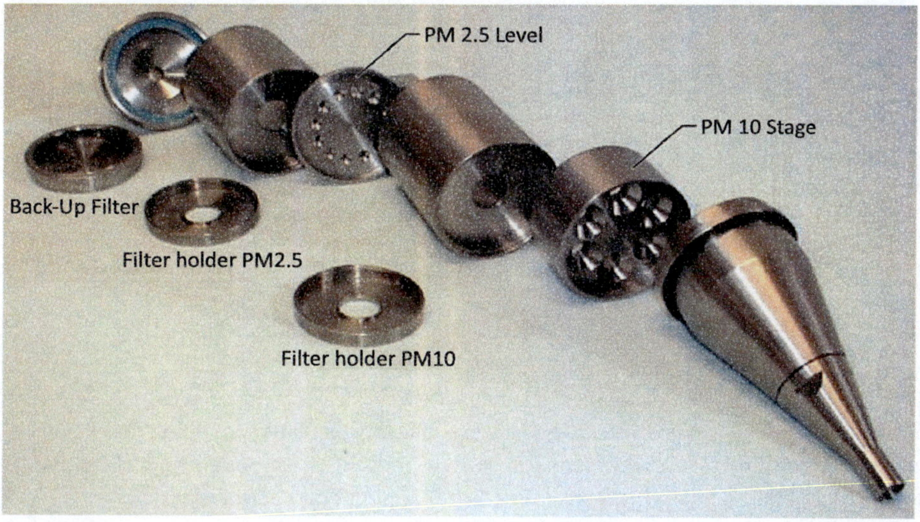

Fig. 13.18 Structure of a cascade impactor for measuring the PM10/PM2.5 dust fractions according to VDI 2066 sheet 10

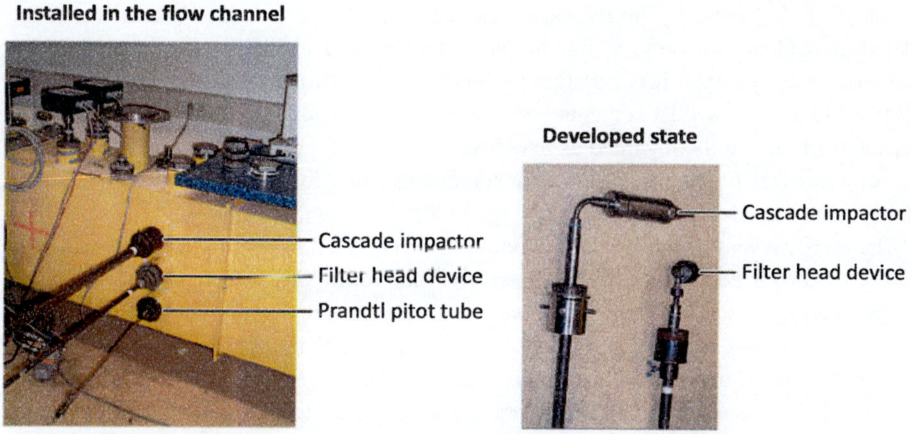

Fig. 13.19 Cascade impactor for measuring the PM10/PM2.5 dust fractions

13.6 Current State of the Art in Fine Dust Measurement

At present, particulate matter measurements are carried out at over 320 stations in Germany as part of the Federal Environment Agency's (Umweltbundesamtes) measurement programme. These are immission measurements, which are carried out continuously according to the principle of beta-absorption or discontinuously by means of high-volume

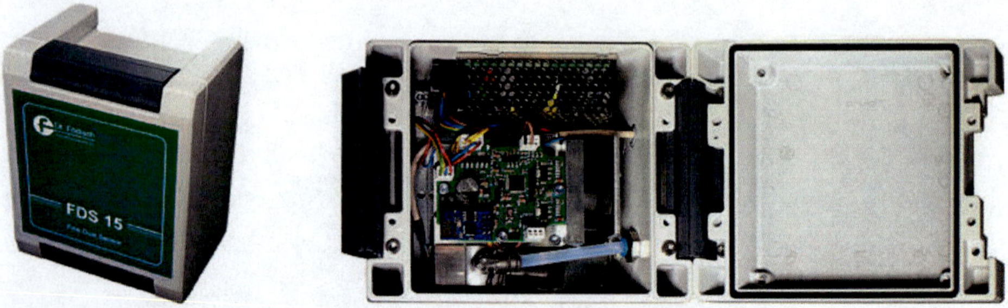

Fig. 13.20 Fine dust sensor FDS 15. (Source: Dr. Födisch Umweltmesstechnik AG)

samplers and gravimetric evaluation. Daily averages are calculated from the measurement results in order to detect possible exceedances of the above mentioned limits. This results in reduction measures for the different source types (e.g. Regulations for industrial plants, soot particle filters for cars) and situations (e.g. Air pollution control plans).

In addition, there are already various portable measuring devices (e.g. from Grimm Aerosol Technik GmbH & Co. KG or TSI GmbH), especially for workplace measurements, with which fine dust can be measured in the lowest concentrations.

A targeted continuous monitoring of particulate matter emissions has not yet been technically implemented. The measuring devices available at industrial plants record the total dust content, whereby, if the corresponding data is available, for example, from impactor measurements, it is possible to conclude on a corresponding fine dust content PM10/PM2.5. At low dust contents (<5 mg/m^3) it can be assumed that the predominant particle fraction is to be regarded as fine dust.

New trends are based on simple sensor solutions for fine dust measurement. An example of this is the FDS 15[2] fine dust sensor (Fig. 13.20).

The fine dust sensor FDS 15 (dimensions130 mm × 160 mm × 90 mm, B × H × T) is an optical sensor for continuous measurement and monitoring of fine dust concentrations. It can be integrated in various applications or used in stationary or mobile applications. Figure 13.21 shows the principle structure of the particulate matter sensor.

The determination of the dust content in the FDS 15 is based on the principle of scattered light measurement. It is factory calibrated and can be used in the measuring range up to μ1400 g/m^3. The device has two conditioned scattered light sensors to improve stability. The air drawn in during operation is preheated at 50 °C. The integrated fan provides a forced flow (2 l/min). An integrated pre-separator is used for the analysis of alveolar particle fractions (PM2.5). The velocity of the measuring gas is selected in such a way that particles are detected representatively.

Furthermore, the FDS 15 carries out a periodic check of zero and reference point. The zero point is set automatically. A high zero point stability is achieved by evaluating the

[2] Dr. Födisch Umweltmesstechnik AG, Markranstädt.

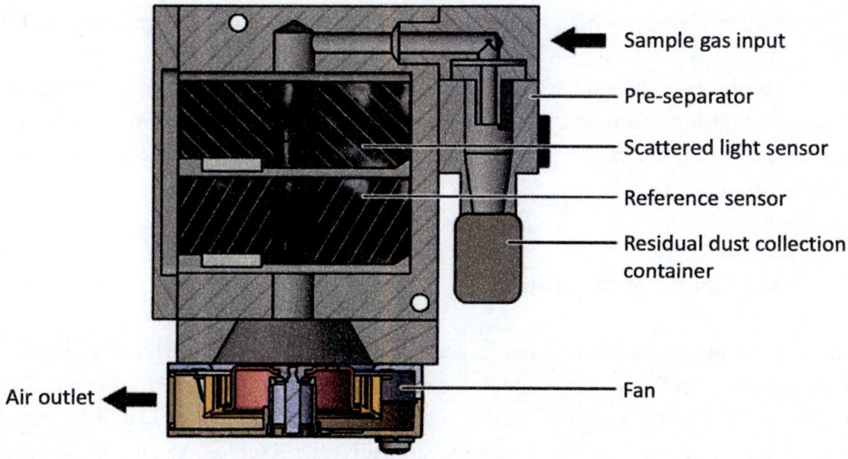

Fig. 13.21 Sensor construction of the FDS 15

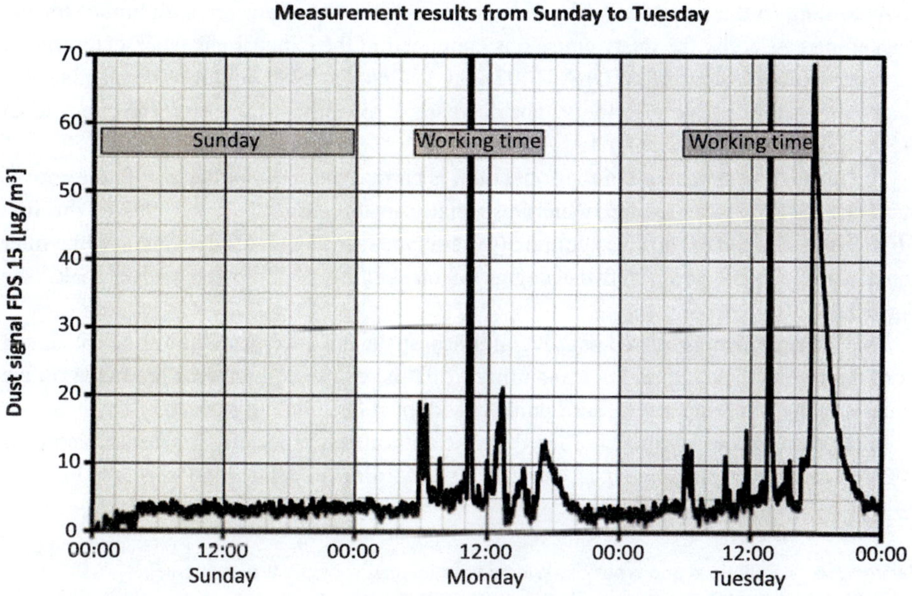

Fig. 13.22 Measurement results of the FDS 15 in a workshop with welding smoke

internal measuring signals. In addition to an RS485 interface for Modbus transmission, there is also the option of an integrated WLAN module.

Figure 13.22 shows measurement results of the FDS 15 over several days in a workshop with welding work. The course of the dust signal can be used to draw clear conclusions about the absence of persons in the room as well as work carried out that generates welding fumes.

The results presented here are exemplary for the good function of the fine dust sensor for workplace monitoring. Further applications are conceivable in combination with weather stations as well as for the establishment of sensor networks for fine dust measurements in particularly polluted areas (e.g. Large cities in emerging markets).

13.7 Soot Count Measurement

For combustion plants for liquid fuels (especially oil firing systems), which fall under the scope of the first BImSchV (Ordinance on Small and Medium-Sized Firing Installations) and partly under the TA Luft (Technical Instructions on Air Quality Control), the soot number in the waste gas shall be measured annually. This serves on the one hand as a measure for air pollution control and on the other hand to save primary energy, as the completeness of the combustion in the plant is monitored at the same time. Table 13.6 shows systems for which measurement of the soot index is mandatory.

According to the first BImSchV, the soot number is the characteristic number for the blackening caused by the dusty emissions contained in the exhaust gas when determining the soot number according to DIN 51402 Part 1, October 1986 edition. The standard for blackening is the optical reflectivity. The decrease in reflectivity by 10% corresponds to an increase in the soot number by 1.

The manual measurement of the soot count is carried out in accordance with the above-mentioned DIN 51402-1, after which, especially in the case of the systems of the first BImSchV, a defined exhaust gas volume (generally ten strokes) is sucked onto a white filter paper using a hand pump and the degree of blackening is compared with a scale (see Fig. 13.23).

This measurement is carried out several times, so that an average value can be calculated from the results. The method is standardized, that is, exact requirements are made on the pump and the filter paper. The soot pumps used must be suitability-tested.

In the case of measurements ordered by the authorities, especially for plants according to TA Luft, the measurement is carried out according to VDI 2066 sheet 8. For this

Table 13.6 Regulations and types of systems with manual soot count measurement

Legal regulation	System type	Regulation
1st BImSchV §7	Oil firing plants with vaporizing burners	Soot count 2 must not be exceeded
1st BImSchV §8	Oil firing systems with atomizing burners	Soot count 1 must not be exceeded
TA Luft 2002 No. 5.4.1.2.2	Single firing plants for fuel oil according to DIN 51603-1 with a thermal output of<50 MW	Soot count 1 must not be exceeded

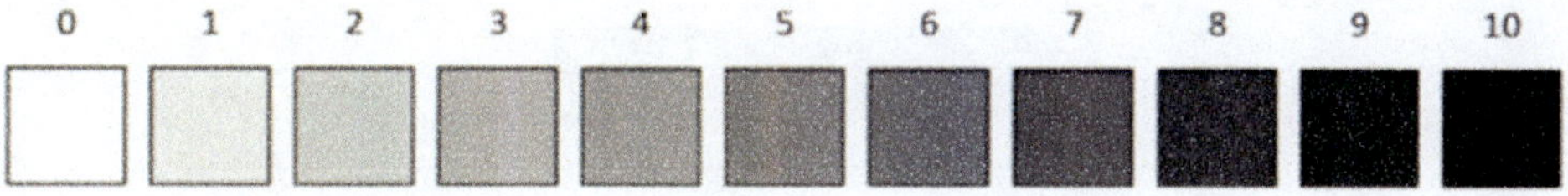

Fig. 13.23 Bacharach scale for determining the degree of blackening

purpose, a measuring filter is coated with soot particles via the analog measuring set-up for gravimetric measurement with plane filter head according to VDI 2066 sheet 1 and then measured with a reflection photometer.

For oil and gas furnaces with a thermal capacity of 10–20 MW (according to 1. BImSchV §11 para. 1) as well as in gas turbine plants using liquid fuels with a thermal capacity <50 MW (according to TA Luft No. 5.4.1.5), continuous monitoring of the soot count is mandatory. The measuring instruments used shall be calibrated by means of conventional methods. Measurement according to DIN 51402-1 is not suitable for this purpose, as too many sources of error can occur. The procedure according to VDI 2066 sheet 8 is used here.

13.8 Continuous Emission Monitoring for Dust

In the broadest sense, the devices used for continuous dust measurement register physical changes caused by the particles in the measuring system and convert them into electrical signals. For this purpose, the material to be measured can be examined directly in the exhaust duct (in-situ measurement) or a partial volume flow can be taken and fed to a measuring device (extractive sampling).

In *in situ* techniques, the measurement signals result from the direct interaction of light or a triboelectric probe with the dust particles in the exhaust duct. The scattered light or transmission/extinction of an emitted light beam or triboelectricity can be used for evaluation.

As with gravimetric measurement, the basis of the extractive methods is a partial flow extraction from the main gas flow as isokinetically as possible. The measuring principles used are scattered light measurement, radiometry or triboelectric measurement.

The process engineering boundary conditions have a considerable influence on the choice of the measuring method. In the case of droplets or aerosols in the exhaust gas flow, for example, extractive sampling with subsequent conditioning of the sample gas flow is used exclusively. Figure 13.24 provides an overview of all methods for continuous dust measurement.

In the following, the individual methods are presented in more detail and their advantages and disadvantages for different measurement tasks are compared.

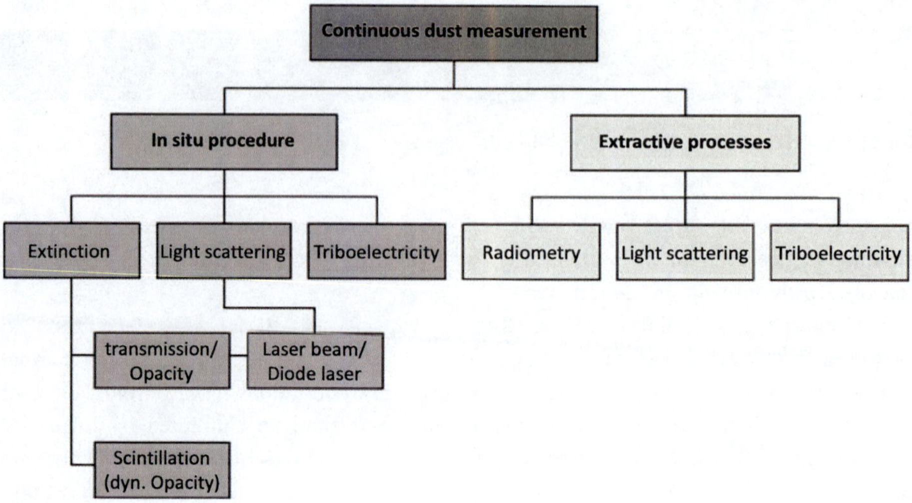

Fig. 13.24 Method of dust measurement

13.9 Transmissiometry

When measuring dust according to the transmission principle, the dust loading of the exhaust gas is measured via the auxiliary variables transmission or extinction. A light beam passes through the exhaust gas flow in a defined cross-section and is attenuated by absorption at the particles (Fig. 13.25).

This method can be used for both qualitative (smoke density determination) and quantitative measurements (dust content measurement). Figure 13.26 shows the basic structure of a transmission measurement.

A transmission measuring device (e.g. Dusthunter T50 from SICK AG) usually consists of an optical measuring unit with a light source and a reflector, which is installed opposite the optical measuring unit. The light emitted by the light source passes through the measuring section twice. The light beam is divided into a measuring light beam and a reference light beam. The intensity of the measuring and reference light beam is opto-electrically converted by a photodetector and transmitted to the evaluation electronics. The dust load can be determined by comparing the received back (attenuated) light and the undamped reference light. The ratio of received light to emitted light gives the value of the transmission:

$$T = \frac{I}{I_0} \tag{13.13}$$

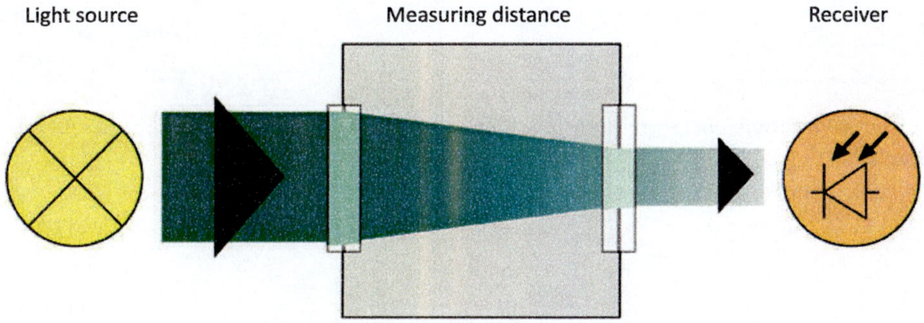

Fig. 13.25 Principle of transmission measurement

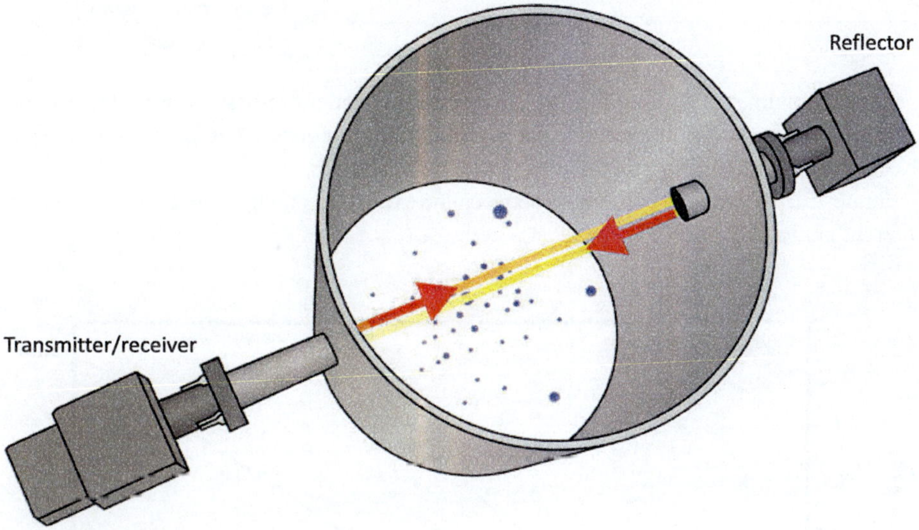

Fig. 13.26 Structure of a transmission measurement

T	Transmission [%]
I	Measured radiation intensity
I_0	Emitted radiation intensity

The lower the transmission values, the higher the solids content or smoke density in the exhaust gas. The logarithm of the reciprocal of the transmission value is called extinction:

$$E = \ln \frac{1}{T} \tag{13.14}$$

E	Absorbance
T	Transmission [%]

The connections between transmission and extinction are described by Lambert-Beer's law:

$$I = I_0 \cdot e^{(-k \cdot c \cdot l)} \tag{13.15}$$

$$E = \ln\left[\frac{1}{T}\right] = k \cdot c \cdot l \tag{13.16}$$

k	Extinction coefficient
c	Concentration [mg/m^3]
l	Measuring section length [m]

After a gravimetric comparison measurement, the dust concentration in mg/m^3 can be determined from the absorbance. The extinction depends on material properties such as size or colour, the dust content and the measurement path length.

Figure 13.27 shows the correlation between the extinction and the dust concentration for different particle diameters.

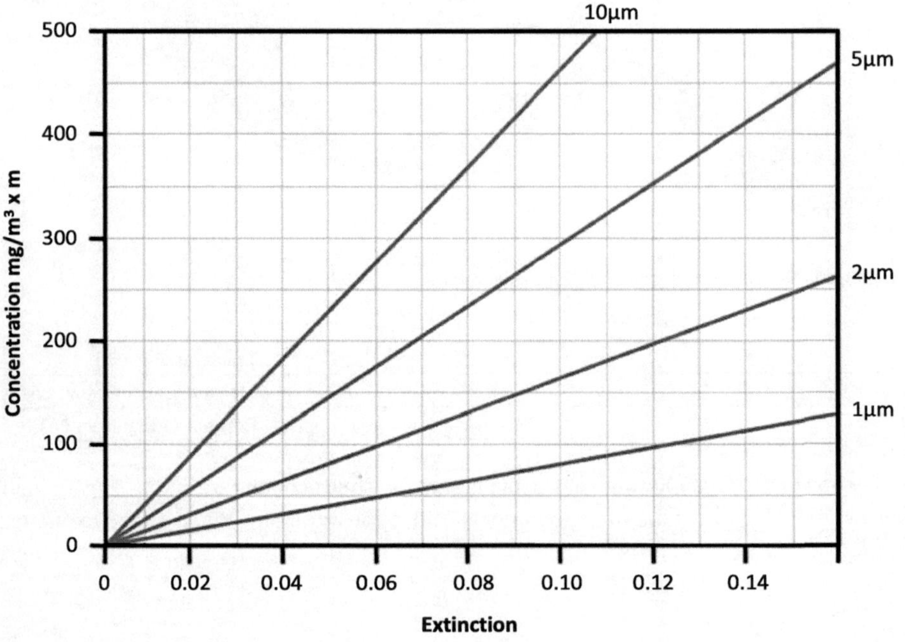

Fig. 13.27 Extinction and dust concentration for different particle diameters

The transmission measuring principle is particularly suitable for medium to high dust concentrations. Meaningful measurement results generally require sufficient measuring path lengths. If the attenuation of radiation (turbidity) is considered, it is the amount of opacity:

$$\mathrm{Op} = 1 - T \tag{13.17}$$

Op	Opacity [%]
T	Transmission [%]

Figure 13.28 shows the relationship between extinction (*E*), transmission (*T*), and opacity (*Op*).

The disadvantage of the opacity measurement (e.g. D-R 290 of the DURAG Group) is the dependence of the measurement result on the measuring point. Even if a conversion to the chimney mouth is made, material properties such as colour, size and surface condition of the dust particles are not taken into account. Therefore, the use of these devices is limited. They are mainly sold on the American and Asian markets.

Conventionally, as described, transmission or absorption is used to determine the measured value. A special type of measurement is dynamic transmission - scintillation (see Fig. 13.29). Here, the variation of the received light intensity, which is caused by the time distribution of the particles in the light beam, is determined. The measurement signal is derived from this variation.

The dynamic transmission reflects the ratio of light variation to light intensity. This method makes it possible to eliminate changes in transmission due to a reduction in the

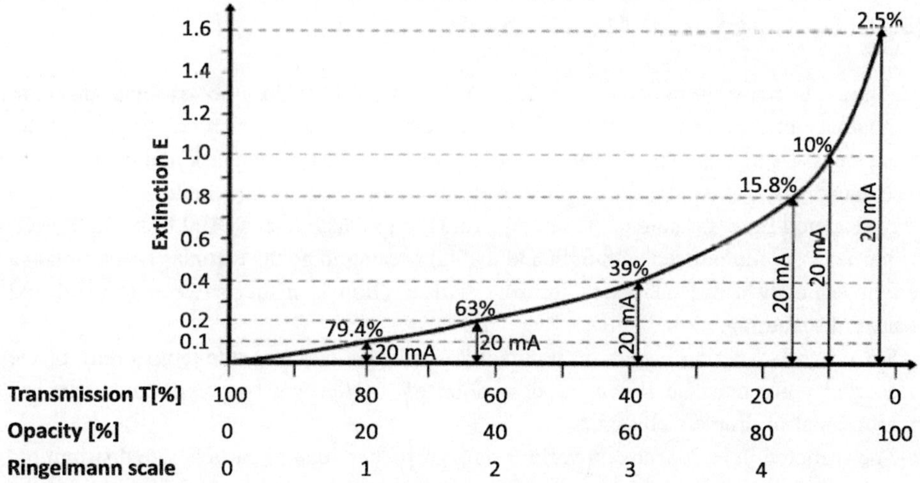

Fig. 13.28 Relationships of radiation absorption

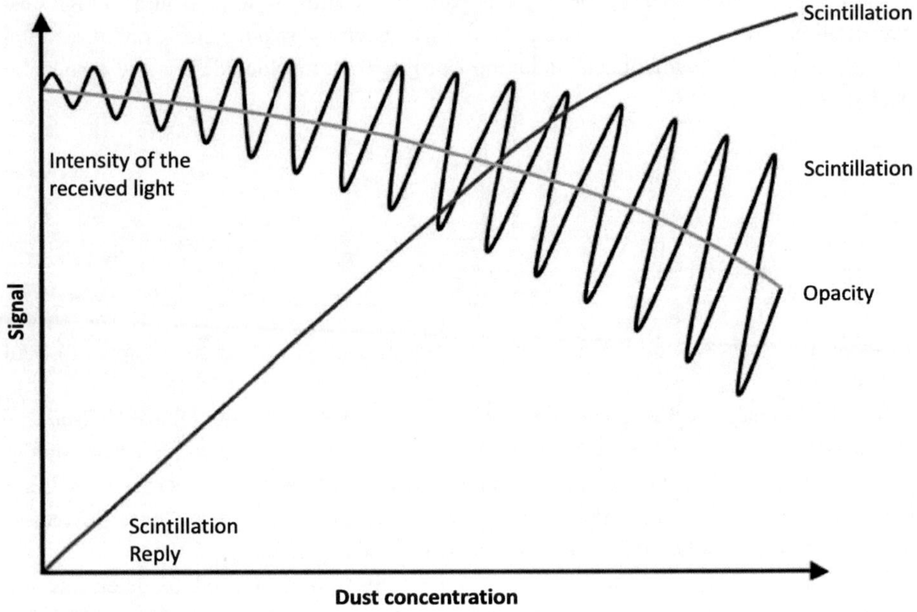

Fig. 13.29 Principle of scintillation measurement

efficiency of the optical transmitter, contamination of optical parts or flow without affecting the measurement signal.

13.10 Scattered Light Measurement

In contrast to transmissiometry, scattered light measurement does not evaluate the light attenuation but the scattering at the dust particles when a light beam passes through an exhaust gas stream (see Fig. 13.30). Depending on the measurement task and the nature of the exhaust gas, this can be done either as an in-situ or extractive method.

In scattered light measurement (see Fig. 13.31, e.g. Dusthunter SF100 from SICK AG) the particles are illuminated with light and the light scattered by the particles is detected at a certain angle by a special optical sensor. A distinction is made between forward and backward scattering.

Since light scattering occurs on both solid and liquid particles, the temperature of the measuring point must be above the dew point of the flue gas to prevent simultaneous measurement of droplets present.

The scattered light depends on various dust properties such as particle size distribution, colour and shape of the particles, refractive index as well as on the optical design of the

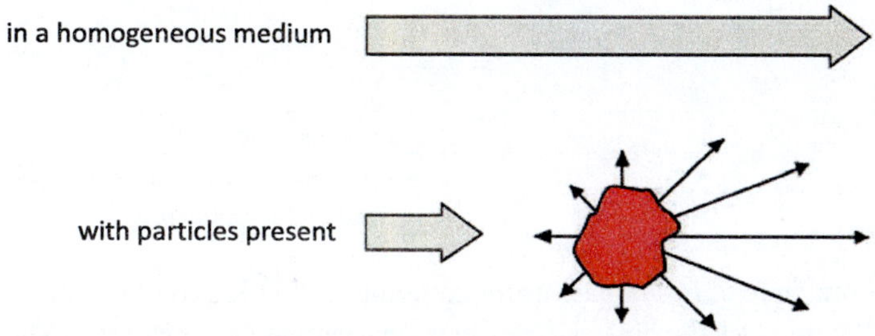

Fig. 13.30 Light propagation in a homogeneous medium and with particles present

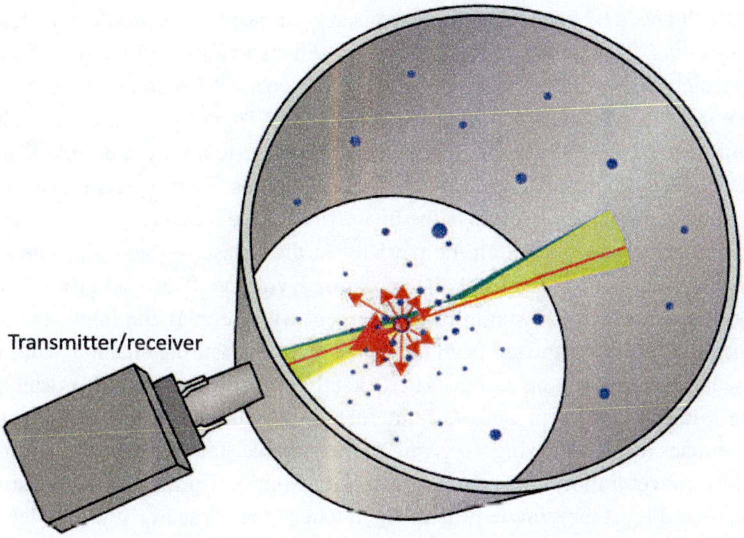

Fig. 13.31 Structure of a scattered light measurement (*in situ*)

measuring device used such as measuring angle, bundle opening and wavelength of the measuring light.

The elastic light scattering on particles with a radius of the order of the wavelength of light is described by the Mie theory (Mie scattering). The Mie theory shows, among other things, the dependence of scattered light on particle size. A significant statement of the Mie theory is that the radiation is scattered more and more in the forward direction as the particle radius increases. This effect is called the Mie effect. After gravimetric calibration, the following relationship can be established between the scattered light intensity and the dust concentration in mg/m^3:

$$c = a_0 + a_1 \cdot I_{ST} \tag{13.18}$$

c	Dust concentration
a_0, a_1	Calibration factors of gravimetric calibration
I_{ST}	Scattered light intensity

Using the principle of *in situ* backward scattering (e.g. D-R 320 of the DURAG Group) a laser diode illuminates the dust particles in the measuring volume of the exhaust duct. A highly sensitive detector detects the light scattered by the dust particles in backward direction. The measuring volume is defined by the intersection of the laser beam and the field of view of the detector. There are various solutions to ensure that the measurement result is not influenced by background light caused by multiple reflections of the laser beam in the exhaust duct. The common practice is to install a so-called light trap in the opposite duct wall, which is associated with high installation costs. Also common is the use of a separate control receiver to detect the background light. This requires a precise alignment of the beam paths. The use of a single imaging optical system with a double detector for background light compensation is particularly advantageous. Two adjacent areas in the flue gas duct are imaged on two detectors in the image plane. The measuring channel detects the light scattered by the laser beam on the particles in the measuring volume. The reference channel looks at an area adjacent to the measuring volume. The background, which is blurred for both detectors, is brought into alignment with internal apertures. The signal of the reference channel is subtracted from the signal of the measuring channel. Thus the light input of the background can be compensated exactly. The advantage of this principle is an installation without light trap and without manual alignment of the beam paths. The principle of backward scattering is particularly suitable for detecting the smallest to medium dust concentrations. In contrast to the transmission principle, the measurement result of scattered light measurement is independent of the measurement path length.

The principle of forward scattering with a measuring probe is shown in Fig. 13.32. The bundled and modulated light of a laser diode illuminates the dust particles in the gas stream. The light scattered in the forward direction is bundled by an objective at the end of the lance and reaches a highly sensitive detector via an optical fibre. The scattered light intensity corresponds to the particle density and is measured by the measuring instrument as a scattered light unit. This measuring arrangement also allows easy installation without complex adjustment. The principle of forward scattering is used for the detection of small to medium dust concentrations.

However, the *in situ* measuring instruments shown so far are only suitable for dust measurement in dry gases. Water droplets or aerosols present in wet or water vapour saturated gases also produce scattered light effects, which falsify the measurement. Therefore, in such cases the extractive measurement method should be chosen.

A partial flow is taken from the exhaust gas flow at a defined velocity (slightly overisokinetic). Before the partial gas flow is fed to a scattered light measuring cell, the moisture contained in the sample gas must be evaporated. One possibility is to supply a

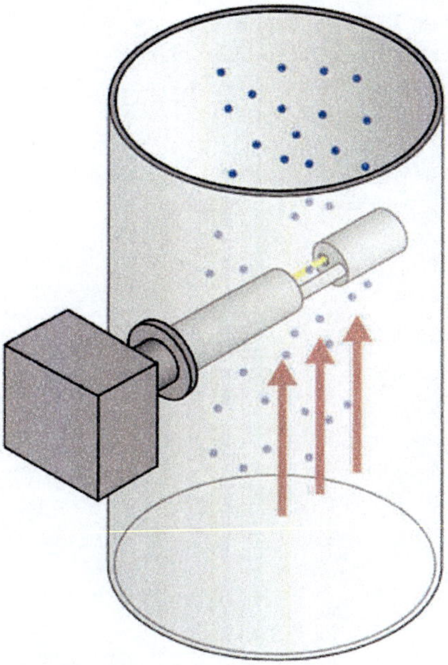

Fig. 13.32 Measuring lance for laser scattered light measurement (e.g. D-R 800 of the DURAG Group)

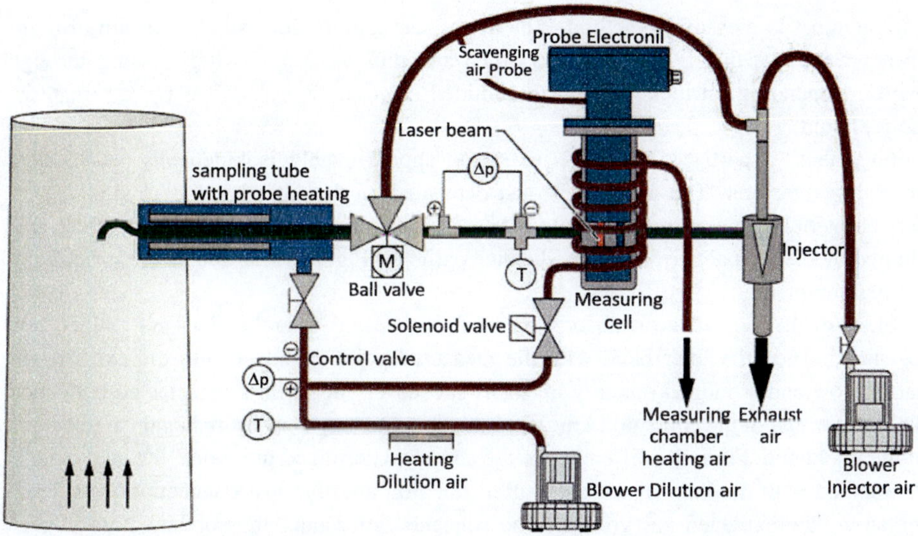

Fig. 13.33 Extractive scattered light measurement (e.g. D-R 820 F of the DURAG Group)

defined (heated) dilution air flow with simultaneous heating of the gas mixture (see Fig. 13.33). Alternatively, water droplets and aerosols can be evaporated by superheating the gas in a thermal cyclone. An example of the structure of an extractive scattered light measurement is shown in Fig. 13.34.

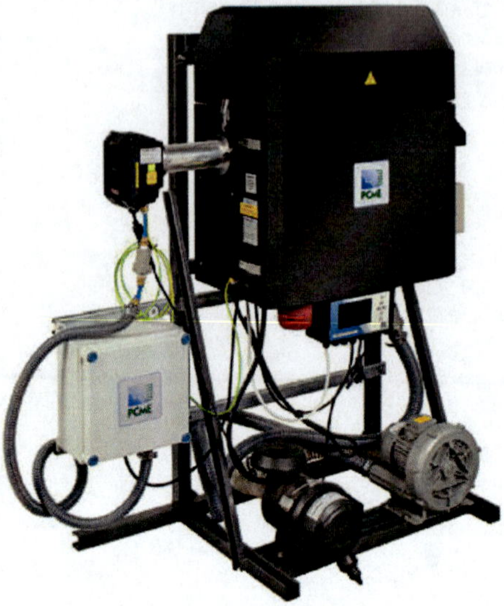

Fig. 13.34 Structure of an extractive scattered light measurement (e.g. STACK 181 WS of PCME Ltd.)

13.11 Beta Radiation Absorption

The radiometric measuring method exploits the fact that the intensity of a beam of high-energy electrons, that is, β rays is absorbed and thus weakened when passing through matter. Depending on their energy, the emitted electrons are absorbed by any material (solid, liquid, gas).

In general, a partial gas flow is extracted approximately isokinetically and sucked through a filter belt. The amount of dust deposited on the filter band is irradiated by a beta-emitter. The measuring principle is based on the absorption of the β rays (electrons) emitted by the radioactive source by the dust particles. Figure 13.35 shows the structure of the measurement.

First of all, the radiation absorption of the undusted filter band is determined and registered. Then the filter band with the measured point is moved into the exhaust gas partial flow and a defined quantity of gas is sucked through it in a preselected collection time. At the end of the collection time, the radiation intensity is measured again at this point of the filter band. From the difference to the already determined pre-value, the dust mass is determined with the help of the calibration function and the dust concentration is determined via the extracted gas volume. The concentration values thus obtained are in each case average values over the sampling period and are output in quasi-continuous steps. By collecting the sample material, this measuring method allows further physical-chemical investigations of the dust. The β radiation absorption depends only slightly on the materials to be investigated, but calibration is necessary in any case.

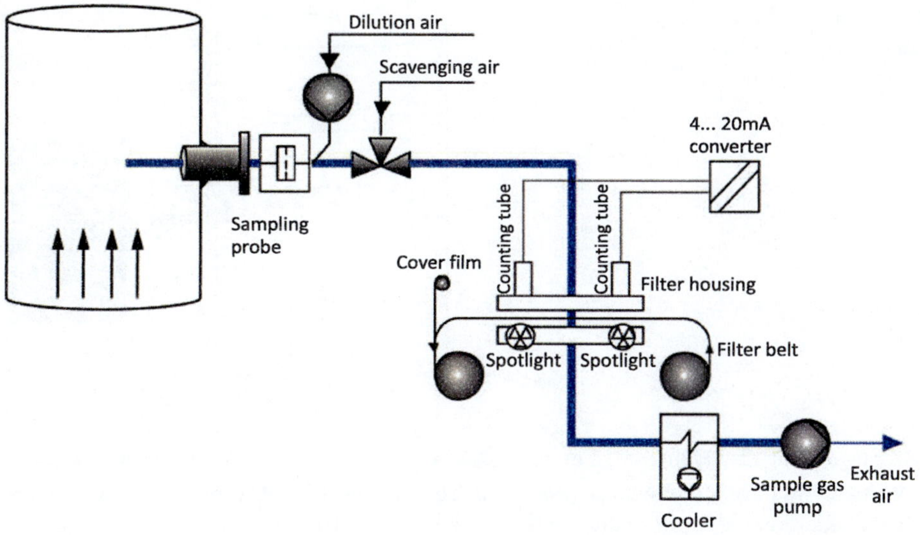

Fig. 13.35 Extractive dust measurement using beta radiation absorption (e.g. F-904-20 of the DURAG Group)

13.12 Triboelectric Dust Measurement

If two bodies are brought into contact with each other by friction or contact, a charge transfer occurs. The charge difference is created when atoms on the surfaces exchange electrons, thus forming a boundary layer with a positive and a negative surface charge at a very small molecular distance from each other. This charge difference, also known as charge fluctuation, is the basis for dust measurement according to the triboelectric principle, whereby the charge exchange between the measuring probe and the dust particles flowing around it as well as impacting on it is used here (Fig. 13.36).

The measuring method is suitable for both in-situ and extractive measurement of dust contents. Qualitative or quantitative measuring tasks are realized. For the *in situ* measurement (e.g. PFM 02 of Dr. Födisch Umweltmesstechnik AG) one or more electrically insulated metal rods are inserted into the gas stream to be investigated, depending on the type of device. The dust particles serve as charge carriers, so that the particles colliding with the rods, rubbing against them and flowing around them cause a measuring current. This measuring current is detected and amplified by the electronic circuit. Depending on the type of dust and gas velocity, the derived signal is a measure of the dust mass fraction in the gas.

Various methods are used to derive the current signal, which are usually characterized as DC (direct current) or AC (alternating current) methods. Both evaluation methods have

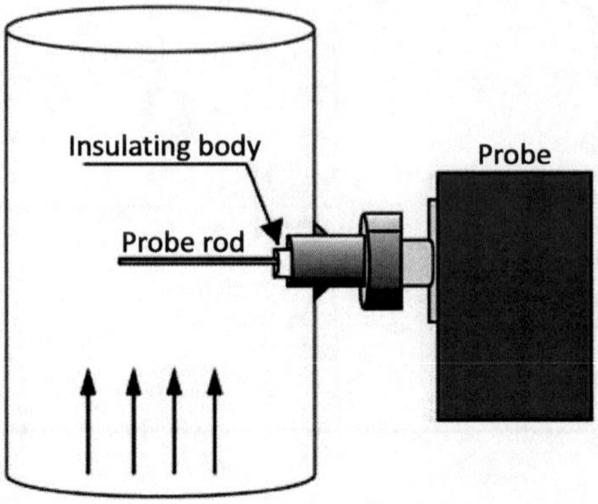

Fig. 13.36 Principle of in-situ measurement using the triboelectric method

advantages and disadvantages. For example, AC-based devices can be used especially when particles have a pronounced precharge. The triboelectric DC probes show a comparatively better linearity, especially in the velocity range >10 m/s.

The triboelectric signal is not only dependent on the dust content. As with other measuring methods, the type of dust, the particle size distribution and particle shape as well as the process-related boundary conditions (gas temperature, humidity, etc.) influence the level of the measuring signal. In contrast to other methods, the speed and degree of turbulence of the flow are also identified as the main influencing factors.

By combining all influencing factors, it can be stated that the most important factors for the charge transfer, that is, the triboelectric measurement signal, are compensated if the influence of the dust concentration is linear and that of the velocity is potentially detected. The mathematical relationship between these variables is as follows:

$$c_{\mathrm{iB}} = A \cdot \mathrm{cal} \cdot v^{\alpha} + D \tag{13.19}$$

c_{iB}	Dust concentration [mg/m^3]
A, D	Constants from the dust calibration
Cal	Measuring signal [V]
v	Velocity of the exhaust gas [m/s]
α	Exponent of speed

The measurement of dust in wet gases requires extractive sampling, as is the case with scattered light measurement. The sample gas flow is sucked in by means of the injector and diluted with clean and appropriately tempered air. In this way, water drops or sticky components are reliably eliminated.

General Factors Influencing the Measuring Principle

First of all, the various factors influencing the triboelectric measuring principle will be considered. Mechanical and electrical dust properties will be distinguished and individual process engineering boundary conditions will be examined in more detail.

Mechanical Dust Properties

The triboelectric measuring signal depends on the intensity, duration and number of contacts (collisions or flow around) between the measuring probe and dust particles. For example, for the grain size of the dust particles, this means that with increasing grain size, the exchanging surface and thus the charge transferred becomes larger. However, as the grain size is approximately a third power of the dust mass, the measurement signal decreases with increasing grain size for the same dust concentrations, as the number of collisions decreases.

Electrical Dust Properties

As a preliminary remark, it should be noted that the use of the triboelectric effect requires the presence of particles with a technological pre-charging. Electrically neutral particles do not cause any charge exchange and therefore cannot be detected.

The triboelectric signal of each dust varies with its electrical conductivity and dielectric constant. Here applies:

The higher the electrical conductivity, the lower the triboelectric interactions. The higher the dielectricity, the higher the triboelectric measurement signal.

However, since the mechanical properties of the dust vary only within narrow limits for the respective measuring task, these do not have to be explicitly taken into account when calculating the dust content. This also applies to the electrical properties. In the case of device calibrations by gravimetric comparison measurements, any variations in the dust properties are "calibrated in".

Properties of the Gas

The dust content has a direct linear influence on the number of contacts between the measuring probe and dust particles and thus on the level of charge exchange. The relation of the mass of the dust to the number of dust particles can be assumed to be constant for the respective measuring task, so that the height of the measuring signal is directly proportional to the dust concentration. The electrical properties of the carrier gas depend on its composition. However, they fluctuate within an order of magnitude that is not relevant for the triboelectric effect.

Exhaust Gas Humidity

The exhaust gas humidity can be considered insignificant as long as the temperature of the exhaust gas is above the dew point. If this is not the case, condensation can occur within the gas and thus, in *in situ* measurements, the formation of a liquid film on the probe surface.

This interferes with the charge transfer between the probe surface and the dust particles and, in extreme cases, leads to a short circuit between the probe and the instrument ground.

Gas Velocity

The speed and the degree of turbulence of the flow directly influence the number, duration and intensity of the contacts between the measuring probe and the dust particles and thus represent the greatest process engineering influencing factors. Approximately, a potency behaviour for the velocity can be assumed. The size of the exponent depends decisively on the value of the velocity.

Gas Temperature

It influences the electrical properties of the dust, probe and gas. In practice, however, it is of little importance due to the small range of fluctuation in the individual measuring tasks. The mathematical relationship between the triboelectric measurement signal and the most important influencing variables can be illustrated as follows:

$$i = \frac{K_1 \cdot c \cdot v^{\alpha}}{d_{50}} + K_2 Q \cdot \frac{c}{d_{50}^3} \tag{13.20}$$

i	Triboelectric probe current
c	Mass concentration
K_1/K_2	Constants
d_{50}	Mean particle size
v	Exhaust gas velocity
Q	Charge of the impinging particles
α	Exponent of speed

In metrological practice, many of the influencing factors are usually constant, so that the triboelectric signal is proportional to the mass concentration.

Influencing Factors Related to the Process

In addition to the already mentioned influences on the triboelectric measuring principle, which can occur to a different extent in every application, there are special influencing factors in some cases.

Isolation

In various plants, the presence of moisture or aerosols can occur for a short time, for example, due to start-up and shut-down processes.. In order to avoid a negative influence on the measuring signal, triboelectric measuring probes can be encapsulated with an

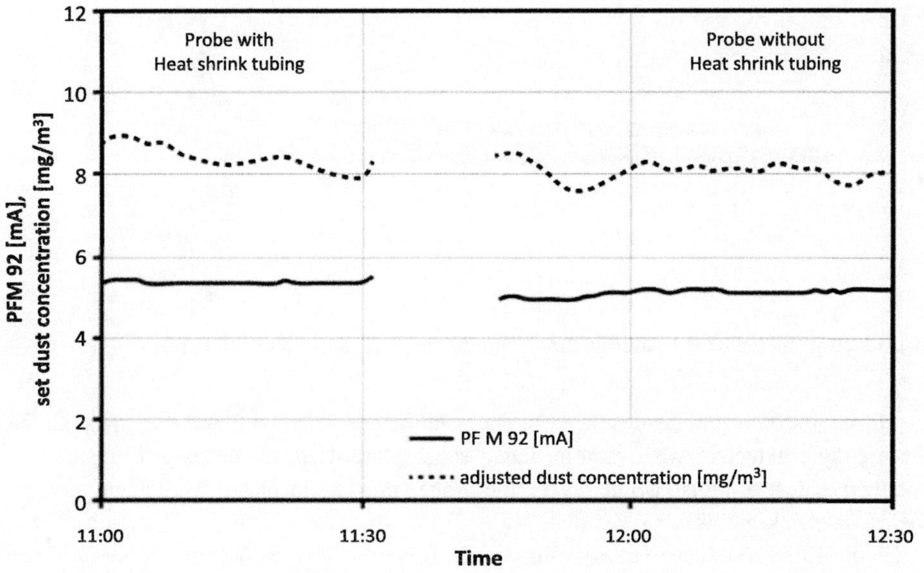

Fig. 13.37 Measurement signals of a triboelectric filter monitor with and without insulation by shrink tubing

insulation, that is, with a non-conductive (high-resistance) material. Tests carried out on the dust channel and many years of practical experience have shown that, despite this insulation layer, the triboelectric measuring effect is virtually unchanged.

The exemplary curves shown in Fig. 13.37 show that with an almost constant dust content of 8 mg/m^3 set at the dust doser, no significant differences in the signal curve with and without probe insulation can be detected.

Contamination of *In Situ* Measuring Instruments

Since during *in situ* measurement the probes are directly exposed to the exhaust gas flow, dust deposits may occur on the sensor surface after a certain operating time. Especially in systems that pass the dew point during start-up and shut-down processes or whose exhaust gas contains (sticky) aerosols, it is not a dust layer that is deposited, but rather a crust of dust, organic coatings and efflorescent oxides and salts.

After it has already been shown that an insulation layer of thin-walled plastic or Teflon does not influence the measurement signal, similar effects could be observed with dust layers up to approximately 3 ... 4 mm. Even a completely dusty probe with relatively thick deposits shows an almost unchanged dust-proportional signal.

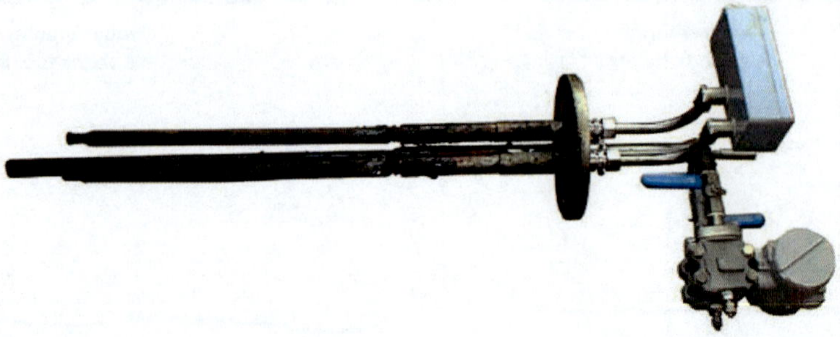

Fig. 13.38 Contaminated measuring probe after use in a chipboard production plant

The condition is that the dust layer or encrustation deposited on the sensors must not be electrically conductive when non-insulated sensors are used. In the case of electrically conductive dust or moist, acidic layers, the probes must be insulated beforehand (e.g. by means of shrink tubing).

Figure 13.38 shows an extremely dirty probe of a triboelectric dust concentration meter. This probe was installed in the exhaust gas of an indirect chip dryer, where aerosols occurred more frequently due to the plant technology. During the entire operating time, the measuring device displayed plausible measured values despite the contamination. This was confirmed by comparative gravimetric measurements.

Experimental investigations on a dust channel, in which the dust concentrations and flow velocities as well as the degree of contamination of the probes were specifically varied, did not show any significant influence of the measurement signal by the deposits. The measurements were initially carried out with dirty measuring probes. In the further course of the investigations, the probes were successively cleaned. Figure 13.39 shows that there is a common calibration function for all contamination levels of the probe for calculating the dust concentration from the triboelectric signal.

Figure 13.40 shows the deviations of the display of a dust concentration meter from a constant dust content with different degrees of contamination of the probes. The results shown prove that the degree of contamination did not significantly influence the measuring signal during the tests.

Electrostatic Precipitators

According to the current state of knowledge, monitoring the clean gas dust content after electrostatic precipitators with triboelectric measuring devices is only possible to a limited extent. The residual electrostatic charge of the dust particles influences the measuring signal immediately after the filter so strongly that no evaluable results are obtained with *in situ* measuring instruments. Installation is therefore only useful if an appropriate dwell time of the dust in the exhaust gas is ensured after the filter or if the dust particles can be discharged by deflection.

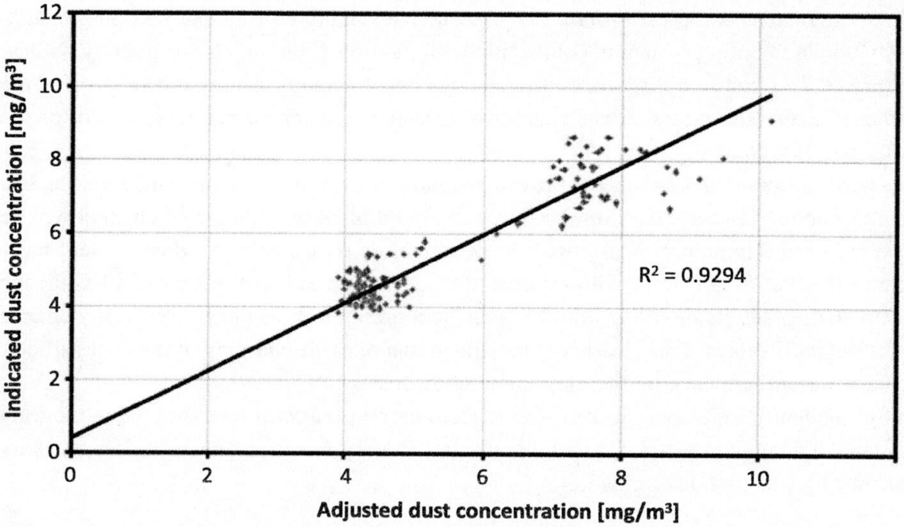

Fig. 13.39 Common calibration function for all pollution levels

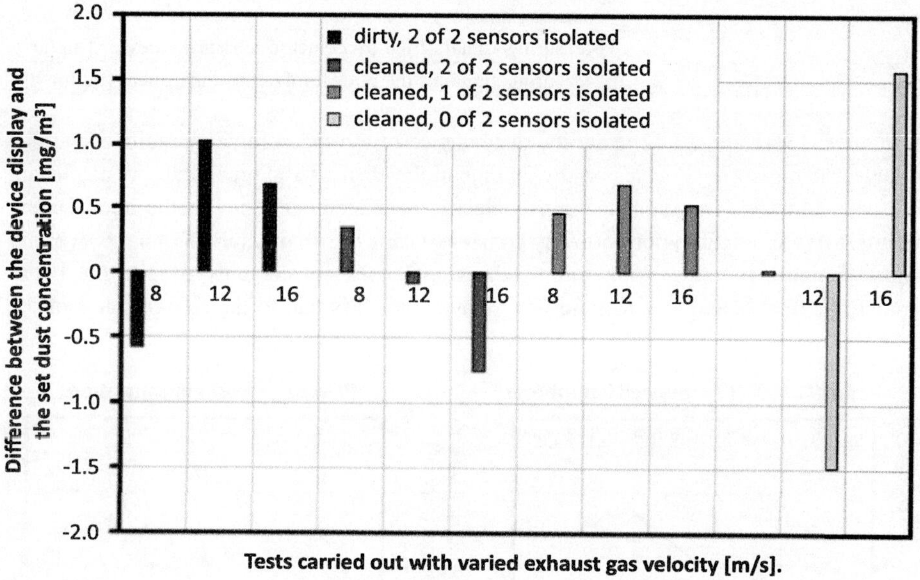

Fig. 13.40 Influence of the degree of contamination on the measuring signal of a triboelectric dust concentration meter

An alternative is the measurement by extractive sampling. This can be done qualitatively with a sampling system in connection with a simple filter monitor or directly with an extractive dust concentration measuring instrument. Sampling causes a discharge of the particles, so that the excess charge after the electrostatic precipitator has no influence on the measured dust content.

These correlations have been proven in various experimental investigations at test facilities and in practical tests. In the following, pilot plant tests on a dust channel will be reported on the influence of electrostatic precipitators on triboelectric dust content measurement. First of all, tests without dust loading of the exhaust gas by increasing the electrostatic precipitator voltage showed that the ionized gas flow did not cause any change at the device displays. This clearly demonstrated that only the charging of the dust particles actually has an influence on the measurement results.

Subsequently, reference measurements were carried out with an extractively working scattered light photometer. For a better presentation of the results, a dimensionless auxiliary quantity Fg was first determined:

$$\mathrm{Fg} = \frac{c_{\mathrm{tr}}}{c_{\mathrm{opt}}} \tag{13.21}$$

Fg	Auxiliary size
c_{tr}	Measuring signal of the triboelectric measuring device [mg/m^3]
c_{opt}	Measuring signal of the optical reference instrument [mg/m^3]

Figures. 13.41 and 13.42 show the dependence of the measurement results of various triboelectric measuring devices on the voltage in the electrostatic precipitator at a constant dust concentration. The ratio of the triboelectric measurement signal to the optical measurement signal remains approximately constant during *in situ* measurement up to a certain voltage point. After this "kink point", the signal rises considerably despite the better separating effect of the electrostatic precipitator, which is due to the charge states of the

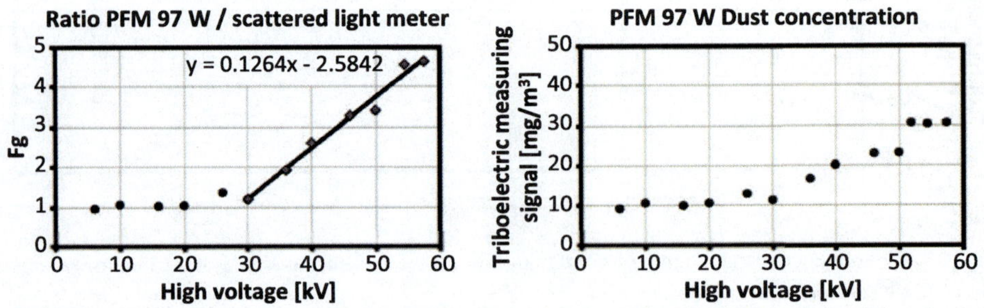

Fig. 13.41 High voltage influence on the measuring signal of a triboelectric dust concentration meter

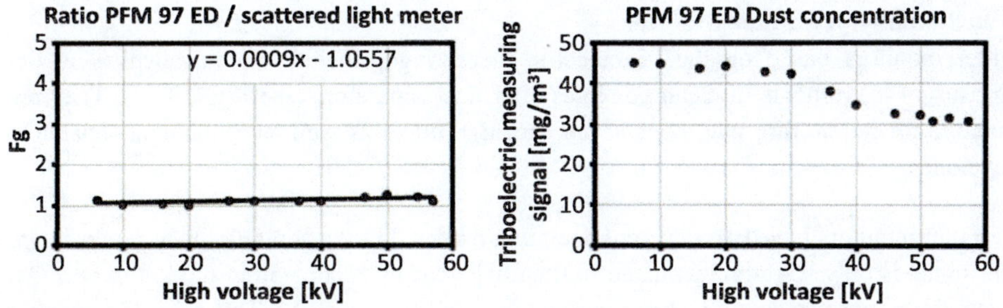

Fig. 13.42 High-voltage influence on the measuring signal of an extractive triboelectric dust concentration meter

dust particles (see Fig. 13.41). In extractive measurement, the pure dust signal follows the actual dust loading of the exhaust gas, while the ratio to the optical reference measurement remains almost constant (see Fig. 13.42).

The measurements were carried out with different types of dust by varying the duct speed and the dust concentration. The reproducible measurement results obtained confirm the statements made. Thus, the suitability of triboelectric, extractive dust measurement even after electrostatic precipitators is clearly proven.

Design and Application of Triboelectric Dust Measurement Technology
In this section we will deal specifically with the devices of Dr. Födisch Umweltmesstechnik AG. Devices from other manufacturers have a comparable operating principle, but differ in their technical design. This will not be considered in the following.

Qualitative Dust Measurement
Qualitative dust measurement includes the monitoring of the clean gas dust content after dedusting plants and at the same time the evaluation of the condition of the exhaust gas cleaning systems. This results in the following advantages for plant operators:

- Emission measurement and filter monitoring with only one device,
- Avoidance of visible waste gas plumes,
- Simplification of the maintenance of filter systems through early detection of beginning filter wear and location of defective filter elements, possibility of targeted maintenance measures,
- Avoidance of product losses.

In order to realize the above-mentioned tasks as effectively and cost-effectively as possible, equipment without major installation and operating costs should be used for this purpose. This requirement is met by measuring instruments based on the triboelectric measuring principle.

Functionality of a Filter Monitor

Filter monitors based on the triboelectric measuring principle are installed as in-situ measuring instruments in clean gas lines after dust separators (see Fig. 13.43). They are suitable for monitoring bag, pocket and cartridge filters as well as centrifugal separators (cyclones).

Filter monitors have two different operating modes. In the operating mode *Integral Off*, the signal is only slightly attenuated in time, whereas in the operating mode *Integral On*, averaging is performed with a time constant of approximately 30 s. This gives the operator the option of initially outputting an average emission value in the standard mode (*Integral On*) that is close to the half-hourly average value relevant to the authorities. Since the dust concentration gradually increases as the filter starts to wear, an alarm can be triggered if a specified limit value is reached (see Fig. 13.44).

In general, two limit values are freely adjustable, so that a lower selected pre-alarm can signal filter wear already beginning. If this is the case, a specific filter diagnosis can be made by changing the operating mode to *Integral Off*. The measuring signal is output continuously for a period of approximately 10 min. As a result, the exact image of the current dust emission can be viewed. Figure 13.45 shows the recording of the dust emissions after a pulse-jet filter for dedusting a cement mill. The dotted line corresponds to an average clean gas dust concentration of 5 mg/m^3 i. N.

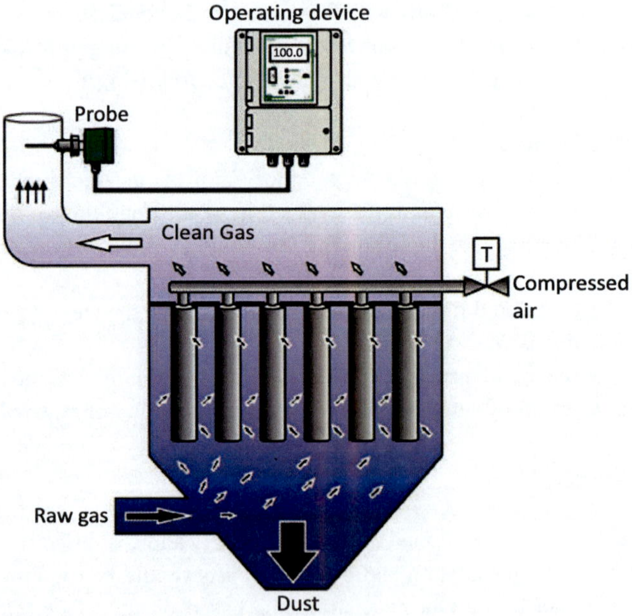

Fig. 13.43 Installation of a filter monitor for monitoring a bag filter

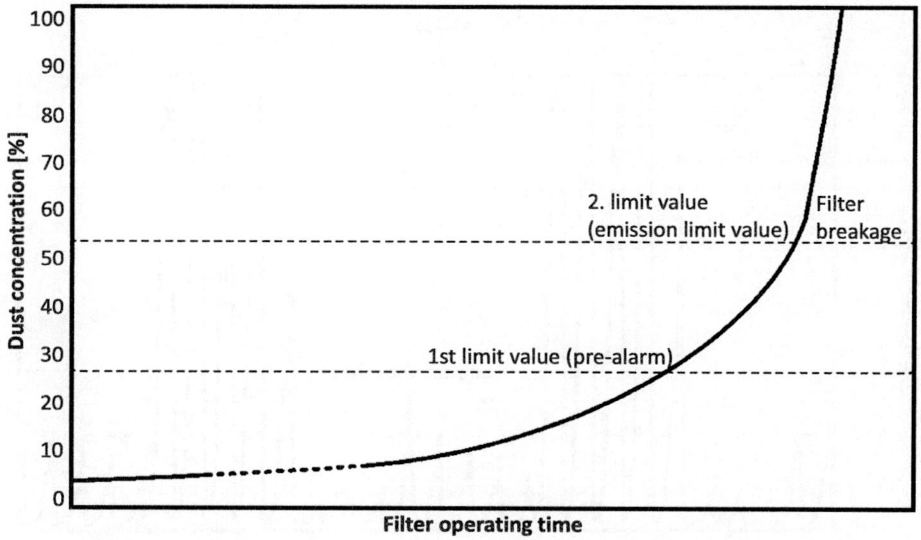

Fig. 13.44 Monitoring a filter in the "Integral On" operating mode

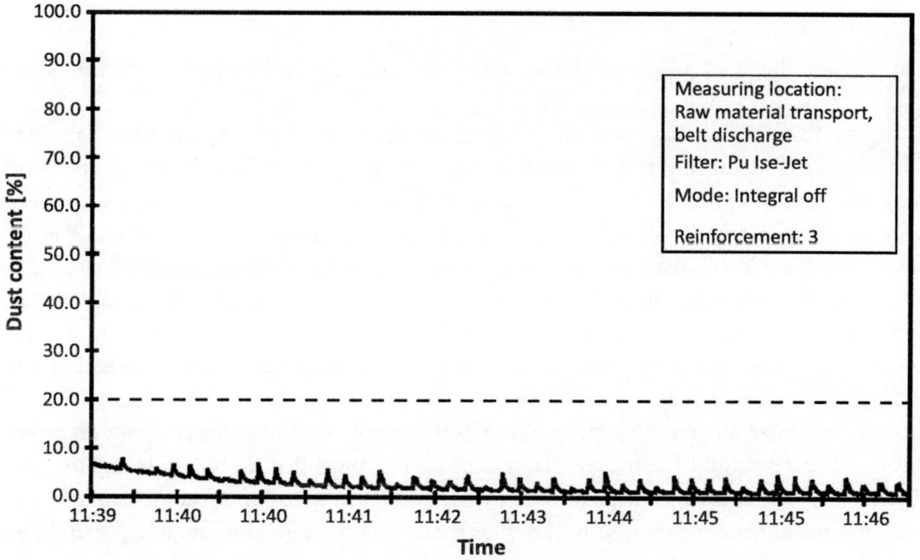

Fig. 13.45 Filter condition diagram with fault-free filter bags

Since the filter hose rows are cleaned at fixed time intervals, cleaning peaks are shown in the diagram. These have a uniform height and steep flanks. From this it can be concluded that the sealing between the raw and clean gas chamber of the filter is optimally secured and

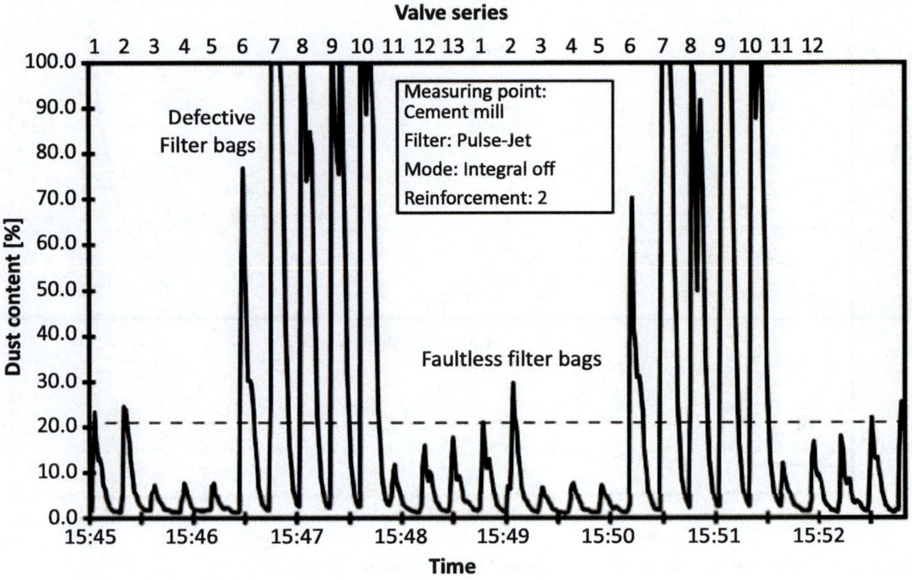

Fig. 13.46 Filter status diagram with defective filter bags

that the filter bags are made of high-quality filter material. If the tips have significantly flatter flanks, it is safe to conclude that the material is of inferior quality.

Figure 13.46 shows the course of the clean gas dust concentration of another bag filter for cement mill dedusting with 13 rows of bags (2 cleaning cycles of the filter). It can be seen that the dust emissions of rows 1–5 and 11–13 are low. On the other hand, a strongly increased dust penetration can be registered for the bag rows 6–10 and a defect can be concluded. As a result, the filter bags of these rows would have to be replaced.

It should be noted that the emissions from these hose lines were not yet or hardly visible at the exhaust outlet. The operator thus receives a warning of the deteriorating filter condition in good time, long before a dust plume is visually perceptible and the authorities or residents take note of it.

Effective filter diagnosis includes not only the localisation of defective filter elements but also the detection of defective diaphragm valves, flaps that do not close tightly, etc. Operators of larger filter systems thus enjoy advantages in terms of effective maintenance. For filter manufacturers, the use of the filter monitors opens up considerable possibilities with regard to their quality assurance. Due to the amplified current signal, even longer distances to the evaluation unit can be overcome without any problems (up to 100 m).

In addition, it is also possible to effectively monitor larger dedusting plants by installing and coupling several devices and the corresponding evaluation software. Figure 13.47 shows the monitoring of the dedusting of a fluidized bed combustion system in a hard coal-fired power plant. The plant consists of 2 filters, each with 6 filter chambers, in which

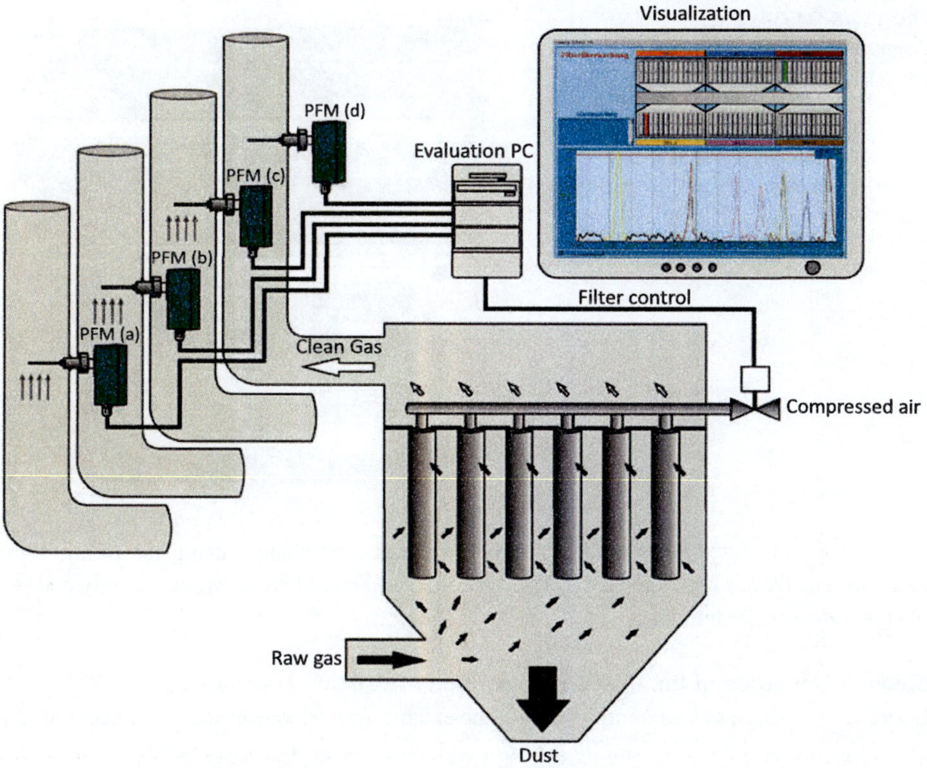

Fig. 13.47 Filter monitoring system

13 rows of 14 hoses are mounted in each of the 13 filter chambers. In the clean gas lines of each filter 4 filter monitors are installed, whose analogue signals are fed to a central evaluation computer together with those of the filter control system. By linking them via a special software it is possible to view the cleaning of the individual rows of hoses with corresponding evaluation. Coloured underlays signal the respective cleaning and any limit value overruns.

After a preset frequency of errors or exceeding of the limit values, an alarm message is generated which describes the type of the last error. Further functions include a retrospective view for viewing the signal curve of a single hose row, the view of all alarm messages and the individual configuration of the software for optimal adaptation to the respective measuring task. Of course, this also includes the free adjustability of the limit values.

The time-consuming troubleshooting after a defect can be significantly limited both by time and personnel. Additional saving effects result from the targeted ordering of filter material, as only defective elements have to be replaced.

However, the filter control does not necessarily require a fixed installation. Mobile filter monitoring systems were developed especially for this purpose (Fig. 13.48). They are easy

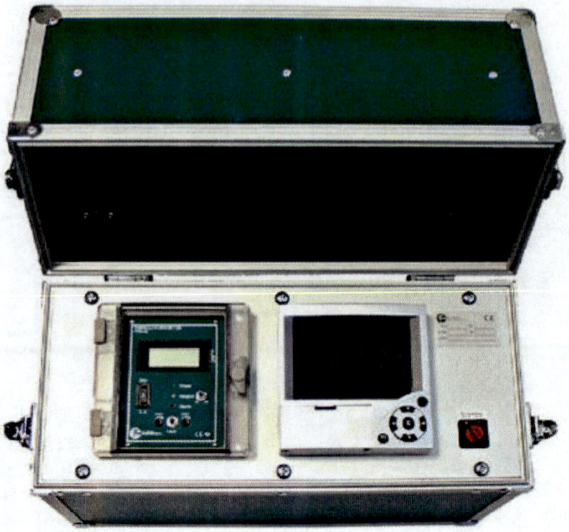

Fig. 13.48 Mobile filter diagnosis device PFM 92 K

to install and operate. Filter defects can be localized immediately using the online line diagram. The data is registered by a digital recorder and stored on an SD card. They can be read out via a serial interface.

Mode of Operation of the Dust Concentration Measuring Instruments

Depending on the type and composition of the exhaust gas, triboelectric dust concentration measurement can be carried out both *in situ* and extractively. Each measuring method has certain influencing factors and physical limits. In triboelectric dust measurement, particle and gas properties influence the measurement signal to a certain extent. The gas velocity exerts the most significant influence. In order to be able to quantitatively determine the dust concentration triboelectrically, this velocity influence must be considered and compensated. For this reason, dust concentration measuring devices for in-situ detection are equipped with an additional recording of the gas velocity or the volume flow and also the exhaust gas temperature. The measuring probe shown in Fig. 13.49 consists of two dust sensors and a dynamic pressure probe with integrated temperature sensor.

The dust sensors, which have a sharp-edged trapezoidal profile, are arranged at the same height in the direction of flow. The selected profile for the dust sensors ensures a low degree of contamination and thus enables a stable measurement over a long period of time. Due to the redundant measurement, occurring errors such as sensor failure, contamination and dust bridges can be detected. The dynamic pressure probe is positioned centrally in front of the dust sensors. The design takes into account that the pitot tube and the dust sensors must not influence each other in terms of flow, but the probe must be compact. The preamplifier for receiving the dust signal, the pressure transmitter and the temperature evaluation integrated in the preamplifier are located in the probe head. In addition to the display and the operating keyboard, the evaluation unit also contains the electronics required for evaluating the

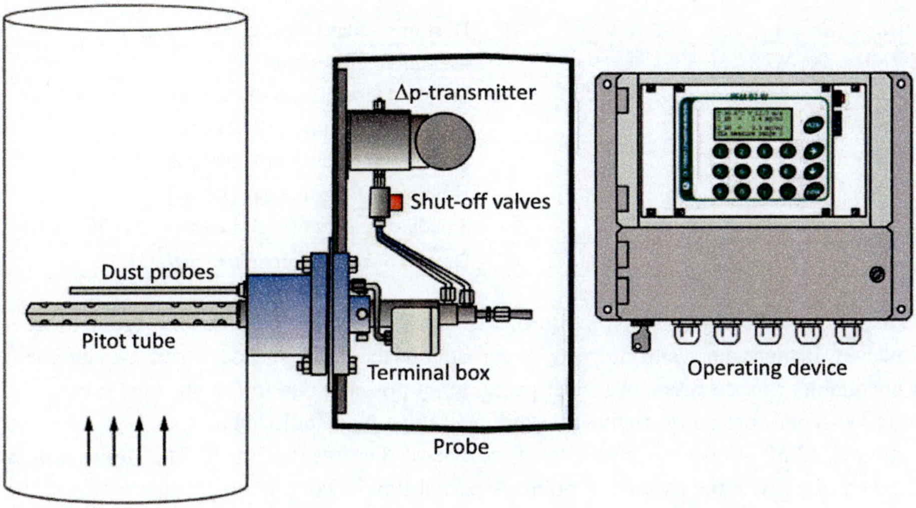

Fig. 13.49 Principle of *in situ* dust concentration measurement

signals. The signals coming from the measuring probe are calculated. All data are output as analogue signals from 4 ... 20 mA. In measuring mode, the calculated data are displayed$c_{\mathrm{i.\,B.}}$ as measured values of the dust concentration in operating condition.

$$c_{\mathrm{iB}} = A \cdot 500 \cdot \mathrm{cal} \cdot v^{\alpha} + D \tag{13.22}$$

$$\mathrm{Exp} = B \cdot v^2 + C \cdot v + E \tag{13.23}$$

c_{iB}	Dust in operating condition [mg/m3]
cal	Raw signal [V]
A, *D*	Calibration constants (determined by gravimetric calibration)
B, *C*, *E*	Constants of velocity compensation
v	Exhaust gas velocity [m/s]
α	Exponent

The constants of the speed compensation are set at the factory and only adjusted if necessary during on-site calibration. The calculation of the dust concentration in the standard state is carried outc_{iN} according to the following correlation:

$$c_{\mathrm{iNtr}} = \frac{T}{T_0} \cdot \frac{p_0}{p} \cdot \left(1 + \frac{f}{\rho_{0\mathrm{H_2O}}}\right) \cdot c_{\mathrm{iB}} \tag{13.24}$$

$c_{\mathrm{iN\,tr}}$	Dust in standard condition dry [mg/m^3]
T	Measured temperature [K]
T_O	Standard temperature (273,15 K)
p_O	Standard pressure (101,325 kPa)
p	Absolute pressure of the flue gas [kPa]
f	Humidity of the flue gas [kg/m^3]
$\rho_{\mathrm{0H_2O}}$	Standard density of water vapour (0.8038 kg/m3)
c_{iB}	Dust in operating condition [mg/m3]

In plants with dry gases, monitoring of dust emissions is generally problem-free. However, if there are water droplets or aerosols in the exhaust gas, in-situ measurement is unsuitable, as measurement in wet gases is not possible due to the physical measuring principle. Therefore an extractive measurement must be installed here.

Figure 13.50 shows the principle of extractive dust measurement. The probe unit is installed directly at the extraction point. A partial flow is continuously extracted from the exhaust gas through an injector. In order to reduce the moisture content of the sample gas and to reduce contamination by sticky exhaust gas components, a defined (heated) dry and dust-free dilution air flow is added to the extracted partial gas flow at the probe inlet. At the same time, the gas mixture is heated to a defined temperature. The sample gas temperature is adjusted in such a way that water droplets are reliably eliminated and the measurement signal is formed exclusively by the solid particles. The elimination of aerosols (e.g. acid, resins, tar) is also achieved by optimizing the sample gas temperature, depending on the task. This sample gas flow is then sucked through the measuring cell with the help of an injector, whereby the volume flow is monitored by measuring the differential pressure over the measuring cell. This is designed as a swirl chamber, on the inner wall of which two

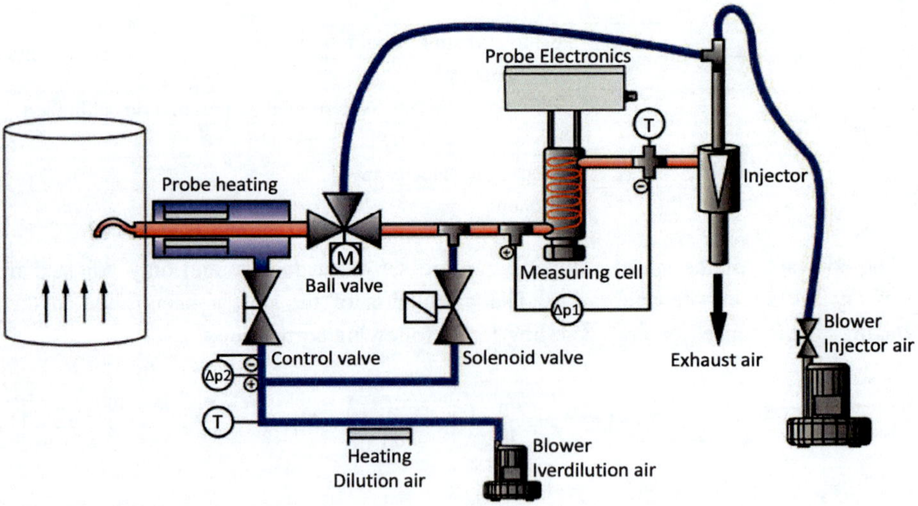

Fig. 13.50 Principle of extractive dust measurement

triboelectric sensors are installed. The measuring gas flow enters the measuring chamber tangentially. This swirling motion makes it possible to generate high speeds and thus corresponding measurement signals at the tribo-sensors.

In triboelectric, extractive dust measurement, a similar mathematical relationship of the physical quantities results in comparison to in-situ measurement. The triboelectric signal *Cal* depends on the dust content of the sample gas, the volume flow in the measuring chamber and the set dilution of the extracted sample gas. The mathematical relationship is therefore as follows:

$$C_{\mathrm{iB}} = A \cdot \left[\frac{\mathrm{Cal}}{F_M} \cdot B \cdot \left(\frac{F}{F_0} \right)^C \right] + D \qquad (13.25)$$

C_{iB}	Dust concentration [mg/m^3]
Cal	Measuring signal [V]
F_M	Extracted sample gas volume flow [m^3/h]
F	Volume flow rate in the measuring chamber [m3/h]
F_0	Flow rate standards (20 m^3/h)
A, D	Calibration constants (determined by gravimetric calibration)
B, C	Constants of the velocity compensation (factory setting, determined device-specifically)

Triboelectric dust concentration meters (e.g. PFM 97 W of Dr. Födisch Umweltmesstechnik AG) continuously measure even the lowest dust concentrations directly in the flue gas stream (*in situ*) (Fig. 13.51). Since neither reflector, light trap nor purge air are required due to the use of the triboelectric effect, the device can be mounted in a flange without counter flange. All assemblies which come into contact with the medium, such as probes, welding sockets and diaphragm of the pressure transmitter, are made of

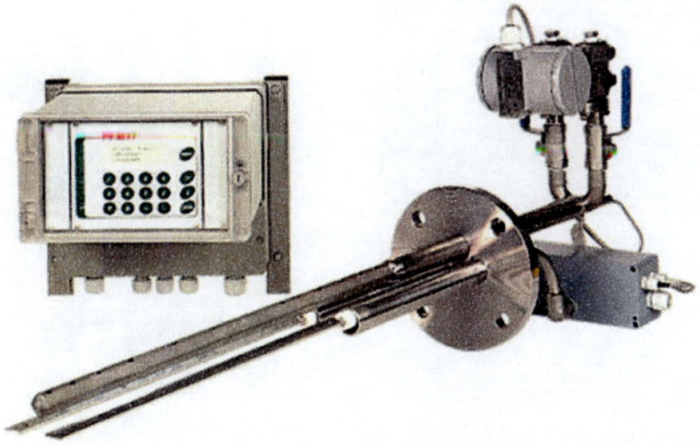

Fig. 13.51 Dust concentration measuring instrument PFM 97

corrosion-resistant materials. In special applications, special materials are used in consultation with the plant operator.

In the case of measurements relevant to the authorities, additional equipment for required exhaust gas boundary parameters can be dispensed with due to the 3 measured variables (dust, volume flow, temperature). The installation effort is reduced to 1 flange.

In addition to the output of the dust concentration values in operating and standard state, the gas velocity or volume flow and the exhaust gas temperature can be displayed separately. The measuring instruments have 4 measuring ranges which can be switched automatically. Over 2 freely selectable limit values, a signal can be given when exceeded. Internally the maintenance and readiness status is constantly checked. The zero point is monitored, so that a reliable measurement can be made even with the smallest dust contents. Zero point deviation and reference point are checked every 4 h.

The dew point of the exhaust gas must not be undercut at the measuring point, otherwise faulty measuring results are to be expected due to condensation. Furthermore, the mounting location for the probe must be sufficiently stable, that is, low-vibration installation should be ensured.

The *in situ* measuring instruments are unsuitable for measuring the dust concentration in wet gases due to the measuring principle. For this reason, continuous measurement of the dust content is carried out extractively with the PFM 97 ED instrument (Fig. 13.52).

This device is, among other things, suitable for dust measurements:

- behind wet separators in waste incineration plants
- in soil remediation plants
- in the woodworking industry (chip dryers)
- in the insulation industry
- behind wet flue gas desulphurisation plants.

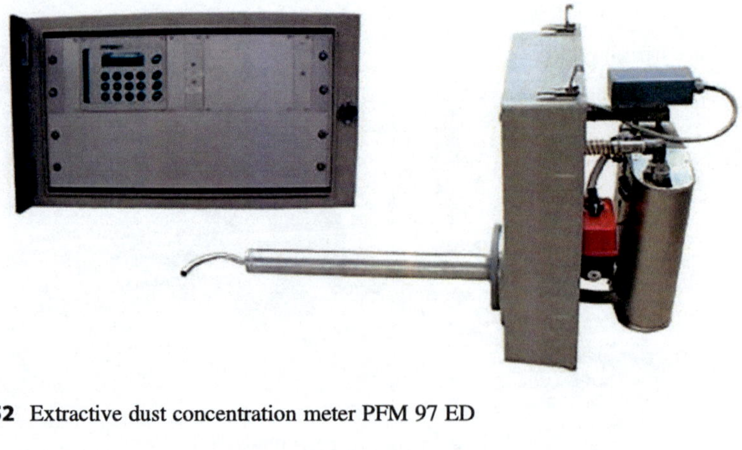

Fig. 13.52 Extractive dust concentration meter PFM 97 ED

The PFM 97 ED consists of a special sampling probe, the triboelectric measuring chamber, gas conditioning (dilution, temperature control), an injector, two side channel compressors and an electronic evaluation unit. For diagnostic and cleaning reasons, the PFM 97 ED automatically carries out a purging operation. This is used to check zero and reference points and to clean the gas paths carrying the sample gas. Due to the extractive measurement the device is independent of a dew point difference and the flow velocity.

13.13 Soot Count Measurement

A special case of optical dust measurement is the determination of the soot number in exhaust gases of various combustion plants. In certain types of plants (e.g. oil or comparable liquid fuels) the continuous measurement of the soot number is mandatory. Scattered light photometers (*in situ* or extractive) or reflection photometers are used to determine the soot number.

In-situ scattered light measuring instruments (e.g. D-R 300 of the DURAG Group, measuring principle see also Fig. 13.31) send light directly into the exhaust gas duct, at the back of which it is collected and absorbed by a light trap. From a defined measuring volume in the duct the resulting scattered light is received and measured at a certain angle (e.g. 90° or backward scattering). Due to the complex installation, this measuring arrangement is particularly problematic in thick-walled chimneys or double-walled pipelines. In these cases, the extractive measuring method must be used.

The extractive soot number measurement is carried out in the same way as the dust concentration measurement, but here only a slight heating of the partial gas flow takes place, since normally no droplets or aerosols occur in the exhaust gases of the plants to be monitored. Therefore the device design is much more compact compared to extractive dust concentration measurement.

Another important difference is the alignment of the sampling probe, which is not placed against the flow as in dust content measurement, but offset by 90°.

In the case of reflection photometry, which is also extractive, a filter band is exposed to the exhaust gas and its blackening is recorded by a photometer (Fig. 13.53). The instrument display and the determined reflection coefficient are in a linear relationship. This method is also described in VDI 2066 sheet 8 (Automatic soot pump devices).

All methods are calibrated according to DIN 51402-1 and are intended to monitor compliance with soot number 1. This soot number is exceeded if the measurement result rounded to a whole number reaches soot number 2 (for soot number $\geq$1.5 as limit value threshold). Figure 13.54 illustrates the recording of measured values for continuous soot number measurement. The measured values of the continuous measuring instruments are recorded on a recorder. Since the soot count measurement also stops when the burner is shut down, operating hour meters are used at the same time, which determine the burner running time on the one hand and the times when the limit value thresholds are exceeded on the other.

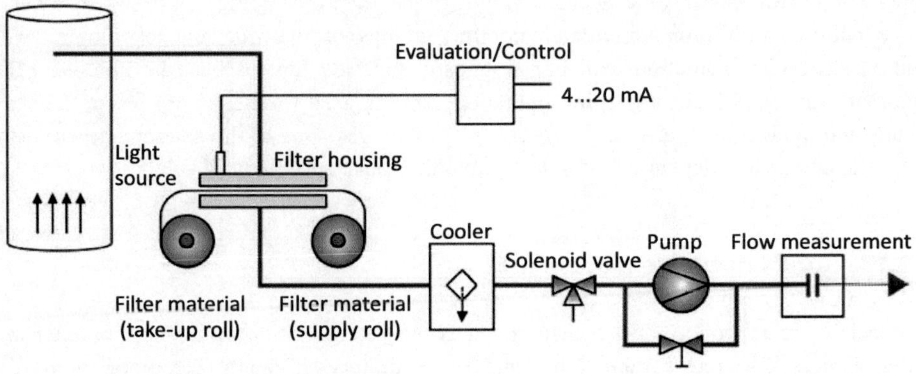

Fig. 13.53 Schematic representation of reflection photometry according to VDI 2066 sheet 8

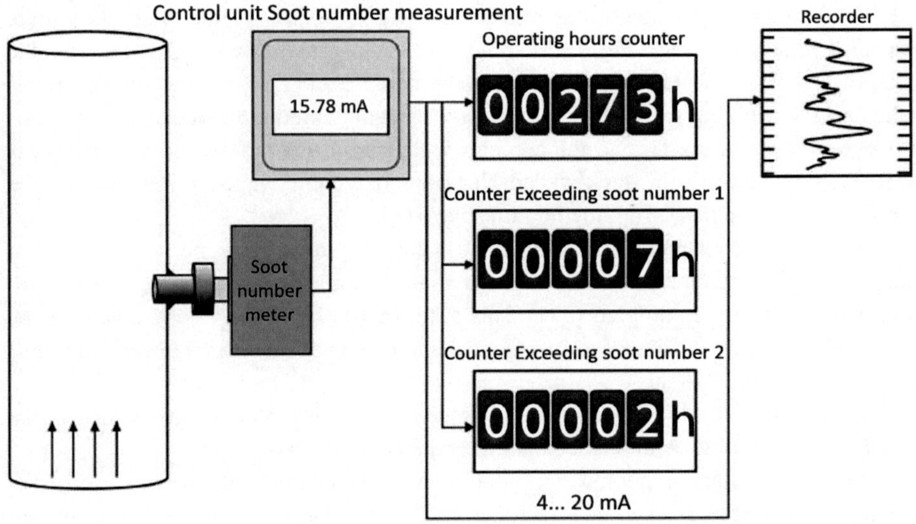

Fig. 13.54 Measured value acquisition for continuous soot count measurement

13.14 Comparison of the Different Continuous Measurement Methods

In summary, the various quantitative measurement methods are compared with regard to their advantages and disadvantages in different categories (see Tables 13.7 and 13.8).

A similar consideration of the qualitative dust measurement methods is not given here. It is known from practice that triboelectric measuring instruments are mainly used here.

Table 13.7 Device arrangement

Device arrangement	*In situ*	Extractive
Transmissiometry	•	
Scattered light measurement	•	•
Beta radiation absorption		•
Triboelectricity	•	•

13.15 Calibration of Dust Measuring Instruments

For the continuous monitoring of dust concentration relevant to the authorities, only suitability-tested measuring instruments may generally be used. Furthermore, it is necessary to have the correct installation of the measuring instruments checked and certified by a measuring point notified in accordance with §29b BImSchG. The proper installation of the dust measuring devices is a fundamental prerequisite for obtaining representative results regarding the dust concentration of plants whose emissions are to be continuously monitored. This also applies to the use of measuring instruments within the scope of process monitoring (e.g. qualitative measurement - monitoring the clean gas dust content of bag filters). Details are regulated in DIN EN 14181 and in the VDI guideline 3950.

When correctly installed, the inlet and outlet sections, shown in Fig. 13.55, are of decisive importance. According to DIN EN 13284-1, DIN EN 15259 as well as VDI 2066 sheet 1, the inlet section must correspond to 5 times the hydraulic diameter. For the outlet section 2 times the diameter is sufficient. If the flue ends behind the measuring plane, 5 times the hydraulic diameter is required.

Furthermore, if possible, a vertical installation location for continuous dust measurement technology should preferably be selected.

After commissioning, continuous dust measuring instruments for official emission monitoring shall be calibrated at periodic intervals by a measuring body notified in accordance with §29b BImSchG and subjected to an annual functional test. The essential requirements for this procedure are defined in the VDI guideline 3950. The time intervals between two calibrations (repeat calibration) are laid down for the respective type of plant in the implementing regulations for the BImSchG or TA Luft. In case of a significant change of the plant to be monitored (e.g. with accompanying change of the dust quality) or the dust measuring device, a repeat calibration is also necessary.

Aim of the Calibration

The aim of the calibration of continuous dust measuring instruments is to determine the analytical function of the complete measuring system. This is carried out among other things, by simultaneous recording of the measurement signal (mA value) of the device to be calibrated and gravimetric determination of the dust concentration in the flue gas duct by means of a standardised measurement procedure or standard reference measurement

Table 13.8 Summary of advantages and disadvantages (The devices in brackets are different for different manufacturers or have application restrictions)

	Transmissiometry	Scattered light measurement *in situ*	Scattered light measurement extractive	Beta radiation absorption	Triboelectricity *in situ*	Triboelectricity extractive
	Investment costs[a]					
Low	•	•			•	
Resources						•
High			•	•		
	Operating costs[b]					
Low					•	
Resources	•	•				
High			•	•		•
	Expenditure for maintenance/service[c]					
Low					•	
Resources	•	•				•
High			•	•		
	Demand for spare and wear parts					
Low	•	•			•	
Resources			•			•
High				•		
	Expenditure for installation/assembly					
Low		•			•	
Resources	•		(•)			•
High			•	•		
	Device structure					
Simple/robust		•			•	
Resources	•		•			•

Complicated				•		
	Special features					
Usability with large exhaust ducts	•	(•)	(•)	(•)	(•)	(•)
Low sensitivity to changes in the measuring medium				•		
Application with wet gases			•	•		•
Collection of samples for follow-up examinations				•		
Combined acquisition of further measured variables (*V*, *T*)					•	
Possibility of combined gas sampling for analysis			•			

[a]The evaluation thresholds here are <10.000€ for *low*, 10.000–20.000 € for *medium* and >20.000 € for *high*

[b]In addition to the normal maintenance costs, for example, for power supply with regard to all equipment aggregates, additional costs, for example, for external purge air etc., are considered in the operating costs

[c] Criteria such as maintenance cycles or the need for specialists and special tools are taken into account when assessing the cost of maintenance/service

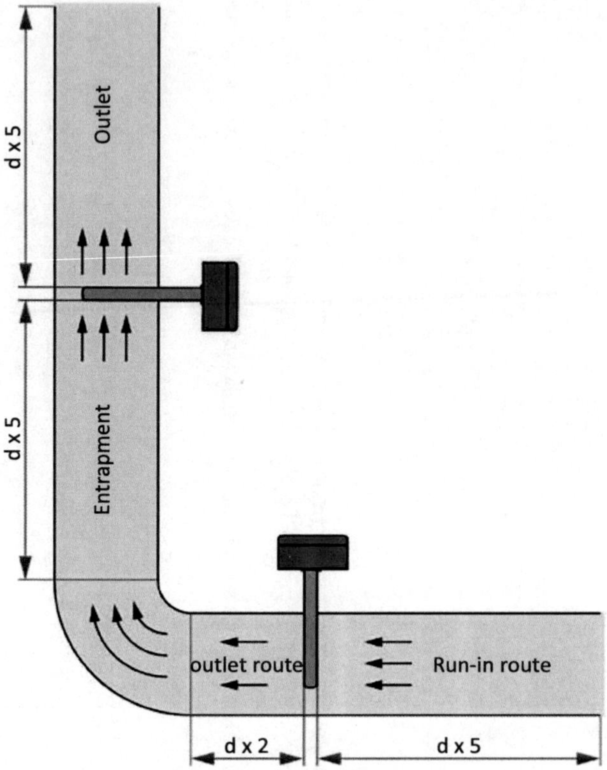

Fig. 13.55 Inlet and outlet section at the measuring point according to DIN EN 13284-1 and VDI 2066 sheet 1

procedure (e.g. according to VDI 2066; DIN EN 13284-1). In consideration of the representativeness, the calibration measurements are carried out as network measurements.

According to VDI 3950, a standard reference measurement procedure is an agreed procedure for the determination of one or more gas characteristics for the determination of which reference materials can practically not be produced. As agreed, the measurement result is a measure of the observed condition. Standard reference measurement methods are generally characterised by the fact that they are based on quantitative chemical or physical reactions whose performance parameters (or reaction parameters) are to be regarded as unchangeable under defined constant conditions. The conventional methods for the individual possibilities of dust monitoring are listed in Table 13.9.

According to VDI 3950, the emission measurement at a plant shall be considered representative if the measurement signal represents the temporal and spatial average value of the dust concentration within given limits. The installation location or sampling

Table 13.9 Overview of Convention measurement procedures

Measured variable	Convention procedure (manual)	Measuring principle autom. measuring equipment
Dust	DIN EN 13284-1 Guideline VDI 2066 sheet 1 Gravimetric determination	Light attenuation Light scattering Radiometry Triboelectricity
Dust (qualitative) Limit value monitoring	DIN EN 13284-1 Guideline VDI 2066 sheet 1 Gravimetric determination	Triboelectricity Dynamic transmission
Soot number	Guideline VDI 2066 sheet 8	Light scattering

point shall be selected such that the waste gas matrix to be measured characterizes the spatial and temporal mean value of the dust concentration (spatial or temporal representativeness). This requires, among other things, a sufficient temporal resolution and an averaging time adapted to the emission course. The analysis function of the complete measuring system is derived by regression calculation from the pairs of values which are usually condensed to half-hourly mean values. The analysis function thus describes the relationship between the dust concentration in the exhaust gas cross section and the measurement signal of the continuous measuring device:

$$c = f(X) \tag{13.26}$$

c	Dust concentration
X	Measuring signal

The installation of the analysis function is the decisive task in the calibration of dust content measuring instruments.

Calibration Procedure

Nearly all dust measuring instruments do not measure the dust concentration directly, but an auxiliary parameter such as the change of optical or electrical signals. However, since the limit value to be monitored is defined as mass per cubic meter (mg/m^3), it is absolutely necessary to calibrate using the reference methods (see Fig. 13.56).

In order to implement the analysis function of the measuring system, 15 comparison measurements are usually to be carried out using the standard reference measuring method (gravimetric dust content measurement) (according to DIN EN 14181, DIN EN 13284-2 and VDI 3950 at least 15 measurements distributed over 3 days). The required number of measurements depends in particular on the possible operating conditions and associated emission conditions at the plant.

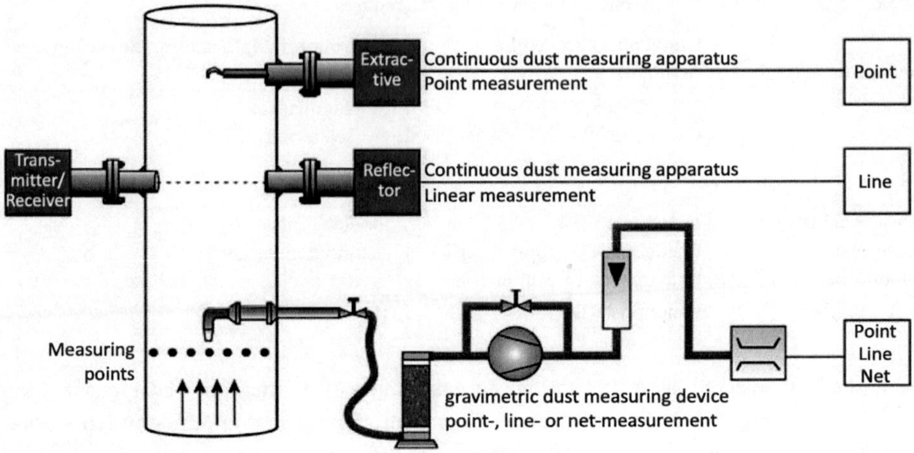

Fig. 13.56 Calibration of continuous dust measuring instruments

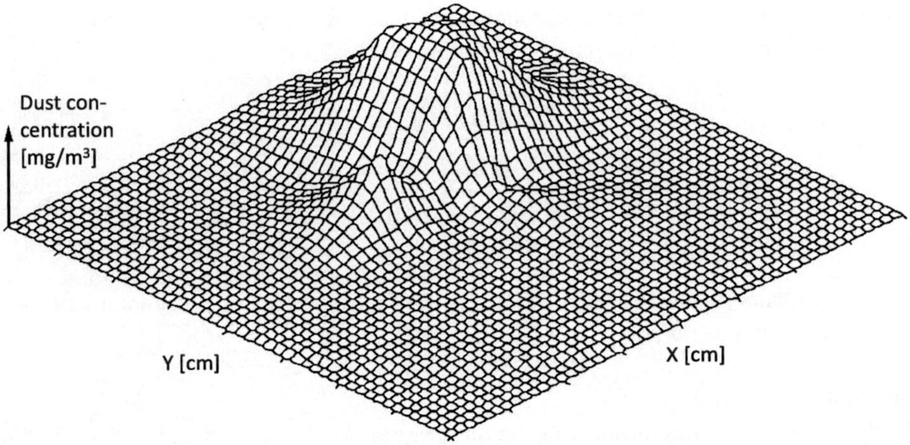

Fig. 13.57 Exemplary inhomogeneous dust distribution in the measuring cross section

In order to achieve sufficient representativeness, the temporal course and spatial distribution of emissions over the measurement cross-section shall be taken into account. For this purpose, the comparative measurements shall be carried out as network measurements. The network-related analysis function is then obtained. In practice a homogeneous dust concentration distribution is often not given. Figure 13.57 shows a typical picture of such an uneven distribution.

The analysis function thus describes the measurement signal of the complete measuring system as a function of the measured object, taking representativeness into account. In contrast, the functional relationship between the measuring signal of the dust monitor and reference materials or adjustment aids is called the device characteristic curve. For example, special filters are used for dust measuring instruments. Within the scope of determining the analysis function, the device characteristic curve is only used to check the function of the measuring instruments.

During calibration, the mass concentration (in mg/m^3) obtained with the conventional method is assigned to the corresponding instrument display. The pairs of values thus obtained are subjected to a regression calculation, whereby a linear analysis function is generally assumed for the emission measuring devices. Only in rare cases is quadratic regression more appropriate.

The general form of the analysis function is therefore

$$c = AX^2 + BX + C \tag{13.27}$$

c	Mass concentration [mg/m^3]
X	Device signal [mA]
A, B, C	Constants

If $A = 0$ is set, the quadratic function changes to the linear function:

$$c = BX + C \tag{13.28}$$

c	Mass concentration [mg/m^3]
X	Device signal [mA]
B, C	Constants

The comparative measurements carried out within the scope of the calibration should cover the entire measuring range of the continuous dust measuring instrument if possible. A further requirement results from the applicable administrative regulations (TA Luft) or implementation regulations of the BImSchG (13./17./27./30. BImSchV). According to these, the measuring ranges of the installed continuous measuring devices should cover approximately 2.5–3 times the plant-specific dust limit value (1.5 times for plants according to 17th BImSchV). For dust concentration measuring devices, however, only limited prior knowledge of the device characteristic curve is available. Whether the above requirement can be met will only be determined after calibration has been completed.

Targeted Variation of the Dust Concentration

During calibration, interventions in the operation of a plant or waste gas purification facility may be necessary with the aim of increasing the dust concentration in waste gas. Typical measures that have been proven in practice are:

- Variation of the current-voltage values for electrostatic precipitators
- Electrical shutdown of complete filter stages for multi-stage electrostatic precipitators
- Variation of the cleaning frequency for bag filters (temporary shutdown of the cleaning, continuous cleaning)
- Targeted increase of the dust concentration due to leaks in the filter material (e.-g. Replacement of individual filter bags with perforated sheets).

However, the latter possibility of a targeted increase in dust concentration should only be used to a limited extent. In order to obtain representative measurement results here, a high degree of operating experience is required. An addition of dust in the clean gas duct with the objective of a defined increase of the dust concentration is not permissible, since serious changes in the particle size distribution or the dispersity state of the dust will result, which, especially in the case of optical measuring instruments, can lead to a considerable influence on the analysis function.

The DIN EN 13284-2 points out that no untypical operating states (e.g. Bypass operation) should be initiated for setting up the analysis function, as a. drastically changed particle size distributions can result (factor 5–10). Figure 13.58 shows a calibration function, in the determination of which precisely those errors were made. After removing a filter hose to increase the dust concentration, the openings in the cover plate were enlarged from 2 mm to 8 mm. This resulted in an extreme change in the measured values, making it impossible to set up a plausible calibration function.

With triboelectric measuring instruments, an addition of dust usually leads to incorrect calibrations, since the charge distribution in the exhaust matrix is permanently changed. In

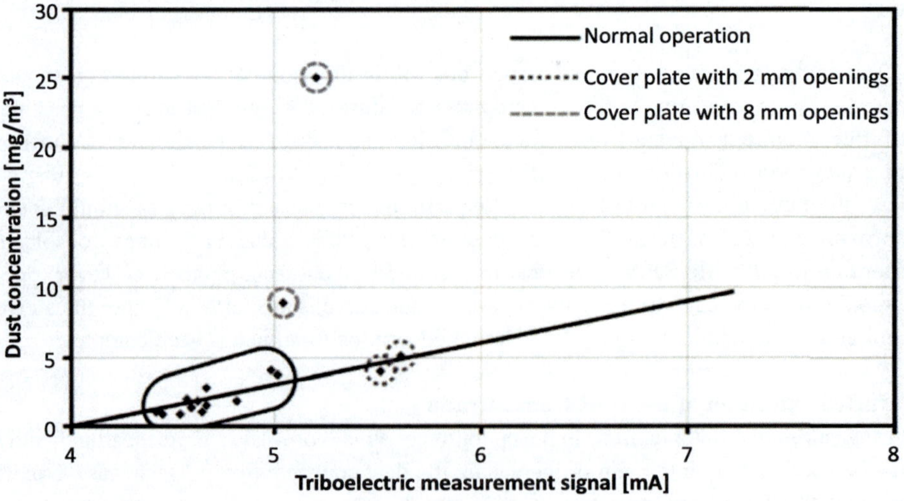

Fig. 13.58 Influence of artificially increasing the dust concentration on the calibration line

some cases even opposite tendencies could be observed, that is, decreasing signal levels when increasing the dust dosage.

Calibration of Optical Dust Measuring Instruments

The correlation between the dust content in the exhaust gas and the measuring signal of the continuous measuring system depends on various influencing variables.

Depending on the dust content, calibration is carried out according to various standard reference measuring methods. A more detailed explanation of the methodology for gravimetric comparison measurement according to VDI 2066 was described in detail in Sect. 13.3. DIN EN 13284-1 contains detailed information on the calibration of optical dust measuring instruments, especially for low dust contents. According to Fig. 13.59, a large number of influencing variables usually influence the measurement signal. Especially with fluctuating operating conditions (e.g. Full load/partial load) or different materials used, different calibration curves can result. In practice, it is therefore recommended that all essential parameters be varied sufficiently during calibration. Fig. 13.60 shows a phenomenon frequently observed with optical in-situ measuring instruments.

It can be clearly seen that the linear regression (1) of the measured values results in a straight line, which shows strong deviations from the coordinate origin and can lead to considerable misinterpretations. The reasons for this are the differences in the particle size distribution, the local distribution in the exhaust gas cross section at different load levels and possible changes in the optical dust properties.

The result of the quadratic regression (2) is a relatively good correlation. However, the curve should never be extrapolated, as otherwise implausible concentration values will be calculated. The calibration lines determined for individual operating conditions provide the best results. If there is a significant change in dust precipitation, for example, due to a

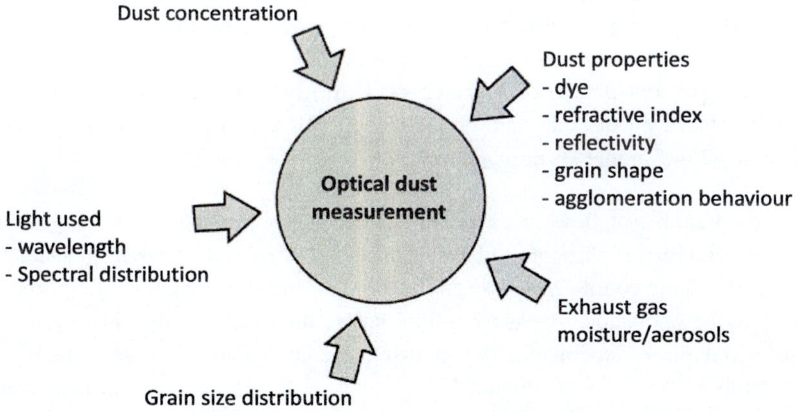

Fig. 13.59 Influencing variables on the optical dust measurement

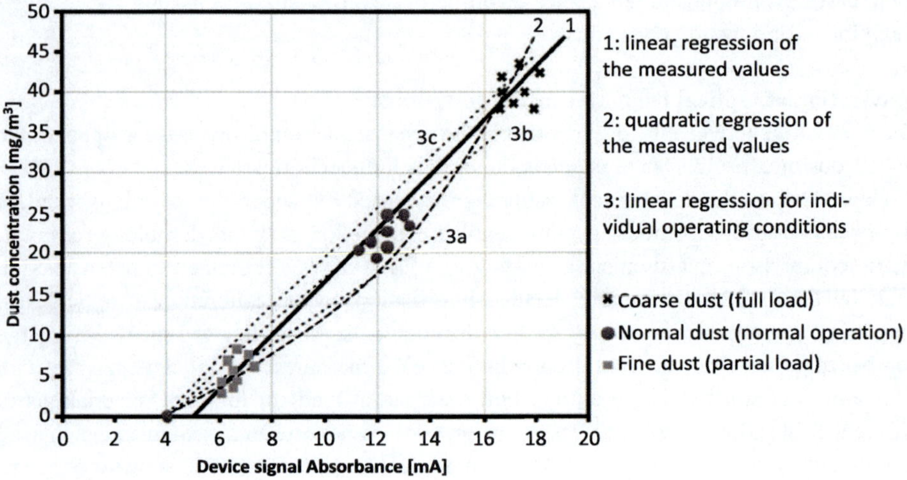

Fig. 13.60 Calibration line of different dusts with different regressions

change in the operating mode or modification of input materials, the correlation between the absorbance and the dust content established by the calibration is no longer valid.

Calibration of Soot Counters

The calibration of continuous soot number measurements is carried out in accordance with VDI 2066 sheet 8, which is a modified standard reference measurement method based on DIN 51402-1. The measurement arrangement is essentially the same as in this guideline, except that the measured value is not determined gravimetrically but by changing the reflectivity of the measurement filter using a special photometer. A fundamental difference in sampling is the arrangement of the sampling probe in the flue. Whereas in gravimetric measurement the opening of the sampling probe is installed against the direction of flow, in soot count measurement the opening is arranged 90° offset. Isokinetic sampling is not necessary because soot particles (diameter <0.1 μm) behave like gases. Relatively low demands are placed on the inlet and outlet sections for the installed continuous devices (inlet section ≥2 × *d*, outlet section ≥1 × *d*).

A defined partial gas flow is extracted via the device arrangement for gravimetric measurement and passed through a plane filter head according to VDI 2066 sheet 1. The evaluation of the soot count is done by a reflection photometer, which was adjusted to the reflectance of the unoccupied measuring filter before the measurement. The soot count is calculated from the reflection coefficient using a conversion factor according to DIN 51402-1. Both are in a linear relationship. A total of 10–15 comparison measurements are made via 3 min. A typical calibration line is shown in Fig. 13.61.

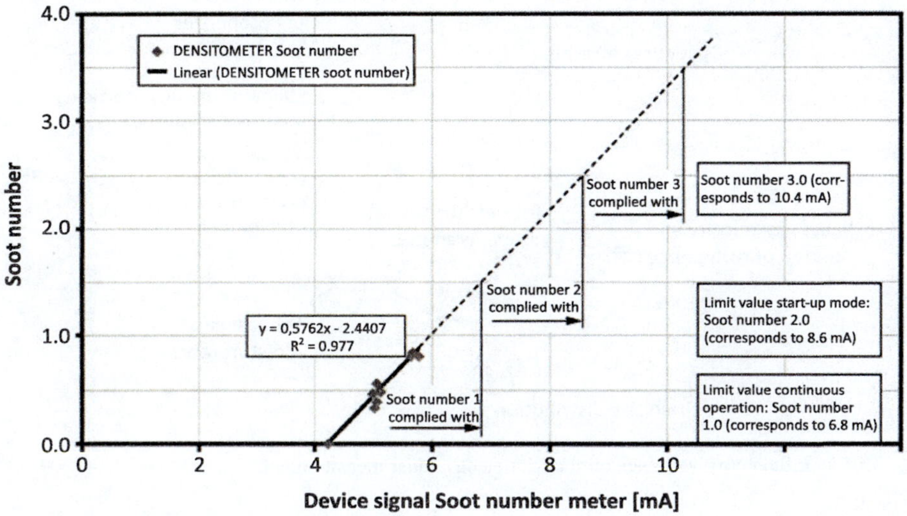

Fig. 13.61 Calibration line of a soot counter

A specific variation of the soot number during calibration is carried out, for example, by changing the combustion air quantity at constant fuel supply or by changing the fuel supply at constant combustion air quantity as well as different load levels.

After calibration, the response thresholds are set in relation to the limit values on the operating hours counters required for continuous soot count measurement. Since the soot count is exceeded when the measurement result rounded to an integer reaches the next soot count, the example in Fig. 13.61 shows the values 6.8 mA for continuous operation (from soot count 1.5) and 8.6 mA for start-up operation (from soot count 2.5).

Calibration of Triboelectric Dust Monitors

Triboelectric dust monitors are used almost exclusively as *in situ* instruments. However, various installations of extractive devices have been carried out so far, especially in wet gases. When calibrating triboelectric dust monitors, certain special features have to be taken into account compared to optical measuring instruments. The triboelectric signal also depends on various influencing variables in the exhaust gas and in the dust to be measured (see Fig. 13.62). If the mechanical and electrical dust properties and the volume flow are almost constant, there is a linear dependence between the dust concentration and the triboelectric measuring signal (see Fig. 13.63).

Influence of the Particle Size

As with all dust measuring instruments, the particle size distribution significantly influences the measuring signal. A typical example is shown in Fig. 13.64. The average particle sizes of the test dusts used are listed in Table 13.10.

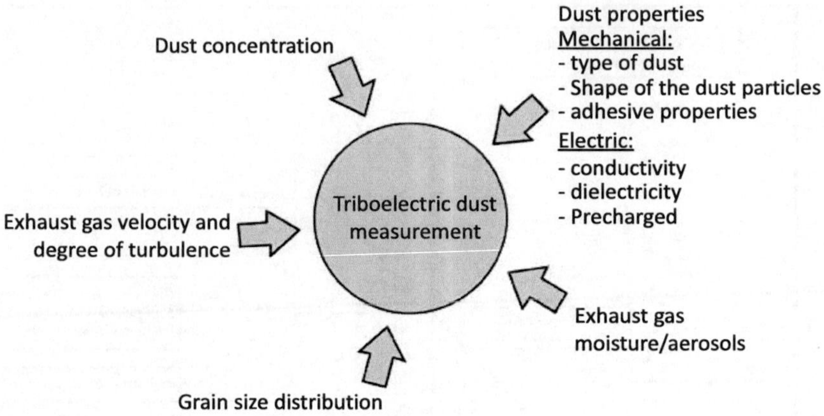

Fig. 13.62 Influencing variables on the triboelectric dust measurement

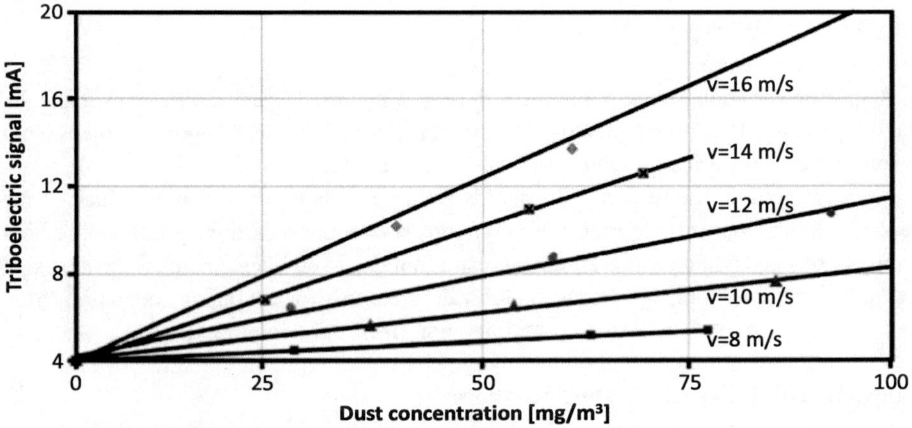

Fig. 13.63 Influence of the flow velocity on the triboelectric dust signal

An exact mathematical description of the grain size influence on the triboelectric measurement signal is not yet possible. Practical experience leads to the following relationship:

$$\mathrm{cal} \approx \frac{1}{\sqrt{d_{50}}} \tag{13.29}$$

cal	Raw signal [V]
d_{50}	Average particle size [µm]

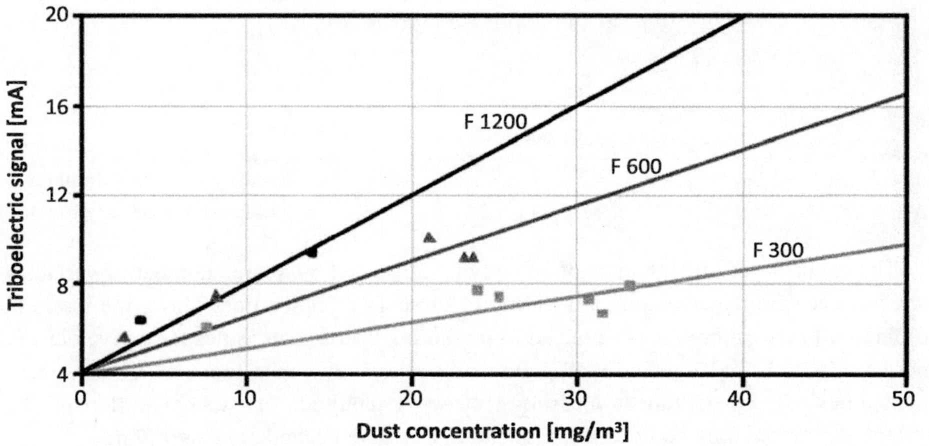

Fig. 13.64 Influence of particle size on the triboelectric dust signal

Table 13.10 Average particle sizes of the test dusts used (determined with Beckman Coulter Laser Granulometer LS 13320)

Test dust	Particle size
Quartz F 300	$d_{50} = 11.00\ \mu m$
Quartz F 600	$d_{50} = 2.75\ \mu m$
Quartz F 1200	$d_{50} = 1.15\ \mu m$

The relevant technical literature also contains information in this respect which represents the following context:

$$\text{cal} \approx \frac{1}{d_{50}} \tag{13.30}$$

cal	Raw signal [V]
d_{50}	Average particle size [µm]

In any case, under otherwise identical conditions (constant dust concentration), finely dispersed particle collectives generate larger measurement signals. The reason for this is the increasing number of particles.

Influence of Speed

Besides the dust concentration, the velocity of the exhaust gas flow has the greatest influence on the triboelectric measurement signal. Experience has shown that this signal depends non-linearly on the gas velocity. The following relationship between the influencing factors applies:

$$c \approx \mathrm{cal} \cdot v^{\alpha} \quad \text{with} \quad \alpha = f(c, d_{50}, v, \ldots) \tag{13.31}$$

c	Mass concentration [mg/m³]
v	Exhaust gas velocity [m/s]

The exponent of the influence of velocity is influenced by a large number of variables and must be determined empirically or within the scope of calibration. This is the essential difference to the calibration of optical dust measuring instruments. Since the gas velocity is necessarily measured anyway in all gravimetric comparative measurements on dust measuring devices, no additional measuring effort is required. The compensation of the velocity influence only requires the application of a given calculation algorithm.

Systems with Constant Exhaust Gas Volume Flow (e.g. Filter Systems)
For systems with small fluctuations of the exhaust gas volume flow (±10%), it is not necessary to consider the exhaust gas velocity during calibration. The evaluation of the gravimetric comparison measurements by means of regression calculation is carried out in analogy to the optical dust measuring devices. Figure 13.65 shows a typical example of a determined calibration curve.

The specification of a confidence interval is omitted, since the filter monitors are usually not connected to emission evaluation computers. The data are recorded by means of

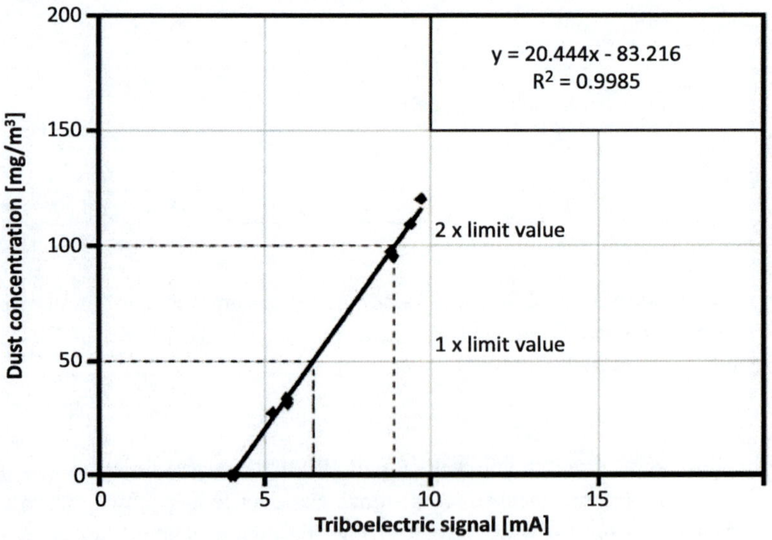

Fig. 13.65 Calibration example of a triboelectric filter monitor

recorders and/or operational process control systems. In this example, the dust limit value of 50 mg/m^3 is reached or exceeded from a signal from ≥6.5 mA. Measuring signals ≥9.0 mA correspond to an exceeding of the twofold limit value.

Systems with Fluctuating Exhaust Gas Volume Flow (e.g.Incinerators)

In systems with fluctuating volume flows, triboelectric measuring devices are only used with online compensation of the influence of velocity. The mathematical relationship between the influencing factors can be described as follows:

$$\begin{aligned} c &= A \cdot 500 \cdot \mathrm{cal} \cdot v^{\alpha} + D \\ \mathrm{Exp} &= B \cdot v^2 + C \cdot v + E \end{aligned} \qquad (13.32)$$

c	Dust concentration [mg/m^3]
cal	Raw signal [V]
v	Exhaust gas velocity [m/s]
A, D	Calibration constants (determined by gravimetric calibration)
B, C, E	Constants of velocity compensation
α	Exponent

With the parameters A to E the calibration function is adapted to the real gas-dust conditions in the measuring cross section. This makes it possible to describe the relationship between dust concentration and velocity in terms of formulae.

During calibration of the devices, in addition to the triboelectric signals, the speed signals must therefore also be continuously recorded (this is required in DIN EN 13284-1 anyway) and usually compressed to half-hourly average values. It must be taken into account that the widest possible range of measured values is generated at the system. The variation of the speeds is only necessary if the fluctuations are more than 10%. As a result of the calibration, the device parameters, among other things, compensate for the influence of speed and the parameters for the emission evaluation computer are determined. The dust concentration can then be determined with the aid of the determined constants. The result of the calibration is thus the representation in Fig. 13.66.

Determination of the Dust Concentration in Standard Condition Dry

The dust concentration in the standard dry state is calculated using the following formula:

$$c_{\mathrm{iNtr}} = \frac{T}{T_0} \cdot \frac{p_0}{p} \cdot \left(1 + \frac{f}{\rho_{0\mathrm{H_2O}}}\right) \cdot c_{\mathrm{iB}} \qquad (13.33)$$

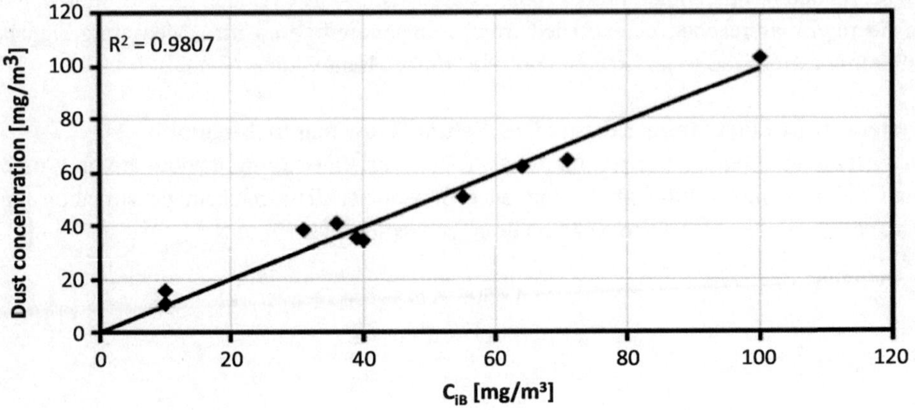

Fig. 13.66 Calibration function of a dust concentration meter

c_{iNtr}	Dust in standard condition dry [mg/m3]
T	Measured temperature [K]
T_0	Standard temperature (273,15 K)
p_0	Standard pressure (101,325 mbar)
p	Absolute pressure of the flue gas [mbar]
f	Humidity of the flue gas [kg/m^3]
ρ_{0H_2O}	Standard density of water vapour (0.8038 kg/m^3)
$c_{\mathrm{i.\,B.}}$	Dust in operating condition [mg/m^3]

Systems with Wet Gases and Sticky Components

The calibration of extractive triboelectric dust monitors differs from the calibration of in-situ instruments only in the compensation of the influence of velocity. In this case, it is not the velocity of the exhaust gas flow that is taken into account, but the volume flow of the sample gas within the measuring device. The mathematical relationship is as follows:

$$c_{\mathrm{iB}} = A \cdot \left[\frac{\mathrm{cal}}{F_M} \cdot B \cdot \left(\frac{F}{F_0} \right)^C \right] + D \tag{13.34}$$

c_{iB}	Dust concentration [mg/m^3]
cal	Raw signal [V]
FM	Sample gas volume flow rate [m^3/h]
F	Volume flow rate in the measuring chamber [m^3/h]
F_0	Flow rate standards (20 m^3/h)
A, D	Calibration constants (determined by gravimetric calibration)
B, C	Constants of the velocity compensation

The calibration procedure is identical to the calibration of optical measuring instruments. The constants of the velocity compensation B and C are preset at the factory and are normally not changed on site. In contrast, the exact determination of the calibration constants A and D is carried out in the course of the gravimetric comparison measurements. After defining the parameters and storing them in the operating device, the calibrated dust measuring instrument displays the values in mg/m^3 or mg/m^3 i. N.

Related References

Baumbach, G.: Luftreinhaltung. Springer, Berlin, Heidelberg (1990)

Bayerisches Landesamt für Umweltschutz: Grundsatzuntersuchungen über die Ermittlung der Korngrößenverteilung im Abgas verschiedener Emittenten (< PM 2,5 und < PM 10), Projekt I, Stand Dezember 2000

Bericht über die Eignungsprüfung der Messeinrichtung PFM 97 ED für Staub der Firma Dr. Födisch Umweltmesstechnik AG, TÜV-Bericht 936/801001/A, Köln: 01.10.2001

BMU: Bundeseinheitliche Praxis bei der Überwachung der Emissionen – Rundschreiben des BMU vom 8.6.1998 – IG I 3–51 134/3, GMBl. Nr. 28, S. 543 (1998)

Födisch, H., Richter, D.: Grundlage Ladungsdifferenz – Triboelektrische Staubmessung unter theoretischen und praktischen Gesichtspunkten. Verfahrenstechnik. **31**(12), 44–49 (1997)

Födisch, H., Schengber, P., Ulbrich, M.: Kontinuierliche Staubgehaltsmessung zur Minderung der Umwelt und Gesundheitsbelastung durch Holzstaubemissionen in der Holz verarbeitenden Industrie. DBU-Reg.-Nr. AZ 16950 (2002)

Födisch, H., Schengber, P., Riemann, U.: Neuartiges Staubmessgerät auf triboelektrischer Basis. Beitrag zum Kolloquium „Neuere Entwicklungen bei der Messung und Beurteilung der Luftqualität". Heidelberg, April 1999

Födisch, H., Schengber, P.: Calibration of triboelectric dust measuring devices. International Conference on Emission Monitoring. Denmark (2002)

Födisch, H.: Staubemissionsmesstechnik. Stand der Technik – Messverfahren – Anwendungsbeispiele. Expert Verlag (2004)

Hoffmann, K.-D.: Entwicklung der Staubemissionsmessungen – Automatische Verfahren. RWTÜV-Schriftenreihe 13, RW TÜV, Essen (1980)

Jockel, W.: Überwachung staubförmiger Emissionen – Messtechnik im Wandel der Zeit. Symposium Emissionsmesstechnik und Feuerungstechnik (1998)

Junk, J., Helbig, A.: Die PM10-Staubbelastung in Rheinland-Pfalz. Gefahrstoffe – Reinhaltung der Luft. **63**(1/2) Springer (2003)

Kaczmarek, H., Kopperschläger, J.: Moderne Filterüberwachung. ZKG Int. **48**(11), 608–614 (1995)

Morawska, L., Salthammer, T.: Indoor Environment. Wiley-VCH, Weinheim (2003)

Paul Gothe Bochum: GMU-Kaskadenimpaktor Johnas II für die Messung von PM10 und 2,5 in strömenden Gasen, Produktkatalog (2003)

Staudinger, G., Obermair, S., Meisl, C.: Treffsicherheit und Zuverlässigkeit der heutigen Zyklonberechnung. VDI-Berichte 1511, 1999

TÜV: Triboelectric and Dynamic Sensors in Innovative Monitoring Systems for Process Integrated Environmental Control. Craft Project SM-S2–7809, Oktober 2000

UBA: Luftreinhaltung: Leitfaden zur Emissionsüberwachung. Umweltforschungsplan des Bundesministeriums für Umwelt, Naturschutz und Reaktorsicherheit, Berichtsnummer UBA-FB 000105, Stand: 2001

UBA: Reinhaltung der Luft/Luftüberwachung in Deutschland, Eignungsgeprüfte kontinuierlich arbeitende Emissionsmesseinrichtungen – Messobjekte, Juni 2003

Vollmer, H.: Entwicklung der manuellen Staubemissionsmesstechnik. RWTÜV-Schriftenreihe. **13** (1980)

Standards

Erste Allgemeine Verwaltungsvorschrift zum Bundes-Immissionsschutzgesetz (Technische Anleitung zur Reinhaltung der Luft – TA Luft), Juli 2002

DIN EN 13284-1 Emissionen aus stationären Quellen – Ermittlung der Staubmassenkonzentration bei geringen Staubkonzentrationen – Teil 1: Manuelles gravimetrisches Verfahren, April 2002 (prEN 2015)

DIN EN 13284-2 Emissionen aus stationären Quellen – Ermittlung der Staubmassenkonzentration bei geringen Staubkonzentrationen – Teil 2: Automatische Messeinrichtungen, Dezember 2004

DIN EN 14181 Emissionen aus stationären Quellen – Qualitätssicherung für automatische Messeinrichtungen, Februar 2015

DIN EN 15259 Messung von Emissionen aus stationären Quellen – Anforderungen an Messstrecken und Messplätze und an die Messaufgabe, den Messplan und den Messbericht, Januar 2008

VDI 2066 Bl. 1 Messen von Partikeln – Staubmessungen in strömenden Gasen – Gravimetrische Bestimmung der Staubbeladung, November 2006 (Überprüfung April 2004)

VDI 2066 Bl. 5 Messen von Partikeln – Staubmessung in strömenden Gasen; Fraktionierende Staubmessung nach dem Impaktionsverfahren – Kaskadenimpaktor, November 1994 (Überprüfung Mai 2015)

VDI 2066 Bl. 8 Messen von Partikeln – Staubmessung in strömenden Gasen – Messung der Rußzahl an Feuerungsanlagen für Heizöl EL, September 1995 (Überprüfung April 2004)

VDI 2066 Bl. 10 Messen von Partikeln – Staubmessung in strömenden Gasen – Messung der Emissionen von PM 10 und PM 2,5 angeführten Quellen nach dem Impaktionsverfahren, Oktober 2004 (Überprüfung Mai 2015)

VDI 3950 Emissionen aus stationären Quellen – Qualitätssicherung für automatische Mess- und elektronische Auswerteeinrichtungen, Dezember 2006 (Überprüfung April 2012)

14 Emission Measurement

Power station - RWE Weisweiler. Source: Gerhard Wiegleb

The fields and applications of gas measurement technology are extremely diverse and even for the expert no longer completely clear. The spectrum ranges from process gas analysis, medical technology and exhaust gas analysis to exotic applications such as odour detection in beer production. In Chaps. 14–18 some important fields of application are pointed out which are currently of special interest. The authors of these 24 application reports have been working for many years in their respective special fields and make an expert contribution to this complex topic. For the sake of clarity, these applications have been divided into *emission measurement technology*, *energy technology*, *life science* and *safety technology*.

Section 14.1 describes in detail the preparation of gas samples, which is of particular importance for the field of emission measurement technology, but is also used in other

G. Wiegleb, *Gas Measurement Technology in Theory and Practice*,
https://doi.org/10.1007/978-3-658-37232-3_14

fields of application. The special features which occur in the different areas of application are described in the corresponding subchapters.

14.1 Gas treatment

The transfer and preparation of the respective gas sample is of great importance in all gas analytical applications. The gas mixture to be analysed is either directly in the ambient air or in a closed process which has to be monitored or controlled. The gas measuring instrument must be adapted to the respective task and supplemented with appropriate peripherals. These peripherals usually consist of pumps, gas lines, sample gas coolers, flame arresters, filters, etc., which are summarized under the term *gas conditioning*. Figure 14.1 shows such a closed process, with the different sampling methods. The process may be a process engineering operation in the chemical industry, the use of natural gas or a combustion process in a large power plant. Basically, there are two possibilities of sampling:

1. The gas sample is brought to the measuring instrument/sensor (→ extractive method, off-line, ...)
2. The measuring instrument/sensor is brought to the process/gas (→ *in situ*, in-line, ...).

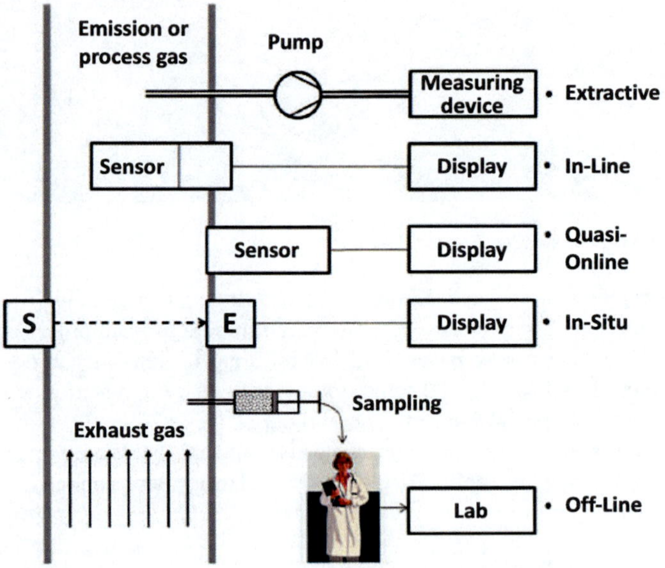

Fig. 14.1 Selection of sampling options in process measuring technology

Both possibilities are used in gas measurement technology. However, the individual methods have advantages and disadvantages which are decisive for the respective application. Therefore, both possibilities are described and compared in the following.

Off-Line Measurement

Gas analytical investigations are best carried out in a laboratory, since the boundary conditions for the measurement are precisely defined. The accuracy can therefore be increased with this procedure. Furthermore, there are various measuring methods available in the laboratory which are not suitable for process gas analysis and only work in a laboratory environment. This type of laboratory analysis is often used for reference measurements to validate the process measuring instruments. For this purpose, however, the gas sample must be brought to the laboratory. For this purpose, so-called gas storage bags are used, which can hold a certain amount of gas. The bags have a pillow shape and have a maximum volume of 0.1–200 L, depending on the design (see Fig. 14.2). The gas storage bag consists of five different layers of the following materials:

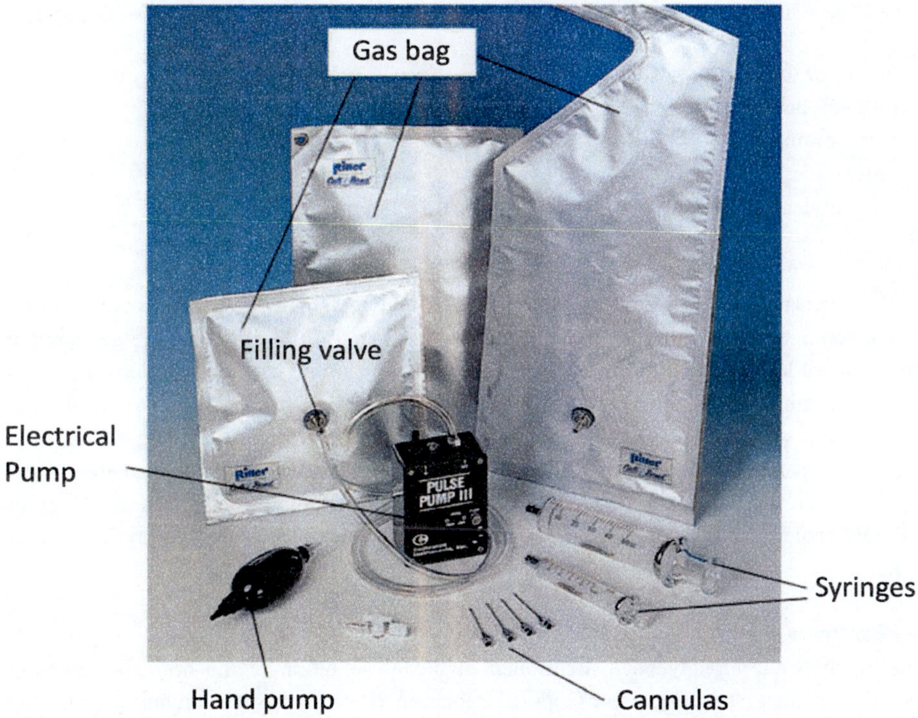

Fig. 14.2 Gas storage bags of the *Cali-5-Bond* ™ type. (Source: Dr.-Ing. Ritter Apparatebau GmbH & Co. KG, Bochum Germany)

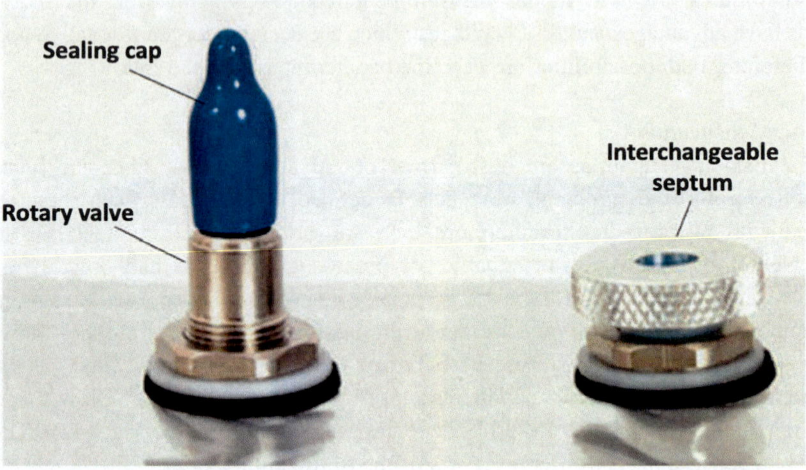

Fig. 14.3 Valve technology for filling and closing the gas storage bag. (Source: Dr.-Ing. Ritter Apparatebau GmbH & Co. KG, Bochum Germany)

- Polyester (outside)
- Polyvinylidene chloride
- Aluminium foil
- Polyamide
- Polyethylene (HDPE) (inside).

The layers are joined to form a single flexible film, with a material thickness of 0.14 mm. This layered structure provides optimal gas tightness and further protects the gas sample from ultraviolet solar radiation. Chemical reactions in the bag are thus prevented and an almost unlimited shelf life of the gas sample is achieved. The filling process is carried out with a pump via a valve (Fig. 14.3). The valve is firmly connected to the bag and ensures that the gas sample remains safely in the bag. The valve is opened by one turn and closed again by one turn after the filling process. The manufacturer offers different valves in different materials. Furthermore, the sample can also be injected with a syringe via a septum.

In Situ Process

This method has already been mentioned in the presentation of the optical measuring methods in Sect. 9.3. The use of optical radiation (IR/UV) has the advantage that the sample is directly detected in a process gas line (e.g. Pipeline) or in a *stack* (*cross-stack*). In this case, the gas concentration c_R is measured under real conditions (pressure p_R and temperature T_R), in the presence of a more or less high water vapour concentration c_{H_2O}. The advantage of this method is that the gas sample is measured directly and no

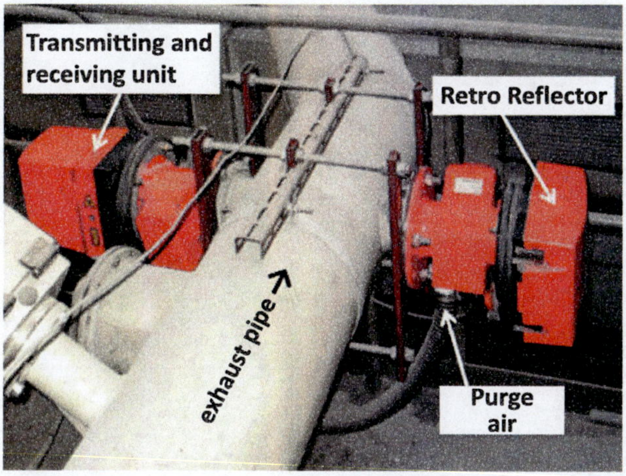

Fig. 14.4 Installation of an optical gas measuring device (Sick AG Reute), which is directly coupled to an exhaust pipe. The necessary purge air is directed to the optical interfaces with appropriate hoses

falsifications by sampling occur. However, it is problematic that the sample also contains other accompanying substances which can have a disturbing influence (e.g. cross-sensitivities) on the concentration measurement. In addition, the optical interfaces must be kept clean with an appropriate flushing device when the radiation is coupled into the measuring chamber. These rinsing devices are mainly required in dust-laden exhaust gases (stack measurement), whereas dust-free gases (e.g. in a natural gas pipeline) do not require any protective devices. Figure 14.4 shows such an *in situ* measuring device on an exhaust pipe.

An alternative possibility is to place the optical measuring section in a gas-permeable protective cover (sintered metal, ceramic) directly in the exhaust gas (Fig. 14.5). This *in-line measurement* eliminates the need for a costly flushing device, as the dust is separated through the porous protective cover. A further advantage of this technique is that the measuring device can be checked with test gas and zero gas under real conditions (c_R, p_R, T_R). A dust coating on the outer wall of the protective cover can be removed by a short air flushing with overpressure, so that a clogging of the pores is prevented. The maintenance effort is particularly low with this procedure.

In addition to the optical methods, other sensor types can also be operated with this *in-line method.* Since the temperature ranges are subject to very strong changes with this method and are usually also at a high level >80°C, the selection of possible sensors is limited, however. A frequently used measuring technique is the use of solid electrolyte sensors (e.g. ZrO_2 probe), which have hardly any restrictions with regard to the ambient temperature T_R. Oxygen measurement in particular can be realized very well with the *in-line method* (see Sect. 4.2).

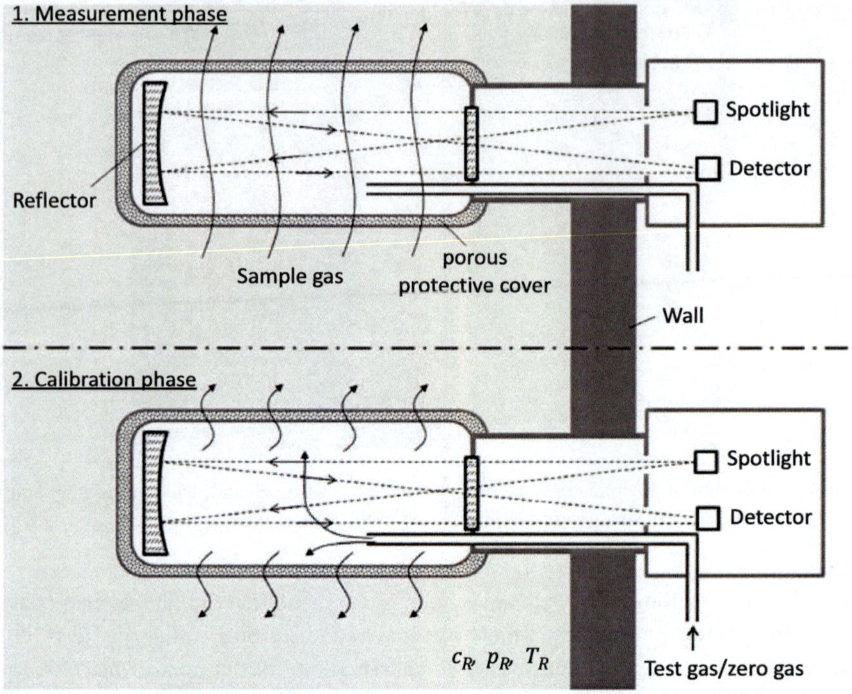

Fig. 14.5 Basic structure of an optical in-line measurement. During the measuring phase (1) the process gas can diffuse from the outside through the porous protective cover into the measuring chamber. In the calibration phase (2) the required zero gas and test gas are switched on via a solenoid valve control. The gases then flow from the inside to the outside, so that defined conditions in the beam path are present in this phase and a corresponding calibration of the measured values can be carried out

Extractive Processes

Often, however, the sample gas is taken from the process and supplied to the corresponding measuring instruments via gas pipes. The advantage of this *extractive method* is that there are no restrictions with regard to the available measuring methods. The measuring instruments usually work under ambient conditions which can be adapted to the respective measuring task. In extreme cases, the measuring instruments are placed in air-conditioned measuring stations where constant conditions always prevail. In order to be able to use the advantages of the extractive method, however, the sample gas must be prepared accordingly. Figure 14.6 shows a typical arrangement of the components of a gas conditioning system, from the sampling point to the gas analyzer.

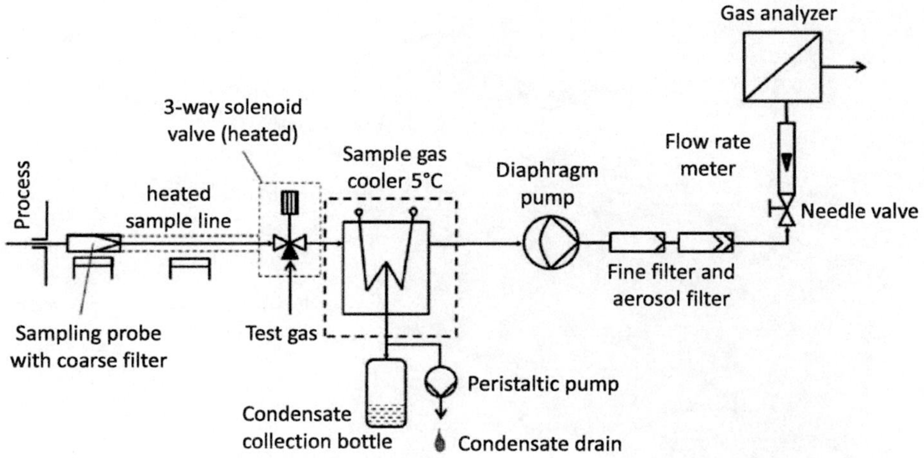

Fig. 14.6 Typical gas flow diagram of an extractive sample preparation with a gas analyzer at the end of the system

Sampling Probe

The gas to be measured is extracted from the process via an extraction probe. In the simplest case, this probe (lance) consists of a thin noble jet pipe (e.g. 6 mm inner diameter), which protrudes into the gas chamber (e.g. Exhaust gas duct, process gas line, ...) via a flange connection. To prevent condensation, these probes are usually electrically heated above the expected dew point of the sample gas. In addition, a coarse filter is installed in this probe to provide pre-filtration. Figure 14.7 shows such a gas sampling probe. The heated assembly is located in an insulated housing with a cover cap which can be opened in case of service, for example to replace the filter element. Furthermore, test gas can also be added to this setup to check the entire gas path from the probe to the analyzer. A non-return valve in the gas supply of the test gas ensures a safe connection in case no test gas bottle is connected.

Gas Pipeline

The outlet of the gas sampling probe is connected to a gas line, which usually consists of a thin hose (e.g. 4 mm inner diameter). Viton is usually used as hose material, which has sufficient chemical resistance and can still be laid flexibly. In chemical process engineering, however, Teflon is preferred, which is even more resistant than Viton. For high-purity gas applications or measurements under high process gas pressure, only complex high-grade jet piping is then considered.

In order to also prevent condensation of water vapour in the gas pipe, the hose is additionally electrically heated and, if necessary, regulated to a constant temperature. Figure 14.8 shows the internal structure of a heated gas pipe. The gas-carrying inner

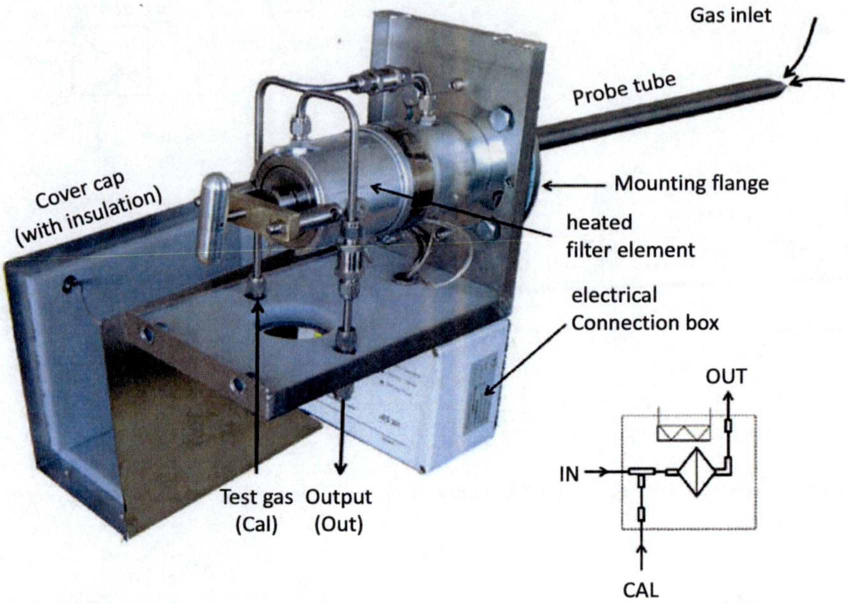

Fig. 14.7 Heated gas sampling probe with an integrated filter element. In addition, this design offers the possibility of feeding test gas via the probe system. (Source: JCT Analysentechnik GmbH, Wiener Neustadt/Austria)

hose usually has an inner diameter of approximately 4 mm. The required heating coil is either wound directly around the hose or an intermediate layer (e.g. metal mesh) is added. If the heating is to take place at a constant temperature, a temperature sensor must still be integrated. Around this structure there is then a thermal insulation of 2–3 cm, so that the heat losses are as low as possible. Heated gas pipes are offered by the various manufacturers in lengths up to 200 m. Figure 14.9 shows the overall view of such a pipe.

Time and Pressure Behaviour of Gas Pipes

The transmission of the measuring gas via a gas pipeline inevitably leads to a time delay. This time response is very important for the practical operation of gas measuring instruments, since subsequent control processes may be influenced by this time response. Figure 14.10 shows such a transmission path, from the sample gas inlet to a gas analyzer. The volume flow $\dot{V}$ is generated by a pump, which provides a defined delivery volume in litres/minute (L/min) or litres/hour (L/h). The time delay t_T (dead time) can be calculated by transporting a *gas package* with the flow velocity w_{gas} over the distance *L*:

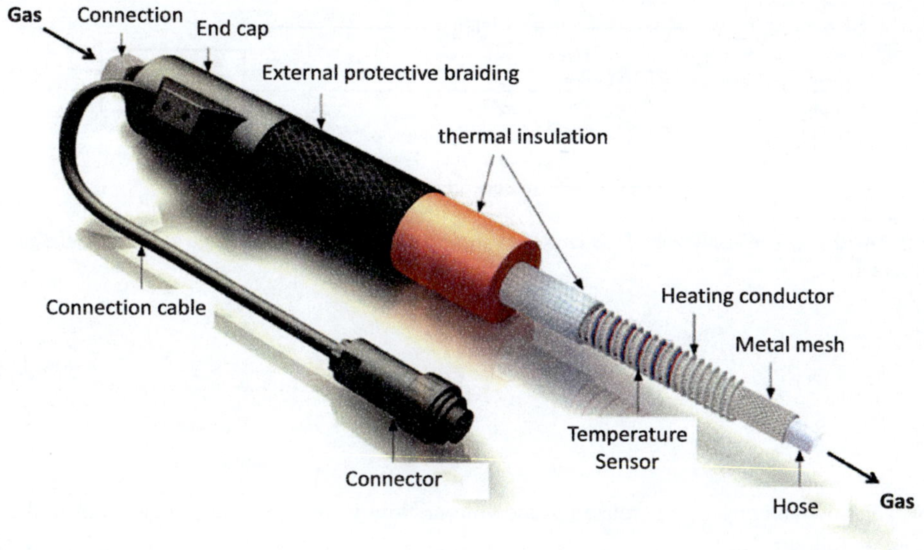

Fig. 14.8 Internal structure of a heated gas pipeline. (Source: JCT Analysentechnik GmbH, Wiener Neustadt/Austria)

Fig. 14.9 General view of a heated gas pipeline. (Source: JCT Analysentechnik GmbH, Wiener Neustadt/Austria)

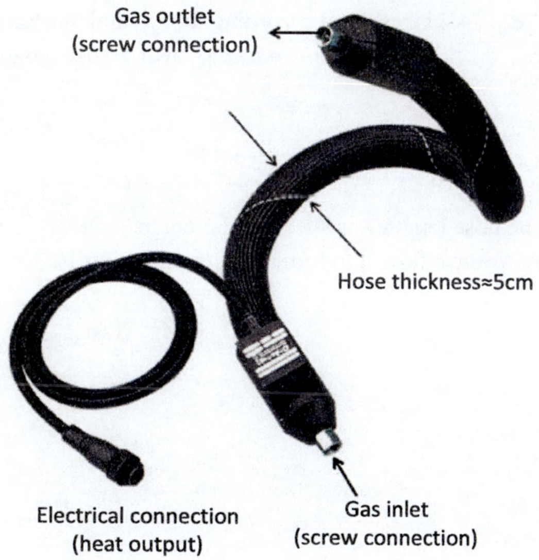

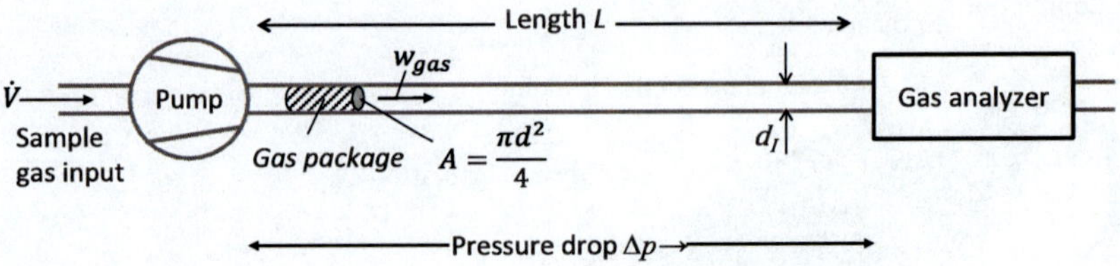

Fig. 14.10 Transmission path of the sample gas from the sampling point (sample gas inlet) to the gas analyzer

$$t_T = \frac{L}{w_{\text{gas}}} \tag{14.1}$$

The flow velocity w_{gas} is related to the volume flow $\dot{V}$ over the inner diameter d_I of the hose as follows:

$$\dot{V} = w_{\text{gas}} \cdot \frac{\pi \cdot d_I^2}{4} \tag{14.2}$$

If Eq. 14.2 is resolved according to w_{gas} and this term is inserted in Eq. 14.1, the theoretical transfer time t_T for a *gas package* over a hose length L is obtained:

$$t_T = \frac{\pi \cdot d_I^2 \cdot L}{4 \cdot \dot{V}} \tag{14.3}$$

The hose length L is specified in meters (m), the inner diameter d_I in millimeters (mm) and the volume flow $\dot{V}$ in L/min or L/h. Equation 14.3 then changes as follows for t_T in seconds:

$$t_T = \frac{\pi \cdot d_I^2 \cdot L \cdot 6}{400 \cdot \dot{V}} \quad \text{for } \dot{V} \, \text{in} \, \text{L/min} \tag{14.4}$$

respectively

$$t_T = \frac{\pi \cdot d_I^2 \cdot L \cdot 3.6}{4 \cdot \dot{V}} \quad \text{for } \dot{V} \, \text{in} \, \text{L/h} \tag{14.5}$$

In this theoretical consideration, mixing influences are neglected, which however always occur due to turbulence in the hose. This process causes the concentration profiles within the tube to become *smoother* and a slow, flowing transition from the concentration level c_O (e.g. zero gas) to c_x (e.g. test gas) is created. This transition is shown in Fig. 14.11. In

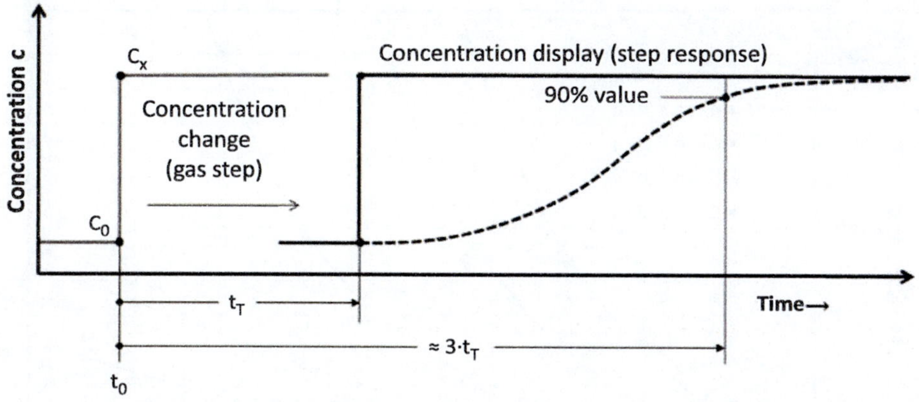

Fig. 14.11 Ideal and real transfer behaviour of an abrupt gas jump from c_0 to c_x at time t_0 and the step response at time t_T

practice, 2–3 times the theoretical value of t_T is usually used for an almost complete gas exchange.

In addition to the temporal behaviour, the pressure drop or pressure build-up in a gas pipeline also plays an important role. This pressure effect must be taken into account, especially in the case of very long gas pipes ($L = 10\,\text{m}\ldots200\,\text{m}$). The pressure build-up Δp, which is shown in Fig. 14.12, can be explained by the *Hagen-Poiseuille law* (Eq. 2.165). According to this law, however, the influence of the volume flow should be linear. However, since gases can be compressed, the behaviour of gases is different from that of liquids.

Sample Gas Cooler

In a further conditioning stage, the sample gas is cooled down to a constant temperature T_K to reach a defined water vapour dew point. The moisture content of the gas sample can vary considerably. Figure 14.13 shows the maximum H_2O concentration of the sample gas as a function of temperature. In this example the sample gas at a temperature of 46 °C has a relative humidity of 33.3%. If the sample gas cools down in the hose at 24 °C after the sampling probe, the condensation process starts at some point in the hose and water droplets are formed which hinder the gas flow. Furthermore, gas components can also dissolve in these water droplets, which are then missing in the gas analysis and thus indicate lower concentration values (lower results). For this reason, the dew point must be undercut at a defined point within the sample preparation to prevent these negative effects. A condensation in the hose would lead to a large contact surface between the condensed water and the dissolving gas and would favour the dissolving process.

As the water vapour can also have an influence on the gas measurement (→ cross-sensitivities), the dew point (TP) should be as low as possible after the cooling process.

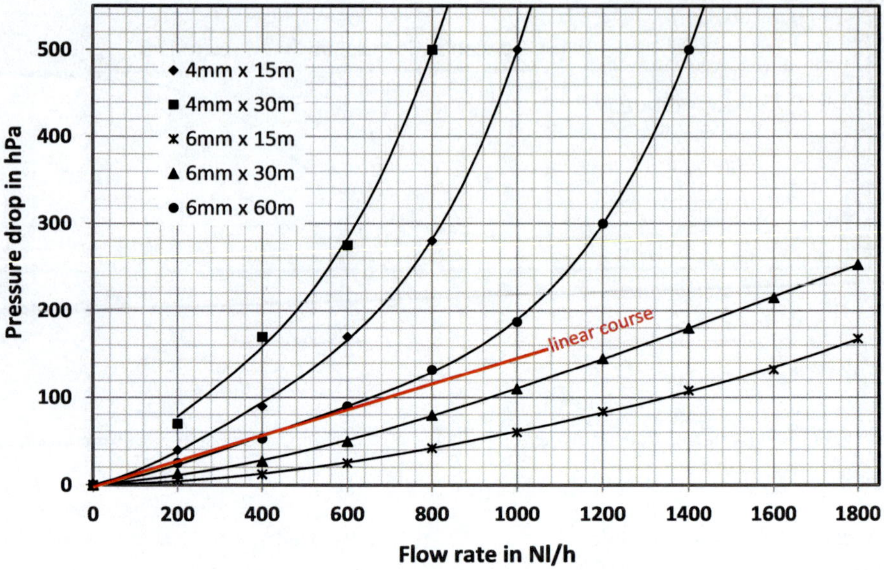

Fig. 14.12 Pressure drop (Δp) for different gas lines (4 mm and 6 mm) and lengths between 15 m and 60 m (data sheet 15.4 of M&C TechGroup GmbH, Ratingen Germany)

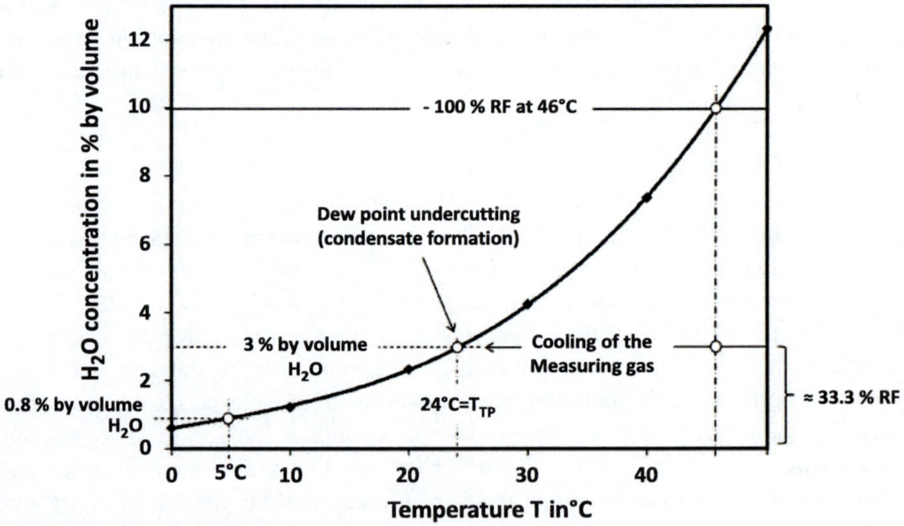

Fig. 14.13 Condensate formation when the temperature falls below the dew point due to cooling of the sample gas in a cooler

However, water freezes at a temperature under 0 °C. As a compromise, a value above the freezing point at $T_K = 5°C$ has become established in gas measurement technology.

The cooling process can take place in different ways. In the simplest case, a so-called *condensate separator* is used to achieve cooling to the ambient temperature T_U. This at least prevents condensation of the water vapour in the gas analyser, since the device temperature T_G is always the same due to the electrical power loss $>T_U$. A vessel can be used as a condensate trap through which the sample gas flows and assumes the ambient temperature due to the dwell time in this vessel. The condensate that accumulates in the vessel is then drained into a collection vessel and disposed of. This cooling process can be forced by water cooling if the gas volume V_{Gas} becomes too large. Figure 14.14 shows such a cooler. The gas sample is taken directly from the process gas line and cooled down to the water temperature by a water cooler. The condensate then precipitates in the inner tube and drips back into the gas pipe due to gravity. To maximize the surface area between the inner pipe and the flowing water, several pipes are connected in parallel (cascaded). The cooling process (→ heat transfer) becomes much better.

For use in other applications, the cascade cooler can also be constructed in plastic (Fig. 14.15). The structure is designed as a wall mounting and the sample gas is passed

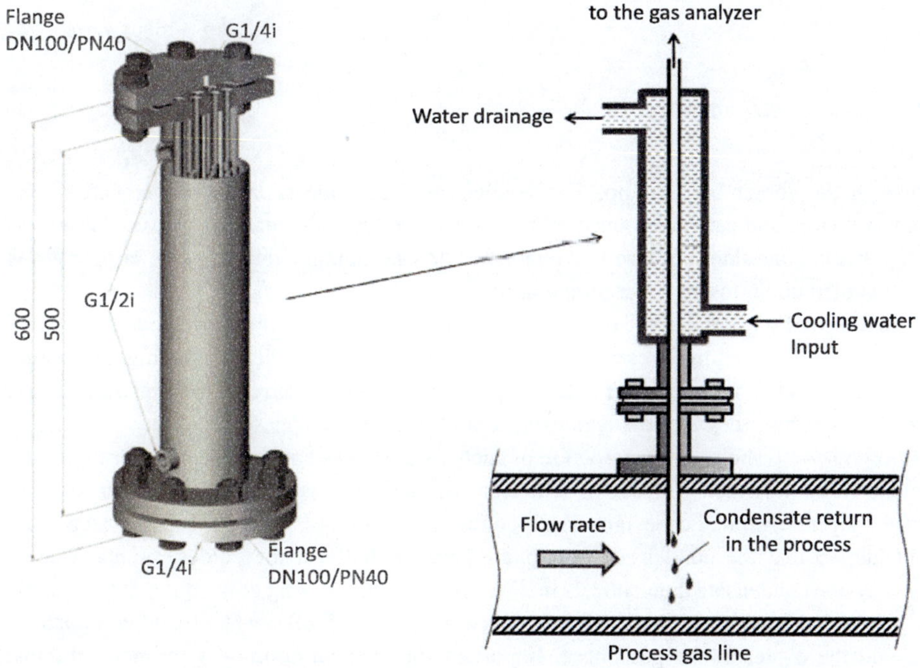

Fig. 14.14 Basic structure of a water cooling system for a process gas extraction. On the *left of the* picture a typical design of a cascade cooler in stainless steel for high cooling capacity is shown. (Source: APM-Technik GmbH, Korschenbroich Germany)

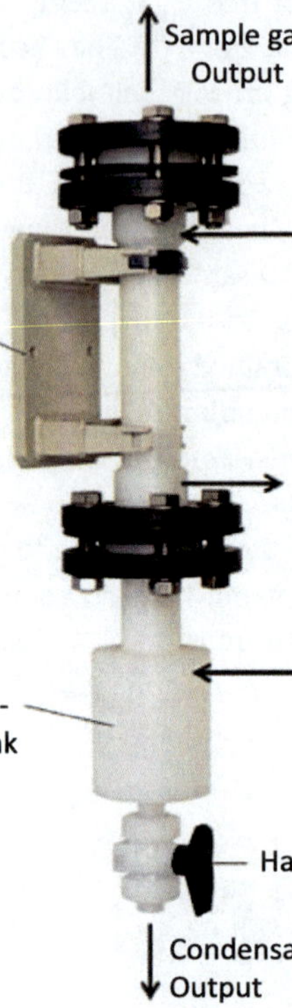

Fig. 14.15 Cascade cooler made of plastic (PVDF) for the use of corrosive process gases. (Source: APM-Technik GmbH, Korschenbroich Germany)

through the cooler by a pump. The condensate then collects in the lower part of the superstructure and can be removed with a hand valve for maintenance purposes. However, the water cooling does not lead to a controlled dew point temperature, as the sample gas is only cooled down to the water temperature.

In a second stage, a defined cooling to the preferred 5 °C must be carried out. In terms of equipment, this cooling is carried out either with a Peltier cooler or, for higher cooling capacities, with a compressor cooler. Peltier coolers are preferred for small sample gas volumes because they are less expensive and also have a longer service life.

Figure 14.16 shows a cross-section of such a sample gas cooler. The humid sample gas enters the cooling area from above, which can be made of glass, plastic (PVDF) or stainless steel, depending on the design. The cooling is transferred into the cooling area from outside, so that the condensate precipitates preferentially at the coldest points. Due to gravity, the condensate then collects in a vessel below the cooling zone. If the entire sample gas is cooled down to 5 °C, there is still a maximum of 0.86 vol% water (residual moisture) behind the cooler in the gas phase. Important for efficient cooling is the good thermal coupling of the condensation vessel with the cooling surface of the Peltier element. Furthermore, the power loss of the Peltier element on the *warm side* must be efficiently dissipated. For this purpose, the *warm side* is equipped with a heat sink, which additionally

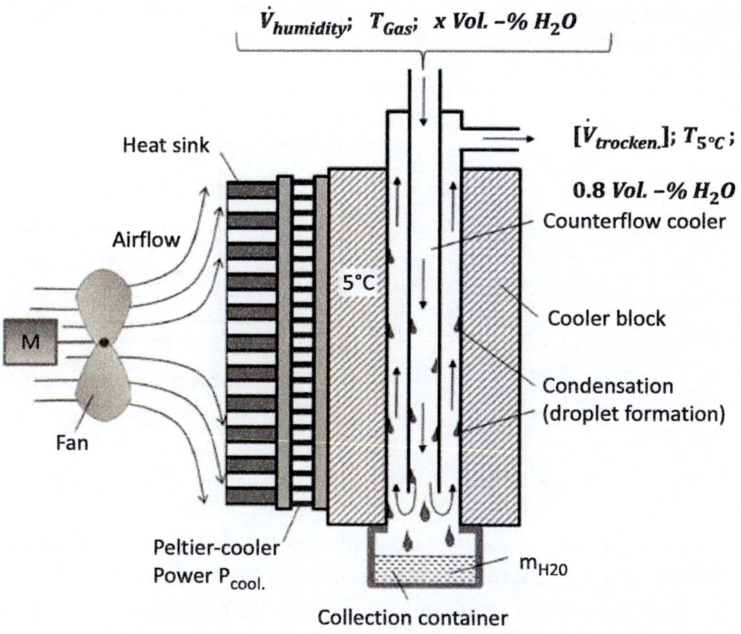

Fig. 14.16 Sample gas cooler with a Peltier element using an additional countercurrent cooling

transports the heat from the cooling unit via a fan. However, the Peltier element can only ensure a certain temperature difference ΔT compared to the ambient temperature T_U. For example, if the cooler is located in a room with a temperature $T_U > 20°C$, the cooling capacity decreases linearly with the ambient temperature and reaches only 1/3 of the original cooling capacity at 40 °C. Figure 14.17 shows this curve of the power decrease.

Furthermore the cooling capacity is influenced by the sample gas. Both the inlet temperature T_{ME} of the sample gas into the cooler and the amount of gas m_{gas} that has to be cooled down to 5 °C are limited by the cooling capacity. The amount of heat Q contained in the sample gas is described by the following equation:

$$Q_{gas} = m \cdot c_p \cdot [T_{ME} - 5\,°C] + \Delta m \cdot r + \Delta m \cdot c \cdot [T_K - 5\,°C] \tag{14.6}$$

Δm is the amount of water that precipitates as condensate. r is the specific heat of evaporation for water (r = 2257 kJ/kg). The specific heat capacity for air is $c_p = 1.0005\,\text{kJ/kg} \cdot \text{K}$ and for liquid water is $c = 4,182\,\text{kJ/kg} \cdot \text{K}$.

With the volume flow $\dot{V} = \dot{m}/\varrho$, one then obtains an expression for the power to be dissipated by the hot sample gas:

Fig. 14.17 Change in cooling capacity of a Peltier cooler at different ambient temperatures T_U

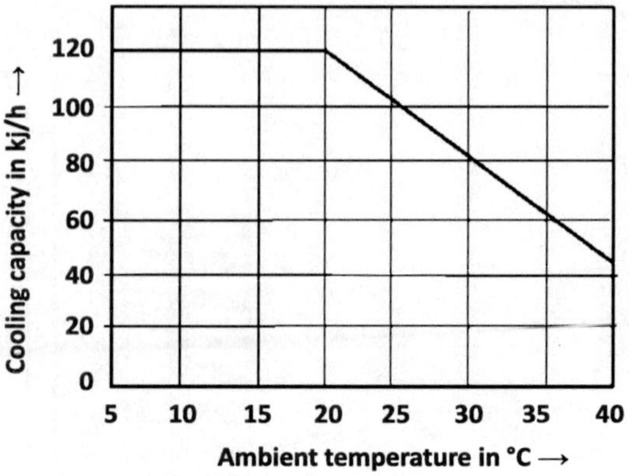

$$P_{\text{gas}} = \dot{V} \cdot \varrho \cdot c_p \cdot [T_{\text{ME}} - 5\,°\text{C}] + \Delta\dot{m} \cdot r + \Delta\dot{m} \cdot c \cdot [T_K - 5\,°\text{C}] \tag{14.7}$$

As the volume flow $\dot{V}$ rate increases, the sample gas cooler must therefore provide ever greater heat output P_{gas} to cool the sample gas down to the required 5 °C. If the volume flow $\dot{V}$ is too large, the cooler can no longer guarantee the temperature of 5 °C at a given cooling capacity and the cooler leaves the control range. The same applies if the sample gas temperature T_{ME} at the inlet of the cooler rises. Figure 14.18 shows the dew point temperature as a function of the sample gas flow (volume flow). It can be clearly seen that this behaviour also depends on the choice of material for the condensation vessel. The stainless steel vessel has, due to its better heat conduction, a more favourable behaviour in this limit area than the glass vessel. A complete sample gas cooler with two separate cooling circuits is shown in Fig. 14.19.

Calculations for the Sample Gas Cooler

Extractive sampling systems often use gas coolers to reduce the temperature T_{ME} of the hot exhaust gas. Usually, the temperature in the cooler falls below the dew point (→ condensate) due to the moisture content of the exhaust gas.

The water vapour content of a gas mixture results from the partial pressure p_D of water vapour. This partial pressure depends on the temperature T and can be described by the *Magnus formula*

$$p_D(T) = 6.11 \cdot \exp\left[\frac{17.08 \cdot T_1}{234.17 + T_1}\right] \tag{14.8}$$

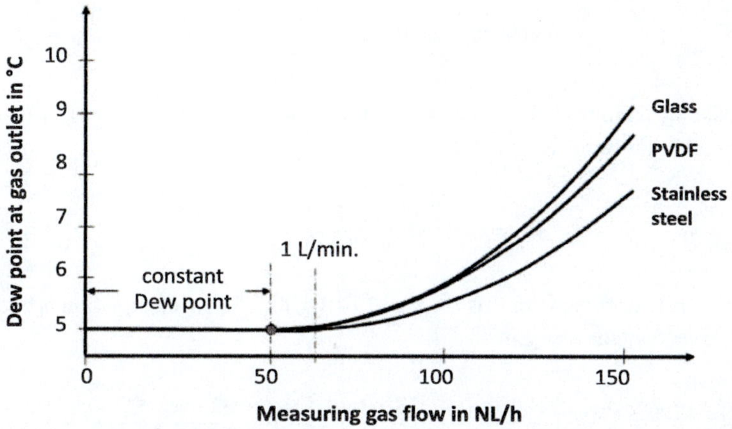

Fig. 14.18 Dew point temperature at the outlet of the cooler for an inlet dew point of 50 °C for different gas flows (Operating Instructions Electric Gas Cooler Series ECP® Version ECP 1000/2000/3000 M&C TechGroup GmbH Ratingen)

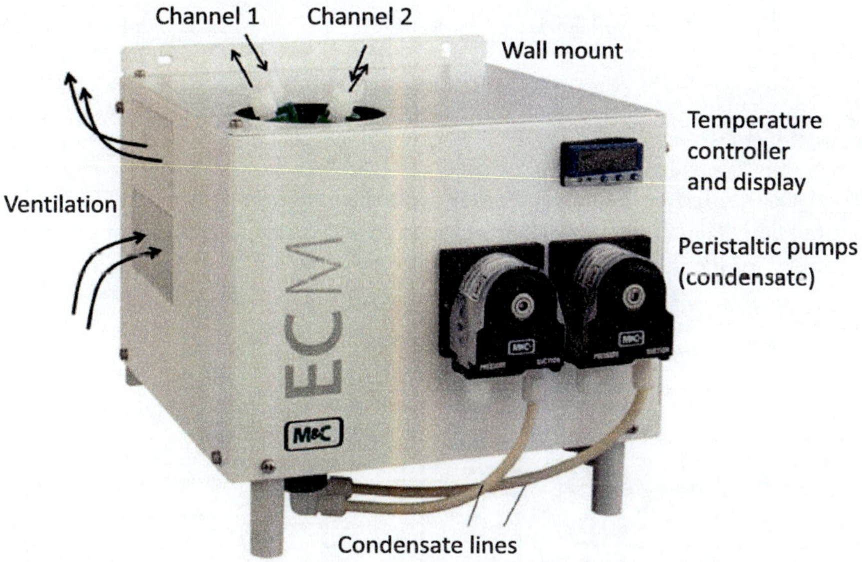

Fig. 14.19 Peltier cooler with two separate gas channels. (Source: M&C TechGroup GmbH Ratingen)

The total amount of water vapour in g/m^3 can be determined by the following approach: The water content in 1 m^3 is the ratio of the saturation vapour pressure to the p_{SD} normal pressure p_0. Expressed in litres, these are then

$$\text{Water content in } 1\,\text{m}^3 = \frac{p_{SD}}{p_0} \cdot 1000\,\text{L} \tag{14.9}$$

18 g H_2O take up a volume of $V_0 = 22.414$ L under normal conditions (1013 hPa, 273.15 ° C). Under real conditions these are then:

$$V_1 = 22.414\,\text{L} \cdot \frac{(273.15\,°\text{C} + T_1)}{273.15\,°\text{C}} \tag{14.10}$$

If one forms a rule of three with Eqs. 14.9 and 14.10, the following equation is obtained for the water vapor content *c* in g/m³:

$$c\left[\text{g/m}^3\right] = \frac{p_{SD}}{p_0} \cdot \left(\frac{1000\,\text{L}}{\text{m}^3}\right) \cdot \frac{18\,\text{g/mol}}{22.414\,\text{L/mol}} \cdot \left(\frac{273.15\,°\text{C}}{273.15\,°\text{C} + T_1}\right) \tag{14.11}$$

The amount of water $\Delta m/t$ which condenses from an exhaust gas in a sample gas cooler (Fig. 14.20) with the temperature $T_2 = 5$°C can be calculated with Eq. 14.11 as follows:

$$\Delta\dot{m} = \left[\frac{(p_{SD} - p_{5\,°\text{C}})}{p_0} \cdot \left(\frac{1000\,\text{L}}{\text{m}^3}\right) \cdot \frac{18\,\text{g/mol}}{22.414\,\text{L/mol}} \cdot \left(\frac{273.15\,°\text{C}}{273.15\,°\text{C} + T_1}\right)\right] \cdot \frac{\dot{V}_1}{1000\,\text{L/m}^3} \tag{14.12}$$

If the exhaust gas has no saturation humidity at temperature T_1 ($p_{SD} \approx 100$ %-RH), the relative humidity *(%-RH)* must be taken into account and the respective vapour pressure p_D must be used for the non-saturated case:

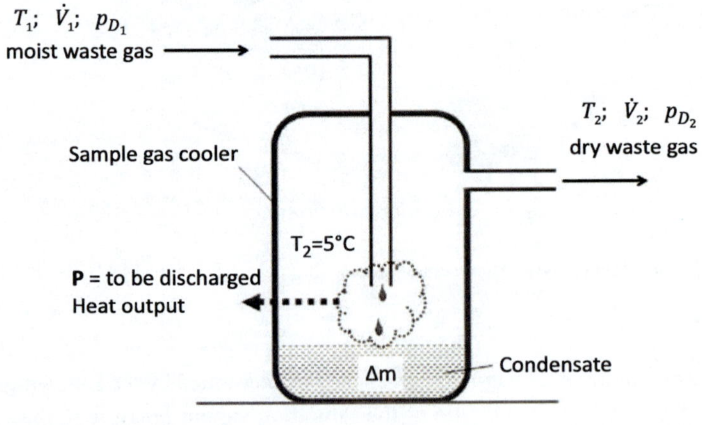

Fig. 14.20 Physical condensation process in a sample gas cooler

$$p_D = p_{SD} \cdot \frac{\% - \mathrm{RH}}{100\%} \tag{14.13}$$

For $p_{5\,^\circ\mathrm{C}} = 8,7$ hPa and using the *Magnus formula* (Eq. 14.8) for p_{SD}, one then obtains the equation for the determination of the condensate quantity $\Delta m/t$ at any dew point temperature T_1:

$$\Delta\dot{m} = \frac{6.11 \cdot \exp\left[\frac{17.08 \cdot T_1}{234.17 + T_1}\right] - 8.7\,\mathrm{hPa}}{1013\,\mathrm{hPa}} \cdot 0.803\,\mathrm{g/L} \cdot \left(\frac{273.15\,^\circ\mathrm{C}}{273.15\,^\circ\mathrm{C} + T_1}\right) \cdot \dot{V}_1 \tag{14.14}$$

Example
An exhaust gas with a volume flow of 3 L/min and a temperature $T_1 = 70^\circ\mathrm{C}$ enters a sample gas cooler. The relative humidity is 85%. The condensate quantity is thus

$$\Delta\dot{m} = \frac{0.85 \cdot 6.11 \cdot \exp\left[\frac{17.08 \cdot 70}{234.17 + 70}\right] - 8.7\,\mathrm{hPa}}{1013\,\mathrm{hPa}} \cdot 0.803\,\mathrm{g/L} \cdot \left(\frac{273.15\,^\circ\mathrm{C}}{343.15\,^\circ\mathrm{C}}\right) \cdot 3\,\mathrm{L/min} \tag{14.15}$$

$$\Delta m = 0.484\,\mathrm{g/\,min} = 29.1\,\mathrm{g/h} = 697.5\,\mathrm{g/d}$$

An estimation of the condensate quantity can also be made from the water vapour concentration in g/m^3 and the volume flow (L/min or m^3/h) (Fig. 14.21). These values can be found in the corresponding tables (e.g. Table 10.2).

Changes in Concentration
The use of a sample gas cooler removes almost the entire water content from the sample gas. At a cooler temperature of $T_K = 5°\mathrm{C}$ and an air pressure of $p_N = 1013$ hPa, only 0.86 vol% H_2O remains in the sample gas as residual moisture after the cooler. Relative to the total volume, the concentration values c^* of the remaining components therefore increase (Clarke 1998). Figure 14.22 shows these ratios graphically. In the sample gas, x vol.-% H_2O and y vol.-% residual gas. It therefore applies $x + y = 100$ %. After the cooling process, the remaining 0.86 vol.% H_2O is still in the sample gas. The percentage calculation then results for the remaining components with the following conversion factor

$$f_{H_2O} = \frac{c^*}{c} = \frac{100\% - 0.86\%\ H_2O}{100\% - x\%\ H_2O} \tag{14.16}$$

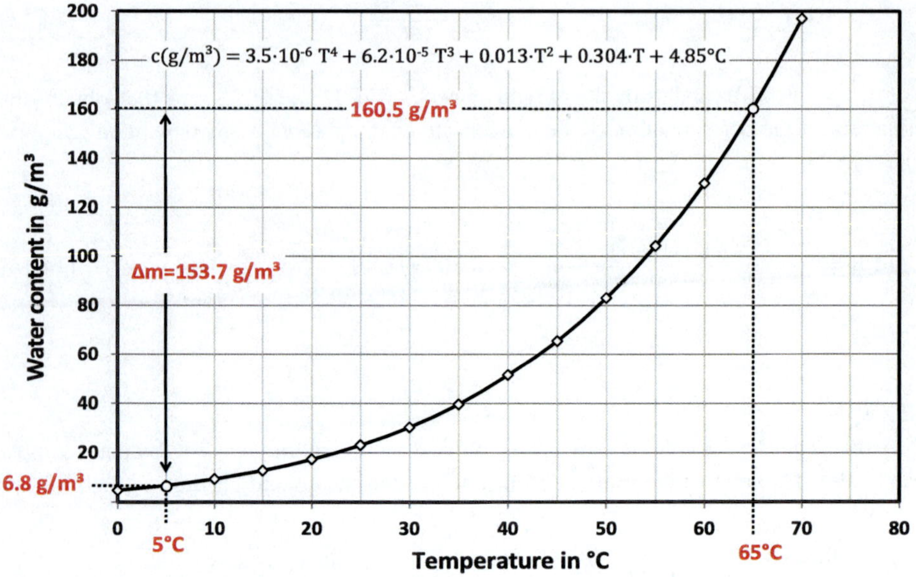

Fig. 14.21 Estimation of the condensate quantity from 1 m^3 Sample gas at a known water content in g/m^3. The curve is derived from the values in the Table 10.2 and was approximated by a trend line $c(g/m^3)$ (sixth degree polynomial)

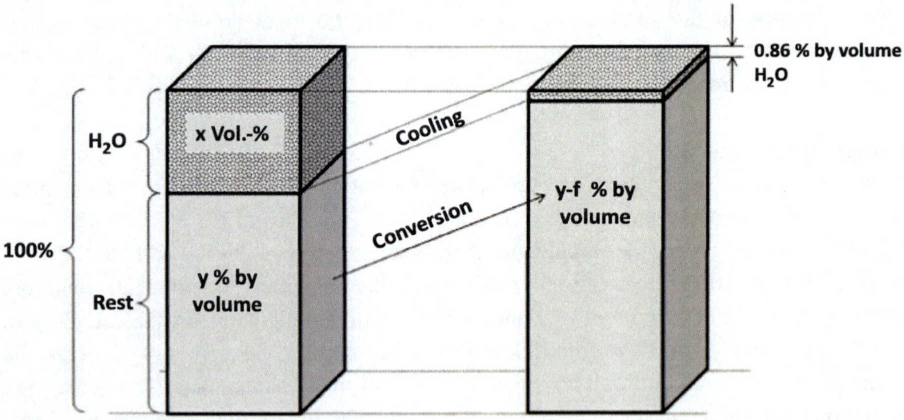

Fig. 14.22 Illustration of the conditions upstream and downstream of the sample gas cooler

In general, this connection can also be described by the partial pressures. With p_D= vapour pressure before the cooler, $p_D(T_K)$= vapour pressure after the cooler (with cooler temperature T_K) and p_{ges} atmospheric air pressure (→ total pressure):

Table 14.1 Vapour pressure (p_D), concentration (c) and conversion factor f_{H_2O} for different dew point temperatures T_{TP} of the sample gas before the sample gas cooler

T_{TP} in ° C	p_D (H_2O)	Vol. % H_2O at 1013 hPa	Conversion factor f_{H_2O}
5	8.7	0.86	1.000
10	12.3	1.21	1.004
15	17.1	1.69	1.008
20	23.4	2.31	1.015
25	31.7	3.13	1.023
30	42.4	4.19	1.035
35	56.3	5.56	1.050
40	73.7	7.28	1.069
45	95.8	9.46	1.095
50	123.3	12.17	1.129
54	150	14.81	1.164
60	199.2	19.66	1.234
66	261.4	25.80	1.336
70	311.6	30.76	1.432
80	473.4	46.73	1.861
90	701.1	69.21	3.220
96	876.7	86.54	7.368

$$f_{H_2O} = \frac{c^*}{c} = \frac{p_{ges} - p_D(T_K)}{p_{ges} - p_D} \tag{14.17}$$

The conversion factor f can be applied to all components in the sample gas and, as a dimensionless quantity, is not linked to a specific concentration specification. The following relationship therefore applies:

$$c^*(\text{before the cooler}) = c\,(\text{after the cooler}) \cdot f_{H_2O} \tag{14.18}$$

With Eq. 14.16 the factor f can be determined for any conditions. Table 14.1 lists the conversion factor f_{H_2O} for a sample gas cooler with temperature T_K = 5°C, at an air pressure of 1013 hPa. In Fig. 14.23 this relationship is shown graphically for different *x-axes.*

Washing Out of Gas Components

The formation of condensate in the sample gas cooler inevitably leads to a more or less large contact of the sample gas with the condensed water. The manufacturers of these cooling units therefore try to keep the contact surfaces as small as possible. However, the contact cannot be prevented. Depending on which gas components are present in the gas mixture, these components can be dissolved in the condensation water. In particular,

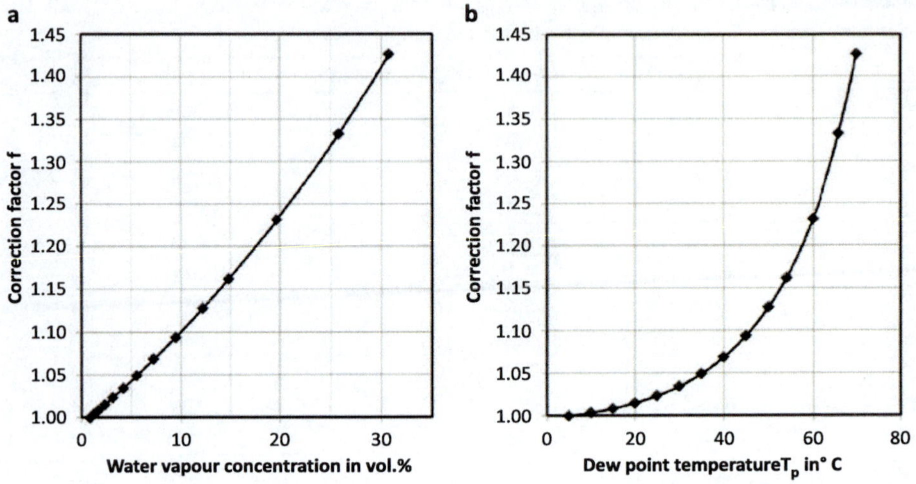

Fig. 14.23 Dependence of the conversion factor f_{H_2O} on the water vapour concentration (volume % H_2 O) (**a**) or on the dew point temperature T_{TP} of the sample gas before the sample gas cooler (**b**) according to Eqs. 14.16 or 14.17

substances such as ammonia (NH_3) and hydrogen chloride (HCl) bind very quickly with water and then form either a lye (NH_3→ ammonia) or an acid (HCl hydrochloric acid), → which then accumulate in the condensate. But also sulphur dioxide (SO $_2$) and nitrogen dioxide (NO_2) show this behaviour, although to a lesser extent. After the sample gas cooler, there is therefore a lower result in the sample gas, which can be considerable. The solution of the substance in water is described by *Henry's law* and increases towards lower temperatures. At the same time the concentration increases with decreasing temperature, so that two opposite effects come into play here.

There is an equation for ammonia with which both the volume effect due to condensation of the water vapour (Eq. 14.17) and the solution of NH_3 in water can be described (Boeker et al. 1999):

$$f = \frac{c^*_{NH_3}}{c_{NH_3}} = \frac{p_{ges} - p_D(T_K)}{[p_D - p_D(T_K)] \cdot M_{H_2O}\Lambda(T_K) + (p_{ges} - p_D)} \tag{14.19}$$

The factor $\Lambda(T_K)$ describes the absorption behaviour of ammonia in water and M_{H_2O} is the molar mass of water (= 18 g/mol). Figure 14.24 shows the course according to Eq. 14.19 for different dew point temperatures (40 °C, 50 °C, 60 °C and 70 °C) at the inlet of the cooler (Boeker 2001). At these temperatures (e.g. 70 °C) the measuring gases are present with saturated water vapour (70% relative humidity). If this gas mixture enters a cooler at a temperature $T_K < 70$°C, the water vapour condenses out and the concentration value rises.

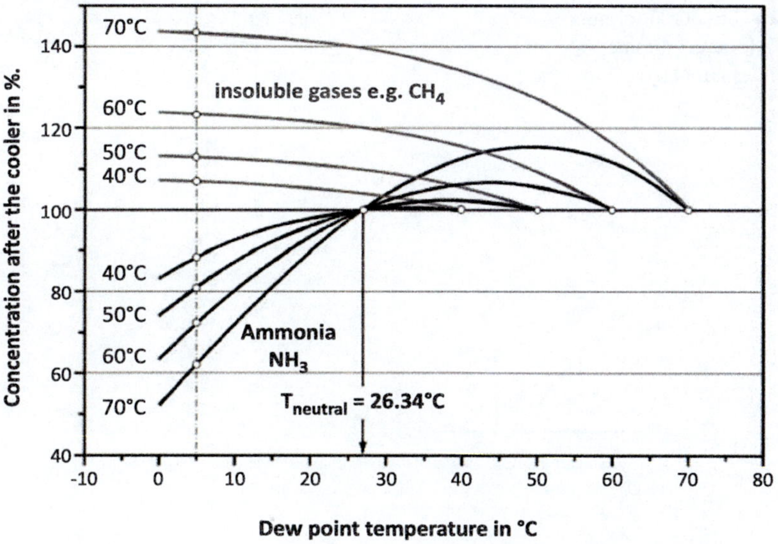

Fig. 14.24 Comparison of the concentration changes for soluble and insoluble gases, using the example of NH_3 and CH_4 in a sample gas cooler for the temperature range from 0 to 40 °C or 70 °C

At the same time, some of the ammonia is dissolved. As the cooler temperature falls, the solubility in the condensed water rises steeply. These two opposite effects result in a maximum for the concentration value at a certain dew point temperature in the cooler. It can also be shown that there is a temperature $T_{\text{neutral}} = 26,34°\text{C}$ for ammonia at which both effects cancel each other out and the original concentration specification (100%) applies. Gases that do not dissolve in water, however, show a different behaviour. With $\Lambda(T_K) = 0$, therefore, you get a gradient which is given by the description according to Eq. 14.17. This behaviour is also shown in Fig. 14.24.

Gas Drying

In order to avoid the effects of water solubility in a sample gas cooler, drying with a water vapour permeable membrane is recommended. *Nafion®*[1] has established itself as a membrane material for gas measurement technology. *Nafion®* is a copolymer of perfluoro-3,6-dioxa-4methyl-7octene-sulfonic acid and tetrafluoroethylene. The structural formula is shown in Fig. 14.25.

This substance spontaneously absorbs water from the gas phase, whereby each sulphonic acid group can bind up to 13 water molecules. The sulfonic acid groups form ion channels within the polymer through which water molecules can be transported.[2]

[1]Trademark of the company E.I. DuPont.

[2]Leaflet Perma Pure measuring gas dryer 97/05 D 9C gasmet GmbH, Karlsruhe Germany

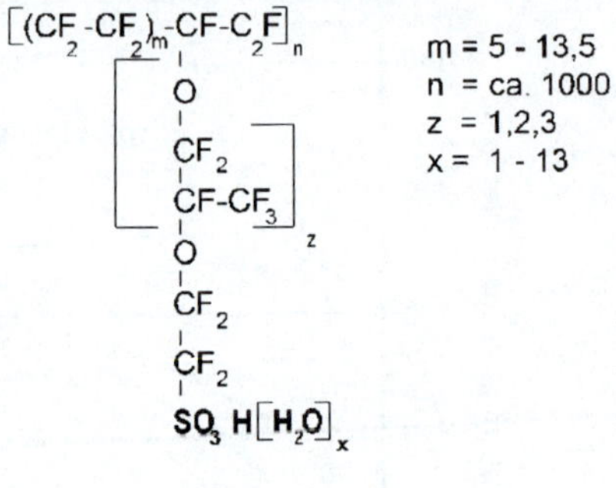

Fig. 14.25 Structural formula of Nafion. (Source: Gasmet Technologies GmbH)

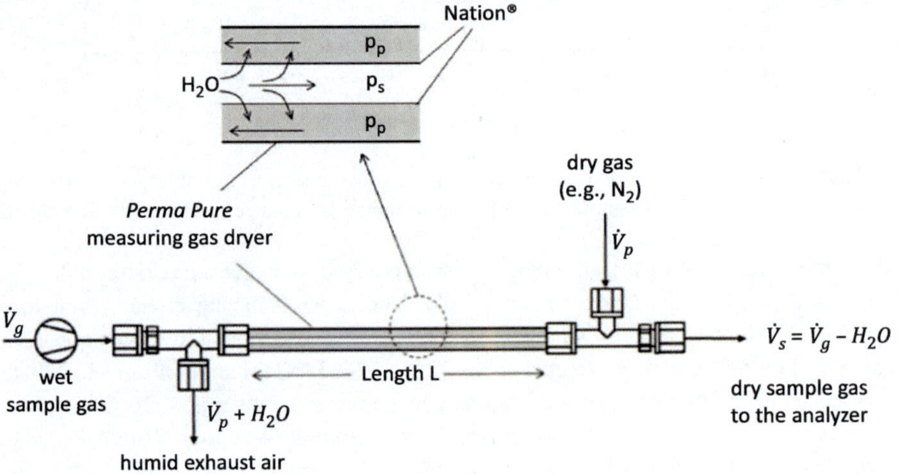

Fig. 14.26 Arrangement of a gas dryer with external drying gas (e.g. Nitrogen)

Figure 14.26 shows the structure of a sample gas drying system with a *Nafion®* membrane. For practical reasons, thin tubes are used which have a large surface area and thus ensure efficient drying. Furthermore, these tubes can be combined to a tube bundle, which allows the drying of larger volume flows (see Fig. 14.27). The sample gas to be dried is passed through these tubes, while a drying gas (e.g. compressed air or N_2) is fed in countercurrent. The required purge gas quantity $\dot{V}_p$ results from the pressure conditions and the volume flow at the inlet of the dryer:

$$\dot{V}_p = F \cdot \dot{V}_g \cdot p_p / p_s \quad (14.20)$$

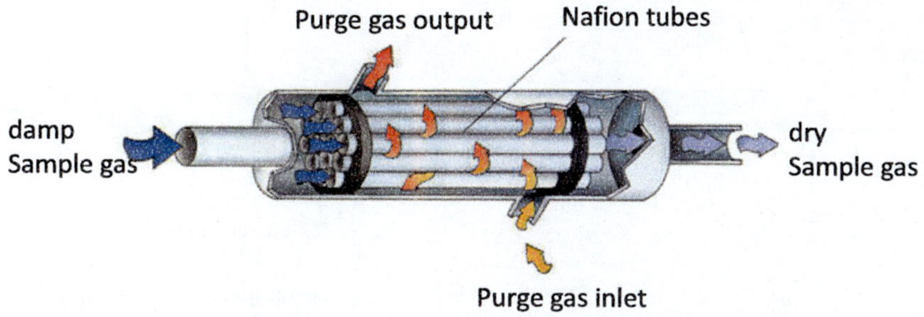

Fig. 14.27 Design of a gas dryer based on Nafion tubes, which are interconnected to form a bundle and thus have a much larger diffusion surface. (Source: Gasmet Technologies GmbH Karlsruhe)

The factor F is an empirical value that can lie between 2–10. The volume flow $\dot{V}_p$ of the dry gas should therefore be at least twice as high as the volume flow of the sample gas. $\dot{V}_s$ For gases containing HCl, the manufacturer recommends 10 times this value (F = 10).

Due to the partial pressure ratio p_p/p_s, the dry gas then absorbs the water vapour from the sample gas and transports it out of the system. The drying gas therefore contains the entire amount of water that was previously removed from the sample gas at the outlet. On the inlet side, however, condensation (falling below the dew point) must be prevented, otherwise the drying effect is lost. In practical cases the sample gas inlet is therefore heated up while room temperature prevails at the end of the drying section (Fig. 14.28). This temperature gradient, which should extend over the entire dryer length L, achieves an optimum drying effect.

For applications in the field of process measuring technology, the Nafion tubes are installed in a stainless steel radiant tube with corresponding pipe fittings for the gas supply (see Fig. 14.29). Although the wall material is of secondary importance with regard to corrosion resistance, because the sample gas is only conducted in the Nafion tube, there are other reasons for using stainless steel:

- because the entire measuring system is equipped with stainless steel piping
- because the dryer should be mechanically self-supporting
- because the heat transfer is better
- because plastic cannot be used for explosion protection reasons
- because a flammable or highly toxic medium has to be dried and the stainless steel pipe offers more protection against kinking/shearing.

Since the degree of drying also depends on the total length of the Nafion tubes, in practice lengths of $L = 0.3$ m, 0.6 m, 1.2 m, 1.8 m, 2.4 m and 3.65 m are used. For better handling, these tube lengths can be formed into a more compact unit by winding them up, which can then also be used in the corresponding equipment.

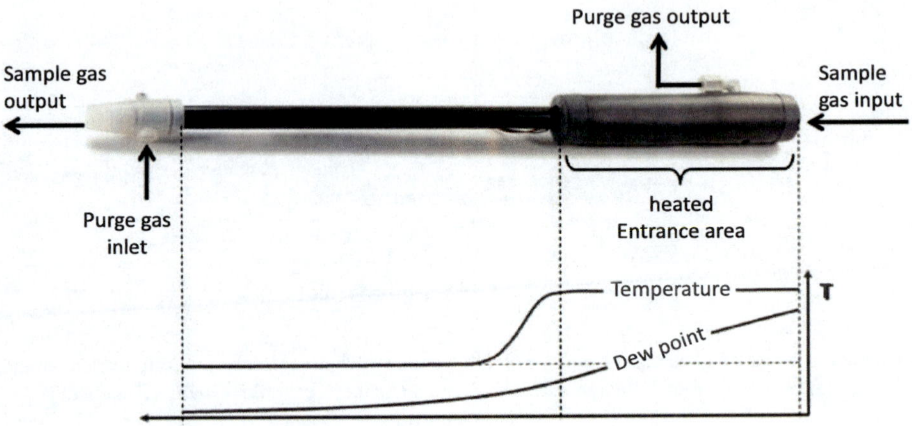

Fig. 14.28 Design of a gas dryer with heated inlet area to prevent condensation from falling below the dew point. (Source: Gasmet Technologies GmbH, Karlsruhe Germany)

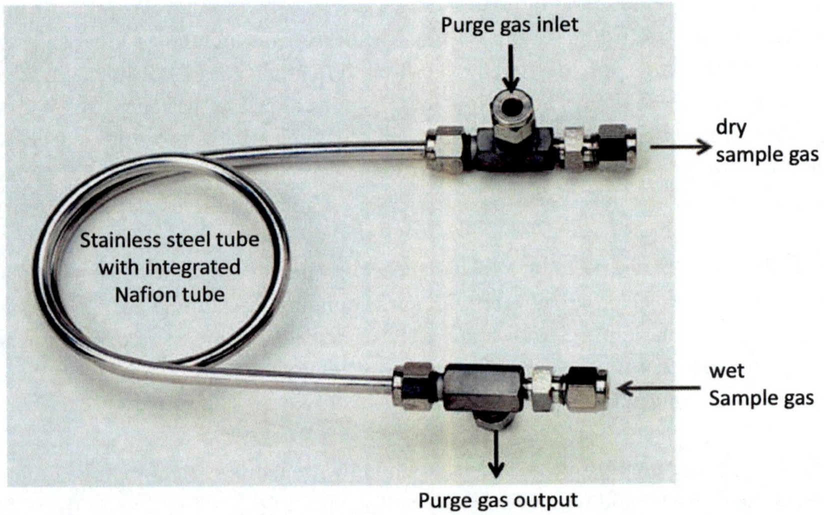

Fig. 14.29 Gas dryer in a stainless steel version for process measurement technology. (Source: Gasmet Technologies GmbH, Karlsruhe Germany)

The degree of drying (dew point temperature) at the outlet of the dryer, depending on the gas flow, is shown in Fig. 14.30. Dew points of −20°C can be reached with very long dryers and low sample gas flows. In special applications dew points of up to −40°C have even been achieved with this technique (Leckrone and Hayes 1997; Kaltenmaier 2015).

If no separate dry gas is available in the application, the dried sample gas can also be used. For this purpose a partial flow (Vp) of the dried sample gas is branched off and passed

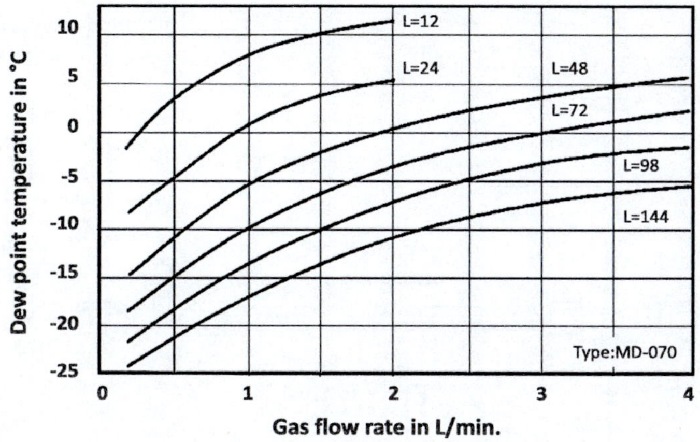

Fig. 14.30 Dew point temperature at the outlet of a *Perma Pure dryer (type: MD-070)* with an inlet dew point of 20 °C

through the gas dryer in countercurrent. To improve the degree of drying, the gas pressure *pp* is reduced by means of a vacuum pump and a needle valve (throttling). To achieve the required partial pressure ratio, the pressure difference must also be at least $\Delta p = p_s - p_p$ 500 hPa. The volume flow is calculated as follows:

$$\dot{V}_p = F \cdot (\dot{V}_p + \dot{V}_s) \cdot p_p / p_s \tag{14.21}$$

The following expression for the purging gas requirement is obtained by transformation:

$$\dot{V}_p = \frac{\dot{V}_s}{(p_s / F \cdot p_p) - 1} \tag{14.22}$$

Alternatively, splitting can be omitted and the entire dried sample gas is first passed through the analyzer and then completely through the dryer. With this measure more dry gas is available and the dew point can be further reduced. Equation 14.22 then changes as follows:

$$\dot{V}_g = F \cdot \dot{V}_g \cdot p_p / p_s \quad \text{it then follows} \quad F \cdot p_p / p_s = 1 \tag{14.23}$$

Figure 14.31 shows such a structure for self-drying. The negative pressure *pp* is set by a needle valve and adjusted with the pressure indicator/pressure sensor.

The use of this drying technique has several advantages, which are particularly useful for the measurement of HCl, Cl_2, SO_2, NO_x, H_2S, etc. With appropriate effort (e.g. central N_2 supply) dew points below $-20^{\circ}C$ can be achieved, which cannot be reached with

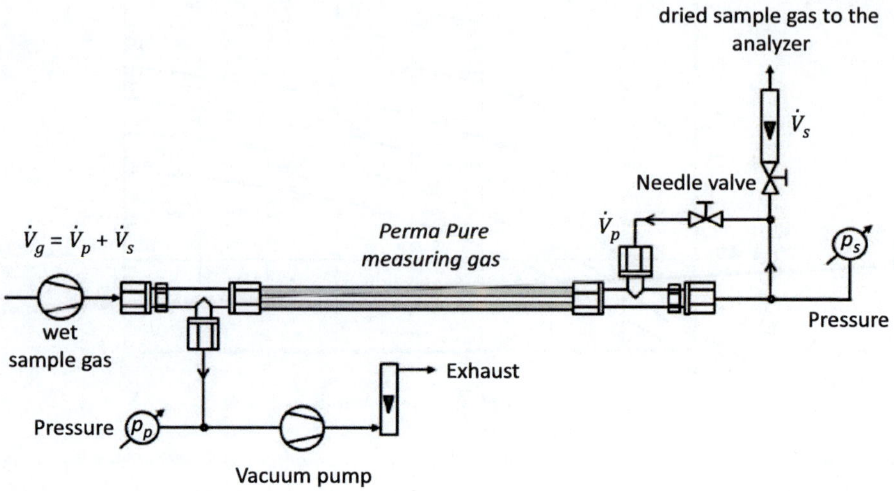

Fig. 14.31 Gas dryer with a return of the sample gas for self-drying

sample gas coolers (→ icing). This results in a very low water vapour concentration in the sample gas and can also reduce possible cross-sensitivities.

However, there is a clear limitation for the measurement of ammonia and other ammonium compounds as well as for polar, organic hydrocarbons. Alcohols, aldehydes and ketones can - like water vapour - penetrate nafion and therefore suffer corresponding losses. Amines and ammonia, on the other hand, even in trace amounts (a few ppm) bring the process to a standstill, that is, drying fails and condensate can get into the device and damage it.

However, ammonia (NH_3) is also a problem for conventional sample gas coolers (→ washing out of NH_3 components), so that this measurement can only be carried out in a completely heated system ($T \approx 180°C$). An alternative to a hot gas measurement is a defined dilution with dry air or dry nitrogen.

In *medical technology* (respiratory gas analysis), a separate dry gas is dispensed with completely and the ambient air is used for this purpose. In this case, the Nafion hose is placed directly on the breathing mask or mouthpiece, thus lowering the dew point to prevent condensation in the downstream gas detector. This dryer is only used once and is exchanged with the next patient (→ *disposable items*).

Gas Dilution

Another possibility to lower the dew point of the sample gas is to mix the humid sample gas with dry zero gas (e.g. Nitrogen/N_2). The diluted sample gas then has a correspondingly lower dew point. Figure 14.32 shows the basic mixing process. Since the diluted sample gas is fed to the gas analyser, the partial flows must be known and must also remain constant during the measurement. The relative humidity after the mixing process depends,

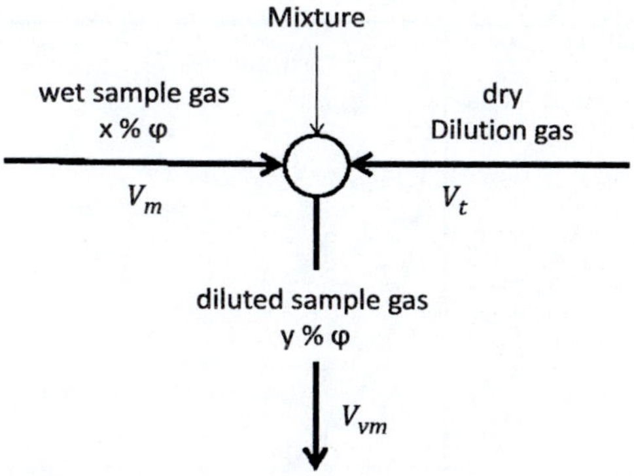

Fig. 14.32 Mixture of a humid sample gas with a dry zero gas (dilution gas) to lower the dew point

like the other gas concentrations, on the mixing ratio *R*. This dilution factor can be determined as follows:

$$R = \frac{V_m + V_t}{V_m} \tag{14.24}$$

This dilution factor is then used to calculate the gas concentration c^* after dilution with an inert gas (Jahnke 2000).

$$c^* = c \cdot R \tag{14.25}$$

The relative humidity (φ) behaves like the gas concentration *c*, so that it can also be determined with the dilution factor *R*:

$$y\%\varphi = x\%\varphi \cdot R$$

respectively

$$y \text{ vol.-}\%H_2O = x \text{ vol.-}\%H_2O \cdot R \tag{14.26}$$

In order to effectively prevent the temperature from falling below the dew point, the dew point of the diluted sample gas must be below the ambient temperature. Figure 14.33 shows an example of the influence of the dilution factor R. In this example, the sample gas has a water vapor concentration of 30% by volume, which corresponds to a dew point of $T_{TP} \approx 70°C$. A dilution by the factor $R = 13$ then leads to a water vapour concentration of 2.3 vol.% ($\rightarrow p_D \approx 23$ hPa). According to the *Magnus formula* (Eq. 14.8), this vapour pressure results in a saturation temperature of $T_S = 20°C$. In practical cases, a dilution

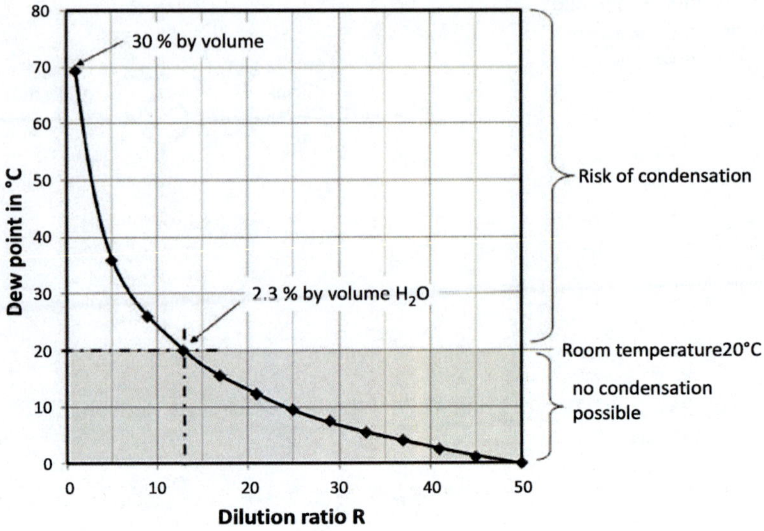

Fig. 14.33 Example of the influence of the degree of dilution on the dew point of the diluted sample gas

wheel R > 30 is chosen to ensure that the water vapour concentration falls safely below the dew point, as in the case of a sample gas cooler ($T_{TP} = 5°C$).

In order to prevent condensation, this dilution process must therefore be carried out as close as possible to the tapping point. The dilution assembly is therefore usually placed directly in the sampling probe. However, the gas path to the dilution unit must be additionally heated in order to prevent condensation on this short section of the path as well. In Figs. 14.34 and 14.35 such a construction is shown. The sample gas is sucked out of the process with the lance of a sampling probe and cleaned through a heated filter. The dilution unit is located directly at the outlet of the filter element. The heating for the filter ensures that the entire gas path is heated above the dew point. The sample gas is then sucked in through a critical nozzle so that a constant volume flow V_m is available. In a critical nozzle the flow velocity is limited to the speed of sound (see Sect. 2.7). The required negative pressure of $p < 0.5$ bar is generated with an injector pump, which is *driven* by the dilution gas with the volume flow V_t. Figure. 14.36 shows a detailed view of the dilution unit.

Since the volume flow V_m of the sample gas is adjusted via the critical nozzle, an almost constant gas mixture must be assumed. If the composition of the main components carbon dioxide (CO_2) and water vapor (H_2O) changes, the velocity of sound in the critical nozzle changes and thus also V_m. The reason for this influence is the molar mass (M) of the substances or the gas mixture. Carbon dioxide (CO_2) has a molar mass of 44 g/mol while water vapour (H_2O) has a molar mass of 18 g / mol. For dry air a value of 28.75 g/mol (Down and Lehr 2005) is given, while a typical exhaust gas is ≈30 g/mol (Jahnke 2000).

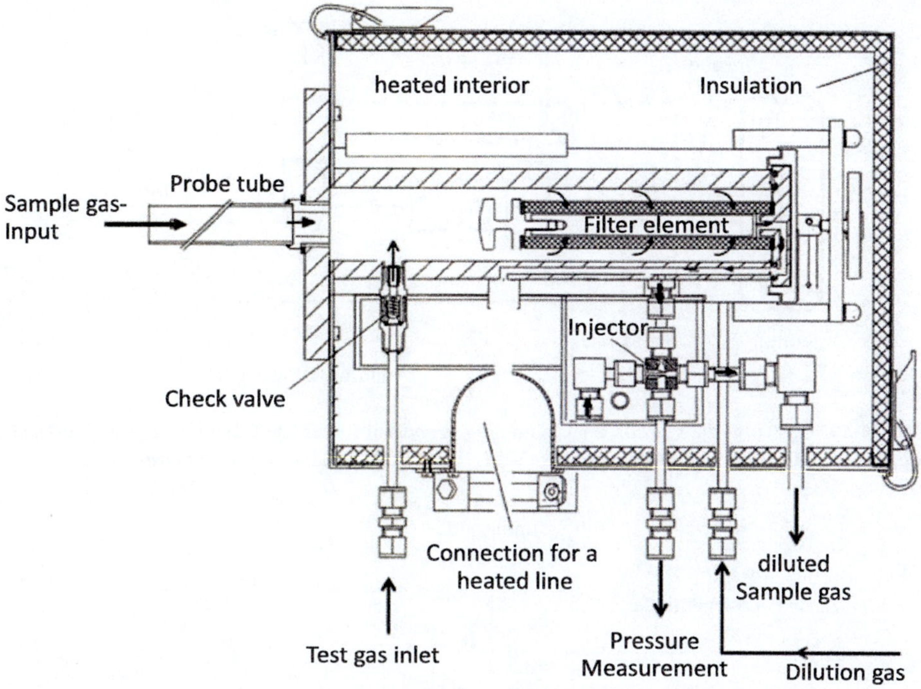

Fig. 14.34 Design of a gas sampling probe with integrated dilution system and test gas supply before the first filter stage. Optionally, this probe type offers the possibility to connect a heated hose to realize a *hot gas measurement.* (Source: M&C TechGroup GmbH, Ratingen Germany)

Furthermore, the actual temperature T and the actual pressure p play a decisive role in the calculation of the sound velocity and thus also the dilution factor. During a calibration with test gas, defined states are present which are marked by p_0, T_0 and M_0. In measuring mode, the values p, T and M are then relevant. The dilution factor R then changes as follows (Jahnke 2000)

$$R = 1 + (R_0 - 1) \cdot \frac{p_0}{p} \cdot \frac{\sqrt{T}}{\sqrt{T_0}} \cdot \frac{\sqrt{M}}{\sqrt{M_0}} \tag{14.27}$$

For the corrected concentration c^* the following expression is then obtained with Eq. 14.25:

$$c^* = c \cdot \left[1 + (R_0 - 1) \cdot \frac{p_0}{p} \cdot \frac{\sqrt{T}}{\sqrt{T_0}} \cdot \frac{\sqrt{M}}{\sqrt{M_0}}\right] \tag{14.28}$$

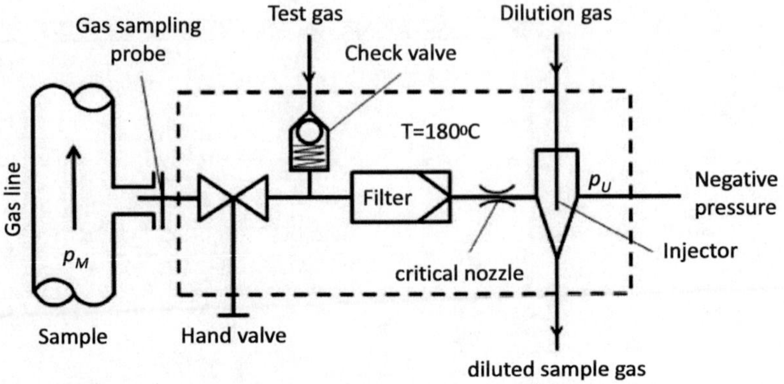

Fig. 14.35 Gas flow diagram of a gas sampling probe with integrated dilution unit and test gas supply before the dust filter. (According to M&C TechGroup GmbH, Ratingen Germany)

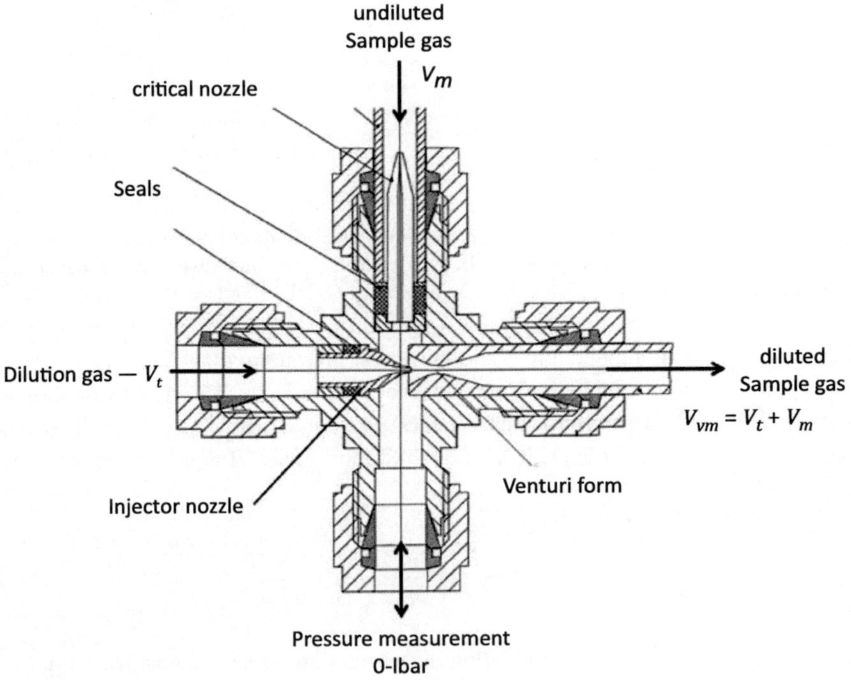

Fig. 14.36 Detailed view of the dilution unit with injector pump and a critical nozzle. (Source: M&C TechGroup GmbH, Ratingen Germany)

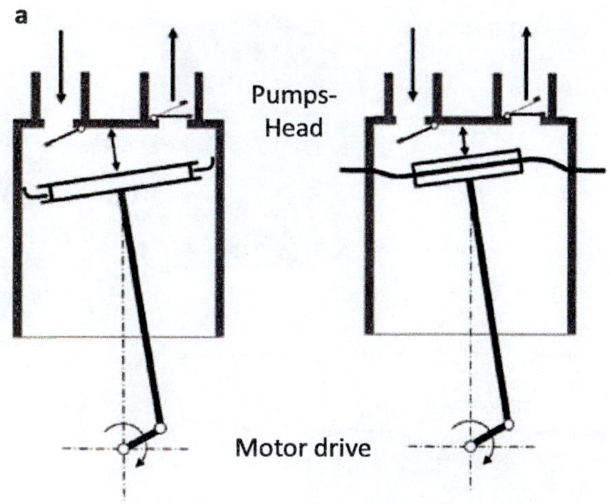

Fig. 14.37 Basic design of swing piston pumps (**a**) and diaphragm pumps (**b**)

Sample Gas Pumps

Gas production can be realized in various ways and is a central component of gas measurement systems. Diaphragm pumps are generally used as gas pumps, which are available with different diaphragm materials. Alternatively, so-called pedal piston pumps are used, which have a similar design. Both types of pumps are shown in Fig. 14.37. An electric motor drives a piston rod via an eccentric, which sets the pendulum piston or the diaphragm in a tumbling motion. If the piston rod moves downwards, the gas is sucked in via the nozzle and the chamber volume fills with the sample gas. After a half turn (180 °) of the motor, the piston or diaphragm moves upwards and the left-hand valve closes, so that the gas then flows out via the right-hand valve. There is therefore one suction and one pumping cycle per revolution. So the gas is not pumped continuously, but is transported in individual *packages*. These pump strokes can interfere with the measuring result in the gas detector, as pressure surges inevitably occur in the system. Some pump manufacturers therefore offer pumps with two pump heads, which operate 180 ° out of phase and thus have a significant reduction in the pressure fluctuations of the individual pump strokes (Fig. 14.38). The gas flow is then conveyed almost continuously, as the gas package from Pump 2 is delivered immediately after the gas package from Pump 1. A prerequisite, however, is that the two pump heads are connected in parallel. The parallel connection also leads to a higher flow rate, as each piston has the same pumping capacity and this is then added in the output. The two pump heads can also be operated in series. The series connection leads to a higher pressure compared to one pump head, because the second pump head compresses the gas from the first pump head again. The pumping capacity (flow rate) remains constant. This connection is therefore also preferred for vacuum operation.

For extremely corrosive gases there are metal bellows pumps, but these are only used in rare cases. Alternatively, pump heads made of Teflon are used, which have sufficient

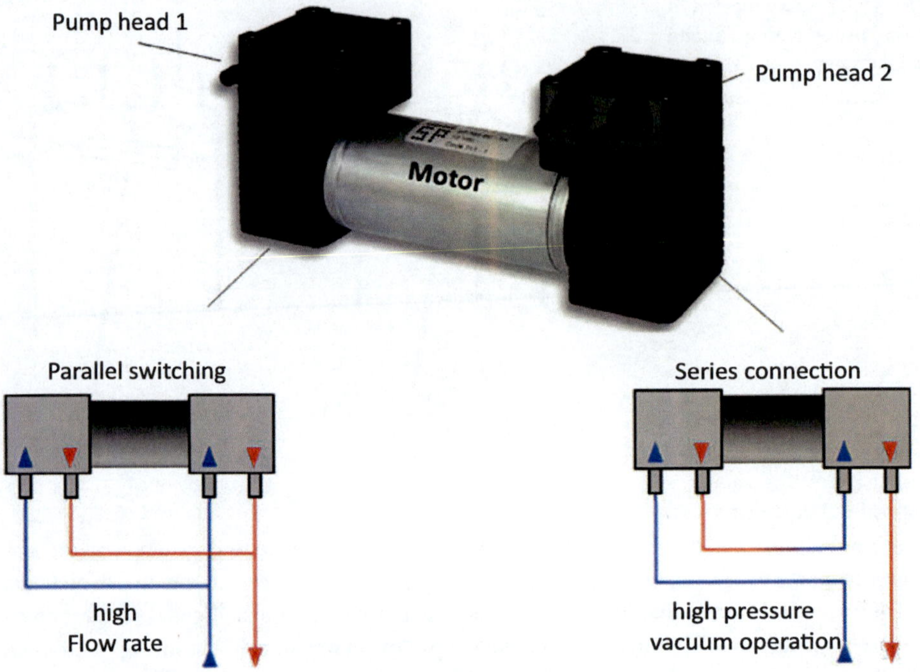

Fig. 14.38 Tandem pump with two pump heads. (Source: Schwarzer Precicion Gmbh, Essen Germany)

corrosion resistance (Fig. 14.39). For portable gas detectors, the pump manufacturers also offer miniaturized versions (Fig. 14.40 and 14.41). With these miniature pumps, the volume flow can also be conveyed in negative pressure and positive pressure operation. However, the volume flows are very low and can be found at 1 L/min. (Fig. 14.42).

Injector Pump

The injector pump works on a pneumatic principle and is driven by compressed air. The compressed air (propellant) enters a chamber via a nozzle and from there is led into a diffuser. The diffuser forms a constriction, similar to a *venturi nozzle,* and a negative pressure is created there due to physical laws (*Bernoulli's equation*) (Fig. 14.43). The measuring gas (suction medium) is then sucked in through this negative pressure. The gas analyzer can be placed alternatively in front of or behind the injector pump (Fig. 14.44). In front of the pump the original concentration in the sample gas is measured. After the injector pump, however, the sample gas is present in diluted form. This application has already been described for the so-called dilution probe. The advantage of the injector pump is that there are no wearing parts (diaphragm, motor, seals, ...) and the pump can also be

Fig. 14.39 Sample gas pump type MP47 230 V for high flow rates up to 6 L/min. (Source: M&C TechGroup GmbH, Ratingen Germany)

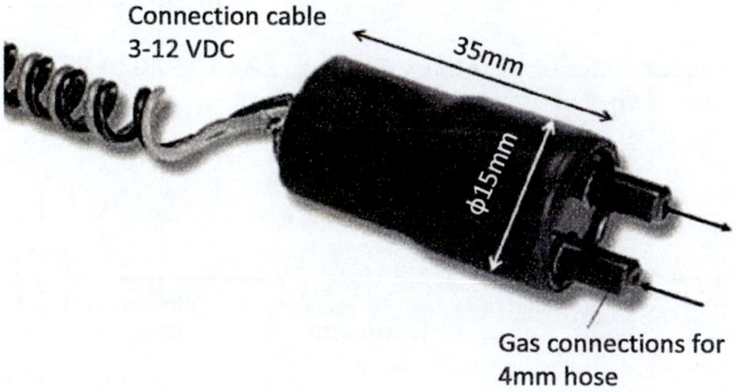

Fig. 14.40 Worldwide smallest micro vane pump type SP 135 FZ/140 FZ. (Source: Schwarzer Precicion Gmbh, Essen Germany)

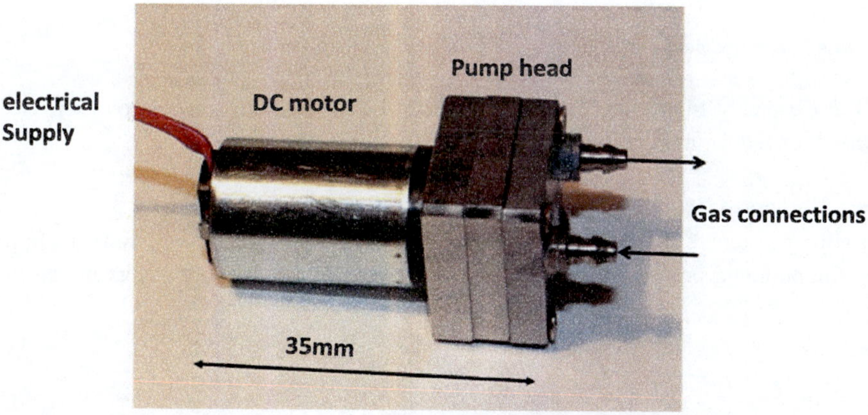

Fig. 14.41 Miniaturised diaphragm pump with a full metal pump head for corrosive gases

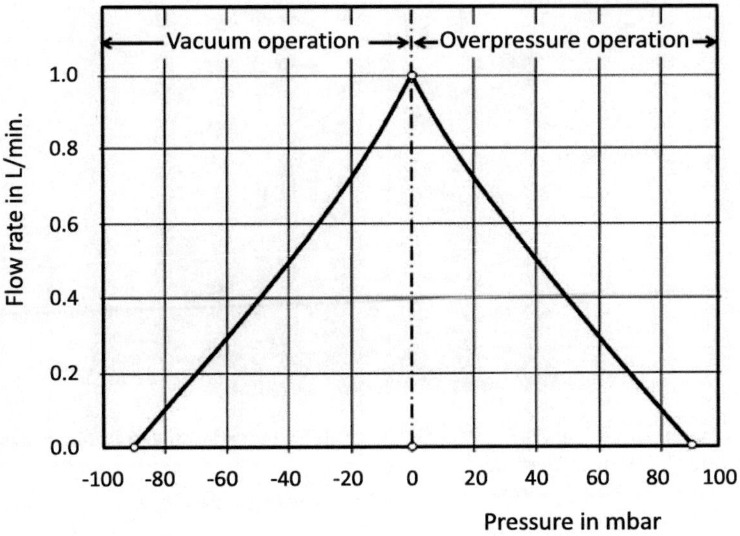

Fig. 14.42 Flow characteristics of the Miro vane pump SP 135 FZ/140 FZ. (Source: Schwarzer Precicion Gmbh, Essen Germany)

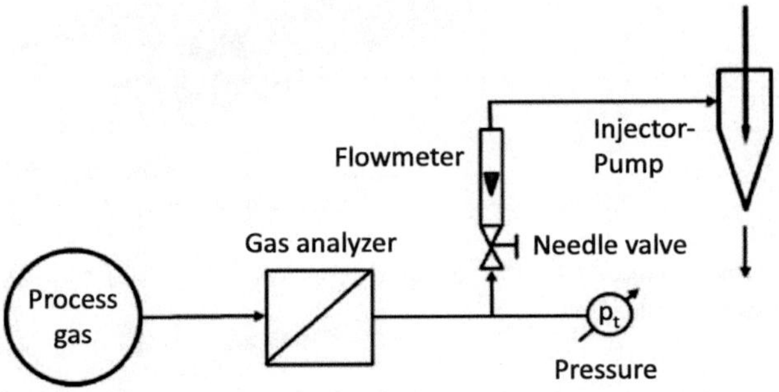

Fig. 14.43 Arrangement of the injector pump in suction operation

used at extremely high temperatures ($T > 200°$ C). The disadvantage, however, is the permanent need for a propellant gas.

Peristaltic Pump (Hose Pump)

Peristaltic pumps are mainly used to pump condensate in gas measuring systems. In this case, the pumping process is carried out directly with a hose by four rollers driven by a

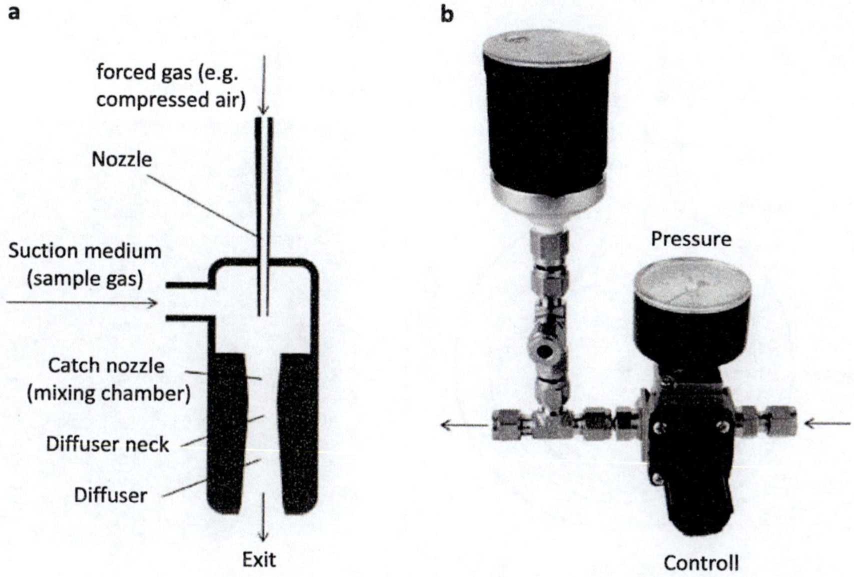

Fig. 14.44 Basic design of an injector pump (**a**) and a commercial version (**b**) for gas measurement technology. (Source: APM-Technik GmbH, Korschenbroich Germany)

motor squeezing a sample volume through the hose (Fig. 14.45). The four rollers are also mounted so that they can rotate within themselves, so that a certain volume is enclosed after the first contact with the tube. This volume is then passed on 360 ° (full rotation) and then conveyed out at the exit. This type of pump works without valves and is therefore very robust. However, the hose material has to be replaced from time to time, as it tires after a certain period of operation and may then leak.

Flow Control

The flow of the sample gas is controlled by a downstream flow meter before it is fed into the gas analyzer. Mechanical floats are preferably used as flow meters, as these operate very reliably and already have an *optical display* (height of the float in the tube). The flow is limited by a needle valve, which is usually integrated directly into the flow meter. Figure 14.46 shows such a control unit. The failure of the pump can be monitored, for example, additionally with a flow control by detecting the position of the float with a light barrier or an inductive sensor.

Gas Filtration

In most of the processes to be monitored, there are more or less particles, dust, soot, etc. in addition to the gas components. These substances must be removed before the gas measurement, as they could possibly disturb the measurement. Especially the optical

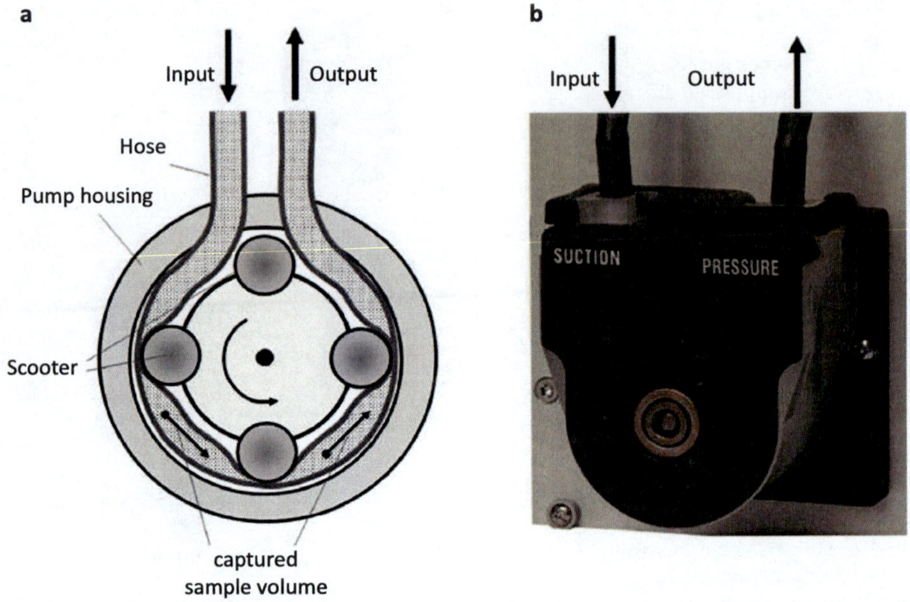

Fig. 14.45 Principle mode of operation of a peristaltic pump (**a**) and a commercial version (**b**) for use in cooling units for condensate transport. (Source: M&C TechGroup GmbH, Ratingen Germany)

measuring methods are very sensitive to contamination. The interfering particles are therefore separated by appropriate filter elements and only the cleaned sample gas reaches the respective gas analyser. Basically, a distinction is made between depth filters and surface filters, which have different filter characteristics.

Depth filters usually consist of a porous material that is permeable to gas, but in which the unwanted particles can settle (see Fig. 14.47a.). The materials used for these filter elements are for example, sintered metal (brass/stainless steel), ceramics, glass, plastic and paper. Whether the different particles are separated in the filter depends on the particle size, the pore size of the filter medium and the filter thickness.

In a surface filter, the filter element usually consists of a thin membrane (e.g. Teflon) or a composite material (fabric and plastic coating) similar to the GoreTex material (see Fig. 14.47b.). The pore size of such membranes is very small, so that these filter materials are mostly used for very small particles. Furthermore, the particles settle directly on the surface and do not penetrate the material (Scott 1995). A so-called *filter cake* therefore forms on the surface, which blocks the further gas passage and causes a pressure increase in a flow system. The absorption capacity of these surface filters is therefore very low compared to a depth filter. These filter types are usually only used for fine filtration or as safety filters.

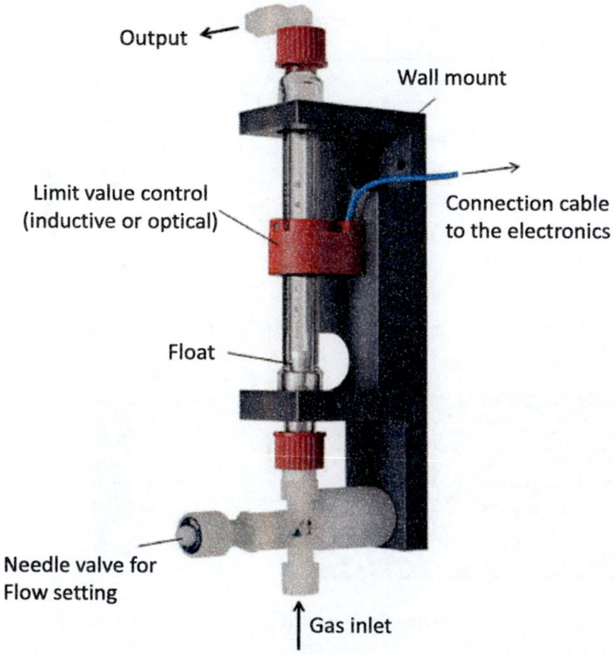

Fig. 14.46 Flow meter with a float to control the flow in a gas analyser. The flow can be adjusted directly with a needle valve. A drop in the flow causes the float to sink, which is monitored by means of an optical or inductive limit value control. (Source: APM-Technik GmbH, Korschenbroich Germany)

Depth filters, on the other hand, have a very high absorption capacity, as the particles can penetrate deep into the filter material. Fine particles move down to the lower layers of the filter, while large particles are trapped by the filter material on the surface or directly below. However, very fine particles can pass through the filter element and are then bound in a second filter stage, which, for example, consists of a surface filter.

Figure 14.48 shows a typical structure of a depth filter. In this case, the sample gas to be cleaned passes through the filter element from the outside and penetrates the entire filter volume. The outer casing of the filter housing is usually made of a transparent material (glass or plastic), so that the degree of contamination of the filter cartridge can also be seen from the outside. Figure 14.49 shows such a filter construction as part of a gas treatment unit.

Filter elements must be replaced after loading, as the gas flow is hindered by the bound particles. The pressure in front of the filter element therefore increases continuously with increasing loading. With a pressure measurement (e.g. Δp -measurement above the filter

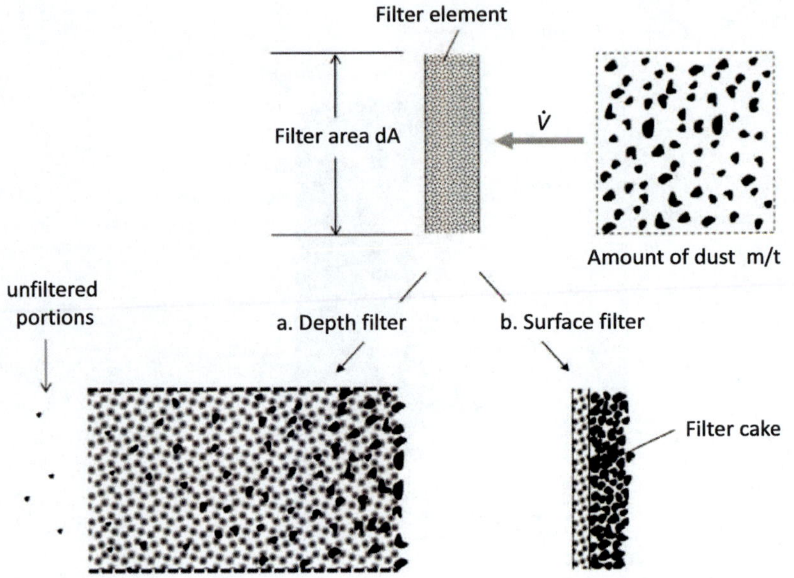

Fig. 14.47 Comparison of the operating principle of a depth filter (*a.*) and a surface filter (*b*)

element) the filter loading can be detected and signalled. To extend the service life of filter elements, the following option is available. Figure 14.50 shows a so-called bypass arrangement. The main flow passes through the inside of the filter element. From the outside, a pump is then used to create a negative pressure and the gas is sucked through the filter element. The coarse particles, due to their own kinetic energy, move through the filter and are not caught by the filter element. It is precisely these large particles that clog the filter pores and thus reduce the service life. Furthermore, the particles that settle on the filter surface are partly carried along again by the main flow. However, this requires a high main gas flow, which should be 10 times higher than the bypass flow (Clevett 1986). The bypass method thus has a significantly longer service life compared to conventional filter elements. A disadvantage of these superstructures is the high gas consumption and the use of 2 pumps.

Gas Absorber

In addition to filtering out particulate accompanying substances, gas absorbers are also used in gas measurement technology to remove certain gaseous substances from the sample gas to be analysed. For example, this measure is used if cross-sensitivities are to be reduced by accompanying components or if undesirable substances have to be removed from the

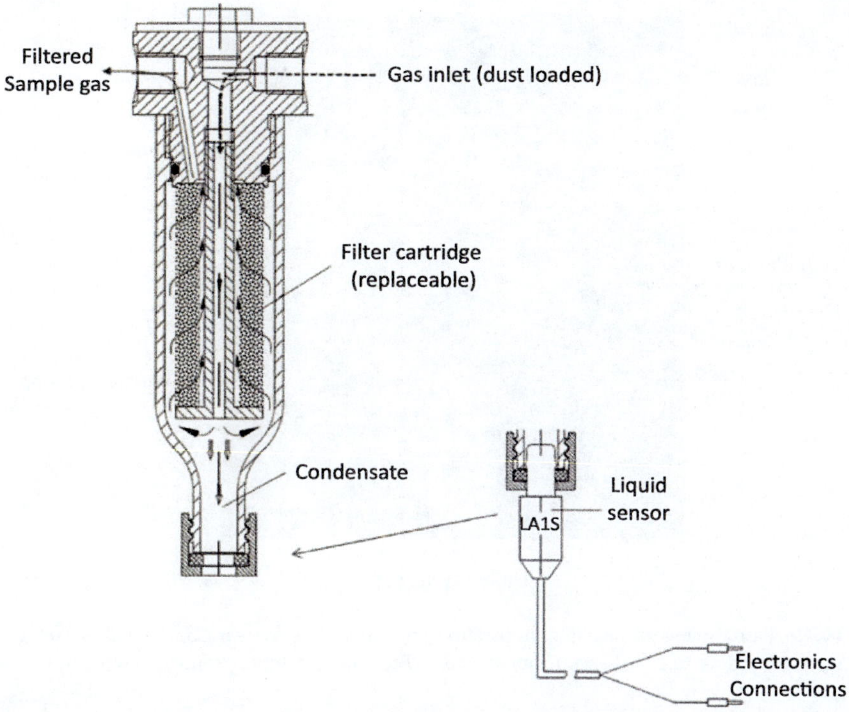

Fig. 14.48 Cross-section through a filter assembly with a replaceable filter cartridge (depth filter). The flow through the filter element is from outside to inside. A possible condensate accumulation then collects in the lower part of the filter structure and can be detected electronically by a liquid sensor (→ conductivity measurement). (Source: M&C TechGroup GmbH, Ratingen Germany)

zero gas (e.g. compressed air). A selection of possible absorber substances is listed in Table 14.2.

One of the most important substances is activated carbon, with which many organic substances can be effectively bound. In particular, activated carbon is used to filter oil vapours or solvents from compressors. This precaution is particularly important when compressed air is used in gas analysis. Figure 14.51 shows the structure of such a filter unit. The absorber material is usually available as granules (1–5 mm) and is filled into a glass tube. The ends are sealed with a filter wadding to prevent the granulate from entering the gas flow.

A widely used absorber material is calcium hydroxide and soda lime, also known as soda *lime*. Breathable lime is used in anaesthetic equipment and in rebreathers and diving rescuers to bind carbon dioxide in order to prevent the harmful effects of too high a CO_2 content in the inhaled air. Breathing lime is also used in space travel. In medicine and

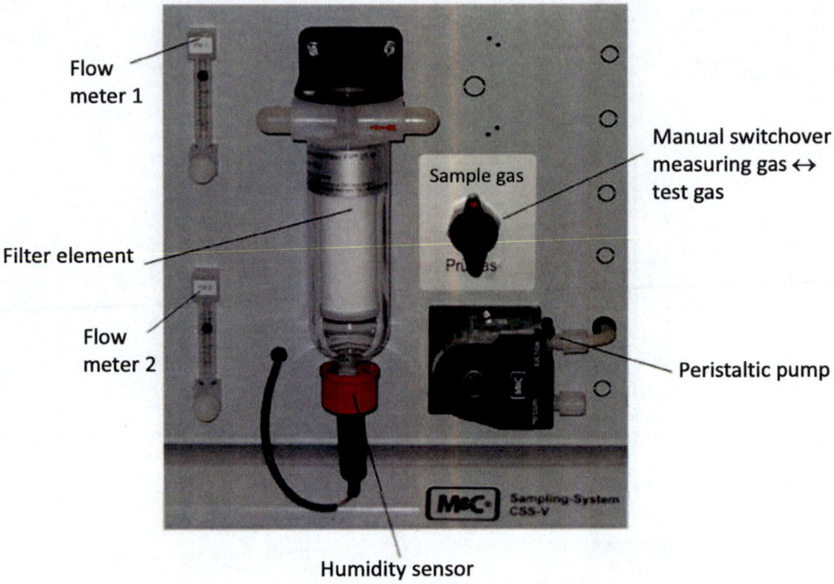

Fig. 14.49 Partial view of a sample preparation unit (*Sampling System CSS-V*) with an integrated filter setup according to Fig. 14.48. (Source: M&C TechGroup GmbH, Ratingen Germany)

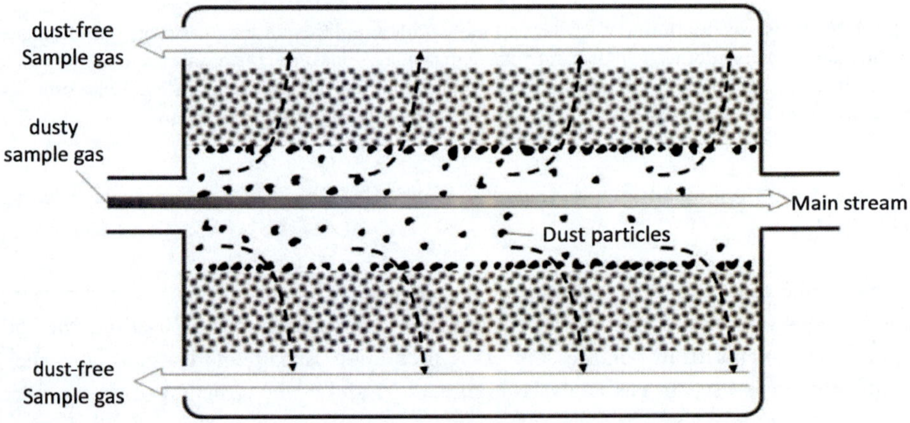

Fig. 14.50 Bypass filter to extend the service life of the filter element

Table 14.2 Common absorber materials for gas analysis

Absorbents	Adsorbed gases
Activated carbon	Oil and solvent vapours
Stuttgart Mass	Aerosols
Silica gel	Steam
Calcium hydroxide	CO_2
Soda lime	CO_2
Purafil II	SO_2, SO_3, NH_3, CS_2, H_2S
Iodine Coal	Mercury vapour (Hg)

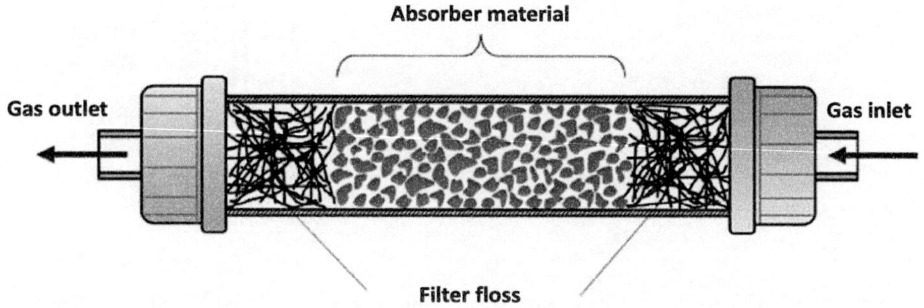

Fig. 14.51 Arrangement of a gas absorber in a filter unit. The granular absorber material is positioned by the filter floss at both ends. The gas flows through the arrangement from one side. The flow direction is not relevant

diving a mixture of calcium hydroxide $Ca(OH)_2$ and sodium hydroxide NaOH is used. The following chemical reactions take place between the soda lime and the carbon dioxide:

$$CO_2 + H_2O \rightarrow H_2CO_3 \tag{14.29}$$

$$H_2CO_3 + 2\,NaOH \rightarrow Na_2CO_3 + 2\,H_2O \tag{14.30}$$

$$Na_2CO_3 + Ca(OH)_2 \rightarrow CaCO_3 + 2\,NaOH \tag{14.31}$$

During the entire reaction, the reaction substances water and sodium hydroxide are constantly renewed. Only calcium hydroxide is consumed and converted to chemically inactive lime (calcium carbonate $CaCO_3$). Theoretically, 1 kg soda lime can therefore bind CO_2 at 20 °C ambient temperature 225 L CO_2. In practice, however, this ability to bind CO depends strongly on the temperature. In fact, the binding capacity decreases steadily with falling temperature, so that it is 1.5 °C at 50%. For this reason, manufacturers of soda lime state a much lower binding capacity. Average absorbers can absorb 10–15 L CO_2 per 100 g. The optimal granulate cross section is 2.5 mm. A pH-indicator is added to the

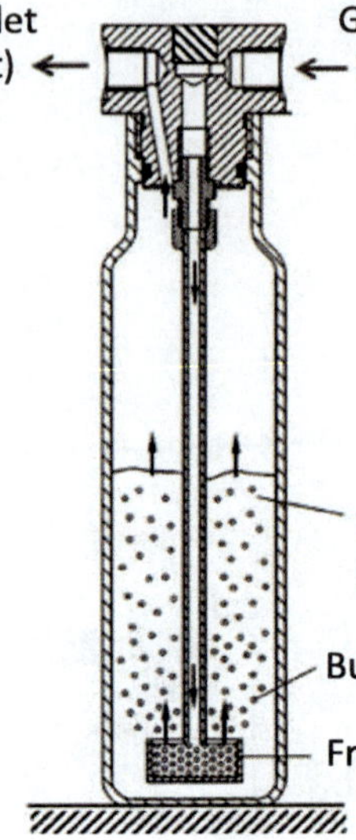

Fig. 14.52 Cross-section through a gas washing bottle/blubber vessel. (Source: M&C TechGroup GmbH, Ratingen Germany)

scrubbing lime, which changes its colour from white to violet at low pH-value, indicating that the absorber is used up.

Gas Humidification/Gas Scrubbing

In certain applications, the sample gas is also humidified in order to achieve a defined dew point or to wash out undesirable substances from the sample gas. Another application is the use of electrochemical gas sensors with a liquid electrolyte. If these gas sensors are exposed to a flow of dry sample gas, they dry out and display incorrect measured values. This drying out process is prevented by the defined humidification.

To do this, the gas is passed through a bottle filled with water (Fig. 14.52). In order to maximize the contact area between the gas and the water, the gas is passed through a frit, which divides the gas flow into many individual bubbles. This measure results in a very large contact area and thus also in good humidification or a good washing effect.

In the procedure shown in Fig. 14.52, the saturation vapour pressure is set at the corresponding water temperature T_W. At the same time, however, other gas components can also dissolve in the water. If the measuring gas is to be humidified without dissolving in the water, a different humidification variant must be selected. The previously described method of gas drying with a *Perma Pure dryer* can also be reversed by passing water through the outer part of the *Nafion tube.* The dry gas flows through the inner part and is surrounded by the outer water. Through the *Nafion wall,* a defined flow of water vapour

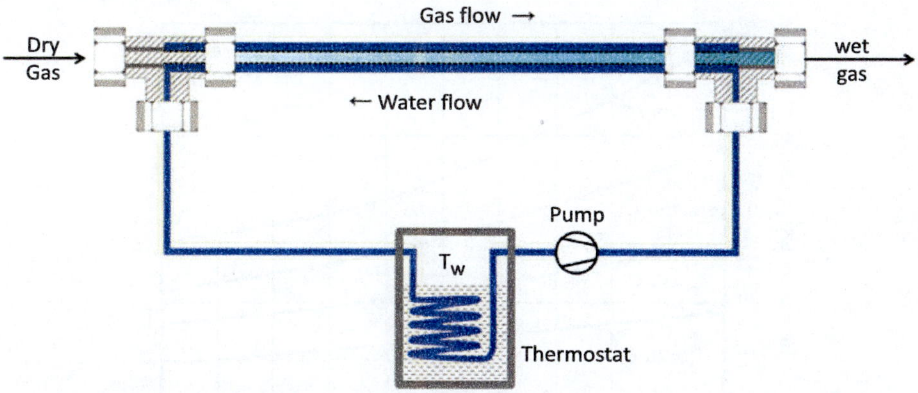

Fig. 14.53 Gas humidifier with a Perma Pure *Nafion tube*. The thermostat temperature is controlled to a constant value T_W and thus ensures a defined moisture content in the gas flow. (According to data sheet of Gasmet Technologies GmbH, Karlsruhe Germany)

then enters the dry gas and humidifies it in this way. In this context one also speaks of a *permeation distillation*.[3]

Figure 14.53 shows such an arrangement. For this purpose, the water is brought to a defined temperature in a thermostat to produce a constant moisture content in the gas flow. Figure 14.54 shows the dew point temperatures for different water temperatures and gas flows. In this example, the air to be humidified is dry (TP = − 40°C) and has a temperature of $T_G = 24°C$. The water is fed into a reservoir with a static pressure of approximately 15 mbar and is permanently circulated. The volume flow of the water is approximately 10% of the gas flow. The curves show an application with gas in overpressure. The gas flows through the *Nafion pipe* and the outer space is filled with water. In applications with gas in negative pressure, (the gas then flows through the exterior and the *Nafion tube* is filled with water) the efficiency of water transport is lower.

Pressure Control

Test gases and zero gas are usually in compressed gas cylinders with an internal pressure of up to 200 bar. In order to be able to use the gas in them at atmospheric pressure (≈1 bar), it must be expanded via a pressure reducer. Figure 14.55 shows a cross-section through such a pressure reducer. It is connected directly to the main valve of the gas cylinder via a screw connection. The main valve is only used to open and close the gas cylinder, so that the contents pressure gauge indicates the current internal pressure of the cylinder. The working pressure is set by an adjusting screw, which compresses a spring. The adjustable spring force F_F acts against the pressure force F_p which is generated by the diaphragm with the output pressure p_A (see Fig. 14.56). In a state of equilibrium, the internal valve opens only

[3]Gasmet Technologies GmbH, Karlsruhe Germany Data sheet 09/06 D 9C.

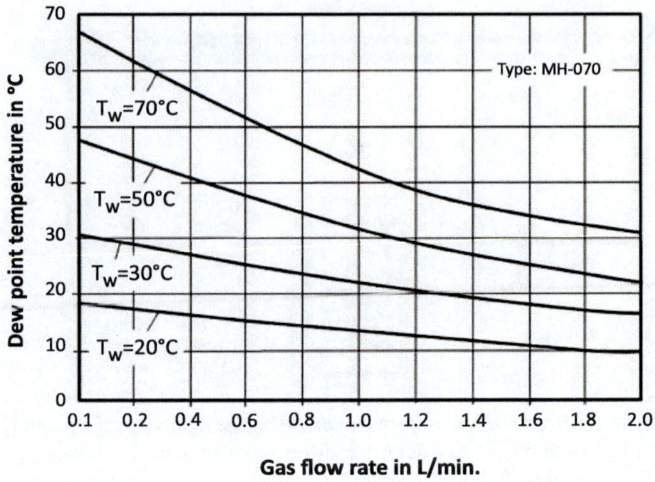

Fig. 14.54 Dew point temperatures of the gas flow at different thermostat temperatures and different gas flow rates. The length of the humidifier was 24″. (According to the data sheet of gasmet GmbH, Karlsruhe Germany)

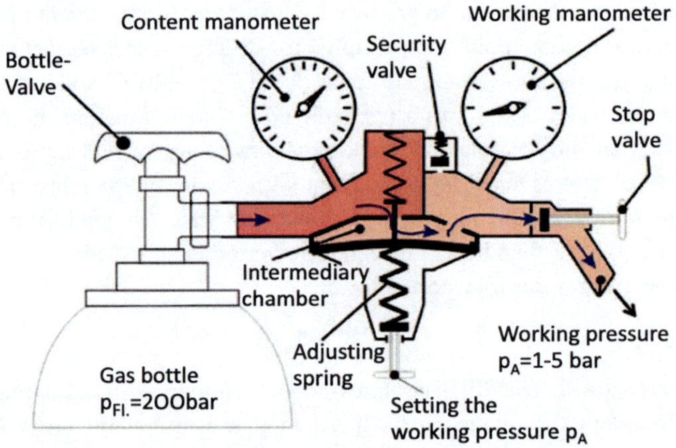

Fig. 14.55 Cross-section through a pressure reducer. (According to Wikipedia)

as far as necessary, until this condition is stable. The working pressure gauge then indicates the outlet pressure. The working pressure can be transmitted to the gas measuring system via a further shut-off valve. The volume flow is usually adjusted by a downstream needle valve and controlled by a flow meter. Depending on the design and application, these pressure reducers are made of brass, nickel-plated brass or stainless steel. The diaphragm can be made of Viton, NBR or stainless steel. For the use of toxic gases, the pressure reducer is equipped with a flushing device so that the remaining gas, inside the pressure

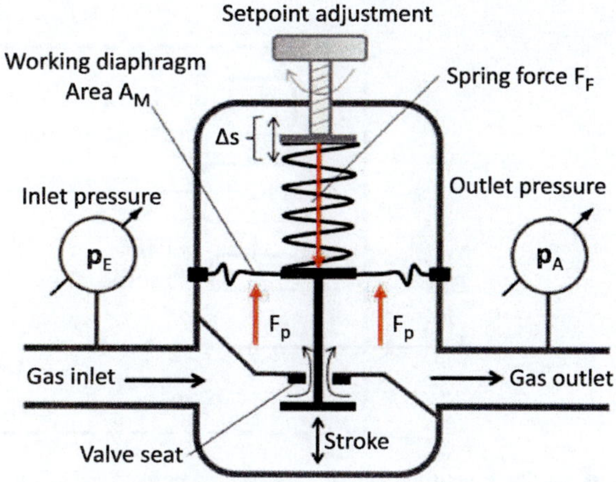

Fig. 14.56 Pressure reducer for general applications in gas application technology

Fig. 14.57 Brass dome pressure regulator for large volume flows up to 5000 m3/h. (Source: WITT-Gastechnik GmbH, Witten Germany)

reducer, does not escape into the environment when the gas cylinder is changed and possibly damage the service personnel.

Figure 14.57 shows a dome pressure regulator for large-scale plants, with a flow rate of up to 5000 m^3. These designs are also used, for example, in the food industry.

It is also possible to carry out an electronic pressure control. For this purpose, the working pressure p_1 must be measured via a pressure sensor (PICA) (see Fig. 14.58). The sensor signal is then transmitted to a control valve which opens or closes to a greater or lesser extent depending on the control action. As a result, a dynamic pressure ($p_1 > p_2$) is produced upstream of the valve which can be constantly controlled in this way. This possibility of pressure control is used in gas measuring systems where several gas analyzers are connected (Fig. 14.59).

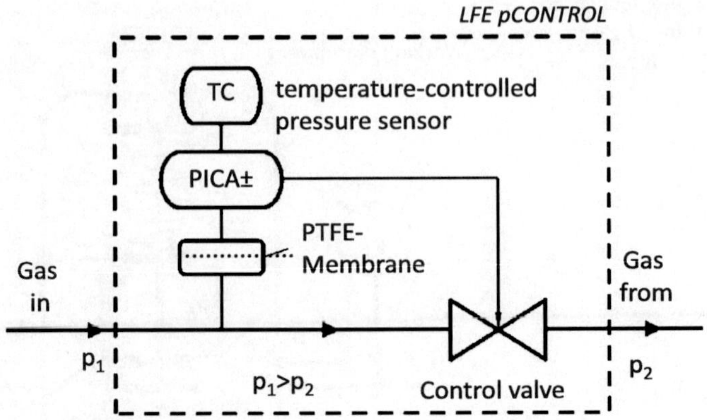

Fig. 14.58 Electronic pressure control (example: *LFE pContol*) with a pressure sensor (PICA) and a proportional control valve

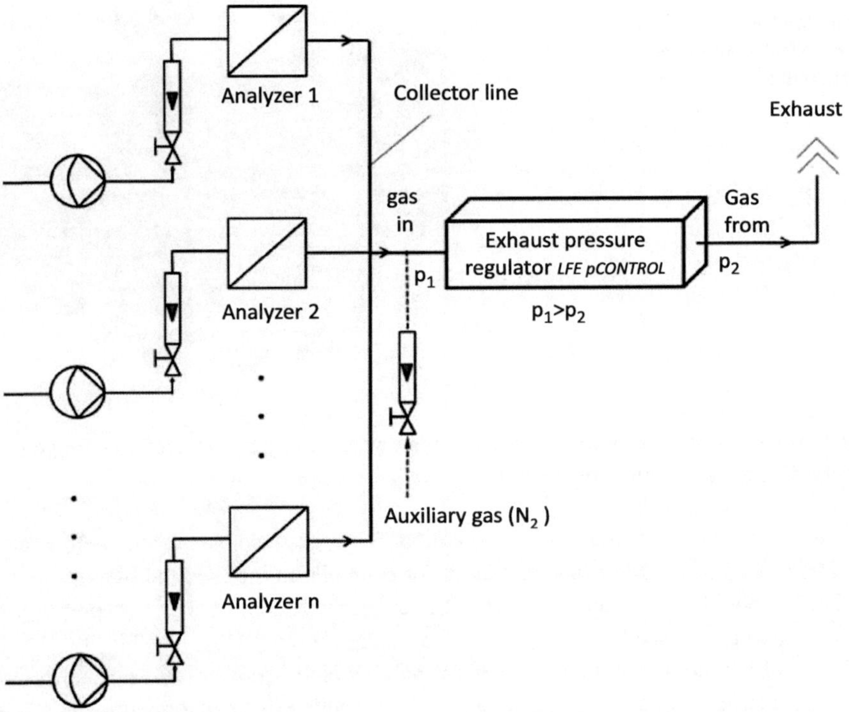

Fig. 14.59 Use of an electronic pressure regulator (*LFE pCONTRL*) in the output of several gas analysers (analysis system)

Analysis Systems

In contrast to single devices, analyzer systems usually consist of several gas analyzers with corresponding peripherals. An essential part of these peripherals is the sample gas conditioning and sampling. In addition to the actual gas measured variables (concentration values c_1, c_2, ..., c_n), other auxiliary variables such as pressure, temperature, flow rate, gas leakage, etc., are also recorded and processed in an analysis system. For this purpose, an analyzer system also contains a control unit which is necessary for the safe operation of the entire system. Figure 14.60 shows a block diagram of a typical analyzer system. The individual modules and devices are either housed in a control cabinet (Fig. 14.61) or mounted on a panel and fixed to the wall. Cabinet mounting is often found in exhaust gas analysis (emission measurement in power plants or in the automotive industry), while wall mounting is used in chemical process engineering and petrochemical/natural gas.

Gas conditioning in analyzer systems is usually also installed in completely independent modules. This has the advantage that these modules can be manufactured separately and can also be replaced if necessary. Figure 14.62 shows such a unit consisting of a sample gas cooler, dust filter, flow meter, etc. The components are mounted on a mounting plate and can be checked and replaced relatively easily during operation. The module shown in Fig. 14.63 is much more compact. In this 1/2–19 $''$ version, only the filter can be changed, while the other components are only accessible by opening the housing. Both variants shown are used for different applications and are thus adapted to the respective application.

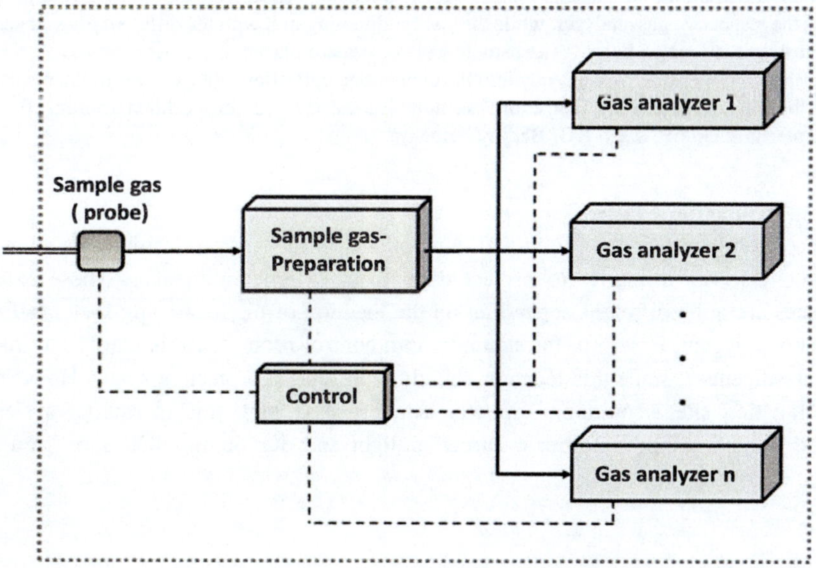

Fig. 14.60 Basic arrangement of modules and devices of a gas analysis system

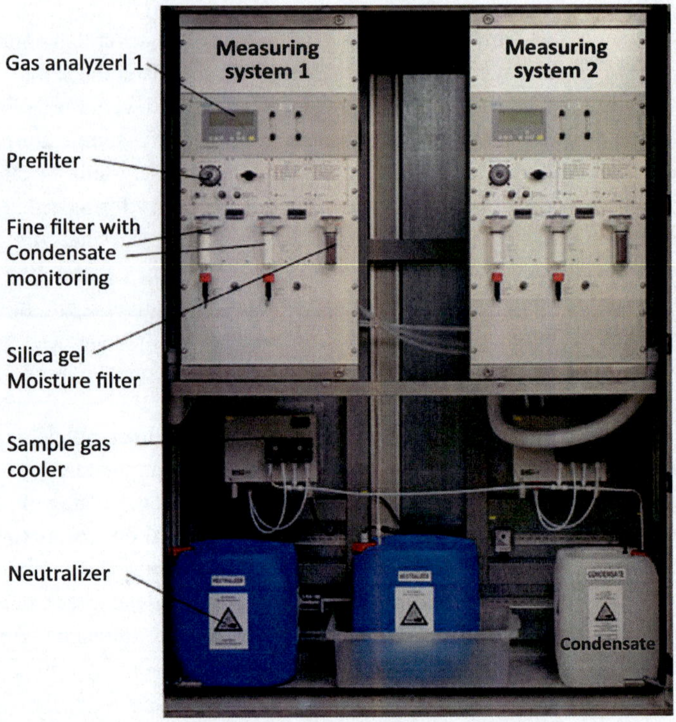

Fig. 14.61 Example of an analysis system with two identical measurement paths. The *upper part* contains the respective gas analyzer, while the gas conditioning unit with the different filter stages and the control unit is located below. The sample gas coolers are placed at the very bottom so that the condensate can flow from there directly into the condensate collection tank. The neutralizer is used to convert the condensation water. The entire assembly is located in a control cabinet. (Source: Pronova Analysentechnik GmbH & Co KG, Berlin Germany)

Housing Protection Classes

The accommodation of the individual components in an analyser cabinet, as shown in Fig. 14.64, serves primarily to protect them from external influences. These external influences are quite different depending on the location of the measuring system. If such a measuring system is set up, for example, in a control room, there is hardly any risk of external influences, since this location already represents a protective space. However, if the installation site is outdoors, directly at a process plant, this measuring system is exposed to all weather conditions. Direct sunlight and dust in summer, snow and frost

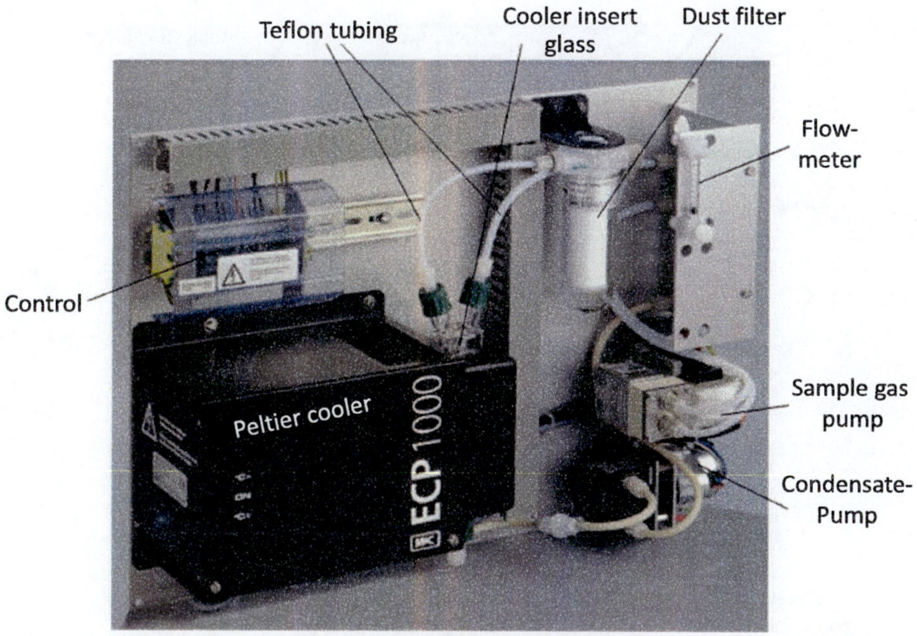

Fig. 14.62 Sample gas conditioning on a mounting plate. (Source: M&C Products GmbH, Ratingen Germany)

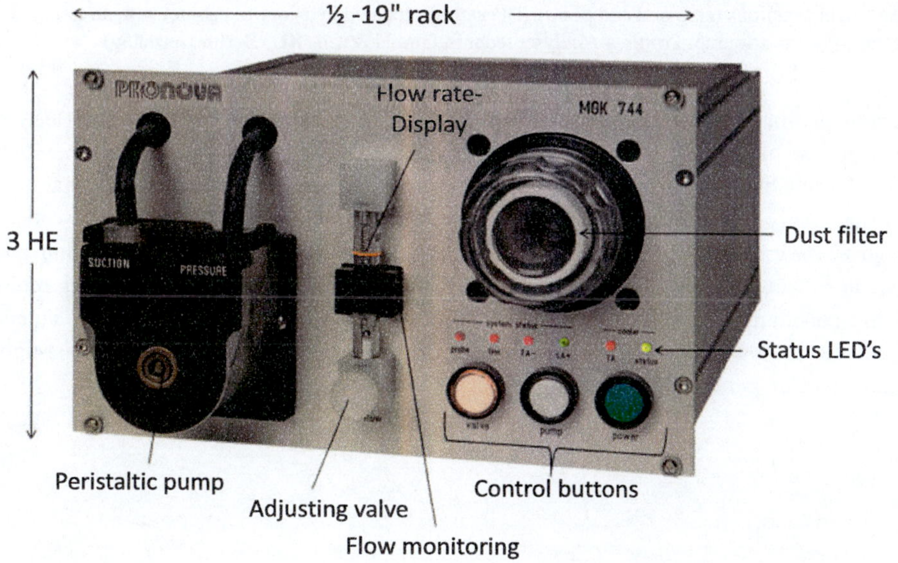

Fig. 14.63 Compact sample gas conditioning in a 1/2–19 ″ housing for installation in a rack system. (Source: Pronova Analysentechnik GmbH & Co. KG, Berlin Germany)

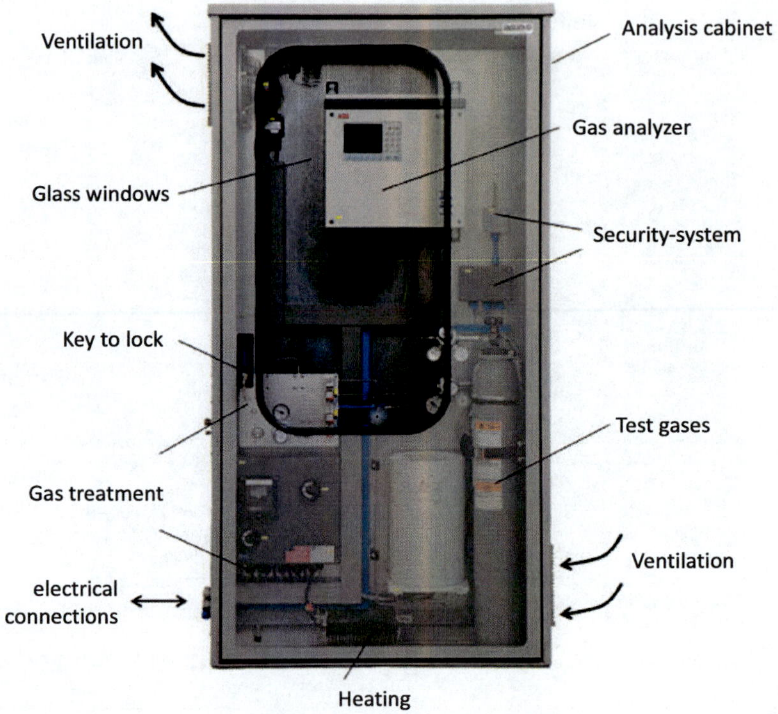

Fig. 14.64 Analysis system with a gas analyzer and an automatic test gas feed to check the instrument functions (zero and end point). The setup is located in a control cabinet with an integrated heating system. (Source: Pronova Analysentechnik GmbH & Co. KG, Berlin Germany)

periods in winter, as well as rain that can fall from all sides. These influences can lead to damage and ultimately to complete failure. The enclosure technology must therefore provide reliable protection against these external influences. In order to obtain an internationally recognised standard for this protective effect of enclosures, the so-called[4] *IP protection class* has been defined. Table 14.3 lists these degrees of protection and specifies them with regard to the possible effects. The first digit of the IP marking refers to the penetration of bodies or dust, while the second digit defines protection against water. According to this table, the designation *IP 54* means that the housing is dust-tight and cannot be penetrated by water.

[4] IP International = Protection.

Table 14.3 Specification of the different IP housing protection classes

First digit			Second digit	
	Touch protection	*Foreign body protection*	*Water protection*	
0	No protection	No protection	0	No protection
1	Touching with the back of the hand	Foreign bodies from 50 mm diameter	1	Vertically falling dripping water
2	Touching with the fingers	Foreign bodies from 12.5 mm diameter	2	Sloping dripping water (75–90 °)
3	Touching with tools	Foreign bodies from 2.5 mm diameter	3	Spray water (60–90 °)
4	Touching with a wire	Foreign bodies from 1 mm diameter	4	Splash water from all directions
5	Touching with a wire	Dust protected	5	Protection against water jets
6	Touching with a wire	Dust proof	6	Jet water from all directions
			7	Temporary immersion
			8	Permanent immersion

14.2 Exhaust Gas Analysis on Motor Vehicles by Means of PEMS

Oliver Franken

Introduction

Exhaust gas analysis on motor vehicles and combustion engines has been carried out for many decades. With the introduction of exhaust emission limits in California, USA, in the early 1960s, emission measurement became mandatory. Uniform exhaust gas regulations in the European Community (EC) first came into force in 1970 and initially limited the emission of carbon monoxide (CO) and hydrocarbons (HC). Later, limits were also added for nitrogen oxides (NO_x) and particulate matter (PM) and particle number (PN) to address air quality requirements. In addition to health effects, fuel consumption and greenhouse gas emissions such as carbon dioxide (CO_2) have become more important in recent years.

In general, a distinction can be made between distance-related limit values for passenger cars and work or load-related limit values for heavy goods vehicles (trucks) and buses. For example, emissions from passenger cars are measured while the vehicle is driving a defined drive cycle on a chassis dynamometers. Truck emissions, on the other hand, are determined by following a dynamic test cycle on engine test benches with exhaust-related components, which includes different load and speed points. Some of these engines can then be installed in different vehicle types. In addition to the above-mentioned car and truck exhaust emission limits, however, exhaust emission regulations have also been developed for other types of vehicles, machines and ships. The classification *non-road mobile machinery* (NRMM), which in this context primarily includes construction machinery and agricultural vehicles, deserves special mention.

Over the years, the pollutant limit values have been continuously reduced in order to improve air quality and to stimulate the optimization of internal combustion and the further development of exhaust aftertreatment systems. In parallel to these improvements measuring technology and measuring procedures as well as test cycles have also been improved continuously in order to simulate real operating conditions as good as possible and to be able to create suitable measures for air pollution control. In addition to type approval procedures described above further test methods were designed to ensure that the exhaust gas behavior of vehicles conforms to legal requirements and will remain so over their useful life. Measures include Conformity of Production (CoP) testing, In-service Conformity (ISC) testing, periodic exhaust emission testing and continuous on-board diagnosis. However, it turned out that these measurements, which are mainly carried out in the laboratory, do not provide an accurate picture of the actual emissions produced by road traffic for various reasons. To close this gap regulators were looking for ways and means to measure real emissions of motor vehicles during normal road operation. This resulted in the requirements for *Portable Emission Measurement Systems* (PEMS). Figure 14.65 shows the use of PEMS measurement technology in different applications.

Fig. 14.65 Portable Emission Measurement Systems (PEMS) in different applications

Problem Definition - Discrepancy Between Air Quality Levels and Emission Targets
Although exhaust gas limits have been continuously tightened over the past decade, road traffic remains the largest source of NO_x and CO emissions. In 2008, road traffic contributed to total emissions of these pollutants in the European Community with 41% and 34% respectively (EEA 2010). In particular, air pollution in cities remained at high levels, so that between 16% and 26% of the urban population in the EU is exposed to higher concentrations of nitrogen dioxide (NO_2) and PM10 (particulate matter of 10 μ m or less) than set by the relevant air quality standards (EEA 2009). Figure 14.66 shows the average annual NO_2 limit value exceedances in the European Community in 2010. The limit values for PM10 and PM2.5 are also regularly exceeded in many cities in the European Community.

It was therefore necessary to find out what caused the persistently high levels of air pollution from road traffic. After the US *Environmental Protection Agency* (US EPA) had already proven at the end of the 1990s that several commercial vehicle manufacturers in the USA were switching between consumption-optimized and emission-optimized engine management programs depending on the application and thus exceeding binding emission

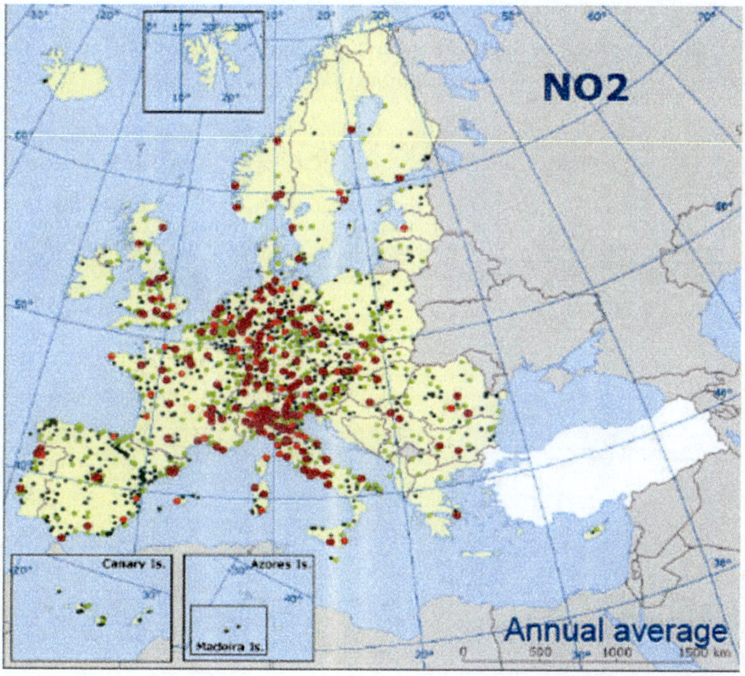

Fig. 14.66 Average annual NO_2 limit value exceedance in the EC in 2010. (According to Cortvriend 2014, p. 3)

limits, the European Commission's Joint Research Center examined the on-road emissions of twelve passenger cars between 2007 and 2010.

Both, vehicles with gasoline engines and with diesel engines exceeded Euro 3–5 standards were examined. The tests were carried out on four different test routes, which were characterized by the following features: country road, motorway, urban traffic and uphill/downhill routes. The average NO_x emissions of all diesel engines tested over the entire test routes were 0.93 ± 0.39 g/km and the average NO_x emissions of the Euro 5 diesel engines tested were 0.62 ± 0.19 g/km. These results indicate that the NO_x emissions of diesel vehicles clearly exceed the Euro 3–5 limits. On average factors of 4–7 were measured over the entire test routes and even a factor of up to 14 for individual average windows were identified compared to the NEDC cycle. It was concluded that the increasing tightening of the limits has not led to a corresponding reduction in on-road NO_x emissions from diesel passenger cars. In comparison, the NO_x emissions of gasoline vehicles as well as the CO and THC emissions of diesel and gasoline vehicles usually remain within the Euro 3–5 levels (Weiss et al. 2011). This is illustrated by an example in Fig. 14.67.

The discrepancy between the NO_2 concentrations determined on a chassis dynamometer in the exhaust laboratory and the actual road traffic emissions has several causes. On the one hand, the currently applied *New European Drive Cycle* (NEDC) is not sufficiently representative to simulate real driving dynamics in real traffic. Strong accelerations, emissions during cold starts and speeds above 120 km/h are not taken into account. In addition, the exhaust emission behavior of motor vehicles was optimized for the NEDC cycle in order to keep pace with the continuously tightening limits.

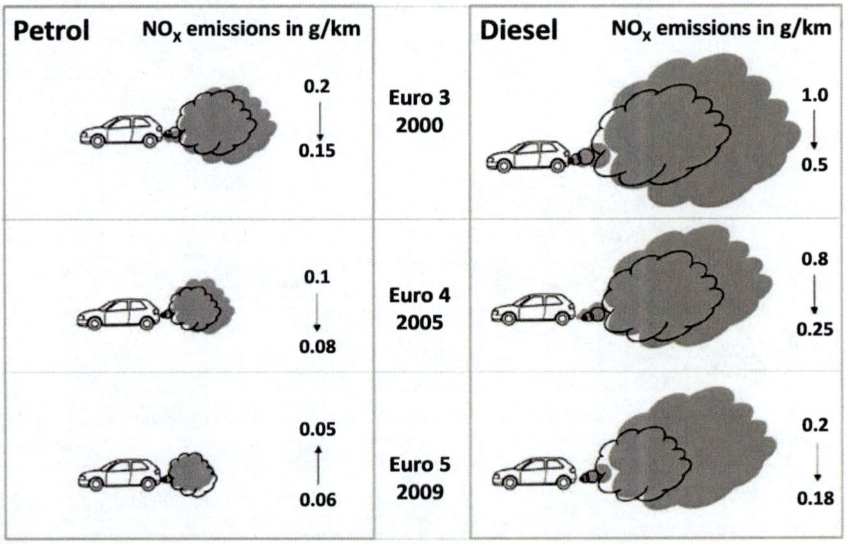

Fig. 14.67 Comparison of limit values and emissions of spark ignition and compression ignition engines between Euro 3 and 5. (According to Cortvriend 2014, p. 10)

Initial, as yet unpublished, investigation results indicate that under certain circumstances particle number emissions from direct-injection spark ignition engines may also exceed the current limit values. Further investigations will take place in this regard and parallel to this, requirements for mobile measuring technology will be defined in order to advance the development of devices. It is therefore possible that by 2017 a binding conformity factor for on-road emissions will also be established for the particle number.

History of Portable Emission Measurement Systems (PEMS) development

As PEMS systems are still a relatively new method for exhaust gas analysis, definitions of PEMS systems are not yet fully established, so that many equipment manufacturers use and claim this term. Mobile or portable emission measurement devices were already commercially available before uniform PEMS system specifications were established in legislation. These devices often originate from the diagnostic and workshop environment and were primarily used for quick verification and diagnostic purposes. Although a further development towards longer service life is also discernible here, the utilized measurement technology does not, however, meet the now specified requirements of the respective legislations. Due to its lower price, however, it represents an alternative for emission measurement in developing countries as well or for fleet screening on a larger scale. Similar applications and restrictions also apply to another subgroup of measuring instruments, which largely consists of on-board sensor technology. These sensors are also installed in production vehicles to provide continuous feedback on the functioning of the exhaust aftertreatment system. Oxygen, nitrogen oxide and PM sensors are primarily used redundantly in aftertreatment systems installed on the vehicle. By using different exhaust gas sampling points in the exhaust aftertreatment system and at the exhaust tailpipe, conclusions can be drawn about the efficiency of these exhaust aftertreatment systems.

The most important group of PEMS devices is characterized by the fact that they meet the regulatory requirements in the USA and Europe. These are complex systems that usually have to meet the same or similar requirements as laboratory measurement technology in order to deliver reliable results. In addition to the integration of vehicle and environmental information, these devices are also equipped with software that allows the analysis and evaluation of the measurement results in accordance with the respective legal requirements. The explanations in this article refer exclusively to these devices.

In summary, PEMS system development has taken place in parallel with legislative programs on *In-Use Compliance* legislation in the USA and *In-Service Conformity* legislation in Europe. The first PEMS systems that met the requirements became commercially available in 2002 and enabled the measurement of gaseous pollutants such as CO, NO, NO_2 and THC. These systems, which were primarily used on and in commercial vehicles, were then expanded in 2005 to include PM measurement technology and proportional dilution of the exhaust gas volume flow.

A further milestone in PEMS development is the trend towards smaller, lighter systems, which should also enable their use in passenger cars. Here, advanced systems have been commercially available since 2014. They are characterized in particular by a compact, highly integrated design and, in addition to installation in the vehicle, also allow time-saving installation on a trailer coupling on the outside of the vehicle.

International Legislation

The legal requirements for PEMS test procedures and for PEMS measuring instruments are described in the following regulations and rules:

- Regulation (EU) No 715/200
- Regulation (EU) No 595/2009
- Regulation (EU) No 582/2011
- Regulation No 49 of the Economic Commission for Europe of the United Nations (UN/ECE)
- USA, Code of Federal Regulations (CFR) 1065, subpart J.

This list concerns primarily Europe and the United States of America. Regulations are also being prepared in South Korea and the People's Republic of China. It is also expected that Japan will include a PEMS element in its legislation.

In addition to various country specific requirements additional laws and regulations will be created for the *Non-road Mobile Machinery* (NRMM) applications.

Measurement Parameters and Requirements for PEMS Systems

In mobile exhaust gas measurement different parameters are recorded depending on the application. For commercial vehicles, the following parameters and values are prescribed: CO_2, CO, NO_x or NO+ NO_2, THC, CH_4 if necessary, and particulate mass. In addition to the pollutants, ambient temperature, pressure, relative humidity and position data are recorded. Furthermore, the exhaust mass flow rate is measured and information from the vehicle's engine control unit (ECU) is collected via on board diagnostic (OBD) interface. The time synchronization of these data is of great importance, since even small time shifts lead to errors in the calculation of emission values. The requirements for recording these measured values are identical in the United States of America, in Europe and in the People's Republic of China, so that there are no major differences in the PEMS hardware legislation existing today. Deviations and in some cases even significant differences lie in the further processing of the collected data and above all in the evaluation and assessment.

In contrast to commercial vehicles, different requirements are emerging for passenger cars. These requirements have been developed in recent years by the European Commission's *Real Driving Emissions-Light Duty Vehicles* (RDE-LDV) working group. A final specification is not yet available at the time of writing this text. However, it is expected that the measurement of THC will be abandoned in order to respect safety requirements during real driving. Furthermore, there is great interest in replacing the measurement of particle mass (PM) by particle number (PN). Since 2014, an evaluation of commercially available devices has been carried out under laboratory conditions. These laboratory tests give reason to test these measuring instruments, some of which are still in pre-series status, under real conditions in driving operation. If a sufficiently good correlation with laboratory procedures can be achieved here, the introduction of a limit value for particle number in driving operation can be expected. Europe is the pioneer here. Whether this test method will be adopted in other countries remains to be seen.

The measuring procedures are also laid down in the above-mentioned laws and regulations and correspond in many respects to the procedures used in exhaust gas laboratories. For example, a non-dispersive infrared (NDIR) analyzer is used for measuring CO_2 and CO. Unlike in exhaust gas laboratories, however, PEMS applications do not use the low CO measuring range in order to save weight and reduce the space required for PEMS installations in the vehicle.

Two measuring methods are permitted for measuring nitrogen oxides: The chemoluminescence detector (CLD) for the selective measurement of NO in combination with a converter which converts NO_2 into NO and is thus able to measure NO_x. Alternatively, *non-dispersive ultraviolet* (NDUV) analyzers can be used, which measure NO and NO_2 simultaneously and add them to NO_x. An example for such analyzer is shown in Fig. 14.68.

NDUV analyzers are used by the majority of PEMS manufacturers because they want to avoid the use of operating gases and they see advantages in direct NO_2 measurements. A special feature is the simultaneous use of two radiation sources which, as shown in Fig. 14.69, emit UV radiation at different wavelengths. An electrodeless discharge lamp is used for NO measurement. A UV-LED also emits in the ultraviolet spectrum, but at a much higher wavelength around 400 nm.

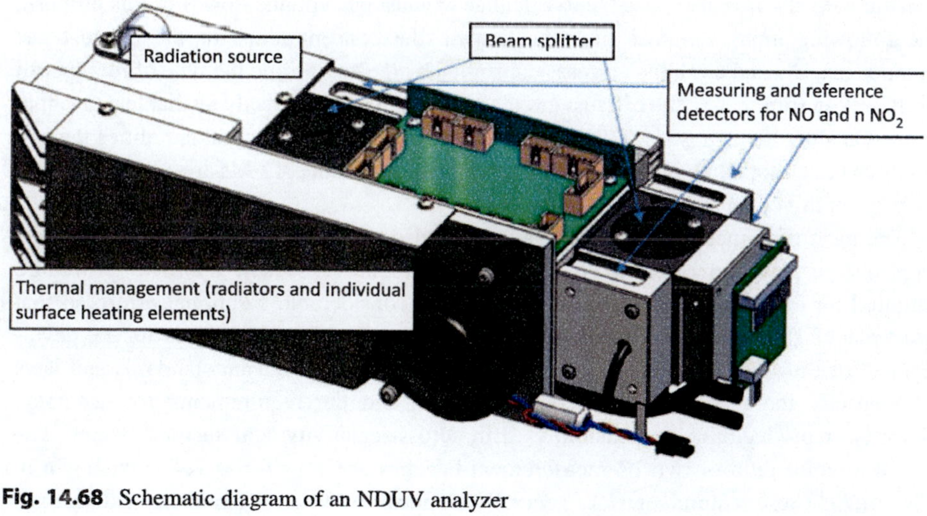

Fig. 14.68 Schematic diagram of an NDUV analyzer

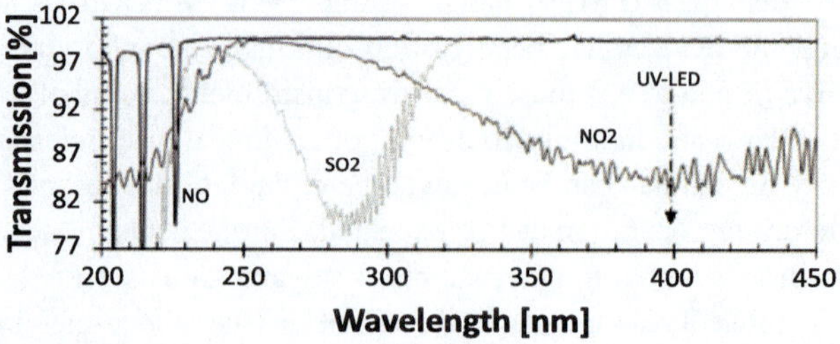

Fig. 14.69 UV transmission spectrum for NO, NO_2 and SO_2

A flame ionization detector (FID) is required to measure total hydrocarbons (THC). Here, the unburned or partially burned carbon fractions of the exhaust gas are completely burned in a hydrogen/helium flame. The ions produced during combustion are proportional to the organically bound carbon content and are measured as current between two electrodes (Wiebrecht 2005). The FID is heated and requires an operating gas (e.g. H_2 / He). For safety and practicability reasons, it is not intended for use in passenger cars until further notice.

The exhaust gas volume flow is measured by means of an exhaust flow meter (EFM). The predominant measuring method is differential pressure measurement in a Pitot tube, according to Henri Pitot. This method enables the direct measurement of exhaust flow independent from information provided by the vehicle's engine control unit. The disadvantage, however, is that a gas-tight piping from the tailpipe to the EFM, which is mounted on the outside of the vehicle, must be created. Depending on the vehicle's exhaust system, a suitable set-up requires an installation effort that should not be underestimated.

Alternatively, it is also possible to calculate exhaust gas volume flow. For this purpose, the following input variables must be known: the concentrations of the exhaust gas components CO and CO_2, the air mass consumed by the engine and the type of fuel (petrol or diesel). In some cases, the exhaust gas mass flow rate is also directly available as a signal at the On Board Diagnostics (OBD) interface. This calculation method simplifies the test setup, but does not allow independent measurement. In practice, EFMs are not always used during test drives for development purposes.

The measuring methods described above can also be replaced by alternative methods. For this purpose, however, a comprehensive alternative system approval procedure required for each measurement. This represents a considerable additional effort, so that most manufacturers of measuring instruments rely on established methods for the development of PEMS. In addition to the measurement methods, relevant standards and laws also specify the respective measurement ranges and the requirements for accuracy, linearity, reproducibility, repeatability, drift, cross-sensitivity and response time. The signal-to-noise ratio of two IR measurement benches at CH_4, 0–100 vol.% is shown in Fig. 14.70. These requirements are generally identical to those for laboratory exhaust gas measurement technology.

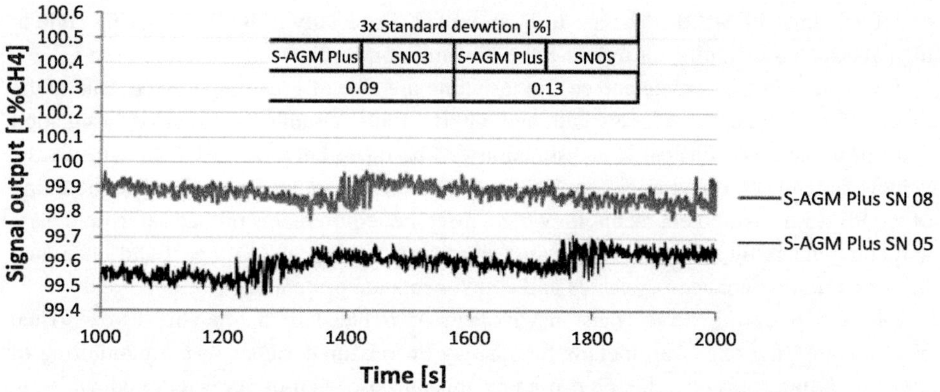

Fig. 14.70 Exemplary signal-to-noise ratio (three-fold standard deviation) of two NDIR measuring benches (type S-AGM Plus) (manufacturer Sensors Europe GmbH Erkrath) at CH_4, 0–100 vol.%

A detailed description of the procedures and performance characteristics is not given here, as these topics have already been dealt with in other chapters. Requirements and specifications for laboratory grade measurement instruments now have to be met even under severe operating conditions while driving. This represents a major challenge for manufacturers of measurement equipment. The extended ambient temperature range T_U from −7 to 40 °C, vibrations and shocks caused by uneven road surfaces and altitudes of up to 2500 meters above sea level have a particular influence on measurement technology. Any effects that occur are compensated by heated enclosures as well as by temperature and pressure compensation. Adequate shock damping is an integral part in the design of the analyzers and moving parts in the measuring benches are eliminated in order to avoid or minimize position and motion dependency effects.

In addition to the legislative requirements for PEMS measurement technology, there are a number of other criteria to be respected. Meeting these criteria is essential for the successful use of PEMS measurement technology in vehicles. Additional criteria are:

1. Safety and occupational health

Within the scope of the risk analysis, several potential hazards and the corresponding countermeasures have been identified. Particularly noteworthy are weight, fastening, exhaust gas sampling and fuel gas supply in the vehicle interior. One of the main development goals is to keep the system weight as low as possible. On the one hand, low weight in combination with modular design should also allow installation by one person, so that no component may weight more than 20 kg. This results in savings in personnel deployment and older workers or women are not excluded from installing and operating PEMS. Since fuel consumption and emission values of the vehicles also increase with additional weight and increased driving resistance, this influence should also be kept as low as possible. A further limitation is partly low payloads as well as low support loads of the trailer coupling on which some PEMS devices are to be mounted. Especially a dynamic driving style at high speeds has a significant influence on the emission level.

Installation in the vehicle and secure fastening are also of great importance. Due to the many different types of vehicles with and without trailer coupling, a versatile installation concept is required. Basically, a distinction can be made between installation inside the vehicle and outside on a trailer coupling for cars or on the trailer for trucks. The mounting of the PEMS measurement technology must meet the requirements of load safety and parts and components must be secured against slipping or falling off. Compact and integrated devices with few connecting cables and components are preferable.

Distraction of the driver must be avoided or reduced to a tolerable level. Visual distraction on the one hand include the display of measured values or the monitoring of the system status. If only a laptop is used for this purpose, a co-driver is required, which in turn has a negative effect on the vehicle test mass. Some PEMS measurement devices offer the possibility of reduced status monitoring and automated test sequences, so that no visual stimuli cause distraction. Ambient distractions on the other hand by temperature and noise inside the vehicle must be taken into account. For example, partially opened windows or tailgates can impair the driver's well-being.

There are two other potential sources of danger when PEMS is installed inside the vehicle: The exhaust gas must be conducted into the vehicle interior to the measuring device and then out of the vehicle interior. Visual checks of the inlets and outlets should help to detect blocked or kinked lines. In addition, some PEMS measuring devices allow the flow in the device to be checked, which provides information on whether the target pressures have been reached and whether pumps are functioning properly. Especially for commercial vehicles, the use of an FID for measuring the total hydrocarbons is also mandatory. FID use requires an operating gas in the form of a helium-hydrogen mixture. In this case, rapid pressure changes in the gas cylinder are usually detected and displayed. However, it is recommended to carry a gas warning sensor for CO_2, NO_x and, if necessary, H while driving.

2. Dimensions and space requirements

Compact design make some PEMS devices ideal for use in RDE-LDV programs. It is a result of an optimization process lasting several years. Especially when no trailer hitch is available for installation outside the vehicle and when small vehicles or sports cars are involved, it is essential to use such an optimized system. Installation in the vehicle should be possible, quick and without major conversion measures, such as removing seats.

3. Practicability

The power consumption of the PEMS device is also an important development and selection criterion, since additional batteries for power supply contribute to the overall weight. The use of small and lightweight LiFePo batteries in combination with an optimized instrument design allows low power consumption has led to significant space and weight reductions in recent PEMS developments.

Another important aspect of practicability is the set-up time, which can be divided into test preparation time of the measuring system and mechanical installation time. In order to guarantee the accuracy requirements of the measuring system, a warm-up time of up to 1 h is necessary. This time is usually used to select the test settings. This includes the selection of the applicable legislation with the corresponding limit values as well as the input of the vehicle-specific parameters. This entry usually needs to be made only once and can be saved for further tests. After warm up, the measuring instrument is calibrated using calibration gases. Almost all PEMS manufacturers now offer automated warm-up and calibration routines, so that measurement readiness is reached at a pre-determined time.

The mechanical installation includes the safe placement of the PEMS system in or on the vehicle and the connection of the vehicle's exhaust system to the measuring device. Due to the lack of standardized connecting pieces, single or multi-flow design, and left, right or double-sided concepts, there is a wide range of different design forms that make the adaptation of the exhaust gas combination a challenge. However, with suitable installation material and preparation or knowledge of the vehicle, great time savings can be achieved here, too. If the aids described above are used, the set-up time can be reduced to approximately 15–30 min (Nagel et al. 2014).

Outlook

The introduction of PEMS about 10 years ago is a very recent development that has only just begun. It poses great challenges for the automotive industry as internal processes and the development strategies for the powertrain have to be reconsidered or redefined. The looming expansion to passenger cars and later to NRMM, as well as a possible inclusion of particle mass and particle number and the resulting modified requirements will accompany both the measurement instrument manufacturers and the automotive industry for the next 10–15 years. Measurement technology, evaluation software, route requirements, emission factors and limit values as well as the definition of environmental and test conditions will continue to change continuously. An example of a summary test report within the framework of RDE-LDV is shown in Fig. 14.71.

In addition to checking compliance with exhaust emission limits in real driving conditions and the resulting improvement in air quality, there are, however, other PEMS applications. For example, conclusions and forecasts on the emission levels of specific geographical zones can be drawn from the measurement data obtained from individual vehicles or specific vehicle fleets. Figure 14.72 shows the nitrogen oxide emissions of a commercial vehicle when crossing the Alps over the San Bernadino Pass. The nitrogen oxide load is proportional to the size of the points.

If this information is combined with findings from traffic censuses and traffic observations, environmental simulations can be generated from which recommendations for traffic guidance, driving behavior and vehicle selection can be derived.

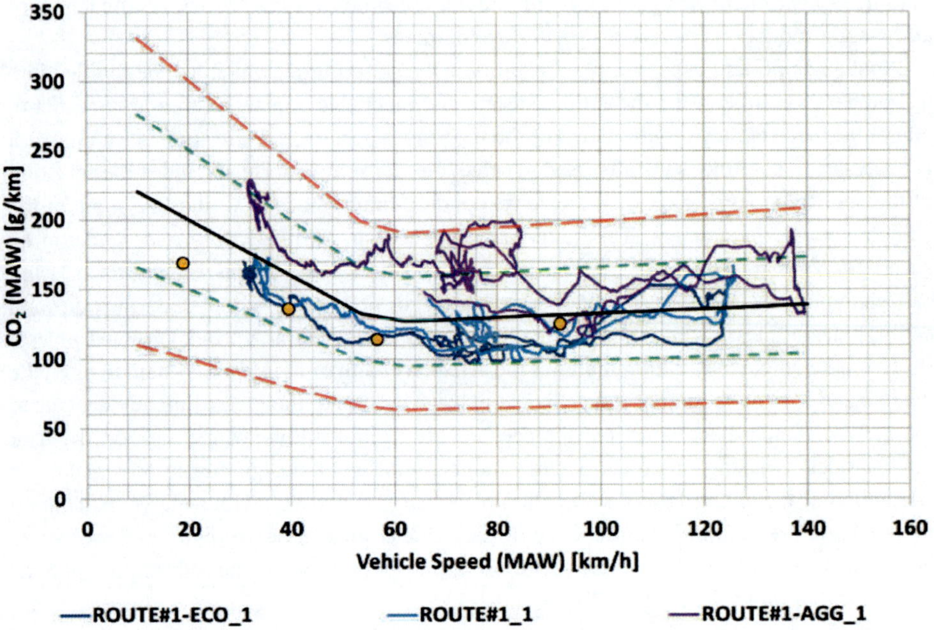

Fig. 14.71 Test evaluation of three measurement runs using EMROAD

Fig. 14.72 Nitrogen oxide emission of a commercial vehicle during a crossing of the Alps on the San Bernadino Pass on August, 5 2003

A further use results from the combination of Portable Activity Monitoring Systems (PAMS) and PEMS. Low-cost PAMS record geostationary and OBD data from many vehicles. This allows representative snapshots of traffic and traffic behavior to be collected from larger vehicle fleets or geographical regions, which can then be backed up with PEMS information. This so-called emission mapping as a simulation method can enable the improved design of environmental zones in actual problem areas.

The increasing limitation of individual traffic through environmental zones, driving bans and car-free inner cities is increasingly restricting the mobility of citizens. Often these measures were introduced as partly exaggerated reactions to poor air quality without considering alternatives such as modified traffic routing or optimized traffic flow. It is therefore hoped that the use of PEMS can demonstrate that cars in actual driving conditions comply with the prescribed limit values. If the obtained measured values provide information for optimized traffic planning by means of simulation and emission mapping, it will be possible to continuously enable individual traffic without traffic-related impairments of air quality and their negative health effects on citizens.

14.3 Continuous Emission Monitoring of Special Compounds by Long-Term Sampling

Jürgen Reinmann

In general, emissions of pollutants such as dust and harmful gases are monitored with analyzers that can measure the corresponding concentrations quasi in real time. For dust and most gases this can also be realized very well with today's proven measuring methods, although it requires very good experience and measurement technology already for the emission measurement of gaseous substances such as HCl or HF. The limit values of these pollutants are in the ppm range or mg/Nm ^{3}. For example, for the monitoring of mercury emissions the limit values are already in the ppb range or μ g/Nm ^{3}. This places very high demands on the applied measurement technology, installation and operation of such Hg analyzers.

There are other pollutants, such as dioxins, furans, PCB's as well as PAH's and heavy metals which cannot be measured in real time with the help of an analyzer. Therefore, the emissions of these substances are generally recorded by manual sampling via 30 min up to a few hours. With annual plant operating times of more than 8000 h such manual sampling represents less than 0.1% of the annual emissions of these substances. Repeated, automated long-term sampling over several days to weeks provides much more information about the actual annual emissions. For the recording of Hg emissions, long-term sampling is one way of recording the annual emission load in the lower concentration range of <5 μ ^{3} g/Nm more easily and also more accurately compared to Hg analyzers.

Dioxins/Furans, PCBs, POPs and PAHs

The monitoring of dioxin and furan emissions from incineration plants is a topic that has been the subject of worldwide discussion for years. General practice to date has been to determine these emissions by extracting a part of the flue gas and collecting the dioxins on an adsorber and then analyzing the adsorber extract in the laboratory. This is usually done by taking 1–3 samples per year for several hours (4–8 h, according to EN 1948–1 or *US EPA method 23A*). This type of monitoring only leads to a low acceptance by the population. A sampling of 8 h at 8000 operating hours per year represents just 0.1% of the operating time. It is generally agreed that continuous dioxin monitoring would be necessary for optimal monitoring of incinerators. So-called online dioxin measurement devices, which can output a dioxin/furan value in ng TEQ/Nm ^{3} almost in real time, will, however, not be available as instruments suitable for industrial use for the foreseeable future. An alternative to close this gap between desire and reality is continuous monitoring by means of long-term sampling. In the meantime, several long-term sampling systems for dioxins/furans and PCBs have been developed.

In Belgium and France and in some Italian regions, the monitoring of dioxin emissions, for example, from waste incineration plants, hazardous waste incinerators, cementkilns, etc., by means of long-term sampling has been made mandatory by law or by implementation in the operating licenses. In several other countries these systems have been installed due to requirements of local authorities, so that today approximately 450 installed systems can be expected worldwide. Long-term sampling is therefore a recognized method for recording emissions of these organic substances, which to date cannot be measured using any online measuring procedure.

Due to the broad application and the increased interest within the EU, a technical specification for long-term sampling of dioxins/furans and dioxin-like PCBs has been developed in recent years by the WG1 of TC264 as CEN TS/EN 1948–5. This technical specification was published in April 2015 and is therefore the first standard worldwide for continuous monitoring of dioxin emissions.

One of the long-term sampling systems available on the market is the AMESA® D System. With this system, a suitability test was carried out by TÜV-Rheinland in 1996/1997 according to the EU notified minimum requirements for long-term sampling systems and announced in the joint ministerial gazette. In 2013, the system was certified in the second generation by TÜV-Rheinland according to the QAL1 requirements of EN 14181 and announced on April 1, 2014 in the Federal Gazette by the Federal Ministry of Justice and Consumer Protection (Bundesanzeiger vom Bundesministerium der Justiz und für Verbraucherschutz) as suitable for long-term sampling of dioxins and furans. The system has been installed in approximately 300 applications since 1996. The following article focuses on the measurement data and experiences gained with this system.

Measurement Method

With *AMESA® D* a continuous sample is taken over a period of time between 6 h and 6 weeks (Fig. 14.73). This sampling is done by isokinetically extracting a partial stream from the flue gas. Dioxins and furans are adsorbed in a cartridge filled with XAD-II adsorber resin including a dust filter (Fig. 14.74). By constantly repeating this measuring cycle, a complete emission monitoring with regard to dioxins and furans is ensured. After changing the cartridge, the sample is analysed in the laboratory to determine the dioxin/furan concentration in the flue gas. The measured value corresponds to the average value of dioxin/furan emissions in the flue gas over the measured sampling period.

The possible measuring range for dioxin and furan is 0.0001–10 ng TE/m ^{3}. In addition, *AMESA® D* measures main flow velocity, partial gas volume flow, flue gas temperature as well as cartridge temperature and continuously records the measured values as 1/2-hour average values (Fig. 14.75).

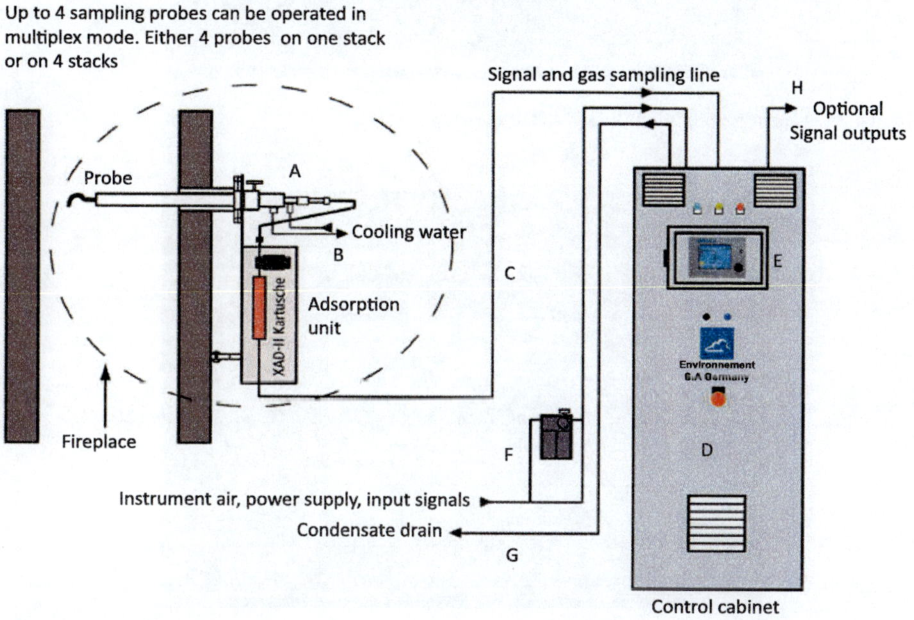

Fig. 14.73 Schematic structure *AMESA® D*. (**a**) Cooled (<70° C) sampling probe for isokinetic sampling of the partial gas flow. (**b**) Sample gas and condensate are sucked through the cartridge filled with adsorber resin (quartz cotton wool filter as pre-filter). (**c**) *Sample* gas and condensate are sucked through the sample gas line to the control cabinet. (**d**) Control cabinet with condensate separation by cooling (<5° C) and infinitely variable regulation of the isokinetic sampling. (**e**) Operation of the *AMESA® D* via central computer with menu-driven software. Data input for system-specific parameter setting and operation via jog-dial and LCD screen. Determination of emission values using external memory card and analysis results. (**f**) Compressed air, supply voltage and signal input lines. (**g**) Condensate and drain lines. (**h**) Signal output line (optional). (**i**) Cooling water connection probe (at $T_{\text{Flue gas}}$ > 70°C)

Fig. 14.74 Cartridge box including XAD-II cartridge

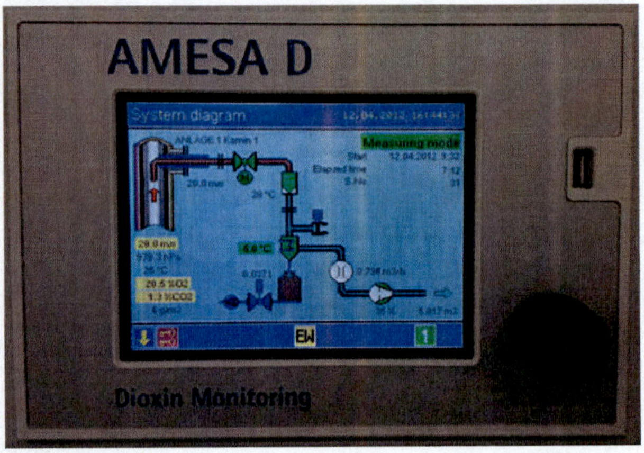

Fig. 14.75 AMESA® D control

Besides dioxins and furans, other POPs (persistent organic pollutents) can be collected and analysed (in combination with PCDF/D analysis)

- PBDF/D, PBCDF/D,
- PCB/PCBz analysis (sum of Tri- to DecaCB, 6 DIN- 12 + WHO congeners; Penta- and HexaCBz)

- PCPh analysis (10 tri- to pentachlorinated congeners)
- PAH analysis (21 PAH components).

Special Features

- Use of the known adsorption process (Funke and Linnemann 1994)
- Continuous automatic measurement (Fig. 14.76)

- Fully automatic registration and taking of a composite sample over a period of up to 6 weeks
- Ensuring a complete monitoring of emissions of dioxins, furans and other POPs (persistent organic pollutents) such as PCBs.
- Depending on the sampling period, information on, for example, the weekly or monthly mean value of the dioxin/furan emission by analysis of the composite sample (XAD-II cartridge) in the laboratory
- Additional information on flue gas velocity, partial gas volume flow, oxygen content, flue gas temperature and cartridge temperature as 1/2 h average value
- No collection of condensate and no associated handling problems when monitoring over long periods of time
- Wide measuring range of dioxin/furan concentration measurement, between 0.0001 and 10 ng/m^3 (as TE according to NATO CCMS model) The measuring range of 0–0.5 ng/m^3 is certified.

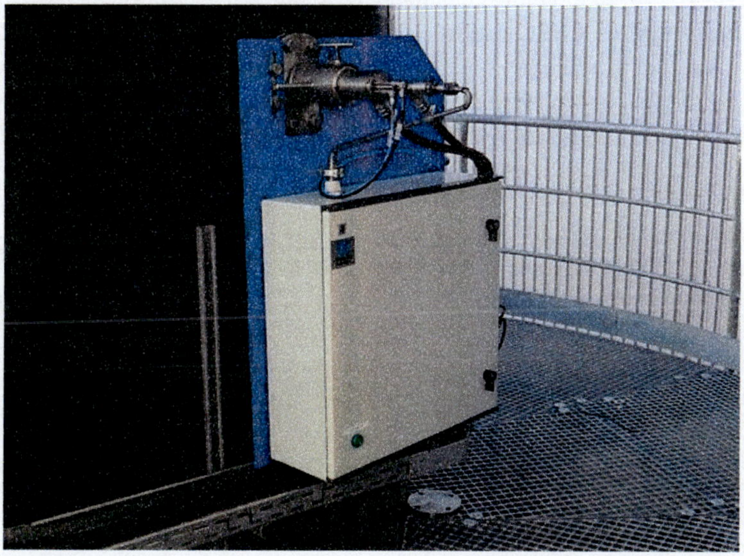

Fig. 14.76 Sampling probe with cartridge box mounted on the stack

All these features offer the following advantages for the plant operator:

- Reduced operating costs through minimization of additive additions
- Cost reduction through optimisation of the built-in dioxin reduction stages
- Minimization of operating and maintenance costs by monitoring the reduction of the furnace temperature
- Higher utilisation of the plant, through possible use of other fuels (e.g. in waste incineration)
- Realisation of the *glass stack* and the associated environmental acceptance.

European Directive and Standards
In Europe, the incineration of waste is regulated by Directive 2010/75/EU.[5] As no CEN or international standard for long-term sampling of dioxins and furans has existed so far, a technical specification for continuous sampling of dioxins and furans was elaborated in WG 1 of TC 264 in the last years as CEN TS/EN 1948–5 and published in April 2015.

The most commonly used international standards for manual monitoring of dioxin emissions in Europe are:

- EN-1948-1 Sampling
- EN-1948-2 Extraction and cleaning of PCDDs/PCDFs
- EN-1948-3 Identification and quantification of PCDDs/PCDFs
- EN-1948-4 Sampling and analysis of dioxin-like PCBs and in the USA
- EPA Method 23A Method for sampling PCDDs and PCDFs from stationary sources.

In many other countries such as Japan, Korea, Taiwan, etc., there exists own standards. Most of these standards are identical to US EPA method 23A except for a few details. All standards are valid for manual sampling and are only validated for sampling periods of a few hours (<16 h). In EN 1948-1, three sampling methods are described (Figs. 14.77 and 14.78).

US EPA Method 23A
The main difference between the US EPA method 23A (Fig. 14.79) and the filter-cooler or cooled probe method of EN 1948–1 is that the condensate is collected after the dioxin/furan adsorbent. According to the US EPA method 23A (§7.2.5) only the amount of condensate is collected to determine the flue gas humidity. However, the condensate is not analysed for dioxins/furans. The AMESA D system uses the so-called simplified method. This method is basically the same as the cooled probe method, with the difference that the condensate trap is constructed according to the XAD II adsorbent, just like the US EPA method 23A. For this reason the condensate produced does not have to be analysed for dioxins and

[5]DIRECTIVE 2010/75/EU OF THE EUROPEAN PARLIAMENT AND OF THE COUNCIL of 24 November, 2010 on industrial emissions (integrated pollution prevention and control) (recast), Official Journal of the European Union, pages L 334/17 to L 334/119

DIRECTIVE 2010/75/EU OF THE EUROPEAN PARLIAMENT AND OF THE EUROPEAN COUNCIL RATES
from 24. November 2010
on industrial emissions (integrated pollution prevention and control pollution)
(rewrite)

Definition of the limit value for dioxins/furans of 0,1 ng/m3 TEQ with a minimum sampling time of 6 hours and a maximum of 8 hours

ANNEX VI

Measurement techniques

2. The sampling and analysis of all pollutants, including dioxins and furans, and the reference measurement procedures for the calibration of automated measurement systems shall be carried out in accordance with CEN standards. Where CEN standards are not available, ISO standards, national standards or international standards ensuring that data of equivalent scientific quality are obtained shall be applied.

Fig. 14.77 Directive 2010/75/EU (extracts)

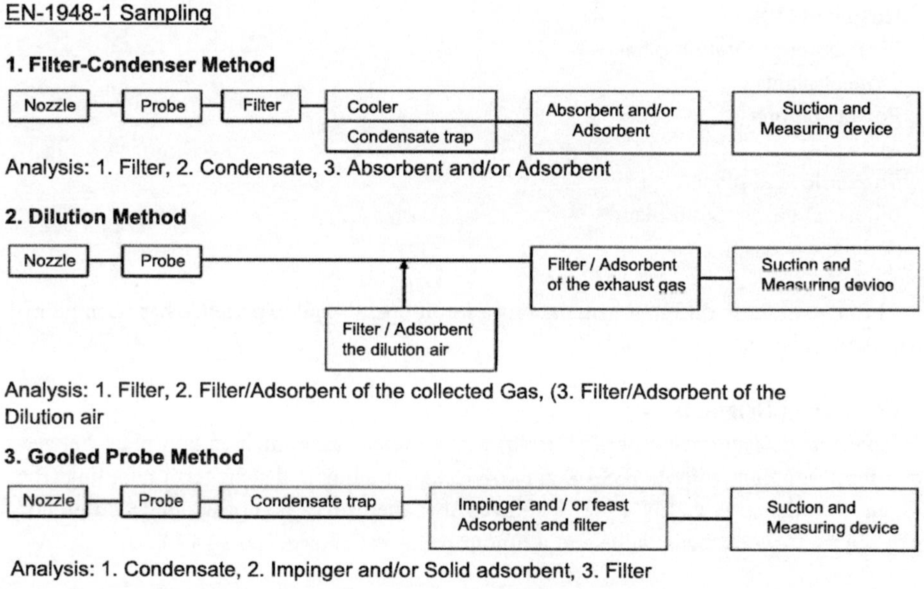

Fig. 14.78 Sampling procedure of EN 1948–1

furans. In this way, routine long-term sampling over several weeks is possible. During a 4-week sampling period, condensate can accumulate up to 200 L. If the condensate would have to be collected for subsequent analysis, this would not be possible in routine operation or would require considerable efforts.

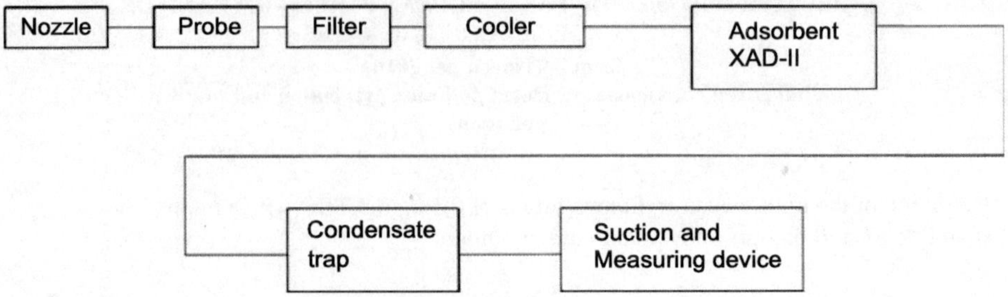

Fig. 14.79 Sampling procedure of US EPA method 23A

Applications and Results
Since 1996 *AMESA®* has been installed in about 300 applications. The installations have been realised in more than 15 different countries.

The installations were carried out in the following plant types:

- Domestic waste incineration plants (MVA)
- Hazardous waste incinerstion plants
- Hospital MVA's
- Biomass cogeneration plants
- Cement plants
- Power stations
- Melting
- Sulphuric acid plant
- Industrial gas disposal plant
- Animal carcass disposal.

Through all these different installations a lot of operational experience has been gained over the last years.

Experience in Belgium
Comparison measurements (de Fré 1998) at a Belgian waste incineration plant between long-term sampling with AMESA® at two-weeks sampling and short-term sampling from 6 h in accordance with EN 1948–1 showed that there may be considerable differences between the 2-week mean value and a 6-hour mean value (see Table 14.4).

The results found were evidence that short-term increased dioxin emissions are not detected by short-term sampling. In the meantime, continuous monitoring of all waste incineration plants in Belgium has been required by law. A two-week sampling period and a limit value of 0.1 TEQ ng/Nm^3 was set. Due to this legal requirement, a network for the continuous monitoring of dioxin/furan emissions from waste incineration plants was set up throughout Belgium between 1999 and 2002 with the AMESA® systems (Becker et al.

Table 14.4 Results of continuous and 6-hour sampling at the IVOO waste incineration plant

Measuring period	Measured dioxin concentration as ng I-TEQ/Nm ^{3} at 11% O_2	
(day-month year)	Continuous monitoring	6 h Sampling
29.12.97–12.01.98	13.4 (14.3)	
12.01.98–26.01.98	8.2 (12.9)	0.25
26.01.98–30.01.98	12.6 (10.1)	
09.02.98–23.02.98	2.11 (2.12)	
23.02.98–09.03.98	0.44	
09.03.98–23.03.98	0.33	0.12
23.03.98–06.04.98	0.8	

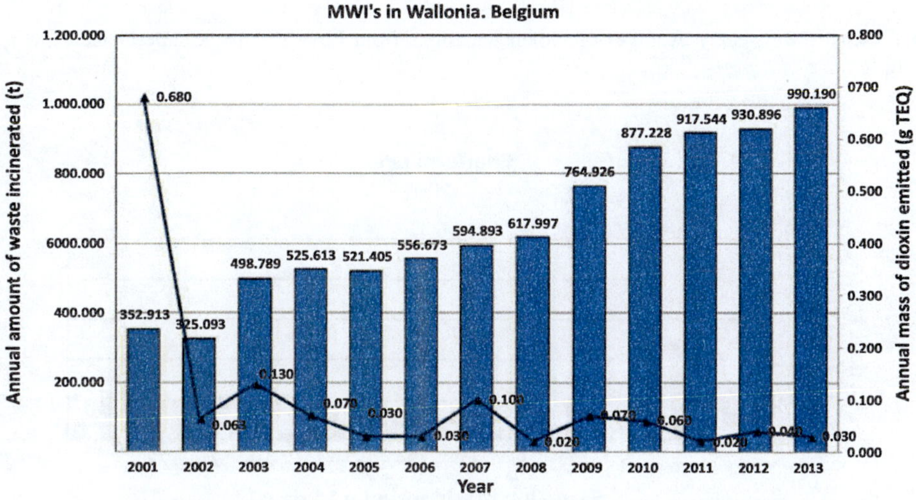

Fig. 14.80 Annual dioxin emissions and waste incineration, Wallonia, Belgium

2000). The results in Wallonia have been published on the Internet since January 1, 2001. Over the years, the total dioxin emissions have been reduced by a factor of about 30, although the total amount of waste incinerated per year has almost tripled. (Fig. 14.80).

During the first years of continuous dioxin monitoring, several exceedances of limit values were observed. The reasons for the limit value exceedances were different and it was important to get to the bottom of them. By recording the operating parameters every 1/2 h it is possible to check the operating conditions during the sampling period and to detect any irregularities. In the future, it will be necessary to avoid operating conditions that have led to increased dioxin/furan emissions or to detect faults more quickly in order to rectify them.

As can be seen in Fig. 14.81, the emission values for this plant were OK for 30 weeks. In period 16, however, the limit value was exceeded by a factor of 39. The reason for this was a filter clogging. As a result, the bypass of the system opened and the flue gas entered the

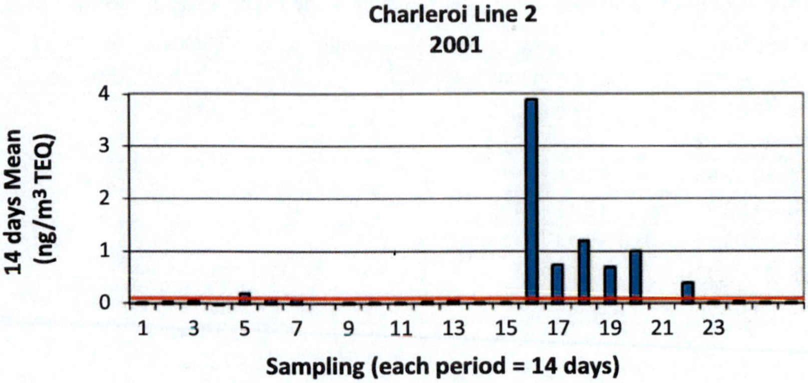

Fig. 14.81 Two-weekly averages of dioxin emissions from Line 2 of the Charleroi incinerator in 2001

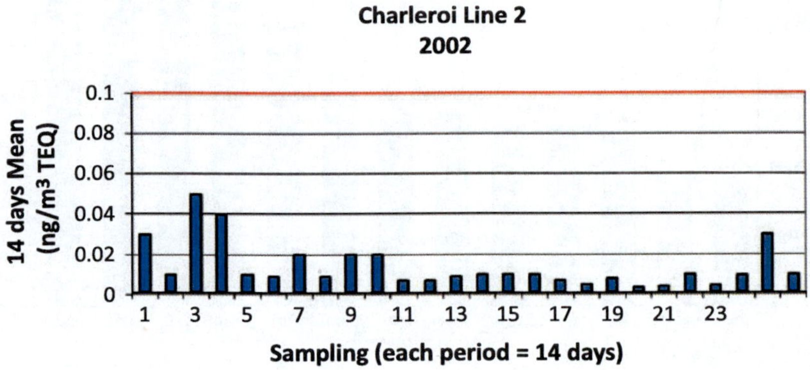

Fig. 14.82 Two-weekly averages of dioxin emissions from Line 2 of the Charleroi incinerator in 2002

stack unfiltered. These increased dioxin emissions were accepted by the authorities due to the operational disturbance. However, after the problem was solved, the emission value in period 17 was still 0.75 ng/m ^{3} TEQ. Defective seals of the bypass system were found to be the cause of this renewed exceeding of the limit value. Part of the flue gas thus continued to enter the stack unfiltered. After the operator was unable to run the plant below the limit value for several weeks, it was shut down by order of the authorities. After major maintenance work on the flue gas cleaning system, the plant was put back into operation in period 23. The slightly increased value in period 22 was due to a 24 h test run. This value therefore only corresponds to an increased start-up value. After this incident the same plant was operated for years with low limit values (Fig. 14.82). This example shows once again how important continuous monitoring of dioxin/furan emissions is in order to quickly detect and eliminate any faults that occur.

If, however, an operator can show by means of continuous monitoring of dioxin/furan emissions that his plant has permanently low emission values, such as the values in Fig. 14.82, the publication of such emission values usually leads to a higher acceptance by the population.

Experience in Germany

In accordance with Directive 2010/75/EU, a minimum temperature of +1100° C for at least 3 s must be maintained in the combustion chamber when burning waste with a chlorine content of more than 1% by weight. In the former 17th BImSchV even a minimum temperature of +1200° C was prescribed. However, this rather high temperature also generates higher operating costs, due to increased corrosion, higher energy consumption and lower waste throughput per unit of time. Lowering the temperature from, for example, 1200 ° C to 1000 ° C can therefore reduce the operating costs by approximately 10–15%. In principle, manual dioxin sampling according to the standard method over a period of 6–8 h can be used to monitor compliance with the limit value even when the combustion chamber temperature is lowered. Such confirmation does not, however, provide the certainty that the specified limit value is permanently complied with. However, continuous monitoring of dioxin emissions by means of long-term sampling can provide a reliable and lasting demonstration of compliance with the limit value even after the combustion chamber temperature has been lowered. In 1998/1999, it was demonstrated at the hazardous waste incineration plant of Infraserv GmbH & Co. Höchst KG using the AMESA® system that lowering the combustion chamber temperature from 1200 °C to 1000 °C had no effect on dioxin emissions (Fig. 14.83) (Mayer et al. 1999).

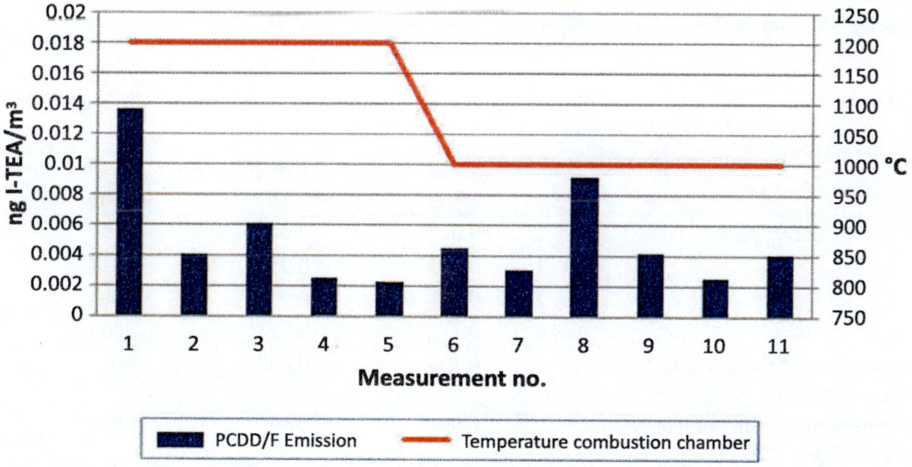

Fig. 14.83 Combustion chamber temperature and dioxin emissions in Infraserve's special waste incinerator

Dioxin emissions from waste wood combustion plants have already been studied in detail. Especially the incineration of treated or PVC-coated wood can lead to increased dioxin emissions. For this reason, larger waste wood incineration plants in Germany must comply with the same limit values as all other waste incineration plants.

In order to establish a better relationship between operating conditions and the resulting dioxin emissions, the local authority prescribed long-term sampling with weekly cartridge changes in a waste wood incineration plant in Germany, which burns treated waste wood as well as packaging material of furniture etc. The weekly cartridge change was chosen to achieve a high time resolution. However, in order to kept the analysis costs for the operator within an acceptable range, only the mixed samples of the 4 weekly extracts of the cartridges were analysed.

During the start-up of the plant, 12 long-term samplings of one week each were carried out between April 2001 and June 2001. Subsequently, the cartridges were extracted and three monthly analyses of the composite samples were performed. To confirm the reliability of the mixed sample results, the extracts of the weekly individual samples were also analysed during this period.

The emission values of the weekly samples varied from 0.0031 to 0.151 ng I-TEQ/m ^{3} and those of the monthly composite samples varied from 0.043 to 0.099 ng I-TEQ/m ^{3}. As can be seen in Table 14.5, the calculated averages of the weekly samples are in good agreement with the analysis results of the monthly composite samples, including in period C when the weekly values varied from 0.003 to 0.1 ng I-TEQ (Fig. 14.84).

This result shows that weekly sampling combined with monthly analysis is a good method for dioxin monitoring. For normal dioxin levels below the limit value, the analysis of monthly mixed extracts allows to keep the costs of analysis low. In case of elevated dioxin levels, it is possible to analyse the individual weekly samples in order to draw more precise conclusions on the increased dioxin emissions due to the higher time resolution. In the future, it will then be necessary to avoid the causes of the increased emissions, such as special operating conditions, mixing of fuels, etc.

Table 14.5 PCDD/F concentration in the clean gas of a waste wood combustion plant in ng I-TEQ/m ^{3} based on 0 ° C, 101.3 kPa, dry, 11% O_2

		Period A *(April 2001)*	Period B *(May 2001)*	Period C *(June 2001)*
Weekly sample [ng I-TEQ/m ^{3}]	1	0.073	0.046	0.103
	2	0.092	0.050	0.034
	3	0.071	0.038	0.013
	4	0.151	0.073	0.003
Mean value of the weekly samples [ng I-TEQ/m ^{3}] [a]	1–4	0.100	0.052	0.039
Monthly composite sample [ng I-TEQ/m ^{3}].		0.094	0.051	0.043

[a] Weekly sample values are weighted according to the sampling volume

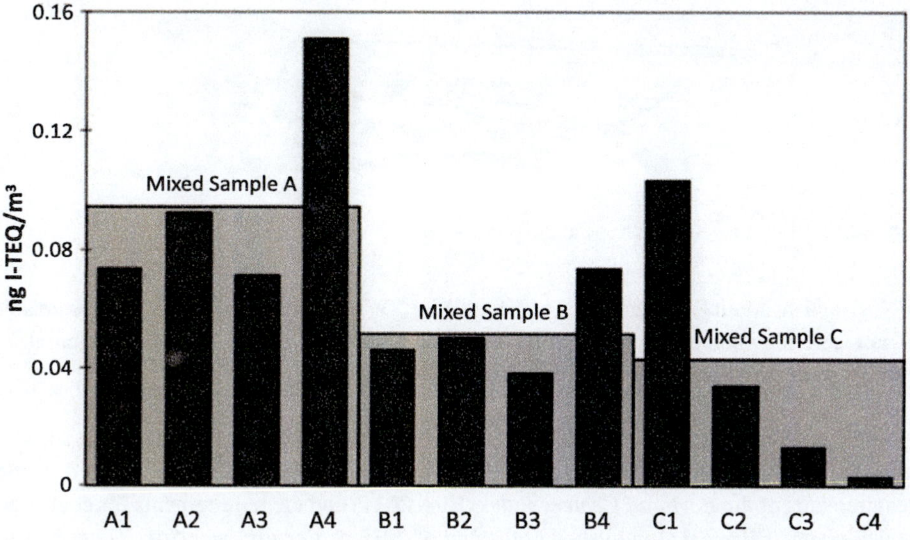

Fig. 14.84 PCDD/F concentrations in weekly (1–4) and monthly composite samples (A-C) from the waste wood combustion plant. ng I-TEQ/m ^{3} based on 0 ° C, 101.3 kPa, dry, 11 vol% O_2

In total, the AMESA® system has been used in about 20 plants in Germany so far. This was either due to the inclusion of continuous monitoring in the notice of approval as a compromise for the concerns of the local residents or on a voluntary basis to carry out own tests. At present 8 plants are in continuous operation.

France

In France, approximately 230 long-term sampling systems have been installed to date. On 21 August, 2010 a new law was published (Journal Officiel de la République Francaise 2010), which requires continuous monitoring of dioxin emissions from waste incineration plants. A sampling period of 4 weeks was determined. The law came into force on November 1, 2010 and had to be implemented by all operators by July 1, 2014.

Italy

To date, about 60 long-term sampling systems have been installed in Italy. To date there is no national legislation requiring the installation of such systems. The installations are prescribed here by the regional authorities (e.g. ARPA Lombardia).

Mercury

Several systems for measuring mercury emissions are already available on the market. However, during the operation of these systems it has been found that these systems require a high level of maintenance, otherwise the measurement errors are too great. According to the 13th BImSchV, the current limit value in Germany is a daily average value of 30 µ

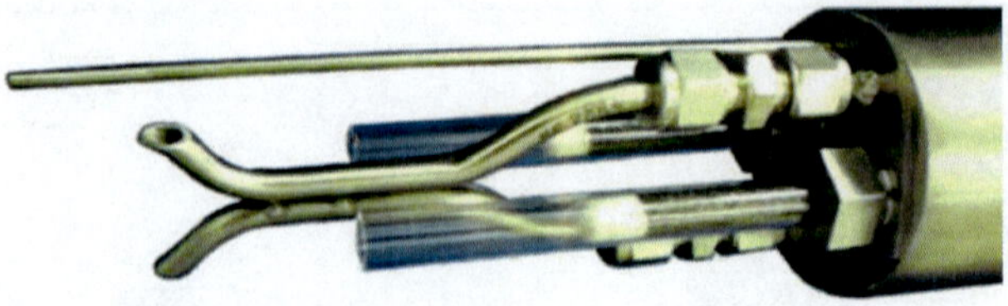

Fig. 14.85 Paired Hg sorbent traps at the probe input

g/Nm^3 and an annual average value of 10 μg/Nm^3. With the currently available systems, it is not certain whether concentration values can be permanently measured accurately enough to demonstrate the new annual mean value.

In the USA, in addition to online Hg measurement, long-term sampling using sorbent traps to record Hg emissions is also accepted. The sampling method is described in the requirements of the *Portland Cement Rules* (EPA 2011) and the requirements for coal-fired power plants (EPA 2013) published at the end of 2013 as *Performance Specification 12B.* A probe equipped with 2 Hg sorbent traps (Fig. 14.85) extracts a partial volume of the flue gas under volume proportional conditions. The gaseous mercury (elemental and ionised) is collected in the sorbent traps. Due to the arrangement of the sorbent traps at the probe entrance, they are the first elements through which the sample gas flows. Losses in the sampling line are thus avoided.

After the sampling period (usually 1–2 weeks), the sorbent traps are replaced and analysed in a certified laboratory or with a mercury analyser. The result is, similar to dioxin sampling, the average mercury concentration over the sampling period.

Based on the proven AMESA system, the AMESA®-M system was designed according to the requirements of PS 12B. In first validation measurements against the standard reference method, very good results were achieved in the concentration range of 1–3 μg/Nm^3. The availability for continuous operation over a period of several months was always above 99%.

Due to the positive experience in the USA, this procedure is also attracting interest in Europe. For this reason, the work of WG8 was resumed in TC264 for standardization work. The aim is to develop a European standard for the measurement of mercury emissions by means of sorbent traps for both short-term and long-term sampling.

Summary

The current results from Belgium and other countries show that it can be beneficial for the environment, the public but also for the operator to monitor dioxin/furan emissions on a permanent basis.

Permanent control and associated improvement of flue gas cleaning are in any case useful to preserve the environment. By retrofitting or, in the case of already existing low emission values, operators have the possibility to achieve a higher acceptance by the population.

A process of rethinking is currently beginning in other countries. This is currently demonstrated by the new legislation in France and the regional regulations in Italy. CEN TS/EN 1948–5 was published in April 2015. This is the first standard worldwide for long-term sampling of dioxins, furans and dioxin-like PCBs. The preparation and publication of this standard also demonstrates the transnational European interest in this type of emission monitoring.

In general, long-term sampling systems such as *AMESA® D* can also be used to detect other POPs, for example, PCBs, PAHs, etc., so that such instruments can also be helpful in the implementation of the Stockholm Convention.[6]

In addition to the application for monitoring dioxin emissions, the modified AMESA® M system also demonstrated its successful application for monitoring Hg emissions. In another variant, the system has also been used for the continuous measurement of arsenic emissions for about 5 years.

In particular for the continuous measurement of heavy metal emissions, it may be worth considering the use of long-term sampling systems, as it is highly probable that there will be no online analysers for a foreseeable period of time.

14.4 In Situ Gas Analysis with Laser Technology

Thomas Beyer, Julian Edler, and Rolf Schiffler

In gas analysis, a whole range of different methods are used in industrial processes. In many applications a direct or even contactless measurement is advantageous. Thus, gas concentrations can be reliably measured under process conditions with short response time in a robust and simple design without falsification. Extractive measurement methods, which require complex sample preparation, cannot offer these advantages. Optical measuring methods have proven to be particularly suitable for *in situ* or *in-line measurements* (Ebert et al. 2005; Allen 1998). These include absorption spectroscopy, in which light with a defined wavelength λ shines through a volume of gas and the gas molecules absorb a certain portion of the light. The TDLS (*tunable diode laser spectroscopy*) method is particularly suitable for fast, continuous in-situ measurements without time delays. This non-dispersive, spectrometric method is highly accurate and highly selective and is used in the GM700 laser spectrometers from SICK AG.

A laser with a defined wavelength (*single mode*) serves as the light source. This wavelength λ can be varied very quickly by changing the laser current and the associated

[6] Stockholm Convention (SC), **2001**. http://www.pops.int

immediate heating or cooling of the laser material. By detecting the radiation simultaneously with the laser control, a high-resolution absorption spectrum can be recorded. In commercially available laser spectrometers, this TDLS technique allows individual spectra to be recorded in less than a millisecond at repetition rates greater than one kilohertz. The spectral resolution of the λ TDLS measurement technique is only limited by the width of the laser emission, which for diode lasers in the near infrared (NIR) range is less than 10^{-3} nm. TDLS measuring instruments do not require wavelength-selective elements. As a further advantage, moving parts such as prisms, gratings or filter cuvettes are omitted, which results in good long-term stability and high availability of the measuring instruments.

For TDLS, robust single emitters in small TO packages are typically used in industrial measuring devices. Today, such continuously radiating semiconductor diode lasers are available in the visible wavelength range (Wiedenmann et al. 2006) and NIR (Totschnig et al. 2003), as well as in the mid-infrared (Seufert et al. 2004) beyond 10 μm (Faist et al. 1994). This allows to exploit the overtones and combination modes of the rotational vibrational states of the molecules in the NIR as well as the significantly stronger fundamental modes in the mid-infrared for the TDLS technique. While relatively expensive quantum cascade lasers or interband cascade lasers (Hildebrandt et al. 2014) are still necessary for measurements in the mid-infrared range, much cheaper laser diodes based on semiconductor materials can be used in the near-infrared range, as they are used in telecommunications. In laser spectroscopy mainly two different types of lasers are used: DFB (*distributed feedback*) lasers (Weih et al. 2014) and VCSEL (*vertical cavity surface emitting laser*). In a DFB laser, a periodically applied grating on the active medium acts as a longitudinal mode filter. The single mode operation required for TDLS is achieved by this filter. In contrast to DFB lasers, a VCSEL (*Vertical Cavity Surface Emitting Laser*) emits the laser light perpendicular to the plane of the semiconductor material. Due to the extremely short resonator and the resulting large distance between the resonator modes, VCSELs intrinsically emit in single mode operation.

In addition to lower manufacturing costs, VCSELs are distinguished from DFB lasers by a significantly lower laser current, but also a lower output power. The lower laser current is often a decisive advantage for industrial applications, since the current consumption and laser power must be limited for measuring devices in hazardous areas. In addition, the eye-safe laser class 1, which is often aimed for measuring instruments, can be achieved with little technical effort. Another significant advantage of VCSEL diodes, compared to DFB lasers, is the faster change of wavelength with the laser current, the so-called laser tuning. The wavelength change is primarily a thermal effect triggered by the laser current. In VCSELs, the active volume is significantly smaller and the thermal coupling of the active region is better, making higher measurement frequencies possible.

However, the wavelengths of VCSEL and DFB lasers can only be changed by current modulation over a very small wavelength range. At 1500 nm, for example, a VCSEL can only achieve tuning up to 5 nm. Since the absorption lines themselves have a line width of only 0.02 nm, several absorption lines can be located in this range. By an additional optimization of the laser temperature controlled by a Peltier element, the wavelength of the laser can be tuned over a somewhat larger range (at 1500 nm approx. 1 to 2 nm) to a specific absorption line of the gas components to be detected. Since only a very small spectral window, in which the laser emission and the absorption lines coincide, is used for gas analysis, an absorption line that is optimally suited for the intended application range must be selected. For this purpose, the temperature dependence of the linestrengths and the adjacent absorption lines of other gas components are taken into account. Thus, even gases in complex mixtures can be detected with high sensitivity and selectivity with only very small cross-sensitivities to other gas components.

To record such spectra, the laser current and thus the wavelength is continuously varied at a fixed laser temperature, often in the form of a ramp. Therefore, in contrast to Fig. 14.86a and b, the detected extinction spectra are overlaid and distorted by a ramp that is orders of magnitude stronger, caused by the change in intensity of the laser. For a simple spectrum with only one absorption line, as shown in Fig. 14.86a, the background can be easily corrected with a simple polynomial fit. Since this is much more difficult for complex spectra (Fig. 14.86b), in the GM700[7] part of the light is reflected at the inside of the exit window onto an additional detector to simultaneously record a spectrum without absorption lines. By dividing the two measured spectra and logarithmizing the result, the change in intensity of the laser is corrected according to Lambert-Beer's law and the absorbance spectrum is calculated.

For the detection of diatomic molecules, such as HF, the area of individual absorption lines can be determined. This can be converted directly into a gas concentration if the measuring distance, pressure and temperature are known (see Fig. 14.86a). The only influence that other gases have on such an absorption line is the effect of extraneous broadening. This means that the width of the observed absorption line varies according to the concentration ratios of the other gases in the gas mixture. With such a simple spectrum, where no other gas has an absorption line in the scan area, this effect can be completely suppressed in the analysis. Multiatomic molecules, on the other hand, often show more complicated spectra. For the measurement of NH_3 and H_2O, a variant of the GM700 can be used, for example the region shown in Fig. 14.86b, in which a larger number of lines overlap. This allows the simultaneous acquisition of both components with very low cross sensitivities.

Due to the extremely narrow line width of the single-mode lasers used ($\lambda < 100$ MHz for Fig. 14.86a and b), the measured spectrum is not influenced by the width of the laser emission. A physically correct extinction spectrum is obtained, which can be compared with spectra calculated with absorption line data from databases (e.g. HITRAN[8]).

---

[7] Laser spectrometer from SICK AG.

[8] HITRAN: THE HITRAN database (**hi**gh-resolution **tran**smission molecular absorption database) http://www.cfa.harvard.edu/hitran/

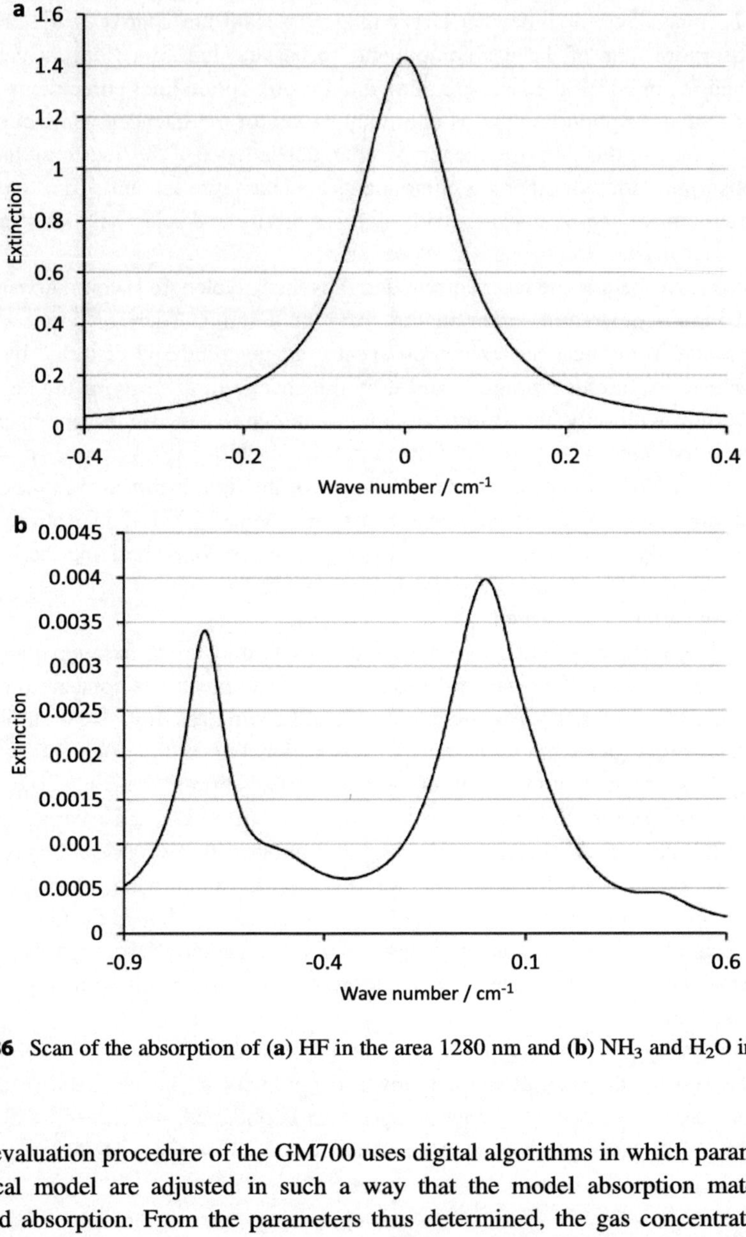

Fig. 14.86 Scan of the absorption of (**a**) HF in the area 1280 nm and (**b**) NH_3 and H_2O in the area 1510 nm

The evaluation procedure of the GM700 uses digital algorithms in which parameters of a physical model are adjusted in such a way that the model absorption matches the measured absorption. From the parameters thus determined, the gas concentrations can be calculated, taking into account the pressure and temperature of the gas mixture and the path length of the measurement path.

For the physical model, suitable shapes for the absorption lines must be defined in the evaluation. Complex line shapes lead to significantly higher computing power for the fit routines. In order to achieve the shortest possible update rates for continuous measurements, ideally the maximum time of the ramp is used for the entire signal processing, evaluation and control. Often a compromise between complex line shape and fast evaluation has to be found. For most applications, so-called *voigt profiles* are used for the absorption line shape. These profiles consider two different types of broadening. On the one hand, the Doppler broadening, which describes the movements of the molecules during the absorption of light, and on the other hand the Lorentz or pressure broadening, which is caused by the broadening of the absorption line by collisions with other molecules in the gas mixture. In some cases, it can even be advantageous to consider only the pressure broadening, since this is mathematically much easier to represent and calculate, and thus small, low-cost processors with low power consumption can be used. Alternatively, a powerful processor can be used, with which a high temporal resolution can be achieved through faster calculation.

The gas component to be measured in industrial applications is often not present in the gas flow over a long period of time - especially when monitoring toxic or explosive gases. At the same time, it must always be ensured that, even after months of operation, the spectral window of the laser emission still corresponds to the absorption line that is to be monitored but is not visible. This poses a particular challenge for industrial TDLS measurement technology. The GM700 from SICK achieves this via a so-called *line locking channel*. A small portion of the laser radiation is reflected at a wedge-shaped window and guided through a gas cuvette, which is permanently installed in the spectrometer and filled with the gas to be measured, to another detector. The spectrum of the gas in the cuvette is thus measured simultaneously with each measurement in the gas flow. A change in the emission wavelength, for example due to aging, can be detected immediately and compensated by adjusting the laser diode temperature. This ensures that the correct gas absorption is always monitored.

In addition, a spectral interference pattern can be generated at regular intervals using a simple interferometer (e.g. a glass pane). For example, this allows to detect and compensate for changes in the tuning of the laser caused by aging effects, which lead to a distortion of the spectra.

Instead of the so-called direct spectroscopy described above, wavelength modulation spectroscopy (WMS) can be used as a further possibility of diode laser spectroscopy. Here, the laser is not only driven by a current ramp, but the current ramp is also overlaid with a fast sinusoidal modulation. The sinusoidal modulation of the current leads to a sinusoidal modulation of the intensity and the laser wavelength. In a typical detector signal, an absorption line with a superimposed, modulated intensity can then be seen. The observed absorption line is clearly extended by the modulated wavelength (see Fig. 14.87). This detector signal can be analyzed with the aid of a phase-sensitive lock-in amplifier, with a short integration time or by a sectional Fourier transformation. As shown in Fig. 14.87 using the example of the O_2 spectrum, the absorption line causes a change in the sinusoidal modulation. While the oscillations are unaffected in the edge regions of the spectrum, they are significantly deformed in the region of the absorption line. In the associated Fourier analysis, corresponding harmonics of the sinusoidal modulation can be seen.

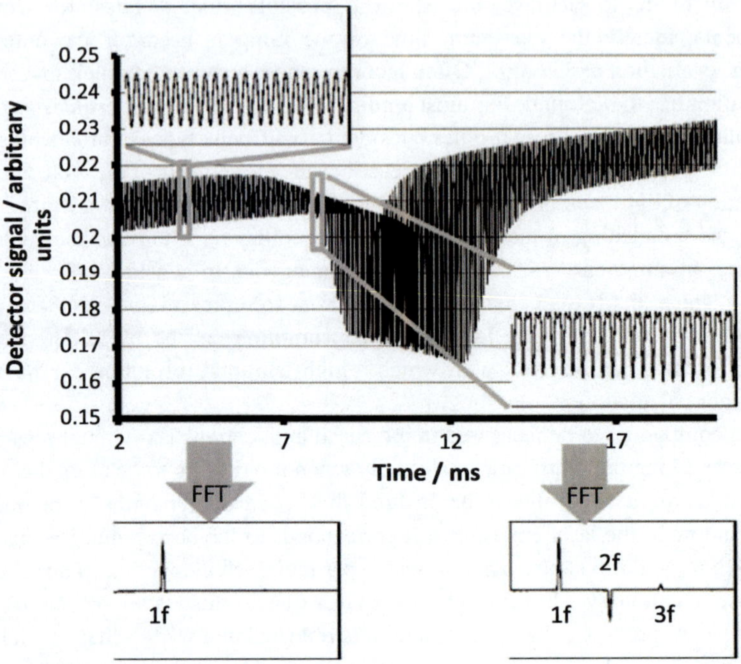

Fig. 14.87 Analysis of the WMS detector signal

Typically, the first harmonic of the sine wave, that is, the signal at twice the modulation frequency, is analyzed. Since the overtones only occur in the area of the absorption line, the intensity of this signal is only non-zero when absorption is present. This suppresses the high offset of the laser power at which the changes caused by the absorption of the gas must be detected in direct spectroscopy. Plotted against the laser ramp, or the laser wavelength, the so-called 2*f* signal is obtained (in Fig. 14.87). By selecting the first overtone, not only the offset of the laser power is suppressed, but also the slope of the spectra caused by the current ramp and the associated increasing laser power. The height of the 2*f*-signal is proportional to the gas concentration after a suitable intensity normalization.

A further advantage of the WMS is the shift of the evaluation to the higher modulation frequency and the associated reduction of the 1/*f*-noise.

Since the Fourier evaluation of a modulated signal is very computationally demanding, it may be sufficient to concentrate on the decisive areas of a 2*f* signal. In the TRANSIC100LP laser spectrometer from SICK AG, therefore, only the two minima and the central maximum of the 2*f* signal are scanned (Fig. 14.88). This makes it possible to use a low-cost processor, achieve a very high measuring rate and offer the advantage of the very low detection limits that wavelength modulation spectroscopy has in a low-cost laser spectrometer.

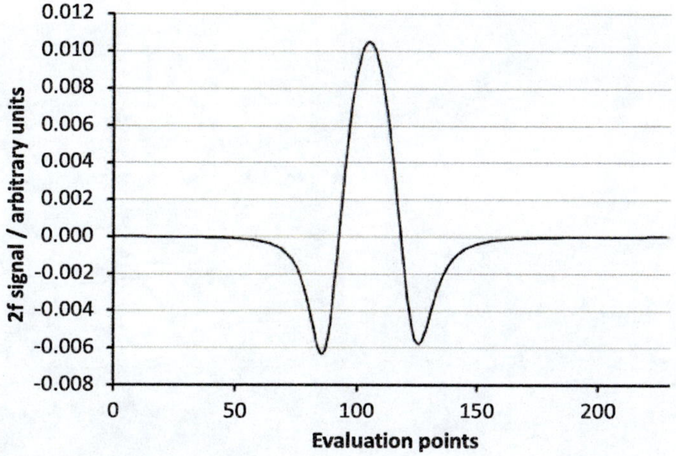

Fig. 14.88 Typical WMS signal at double modulation frequency

Challenge of the Applications

Analysers with TDLS technology can be used in a wide range of applications. In gas measurement technology, these are often applications in large industrial plants, such as power stations, cement works, waste incineration plants or process measurement in chemical plants or refineries (Fig. 14.89). The application conditions vary greatly and with gas temperatures up to over 400 ° C, duct diameters from about 0.5 to over 10 m and dust contents from a few milligrams up to several 10 grams per cubic meter, they are a challenge for any measuring system.

One way to take into account the different application conditions is the use of extractive measurement technology. In this case, sample gas is extracted from the flue gas duct via a probe and fed to a separate analysis system for analysis. However, this method has limitations, both in terms of the effort required for such a system and the range of measurable components available. Especially highly reactive gases, such as NH_3, HCl, HF, will not survive transmission through a gas pipe and gas treatment without loss. The measured values are falsified by surface effects or washing out in the cooler, so that a special form of extractive measurement technology with a heated sample extraction and measuring cell must be used (*hot-wet*).

There are various versions of in-situ measurement technology, but it is basically *contactless* and therefore particularly suitable for problematic measurement components such as HF, HCl or NH_3, high dust content and for measurements without delay (Table 14.6).

Fig. 14.89 Application of the GM700 laser gas analyzer from SICK on an exhaust duct

Table 14.6 Metrological properties of GM 700

Features of the GM700	
Measuring components	O_2, NH_3, HF, HCl, H_2O
Channel diameter	To 6 m or independently when using measuring probes
Detection limit for NH_3	0.35 mg/m ^{3}
Measuring range NH_3	Dependent on measuring distance, up to 3000 mg/m ^{3}
Exhaust gas temperature	−40 to 430 ° C
Design	For use in large industrial plants

Constructive Solutions

Typically, *in situ* analyzers are mounted on an exhaust duct in a *cross-duct configuration*, that is, transmitter and receiver are installed directly opposite each other on the duct (see Fig. 14.90). The light path available for the measurement depends on the diameter of the flue or stack. For such a *cross-duct installation,* connecting flanges with very tight tolerances must be installed in the flue, so that the emitted light beam arrives at the receiver and is not cut by the flange pipe. In addition, this alignment must be maintained over various operating states of the system and the associated duct movements.

In many cases, the ash or dust content of the gas weakens the transmitted laser signal. The closer the measurement is to the furnace, for example, for a control system, the higher the dust content. For coal-fired power plants, 10 ... 20 g/m ^{3} are the rule. Depending on the

Fig. 14.90 Installation of a measuring system in a *cross-duct arrangement*

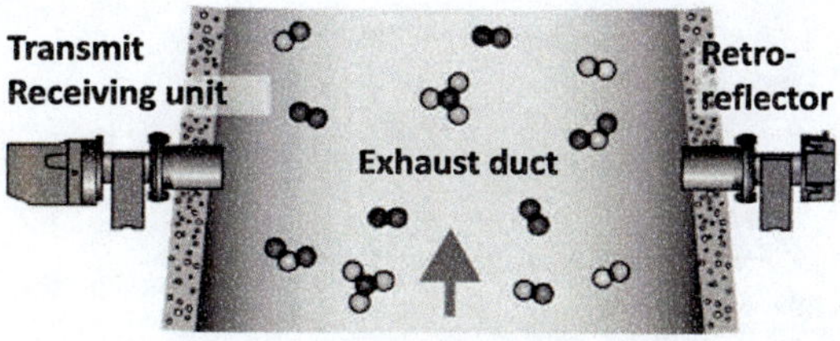

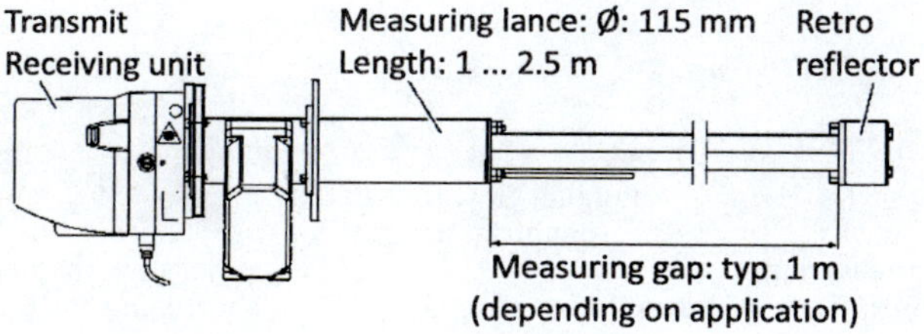

Fig. 14.91 TDLS analyser GM700 with measuring probe

type of coal, however, values from 50 ... 70 g/m ^{3} can also occur. With these enormous quantities, optical measurement is no longer possible or requires a very short measuring distance. However, since the sensitivity (detection limit) depends on the length of the available measuring distance, a very short measuring distance is rarely a sensible solution, depending on the measuring task.

The measuring probes offered for the GM700 from SICK offer innovative solutions precisely here, which allow installation from one side of the channel on the one hand, and simpler handling (alignment, inspection) on the other (Fig. 14.91). A measuring probe connected to the analyzer (transceiver unit) is used to keep the reflector in the correct position at all times. The measuring section through which the sample gas flows is located between the transmitter/receiver unit and the reflector.

It is also true for the use of measuring probes that dust content in the gas weakens the signal. Compared to the alternative of drawing air-purged, measuring distance-shortening tubes into a 10 m channel, however, the choice of a suitable measuring distance with a probe is much easier to implement, as Fig. 14.92 shows.

Today, power plants often use imported coal of various origins and compositions. At the time of the design of a measuring system, the resulting differences, for example, in ash content, can hardly be considered. For measurements in heavily dust-laden gases, the measuring gap of a measuring probe can be equipped with a ceramic filter. This makes the measurement completely independent of the level of the dust concentration and also independent of strong fluctuations, as they occur with different coal types or with periodically running cleaning cycles of the heat exchanger surfaces (soot bubbles). These filters work purely passively by diffusion of the gas to be measured and therefore do not clog up even over long operating periods.

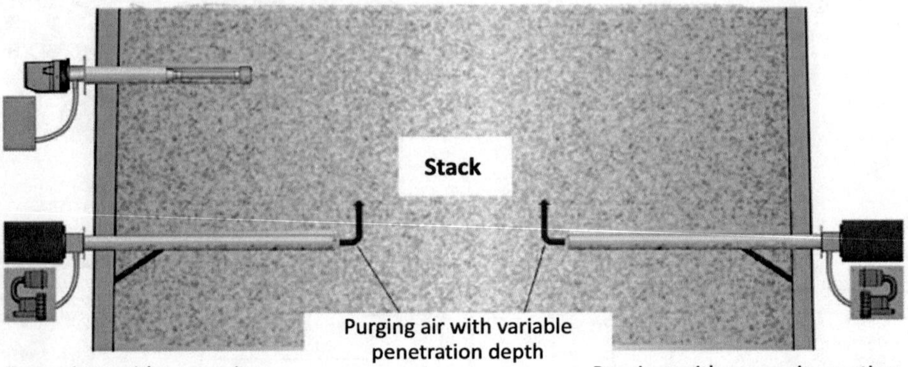

Fig. 14.92 Installation comparison cross-duct and measuring lance

Example of an Application: Measurement of NH_3 Slip in DeNOx[9]-Processes
Industrial plants often generate waste gases during the process. These are subject to certain regulations for the discharge of the gases into the atmosphere. In order to comply with specified limit values, exhaust gas purification systems must often be used. A denitrogenation process (DeNOx) is used for the removal or reduction of nitrogen oxides, as they occur in combustion processes (e.g. coal-fired power plant) or chemical plants (e.g. nitric acid production) (Fig. 14.93). In addition to compliance with emission limit values, challenges for plant operation arise from the possible inclusion of excess NH_3 in fly ash, gypsum and waste water from the desulphurisation plant, as this prevents further use of the power plant ash as an aggregate in cement production and gypsum in the construction industry. In addition, polluted waste water must be treated at great expense. The formation of ammonium salts with other waste gas components such as SO_2 or CO_2 can also cause considerable problems through deposits on heat exchangers or sewer installations and should therefore be avoided.

The Process
Nitrogen monoxide is formed in combustion gases from various sources:

- nitrogenous substances in the fuel (1 g N/kg fuel gives up to 370 mg/m^3 NO_2 for ashless coal and 0% Rest-O $_2$)
- prompt-NO from atmospheric nitrogen with participation of hydrocarbon radicals (Baumbach 1993)
- thermal NO from atmospheric nitrogen at temperatures >1200° C.

[9]DeNOX = Processes for the reduction of nitrogen oxide concentrations in gases.

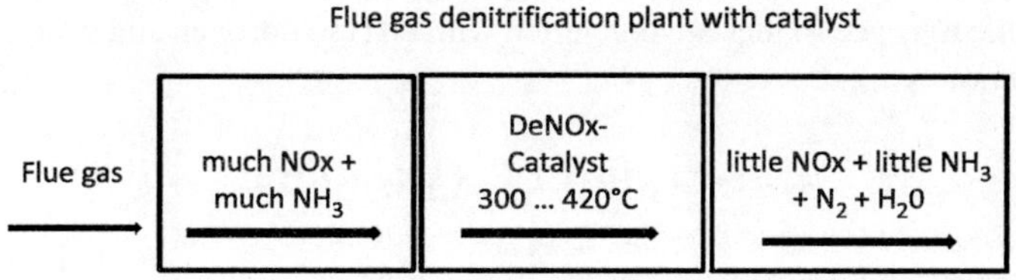

Fig. 14.93 DeNOx plant with catalyst

At temperatures under 600 ° C, sufficient residual oxygen and corresponding residence times, part of the NO can be converted to NO_2. Within the plant, this proportion is negligible in most cases. In the atmosphere the complete conversion takes place, which is why the nitrogen oxide emissions are converted to NO_2, although the component nitrogen monoxide is measured. As an example, for hard coal-fired plants approximately 1200 ... 1700 mg/m ^{3} in old plants, or approximately 600 ... 800 mg/m ^{3} in NO_x upgraded plants, calculated as NO_2.

Large industrial plants therefore have to take measures to comply with the emission limit values for nitrogen oxides (in the above example 200 mg/m ^{3} calculated as NO_2) for their waste gases. While for some plants so-called primary measures such as the use of low NO_x burners, air staging and exhaust gas recirculation already lead to success, others have to apply active DeNOx measures:

- SNCR[10]: in this process urea (splits off NH_3 to form biuret) or ammonia solution is injected into the combustion chamber at temperatures above 800 °C

$$2\,H_2N-CO-NH_2 \rightarrow H_2N-CO-NH-CO-NH_2+NH_3$$

- SCR[11]: here ammonia (solution) is added to the flue gas. The reaction then takes place in a catalyst at temperatures which are generally between 340 and 400 ° C, or even below this temperature in the case of special types of plant or catalyst. The plant can be designed with a catalyst directly behind the boiler before dedusting. LowDust systems are usually located at the end of the flue gas duct if there is no space available behind the boiler when retrofitting in existing systems. In this case it is necessary to heat the flue gases before the catalyst.

[10] SNCR = Selective non catalytic reduction, DeNOx process with direct reduction of nitrogen oxides without the involvement of a catalyst.

[11] SCR = Selective catalytic reduction, DeNOx process with reduction of nitrogen oxides with catalyst support.

In both cases the nitrogen oxides are converted with NH_3 to nitrogen and water. Mainly the following reaction is used:

$$4\,NO + 4\,NH_3 + O_2 \rightarrow 4\,N_2 + 6\,H_2O$$

But other nitrogen oxides are also recorded in this process

$$6\,NO_2 + 8\,NH_3 \rightarrow 7\,N_2 + 12\,H_2O$$
$$2\,NO + 2\,NO_2 + 4\,NH_3 \rightarrow 4\,N_2 + 6\,H_2O$$

The amount of NH_3 required for the reaction is determined by measuring the nitrogen oxides in the raw gas. Besides the determination of the DeNOx efficiency based on the nitrogen oxide concentration on the clean gas side of the plant (behind the catalyst), the knowledge of a possible ammonia excess (*slip*) is interesting for the control of the feed rate. This value, from a continuous long-term measurement, can also be used for planning the replacement of a used catalyst.

The Challenge
The direct reaction of nitrogen oxides with NH_3 (SNCR), as well as the use of catalytic converters (SCR) require certain minimum temperatures. For this reason, a DeNOx process is usually integrated into the raw gas flow before the heat exchanger (SNCR, 850 ° C ... 1050 ° C), or after it (SCR, 340 ° C ... 420 ° C). In addition to the high temperature, the gas flow in plants with solid fuels also contains a high amount of dust/ash. Cleaning processes of the heat exchanger surfaces to remove deposits cause periodic increases due to ashes released into the gas flow. Temperature and abrasion resistant materials, suitable filters and/or measuring methods that can cope with the conditions are required here. The task is to ensure that the emission limit values are reliably maintained even with changing NO concentrations on the raw gas side, while at the same time detecting NH_3 slip caused by overrunning the plant as well as catalyst defects and aging in good time. This requires measurements that allow the NH_3 quantity to be controlled under the given conditions.

Solution
The amount of ammonia required for the reaction is determined by measuring the nitrogen oxide load between the boiler and the catalyst inlet. Due to short distances to the NH_3-dosage point and high gas velocities, only *in-situ measurement technology* is suitable, which guarantees a measurement without delay. In plants with typical dust contents of more than 10 g/m ^{3}, active measuring distances between 500 mm and a maximum of 1500 mm result for optical measuring methods, whereby the measurement should continue to function even with the dust content increased by soot blowing. For reliable compliance with the emission limits, NO is also measured on the clean gas side and enters the control system as a correction. The UV measurement technology advantageously used for nitrogen

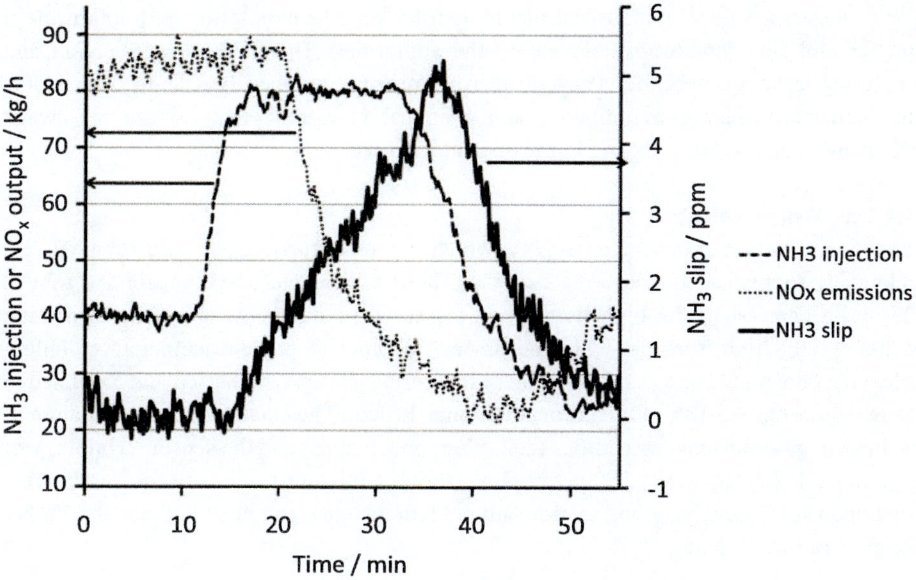

Fig. 14.94 NH_3 measurement curve with increase of the added quantity

oxide measurement is often not suitable for recording the ammonia slip at this point because high SO_2 concentrations superimpose the spectrum. With near-infrared lasers, suitable wavelengths are available for NH_3 measurement at which SO_2 shows no absorption and therefore does not interfere with the measurement. With the use of measuring probes with dust filters, the measurement becomes completely independent of the dust content and the measurement remains unrestrictedly available even when coal types with strongly varying ash contents arc used, as well as during cleaning phases. Figure 14.94 shows a curve based on real measurement data from the GM700, which has been reproduced for clarity.

Summary

TDLS laser spectroscopy provides an extremely selective, highly accurate method for the analysis of a large number of different gas components. In *in situ* technology it is particularly suitable for demanding measurement tasks, such as NH_3 measurement in DeNOx plants. The gas concentration is determined directly in the process in order to provide an unbiased picture of the process flow.

14.5 Modern Hot Gas Analysis

Stefan Roman

Gas analyzers are used in industrial plants both for process monitoring and optimization and for emission monitoring relevant to the authorities. The fields of application and measuring tasks are versatile. They range from use in combined heat and power plants and incineration plants, in refineries and chemical industrial plants to use in sewage treatment plants, paper mills and in the cement industry.

Hot Gas Measurement

Continuous multi-component analyzers, which are used for hot gas measurement, are technically sophisticated measuring systems. These devices must be compact and robust. This is in contrast to the high number of measuring components and their measuring accuracy, which are recorded with the device. In order to prevent temperatures falling below the dew point (water and acid), these measuring systems are heated and regulated to temperatures of over 185 ° C. The constructional difficulty lies in the thermal separation of the heated, gas-contacted assemblies (cell, filter, pneumatics) and the sensor technology or electronics (Wiegleb 1984). After sampling, the gas mixture to be measured is filtered, extracted via a heated sampling section and fed into the measuring cell without any further complex preconditioning.

System Structure

As an example of a multi-component analyzer, Fig. 14.95 shows the MCA 10 HWIR[12] of Fig. 14.95a and this within a system application using an analyzer cabinet (Fig. 14.95b).

The MCA 10 HWIR is a hot gas analyzer for the continuous measurement of gas components, approved according to DIN EN 15267 or the *MCERTS*[13] certificate, which is recognized throughout Europe. Table 14.7 lists the measuring components with the smallest and largest certified measuring ranges.

The system structure above the analyzer cabinet is shown in Fig. 14.96. The MCA 10 HWIR with the filter wheel photometer is located in the air-conditioned sheet steel cabinet with the control unit mounted in the door. Pneumatic gas delivery is provided by a low-maintenance air jet pump integrated in the unit. The sampling probe, the sample gas line and the gas path in the analyser are thermostatically controlled at 185 ° C.

The pneumatic connections are made via a heated gas distribution block integrated in the photometer. This is accessible from the side of the analyzer and contains the air jet pump, the connections for the exhaust gas and for up to three sample gas lines, which are connected directly. Cold bridges, which lead to salt formation and ultimately to deposits and blockages in the system, are thus eliminated. This makes it possible, for example, to easily implement a measuring point switchover of up to three sampling points. In the case of flue gas cleaning systems, such a switch-over also allows the separation efficiency of the

[12]Manufacturer: Dr. Födisch Umweltmesstechnik AG.

[13]MCERTS is the Environment Agency of England & Wales (EA) Monitoring Certification Scheme.

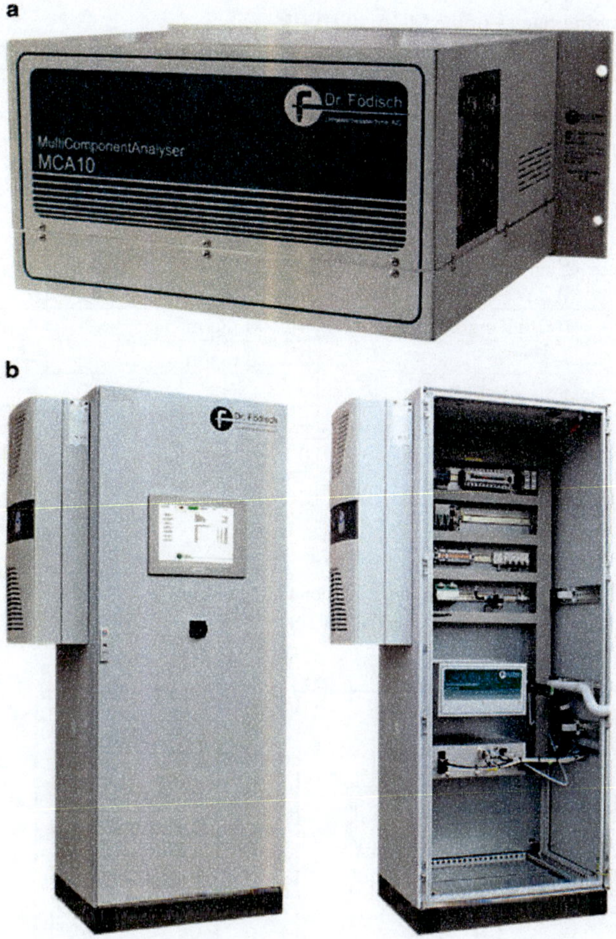

Fig. 14.95 (**a**) MCA 10 HWIR multi-component analyser, (**b**) MCA 10 HWIR analysis cabinet

individual cleaning stages to be determined with a measuring system. A further possibility offered by such a pneumatic distribution system is the extension of the system by additional gas analysers. Often total hydrocarbon analysers or mercury measuring devices are connected. Costs for the plant operator can be reduced because all devices use the same sampling system.

Thermal stability is essential for the accuracy of the measurement results, as the pyroelectric detector is sensitive to any change in temperature. Such analysis systems must therefore be air-conditioned. To ensure the precision of the measurement results, the installation site must also be selected accordingly with regard to temperature and vibrations. Analysis cabinets are often installed in air-conditioned analysis containers.

Figure 14.97 shows the gas flow diagram of a possible system design.

Table 14.7 Measuring ranges of the MCA 10 HWIR

Component	Certification area	Measuring range 2	Measuring range 3
CO	0 ... 75 mg/m^3	0 ... 300 mg/m^3	0 ... 5000 mg/m^3
CO_2	0 ... 25 Vol.%	0 ... 50 Vol.%	–
NO	0 ... 200 mg/m^3	0 ... 400 mg/m^3	0 ... 3000 mg/m^3
NO_2	0 ... 50 mg/m^3	0 ... 500 mg/m^3	–
N_2O	0 ... 50 mg/m^3	0 ... 3000 mg/m^3	–
NH_3	0 ... 10 mg/m^3	0 ... 50 mg/m^3	0 ... 500 mg/m^3
SO_2	0 ... 75 mg/m^3	0 ... 300 mg/m^3	0 ... 2500 mg/m^3
HCl	0 ... 15 mg/m^3	0 ... 90 mg/m^3	0 ... 5000 mg/m^3
HF	–	0 ... 20 mg/m^3	–
H_2O	0 ... 40 vol.%	–	–
CH_4	0 ... 50 mg/m^3	0 ... 500 mg/m^3	–
C_{ges}	0 ... 15 mg/m^3	0 ... 30 mg/m3	–
O_2	0 ... 25 vol.%	–	–

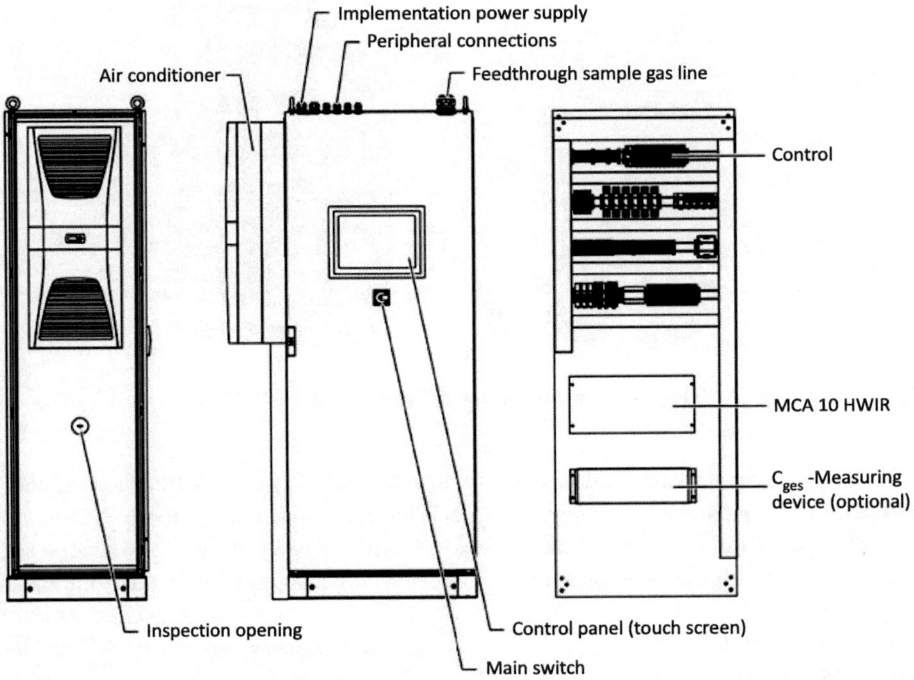

Fig. 14.96 MCA 10 HWIR analysis cabinet

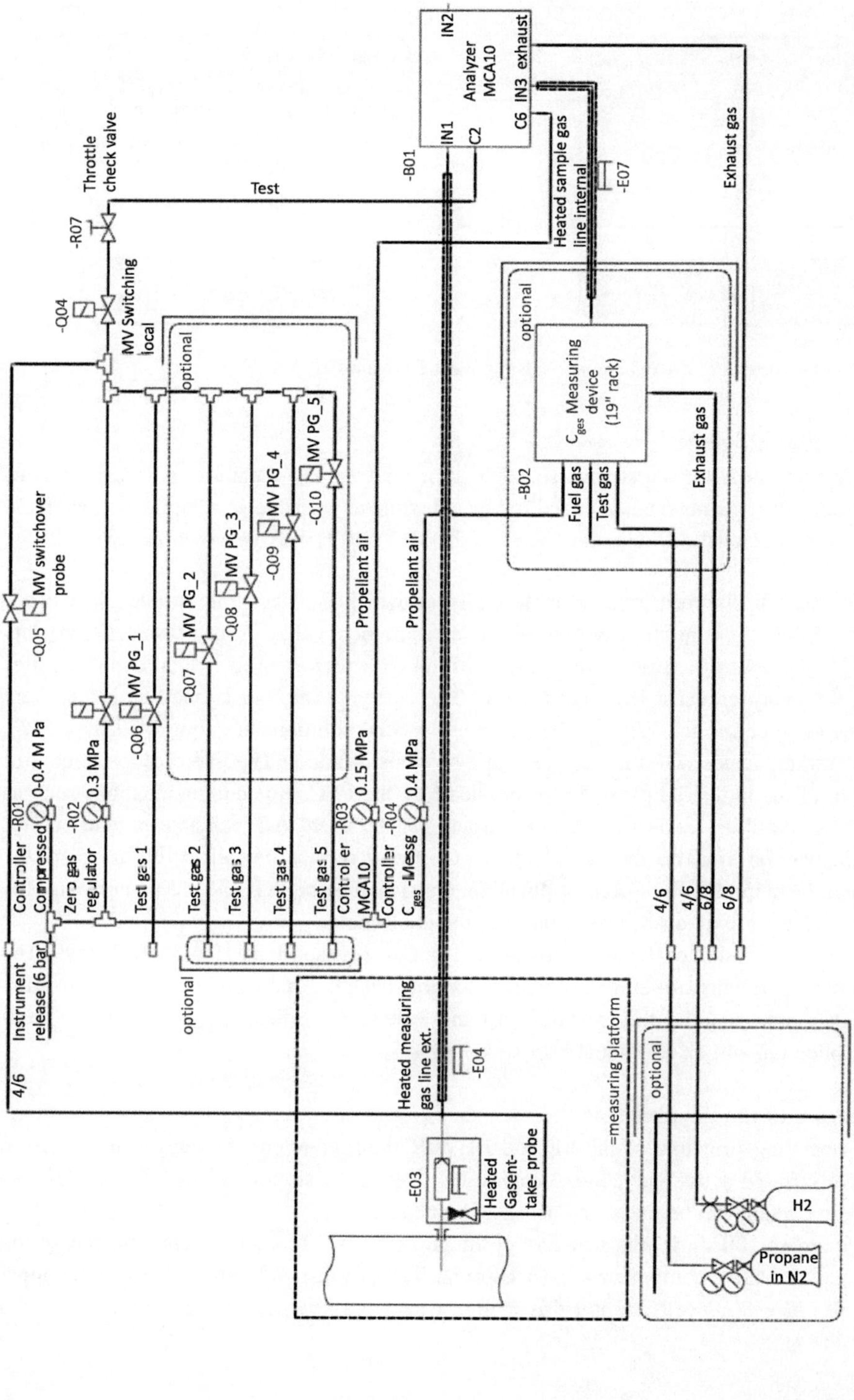

Fig. 14.97 Gas flow chart of a 17th BImSchV system with optional test gas connection

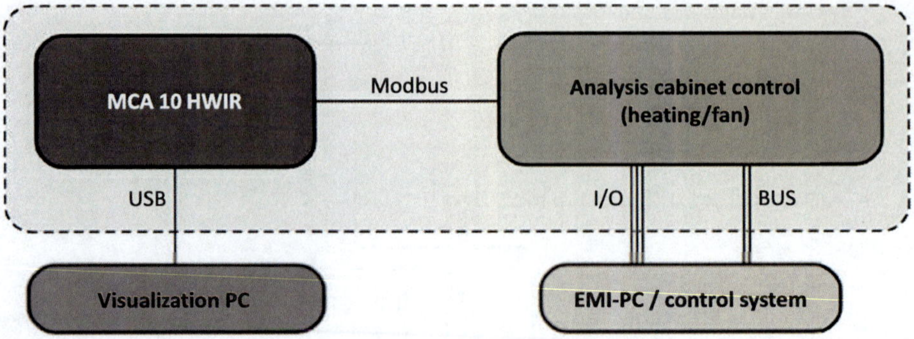

Fig. 14.98 Structure of CPU components (application example)

Automation Concept

In order to meet the most diverse application and environmental conditions, modern analytical systems must be designed to be as flexible as possible. An application example of the structure of the CPU components in the MCA 10 HWIR is shown in Fig. 14.98.

The heart of the measurement is the analyzer, which evaluates the sample gas concentration that flows continuously through the measurement cell. For user visualization, data storage and control, an operating panel with touch screen is used, which communicates with the analyzer via a USB connection. The system components such as heaters and valves, time-controlled zero setting and other special solutions are controlled by a PLC. This communicates with the analyzer via a Modbus protocol. The interface to the control system of the industrial plant is also provided by the PLC. Bus communications such as Modbus, Profibus, CAN-Bus or mA outputs can be used. All available remote control options can be used via the visualization PC, giving service personnel full access to the system. For a minimalist system application, the visualization PC is not necessary, since the entire calculation algorithm takes place in the photometer.

A mobile version of the multi-component analyzer is the MCA 10 m in Fig. 14.99. The system components are set in the smallest possible design. A tablet is used for visualization and data storage. The thermostatting of the system is realized by means of a speed-controlled fan and an additional heat sink.

Structure of the Photometer

The operating principle of the MCA 10 HWIR multi-component analyzer allows up to twelve infrared active gas components to be determined simultaneously. Optionally, the oxygen content can be measured using a zirconia cell.

Figure 14.100 shows the structure of the photometer. The optical bench for measuring the infrared active components consists primarily of an infrared light source with chopper wheel, a measuring cell, a motor-driven filter wheel and a pyroelectric detector. The heated

Fig. 14.99 Mobile hot gas analyzer MCA 10 m

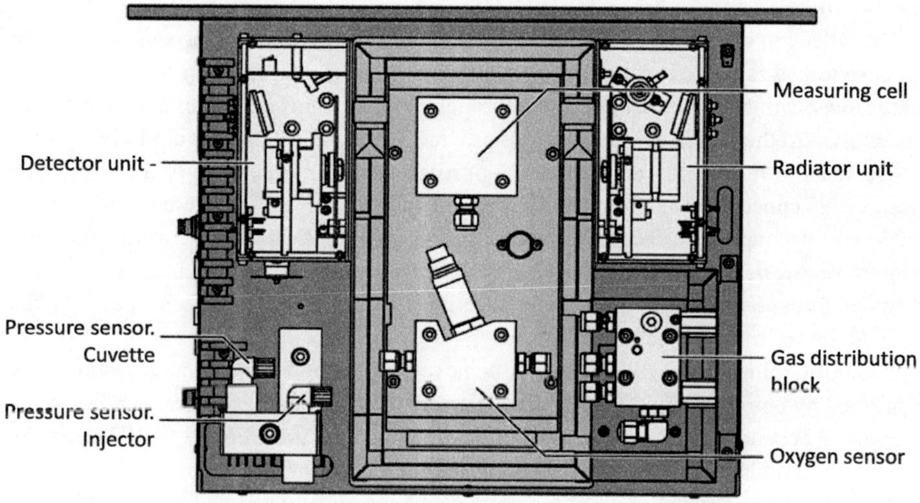

Fig. 14.100 Design of the MCA 10 HWIR photometer

components, such as the gas distribution block and measuring cell, are surrounded by an effective insulating material for thermal separation from the rest of the photometer.

The infrared light source contained in the emitter unit emits a broadband, constant light beam. This radiation is collected by a focusing mirror, modulated by the chopper wheel and directed into the measuring cell, through which the sample gas flows continuously. Pressure sensors are used to monitor the cell pressure and the injector pressure on the suction side. In order to compensate for the negative pressure in the cell, all measured

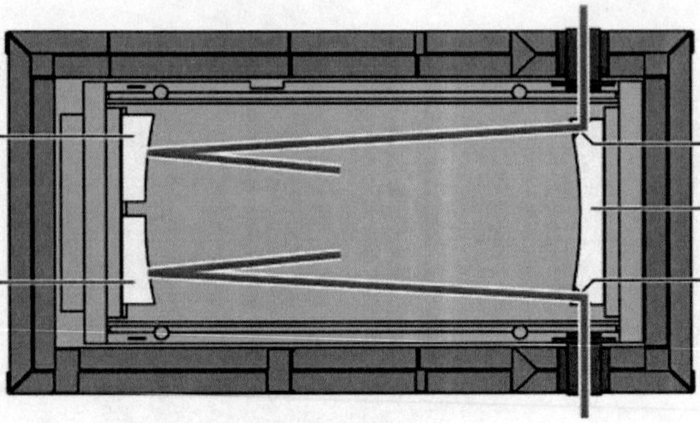

Fig. 14.101 Analysis cuvette with one beam folding (white cell)

values are subjected to pressure normalization. The structure of the cell is shown in Fig. 14.101.

The reflection of the infrared light begins at the coupling surface at the sphere mirror in the cuvette and is passed on to the second mirror. A path length of up to 10 m can be achieved by variable multiple reflections through the three mirrors. By default the cuvettes are adjusted to the length of 7.2 m. These provide a sufficiently high signal-to-noise ratio for the smallest measuring ranges and do not reach the saturation range of the absorption even at high concentrations. Infrared radiation is guided from the cuvette into the detector unit via the decoupling surface on the sphere mirror. The filter wheel, a focusing mirror and the pyroelectric detector with its control electronics are located there. The infrared radiation is guided through the measuring or reference filters located on the filter wheel and via the focusing mirror to the detector.

The positioning of the filter wheel is achieved by a stepper motor and continuously monitored by an optical light barrier. The interference filters and gas filters applied to the filter wheel depend on the type of gas components to be measured. Figure 14.102 shows an example of the filter wheel configuration for the MCA 10 HWIR.

After the infrared radiation has been component-specifically selected by the filters, it is directed via a further focusing mirror to the broadband pyroelectric detector. The optical bench is of simple design to meet the adverse conditions in industry. For example, the light beam is focused without complex optics and the light is only parallelised via the elliptical focusing mirror.

Measurement

As usual with filter wheel photometers, the MCA 10 HWIR also uses gas filter correlation and the bifrequency measuring method. The oxygen measurement is performed with the aid of a zirconium dioxide cell. In the bifrequency measuring method the components SO_2, CO_2, NO_2, H_2O and HF are measured. Figure 14.103 shows a real spectrum of an SO_2

Fig. 14.102 Filter wheel MCA 10 HWIR

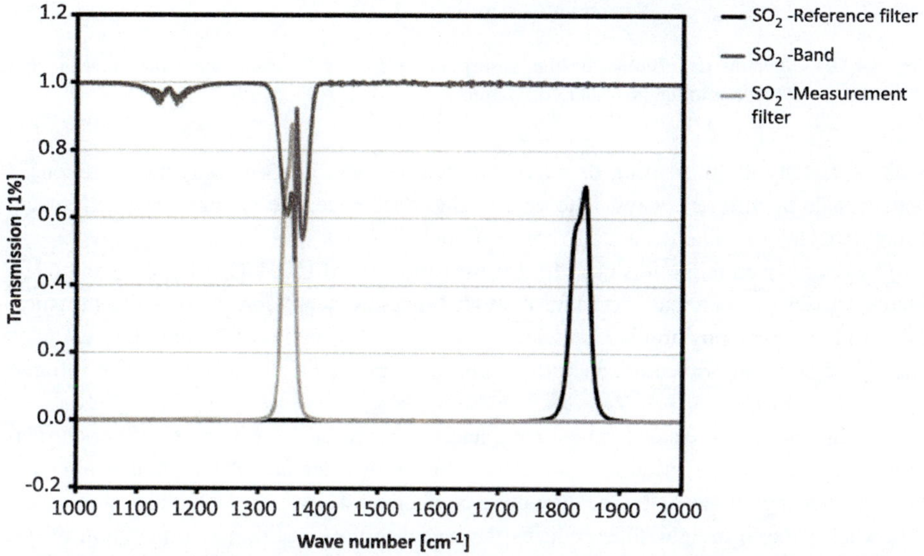

Fig. 14.103 Spectral distribution of the absorption band and the measuring and reference filters for an SO_2 measurement using the bifrequency measurement method

measurement. The measurement filter is located in the active band of SO_2, whereby the reference filter is not affected by SO_2.

A disadvantageous property of bifrequency measurement is the sometimes strong cross-sensitivities to other components in the gas mixture. If a gas component in the spectral band

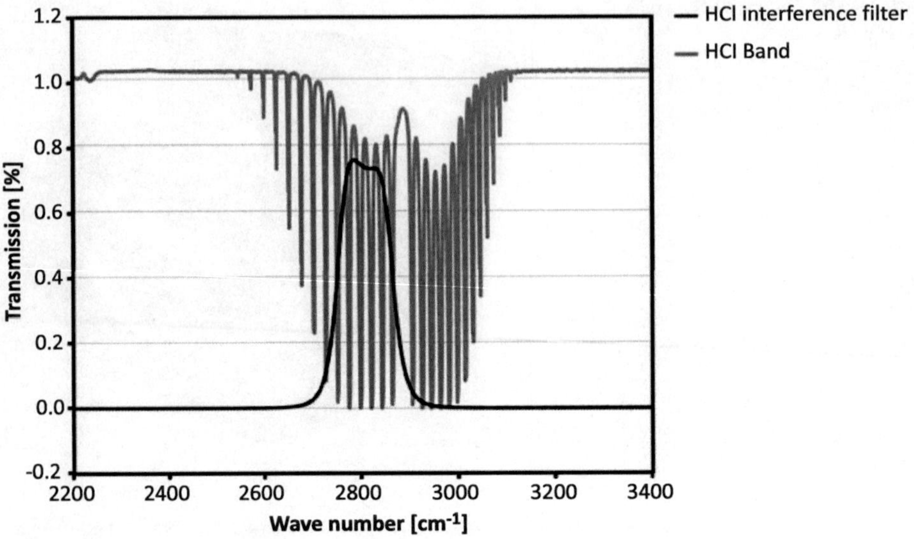

Fig. 14.104 Spectral distribution of the absorption band and measurement filter for an HCl measurement according to the gas filter correlation

is absorptive by the measuring or reference filter, the interference component influences both signals in different ways. The greater this difference, the greater the interference, which must later be mathematically compensated.

The gas filter correlation is used for the measurement of CO, NO, HCl, CH_4 and NH_3. Figure 14.104 shows a real spectrum of an HCl measurement. Here the spectral measurement and reference position is identical. By swinging the HCl gas filter into the reference signal, the substance-specific spectrum is almost completely blanked out of the infrared light.

The disruptive influence is almost completely compensated for in the gas filter correlation by the measuring principle, since the measuring filter and the reference filter are influenced in the same way by the disruptive component. However, the additional radiation loss when passing the gas filter reduces the signal-to-noise ratio. The production of gas filters is complex and cost-intensive. High safety requirements must be observed during use.

The synchronization between filter wheel, chopper wheel and the detector evaluation is essential for the stored calculation algorithm. A measurement cycle includes a measurement of each filter position on the filter wheel. The absorption A is calculated according to Eq. 14.32 from the energy values of the measuring filter E_{Mess} and reference filter E_{Ref}, taking into account the zero point N.

$$A = \frac{E_{\mathrm{Ref}}}{N \cdot E_{\mathrm{Mess}}} - 1 \tag{14.32}$$

The zero point N is determined at the time of inerting the cuvette according to Eq. 14.33.

$$N = \frac{E_{\mathrm{Ref}*}}{E_{\mathrm{Mess}*}} \tag{14.33}$$

With the calculated absorption the first additive calculation loop is run through. The previously calibrated cross-sensitivities are added to the absorptions of the other measuring channels in several iteration steps. Table 14.8 lists the cross-sensitivities to be compensated to the measuring components. The concentration value is calculated from the iteratively determined absorbance using an allocation table.

Oxygen Measurement

Oxygen is measured with the aid of a zirconium dioxide cell (Fig. 14.105). Inside the cell, the sample gas is separated from the reference gas (ambient air) by a zirconia membrane. Depending on the partial pressure of the oxygen, an oxygen ion movement through the membrane occurs. This leads to an electrical potential difference.

The oxygen sensor used consists of a measuring cell and a pump cell, which ensures a constant oxygen concentration in the measuring cell. The energy expended is a measure of the oxygen concentration.

The proven proportionality of the raw signal to the oxygen concentration ensures consistent accuracy in all concentration ranges.

Table 14.8 Compensation table

Measuring	Cross sensitivity component											
component	HF	CO_2	HCl	H_2O	CH_4	CO	NO_2	SO_2	N_2O	NO	NH_3	O_2
HF	–	•		•	•							
CO_2		–		•	•							
HCl		•	–	•	•							
H_2O		•		–	•							
CH_4		•		•	–							
CO		•		•		–						
NO_2				•	•		–			•		
SO_2		•		•	•			–		•		
N_2O		•		•		•			–	•		
NO		•		•						–		
NH_3		•		•							–	
O_2				•								–

•Compensation required

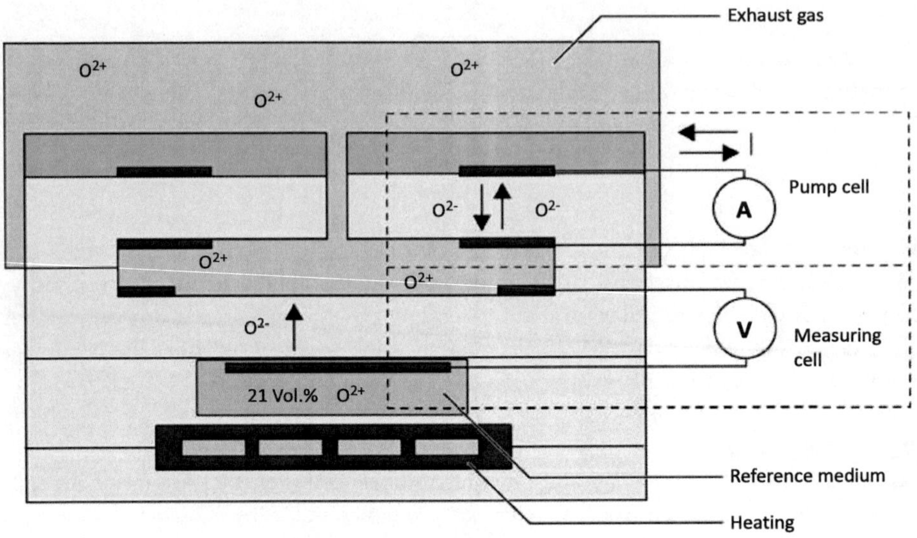

Fig. 14.105 Operating principle of the oxygen measurement

The MCA 10 HWIR multi-component analyzer calculates internally all concentrations required according to the specification with all necessary compensations and standardizations. The main board performs all tasks of photometer control, sensor evaluation, concentration calculation and interface communication.

Emission Monitoring Relevant to Authorities

In order for an analysis system to be used for emission monitoring relevant to the authorities in large combustion plants, waste incineration plants or solvent-using industrial plants, it must meet the first quality assurance level (QAL1) according to DIN EN 14181. The requirements for this are laid down in DIN EN 15267-3 "Air quality - Certification of automatic measuring systems - Part 3: Minimum requirements and test procedures for automatic measuring systems for the monitoring of emissions from stationary sources" and must be confirmed by an authorised testing institute by means of a complex and lengthy suitability test. The test procedure provides for a laboratory test and a field test of two identical automatic measuring systems (AMS). Due to the depth of DIN EN 15267-3, the scope of testing is only partially explained below.

The scope of the equipment at the time of the suitability test is binding for the measuring equipment used later and relevant to the authorities. This includes not only the photometer design and software version, but also all system elements such as probe, sample gas line, cabinet air conditioning, cabinet control, etc. Any changes to the system must be notified to the testing institute commissioned and will be assessed by the latter. Depending on the

Table 14.9 Categorization for device changes

Category	Description	Scope of the audit
Type 0	Changes that have no measurable impact on the performance of the AMS	Notification of change
Type 1	Changes that could have an impact on the performance of the AMS, although subsequent tests show that the impact of such changes is not significant	Supplementary test
Type 2	Changes that have a significant impact on the performance of the AMS	Supplementary test or new test

extent of the change, the results of the original suitability test must be verified by supplementary tests. Modifications are classified in categories listed in Table 14.9. For a multi-component analyser such as the MCA 10 HWIR, these minimum requirements apply to each of the thirteen measuring components. In addition, the response factor determination of the total hydrocarbon measurement is also required. Both proficiency testers must meet the minimum requirements equally. The deviations of the AMS determined in the process are evaluated individually and are included in the predicate of the overall measurement uncertainty for each component.

Laboratory Test

In a laboratory test, in addition to general minimum requirements, the properties listed in Table 14.10 are examined and evaluated in detail. The table is an excerpt from DIN EN 15267-3 and shows the individual limit values for the general minimum requirements of the AMS.

Test Setup Laboratory

The test setup in the laboratory is shown in the application example in Fig. 14.106. Two identical analyzer systems (suitability test apparatus 1 and 2) provided by the instrument manufacturer measure the same gas mixture. In order to avoid mutual interference between the analyzer systems and feedback to the mass flow controllers, the line after the mixing section is opened (atmospheric pressure). The offered volume flow of the gas mixture must be large enough so that no false air is sucked through the systems. The gas mixture is fed in a defined way via various mass flow controllers, a pump and an evaporator. The data are then compared by use of computer-based evaluation software.

Field Test

The field test of an AMS is normally carried out on an industrial plant which shows the most difficult boundary conditions (DIN EN 15267-3) for the application area for which the manufacturer is seeking certification. In addition to general minimum requirements, the properties listed in Table 14.11 are examined in detail. The field test duration must be at least three months and may not be interrupted during the entire test period.

Table 14.10 Minimum requirements for AMS, laboratory test (DIN EN 15267-3)

Process parameter	Minimum requirements Gases except O_2		Examination according to Subsection
Setting time	≤ 200 s ≤ 400 s for NH_3, HCl and HF	≤ 200 s	10.9
Repeated standard deviation at zero	≤2.0 %[a]	≤0.20 %[b]	10.10
Repeat standard deviation at reference point	≤2, 0 %[a]	≤0.20 %[b]	10.11
Lack of fit	≤ 2.0% [a]	≤0.20 %[b]	10.12
Influence of temperature at zero point	≤5.0 %[a]	≤0.50 %[b]	10.14
Influence of the temperature at the reference point	≤5.0 %[a]	≤0.50 %[b]	10.14
Influence of the sample gas pressure at the reference point for Δp from 3 kPa	≤2.0 %[a]	≤0.20%[b]	10.15
Influence of the sample gas volume flow	≤2.0%[a]	≤0.20 %[b]	10.16
Influence of the mains voltage −15%, +10%	≤2.0 %[a]	≤0.20 %[b]	10.17
Influence of vibrations	≤2.0 %[a]	≤0.20 %[b]	10.18
Cross-sensitivity	≤4.0 %[a]	≤0.40 %[b]	10.19
Emigration of the measuring beam for in-situ AMS	≤2.0 %[a]	–	10.20
NO_x Converter efficiency	≥95.0 %	–	10.21

[a]Percentage value related to the upper limit of the certification range
[b]Percentage value as oxygen volume concentration (volume fraction)

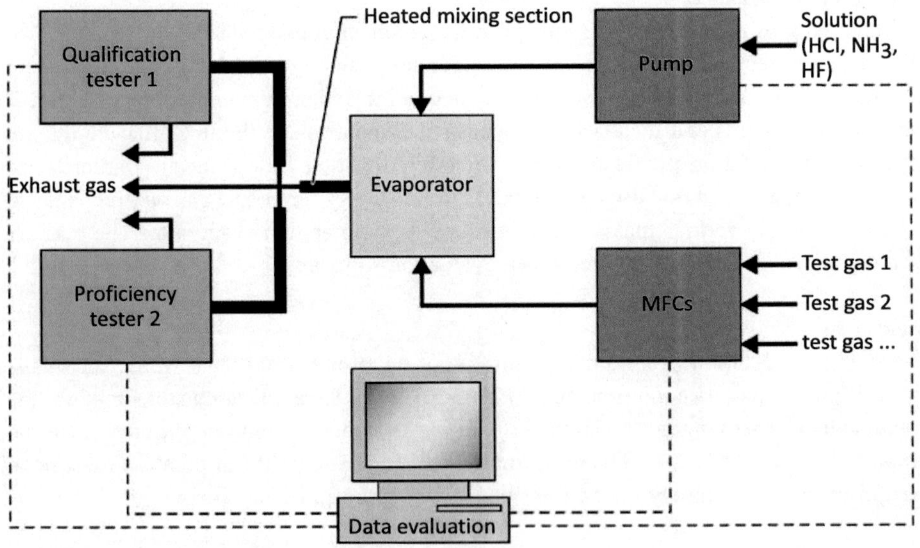

Fig. 14.106 Test setup laboratory (application example)

Table 14.11 Minimum requirements for AMS, field test (DIN EN 15267)

Process parameter	Minimum requirements Gases except O_2		Examination according to Subsection
Correlation coefficient of the calibration function, R^2	≥ 0.90	≥ 0.90	12.1
Setting time	≤ 200 s ≤ 400 s for NH_3, HCl and HF	≤ 200 s	12.2
Lack of fit	$\leq 2.0\%$ [a]	$\leq 0.20\%$ [b]	12.4
Shortest maintenance interval	8 days	8 days	12.4
Zero point drift during maintenance interval	$\leq 3.0\%$ [a]	$\leq 0.20\%$ [b]	12.5
Reference point drift during maintenance interval	$\leq 3.0\%$ [a]	$\leq 0.20\%$ [b]	12.5
Availability	$\geq 95.0\%$	$\geq 98.0\%$	12.6
Comparison precision, R_{field}	$\leq 3.3\%$ [a]	$\leq 0.20\%$ [b]	12.7

[a]Percentage value related to the upper limit of the certification range
[b]Percentage value as oxygen volume concentration (volume fraction)

Test Setup Field

The test setup in the field is shown in the application example in Fig. 14.107. The setup is not fundamentally different from that in the laboratory. In order to obtain comparable results, the two proficiency testers are installed on the same stack in the industrial plant. To validate the minimum requirements listed in Table 14.11, the field test includes various test gas tasks. For test gas experiments, both measuring systems can be guided over a sampling probe and draw in the same test gas mixture via the test gas inlet.

At the beginning and end of the field test, comparative measurements are performed to determine the calibration functions for the measured components. At the same time, the response times and linearities in the field are also determined. To determine the zero and reference point drift, a test gas application and a zero point check must be performed ten times during the test period, and at a minimum every 4 weeks.

The availability must be higher than 95% for all components and higher than 98% for oxygen. It should be noted that interruptions caused by malfunctions, maintenance work and zero and reference point checks have a negative effect on the availability calculation.

The measurement data of the two systems from a simultaneous, continuous and three-month field test interval are used to calculate the reproducibility. This includes the synchronization of the measurement results. All half-hourly mean values between 30% and 100% of the limit value of the measurement component to be monitored are included in the analysis. According to DIN EN 15267-3, the reproducibility R_f is calculated according to Eqs. 14.34 and 14.35.

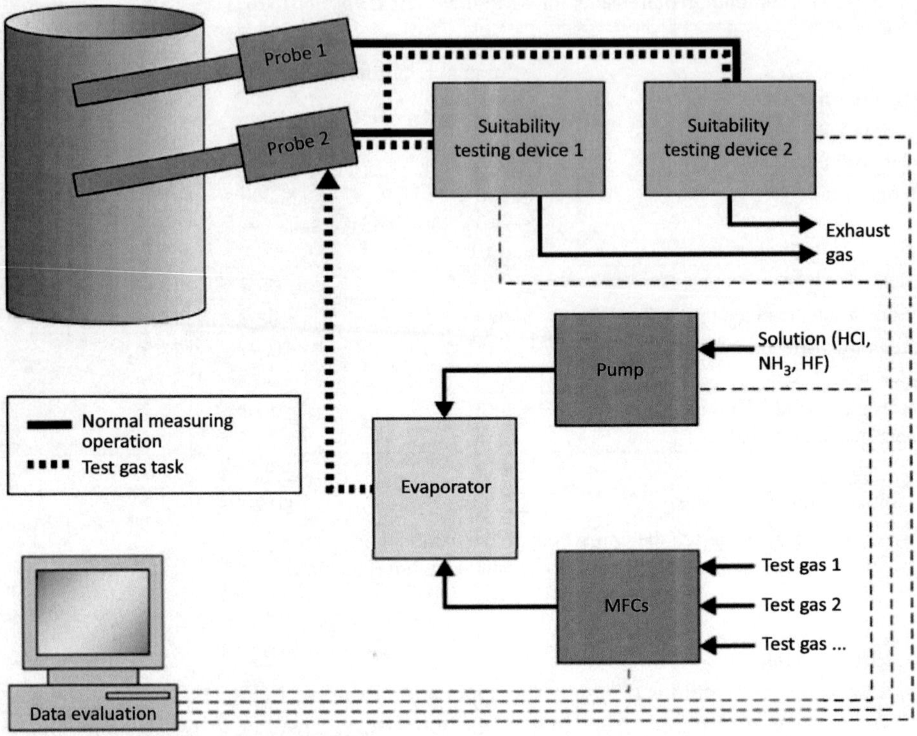

Fig. 14.107 Test setup field (application example)

$$R_f = t_{n-1;0{,}95} \cdot s_D \tag{14.34}$$

$$s_D = \sqrt{\frac{\sum_{i=1}^{n} (x_{1,i} - x_{2,i})^2}{2n}} \tag{14.35}$$

The following applies to the calculation:

R_f	Reproducibility under field conditions
$t_{n-1;0,95}$	Student *t-factor* for bilateral segregation and confidence level of 95% and a number of degrees of freedom of $n-1$
s_D	Standard deviation of the differences determined from duplicate determinations
$x_{1,i}$	*i-th* measurement result of the first measuring device
$x_{2,i}$	*i-th* measurement result of the second measuring device
n	Number of duplicate determinations.

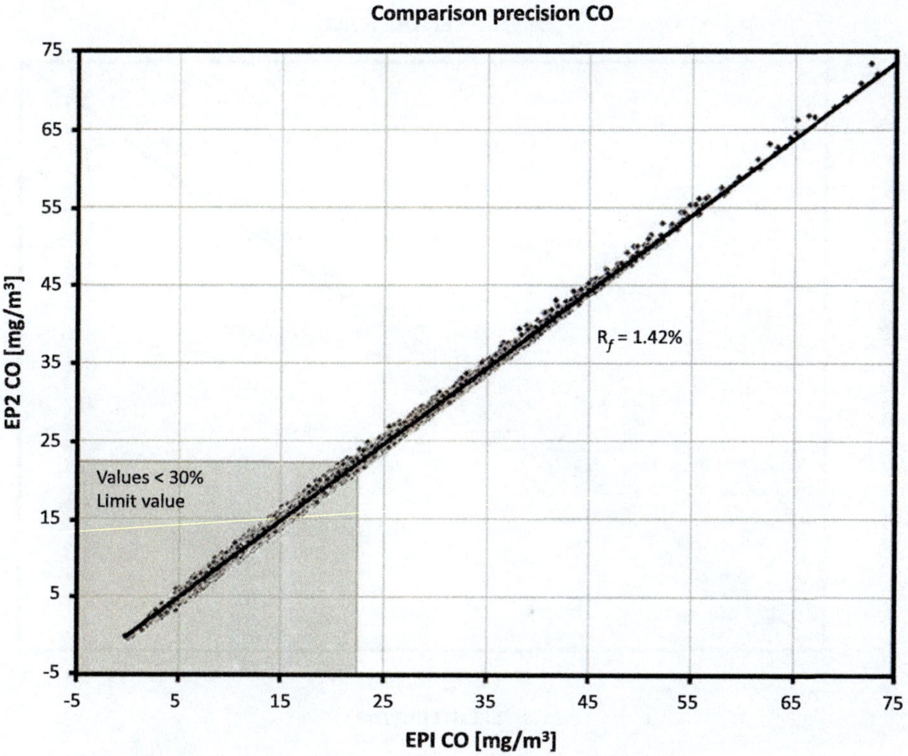

Fig. 14.108 Comparison precision from double determination, measuring range 0–75 mg/m^3 CO

Since the number of duplicate determinations is sufficiently high for a period of three months, a student factor of 1.96 can be assumed. The smaller the difference between the pairs of values, the smaller the standard deviation. The minimum requirement for reproducibility is 3.3% of the certification range or 0.2% by volume for oxygen.

A clear illustration for the significance of the reproducibility from the double determination can be seen in Figs. 14.108 and 14.109.

After the report has been prepared by the testing institute, it is submitted to the Länder Committee for Immission Control (LAI). The members of this committee are representatives of the highest authorities responsible for immission control in the Länder (LAI Homepage, 2015). Afterwards, the approval will be published in the Federal Gazette.

It is becoming increasingly difficult for device manufacturers to achieve QAL1 certification. On the one hand, the required measuring ranges of the components are approaching the technically feasible limits of the currently existing measuring methods. On the other hand, the increased requirements lead to additional testing costs, so that suitability tests are becoming increasingly less attractive. Especially for smaller equipment manufacturers, the testing of such a complex multi-component analyzer is a high economic risk.

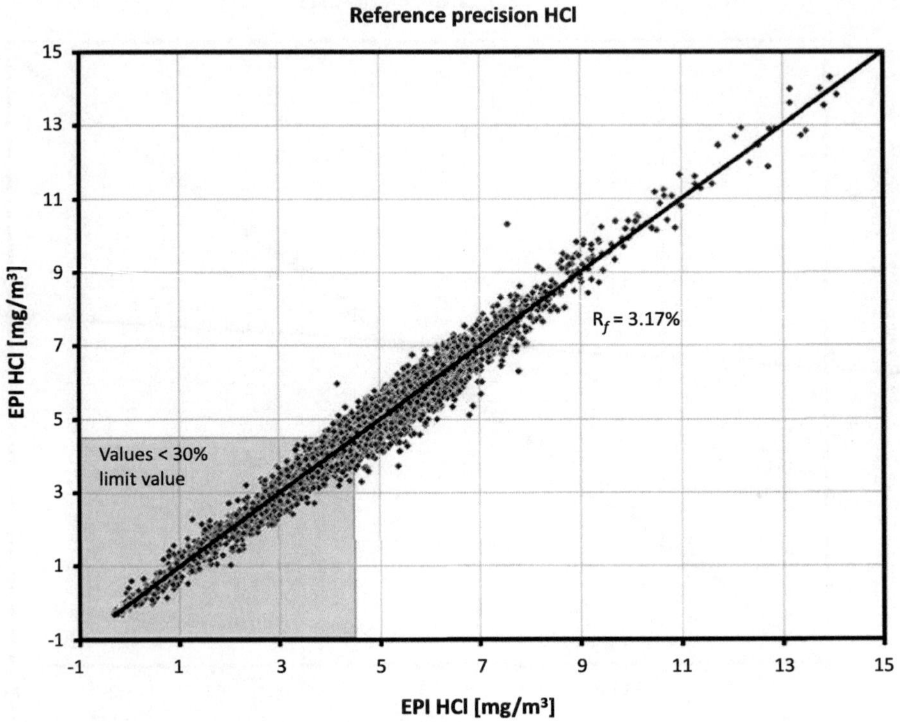

Fig. 14.109 Comparison precision from double determination, measuring range 0–15 mg/m^3 HCl

14.6 Official Emissions Monitoring of Incineration Plants

Roland Bianchin, David Triebel, and Thomas Eisenmann

Operators of industrial and commercial plants are obliged by law to have the air pollution emitted by their plants regularly inspected. The Federal Immission Control Act (BImSchG) with its associated ordinances (BImSchV) and administrative regulations serves as the basis for emission monitoring. In principle, the BImSchG offers the administration a wide range of instruments for enforcing the basic obligations of the operator and thus ensuring the safety of installations *requiring a permit with* regard to human health, material goods and the environment.

Principles of Emission Monitoring in Germany

The Technical Instructions on Air Quality Control (TA Luft) is the administrative regulation for the Federal Immission Control Act of the German Federal Government. It is addressed to the approval authorities for industrial and commercial plants requiring approval. Based on the general requirements of TA Luft, the respective authorities draw up adapted requirements which must be fulfilled by the plant operator. Even existing, old plants must reach the state of the art and reduce pollutant emissions within certain transition periods. The administrative regulation *TA-Luft* is binding for various plants according to BImSchG including incineration plants up to 50 MW. Based on the pollutant mass flow emitted by the industrial plant, the competent authority requires the operator to continuously measure the pollutant concentration. For power plants, heating plants and refineries with services via 50 MW, waste and co-incineration plants as well as crematoria and biomass plants, the monitoring equipment to be provided by the operators, the limit values to be observed and exemptions are laid down in the relevant Federal Immission Control Ordinances (BImSchV). The operator must ensure that the plants are constructed and operated in such a way that the requirements arising from these regulations are met. The competent authority is obliged to punish any infringements of these regulations and is entitled to shut down the plant.

TA-Luft
The purpose of this Technical Guidance Document is to protect the general public and the neighborhood from harmful effects of air pollution on the environment and to take precautions against harmful effects of air pollution in order to achieve a high level of protection of the environment as a whole.

First BImSchV (Ordinance on Small and Medium-Sized Firing Installations)
This Ordinance shall apply to the construction, condition and operation of combustion plants which do not require a permit in accordance with §4 of the Federal Immission Control Act.

Second BImSchV (Ordinance on the Limitation of Emissions of Volatile Halogenated Organic Compounds)
This Regulation shall apply to the construction, nature and operation of installations containing solvents or other volatile halogenated organic compounds.

13th BImSchV (Ordinance on Large Combustion and Gas Turbine Plants)
This Regulation shall apply, with certain exceptions, to the construction, characteristics and operation of combustion plants, including gas turbine plants and gas turbine plants designed to drive machinery with a rated thermal input of 50 megawatts or more and using solid, liquid or gaseous fuels.

17th BImSchV (Ordinance on the Incineration and Co-incineration of Waste)
This Regulation shall apply to the construction, characteristics and operation of incineration or co-incineration plants. For exceptions, please refer to the Regulation.

27th BImSchV (Ordinance on Installations for Cremation)
This ordinance contains immission control provisions on air pollution control during the operation of crematoria. The aim of the regulation is to reduce the emission of air pollutants from crematoria.

30th BImSchV (Ordinance on Installations for the Biological Treatment of Waste)
This Regulation shall apply to the construction, nature and operation of installations in which municipal waste and waste which can be disposed of as municipal waste is treated by biological processes or a combination of biological and physical processes. For exceptions, please refer to the Regulation.

31st BImSchV (Ordinance on the Limitation of Emissions of Volatile Organic Compounds Due to the Use of Organic Solvents in Certain Installations)
This Regulation shall apply to the construction and operation and where activities involving the use of organic solvents are carried out, in so far as solvent consumption in the activities concerned exceeds the thresholds specified. The areas of application, thresholds and exceptions can be found in the Regulation. In addition to the above-mentioned administrative regulations and laws, the following guidelines also form the basis for continuous monitoring in Germany:

DIN EN 14181 Quality Assurance for Automatic Measuring Equipment
This standard describes the necessary quality assurance procedures to ensure that an automatic measuring system (AMS, commonly referred to as CEMS, Continuous Emission Monitoring System, see Fig. 14.110) for the measurement of emissions is capable of meeting specified requirements for the uncertainty of measured values.

Uniform National Practice in Monitoring Emissions
RdSchr. d. BMU from June 13, 2005 - IG I 2 - 45053/5 and from August 8, 2010 - IG I 2 - 51134/0. 2010 as well as the status identification and classification in the valid version from August 01, 2012. Directive on:

- the suitability testing of measuring and evaluation facilities for continuous emission measurements, the continuous recording of reference or operational parameters and for the continuous monitoring of emissions of special substances
- the installation, calibration and maintenance of continuously operating measuring and evaluation equipment,
- the evaluation of continuous emission measurements.

DIN EN 15267
Certification of automatic measuring equipment.

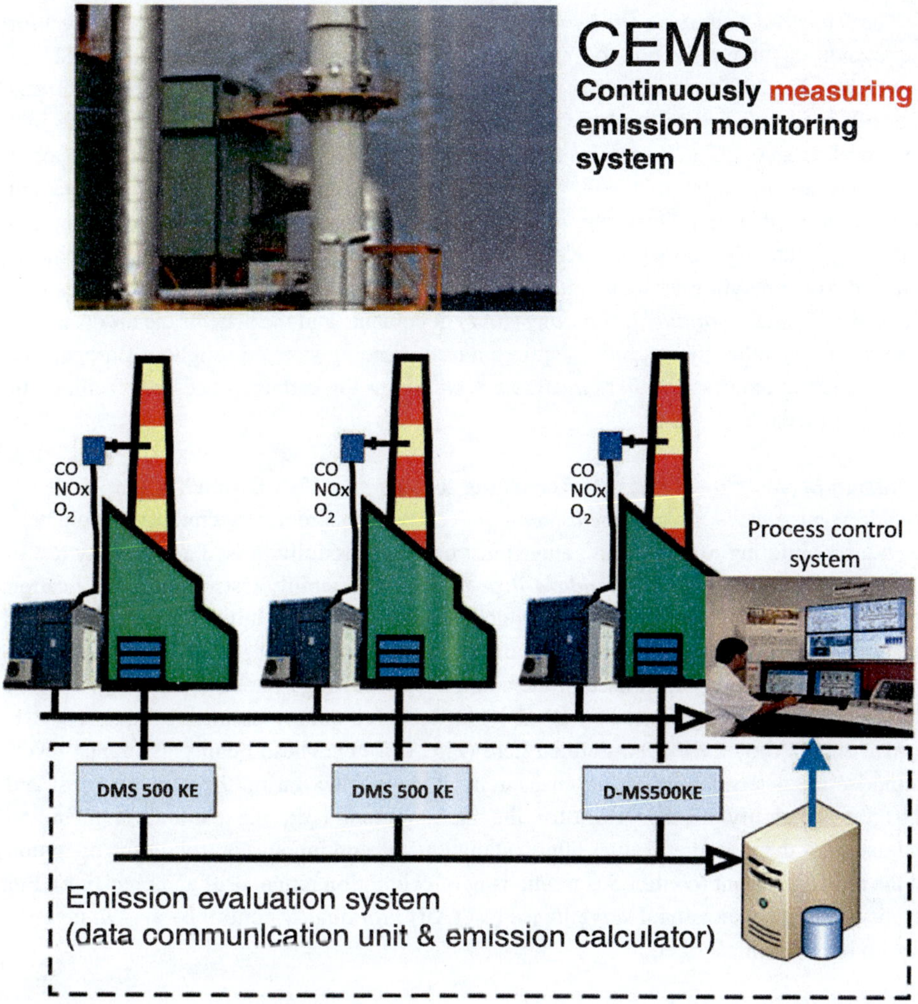

Fig. 14.110 Standard configuration of an emission monitoring system with continuous measuring systems and emission calculator. RobertRED - Cc-by-3.0-en

Part 1:	Basics
Part 2:	Initial assessment of the manufacturer's quality management system and monitoring of the manufacturing process after certification
Part 3:	Minimum requirements and test procedures for automatic measuring systems for the monitoring of emissions from stationary sources

The *Industrial Emissions Directive (IED)* is particularly important at European level for the avoidance and reduction of emissions. This European Directive 2010/75/EU of November 24, 2010 on industrial emissions IED (integrated pollution prevention and control) was published in the EU Official Journal on December 17 2010 and came into force on January6, 2011. It contains an integrated approach to the prevention and control of emissions to air, water and soil, waste management, energy efficiency and accident prevention. In this new Directive, seven different existing Directives have been revised and merged (among others Large Combustion Plants Directive, Directive on the incineration and co-incineration of waste, old IPPC Directive). With the entry into force of this IED Directive, the *best available technology (BAT)* documents and the state of the art described therein, among other things, have acquired a more binding status throughout Europe. The IED Directive requires the *best available technology* for certain processes, facilities or operating methods.

Emission Monitoring for Plants According to 13th and 17th BImSchV

Requirements on the measurement technology: The measurement technology equipment used for monitoring official limit values must have successfully passed a suitability test in accordance with the *Federal Standard Practice.* Such suitability tests require certification in accordance with DIN EN 15267, which, in addition to an initial assessment of the manufacturer's quality management (initial audit), also ensures regular monitoring during production. The quality assurance of the automatic measuring systems (AMS, Fig. 14.110) during operation is regulated in DIN EN 14181. Three quality assurance levels (QAL1, QAL2 and QAL3) have been defined here (Fig. 14.111). These quality assurance levels comprise the determination of the measurement inaccuracy during the suitability test and thus the suitability of the AMS for the measurement task, the calibration including validation of the measured values after installation and continuous control during operation at the industrial plant (continuous monitoring of calibration range, drift and precision). For quality assurance, an annual surveillance test (AFT) for quality control by a §29b measuring body is also prescribed.

The last amendment of this standard was made in February 2015 and *Predictive Emissions Monitoring Systems (PEMS)*[14] are used as an alternative to *Automatic Measurement Systems (AMS)*, especially outside the EU. To be suitable for this task, PEMS must be fully compliant with the applicable, valid regulations for emission monitoring. In the past, PEMS was mainly used according to the regulations of the US Environmental Protection Agency (US EPA). Therefore, there is no suitable German term for PEMS and PEMS is also used in German. PEM systems (PEMS) are software-based; therefore, they do not require gas analyzers and associated components such as sampling equipment, heated pipes or measuring containers such as an automatic measuring system (AMS). PEMS are

[14]The term is used in the automotive sector for Portable Emission Monitoring **Systems.**

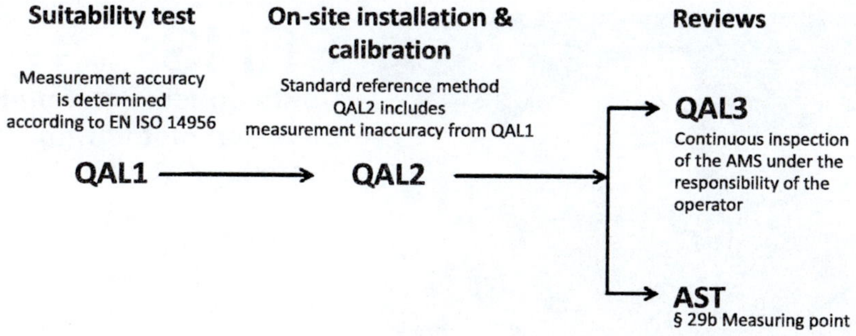

Fig. 14.111 Simplified representation of quality assurance according to DIN EN 14181

connected to the process control system (see Fig. 14.112) and use process variables provided by the process control system to enable continuous determination of air pollutants such as NO_x, SO_2, CO, NH_3, HC, etc., or O_2 and CO_2. PEMS are primarily suitable as an alternative to AMS for all emission sources that are gas or oil fired. When used correctly, PEMS provide results that are fully comparable to AMS in terms of accuracy and data quality. PEMS are often used together with emission calculators and in this combination, they represent self-sufficient monitoring solutions, although PEMS can of course also work independently. In order to be recognized as a certifiable, accepted alternative to AMS, PEMS must also be based on a solid foundation of regulations and strictly comply with their requirements, whereby regulations of the *U.S. Environmental Protection Agency (U.S. EPA)* have so far been the most widely used: Performance Specification (PS)16 as part of the so-called New Source Performance Standards (NSPS), also known as the 40 *Code of Federal Regulation Part 60.* Also, in 40 CFR Part 75 there are alternative monitoring requirements in Subpart E that PEMS fulfills. In Europe, a working group (Working Group 37) developed a PEMS standard CEN/TS 17198 within the CEN/Technical Committee TC 264 *Air Quality*, which will consider the relevant European standards EN 14181 and EN 15267.

Requirements on the Evaluation Technology

The minimum requirements for electronic evaluation equipment, including system-related classification regulations, are specified in detail in *national practice.* Below is an overview of the basic requirements for electronic evaluation systems (emission calculators).

- The evaluation facility must completely fulfil the registration, averaging, validation, classification, and evaluation according to the minimum requirements of the *Federal Uniform Practice.*
- The validated half-hourly average values are determined at the end of the integration interval from the integral values of the raw measured value after appropriate standardization (temperature, pressure, humidity) and oxygen reference value

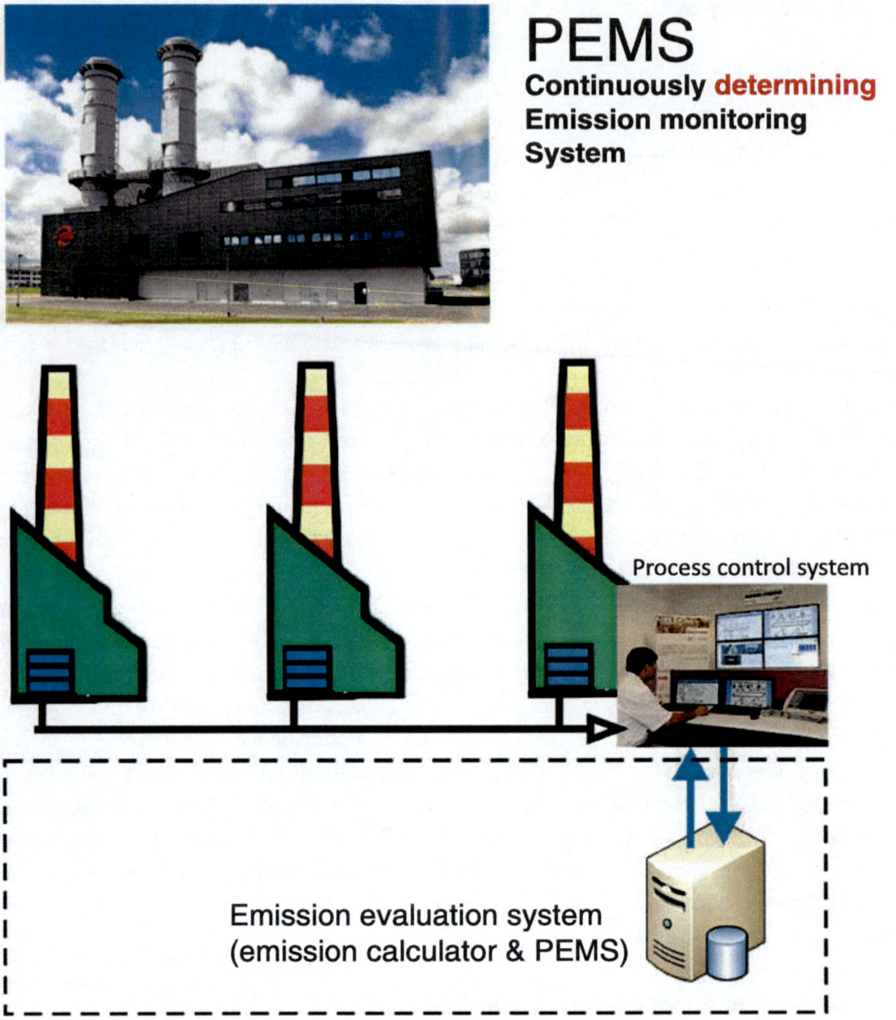

Fig. 14.112 Emission monitoring with PEMS and emission calculator. RobertRED - Cc-by-3.0-en

calculation after deduction of the measurement uncertainty (confidence interval) determined during calibration.

- Negatively validated mean values are set to zero.
- Validated mean values outside the valid calibration range (No. 6.5 DIN EN 14181) must be stored with the associated time and status.
- The daily mean values shall be calculated as the arithmetic mean of the validated half-hourly mean values.

- Any day on which more than 6 (13th BlmSchV) or 5 (17th BlmSchV) half-hourly mean values cannot be formed due to maintenance or malfunction of the continuous measuring system is declared invalid.
- A maximum of 10 such invalid days are permitted per year.
- All average values must be rounded and classified according to DIN 1333 (see also TA-Luft 2.9). In addition to this frequency distribution, the measured values relevant to the authorities shall be stored as individual values with the associated time (date, time), status and a characteristic value for the operating mode and shall be provided in table form with a reasonable time grid as a pdf file.
- Classification must take place from 00 hrs.
- The daily mean value shall be determined for the interval 00–24 h if at least 12 valid half-hourly mean values are available.
- With redundant systems (second, independent and spatially separated recording system for data), the daily protocol printout can be dispensed with.
- In redundant systems, the raw value recorders can be omitted if the electronic recording does not exceed an averaging time of 5 seconds.
- The availability of the evaluation facility must be at least 99%.
- Evaluation systems are to be equipped with a radio clock DCF77 and adjusted daily.
- Measurement inputs and inputs for status signals can be combined in a suitable digital interface between the measuring instrument and the evaluation unit (for details see DIN VDI 4201).
- It must be possible to test/simulate the digital interface.

According to these requirements, the measured pollutant values, the reference values and the associated status information (maintenance, malfunction, start-up, etc.) shall be continuously recorded by the emission computers and the standardized oxygen-related concentrations shall be calculated as follows (Fig. 14.113).

The measured values of the pollutants and the reference values are to be recorded by the Emissions Calculator with a resolution of <5 seconds and integrated over the period of time of 30 min as prescribed in Germany. The *half-hourly average values* thus obtained are the basis for the standardization with pressure, temperature and humidity, as well as the oxygen reference value calculation (depending on the respective plant type). For a correct standardization calculation, the principle of AMS (extractive, *in situ*) and the respective conditions of the plant have to be considered. If humidity and pressure conditions are constant, measurement of these parameters is not necessary. These reference values can then be taken into account via calibration or a correction invoice. The validation of the standardized oxygen-related measured values is only carried out for quantities that must comply with an official limit value. Freight is used for pollutant balancing and is determined from standardized values (i.e. not validated). Exception: Plants where official limit values for freight have to be observed (e.g. Plants according to 30. BImSchV). The

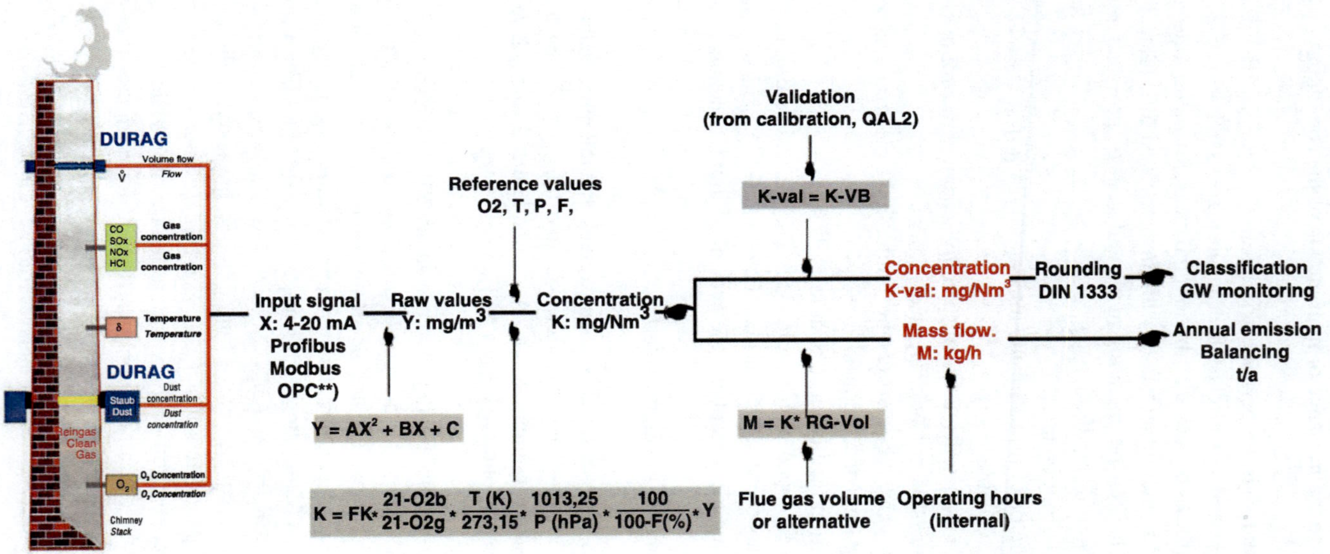

Legend: O2 - Oxygen X - Current value O2b- Oxygen reference value Fk -correction fact
T - temperature Y - Raw value O2g- measured oxygen
P - Pressure K - Concentration RG-Vol - Flue gas volume flow
F - Humidity M - mass flow N- in standard condition (dry, O 2 related*)

* - O2 reference value calculation for pollutants not applicable for measured values below the reference oxygen content for plants with ARE. (for boilers with DENOX equipment for NOx, for boilers with FGD's for S02 before and after FGD, as well as SAG calculations, for boilers with dust filters for dust).
** - digital interfaces according to DIN 4201 sheet 1-4

Fig. 14.113 Overviews of the calculation rules for emission monitoring validated half-hourly average values are to be rounded according to the current classification regulations before classification according to DIN 1333 (see also TA-Luft 2.9).

From the validated (unrounded) half-hourly mean values, an arithmetic calculation of the daily mean values is performed after the end of the day (quorum: at least 12 valid measured values). The determined result must be rounded according to DIN 1333 (see also TA-Luft 2.9). On days on which more than 5 (17th BImSchV) or 6 (13th BImSchV) half-hourly mean values are in maintenance or malfunction, an additional class (TS3) parallel to the daily mean value classification (T1 to T10, TS1) is used. For plants in accordance with 17th BImSchV, the temperature of the post-combustion zone shall also be recorded and stored as 10-minute mean values in the emission calculator. In accordance with the current regulations, the calculation of half-hourly, daily and 10-minute averages as well as annual averages will be mandatory in the future. These are arithmetically calculated from the valid daily mean values of the year and, together with the number of daily mean values used for calculation, must be stored in the Emissions Calculator System over a period of 6 years, like all other emissions-relevant measured and calculated values. As required by federal practice, all official measured values and plant statuses are stored in classes according to the plant type. Below is the classification regulation for plants according to 13th BImSchV (Fig. 14.114). Here, the half-hourly mean values are stored in 20 classes. The emission limit value occupies the end of class 10, the double limit value the end of class 20. The daily mean values are to be classified in 10 classes, here the end of class 10 corresponds to the emission limit value. The special classes provide information on the plant and measurement status. The classes S11 to S13 reflect the failures of the waste gas purification plant. It should be noted that the class memory for the cumulative failure of waste gas purification plants is not deleted at the turn of the year like all other classes but is maintained over a period of 12 months. Classes S15 and S16 are only activated in systems according to 17th and 30th BImSchV.

For plants according to 17th BImSchV, the half-hourly mean values are also classified into 20 classes (Fig. 14.115). However, the half-hourly limit value for these plants occupies the end of class 20. In addition, the 10-minute average value of the temperature of the afterburning zone must be classified in 20 classes. This classification is inverse, that is, temperatures above the minimum temperature are good (< Class 10), temperatures below the minimum temperature (>Class 10) are bad and must lead to reactions on the part of the plant (e.g. Setting the waste feed). Classes S13 and S14 are not assigned for systems according to 17th BImSchV. All start-up/shut-down processes must be classified in the *normal classes*.

According to DIN EN 14181, monitoring of the valid calibration range is required for all systems. This is determined by the calibration institute during QAL2 and has a range of 0–110% of the highest calibration value (Fig. 14.116). Classes S09 and S10 provide information on compliance with the valid calibration range.

The calibration function is only valid within the calibration range. Average values outside the valid calibration range must be stored by the emission calculator with the corresponding time and status. A new calibration QAL2 becomes necessary if within 1 year >5 % of all values per week of more than 5 weeks or >40 % of all values of at least 1 week

>4h-mean values	
M1	-Zero
M2	
M3	
M4	
M5	
M6	
M7	
M8	
M9	
M10	-GW
M11	
M12	
M13	
M14	
M15	
M16	
M17	
M18	
M19	
M20	- 2* GW
S1	GW Exceedance: >2* GW

Daily mean values	
T1	-Zero
T2	
T3	
T4	
T5	
T6	
T7	
T8	
T9	
T10	-GW
TS1	GW Exceedance: >GW
TS2	Formation of a valid TMW not possible
TS3	Day on which the AMS > 6 HMW was not iB due to WA orST
TS4	GW forSAG complied with
TS5	GW Excess: < GW for SAG

Special classes	
S1	GW Exceedance: >2* GW
S2	2/3 criterion not met
S3	with replacement value for reference value
S4	2/3 criterion not met because of STAMS
S5	2/3 criterion not met due to WA of the AMS
S6	regulated plant operation
S7	2/3-criterion not fulfilled due to the nature of the plant
S8	implausible values that do not fall into classes S2 to S5 and S7 as well as integration values that are not classifiable
S9	Short-time counter for 1 ntegration values outside the calibration range
S10	Long-term counter with regard to integration values outside the calibration range
S11	ARE failure
S12	current ARE failure, also beyond day change
S13	Default of ARE per 12 months, rolling
S14	GW overrun (> 2* GW) during start-up or shut-down operation

Fig. 14.114 Classification of plants according to the 13th BImSchV

>4h-mean values	
M1	-Zero
M2	
M3	
M4	
M5	
M6	
M7	
M8	
M9	
M10	
M11	
M12	
M13	
M14	
M15	
M16	
M17	
M18	
M19	
M20	_HGW
S1	GW exceedance: > HGW

10'-mean values	
$T_{NBZ}1$	$>T_{min}$
$T_{NBZ}2$	
$T_{HBZ}3$	
$T_{N3Z}4$	
$T_{NBZ}5$	
$T_{NBZ}6$	
$T_{NBZ}7$	
$T_{NBZ}8$	
$T_{NBZ}9$	
$T_{NBZ}10$	$\_T_{min}$
$T_{NBZ}11$	
$T_{NBZ}12$	
$T_{NBZ}13$	
$T_{NBZ}14$	
$T_{NBZ}15$	
$T_{NBZ}16$	
$T_{NBZ}17$	
$T_{NBZ}18$	
$T_{NBZ}19$	
$T_{NBZ}20$	$<T_{min}$
$T_{NBZ}21$	WA orST of the AMS

Daily mean values	
T1	-Zero
T2	
T3	
T4	
T5	
T6	
T7	
T8	
T9	
T10	_TGW
TS1	GW exceedance: >TGW
TS2	Formation of a valid TMW not possible
TS3	Day on which the AMS >5 HMW was not iB due to WA orST

Special classes	
S1	GW exceedance: >HGW
S2	2/3 criterion not met
S3	with replacement value for reference size
S4	2/3 criterion not met because of STAMS
S5	2/3 criterion not met because of WA ofAMS
S6	plant operation subject to supervision
S7	2/3-criterion not fulfilled due to the system
S8	implausible values that do not fall into classes S2 to S5 and S7 as well as integration values that do not have to be classified
S9	Short-time counter for integration values outside the calibration range
S10	Long-term counter with regard to integration values outside the calibration range
S11	ARE failure
S12	current ARE failure
S15	for dust HMW < 150 mg/m^3 and failure of the ARE
S16	for dust HMW > 150 mg/m^3 and failure of the ARE

Fig. 14.115 Class divisions for plants according to the 17th BImSchV

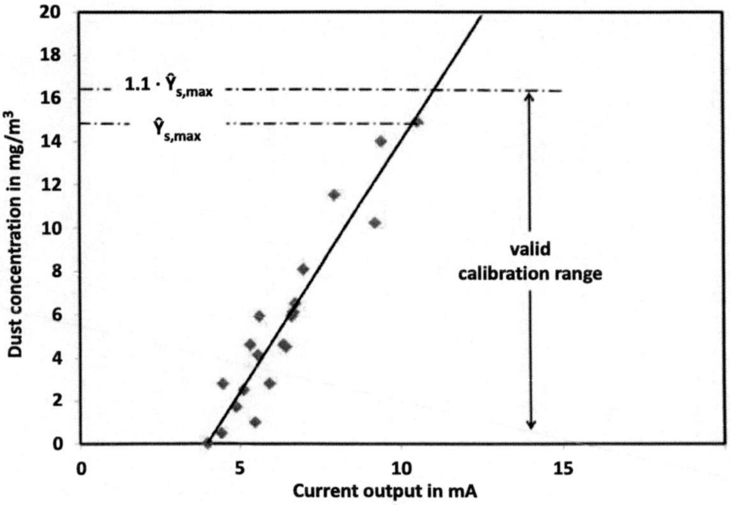

Fig. 14.116 Schematic representation of the valid calibration range

are outside the valid calibration range. If the plant is not operated continuously, the percentages may be calculated on the basis of the last 168 operating hours, which represent 1 week of plant operation.

Provision of Emission Data for the Authority

For online monitoring of emission values by the authority, there is the possibility to oblige the operator to regularly transfer the classified values to the authority. Two systems have been established for this purpose in Germany in recent years. The emission data transmission (EFÜ) in accordance with the *Federal Standard Interface Definition* and the data provision via a server on the Internet (EFÜ.www). According to U.S. Environmental Protection Agency (U.S. EPA, this is called Electronic Data Reporting (EDR).

Features:

EFÜ	serial transmission of the half-hourly average values via telephone modem in a specified time window daily after midnight The classification of the measured values and the calculation of daily mean values is carried out on the authority computer in parallel to the classification in the operator's emission computer. The authority has the possibility to receive immediate reports and to actively query current data if required.
EFÜ.www	on HTML sites on the Internet, the discrete 10-minute, half-hourly and daily mean values are available in line diagrams in both current and historical form and are provided, like the original classification reports from the operator's emissions calculator, as tables in pdf format. The values are updated after each classification. Using the immediate report function, current events can be automatically displayed on the Internet platform or sent by e-mail. The data transfer takes place via s FTP. Access to the data is secured by username and password.

Operators of installations pursuant to the 13th BImSchV are exempted from the obligation to prepare an annual report if data are transmitted continuously to the authority.

Basic Structure of an Emission Evaluation System

Figure 14.117 shows the modern, modularly structured emission evaluation system D-EMS 2000. The current version of the system developed by the Hamburg-based company DURAG data systems GmbH has been suitability-tested by TÜV Rheinland for plants according to TA-Luft, first, second, 13th, 17th, 27th, 30th and 31st BImSchV and announced as such in the Federal Gazette March 02, 2012, No. 36, page 920, Chapter III No. 1.2 by the *Länder Committee on Immissions*.

The certification was based on the requirements of the following standards and directives:

- Uniform national practice in monitoring emissions; RdSchr. d. BMU of 13.6.2005 - IG I 2 - 45053/5 and of 04.8.2010 - IG I 2 - 51134/0,
- Remote emission monitoring (EFÜ) Interface definition in the version of the LAI decision of September 28, 2005 (corrected version of November 15, 2006)
- DIN EN 14181,
- DIN EN 15267,
- VDI 4201.

The D-EMS 2000 system meets the current requirements for official evaluation systems, can be used for plants of any size due to its modular structure and is optimally prepared for future requirements. It enables operators to carry out transparent local evaluation of the emission values to be monitored and the associated process data in addition to an evaluation in conformity with the authorities. For the operator to comply with the legally required availability of 99% for such systems, modules with the latest fixed storage technology, raid components and external backup systems are provided, depending on the level of equipment. This also makes it possible to dispense with conventional raw value recorders and daily log printouts. To meet further requirements such as the remote transmission of emission data to the authority, the modules EFÜ, in accordance with the standard national interface definition, and EFÜ.www for data provision to the authority via the Internet are available. The system has the conventional standard interfaces (4–20 mA, binary contacts) required by the *national standard practice* and supports communication via digital interfaces (PROFIBUS, Modbus and OPC, according to VDI 4201) with measuring devices in the field. The interface to the process is provided by the D-MS 500 KE data communication units, which have internal data buffer memories of up to 96 days. In the event of problems with data transmission to the evaluation PC, the system workstation, or malfunctions of the system workstation itself, no data is lost. After the functionality has been restored, all raw data is automatically supplied, all calculations relevant to the

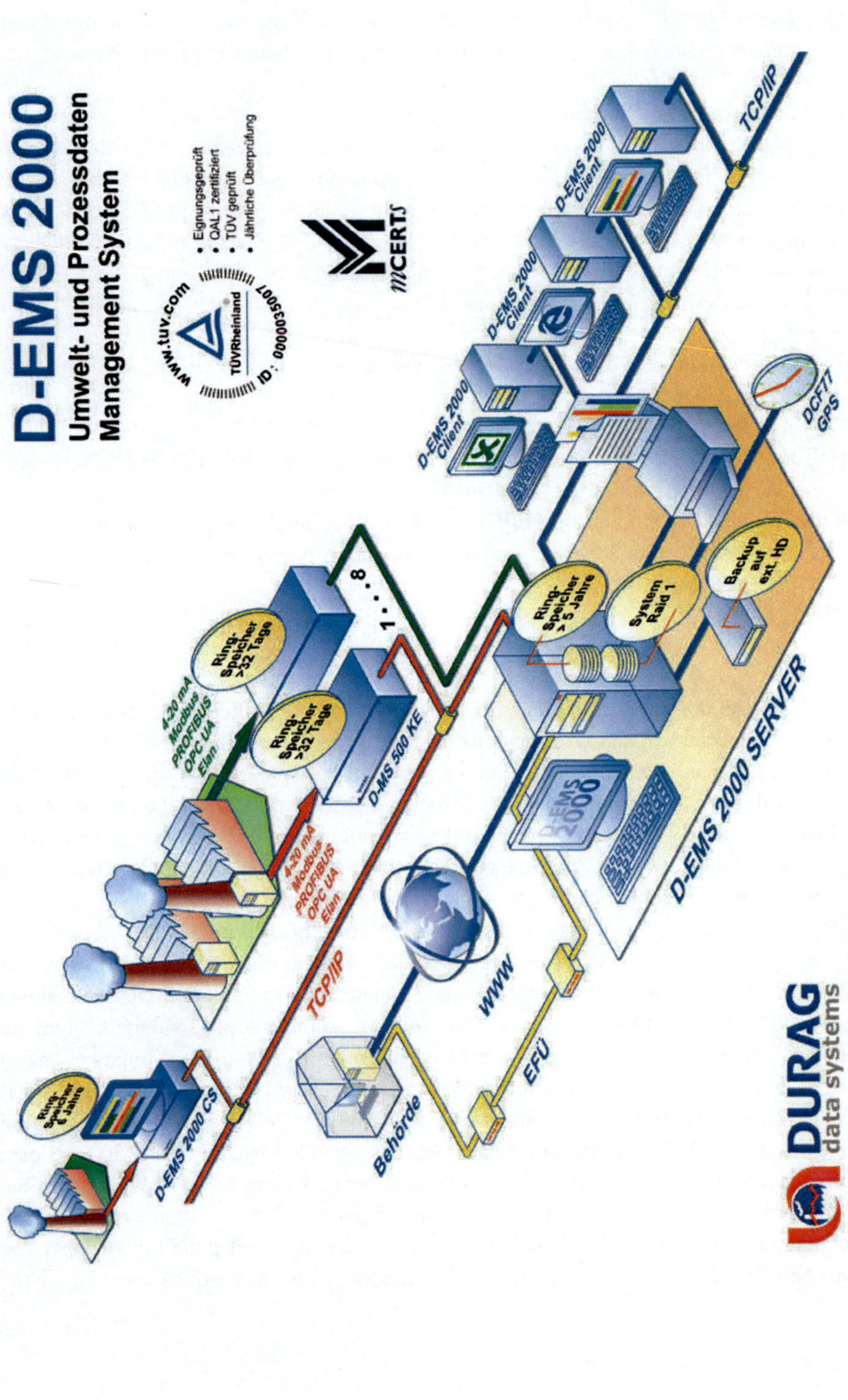

Fig. 14.117 Emission evaluation system D-EMS 2000

authorities, including the creation of classification reports, are carried out in the correct sequence and the data provision for the authority (EFÜ) is automatically completed. All data relevant to the authorities, including the daily classification protocols, are stored securely in the D-EMS 2000 system and are available for display on the screens or output on a color printer for the legally required period of 6 years. The screens of the system workstation and the PCs in the customer's data network can be used to display current data such as

- Raw values (second values),
- Instantaneous values (minute integrals),
- Prognostic trends for half-hourly values,
- Free loads for half-hourly values,
- Current daily mean values,
- Prognostic trends for end-of-day values,
- Free loads for daily mean values,
- Operating times for various operating modes,
- Incoming messages (including comment fields).

The following historical data (storage depth: local up to 6 years or unlimited from the external HD) can be freely selected by the user for display in line charts and/or tables:

- Secondary values,
- Minute readings,
- Half-hour averages,
- Daily mean values,
- Operating hours,
- Monthly averages and annual emissions,
- Comments and related messages,
- Protocols of authorities (classification protocols),
- Availability analyses.

In accordance with the new minimum requirements, which allow paperless data storage, the raw values and daily classification protocols are recorded electronically and stored securely on a redundant, spatially separated system (external hard disk with automatic restart). The D-EMS 2000 system is completed by the following additional modules:

D-EMS 2000 QAL
Module for the computer-aided fulfilment of the requirements of QAL3 (DIN EN 14181) optionally via CUSUM, Shewhart or EMWA cards including the corresponding documentation.

D-EMS 2000 BUBE

Module for automated data communication with the authority software *BUBE-online* and thus the fulfilment of the 11th BImSchV *annual emission declaration* as well as the information obligation for the PRTR. For additional information on existing client systems or existing PCs in the operator's data network, the modules D-PM.www for web-based data provision and D-PM.ms, which enables database access from MS-Excel, are available. By means of the client software *WinDeva* the complete functionality of the system workstation can be mapped on existing workstations of the operator.

Outlook

As described in the previous chapters, the recording and evaluation of controlled emissions from stationary sources in Germany and Europe is subject to a detailed set of regulations which are being amended and further developed at both EU and state level. This can be seen particularly clearly in the introduction of Directive 2010/75/EU at the end of 2010 or the revision of the 13th BImSchV for large combustion plants and the 17th BImSchV - the Ordinance on the Incineration and Co-incineration of Waste - in 2013.

Currently, both the TA Luft and the Federal Uniform Practice for Monitoring Emissions, which is important for the continuous evaluation of emission measurements, are under revision. Furthermore, the working group WG9 in CEN / TC 264 *Air Quality is* developing a European standard for quality assurance for data of emissions from stationary sources obtained with automatic measuring devices. This standard will be of particular importance for emission calculators. Working group WG37, which is currently working on a European standard for PEMS, was already mentioned on page 7.

Both the measuring systems including PEMS and the data acquisition and evaluation systems must adapt to this changing landscape of regulations through continuous development work in order to provide a secure data basis for the assessment of emissions and their reduction in the future. It is essential that high minimum requirements are set and adhered to in terms of functionality as well as quality assurance of the results and that this is ensured by independent testing.

References

Section 14.1

Boeker, P., Rechenbach, T., Schulze Lammers, P.: Messung biogener Gase-Konzentrationsverschiebungen durch Auskondensation von Gasfeuchte. Agrartechnische Forschung. **5**(1), 1–8 (1999)

Boeker, P.: Absenkung des Messgastaupunktes-Konzentrationsänderungen und Auswaschungen von Schadgasen und Geruchsstoffen. Agrartechnische Forschung. **7**(3), 72–76 (2001)

Clarke, A.G.: Industrial Air Pollution Monitoring. Chapman & Hall (1998)

Clevett, J.C.: Process Analyzer Techonology. Wiley & Sons (1986)

Down, R.D., Lehr, J.H.: Environmental Instrumentation and Analysis Handbook. Wiley-Interscience (2005)
Jahnke, J.A.: Continuos Emission Monitoring. John Wiley & Sons, New York (2000)
Kaltenmaier, K.: private Mitteilung (2015)
Leckrone, K.J., Hayes, J.M.: Efficiency and Temperature Dependence of Water Removal by Membrane Dreyers. Anal. Chem. **69**(5), 911–918 (1997)
Scott, K.: Handbook of Industrial Membranes. Elsevier (1995)

Section 14.2

Cortvriend, J.: Präsentation „Emission Reductions Resulting from the Implementation of Euro Standards" anläßlich der 2. International Conference Real Driving Emissions, DG Environment, Europäische Kommission, Düsseldorf, 17.9.2014
EEA: Exceedance of air quality limit values in urban areas (CSI 004). European Environmental Agency (2009). http://www.eea.europa.eu/dataand-maps/indicators/exceedance-of-air-quality-limit-1/exceedance-of-air-quality-limit-1. Zugegriffen am 11. Oktober 2010
EEA: European Union emission inventory report 1990–2008 under the UNECE convention on Long-range Transboundary Air Pollution (LRTAP). EEA Technical report No 7/2010. EEA –European Environmental Agency. Kopenhagen, Dänemark (2010)
Nagel, C., Fähnle, M., Draht, J., Thiel, M., Janott, F. – Audi AG, Neckarsulm, Gromann, M. – Bertrandt Ingenieurbüro GmbH, Neckarsulm: PEMS 2.0 – Einsatz eines anwendungsoptimierten, mobilen Abgasmesssystems im Pkw, Vortrag im Rahmen der AVL Abgas- und Partikelkonferenz, Ludwigsburg, Deutschland (2014)
Weiss, M., Bonnel, P., Hummel, R., Manfredi, U., Colombo, R., Lanappe, G., LeLijour, P., Sculati, M.: Analyzing on-road emissions of light-duty vehicles with Portable Emission Measurement Systems (PEMS). Institute for Energy, DG Joint Research Center, Europäische Kommission, EUR 24697 EN (2011)
Wiebrecht, J.: Einführung in die Abgasmesstechnik – Erläuterungen zur Abgasmesstechnik in der Automobilindustrie am Beispiel der AUDI AG (2005)

Section 14.2: Further Reading

Nietsch, I., Wiegleb, G.: NO^x Gassensor auf der Basis der UV-Resonanzabsorption, 6. Dresdner Sensor-Symposium 9.12.2003
Nietsch, I. Wiegleb, G.: A Novel NO^x Gas Sensor based on Resonance Absorption technique in the UV-Range. SENSOR 2005 conference proceedings II. B7.4, 12. Mai 2005

Section 14.3

Becker, E., Reinmann, J., Rentschler, W., Mayer, J.: Continuous Monitoring of the Dioxin/-Furan Emission of all Waste Incinerators in Belgium. Organhalogen Compounds. **49S** (2000)
De Fré, R., Wevers, M.: Underestimation in dioxin emission inventories. Organhalogen Compounds. **36**, 17–20 (1998)

Environmental Protection Agency (EPA): National Emission Standards for Hazardous Air Pollutants From the Portland Cement Manufacturing Industry and Standards of Performance for Portland Cement Plants (2011)

Environmental Protection Agency (EPA): National Emission Standards for Hazardous Air Pollutants From Coal and Oil-Fired Electric Utility Steam Generating Units and Standards of Performance for Fossil-Fuel-Fired Electric Utility, Industrial-Commercial-Institutional, and Small Industrial-Commercial-Institutional Steam Generating Units (2013)

Funke, W., Linnemann, H.: Messen von PCDD, PCDF, sowie organischen Substanzen mit vergleichbarer Flüchtigkeit und Polarität in Abgasen von Feuerungsanlagen bei Anwendung des „Adsorptionsverfahrens". GfA Report, Januar 1994

Journal Officiel de la République Francaise, Texte 10 und Texte 11 sur 126, 21.08.2010

Mayer, J., Rentschler, W., Sczech, J.: Long-Term Monitoring of Dioxin Emissions of a Hazardous Waste Incinerator during Lowered Incineration Temperature. Organhalogen Compounds. **41** (1999)

Section 14.3: Further Reading

Mayer, J., Grümping, R.: Continuous Monitoring of Dioxin Emissions from a Waste Wood Combustion Plant. Organhalogen Compounds. **59**, 81–83 (2002)

Section 14.4

Allen, M.G.: Diode laser absorption sensors for gas-dynamic and combustion flows. Meas. Sci. Technol. **9**(4), 545 (1998)

Baumbach, G.: Luftreinhaltung. Springer (1993)

Ebert, V., Teichert, H., Giesemann, C., Saathoff, H., Schurath, U.: Fasergekoppeltes In-situ-Laserspektrometer für den selektiven Nachweis von Wasserdampfspuren bis in den ppb-Bereich (Fibre-Coupled In-situ Laser Absorption Spectrometer for the Selective Detection of Water Vapour Traces down to the ppb-Level). tm–Technisches Messen/Plattform für Methoden. Systeme und Anwendungen der Messtechnik. **72**(1), 23–30 (2005)

Faist, J., Capasso, F., Sivco, D.L., Sirtori, C., Hutchinson, A.L., Cho, A.Y.: Quantum cascade laser. Science. **264**(5158), 553–556 (1994)

Hildebrandt, L., von Edlinger, M., Scheuermann, J., Nähle, L., Fischer, M., Koeth, J., Höfling, S.: Distributed Feedback Interband Cascade Lasers and their Spectroscopic Applications in Gas Sensing. In Laser Applications to Chemical, Security and Environmental Analysis (pp. LTu3D-4). Optical Society of America (Juli 2014)

Seufert, J., Fischer, M., Legge, M., Koeth, J., Werner, R., Kamp, M., Forchel, A.: DFB laser diodes in the wavelength range from 760 nm to 2.5μm. Spectrochim. Acta A Mol. Biomol. Spectrosc. **60**(14), 3243–3247 (2004)

Totschnig, G., Lackner, M., Shau, R., Ortsiefer, M., Rosskopf, J., Amann, M.C., Winter, F.: High-speed vertical-cavity surface-emitting laser (VCSEL) absorption spectroscopy of ammonia (NH3) near 1.54 μ m. Applied Physics B. **76**(5), 603–608 (2003)

Weih, R., Nähle, L., Höfling, S., Koeth, J., Kamp, M.: Single mode interband cascade lasers based on lateral metal gratings. Appl. Phys. Lett. **105**(7), 071111 (2014)

Wiedenmann, D., Grabherr, M., Jäger, R., King, R.: High volume production of single-mode VCSELs. In: Integrated Optoelectronic Devices. Proceedings of SPIE – The International Society for Optical Engineering. (Februar 2006) DOI: https://doi.org/10.1117/12.657655

Section 14.4: Further Reading

Fitzer, E., Siegel, D.: Stickoxid-Emissionen industrieller Feuerungsanlagen in Abhängigkeit von den Betriebsbedingungen. Chemie Ingenieur Technik. **47**(13), 571–579 (1975)

Köbel, M., Elsener, M.: Entstickung von Abgasen nach dem SNCR-Verfahren: Ammoniak oder Harnstoff als Reduktionsmittel? Chemie Ingenieur Technik. **64**(10), 934–937 (1992)

Schäfer, G., Riedel, F.N.: Über die Bildung von Stickoxiden in Großfeuerungsanlagen, deren Einfluß auf die Umwelt, ihre Verminderung sowie ihre Entfernung aus den Abgasen der Kraftwerke. Chemiker-Zeitung. **113**(2), 65–69 (1989)

Section 14.5

DIN EN 15267-3: „Luftbeschaffenheit – Zertifizierung von automatischen Messeinrichtungen – Teil 3: Mindestanforderungen und Prüfprozeduren für automatische Messeinrichtungen zur Überwachung von Emissionen aus stationären Quellen“ (2007a)

Bund/Länder-Arbeitsgemeinschaft für Immissionsschutz.: http://www.lai-immissionsschutz.de (2015) Zugegriffen: 6.11.2015

Wiegleb, G.: Gasanalysesystem für erhöhte Prozesstemperaturen. Tech. Mess. **51**, 385–393 (1984)

Section 14.5: Further Reading

Dr. Födisch, H.: Staubemissionsmesstechnik. Expert (2004)

Paragraph 14.6: Standards and Guidelines

BImSchG – Bundes-Immissionsschutzgesetz. Gesetz zum Schutz vor schädlichen Umwelteinwirkungen durch Luftverunreinigungen, Geräusche, Erschütterungen und ähnliche Vorgänge (2002)

Erste Allgemeine Verwaltungsvorschrift zum Bundes-Immissionsschutzgesetz (Technische Anleitung zur Reinhaltung der Luft – TA Luft) (2002)

Bundeseinheitliche Praxis bei der Überwachung der Emissionen; RdSchr. d. BMU vom 13.6.2005 – IG I 2–45053/5 und vom 04.8.2010 – IG I 2–51134/0. sowie der Statuskennung und Klassierung in der gültigen Fassung vom 01.08.2012

Erste Verordnung zur Durchführung des Bundes-Immissionsschutzgesetzes (1. BImSchV – Verordnung über kleine und mittlere Feuerungsanlagen – 1. BImSchV) (2010)

Zweite Verordnung zur Durchführung des Bundes-Immissionsschutzgesetzes (2. BImSchV – Verordnung zur Emissionsbegrenzung von leichtflüchtigen halogenierten organischen Verbindungen – 2. BImSchV) (1990)

Dreizehnte Verordnung zur Durchführung des Bundes-Immissionsschutzgesetzes (13. BImSchV – Verordnung über Großfeuerungs-, Gasturbinen- und Verbrennungsmotoranlagen – 13. BImSchV) (2013)

Siebzehnte Verordnung zur Durchführung des Bundes-Immissionsschutzgesetzes (17. BImSchV – Verordnung über die Verbrennung und die Mitverbrennung von Abfällen – 17. BImSchV) (2013)

Siebenundzwanzigste Verordnung zur Durchführung des Bundes-Immissionsschutzgesetzes (27. BImSchV – Verordnung über Anlagen zur Feuerbestattung – 27. BImSchV) (1997)

Dreißigste Verordnung zur Durchführung des Bundes-Immissionsschutzgesetzes (30. BImSchV – Verordnung über Anlagen zur biologischen Behandlung von Abfällen – 30. BImSchV) (2001)

Einunddreißigste Verordnung zur Durchführung des Bundes-Immissionsschutzgesetzes (31. BImSchV – Verordnung zur Begrenzung der Emissionen flüchtiger organischer Verbindungen bei der Verwendung organischer Lösemittel in bestimmten Anlagen – 31. BImSchV) (2001)

DIN EN 14181:2015–02 Emissionen aus stationären Quellen – Qualitätssicherung für automatische Messeinrichtungen. Deutsche Fassung EN 14181:2014

DIN EN 15267-1:2009–07 Luftbeschaffenheit – Zertifizierung von automatischen Messeinrichtungen – Teil 1: Grundlagen. Deutsche Fassung EN 15267-1:2009

DIN EN 15267-2:2009–07 Luftbeschaffenheit – Zertifizierung von automatischen Messeinrichtungen – Teil 2: Erstmalige Beurteilung des Qualitätsmanagementsystems des Herstellers und Überwachung des Herstellungsprozesses nach der Zertifizierung. Deutsche Fassung EN 15267-2:2009

DIN EN 15267-3:2008–03 Luftbeschaffenheit – Zertifizierung von automatischen Messeinrichtungen – Teil 3: Mindestanforderungen und Prüfprozeduren für automatische Messeinrichtungen zur Überwachung von Emissionen aus stationären Quellen. Deutsche Fassung EN 15267-3:2007b

Emissionsfernüberwachung (EFÜ) Schnittstellendefinition in der Fassung des Beschlusses des LAI vom 28.09.2005 (korrigierte Fassung vom 15. November 2006)

Richtlinie 2010/75/EU des Europäischen Parlaments und des Rates über Industrieemissionen (integrierte Vermeidung und Verminderung der Umweltverschmutzung) (Neufassung) (RL 2010/75/EU: Industrieemissionen (integrierte Vermeidung und Verminderung der Umweltverschmutzung) (Neufassung) – 2010/75/EU) (2010)

U.S. Code of Federal Regulations, 40 CFR PART 60—STANDARDS OF PERFORMANCE FOR NEW STATIONARY SOURCES (NSPS)

U.S. Code of Federal Regulations, 40 CFR Part 60, „Performance Specification 16 for Predictive Emission Monitoring Systems and Amendments to Testing and Monitoring Provisions“. Federal Register, Vol. 74, No. 56, FR 40297 (2009)

U.S. Code of Federal Regulations., 40 CFR PART 75—CONTINUOUS EMISSION MONITORING

U.S. Code of Federal Regulations., 40 CFR Part 75, Subpart E. „Alternative Monitoring Systems“, Federal Register, Subpart E

VDI 4201 Blatt 1–4 Mindestanforderungen an automatische Mess- und elektronische Auswerteeinrichtungen zur Überwachung der Emissionen – Digitale Schnittstelle (2010–2014)

Power Engineering

15

Petrochemical plant. (Source: fotolia)

15.1 Energy Metering and Other Metering Tasks in the Gas Industry

Joachim Kastner

The Importance of Gas in the Energy Industry
Natural gas is produced in geological sediments under exclusion of air, at high pressure and elevated temperature from incompletely decomposed sunken marine biomass. Natural gas is often produced together with crude oil. In conventional natural gas reservoirs, the natural gas is trapped in porous reservoir rocks under an impermeable cap rock and can therefore

G. Wiegleb, *Gas Measurement Technology in Theory and Practice*,
https://doi.org/10.1007/978-3-658-37232-3_15

flow freely to a downhole. This type of conventional reservoir has dominated natural gas production until now because of its yield and economic efficiency. In recent years, however, natural gas has increasingly been produced from unconventional reservoirs. These include *tight gas,*[1] *shale gas*, coal bed methane and methane hydrates. In the case of *tight gas* and shale gas, the gas is trapped in non-porous rock and must first be made available using complex production methods, *hydraulic fracturing* (or *fracking* for short).

Among the fossil primary energy sources, natural gas is particularly attractive because of its flexibility, efficiency and cleanliness. Compared with other fossil primary energy sources, natural gas produces fewer local pollutant emissions when burned and the lowest CO_2 emissions per energy unit, since the ratio of hydrogen to carbon atoms in natural gas is relatively high. The high flexibility of natural gas allows a wide range of applications:

- One of the main applications is heating of buildings, where compact, powerful, efficient and clean heating devices are available.
- Another significant use is the generation of electricity from natural gas. Combined gas and steam turbines are increasingly being used in modern gas-fired power plants; they achieve electrical efficiencies of up to approximately 60% and are easily controllable. Gas-fired power plants thus represent an ideal complementary technology to the volatile renewable energy sources.
- In cogeneration, electricity and heat generation are combined with a high overall efficiency. Ideally, it is used in a decentralised way to enable efficient heat utilisation. Natural gas is also an advantageous fuel for this purpose, either with classic combustion engines or with modern fuel cells.
- Natural gas is also used as fuel for vehicles. *Compressed natural gas* (CNG) is commonly used in passenger cars. It is stored at a pressure of several hundred bar and at ambient temperature in pressure vessels. Current projects deal with the replacement of diesel or heavy oil by *liquefied natural gas* (LNG) in heavy duty transport, especially for ships, trucks and special vehicles. The gas is stored in cryotanks at a temperature of about $T \approx -164°C$ and near atmospheric pressure ($\approx$1 bar).
- The chemical industry recycles about 11% of the natural gas consumed in Germany, 8% of which is energetic and 3% material. Compared to oil, natural gas still plays a subordinate role in material recycling, but it is expected that the share of natural gas will grow in the future due to the greater reach of resources.
- Finally, natural gas serves as an energy source for thermal processes, especially in the glass and ceramics industry and in metal processing. Natural gas is often indispensable here, if the flame is used as a tool or if a special combustion atmosphere is essential for the process.

[1] Natural gas from unconventional deposits.

- The future potential of natural gas has further increased in the recent years through diversification of gas procurement, mainly through exploration of new unconventional fossil gas sources such as *tight gas*, shale gas and coal seam gas, but also through regenerative gas production. At present, oil accounts for the largest share of primary energy sources, followed by coal and gas. The share of natural gas is growing steadily while the shares of oil are shrinking and those of coal are tending sideways. Current studies forecast that by 2035 the share of fossil primary energy sources will be about the same as that of coal at around 27%.
- The diversification of gas procurement is not only achieved through new gas sources, but also through growing global gas trading. The expansion of transport infrastructure, especially for LNG, is increasing the number of potential export–import relationships and thus competition and security of supply.
- The natural gas industry with its technology and infrastructure can also make valuable contributions to the energy turnaround. It integrates fossil and regenerative energy sources and offers high transport and storage capacities and an extensive distribution network.
- After processing to pipeline quality, the regenerative primary energy biogas can be fed into the existing infrastructure of the natural gas industry and used more efficiently on a decentralized basis, ideally in combined heat and power generation.
- Currently, the so-called *power-to-gas concept* is being discussed, in which the flow of energy is from the electricity to the gas network. Electrical energy, ideally from surpluses from renewable power generation, is stored in the gas grid via electrolysis as hydrogen or synthetic methane. This would allow large amounts of energy to be stored relatively efficiently over long periods of time. Through a bidirectional coupling with the power grid, the natural gas industry could help to manage the volatility of the growing renewable energy share.

The described trends, new fossil and regenerative gas sources combined with a worldwide gas trade, lead to a growing range and dynamics of gas composition. Regenerative gas sources can even bring new gas components, especially hydrogen, into the public supply networks. Liberalisation and unbundling in the gas industry strengthen the trend towards greater fluctuations in gas composition and extend the variations right through to the gas user. The variation of gas composition creates new and growing challenges in gas metering and especially in industrial and domestic gas use. Innovations in measurement and control technology are therefore in demand to develop the potential of the gas industry even under these volatile conditions.

Measurement Tasks in the Gas Industry

The primary measurement task in the gas industry is energy measurement. In addition to volume measurement, gas quality measurement plays a decisive role, it also performs various secondary measurement tasks in the natural gas industry, in gas transport but also in gas use. The main measurement tasks in the gas industry will be explained in the following.

Energy

Gas trading is based on energy billing, as the energy content of the gas represents the decisive utility value for the consumer. In Germany, for example, thermal energy billing is regulated in a standard of the German Technical and Scientific Association for Gas and Water (Deutsche Vereinigung des Gas- und Wasserfachs – DVGW). The basic task for gas metering is therefore the determination of energy. The energy E is calculated as the product of the gas volume V with the volumetric calorific value H_s (Eq. 15.1). The volume and calorific value must refer to the same gas state (pressure p, temperature T).

$$E = V \cdot H_s \tag{15.1}$$

In practice, the volume V is measured under operating conditions (p, T) of the pipeline, while the calorific value$H_{s,\,0}$ refers to standard conditions (p_0, T_0). The index 0 here refers to standard conditions, $p_0 = 1013.25$ mbar and $T_0 = 273.15$K. For the energy calculation, therefore, the operating volume V must first be converted to standard volume V_0. Natural gas is not an ideal gas; its real gas behaviour is taken into account in the conversion by the compressibility index K. The equation for calculating the energy is then

$$E = V_0 \cdot H_{s,0} = V \cdot \frac{p \cdot T_0}{p_0 \cdot T} \cdot \frac{1}{K} \cdot H_{s,0} \tag{15.2}$$

In addition to the state (p, T), the compressibility coefficient K depends significantly on the gas quality, that is the gas composition and its physical parameters. In the gas industry, standardised equations of state, for example according to ISO 12213, are used to calculate the compressibility coefficient. These equations require as input data either the molar gas composition or characteristic gas parameters, such as calorific value, standard density and various gas components. The determination of energy therefore requires not only volume measurement but also, and above all, an analysis of the gas quality.

Gas Quality

Pipeline-quality natural gas mainly consists of hydrocarbons (typ. 75–100 mol%), as well as the inert gases nitrogen (typ. 0–20 mol%) and carbon dioxide (typ. 0–5 mol%). The hydrocarbon content consists mainly of methane CH_4 and the series of alkanes (molecular formula C_nH_{2n+2}, order n), the representatives are ethane C_2H_6, propane C_3H_8, butane C_4H_{10}, pentane C_5H_{12}, hexane C_6H_{14}, etc., and their isomers. The concentration of the alkanes typically decreases with increasing order n. Other components such as helium, oxygen, sulfur components and water vapor occur in transport networks only in traces, typically in the concentration range from a few 1 ppm to a few 100 ppm (ppm parts= per million). By feeding in biogas and hydrogen from renewable energy sources, the new gas components oxygen and hydrogen can occur in significant concentrations.

The physical parameters of the gas vary with the gas composition. The gas quality thus plays an important role in energy billing, first directly as the calorific value of the gas and second indirectly as compressibility in volume determination. In addition, there are numerous other gas parameters that play an important role in gas production, transport, trade and especially in gas use.

Due to its importance for the gas industry, the gas quality is defined in various national and international standards. In Germany, gases in public supply must comply with the specifications of DVGW Code of Practice G260. DVGW Code of Practice G262 applies to the feed-in of gases from renewable sources.

On the European level, there is the industry agreement EASEE-gas with coordinated business practices (Common Business Practices) for H-gas in cross-border gas trading. Currently, CEN working groups are working on the standardisation of H-gas (Mandate M400) and biogas (Mandate M475).

At international level, there is no standard for gas composition, but there are various standards for measuring gas composition and calculating gas parameters. In the following, some further gas parameters and their meaning are presented:

Wobbe Index

A decisive parameter of the gas composition is the Wobbe index $W_{s,\,0}$, which is calculated from the calorific value $H_{s,\,0}$ divided by the square root of the relative density d (Eq. 15.3, see also Sect. 2.6). The Wobbe index is a measure of the thermal output of a gas burner, that is two different gases with the same Wobbe index deliver the same burner output at otherwise identical settings. The Wobbe index is therefore important for the majority of gas applications, especially in heating. The range of fluctuation of the Wobbe index is therefore locally limited in the gas networks to avoid problems in the use of gas. Due to the trends in the gas industry described above, limiting the fluctuation bandwidth will become more difficult in the future and require increased use of measurement and control technology for gas conditioning and process control.

$$W_{s,0} = \frac{H_{s,0}}{\sqrt{d}} \tag{15.3}$$

Methane Number

The anti-knock properties of the fuel are important for the combustion of gas in engines. For high efficiency, high compression is aimed for in gas engines, but on the other hand, this involves the risk of wear-promoting knocking due to premature self-ignition of the fuel. Knock resistance is described by the so-called methane number MZ, which corresponds approximately to the octane number of petrol (see also Sect. 2.6). The scaling of the methane number is based on the methane content of a binary methane/hydrogen mixture, that is MZ = 0 the methane number of pure hydrogen and, MZ = 100 for pure methane, the intermediate values are defined by the methane content of the corresponding binary methane/hydrogen mixture. The methane number of a natural gas mixture is obtained by comparison with the binary methane/hydrogen mixture of the same knock resistance. Inert gases such as nitrogen and carbon dioxide increase the knock resistance, hydrocarbons such as ethane, propane, butane, etc. reduce it. Current engine control systems usually work with knock sensors that detect the occurrence of early ignition and take appropriate measures such as ignition timing shift, mixture adjustment or power reduction. By quickly and continuously measuring the methane number of the fuel, engine operation could be proactively optimized before knocking occurs in the first place.

In addition to the primary combustion parameters, such as calorific value, density, air demand, Wobbe index and methane number, there are other gas parameters that are relevant to integrity and safety in the transport and use of gas. There are therefore good reasons for all parties involved in the production, transport and consumption chain to specify these gas parameters, to define them in contracts and to monitor them by measurement.

Hydrocarbon Dew Point

The higher hydrocarbons in natural gas can condense at pressures and temperatures typical of the process. The hydrocarbon dew point is the temperature at which, at a given pressure, the hydrocarbons in the gas begin to condense. The dew point temperature increases with increasing concentration of higher hydrocarbons and their hydrocarbon order. Figure 15.1 shows a typical phase diagram of natural gas, the curve describes the phase boundary between the liquid, gaseous and two-phase state. In typical natural gases, the hydrocarbon dew point first increases with pressure and reaches a maximum, the so-called *Cricondentherm*. With further increasing pressure the dew point temperature decreases again, this effect is called retrograde condensation. The *Cricondentherm* for natural gases is typically in a pressure range between about 25 and 45 bar.

Hydrocarbon condensation can have a negative effect on the function of gas systems, especially of controllers, valves and measuring systems. The operation of gas turbines can be disturbed by hydrocarbon condensate. Accumulations of condensate in the gas transport system can occur in surges and lead to temporary overheating of the machine. However, a uniform concentration of condensate can also have a negative effect on the operation of a gas turbine by causing delayed ignition or after ignition.

Because of this importance, the hydrocarbon dew point is defined in a number of technical regulations. The DVGW regulations require that the condensation point of hydrocarbons at line pressure is lower than the ground temperature. The industry agreement EASEE-gas specifies a limit of −2°C for the pressure range 1–70 bar.

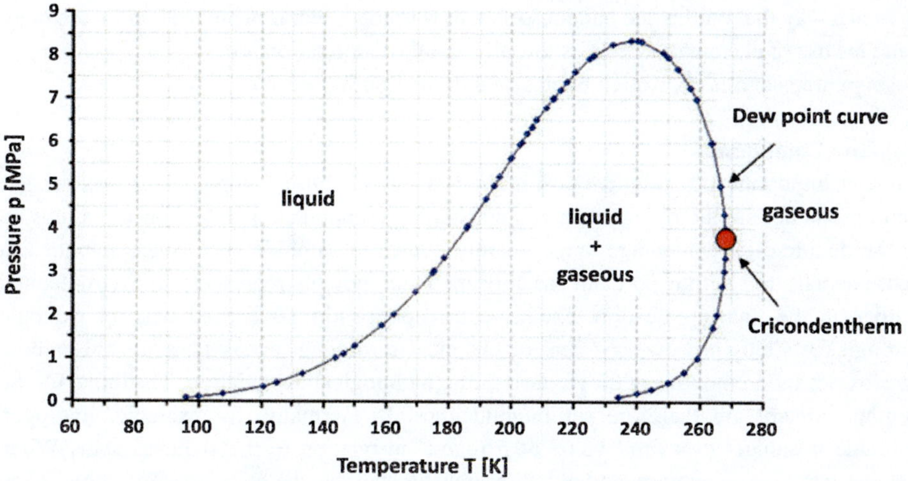

Fig. 15.1 Phase diagram of a typical natural gas. The envelope curve covers the two-phase range and separates the ranges of the liquid and gaseous phases. When the gas state exceeds the dew point curve from the gas phase to the two-phase region, hydrocarbon condensation occurs. The dew point temperature depends on the pressure. The maximum dew point temperature is called *Cricondentherm*

Water Dew Point

In addition to the hydrocarbon dew point, the water dew point is of course also important. Condensing moisture in combination with small amounts of sulphur compounds or oxygen leads to corrosion. In addition, methane forms solid methane hydrates with water, which can lead to mechanical disturbances or damage to the transport system. The network operators therefore take great care to ensure that no moist gas enters the transport pipeline. The DVGW regulations limit the water content to 200 mg/m^3 at design pressures up to 10 bar, above that even only to 50 mg/m^3. The industry agreement EASEE-gas specifies a limit of -8°C at design pressures of 80 bar.

Odorization

Pipeline-quality natural gas is generally odorless. Odorants are added to the natural gas to make the danger of possible leaks perceptible to humans by means of a warning odor. Usually, this odorization is only carried out in the distribution networks, and in some countries also in the transport networks. The odorants commonly used are based on organic sulphur compounds such as methanthiol, ethanthiol, propane-1-thiol, 2-methyl-propane-2-thiol (tert-butylmercaptan, TBM) and tetrahydrothiophene (THT). Due to their strong odor, concentrations in the ppm range are already sufficient for adequate odorization. Nevertheless, the odorants contribute significantly to the sulfur content of natural gas. Since the general aim is to reduce the sulphur content of fuels, sulphur-free odorants have also been developed. They contain a mixture of ethyl acrylate, methyl acrylate and 2-ethyl-3-methylpyrazine. During odorization, the concentration in the ppm range must be adjusted in such a way that on the one hand a sufficient warning effect is achieved and on the other hand the use of chemicals is kept as low as possible. Odorization is carried out by means of dosing pumps and is controlled by gas analysis measuring devices.

Sulphur Compounds

Another important measurement task is the analysis of sulfur compounds in natural gas. During production, the raw gas can contain sulphur compounds in high concentrations up to the double-digit percentage range, mainly hydrogen sulphide H_2S. Since sulphur can cause damage to gas transport and gas utilisation facilities and represents an environmental burden in the final application, extensive desulphurisation is carried out, for example through the Claus process, physical or chemical scrubbing, catalysts, activated carbon, zinc oxide, and in the case of biogas also through biological degradation. The limits for the sulphur content are therefore set in guidelines. In Germany, for example, hydrogen sulphide is subject to a limit value of 5 mg/m^3 in relation to the standard state. When feeding into the gas network and at border transfer stations, the concentration of hydrogen sulphide is therefore continuously monitored with process measuring instruments.

Gas Usage

Gas utilisation processes are dependent on the gas composition and the associated gas parameters. Up to now, however, the composition in Germany but also in other countries at one location only varies within relatively narrow bandwidths, so that building heating systems can be adjusted to this and industrial gas users can design and optimize their processes accordingly. Current trends show, however, that these historical conditions will not necessarily remain valid into the future. Current committee activities are aimed at a European standardization for H-gas with a wide bandwidth. Although the existing national gas standards already allow a wide range of gas, this range was not controlled locally in a supply area. In the context of the creation of a European internal energy market, however, it is expected that the permitted bandwidth of gas composition will be increasingly used by energy suppliers. It can therefore be expected that in the future, measurement tasks for process gas analysis will also arise in sensitive processes of gas use. The aim here is to maintain the requirements for safety, environmental protection, energy efficiency and process quality in an environment of greater gas quality fluctuations. The need to handle gas variations in gas usage is not fundamentally new. Dynamic gas composition regulations have already been developed for certain gas applications.

One example is the *SCOT* combustion system for premix burners from Elster Kromschröder (Fig. 15.2). The system evaluates the ionisation signal of an electrode in the gas flame in order to control the gas/air mixture and thus the combustion quality. The physical relationship between the ionization current and the air ratio λ is exploited. The system is suitable for premix burners in the output range from 3 to 60 kW. Other dynamic methods measure the exhaust gas composition and thus optimally adjust the combustion in the event of varying gas composition (Giese 2013).

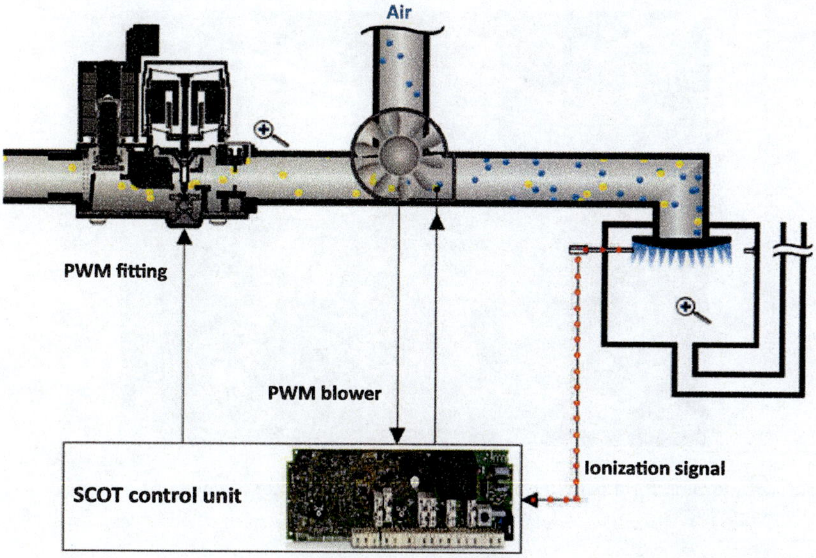

Fig. 15.2 Elster-Kromschröder *SCOT* combustion system for premix burners, consisting of an electrode for the flame ionisation current, control unit, adjustable fan and electronic gas fitting. (Source: Elster GmbH)

However, the majority of industrial gas applications do not have dynamic control capabilities and must be optimized for a given gas quality, which is assumed to be constant. In a future scenario of fluctuating gas composition, dynamic measurement and control solutions must be developed for sensitive gas use processes. The topic is complex due to the variety of application processes and requirements, some of which are conflicting objectives.

For many gas application technologies, especially for domestic building heating systems, the Wobbe index is the decisive parameter of the gas quality. Its stability is therefore usually a priority. However, there are also numerous processes and effects in industrial gas use where the Wobbe index of the fuel is secondary or even insignificant. A consideration of the possible effects of gas quality fluctuations on industrial applications can be found in Kastner (2013). In the following, some examples of the relationships between gas composition and gas use processes are described.

In order to meet the requirements for energy efficiency and pollutant emissions, auxiliary units are used for air preheating and heat recovery, and the aim is to achieve near-stoichiometric combustion. Such multi-dimensionally optimised processes can, however, react more sensitively to fluctuations in gas quality.

Effects caused by variations in gas composition on an optimally adjusted burner affect the following variables, for example: flame temperature, stoichiometric air demand, flame geometry in relation to the combustion chamber geometry, heat transfer (shifting of the

Fig. 15.3 Gas turbine for a combined cycle power plant. (Source: Siemens AG)

reaction zone), pollutant formation. In the case of a glass melt, for example, heat transfer is mainly affected by infrared radiation. It depends on the composition of the combustion products of CO_2 and H_2O and thus again on the composition of the fuel gas. The composition of the exhaust gas also has a sensitive effect on the quality of the glass through chemical reactions; colour defects or streaks can be the result.

Another very important use of gas is in gas turbines for electricity generation, which account for about one third of German gas consumption (Fig. 15.3). Modern gas-fired power plants work with combined gas and steam turbines and achieve total electrical outputs of almost 600 MW and an electrical efficiency of more than 60%. Of course, systems optimised in this way are also sensitive to variations in gas composition. Once again, it is a matter of optimising energy efficiency, pollutant emissions and plant wear.

The described trends in the energy industry and their effects on the use of gas make innovations necessary to control the changed operating conditions. In addition to the safety and environmental aspects, the effects in energy efficiency and product quality in industrial gas use mean hard cash.

Most approaches to process optimization for gas variations require a powerful process gas analysis. While accuracy and sensitivity are the most important factors in gas quality measurement for energy billing and monitoring of gas specifications, the dynamics of the measurement also play a major role in the measurement tasks for gas usage.

Process Gas Analysis Equipment in the Gas Industry

As previous observations show, the natural gas industry is becoming increasingly important in the energy industry. The current trends of diversification and global gas trading require efficient gas measurement technology, especially for gas quality measurement. We have learned about the variety of measurement tasks in the gas industry. In the following, the most common measurement technologies for gas quality measurement in the natural gas industry will be presented.

Calorimeter

Calorimeters are measuring devices for heat quantities, they can be used to determine the calorific value of gases, they are therefore among the oldest measuring devices in the gas industry. Process calorimeters usually work continuously by burning the sample gas and, depending on the design, directly measure the calorific value, heating value or Wobbe index. In classic combustion calorimeters, the gas is burnt and the heat released is transferred to a stream of coolant, its temperature rise is a measure of the heat output released during combustion. Depending on the design, the measurement directly provides the Wobbe index or the calorific value. If the exhaust gases are cooled down during heat transfer to such an extent that the moisture in the exhaust gas condenses, the upper Wobbe index or the corresponding calorific value is determined. If the moisture does not condense, the lower Wobbe index or the calorific value is obtained. Often a gas density measurement is integrated in order to calculate the missing variables calorific value or Wobbe index. Calorimeters work continuously and are therefore well suited for measurement and control technology. However, the finite heat capacity of the measuring system results in a certain inertia of the measurement. The time t_{90}, that is the time until 90% of a step response is reached, is typically about 1 min.

A more modern version of the calorimeter burns the sample gas in excess air in a catalyst. A lambda sensor measures the oxygen content of the exhaust gas and determines the air requirement for stoichiometric combustion. Depending on the design, the gas parameter Wobbe index or calorific value can be determined from this again, with an additional density measurement determining the missing parameter in each case. The correlation between oxygen demand and gas parameters depends on the gas components, such as hydrocarbons, nitrogen, carbon dioxide or hydrogen, so the correlation parameters must be adapted to the type of gas. The advantage of this version of the calorimeter is that the measurement is not based on a slow heat transfer between heat capacities. Therefore such calorimeters achieve very short response times of about 5 s (t_{90}). The advantage of calorimeters is that they can process practically all fuel gases, react continuously and quickly to very quickly. This makes them particularly suitable for measurement and control technology. However, calorimeters are relatively large and sensitive to ambient conditions, and there is also extensive explosion protection. In billing measurement, calorimeters have therefore been largely replaced in recent decades by other methods, especially by gas chromatographs (GCs).

Gas Chromatography

The supreme discipline of gas analysis is gas chromatography (GC), which provides detailed information about the composition of the sample gases with high sensitivity and accuracy. Process gas chromatographs (PGCs) in the natural gas industry work discontinuously with a cycle time of a few minutes, making them fast enough for fiscal gas measurement, but only conditionally suitable for fast measurement and control tasks. The equipment required for GC is relatively high, and high-quality calibration gases are needed in addition to the carrier gas.

How GC works is explained in Chap. 5. In the following, the most important functions are summarized once again and related to the application in the natural gas industry. The main components of GC are injector, separation column and detector. Since their functions are very strongly temperature dependent, the individual components are very precisely temperature controlled. A carrier gas, often helium, flows from the injector through the separation column system to the detector. At start-up, a small volume of sample gas is injected into the carrier gas flow in the injector and carried through the system. The gas components of the sample gas interact with the stationary phase of the separation column to varying degrees and are therefore transported to the detector at different rates. The detector thus delivers a time-dependent signal that represents the detector response to the individual gas components, the so-called chromatogram. GC therefore works discontinuously. In chromatogram evaluation, the peak areas or peak heights are determined, assigned to the respective gas components and quantified by comparison with defined calibration gases. There is a variety of column types and separation materials. I natural gas analysis typically micropacked columns and capillary columns are used. There is also a variety of detectors, such as the thermal conductivity detector (TCD), the flame ionization detector (FID), the photoionization detector (PID) and the flame photometric detector (FPD). A particularly high sensitivity and further selectivity can be achieved if mass spectrometers or ion mobility spectrometers are connected downstream of the separation column as detectors. PGCs in natural gas analysis can solve most measuring tasks satisfactorily with the TCD.

Classical GCs consist of discrete modules, which are integrated into field-suitable housings in PGCs. Some modern PGCs are based on microsystem technology, so-called micro-GCs. The assemblies are designed as microelectromechanical systems (MEMS). The advantages are a compact modular design, lower media consumption, higher sensitivity and linearity. Due to the high integration of GC technology, on-site service is also simplified by module replacement. In the following, an established as well as a novel micro-GC technology will be presented.

Classical Micro-Gas Chromatography

In the Agilent 490 micro-GC, the key components injector for sample dosing and detector for signal generation are designed in microsystem technology and integrated into a GC module together with an electronic pressure regulator (Fig. 15.4). This results in low media consumption, high sensitivity and linearity as well as a modular design with high spatial integration. The injector is made of silicon and glass and has diaphragm valves that are pneumatically actuated by the carrier gas pressure. The electromagnetic control valves are located outside the actual injector, thus avoiding the interference caused by the waste heat of the solenoid coils. The detector is also manufactured in silicon technology, with low

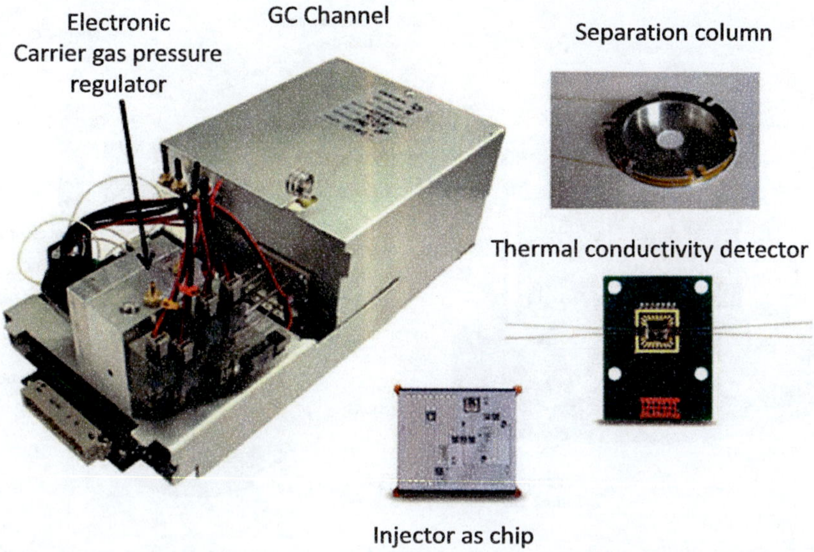

Fig. 15.4 Micro-GC module with the components injector, separation column, thermal conductivity detector (TCD) and electronic pressure regulator for carrier gas. (Source: Agilent)

dead volume and good thermal coupling of the filament with the gas flow, resulting in very high sensitivity, repeatability and linearity. Matching the small geometric cross-sections of injector and detector, separation columns with very small internal diameters can be used, thus achieving high separation performance. The coordination of the fluidic parameters of the key components is a prerequisite for the high analytical performance of the overall system. A precise carrier gas pressure control is essential for a stable and reproducible gas chromatographic separation. In this micro-GC technology, this is achieved by an electronic pressure regulator which is directly integrated into the GC module. Each GC module of a multi-channel system thus has an individual carrier gas pressure control.

Based on Agilent micro-GC technology, Elster has developed the EnCal 3000 PGC for the natural gas industry. This enables solutions to be presented for numerous measuring tasks in the gas industry. For the fiscal measurement of natural gas, it is initially sufficient to separate the sample gas into its main components methane, nitrogen, carbon dioxide and a detailed hydrocarbon analysis up to hexane plus higher components (C6+). In an extended version, the higher hydrocarbons up to nonane (C9) are also individually recorded. In both cases the carrier gas is helium. For the analysis of biogas, the components methane, carbon dioxide, nitrogen, oxygen and hydrogen are recorded, for the conditioning of biogas additionally propane and butane. Two carrier gases, helium and argon, are used here for optimal measurement. These important applications can be demonstrated with two separation columns (GC modules) (Fig. 15.5).

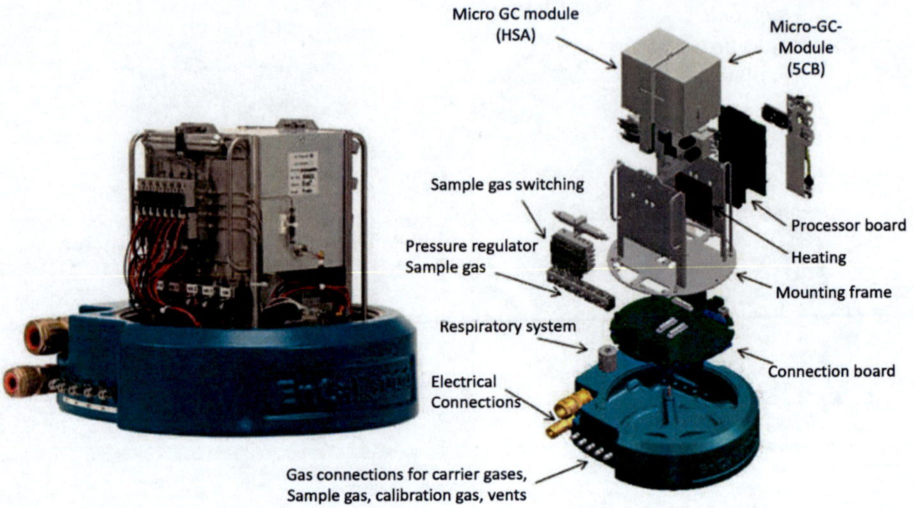

Fig. 15.5 EnCal 3000 process gas chromatograph Standard housing for 2-micro-GC modules, 6-channel sample gas switching and processor board for autonomous signal processing and communication. (Source: Elster GmbH)

The growing variation and complexity of the gas composition increases the need for information about the transported gases. The introduction of regenerative gases such as biogas and electrolysis gas (*power-to-gas*) also requires the further development of process measurement technology. New measurement tasks are: substance concentrations of hydrogen, oxygen, sulphur, odorants, dew points of water and hydrocarbons, etc.

The abundance of applications requires more complex GC systems, which can be easily implemented thanks to the modularity of micro-GC technology. Current product developments therefore concern complex measuring systems, such as the EnCal 3000 Quad, which can operate up to 4 GC modules and supply them individually with carrier gases (Fig. 15.6).

Novel Micro-Gas Chromatography

As a consistent further development of the established micro-GC, the separation columns of a new micro-GC technology were also designed in microsystem technology (Fig. 15.7 left). For this purpose, the separation columns are manufactured by anodic bonding from silicon-Pyrex discs in dimensions of a few centimeters. They are available as thin-film versions and as micropacked versions. Before bonding, the thin-layer column is coated in a plasma process with a stationary phase for the gas chromatographic separation of higher hydrocarbons. A variety of commercial packing materials are available for the packed columns, they allow the separation of the main components and permanent gases. Thanks to their small dimensions and thermal masses, the separation columns can be programmed very easily in terms of temperature, which in turn makes powerful analysis programs

Fig. 15.6 Process gas chromatograph EnCal 3000 Quad. Double housing for up to 4-micro-GC modules. Mounting frame with EnCal 3000 Quad, carrier gas and calibration gas. (Source: Elster GmbH)

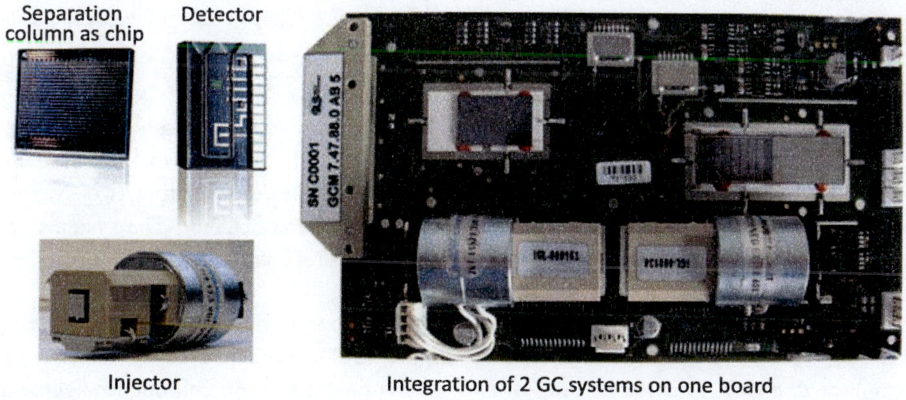

Fig. 15.7 Novel micro-GC technology with the key components injector, separation column and detector. The separation column is designed as a chip with direct column heating and allows powerful temperature programming. Up to 2 complete GC systems can be integrated on one fluidic-electronic board in Eurocard format. (Source: Elster GmbH)

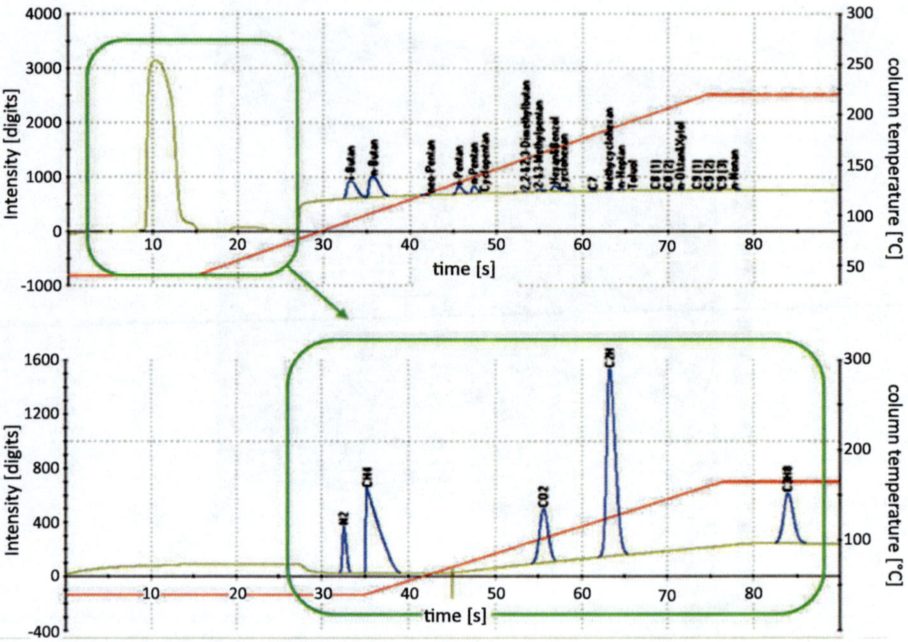

Fig. 15.8 Analysis of an L-gas sample using a micro-gas chromatograph with temperature-programmed chip separation columns in column configuration. The *upper* chromatogram shows the signal of column 1: The gas components nitrogen, methane, CO_2, ethane and propane are separated as a sum peak from the higher gas components and passed on to column 2 by column switching. This is followed by the separation of the higher hydrocarbons to nonane. The *lower* picture shows the signal from column 2. The sum peak from column 1 is separated into its components. (Source: Elster GmbH)

possible. These novel separation columns are integrated together with microsystem injectors and detectors on fluidic electronic boards to form independent GC systems. Figure 15.8 on the right shows, for example, a GC module representing two systems each consisting of an injector, a chip separation column and a detector. The systems can each analyze independently in parallel or work together in a column circuit.

This novel micro-GC technology enables a wide range of applications in the energy industry, but also in the chemical industry. Figure 15.8 shows the analysis of an L-gas sample with a GC system consisting of two separation columns in column arrangement. Figure 15.9 shows the analysis of biogas.

The analytical specifications of this new micro-GC technology are certainly not yet at the high level of established process GC, but it can already be used to solve significant measurement tasks in the gas industry. This new dimension in micro-GC technology opens up the potential for the development of more cost-effective PGCs with lower maintenance

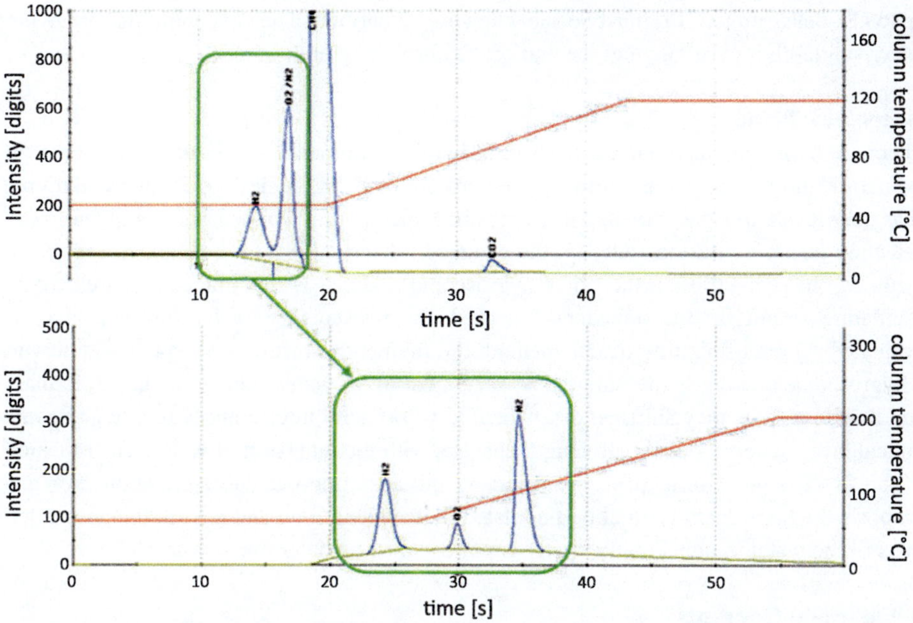

Fig. 15.9 Analysis of a biogas sample using a micro-gas chromatograph with temperature programmed chip separation columns in column configuration. The upper chromatogram shows the signal of column 1 with the components hydrogen, nitrogen + oxygen, methane, CO_2. The lower chromatogram shows the signal from column 2: the sum peak of column 1 is separated by a molecular sieve column into its components oxygen and nitrogen. (Source: Elster GmbH)

costs, which can also reach applications that were previously closed to sophisticated measurement technology.

Hydrocarbon Dew Point Measurement

The measurement of the hydrocarbon dew point is carried out classically by dew point mirrors. A mirror is exposed to sample gas and cooled down in a controlled manner until condensate is deposited. The dew point temperature is usually detected optically by changing the reflection behaviour of a light beam; this can be detected visually or automatically.

An alternative determination of the hydrocarbon dew point is made from the amount of substance analysis of the gas and subsequent phase equilibrium calculation. This requires a very sensitive and detailed molar mass analysis, as provided by laboratory chromatographs and more powerful process chromatographs. The individual boiling points of the alkanes and their isomers as well as the relevant cyclic hydrocarbons must be taken into account in the modelling. The GC analysis is very demanding, as the hydrocarbon isomers mentioned up to C9, better up to C12, must be separated and quantified. The concentrations are in the lower single-digit ppm range at the end of the chromatogram.

Water Dew Point

There are numerous methods for measuring humidity in media and in the atmosphere. In the natural gas industry, the following methods are used. In the dew point mirror, a mirror charged with sample gas is cooled in a controlled manner until condensate precipitates. The detection is carried out optically by the changed reflection of a light beam. In capacitive methods, the dielectric constant of a hygroscopic material is determined in an electrical oscillating circuit. In the impedance method, the resistance of a hygroscopic layer is evaluated. In the oscillating quartz method, the frequency shift of a quartz provided with a hygroscopic coating is measured. The accumulation of water leads to a change in mass and thus to a frequency shift. Another method works with interferences in a hygroscopic optical layer, whose optical path length changes with the inclusion of water. The moisture in the process gas can also be measured by the absorption of laser radiation. For this purpose, the laser beam is quickly modulated in wavelength on and next to the absorption lines of the water vapor. Finally, water vapor can also be determined with GCs.

Sulphur and Odorants

Sulphur components and odorants are determined in the gas industry by various measuring methods. In the field of gas production, the concentrations can be very high, in the case of hydrogen sulphide H_2S, for example, up to the double-digit percentage range. In gas transport and distribution networks, on the other hand, the concentrations are in the single-digit ppm range. The measuring instruments used must therefore be very sensitive. PGCs with different detector types are usually used to monitor the concentration of sulphur components and odorants. TCDs are very versatile and, with suitable chromatographic separation, allow the measurement of sulphur components and odorants in combination with calorific value determination. However, their sensitivity is just sufficient for limit monitoring of trace gases. Special sulphur GCs work with electrochemical detectors that react selectively to sulphur and are very sensitive. In the FPD, the eluate of a GC is burnt in an oxyhydrogen flame and the light emission released is detected at a specific wavelength, for sulphur 394 nm. Another very sensitive and specific detection method is ion mobility spectroscopy, where the eluate is ionized and a spectrum of the ion mobility is recorded in an electric field. Mobile measuring instruments for randomly checking the odorant concentration work with electrochemical detectors. Lead acetate paper is also used for measuring hydrogen sulphide; it turns dark under its influence. Automatic process measuring instruments using this method contain lead acetate paper strips on coils. The paper is exposed to the sample gas at certain points and the blackening is evaluated optically.

The standards for natural gas define not only a limit value for H_2S, typically 5 mg/m^3, but usually also a limit value for total sulphur, typically 30 mg/m^3. It is measured approximately by discretely measuring the concentrations of the most important sulphur compounds and calculating the total sulphur from this. In a procedure for the genuine determination of total sulphur, all sulphur components are reduced to hydrogen sulphide on a catalyst using hydrogen, which is then measured using the usual methods.

Correlative Sensor Systems

GCs are relatively expensive and complex, they also do not provide a continuous signal for measurement, control and regulation purposes. Calorimeters, on the other hand, operate continuously, but are demanding in terms of maintenance and service. For this reason, numerous alternative methods for gas quality measurement based on sensor systems have been developed over the last 20 years (Schley et al. 2003). It must be checked in detail whether the required gas parameters can be determined and whether the sample gases are suitable for the sensor measurement method. The aim was and is to develop simpler, more robust measuring systems for measurement and control purposes, but also for energy billing.

What is the idea behind these alternative measurement methods? Natural gas is a complex gas mixture with numerous components, yet it has a certain regularity. As described above, classical natural gas consists essentially of the three groups of substances hydrocarbons (CH), carbon dioxide (CO_2) and nitrogen (N_2), whereby the hydrocarbons are essentially methane and alkanes, whose concentrations decrease with increasing order. From this typical composition, it follows that natural gas mixtures have considerably less degrees of freedom than gas components. However, this also means that important gas parameters such as calorific value, density, Wobbe index, methane number can be correlated from relatively few characteristic physical quantities without the need for a detailed gas component analysis. Typical measured variables for correlative sensor systems are for example the infrared absorption, heat conductivity, heat capacity, viscosity and sound velocity of the sample gas. In some methods, the same parameter is also measured at different conditions to obtain additional information about the gas mixture. The measurement is usually carried out continuously, with reaction times ranging from several seconds to minutes. Sensor measurement methods are therefore advantageous for process control. Correlative systems do not require a carrier gas and usually only simple calibration gas mixtures. The correlative sensor measurement method *GasLab* from the Elster company is presented here as an example (Kastner 2013).

This measuring method is also based on the characteristic composition of natural gas from hydrocarbons (CH), carbon dioxide (CO_2) and nitrogen (N_2). The infrared absorption of the carbon dioxide, the hydrocarbon mixture and the thermal conductivity of the total gas are measured values. The absorption band of carbon dioxide is free in the spectrum of natural gas without overlap with other absorption bands. Therefore the carbon dioxide concentration can be determined directly from it. The molar calorific value of the hydrocarbon fraction correlates with the infrared absorption of a specific IR band with defined wavelength position and bandwidth in the spectral range of about 3.5 μm. The thermal conductivity is modelled as a function of the concentrations of the groups of substances and

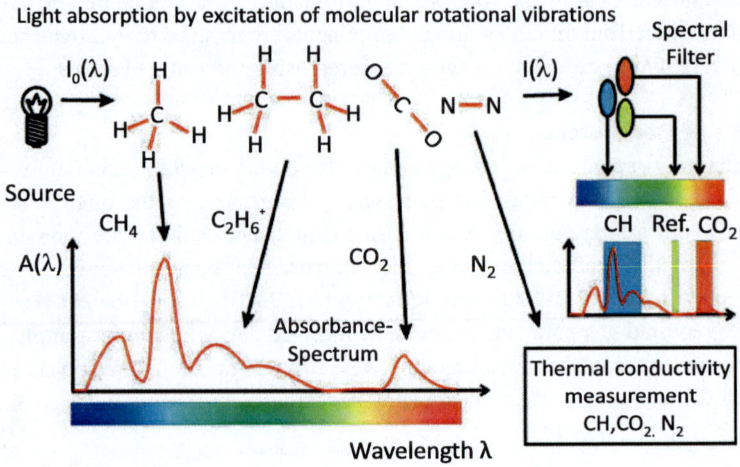

Fig. 15.10 Measuring procedure of the correlative sensor system *GasLab*. The infrared absorption of hydrocarbons provides the molar calorific value of the hydrocarbon mixture, the infrared absorption of carbon dioxide provides its molar fraction. As nitrogen is not infrared-active, it is determined indirectly from a thermal conductivity measurement. (Source: Elster GmbH)

the thermal conductivity of the hydrocarbon fraction. The concentration of nitrogen, which is not directly accessible by measurement, is determined by iterative adaptation of the model value to the measured values. Thus the sample gas is sufficiently described and calorific value, Wobbe index, density, as well as CO_2concentration can be determined (Figs. 15.10 and 15.11).

The *GasLab* sensor system has an explosion-proof housing with gas connections for sample gas and calibration gas (Fig. 15.12). The sample gas flows continuously through the sensor system. By means of a bypass the sample run time from the sampling probe to the measuring instrument can be shortened. The instrument has an integrated pressure regulator and a display with capacitive control elements. A typical application of the *GasLab* is the fast and continuous measurement of gas quality variations for the control of gas turbines. The measuring rate is 1 Hz with a response time of about 5 s (t_{90}).

The evaluation algorithms of the correlative sensor measurement methods are based on individual model assumptions and measurement variables. They therefore cover different application areas and have individual cross-sensitivities. Due to the constant changes in the gas industry and the introduction of new gas types and components, the model assumptions of the correlative measurement methods are constantly being questioned. Currently, the feeding of hydrogen into the public gas distribution system is being discussed in the German gas industry. Due to these new developments, correlative sensor methods require fundamental investigations regarding suitability, sensitivity and cross-sensitivity. In principle, however, sensor measurement methods can also be further developed for this extended area of application (Kastner and Porsch 2011).

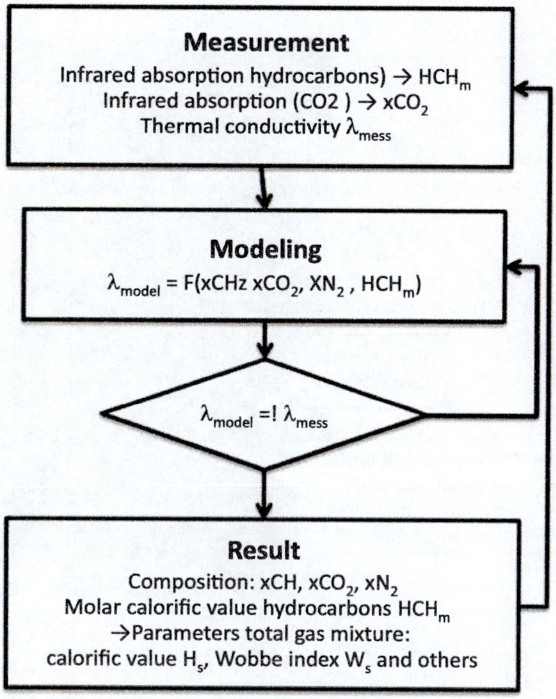

Fig. 15.11 Measuring procedure of the correlative sensor system *GasLab*. The molar calorific value of the hydrocarbon mixture HCH_m and the molar fraction of carbon dioxide xCO_2 are determined by infrared absorption. The molar proportions of hydrocarbons xCH and nitrogen xN_2 are determined by adapting the modelled thermal conductivity λ_{modell} and the measured thermal conductivity λ_{mess}. This allows the calorific value, Wobbe index and other characteristic values of the gas mixture to be determined continuously and quickly. (Source: Elster GmbH)

Gas Measurement Systems

Gas measuring systems are available in different size classes. While gas measuring stations in the distribution network fit into a switch cabinet or container, systems on the transport level can take on the dimensions of machine halls. As a rule, a gas measuring station has a volume measurement system; in the case of large gas flows, it is often redundant and equipped with different measuring methods. On the one hand, this reduces the probability of failure, and on the other hand, systematic errors can be detected between the measuring methods. A gas quality measurement is used when the gas quality can change in relation to other gas quality measurements in the network, for example by mixing in other gas qualities or in networks with multi-side feed and pendulum zones where the flow direction can be reversed. Gas quality measurement is also necessary if energy billing must be carried out between network operators or with a consumer. Billing is based on a gas volume-weighted calorific value, that is the average calorific value weighted with the volume flow in the measuring period. Especially at border crossings and underground storage facilities, the gas quality is monitored with regard to limit values of gas components such as sulphur and dew points in addition to energy measurement. Figures 15.13 and 15.14 show a gas measuring station at a national border with redundant volume measurement as well as gas quality measurement for billing and gas quality monitoring.

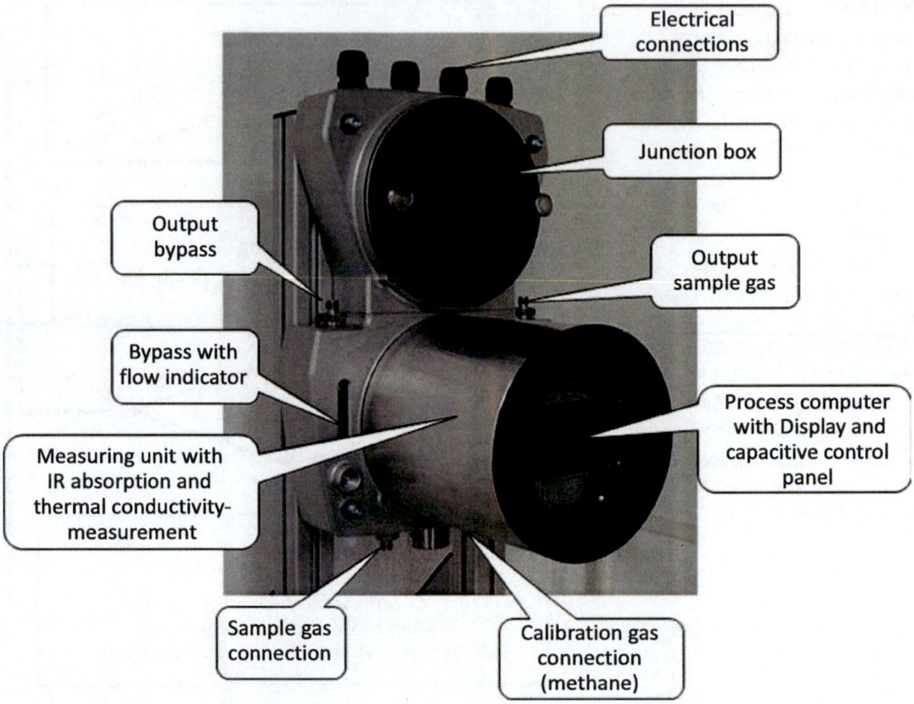

Fig. 15.12 *GasLab* sensor measuring system. (Source: Elster GmbH)

Fig. 15.13 Gas measuring station at a border crossing. The volume measurement is carried out redundantly with different measuring methods, ultrasonic and turbine. In the background, the gas

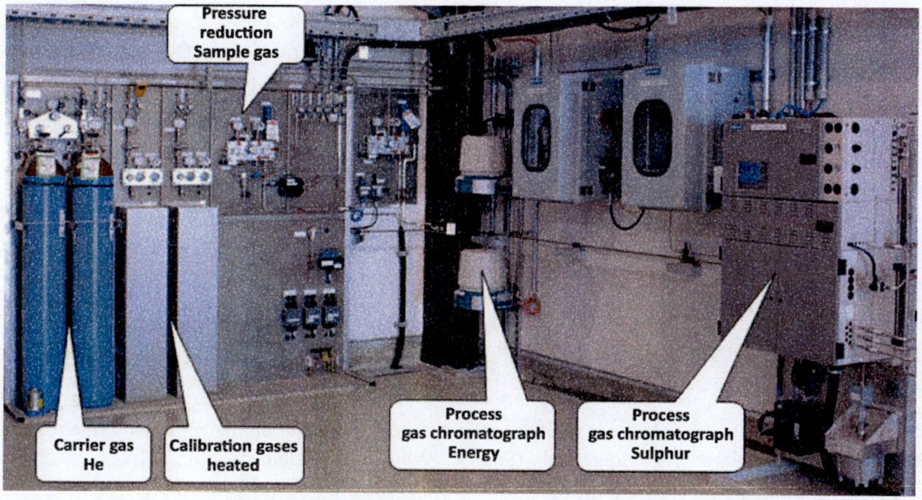

Fig. 15.14 Gas quality measurement of a gas measuring station at a border crossing. Several process gas chromatographs are used for energy billing and monitoring gas quality, such as sulphur concentration and dew points. (Source: Elster GmbH)

15.2 Energy Measurement in the Biogas Application

Ernst Murnleitner

Biogas and sewage gas are valuable energy sources that can be used to generate electrical energy and heat. Therefore, the determination of the energy flow E, for the trade with this raw biogas, is of special interest. Due to the fluctuating composition, however, it is not sufficient to measure only the gas volume flow, the gas composition must also be determined. The energy value is then calculated from both quantities. In addition to the goal of determining the energy value, the gas measurement technology in biogas plants (Fig. 15.15) is also used for process optimization and process control (Murnleitner 2001; Neumann 2012) as well as for the control of the subsequent gas treatment.

Before the measuring methods can be discussed, the special features of biogas measurement that result from the properties of biogas must be considered. Biogenic gases differ from many other gases to be measured by their fluctuating composition, the relatively large number of gas components and the corrosive properties. This has an influence on the flow measurement as well as on the gas composition measurement and on the combination of the

Fig. 15.13 (continued) quality measurement for energy billing and monitoring of the gas quality with several gas chromatographs. (Source: Elster GmbH)

Fig. 15.15 View of a typical biogas plant. *On the left* the fermenters, on the *right* the fermentation residue storage with large gas storage tank. (Source: Awite GmbH Langenbach)

two, the determination of the energy flow. The most common and important biogenic gas is biogas. Depending on its origin, it is also called as sewage gas or landfill gas. As a differentiation, biogas processed to natural gas quality is called biomethane.

Generation of Biogas

Biogas is produced in a multi-stage process during anaerobic fermentation by microorganisms (Schlegel 1992). Several groups of organisms are involved in this process (Fig. 15.16).

In a first step, long-chain organic compounds of enzymes excreted by bacteria are broken down into smaller building blocks such as glucose, which are then taken up by the acidogenic bacteria in a second step and used to generate energy in the cells. Due to the lack of oxygen, the microorganisms cannot generate energy directly by producing carbon dioxide. Instead, the substrate is broken down into oxidized (carbon dioxide, organic acids) and reduced compounds (hydrogen, hydrogen sulfide, alcohols). In a third and fourth step, the products excreted by a group of microorganisms are split up again by specialized microorganisms, the acetogenic bacteria and the methanogenic archaeae,[2] leaving at the end the most oxidized low-energy carbon dioxide (CO_2) and the most reduced high-energy methane (CH_4). The energy rich, gaseous methane can no longer be further utilized anaerobically. This disadvantage for the microorganisms becomes an advantage for the use, because the biogas produced, in combination with the previously missing oxygen, can finally be used for energy production. Depending on how highly oxidized the carbon in the initial substrate is, different amounts of methane are produced. Carbon in carbohydrates

[2] Archaeae are the third domain of living organisms, next to bacteria and eukaryotes (fungi, plants, animals).

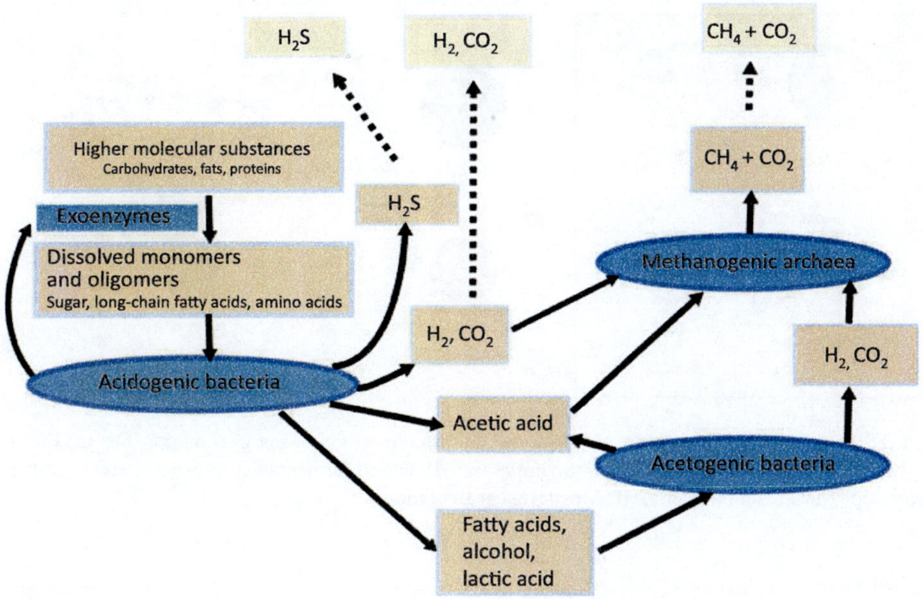

Fig. 15.16 Development of biogas. *Left side*: hydrolysis and acidogenesis, *right side*: acetogenesis and methanogenesis

(oxidation state +2) 1:1 is split into methane (oxidation state 0) and carbon dioxide (oxidation state +4) (Fig. 15.17a). Fat contains little oxygen and is therefore little oxidized. More methane is formed here (Fig. 15.17b). In addition to carbon, oxygen and hydrogen, protein also contains the elements sulfur and nitrogen. Therefore, hydrogen sulfide and ammonia are also formed (Fig. 15.17c). Other fermentation processes in which no archaea are involved (Fig. 15.16: acidogenic and acetogenic bacteria) produce carbon dioxide and hydrogen (Fig. 15.17d). Air for microbiological desulpurisation (Fig. 15.17e) and water vapor (Fig. 15.17f) complete the biogas composition (Fig. 15.17g).

Microbiological Desulfurization

When oxygen is added to the system, it is consumed by microorganisms. If this oxygen (O_2) in the form of air (Fig. 15.17e) is only introduced into the gas space, so-called sulfur bacteria consume the oxygen, because they convert the hydrogen sulfide (H_2S) produced under anaerobic conditions into elemental sulfur by generating energy. This is used for microbiological desulfurization. The oxygen is not completely consumed here but remains in the biogas at about 0.5 vol.%. The air supplied results in a nitrogen content in the biogas, which can limit the attainable methane content in a later gas treatment.

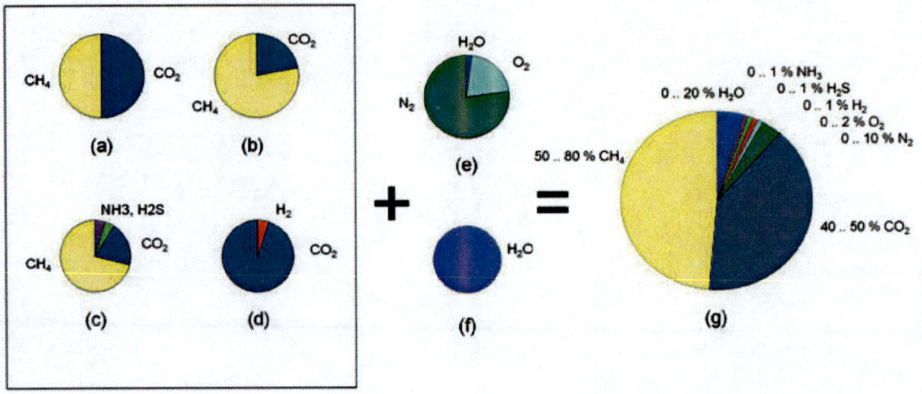

Fig. 15.17 Variable composition of biogas due to different origin of the gas components. (**a**) From carbohydrates. (**b**) From fat. (**c**) From proteins. (**d**) Other fermentation processes. (**e**) Air for microbiological desulpurisation. (**f**) Moisture. (**g**) Real biogas

Moist Gas

Since biogas is produced in the liquid substrate, the gas is initially saturated with water vapor (H_2O). Depending on the temperature, the water content (Fig. 15.17f) can be very high, for example 16 vol.% at 55 °C. Figure 15.18 shows the water vapor content in the biogas at saturation, calculated according to Eq. 15.9.

Importance of Flow and Gas Composition Measurement

Figure 15.19 schematically shows a biogas plant with gas utilisation. The gas composition and the gas quantity are recorded at several points (a–k) in corresponding large plants. In the case of gas from the fermenter (a), the methane content and the hydrogen content are of particular interest, since both represent a parameter for process control, as well as the gas flow, if several fermenters are operated in parallel or one after the other. In addition, the hydrogen sulfide concentration (H_2S) is also measured in order to assess the desulpurisation performance. Air is introduced into the head space of the secondary fermenter/final storage tank and the oxygen content is controlled to setpoints of max. 1 vol.%. The oxygen setpoint value can be made dependent on the sulfur content achieved. For this reason, the oxygen and hydrogen sulfide content is of particular interest in (b). Depending on the gas treatment process, the hydrogen sulfide content must be reduced even further. This gas purification takes place between (c) and (d). Therefore, the hydrogen sulfide content is measured at these points and, depending on the desulphurization process, the oxygen content at (d) is also measured. The purified biogas can either be used directly in combined heat and power plants or fuel cells (e), upgraded to natural gas quality biomethane (f) or fed into a local biogas network (g). At these points, the gas flow and the methane content are recorded and the energy flow is calculated from this. In addition, the hydrogen sulfide and oxygen content must be measured at these points, because corresponding upper limits must always be observed for the subsequent units of use. In

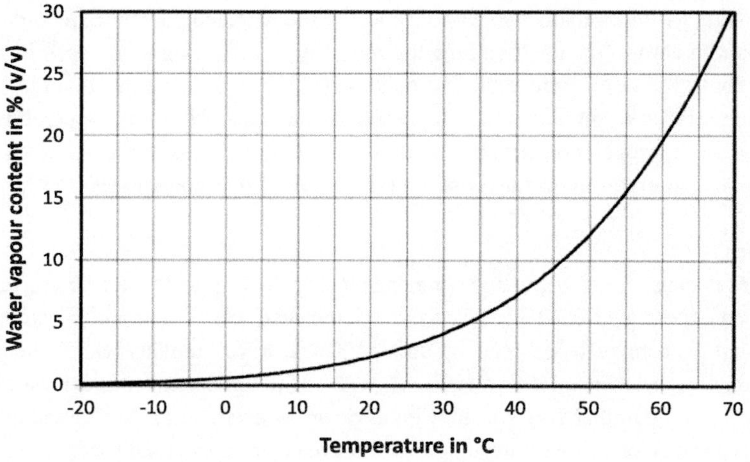

Fig. 15.18 Water vapor content (vol.%) at saturation as a function of temperature at a pressure of $p = 1013$ mbar

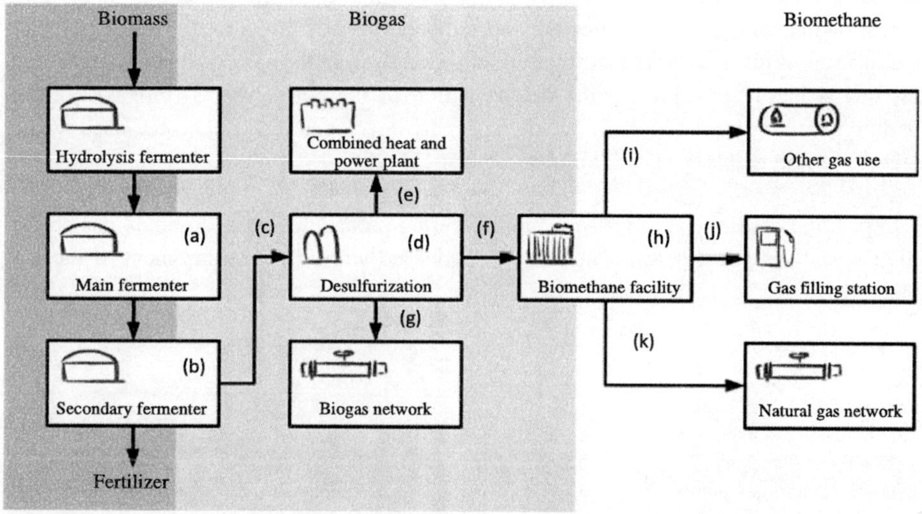

Fig. 15.19 Biogas plant and gas utilization. The letters indicate the positions where the gas composition and gas quantity is measured: (**a**) fermenter, (**b**) microbiological desulphurization (integrated in the fermenter), (**c**) input gas purification/raw biogas sales, (**d**) desulphurization, (**e**) upstream of the CHP, (**f**) input biomethane plant, (**g**) feed into a biogas network, (**h**) gas treatment in the biomethane plant, (**i**) other use of the biomethane, (**j**) gas filling stations, (**k**) feed into the natural gas network

gas processing (h) to biomethane (98 vol.% CH_4), the methane content is recorded as the most important control parameter either continuously or according to the specifications of the plant control system. Before the biomethane (i,j,k) is used, the gas quantity and gas quality are recorded again and the energy value is calculated from this. Depending on the PGC used, the hydrogen content (H_2) is also measured at (k), as this must not be higher than 0.2 vol.% in order not to falsify the energy value of the biomethane.

Challenges

Raw biogas is moist, has a fluctuating composition (Fig. 15.17g) and corrosive properties. All three properties pose great challenges for the measurement of gas volume and gas composition. The moisture content of biogas can be up to 100% relative humidity. Since thermophilic fermentations can be obtained from a fermenter with 55 °C, this corresponds to a water content of 16 vol.%. On the further way it usually cools down, whereby water condenses out. Part of the hydrogen sulfide contained in the biogas is dissolved in the condensed water. When oxygen enters the gas or is even added in a targeted manner in the form of air for microbiological desulfurization, the hydrogen sulfide can be oxidized to sulfurous acid and sulpuric acid.

Methods of Gas Composition Analysis

The gas composition is determined by the physical and chemical properties of the gas. Corresponding measured variables are the temperature and pressure as well as the gas composition. While PGCs are mainly used for the analysis of the gas composition in natural gas, this is hardly possible for the determination of the corrosive raw biogas. For the condition measurement of biogas, mostly specially developed instruments are used which meet the requirements regarding robustness (Fig. 15.20).

Depending on the type of sensors and measuring procedures used, complex gas preparation in the analytical instruments can be dispensed with. The measurement of the gas

Fig. 15.20 Opened process analyzer for biogas. *Top*: Signal converter and digital modules. *Left*: Infrared sensors and electrochemical sensors, *including* sample gas pumps. *Right*: Valves. (Source: Awite GmbH Langenbach)

composition of raw biogas is usually carried out discontinuously on an hourly basis, as the biogas generation process is slow and thus considerably longer sensor lifetimes can be achieved. Typical measuring intervals of 30 min are usual for monitoring gas purification. The measurement is often carried out continuously during the gas treatment process, as the gas is already cleaned and causes only little wear on the sensors. Table 15.1 shows the measuring principles used in practice for the gas analysis of biogas.

NDIR Procedure https://www.witec-sensorik.de/en/products/infra-sens/
Infrared absorption measurement is very often used to determine the methane and carbon dioxide content of biogas. Most of the devices work according to the so-called two-beam method, which can largely compensate for the effects of ageing and pollution. With infrared measurement, small measurement uncertainties can be realized, depending mainly on the calibration interval. While the standard measurement uncertainty is only 0.2 vol.% directly after calibration, when using high-precision calibration gases, it increases to about 2 vol.% over the course of a year, taking into account temperature fluctuations of $\Delta T = \pm 10°$ C.

Thermal Conductivity
The thermal conductivity for the determination of methane and carbon dioxide can be realized relatively inexpensively. However, since other gas components contained in biogas have a similar thermal conductivity to biogas (Fig. 15.21: comparison of air and 50–55% methane in carbon dioxide), this measuring principle is only suitable for two-component mixtures. If the presence of air and water vapor can be excluded for biogas, then the measuring method is also suitable. Under no circumstances, however, should the zero point be set at ambient air, as this can cause a measurement error of >10%. The cross-sensitivity to air is sometimes corrected mathematically via the oxygen content. One uncertainty in this correction, however, is the possibility that some of the oxygen has been consumed and thus the actual nitrogen concentration is not known.

Ultrasound
Similar to the thermal conductivity, the measurement of the ultrasonic velocity is used to determine the methane and carbon dioxide content. Figure 15.22 shows the sonic velocity of biogas with different compositions compared to air, nitrogen, oxygen and water vapor. For estimating the methane content, ultrasonic measurement in combination with ultrasonic flow measurement is suitable since the ultrasonic transit times are recorded here anyway. However, if air is added for biological desulphurization, but the gas is not dry or residual air gets into the gas, this measurement is less suitable.

Oxygen Measurement
Either the electrochemical or the paramagnetic measuring principle is used to measure the oxygen content in biogas. Both methods are suitable. Paramagnetic sensors have a longer

Table 15.1 Advantages and disadvantages of the measuring principles used in raw biogas analysis. Measuring principles that are either not possible or are not used in practice are marked with –

Measuring principle	CH_4	CO_2	O_2	H_2S	H_2	Humidity
Infrared	Selective	Selective	–	–	–	–
Electrochemical	–	–	Selective; cost-effective	Cost-effective; cross-sensitivities	Cost-effective; selective	–
Paramagnetic	–	–	Selective; sensitive	–	–	–
Thermal conductivity	Low cost; non-selective	Low cost; non-selective	–	–	Only at high concentrations	–
Ultrasound	Non-selective	–	–	–	–	–
Electrical-capacitive	–	–	–	–	–	Selective

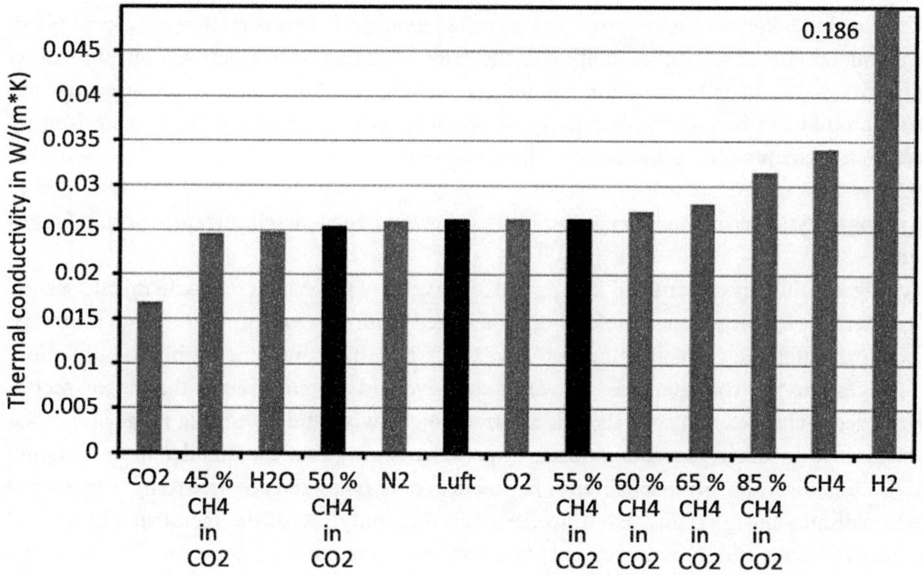

Fig. 15.21 Thermal conductivity values of biogas and its gas components, sorted in ascending order. The thermal conductivity of air is similar to that of biogas

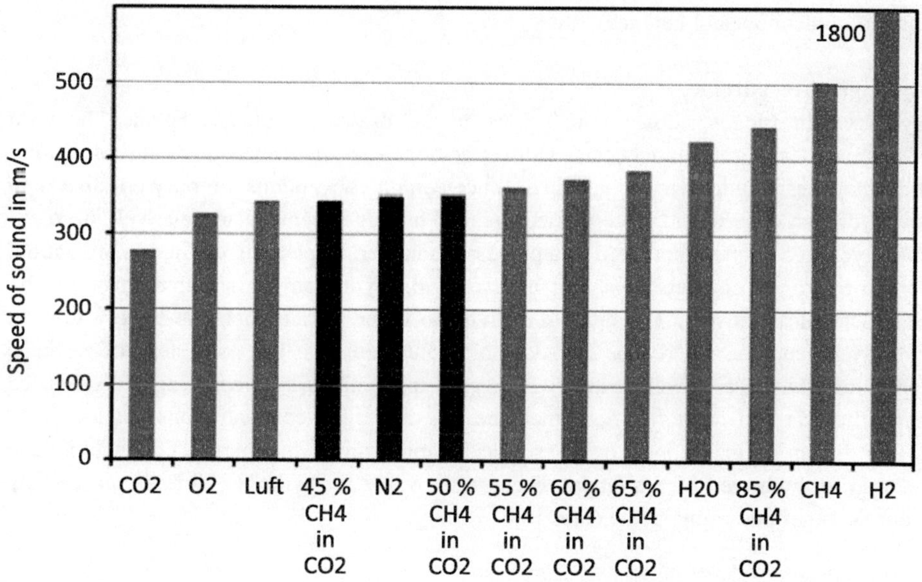

Fig. 15.22 Sound velocity of different gases sorted in ascending order. The speed of sound of air is similar to that of biogas

life span, but contain a moving part, the so-called *dumbbell*. This is relatively susceptible to contamination. Therefore, in addition to the price advantage, the electrochemical measuring principle is usually used for oxygen measurement. These sensors are used up and therefore have to be replaced regularly. A possible failure can be predicted quite well by test measurements with pure air.

Hydrogen Sulphide Measurement https://www.witec-sensorik.de/en/products/ultra-sens/
Hydrogen sulfide is determined in biogas exclusively by means of electrochemical sensors. In other areas, hydrogen sulfide is also realized with UV absorption or by means of chemical automated precipitation of silver salt and measurement of black coloration. However, the last two methods are very expensive and are not used in the biogas sector. With electrochemical sensors, there is a correlation between the measuring range and cross-sensitivities to other gas components. Experience shows that the smaller the measuring range, the lower the cross-sensitivity. In practice, only cross-sensitivities to hydrogen exist for small measuring ranges. Even if this is less than only 1% of the measuring range end value (full scale, FS), a hydrogen sulfide sensor with a measuring range of 10 ppm, with a hydrogen content of 1000 ppm, can no longer perform an H_2S measurement, since the FS has already been reached by the H_2 content. At very high measuring ranges, of several thousand ppm, cross-sensitivities also exist to other chemical compounds, such as alcohols and ketones. This can cause problems in biowaste fermentation plants due to an increased content of alcohols and ketones.

Hydrogen Measurement
Hydrogen in biogas allows conclusions to be drawn about the balance between acidogenesis and methanogenesis. During acidogenesis hydrogen is produced, during methanogenesis hydrogen is consumed. Since certain conversions are only possible with low hydrogen content and since electrochemical hydrogen sensors are relatively inexpensive, hydrogen is often measured as a process parameter. Depending on the gas utilization, certain hydrogen contents must not be exceeded. Hydrogen is measured in biogas by electrochemical sensors. The cross-sensitivity to other gas components is very low. In hydrolysis reactors, where due to a low pH value only the first two steps of methane formation take place, as well as in *power-to-gas*[3] applications, where hydrogen is produced by hydrolysis and converted into methane, the hydrogen concentrations can be much higher than in normal biogas. For these special applications, the thermal conductivity can be used for hydrogen measurement, because the value for hydrogen differs considerably from the other gas components (Fig. 15.21).

[3] Power-to-Gas: electric hydrolysis of water and production of hydrogen.

Humidity Measurement https://www.witec-sensorik.de/en/products/options/humisens/

The gas composition includes the measurement of the concentration as well as the detection of the humidity. Gas analysis devices can measure either moist gas or dry gas (after dehumidification by means of a gas cooler). The water vapor content is required for the conversion between both or for the energy value calculation. For biogas, the electric-capacitive measuring principle is used almost exclusively. Service lives of several years in raw biogas have already been proven. The relative humidity is converted into the water vapor content according to Eq. 15.9 using pressure and temperature.

Methods of Flow Measurement

The flow measurement methods can be divided into volume measurement and mass measurement. For biogases, volumetric measuring methods such as turbine wheel, vortex and ultrasound are often used. Of the mass flow methods, thermal mass flow measurement, Coriolis and differential pressure measurement are often used. With volumetric flow measurements, a signal is obtained which is proportional to the operating volume. Pressure and temperature change the operating volume and thus also the measurement signal. The relationship can be described with the equation of state of ideal gases.

$$\frac{p \cdot V}{T} = \text{const.} \tag{15.4}$$

From Eq. 15.4, it can be directly derived that a doubling of the pressure p leads to a doubling of the measurement signal, just as a corresponding temperature increase T, for example from 273 K (0 °C) to 546 K (273 °C). In practice, pressure fluctuations of ±5% are caused by the weather alone. Depending on the season, the temperature in the gas pipeline can vary between summer and winter when laid outdoors around 50 °C. Therefore, with these measuring methods, the measurement and mathematical correction of pressure and temperature is necessary. With the mass flow methods, a signal is obtained which is directly proportional to the mass flow, that is there is no dependence on pressure and temperature as with the volume measurement methods. For a given gas composition, the mass flow can be converted directly to a volume flow at reference conditions. A difficulty in practice, however, is the determination of a sufficiently accurate gas composition. Each flow measurement principle has advantages and disadvantages when measuring biogas (see Table 15.2).

Energy Value Calculator

There are currently no officially approved devices[4] for determining the energy value of raw biogas. The energy flow of raw biogas is calculated from the flow rate multiplied by the

[4]Exception in the calibration law for fuel gases <6, 5 kWh/m^3.

Table 15.2 Advantages and disadvantages of flow measurement methods for measuring biogas

Measurement method	Advantages	Disadvantages
Ultrasound	High measuring accuracy No moving parts Insensitive to dirt Low pressure drop No influence of the gas composition	Installation of pipe section necessary Not possible with high CO_2 content Not possible with dry biogas Pressure and temperature measurement necessary
Turbine wheel	High measuring accuracy Low pressure loss No influence of the gas composition	Wear of the bearings (service life) Sensitive to contamination Pressure and temperature measurement necessary
Vortex	No moving parts No drift High measuring accuracy Compact design Drops of water no problem	Turbulent flow necessary Pressure and temperature measurement necessary
Thermal dispersion	No moving parts (particles in gas) Compact design, easy installation Different nominal sizes with the same sensor Low pressure loss No pressure and temperature compensation necessary	Further dependence on the gas composition Uncertainty of correction factors
Coriolis	No moving parts (particles in gas) Insensitive to vibration No pressure and temperature compensation necessary	Depending on the gas composition Expensive High pressure drop?

methane content and the energy content of methane (calorific value). It must be ensured that the gas composition and gas quantity are based on the same conditions. In practice, this often leads to problems with the gas humidity. Table 15.3 lists the possible scenarios where humidity is considered.

Raw biogas is moist. The flow rate from the fermenter is therefore measured in the moist state. If the gas analysis is also carried out in the humid gas, no measurement and correction of the humidity is necessary (Case 1). However, this requires that no moisture can condense out between flow rate measurement and gas analysis. Sample gas lines may therefore have to be heated or kept frost-free. If the flow rate from the fermenter is determined in the moist state and the analysis is performed with dried gas (analysis gas drying), then a correction of the moisture is necessary (Case 2). Here it must also be considered that calibration gases are dry and dried sample gases still have residual humidity of 0.7 vol.% (Eq. 15.9 and Fig. 15.18). In this case, the almost dry gas analysis is corrected by the 0.7 vol.% residual

Table 15.3 Consideration of gas humidity in the calculation of energy value

Case	Gas flow	Gas analysis	Humidity
1	Humid	Humid	Do not consider
2	Humid	Dry	Correct
3	Dry	Dry	Do not consider
4	Dry	Humid	Correct

moisture and the gas flow is converted to dry gas. For the conversion, the moisture content is required for the flow measurement. If the gas flow measurement is installed immediately after the coldest point of the gas drying process, then it may be possible to dispense with a moisture measurement and instead calculate the moisture content from the saturation curve at the corresponding temperature according to Eq. 15.9. For this purpose, a constant value is used instead of the real measured humidity value. Otherwise, the relative gas humidity is measured by means of a sensor.

In Case 3, the gas is dry both in the gas analysis and in the gas flow measurement. If both residual moisture contents are different, this will cause a deviation.

Case 4 has no relevance in practice.

Hydrogen, hydrogen sulfide and ammonia also have a calorific value, but their concentration in biogas is usually very low. For this reason, only the methane content is considered below.

$$E = F_{CH_4} \cdot H_S \tag{15.5}$$

$$F_{CH_4} = F_n \cdot \frac{c_{CH_4}}{100} \tag{15.6}$$

F_{CH_4} = flow rate for pure methane (m^3/h) at 0 °C and 1.013 bar,[5] F_n= flow rate at 0 °C and 1.013 bar based on dry gas (m^3/h).c_{CH_4} = methane content in volume %, energy flow E= (kW), calorific value H_S= (1.013 bar, 0 °C) at a condensation temperature of 25 °C (kWh/m^3).[6]

While the mass flow meter already provides a signal proportional to the mass flow or the flow rate under reference conditions, when using a volume flow meter, the correction of the moisture content still has to take temperature and pressure into account (Eq. 15.7). When using a mass flow sensor, only the humidity must be taken into account (Eq. 15.8).

$$F_n = F_b \cdot \frac{p}{1013} \cdot \frac{273.15}{273.15 + t} \cdot \frac{100 - c_{H_2O}}{100} \tag{15.7}$$

[5]Reference conditions vary by industry and country.

[6]Calorific value combination I according to DIN 51857.

$$F_n = F_{n,\text{wet}} \cdot \frac{100 - c_{H_2O}}{100} \tag{15.8}$$

F_b= flow rate under operating conditions (at p bar and t °C), p= pressure in bar, t= temperature in °C, c_{H_2O} = water vapor content in vol.%.

The water content in the gas is calculated by means of a modified Magnus formula from the temperature and the relative water content, taking pressure into account (Eq. 15.9 and Fig. 15.18).

$$c_{H_2O} = \frac{h}{p} \cdot 0.0061078 \cdot \exp\left(\frac{17.08 \cdot t}{234.18 + t}\right) \tag{15.9}$$

h= relative humidity (in %)

In summary, Eq. 15.10 shows the energy value calculation using a volume flow sensor and Eq. 15.11 the energy value calculation using mass flow measurement.

$$E = \frac{F_{b,\text{wet}} \cdot c_{CH_4} \cdot H_s}{100} \cdot \left[1 - \frac{6.1078 \cdot h \cdot e^{17.08 \cdot t/(234.18+t)}}{p \cdot 10^5}\right] \cdot \frac{p \cdot 273.15}{1.013 \cdot (273.15 + t)} \tag{15.10}$$

$$E = \frac{F_{n,\text{wet}} \cdot c_{CH_4} \cdot H_s}{100} \cdot \left[1 - \frac{6.1078 \cdot h \cdot e^{17.08 \cdot t/(234.18+t)}}{p \cdot 10^5}\right] \tag{15.11}$$

Figure 15.23 shows the course of the measured values at a biogas plant over 1 year.

In summer, the gas temperature is significantly higher, which also leads to a higher water vapor content in the gas. On day 281, there was a heavy storm with flooding, which also affected the operation of the biogas plant. The collapse of gas production due to substrate shortage is clearly visible. An increase in the methane content in such a situation is normal. This is explained by a relatively stronger decrease of carbon dioxide production (Fig. 15.16 left side) compared to methane production (Fig. 15.16 right side) and a faster limitation of the rapidly degradable carbohydrates compared to the more difficult to degrade protein compounds (higher methane yield of protein degradation in Fig. 15.17c compared to carbohydrates in Fig. 15.17a). The incident on day 281 shows that the gas measurement can also be used for condition detection. While the methane content increases when there is a substrate deficiency, the hydrogen content increases when there is an overload because a lot of hydrogen is produced by the acidogenic bacteria (Fig. 15.16 left), which cannot be consumed quickly enough by the methanogens (Fig. 15.16 right).

Measurement Uncertainty of the Energy Value

In Great Britain, the determination of the measurement uncertainty of the energy value is required for combined heat and power plants (CHPQA 2013). Independent of this, measured values should always be stated together with the measurement uncertainty (GUM 2008; Barwick and Prichard 2011). The measurement uncertainty limits the value

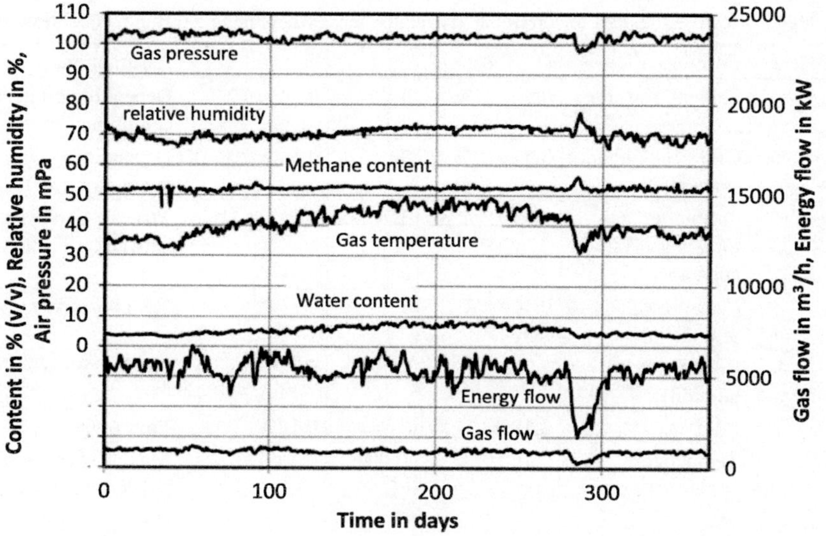

Fig. 15.23 Course of the measured values at a biogas plant over 1 year (daily mean values)

range within which the measured value lies with a certain probability. The range covering approximately 68% (corresponding to ±standard deviation) is called the standard measurement uncertainty. An expanded measurement uncertainty is often specified with a coverage factor of 2, which covers approximately 95% of the measured values. The total measurement uncertainty of the energy flow must take all relevant factors into account. It is calculated according to Eq. 15.12.

$$U(k=2)_{\text{Overall}} = 2 \cdot u_{\text{Overall}} - 2 \cdot \sqrt{\sum ue_i^2} = 2 \cdot \sqrt{\sum (S_i \cdot u_i)^2} \tag{15.12}$$

$U(k=2)$= expanded total uncertainty with coverage factor 2 (95% confidence interval); u_{Overall}= standard uncertainty (68% confidence interval), ue_i= effective standard uncertainty of a sensor or parameter, u_i= standard uncertainty of a sensor or parameter, S_i= sensitivity factor for the uncertainty of a sensor or parameter.

The sensitivity factors were derived from the Eqs. 15.10 and 15.11. Table 15.4 shows the measurement uncertainties as estimated using a practical example with a turbine wheel flow meter and a process analyzer. The flowmeter used was approved by the weights and measures office with 1% calibration error limit for natural gas with air as calibration medium. The traffic error limit during the period of use may be doubled. There is currently no custody transfer approval for raw biogas. However, the same uncertainty was applied. Since the standard uncertainty has a confidence interval of 68%, the calibration error limits have been halved.

Table 15.4 Effective standard uncertainties (typical values) when using a turbine wheel flow meter

Parameters	Description	*U*	*S*	*ue*	Assumptions
$F_{n,\ wet}$	Flow rate (wet)	0,5 ... 1%	1	0,5 ... 1%	Depending on general conditions
c_{CH_4}	CH_4 concentration	0,5 ... 2%	1	0,5 ... 2%	Depending on the calibration interval[a]
Calgas	Calibration gas for gas analysis apparatus	0,05 ... 1%	1	0,05 ... 1%	Traceable analysis between 0.1 and 2%[b]
Hs	Calorific value of CH_4	0	1	0	Not considered
H	Relative humidity	1,2%	0,1	0,12%	100% at 46 °C
P	Print	0,18%	1	0,18%	Environment ±10%
t	Temperature	0,16 °C	0,006/°C	0,10%	Max. 46 °C; +1 °C increases the water content by 5%
Total standard uncertainty *u*				0,7 ... 2,5%	
Total expanded uncertainty $U(k = 2)$				1,4 ... 5%	

[a]55 vol % methane, 2 weeks calibration interval; ±2 ... 10 °C ambient temperature
[b]Extended measurement uncertainty ($k = 2$) available on the market

The uncertainty of the methane concentration measurement depends to a large extent on the fluctuations of the ambient temperature and the calibration interval. In addition to the flow rate and the methane concentration, the uncertainty of the calibration gas also has an influence on the calibration of the gas analyzer, as do the gas humidity, the gas temperature and the gas pressure. Even the calorific value is subject to uncertainty because biogas is not an ideal gas and the calorific values themselves are subject to uncertainty (DIN 51857). However, this uncertainty is assumed to be 0 here, as comparative measurements are based on the same values. This results in an overall standard uncertainty of 0.7–2.5%. Depending on the industry, this standard measurement uncertainty is converted into an expanded measurement uncertainty by multiplication with a coverage factor; the usual factor is $k = 2$. The actual measurement deviation is then very probably within the confidence range of the expanded measurement uncertainty (95%) from ±1.4 to 5%.

In thermal mass flow measurement, the gas composition also has an influence on the measurement uncertainty.

Table 15.5 shows how the measurement uncertainty of the thermal mass flow measurement for biogas was estimated using a practical example. This example is a gas measurement, upstream of the cogeneration plant, with relatively constant conditions, the resulting standard measurement uncertainty is 1%. The classification of the combined measurement uncertainty was made according to CHPQA17 (2008).

Table 15.6 shows typical uncertainty values when using a mass flow sensor with thermal dispersion; here the standard measurement uncertainty of thermal flow measurement was

Table 15.5 Estimation of the measurement uncertainty of thermal gas flow measurement for biogas. Measurement before the cogeneration unit

Standard uncertainty		Description
$u_{CalFlow}$	0,25%	Calibration stand
u_{CalGas}	0,25%	Composition of the synthetic calibration gas
u_{FRep}	0,59%	Measuring error according to manufacturer's specifications (for measuring range 500 m^3/h)
$u_{FNomTemp}$	0,15%	Temperature deviation
u_{FNomCH_4}	0	Deviation calibration medium real biogas
u_{FTrans}	0,21%	Signal transmitter
u_{TGas}	0,21%	Actual gas temperature
u_{TAmb}	0	Actual ambient temperature: Contained in $u_{FNomTemp}$
u_{Press}	0	Pressure dependence
u_{CH_4}	0,44%	7 m/s flow velocity (±10% rel. CH_4)
u_{CH_4Meas}	0,17%	Measured methane content
u_{Air}	0,15%	5 ± 1% (v/v) air in biogas
u_{Hum}	0,22%	Gas humidity: 10 ± 1°°C, saturated
u_{Gas}	0,14%	Gas composition: Other gas components
u_{FDrift}	0	No drift of the flow sensor (CHPQA 2013: GN17-4)
u_{TDrift}	0	No drift of transmitter and auxiliary sensors (CHPQA 2013)
Combination		
u_{FStand}	0,35%	*Calibration stand:* $u_{FStand} = \sqrt{u_{CalGas}^2 + u_{CalFlow}^2}$
u_{FNom}	0,15%	Calibration conditions: $u_{FNom} = \sqrt{u_{FNomCH_4}^2 + u_{FNomTemp}^2}$
u_{FPrim}	0,70%	Reference conditions: $u_{FPrimary} = \sqrt{u_{FStand}^2 + u_{FRep}^2 + u_{FNom}^2}$
u_{FFluct}	0,60%	$u_{FFluct} = \sqrt{u_{CH_4}^2 + u_{CH_4Meas}^2 + u_{Air}^2 + u_{Hum}^2 + u_{Gas}^2 + u_{TGas}^2 + u_{TAmb}^2 + u_{Press}^2}$
u	0,94%	Eq. 15.12
$U(k = 2)$	1,89%	

Table 15.6 Effective standard uncertainties (typical values) when using a mass flow meter with thermal dispersion

Parameters	Description	*u*	*S*	*ue*	Assumptions
$F_{n,\ \mathrm{wet}}$	Flow rate (wet)	1 . . . 3%	1	1 . . . 3%	Depending on general conditions
$c_{\mathrm{CH_4}}$	CH_4 concentration	0,5 . . . 2%	1	0,5 . . . 2%	Depending on general conditions[a]
Calgas	Calibration gas for gas analysis apparatus	0,05 . . . 1%	1	0,05 . . . 1%	Traceable analysis between 0.1 and 2%[b]
Hs	Calorific value of CH_4	0	1	0	Not considered
h	Relative humidity	1,2%	0,1	0,12%	0% at 46 °C
p	Print	0,18%	0,1	0,02%	Environment ±10%
t	Temperature	0,16 ° C	0,005/°C	0,08%	Max. 46 °C;+1°C increases the water content by 5%
Total standard uncertainty *u*				1 . . . 4%	
Total expanded uncertainty $U(k = 2)$				2 . . . 8%	

[a]The calibration interval; 55 vol.% methane, 2 weeks calibration interval; ±2 ... 10 °C ambient temperature
[b]Extended measurement uncertainty ($k = 2$) available on the market

set at 1–3%, since in practice, depending on the system and installation location, correspondingly larger uncertainties can also occur. The increased measurement uncertainty here is mainly due to the fact that in this practical example oversized pipes were used, which lead to a low flow rate and the sensor is therefore operated at the lower end of the measuring range (data not shown).

Compared to volumetric flow methods, such as the turbine wheel, the influence of pressure and temperature is less significant, since these two parameters mainly affect only the water content with this measuring principle. However, the uncertainty of the flow measurement is significantly higher because the gas composition has an influence on the measurement signal (Eq. 15.13). A mathematical correction of the gas composition is already taken into account in the estimation of the measurement uncertainty. An energy value calculator functionality is stored in the gas analyzer used, which corrects the value of the flow sensor accordingly. However, this correction itself is itself subject to an uncertainty, due to the uncertainty of the correction factors and the measurement of the gas composition. Without this correction, systematic deviations in the two-digit percentage range occur.

Equation 15.13 (Endress and Hauser 2003) shows that the flow velocity v is directly proportional to the density and heat capacity of the gas. However, the measurement signal also depends on the thermal conductivity λ. Depending on the flow velocity, the correction is 1–3% for a change in biogas composition from 45 to 50 vol.% methane (in carbon dioxide)

$$Q_L = \Delta T \cdot \sqrt{K + (2 \cdot \pi \cdot \lambda \cdot c_v \cdot \rho \cdot v \cdot d)} \tag{15.13}$$

Q_L= heat dissipation wire, ΔT= temperature difference between wire and gas, λ= thermal conductivity of the gas, c_v= specific heat capacity at constant volume, ρ= density of the gas, v= flow velocity, d= diameter of the wire.

Despite the higher measurement uncertainty, thermal dispersion is used relatively frequently in practice in biogas plants because this measurement method is robust and the investment costs are comparatively low.

Summary

Due to the gaseous nature of biogas, as well as its changing composition and moisture, the energy content of the raw biogas is not measured by natural gas analysis methods, but rather using specialized equipment. In addition to the energy content, the gas composition of the biogas is also of interest during the individual production and treatment steps and is used to support process control. The methane produced with the biogas determines the energy content of the gas. Other gas components have only a minor influence on the energy content due to their non-existent calorific value or their low concentration. Future *power-to-gas* applications could also lead to higher hydrogen contents, which must then be considered accordingly in the determination. In addition to the gas composition, including humidity, temperature, pressure and glass flow have different influences on the uncertainty of the resulting energy value depending on the general conditions. A measured value can only be used in connection with its uncertainty. Overall, the measurement uncertainty of the energy value determination of the biogas is relatively high. Since many parameters influence the measurement uncertainty, in practical use the effort required to improve the measurement uncertainty of a partial component must be weighed against the overall measurement uncertainty. Computational correction methods can significantly reduce the measurement uncertainty, which can make it possible to use more cost-effective measurement methods.

15.3 SF_6 Leakage Monitoring in Switchgear

Roland Kurte, Alexander Hein, and Thomas Heckler

Sulphur hexafluoride (SF_6) is known for its high dielectric strength and has its primary application worldwide as an insulating and extinguishing agent in electrical switchgear. It is non-flammable, chemically inert, non-toxic, colourless and odourless. With a molecular weight of 146.06 g/mol it is a heavy gas and can lead to suffocation by air displacement in closed rooms (e.g., maintenance tunnels) at high concentrations. Due to its high global warming potential of 22,800 compared to CO_2, emissions are monitored worldwide and should be reduced to a minimum. In Europe, this is regulated by the fluorinated greenhouse gas regulation (EU-Verordnung Nr. 517/2014).

Insulating Gas for Switchgear

Most electrical switchgear requires the use of SF_6 gas as an insulating and arc-quenching medium, especially at higher voltage levels. SF_6 is the substance with ideal properties for these two functions. To ensure the insulating and arc-quenching properties, a defined amount of gas per volume, the gas density of SF_6, is required. Because the gas density as a derived measurand cannot be measured directly, the pressure measurement normalized to 20 °C is often used as a substitute measurand for the gas density. When monitoring the SF_6 density 1 K deviation can lead to a pressure deviation of 3 kPa SF_6 at 20 °C.

If the SF_6 gas density falls, the insulation properties are reduced, so that below a certain value safe operation of the switchgear cannot be guaranteed. Therefore it must be constantly monitored.

Figure 15.24, shows a simplified gas insulated switchgear. The temperature, and hence the density, in the vessel can normally be assumed to be uniformly distributed with certain accuracy as long as the gas is not present in the liquid phase.

Real Gas Behaviour of SF_6

The real behaviour of SF_6 in the three dimensions pressure, density and temperature is not directly proportional. This requires the use of the "real gas" equation in order to get an accurate result. Simplified calculations, considering SF_6 an "ideal gas", would be convenient, but would not sufficiently describe the behaviour of SF_6. SF_6 density can be measured indirectly with temperature compensated pressure gauges or sensors.

$$\frac{p}{\varrho RT} = Z = 1 + B(T) \cdot \varrho + C(T) \cdot \varrho^2 \tag{15.14}$$

$$\text{with} \quad B(T) = B_0 + \frac{B_1}{T} + \frac{B_3}{T^3} + \frac{B_5}{T^5}$$

$$\text{and} \quad C(T) = C_0 + C_1 \cdot T + C_2 \cdot T^2$$

Virial Eq. 15.14 describes the relationship between density (ϱmuss durch ϱ auch in den Formeln wieder ersetzt werden), pressure (p) and temperature (T). The constant R, B_x and C_x have been determined empirically.

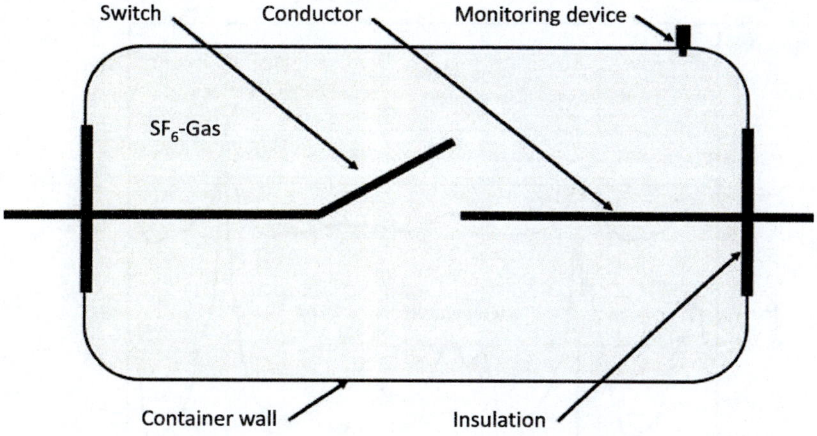

Fig. 15.24 Schematic diagram of a SF_6 gas-insulated switchgear

SF_6 Gas Density Instruments Versus Standard Pressure Gauges

The ambient temperature can change significantly over different time periods, such as daily or seasonal fluctuations (see Fig. 15.25). If such temperature changes are not compensated directly at the instruments, the uncompensated pressure measurement must be corrected using external temperature measurement and a temperature compensation table to obtain a density value.. Most *p*/*T* charts can only be read in steps of 5 or 10 degrees, so that accurate compensation is very time consuming and often inaccurate. This is an important argument why the use of standard pressure gauges for SF_6 density measurement is not recommended.

Figure 15.25 shows an example of the pressure and temperature curve of an SF_6 gas-filled vessel outdoors over 5 days. The recordings were made in Central Europe in spring 2013 with a GDHT-20 sensor from WIKA. The temperature fluctuates between 10 °C at night and almost 40 °C during the day. The resulting pressure change is between approximately 3.1 and 3.5 bar abs. It is clearly visible how the pressure increases with rising temperatures.

Another factor that influences plant safety is the humidity in the gas. The two main advantages of SF_6, the excellent arc quenching properties and the following recombination of the SF_6 molecules, depend on the degree of purity. With higher humidity levels, more decomposition products, such as HF and SO_2, are formed during a switching operation or electrical fault (Heise et al. 1997). Extreme high humidity levels may lead to condensation and thus to liquid and conducting water on the surfaces.

Quality Requirements for Gas Density Measuring Instruments

Electrical switchgears can be installed outdoors, under the harshest conditions with lifetimes measured in decades. Therefore SF_6 gas measurement technology must be developed for the harshest environments and with the appropriate reliability and accuracy.

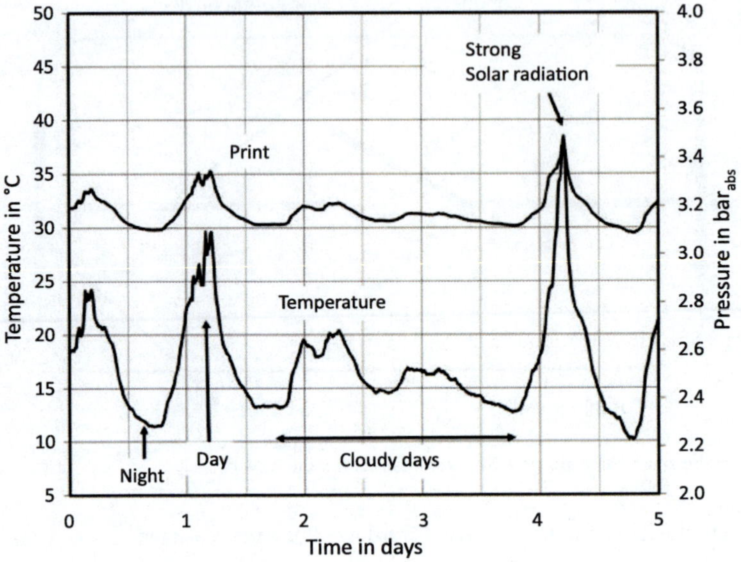

Fig. 15.25 Temperature signal and pressure signal over a 5-day period

Quality criteria for SF_6 gas density indicators are:

- For outdoor installations: high resistance to environmental conditions such as temperatures of −40°C... +80° C, continuous strong UV radiation.
- High accuracy (depending on application and filling pressure).
- High vibration and shock resistance.
- No false alarms due to temperature fluctuations or vibrations during switching operations.
- Measuring systems made of stainless steel and tested for leaks ($<1 \cdot 10^{-8}$ mbar l/s)
- Hermetically sealed housing – no penetration of water or moisture, error-free operation at high altitudes.

For mechanical gas density measuring instruments the compensation of the temperature-related pressure fluctuations, as shown in Fig. 15.26, is usually achieved with a bimetal. The bimetal deforms according to the ambient temperature and thus causes a change in length. This behaviour counteracts with the movement of the *Bourdon pressure measuring spring* and the pointer consequently indicates a pressure referenced to 20 °C .

Since the temperature has a great influence on the signal, it is important to install the instrument at a location with conditions which are representative for the whole system, in order to obtain reliable measurements.

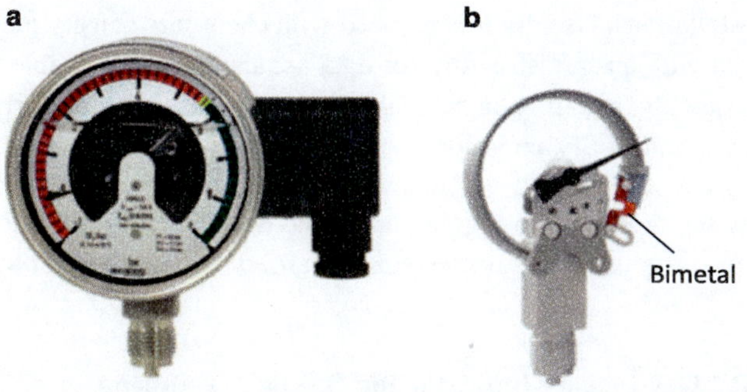

Fig. 15.26 (**a**) Gas density monitor with bimetal temperature compensation. (**b**) Measuring unit with the bimetal

Typical problems are:

- Remote installations for better readability of the display and therefore, potentially significantly different temperatures in the tank and at the instrument.
- Direct sunlight on the device heats it up faster than the gas tank, with it's much larger volume.
- Wind, rain and/or snow cool the unit faster than the tank.
- Non-homogeneous temperature distribution in the tank due to operating conditions or changes in ambient conditions (temperature rise in the morning /drop in the evening).

Typical solutions are: covers or insulation accessories for the instruments. The non-homogeneous temperature distribution in the tank must be taken into account when planning the density sensor measuring point.

Device Maintenance

The devices should be designed to require as little maintenance as possible. In some Asian countries, annual calibration is mandatory. In these cases, the measuring instruments should be installed on special maintenance-friendly valves or couplings which are self-sealing in order to prevent the loss of SF_6 into the atmosphere during the calibration process. This guarantees easy, quick and environmental friendly maintenance of the installed measuring systems.

Developments in SF_6 Monitoring

In addition to its excellent insulation and arc quenching properties, SF_6 is considered a very potent greenhouse gas. National and international regulations impose more restrictive conditions on the SF_6 applications. Global trends also show an increased demand for continuous leakage monitoring (and reporting to the authorities) for individual SF_6 gas compartments. As a part of this development, online monitoring has become the standard for many new installations. Usually it is realized with electronic density meters, which are specially equipped with an infrastructure for data acquisition. This enables the system to calculate losses quickly and to generate alarm signals resulting in early detection of leakages and prevention of SF_6 emissions.

Online density monitoring with logging and trend detection can be used for preventive maintenance on all systems, as retrofitting can be carried out without problems on existing systems.

Example of a SF_6 Gas Density Indicator for Online Monitoring

One example for an SF_6 gas density indicator which is ideally suited for online monitoring is the GDHT-20 from WIKA (see Fig. 15.27). The very accurate pressure measurement with an error of less than 0.06% of the span (in the positive temperature range) and a max. Temperature error of 1 K enables this sensor to precisely calculate the prevailing density using real gas equations. In addition to pressure, temperature and density, the sensor also provides a measured humidity. It is very robustly designed and communicates digitally via the MODBUS_RTU protocol, which has become established as a standard in a large part of the industry.

Figure 15.28 represents the density signal calculated directly in the GDHT-20 transmitter from the data shown in Fig. 15.25. This signal is approximately 20 g/l for this vessel and is very stable over time, which indicates a tight vessel. There are only upward deflections on days 1 and 4. These are due to the fact that there was a disbalance between the gas temperature and the sensor temperature at this time due to the high temperature gradient. This imbalance leads to an incorrect density calculation and has nothing to do with the accuracy of the sensor technology. With today's filter algorithms, however, it is easy to eliminate these effects for a long-term trend analysis.

An example of the above-mentioned long-term trend analysis is shown in Fig. 15.29. Here, three different SF_6 gas compartments in one of the largest gas-insulated switchgear in the world were equipped with GDHT-20s. With a simple compensation line, the temperature-related density fluctuations in the signal can already be filtered out and thus a clear statement about the tightness of the compartment can be made after a few days. Both compartments, 1 and 2, show a constant density curve over the entire period of 30 days. In compartment 3, however, a clear decrease in density can be observed already after 10 days and as a result actions can be taken early to prevent further leakage of SF_6 and failure of the plant.

The lower the leak-rate, the longer the time required to be able to make a qualified statement. This is due to the fact that in the case of particularly small leakages, the loss of density is initially overlaid by the daily fluctuations and a clear trend can only be identified over a longer period of time. Individual control measurements are only poorly suited here in comparison to online monitoring, as one would always have to measure under the same environmental conditions, which is practically impossible to realize.

Fig. 15.27 Product image of the WIKA_GDHT-20 transmitter

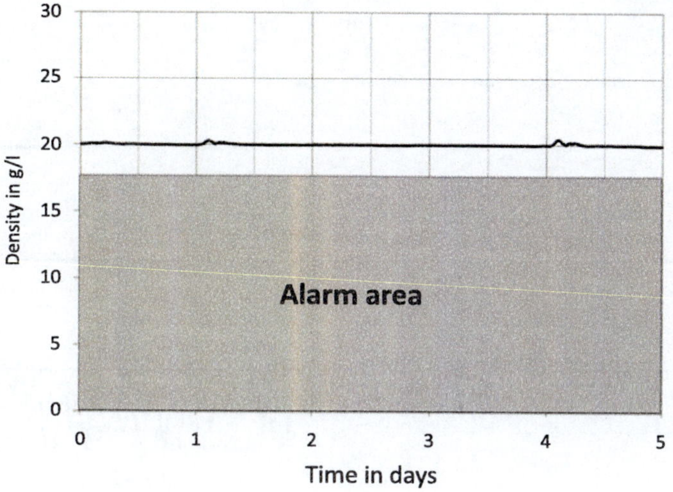

Fig. 15.28 Resulting density signal

In addition, the modern sensor is able to measure the humidity in the corresponding gas compartment (Mitter and Kurte 2015). In field tests on a switchgear in a hydroelectric power plant, a dew point mirror served as a reference for humidity measurement. Figure 15.30 shows a 24-h measurement sequence with the GDHT-20 on a gas compartment. The reference curve of 93.74 ppm_w of the dew point mirror is an averaged value from several measurements. The humidity sensor integrated in the GDHT-20 can determine the frost point exactly according to the specification up to 3 K. As can be seen, the frost point could even be determined to an average accuracy of 0.7 K in the real measurement.

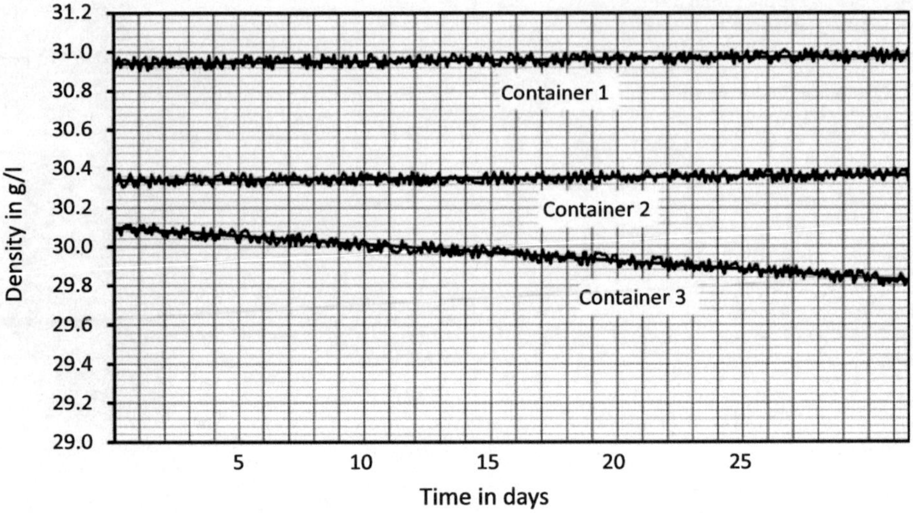

Fig. 15.29 Early leakage detection using density signal

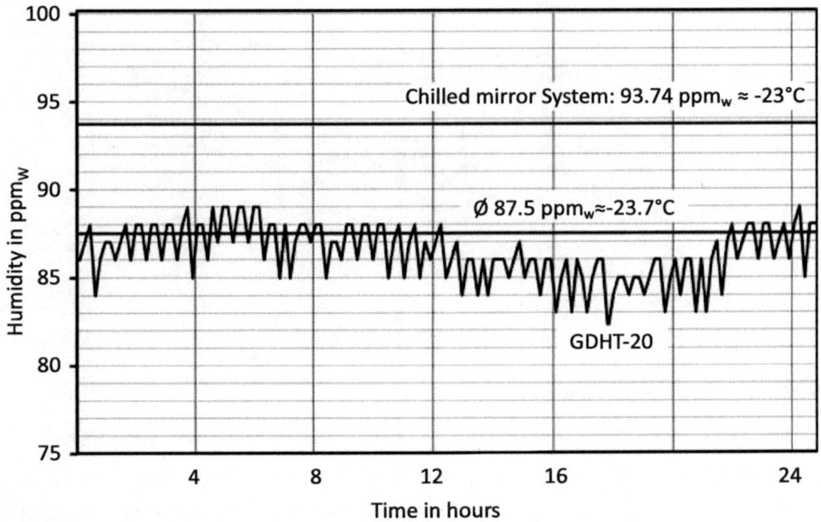

Fig. 15.30 Comparison of measured value GDHT-20 and chilled mirror system

The smaller fluctuations in the signal of ±1 ppm$_w$ correspond to the minimum resolution of the sensor. Higher fluctuations are caused by thermal effects. For example, the moisture bound to the compartment walls tends to pass into the gas at higher temperatures and vice versa, as can be seen in the time period of 12–22 h in Fig. 15.30.

These temperature-dependent fluctuations can lead to false conclusions being drawn about the total humidity in the compartment if, as with a dew point mirror, measurements are only taken sporadically and at certain times of the day. If, for example, measurements had been taken only at hour 16, a supposedly lower measured humidity value would have resulted. This problem is avoided with continuous monitoring.

Conclusion
The density measurement of SF_6 gas depends on many factors and environmental influences. A prerequisite for accurate measurement is therefore to know these influences and to interpret the signals accordingly. With the increasing demands on the tightness of gas-insulated systems, the demands on the measurement technology also increase. It is therefore important to use a sensor which, like the GDHT-20, filters out most of the disturbing factors and thus makes it considerably easier to evaluate the measurement results.

15.4 Gas Quality Tracking (*SmartSim*)

Peter Schley

In this chapter, a new method for the simulation of gas networks – in the following called *SmartSim* – is presented. Based on this method, the calorific values at the exit points in supply areas with several natural gas or biomethane feed-in points can be determined and used for the billing of end customers. In this way, conditioning with LPG can be avoided when biomethane is injected into the gas network and thus, costs can be reduced significantly.

The procedure can be applied to both transmission and distribution networks. Distribution networks are often characterized by a complex network topology and an incomplete measurement infrastructure. The calculation model is explained and a validation of the method is presented using the example of a regional distribution network.

Background
In the context of the "energy transition", natural gas will make an important contribution to a more efficient and environmentally friendly energy supply. The integration of renewable energy sources, such as biomethane or hydrogen generated from surplus *wind power*, into the natural gas system offers enormous potential for reducing CO_2 emissions and increasing security of supply. At the same time, however, these developments pose a major challenge for gas infrastructure and gas applications. Fluctuations in the gas composition, as we have already been observing in Germany for some years due to the increasing diversification of the supply of natural gas/LNG (Altfeld and Schley 2011), will continue to increase as a result.

There are currently about 7500 biogas plants in Germany, of which about 100 annually feed about 600 million standard cubic metres of biogas processed to natural gas quality into the gas network – usually into the regional distribution network. According to the objectives of the Federal Government, this value is to increase to 6 billion standard cubic metres by 2020 and to 10 billion standard cubic metres by 2013 (Gasnetzzugangsverordnung vom 3. September 2010). In future, *power-to-gas* technology will make it possible to convert surplus electricity from wind or photovoltaic plants into hydrogen by means of electrolysis and then feed it directly into the natural gas grid. The overall efficiency (electrolysis → storage → conversion into electricity) is $\eta \approx 50\%$. According to DVGW Code of Practice G260, which specifies the gas quality required for feeding into gas networks in Germany, or according to DVGW Code of Practice G262 for gases from regenerative sources, concentrations of up to 10 vol.% of hydrogen are currently theoretically possible. In practice, however, further restrictions must be considered with regard to natural gas vehicles (DIN 51624), gas turbines or underground storage facilities, so that this limit cannot usually be exhausted today. An alternative is the methanation of hydrogen to synthetic methane, whereby a further reduction in efficiency must be accepted. Currently, there are about 20 pilot projects in Germany to test the *power-to-gas* technology. The projects, some of which are in the planning and some of which are already in the implementation phase, are pursuing both the direct feed-in of hydrogen and the methanisation approach. The electrolysis plant built by E.ON in Falkenhagen (Brandenburg), which will produce up to 360 standard cubic meters of hydrogen and feed it into the Ontras gas network, has been in trial operation since July 2013.

In addition to compliance with DVGW worksheets G260 and G262, the gas quality is also relevant with regard to correct energy billing. According to the calibration law, the calorific value at the end customer must be determined with a maximum uncertainty of 2% (corresponds to the traffic error limit). From this requirement, the so-called *2% rule* was derived in DVGW Code of Practice G685 (gas billing), which states that when different gases are fed into a supply area, the calorific values of the fed-in gases, averaged over the billing period, may not deviate by more than 2% from the volume-weighted calorific value for the supply area (Fig. 15.31). Therefore, when the biomethane is fed into the grid, it is usually conditioned to the calorific value prevailing in the grid by adding LPG. However, this is associated with considerable costs and a deterioration of the CO_2 balance. In the case of feeding hydrogen into the grid, on the other hand, an adjustment of the calorific value would hardly be conceivable for technical and economic reasons.

Against this background, as well as the general trend of increasing fluctuations in natural gas quality, there is an urgent need for intelligent solutions for simple and cost-effective energy determination, especially for gas distribution networks. One possible solution is the computational tracking of gas quality, which has been established on the transport network level for several years (Altfeld et al. 2008). These so-called calorific value reconstruction systems allow the calorific value to be determined mathematically at any time and any place in the entire network. A prerequisite for the use of such systems is that, in addition to topological information about the network, calibrated measured values of the calorific

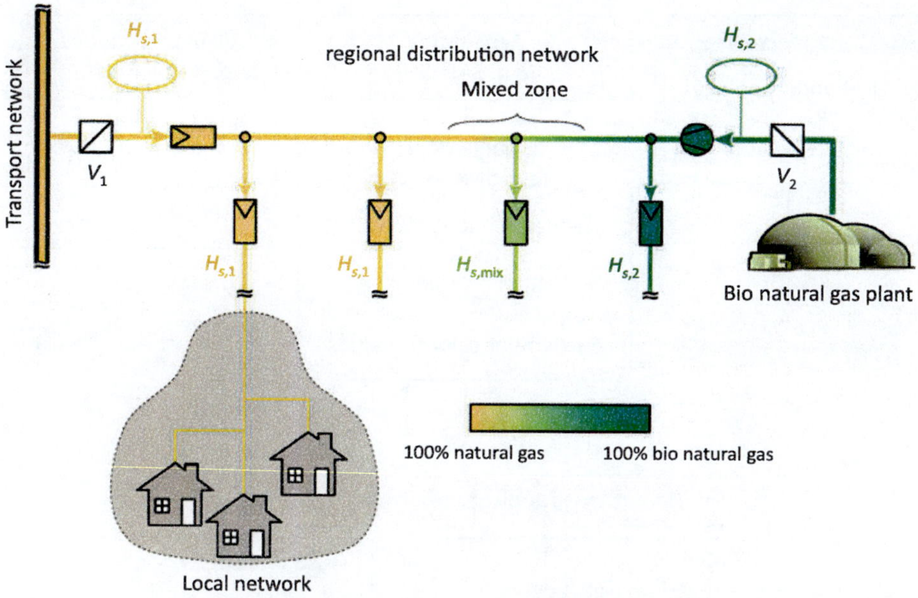

Fig. 15.31 Simplified representation of a regional network with natural gas/bio natural gas feed-in

value at the entry points and the volumes at the entry and exit points are available. For distribution networks, it has not been possible in the past to track calorific values mathematically with sufficient accuracy for gas billing. The problem here is that insufficient measured values of the volumes are often available, especially at the exit points. With *SmartSim*, a new method has been developed in the recent years that can also be used for distribution networks with an inadequate measuring infrastructure. The procedure is explained in more detail below. A detailed description of the algorithms is given in Hielscher et al. (2014).

Calculation Model

In contrast to transport networks, distribution networks generally have more complex network structures, lower flow velocities and lower pressures. With *SmartSim*, a novel method has been developed that can be used for both transport and distribution networks. Due to efficient algorithms, the calculation model described in the following is characterized by a high calculation speed, so that even very complex networks can be simulated in a reasonable calculation time. Furthermore, the integration of the calculation kernel into the *SmartSim software* leads to a high flexibility in terms of handling and connection to other IT systems.

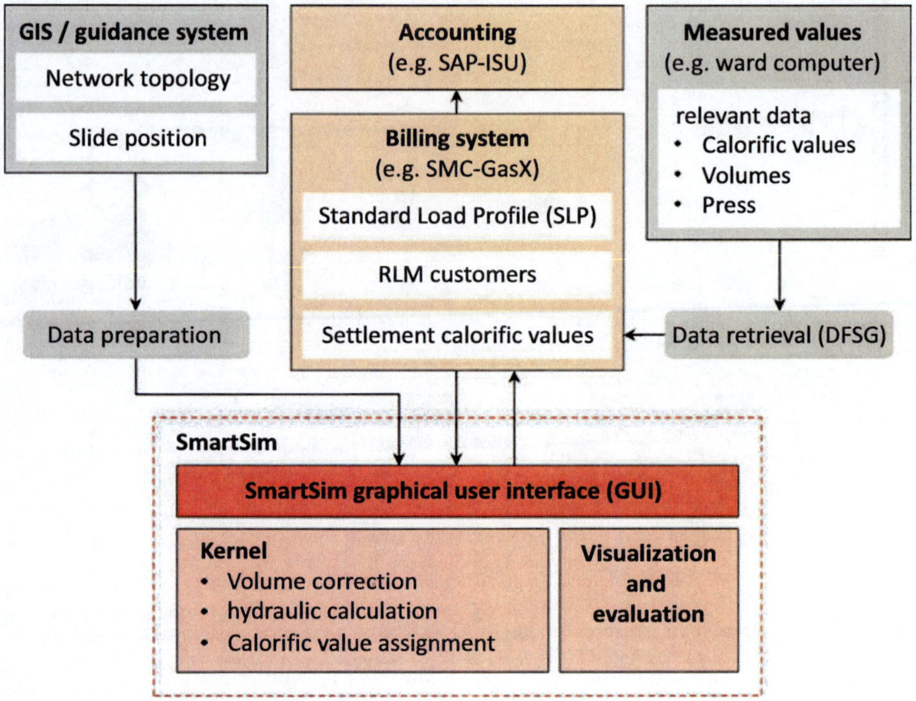

Fig. 15.32 *SmartSim* data flow

The input data of the simulation as well as the data flow are schematically sketched in Fig. 15.32. The calculation is performed in the following steps, which are explained in detail below:

- Determination of exit volumes/volume correction.
- Flow calculation.
- Calorific value assignment.

(i) Determination of Exit Volumes/Volume Correction

For gas quality tracking in a regional network, the volumes at the entry and exit points of the network under consideration are required in addition to the entry calorific values. Calibrated measured values based on hourly averages are usually available for the calorific values and volumes at the entry points. The exit volumes, which are usually not measured directly, are the sum of all end customers in the downstream local network. A distinction is made here between

- consumers with registered measurement – so-called RLM customers, and
- consumers whose current energy consumption is determined using standard load profiles – so-called SLP customers.

A registerred measurement is generally used for consumers with an hourly power output of more than 500 kW or an annual consumption of at least 1.5 million kWh. For smaller offtake volumes, simplified methods – so-called standard load profiles – are used, which were developed at the Technical University of Munich on behalf of BGW (Federal Association of the German Gas and Water Industry) and VKU (Association of Municipal Enterprises) (Hellwig 2003).

The sigmoid function $h(\vartheta)$ displays the relative consumption as a function of the weighted daily mean temperature.

$$h(\vartheta) = \frac{A}{1 + \left(\frac{B}{\vartheta - 40^\circ\mathrm{C}}\right)^C} + D \tag{15.15}$$

The parameters A, B, C and D were defined for different load profile types (e.g., single-family house, apartment building, local authorities, retail trade).

Furthermore, if a previous reading interval – the so-called period consumption Q – is known, a customer value KW (in kWh) can be determined, which reflects the specific standardized consumption of the customer. Finally, if the temperature is known, the expected hourly energy consumption (in kWh) of a customer can be determined as follows

$$Q_{\mathrm{SLP}}(h) = \mathrm{KW} \cdot h(\vartheta) \cdot F(d) \cdot \mathrm{SF}(h, \vartheta). \tag{15.16}$$

Where $F(d)$ is the weekday factor and $\mathrm{SF}(h, \vartheta)$ is the hour factor. The conversion of the energy consumption to the gas volume is carried out via the volume-related calorific value *Hs* and results in

$$V_{\mathrm{SLP}}(h) = \frac{Q_{\mathrm{SLP}}(h)}{H_s}. \tag{15.17}$$

The use of standard load profiles with regard to gas network simulation is suitable in that it avoids upgrading the metering infrastructure, which is associated with considerable costs. However, the accuracy of the volumes determined by SLP does not initially correspond to the quality that would be required for billable calorific value tracking. For this reason, the correction procedure described below was developed, on the basis of which the accuracy of the volumes determined can be significantly improved.

First of all, it is assumed that in the gas network under consideration the standard volumes at all entry points are measured on the basis of hourly values, so that the sum of the entry volumes is known exactly.

$$V_E(h) = \sum_i V_{E,i}(h) \tag{15.18}$$

The total exit volume is the sum of the measured volumes of RLM customers and the volumes of SLP customers determined in accordance with Eq. 15.17.

$$V_A(h) = V_{A,\mathrm{RLM}}(h) + V_{A,\mathrm{SLP}}(h) = \sum_i V_{A,\mathrm{RLM},i}(h) + \sum_i V_{A,\mathrm{SLP},i}(h) \tag{15.19}$$

Since pressure fluctuations can occur in a gas network, the change over time of the gas volume present in the network – the so-called line pack – is determined by the following equation.

$$\Delta V_{\mathrm{Netz}}(h) = V_{\mathrm{Netz}}(h) - V_{\mathrm{Netz}}(h-1) = \frac{p(h) - p(h-1)}{p_n} \cdot \frac{T_n}{T} \cdot V_{\mathrm{Geo}} \tag{15.20}$$

The mean network pressure can be determined by measuring the pressure at representative points in the network. The gas temperature can be assumed to be constant (e.g., 8 °C). Taking network respiration into account and assuming that the gas density in the gas network or network section under consideration can be assumed to be constant, the following volume balance can be drawn up.

$$\Delta V_{\mathrm{korr}}(h) = V_E(h) - V_{A,}(h) - \Delta V_{\mathrm{Netz}}(h) \tag{15.21}$$

The uncertainty contributions of network respiration and the measured volumes *VE* or $V_{A,\mathrm{RLM}}$ are comparatively low, which is why possible balance sheet deviations are mainly due to inaccuracies in the determination of the volumes of SLP customers. As a result, the difference volume determined according to Eq. 15.21 can be used to correct the volumes of SLP customers in proportion to consumption.

$$\widehat{V}_{A,\mathrm{SLP},i}(h) = \left(1 + \frac{\Delta V_{\mathrm{korr}}(h)}{\sum_i V_{A,\mathrm{SLP},i}(h)}\right) \cdot V_{A,\mathrm{SLP},i}(h) \tag{15.22}$$

Finally, the total volume at an exit point is the sum of the consumers behind this node.

(ii) Flow Calculation

A complete description of the flow state is based on the three conservation equations of mass, momentum and energy. In this context, conservation of momentum and energy are used to describe dynamic changes of state such as pressure surges that propagate in the natural gas network at the speed of sound within a few seconds or minutes. In contrast, the temporal resolution of the input data is usually 1 h. Consequently, the influence of the dynamic state changes described by pulse and energy conservation can be neglected. Thus the flow calculation is reduced exclusively to the mass conservation equation:

$$\left(\frac{\partial(\rho A)}{\partial t}\right)_x + \left(\frac{\partial(u\rho A)}{\partial x}\right)_t = 0 \tag{15.23}$$

Differences from the injected or withdrawn natural gas mass are called linepack and result from pressure changes. The simulation of these pressure changes leads to the fact that the time derivative is not equal to zero. To determine the pressure change of the natural gas network, the following mass balance is calculated hourly for the entire natural gas network. The injected mass is marked with m_{E} and the withdrawn mass with m_{A}. The mass difference Δm then results:

$$\Delta m = \sum m_{\mathrm{E}} - \sum m_{\mathrm{A}} \tag{15.24}$$

Taking into account the pressure distribution of the natural gas network, the difference in mass Δm is distributed to the individual pipes. The mass change of a pipe r_i depends on the average density $\rho_m(p_m)$, the geometric volume V_{geo} and the mass $m_{r,\,i}$ of the pipe i at the beginning of the hour:

$$\Delta m_{r,i} = \Delta m \cdot \left(\frac{\rho_{m,i}(p_{m,i}) \cdot V_{\mathrm{geo},i} - m_{r,i}}{\sum_j \rho_{m,j}(p_{m,j}) \cdot V_{\mathrm{geo},j} - m_{r,j}}\right) \tag{15.25}$$

To determine the mean pressure p_m, the integral mean value according to Mischner et al. (2011) is used in a numerically stable form. The pressure p_1 is the pressure at the pipe inlet, p_2 the pressure at the pipe outlet:

$$p_m = \frac{2}{3} \cdot \left(p_2 + \frac{p_1 \cdot p_1}{p_1 + p_2}\right) \tag{15.26}$$

The mean density $\rho_m(p_m)$ is determined by means of a simplified virial equation taking into account the real gas behaviour. To calculate the virial coefficients B_i as well as the molar mass M_i, the SGERG equation (ISO 12213/2 2006) is used once for all feed-in nodes, then the material properties are propagated volumetrically weighted through the natural gas network. The gas temperature T and the general gas constant R_m are constant for the entire network:

$$\rho_m = -\frac{1}{2 \cdot B_i} - \sqrt{\left(\frac{1}{2 \cdot B_i}\right)^2 + \frac{p_m \cdot M_i}{R_m \cdot T \cdot B_i}} \tag{15.27}$$

The mass flows $\dot{m}_{r,i}$ are calculated taking into account the mass changes $\Delta m_{r,\ i}$(Eq. 15.25) of the individual pipes by solving a linear system of equations and then to determine the flow velocities u_i and the associated pressure losses Δp_i used:

$$u_i = \frac{\dot{m}_{r,i}}{\rho_{m,i}\left(p_{m,i}\right) \cdot A_i} \tag{15.28}$$

The calculation of the pressure losses Δp_i is carried out according to Darcy-Weisbach (Mischner et al. 2011; DVGW GW 303-1 2006) using the approach of Zanke (1993, 1996) to determine the pipe friction coefficient $\lambda_{m,\ i}$. Index 1 indicates the entry into pipe i:

$$\Delta p_i = p_{1,i} \cdot \left(1 - \sqrt{1 - \lambda_{m,i} \cdot \frac{L_i}{D_i} \cdot \frac{\rho_{1,i}}{p_{1,i}} \cdot u_{1,i}^2}\right) \tag{15.29}$$

Zanke's approach provides a consistent method for describing the pipe friction coefficient over the entire laminar and turbulent area with validity for both hydraulically smooth and hydraulically rough pipes. Depending on the topology description used, both the roughness k and the integral roughness k' can be used:

$$\lambda = \lambda_{\text{turb}} \cdot \alpha + \lambda_{\text{lam}} \cdot (1 - \alpha) \tag{15.30}$$

with:

$$\lambda_{\text{turb}} = \left[-2\lg\left(2.7\frac{(\lg(Re))^{1,2}}{Re} + \frac{k}{3.71D}\right)\right]^{-2} \tag{15.31}$$

$$\lambda_{\text{lam}} = \frac{64}{Re} \tag{15.32}$$

$$\alpha = \exp(-\exp(-(0.0025 \cdot Re - 6.75))) \tag{15.33}$$

Evaluations of various regional distribution networks have shown that approximately 90% of all flow velocities u are in the range between 0.1 and 2 m/s. Despite the relatively low flow velocities, only about 10% of the flow states are laminar, that is the Reynolds number Re is less than or equal to 2300. Consequently, about 90% of the flow states are turbulent.

$$Re = \frac{u \cdot \rho \cdot D}{\eta} \tag{15.34}$$

The calculated pressure losses Δp_i are used to determine the node pressures p_k in the natural gas network. For a given pressure distribution, the mean network pressure is iterated until the mass $m_{r,\,i}$ of all pipes at a given mass change $\Delta m_{r,\,i}$ is equal to the mass resulting from the product of mean density $\rho_m(p_m)$ and geometric volume V_{geo}. Since the pressure and mass distribution are mutually dependent, an iterative procedure for solving the Eqs. 15.24, 15.25, 15.26, 15.27, 15.28, 15.29, 15.30, 15.31, 15.32, 15.33 and 15.34 is required. After reaching the abort criterion of the iteration, the flow velocities, mass flows and pressures are used to calculate the new distribution of the material data such as H_s, ρ_n and B_i in the gas network.

(iii) Calorific Value Assignment

After the flow calculation has been carried out, the flow velocities as well as mass and volume flows in the individual pipes are known. In the last step, the exit volumes for each exit point are now tracked backwards through the network to the entry points. This is done using a special gas package model, the so-called *back-propagation algorithm*. This enables the system to show the proportion of gas fed into the network at each exit point, including the transit times. This procedure has the advantage that the exit values are always assigned to the calibrated measured entry values. In addition, further gas components or relevant gas characteristics such as Wobbe index, *K*-number or CO_2 emission factor can be derived from the entry points in one calculation step. As a result, *SmartSim* provides the calorific values for all exit points of the network, which can then be used for billing. Usually, the settlement calorific value is determined as a volume-weighted monthly average.

Validation of the Process Using a Regional Distribution Network

In the following, the procedure is validated using the example of a regional distribution network in two different ways. First, the uncertainty is determined theoretically on the basis of a *Monte Carlo simulation (MCS)*. Afterwards the verification of the measurement accuracy is carried out experimentally, using a mobile PGC.

The Net Lüchow of Avacon

The topology of the examined network is shown in Fig. 15.33. The network, with a total pipe length of approximately 80 km, comprises 40 exit points (which are followed by local networks), three natural gas entry points (LA24, LG01 and LV40) and a bio natural gas entry point (LA45). Each injection point has a pressure sensor that provides the pressure information for the volume correction procedure. Four valves (VEN_1–VEN_4) are used for network control. The three natural gas injection points are supplied by a common transport pipeline, which is also shown in Fig. 15.33.

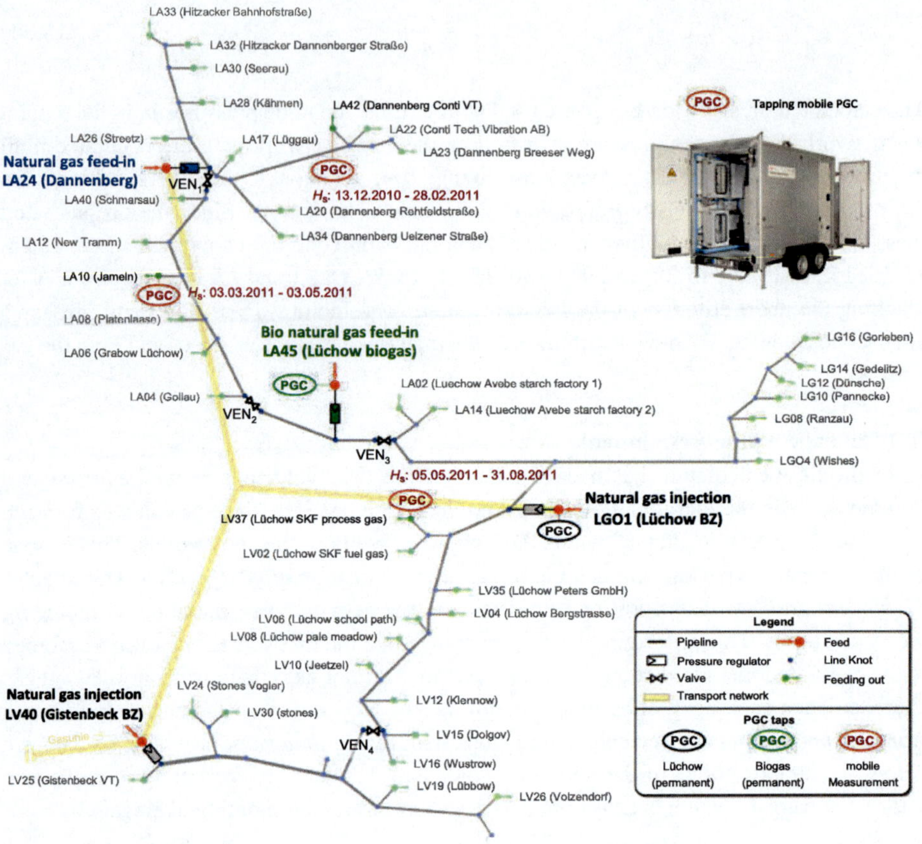

Fig. 15.33 Schematic representation of the *Lüchow-Dannenberg* topology

The calorific value of the conditioned biomethane is measured at the LA45 injection point with a PGC; another PGC is used to determine the calorific value of the natural gas at the Lüchow injection point (LG01). Since the three natural gas injection points are supplied via a transport pipeline, the PGC measurement at LG01 is also used as a basis for the injection points LA24 and LV40. Due to comparatively low flow velocities in the transport pipeline and the resulting differences in transit times, the calorific values at the feed-in points do not match in time. Therefore, the relevant sections of the transport network are included in the simulation calculation.

Uncertainty Determination Via a Monte Carlo Simulation

Within the scope of this work, a comprehensive uncertainty calculation for a calorific value tracking according to the *Guide to the Expression of Uncertainty in Measurement* (ISO/GUM 2008) is carried out for the first time. Due to the high complexity of distribution networks and the often insufficient measurement infrastructure, the method of MCS according to GUM is suitable. Here, the relationship between the uncertainties of the input variables and the resulting uncertainty of the calorific value is determined empirically. The variation of all input variables is performed in parallel and randomly distributed in so-called scenarios. The scenario with unchanged input variables is referred to in the following as the reference scenario (Ref). At least 1537 scenarios are required for the uncertainty determination on a 95% confidence interval, so that the number of scenarios is set at $N_{\mathrm{MC}} = 1600$. The volume-weighted settlement calorific values $\overline{H}_s$ of scenario j at node i in month T are each compared with the settlement calorific value of the reference scenario $\overline{H}_s^{\mathrm{Ref}}$:

$$\delta\overline{H}_{s,j,i,T} = \frac{\overline{H}_{s,j,i,T} - \overline{H}_{s,i,T}^{\mathrm{Ref}}}{\overline{H}_{s,i,T}^{\mathrm{Ref}}} \tag{15.35}$$

The standard uncertainty $u\left(\delta\overline{H}_{s,i,T}\right)$ is then calculated according to Eq. 15.36 and is converted to the expanded measurement uncertainty $U\left(\delta\overline{H}_{s,i,T}\right)$ with a coverage factor $k = 2$:

$$u\left(\delta\overline{H}_{s,i,T}\right) = \sqrt{\frac{1}{N_{\mathrm{MC}} - 1} \cdot \sum_{j=1}^{N_{\mathrm{MC}}} \left(\delta\overline{H}_{s,j,i,T} - E\left(\delta\overline{H}_{s,i,T}\right)\right)^2} \tag{15.36}$$

For the mathematical investigation of the method, the standard uncertainties u shown in Table 15.7 are applied for the input variables, for which a rectangular distribution is assumed. After execution of the MCS the expanded measurement uncertainty $U\left(\delta\overline{H}_{s,i,T}\right)$ is visualized over the topology. The results are shown in Fig. 15.34 as examples for the months January 2011 and June 1011. In addition, the corresponding flow situations are shown in the upper pictures. On the one hand the distribution of the injected gases is marked; on the other hand the volume flow is characterized by the thickness of the envelopes drawn around the pipes.

It is shown that all expanded uncertainties remain well below the required limit of 2%. In January, the maximum expanded uncertainty is below $U\left(\delta\overline{H}_s\right) \leq 0.6\%$, in June below $U\left(\delta\overline{H}_{s,}\right) \leq 0.3\%$. The uncertainty here depends on the respective flow situation. In January, for example, in the section between the biogas plant and the Dannenberg injection point, a swing zone of natural gas and a biogas–natural gas mixture occurs, which results in relatively high uncertainties. In June, on the other hand, the upper part of the network is almost completely supplied with bio–natural gas; only the right-hand section contains a mixing zone. It is in this mixing zone that the greatest uncertainties occur.

Table 15.7 Standard uncertainties of input variables for Monte Carlo simulation (MCS)

Size	Uncertainty
Pipe roughness	[20% ... 200%]
Pipe length	[85% ... 115%]
Pipe diameter	[94% ... 106%]
SLP energy 1 (systematic, all nodes)	[70% ... 130%]
SLP energy 2 (systematic, single nodes)	[70% ... 130%]
SLP energy 3 (hourly value, all nodes)	[70% ... 130%]
Measured decrease	[97% ... 103%]
Feed-in volume	[97% ... 103%]
Feed-in pressure	[94% ... 106%]
Gas temperature	[98.2% ... 101.8%]

Validation by Comparison Measurements with a Mobile PGC

The developed *SmartSim* calculation kernel will be validated with the data of the field test conducted in 2011 in the Lüchow network of Avacon AG (Schley et al. 2011, Schenk et al. 2011). Exemplary measurements were carried out with a mobile PGC at different exit points. A representation of the topology as well as the time and location of the mobile PGC is shown in Fig. 15.34. Table 15.8 shows the comparison between the measured calorific values $H_{s,\text{ gem}}$ and the calorific values calculated $H_{s,\text{ ber}}$ with SmartSim in relation to the monthly mean values. It can be seen that the deviations over the entire test period are less than 0.1%.

Conclusion

In this chapter, a new method for the simulation of gas distribution networks is presented. It is characterized by high accuracy and extremely short computing times. The integration of the calculation core into the *SmartSim software* developed by E.ON Technologies for tracking calorific values enables complex networks to be calculated very efficiently. The validation of the system is carried out both by an uncertainty analysis based on a MCS and by measurements. The results are presented exemplarily using a regional distribution network of Avacon AG. Despite conservative estimation of the input uncertainties, the expanded uncertainties remain below 0.6%. The deviations from the measurements are even significantly smaller.

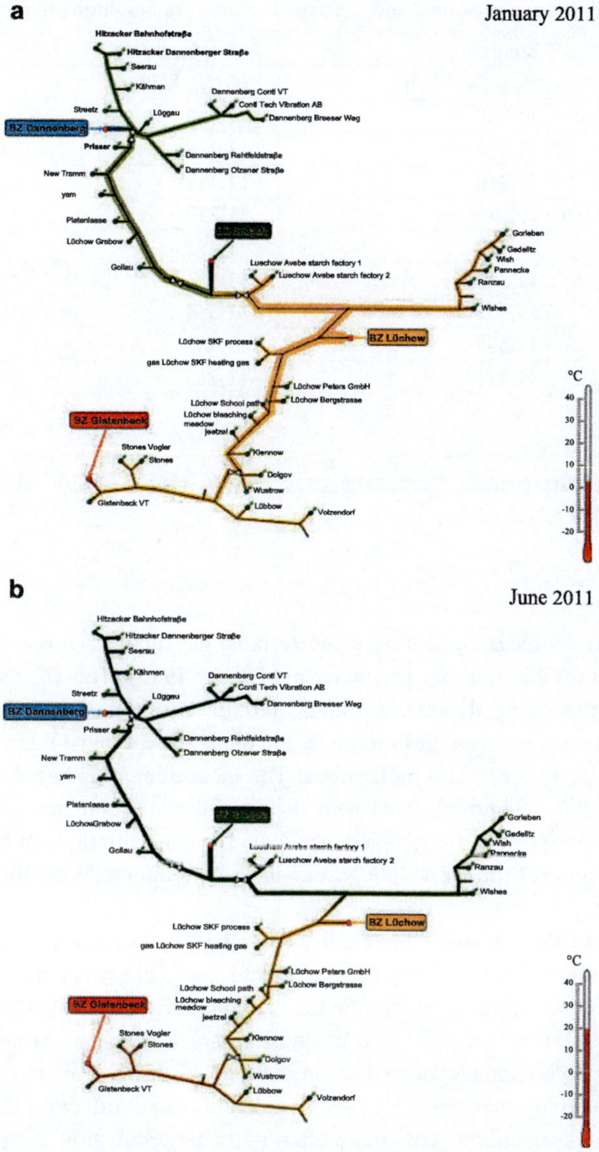

Fig. 15.34 Visualization of the flow conditions for the months January (**a**) and June 2011 (**b**). The fed-in gases (natural gases: blue, yellow, red; bio natural gas: green) or their mixture proportions are marked in colour

Table 15.8 Comparison of measured and calculated calorific values during the field test

	SmartSim	Measurement	
Month	$H_{s,\ neu}$ kWh/m³	$H_{s,\ gem}$ kWh/m³	$\Delta H_{s,\ gem}$ %
12.2010	11,368	11,369	0.01
01.2011	11,334	11,338	0.04
02.2011	11,338	11,337	−0.01
03.2011	11,275	11,277	0.02
04.2011	11,273	11,283	0.09
05.2011	11,353	11,353	0.00
06.2011	11,355	11,354	−0.01
07.2011	11,308	11,307	−0.01
08.2011	11,211	11,213	0.02

15.5 Multicomponent Gas Analysis Using the FTTCA[7] Analysis Method

Axel-Ulrich Grunewald

Thermal conductivity measurement for characterizing gas mixtures is one of the oldest gas analysis methods used in industry (Siemens and Halske 1913). The TCD was the first gas analyzer used in the chemical industry around 1920 for process measurements to determine the quantitative composition of gas mixtures. Each gas has a typical thermal conductivity that depends on its viscosity and molar mass. The measurement is based on the principle that when a mixture of known gases with different thermal conductivities is used, the mixture has a thermal conductivity that depends on the concentrations of the components. Thus, the proportions of the individual gases can be determined (Wassilijewa 1904).

Gas Analysis with the Classical Thermal Conductivity Measuring Method
The principle of thermal conductivity measurement is particularly useful when the gases to be measured differ significantly in their thermal conductivity and the gas mixture is binary or quasi-binary. Mixtures are described as quasi-binary if they consist of more than two components, but the component to be measured differs so clearly from the others regarding its thermal conductivity that changing it has a tolerably small effect on the concentration determination. This situation occurs more often in the determination of hydrogen or even helium, since these two gases have a significantly higher thermal conductivity than all the others (see Table 15.9).

[7] Fourier transformed thermal conductivity.

Table 15.9 Thermal conductivity λ, thermal conductivity gradient with respect to temperature dλ/dT, determined between 25 °C-100 °C, and the resulting temperature coefficient TK = (dλ/dT)/λ

Gas	λ at 100 °C mW/(m·K)	dλ/dT mW/(m·K²)	TK = (dλ/dT)/λ 10^{-3}/K
He	174	0.32	1.84
Ar	21	0.04	1.90
H_2	211	0.40	1.90
N_2	31	0.07	2.15
Methane	44	0.13	3.03
CO_2	22	0.08	3.64
Propane	27	0.12	4.44
Butane	25	0.12	4.80
Ethane	34	0.17	5.10

Source: VDI Wärmeatlas, sixth edition 1991

Today, the determination of concentrations by measuring the thermal conductivity is preferably used in process gas analysis to measure gases that are not accessible to infrared-optical measurements due to the lack of permanent dipole moment. These include homo-nuclear gases, such as hydrogen and nitrogen, and noble gases, such as helium and argon. Another advantage is the quite simple and robust instrument design and the purely physical measuring principle without consumables.

With the classic macroscopic setups, a comparison side, which is as identical as possible to the measuring side and which is filled with gas, is necessary to compensate for the influence of the fluctuating ambient temperature on the measuring effect in a bridge circuit. In addition, due to the relatively large thermal mass of the measuring element, the display follows the concentration changes in the measuring gas only with a significant delay.

Further Development of Sensor Technology Through the Emerging Microsystems Engineering

Developments in microsystems technology make it possible to manufacture the sensor elements reproducibly in small dimensions. On the one hand, this has reduced production costs, and on the other hand, miniaturization also has metrological advantages. In the case of sensor elements from micromechanical production, heat conduction through the gas dominates by far the undesired parasitic effects, such as the dissipation of heat via the solid state structures. Calculations showed that with nitrogen as the measuring gas 90% of the heat is dissipated by the gas (Völklein 2007). This means that the gas-filled comparison side is dispensable. In addition, the microstructure of the measuring element has such a small thermal mass that the time constant for changes in concentration is less than 0.1 s (Messkonzept 2004).

Figure 15.35 shows a micromechanical sensor element. The temperature-dependent measuring resistors, which heat and measure simultaneously, are applied to a silicon oxynitride membrane about 1 μ m thick. The temperature of the membrane is adjusted so

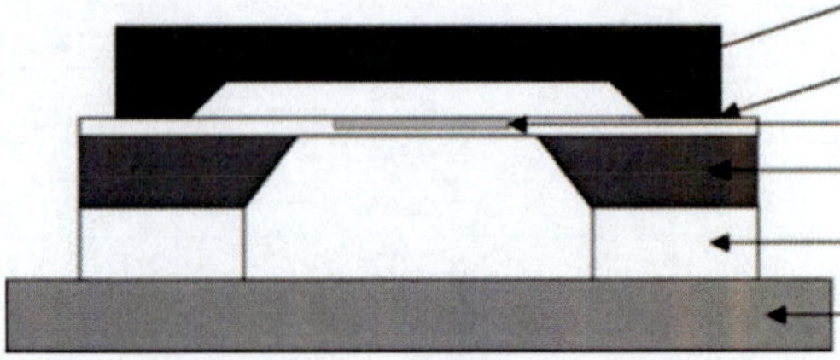

Fig. 15.35 Schematic diagram of the micromechanical thermal conductivity sensor TCS208 (Wagner 1988)

Fig. 15.36 FTC 300 thermal conductivity meter of Messkonzept GmbH, Frankfurt am Main

that it is, for example, 70 K above the temperature of the thermostatted overall structure. The measured variable is the electrical voltage at the measuring resistors, which is necessary to maintain the 70 K overtemperature. Figure 15.36 shows a compact thermal conductivity analyzer equipped with a micromechanical sensor.

The Determination of Complex Gas Mixtures

These new properties of micromechanical sensor elements led in the 1980s to new metrological approaches that were intended to eliminate the disadvantages of classical heat conduction measurement – in particular the restriction to binary or quasi-binary gas mixtures (Goeldner et al. 1988). The underlying idea is to determine the thermal conductivity of gas mixtures at different temperatures in order to use this additional information to break down the composition of more complex mixtures. Table 15.9 shows some examples of relevant gases with their thermal conductivities λ and how they change with temperature $d\lambda/dT$. The temperature coefficient TK shown in the third column indicates that the gases can be roughly divided into three groups with respect to this property. Elementary gases have the smallest TK, multi-atomic gases are in the middle range, and the largest TK is found in higher hydrocarbons.

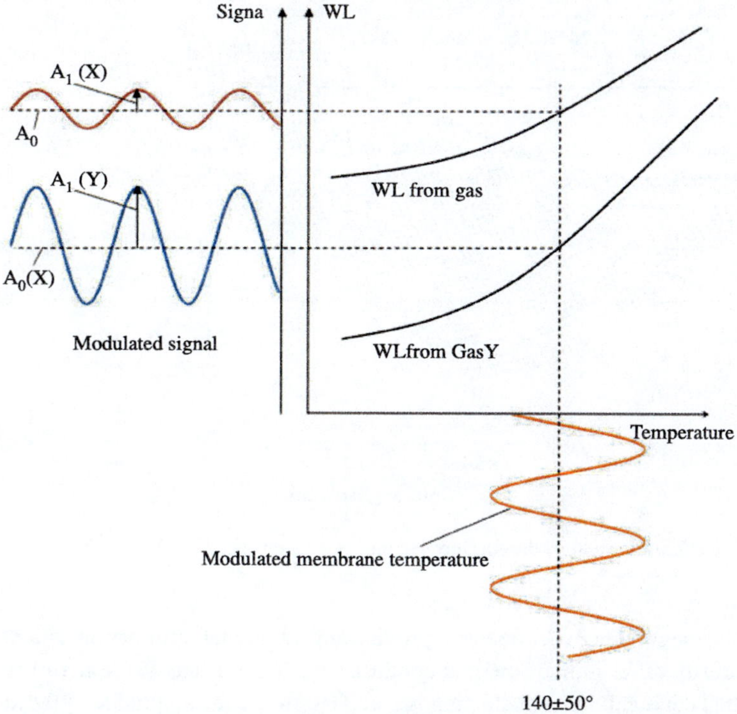

Fig. 15.37 Influence of the modulated membrane temperature on the measuring signal

A further development of these ideas and their concrete physical implementation took place at the beginning of this century (Grunewald 2001a, 2001b). This new analytical method aims to determine simultaneously the thermal conductivity and its change with temperature with a single measuring element. For this purpose, a sinusoidal voltage curve is applied to the Ni measuring resistors (see Fig. 15.35). This results in a periodic temperature change of the membrane. The circuit is designed as a closed-loop temperature control with cyclical setpoint control. In this way, a constant temperature cycle is achieved. The voltage cycle required for this is then essentially a function of the thermal conductivity of the sample gas and its change with temperature. This principle is shown schematically in Fig. 15.37. With the circuit described, the constant membrane temperature of 140 °C is superimposed on an alternating variable of ±50° C with the frequency of 10 Hz, as shown below the temperature axis. The thermal conductivity (TC) of the gases X and Y as well as the resulting measurement signals are plotted on the ordinate. If the gas Y is located in the sensor element, the direct signal $A_0(Y)$ is generated and superimposed by the alternating signal $A_1(Y)$. For the gas X with its higher thermal conductivity, the direct signal $A_0(X)$ increases to the value. However, as the gas X has$d\lambda/dT$ a smaller gradient of thermal conductivity, the amplitude of the alternating signal $A_1(X)$ will decrease to the value. In this

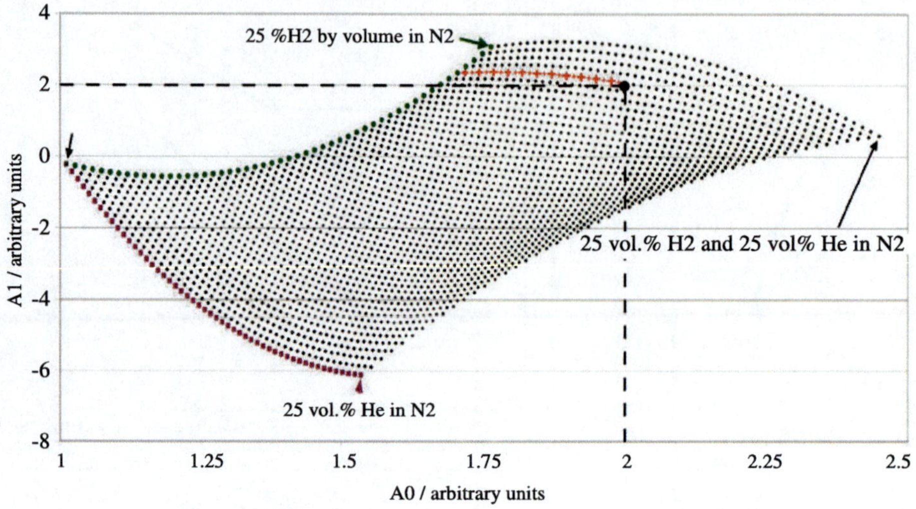

Fig. 15.38 Calibration field of the mixture He and H_2, each 0 ... 25 vol.%, in N_2

example, the magnitude $A_1(X)$ does not provide any additional information, since the gases X and Y already differ in their thermal conductivity. If gas Y and gas X at 140 °C had the same thermal conductivity, a distinction is still possible via the amplitude of the alternating signal. In the case of ternary mixtures of unknown composition, this additional information often allows all three concentrations to be broken down.

For quantitative analysis, the waveform must be characterized. For this purpose, the series expansion according to Fourier is suitable for periodic functions. In the real form it is

$$f(t) = \frac{a_0}{2} + \sum_{k=1}^{\infty} (a_k \cos(kt) + b_k \sin(kt)) \tag{15.37}$$

The element a_0 in Eq. 15.37 represents the thermal conductivity A_0. The coefficients a_1 and b_1 give the amplitude $A_1 = \sqrt{a_1^2 + b_1^2}$ of the fundamental oscillation and contain primarily the information about the slope of the thermal conductivity in the considered temperature interval 140 ± 50°C. The higher coefficients are a function of the curvature of the thermal conductivity in the considered temperature interval. In practice, the series development is broken off after the first or second link, depending on the requirements. To distinguish the new method from the classical thermal conductivity measurement, the acronym FTTCA (Fourier Transformed Thermal Conductivity Analyzer) was coined (Grienauer 2006).

A clear link between the measuring effect and the gas concentrations is called – not quite correctly – a calibration. Figure 15.38 shows a calibration field, the thermal conductivity A_0

normalized to nitrogen is plotted on the abscissa. The ordinate shows the travel of the fundamental oscillation A_1. The value for nitrogen is arbitrarily set to 0. The squares leading from the left-hand corner marked N_2 to the point at 25 vol.% He are formed when only He is added to the nitrogen in 0.5 vol.% steps. The small circles show the course which is created when only hydrogen is added to the nitrogen in 0.5 vol.% steps. All other points represent ternary mixtures of N_2, He and H_2. The distance is 0.5 vol.%, the transitions between the points are continuous. Thus, the resolution can be increased by interpolation.

Once such a calibration field has been established, the Fourier coefficients A_0 and A_1 determined in the measurement can be converted into gas concentrations. To do this, a perpendicular is dropped through A_0 onto the abscissa and through A_1 a perpendicular onto the ordinate. The concentration point closest to the point of intersection of the perpendiculars indicates the composition of the ternary mixture. If, for example, the measurement as shown in Fig. 15.38 by the dashed lines $A_0 = 2$ and $A_1 = 2$, then the point of intersection of the perpendiculars is obtained by following the upper circles from the N_2 point to the 23 vol.% H_2 point and then moving forward on the points represented by crosses to 11.5 vol.% He. The concentration of nitrogen is 65.5 vol.% and results from the addition to 100 vol.%.

As the nicely spread out calibration field already indicates, a complete determination of the mixture N_2–He–H_2 is possible, which is unthinkable with the classical thermal conductivity measurement. However, it should not be concealed at this point that the change in thermal conductivity in this example does not explain the shape of the calibration field, but obviously other gas properties are also responsible for this.

The Determination of the Composition of Ternary Mixtures

A process-relevant mixture can be approximated quite well by the gases H_2, CO_2 and N_2. Among other things, it occurs in gas treatment for fuel cells. In addition to the determination of the hydrogen concentration, the monitoring of the carbon dioxide concentration is often desired from a process engineering point of view. Figure 15.39 shows the results of comparative measurements with a classical thermal conductivity meter (dashed lines) and with the FTTC method (solid lines).

The H_2 concentration is kept constant at 2 vol.% (squares), 10 vol.% (circles) and 18 vol.% (triangles) in three series of measurements. Then 10, 20 and 30 vol.% CO_2 is added to the H_2 in N_2 mixture step by step, to 100 vol.% is always added N_2. While the measured values of the FTTC method lying on the solid lines show that this H_2 measurement is almost completely undisturbed by the change in the associated gas, the classic measured values lying on the dotted lines show the error in the H_2 measurement, which increases approximately in proportion to the CO_2 concentration. For the CO_2 determination of this application example, the classic thermal conductivity measurement no longer provides a meaningful result if there is H_2 in the sample gas in the volume % range. In Fig. 15.40, therefore, only the CO_2 signal obtained with the FTTC method is plotted against

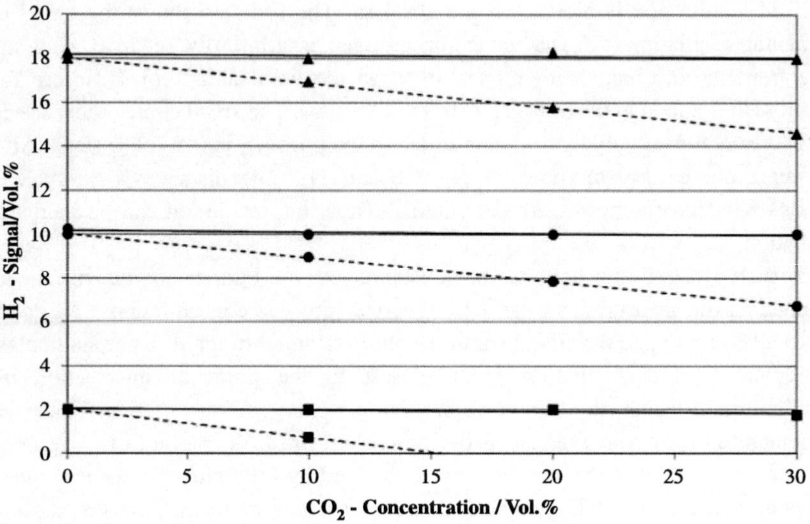

Fig. 15.39 The influence of increasing CO_2 concentration on the measurement of H_2 in the mixture H_2–CO_2–N_2

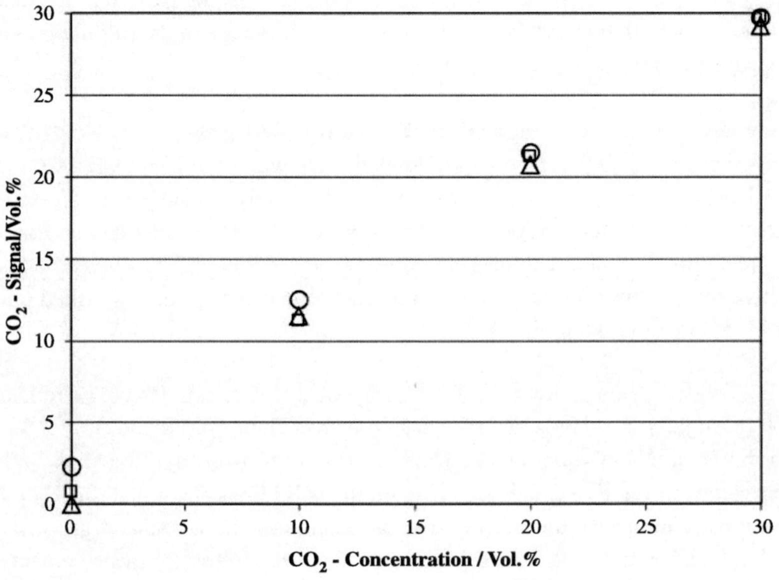

Fig. 15.40 CO_2 measurement in the mixture H_2–CO_2–N_2 with the FTTC method

the actual CO_2 concentration. The triangles represent the measured values at 2 vol.%, the circles at 10 vol.% and the squares at 18 vol.% H_2. The CO_2 concentration can be determined accurately to about 2 vol.%, although the H_2 concentration changes from 0 to 18 vol.%. This is surprisingly good, considering that the thermal conductivity of H_2 according to Table 15.9 is almost ten times higher than the thermal conductivity of CO_2.

The measurements in the mixture H_2–CH_2–N_2 provide in principle comparable results and are therefore not listed here. The biogas, which is interesting from a practical point of view, approximated by the mixture CH_2–CO_2–N_2, can also be determined quite well with the FTTC method.

Differentiation from Natural Gases

The FTTC method can also be used to measure other properties of gases. In this case, the decoding of the composition of the gas is omitted and the property of interest is directly correlated with the measured variables. Figure 15.41 shows various natural gases commonly used in Germany. The thermal conductivity A_0, standardised to methane, is plotted on the abscissa. The ordinate shows the change in thermal conductivity with temperature A_1, which was set equal to 0 for methane. The four natural gases, with the corresponding calorific values in brackets, are clearly separated from each other in the diagram. Each symbol in the diagram represents the result of an entire series of measurements. This allows the natural gases and thus their calorific values to be identified by the measuring method.

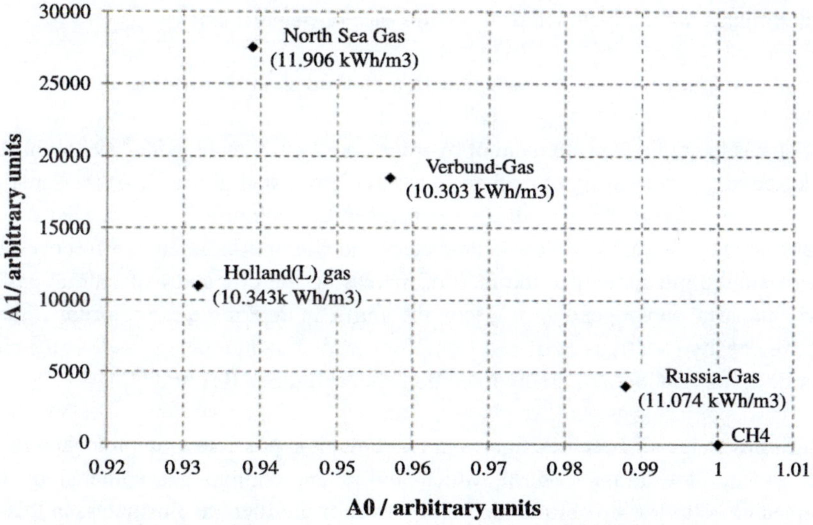

Fig. 15.41 Identification of different qualities of natural gas

A qualitative explanation for the arrangement of the gases in Fig. 15.41 is provided by the following simple observation. The Russia gas owes its high calorific value to the methane content of almost 98 vol.%. Thus, it is close to the point for methane. In the case of North Sea gas, the high calorific value is due to the high proportion of about 11 vol. % hydrocarbons, longer chain than methane. According to Table 15.9, this results in a lower thermal conductivity A_0 and a high amplitude A_1. In the case of Dutch and "Verbund" gas, higher N_2 contents of around 10 vol.% lead to smaller calorific values. A quantitative approach to calorific value determination is described below.

Concentration Calculation According to the Matrix Method

This new abstract approach to solving the multidimensional calibration task was developed and implemented by Grienauer (2012). The first five coefficients of the Fourier series a_0, a_1, a_2, b_1, b_2 given in Eq. 15.37 are used as measured variables, together with the square $(a_0)^2$, to form the columns of a square matrix, the so-called coefficient matrix. This is compared with the so-called concentration matrix, whose six columns consist of the calibration gas concentrations to be set and some of their squares and products. During calibration, the six gas mixtures used to create the concentration matrix are set one after the other. The Fourier analysis of the time-dependent signal curve provides the elements of the coefficient matrix. Using a well-known matrix operation, it can be shown that the product of the inverse coefficient matrix and the concentration matrix results in the transformation matrix, which converts a vector formed from the measured coefficients (a_0, a_1, a_2, b_1, b_2, $(a_0)^2$) into the vector whose elements are formed from the gas concentrations. In this sense, the transformation matrix contains the calibration of the measuring instrument. Computer-based algorithms were developed to solve this task (Grienauer 2014).

Two of the applications with this evaluation method shall be mentioned here:

- One task is the rapid measurement of hydrogen purity in the range 98.4–99.5 vol.% in a background gas consisting of methane, carbon dioxide and nitrogen. At the same time, the required determination of the gas density in this mixture is also carried out. The measurement is described as low maintenance and with low calibration effort over years.
- The second application is in the field of increasing the efficiency of natural gas-fired large industrial steam generators. Here, the aim is to determine the calorific value, the relative density (relative to air) and from this the Wobbe number, as well as the carbon dioxide concentration and the hydrocarbon concentrations (C1 and C2).

With the knowledge of these key figures in real time, it is possible to implement a forward control strategy for burner control, which allows the optimal gas composition to be maintained close to the stoichiometric point – even if the fuel gas fluctuates. In this way, energy losses can be minimized.

15.6 Combustion Optimisation

Frank Hammer

One of the largest anthropogenic air pollutants is emissions from combustion processes. These processes mainly produce dust, soot, NO_x and CO_e[8] as air pollutants. The most sources of these emissions are combustion processes in industrial and power plant combustion systems, motor combustion for vehicle drives and building heating systems.

The main factors influencing the amount of pollutants in the flue gas of combustion plants are shown in Fig. 15.42. Apart from the type of combustion plant and the fuels, the combustion process is the main factor determining the combustion products produced. Various emission reduction measures are available. An important element in assessing success is final control measures to ensure fault-free burner operation and compliance with emissions. An example of this is the inspection of small to medium capacity combustion plants prescribed by law and regularly carried out by the chimney sweep trade (first BImSchV 2010). Industrial and power plant firing systems of 50 MW and above are usually continuously monitored (13th BImSchV 2013).

The legal regulations for air pollution control in Germany result from the Federal Immission Control Act (BImSchG), its ordinances (BImSchV) and administrative regulations, such as the Technical Instructions on Air Pollution Control (TA Luft) and on Protection against Noise (TA Lärm).

Combustion

The aim of any combustion process should be to maximise efficiency while minimising pollutant emissions. This is not always easy to achieve. Combustion is the chemical reaction of a fuel with an oxidiser, usually atmospheric oxygen, resulting in heat generation. Ideally, all fuel molecules react with the oxygen in the air, without oxygen being missing or remaining. This is referred to as stoichiometric combustion with the combustion air ratio of $\lambda = 1$.[9] In reality, however, a certain amount of excess air ($\lambda > 1$) is necessary for the complete combustion of any fuel.

The influence of the air ratio λ or the residual oxygen content on the efficiency and the pollutant emissions of a combustion plant is shown in Fig. 15.43. Excessive excess air leads to *exhaust heat losses*. The excess air does not actively contribute to combustion and escapes from the chimney of the combustion plant as heated ballast air. A reduction in the air ratio should therefore be aimed for. However, not too far, because a lack of air leads to *losses due to incomplete combustion*. The so-called *emission edge* marks the operating state

[8] CO_e: the CO equivalent is the sum of all unburnt components in the exhaust gas, for example CO, H_2, C_xH_y, etc.

[9] λ: air number, air ratio, combustion air ratio.

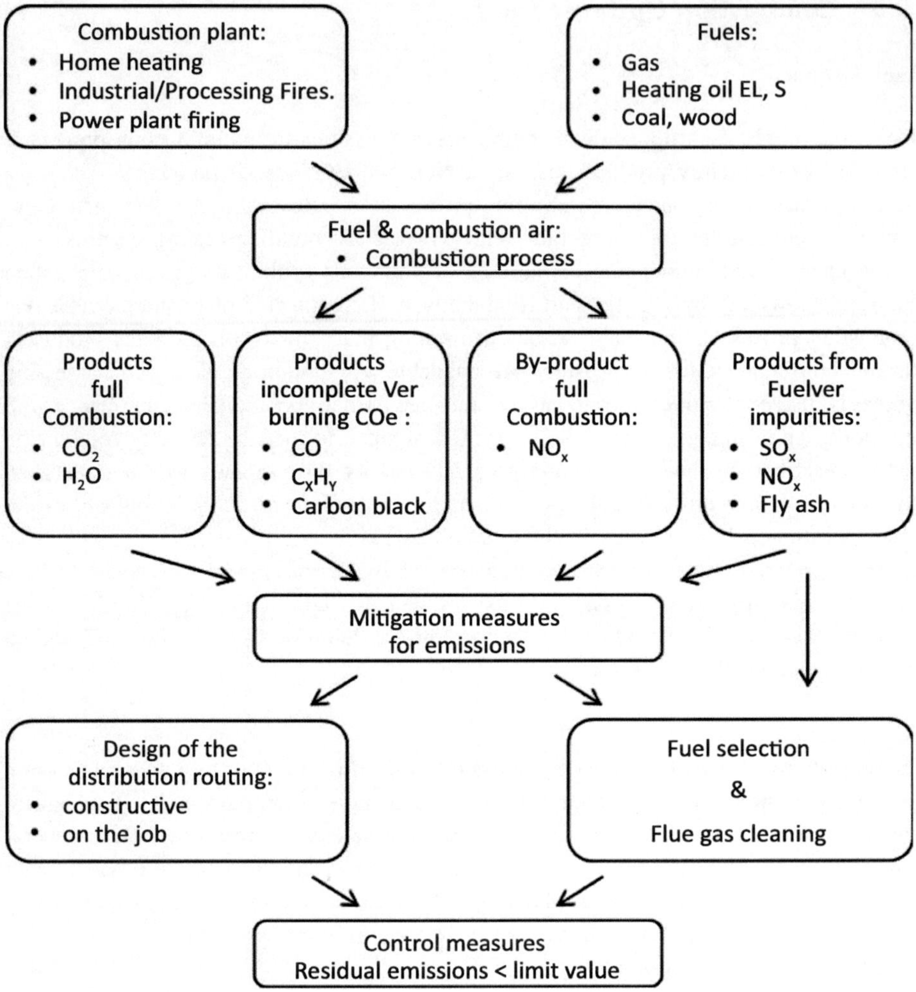

Fig. 15.42 Influencing variables on pollutant emissions in the flue gas of combustion plants. (Source: According to Baumbach 1993)

of a combustion plant in which – with further reduction of the air ratio λ – too little oxygen is available for complete combustion of the fuel and this, together with its energy content, escapes from the chimney unburned and unused. Ideally, the plant should be operated at the optimum air ratio shortly before the so-called emission edge, which can certainly be the case $\lambda_{opt} = 1.03$ in today's plants. However, the question λ_{opt} of this is not easy to answer, because the position of the emission edge varies from plant to plant and depends on burner design and on the state of combustion.

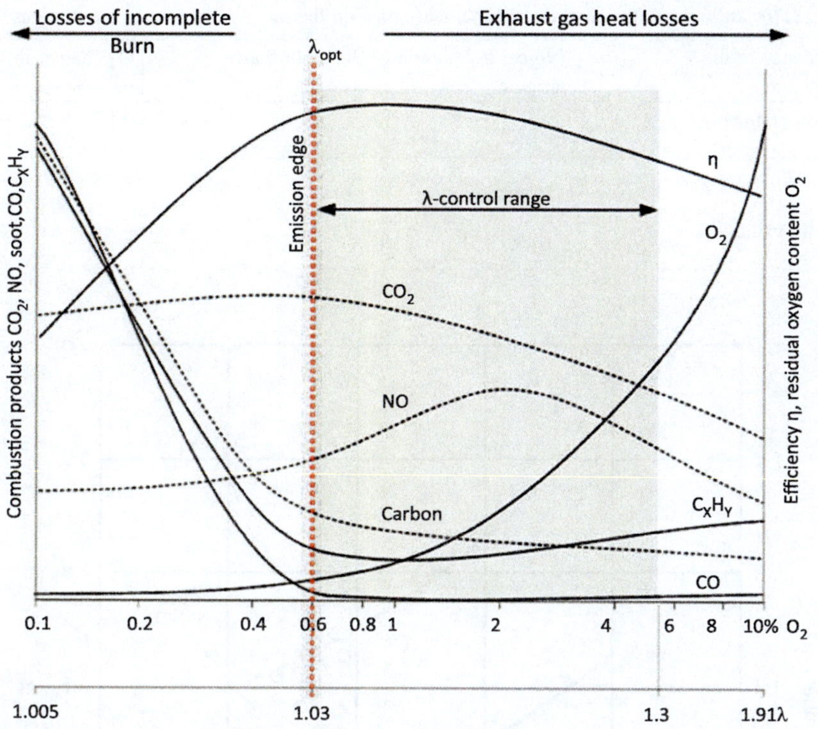

Fig. 15.43 Course of the combustion products and efficiency as a function of the air ratio λ

Disturbance Variables on the Combustion

The challenge for every combustion process is gradually changing conditions and rapid external combustion disturbance variables such as

- combustion air (temperature, pressure, humidity),
- fuel (calorific value, temperature, viscosity),
- pollution (burner, combustion chamber, boiler, flue),
- chimney draught (wind, temperature, barometric air pressure) and
- mechanics (play, hysteresis, wear).

Typical fluctuations of the combustion air temperature of ±20 °C cause O_2 changes of ±1.5 vol.% O_2. Table 15.10 shows the influence of typical disturbance variables on the O_2 content in the flue gas of a combustion plant.

A permanently set combustion process without sensory control is completely exposed to these O_2 fluctuations. Figure 15.44 shows an example of the scatter band of around $\pm 2.5\sigma$

Table 15.10 Influence of typical disturbance variables on the oxygen content in the flue gas

Disturbance to the combustion	Typical fluctuation of the disturbance variable	O_2 change in vol. %
Air temperature	±20 °C	±1.5
Air pressure	±25 mbar	±0.8
Calorific value	±10%	±2.0

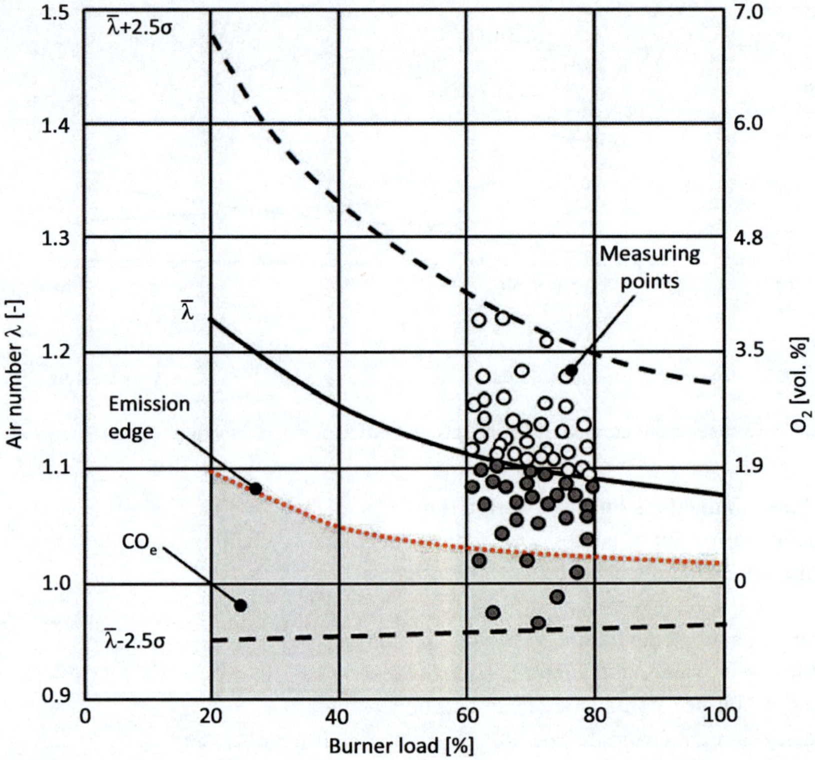

Fig. 15.44 Scattering of the air ratio λ over the burner firing-rate of an uncontrolled combustion plant

the mean value of the air ratio λ over the burner firing-rate, measured on an uncontrolled 4 MW gas firing system. If, for example, the air ratio or the O_2 content decreases due to typical fluctuations, there is a risk of incomplete combustion with high pollutant emissions of CO_e, as the emission edge is undercut. In order to avoid this critical state, an uncontrolled system must be operated at a *safety distance* from the emission edge – that is with a large excess of air. However, this is equivalent to low efficiency due to exhaust heat losses. Monitoring/control and safe adjustment of combustion to compensate for such disturbances and fluctuations is indispensable for both environmental and safety reasons.

The prerequisite for any control system is suitable sensor and actuator technology, in particular a freely adjustable fuel/air ratio. In the past, it was difficult to find a way to intervene in rigid mechanic fuel/air ratio solutions.[10] Only the electronic fuel/air ratio control[11] has opened up cost-effective intervention possibilities in the recent years, with which the firing system can now be flexibly adjusted to the changed conditions. Since then, the O_2 control on industrial combustion plants has been state of the art. The sensors required for control, as well as two different control strategies, are presented below.

Exhaust Gas Sensors

To monitor/control dynamic combustion and to compensate for disturbances on combustion, fast-response sensors must be placed, ideally directly in the flue gas duct of the combustion system. These in-situ exhaust gas sensors are exposed to high loads in the flue gas. In addition to the known combustion products (see Fig. 15.42), these include temperature, pressure, humidity, steam, additives, HF, SO_2, SO_3, H_2SO, ash, dust, heavy metals, boiler abrasion, vibrations, etc. Therefore, robust, highly dynamic gas sensors based on solid state electrolytes are used for this task. The best-known example of a solid electrolyte sensor is the λ probe, which is mainly used in the automotive sector. In industry, solid electrolyte sensors are mainly used for measuring O_2 and detecting CO_e. Figure 15.45 shows an example of a combination probe for simultaneous measurement of both relevant process variables O_2 and CO_e.

Table 15.11 shows some relevant technical data of the combi-probe KS1D (LAMTEC 2015), which has been tested according to the new standard EN 16340 "Safety and control devices for burners and appliances burning gaseous or liquid fuels- Combustion product sensing devices".

Figure 15.46 shows the principle of the thimble-like KS1D probe. It is located in the flue gas duct of the firing system. The zirconium oxide ceramic separates the reference gas chamber gas-tight from the measuring gas chamber. The platinum reference electrode is located on the inside of the functional ceramic in contact with the reference gas (ambient), while the two measuring electrodes for O_2 and CO_e are located on the outside of the ceramic in contact with the measuring gas. The O_2 electrode 1 made of platinum and the CO_e electrode 2 made of a platinum/precious metal mixture differ only in the material. The different catalytic and electrochemical properties of this material enable the detection of

[10] Mechanic ratio control: actuators (e.g. valves) for gas and air are mechanically rigidly coupled via rods.

[11] Electronic ratio control: actuators (e.g. valves) for gas and air are electrically operated independently of each other.

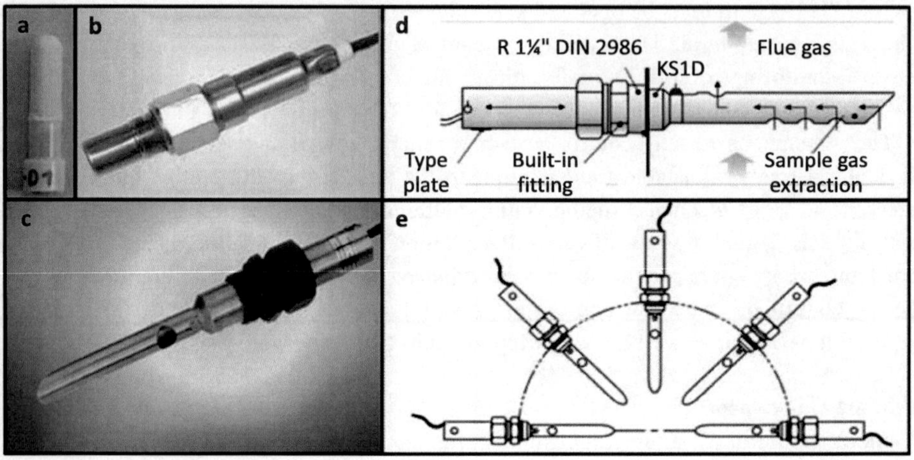

Fig. 15.45 Combi-probe KS1D: (**a**) solid electrolyte sensor element; (**b**) probe with integrated sensor element; (**c**) probe in standard housing; (**d**) probe installation; (**e**) installation position in flue gas duct

Table 15.11 Extract from the technical data of the combi-probe

Measuring range	O_2	0 . . . 21% O_2
	CO_e	0 ...1.000 ppm
Measurement accuracy	O_2	±5% of measured value, not better than ±0.3 vol.%
	CO_e	±25% of measured value, not better than ±20 ppm
Response time T_{60}	O_2	<3 s (unfiltered <3 s)
	CO_e	<9 s (unfiltered <3 s)

Fig. 15.46 Basic structure of the KS1D CO_e/O_2 probe

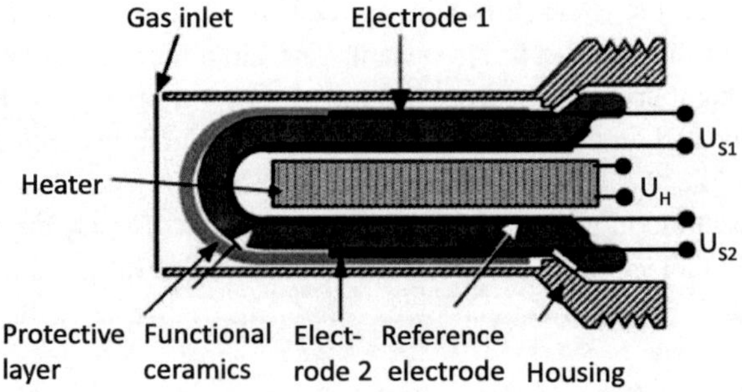

CO_e. The probe is heated and controlled to temperatures of $T_i \approx 650°C$ by an integrated heater. At this temperature, the solid electrolyte ceramic is oxygen ion conductive and the two sensor signal voltages U_{S1} (between electrode 1 and reference electrode) and U_{S2} (between electrode 2 and reference electrode) are formed and can be measured.

The sensor voltage at both electrodes U_{Si} with $i = 1, 2$ initially corresponds to the known Nernst voltage, which is measured according to

$$U_{Si} = U_{0,i} + \Re T_i / 4F \cdot \ln\left(p_{O_2,\text{ref}} / p_{O_2,\text{mess}}\right) \tag{15.38}$$

depends on the oxygen partial pressure $p_{O_2,\text{mess}}$ to be measured in the exhaust gas. The O_2-partial pressure of the ambient as a reference is known and is constant at $p_{O_2,\text{ref}} = 21$ vol. % The universal gas constant $\Re$ and the Faraday constant F are also known. A simple one-point calibration in ambient air with $p_{O_2,\text{mess}} = p_{O_2,\text{ref}} = 21$ vol.% provides $U_{Si} = U_{0,\,i}$ and thus directly the sensor-specific offset voltages $U_{0,\,i}$ at the set heater temperature and the resulting temperatures T_i of both electrodes.

In the presence of combustible gases CO_e, an additional non-Nernst sensor voltage U_{COe} is formed at the CO_e measuring electrode 2, which adds to the pure Nernst oxygen signal. The resulting sensor signal at electrode 2 is thus

$$U_{S2} = U_{S1} + U_{CO_e}, \tag{15.39}$$

and resolved by U_{CO_e}

$$U_{CO_e} = U_{S2} - U_{S1} \tag{15.40}$$

where U_{COe} is a measure of the combustible components CO_e.

Figure 15.47 shows the two signals U_{S1} and U_{S2} KS1D above the O_2 content in the flue gas of a combustion plant. In addition, the concentration of unburned components CO_e in ppm is also plotted on the second *y*-axis. A typical curve of CO_e concentrations with gradually reduced air ratio λ or O_2 content shows a significant increase in CO_e concentration due to combustion air shortage when approaching incomplete/poor combustion (see also Fig. 15.43).

In the excess air area, with clean CO_e-free combustion, both sensor signals U_{S1} and U_{S2} are identical and show the current oxygen content in the flue, according to the Nernst equation. Near the emission edge, however, the sensor signal of the second electrode U_{S2} increases overproportionally due to the added non-Nernst CO_e signal.

O_2 Control

The aim of the λ or O_2 control is to set the combustion plant to a fixed λ value with sufficient safety distance from the emission edge to prevent the risk of incomplete combustion. In Fig. 15.43, the λ control range of today's plants is highlighted in grey. Typical values are λ=1.1–1.3 and sometimes even higher. The greater the measurement uncertainty and the measurement error, for example due to drift of the O_2 measurement or ingress air,

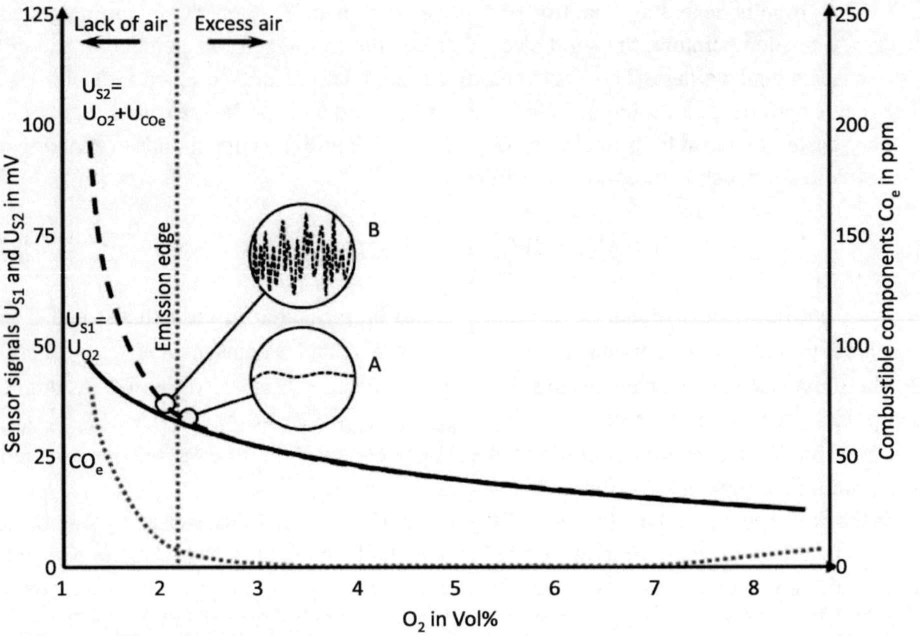

Fig. 15.47 Typical signal curve of the electrode voltages U_{S1} and U_{S2} over the residual oxygen content in the flue gas duct of a combustion plant

and the greater and more dynamic the fluctuations on combustion described above, the greater the safety distance to the emission edge must be selected. This safety margin is necessary due to the process, but has an unfavourable effect on the efficiency, since the optimisation potential is wasted up to the plant- and fuel-specific combustion optimum near the emission edge.

The classic O_2 control with firing-rate dependent O_2 setpoints compensates for these fluctuations to a large extent. However, firing with O_2 control does not achieve the maximum possible energy savings, since a "safety margin" to the firing system optimum is still necessary. In addition to O_2 control, advanced CO/O_2 control for combustion optimisation – also known as emission edge control – enables a further approach to the emission edge up to the efficiency maximum.

CO/O_2 Optimisation

In order to locate the emission edge, the fuel/air ratio is reduced dynamically by means of suitable actuators from the excess air area towards a lower air ratio λ without influencing the power control of the combustion plant until the CO_e sensor signal U_{S2} spreads from the O_2 signal U_{S1} at the emission edge (see Fig. 15.47). According to Eq. 15.40, the difference between the two signals $U_{S2} - U_{S1} = U_{CO_e}$ is a measure of CO_e. If a predetermined threshold value of CO_e is exceeded, the emission edge is considered to be found. Now the control elements are moved slightly in the direction of the excess air again; the optimum operating point λ_{opt} of the system "before" the emission edge with maximum efficiency is thus found. This procedure is repeated continuously in order to be able to ensure operation close to the optimum combustion level even under changing conditions or burner firing-rates that lead to a shift in the emission edge.

Rapid changes or disturbances to the optimum operating point λ_{opt} are immediately detected during CO/O_2 optimisation through permanent monitoring of CO_e emissions and information on the current O_2 content in the exhaust gas. The system is then moved to a safe operating point with the minimum excess air required for safe operation and, starting from this point, the system is brought back to its new optimum operating state under the changed conditions using the procedure described above.

The dynamic sensor signal behaviour of the CO_e electrode can be used as an extremely advantageous criterion for finding the emission edge. Zoom A in Fig. 15.47 shows the relatively quiet sensor signal $U_{S2}(t)$ behaviour in the area of complete, clean combustion. Zoom B, on the other hand, shows the highly dynamic signal behaviour in the area of incomplete, poor combustion behind the emission edge. By considering the probe dynamics (amplitude, frequency, etc.), the detection of the emission edge has thus been reduced to a simple and clear yes/no statement.

The use of signal dynamics, instead of a fixed sensor signal threshold value, is an age- and drift-tolerant method, since the dynamics and thus the emission edge can still be clearly recognized even with an aged or contaminated sensor element. Drift compensation or calibration in the field is therefore not necessary. Only the function of the sensor must be ensured from time to time. As a plausibility check, for example, the CO increase with each ignition of the burner and the associated increase in the sensor signal can be used.

CO/O_2 optimization has been successfully in use since about 2002. The most important advantages of CO/O_2 optimisation over O_2 control are summarized below:

- higher energy savings through permanent optimisation at each firing-rate point,
- better control behaviour due to significantly shorter response time,
- ingress air independent,
- failsafe,
- robust and
- low-maintenance.

Savings Potential

For combustion control, the complete range of electronic burner control units, flame monitors, fuel/air ratio controllers and CO_e/O_2 measuring devices with associated sensor technology is available on the market today.

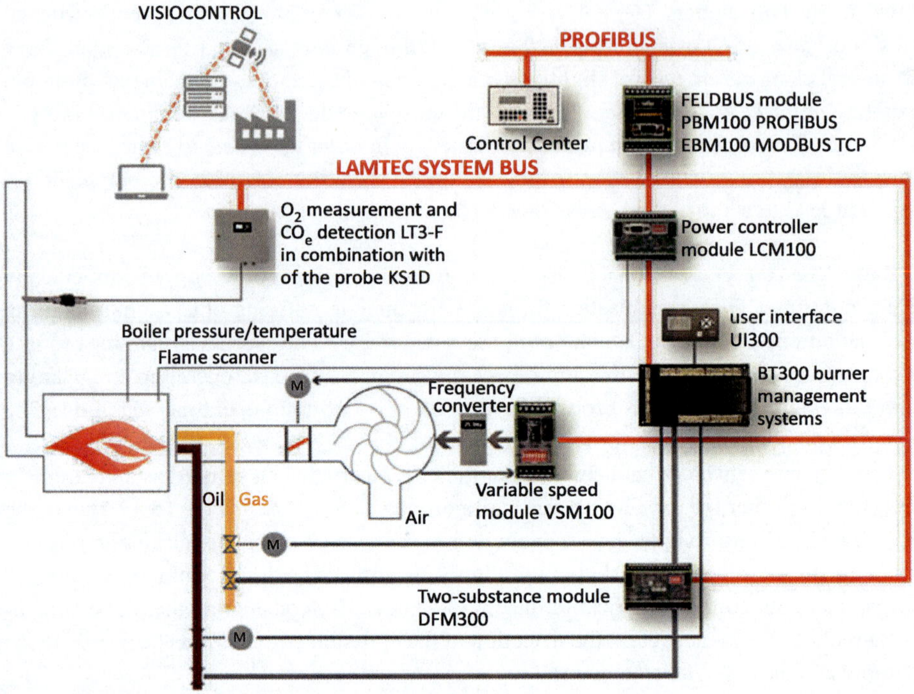

Fig. 15.48 Main components of a combustion control system

The main components of a combustion control system are shown in Fig. 15.48. While these modules are already integrated in new plants, old plants can be retrofitted with the corresponding components and thus upgraded to electronic ratio control. In the following, the conversion of an existing plant is described as an example and the resulting improvements and savings are presented.

Figure 15.49 shows the high-pressure steam boiler of a thermoprocessing plant for the food industry with a 10 MW gas burner to be upgraded with CO/O_2 optimisation and firing-rate dependent speed control (Fig. 15.49a). During the conversion, the control and safety technology of the burner, such as the burner control unit, the valve leakage test device and the burner firing-rate controller were first dismantled. The rigid mechanical ratio control was dissolved, that is the gas/air linkage was dismantled (Fig. 15.49b). The gas and air dampers were fitted with suitable electric actuators (Fig. 15.49c). The frequency converter for the 30 kW combustion air fan was installed and its motor was equipped with a NAMUR sensor for speed measurement. The CO_e/O_2 probe (Fig. 15.49d) was placed in the flue gas duct and the associated probe electronics (Fig. 15.49e) was mounted on the wall.

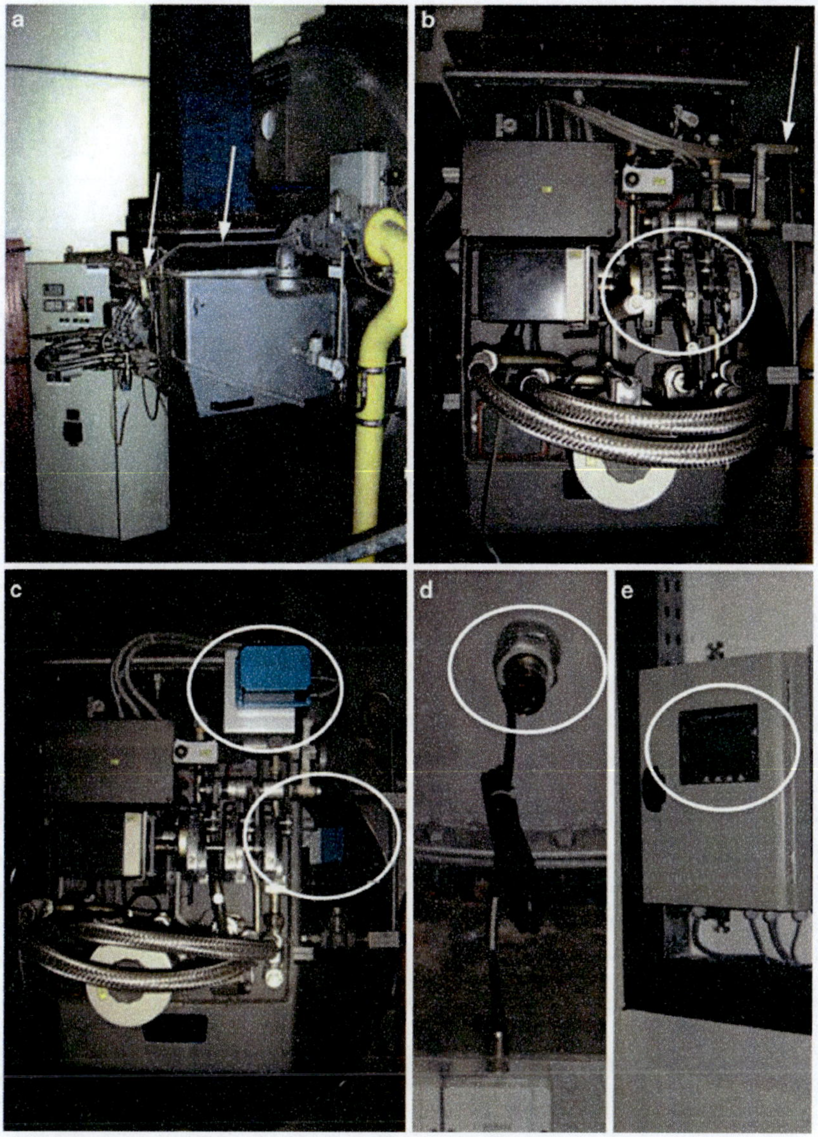

Fig. 15.49 (**a**) Burner with gas/air linkage (*arrows*). (**b**) Cam discs of the rigid mechanical fuel/air ratio control – before conversion. (**c**) Electric actuators of the electronic fuel/air ratio control – after conversion; (**d**) Probe in the flue gas duct; (**e**) Probe electronics in the wall mounting housing

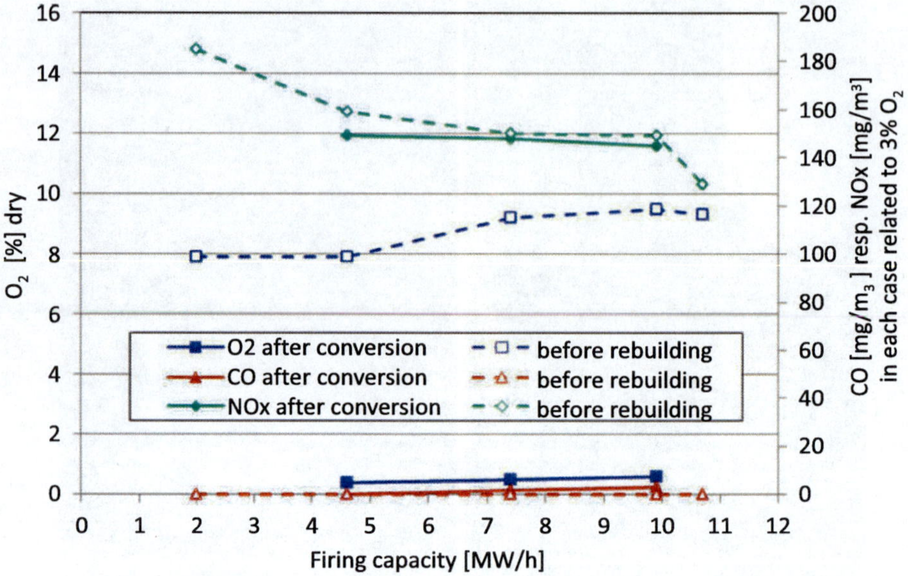

Fig. 15.50 Comparison of the relevant exhaust gas components before and after the conversion

Conversion and commissioning including electrical installation were completed after 3 days (Witte et al. 2009).

The relevant flue gas components, in comparison before (dashed) and after the conversion (solid), are shown in Fig. 15.50. As a result of the *emission edge strategy,* the oxygen content in the exhaust gas could be significantly reduced over the entire firing capacity range of the burner from approximately 8–9.5 vol.% O_2 to below 1 vol.% O_2. At the same time, the CO concentration, the indicator of poor combustion, always remained at 3 mg/m^3 and is therefore negligible. A review of the NO_xcontent also showed that the conversion and CO/O_2 optimisation in no way had a negative effect on these emissions, but rather that the level could be lowered somewhat due to the near stoichiometric combustion.

The efficiency improvements were verified by means of two methods; directly via the gas quantity fed in, taking into account the gas parameters and the steam quantity released, and via *Siegert's formula* according to

$$q_A = (\theta_A - \theta_A) \cdot \left(\frac{A_2}{21 - O_2} + B \right) \tag{15.41}$$

with the flue gas loss q_A, in %, the flue gas temperature θ_A or the combustion air temperature θ_B in °C, the oxygen content in the flue gas O_2 in vol.%. and the two fuel-

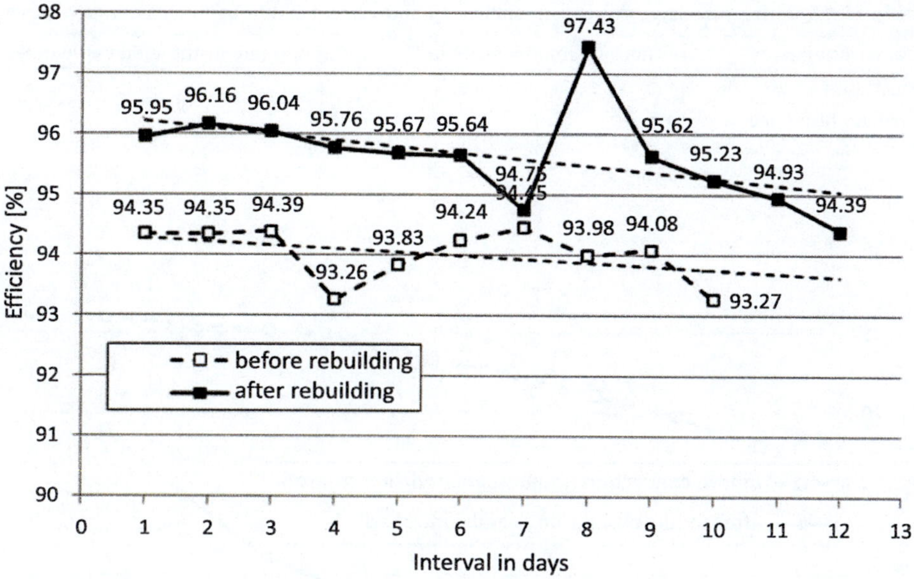

Fig. 15.51 Comparison of the efficiency before and after the conversion

specific parameters A_2=0.66 and B = 0.009 for natural gas. All parameters required for this were measured and recorded.

The calculation according to Eq. 15.41 resulted in an average value of 1.5% efficiency improvement. The values for the calculation according to the direct method were recorded over a period of 10 days and resulted in an average value of 1.6% according to Fig. 15.51. Both methods are a conservative estimation of the efficiency gain, which is actually even higher. Table 15.12 shows in detail the firing-rate range dependent efficiency improvements with associated firing-rate distribution curve, that is the percentage operation of the burner in this firing-rate range.

Thus, an annual energy saving of approximately 1,000,000 kWh was determined. By equipping the combustion air fan with a frequency converter, a demand-oriented, firing-rate dependent variable speed control was possible, which led to an additional saving of electrical energy of 125.000 kWh per year (see Fig. 15.52). Assuming fuel costs of 5 ct/kWh$_{Gas}$ and electricity costs of 6 ct/kWh$_{el}$, this results in annual gas savings of approximately 50,000 Euros. The savings from variable speed control amount to another approximately 11,250 Euro per year. This results in a total saving of approximately 61,250 Euro per year, which leads to a payback period of well under 1 year. In addition to these fuel and cost savings, there is also an annual CO_2 reduction of approximately 270 t per boiler.

Table 15.12 Firing-rate range dependent efficiency improvement and firing rate distribution curve

Burner firing rate	Efficiency improvement in %	Firing-rate distribution curve in %
Small firing rate	0.9	7
Medium firing rate	1.1	64
Full firing rate	2.5	29

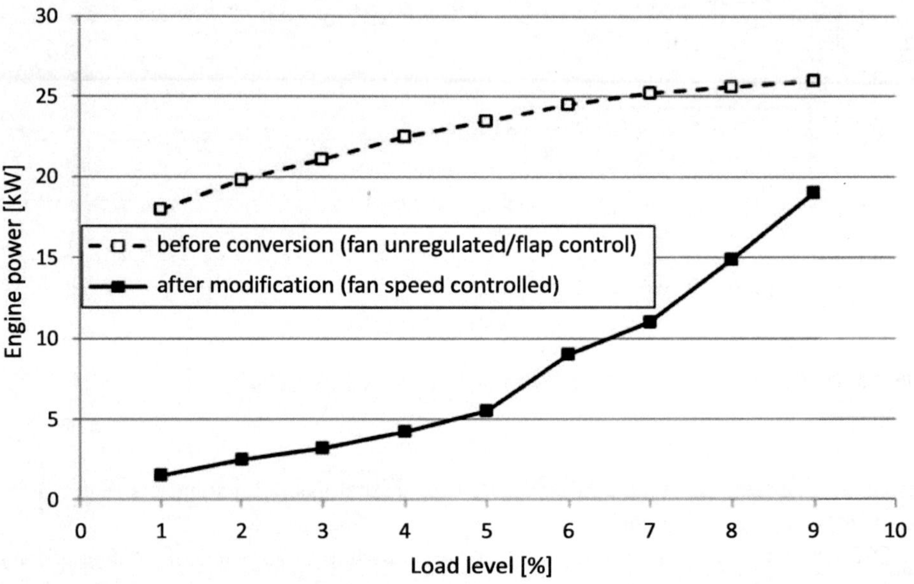

Fig. 15.52 Comparison of the required motor power for the speed fan before and after the conversion

Excursion into the World of Standards

EN 298 applies to burner controls, while EN 12067-2 applies to electronic fuel/air ratio controllers. In addition, the relevant burner standards, such as EN 676 for gas burners and EN 267 for oil burners with fan, must be considered. These burners, in turn, are operated in combustion systems whose standards must also be complied with (see Table 15.13). For industrial thermoprocessing plants, this is for example EN 746-2 and for steam boilers or thermal oil heaters EN 12952/12953 or EN 4754. The electrical requirements for pressure equipment are described in a special standard EN 50156.

Table 15.13 Standards and regulations for combustion plants

Plant	Combustion plant		
Type	Industrial thermoprocessing plant	Steam Boiler	Thermal oil heater
Definition	Machine	Pressure equipment	
Applicable directive	2006/42/EC	2014/68/EU	
Applicable standard	EN 746-2	EN 12952/ 12953	EN 4754
Electrical requirements according to	EN 746-2	EN 50156-1	
Equipment must be	Type-tested SIL PL	Type-tested Loop must be suitable for use in SIL and meet additional requirements	

Table 15.14 Reference to product standards for the provision of safety certification for safety devices/equipment on industrial thermoprocessing equipment according to EN 746-2

Safety device according to EN 746-2	Refers to product standard or solution approach in
Automatic burner control	EN 298
Gas/air pressure sensing devices	EN 1854
Automatic shut-off valves (gas)	EN 161
Valve proving systems	EN 1643
Electronic fuel/air ratio control	EN 12067-2
Shut-off valves (oil)	DIN EN ISO 23553-1

The most important functions of a control system for combustion plants must be performed in a fail-safe manner. Faults in the device or in the sensors and actuators must be detected and always lead to a safe state. In addition, the equipment parts used must have a type approval in accordance with the applicable standards. In addition, there may be requirements for a specific safety integrity level (SIL) according to EN 61508/61511 or performance level (PL) according to EN 13849. EN 746-2 refers here to proven product standards and – if these are not available – to standardised and proven approaches to solutions for providing the required safety certificates (see Table 15.14).

The system presented here for CO/O_2 optimisation with the fail-safe probe electronics Lambda Transmitter LT3-F in combination with the combination probe KS1D takes this development into account. For the first time worldwide, the system meets the requirements of EN 61508 up to SIL2 – from the sensor to the control function. For this purpose, the probe has been tested according to EN 16340 "Combustion product sensing devices". With this new standard, exhaust gas sensors can be tested and certified for the first time in a uniform and thus comparable manner. They thus meet a safety standard that has been harmonised at European level, which helps to reduce potential distortions of competition.

As safety-related components, the exhaust gas sensors increase the already high safety level of gas applications in combustion technology as well as plant availability and plant service life, even in complex thermoprocessing plants.

15.7 Gas Analysis of Metal Samples (C, S, H, N, O)

Peter Paplewski

Traditionally, the term *quantitative elemental analysis* in organic chemistry is understood to mean the quantitative determination of the elements carbon (C) and hydrogen (H) as the main constituents of all organic substances, as well as the most important of the so-called "heteroatoms": nitrogen (N), oxygen (O), sulfur (S), more rarely also phosphorus (P) and the halogens (F, Cl, Br, I). The development of elemental analysis is inseparably linked to names such as Antoine Laurent de Lavoisier, Joseph Louis Gay-Lussac, Johan Kjeldahl, Jean Baptiste Dumas, Justus von Liebig (Liebig 1831) and Fritz Pregl. As Fig. 15.53 shows, the first available apparatuses resembled more a laboratory experiment than a commercial analyzer of instrumental analysis. After digestion or combustion of the sample,

Fig. 15.53 Historical photograph of one of the first apparatuses for carbon determination from Ströhlein GmbH & Co KG (Thuringia, around 1895). (The Ströhlein company was founded in Thuringia in 1892, initially as a pure glassblower's business, and after the turmoil of the Second World War moved its headquarters first to Düsseldorf, later to Kaarst. After the company went bankrupt in 1997, the field of elemental analysis was continued as JUWE Laborgeräte GmbH (Viersen) before becoming part of Bruker Corp. in 2008)

the actual quantitative determination of the collected combustion gases was carried out using considerable amounts of wet chemistry (e.g., absorption solutions) by back titration or gravimetry.

Current commercially available devices of the instrumental organic CHNS elemental analysis work on the principle of the so-called combustion analysis. A precisely weighed sample is transferred quantitatively via a sample lock into a combustion tube and combusted at continuous temperatures of up to ~1400 °C C (due to exothermic reactions temporary up to 1800 °C) with the addition of pure oxygen (O_2). The released gaseous oxidation products (analysis gases) are then carried through the complete analyzer with the aid of an inert carrier gas (usually pure helium). In order to reduce the nitrogen oxides (NO_x) to nitrogen (N_2) and to remove any excess O_2 from the analysis gas, the analysis gas is then passed through a packing of hot copper (chips or granules). Only the analysis gases (hereinafter referred to as *analytes*) remain in the He carrier gas stream: N_2, CO_2, H_2O and SO_2. From this point onwards, two equivalent approaches are chosen for the selective and quantitative determination of the individual elements carbon (as CO_2), hydrogen (as H_2O), nitrogen (as N_2) and sulphur (as SO_2):

1. Chromatographic separation of the gas mixture into the individual analytes, which elute from the separation column with a time delay (see Sect. 15.5). The detection of the individual components can then be carried out cost-effectively with a single, non-specific but sensitive detector with a high dynamic range, such as a TCD or FID (see Sect. 15.3), since the selectivity is given in the form of the different retention times of the analytes. In a modification of the chromatographic principle, the gas mixture is passed successively through different packed adsorption columns (one for each analyte, except N_2, which is not adsorbed by any stationary phase). The packing material (stationary phase) of the columns is generally carbonized molecular sieve with pore sizes between 2 and 5 Å. By different tempering or selection of the stationary phase, different selectivities and capacities for different analytes can be achieved. In the current example, the gas mixture passes through three adsorption columns A, B and C successively, whereby column A selectively adsorbs H_2O, column B selectively SO_2 and column C selectively CO_2. N_2 passes all three columns and is directly determined by a TCD. The carrier gas stream is then directed through a valve circuit exclusively through column C and the adsorbed analyte CO_2 is desorbed by rapid heating of the column and then also measured with the TCD. The same procedure is carried out with the separation columns B and A. The result of this analysis sequence is a time/intensity diagram with four baseline separated peaks (one for each analyte). The adsorption of analytes onto a stationary phase followed by desorption is an established procedure in GC known as "purge & trap". In addition, the analytes are enriched within the separation columns and can be thermo-desorbed again in a targeted manner by rapid heating of the columns. This method is also known from GC under the name "Cryo-(Re)Focusing". A major disadvantage of the separation methods discussed here is their time requirement, since

they are purely sequential, discontinuous processes. Another complication is the limited capacity of each separation or adsorption column with respect to the respective analyte.
2. C, S and H are determined by using gas-specific detectors (usually NDIR detectors, see Sect. 15.7) which selectively detect only one of the analytes CO_2, SO_2 and H_2O of the analysis gas mixture at a time. Since nitrogen, like all diatomic homonuclear molecules with $C_{\infty v}$symmetry (Atkins 1988), is infrared inactive, the determination of the nitrogen content is usually carried out with a TCD. Due to the non-specific detection with a TCD, all components except the analyte N_2 and the carrier gas itself must be removed beforehand. For this purpose, the analysis gas mixture is passed through various packed beds of suitable solid absorbents/reagents (glass tubes filled with reagents, *scrubber* or *trap*). Table 15.15 provides an overview of some of the absorbents and reagents commonly used in elemental analysis.

At this point, the attentive reader will have noticed that a (simultaneous) determination of oxygen and the halogens (F, Cl, Br, I) in addition to C, H, N, and S is not possible using the methods described above. The determination of the halogens is carried out after combustion by means of electrochemical cells or after absorption in a suitable buffer solution by back-titration with coulometric end-point detection (coulometry). The determination of oxygen, must be carried out in pure inert gas (high-temperature pyrolysis with exclusion of oxygen) under reductive conditions (graphite contact) at temperatures >1000°C. According to Eq. 15.42, the so-called "*Boudouard equilibrium*[12]" (Hollemann and Wieberg 1985), carbon monoxide (CO) is quantitatively produced under these conditions.

$$\{C\}_{\text{solid}} + CO_2 \underset{T<400\,^\circ C}{\overset{T>100\,^\circ C}{\rightleftharpoons}} 2CO \quad (\Delta_R H = +172{,}6\,\text{KJ} \cdot \text{mol}^{-1}) \tag{15.42}$$

Besides the described methods of organic CHNS elemental analysis there is a, generally less known, branch of elemental analysis which deals exclusively with the analysis of inorganic solids, especially metals and other refractory compounds. Since especially the light elements C, S, O, N, H have a significant influence on the material properties and alternative analytical measurement methods such as ICP-OES/MS, S-OES or XRF have deficits exactly here, the applicability of this special form of elemental analysis extends over the entire value chain, from the raw material to the finished component, across almost all industries. For several reasons, the described conventional methods of organic elemental analysis cannot be applied adequately to these inorganic solids. Two of these reasons are explained in more detail here:

[12]The Boudouard equilibrium plays an important role in several large-scale processes such as the blast furnace process or the production of generator gas. As an endothermic, volume-increasing reaction, the equilibrium is completely on the CO side at temperatures above 1000 °C and pressures around 0.1 MPa.

Table 15.15 Some common absorbents and reagents used in elemental analysis

Absorbent/reagent	Effective component	Use	Temperature
Magnesium perchlorate (as dihydrate), on carrier or pure substance	$Mg(ClO_4)_2$	Gas drying (irreversible)	RT
Phosphorus pentoxide on carrier (possibly with indicator)	P_4O_{10}	Gas drying (irreversible)	RT
Sodium hydroxide on carrier	NaOH	Removal of CO_2 (irreversible)	RT
Manganese dioxide (granules)	MnO_2	Removal of SO_2 (irreversible)	RT
Schütze reagent (sometimes called: Schuetze, or Schutze reagent), acid-stabilized iodine pentoxide on carrier	I_2O_5	Selective oxidation of CO to CO_2 at RT (H_2 is not oxidized to H_2O)	RT
Hopcalite	MnO_2/CuO/ MO_x	Mixed catalyst that selectively oxidizes CO to CO_2 at RT (H_2 is not oxidized to H_2O)	RT
Platinum on carrier (silica or alumina); e.g. platinized silica	Pt	Catalytic oxidation of CO to CO_2 in the O_2carrier gas stream	~350°C
Copper(II) oxide (granules), copper(II) oxide with rare earths	CuO	Oxidation of CO to CO_2 and H_2 to H_2O. oxidation of hydrocarbons, e.g., CH_4, to CO_2 and H_2O (catalytic in oxygen stream)	~600°C
Cellulose wool	$[C_{12}H_{20}O_{10}]_n$	Removal of SO_3	RT
Copper (shavings)	Cu	Removal of O_2, reduction of NO_x to N_2 in inert gases	~650°C
Antimony (granules), potassium iodide (granules), 2,4-Dinitrodiphenylamine	Sb, AI, $C_{12}H_9N_3O_4$	F/cl and general halogen absorbers	RT

1. *Maximum temperature*: at ~1400–1500° C the maximum permissible continuous operating temperature of conventional ceramic high temperature materials (e.g., -α-Al_2O_3, corundum, MgO magnesia) is reached. However, these temperatures are not sufficient to safely and completely *combust* refractory elements, compounds or materials such as nitride and oxide ceramics, metals such as tungsten, hafnium, osmium, etc., nor simple steels, ferro-alloys and difficult-to-dissolve ores. Temperatures beyond 2000 ° C (partly ~3000°C) are required for this.
2. *Time required*: if a cost-intensive process is controlled on the basis of the analysis results, the factor time is cash money and analysis results should be available within the shortest possible time (<2 min). Due to this requirement, time-consuming, discontinuous processes such as chromatographic separation or *purge & trap* are virtually ruled out by themselves.

Although the underlying principles such as detection by NDIR or TCD and analytes (CO, CO_2, N_2, H_2O, H_2, SO_2) of the organic (solid, liquid, gaseous) elemental analysis are comparable to those of the inorganic solid elemental analysis, in practice there are significant differences with regard to the suitability of the two related methods. In the following, only inorganic solids analysis will be discussed, whereby the common, physical principles and challenges with regard to gas measurement technology will also be addressed.

Gases in Metals

It has long been known that almost all metals contain small amounts of oxygen, nitrogen and hydrogen. In this context, the term *gases in metals* has been established for historical reasons, regardless of whether these elements are actually present in the metal as oxides, nitrides, hydrides or in other, unbound forms. The use of the word "gas" is due to the fact that these elements are present in their natural, unbound state as gases and are in turn released as gases during analysis (Jordan and Eckman 1925). In fact, the presence and content of these gas components in metallic materials[13] has a great influence on the quality of the material. However, the obviously different qualities could not be explained by the usual variation of the alloy components (at that time detectable by (wet) chemical analysis, today by XRF[14] or S-OES[15] spectrometry) or by different mechanical or thermal process steps. In these cases, due to the lack of suitable analytical methods, the gases in the metals were initially held responsible for the different qualities. After the development of suitable analytical methods, this assumption was confirmed (Jordan and Eckman 1925).

Oxygen in molten metals, forms insoluble oxidic inclusions with additives or alloying elements such as Al, Si and Ca. These inclusions impair the mechanical properties such as impact strength and ageing embrittlement of the materials. Oxygen that reaches the steel during forging or other hot forming processes can cause *red shortness*. Nitrogen also forms inclusions in steels and other non-ferrous metals (e.g., Al and Ti), which reduce the ageing stability of the material and, above a certain content (~120 ppm), increase the risk of *segregation* of the phases during the solidification process of the melt. On the other hand, nitrogen is used as an alloying component (contents up to 0.4%) in CrNiMn steels, as it stabilizes their austenitic microstructure and, for example, also increases the hardness of manganese steels.

Hydrogen plays a special role in this consideration and is a highly undesirable element in all ferrous and non-ferrous metals, since it causes cracking, pores, embrittlement (hydrogen-induced brittle fracture) and a large number of other damage patterns. Since hydrogen for example surrounds us omnipresent in the form of (air) moisture and many

---

[13] Both ferrous and non-ferrous metals are generally affected (Jordan and Eckman 1925).

[14] XRF: (English: **X-Ray** Fluorescence), X-ray fluorescence spectrometry.

[15] S-OES: (Spark **Optical** Emission Spectrometry), optical emission spectrometry with excitation by electric arc excitation (Note: "Spark": excitation in Ar atmosphere, "Arc": excitation in air).

metals catalyze the dissociation of hydrogen-containing compounds, the hydrogen input not only plays a role in the production of the raw material, but certainly also in almost every process of the production chain up to the finished product (hardening, welding, galvanizing, pickling, hot or cold forming) and beyond that also in the use of the finished workpiece (hydrogen corrosion). Due to its small size[16] and the resulting small cross-section, hydrogen has an extraordinarily high mobility, speed of diffusion and permeation rate in all materials and especially in metals. *Ab initio* calculations show that hydrogen is not present atomically in the metal lattice, but that the H atom immediately allocates its valence electron to the electron gas of the metallic bond[17] as soon as it enters the structure[18] (Katzarov and Paxton 2014). This would mean that not a H atom but a free proton moves (*diffuses*) around in the metal structure. However, a proton is smaller by a factor of ≅42.500 (see Footnote 17) compared to the diameter of the already tiny H atom. As a result, once hydrogen is inside a metal, there are almost no barriers for it. The term *diffusible hydrogen*, which describes the high diffusion rate of hydrogen in metals, has become generally accepted. The only known way to reduce speed of diffusion and to preserve the diffusible hydrogen in the material for a subsequent measurement after the entry is to *freeze out* at the lowest possible temperatures (generally in liquid nitrogen at ~ −196°C), which corresponds to a kinetic inhibition of diffusion. For example, the two valid standards for the measurement of diffusible hydrogen in welds (ISO 3690 and AWS A4.3, see Table 15.16) require *freezing* the test specimen immediately after its preparation (ISO 3690: within 4 s). Any further specimen preparation takes place exclusively on the cooled specimens. The high diffusion rate already at RT also explains the non-existence of certified reference material for diffusible hydrogen (the hydrogen content would either not be stable over time or not diffusible).

At this point, the reader may be forced to ask the question: "*Why does hydrogen damage the material at all or why is hydrogen, which leaves the material, a problem at all?*" In preparation for answering this question, it should be remembered that steel is a very inhomogeneous material with various microstructures, including areas with more crystalline or amorphous characteristics and grain boundaries. Furthermore, there are the usual

[16]Hydrogen is the lightest known element with the smallest covalent radius ($r_c = 37$ pm) of all known elements. The naked nucleus of the H atom consists of a single proton ($r = 0.00087$ pm), the H_2 molecule has a covalent radius of 276 pm. For comparison: Fe ($r_c = 125$ pm), Fe^{3+} ($r_c = 68$ pm), distance of the Fe^{3+} atomic cores in the iron lattice (bbc lattice):~287 pm. All data from Huheey (1983) and TK Steel Research (2013).

[17]A simplified model of the metallic bond is based on the idea that not the neutral metal atoms form the crystal lattice but the atomic bodies (cations such as Fe^{3+}), which make their valence electrons freely available to the entire crystal compound. These free, mobile valence electrons penetrate the entire material compound as a cloud and are also called *electron gas*. The free moving electron gas explains the good electrical conductivity of the metals compared to covalently or ionically bound compounds.

[18]Admittedly, the *ab initio* calculations were based on a perfect single crystal and the question arises to what extent such conditions actually exist in the very inhomogeneous material steel.

defects of a lattice: voids, interstitial and substitutional atoms, dislocations and precipitates (e.g., metal carbides). There are several theoretical models (e.g., *Decohesion model* or the HELP process (***Hydrogen Enhanced Local Plasticity***) that try to explain certain aspects of the problem. The most obvious theory, however, is the *pressure theory*, which is also called cavity model: In sufficiently large voids in the lattice, the H atoms accumulated in the material can recombine to form molecular hydrogen (H_2). Due to the many times larger volume of the molecule compared to the atom, there is a pressure increase up to 1000 bar within the lattice defect. The high internal gas pressure causes stresses which overlap with the existing internal stresses as well as externally applied stresses and lead to premature crack formation or crack growth. Such *hydrogen induced cold cracks* can occur even without an external load and, due to the temperature dependence of diffusion, with a delay.

Due to the inhomogeneity of steel, a distinction is made from an empirical point of view not only between the diffusible hydrogen, which has a high mobility in the material and is mentally present in atomic form (or as a proton), and the bound hydrogen, which is present as a chemically stable hydride. These two hypothetical forms represent only the two extremes in which hydrogen can occur with respect to mobility in the structure. In addition, there is the *residual hydrogen*, which is adsorbed/bound to a greater or lesser extent to various energy sinks (called *traps*) within the material. It does not diffuse at RT, but requires a certain kinetic activation energy (in the form of heat or mechanical tension) to become mobile again in the material (to be desorbed from its trap). Depending on the amount of activation energy required, one can distinguish, relatively arbitrarily, between reversibly or irreversibly bound residual hydrogen. The total hydrogen is therefore the sum of diffusible, residual and bound hydrogen and its determination requires a fusion of the complete sample to release all hydrides with certainty. The analytical methods described below allow not only the determination of the total hydrogen content, but also the determination of the diffusible and residual components through specific temperature control during the analysis. The kinetic information gained provides the metallurgist with insights into the microstructure of the steels (Bergers et al. 2010). The high-strength steels, which are used for energy efficiency reasons, especially in the automotive sector, are particularly susceptible to hydrogen embrittlement. Not only steels, but also non-ferrous metals such as copper or titanium tend to become hydrogen-induced brittle. In German-speaking countries, the term hydrogen sickness of metals (Baukloh and Stromburg 1937) was used for this.

It is not only in the determination of hydrogen that one would like to know more about the different binding forms of the gases in the metals, but also in the determination of oxygen such analytical methods have become a standard procedure (Prumbaum and Orths 1979). By specific heating rates an oxide separation is achieved, that is a separation of different oxidic forms such as the surface oxides from the bulk oxides. As Table 15.16 shows, there are a number of standards according to which metallic materials are tested for O, N and H. However, the applicability of these methods is not limited to metals but is generally applicable to inorganic solids. For example, refractory oxide or nitride ceramics

Table 15.16 Overview of some standard methods for testing metals for the "gases" O, N, H

Standard	Title
ASTM E 1019	Standard test methods for determination of carbon, sulfur, nitrogen and oxygen in steel and in iron, nickel, and cobalt alloys
ASTM E 2575	Standard test method for determination of oxygen in copper and copper alloys
ASTM E 1409	Standard test method for determination of oxygen and nitrogen in titanium and titanium alloys by the inert gas fusion technique
ASTM E 1447	Standard test method for determination of hydrogen in titanium and titanium Alloys by the inert gas fusion thermal conductivity method
ASTM E 1937	Standard test method for determination of nitrogen in titanium and titanium alloys by the inert gas fusion technique
ASTM E 1569	Standard test method for determination of oxygen in tantalum powder
ASTM E 1587	Standard test methods for chemical analysis of refined nickel
ASTM E 2792	Standard test method for determination of hydrogen in aluminum and aluminum alloys by inert gas fusion
DIN EN 10276-2	Chemical analysis of ferrous materials – determination of oxygen content of steel and iron – part 2: Measurement of infrared absorption after melting under inert gas
ISO 10720	Steel and iron – Determination of nitrogen content – Thermal conductimetric method after fusion in a current of inert gas
ISO 15351	Steel and iron – Determination of nitrogen content – Thermal conductimetric method after fusion in a current of inert gas (routine method)
ISO 17053	Steel and iron – Determination of oxygen – Infrared method after fusion under inert gas
ISO 22963	Titanium and titanium alloys – determination of oxygen – Infrared method after fusion under inert gas
DIN 25710	Determination of the equivalent hydrogen content in uranium dioxide powders and tablets of UO_2, $(U,Gd)O_2$ and $(U,Pu)O_2$ by the carrier gas method
EN ISO 4491	Metallic powders – Determination of oxygen content by reduction methods – Part 4: Total oxygen by reduction–extraction
DIN EN ISO 3690	Welding and allied processes – Determination of hydrogen content in arc weld metal
ANSI AWS A4.3	Standard methods for determination of the diffusible hydrogen content of martensitic, bainitic and ferritic steel weld metal produced by arc welding

with high melting points, glasses, ores, slags and catalysts for the petrochemical industry are routinely tested for O, N and H.

Carbon and Sulphur in Metals

The importance of the element carbon in iron and steel production is well known. The properties of the finished material (such as hardness, hardenability, cold and hot machinability, etc.) depend to a large extent on the carbon content of the material. Even before silicon, carbon is the element that is present in iron as the main alloy component. Typical concentration ranges in low-alloy steels range from~0.05–2% C, in cast iron at ~2–3% C to

~6.1% C or 29.9% C in the ceramic materials tungsten carbide (WC) or silicon carbide (SiC). Sulphur occurs in nature in the form of sulfide ores or minerals (e.g., minerals containing FeS_2, PbS, ZnS and sulphates) as well as in all fossil fuels. In addition to the undesirable SO_2 emissions during the combustion of fossil fuels or the roasting of sulfide ores, sulfur is also a strong catalyst poison. Similar to the gases N, O and H, sulfur can be dissolved (*interstitially*) in the cavities of the *bcc* and *fcc iron lattice* (Bergers et al. 2010). In most cases, these interstitially dissolved elements exert negative influences on the mechanical properties of the material. Sulfur is therefore an undesirable element in metals as well as in other inorganic materials such as cement, lime, glass, slags and ores. Table 15.17 provides an overview of a selection of standards according to which different materials are analyzed for carbon and sulfur.

Despite certain analogies, for example with the detectors used, there is an essential difference between CS analysis and ONH analysis, which will be pointed out here at the beginning: CS analysis is a pure combustion technique whereas ONH analysis is a reductive melt digestion under inert gas – the exact opposite of combustion.

Determination of C and S by Combustion in an HF-Induction Furnace

In contrast to the organic CHNS elemental analysis analyzers mentioned at the beginning, the simultaneous determination of carbon and sulfur is carried out by high-temperature combustion in an HF-induction furnace with pure oxygen as carrier gas[19] and under increased pressure. For the analysis, a solid sample is weighed into a ceramic crucible and then a certain amount (~0.7–1.5 g) of a high-purity, metallic surcharge material, so called *accelerator* is added. Tungsten granules, copper chips, pure iron chips and tin granules are used as accelerator. The purpose of these aggregates is derived from the English term *accelerator.*

- Due to their highly exothermic reaction with oxygen, these additives accelerate the combustion process and allow higher combustion temperatures to be reached.
- All the above-mentioned accelerators are good electrical conductors and couple with the high-frequency alternating field of the induction furnace. If the sample consists of a non-conductive material (e.g., metal oxides), the use of accelerators is absolutely necessary to achieve any interaction with the HF field. However, accelerators are also used when a metallic sample would provide a sufficiently good RF coupling, for the reason given above.

[19] Pure oxygen acts as a carrier gas (transport medium of the analysis gases) as well as a reaction partner during combustion (oxidation).

Table 15.17 Overview of standard methods for testing for total carbon and sulfur

Standard	Title
ASTM E 1019	Standard test methods for determination of carbon, sulfur, nitrogen and oxygen in steel and in iron, nickel, and cobalt alloys
ASTM E 1587	Standard test methods for chemical analysis of refined nickel
ASTM G 79	Standard practice for evaluation of metals exposed to carburization environments
ASTM C 1494	Standard test methods for determination of mass fraction of carbon, nitrogen, and oxygen in silicon nitride powder
ASTM E 1941	Standard test method for determination of carbon in refractory and reactive metals and their alloys
ASTM E 1915	Standard test methods for analysis of metal bearing ores and related materials by combustion infrared absorption spectrometry
ASTM C 25	Standard test methods for chemical analysis of limestone, quicklime and hydrated lime
ASTM C 1408	Standard test method for carbon (total) in uranium oxide powders and pellets by direct combustion-infrared detection method
UOP 703	Carbon on catalysts by induction furnace combustion and infrared detection
DIN EN ISO 9556	Steel and iron – determination of total carbon content – method using infrared absorption after combustion in an induction furnace
ISO 10719	Steel and iron – Determination of non-combined carbon content – Infrared absorption method after combustion in an induction furnace
ISO 21614	Determination of carbon content of UO2, (U, Gd)O2 and (U, Pu)O2 powders and sintered pellets – Combustion in a high-frequency induction furnace – Infrared absorption spectrometry
DIN EN ISO 15349-2	Non-alloy steel – determination of low carbon content – Part 2: Method using infrared absorption after combustion in an induction furnace
ISO 4935	Steel and iron; determination of sulfur content; infrared absorption method after combustion in an induction furnace
ISO 13902	Steel and iron – Determination of high sulfur content – Infrared absorption method after combustion in an induction furnace
ISO 4689-2	Iron ores – Determination of sulfur content – Part 2: Combustion/titration Method
DIN EN ISO 15350	Steel and iron – determination of total carbon and Sulphur content – infrared absorption method after combustion in an induction furnace (standard method)
ISO 7524	Nickel, ferronickel and nickel alloys; determination of carbon content; infra-red absorption method after induction furnace combustion
ISO 7526	Nickel, ferronickel and nickel alloys; determination of sulfur content; infra-red absorption method after induction furnace combustion

- Tin and pure iron, as accelerator material, also have a flux/melting agent effect, which facilitates homogeneous melting of the sample. For this reason, combinations of different materials such as W/Sn or W/Fe are often used as accelerator.

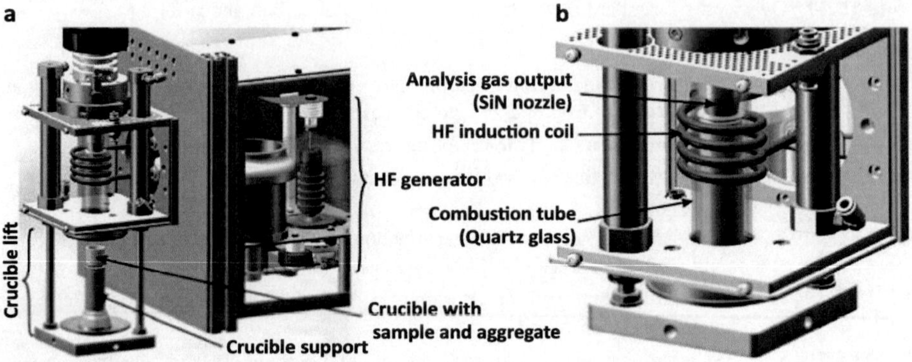

Fig. 15.54 Components of the high-frequency induction furnace of a simultaneous CS analyzer (Model G4 ICARUS HF AC from Bruker Elemental). (**a**) Overview of the HF generator, HF furnace (open state) with crucible lift and part of the fully automatic cleaning device with integrated particle filter. (**b**) Details of the HF furnace chamber in closed state with combustion tube (quartz glass), surrounding HF induction coil and analysis gas extraction nozzle

The heart of a CS combustion analyzer, the high-frequency induction furnace, is shown in Fig. 15.54. The actual combustion chamber is formed by a so-called combustion tube made of quartz glass. It sits concentrically inside an induction coil, which is fed with high-frequency energy (2.2–4 kW) by an HF generator. The crucible with sample and accelerator is placed on the crucible support and when the furnace chamber is closed, the crucible lift moves the crucible to the center of the induction coil and simultaneously seals the furnace chamber.

During heating, the HF-tube of the HF-generator is operated with a high-frequency high voltage and the alternating electromagnetic alternating field is transmitted to the sample via the induction coil. The alternating field induces eddy currents in the metallic sample itself – or in the metallic accelerator – which lead to rapid heating of the sample and thus to a rapid increase in the electrical resistance of the material (resistance heating in the material itself). The method of inductive heating enables highest power flux densities, which are only exceeded by laser pulses. Overall, this direct type of heating (comparable to a microwave oven) results in very short heating phases. Since the quartz of the combustion tube as well as the ceramic of the crucible do not interact with the HF-radiation, it is even possible to reach temperatures in the sample which, at 2100–2200 °C, are far beyond the melting point of the crucible material itself. The crucible is heated relatively slowly and indirectly by the liquid melt of the sample and survives the short heating phases of approximately 40 s undamaged.

After placing the crucible with the prepared sample, the further analysis procedure is as follows: at the push of a button, the crucible lift closes and moves the crucible with the sample into position. The entire analyzer including the furnace chamber is flushed with oxygen for a few seconds to purge the ambient air out of the system. Then the furnace chamber is supplied with a constant flow of oxygen (2–4 l/min at an abs. Pressure of

3–4 bar) and the HF generator is switched on. Due to the inductive heating described above, the sample melts, the accelerators react spontaneously with the oxygen to form metal oxides and the carbon- or sulfur-containing components of the sample are oxidized to the thermodynamically stable end products CO_2 and SO_2. The analysis gases formed are purged out of the furnace chamber by the carrier gas in a flow-optimized manner and initially freed from particles by highly efficient, heated fine filters. To avoid problems with delayed SO_2 purging,[20] the analysis gas mixture then passes through a trap with $Mg(ClO_4)_2$ (see Table 15.15) to remove any traces of moisture[21] that may have formed during combustion.

At this point, the analysis gas consists only of the analytes CO_2, SO_2 and possibly CO and first the sulfur content of the sample (as SO_2) is quantitatively determined by means of NDIR detectors https://www.witec-sensorik.de/en/applications/elemental-analysis/. Carbon monoxide (CO), is possibly present in traces, due to incomplete combustion or a short residence time in the hot zone (the crucible). In order to convert CO quantitatively into CO_2, after the sulfur measurement has been carried out, the analysis gas undergoes a catalytic oxidation process with platinized silica gel (see Table 15.15). At temperatures above 350 °C, the platinum in the O_2 stream catalytically and quantitatively oxidizes CO to CO_2. As an undesirable side reaction, SO_2 is oxidized to SO_3, but only partially. Since SO_3 is a solid (corrosive) under normal conditions, it is subsequently removed by an SO_3 trap consisting of cellulose wool (see Table 15.15). At this point the analysis gas consists only of CO_2,[22] remaining SO_2 and of course oxygen. In the last step of the simultaneous analysis the carbon content of the sample (in the form of CO_2) is determined by NDIR detectors before the analysis gas leaves the analyzer. Throughout the analysis the detector signal

[20] It is known that SO_2 in combination with moisture not only forms highly corrosive compounds, but that these are also easily adsorbed on the walls of pipes. The consequences are delayed rinsing and even reduced results due to partially adsorbed SO_2 and increased corrosion. This phenomenon can only be mitigated by tempering the pipes. The immediate drying of the analysis gases is a remedy. In addition, care should be taken to ensure that all parts in contact with the medium are compatible in terms of inertness to the corrosive analysis gases. For example stainless steel lines (high Mn content) in the petrochemical industry are known to absorb various sulphur components such as SO_2, COS and thiophene. MnO is even used for SO_2 absorption at RT (see Table 15.15). The use of stainless steel pipes is remedied by inert coatings such as Sulfinert® or Silicosteel® or fully fluorinated, diffusion-proof polymers such as PFA, PVDF.

[21] In cases in which minerals or natural substances of unknown composition are analysed, the organic components contained in them or water of crystallization during combustion can cause the release of large quantities of water. In these cases the use of an upstream inline membrane dryer (e.g. from Nafion®) is recommended. If the sample contains halogens, a halogen trap (see Table 15.15) is usually connected upstream of the $Mg(ClO_4)_2$ moisture absorber.

[22] Although the probability of partial CO formation under the existing conditions can be classified as low, ASTM E 1019 prescribes the use of an oxidation catalyst. The also possible approach of determining CO and CO_2 in parallel by suitable NDIR detectors and "calculating" the two signals to a common carbon value would therefore not be in conformity with the standard.

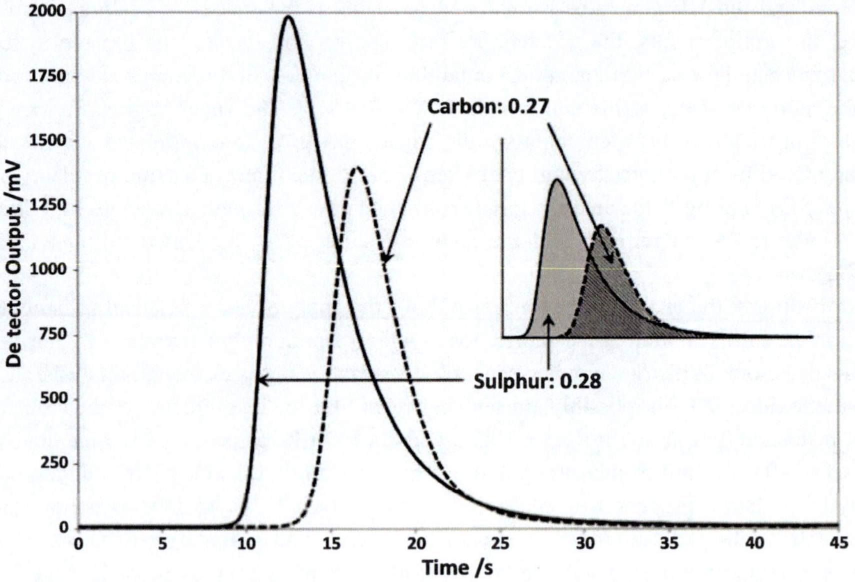

Fig. 15.55 Time-resolved signal (*x*-axis: time/s, *y*-axis: detector signal/mV) of the CO_2 and SO_2 NDIR detectors during an analysis of ~0.5 g of a certified steel standard material *Dashed curve*: SO_2 signal, *solid curve*: CO_2 signal. *Top right*: Integration of the area under the signals, indicated by the *filled or hatched areas* under the peaks

from the NDIR detectors is recorded at a sampling rate of at least 5 Hz and then evaluated. Figure 15.55 shows an example of the signals during the analysis of a steel sample.

Figure 15.55 shows a detector baseline that initially runs near zero for ~9 s. After this so-called dwell time, the first analysis gas reaches the SO_2-NDIR detector and its signal rises very fast (within approx. 3 s) to its maximum value. After exceeding the maximum height, the signal drops rapidly within 15 s to slowly approach the original baseline height. The signals of this time-resolved detector signal are called "peaks" in analogy to the GC. The slightly slower decay after the peak maximum (compared to the rise before the maximum) is called *tailing*.[23] Due to *tailing*, a peak is not perfectly symmetrical, but has a certain asymmetry, which is basically caused by the purging behavior including a longitudinal diffusion along the carrier gas flow. After about 12 s the peak appears in the signal of the CO_2 detector and after ~45 s both signals have returned to the original height of the baseline and the analysis is finished. The quantitative evaluation of the peaks is computer-

[23] If a signal rises more slowly than it falls after the maximum, the peak is called "fronting".

aided based on their peak area (integration of the signal over time). The peak area (*I*) corresponds directly to the absolute mass of the element to be determined via the so-called *response factor*, which corresponds to the slope of the calibration curve (*b* in Eq. 15.43 in µg element/unit of area). Like most methods, IR absorption spectrometry is not an absolute method, but requires calibration. This is done by analyzing certified reference material with a precisely known element concentration. The contribution of the accelerators used and the crucibles to the area of an element signal are determined separately in the form of a so-called *blank value, a* in Eq. 15.43) and subtracted from the resulting peak area of the analysis.

$$c[\mathrm{ppm}] = \frac{b\left[\frac{\mu\mathrm{g}}{\mathrm{area}}\right](I[\mathrm{area}] - a[\mathrm{area}])}{E[\mathrm{g}]} \tag{15.43}$$

With regard to gas measurement technology, the technology presented places special demands on the detectors and flow constancy:

- The course of the signal over time must be recorded as accurately as possible, especially on sections with high gradients and turning points, in order to minimize errors due to signal *drift*. In other words, the sharper the peak, the more measurement points per time unit are required and the higher the sample rate must be.
- Fluctuations of the (mass) flow (at constant total pressure → particle density) must be avoided. Fluctuations in flow cause variations in the macroscopic linear velocity of a hypothetical volume element and thus different residence times in the absorption path of the detectors.
- Due to the rapid transient change of analyte concentrations during analysis, the measuring range of the detectors used does not correlate in any way with those from a static experiment. Along the entire flow path, and thus also within the absorption path of the detectors, there is a transient concentration gradient of the analytes. Consequently, a detector only sees the average of the analyte concentration along its absorption path for each measured value. The linearization of the NDIR detectors is carried out using known gas mixtures either statically or by continuous flow. Due to the transient change of the mixture in the real measurement, however, the maximum measuring range does not correlate with the concentration values during linearization but must be determined empirically.

Example

The signal shown in Fig. 15.55 shows the analysis of a steel sample with a sulfur content of 0.280% (2800 ppm) with a weight of 0.0840 g. This corresponds to a mass of 235.2 µg sulfur (469.9 µg SO_2), which passes through the detector within a time of ~20 s (peak width). The concentration of the analyte in the gas phase (in vol.%) in the detector, averaged per unit of time, results from the signal height alone: The S-signal in Fig. 15.54

has a maximum height of 0.2 V. The detector used shows a signal height of 10 V at a constant 500 ppm SO_2. During the peak maximum, the averaged volume concentration in the detector ~10 ppm SO_2.

Carrier Gas Hot Extraction

The quantitative determination of oxygen, nitrogen and hydrogen is carried out according to the so-called carrier gas hot extraction method. In the English-speaking world this method is known as *inert gas fusion* (IGF), *gas fusion analysis* (GFA) or *carrier gas hot extraction.*[24] The sample is melted in a disposable crucible made of high-purity graphite at high temperatures (up to 3000 °C) in a carrier gas stream of an inert gas. The graphite crucible has three main functions in this process:

- The graphite crucible is part of the electrode furnace. Due to its electrical resistance, the graphite represents the actual heating element of the furnace.
- The carbon of the crucible material itself represents the reagent with which oxidic sample components react to form carbon monoxide (CO). Any CO_2 released (in traces) would also be converted into CO by reaction according to Eq. 15.42 with the graphite of the crucible.
- The graphite crucible also functions as a sample vessel for the liquid metal melt.

For analysis, a solid sample is weighed in (approx. 100 mg to 1.0 g depending on the content of the elements) and transferred via a sample introduction funnel into the sample lock (see Fig. 15.56) and a new graphite crucible is placed on the crucible tip. Fine powdered samples are weighed into a capsule made of nickel or tin. To start the analysis the furnace closes by pneumatically raising the lower electrode and pressing the graphite crucible against the upper (hollow) electrode. After a programmable purging time with carrier gas, during which ambient air is expelled from the furnace chamber, the so-called degassing process starts: While further purging with carrier gas, a continuously controlled electric current (max. 1100 A, corresponding to 7 kW) is passed through the electrodes and the crucible heats up. The temperature set during degassing is approximately 200 °C above the actual analysis temperature and in order to remove any impurities (e.g., traces of hydrocarbons) from the graphite of the crucible. After outgassing, a short stabilization phase is carried out to reach the actual analysis temperature[25] and the correct carrier gas flow. At the actual start of the measurement, the sample is automatically dropped through the sample lock into the heated graphite crucible and melted. The oxygen in the sample, which is usually in oxidic form, reacts with the excess carbon in the graphite crucible to

[24] The use of the word hot is unfortunate in this context. Since the sample must be completely melted for the overall determination, carrier gas melt extraction would be the more appropriate term.

[25] O, N in steel: ~2100 – 2200 °C; O, N in Ti and refractory metals ~2800 – 3000 °C (with Ni flux in ratio 10 : 1); H in all materials: 1600–1650 °C (1800 °C when using Ar as carrier gas).

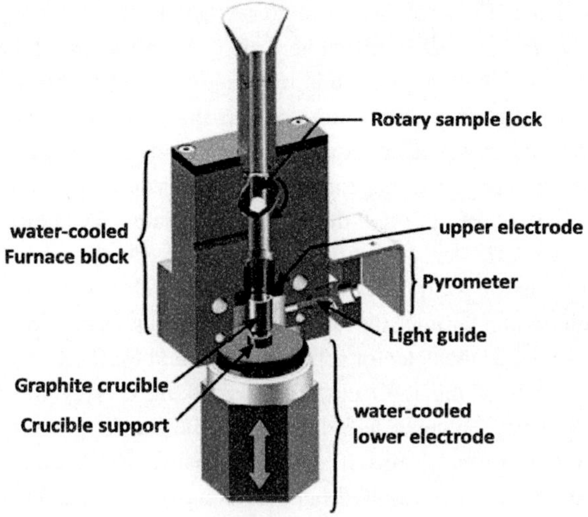

Fig. 15.56 Diagram of an electrode furnace in closed condition. The entire upper furnace block and the vertically movable lower electrode are water cooled. When the furnace is closed, the graphite crucible is pressed against the upper electrode by the lower electrode. The sample is transferred to the sample lock via a funnel. By turning the sample lock the sample drops into the crucible. The freely programmable temperature of the furnace is monitored by a non-contact optical sensor (calibrated pyrometer) (calibration to ensure that the temperature of the crucible wall detected by the pyrometer matches the temperature of the liquid sample in the crucible)

form CO. At the same time nitrogen is released from the sample in the form of N_2, for example, by decomposition of nitrides and hydrogen in the form of H_2. Together with the carrier gas, the analysis gas mixture is first passed through particle filters and downstream fine filters before it is passed at a constant flow through the detector systems for the actual measurement.

For the actual detection or quantification of O, N and H, there are several possible procedures, which differ in the detection system used (NDIR or TCD) and the actual analyte molecule.

- *Nitrogen*: there is actually no (cost-effective) choice here, since the N_2molecule is the thermodynamic low sink of nitrogen chemistry and cannot be easily, selectively and quantitatively converted into other compounds. Due to its low boiling point, it also cannot simply be frozen out or removed from the carrier gas stream by absorbents. In all commercially available devices, nitrogen is always detected as N_2molecule by means of TCD, since due to the IR inactivity mentioned above, vibrational spectroscopic detection methods such as NDIR are not applicable. Since a TCD is a non-specific detector, the analysis gas may only contain the analyte (N_2) and the carrier gas itself before entering the measuring cell of the TCD, and the TCD is therefore always the last detector in the measuring chain.

- *Oxygen*: always detected as the first analyte using NDIR, preferably in the form of CO. Alternatively, it is possible to convert all CO by means of CuO or I_2O_5 (see Table 15.15) into CO_2 and to detect this by means of NDIR. However, this method has the disadvantage that even with a pure oxygen analyzer there is a high consumption of these oxidation reagents. On the other hand, CO and CO_2 detectors can be used in parallel to extend the measuring range and for samples with very high oxygen contents.
- *Hydrogen*: here, there is the option of detection either with a TCD (possibly also a mass spectrometer) as H_2 or after oxidation with CuO in the form of H_2O via NDIR. However, detection in the form of water has some basic disadvantages: To avoid losses due to condensation, transfer lines that are as short and inert as possible shall be used and the transfer line as well as the detector cell must be heated Furthermore, the sensitivity as well as the selectivity of the H_2O detection by NDIR is very limited. A TCD, for example, has a linearity range of 5–6 decades, while the IR absorption has only two decades linearity (Colón and Baird 2004). The sensitivities or detection limits of the detectors behave similar, so that hydrogen is usually detected by means of TCDs, especially since the critical contents in most materials are in the lower, single-digit ppm range or below.
- For the reasons given above and those discussed in Sect. 3.1 (quantitative determination in multicomponent systems), it follows that when using a TCD for H and N, both elements cannot be determined simultaneously in one analysis. Since lower temperatures have to be maintained[26] for H analysis than are often required for the decomposition of refractory nitrides, samples are usually analyzed once for ON and once for H. Due to the thermal conductivity of the gases (see Table 3.1), helium is used as the carrier gas for the nitrogen determination, whereas nitrogen is used for the hydrogen determination, since the difference in thermal conductivity between the analyte and the carrier gas is biggest in this case. However, the use of helium is becoming more and more of a problem for reasons of availability and price, so that there is a general desire to switch to argon as carrier gas. With their programmable amplification circuits, polarity inversion of the bridge current (always positive peaks), heat exchangers that bring reference and analysis gas to exactly the same and constant temperature and other developments, current TCDs allow N determination using Ar as carrier gas with almost no loss of sensitivity. However, when using Ar as carrier gas for the H-determination an additional separation column must be used to separate nitrogen (from the sample) from the analyte hydrogen.

The detection sequence in *ON mode* is as follows: oxygen determination by NDIR of CO followed by an oxidation furnace with CuO. CuO oxidizes H_2 to H_2O and CO to CO_2 (in simultaneous instruments hydrogen is then determined as H_2O by NDIR). The analysis gas then passes through a trap of $Mg(ClO_4)_2$ to remove H_2O. After a possible subsequent

[26] At temperatures above 1650 °C, hydrogen depletion occurs, probably due to the formation of methane (CH_4) by reaction of H_2 with the graphite of the crucible.

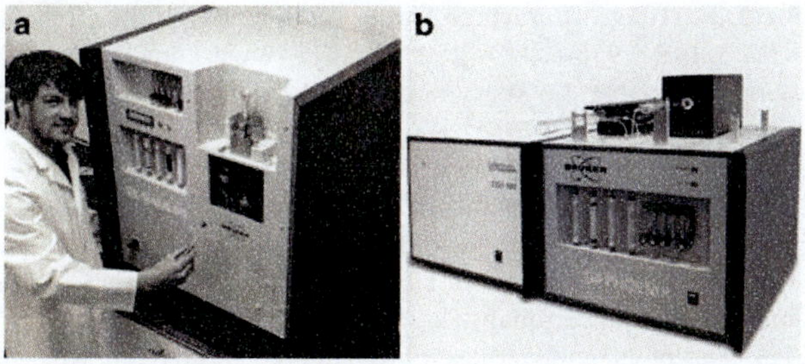

Fig. 15.57 (**a**) Bruker G8 GALILEO ONH in the laboratory. (**b**) Bruker G4 PHOENIX analyzer for diffusible hydrogen according to the principle of artificial ageing with coupling to a mass spectrometer (thermodesorption mass spectrometry)

determination of CO_2 by NDIR to extend the measuring range/control the CO_2 is quantitatively removed by a CO_2 trap (NaOH on carrier, followed by further drying). At this point, the analysis gas contains only the analyte N_2 and the carrier gas. The last measuring element is then a TCD for nitrogen determination.

The *H mode* (or OH mode) differs from the ON mode by the carrier gas used, which is switched automatically by valve switching, by lower temperatures in the electrode furnace and by the use of contactor reagent (I_2O_5, see Table 15.15) instead of CuO. The analysis gas also passes through the contactor reagent by valve switching, $which_2$ oxidizes CO to CO but does not convert H_2. After a possible oxygen determination using NDIR, CO_2 is oxidized to CO, CO_2 and moisture are removed and the remaining analyte H_2 is determined with the TCD. The determination of the diffusible hydrogen can be achieved by combining other furnaces (e.g., infrared heated furnace) by coupling to the TCD of an ONH instrument. For contents below 1 ppm or very long measurements mass spectrometers are used.

At the end of this chapter and in contrast to Figs. 15.53 and 15.57 shows the current series of Bruker instruments regarding ONH (G8 GALILEO) and diffusible hydrogen (G4 PHOENIX).

References

13. Verordnung zur Durchführung des Bundes-Immissionsschutzgesetzes (Verordnung über Großfeuerungs-, Gasturbinen- und Verbrennungsmotoranlagen – 13. BImSchV). Bundesministeriumder Justiz und für Verbraucherschutz. http://www.gesetze-im-internet.de/bundesrecht/bimschv_13_2013/gesamt.pdf. Zugegriffen: 02.05.2013

Altfeld, K., Schley, P.: Entwicklung der Erdgasbeschaffenheiten in Europa. gwf-Gas/Erdgas. **152**, 544–550 (2011)

Altfeld, K., Bödeker, J., Frieling, H., Schley, P., Uhrig, M.: Modelling of Gas Flow in Pipelines Tracking Gas Quality. Proceedings of the International Gas Research Conference, Paris, 8–10. Oktober 2008

Atkins, P.W.: Physikalische Chemie. 1. Aufl. S. 421, 435–439. VCH, Weinheim (1988)

Barwick, V.J., Prichard, E. (Hrsg.): Eurachem Guide: Terminology in Analytical Measurement – Introduction to VIM 3. ISBN 978-0-948926-29-7 (2011) www.eurachem.org

Baukloh, W., Stromburg, W.: Über die Wasserstoffkrankheit einiger Metalle. Z. Metallkunde. **29**, 427–430 (1937)

Baumbach, G.: Luftreinhaltung. Springer-Verlag (1993)

Bergers, K., Flock, J., et al.: Determination of hydrogen in steel by thermal desorption mass spectrometry. Steel Res. Int. **7**(81), 499–507 (2010)

CHPQA: CHPQA Guidance. United Kingdom Department of Energy and Climate Change https://www.gov.uk/combined-heat-power-quality-assurance-programme (2013)

Colón, L.A., Baird, L.J.: Detectors in modern gas chromatography. In: Grob, R.L., Barry, E.F. (Hrsg.) Modern Practice of Gas Chromatography, 284, 289–298, 4. Aufl. Wiley Interscience (2004)

DIN EN 16726 (Entwurf): Gasinfrastruktur – Beschaffenheit von Gas – Gruppe H; Deutsche Fassung prEN 16726 (2014)

DVGW: Berechnung von Druckverlusten bei der Gasverteilung. Deutsche Vereinigung des Gas- und Wasserfaches. GW 303-1. DVGW, Bonn (2006)

DVGW Arbeitsblatt G260. DVGW Deutsche Vereinigung des Gas- und Wasserfaches e.V., Bonn, Mai 2008 (2008)

DVGW Arbeitsblatt G262. DVGW Deutsche Vereinigung des Gas- und Wasserfaches e.V., Bonn, September 2011 (2011)

DVGW Arbeitsblatt G685. DVGW Deutsche Vereinigung des Gas- und Wasserfaches e.V., Bonn, November 2008 (2008)

Endress & Hauser Flowtec AG: Durchfluss Handbuch (2003)

Erhebungen des Schornsteinfegerhandwerks in der Bundesrepublik Deutschland für das Jahr 2013. Bundesverband des Schornsteinfegerhandwerks – Zentralinnungsverband (ZIV) – http://www.schornsteinfeger.de/bilder_ziv/files/erhebungen2013.pdf

Erste Verordnung zur Durchführung des Bundes-Immissionsschutzgesetzes (Verordnung über kleineundmittlere Feuerungsanlagen – 1. BImSchV). Bundesministerium der Justiz und für Verbraucherschutz. http://www.gesetze-im-internet.de/bundesrecht/bimschv_1_2010/gesamt.pdf. Zugegriffen: 26.01.2010

EU-Verordnung Nr. 517/2014 über fluorierte Treibhausgase

Gasnetzzugangsverordnung vom 3. September 2010 (BGBl. I S. 1261)

Giese, A.: Gasbeschaffenheitsschwankungen – Mögliche Auswirkungen auf industrielle Anwendungen. GASWÄRME Int. **02**, 70–75 (2013)

Goeldner, H.D. et al.: US Patent 4,902.138 (1988)

Grienauer, H.: FTTCA Fourier-Transform-Thermal-Conductivity Analysis. Paper presented at themeeting of the Analysis Division of ISA, Anaheim (2006)

Grienauer, H.: FTTCA – Gasanalyse mit thermisch modulierten Wärmeleitfähigkeits-Sensoren mit Fourier-Analyse des Messsignals. 16. GMA/ITG-Fachtagung Sensoren und Messsysteme (2012)

Grienauer, H.: persönliche Mitteilung (2014)

Grunewald, A.U.: Modern Thermal Conductivity Technology. Paper presented at the Emerging Technology Conference, Houston (2001a)

Grunewald, A.U.: Method and Device for Determining Gas Concentrations in a Gas Mixture. Patent WO2001027604 A1 (2001b)

GUM: Uncertainty of measurement – Part 3: Guide to the expression of uncertainty in measurement. ISO/IEC Guide 98-3, International Organization for Standardization (ISO), Geneva http://www.bipm.org/utils/common/documents/jcgm/JCGM_100_2008_E.pdf (2008)

Heise, H.M., et al.: Gas analysis by infrared spectroscopy as a tool for electrical fault diagnostics in SF6 insulated equipment. Fresenius J. Anal. Chem. **358**(7), 793–799 (1997)

Hellwig, M.: Entwicklung und Anwendung parametrisierter Standard-Lastprofile. Dissertation, TU München (2003)

Hielscher, A., Fiebig, C., Schley, P., Span, R., Schenk, J.: Brennwertverfolgung mit SmartSim – einneuer Rechenkern zur Strömungssimulation. gwf-Gas/Erdgas. **155**, 736–743 (2014)

Hollemann, A.F., Wieberg, E.: Lehrbuch der Anorganischen Chemie. S. 720, 1128. 91–100. Aufl., Walter de Gruyter (1985)

Huheey, J. E.: Inorganic Chemistry. 3rd ed., S. 258–259, Tab. 6.1 und die in Fußnoten erwähnten Referenzen. Harper International SI Edition (1983)

ISO 12213: Natural Gas – Calculation of compression factor. Part 1: Introduction and guidelines; Part 2: Calculation using a molar composition analysis; Part 3: Calculation using physical properties. International Standard ISO 12213, 2. Aufl. (2006)

ISO 12213: Natural Gas – Calculation of Compression Factor. Part 1: Introduction and Guidelines; Part 2: Calculation Using a Molar Composition Analysis; Part 3: Calculation Using Physical Properties

ISO 23874:2006(E) Natural Gas – Gas chromatographic requirements for hydrocarbon dewpoint calculation

ISO 6974-1:2012: Natural gas – Determination of composition and associated uncertainty by gas chromatography – Part 1: General guidelines and calculation of composition (2012)

ISO 6976:1995: Natural gas – Calculation of calorific values, density, relative density and Wobbe index from composition (1995)

ISO/GUM: Evaluation of measurement – Guide to the expression of uncertainty in measurement

JCGM 2008; Supplement 1 to the "Guide to the expression of uncertainty in measurement" – Propagation of distributions using a Monte Carlo method

Jordan, L., Eckman, J.R.: „Gases in Metals: II. The Determination of Oxygen and Hydrogen in Metals by Fusion in Vacuum". NIST: Scientific papers of the Bureau of Standards, USA (1925)

Kastner, J.: Neue Aufgaben für die Gasbeschaffenheitsmessung in der industriellen Gasverwendung. gwf-Gas Erdgas. **11**, 834–842 (2013)

Kastner, J., Porsch, T.: Neues Sensormessverfahren zur Gasbeschaffenheitsmessung von Biogas. gwf-Gas Erdgas. **10**, 704–711 (2011)

Katzarov, I.H., Paxton, A.T.: Vortrag F01 auf der Steely 2014 (2nd International Conference on Metals and Hydrogen, OCAS, Ghent 2014) King's College, London. Conference Homepage: http://www.steelyhydrogen2014.be/

Liebig, J.: Über einen neuen Apparat zur Analyse organischer Körper, und die Zusammensetzung einiger organischer Substanzen. Annalen der Physik. **21**, 1–47 (1831)

Messkonzept GmbH, Frankfurt, interner Bericht (2004)

Mischner, J., Fasold, H.-G., Kadner, K.: gas2energy.net mit DVD: Systemplanerische Grundlagen der Gasversorgung. Oldenbourg Industrieverlag, München (2011)

Mitter, H., Kurte, R.: Enhancement Factors for Moist Sulphur Hexafluoride (1–38). Int. J. Thermophys., eingereicht (2015)

Murnleitner, E.: State Detection and Feedback Control of the Anaerobic Waste Water Treatment. Dissertation, TU-München (2001) http://mediatum.ub.tum.de/doc/603312/603312.pdf

Neumann, H.: Biogas.Was die Gasanalyse leisten muss. Top Agrar Energiemagazin. **4**, 14–17. http://www.topagrar.com/archiv/Gasanalyse-Das-bietet-der-Markt-992930.html (2012)

Prumbaum, R., Orths, K.: Verfahren und Einrichtung zur Schnellbestimmung von Gasen, insbesondere des Sauerstoffes, und der vorliegenden Bindungsformen in Metallen. Gießerei-Forschung. **31**(2/3), 71–82 (1979)

Schenk, J., Schley, P., Hielscher, A.: Brennwertverfolgung in Verteilnetzen: Teil 2 – Auswertung Feldversuch und Implementierung. GWF/Gas, Erdgas. **10**(10), 676–683 (2011)

Schlegel, H.G.: Allgemeine Mikrobiologie. Verlag Thieme, Düsseldorf (1992)

Schley, P., Kastner, J., Wiegleb, G.: gas-lab Q1 – ein neues Messgerät zur Bestimmung von Gasbeschaffenheitskenngrößen. gwf-Gas Erdgas. **4**, 238–242 (2003)

Schley, P., Schenk, J., Hielscher, A.: Brennwertverfolung in Verteilnetzen: Teil 1 – Entwicklung und Validierung des Verfahrens. GWF/Gas, Erdgas. **152**(9), 552–556 (2011)

Sicherheits- und Regeleinrichtungen für Brenner und Brennstoffgeräte für gasförmige oder flüssige Brennstoffe – Abgasfühler. Europäische Norm DIN EN 16340:2014-10 (D), Oktober 2014

Siemens & Halske: Patent DRP 283677 (1913)

SigmaAldrich/SupelcoWebsite:CarbosieveTMS-II (. . . analyzingmixtures of permanent gases and C1-C2 Hydrocarbons . . .) http://www.sigmaaldrich.com/analytical-chromatography/analyticalproducts.html?TablePage=14540719

Technische Daten Kombi-Sonde KS1D im Gehäuse. Druckschrift DLT3033-15-aDE-002, LAMTEC Meß- und Regeltechnik für Feuerungen GmbH & Co. KG http://www.lamtec.de/produkte/feuerungsoptimierung/ks1d.html (2015)

ThyssenKrupp Steel Research, Duisburg, persönliche Mitteilung (2013)

Völklein, F.: Fachhochschule Rhein Main, persönliche Mitteilung (2007)

Wagner, G.R.: TCS208 Wärmeleitfähigkeitssensor (1988)

Wassilijewa, A.: Wärmeleitung in Gasgemischen. Physikal. Zeitschr. **5**(22), 737–742 (1904)

Witte, G., Schmitz, U., Schütt, T.: O2- und CO-Reduktion an Feuerungsanlagen. Fachbericht. GASWÄRME International. Vol. 58. Ausgabe. **1**(2), 30–3s (2009)

Zanke, U.: Zur Berechnung von Strömungswiderstandsbeiwerten. Wasser & Boden. **1** (1993)

Zanke, U.: Zum Übergang hydraulisch glatt- hydraulisch rauh. Wasser & Boden. **10** (1996)

Life Sciences 16

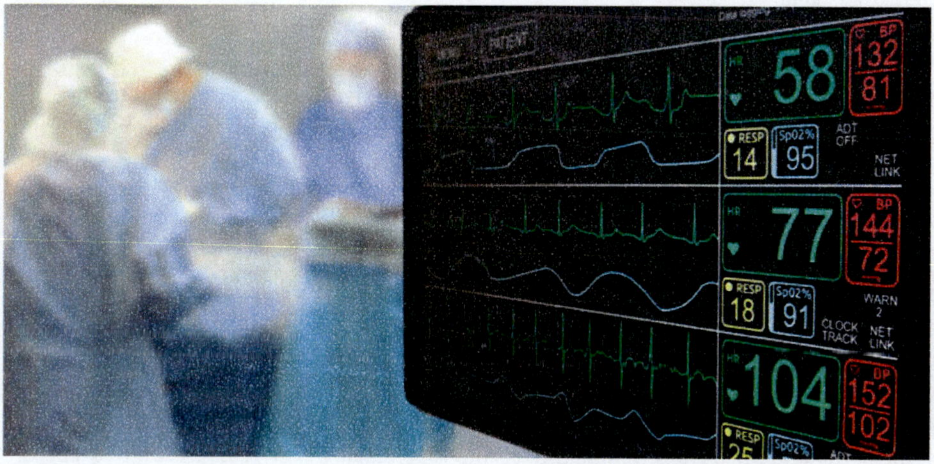

In many areas of biology and medicine, the use of gas analytical methods is of great importance. With the help of these technical possibilities, important conclusions can be drawn about metabolism, respiration and also about diseases. Gas measurement technology has therefore also opened up completely new diagnostic possibilities in this area of application. This aspect is described in this chapter on the basis of very different areas of application.

G. Wiegleb, *Gas Measurement Technology in Theory and Practice*,
https://doi.org/10.1007/978-3-658-37232-3_16

16.1 Ammonia Gas Analysis in Animal Husbandry

Michael Unruh

Solving a measurement task in gas analysis solely by using a sensor element available on the market in combination with electronics for sensor signal amplification will usually not be successful. The task of an equipment manufacturer is rather to determine and consider the general and operating conditions of the respective application as well as the customer's objectives right at the beginning of the design process. Only on this basis can a measuring system be created that meets the customer's requirements, in many cases in which the sensor technology must be supplemented by components for sample gas conditioning and transport as well as the control of all system components. However, the gas measuring system will ultimately only become a complete device that can be offered as a product on the market if suitable interfaces for the user and peripherals are added. These steps of product development are explained below using the example of "Ammonia gas analysis in animal husbandry".

Ammonia Contamination in Animal Stables

When animals are kept in stables, excrements inevitably arise. Due to the fermentation processes taking place in the excrements, gas is generated. These gases first mix with the air in the animal stable and are then discharged outside via the stable ventilation system. One of the fermentation products is ammonia.

If ammonia concentrations in the stable are too high, the well-being and health of the animals is impaired. In addition to the aspect of animal welfare in general, this also entails risks in terms of financial losses for the animal owners. Both yield and meat quality are negatively affected. First measures to reduce the ammonia problem can entry into force even before the gas is formed and then include feed additives, feeding and treatment of litter and excrements. The amount of ammonia produced in spite of these measures must be sufficiently diluted by the ventilation of the stable. However, the use and extent of ventilation is limited due to animal husbandry requirements. For example, narrow temperature ranges must be maintained in the barn and the air flow velocity must not exceed specified limits. There are always costs associated with the implementation of ammonia reduction measures. Since the worst-case conditions must currently be covered, these costs are correspondingly high. A (quasi)continuous ammonia measurement is an essential prerequisite for optimisation. It is the only way to assess the effectiveness of the measures during operation and to continue to ensure compliance with the ammonia limit values.

Since ammonia also has a damaging effect on the environment, emissions from the stable must be kept within reasonable limits. Not least the odour nuisance in the vicinity of the stables is becoming increasingly important. For this reason, the use of filter and/or gas scrubber systems for exhaust air purification is mandatory. An ammonia measurement, this time comparing the concentrations in the raw gas from the stable in front of and in the clean gas behind the exhaust air purification system, serves here to monitor the effectiveness of the purification system. The incentive for the use of gas measurement technology is the reduction of costs for operation, material usage and the removal of contaminated consumables with an optimal operation of the plants. In addition, the use of ammonia measurement may also be required by official regulations.

The following information gives an impression of the importance of the animal husbandry field of application. According to the Federal Statistical Office, there were a total of approximately 200,000 farms with livestock in Germany in 2013. Approximately a total of 171 million livestock are recorded, including 115 million chickens, 13 million cattle and 28 million pigs (DeStatis 2014).

Despite these impressive figures and even already existing requirements regarding the ammonia problem in European and national regulations (e.g., Directive 2001/81/EC, Animal Husbandry Decree), no significant use of gas measurement technology was recorded for the applications described above until the beginning of 2015. Rather, there are occasional indications that implementation must be suspended because no suitable measurement technology is available. What is the reason for this statement?

State of the Technology

In fact, (quasi)continuous ammonia measurements are already frequently carried out in animal stables, for example as a part of the certification of exhaust air purification systems. However, special laboratory analysis technology based on optical or optical-acoustic absorption methods is used. This is mounted on a test plant at great expense. The measurement is supplemented by control measurements with sample collector and subsequent chemical analysis. For the temporary test operation, a permanent supervision of the measurement by qualified personnel is necessary. The use in a wide field is therefore excluded.

In the form of gas detector tube measurement and hand-held measuring equipment equipped with an electrochemical sensor (EC) for ammonia, there are also simple methods that can be used to sporadically determine the ammonia concentration in the barn or its surroundings.

The use of the EC measuring method for continuous ammonia monitoring with a stationary mounted system is initially suitable. After all, this sensor technology has been used very successfully on a large scale and for a long time in the field of ammonia refrigeration plants. There it is used to detect leaks in the refrigeration system. However, in this application, the measuring component ammonia is normally not present at all or only in extremely low concentrations in the ambient air in which the sensor is operated (usually far below 1 ppm). In animal stables, however, a basic level of ammonia is always to be expected, ranging from a few to several tens of ppm. Stable occupancy, feeding, climatic environmental conditions and ventilation are of decisive importance for the formation of ammonia contamination and its temporal trend. However, this basic load makes a disadvantage of all electrochemical ammonia sensors significant. The electrochemical cells have a dose-dependent life time, that is the more and longer they are exposed to ammonia, the quicker they are consumed. Even at a continuous level of 20 ppm ammonia, according to theoretical calculations, the life span of even the best cells is shortened to about 750 h. A value that can be verified in practical tests. This period corresponds to only 1 month. A first phase of stable measuring performance, the duration of which is in the range of the period just mentioned, is followed by a second phase during which the sensitivity of the sensor decreases steadily and comparatively quickly. Since a device permanently mounted in the barn is difficult to reach for calibration and adjustment, the sensor should actually always be replaced before this second phase is reached. A service life of less than 1 month would therefore be realistic. Changing the sensor element at such short intervals is ruled out for cost reasons alone and cannot be reconciled with the requirements of operational procedures in animal husbandry.

The use of other ammonia-sensitive measuring methods, for example metal oxide semiconductors or catalytic combustion cannot be used due to strong cross-sensitivities, susceptibility to poisoning and/or insufficient measuring accuracy for this application. Even the new development of an opto-chemical process (Winkler et al. 2011) has not reached product maturity.

Determination of the General Conditions to Be Observed

The first important considerations are to be made with regard to the expected ammonia concentrations and the requirements for the measuring accuracy to be achieved. The limit values to be observed are in the range up to 20 ppm. In order to be able to make a statement about the level of the actual load and to observe a trend, even if short-term concentration peaks occur, the measuring range should go well beyond the limit values mentioned. Following the experience with the measurement of toxic gases in working areas, a target value of 100–200 ppm as the upper limit of the measuring range seems to be suitable. With regard to the measuring accuracy, the absolute measuring tolerance must first be taken into account. It is particularly important when monitoring limit values. In order to ensure sufficient safety here and at the same time avoid too early alarm messages, the deviations from the real concentrations in the range of the alarm thresholds should not exceed a value of 1–2 ppm. If the measurement signal is to be used in control loops, for example in the ventilation system, resolution, linearity and reproducibility of the measurement must also meet additional requirements. Even small changes of the ammonia concentration in the stable of about 0.2 ppm should be reliably traceable by means of the measurement technology.

The climatic conditions in the barn make it easier to comply with these conditions. Their range of fluctuation must in any case be severely restricted to protect the animals. However, unlike gas detection in industrial environments, outdoor use can be excluded. The unavoidable disturbing influences due to changes in temperature, humidity and pressure will be relatively small. A prerequisite for this, however, is that calibration and adjustment are carried out under conditions that take into account the average conditions during the occupation of the stable with animals.

In contrast to the use of gas detection in the context of explosion protection or health protection, the occurrence of a hazard due to sudden release of the flammable or toxic gas, for example due to a leakage, is not to be expected. Continuous monitoring for the immediate detection of a very rapidly increasing hazard is not necessary. The formation of the ammonia concentration in the stable is rather a process with relatively slow changes. A quasi-continuous measurement, which only provides an actual measured value at longer intervals, is therefore completely sufficient for such process dynamics. Measuring rates of ≥ 1 per hour are sufficient for this measuring task.

When designing the devices, a high dust load in the air must be taken into account, which can cause blockages of the gas inlets of sensors or the sample gas lines when using aspirated systems. Furthermore, aggressive gases, for example hydrogen sulphide, together with the moisture loading of the stable air can cause corrosion on the housings and components of the measuring equipment.

All solutions that would meet the requirements mentioned so far should always be evaluated under the reservation that operation is possible with a reasonable expenditure of time and money. Service lives of at least 12 months for sensors and consumable material, such as filters, should therefore be achieved. As access to the stables is already limited for hygienic reasons, inspections and maintenance should only be necessary at intervals of at least 3 months or during periods of stabling.

Solution Approach

The direct use of gas detection technology available on the market is ruled out for the reasons mentioned above. The appearance of new sensor technologies does not seem realistic, at least not in the medium term. A promising solution can therefore only be based on the appropriate adaptation of an already available technology. After analysis of the basic conditions, one quickly comes to the conclusion that all basic conditions can be fulfilled with the exception of the required service life when using an electrochemical cell. As in ammonia sensors the lifetime of the sensor is coupled to the dose to which it is exposed, this problem can be solved. An operating mode for the sensor element must be found in which the sensor only comes into contact with ammonia-loaded sample gas for very limited periods of time. For example, if the sensor is exposed to stable air for only 5 min every hour, the dose exposure will be reduced in a ratio of 1:12 proportionally and the lifetime will increase by a factor of 12. If the above mentioned service life of 1 month is taken as a basis for continuous measuring operation, the change to intermittent operation would already achieve the target sensor life time of at least 1 year. However, this solution can only be pursued because the dynamics of the "animal stable" system to be monitored is relatively slow. On the other hand, such a mode of operation would be completely unsuitable in the safety monitoring of a refrigeration plant, where ammonia leakage can occur at any time and must be detected as quickly as possible.

The transition to intermittent operation means that the sensor is not used in diffusion operation directly in the stable air. Instead, an aspirated measuring system must be used with which the sensor can be alternately supplied with the sample gas or ammonia-free atmosphere. This can be realized quite easily by a sampling system with a switchover option for the gas supply.

The provision of an ammonia-free atmosphere for the purging cycle proves to be a greater challenge. At first glance, it seems appropriate to extract this from the uncontaminated air outside the barn. Two variants are still conceivable. In the first case, the measuring device with sensor is located outside the stable and the measuring gas is sucked in from there, in the second case the measuring device with sensor remains inside the stable and the air is supplied from outside. With both variants, however, no satisfactory results could be achieved in practical tests. The main problem was the formation of condensate in the sample line hoses. Due to the very high water solubility of ammonia, even small water occupancy of the gas paths caused significant effects on the ammonia concentration still reaching the sensor. The absorption effects strongly falsify the measuring results. In any case, a compact design of the system with suction points for stable air and ammonia-free atmosphere mounted locally at a greater distance prevents a compact construction of the measuring equipment. The effort for the installation of external components and their maintenance is not insignificant and is in contrast to the goal of making the equipment easy to use in agricultural environments.

Only with the integration of the next development step, a long-term stable measuring system could be realized. This step represents the transition to the generation of ammonia-free atmosphere from the stable air itself. For this purpose, a suitable filter element is inserted into the gas path for the purge air, which completely absorbs ammonia. Its effectiveness has been proven by tests in which test gases containing 15,000 ppm ammonia were passed through the filter. On a downstream ammonia sensor, no reaction whatsoever could be detected in the ppm range. The filter construction is very simple, the filter materials used are very inexpensive, non-toxic and can be disposed of without any problems. Despite the relatively small space requirement, the desired service life of more than 12 months is achieved.

The prototype shown in Fig. 16.1a was built in small numbers and was used for testing under real operating conditions. The steel wall-mounted housing contains the actual ammonia transmitter, the sample gas pump, the filter element and the switchover for sample gas and purge air as components. The electronic board takes over the cyclic control of the gas path switching and the pump. In addition, the measuring signal of the transmitter is "frozen" outside the actual measuring phases. This provides a stable 4–20 mA standard signal that can be further processed by a gas detection control unit, a PLC or a data logger. Figure 16.1b shows the associated gas path of the system.

Over a period of about 3 years, the prototypes were continuously used in different stables. After some optimization steps, which still proved necessary in the first year with regard to the type of sensor used and additional protective measures against ammonia contamination, the previously formulated goals with regard to measurement quality and

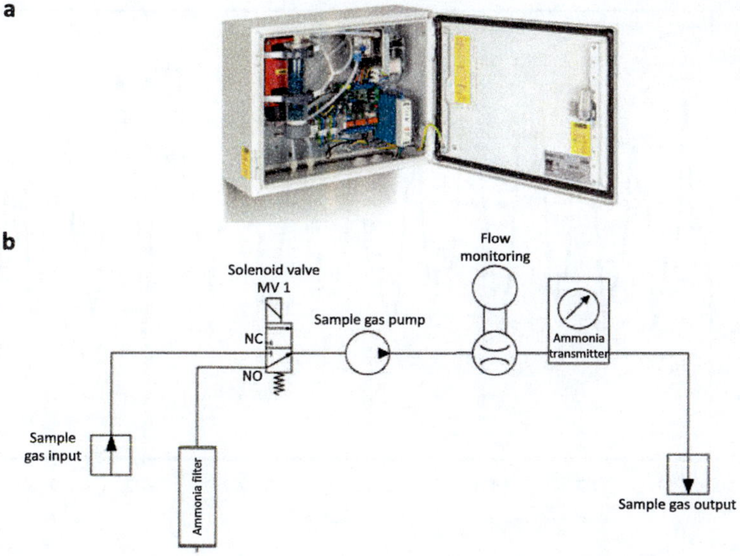

Fig. 16.1 (**a**) Test device for ammonia measurement in an animal stable. (**b**) Scheme of the gas path in the test device

long-term stability were met. Patent applications for the process were subsequently filed with the German and European Patent Office.

Figure 16.2 shows the results of a trial use at a chicken house. The aim was to investigate the effectiveness of the exhaust air treatment. For this purpose, the ammonia concentration was measured before and after the scrubber. Both units were mounted in the raw gas area of the plant and were exposed to very high levels of dust, large air flows, humidity and acidic spray water. The first device determined the gas concentration in the raw gas, while the second device sucked its sample gas on the clean gas side from the area of the scrubber exhaust air. The figure shows the course for the raw gas measurement. It can be seen very clearly that the ammonia load, which is relatively stable on average over the period shown, is clearly superimposed on a daily cycle. This is generated by the influences of the day/night temperature curve, the existing time control of the ventilation and cyclical processes in animal husbandry, for example feeding. The measurement on the clean gas side correlated with this process. However, the ammonia concentration is already greatly reduced by the washing process to values at 2 ppm NH_3. Since the data recording in this area contains significant disturbing influences due to noise, we will not present it here.

Figure 16.3 gives an impression of the operating conditions in an animal stable, where the device measures directly in the air of the stable. In addition to the ammonia concentration, the carbon dioxide concentration, the temperature and the relative humidity are also recorded in this test. In this trial application, two measuring devices were installed in parallel in two stables under almost identical operating conditions. In one barn only one

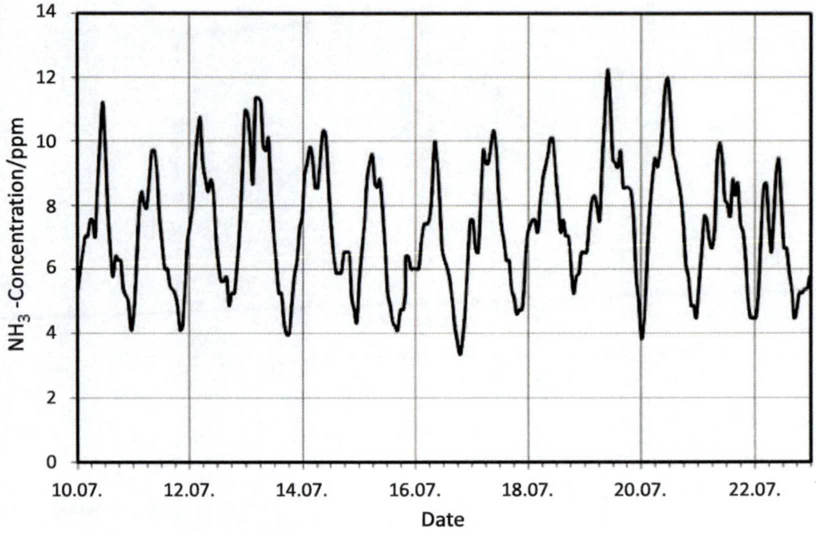

Fig. 16.2 NH_3-concentration curve in the raw gas exhaust air flow of a chicken house

Fig. 16.3 Use of the test device in a pigsty

feed additive was used. With the ammonia measurement its influence on the ammonia formation should be investigated. Figure 16.4 shows the ammonia and carbon dioxide values determined over a period of almost 3 months for the barn with feed additive. Temperature and humidity were almost constant during the measuring period. The second

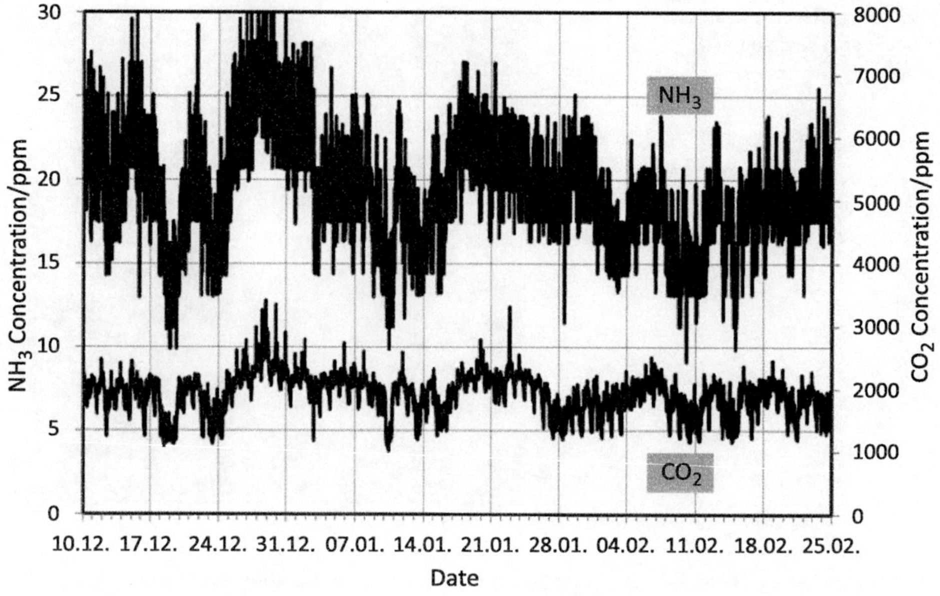

Fig. 16.4 NH_3- and CO_2-concentration curve in a pigsty

measuring device in the barn without feed additive delivered almost the same measured values and curves. Only the ammonia concentration was about 50% higher.

As already with the measurements in the henhouse described above, the superimposition of a daily course can be seen again. Also the correlation between ammonia and carbon dioxide, which could be observed in a wide range, was to be expected in case of increasing and decreasing concentrations. This correlation should exist as long as the common cause is the change in ventilation conditions.

Thus the continuous load of the stable air was over 20 or 30 ppm, which would have led to a clearly recognizable loss of sensitivity after only 1 month of normal sensor operation. As such a drop in the measurement signal is not even remotely detectable in either of the two measuring devices, the effectiveness of the developed measuring method is clearly evident.

Further Development to a Product

After the successful trial use, the transfer of the device technology would have been possible directly into a marketable product. However, this would only have provided an 4 . . . 20 mA interface for the measured value of the ammonia concentration. Intensive discussions with the operators during the test runs soon showed that the potential customer base is not to be seen in the operators of the barn. Rather, it must be possible to integrate the measurement into the systems of the stable construction companies or manufacturers of the

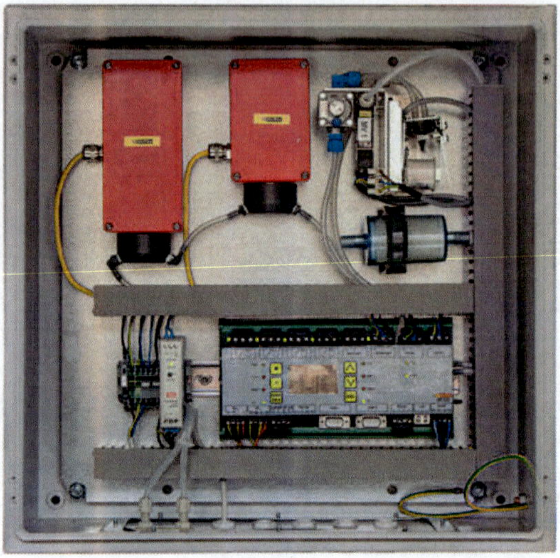

Fig. 16.5 NH_3-stable unit concept after revision and expansion (pre-series)

air purification systems. For this purpose, it was essential to be able to expand the existing system with individual options according to customer requirements. Some examples of customer requirements are listed below:

- Acquisition of additional measured values: carbon dioxide concentration, temperature, pressure, humidity.
- Corrosion resistant version.
- Measuring point switching.
- User interface with measured value display and integrated parameterization option for system functions.
- Adjustable alarm thresholds with associated relay outputs.
- Data interfaces: analogue current/voltage outputs, digital (bus) transmission.
- Spatially separated remote measured value display.
- Data logger.

Based on the experience with modular concepts for aspirated systems including sample gas conditioning, for example in the field of biogas or landfill applications, these requirements could be integrated with relatively minor modifications. Besides an adaptation of the housing design, the addition of a powerful control and evaluation unit proved to be the most important measure to achieve the desired system flexibility. Only in this way was it possible to achieve simple adaptation to the respective customer-specific equipment requirements while at the same time keeping within a market-oriented cost framework. Figure 16.5 shows the final NH_3-stable device concept after revision.

16.2 Gas Measurement Technology in Emergency Ventilation

Frank Herrmann

In order to better understand the necessity of gas measurement technology in emergency ventilation, it is important to go into the basics of natural respiration and finally into ventilation technology. In natural respiration, also known as spontaneous breathing, ambient air is drawn in by the formation of negative pressure in the lungs. The diaphragm, a large muscle that separates the heart-lung area from the abdominal area, is mainly responsible for this. The diaphragm moves almost like a piston from top to bottom during inhalation (inspiration) and moves upwards again during exhalation (expiration). Due to their spring action, the ribs form the restoring forces for exhalation, which is passive during resting breathing. The aim of the breathing effort is always the supply of fresh breathing air during inspiration and the expulsion of used air during expiration. In simplified terms, oxygen is supplied to the lungs and carbon dioxide is removed from the lungs.

Requirements on the Sensor Technology

Breathing is therefore, from a purely technical point of view, the cyclical movement of gas to or from the lungs with varying gas pressures and gas flows, also simply called flow. The example of natural respiration is shown in Fig. 16.6. The upper curve shows the pressure curve in the lungs. In the inspiration phase, the pressure in the lungs is lowered, whereas in expiration it rises slightly above ambient level. Associated with the pressure reduction is an

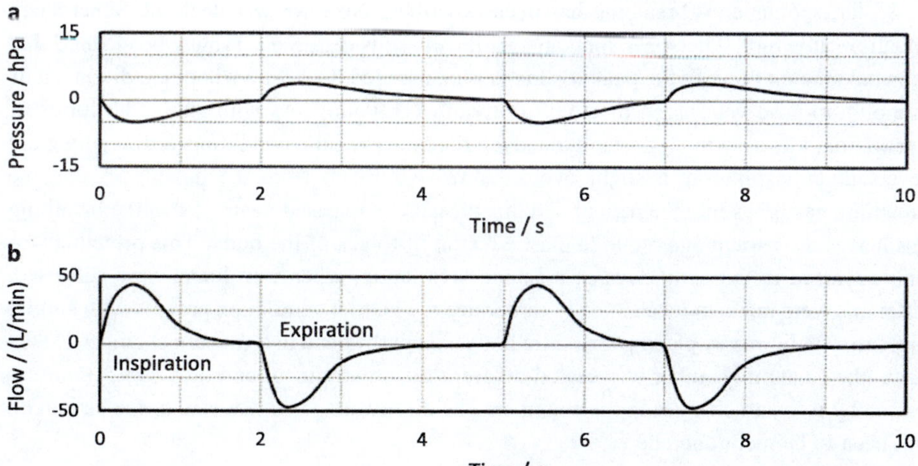

Fig. 16.6 Qualitative pressure (**a**) and flow (**b**) curve of an adult during spontaneous breathing in the lungs

Table 16.1 Orders of magnitude for respiratory rate and respiratory volume as well as mean flow and peak flow of a healthy person at rest during spontaneous breathing (Jacobsohn 2015)

	Respiratory rate (RR)	Respiratory volume (TV)	Medium flow (Q_{medium})	Peak flow (Q_{max})
Adult	10–15/min	350–800 mL	25 L/min	Until 150 L/min
Child	15–20/min	150–350 mL	15 L/min	30 L/min
Infant	20–30/min	50–150 mL	7 L/min	15 L/min
Baby	30–40/min	20–50 mL	4 L/min	9 L/min

inflow of air into the lungs (positive flow) and with the pressure increase an outflow from the lungs (negative flow). This is shown in the lower curve.

The qualitative pressure and flow curve of an adult during spontaneous breathing in Fig. 16.6 serves as an example. Even if persons with respiratory diseases are not considered here, the variance in flow and respiratory rate (RR) (respiratory cycles per minute) from infant to adult is very large. The pressures generated correspond to the vehemence of the breathing effort: low at rest and correspondingly high during exertion. To estimate the tidal volume (TV), the volume moved during a breath, science has found a factor that describes the amount of air required per kilogram of ideal body weight, independent of age. A value between 6 and 8 mL/kg body weight can be assumed (Oczenski 2012).[1] Looking at body weights from approximately 3 to 100 kg, the result is TVs as shown in Table 16.1. Since the shape of the flow curve differs only slightly in healthy people, differences in the peak flows are correspondingly apparent in addition to differences in the TVs. As will be shown later, this variance is a challenge for flow measurement technology.

So far, spontaneous breathing has been described. Now we will deal with emergency situations in which a person's spontaneous breathing is disturbed. It must be ensured that the vital amount of air in the patient's lungs is exchanged. One way without additional aids is mouth-to-nose ventilation or mouth-to-mouth ventilation. As with almost all forms of ventilation, it is easy to understand that even here the pressure conditions for breathing are reversed. In inspiration, a slight overpressure is built up from the outside so that the breathing gas flows into the patient and this pressure is released again so that the breathing gas leaves the patient again due to elastic restoring forces of the body. This principle was only reversed in the early days of machine ventilation, when iron lungs were still used. Attempts were made to mechanically reproduce the natural ventilation process with similar pressure conditions by placing the entire body except the head in a vacuum chamber. At the neck, the vacuum chamber was sealed off from the environment and thus ventilation was forced by pressure changes in the chamber. The disadvantages in the care of the patient do not need to be mentioned here.

[1]For general and detailed questions about breathing and ventilation (Oczenski 2012) is highly recommended.

For further solutions to the ventilation problem, they resorted to the techniques that already existed. With the knowledge gained from Table 16.1, an attempt was made to administer the necessary TV to the patient. A piston-like system, which is still common in anaesthesia technology, ensures that the patient receives the necessary volume of respiratory gas. Ventilation was carried out in this way for decades and this is still quite common today. This is called *volume-controlled ventilation*, as the patient is given a fixed volume. The decisive advantage is that the patient receives a fixed volume of breathing gas. This form of ventilation becomes problematic if there are blockages or there is an oversized amount of gas in the respiratory tract. Since the volume is continuously pushed further into the patient, if there is insufficient control, the pressure in the lungs can rise to such an extent that lung damage can occur. Even the simplest machine ventilation with a resuscitator bag is therefore usually equipped with a pressure relief valve on the bag to prevent lung damage. Modern ventilators monitor this with appropriate sensor technology, with which the ventilation controller can limit the maximum pressure. Another problem is leakage between the ventilator and the patient. Especially, when ventilating through a ventilation mask, leakages in the area of the mask are normal. The volume supplied to the patient therefore only partially reaches the patient, so that the advantage of volume control is lost.

Another form of *volume-controlled ventilation* can also be achieved with a compressed gas-powered system. The basic element is a proportional valve, which is supplied on the input side by a ventilation gas at high pressure. Typically, here is an overpressure of more than 3 bar compared to the environment. A critical flow (flow at the speed of sound) occurs within the valve, which determines the mass flow through the valve. The opening width of the valve is thus proportional to the initial mass flow of the valve, so that a constant flow is generated with a constant valve position. With this *flow-controlled ventilation,* the applied volume can thus be dosed over the opening time of the valve.

Since the seventies of the last century, the electrically driven radial blower has also been used as an alternative source of ventilation. With the help of the blower, ventilation independent of compressed gas is possible. Inside the radial blower, air is sucked in near the axis and expelled outwards by means of radially arranged blades. At the end of the blades, the air has approximately approached the circulation speed of the fan wheel, so that the kinetic energy of the air is noticeably increased. When this accelerated air is decelerated again, the static pressure increases, which can then be used for ventilation. The achievable ventilation pressure is therefore dependent on the peripheral speed of the radial blades. This results in a *pressure-controlled form of ventilation,* the pressure of which can be adjusted via the rotation speed. The decisive advantage of blowers compared to piston-like systems is that blowers are only slightly larger or even smaller than a fist, which means that any large TV can be administered. Now the ventilation pressure can be adjusted in a targeted manner, whereby the administered volume remains unknown without additional measuring technology.

Now that machine ventilation techniques have been described, the effects of ventilation modes on the ventilation process will be presented below. Figures 16.7 and 16.8 show the typical pressure and flow curve of these ventilation modes. Volume-controlled ventilation

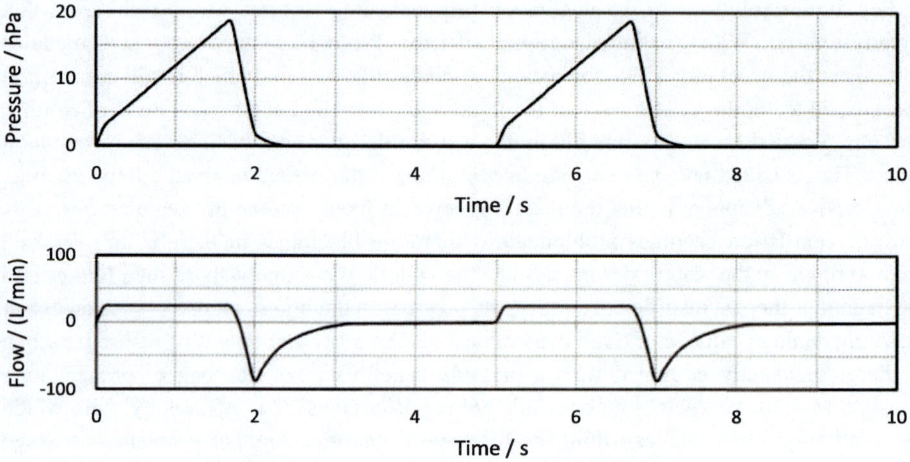

Fig. 16.7 Typical volume-controlled ventilation in an adult (Jacobsohn 2015)

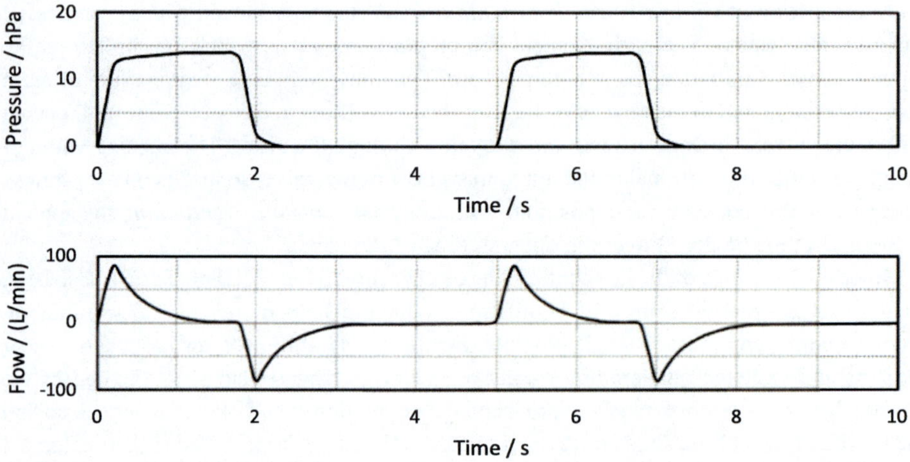

Fig. 16.8 Typical pressure-controlled ventilation in an adult (Jacobsohn 2015)

in Fig. 16.7 represents ventilation with constant flow in inspiration. If the flow is known, the administered volume can be calculated over the inspiratory time, so that ventilation can be performed in volume-controlled mode. There are other flow profiles for volume-controlled ventilation in ventilation technology, but they will not be discussed in detail here.

Figure 16.8 shows pressure-controlled ventilation. Due to the rapid increase in pressure, the ventilation flow rises almost immediately to its maximum value and then drops asymptotically to zero. Common to both figures (Figs. 16.6 and 16.7) is that the pressure is always kept constant during the expiratory phase.

The illustrations show steep flow or pressure ramps. These fast ramps are a challenge for the gas sources, the actuators, as they usually require high accelerations of the moving parts. However, these fast ramps are needed to relieve spontaneously breathing patients during their breathing efforts. Spontaneously breathing patients also place high demands on the gas sensor technology during phase changes, because here it is the patient who triggers the change from inspiration to expiration and vice versa and not the device which forces the change in a time-controlled manner. The gas sensors must detect the patient's breathing efforts early so that the ventilator does not work against the patient thus causing the benefit of the respiratory support provided by the ventilator to be lost.

In the following, the requirements for gas sensor technology will be discussed in more detail. The basic task of sensors in ventilation is to detect safety-critical situations and to record the ventilation status so that the patient can be given the best possible respiratory assistance. The detection of the breathing phase (inspiration or expiration) is of great importance for good respiratory support. If we look at the phase transition from expiration to inspiration, the ventilator initially delivers a constant expiratory pressure so that the patient's breathing efforts become noticeable by a change in flow. At the end of expiration, the patient's flow has dropped to about zero. The patient's effort to breathe in is reflected in an almost immediate change in flow towards the patient. For good respiratory support, the flow sensors must detect the attempt to inhale within a maximum of 50 ms and control the actuators very quickly to provide the patient with the necessary breathing gas. Flow changes from 1 l/min and below must be detected for good triggering. Particularly in emergency situations where an ambulance or helicopter is moving, air movements in the hose system alone can create such flow changes. Considering the peak flow of an adult of 150 L/min (see Table 16.1), which the flow measurement system must also be able to detect exactly, these small flow changes are a great challenge.

At the transition from inspiration to expiration, the ventilation mode determines the trigger condition. In pressure-controlled ventilation, the flow change must again serve as the criterion, whereas in flow-controlled ventilation the pressure change is used as the trigger value. In general, it can be seen that the change from inspiration to expiration is easier to detect than from expiration to inspiration.

For ventilation, gas pressure and gas flow have so far been considered. However, the gas composition also plays an important role in ventilation. A healthy person normally breathes in ambient air, which consists of approximately 21% volume fraction of oxygen. The organism absorbs part of the oxygen in the lungs and releases carbon dioxide in return. Exhaled air typically contains about 17% volume fraction of oxygen and 4% volume fraction of carbon dioxide. Some disturbances in the organism show up early as a deviation in the CO_2 content at the end of expiration, which is why the CO_2 content plays an important role in diagnostics. Even if the CO_2 content is not directly relevant for ventilation control, emergency ventilators often contain CO_2 sensors for diagnostic purposes.

In case of a disturbance in the respiratory system and especially during cardiopulmonary resuscitation, the oxygen content in the inspiration is increased for the patient. Depending on the condition, oxygen concentrations of up to 100 vol.% are administered. This is done by mixing ambient air and oxygen. An oxygen sensor monitors the concentration and influences the oxygen mixer. During the technical implementation of the oxygen mixer, compromises often have to be made, which are discussed in the section on oxygen measurement technology.

As in many other technical areas, the general conditions in medical technology are defined by mostly international standards. For example, gas measurement technology is currently described in the standard DIN EN ISO 80601-2-55. This includes, among other things, oxygen and carbon dioxide measurement in emergency ventilation. Requirements for anaesthetic gases are also described, but these are not relevant for emergency ventilation. Many other standards are also applied, for example concerning safety, such as the standards on shock and vibration resistance or aviation standards if the ventilator is to be used in aircraft. All in all, there are demanding boundary conditions for sensor technology in emergency ventilation.

Pressure Measurement Technology

Since a patient is given compressed gas during ventilation, it is obvious that the gas pressure must be monitored. Usually in this technology, absolute or differential pressures in the range from a few hundred hectopascals up to several hundred bars are measured. In ventilation, however, 60 hPa already represents a pressure limit that should not be exceeded, so that very sensitive sensors are required for ventilation.

Pressure sensors work indirectly. A membrane deforms as a result of a pressure difference between the front and back of the membrane. This deformation is measured and the resulting pressure difference is calculated. The spring constant of the membrane and the sensitivity of the sensor elements determine the pressure measuring range. Modern pressure sensors today are constructed as silicon MEMS (*Micro-Electro-Mechanical-Systems*) elements, as shown in Fig. 16.9. The membrane and the sensor elements are combined on a tiny piece of silicon, only a few square millimetres in size. The deformation of the membrane is correspondingly small, so that the sensitivity of the measuring resistors must be very high. The figure shows that the silicon wafer (the *die*) is back-etched in the area of the membrane to achieve the necessary flexibility of the membrane. Shown here is a differential pressure sensor where the pressures are applied to the top and bottom side. For absolute pressure measurement, the lower hole is closed and the space under the membrane is evacuated. Thus the pressure is measured in relation to the vacuum, that is the absolute pressure. In the example shown, the mechanical stress in the diaphragm is measured via piezoresistive resistors. This is the most widely used method because piezoresistive resistors can be integrated very easily into the semiconductor manufacturing process. The sensors are inexpensive and can be purchased in a pressure range from a few hectopascals to about ten bar.

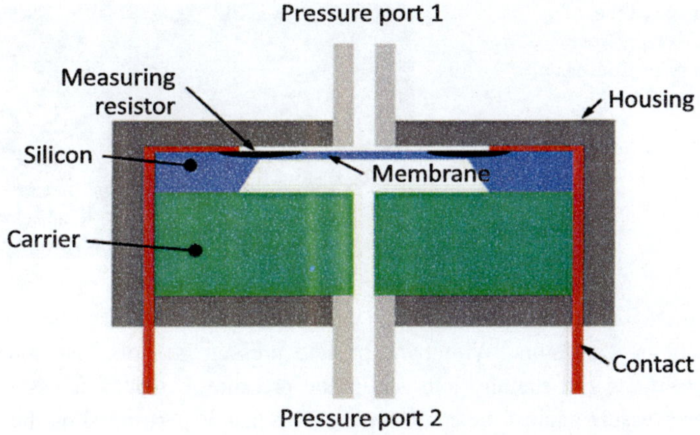

Fig. 16.9 Principle of the piezoresistive pressure sensor

Important for this sensor type is the high measuring speed, which is achieved by the small size and thus the low mass inertia. Above, in connection with the phase change of inspiration and expiration, the time for trigger recognition was mentioned by 50 ms. Piezoresistive pressure sensors achieve response times of less than 1 ms and are therefore very well suited for ventilation technology. However, the disadvantage of zero point drift, to which piezoelectric systems are strongly prone, should not be concealed here. This is usually only really noticeable in the lower measurement range. If these sensors are used for flow measurement in a ventilator, as described later, the drift must often be measured and compensated. For this purpose both inputs of a differential pressure sensor are short-circuited and an offset measurement is carried out. This offset is then used to compensate the subsequent measurements.

A piezoresistive pressure sensor only functions reliably over a larger temperature range if extensive compensation measures have been taken within the sensor. For this reason, the final sensor consists not only of the membrane with sensors but also of an *ASIC (Application Specific Integrated Circuit)*, an additional silicon chip which, in addition to linearization of the characteristic curve, also performs temperature compensation. The membrane and ASIC are then installed in a sensor housing. Such sensors, if well compensated, cost well over 15 € in thousands of units. Considering that in a ventilator five pressure sensors and more are sometimes used, these *inexpensive* sensors represent a significant cost factor.

Falling prices for electronic products are mostly caused by the demand in the consumer sector, where large quantities are required. Absolute pressure sensors are used here as altimeters, but are also very important in ventilation technology. Many sensor values have to be calibrated with the absolute pressure in order to obtain comparable measured values. Especially for gas concentrations but also for *flow* measurement, absolute pressures are required. Nowadays, absolute pressure sensors with excellent measuring accuracy are available for less than 2 €. These sensors are optimized for extreme resolution, so that

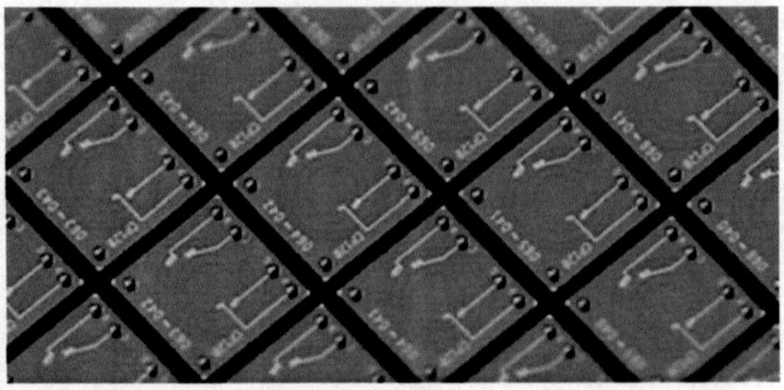

Fig. 16.10 Image of the CPS120 capacitive pressure sensor. (Courtesy of Consensic, Inc., Burlingame, CA)

the sensor can even detect step-by-step stair climbing. Such sensors can be used well and cost effectively in ventilators. With two absolute pressure sensors, one measuring the ambient pressure and one the absolute ventilation pressure, it would be easy to measure the ventilation pressure against the environment. This task is performed by the differential pressure sensors described above, which are considerably more expensive. However, this would require a higher measuring rate with speeds under 5 ms, which current sensors with measuring rates of a few Hertz cannot yet achieve. However, there seems to be a trend that sensor manufacturers are designing their absolute pressure sensors to be adjustable, so that the measuring rate can be further accelerated at the expense of resolution.

In the area of larger pressure transmitters, capacitive sensors are used to keep the long-term drift of the pressure sensor low. In MEMS technology, chip manufacturers have extensive experience with capacitive sensors such as accelerometers and gyrometers, as well as with capacitive microphones, which are ultimately highly sensitive dynamic pressure transducers. Capacitive MEMS sensors are very inexpensive (often available at 1 €) and yet fast and accurate. In the consumer sector, the first capacitive MEMS pressure sensors are now available that seem to combine all the desired advantages (Fig. 16.10).

In addition to these silicon MEMS pressure sensors, a completely different pressure measurement technology has become established over the last 15 years, which requires a low gas flow over the sensor. In contrast to the method of determining the effect of pressure as the deflection of a membrane, here the effect of pressure as the driving force for a gas flow is measured. The gas flow is again measured with a MEMS element consisting of a heater, two temperature sensors and an electronic evaluation unit, as shown in Fig. 16.11. A heat cloud is formed around the heater, the temperature of which is measured by the temperature sensors. Without gas flow, the temperature at both sensors is identical. Even a small flow leads to a noticeable temperature difference at both sensor elements. The special feature of this principle is the very high sensitivity and stability of the sensor at small gas flows, which decreases significantly for increasing flows. In addition, these sensors show practically no drift behaviour and thus an enormous long-term stability with regard to zero-point drift and sensor drift. Strictly speaking, this is therefore a flow sensor that is used for pressure measurement with the aid of series-connected pneumatic resistors. Since all production steps, except for the etching of the membrane, can be carried out in a standard

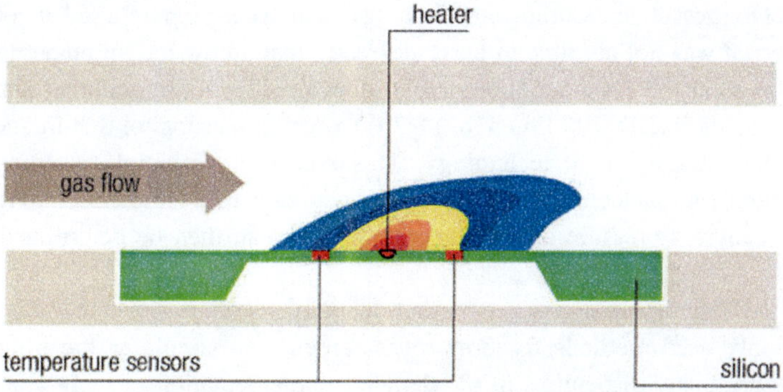

Fig. 16.11 CMOSens® principle. (Courtesy of Sensirion AG, CH-Stäfa). The depicted heat cloud spreads from the heater to the temperature sensors. The temperature difference is a measure for the gas flow

CMOS process (process for the production of computer chips), the sensor can be manufactured at a very reasonable price.

One of the first manufacturers to use this technology was the Swiss company Sensirion. Today other companies still use this heat balance method. The fact that a flow from the measuring gas is necessary here is usually less of a problem, as it is a small flow of about 0.1 L/min or much less. To limit these flows, however, very small, usually capillary resistors containing tiny tubes are also required. Even with the best possible filtering of the gas, it cannot be ruled out that these tubes could become clogged with dirt, which would lead to an unnoticed failure of the sensor. A similar problem would arise if water vapour were to condense. This could be the case, for example if a ventilator is taken from a cold vehicle and then used in a warm, humid environment. Since this warm, moist air is drawn in for ventilation, condensation is almost impossible to avoid. Another case is the use of the sensor close to the patient, where condensation occurs during exhalation. Here such differential pressure sensors cannot be used.

Flow Measurement Technology

The volume delivered to the patient is one of the most important therapeutic parameters in ventilation technology. In most cases, the ventilation volume is calculated by integrating the ventilation flow over the inspiratory time. Since the inspiration time, for example for infants, is completed after just under a second, the flow measurement must be fast so that the error of the calculated ventilation volume does not become too large. Considering that in infants the average flows are sometimes below 4 L/min, but that peak flows of up to 150 L/min are possible in adults, one can surmise the challenge for flow measurement technology.

In contrast to pressure measurement technology, which can be purchased at good quality for ventilation, it was not possible to purchase ready-made flow measurement technology for ventilation about 10 years ago. For circular duct sizes from 50 mm inner diameter, for example the standard (DIN EN ISO 5167) helps to design resistors of this frequently used method of flow measurement technology. This measuring principle is comparable to current measurement in electrical engineering: a pneumatic resistor is built into the volume flow, which causes a pressure drop. The volume flow can then be determined from the pressure drop.

The method of differential pressure measurement over a resistance element is one of the most commonly used methods for flow measurement. As simple as the principle may sound, there are many difficulties in the detailed implementation of it. In contrast to an ohmic resistor in electrical engineering, which gives a linear relationship between the voltage drop across the resistor and the flowing current, the relationship between pressure drop and flow rate is non-linear. In the case of turbulent flow, which can be realised for example by means of a sharp-edged orifice in the flow channel, there is a quadratic relationship between the volume flow $\dot{V}$ and the pressure drop Δp, whereby the constant c reflects the effect of the orifice.

$$\Delta p = c \cdot \dot{V}^2 \tag{16.1}$$

The decisive disadvantage of the sharp-edged orifice is that at very small flows the change in pressure drop is close to zero. At these flows the useful signal is almost negligible, so that small flow changes can hardly be detected. However, the detection of small flows is important for the inspiration trigger, which triggers the change from expiration to inspiration. Sensors based on the heat balance method described above can almost compensate for the disadvantage of the quadratic characteristic curve. Therefore, various sensor manufacturers offer differential pressure sensors with a root characteristic for this problem. For a few years now, complete flow sensors with a large cross section have also been offered for ventilation technology, which usually operate according to the heat balance method and can measure small as well as large ventilation flows properly. So much for the theory.

In practice, flow profiles often form within a ventilator that vary with the flow due to the limited space available, the small cross sections and the necessary changes of direction. This means that a clean flow to the flow sensors is no longer possible, so that the sensors must be calibrated and compensated for each application by the ventilator manufacturer. On closer inspection, the dependence of the sensors on ambient pressure, temperature and gas composition can be seen. The ambient pressure has usually only a linear dependence and can be compensated by measuring the absolute pressure in the ventilation flow. The temperature compensation within the sensors works very well. It measures the temperature of the gas in the pressure sensor. In a setup consisting of a pneumatic resistor and a pressure sensor as described above, the gas temperature in the main flow is therefore not measured by the pressure sensor. The temperature sensor within the pressure sensor is therefore only used to compensate for the temperature-dependent errors of the pressure sensors themselves. Since the flow or volume display of the ventilator must always refer to a reference temperature, for example 37 °C (conditions on the patient), the measured ventilation flow and volume must be corrected with the current gas temperature, which must be measured separately in the ventilation gas. Sensirion, as one of the pioneers of flow sensor technology in ventilators, has recognized this problem and has been offering flow sensors with additional temperature measurement in the main flow channel since 2015.

However, the heat balance principle is also suitable for direct flow measurement in the main flow, since the sensor element itself already performs a mass flow measurement. The sensor element is very small due to the low-cost MEMS production and can therefore only measure the velocity on one flow filament. To determine the flow, the velocity profile in the flow channel must be known or the profile for the different flows and ventilation pressures must be reproducible. However, not only the main flow measurement with small sensor elements has this problem of punctual measurement, but also the pressure measurement via a resistance element. For pressure measurement, pressure or bypass flow taps must be attached to the flow channel, which must work reproducibly in all operating positions.

The flow sources are responsible for possible disturbances of the flow within a ventilator. In proportional valves when dosing gas from a pressure supply, sound velocity occurs in the narrowest cross-section. This is associated with turbulences, which usually severely disturbs the flow profile, since long calming distances are hardly possible due to the spatial restrictions. Respiratory blowers cause similar problems. The pressure is dependent on the fan speed and almost independent of the ventilation flow. Depending on the speed, variable flow profiles are created with the same ventilation flows. A reliable flow measurement is therefore almost impossible without suitable laminarization measures.

The range of other flow measurement methods is wide and all methods struggle with similar problems. The hot-wire methods with all their variations (one or more wires, modulated heating currents, transit time measurements, etc.) all measure at discrete locations and are usually expensive to manufacture because of the individual production. The ultrasonic methods are ideal scanners, that means they always measure only a current and not an average speed. To avoid undersampling for ventilation, sampling rates of several kilohertz would be necessary. Such fast and compact ultrasonic sensors are not yet available.

From these considerations, it can be seen that flow measurement and thus volume determination is one of the greatest difficulties in gas measurement technology for ventilation systems. The possible errors caused by gas mixtures have not been mentioned so far. For example, the pressure drop at a pneumatic resistance is approximately 10% greater for oxygen than for air. When measuring the flow of air–oxygen mixtures, the gas composition must therefore also be known. The air humidity must also be taken into account. Relevant standards for ventilation therefore only require an accuracy of ±20% of the measured value for volumes above 100 mL for a volume measurement. If we take into account sensors close to the patient, which must be regularly sterilized, these requirements represent an additional hurdle.

Oxygen Measurement Technology

Oxygen is administered as a medicine in the respiratory system. Depending on the state of health, the oxygen concentration is increased from volume fraction of 21% in ambient air to up to volume fraction of 100% oxygen. To ensure that the dosage of oxygen can be carried out reliably, the above-mentioned standard DIN EN ISO 80601-2-55 defines the accuracy limits for oxygen measurement, which are shown in extracts in Table 16.2. Some standard requirements have been extended by further standards and through practical knowledge gained by Weinmann Emergency for the emergency sector. The requirements of *Weinmann Emergency* are thus a superset of the standard requirements. If you look for example at the requirement for the T_{90} time in the table, you will see that the standard does not describe a minimum speed. The only requirement is that the rise time be specified in the instructions for use. For this purpose, the standard describes the exact test procedure for determining the rise time. This enables the device manufacturer to freely select the sensor principle.

In ventilation technology, oxygen sensors are primarily used to monitor the oxygen concentration in the ventilator. The gas mixture is measured only during inspiration and not during expiration. As described below, expiration is monitored with CO_2 sensors. To create a respiratory gas mixture of air and oxygen, typically two methods described below are used.

Method 1 is shown in Fig. 16.12. A gas mixing and storage chamber before the actual ventilation control is fed by an air and an oxygen source. In the chamber is the oxygen sensor, which controls the gas sources. If there is a sufficient amount of gas in the chamber, the gas sources can work almost statically and the measuring speed of the oxygen sensor may be slow.

Method 2 in Fig. 16.13 shows two gas sources whose flow is dynamically controlled by a flow controller in each gas path so that the required gas mixture is correctly set for all possible ventilation flows and for each gas concentration. Since the existing oxygen sensor technology is not able to realize measuring frequencies of at least 50 Hz to control the gas composition in real time during a breath, the oxygen control is implemented with a superimposed controller. The gas mixture is passed over the oxygen sensor, which detects an average oxygen concentration of the respiratory flow. From this measurement, a fine tuning for the two flow controllers can then be made.

As always, each method has its advantages and disadvantages. Method 1 can be equipped with comparatively slow flow regulators for the two gas paths. Only the flow controller at the chamber outlet must be able to control the ventilation flow with high temporal resolution. This also requires a fast flow measurement at the outlet. The oxygen concentration in the chamber can be easily adjusted by using the oxygen measurement cell and adjusting the flow regulators for air and oxygen. The disadvantage is the required size of the mixing chamber. Depending on the patient, TVs of up to 2 L are possible, which

Table 16.2 Extract from DIN EN ISO 80601-2-55:2011 for measuring oxygen concentration and extended requirements by Weinmann Emergency according to other standards or practical knowledge from the emergency sector

Title	Constraints
Measuring range	Gas concentration 15%–100% O_2 volume fraction
Accuracy	± (volume fraction of 2.5% + 2.5% of gas concentration) for all operating conditions, independent of flow, can be compensated with external parameters such as ambient pressure. To be checked at gas concentration 15, 21, 40, 60, 100% O_2 volume fraction, rest N_2 (DIN EN ISO 80601-2-55)
T_{90} *time*	To be named at gas concentration 60% O_2 volume fraction and 60 L/min flow (DIN EN ISO 80601-2-55) <10 s
Drift	Accuracy must be maintained at least 6 h without recalibration (DIN EN ISO 80601-2-55)
Operating temperature	−20 to 50 °C (−20 to 60 °C internal)
Ambient pressure	540–1060 hPa
Ambient humidity	5–95% RH, non-condensing or up to pH_2O = mbar (partial pressure of water vapour)

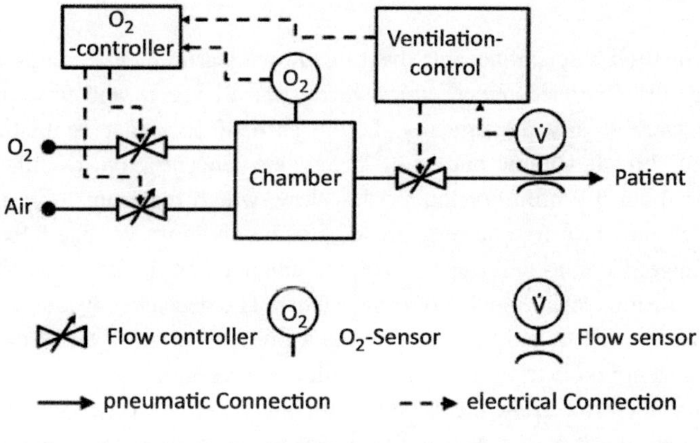

Fig. 16.12 Oxygen mixer according to method 1 with one mixing chamber

must be kept safely in the chamber, otherwise the flow regulators at the chamber inlet would have to work very dynamically again. A further point is that the chamber must be operated with overpressure in order to drive the downstream flow controller. If it is further assumed that the air, as is usual in emergency ventilation, is obtained from the environment and not made available via a compressed gas system, the flow regulator on the air side must be implemented with a pump. Although this principle makes gas mixing easy, the design is too large for compact mobile devices in emergency ventilation.

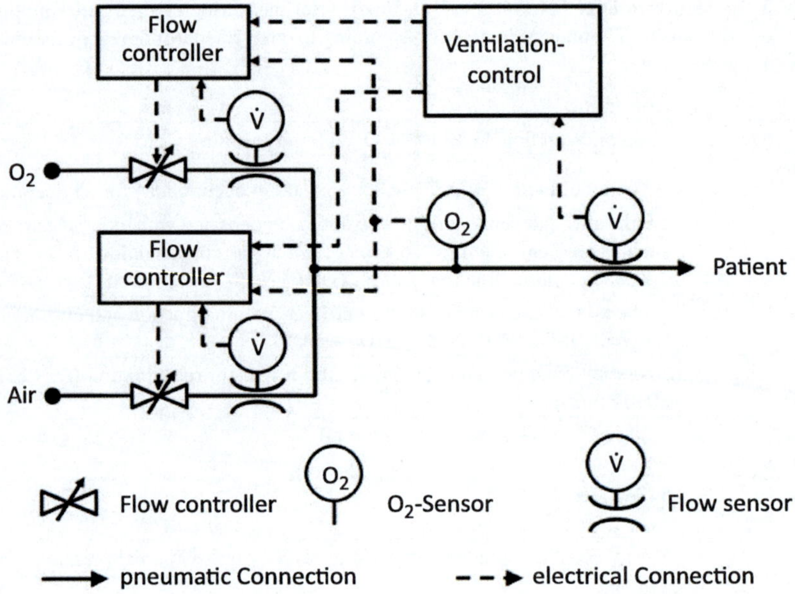

Fig. 16.13 Oxygen mixer according to method 2 with two flow controllers

Looking at method 2, it is noticeable that three flow measurement systems are required. This eliminates the chamber and the third valve, but all valves and flow measurement systems must work highly dynamically. In the case of an infant ventilation with for example 50 mL breath volume and at 40% oxygen concentration, average ventilation flows result of about 4 L/min. For the oxygen flow, which represents only a fraction of the total flow, it must therefore be possible to measure and dose volume flows down into the millilitre range. Here, as described above, the demands on the flow sensor technology but also on the dosing technology become very clear. The decisive advantage, however, is that the flow source for air only has to provide a pressure at the level of the ventilation pressure and does not have to fill a chamber with overpressure.

Both methods therefore allow the use of comparatively slow oxygen sensors. However, method 2 requires the use of a sensor that measures an average oxygen concentration. Sampling sensors, such as optical absorption measurement, are less suitable. For this reason, Table 16.3 provides an overview of the current oxygen sensor technology.

In the following, the overview provided by the table will be supplemented by detailed explanations. Galvanic cells consist of an electrolyte that emits a voltage in the mV range depending on the oxygen concentration present. The voltage can be directly converted into an oxygen concentration by the device. These sensors are consuming as soon as they come into contact with oxygen. Due to the consumption of the sensors, they must be recalibrated at least daily. This type of sensor is standard in ventilation and very widespread due to its low price. All other sensor principles listed in Table 16.3 are non-consuming. The second

Table 16.3 Examples of oxygen sensors and their suitability for emergency ventilation

	Galvanic cells	O_2 flux (zirconium)	Para-magnetic	Fluorescence quenching	Visual
Cross-sensitivity	+	++	++	++	++
Drift	–	+	+	+	+
Lifetime	o	++	++	+	++
Measuring speed	o	+	o	++	++
Energy demand	++	– –	+	++	+
Size	++	++	+	++	o
Robustness	++	+	–	++	++
Price	+	+	– –	–	–

sensor in the list, the zirconium sensor, has been used for years in automotive engineering as a so-called Lambda oxygen sensor. Zirconium or more precisely zirconium dioxide (ZrO_2) is a ceramic that conducts oxygen ions at temperatures above 600 °C. The oxygen content can be deduced from the ion current. The main disadvantage of this technique is the very high operating temperatures, which lead to high energy consumption. Paramagnetic sensors exploit this special paramagnetic property of oxygen. Similar to zirconium sensors, they work very selectively and show practically no cross-sensitivity to other gases. In the case of paramagnetic sensors, a dumbbell-shaped test device filled with nitrogen is suspended from a torsion spring in a strong magnetic field. Depending on the oxygen concentration within the arrangement, a force is generated which causes the dumbbell to turn away from the magnetic field. This force is a measure of the oxygen concentration. The system is mechanically very filigree, so that it seems less suitable for rough emergency use. It is also very expensive. Its use can be justified in a hospital, where it pays off in comparison to galvanic sensors which have to be changed frequently.

The last two methods work optically. Fluorescence quenching uses the effect that certain colour pigments embedded in plastic change their fluorescence behaviour depending on the oxygen concentration. Optical absorption uses the absorption behavior of oxygen at 762 nm wavelength. For this purpose, a tunable laser must cyclically pass through a range of wavelengths while simultaneously measuring the absorption behavior of the gas in the sample chamber. The optical path length in the sample gas essentially determines the size of the setup. Due to the relatively expensive laser, the principle is quite expensive and cannot be used everywhere because of its sampling principle.

Typically, all measuring methods record the partial pressures of the gas component, so that the absolute pressure must be taken into account for the calculation of the gas concentration. Since practically all gas sensors drift more or less strongly, they require a reference gas for calibration at regular intervals. The ambient air is often used for this. The oxygen volume fraction in dry air is 20.95%. For humid air the water vapour pressure has to be subtracted from the air pressure and then the actual oxygen content has to be calculated for the calibration. This approach also applies to CO_2 measurement.

Carbon Dioxide Measurement Technology

From the carbon dioxide content at the end of expiration, important information about the body's metabolic activity can be deduced. For this reason, the CO_2 content is often used in medical technology as a non-invasive diagnostic parameter and the measuring technique is called capnometry. In capnometry, the CO_2 content is determined with high temporal resolution during exhalation. In addition to the progression of the gas content, the concentration at the end of the exhalation phase is of importance. Capnometry is also used to check the position of the intubation tube. Especially in emergency medicine, due to the often poor accessibility to the patient, it often happens that the ventilation tube has been placed in the esophagus instead of the trachea by mistake. Using the CO_2 data from the first ventilation cycles, the incorrect position of the intubation can be quickly detected and corrected. Analogous to the oxygen measurement technology, the requirements for the sensor technology will first be shown. Here, too, the standard DIN EN ISO 80601-2-55 will be applied, which is shown in Table 16.4 as a supplement to Table 16.2 for CO_2 measurement.

Compared to oxygen measurement, capnometry must be much more responsive in order to measure the CO_2 content during a respiratory phase. The standard again only stipulates that the measuring speed must be specified in the instructions for use. In practice, however, high measuring rates are required. Since the expiratory gas must be measured during CO_2 measurement, either the patient's respiratory flow is measured in the mainstream close to the patient or a secondary flow of respiratory gas is led to a sidestream so that it can be examined away from the patient. The mainstream measurement delivers measured values without delay, whereas the sidestream measurement causes a delay. However, this delay is not problematic, since the CO_2 trend is decisive for diagnosis. With CO_2 measurement, expiratory gas is always examined, which can contaminate the sensor channel, so that hygiene measures must be observed when changing patients.

Table 16.4 Supplements to Table 16.2 for measuring the CO_2 concentration

Title	Constraints
Measuring range	Gas concentration 0–10%-CO_2 volume fraction
Accuracy	± (volume fraction of 0.43% + 8% of gas concentration) for all operating conditions, independent of flow, can be compensated with external parameters such as ambient pressure, to be checked at gas concentration 0%, 2.5%, 5%, 10% CO_2 volume fraction, rest N_2 (DIN EN ISO 80601-2-55)
T_{90} time	To be named at gas concentration 5% CO_2 volume fraction and 60 L/min flow (DIN EN ISO 80601-2-55) <250 ms

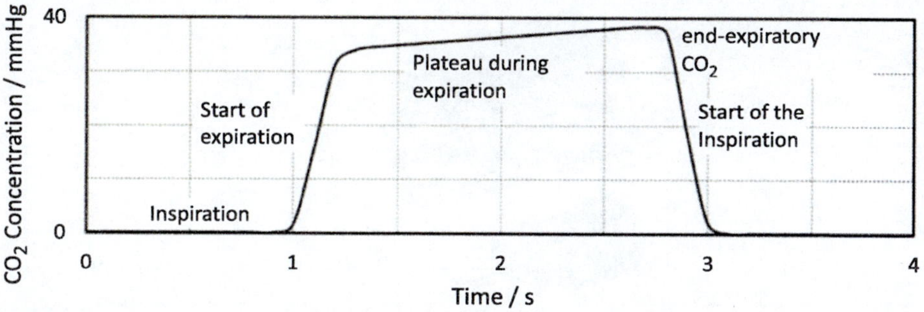

Fig. 16.14 Normal capnogram with markings of the respiratory phases

The progression of a normal capnogram is shown in Fig. 16.14. It shows the respiratory phases and the associated change in CO_2 concentration. If no capnogram can be displayed, only the expiratory CO_2 is shown as a meaningful value.

The previous remarks on capnometry only describe the external conditions that the gas measurement technology has to cope with. Considering the measuring principles used in capnometry, only the optical absorption method has been able to establish itself due to the high measuring rate required. For this method, the absorption behavior of CO_2 in the infrared (IR) at a wavelength of 4.26 µm is used. For exact concentration measurement, the absorption must be measured in a CO_2-free reference gas and in the sample gas. As the absorption lines of water vapour lie in a near wavelength range, the air humidity must also be included in the measurement or the sample gas must be dried before the measurement.

For measurement close to the patient, the measuring device is placed directly at the end of the ventilation hose. Thus, size and weight but also robustness play an important role. Large systems make access to the patient difficult, while heavy systems can pull the tube out of the patient's throat due to their weight. The robustness of the measuring device must also be given the same high priority so that prudent action benefits the patient and not the equipment. The requirements of filigree and robustness often contradict each other, which makes it difficult to set up a sensor close to the patient. Since the measurement is optical, hygiene requirements can easily be met with a measuring cuvette that is changed with each patient. However, the patient-close sensor system must be able to cope with condensation, as condensation can only be avoided with great effort when the sensor is so close to the patient.

The capnometry remote from the patient is mostly used in ventilators but often also in defibrillators. Typically, a small pump is used to draw a constant bypass flow from the respiratory gas stream, which is then passed through a measuring cell. Such a setup is shown as an example in Fig. 16.15. Due to the external suction hose for the bypass flow there are further requirements for the CO_2 measuring system. The system must be able to detect whether a hose is present and whether it is snapped off or blocked. Since the suction rates are very low (about 100 mL/min), the hose cross-sections are also small. As a result,

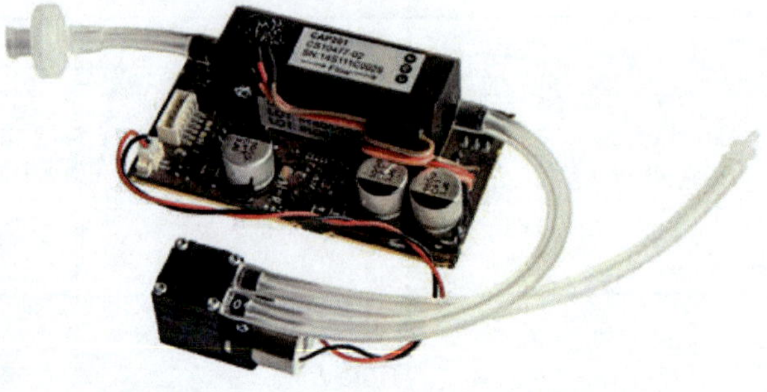

Fig. 16.15 Module for sidestream capnometry from the company Corscience (with the kind permission of Corscience GmbH& Co. KG, Erlangen). The figure shows the optical measuring system and the evaluation electronics on the PCB (*top*) and the external micropump (*bottom*)

the extracted gas reaches the measuring cell with a delay of about 5–10 s. The exhaled gas is saturated to 100%, so that condensation occurs in the suction hose, which can clog the hose. For this reason, the suction hose usually contains moisture absorbers. To prevent suction pump disturbances from being transmitted to the measuring cell too often, the pump is located downstream and therefore sucks through the measuring cell.

Example of an Emergency Ventilator

The MEDUMAT Standard² from Weinmann Emergency will serve as an example of how the gas sensors described above can be installed in an emergency ventilator. This is a compressed gas-powered ventilator that uses an oxygen supply with pressures from 3 to 6 bar as its pressure source. Figure 16.16 shows the simplified block diagram of the device. The oxygen concentration can be changed with the Air Mix Valve 10 by drawing in ambient air via the injector 15 according to the jet pump principle. Thus the oxygen concentration can be reduced up to 55% oxygen.

As can be seen in the description of Fig. 16.16, the MEDUMAT Standard² uses five pressure sensors 2–6 in addition to the CO_2 module 1. All pressure sensors contain an internal temperature sensor which gives a good indication of the gas temperature, due to the pneumatic channel made of aluminum, as can be seen in Fig. 16.17. Figure 16.17 shows the upper housing shell with the electronics on the left and the lower housing shell with the pneumatics and most of the sensors on the right. Both halves of the housing are folded and screwed together for mounting, so that the CO_2 module 1 folds into the gap in the lower shell. Due to its compact design, the device has a size of 206 mm × 137 mm × 130 mm and weighs only 2.5 kg with battery. Figure 16.18 shows the ready-to-use unit with oxygen cylinder and ventilation hose.

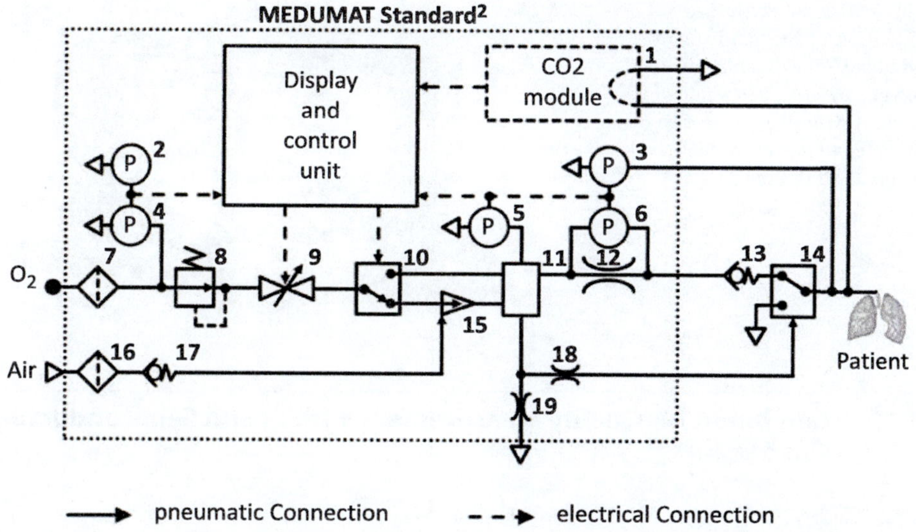

Fig. 16.16 Simplified block diagram of the MEDUMAT Standard² from Weinmann Emergency. 1: CO_2 module according to the NDIR method; 2: Ambient pressure sensor; 3: Patient pressure sensor; 4: Oxygen inlet pressure; 5: Mixing chamber pressure; 6: Differential pressure sensor for flow measurement; 7: Oxygen filter; 8: Pressure reducer; 9: Proportional valve for patient gas dosing; 10: Air Mix valve for 100% oxygen or for mixing ambient air to reduce oxygen; 11: mixing chamber; 12: pneumatic resistor for flow measurement; 13: check valve; 14: pneumatically controlled exhalation valve for the patient; 15: injector for enrichment with ambient air; 16: ambient air filter; 17: check valve; 18, 19: pneumatic resistors for exhalation valve control

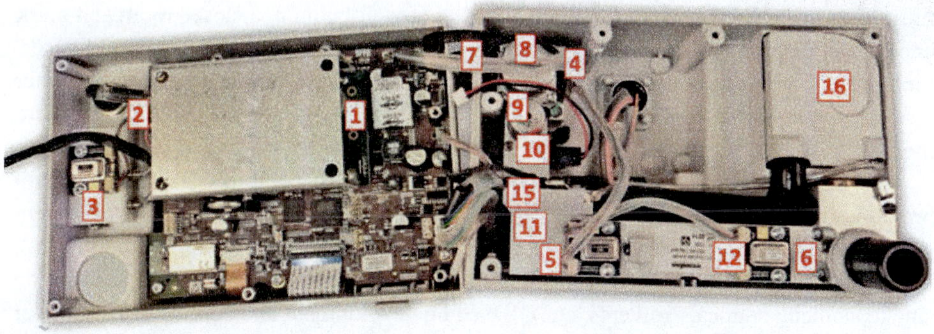

Fig. 16.17 Internal structure of the MEDUMAT Standard² from Weinmann Emergency. The numbering corresponds to the block diagram in Fig. 16.16

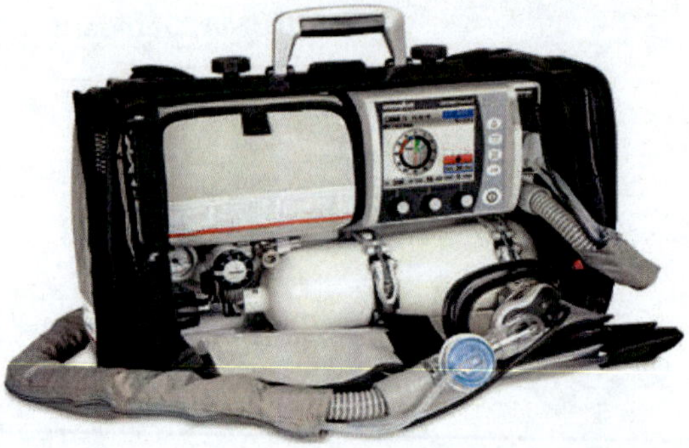

Fig. 16.18 Ready-to-use MEDUMAT Standard[2] (here without CO_2 extraction) with oxygen cylinder and ventilation hose. A defibrillator can also be mounted to the left of the ventilator if desired

16.3 Combined Air Quality Measurement with IR and Semiconductor Gas Sensors

Olaf Kiesewetter, Jürgen Müller, and Matthias May

The quality of indoor air has a significant influence on the well-being and health of the people who spend time in the room. Climate and air quality also influence in particular the ability to concentrate and judge as well as reaction time and condition. Typical sources of indoor air pollution are furniture, floor coverings, equipment used, cleaning agents and outdoor air emissions. But also the people themselves who are in the room unconsciously have a great influence on the air quality. Primarily, oxygen is consumed by breathing and carbon dioxide is exhaled. In addition, there are body odors, perfumes, tobacco smoke, clothing, etc., which affect the indoor air. Depending on the type of input, the smell is perceived as either disturbing or pleasant. Perfume or air fresheners are deliberately applied substances that are intended to increase the sense of well-being in a closed room and mask unpleasant, disturbing odors. Other outgassing's from mainly older electrical appliances, furniture or carpets may be odorless but are undesirable because they are toxic in higher concentrations. These include in particular *volatile organic compounds* (*VOCs*), which are found in solvents, adhesives, paints, varnishes and plastics.

This article deals with an innovative sensor system that combines gas measurement principles of photoacoustic and MOX[2] semiconductor gas sensor technology. This system will be integrated in buildings and vehicles for the detection of carbon dioxide (CO_2), selected VOC key substances and for the detection of hazardous situations (e.g., cigarette smoke, spilled chemicals/solvents, etc.). Based on development and application results for gas sensors for demand-based ventilation in buildings, the measuring characteristics of a demonstrator for an innovative patented VOC–CO_2 sensor element for the detection of air quality in rooms are presented. Under certain conditions, this method can also be used to

[2]Metal oxide.

differentiate between pleasant and unpleasant odors. This advantageous sensor concept is currently used in buildings to optimize air exchange rates from an odor point of view and with the aim of saving energy.

For the climatic conditions in interiors, for example also in vehicles, further parameters have to be evaluated: temperature, humidity, air flow or direct sunlight. Special features that prevail in the respective locations must be considered additionally. Thus, at workplaces, in laboratories, hospitals or even offices, a targeted air quality monitoring with controlled ventilation concepts is necessary. By intelligent screening of the ambient conditions, efficient energy savings can be achieved. Emissions from machines, plants or technological processes, chemical entry into the ambient air, spreading of aerosols or gases are detected with the patented VOC–CO_2 sensor element. These signals can be used for further evaluation and control measures. Thus, for example, vehicle interiors or passenger compartments in public transport can also be optimally ventilated or air conditioned. The indoor air quality in vehicle cabins is essentially determined by the occupants (breathing, tobacco smoke, clothing, perfumes, etc.), transported objects (food, plants, luggage, etc.) and, if applicable, animals as well as by inputs from the outside air (exhaust gas components, dust, pollen, odors, etc.). Typical components of the vehicle interior air are for example soot particles, nitrogen oxide/NO_x and carbon monoxide/CO from exhaust gases of particularly older vehicles (Wiegleb and Zöchbauer 1994), carbon dioxide/CO_2 from exhaust gases and as a natural metabolic product of humans and animals (Köhne and Wiegleb 2008), ozone/O_3 (from solar radiation), abrasion and dust particles as well as VOCs. As shown above, an increase in CO_2 via 1000 ppm and the accumulation of certain VOCs, nitrogen oxides and carbon monoxide already cause symptoms of fatigue and a decrease in concentration and judgement.

MOX Semiconductor Gas Sensors

MOX semiconductor gas sensors are very well suited for investigations of indoor air quality with regard to selected target substances, such as flue gases, carbon monoxide, ammonia, nitrogen oxides, alcohols, benzene and VOCs. Due to their high sensitivity in the ppm-/ppb-range, the wide measuring range and the high stability a long lifetime can be guaranteed.

The operating principle of semiconductor gas sensors is based on the change in conductivity of a sensitive semiconductor layer when exposed to gases. These sensitive layers react to reducing gases by reducing the resistance and to oxidizing gases by increasing the resistance. The reaction takes place in the concentration range from a few ppb up to 5 vol.%. The respective detection range of the gas sensor element depends on the sensor type used and the type of gas to be detected.

For an optimal reaction of the gas on the semiconductor the sensitive layers are heated. A heater is integrated in the sensor element for this purpose. The resistance of the platinum heater has a defined temperature coefficient, which allows the working temperature of the sensor to be controlled. This compensates for fluctuations in the ambient temperature as far as possible and specifically influences the sensitivity of the sensor.

The selectivity of a gas sensor with respect to the target substances can be deliberately controlled by the properties of the respective gas-sensitive layer(s) (layer materials and their structure, doping, etc.). Further possibilities arise for example by using specific filters and/or by controlling the operating modes. Due to a compact design of multiple gas sensors (gas sensor arrays and/or gas sensor elements with several sensitive layers) this sensor type is suitable for a variety of detection applications of different gases.

The majority of target substances can be detected with gas sensors with several sensitive layers via pattern recognition. It is advantageous to arrange the sensitive layers on a chip in order to minimize location and concentration-related artifacts. Depending on the sensor design and the manufacturing process used (substrate/chip technology, etc.), such multiple gas sensors can also be produced comparatively inexpensively. Advantages of MOX-semiconductor gas sensors, compared to other gas sensor principles, are especially the high sensitivity to the respective target gases, the comparatively simple and inexpensive miniaturization, long lifetime and the cost-effective production in small as well as in large series.

Figure 16.19 shows a triple semiconductor gas sensor with which a wide variety of gases and gas mixtures can be measured from the lower ppb to the volume percent range (Kiesewetter et al. 2004, 2005; Kittelmann et al. 2006). The sensor module consists of a MOX semiconductor gas sensor element with three different gas-sensitive layers for reducible, easily and hardly oxidizable gases on a microstructured ceramic substrate in a TO39 housing with four connections as well as an electronic module for controlling the sensor element, calibration data storage, signal preprocessing and data communication (Fig. 16.19). The application-specific control and signal processing of the sensor element enables the highly sensitive detection of specific gases, gas mixtures and selected VOC lead substances for their subsequent external classification and quantitative analysis (e.g., using PCA[3]).

In addition to the target substances described above, such as selected VOC markers, the detection of CO_2 is also necessary for the application. CO_2 as an inert gas reacts only with difficulty or hardly with the surfaces of the sensitive metal oxide layers of MOX semiconductor gas sensors. For the detection of CO_2, photoacoustic measuring methods can be used.

As described above, the semiconductor gas sensor is heated. Due to the usual sensor temperatures of 300–450 °C, it is possible to use the semiconductor gas sensor simultaneously as an IR emitter/transmitter for the photoacoustic measuring principle described below for CO_2 detection (Kiesewetter et al. 2014).

[3]Principal Component Analyses.

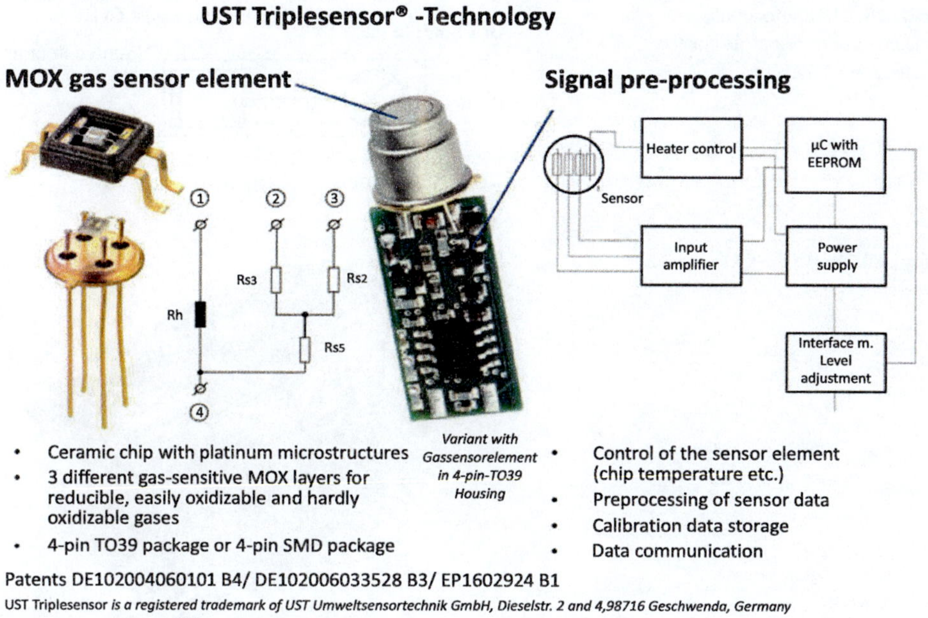

Fig. 16.19 UST Triplesensor®-gas sensor with 3 sensitive layers and sensor electronics

Photoacoustic Measuring Principle for the Detection of CO_2
Figure 16.20 shows the patented photoacoustic measuring principle for CO_2 detection (Kiesewetter and Riesenberg 1999).

Detection is achieved by recording and evaluating the change in the radiation emitted by the IR transmitter, which is attenuated by the existing CO_2 concentration in the gas. Reference measurements are used for continuous zero adjustment. The advantages of this measuring method are the high selectivity, the detection of almost all energy bands of the sample gas component as well as the lowest drift and ageing of the sensor.

Sensor Element for Detection of CO_2 and VOCs
Figure 16.21 shows an implemented sensor sample for simultaneous detection of CO_2 and VOCs. The patented VOC–CO_2 sensor element (Kiesewetter 2010) consists of a photoacoustic gas sensor system for the detection of CO_2 and a MOX semiconductor gas sensor system (UST Triplesensor® Technology) for the simultaneous detection of room climate- and ambient-relevant measured variables (e.g., selected VOC lead substances) as respective reference variables for air quality (Fig. 16.22). If required, hazards can also be detected with the aid of selected specific substances (e.g., combustion gases, etc.), which can arise in the event of smoldering or cable fires (Kiesewetter et al. 2014).

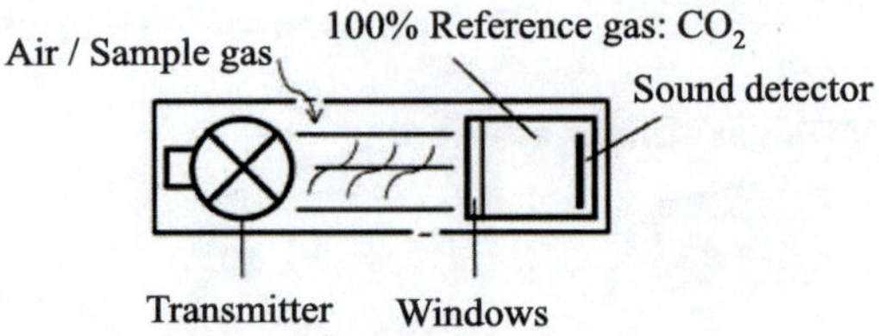

Fig. 16.20 Photoacoustic measurement principle for the detection of CO_2

Fig. 16.21 VOC–CO_2 sensor element

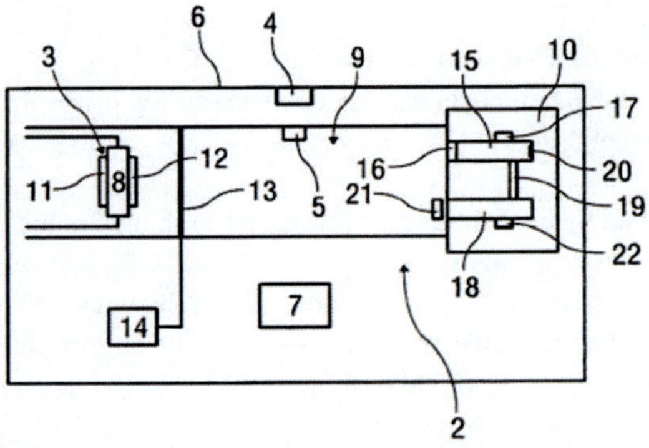

Fig. 16.22 VOC–CO_2 sensor element – principle (Kiesewetter 2010) – 2. optical gas sensor, 3. chemical gas sensor, 4. ambient temperature sensor, 5. humidity sensor, 6. housing, 7. evaluation unit, 8. radiation source, 9. measuring cell, 10. detector device, 11. semiconductor metal oxide layer, 12. coating, 13. shutter, 14. motor, 15. first detector cell, 16. window, 17. first microphone, 18. second detector cell, 19. flow sensor, 20. first infrared radiation sensor, 21. second infrared radiation sensor, 22. second microphone

In addition to the photoacoustic sensor for CO_2 detection, application-specific MOX gas sensors are used for different gases and gas mixtures. VOC semiconductors are used, as already described, for the detection of measured variables relevant to well-being and health.

Other specific semiconductor materials are used for example for the detection of hazardous situations by reaching explosion limits or fire situations (smoldering, cable fires), chemical accidents or continuous enrichment of hazardous gases (e.g., cellar rooms: sewage gases, digester gases). A wide range of applications is possible depending on the equipment and configuration. The overall presentation of environmentally sensitive data also includes the recording of the current temperature and humidity. The detection of this climate data can be used to control air conditioning and ventilation systems or even just shown on a display. All relevant data of the VOC–CO_2 sensor element can be output via an interface, for example to a control system for air conditioning and ventilation systems or to a PC for data analysis.

In the following, two typical application scenarios with selected measurement results for air quality measurement are presented. The VOC–CO_2 sensor elements were installed in test units of the UST Air Quality Monitor, a device with touch screen display. The UST Air Quality Monitor detects relevant parameters of indoor air, such as the concentrations of carbon dioxide and selected VOC key substances as well as room temperature and humidity. The basic design of the UST Air Quality Monitor is shown in the block diagram Fig. 16.23.

The data transfer is carried out via an integrated RS232 interface. The current air quality data can be read directly from the display of the UST Air Quality Monitor (Fig. 16.24).

A UST air quality monitor has been located in a consultation room for some time. As will be seen later, the decisive air quality feature here is the measured CO_2 value. In this room, the number of people present varies between 0 and 18. The second monitor is installed in "Cycle 1", a production area for temperature and gas sensors. Here, substances such as screen printing oils and solvents worked. The VOCs released in the room are decisive for the air quality assessment.

Results of the Long-Term Study in the Consulting Room

The UST air quality monitor in the meeting room has been used as an internal test device for over 2 years. The data are continuously read out and checked for plausibility. For this purpose, the activities (ventilation cycles) and the number of persons in the room are recorded in a protocol. Thus, it is compared whether and how particularly the carbon dioxide signal is influenced by the people present. The integrated semiconductor gas sensors react to changes in the VOCs in the room air caused by the people present, furniture, flooring and electrical equipment. The daily diagram shown in Fig. 16.25 shows the signal characteristics on a typical Monday. Clearly visible is the increase in CO_2 concentration from approximately 400 ppm to via 1400 ppm in the room at 9.00 a.m. According to the protocol, 18 people took part in a consultation at this time. The VOC2 value also rose slightly from 0 to 10 (without unit). Here, it can be assumed that this value – especially in the morning – was caused by perfume or other body care products. This VOC entry into the air is not perceived as unpleasant. On the other hand, the high CO_2 value of 1400 ppm will have affected the concentration of the employees and will have had a tiring effect. For this reason, the window was tilted open for ventilation after the end of the staff

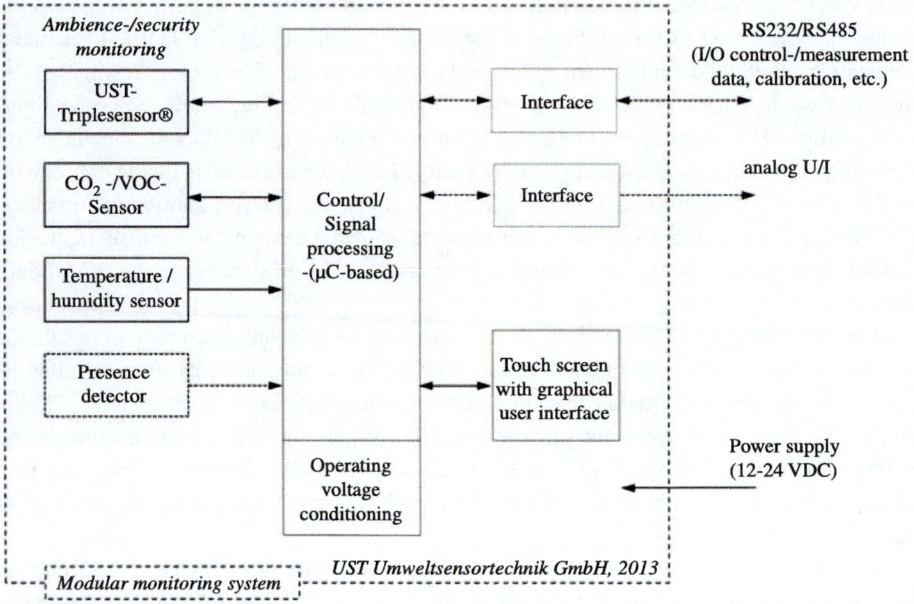

Fig. 16.23 UST Air Quality Monitor – block diagram

Fig. 16.24 UST Air Quality Monitor – history display

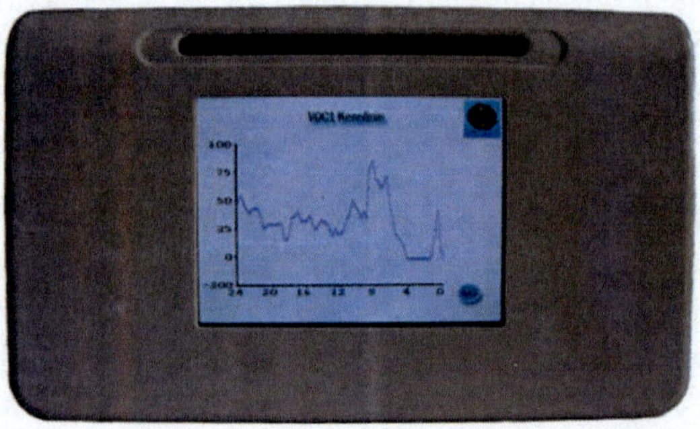

meeting. The carbon dioxide concentration is visibly decreasing. At noon at 12.00 noon, the window was closed and the door to the corridor was left open for ventilation. As a result, the CO_2 concentration has not changed significantly. However, the VOC1 value (without unit) has increased noticeably. Through the open door, substances from outside have entered the meeting room (e.g., exhaust fumes due to the proximity of the motorway, etc.). The meeting from 4.30 p.m. to approximately 5.30 p.m. with five participants caused the carbon dioxide concentration and the VOC1 value to rise again. The VOC2 value remained largely unaffected by this event. At the end of the meeting, the values slowly returned to the initial values when the room was closed.

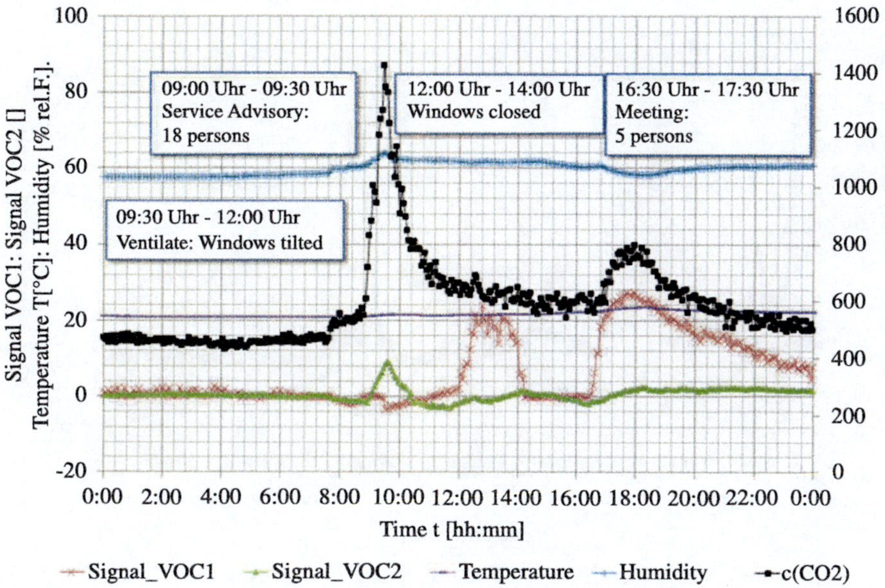

Fig. 16.25 UST air quality monitor with VOC–CO_2 sensor element – measurements in the conference room of UST Umweltsensortechnik GmbH – daily diagram from September 15, 2014

Figure 16.26 shows the course of measured values for individual days of the entire week. The principle of the detection of carbon dioxide and VOCs is shown. At this point, we will not go into the individual events any further. The increases in CO_2 concentration are easy to recognize when people are present in the meeting room. The decay of the individual signals in the evening and at night is also clearly visible. There are no activities in the closed meeting room at weekends. The CO_2 and VOC2 values do not change.

Results of the Long-Term Study in Production Cycle 1

In production cycle 1, the basic construction of the temperature and gas sensors takes place. The platinum-coated sensor chips are laser structured and prepared for further processing. This includes, for example, screen printing processes and cleaning processes. Accordingly, screen printing oils and harmless solvents are used here. A varying number of people work in the rooms from 6.00 to 22.00 h. In the time between the start of the shift and the change of shift, an increase in CO_2 concentration from 400 ppm to approximately 1200 ppm is measured, which fluctuates greatly depending on the ventilation. After the shift change at 2 p.m., the carbon dioxide value decreases continuously and the total VOC value increases slightly. Ventilation control compensates for the input of volatile organic substances from screen printing oils and harmless solvents as well as additional substances from perfume and personal care products. These substances can be easily separated by the VOC1 (sum

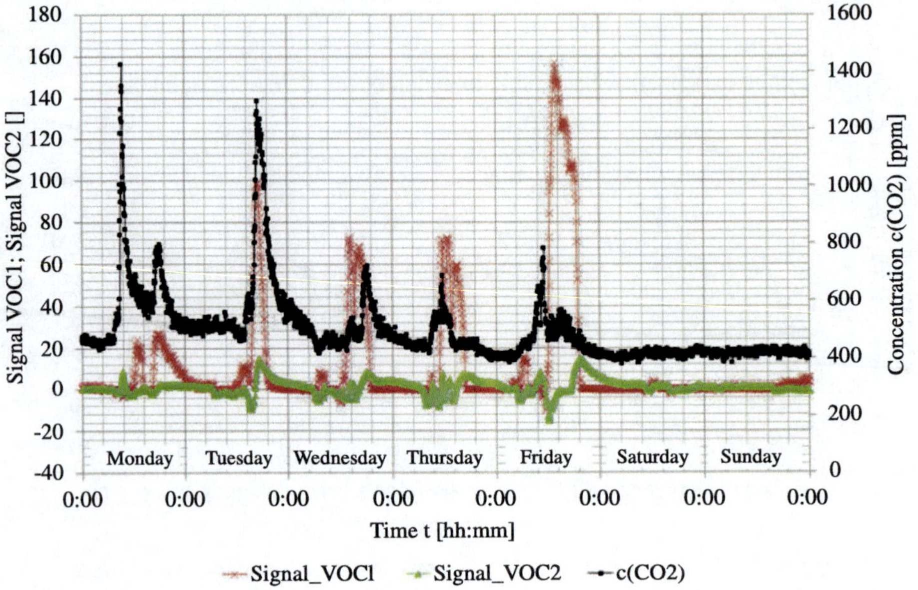

Fig. 16.26 UST Air Quality Monitor with VOC–CO_2 sensor element – measurements in the conference room of UST Umweltsensortechnik GmbH – weekly diagram: September 15–21, 2014

signal) and VOC2 signals. Figure 16.27 shows the change in air quality caused by the input of VOC substances. The jumps between the individually displayed measuring points are caused by the automatic ventilation system. The change in ventilation intensity changes the VOC and CO_2 concentration in the room air. This can be seen very clearly on Monday morning before 6.00 a.m., as the ventilation is in a stationary or balanced state during this period.

Figure 16.28 shows the measurements in production cycle 1 for 1 week. As in the meeting room, the principle of detecting carbon dioxide and VOCs without making individual evaluations is shown here only once more. Very good to see is also here the decay of all relevant air quality data on the weekend after end of shift on Saturday noon.

Cycle 1 is the production area in which, for technological reasons, harmless solvents and other VOC-containing substances are used. This has already been described in the diagrams above. The following diagrams show the effect of a regulated ventilation control on this air pollution. Figure 16.29 shows the raw signals of all air quality parameters from a Friday 6.00 a.m. to the following Monday 9.00 a.m.

The extract from the course of the sensor raw signals shows that in the period shown, with 10 times the number of air changes (Fig. 16.32) and two to three people, comparatively small changes were recorded with regard to CO_2 concentration (Kiesewetter et al.

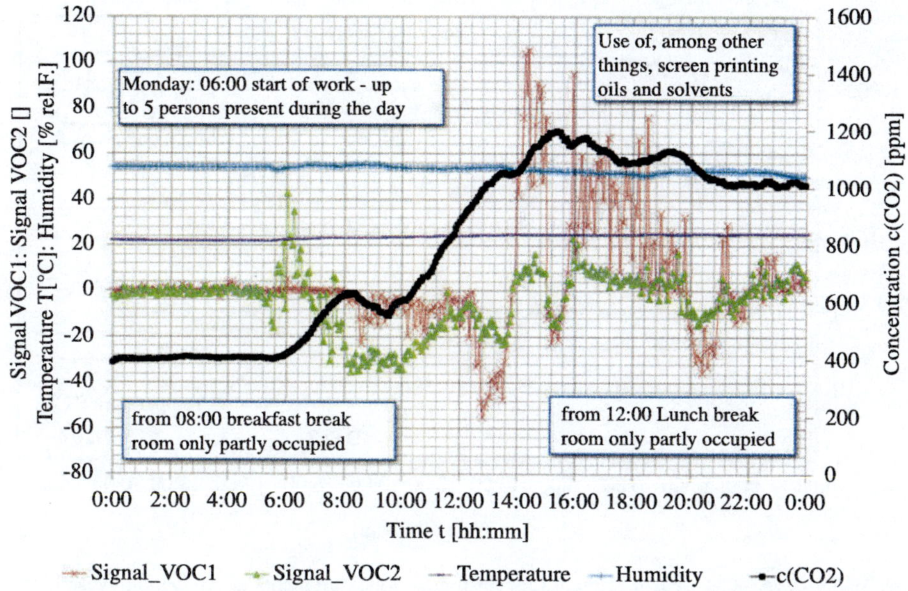

Fig. 16.27 UST air quality monitor with VOC–CO_2 sensor element – measurements in production cycle 1 of UST Umweltsensortechnik GmbH – daily diagram of September 15, 2014

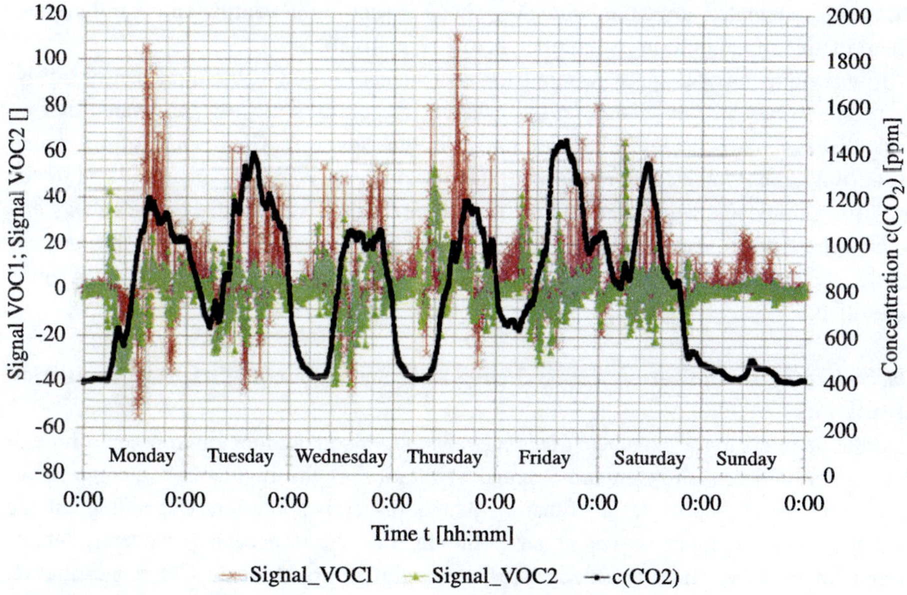

Fig. 16.28 UST air quality monitor with VOC–CO_2 sensor element – measurements in production cycle 1 of UST Umweltsensortechnik GmbH – weekly diagram: September 15–21, 2014

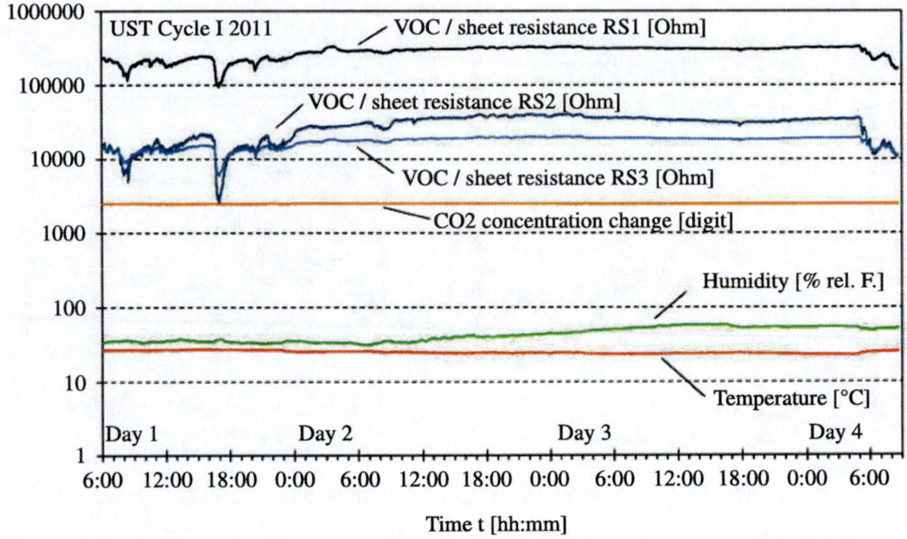

Fig. 16.29 VOC–CO_2 sensor element (demonstrator) – measurements in production cycle 1, raw signals from the sensors (excerpt)

2014). By means of a calibration stored in the UST Air Quality Monitor, the number of air changes is calculated and a corresponding signal is sent to the ventilation control system. The principle of calibration is described below (Fig. 16.30).

In three-dimensional space, where each axis stands for one of the three sensor layers, occurring events in which the sensitive layers react differently to gas events can be evaluated separately by their direction (angle in the diagram). The amount of the spatial expansion represents the concentration of the detected gas. With the data calculated by the PCA (magnitude and angle), the VOC–CO_2 sensor element is calibrated in such a way that when events occur, gas types and concentrations are detected accordingly. From these results, the air exchange rate is calculated and the air conditioning or ventilation system is controlled.

Derivation of an Optimized Air Exchange Rate for a Demand-Oriented Ventilation of Interiors

Ventilation refers to the exchange of air between indoor and outdoor environments through the supply of fresh air. Demand-oriented ventilation is the regulated exchange of air depending on the concrete air situation in the respective interior. Depending on the dominance of the main source of air pollution, one or, if necessary, several suitable command variables must be selected for automatic control (human: CO_2 concentration and/or other sources: concentrations of selected VOC markers).

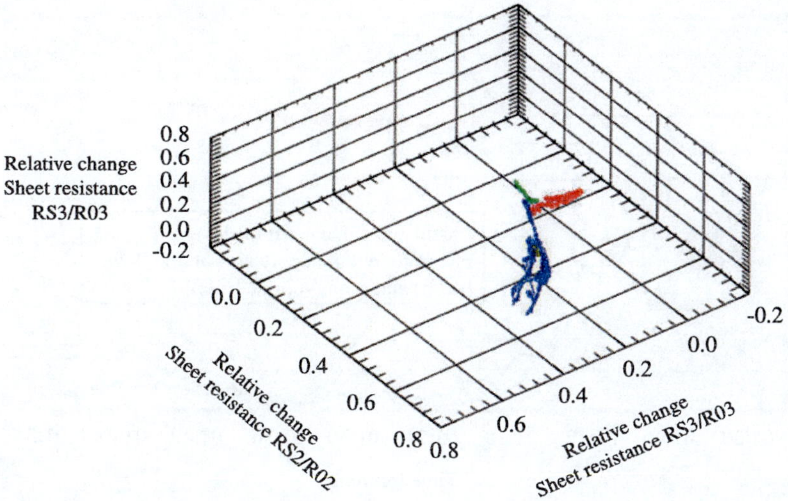

Fig. 16.30 3D representation of the three evaluated sensor layers for evaluation using PCA (Principal Component Analysis)

With the measured values of the air quality measurement, the number of air changes currently required can be calculated. An example of the tests in the aforementioned production cycle 1 is shown in Fig. 16.31.

By optimizing the number of air changes, the energy consumption for the ventilation system can be reduced (Fig. 16.32). For the example of the aforementioned production cycle 1, the energy consumption of the ventilation system without optimization of the number of air changes was approximately 4 kWh, with optimization the energy consumption of the ventilation system was approximately 2 kWh (Kiesewetter et al. 2014).

Detection and Separation of Pleasant and Unpleasant Odors

The majority of air quality sensor elements currently available on the market only detect VOC sum signals, including nitrogen oxides, gaseous sulfur compounds and ammonia, and it is not possible to separate pleasant and unpleasant odours or even toxic air constituents. The present sensor concept, based on the UST Triplesensor® Technology, enables group-selective separation by arranging three gas-sensitive active layers on one chip. Figure 16.33 shows the sensitivities to “bad smelling” substances (ammonia, sulphur dioxide), to a process-related VOC and to a good smelling VOC (perfume) (Kiesewetter et al. 2014).

Conclusion

The results presented show that the combined VOC–CO_2 sensor technology can be used to detect air quality and user-specific air compositions in building and vehicle interiors. Based

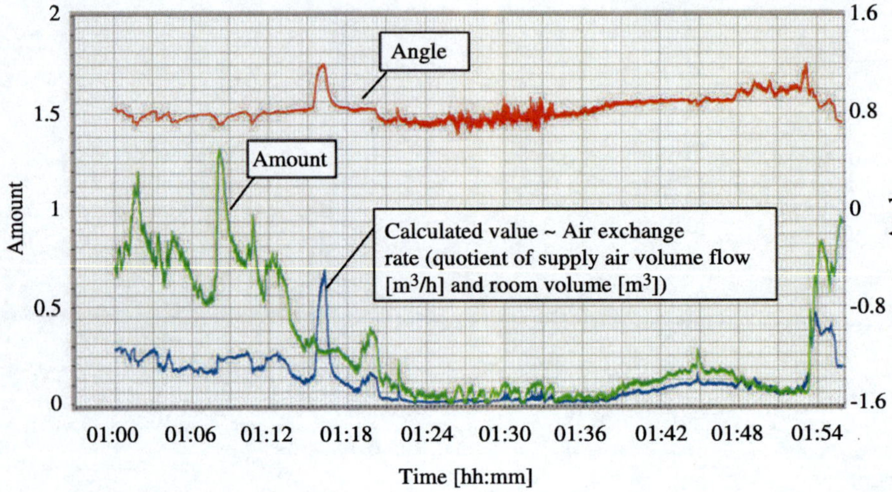

Fig. 16.31 VOC–CO_2 sensor element (demonstrator) – measurements in cycle 1, signal evaluation UST Triplesensor® by means of PCA and calculated number of air changes (section of measurement cycle)

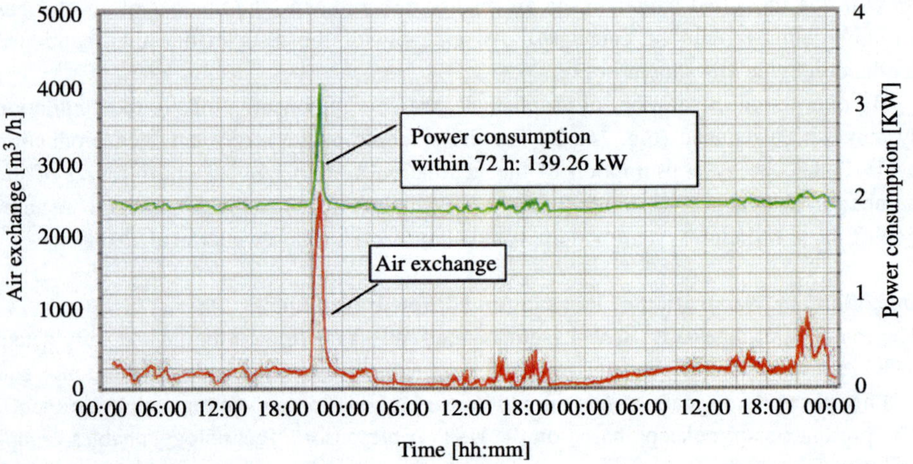

Fig. 16.32 VOC–CO_2 sensor element (demonstrator) – air exchange and power consumption over a measurement cycle from 72 h

on these data, the air quality in these rooms can be improved by targeted control of the air conditioning and ventilation systems, while at the same time saving energy.

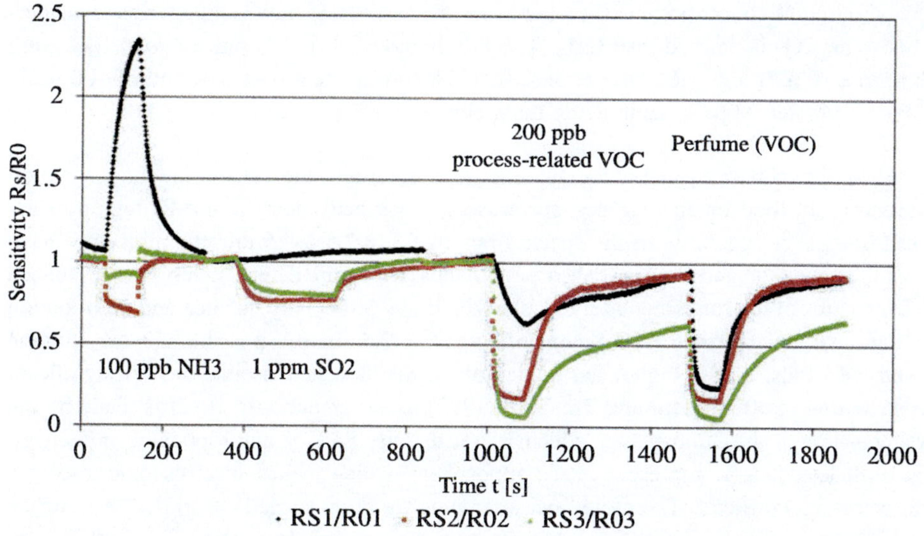

Fig. 16.33 UST Triplesensor® – sensitivities when exposed to ammonia (NH_3), sulfur dioxide (SO_2), process-related VOC and aromatics/VOCs

16.4 Measurement of Dissolved Gases in Water

Peer Fietzek

This chapter discusses the motivation for measuring dissolved gases such as carbon dioxide (CO_2), methane (CH_4) and oxygen (O_2) in different waters and gives a qualitative presentation of the measurement techniques and methods used.

Greenhouse Effect

The gases present in the atmosphere have a significant influence on important climatic relationships. These are in particular the so-called greenhouse gases, which are transparent for the short-wave electromagnetic radiation of the sun, but are an absorber for the long-wave *reflection of* the earth. The absorbed energy is emitted as thermal radiation and thus increases the temperature of the atmosphere. This natural greenhouse effect leads to an increase in the mean land and ocean surface temperature of $\Delta T_N \approx 33°C$ (Schönwiese 2008). The metrologically proven temperature increase of the last 100 years is linearly approximated $\Delta T \approx 0.9°C$ (Stocker et al. 2013). This is mostly explained by the increase in atmospheric concentrations of greenhouse gases, especially carbon dioxide (CO_2) and methane (CH_4). Anthropogenic processes, that is processes influenced by human activity, such as the burning of fossil fuels or changes in land use, contribute to an increase in greenhouse gas concentrations (Stocker et al. 2013). Even if, in absolute terms, considerably more CO_2 is emitted than CH_4, it should be noted that CH_4 has a *Global Warming Potential* (*GWP*) of 28 (Stocker et al. 2013). This means that over a period of 100 years CH_4 contributes 28 times as much to the greenhouse effect as CO_2.

The physical background of the greenhouse effect is also the basis of the most common measuring method for these gases, spectroscopy or spectrometry in the IR region of the electromagnetic spectrum. In the spectral range of $\lambda \approx 2.5 - 25$ µm, the molecules have their characteristic and strongest absorption bands, which are determined by the resonance wavelengths of the fundamental molecular vibrations (especially valence and deformation vibrations). In the near-IR (NIR) range of $\lambda \approx 0.8 - 2.5$ µm, it is possible to work with the overtone bands, that is higher harmonics of the fundamental oscillations. A significant contribution (≈60%; Kiehl and Trenberth 1997) to the greenhouse effect is made by the water vapor in the atmosphere, which is not directly part of anthropogenic influences. Nevertheless, it is omnipresent in the atmosphere, which makes spectroscopic trace gas measurements difficult. Therefore, measurements are often carried out in the range of the absorption windows of water vapor. These are for example at $\lambda \approx 1.6 \pm 0.05$ µm, $\lambda \approx 4 \pm 0.5$ µm and between $\lambda \approx 8 - 14$ µm.

Solubility of Gas in Water

All gases that occur in the atmosphere also dissolve to a greater or lesser extent in the ocean and inland waters. The physical relationship between the partial pressure p_i of a gas in the gas phase and the quantity L_i of dissolved gas in an adjacent liquid is described in chemical equilibrium by *Henry's law* (see Sect. 2.4). In some cases, geological processes on the sea floor or biological processes in the water column also produce corresponding gases, which can then be released into the atmosphere. In addition to the measurement of greenhouse and trace gases in the atmosphere, the measurement of dissolved gases in waterbodies is also receiving considerable scientific and industrial attention. Different measurements are of interest depending on the research question pursued:

- Measurements at the water/atmosphere interface to determine for example gas fluxes.
- Measurements in the water column to identify and characterize water masses or to investigate processes in this volume.
- Measurements near the ground, the benthos, to investigate gas emissions or biogeochemical processes taking place there.

Marine Carbon Cycle and Marine CO_2 System

The measurement of dissolved CO_2 in surface water, for the identification of aquatic sources and sinks, is scientifically widespread and has been significantly influenced since the 1960s (Takahashi 1961). A regularly updated climatology of the CO_2 flux (Takahashi et al. 2014) as well as freely accessible databases (Pfeil et al. 2013) are the results of these actions. In the meantime, such an advanced stage has been reached that corresponding measurements are in the process of being carried out routinely by public authorities (see the implementation of the ICOS infrastructure, *Integrated Carbon Observation System*[4]).

However, carbon dioxide is of interest not only because of its influence on the greenhouse effect, but also because of its key contribution to the marine carbon cycle. To understand these complex processes, from local to global scale, measurements in the water column, at the water surface and increasingly also in coastal areas are necessary. The three essential and global processes of the marine carbon cycle are the physical, the biological and the calcium carbonate cycle. The physical cycle is driven by the increasing solubility of CO_2 in cold water and the decreasing solubility of CO_2 in warm water masses. The biological cycle comprises the aspects of primary production and remineralization as well as the conversion of inorganic to organic carbon. Finally, the calcium carbonate cycle describes the precipitation or biological formation of calcium carbonate (calcite and aragonite) and its dissolution.

The fact that about 30% of anthropogenic CO_2 emissions are absorbed by the ocean (Stocker et al. 2013) causes another important driver for measurements of dissolved CO_2: ocean acidification. The more CO_2 is dissolved in seawater, the more the pH decreases, the more acidic the water becomes (Caldeira and Wickett 2005). In particular, but not exclusively, this fundamental change makes it difficult for calcifying organisms to build their skeletons or shells (Riebesell et al. 2000). The impact of this change on the marine biota is not yet fully understood and is currently being studied in many research projects worldwide. The reason for the close connection between CO_2 and pH value is the fact that dissolved CO_2 in water forms carbonic acid (H_2CO_3). This in turn immediately dissociates into hydrogen (H^+), bicarbonate (HCO_3^-) and carbonate ions (CO_3^{2-}). In fact, only about 1% of inorganic carbon is present in the ocean as physically dissolved CO_2 and carbonic acid; about 90% is hydrogen carbonate ions and the rest carbonate ions. Since the pH value is by definition directly related to the amount of hydrogen ions, it is influenced by the equilibria that are established within this carbonic acid dissociation system, the marine CO_2 system (Millero 2007). The more CO_2 dissolves, the more the equilibrium tends towards the dissolved gas and a low pH, resulting in a reduced amount of free carbonate ions. Besides the CO_2 partial pressure (pCO_2) or CO_2 fugacity[5] (fCO_2), the pH, total alkalinity and dissolved inorganic carbon are measurable variables of the marine CO_2 system. At least two of these variables must be present in order for the CO_2 system to be completely determined using thermodynamic constants (Millero 2007). Commercial underwater sensors suitable for field use so far only exist for the parameters CO_2 and pH (see Byrne et al. 2010).

[4] www.icos-infrastructure.eu

[5] In contrast to the partial pressure, the fugacity of a gas also takes into account real gas effects, which are described by its equation of state.

Further Applications for Dissolved CO_2 Measurement
Due to the outlined relevance of CO_2 for a multitude of biogeochemical processes, the increasing number of climate and environmental models relating to CO_2 also benefit from an improved spatio-temporal coverage of marine and atmospheric measurement data. A further, currently still rather scientific field of application for pCO_2 measurements is in connection with *carbon capture and storage* (CCS). In for example power plants, CO_2 is captured and stored into reservoirs. The latter can be located in depleted natural gas storage sites on land or on or under the seabed (Metz et al. 2005). Since 1996, for example, CO_2 has been captured from the natural gas produced in the Sleipner field in the North Sea and stored in the sediment through an additional borehole. In this context, the measurement of dissolved CO_2 is crucial for background measurements before injection or for leakage measurements in the vicinity of underwater structures. Long-term measurements are necessary to verify the planned emission-free storage. In areas such as the region around the island of Panarea in the Mediterranean Sea (north of Sicily), volcanic processes cause gas to escape at the sea floor. This consists of 94% CO_2 (Caramanna et al. 2011). Such areas are of great scientific interest to investigate the biological effects of CO_2 (see ocean acidification) and to carry out physical investigations of the solubility behavior and the reach of gas plumes. These are natural analogues to leakage scenarios that could occur in the environment of marine CCS. For the same reason, gas releases at the sea floor are also artificially induced, which also requires adequate sensor technology (Blackford et al. 2014).

In combination with the measurement of dissolved oxygen (O_2), pCO_2 is also an important parameter in fish farming and aquaculture, although the accuracy requirements sometimes differ significantly from oceanographic applications. In the marine environment, the combination of O_2 and CO_2 sensors is crucial to study oxygen-based primary production and to perform photosynthesis and respiration measurements. The measurement of O_2 is also essential for the investigation of oxygen minimum zones in seas and rivers, that is areas in which only those organisms can survive whose metabolism functions without oxygen (anaerobically).

The so-called Argo Network is a successful, international joint effort that has meanwhile deployed almost 4000 profiling deep drifters (English *floats*) in the world's ocean (JCOMMOPS, n.d.). These platforms can change their buoyancy with the help of an oil-filled swim bladder and thus take measurements along vertical profiles with a very low energy input. At regular intervals, the floats emerge to such an extent that an antenna protrudes from the water and the recorded measurement data can be transmitted via satellite to the control facilities on the mainland. They have no active propulsion and therefore drift with the currents, especially horizontally. Originally, primarily equipped with sensors for physical measurement parameters such as temperature, conductivity and pressure, they are now increasingly being expanded to include biogeochemical measurements (see SOCCOM, n.d.). Currently, about 250 of the Argo floats are equipped with O_2 sensors.

Motivation for Measuring Dissolved CH_4
The measurement of CH_4 plays a particularly important role in dammed, non-flowing inland waters (e.g., at dams), since large quantities of the greenhouse gas are produced here due to decomposition of organic material (*methanogenesis*) and can be released into the atmosphere (Maeck et al. 2013). Methane is formed in the sediment under anoxic conditions, mostly by microbial conversion of CO_2 and hydrogen to CH_4, which in turn is relates to CO_2 and O_2 measurements. As with CO_2 and investigated with a similarly strong scientific interest, there are also natural, volcanic or geothermal gas seeps in different regions of the ocean, which contain large amounts of hydrocarbons and thus also CH_4.

Trapped in a cage of water molecules, methane can be present as a stable clathrate, also known as methane hydrate or methane ice, under certain thermodynamic conditions (high pressures, low temperatures). Since these conditions occur in many places in the ocean, there are enormous deposits of this natural, fossil energy source worldwide. It is estimated that there is significantly more carbon in the form of methane hydrate worldwide than in the form of oil, gas and coal combined (Stocker et al. 2013). This statement, in combination with the realization that 1 m^3 methane hydrate can theoretically contain gaseous methane under standard conditions up to 164 m^3 (Kvenvolden 1993), makes it clear why methane ice is under discussion as a potential energy source. In 2013, Japan was the first country in the world to extract methane from a hydrate deposit for over 6 days (Anon, n.d.). Further projects of this kind are being worked on and planned worldwide. Particularly against the background of a further global warming, the conditions of hydrate deposits can leave the stability region, which can result in enormous CH_4 emissions into the water column and finally also into the atmosphere over longer periods of time (Stocker et al. 2013). Investigations of methane hydrates are therefore in the focus of research projects worldwide and represent an important application for appropriate gas sensor technology, not least because of the potential industrial perspective.

Offshore Deployments
One of the largest fields of applications for dissolved gas measurement is the offshore oil and gas market. The detection of leaks at pipelines and underwater structures can be carried out at a very early stage using gas sensors or more complex leak detection systems (Fig. 16.39). Potential follow-up costs and damage can thus be minimized. The fact that only 70% of the hydrocarbons produced reach the private or industrial consumer and 30% are lost during extraction, transport, refining and distribution (Devold 2013) underlines the need. In addition to the early detection of and search for leaks, background measurements prior to production or after decommissioning are also potential areas of application. The development of deeper and deeper deposits and an increase in automation go hand in hand with a growing environmental awareness, which is being driven forward by events such as the *Deep Water Horizon disaster* in the Gulf of Mexico in 2010. Since the predicted global

reserves of natural gas or methane hydrates are much greater than those of oil, a larger field of application for gas sensors can be expected in the coming decades.

In addition to the prominent greenhouse gases CO_2 and CH_4 presented here, other gases are also receiving increasing attention. The measurement of nitrogen or nitrogen compounds such as the greenhouse gas nitrous oxide (N_2O) is relevant in connection with nitrification and denitrification processes and generally for the investigation of the nitrogen cycle.

Measurement Technology

The direct measurement of dissolved gas molecules in liquids is not common in marine and limnological environments. Direct absorption measurements in the NIR and IR are considerably complicated due to the strong absorption of liquid water. The absorption coefficient in water of radiation with a wavelength $\lambda > 2$ μm is already 20 cm^{-1} and increases with longer wavelength. For underwater measurements at depths of up to 4000 m, Raman spectrometers have been used occasionally in the scientific environment, which operate with an excitation wavelength of $\lambda = 532$ nm and thus have sufficient transmission in seawater (Pasteris et al. 2004). The aim was to measure CH_4 and CO_2 in the vicinity of hydrothermal springs and gas vents. Also mass spectrometers (MS) equipped with a membrane inlet, referred to as a *membrane inlet mass spectrometer* (MIMS), are used for the measurement of dissolved gases (Schlüter and Gentz 2008; Wenner et al. 2004). The advantage of a multiparameter measurement is contrasted by a high technical complexity. Especially problematic is the reliable calibration of a MIMS instrument.

Due to the motivation discussed at the beginning, the measurement of greenhouse gases is very common in the atmosphere (see Sect. 2.2). The accuracies and precision achieved by optical absorption measurements in the NIR and IR range meet the highest requirements in this area. NDIR[6]-based systems (see Chap. 7) are increasingly complemented by CRDS[7] or OA-CEAS[8] instruments, the realization of which has been pushed forward by the rapid progress in the field of narrow-band laser diodes or QCL[9] light sources. Major advantages of optical absorption measurement are traceability through regular measurement of certified gas standards, no chemical alteration of the target molecules and high selectivity. These techniques are also very common for the measurement of dissolved gases or, as in the case of CRDS and OA-CEAS, are becoming more and more common for this application. However, in order to be able to use corresponding methods for the measurement of dissolved gases, a gas space (*headspace*) must first be created. The gas partial pressures in

[6] **N**on **D**ispersive **I**nfra**R**ed.

[7] **C**avity **R**ing **D**own **S**pectroscopy.

[8] **O**ff **A**xis **C**avity **E**nhanced **A**bsorption **S**pectroscopy.

[9] **Q**uantum **C**ascade **L**aser.

this headspace (gas phase) are in *equilibrium* with the corresponding concentrations in the liquid phase.

In the area of surface water measurements, this is possible through a variety of different equilibrator designs, such as spray, marble or column equilibrators, as well as combinations of the previously mentioned designs. All these techniques have in common that a water flow should be brought into partial pressure equilibrium as quickly as possible with regard to all gas molecules dissolved in it with a sample gas circuit, usually pumped through the equilibrator in countercurrent. A precise measurement of the target gas concentration in the equilibrated gas circuit can then be used to determine the amount of dissolved gas present in the water (→ Henry's law). Often, however, the measurement of the gas partial pressure is sufficient, since this is the essential and the driving parameter of for example exchange processes. For the measurement of CO_2 in surface water and for the determination of the gas flux between atmosphere and sea, complex flow-through systems are used, which usually consist of a wet box containing the equilibrator and a dry box containing the electronics and the gas analyzer (see Fig. 16.34). Usually four standard gases are used to completely characterize the NDIR sensor at regular intervals (Pierrot et al. 2009). While systems of this kind were initially only used on research vessels, corresponding measurements can now be carried out largely unattended on merchant ships or ferries. Recent advances in gas sensor technology now also allow for the measurement of CO_2 isotopologues in surface water (Friedrichs et al. 2010), the combined measurement of CH_4 and CO_2 (Gülzow et al. 2011) or also of N_2O, CO and CO_2 (Arévalo-Martínez et al. 2013).

An essential tool in marine research and a standard oceanographic technology is a so-called water sampling rosette (see Fig. 16.38), with the aid of which water samples can be taken in the water column and brought on the deck of a research vessel. Discrete water samples are still the only way to obtain field data for many laboratory-based measurements. A common reference method for measuring dissolved CH_4 in discrete samples is based on measurement using a gas chromatograph (GC). The water sample is filled bubble-free into a sample vial. Then a gas space is created in the vial with an inert gas and a partial pressure equilibrium between *headspace* and liquid is achieved by sufficient shaking. A sample from the *headspace* is now measured in the GC and the CH_4 concentration is determined. The amount of dissolved CH_4 originally present in the water can be deduced (see Capasso and Inguaggiato 1998). The determination of CO_2 of a discrete seawater sample is usually done by characterizing the carbonate system by measuring the total alkalinity and dissolved inorganic carbon of the sample. As written in the motivation, by measuring these two parameters (including temperature and salinity) the CO_2 system is completely defined and pCO_2 can be determined (Dickson et al. 2007). The discrete reference method for the determination of O_2 still goes back to the Winkler titration first described in the 1880s (Winkler 1888). In this method, the oxygen in the sample is first fixed as manganese hydroxide and later determined titrimetrically.

The two principles proven from the flow systems, optical absorption measurement in the IR and partial pressure equilibration, are also used in submersible sensors. The key element here is a membrane equilibrator which, by means of a semi-permeable membrane, enables

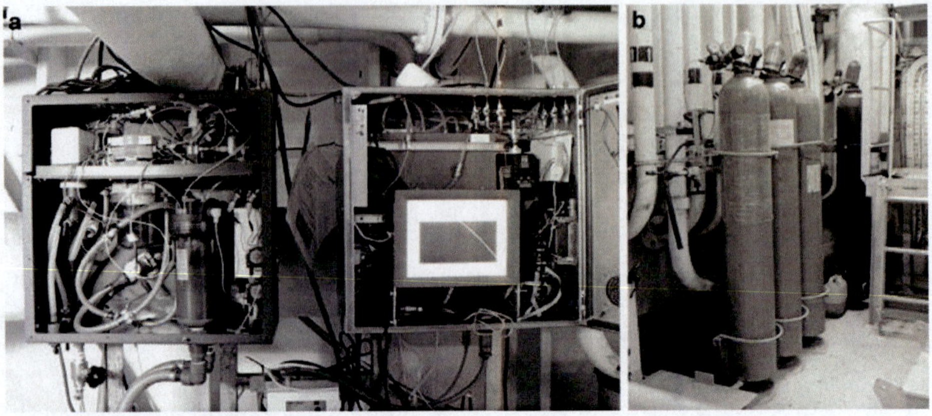

Fig. 16.34 Picture (**a**) Shows the wet (*left*) and dry box (*right*) of a flow-through system for measuring CO_2 in surface water and in the atmosphere (General Oceanics Inc., Miami, USA). The boxes have a size of approximately 600 × 600 mm. The central, cylindrical element of the wet box is the spray equilibrator. The NDIR analyzer is located in the dry box behind the monitor. The system is installed on a merchant ship, which collects valuable data for scientific purposes. The gas cylinders for regular calibration of the NDIR sensor are located in an extra room (**b**). (By courtesy of Arne Körtzinger, GEOMAR)

phase separation between liquid water on the one hand and an internal gas circuit on the other. At the same time, the membrane material is selected to have the highest possible permeability for the gases dissolved in the water. This is the case, for example, for polydimethylsiloxane (PDMS, *silicone*) or Teflon® AF. In such a setup, the gas molecules from the water also dissolve in the membrane, diffuse through the separation layer and desorb into the gas circuit of the sensor. This process continues until a partial pressure equilibrium is established between the dissolved gas in the water on the *front side* and the gas circuit on the *back side* of the membrane. If the partial pressure drops on the water side, the described process also runs in the other direction. Figure 16.35 shows the schematic structure of such an underwater sensor. A detailed description can be found in (Fietzek et al. 2014). The gas concentration in the circuit is measured either by a NDIR sensor or a TDLAS[10] module (Schmidt et al. 2013; see Fig. 16.36). Figure 16.37 shows an exemplary calibration procedure for such a CO_2 underwater sensor.

Other methods for measuring dissolved CO_2 are also used. Their measuring principle is not based on the CO_2 measurement in an equilibrated headspace as described above, but on the pH determination of a CO_2-equilibrated liquid in the sensor. This is based on the relationship between pH and pCO_2 defined by the carbonate system. The first measurements of dissolved CO_2 were developed for medical applications in the 1950s and were based on a wet-chemical pH measurement behind a Teflon® membrane

[10]**T**unable **D**iode **L**aser **A**bsorption **S**pectroscopy.

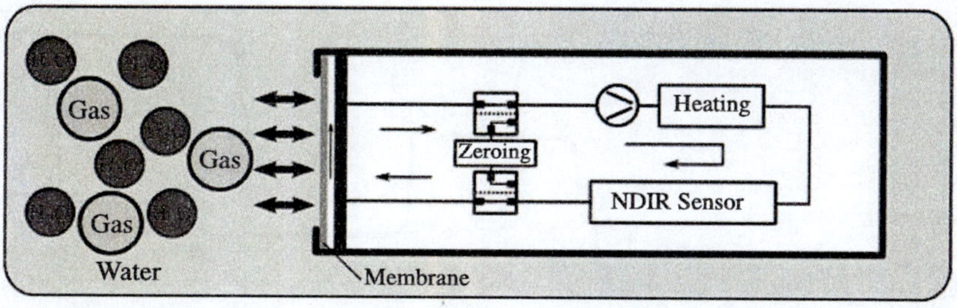

Fig. 16.35 Schematic drawing of a HydroC® sensor from -4H-JENA engineering GmbH, Jena, Germany (formerly from KM Contros GmbH, Kiel, Germany). The gases dissolved in the water diffuse through a flat membrane, mechanically reinforced against the hydrostatic pressure on the rear side, into the gas circuit of the sensor. A pump circulates the air between the membrane equilibrator and an NDIR sensor. A CO_2 sensor can be zeroed at regular intervals by passing the air through a CO_2 absorber and not past the membrane

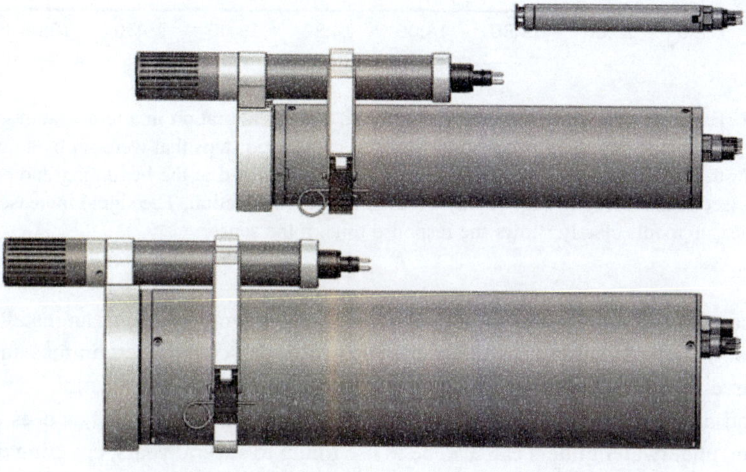

Fig. 16.36 Shown are a HydroFlash O_2 oxygen optode, a HydroC® CO_2 NDIR based pCO_2 sensor and a HydroC® Plus CH_4 TDLAS based CH_4 sensor in size comparison (*from top to bottom*). The titanium housing of the HydroC® CO_2 has a length of 380 mm and a diameter of 89 mm. The heads of the sensors are located on the left, the connections for power and data transmission on the right. The two HydroC® sensors are equipped with a submersible pump in the head area to ensure an even water flow onto the membrane and thus a fast response time of the sensor

(Severinghaus and Bradley 1958; Stow et al. 1957). Contemporary variants of this measurement method use spectrophotometric pH measurement in an equilibrated dye solution with known characteristics (DeGrandpre et al. 1995; Lefèvre et al. 1993). The deployment time of such sensors is limited by the amount of dye solution carried. In contrast to the optical measurement in an equilibrated headspace, no continuous measurements are possible with this method.

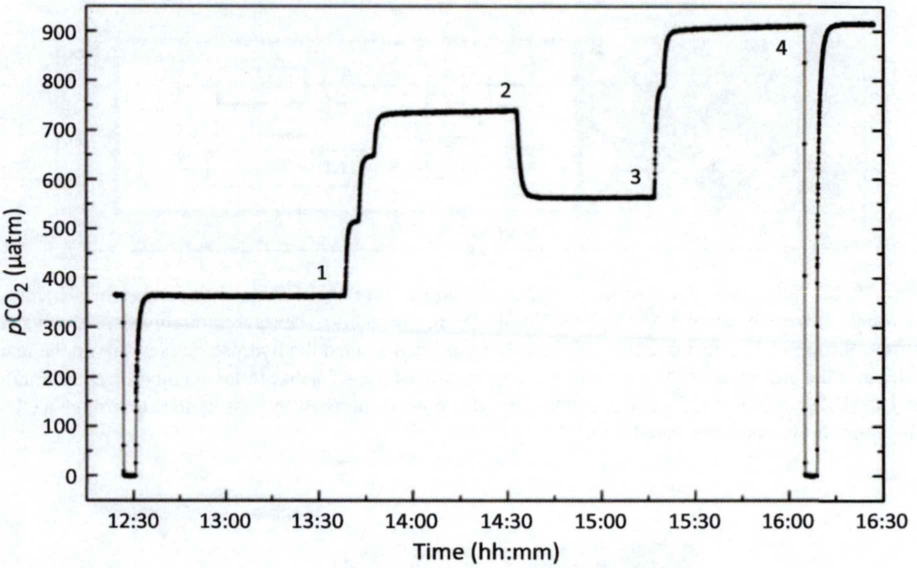

Fig. 16.37 Exemplary signal curve during a HydroC® CO_2 calibration in a temperature-stabilized water tank. Numbers 1–4 indicate the individual CO_2 calibration steps that were set by adjusting the pH value in the tank water. The zero gas measurements performed at the beginning and end of the calibration (see sensor setup in Fig. 16.35) are used for drift correction. The signal increase after the zero gas measurements clearly shows the response time of the sensor

The use of sensors in water is always coupled to an integration into or an installation on a platform. Only by an appropriate platform the sensor can cover a certain measuring area for a desired duration (Fietzek et al. 2011). Buoys or moorings, for example, are defined locally and are installed for up to 10 years. Modern mobile platforms such as floats or surface and underwater gliders can also be in use for up to several years, covering distances of sometimes more than 10,000 km. Finally, remotely operated submersible vehicles are capable of carrying out locally very defined work in the cm range on the seabed or of covering distances of up to several km. In doing so, they usually do not remain in operation for more than a few days at a time. Underwater sensors therefore have to be versatile and capable of being integrated into a wide variety of platforms. Figures 16.38 and 16.39 show some examples of this.

Another decisive aspect of underwater sensors for dissolved gases is their response time. The dominant effect in this context is the diffusion of the gas molecules through the semi-permeable membrane and not the response time of the NDIR or TDLAS detector in the gas circuit of the sensor. In order to minimize the equilibration time of the gas circuit, the thickness of the membrane separation layer is chosen as thin as possible (in the µm range) and usually an underwater pump is also used to generate a flow of water onto the membrane (see Fig. 16.36). For CO_2 sensors, time constants ($t_{63\%}$) of a little over 1 min were thus

Fig. 16.38 This compilation shows two different platforms for dissolved gas sensors: a water sampling rosette for measurements in the water column with sample bottles in the upper part of the metal frame and several gas sensors in the lower part (**a**); a profiling depth drifter (English *float*) for autonomous applications with a CO_2 sensor on the lower right (**b**). (By courtesy of **a** Mark Schmidt, GEOMAR, **b** Björn Fiedler, GEOMAR)

achieved (Fietzek et al. 2014). In Fig. 16.37, the response time of such a sensor can be clearly seen in the signal increase after the two zero gas measurements. The response time must be taken into account especially if the sensors are installed on a moving platform and if the localization of measured events should be as accurate as possible. Mathematical methods exist to correct a data set with respect to its response time as described for example in (Miloshevich et al. 2004) or applied to underwater sensor data in (Atamanchuk et al. 2014a; Fiedler et al. 2013).

For the measurement of O_2, more and more optical sensors based on the principle of fluorescence quenching have prevailed in recent years, the so-called *optodes* or *optrodes*. Essential for the measurement are special dyes, so-called fluorophores, which are embedded in a polymer matrix. This dye membrane is in contact with the water from one side while the fluorophores can be optically excited and read from the back. The more oxygen is present in the water and diffuses into the dye membrane, the less fluorescence occurs. Due to their long-term stability in particular, the optodes increasingly displaced the electrochemical Clark cells. The process of fluorescence quenching was already described in Kautsky (1939). However, it was not until the late 1990s that sensors based on this method found their way into marine research (Demas et al. 1999; Klimant et al. 1995). Further important characterization steps followed (e.g., Bittig et al. 2014; Körtzinger et al. 2005;

Fig. 16.39 This picture mosaic shows different application scenarios for dissolved gas sensors: a *working-class ROV* for inspection tasks in an industrial offshore environment (1a) with an installed leak detection system (1b); a FerryBox flow-through system for multi-parameter measurements in surface water (2a) with a flow-through CO_2 sensor (2b); an autonomous *Wave Glider* where the integrated sensors are hidden by solar panels (3). (By courtesy of: (3) Björn Fiedler, GEOMAR, (2) -4H-JENA engineering GmbH)

Tengberg et al. 2006). These small and economical sensors with response times ($t_{63\%}$) in the one or two-digit second range can be integrated into platforms correspondingly well (see Fig. 16.36). There have been initial successes in transferring this measuring method to other target molecules such as CO_2 and in using corresponding sensors in the marine science environment (Atamanchuk et al. 2014b).

A major challenge for all underwater gas sensors is the signal stability during long-term deployments. Signal drifts must be taken into account and compensated for over periods of several months to a few years in order to achieve the best possible measurement quality. In the case of CO_2 sensors, zero gas measurements for drift correction in the field can be carried out regularly as described above (see Figs. 16.35 and 16.37). In general, the best results can always be achieved by an accurate calibration before and after a deployment (e.g., Fietzek et al. 2014). For oxygen optodes that are installed for example on profiling floats, there are two further possibilities for correction during long field missions. On the one hand, measurements in the atmosphere[11] can be used, which are always possible with an optode when the float is at the surface, to send the measurement data via satellite. On the other hand, measurement data in deep water of known oxygen concentration can be used to check the signal stability. Not least with regard to the aspect of long-term stability of

[11]The concentration of oxygen in the atmosphere can be assumed to be sufficiently constant. In this context, a correct assumption of the air humidity is more decisive.

dissolved gas sensors, expectations are high for laser-based measuring methods such as TDLAS.

Summary
The range of applications of sensors for the measurement of dissolved gases is extremely versatile and ranges from lakes and rivers to seas and the ocean. It includes measurements on the water surface as well as measurements in the deep sea at depths of several 1000 m. Depending on the application and motivation for the measurement, the concrete requirements for the measurement quality such as accuracy and precision differ. What unites all these measurements are the generally demanding measurement conditions due to aspects like the following:

- The sensors must always be part of a suitable platform in order to be able to spatially and temporally perform the desired measurements.
- Long autonomous deployments place high demands on the robustness of the measuring methods used.
- Biofouling is always a great challenge when measuring in aquatic environments. This is the plant and animal growth that settles and develops on all bodies that are placed in the light-flooded part of a body of water.
- Especially within profiling measurements, underwater sensors also experience significant temperature gradients that must not affect the measurement technology.
- Furthermore, salt water is a challenging medium in which most metallic materials tend to corrode in the long run.
- After all, any sensor design must always be able to cope with the enormous hydrostatic pressure in the water column.

Despite, or perhaps because of, these challenging measurement conditions and the relevance of many of the applications, dissolved gas measurement represent a growing market. There are many interesting scientific but also promising industrial applications, which will be developed with increasing optimization of dissolved gas sensors; for example waste water treatment or the detection of toxic gases in the sea.

Membrane-based and thus equilibration-based optical sensors predominate in the environment of underwater gas measurements. Since it is possible to create an equilibrated headspace with regard to the dissolved gases even under water, different measuring methods can be used flexibly. In the future, multi-parameter gas and isotopology-selective measurements under water are conceivable and realistic with such a sensor design. This will make it possible to draw conclusions about the formation and reaction history of certain gases such as CO_2 and CH_4 in the water masses of interest. Further advances with dye-based optodes will produce highly autonomous sensors that will be widely used due to their properties.

16.5 Direct Respiration Analysis in Farm Animals

Tobias Lehmann, Scott Zimmerman, and Patrick Zimmerman

Currently, 98% of total global emissions of anthropogenic greenhouse gases are represented by the gases CO_2, CH_4 and N_2O. Looking at the specific, climate-relevant properties of CH_4, it can be seen that CH_4 has a significantly higher relative absorption capacity of thermal radiation (factor ≈28) compared to CO_2 and thus also brings with it a higher warming potential (GWP[12]). The volume fraction of CH_4 in the atmosphere has increased over the past 250 years from 0.7 to 1.75 ppm (IPCC 2007).[13]

Problem

Of the total CH_4 emissions, around 70% are caused by humans and of these, 60% come from agriculture (LEAD 2007). Thus agriculture is the most important single source of CH_4 emissions. The increase in agricultural CH_4 emissions was significantly influenced by a steady intensification of animal husbandry, which has quadrupled since 1950 (Vabitsch 2006). For example, the UN Food and Agriculture Organization expects an increase in CH_4 emissions from livestock farming of up to 60% by 2030. In terms of the greenhouse gas CH $_4$, it can be said that one cow emits approximately 200–400 g CH_4 per day (Flachowsky and Brade 2007). In addition to the environmental influences mentioned above, there are also further problems at the level of the individual animal or the herd association, which are immediately apparent to users, researchers and animal breeders. This is because the measurable CH_4 emission of ruminants reflects a significant loss in the energy balance of the animals, whereby typically 6–8% of the energy supplied by ingested food is lost due to CH_4 emission (Johnson and Johnson 1995; Lassey et al. 1997; McGinn et al. 2004). This of course represents a financial burden for the breeder, as well as an additional emission of greenhouse gases. Studies have shown that the majority of CH_4 emissions (92–99%) are emitted by the *belching* (lat. Ructus) of ruminants (Grainger et al. 2007). The amount of CH_4 emitted, or the frequency and intervals between belching or a change in it, could presumably also provide information about the health condition or a change in the health condition of the cattle under consideration. Furthermore, it is possible to search for individuals with a special genetics that have a better energy balance in order to pass this on to subsequent breeding. Further investigations have shown that the choice of feed, feeding times and quantity of feed has a considerable influence on CH_4 emissions, which may be able to significantly reduce them (Beauchemin and McGinn 2005; Jalc and Certik 2005). Thus, both environmental and economic aspects are taken into account in this

[12] Global Warming Potential.

[13] The current methane content is 1.85 ppm according to measurement data from the Mauna Loa measuring station in Hawaii. Based on these data, a stagnation of the methane increase in the years 1999–2006 could also be observed. A precise explanation for this unexpected phenomenon is still pending.

problem. The focus of this problem will be the consideration of CH_4 emission on the basis of the individual animal with the emphasis on CH_4 emission from the rumen, as well as the associated gas measurement technology and its special character.

State of the Art

Up to now, only isolated respiratory gas analyses have been carried out in studies or pilot plants using complex, large and cost-intensive analysers or GCs which simultaneously monitored several or individual cattle in a defined space, a so-called respiration chamber. Metrological approaches to monitor emissions directly near the head in the form of hoods, masks or probes, which have to be mounted on or in the cattle, proved to be ineffective. Another possibility, even without the use of respiration chambers, would be the SF_6 tracer method (Zimmerman 1994). In this method, a known contingent of SF_6 diffuses out of a permeable reservoir in the rumen to interact with the locally available CH_4. The remaining SF_6 quantity is then measured. However, this method is also time-consuming, very cost-intensive and limited to the observation of individual animals. With this method, too, the SF_6 gas emitted in the lower ppm range would have to be collected for later analysis, that means the test animal would have to be equipped with the appropriate apparatus. So far, there was no precise, compact and cost-optimised analysis system that was easy to integrate and could reliably solve this measuring task individually per bovine animal with measured values in the range of seconds under harsh environmental conditions.

Objective

The object and objective of the development project was to develop an economical, compact and reliable respiratory gas analysis system that can be used to record the CH_4 and CO_2 emissions of individual cattle. The respiratory gas analysis system developed should be simple, robust, as accurate as possible and capable of continuously determining the emissions in the normal everyday life of farm animals.

With the knowledge gained, new feeds and feeding methods can be developed, a targeted search can be made for individuals with optimal, more energy-efficient genetics and, in the long term, this respiratory gas analysis can be used to develop a strategy to reduce emissions. Another aspect is that it could be decided more quickly which cattle can be used for milk production or which cattle is more suitable for meat production. Furthermore, with the help of respiratory gas analysis, changes in animal health can be detected early and countermeasures can be initiated in time. A possible application for emission certificates, following the example of the CO_2 emission certificates of the industry, would be a further future application example.

Solution Approach

Based on this current problem, the company C-Lock Inc.[14] from Rapid City, South Dakota (USA) outlined a breathing gas analysis system. Due to the long-time experience of Sensors Europe GmbH in the field of gas analysis, already available know-how can be used, so that a cooperation between C-Lock Inc. and Sensors Europe GmbH for this breathing gas analysis project could be realized. The innovation of this project lies in the provision of all necessary information such as the individual CO_2 – and CH_4 – concentration, air pressure, wind speed, humidity, flow rate and ambient temperature. Figure 16.40 gives a first approximation of the basic idea of the breathing gas analyzer.

With this respiratory gas analysis system, the CH_4 and CO_2 emissions of cattle can be recorded directly and individually for each cattle in order to actively control feeding times and feed use (stall husbandry) and to obtain an estimate of the average CH_4 emissions from the rumen for each cattle and for the herd. The cow breathes directly into the breathing gas analyser during feed intake or during milking. If the ructus occurs during this process, it is also measured. The measurement data are assigned to each animal individually using its RFID tag and stored in a database for further evaluation. The information and measurement results obtained in this way are then evaluated by the company C-Lock Inc. and made available to researchers and users. Due to its compact design, this analysis system can be installed in milking robots or in automatic feeders on the pasture as well as in the barn in order to measure the ruminants' ructus. The above described respiratory gas analysis system of the company C-Lock Inc. is marketed under the product name GreenFeed™ and has already been applied for a patent (Zimmerman 2012).

Structure of the System

In order to develop a usable device (see Fig. 16.41) from the first ideas sketch (see Fig. 16.40), some adjustments and special features were necessary, because the respiratory gas analysis system has to function smoothly in the rough everyday life in wind and weather on the pasture, as well as in the somewhat better protected stable, also in the long run. The longest possible fault-free operation (*Mean Time Between Failures*, MTBF) is desired, since the systems are partly used in terrain that is difficult to access and the availability of technical experts on site is sometimes difficult. Furthermore, it had to be considered that the intended operational environment is heavily dust-laden and has a high relative humidity. Corrosive materials can also occur in the application environment, as well as a strong external force effect of the cattle (keyword shock and impact resistance of the measurement technology and its periphery). Ultimately, even the smallest living

[14]The company C-Lock Inc. was founded by Dr. Patrick R. Zimmerman in Rapid City. Dr. Patrick R. Zimmerman has been instrumental in the development of the SF_6 Tracer method and was for many years Director of the Institute of Atmospheric Sciences at the University of Mines and Technology in South Dakota.

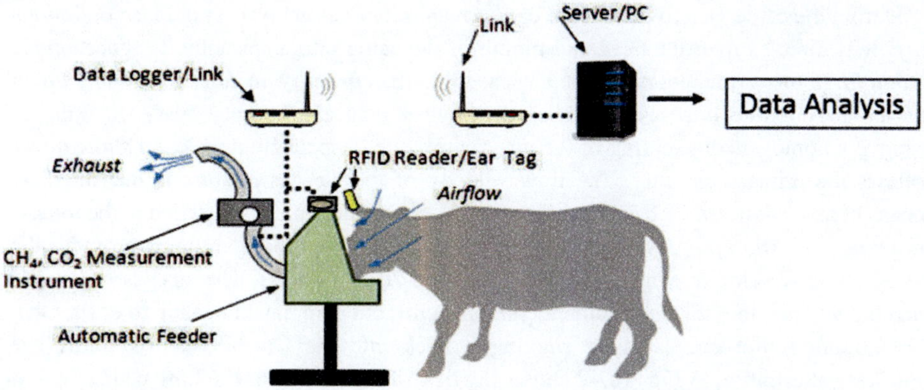

Fig. 16.40 Idea sketch of the breathing gas analyzer to be developed. (Source: C-Lock Inc.)

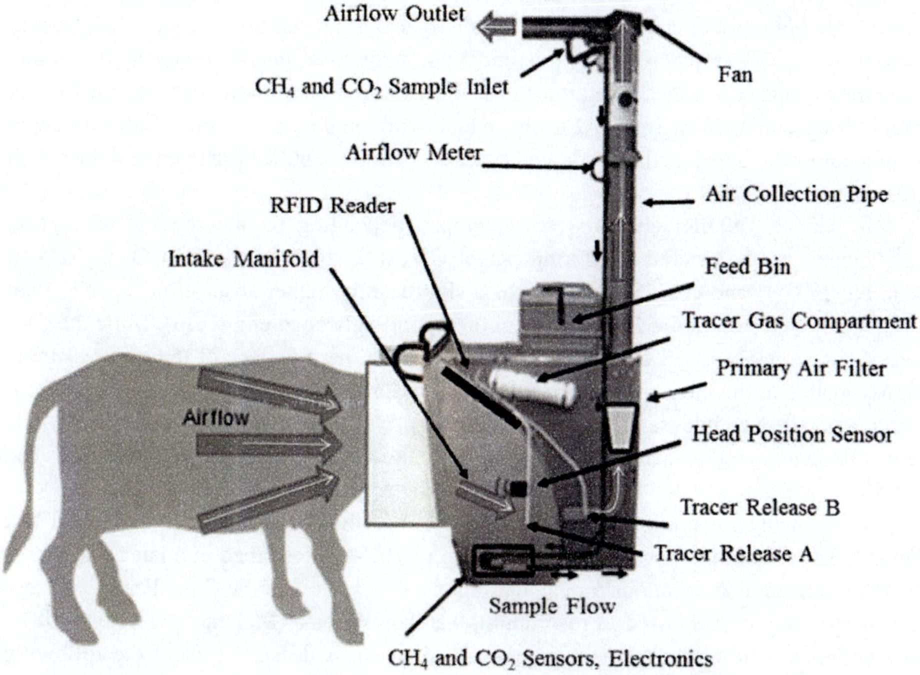

Fig. 16.41 Overview of the GreenFeed™ respiratory gas analysis system. (Source: C-Lock Inc.)

creatures such as mice or insects play a significant role in the long-term error consideration; special protection of cables and hoses against rodent bites had to be implemented.

During the cattle visit to the GreenFeed^{TM} respiratory gas analysis system, a continuous *airflow* is directed over the head and mouth of the cattle into a specially designed *intake manifold.* In the further course of the air–gas mixture, a *primary air filter* follows in which the air–gas mixture is first cleaned, as the environment can be very dusty. In order to achieve a homogeneous mixing of the air–gas mixture, a specially developed *static mixer* follows the primary air filter. The flow velocity of the air–gas mixture is measured by means of an *airflow sensor* based on a hot film anemometer, which is installed in the middle of the *air collection pipe*. A powerful fan (*air fan*) at the top of the chimney provides the necessary flow velocity, which is usually between 20 and 40 L/s. The air–gas mixture is then blown back into the environment (*sample vent*), but care must be taken to ensure that the mixture is not circulated by sucking it back into the GreenFeed^{TM} system. The following illustration in Fig. 16.42 shows the described stations of the flow path.

The actual sampling point of the air–gas mixture to be analysed is located in the upper part of the stack. Behind the fan, a *sample port* is used to continuously take a sample from the air–gas mixture and return it to the lower part of the GreenFeed^{TM} System for further analysis. In order not to contaminate the sensitive analysis technique, the sample is passed through a *secondary sample air filter* and then through a *sample pump* to the actual measuring point. The exact measurement of the CO_2 and CH_4 components is carried out with OEM products from Sensors Europe, which work on the basis of the NDIR two-beam method and were developed and adapted especially for this application in cooperation with C-Lock Inc. (Fig. 16.43)

Two separate NDIR sensors were developed depending on the specific absorption coefficients and the desired measuring range (CH_4 0–2 vol.% in N_2 and CO_2 0–1.5 vol. % in N_2) of CH_4 and CO_2. Since CO_2 has a significantly higher absorption capacity than CH_4, the optical path of the CO_2 sensor is correspondingly smaller (~2 cm), while the CH_4 sensor has a visibly larger optical path (~9 cm). Since the influence of the air pressure is clearly visible in the measurements, a pressure measurement is carried out, on the basis of which a pressure compensation for the respective gas channel can be derived. Each NDIR sensor has a temperature sensor whose signals are used for temperature compensation. The specified accuracy is stated as 1% in relation to the respective *full scale of range* (FS).

The following illustration in Fig. 16.44 describes the flow pattern of the actual test sample. The calibration port also shown in Figure 16.44 is described in a later chapter.

The mathematical equations for calculating the air–gas flow of the GreenFeed^{TM} system are comparable to those used in respiration chamber systems (McLean and Tobin 1987). The volumetric flow rate Q_c, as a function of time (t), is described using the following equation (Eq. 16.2):

$$Q_{c(t)} = C_{P(t)} \cdot \left(\mathrm{Conc}_{(t)} - \mathrm{BkConc}_{(t)}\right) \cdot Q_{\mathrm{air}_{(t)}} \tag{16.2}$$

Here the current gas concentration in ppm related to the volume is described with $\mathrm{Conc}_{(t)}$ and the current background gas concentration in ppm with $\mathrm{BkConc}_{(t)}$. The total volumetric

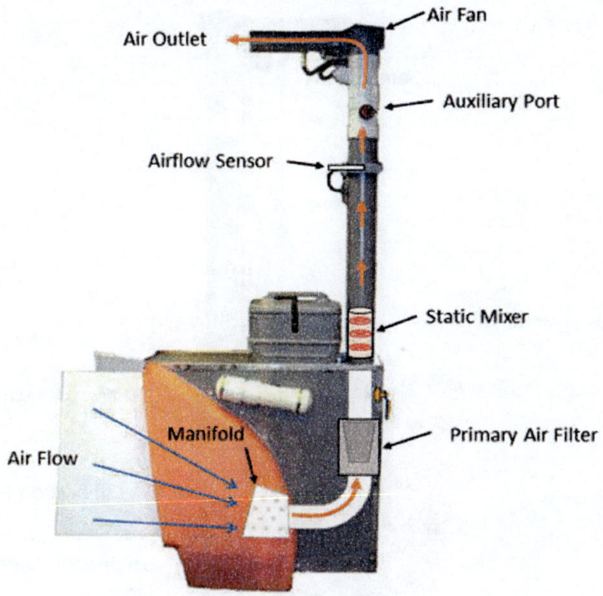

Fig. 16.42 Flow pattern of the air–gas mixture in the GreenFeed™ System. (Source: C-Lock Inc.)

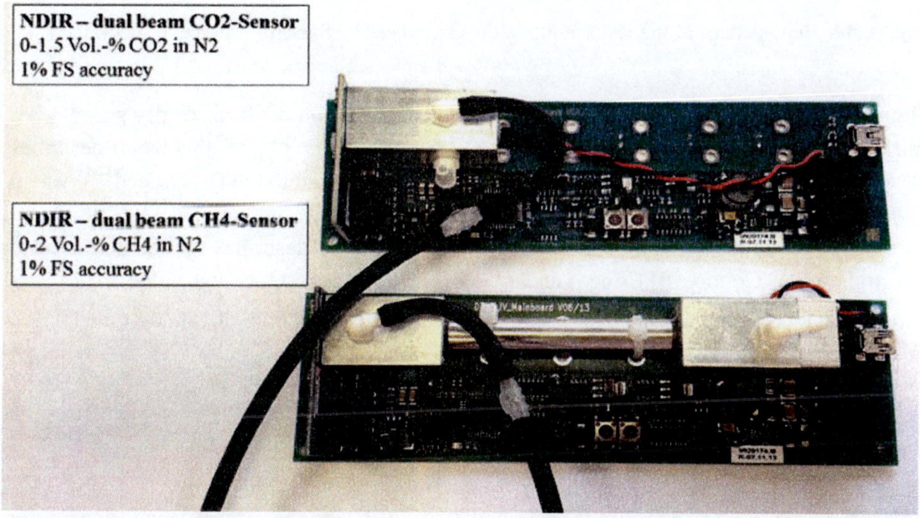

Fig. 16.43 NDIR gas sensors (CO_2 and CH_4) of the Gasmitter™ series from Sensors Europe GmbH (The NDIR sensors specially developed for this respiratory gas analysis project are based on the new Gasmitter™ series from Sensors Europe GmbH, which can be equipped with different NDIR sensors for different gases and measuring ranges and is therefore successfully used in many applications)

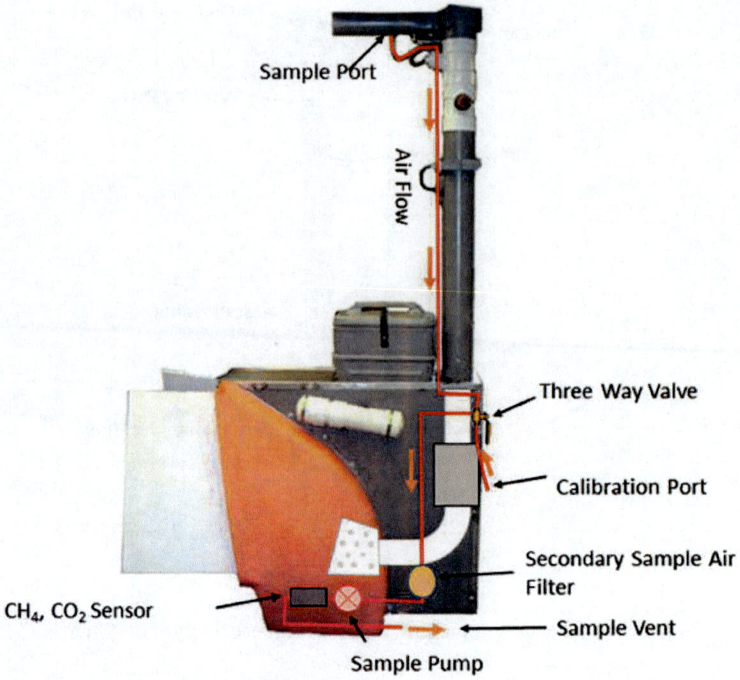

Fig. 16.44 Flow pattern of the test sample in the GreenFeed^{TM} System. (Source: C-Lock Inc.)

flow rate is also represented as $Q_{air(t)}$, which was measured on the basis of dry gas at 1 bar and is indicated with Liter/s. The air intake rate is expressed in $C_{p(t)}\%$, this factor describes the ratio of the efficiency with which the emissions are drawn into the GreenFeed^{TM} system via the intake. The measurements and calculations of the GreenFeed^{TM} System are normalized on a time basis of 1 s. The following paragraph describes the method used to determine $C_{p(t)}$, $Conc_{(t)}$, $BkConc_{(t)}$ and $Q_{air(t)}$ (see Eq. 16.4). Once the value for $Q_{c_{(t)}}$ is known, the ideal gas law can be applied to determine the mass flow Q_m at any time (t) (see Eq. 16.3).

$$Q_{m_{(t)}} = Q_{c_{(t)}} \cdot \frac{273.15}{273.15 + T_{air_{(t)}}} \cdot p_c \tag{16.3}$$

Here, the current air temperature is described $T_{air(t)}$ in °C and the respective gas density for CH_4 and CO_2, based on 1 bar and 273,15 K, is expressed in p_c. The CH_4 and CO_2 concentration measurements show a time delay of approximately 4–5 s due to the transit time from the nose of the cow through the GreenFeed^{TM} system to the NDIR sensor. The values measured in real time (RFID, head position measurements, etc.) are synchronized with the time-delayed values. The background concentrations of CO_2 and CH_4 are determined by measuring the ambient air concentrations before and after the visit of a bovine in the GreenFeed™. During the visit the $BkConc_{(t)}$ is calculated by means of a linear function over the time from start to end point. The most important criterion for an accurate calculation is that the background concentration does not change outside the linear function during the visit.

$$Q_{air_{(t)}} = V_{m_{(t)}} \cdot V_{adj_{(t)}} \cdot \frac{Ap}{1000} \tag{16.4}$$

The cross-sectional area of the chimney pipe was measured at 0.00811 m^2 and is expressed as Ap. The flow velocity measured in m/s in the middle of the chimney pipe is also described as $V_{m_{(t)}}$. Since the flow velocity in the chimney pipe can vary depending on the measuring point, an experimentally determined correction factor of 0.88 is given in order to calculate an average flow velocity. The correction factor of the flow velocity is given as $V_{adj_{(t)}}$. The air intake rate C_p is determined by means of a propantracer system and is calculated from the resulting ratio of the two injection points (*feeder tracer and standard tracer*, see Fig. 16.45). The NDIR CH_4 sensors also react to C_3H_8 concentrations, but it is not possible to differentiate between CH_4 and C_3H_8. Usually the propane tracer system is activated when there is no cattle in the GreenFeed™ system. As the propane sometimes contains odorous substances, a special filter (*tracer de-odorizer*) is used to minimize this odor, otherwise the cattle would avoid the device. Approximately every 2–6 h this propane tracer is automatically inserted.

This method is used to monitor the measuring characteristics over a longer period of time. A change between current measurement results and measurement results of previous tracer tests could indicate a changed calibration of the CH_4 sensors or the flowmeters. With this method, the dead time between suction and measurement can also be determined very precisely. For the general calibration of NDIR gas sensors, gas cylinders or calibration gas bags are usually available. The calibration can be performed automatically (gas cylinders) or manually by the user (calibration gas bag). Due to the not static position of the cattle head clear measuring errors could arise. Therefore only measured values are brought into a further evaluation if the head position sensors detect the head in a firmly defined area. All other measured values are discarded so as not to influence the accuracy of the measurement.

First Measurements

If a bovine animal is now lured into the GreenFeed™ system by a certain amount of feed, it will start to eat. At the same time, at irregular intervals, the Ructus occurs, whose emissions continuously enter the intake tract of the GreenFeed™ System. Figure 16.46 clearly shows the background portion of the ambient methane, as well as the respective emissions from the lungs and the ructus.

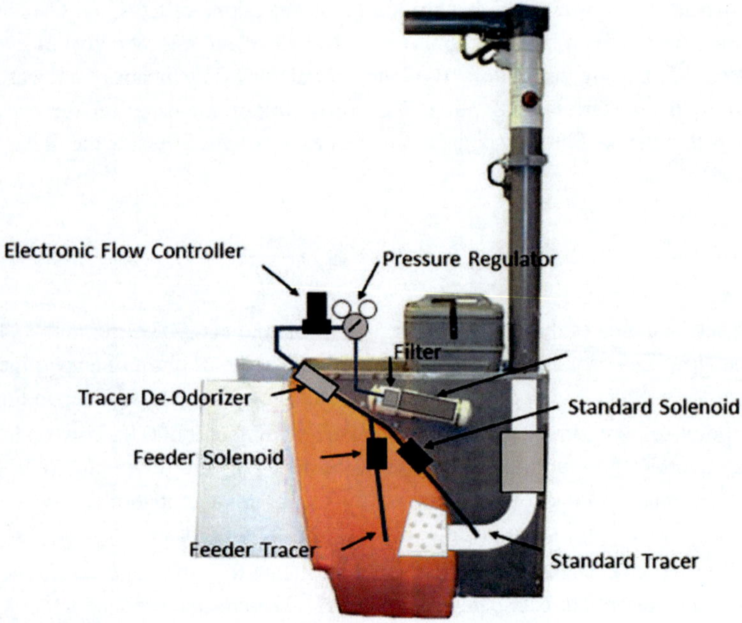

Fig. 16.45 Propane tracer in the GreenFeed™ System. (Source: C-Lock Inc.)

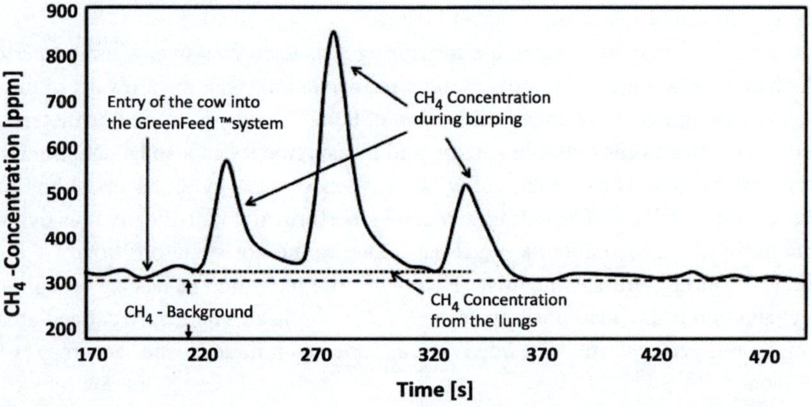

Fig. 16.46 CH_4 emissions from cattle in the GreenFeed™ System. (Source: C-Lock Inc.)

If these raw data are now further processed using the additional criteria or measurements, the emission of CH_4 can be represented in grams per time (see Fig. 16.47).

For researchers working in animal nutrition, the individual CH_4 emission per bovine animal in g/d is an important piece of information. Based on the measurements carried out,

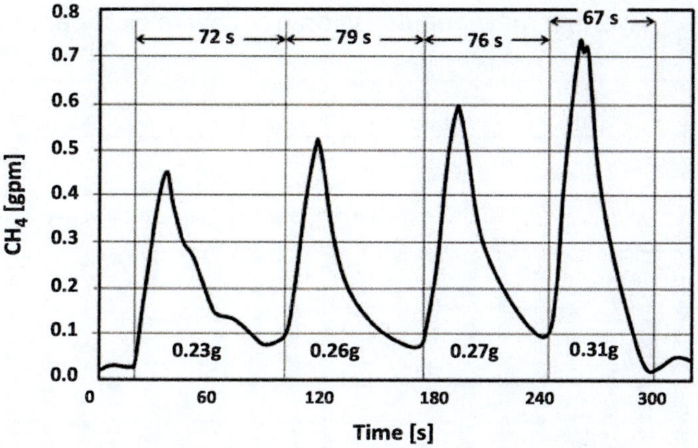

Fig. 16.47 Bovine methane emission in grams per unit time. (Source: C-Lock Inc.)

a value for the average CH_4 emission in grams per day of the respective animal can be derived by summation.

Equation 16.5 shows the summation of the individual CH_4 emissions per ructus in relation to Fig. 16.47. After completion of the summation of the individual ructus, the value is extrapolated from g/s to g/d.

$$\text{Methane Flux} = \frac{0.23\text{g} + 0.26\text{g} + 0.27\text{g} + 0.31\text{g}}{72\text{s} + 79\text{s} + 76\text{s} + 67\text{s}} = 0.00364\frac{\text{g}}{\text{s}} = 13.1\frac{\text{g}}{\text{h}} = 314\frac{\text{g}}{\text{d}} \quad (16.5)$$

This measurement is now repeated with many different cattle, whereby the measurement data are assigned to the individual cattle according to its specific RFID number. How often a cow receives some feed per day on the GreenFeed™ System, and especially the amount of feed per session, can be freely defined by the user per cow. Usually, a cow visits the GreenFeed™ System 2–3 times a day. The relative measurement error is further minimized by continuously repeating the measurement over days, weeks and months. Thus, an average value per cow or herd can be determined (see Fig. 16.48).

If these measured data were now to be put into relation with the data on the amount of water and feed consumed, or the DMI (*dry matter intake*) and the milk quota produced, this would provide a basis of information on the corresponding cattle, which is of great interest to animal nutritionists.

Outlook

The GreenFeed™ System is currently being successfully used on almost all continents, except Antarctica and Africa, and is continuously generating data. Renowned researchers from Sweden, Finland, Brazil, Belgium, the Netherlands, France, Korea, Australia, New Zealand and the USA are already working with the GreenFeed™ System. The

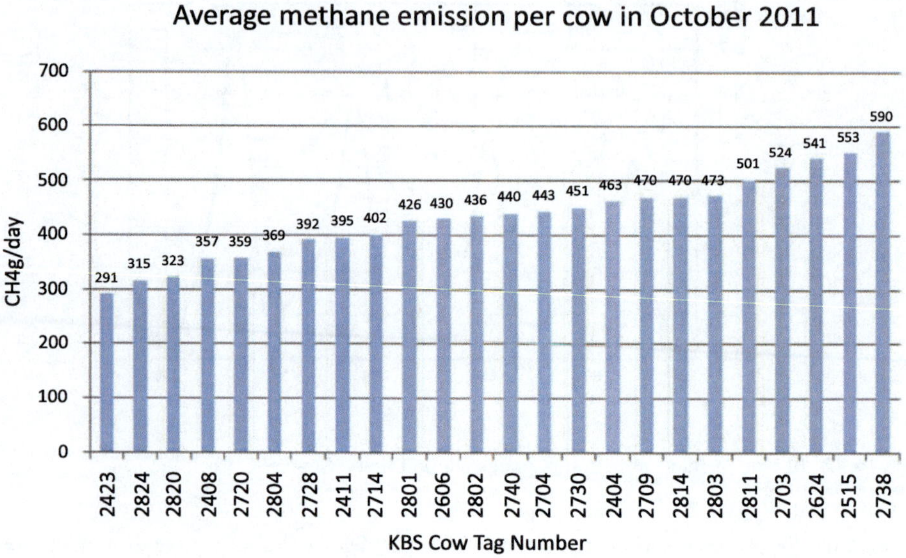

Fig. 16.48 Average methane emission per cow (October 2011). (Source: C-Lock Inc., Kellog Biological Station (KBS))

users and researchers regularly meet at international meetings and seminars, such as the *GreenFeed Users Class* of the company C-Lock Inc. or the *Methagene Training School* of the Leibniz Institute for Farm Animal Biology (FBN) to discuss the latest findings. The first GreenFeed™ System in Germany was put into operation in December 2014 near Rostock at the Leibniz Institute of Farm Animal Biology (FBN). Although the GreenFeed™ System is currently still being used in the form of scientific studies at universities or research institutes, a future use by large-scale breeders or commercial users is planned.

16.6 $^{13}CO_2/^{12}CO_2$ Measurement of Respiratory Gas in Medical Diagnostics

Walter Fabinski and Thomas Weyrauch

This paper describes the use of NDIR[15] measurement technology to determine the isotope ratio of $^{13}CO_2$ and $^{12}CO_2$ in respiratory gas analysis in medical diagnostics. For this purpose, the mode of action is explained using the example of diagnosis of infection of the stomach with the bacterium *Helicobacter pylori* (HP), and the requirements, structure and functionality of NDIR measurement technology are presented. The results are

[15]NDIR = Non-Dispersive Infra Red.

compared with those of the established mass spectroscopy (MS) method. Besides the use for HP detection, the method has potential for further investigations of metabolic processes in human, animal and plant organisms involving carbon compounds while avoiding radioactive tracers.

Background

Barry J. Marshall and J. Robin Warren were awarded the Nobel Prize in Physiology and Medicine in December 2005 for their discovery of the bacterium HP as the main cause of complaints in the stomach and duodenum. This discovery has fundamentally changed the treatment of stomach diseases. Marshall and Warren found that HP lodges in the stomach wall by its flagella and, by increasing the secretion of stomach acid, can cause diseases such as gastritis, ulcers and, in the case of chronic infection, cancer. According to available studies, about half of the world population is infected with HP. HP is considered one of the most common bacterial diseases after caries. In Germany alone, the number of people infected with HP was estimated at 30% in the mid-1990s. In total, 5–10% of these suffer from ulcers in the stomach and duodenum (Haisch 1995; Haisch et al. 1996). Before the discovery of HP, it was believed that no living organism, not even a bacterium, could survive in the extremely acidic environment of the stomach. People who suffered from gastritis or peptic ulcers should avoid stress, smoking and alcohol and were treated with histamine blockers. However, this brought only short-term relief but no cure. In order to find the cause, however, invasive methods such as gastroscopy, which penetrated the patient's body, were still required. From today's point of view, the achievements of Marshall and Warren cannot be overestimated. They brought about a change in thinking and promoted the search for new methods of diagnosis and treatment and led to a significant reduction in the infection rate in industrialized countries. This also applies to the gastric cancer rate (Kazutoshi et al. 2008).

Initially, research focused on its physiology and pathogenicity (i.e., its ability to cause disease). HP is a rod-shaped bacterium (Fig. 16.49), which holds on to its flagella in the protective stomach mucosa. Here, it can feed and reproduce, secreting toxic substances that cause further damage to the stomach lining. As a result, the caustic stomach acid can attack the stomach wall, which can lead to further inflammation and serious complaints. It is estimated that almost all duodenal ulcers are due to HP (Haisch, et al. 1996). Research has shown that the bacterium absorbs urea and converts it into CO_2 and ammonia through an enzymatic process (urease). The basic ammonia neutralizes the acidic gastric juices and forms a survival envelope around the bacterial cell. The CO_2 formed in the process enters the air we breathe via the bloodstream and lungs. This is how the bacterium reveals itself. If it is possible to detect the CO_2 produced by the bacterium, a non-invasive diagnostic method for HP infection is obtained from the breath of the person being examined. But how can the small amount of CO_2 produced by the bacterium be distinguished from the CO_2 constantly present in the air we breathe? In nature, carbon occurs in the form of the stable isotopes ^{12}C, ^{13}C and the radioactive isotope ^{14}C. ^{12}C has the largest share with almost 99%

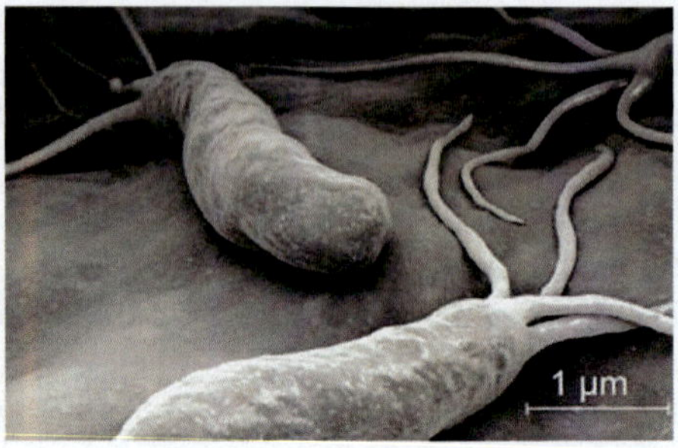

Fig. 16.49 View of the bacterium *Helicobacter pylori*, taken with a scanning electron microscope

of the total global carbon, followed by ^{13}C with about 1%. In contrast to radioactive ^{14}C, ^{13}C and ^{12}C are stable and completely harmless to humans. If the ^{12}C atom in the urea CO $(NH_2)_2$ or CH_4N_2O is replaced by the rare ^{13}C atom, $^{13}CO_2$ is formed under the influence of the urease (Haisch et al. 1996). Now all that is needed is to measure CO_2 isotope-selectively in the air we breathe in order to detect HP. This ratio can be determined for example with a MS. However, this is an expensive procedure that is associated with high operating and maintenance costs. What was missing for clinical use was a simple, robust, inexpensive and yet highly sensitive and selective measuring instrument. The search was on for a device that would meet the requirements of medical practice and completely avoid the use of radioactive material.

Requirements for an Isotope Selective Diagnosis and Treatment Concept

The parameter to be determined for a breath test in clinical application is the $\delta^{13}C$-value as a measure of the isotope ratio (Haisch et al. 1996). It is defined as%:

$$\delta^{13}C = (R_P/R_R - 1) \cdot 1000‰ \tag{16.6}$$

Here, R_P is the isotope ratio of a specific breathing gas sample and R_R is the isotope ratio of a reference. In general, the ambient air is used as a reference. In medical practice, this examination is simplified in such a way that only the difference between the $\delta^{13}C$ values before (zero sample) and after administration of the substrate administered with ^{13}C is used for the evaluation. This is expressed as *Delta over Baseline* (DOB) and expressed in per mil (‰).

The concentrations in the expiratory air are between 3 and 5 vol.% CO_2. The remainder consists essentially of the gases oxygen (approx. 16 vol.%), nitrogen (approx. 78 vol.%) and argon (approx. 1 vol.%). The ratio of $^{13}CO_2$ and $^{12}CO_2$ in the ambient air is (1 ± 0.003) %. If HP is present, this ratio will change in the breathing air, since the substrate marked with ^{13}C is split up into ammonia and $^{13}CO_2$. Medical users demand a detection accuracy of the $\delta^{13}C$-value of $\leq 0.7‰$ (Haisch 1995; Haisch et al. 1996).

Another important point for the design of a measuring instrument is the sample handling. Breath samples are usually collected in aluminum-coated plastic bags of about 1.5 L. To avoid contamination of the samples, they should be analyzed without decanting into special containers and without further preparation. Apart from carbon dioxide, water vapor is the only IR-active gas contained in relevant quantities in the breath samples. If the measuring method has a cross-sensitivity to water vapor, this must either be compensated for or the water vapor must be removed before the measurement.

In addition to the metrological requirements, there are a number of other requirements that must be met for use in clinical practice. These include simple handling and user-friendliness, low maintenance frequency, low maintenance costs and digital communication capability. These points will only be dealt with in passing in this article.

When selecting the appropriate measurement method, it is important to find a procedure that, in addition to the above-mentioned requirements, guarantees a measurement of $^{13}CO_2$ unaffected by $^{12}CO_2$. This is a core problem of measurement. For this purpose, a property is sought in which the two molecules differ. $^{13}CO_2$ and $^{12}CO_2$ have many common properties. However, they differ significantly in their atomic mass. This is expressed in clearly different absorption spectra of the isotope gases. An optical method can therefore form the basis for a separate measurement of the isotopic gases. When determining carbon dioxide concentrations with IR spectrometers, the strong band at 4.3 μm (wave number 2300 cm^{-1}) is usually evaluated. This band corresponds to the excitation of the fundamental tone of the asymmetric stretching oscillation (Schulze 1976). Since in this oscillation the carbon atom alternately oscillates against the two oxygen atoms, the oscillation frequency also depends on the mass of the carbon atom. One obtains between the band centers of $^{13}CO_2$ and $^{12}CO_2$ a shift of the wave number ν of about 66 cm^{-1}. Figure 16.50 shows a small section of the spectrum of $^{13}CO_2$ and $^{12}CO_2$, in which this shift $\Delta\nu \approx 66 cm^{-1}$ is clearly visible.

The spectral-line fine structure is created by superimposing the quantum transitions of vibrational states on the rotational states in the molecule. This rotational vibration spectrum was calculated from literature data. It should be noted that in Fig. 16.50 the proportion of $^{13}CO_2$ has been greatly increased compared to the natural proportion in order to get a better overview. With the natural isotope ratio of 1–100 the $^{13}CO_2$ band would hardly be visible. The decisive factor for an accurate, isotope-selective measurement is that the bands are sufficiently separated. IR spectroscopy offers an approach to this.

The NDIR Method, Function, Structure and Characteristic Features

The Institute for Laser Medicine at the University of Düsseldorf was also involved in the development of the requirements and investigation of boundary conditions. Together with ABB Automation GmbH, Frankfurt (M), possibilities for the use of a suitable measurement technology were discussed. The above-mentioned distinguishing features of isotope gases in their absorption spectra led to the proposal to use the NDIR technology. The NDIR-

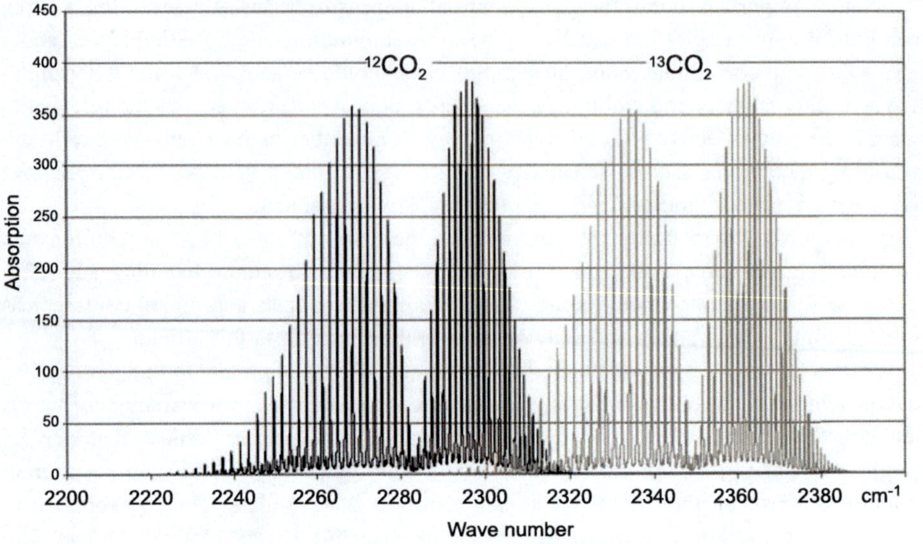

Fig. 16.50 Calculated absorption spectra (rotational vibration spectra) of the fundamental bands of $^{13}CO_2$ and $^{12}CO_2$. For the line shape, the Lorenz distribution was used

photometer URAS[16] was chosen, which had already proven itself over many years in gas analyzers for industrial applications (Fabinski et al. 1993a, b). Main applications are the measurement of CO, CO_2, NO, SO_2, N_2O and CH_4 in production processes, the compliance with environmental regulations and explosion protection.

The NDIR method operates in the mid-IR spectrum, in a wavelength range between $\lambda = 2.5$ µm and $\lambda = 16$ µm. The rotational vibration bands of heteroatomic gases are in this spectral range. Each of these gases has an absorption spectrum with a characteristic fine structure (*fingerprint*) that is individual for the gas (see also example Fig. 16.50). The NDIR method uses this fine structure for a selective detection of the measuring component. The selectivity is produced by the specific absorption of the measuring component. Therefore the detector is filled with the measured component. It is specially manufactured for each application. The method does not use monochromatic radiation (laser) and does not use any dispersive elements (e.g. grating).

The basic design of this NDIR analyzer consists of a thermal radiator whose radiation is mechanically modulated by an aperture wheel and passes through a cuvette into the radiation receiver (see Fig. 16.51). In the URAS photometer, the cylindrical cuvette is divided by a partition into a measuring and a reference side. The aperture wheel lets the radiation fall alternately through the measuring and reference side of the cuvette. Single

[16]Product name of the company ABB Automation GmbH Frankfurt.

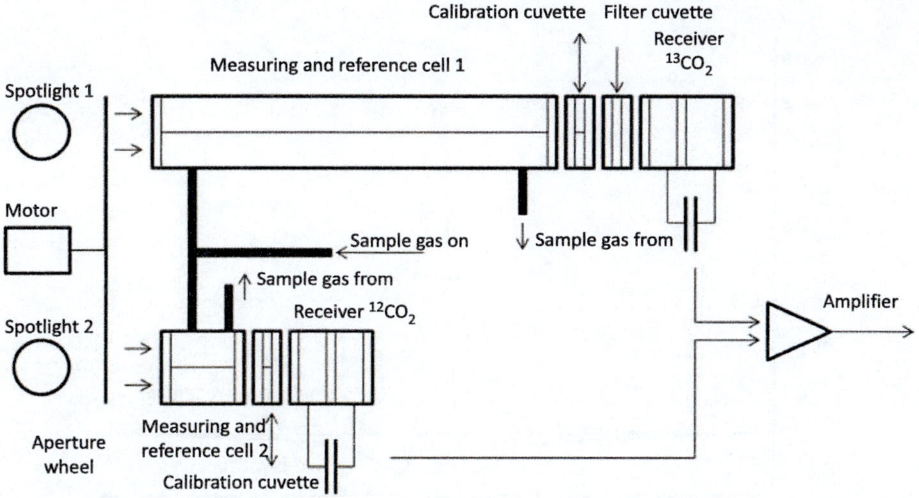

Fig. 16.51 Functional diagram of the NDIR analyzer URAS for measuring $^{13}CO_2$ and $^{12}CO_2$ concentrations

and double layer receivers may be used as receivers (Luft 1955, 1956). Due to its better metrological properties, a two-layer receiver is used here. This contains two chambers one behind the other in the direction of radiation, in which the measurement component (e.g., $^{13}CO_2$) is enclosed. Due to this gas filling, the detector is only sensitive to one specific gas (selective receiver). The radiation energy absorbed in the area of the absorption lines of the measuring component excites internal oscillations and rotations within the absorbing molecules. Molecular collisions result in a conversion into thermal energy in a time that is small compared to the period of the modulation. This results in a pressure signal in phase with the modulation, that is absorbed by a membrane capacitor which is installed between the front and rear receiver chambers, in order to measure their pressure difference Δp. It consists of a movable membrane with a fixed counter electrode connected to the measuring chambers via holes. Due to the non-linearity of the Lambert–Beer absorption law (e-function), the line shape in the front and rear chambers is spectrally different. While in the front chamber the energy occurs largely in the center of a line (positive effect), in the rear chamber more energy is converted in the flanks due to its greater length. Figure 16.52 shows the spectral sensitivity of a rotation line of a receiver built with two chambers behind each other (*Selectivity Function*). While the centers of the lines are formed according to a Lorenz distribution (Fabinski and Ascherfeld 1980; Schaefer et al. 1995), the line flanks contain negative components. Interfering gases, whose line flanks overlap with the lines of the measuring component, have positive and negative effects which partially cancel each other out. This improves the selectivity of the dual-layer receiver considerably. In addition non-selective contaminations, for example in the cuvette, partially compensate each other. This leads to a higher zero-point stability. An *indentation can* be seen at the center of the

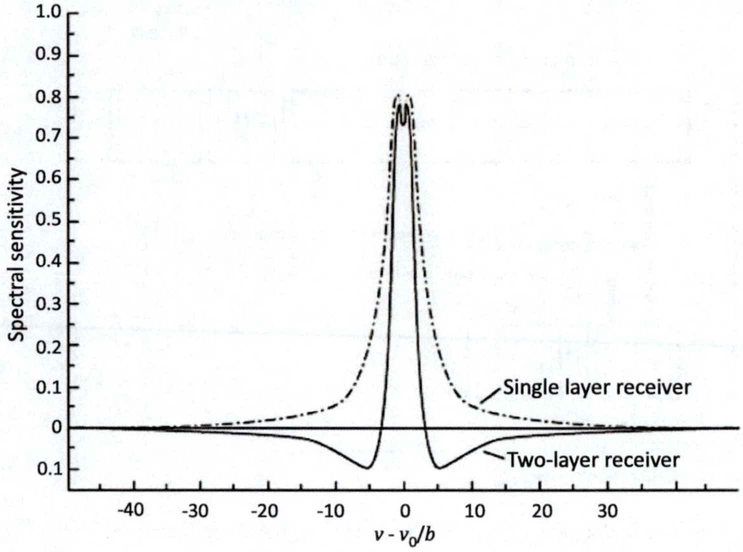

Fig. 16.52 Calculated spectral sensitivity (*selectivity function*) of a rotation line with Loren profile in a 2-layer receiver

line. It is caused by the fact that in the thin layer behind the window a part of the measuring effect is lost through thermal conductivity of the window (Schaefer et al. 1995).

The pressure signals are only a few nanobars (10^{-9}bar) at the detection limit. They are converted into an electrical signal, digitized and displayed. If a gas is introduced into the measuring side of the measuring cuvette, the radiation intensity is weakened by pre-absorption on the measuring side. On the reference side, however, the radiation intensity remains constant. The difference between the two sides is a measure of the concentration of the measuring component.

In summary, the following metrological properties result for the NDIR measuring method. The absorption spectra of the gases filled to the receivers coincide exactly with those of the measuring components in their fine structure, resulting in high sensitivity (resonance absorption). The technique with two receiver chambers, arranged one behind the other, generates lines with negative components in the flanks, which further improve the selectivity of the receiver and the zero point stability. All absorption lines of the gas contribute to the measurement signal. The coincidence of the absorption lines of the measuring component in the measuring cuvette with those of the receiver filling gas is natural and unchangeable. Therefore, no adjustment is necessary. These are the essential advantages compared to all other optical methods which evaluate only one or a few rotation lines. Furthermore, no sensitive optical components or adjustments are required. This gives the NDIR method a high robustness at minimal cost and makes it an alternative to MS for the measurement task at hand.

Practical Implementation, Overall Measuring Equipment
After the first, promising laboratory tests on a modified URAS photometer, the metrological potential was investigated in accordance with the requirements of clinical use (Haisch 1995). Later, other manufacturers[17] of medical devices took over the development for the final product. Important in this context were the adaptation to the practice in the clinical environment and the worldwide distribution of the devices. The result is a measuring technique that is able to compare two breath samples. For this purpose, the patient blows into a breathing bag (*Zero Breath Test*) and then ingests an urea sample marked with ^{13}C, either as a tablet or mixed in a drink. About 10–30 min later, allowing for enzymatic conversion, the patient blows into a second breathing bag. The carbon dioxide content in the two samples is analysed and calculated as $\delta^{13}CO_2$. These values are compared with each other. If there is a significant increase in the $\delta^{13}C$-value, this indicates an infestation with HP (Fabinski et al. 1993a, b). For infected patients, a seven-day so-called *triple therapy* with two different antibiotics and an acid blocker is usually recommended.

In the application described here, a measuring range of 0 . . . 500 ppm $^{13}CO_2$ is recorded in the first IR channel. The second channel is used to measure $^{12}CO_2$ with the measuring range 0 . . . 5 vol.%. Despite the separation of the spectra (Fig. 16.50), the overlapping of the absorption bands in the first channel leaves a residual cross-sensitivity due to the $^{12}CO_2$ present in the 100 times higher concentrations. This undesired effect is reduced by pre-absorption by means of a gas filter filled with $^{12}CO_2$ (positive filtering) and the remaining amount is compensated by a simple mathematical correction. Furthermore, it must be taken into account that the correlation between absorption by the measuring component and the gas concentration follows the Lambert–Beer absorption law. The output signal must therefore be linearized. This is done by electronic means. The different concentrations of the measuring ranges for the $^{13}CO_2$ and $^{12}CO_2$ concentrations are adjusted by different cell lengths so that similar output characteristics are obtained for both channels.

Since the metrological properties of gas analyzers can change over time due to aging of the components, they must be readjusted at specified intervals. For this purpose, the URAS photometer has an adjustment unit in each channel which holds a calibration cuvette filled with the respective measuring component (Fig. 16.53).

For adjustment, the sample gas is replaced by zero point gas and the zero point is adjusted automatically if necessary. The zero point gas is obtained simply from ambient air by passing it through a CO_2 absorber. The calibration cells filled with the isotope gases are then moved into the beam path by an electric motor drive. The gas fillings of the calibration cells are stable over many years, because the cells are made of hermetically sealed metal-glass compounds and no internal chemical reactions can take place. Figure 16.54 shows the

---

[17] Wagner Analysentechnik GmbH, Fischer Analyseninstrumente GmbH.

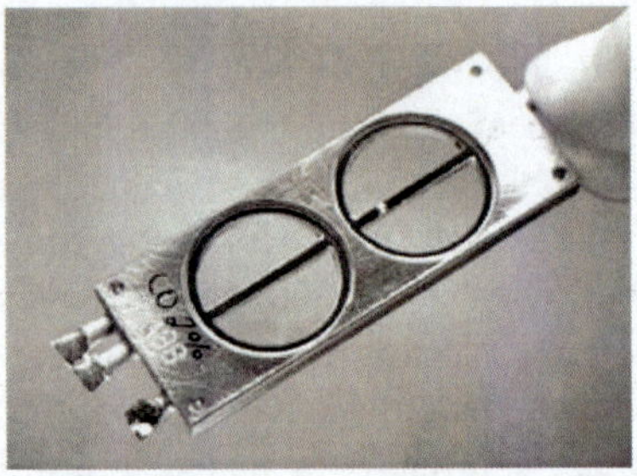

Fig. 16.53 View of the calibration cuvette of the URAS photometer. The cuvette is filled with a defined concentration of $^{12}CO_2$ or $^{13}CO_2$

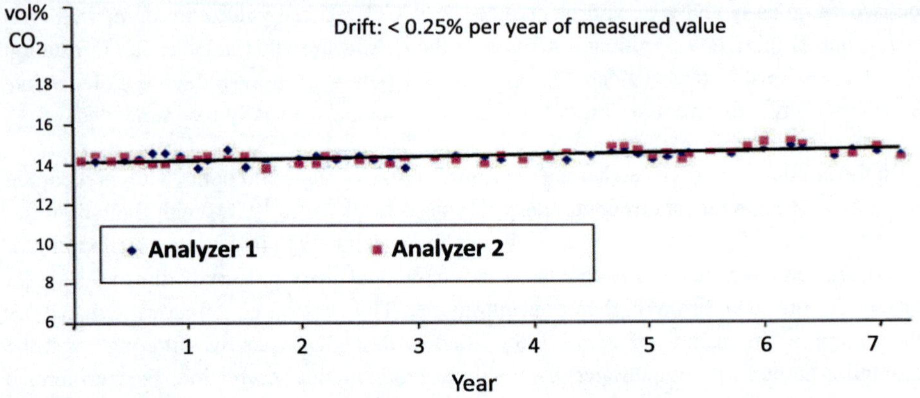

Fig. 16.54 Time course of the sensitivity of two analyzers (URAS) during a regular adjustment with the calibration cuvette

sensitivity curve of a CO_2 measurement with calibration of the adjustment unit over 10 years.

For this purpose, two instruments were adjusted at different time intervals according to the described procedure at zero and calibration point and then tested with a test gas of known concentration. The result was a long-term stability of approximately 0.25% of the measured value/year. Based on a requirement for stability of <1% of the measured value/year, this result means that the use of expensive isotope gases in conventional test gas cylinders can be dispensed with. Operation and handling of the instrument are thus considerably simplified (Fabinski 1992). It is recommended to check the sensitivity annually with a test gas.

To investigate the metrological performance of NDIR measurement technology for its intended use, a number of investigations were carried out, of which two examples are given here. A meaningful investigation for clinical use is the comparison with the established method of MS. For this purpose, mixtures with test gases were prepared and analyzed by a MS and the URAS. Figure 16.55 shows the correlation between the two methods. The equation of the regression line is

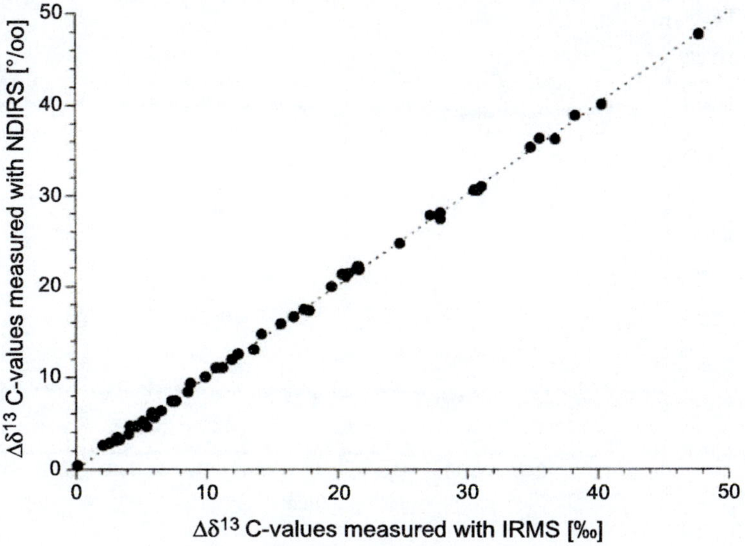

Fig. 16.55 Correlation between a mass spectrometer (IRMS) and the URAS (NDIRS) with synthetic gas samples

$$y = 0.998 \cdot x - 0.01 \qquad r = 0.999, \quad n = 50 \tag{16.7}$$

For this application, this corresponds to a standard deviation of 0.3‰, which is clearly within the requirement of 0.7‰ (Haisch et al. 1996).

A further criteria is the accuracy of the measurement. According to DIN 55350, the term accuracy should not be specified with a numerical value. Here, the term measurement uncertainty is usually given as the standard deviation of repeatability. Figure 16.56 shows the result of the measurement of the repeatability of the carbon isotope ratio in the NDIR analyzer investigated here. Gas with 4.5 vol.% CO_2 and an isotope ratio of 1.0801% and gas with 4.0 vol.% CO_2 and an isotope ratio of 1.0630% were measured alternately. The standard deviation was 0.27‰ with the first gas and 0.26‰ with the second gas for a total test duration of 20 h. These deviations are also well within the required values.

The metrological investigations showed a high potential for the use of URAS for the measurement task presented here. Figure 16.57 shows the view of the NDIR spectrometer module. This is built as an OEM[18] module and delivered to the sales partners. The necessary sensitivity in the practical environment was achieved by optimizing the measurement and system technology so that only a small amount of substrate is required for the investigation (75 mg). Together with the breathing bags, this keeps the costs for the test low. If the results indicate an infestation, the treatment will be accompanied by further

[18]OEM = Original Equipment Manufacturer.

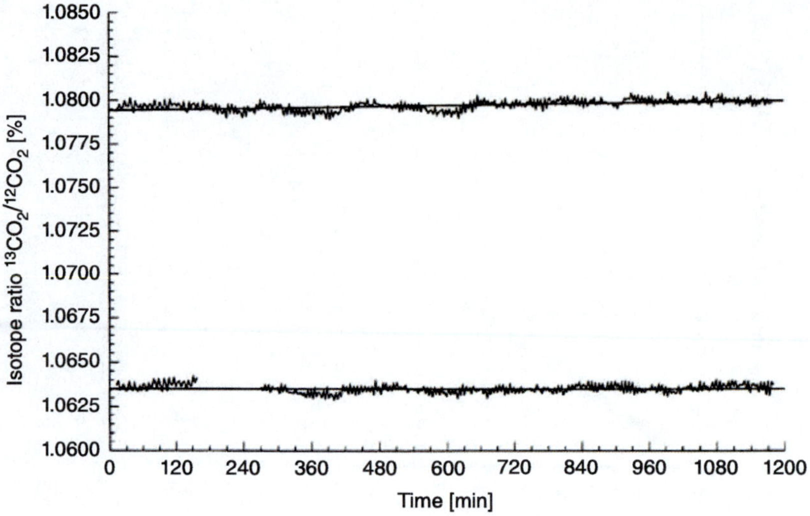

Fig. 16.56 Measuring the repeatability of the URAS. One gas with an isotope ratio of 1.0801% and one with an isotope ratio of 1.06392 were measured alternately over 20 h

breath tests after the start of the therapy in order to assess the effectiveness of the therapy. Figure 16.58 shows the example of the course of the δ^{13}C value in a non-infected and in a patient infected with HP. The samples were taken up to 1 h after ingestion of the urea marked with ^{13}C. The procedure requires the use of invasive methods neither for diagnosis nor for success control.

For clinical application, the medical technology companies have further adapted and optimized the measurement technology for the environment that is common here. For this purpose, the NDIR spectrometer module is supplemented with a gas control and gas feed pump, filters, gas lines, valves, as well as the connection technology for tubes and digital communication. Figure 16.59 shows a view of the FANci 2 from Fischer Analysentechnik GmbH, Leipzig, together with a laptop for control and communication. Up to 16 connection pieces for sample gas bags are provided, so that several examinations can be carried out efficiently and quickly one after the other. The metrological data and control signals are transmitted to the computer with the data output via the digital interface which is common in medical measurement.

The measurement and diagnostic technology has been supplied worldwide since the mid-1990s. Necessary adjustments have been made and national and international approvals have been granted.

The analyses show that the global distribution of HP is very different. For example, the infection rate in Central and Northern Europe is about 30% and in Africa and Asia up to over 90% of the population (Haisch et al. 1996). The findings also show that infections do not necessarily lead to an outbreak of disease. According to estimates, effects are only seen

Fig. 16.57 View of the NDIR module URAS 14

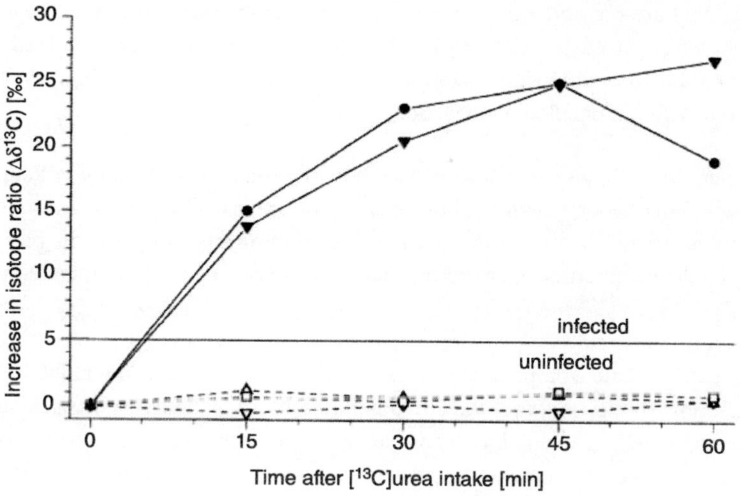

Fig. 16.58 Rise for the course of $^{13}CO_2$ accumulation in the air breathed during the $\delta^{13}C$ urea test. Patients with $\delta^{13}C$ values via 5‰ are considered infected

in about 5% of those affected. In Germany and other European countries, the costs for the breath test are covered by health insurance companies.

Further Applications

The positive experience gained so far in practical use as a breathing gas analyzer led to the expectation that this method could also be used to open up other areas of application. For this purpose, substrates marked with ^{13}C are used, which act exclusively on the organs to be examined. To be recognized as a diagnostic agent, the new applications must also undergo

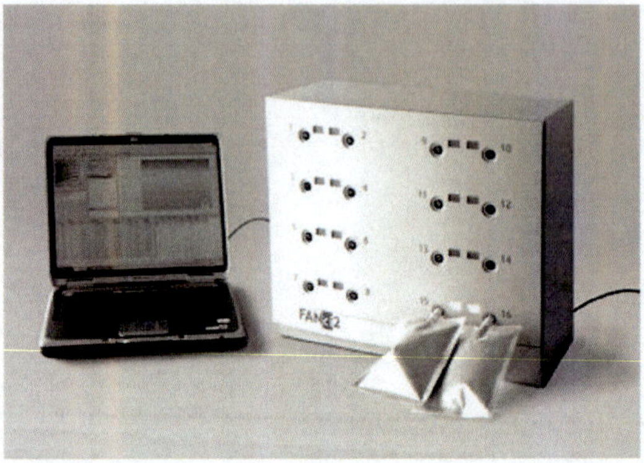

Fig. 16.59 View of the FANci 2 diagnostic system from Fischer Analyseninstrumente GmbH, Leipzig

clinical tests and be approved in accordance with legal regulations. In addition to examinations for stomach complaints, there are already more than 30 diagnostic methods for diseases that are possible through the precise and highly sensitive breath test with the NDIR technology in the system. Many research institutes are now involved in these examinations. For this purpose an example of a liver test is shown in Fig. 16.60, which is carried out with ^{13}C-labeled methacetin.

The upper curve shows the course of metabolism with $\delta^{13}C$ in healthy volunteers. The lower curves show patients with liver cirrhosis of varying severity. Further examples with applications are listed in the table. ^{13}C tracer technology is also used in biomedicine, veterinary medicine, pharmacy, chemistry, plant physiology and environmental protection.

Summary and Future

The investigation of the isotope ratio of elements is usually performed with MS. Since this determination in connection with the administration of ^{13}C-labelled substrates has gained diagnostic importance in medicine, there has been a demand for inexpensive, robust and easy-to-use analytical instruments for the measurement of respiratory gas samples. The application examples listed here show that the NDIR method has the potential for such applications in medical diagnostics due to its metrological properties. It was unclear whether, despite the overlapping of the absorption bands of $^{13}CO_2$ and $^{12}CO_2$, a sufficiently high accuracy and selectivity could be achieved for the separate determination of the concentration of the two isotopes $^{13}CO_2$ and $^{12}CO_2$. In order to be able to measure isotope-selectively, the URAS method was modified in some points and metrologically optimized. Together with a supplementary medical system technology, a measuring accuracy for the ratio of $^{13}CO_2$ to $^{12}CO_2$ of 0.3‰ was achieved with synthetic gas mixtures. The clinical trials resulted in a measurement uncertainty of $\leq$0.5‰. Although this is higher than uncertainties achieved with the MS analysis, it is sufficient for medical applications.

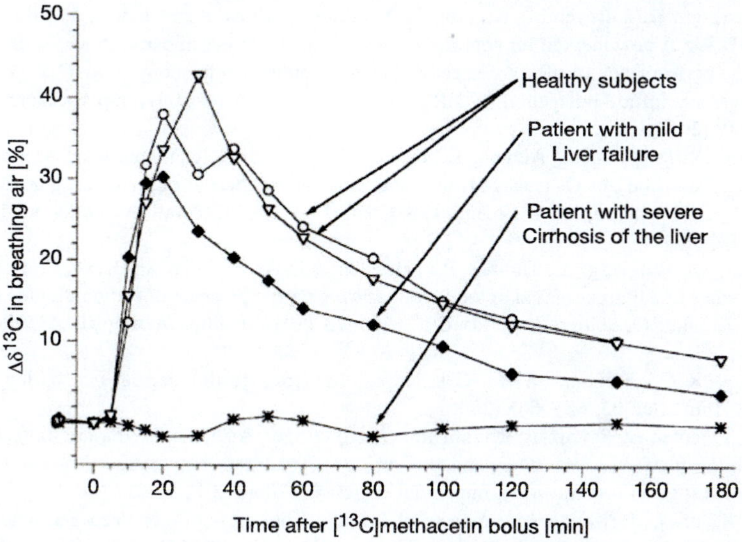

Fig. 16.60 Liver test with ^{13}C-labelled methacethine in patients with different liver cirrhosis

Table 16.5 Application examples for the δ^{13}C testes and use in medical respiratory gas analysis

Test substance	Organ	Process under investigation
Urea	Stomach	Bacterial infestation (HP)
Acetate, octanoate	Stomach	Gastric emptying
Methacetin	Liver, general	Functional liver mass
Keto-isocaproic acid	Liver, mitochondria	Capacity of liver mitochondria
Glucose	All institutions	Glucose utilization under diabetes

The respiratory gas analyzer presented here was developed by an interdisciplinary team from the fields of medicine and analytical measuring technology. It works on the basis of the URAS of ABB Automation GmbH and is characterized by simple operation, compact design and high robustness at low costs for acquisition and maintenance. Together with medical device manufacturers, the device has been adapted to the usual environment in clinical practice. This system enables δ^{13}C breath tests to be performed quickly (Table 16.5). The analyzer module in the measuring system can also be used for other isotope selective tasks in science and industry by simple modifications.

References

Anon: News release – Gas Produced fromMethane Hydrate (provisional) 1. http://www.jogmec.go.jp/english/news/release/news_01_000006.html, n.d.

Arévalo-Martínez, D.L., Beyer, M., Krumbholz, M., Piller, I., Kock, A., Steinhoff, T., Körtzinger, A., Bange, H.W.: A new method for continuous measurements of oceanic and atmospheric N_2O, CO and CO_2: performance of off-axis integrated cavity output spectroscopy (OA-ICOS) coupled to non-dispersive infrared detection (NDIR). Ocean Sci. **9**(6), 1071–1087. http://oceanrep.geomar.de/21900/ (2013)

Atamanchuk, D., Tengberg, A., Aleynik, D., Fietzek, P., Shitashima, K., Lichtschlag, A., Hall, P.O.J., Stahl, H.: Detection of CO_2 leakage from a simulated sub-seabed storage site using three different types of pCO_2 sensors. Int. J. Greenh. Gas Control, in press (2014a). https://doi.org/10.1016/j.ijggc.2014.10.021

Atamanchuk, D., Tengberg, A., Thomas, P.J., Hovdenes, J., Apostolidis, A., Huber, C., Hall, P.O.J.: Performance of a lifetime-based optode formeasuring partial pressure of carbon dioxide in natural waters. Limnol. Oceanogr. Methods. **12**(2), 63–73 (2014b). https://doi.org/10.4319/lom.2014.12.63

Beauchemin, K.A., McGinn, S.M.: CH_4 emissions from feedlot cattle fed barley or corn diets. J. Anim. Sci. **83**, 653–661 (2005)

Bittig, H.C., Fiedler, B., Scholz, R., Krahmann, G., Körtzinger, A.: Time response of oxygen optodes on profiling platforms and its dependence on flow speed and temperature. Limnol. Oceanogr. Methods. **12**(8), 617–636 (2014). https://doi.org/10.4319/lom.2014.12.617

Blackford, J., et al.: Detection and impacts of leakage from sub-seafloor deep geological carbon dioxide storage. Nat. Clim. Chang. **4**(11), 1011–1016 (2014). https://doi.org/10.1038/nclimate2381

Byrne, R.H., DeGrandpre, M.D., Short, R.T., Martz, T.R., Merlivat, L., McNeil, C., Sayles, F.L., Bell, R., Fietzek, P.: Sensors and systems for in situ observations of marine carbon dioxide system variables. In: Hall, J., Harrison, D.E., Stammer, D. (Hrsg.) Proceedings of OceanObs'09: Sustained Ocean Observations and Information for Society. ESA Publication WPP-306, Venice, Italy. http://eprints.ifm-geomar.de/8792/ (2010)

Caldeira, K., Wickett, M.E.: Ocean model predictions of chemistry changes from carbon dioxide emissions to the atmosphere and ocean. J. Geophys. Res. **110**, C09S04 (2005). https://doi.org/10.1029/2004JC002671

Capasso, G., Inguaggiato, S.: A simple method for the determination of dissolved gases in natural waters. An application to thermal waters from Vulcano Island. Appl. Geochemistry. **13**(5), 631–642 (1998). https://doi.org/10.1016/S0883-2927(97)00109-1

Caramanna, G., Fietzek, P., Maroto-Valer, M.: Monitoring techniques of a natural analogue for sub-seabed CO_2 leakages. Energy Procedia. **4**, 3262–3268 (2011)

DeGrandpre, M.D., Hammar, T.R., Smith, S.P., Sayles, F.L.: In-situ measurements of seawater pCO_2. Limnol. Oceanogr. **40**(5), 969–975 (1995)

Demas, J.N., DeGraff, B.A., Coleman, P.B.: Oxygen sensors based on luminescence quenching. Anal. Chem. **71**(23), 793A–800A (1999). https://doi.org/10.1021/ac9908546

der Europäischen Gemeinschaften L 309/22 vom 27.11.2001

Devold, H.: Oil and gas production handbook – An introduction to oil and gas production, transport, refining and petrochemical industry, 3. Aufl., ABB Oil and Gas, Oslo. http://www.texvet.org/sites/default/files/01_Oil%20and%20Gas%20Production%20handbook.pdf (2013)

Dickson, A.G., Sabine, C.L., Christian, J.R. (Hrsg.): Guide to best practices for ocean CO_2 measurements. PICES Special Publication 3. http://aquacomm.fcla.edu/1443/ (2007)

DIN EN ISO 5167: Durchflussmessung von Fluiden mit Drosselgeräten in voll durchströmten Leitungen mit Kreisquerschnitt (2003)

DIN EN ISO 80601-2-55: Besondere Festlegungen für die Sicherheit einschließlich der wesentlichen Leistungsmerkmale von Überwachungsgeräten für Atemgase (2011) 1174 16 Life Science

DLG: Testzentrum Technik und Betriebsmittel, Prüfberichte zu Abluftreinigungssystemen. DLG e. V., www.dlg-test.de

Fabinski, W.: On-line Analysatoren, automatischer Abgleich mit internen Standards. Staub, Reinhaltung der Luft. **52** (1992)

Fabinski, W., Ascherfeld, M.: Funktion und Aufbau des NDIR-Prozeßfotometers Uras 3. Technisches Messen tm. **7(8)**, 257–264 (1980)

Fabinski, W., Fuß, W., Haisch, M., Hering, P., Wiegleb, G.: Verfahren und Vorrichtung zur Bestimmung der Isotopenverhältnisse in Gasen. Deutsche Patentanmeldung DE 4224146.4 (14.7.1993) (1993a)

Fabinski, W., Fuß, W., Haisch, M., Hering, P., Wiegleb, G.: Messen der $^{13}CO_2$-Konzentration in Atemluft mit einem nichtdispersiven IR-Spektrometer. Verhandlungen der DPG-Frühjahrstagung in Berlin. MO 12.5, S. 359 (1993b)

Fiedler, B., Fietzek, P., Vieira, N., Silva, P., Bittig, H.C., Körtzinger, A.: In situ CO_2 and O_2 measurements on a profiling float. J. Atmos. Ocean. Technol. **30**, 112–126. http://oceanrep.geomar.de/16572/ (2013)

Fietzek, P., Kramer, S., Esser, D.: Deployment of the HydroCTM (CO_2/CH_4) on stationary and mobile platforms – Merging the trends in the field of platform and sensor development. In: Proceedings of Oceans 11 MTS/IEEE Conference, 19–22 Sept. 2011, S. 1–9, Kona, Hawaii, USA. http://ieeexplore.ieee.org/xpl/login.jsp?tp=&arnumber=6107129&url=http://ieeexplore.ieee.org/stamp/stamp.jsp?tp=&arnumber=6107129

Fietzek, P., Fiedler, B., Steinhoff, T., Körtzinger, A.: In situ quality assessment of a novel underwater pCO_2 sensor based on membrane equilibration and NDIR spectrometry. J. Atmos. Ocean. Technol. **31**(1), 181–196 (2014)

Fischer Analyseninstrumente GmbH (FAN), Leipzig, Firmenschriften. Heli FAN plus (2015)

Flachowsky, G., Brade, W.: Potenziale zur Reduzierung der Methan-Emissionen bei Wiederkäuern. Züchtungskunde. **79**, 417–465 (2007)

Flachowsky, G., Brade, W., Feil, A., Kamphues, J., Meyer, U., Zehetmeier, M.: Carbon-Footprints bei der Primärerzeugung von Lebensmitteln Tierischer Herkunft. Übersichten zur Tierernährung. **39**(1/2011), 1–45 (2011)

Friedrichs, G., Bock, J., Temps, F., Fietzek, P., Körtzinger, A., Wallace, D.W.R.: Toward continuous monitoring of seawater $^{13}CO_2/^{12}CO_2$ isotope ratio and pCO_2: Performance of cavity ringdown spectroscopy and gas matrix effects. Limnol. Oceanogr. Methods. **8**, 539–551 (2010). https://doi.org/10.4319/lom.2010.8.539

Grainger, C., Clarke, T., McGinn, S.M., Auldist, M.J., Beauchemin, K.A., Hannah, M.C., Waghorn, G.C., Clark, H., Eckard, R.J.: CH_4 emissions from dairy cows measured using the sulfur hexafluoride (SF6) tracer and chamber techniques. J. Dairy Sci. **90**, 2755–2766 (2007)

Gülzow, W., Rehder, G., Schneider, B., Deimling, J.S.v., Sadkowiak, B.: A new method for continuous measurement of methane and carbon dioxide in surface waters using off-axis integrated cavity output spectroscopy (ICOS): An example from the Baltic Sea. Limnol. Oceanogr.Methods. **9**, 176–184 (2011). https://doi.org/10.4319/lom.2011.9.176

Haisch, M.: Quantitative isotopenselektive Infrarotspektroskopie zur Bestimmung des Kohlenstoffisotopenverhältnisses in der Atemluft. Dissertation, Heinrich-Heine-Universität, Düsseldorf (1995)

Haisch, M., Hering, P., Fabinski, W., Zöchbauer, M.: Isotopenselektive Konzentrationsmessung an Atemgasen mit einem NDIR-Spektrometer. Technisches Messen tm. **9**, 322–328 (1996)

IPCC (AR4): Zusammenfassungen für politische Entscheidungsträger. Intergouvernemental Panel on Climate Change IPCC, WMO/UNEP: Klimaänderung 2007 (2007)

Jacobsohn, A.: Rechnungen aus dem Lungensimulationsmodell. Weinmann Emergency Medical Technology GmbH + Co. KG (2015)

Jalc, D., Certik, M.: Effect of microbial oil, monensin and fumarate on rumen fermentation in artificial rumen. Czech J. Anim. Sci. **50**(10), 467–472 (2005)

JCOMMOPS: Argo – part of the integrated global observation strategy. http://www.argo.ucsd.edu (Abgerufen am 15. März 2015), n.d.

Johnson, K.A., Johnson, D.E.: CH_4 emissions from cattle. J. Animal Science **73**, 2483–2492 (1995)

LEAD (LIVESTOCK, ENVIRONMENT AND DEVELOPMENT INITIATIVE): Livestock and Environment toolbox. www.lead.virtualcenter.org/dec/toolbox/indust/ghgas.html. Accessed: 13.03.2007

Kautsky, H.: Quenching of luminescence by oxygen. Trans. Faraday Soc. **35**, 216–219 (1939). https://doi.org/10.1039/tf9393500216

Kazutoshi, F., et al.: Effect of eradication of Helicobactor Pylori of metachronous gastric carcinoma after endoscopic resection of early gastric cancer: an open-label, randomised controlled trial. The Lancet. **372**(9636), 392–397 (2008)

Kiehl, J.T., Trenberth, K.E.: Earth's Annual Global Mean Energy Budget. Bull. Am. Meteorol. Soc. **78**(2), 197–208 (1997). https://doi.org/10.1175/1520-0477(1997)078<0197:EAGMEB<2.0.CO;2

Kiesewetter, O.: Sensoranordnung. Deutsches Patent, Nr. DE102010003966 B3 (02.01.2010)

Kiesewetter, O., Riesenberg, R.: Verfahren und Vorrichtung zur Ermittlung der Konzentration von Gasen. Deutsches Patent, Nr. DE19957364 B4 (29.11.1999)

Kiesewetter, O., Ewert, A., Melchert, V., Kittelmann, S.: Anordnung zur Detektion von Luftinhaltsstoffen und Verfahren zum Betreiben der Anordnung. Deutsches Patent, Nr. DE102004060101 B4 (13.12.2004)

Kiesewetter, O., Ewert, A., Melchert, V., Kittelmann, S.: Assembly for detecting air components. European Patent, No. EP 1602924 B1 (02.06.2005)

Kiesewetter, O., Kiesewetter, N., Kraußer, A., Müller, J., May, M.: InnovativeCO_2-/VOC Sensorelemente zur Detektion der Luftqualität in Fahrzeuginnenräumen. In: T. Tille et al. (Hrsg.) Sensoren im Automobil V, 77–91, Haus der Technik Fachbuch, Bd. 132, ISBN 978-3-8169-3207-9, expert (2014)

Kittelmann, S., Ewert, A., Kiesewetter, O.: Anordnung zur Detektion von Luftinhaltsstoffen. Deutsches Patent, Nr. DE 102006033528 B3 (20.07.2006)

Klimant, I., Meyer, V., Kühl, M.: Fiber-optic oxygen microsensors, a new tool in aquatic biology. Limnol. Oceanogr. **40**(6), 1159–1165 (1995). https://doi.org/10.4319/lo.1995.40.6.1159

Köhne, S., Wiegleb, G.: Gassensorik in der Fahrzeugtechnik. Design & Elektronik. **7**, 62–63 (2008)

Körtzinger, A., Schimanski, J., Send, U.: High quality oxygen measurements from profiling floats: A promising new technique. J. Atmos. Ocean. Technol. **22**(3), 302 (2005). https://doi.org/10.1175/JTECH1701.1

Kvenvolden, K.A.: Gas hydrates – geological perspective and global change. Rev. Geophys. **31**, 173–187 (1993). https://doi.org/10.1029/93RG00268

Lassey, M.U., Martin, R.J., Walker, C.F., Shelton, I.D.: CH_4 emissions measured directly from grazing livestock in New Zealand. Atmos. Environ. **31**, 2905–2914 (1997)

Lefèvre, N., Ciabrini, J.P., Michard, G., Brient, B., DuChaffaut, M., Merlivat, L.: A new optical sensor for pCO_2 measurements in seawater. Mar. Chem. **42**(3–4), 189–198 (1993). https://doi.org/10.1016/0304-4203(93)90011-C

Luft, K.F.: Vorrichtung zum Analysieren von insbesondere gasförmigen Gasgemischen durch Absorption von Strahlung. Deutsches Patent: DBP1 017 385 (1955)

Luft, K.F.: Physique de L'Infrarouge Analyse des gaz par absorption infrarouge non dispersive. C.R. Acad. Sci. **242**, 482–484 (1956)

Maeck, A., Delsontro, T., McGinnis, D.F., Fischer, H., Flury, S., Schmidt, M., Fietzek, P., Lorke, A.: Sediment trapping by dams creates methane emission hot spots. Environ. Sci. Technol. **47**, 8130–8137 (2013). https://doi.org/10.1021/es4003907

McGinn, S.M., Beauchemin, K.A., Coates, T., Colombatto, D.: CH_4 emissions from beef cattle: effects of monensin, sunflower oil, enzymes, yeast, and fumaric acid. J. Anim. Sci. **82**, 3346–3356 (2004)

McLean, J.A., Tobin, G.: Animal and Human Calorimetry. Cambridge University Press, Cambridge (1987)

Metz, B., Davidson, B.O., De Coninck, H., Loos, M., Meyer, L. (Hrsg.): IPCC: IPCC special report on carbon dioxide capture and storage. Cambridge University Press. http://www.ipcc.ch/pdf/special-reports/srccs/srccs_wholereport.pdf (2005)

Millero, F.J.: The marine inorganic carbon cycle. Chem. Rev. **107**(2), 308–341 (2007). https://doi.org/10.1021/cr0503557

Miloshevich, L.M., Paukkunen, A., Vömel, H., Oltmans, S.J.: Development and validation of a time-lag correction for vaisala radiosonde humidity measurements. J. Atmos. Ocean. Technol. **21**(9), 1305–1327. http://journals.ametsoc.org/doi/abs/10.1175/1520-0426(2004)021<1305:DAVOAT>2.0.CO;2 (2004)

Oczenski, W.: Atmen – Atemhilfen: Atemphysiologie und Beatmungstechnik. 9. Aufl. Thieme (2012)

Pasteris, J., Wopenka, B., Freeman, J.J., Brewer, P.G., White, S.N., Peltzer, E.T., Malby, G.E.: Raman spectroscopy in the deep ocean: Successes and challenges. Appl. Spectrosc. **58**(7), 195–208 (2004). https://doi.org/10.1366/0003702041389319

Pfeil, B., et al.: A uniform, quality controlled Surface Ocean CO_2 Atlas (SOCAT). Earth Syst. Sci. Data. **5**, 125–143 (2013). https://doi.org/10.5194/essd-5-125-2013

Pierrot, D., Neill, C., Sullivan, K., Castle, R., Wanninkhof, R., Lüger, H., Johannessen, T., Olsen, A., Feely, R.A., Cosca, C.E.: Recommendations for autonomous underway pCO_2 measuring systems and data-reduction routines. Deep. Res. Part II Top. Stud. Oceanogr. **56**(8–10), 512–522 (2009). https://doi.org/10.1016/j.dsr2.2008.12.005

Richtlinie 2001/81/EG: Richtlinie 2001/81/EG des Europäischen Parlaments und des Rates vom 23. Oktober 2001 über nationale Emissionshöchstmengen für bestimmte Luftschadstoffe, Amtsblatt

Riebesell, U., Zondervan, I., Rost, B., Tortell, P.D., Zeebe, R.E., Morel, F.M.: Reduced calcification of marine plankton in response to increased atmospheric CO_2. Nature. **407**(6802), 364–367. http://www.ncbi.nlm.nih.gov/pubmed/11014189 (2000)

Schaefer, W., Fabinski, W., Zöchbauer, M.: How the thermal conductivity of the enclosed gas influences the spectral response of a Luft-type detector. J. Mol. Struct. **349**, 247–250 (1995)

Schlüter, M., Gentz, T.: Application of membrane inlet mass spectrometry for online and in situ analysis of methane in aquatic environments. J. Am. Soc. Mass Spectrom. **19**, 1395–1402 (2008). https://doi.org/10.1016/j.jasms.2008.07.021

Schmidt, M., Linke, P., Esser, D.: Recent development in IR sensor technology for monitoring subsea methane discharge. Mar. Technol. Soc. J. **47**(3), 27–36. http://oceanrep.geomar.de/21452/ (2013)

Schönwiese, C.-D.: Klimatologie. 3. Aufl. UTB, Stuttgart (2008)

Schulze, W.: Molekülbau. Bd. 786. Sammlung Göschen (1976)

Severinghaus, J.W., Bradley, A.F.: Electrodes for blood pO_2 and pCO_2 determination. J. Appl. Physiol. **13**(3), 515–520 (1958)

SOCCOM: Southern Ocean Carbon and Climate Observations and Modeling (SOCCOM). http://soccom.princeton.edu. Accessed: 30. März 2015

Statistisches Bundesamt (DeStatis): Fachserie 3 Reihe 2.1.3; Land- und Forstwirtschaft, Fischerei; Viehhaltung der Betriebe; Agrarstrukturerhebung 2013; Artikelnummer: 2030213139004; Wiesbaden (2014)

Stocker, T.F., Qin, D., Plattner, G.-K., Tignor, M., Allen, S.K., Boschung, J., Nauels, A., Xia, Y., Bex, V., Midgley, P.M. (Hrsg.): IPCC: Climate Change 2013: The Physical Science Basis. Contribution of Working Group I to the Fifth Assessment Report of the Intergovernmental Panel on Climate Change. Cambridge University Press, Cambridge, United Kingdom and New York, NY, USA. http://www.climatechange2013.org/ (2013)

Stow, R.W., Baer, R.F., Randall, B.F.: Rapid measurement of the tension of carbon dioxide in blood. Arch. Phys. Med. Rehabil. **38**(10), 646–650. http://europepmc.org/abstract/MED/13471145 (1957)

Takahashi, T.: Carbon dioxide in the atmosphere and in Atlantic Ocean water. J. Geophys. Res. **66**, 477–494 (1961)

Takahashi, T., Sutherland, S.C., Kozyr, A.: Global Ocean SurfaceWater Partial Pressure of CO_2 Database: Measurements Performed During 1957–2013 (Version 2013), ORNL/CDIAC-160, NDP-088(V2013) (2014). https://doi.org/10.3334/CDIAC/OTG.NDP088(V2013)

Tengberg, A., Hovdenes, J., Andersson, H.J., Brocandel, O., Diaz, R., Hebert, D., Arnerich, T., Huber, C., Körtzinger, A., Khripounoff, A., Rey, F., Rönning, C., Schimanski, J., Sommer, S., Stangelmayer, A.: Evaluation of a lifetime-based optode to measure oxygen in aquatic systems. Limnol. Oceanogr.Methods. **4**(1964), 7–17. http://citeseerx.ist.psu.edu/viewdoc/download?doi=10.1.1.126.3184&;rep=rep1&;type=pdf (2006)

Tierhaltungserlass: Erlass „Immissionsschutzrechtliche Anforderungen an Tierhaltungsanlagen“ des Ministeriums für Klimaschutz, Umwelt, Landwirtschaft, Natur- und Verbraucherschutz des Landes Nordrhein-Westfalen vom 19. Februar 2013

TierSchutzNutztV: Verordnung zum Schutz landwirtschaftlicher Nutztiere und anderer zur Erzeugung tierischer Produkte gehaltener Tiere bei ihrer Haltung (Tierschutz-Nutztierhaltungsverordnung – TierSchNutztV), Fassung vom 22.08.2006

Vabitsch, A.M.: Qualitativer Vergleich von Modellen zur Bewertung von Klimaschutzmaßnahmen in Europaunter besonderer Berücksichtigung der Landwirtschaft. Disseration, Universität Hohenheim (2006)

Wagner Analysen Technik – Kibion GmbH, Bremen: Firmenschriften IRISTM (2015)

Wenner, P.G., Bell, R.J., Van Amerom, F.H.W., Toler, S.K., Edkins, J.E., Hall, M.L., Koehn, K., Short, R.T., Byrne, R.H.: Environmental chemical mapping using an underwater mass spectrometer. TrAC – Trends Anal. Chem. **23**, 288–295 (2004)

Wiegleb, G., Heitbaum, J.: Semiconductor gas sensor for detecting NO and CO traces in ambient air of road traffic. Sens. Actuat. B. **17**, 93–99 (1994)

Wiegleb, G., Zöchbauer, M.: New optical methods for applications in process analysis and medical instruments. Exp. Tech. Phys. **40**(2), 241–257 (1994)

Winkler, L.W.: Die Bestimmung des im Wasser gelösten Sauerstoffes. Berichte der Dtsch. Chem. Gesellschaft. **21**(2), 2843–2854 (1888). https://doi.org/10.1002/cber.188802102122

Winkler, N. et al.: Opto-chemische NH3-Sensoren im ppm und ppb-Bereich für den Einsatz in der Landwirtschaft. 10. Dresdner Sensor Symposium (2011)

Witzke, H.V., Noleppa, S.: Methan und Lachgas, die vergessenen Klimagase. WWF Deutschland (2007)

Zimmerman, P.: Measurement of methane emissions from ruminant livestock using a SF6 tracer technique. Environ. Sci. Technol. **28**(2), 359–362 (1994)

Zimmerman, P.: Method and system for monitoring and reducing ruminant methane production. US Patent 8307785 B2. (2012)

Biotechnology 17

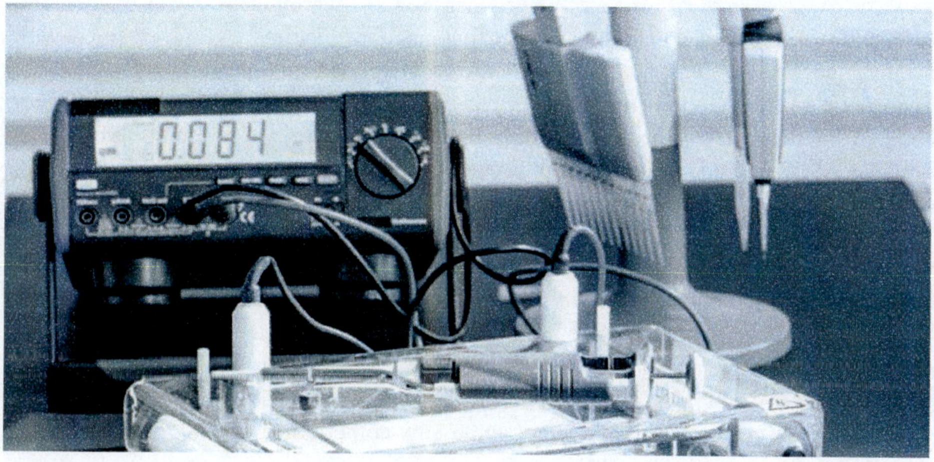

17.1 Gas Analysis for Fermentation Processes

Holger Müller and Udo Schmale

Fermentation is a process that was already used in the Mesolithic period, initially only for food production. Among the best known and oldest applications are wine preparation, vinegar production and beer brewing (Salomon 2010).

Today, in addition to these applications, vaccines, tissues, therapeutic proteins and food supplements are also produced in fermenters. Biogas is also produced in a fermenter with

G. Wiegleb, *Gas Measurement Technology in Theory and Practice*,
https://doi.org/10.1007/978-3-658-37232-3_17

the help of a wide variety of microorganisms. Basically, a distinction is made between fermentations that are carried out with the addition of oxygen (aerobic fermentation) and those that take place without oxygen (anaerobic fermentation).

Even today, for most biotechnological processes only a few process data are recorded during fermentation. Instead, samples are taken at intervals and analysed in a laboratory. Even small changes to the process can mean that it has to be completely re-sampled. The PAT (Process Analytical Technology) initiative of the FDA (Food and **Drug** Administration) is expected to change this for future processes. From the FDA's point of view, PAT should be a system for the design, analysis and control of the production process, which ensures product quality through continuous measurements of so-called critical quality and performance attributes during the process (FDA 2004). Some of these attributes can also be determined by means of gas analysis.

Aerobic Processes

Aerobic fermentations use microorganisms that consume oxygen (O_2) and produce carbon dioxide (CO_2) to grow or produce the product. For this reason, sensors for oxygen and carbon dioxide are most common in aerobic fermentations. For many years, analysers with paramagnetic oxygen sensors and IR photometers were used in fermentation gas analysis. These were sensitive to moisture and were associated with high acquisition costs. Therefore, several fermenters were usually connected to one analyser via gas multiplexers. The sample gas was additionally dried to protect the analyser. Figure 17.1 shows the structure of a system with several fermenters connected to one .

A positive side effect of drying the sample gas was that the measured gas concentrations could be directly compared with the values of the input gas of the fermenter. Due to the enrichment of the gas with moisture from the fermentation broth, the gas concentrations at the fermenter outlet are lower than at the fermenter inlet, even if the fermenter contains only water without any biological activity. The gas concentrations are diluted by the moisture in the exhaust gas. At the same time, the volume flow at the fermenter outlet is thus higher than at the inlet (Sahm et al. 2014).

The water content of the fermenter exhaust air can be as high as 5 vol.%, depending on whether or not the fermenter has a gas cooler to minimize the loss of liquid volume. In principle, the dilution depends on the absolute humidity of the sample gas. This can be read from Fig. 17.2 using the dew point, that is the temperature at which the water in the sample gas condenses.

Often, however, the measured concentrations of these two gases are not of immediate interest to users. Rather, the measured values are used to calculate other values that cannot be determined directly. The values of interest for microbiologists, which can be calculated from the measured gas concentrations of O_2 and CO_2, are the *oxygen uptake rate* (*OUR*), the *carbon dioxide production rate* (*CPR*) and the quotient of both, the so-called respiratory quotient (RQ). The OUR indicates how much oxygen per litre fermenter volume and hour was consumed by the microorganisms, the CPR how much CO_2 was produced by the

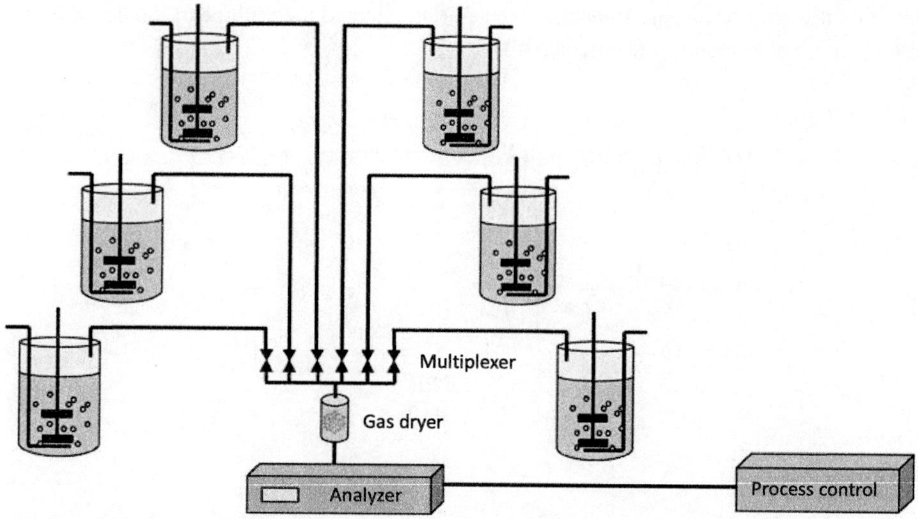

Fig. 17.1 Fermentation system consisting of 6 stirred fermenters with a gas multiplexer, a gas cooler for drying the sample gas and a gas analyser

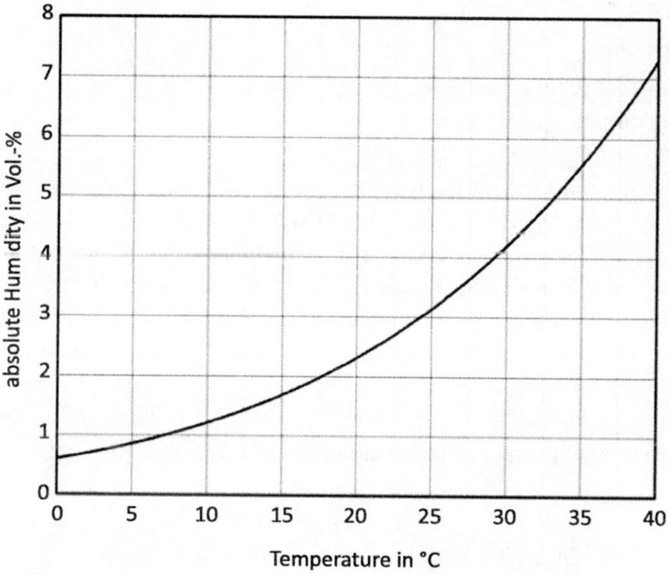

Fig. 17.2 Dependence of the absolute humidity (water vapour concentration in vol.%) of a gas on its dew point temperature

microorganisms. For its determination, the aeration rate and the volume of the liquid in the fermenter are also required (Storhas 2013).

$$\mathrm{OUR} = \frac{\dot{V}_{G,n}}{V_{M,n} \cdot V_{R,L}} \cdot \left[Y_{O_2}^{\alpha} - \frac{\left(1 - Y_{O_2}^{\alpha} - Y_{CO_2}^{\alpha}\right)}{\left(1 - Y_{O_2}^{\omega} - Y_{CO_2}^{\omega}\right)} \cdot Y_{O_2}^{\omega} \right] \tag{17.1}$$

$$\mathrm{CPR} = \frac{\dot{V}_{G,n}}{V_{M,n} \cdot V_{R,L}} \cdot \left[\frac{\left(1 - Y_{O_2}^{\alpha} - Y_{CO_2}^{\alpha}\right)}{\left(1 - Y_{O_2}^{\omega} - Y_{CO_2}^{\omega}\right)} \cdot Y_{CO_2}^{\omega} - Y_{CO_2}^{\alpha} \right] \tag{17.2}$$

$$\mathrm{RQ} = \frac{\mathrm{CPR}}{\mathrm{OUR}} \tag{17.3}$$

with

$\dot{V}_{G,n}$ = flow rate in $\frac{\mathrm{L}}{\mathrm{h}}$;
$V_{M,\,n}$ = molar volume of the (ideal) gas $\left(22,41\,\frac{\mathrm{L}}{\mathrm{mol}}\right)$;
$V_{R,\,L}$ = volume of liquid assets;
$Y_{O_2}^{\alpha}$ = O_2 at inlet (0,2097 for air);
$Y_{O_2}^{\omega}$ = O_2 at output;
$Y_{CO_2}^{\alpha}$ = CO_2 at inlet (0.04 for air);
$Y_{CO_2}^{\omega}$ = CO_2 at output.

The input concentrations are normally known, since either dried and filtered air with the concentrations given above is used, or defined gas mixtures are adjusted by means of mass flow controllers. Therefore a gas measurement usually takes place only in the exhaust gas flow of the fermenter. Depending on the organisms used, the RQ value often allows direct conclusions to be drawn about the metabolic process currently taking place.

This can be explained very well using the example of a yeast fermentation with glucose as substrate. In this process, there are different metabolic processes that require different amounts of O_2 and produce different amounts of CO_2 (Höra 2002). If enough oxygen is available, the (aerobic) process of respiration proceeds as follows:

$$\begin{array}{c} C_6H_{12}O_6 + \ 6\,O_2 \rightarrow 6\,CO_2 + 6\,H_2O \\ \text{Glucose} + \text{Oxygen} \rightarrow \text{Carbon dioxide} + \text{Water} \end{array} \tag{17.4}$$

A lot of biomass (yeast cells) is produced in this process. If, on the other hand, no oxygen is available, ethanol is produced during (anaerobic) fermentation:

$$\begin{array}{l} C_6H_{12}O_6 \rightarrow 2\, C_2H_5OH + 2\, CO_2 \\ \text{Glucose} \rightarrow \text{Ethanol} + \text{Carbon dioxide} \end{array} \tag{17.5}$$

Hardly any new cells are formed. If glucose is no longer available, ethanol is used as substrate:

$$\begin{array}{l} C_2H_5OH + 3\, O_2 \rightarrow 2\, CO_2 + 3\, H_2O \\ \text{Ethanol} + \text{Oxygen} \rightarrow \text{Carbon dioxide} + \text{Water} \end{array} \tag{17.6}$$

In addition, the so-called *crabtree effect* occurs with sufficient oxygen supply and high glucose concentrations (greater than 0.1 g/l). In addition to biomass, ethanol is also produced. The corresponding sales equation is:

$$\begin{array}{l} C_6H_{12}O_6 + 0,25\, O_2 \rightarrow 1,92\, C_2H_5OH + 2,16\, CO_2 + 0,25\, H_2O \\ \text{Glucose} + \text{Oxygen} \rightarrow \text{Ethanol} + \text{Carbon dioxide} + \text{Water} \end{array} \tag{17.7}$$

This means that during aerobic respiration an RQ of 1 should be achieved, if there is too much glucose the RQ should be set to a value of about 8.6. As soon as the glucose is used up, the RQ will adjust to a value of about 0.7. Since in a real fermentation never only one process takes place, the real RQ will be between 0.7 and 8.6 if there is sufficient oxygen supply. This also results in a simple possibility for process control. Depending on whether one wants to grow as many yeast cells as possible ($RQ = 1$) or whether one wants to produce ethanol ($RQ > 1$) with their help, the process parameters such as fumigation (flow rate and gas concentration) and feeding can be controlled in such a way that a previously determined RQ is set and maintained.

For illustration purposes, the processes just described will appear in an example of measured values in the following. At the beginning of the development of a new bioprocess, shaken reaction systems such as shake flasks or microtiter plates are usually used in laboratory experiments. Rohe (2013) Currently, only shak flasks are suitable for gas analysis because the gas volumes in microtiter plates are too small. A possible setup for such shake flasks is shown in Fig. 17.3, where each shake flask has two connection options for gas sensors and an additional opening for gas exchange with the environment. This exchange takes place by diffusion. Of course, with such a system the calculation of OUR and CPR is not possible according to the above formulas, because there is no defined gas flow through the fermenter (here shake flask). Instead, the gas exchange between the shake flask and the environment takes place through a sterile filter. Nevertheless, there are also possibilities for determining the uptake and formation rates for such systems. However, these are always system-specific and therefore dependent on the manufacturer. Since such small-scale systems are almost exclusively parallelized systems for the rapid evaluation of several only slightly modified approaches, the required formulas are already included in the

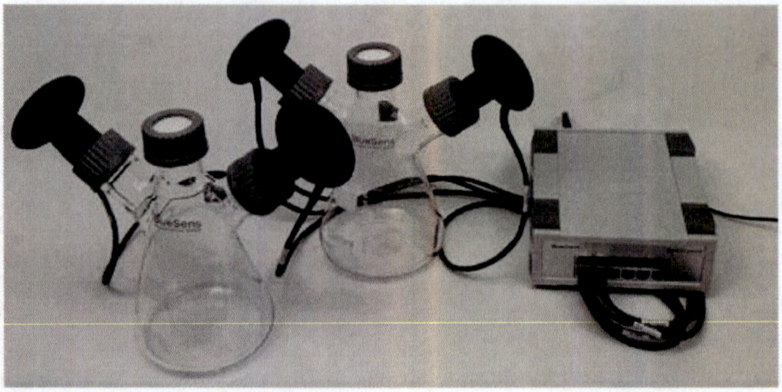

Fig. 17.3 Shake flask with one O_2 and CO_2 sensor each and a multiplexer box for connecting up to 12 sensors. (Source: BlueSens GmbH, Herten)

corresponding evaluation software of the manufacturers. An example of a measurement in a shake flask is shown in Figs. 17.4 and 17.5, where yeast was cultivated in a shake flask.

During the first approximately 5 h after the start of measurement, the consumption and formation rates were very low, they were below the calculation limit of the software used during this time. Figure 17.5 shows the stages of yeast fermentation described above. At the beginning of the measurement there is an excess of glucose. This causes the RQ to rise to about 7. After approximately 20 h, the yeast cells begin to use the ethanol. After experiments with microtiter plates or shake flasks, the process is then transferred to mostly stirred fermenters. Such systems are also constructed in parallel (see Fig. 17.1).

Today, the exhaust gas streams from the fermenters are often no longer routed through a central analyser. Miniaturisation in gas measurement technology has made it possible to monitor each fermenter with its own sensors. Either a single sensor can be used for each gas or, as shown in Fig. 17.6, sensors with several measured variables (here CO_2, O_2, pressure and humidity).

In the process just described, the importance of RQ for the fermentation of yeast was demonstrated. In addition to the determination of the oxygen and carbon dioxide content, ethanol measurement is particularly meaningful in this process. In yeast fermentation, there are different objectives. On the one hand, large quantities of yeast are produced in the baking and food industry, on the other hand, ethanol is produced by yeast fermentation as a part of biofuels. Depending on the application, different ethanol sensors are sometimes used. On the one hand, the ethanol content in the liquid can be determined directly, on the other hand, the ethanol is measured in the gas phase. Both methods have their advantages and disadvantages. In principle, gas measurement is simpler, as the sensor does not come into contact with the fermenter broth and therefore there is no danger of the sensor being covered with growth/biofilms. A disadvantage is that the concentration of ethanol in the gas phase does not depend exclusively on the ethanol concentration in the liquid. In addition, there is also a connection via the vapour pressure with the temperature of the measuring gas; more precisely with the dew point of the measuring gas. On the way from the

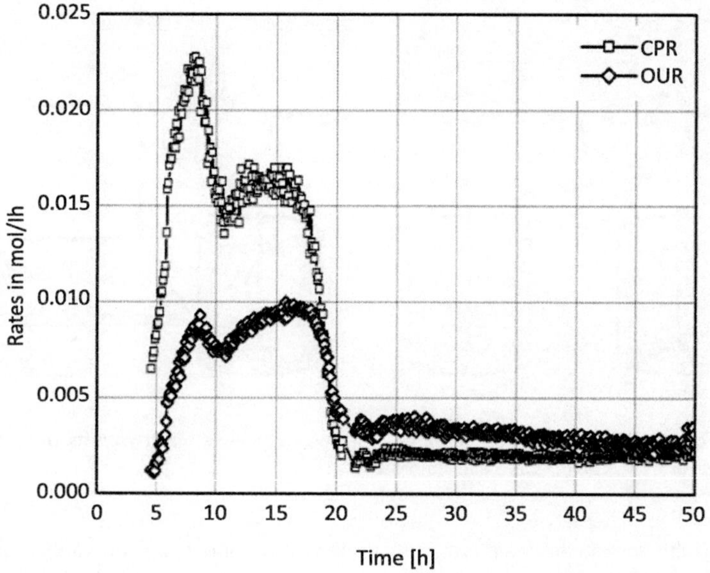

Fig. 17.4 Measured formation and consumption rates of a yeast fermentation in shake flasks. Between hour 5 and hour 18, the production rate for CO_2 is higher than the uptake rate for O_2. From hour 20 onwards, the ratio is reversed

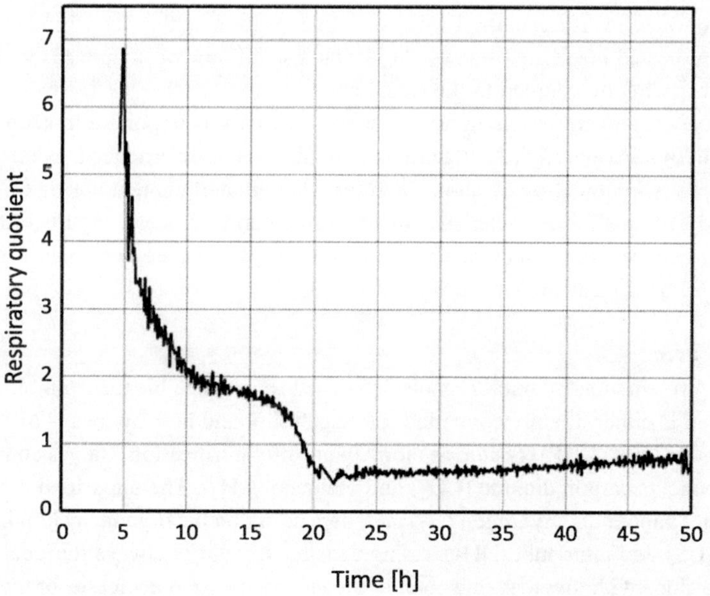

Fig. 17.5 Measured respiratory quotient of a yeast fermentation in a shake flask

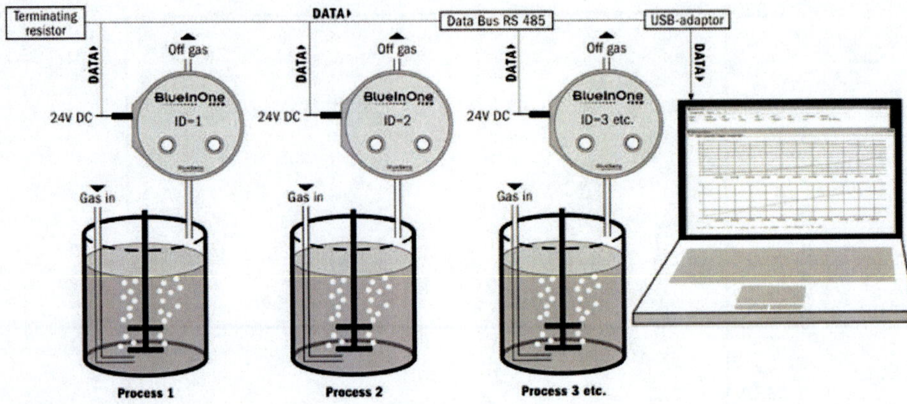

Fig. 17.6 Fermentation system consisting of 3 stirred fermenters, each with its own exhaust gas measurement for CO_2, O_2, pressure and humidity

fermenter to the sensor the water/ethanol mixture can condense. The vapour pressure of ethanol depends on the lowest temperature on this path. Figure 17.7 shows the displayed concentration of a sensor calibrated at 37 °C for an ethanol concentration of 15 vol.% in the liquid at different temperatures of the liquid. For example, if a fermenter has a reflux condenser in the flue gas line with a temperature of 18 °C, the sensor will display only 4 vol.% instead of 15 vol.%. Therefore, knowledge of the dew point is particularly important for these measurements. To prevent condensation on the way from the fermenter to the sensor, heated hoses are often used to bring a small flow of sample gas to the sensor. These are then directly coupled to the fermenter.

In a process designed for maximum yeast production, it is important to keep the RQ at 1 if possible. In addition, as little ethanol as possible should be produced, as the growth of the yeast cells is inhibited by ethanol. Therefore, the ethanol content during fermentation can be used as a quality characteristic for the fermentation process. Figure 17.8 shows a comparison between the values of an online gas phase measurement with a NDIR gas sensor and a liquid measurement.

Anaerobic Processes

Anaerobic fermentation by microorganisms is used to produce biogas from biomass. The gas produced is either directly converted into electricity and heat by means of combinded heat and power units (CHP)) or sold as biomethane after purification. Biogas consists of the main components carbon dioxide (CO_2) and methane (CH_4). The unpurified gas may also contain water vapour and hydrogen (H_2) and, depending on the organic material, hydrogen sulphide (H_2S) and ammonia (NH_3). A high methane yield is always the goal of biogas production, since with low methane content the electricity yield decreases or the CHP unit

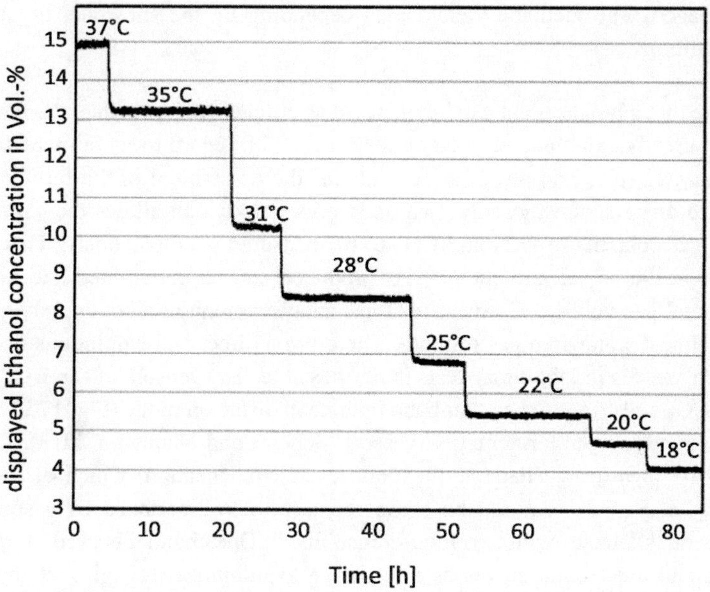

Fig. 17.7 Display of an ethanol sensor calibrated on 37 °C for a mixture of 15 vol.% ethanol in water at different dew points

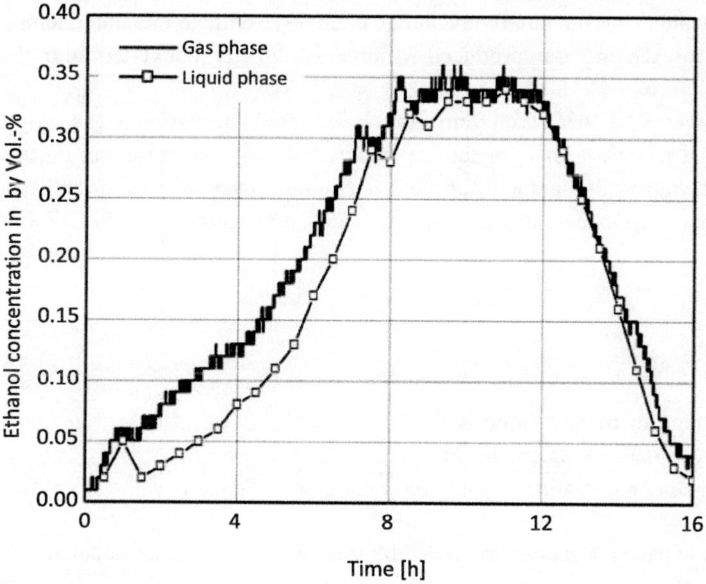

Fig. 17.8 Comparison of a gas phase and a liquid phase measurement of ethanol in a yeast fermentation for yeast production

can be damaged. The methane yield varies depending on the substrate, organisms and process conditions.

To ensure that a biogas plant can be operated as efficiently as possible, the fermentation substrate is already examined on a laboratory scale with regard to its fermentability. VDI 4630 summarises rules and test arrangements for the assessment of the fermentability of substrates. In this chapter, the only method is presented which allows the gas volume as well as the gas contents of CO_2 and CH_4 to be measured simultaneously. This offers the advantage that the actual amount of CH_4 produced can be determined and not just the unspecific total gas volume. Conventional gas analysers require a continuous gas flow as well as an almost water-free gas mixture. Therefore a direct and continuous use on 1–2 l fermentation vessels in laboratory tests is not possible. The sensors of the BCP series for measuring CO_2 and CH_4 on the other hand operate in diffusion mode (Fig. 17.9) and can be connected directly to the fermentation vessels (Scherer and Neumann 2013). An internal heating of the measuring chamber prevents water condensation, which also eliminates corrosion by H_2S. Furthermore, the sensor consists of a broadband light source and a pyroelectric dual detector with two interference filters. One channel is used to measure the desired gas and the second serves as a reference to minimize the aging of the radiation source (Scherer et al. 2004). The sensor is hermetically sealed and can therefore be connected directly to fermentation vessels or even pipelines.

The BCP sensors in combination with BlueVCounts for volume measurement result in the Yieldmaster measuring system (Fig. 17.10), which continuously and online records the produced methane gas quantitatively and qualitatively. With an integrated temperature and pressure measurement, the produced volume is directly converted into the standard volume.

According to VDI 4630, each fermentation test should be carried out as a triplet, as these are biological processes that are subject to a certain range of variation. Furthermore, the inoculum is examined in parallel for the *fermentation potential* (Fig. 17.11).

The amount of methane produced can be calculated according to Eq. 17.8:

$$V_{CH_4} = V_B \cdot C_{korr}^{tr} \tag{17.8}$$

There is

V_{CH_4} volume of methane produced in mL
V_B volume of biogas produced in mL
C_{korr}^{tr} corrected concentration of the biogas component in the dry gas in vol.%

If only the methane concentration should be measured, C_{korr}^{tr} can be calculated according to Eq. 17.9:

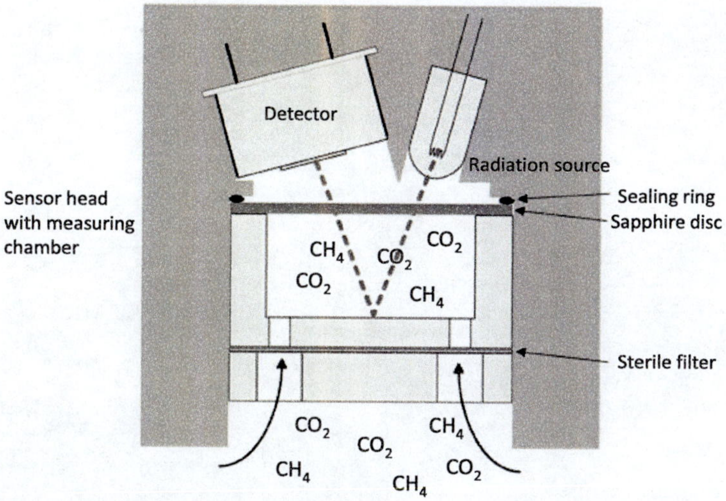

Fig. 17.9 Schematic diagram of an infrared gas sensor for measuring CO_2 or CH_4. In this setup, the optical components (detector and radiation source) are located in the sensor head and separated from the sample gas by a sapphire disc. The sample gas enters the measuring chamber by diffusion and absorbs the IR radiation introduced. The dual detector analyses this radiation at the wavelength of the gas to be examined and at a reference wavelength at which ideally no other gas absorbs radiation. Opposite the fermentation vessel/tube, the measuring adapter is equipped with a PTFE filter to keep particles and water away from the measuring chamber

$$C_{\text{korr}}^{\text{tr}} = C_{t2}^{\text{tr}} + \left(C_{t2}^{\text{tr}} - C_{t1}^{\text{tr}}\right) \cdot \frac{V_K}{V_B} \tag{17.9}$$

with

$C_{\text{korr}}^{\text{tr}}$ corrected concentration of the biogas component in the dry gas in vol.%
C_t^{tr} measured concentration of the biogas component in the dry gas in vol.%
V_K headspace volume in mL
V_B volume of biogas produced in mL
t time of measurement ($t_2 > t_1$).

If CO_2 and CH_4 are present, Eq. 17.10 applies:

$$C_{\text{korr}}^{\text{tr}} = C_{CH_4(CO_2)} \cdot \frac{100}{C_{CH_4 + C_{CO_2}}} \tag{17.10}$$

There is

$C_{\text{korr}}^{\text{tr}}$ corrected concentration of the biogas component in the dry gas in vol.%

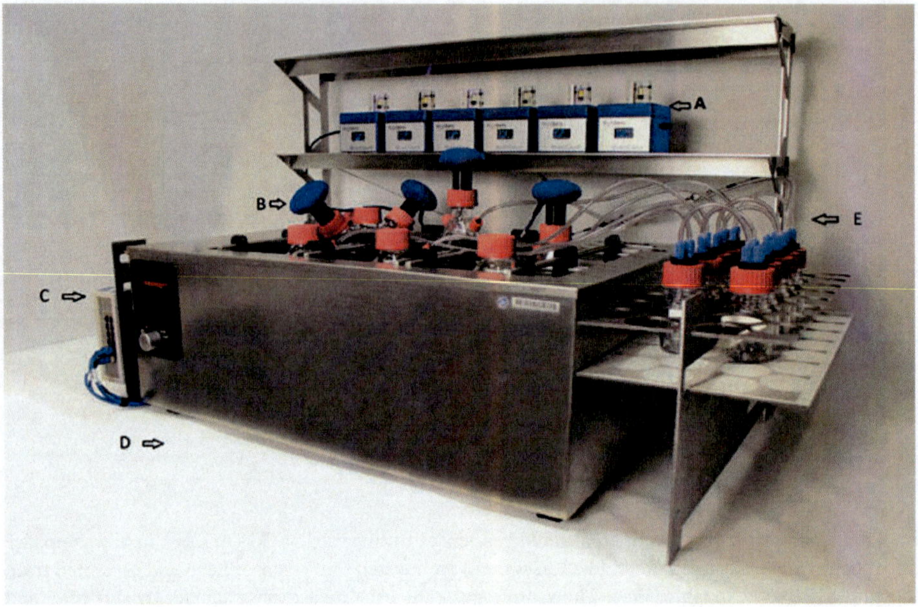

Fig. 17.10 Yieldmaster measuring system for simultaneous determination of the produced gas volume and the methane content during the fermentation of organic substances. A Volume meter BlueVCount, BB BCP-CH_4 sensor, C Electronic multiplexer for data recording, D water bath, E water trap and several glass vessel (Source: BlueSens gas sensor GmbH, Herten)

$C_{CH_4(CO_2)}$ measured gas concentration of methane (or carbon dioxide) in the gaseous medium measured in vol.%

C_{CH_4} measured methane concentration in the gas in vol.%

C_{CO_2} measured carbon dioxide concentration in the gas in vol.%.

The average methane volumes thus calculated are then obtained from the triple measurements (Fig. 17.12). Subtracting the volume produced from the inoculum gives the methane yield from the respective substrate and thus serves for better comparability.

Hydrogen Production

Hydrogen is still being traded today as the energy carrier of the future. Cars, buses, forklifts, boats and even airplanes are now moved with hydrogen (DWV 2014). Since hydrogen simply turns into water when burned, it is apparently a very environmentally friendly energy carrier. Of course, the hydrogen must also be produced in a climate-neutral way. Furthermore hydrogen is considered to be better to store than “electricity”.

For example, with wind power the electrolysis of water to hydrogen can be done, if only little electricity is needed, but there is a lot of wind.

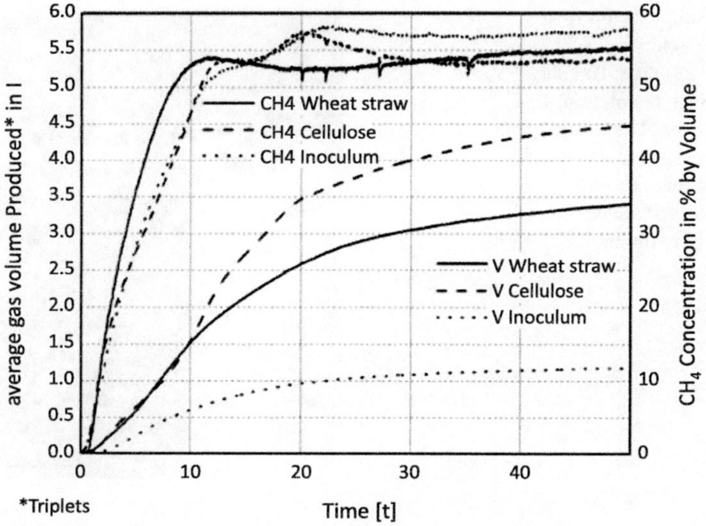

Fig. 17.11 Temporal course of a fermentation of wheat straw, cellulose and the inoculum (fermentation residue from biogas plant). The produced gas volume and the methane gas concentration were recorded

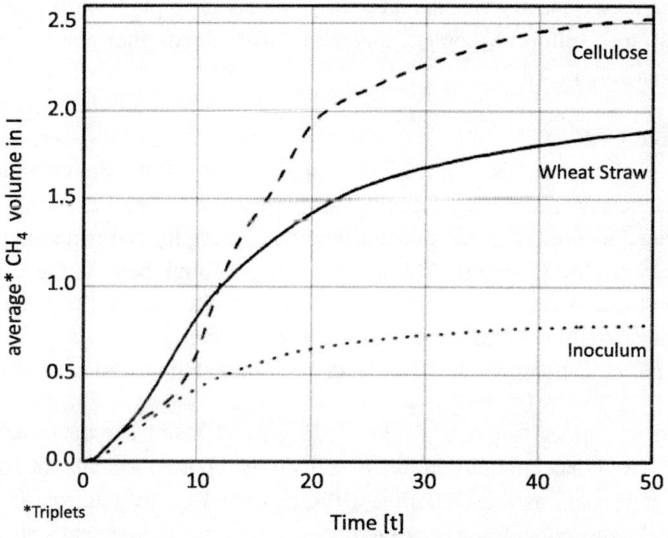

Fig. 17.12 Average (triple approaches) volume of methane produced for cellulose, wheat straw and the inoculum used (sludge from a biogas plant) after applying eqs. 1 and 2 according to VDI 4630

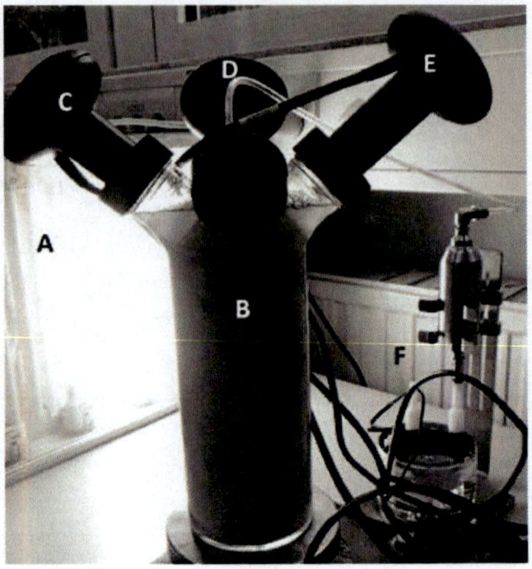

Fig. 17.13 Simple setup for hydrogen production with algae. A: Light source, B: Glass vessel with algae in solution, C, D, E: Gas sensors for measuring H_2, O_2 and CO_2concentrations, F: Milligascounter for measuring the resulting gas volume

Other ways of producing hydrogen are the production with algae (Fig. 17.13) such as *Chlamydomonas reinhardtii*. This algae produces H_2 under sulfur deficiency and saturated light conditions to balance the internal electron balance (Müllner 2008).

The following diagram shows a typical measurement curve for oxygen and the produced hydrogen volume. It is clearly visible that due to the lack of sulphur the oxygen production decreases and the culture becomes anaerobic. The algae then produces hydrogen (Fig. 17.14; Melis et al. 2000).

Conclusion and Outlook

Fermentation processes are used in numerous applications for the production of pharmaceutical products, raw materials, biofuels, and luxury foods and beverages. At present, often only a small amount of process data is recorded during the fermentation process. The continuous, decentralized measurement of important gas components in the fermenter or in the exhaust gas stream can provide important indicators for ensuring product quality and process optimization. Non-contact gas analysis offers advantages over measurement in the liquid phase. Fermentation processes can be divided into aerobic and anaerobic processes.

For aerobic processes, a gas analysis of oxygen (O_2) and carbon dioxide (CO_2) is suitable. The gas concentrations allow a software-supported calculation of important process indicators such as the OUR, the CPR and the RQ. In addition to oxygen and carbon dioxide measurement, the measurement of other gas components, such as ethanol measurement in the gas phase of yeast fermentations, may also be useful in aerobic processes.

In anaerobic digestion processes, biogas is usually produced from biomass, whereby the aim of effective biogas production is to achieve a maximum methane yield. Depending on

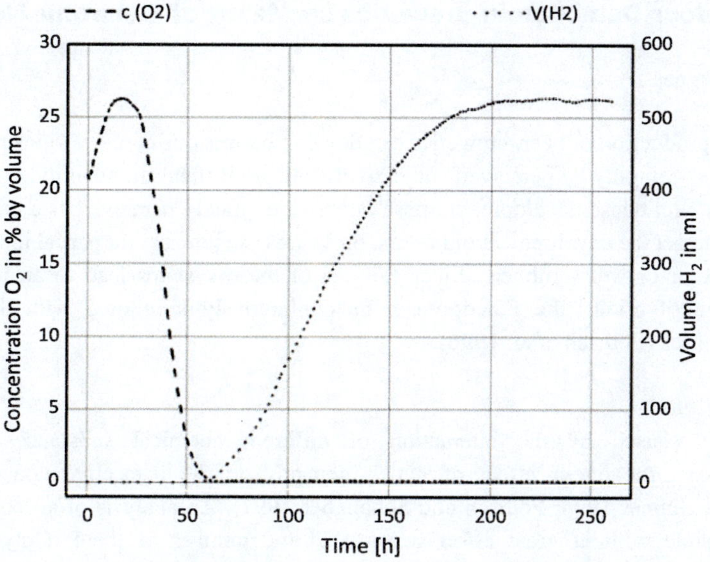

Fig. 17.14 Cultivation process of a *Chlamydomonas reinhardtii* culture based on the oxygen concentration and the produced hydrogen volume. O_2 is measured with a BCP-O_2 sensor. The produced hydrogen volume is calculated with a H_2 gas sensor (BCP-H_2) and a volume measurement (milligascounter) (Hippler and Weingarten 2015)

the substrate and process conditions, the methane yield varies and preliminary tests in the laboratory can provide information about ideal process conditions and substrate compositions. VDI4630 summarizes the rules for a standardized measurement of substrate fermentability. Based on these rules, an additional measurement of the actually produced methane quantity in real time is desirable.

The use of H_2-sensors in the biotechnical production of hydrogen also makes sense.

The gas analysis of fermentation processes offers numerous advantages and will in future play a more important role in bioprocess technology. It can be assumed that different industries will achieve different speeds in the introduction of gas analysis technology for their fermentations. Biotechnical production processes that rely on high quality or high production efficiency will switch to gas analysis more quickly. At the same time, production with minimal unwanted by-products also means greater sustainability. With regard to quality assurance and process control, a decentralized, continuous gas measurement makes more sense than a central analysis technique. Efficient gas analysis in the fermentation process can provide a decisive competitive advantage in highly competitive markets.

17.2 Odour Detection in Breweries by Means of Electronic Nose

Michael Wagner

During the production of beer, breweries can develop odours during the various processes. These odours can only be perceived subjectively and are difficult to quantify. Particularly in locations with dense development, these odours are quickly perceived as a nuisance.

To counteract the development of odours, breweries use ionizers for partial ionization of the exhaust air or wet scrubbers. Since failures of these systems lead to an immediate development of odour, the function can be continuously monitored with the aid of electronic noses (Födisch et al. 1996).

Odour Detection

Odours are caused by the interaction of different chemical substances (volatile hydrocarbons), the odour effect of which depends on the respective concentrations (Schön and Hübner 1996; Födisch and Schengber 1997). An analysis of all components is not possible without great effort because of the number of them. Only identical compositions and concentrations are comparable in their odour intensity (Födisch and Schengber 1998).

Olfactometry (definition according to VDI 3881 Bl. 1) offers a possibility to determine and record odours. This procedure can only be carried out discontinuously and is also very time and labour intensive. A technical, automated solution would be preferable here.

Conventional Odour Sensor Technology

Semiconductor sensors based on metal oxides, for example, can be used for the selective measurement of odours (Wiegleb 2001). Among other things, tin dioxide (SnO_2) with different doping is used as a sensitive layer. The principle structure of a metal oxide semiconductor gas sensor is shown in Fig. 17.15.

The sensors are heated to a constant temperature to achieve selectivity towards different hydrocarbons. They change their conductivity when reducing (volatile hydrocarbons) or oxidizing (oxygen) different gases. This signal can be detected by appropriate electronics.

In order to be able to distinguish odours, several sensors, each of which reacts to different chemical compounds, must be arranged in an array. The number of sensors indicates the resolution of the selective odour pattern recognition.

Metal oxide semiconductors have the disadvantage that they have a cross-sensitivity to air humidity. The influence can be different on each sensor of the array and requires selective compensation. In addition, if a gas is exposed to the sensor for a longer period of time, saturation effects can occur, which can only be counteracted by regular baking.

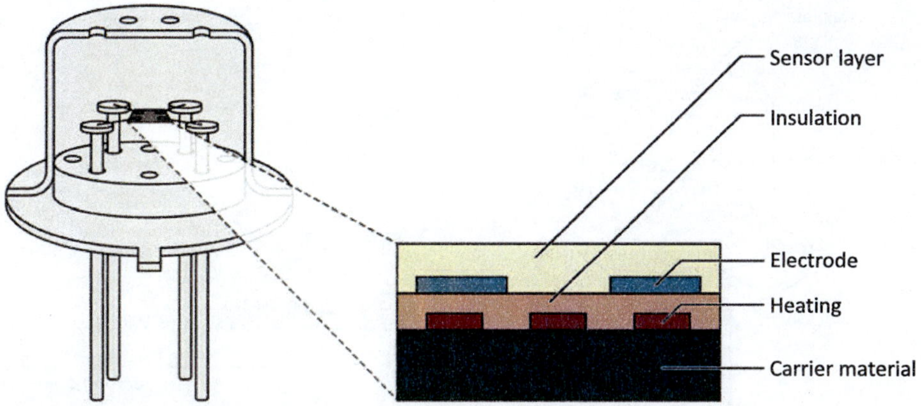

Fig. 17.15 Gas sensor – scheme

Odour Sensing by Means of Virtual Gas Sensor Array (VGSA)

To reduce the number of sensors, a VGSA was developed. By modulating the temperature of the sensor element in several steps the optimal sensitivity to the different gases can be achieved selectively. Each individual temperature step thus corresponds to the measured value that would be determined by a single metal oxide sensor. The signal is not measured via the resistance of the sensor, but via the electronic charge transfer. External influences, such as different saturation effects and cross-sensitivities to water vapor, are minimized by this method. By means of 48 temperature steps an array of 48 sensors can be simulated, which react to different gas compositions.

Figure 17.16 shows the structure of the virtual gas sensor array.

Design and Integration into an Industrial Multi-Gas Analyzer

The multi-gas analyzer MGA 12[1] is used for the continuous measurement of pollutants in flue gas (e.g., CO, CO_2, SO_2, NO) and the measurement of O_2 as well as for continuous process monitoring. The analyzer is suitability-tested according to DIN EN 15267-3 and certified according to QAL1. As a component of the MGA 12 analysis system, it is suitability-tested and certified for systems according to TA Luft, 13th and 27th BImSchV in accordance with DIN EN 15267-3. Two independent, selectively operating measuring principles are used in the MGA 12: infrared absorption and electrochemical cell. Due to the functional principle of infrared absorption, up to five infrared active gas components can be determined simultaneously. This spectroscopic method is based on the absorption of non-dispersive infrared radiation (NDIR photometry).

[1] Dr. Födisch Umweltmesstechnik AG.

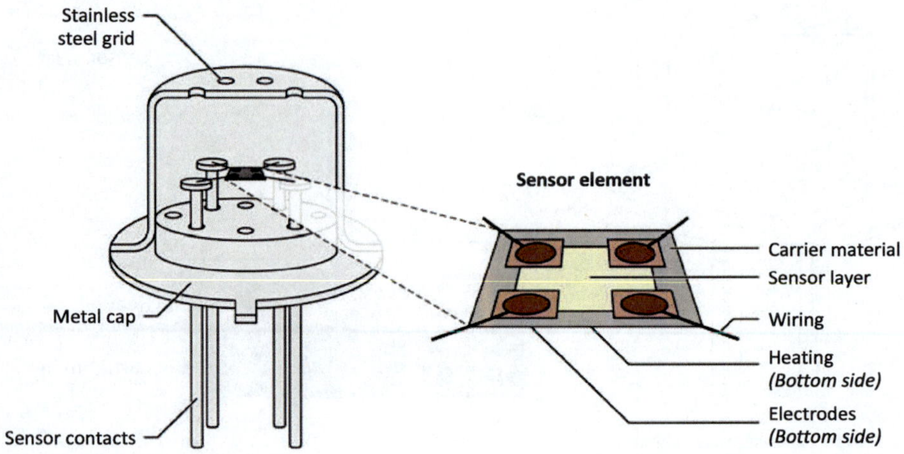

Fig. 17.16 Virtual gas sensor array

Fig. 17.17 Gas flow chart MGA12 with VGSA – scheme

For practical application, the VGSA is integrated into this proven and industrially suitable multi-gas analyzer. The internal gas flow chart of the instrument is shown in Fig. 17.17.

The sample gas is fed into the analyser. Based on the infrared absorption, CO_2, CH_4 and CxH_y are measured. Finally, the water content and temperature are determined. At the outlet of the photometer, the sample gas is led into the gas block. The gas block is shown in Fig. 17.18. All measurement data are recorded and calculated by the internal electronics of the analyzer. Figure 17.19 shows the analyzer design with integrated gas block.

Odour Pattern Recognition and Evaluation Algorithm

For the evaluation of the sensor signals, a software for the recognition of odour patterns is necessary. When exposed to a gas, the individual sensors output specific measured values that can be recorded in a pattern. The number of sensors thus corresponds to the resolution of selective odour pattern recognition. The sensor array cyclically outputs 48 measured values relative to the stored zero point, which was previously set with odour-free ambient air.

The odour pattern is stored in the electronics and serves as a reference for recognition. The diagram in Fig. 17.20 illustrates the application of methane in different concentrations to the sensor.

Fig. 17.18 Gas block for the integration of the air quality sensor and VGSA

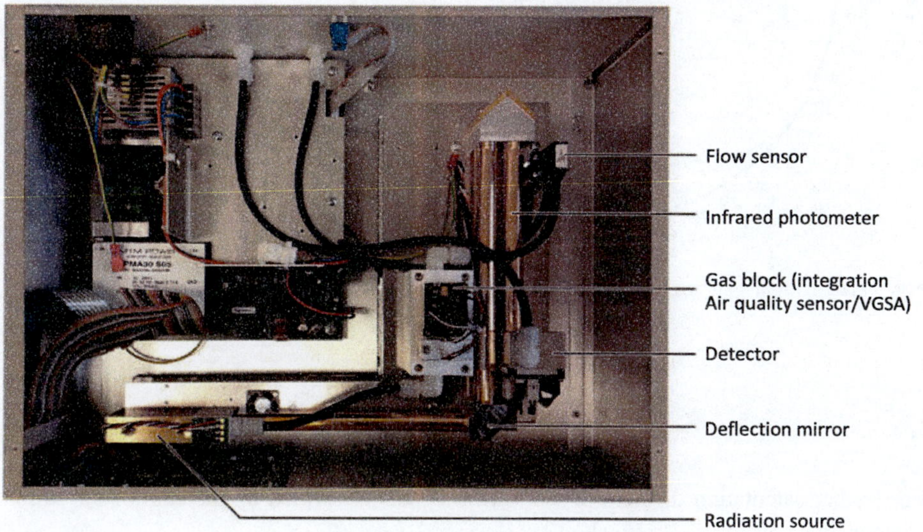

Fig. 17.19 Analyzer with integrated gas block

The measured values from the feeding of methane result in a typical pattern for the detection, which shows the different sensitivities of the virtual sensors.

Each individual measured value can be evaluated by a factor *f*. This makes it possible to adjust the signals of the individual virtual sensors to each other or to remove individual sensors from the calculation. A sum signal is formed from the measured values multiplied by the factor *f* according to Eq. 17.11.

$$\text{SUM} = \sum_{0}^{n} |\text{VGSA}_n \cdot f_n| \tag{17.11}$$

Pattern recognition is achieved by comparing the currently measured values of the individual virtual sensors with the values of the stored pattern. For this purpose, the sum of the difference of the deviations is calculated as mean value according to Eq. 17.12.

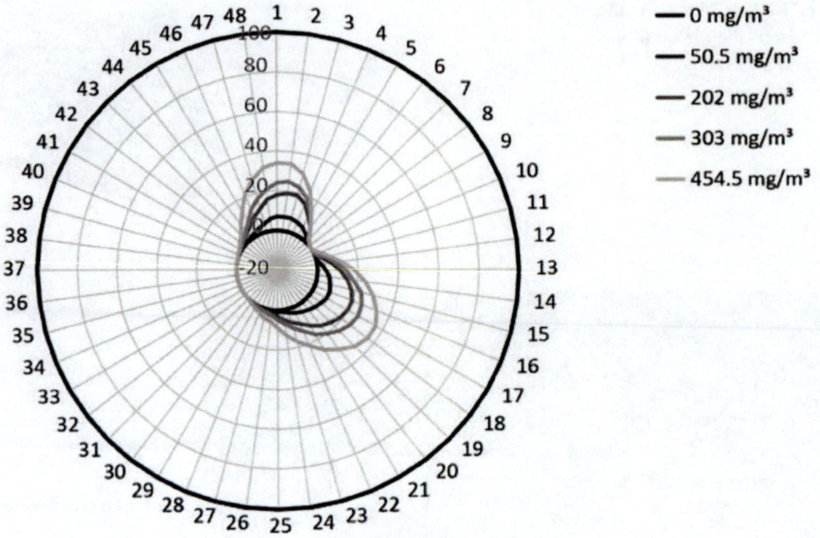

Fig. 17.20 Odour patterns of methane in different concentrations

$$\mathrm{Abw_{abs}} = \frac{\sum_{0}^{n} |\mathrm{VGSA}_n - \mathrm{Pattern}_n|}{n} \tag{17.12}$$

For further calculation the sum of the sensor signals is divided by the sum of the stored sample signal according to Eq. 17.13.

$$\mathrm{Abw_{rel}} = \frac{\mathrm{Sum}}{\sum_{0}^{n} \mathrm{Pattern}_n} \tag{17.13}$$

The factor resulting from Eq. 17.13 indicates the relative deviation ($\mathrm{Abw_{rel}}$) of the current measurement to the stored sample. To obtain a common value for further calculation in the algorithm of the multi-gas analyzer, the measurement result is calculated by a cubic regression from the absolute deviation ($\mathrm{Abw_{abs}}$) and the relative deviation ($\mathrm{Abw_{rel}}$) according to Eq. 17.14.

$$\mathrm{Erg} = \mathrm{Abw_{rel}} \cdot \left(a \cdot \mathrm{Abw_{abs}^3} + b \cdot \mathrm{Abw_{abs}^2} + c \cdot \mathrm{Abw_{abs}} + d\right) \tag{17.14}$$

Through this evaluation, it is possible to recognize the pattern and the degree of approximation (concentration). The signal generated from Eq. 17.14 can then be provided with a calibration factor and output as a percentage signal for recognition.

Plausibility Check by Infrared Photometer

The measured values of the infrared photometer are used to check the correct pattern recognition. In the photometer of the gas analyzer the group of hydrocarbons (C_xH_y), methane (CH_4) and carbon dioxide (CO_2) are measured as infrared components. The influence of these components on the VGSA can thus be minimized. Multi-stage iteration methods are used to account for other cross-sensitivities, for example the water vapor concentration. In the infrared photometer, no saturation effects can occur due to exposure to gases. This property is used to perform a plausibility check of the measurement results of the VGSA. Thus it can be distinguished whether the measurement signal is a real change in the gas composition or whether the VGSA is drifting due to a saturation effect.

Characteristics of the Tapping Point

The exhaust air from the boiling pans is collected by a collecting pipe and passed through an ionisation system. The voltage discharges break down oxygen and water molecules into charged ions. The highly reactive ions then form bonds with the organic molecules of the exhaust gas. Furthermore, ozone is formed during this process, which has a germicidal effect and splits long-chain hydrocarbons. The tapping point is located between the ionisation system and the wet scrubber. At this point, it is possible to monitor the operation of the ionisation system. Figure 17.21 schematically shows the tapping point in the process of exhaust air purification. In a connected wet scrubber, additional odours are bound by sprinkling the exhaust gas with water. The odours are then discharged into the open air through an exhaust chimney.

Gas Conditioning

The water content at the tapping point is approximately 80 vol.%. The gas is preconditioned by a special withdrawal probe and a proportion of the condensing water is returned to the process. In the analysis system, the dew point is lowered to 5 °C by a multi-stage gas cooling system. This minimizes the influence of the water vapor on the measured values and prevents condensation in the measuring instrument. To counteract any drift, the analyzer regularly sets a zero point for all sensors with clean and odor-free ambient air. For this purpose, the sample gas path is closed by a solenoid valve and zero air is supplied instead. Figure 17.22 shows the basic gas flow diagram of the entire analyzer system.

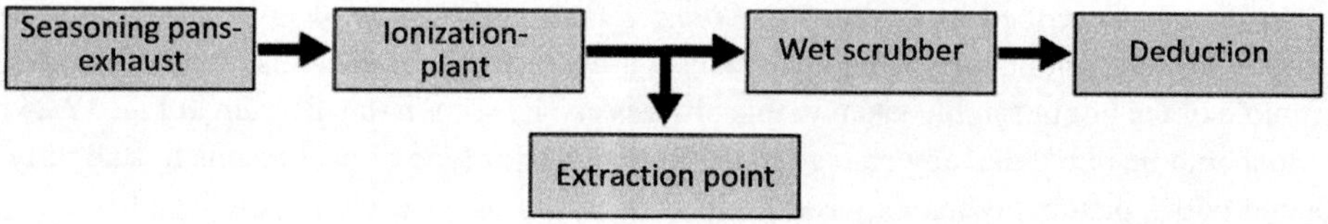

Fig. 17.21 Exhaust air treatment with extraction point – scheme

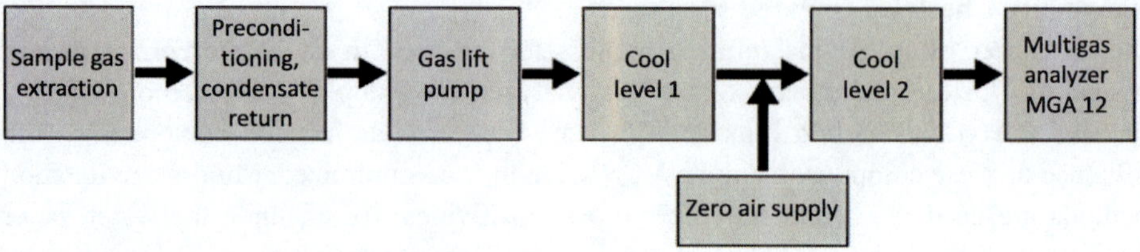

Fig. 17.22 Gas flow chart analyzer system – scheme

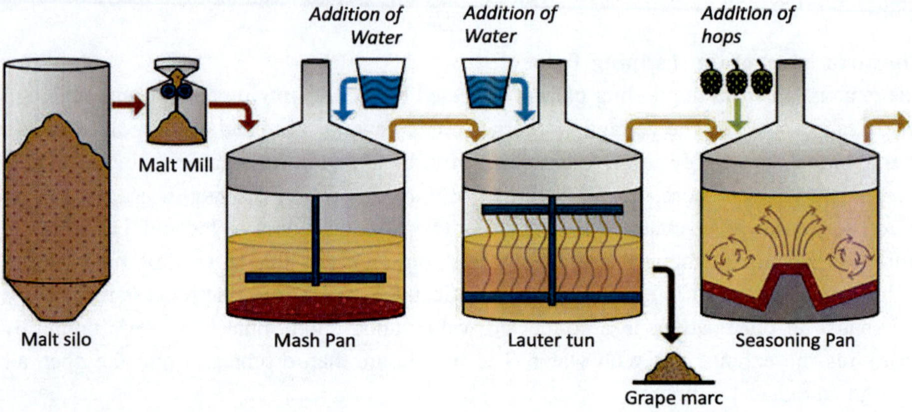

Fig. 17.23 Possible sources of odour development during the brewing process

Measurements in the Real Process

The brewing process consists of several sub-processes, which result in different odours. The malt is stored in a silo and passes through a malt mill into the mash pan. By adding water and slowly increasing the temperature, enzymes are activated which convert the starch into sugar. As this process takes place below the boiling temperature of water, a low odour development is to be expected. The solid components are then separated from the mash in the lauter tun. By far the greatest odour development occurs during boiling in the wort pan. The wort is boiled over a longer period of time. Depending on the recipe, further ingredients are added in this partial step. Figure 17.23 shows the possible sources of odor development during the brewing process.

As already described in the section *Characteristics of the sampling point*, the odour sample is taken continuously from the exhaust air downstream of the ionisation system. An example of the odour sample taken in this process can be seen in the diagram in Fig. 17.24.

It should be noted that any change in the ingredients or type of beer results in a slightly altered odour pattern. When exposed to different types of beer, the sensitivity of sensors located close together changes. The overall odour pattern nevertheless remains

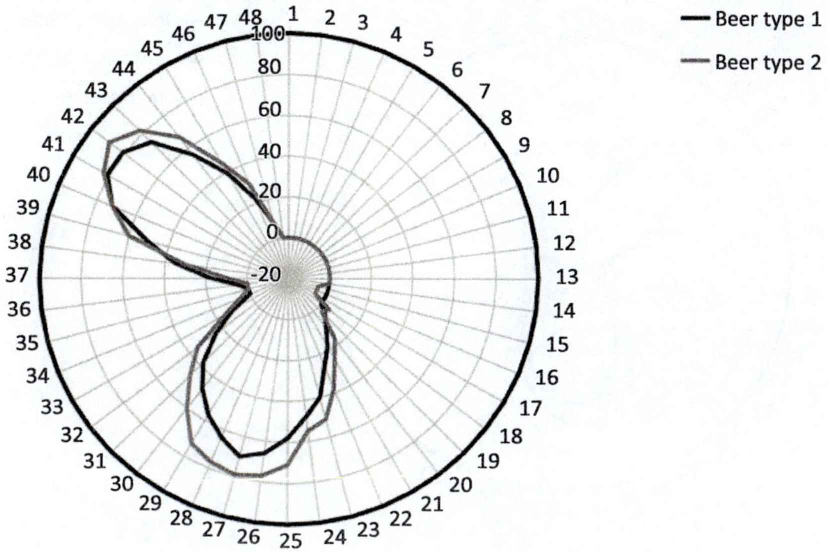

Fig. 17.24 Odour patterns of different beer types during boiling in the seasoning pan

comparable. Several patterns can be stored in the multi-gas analyzer, which must be selected prior to sampling.

In further investigations, different performance levels in the range from 0 to 150% of the ionization system were set and the sensor signals were simultaneously recorded. Figure 17.25 illustrates the relationship between a high sensor signal with ionization switched off and a low sensor signal at 150% ionization power. The basic pattern for the typical odour is always maintained.

The odour pattern with 0% ionisation performance from Fig. 17.25 was stored in the software, which corresponds to a recognition of 100%. The chronological sequence of a brewing process without an ionizing system connected is shown in the diagram in Fig. 17.26. The result of the measurement is calculated according to Eq. 17.14 and then scaled with a factor to the range 0–100%.

The highest odour concentration is achieved during cooking. The pattern is briefly detected at 100%. This means that at this point in time the same odour load exists in intensity and type as before when the sample was recorded or calibrated. After adding the ingredients, the temperature of the wort drops below boiling point and the odour development decreases. As soon as the wort begins to boil again, the odour concentration also rises again. In comparison, Fig. 17.27 shows another boiling process. In order to reduce the odour, the ionisation system was switched on during this process. Odour development is also visible, but is detected at a lower level.

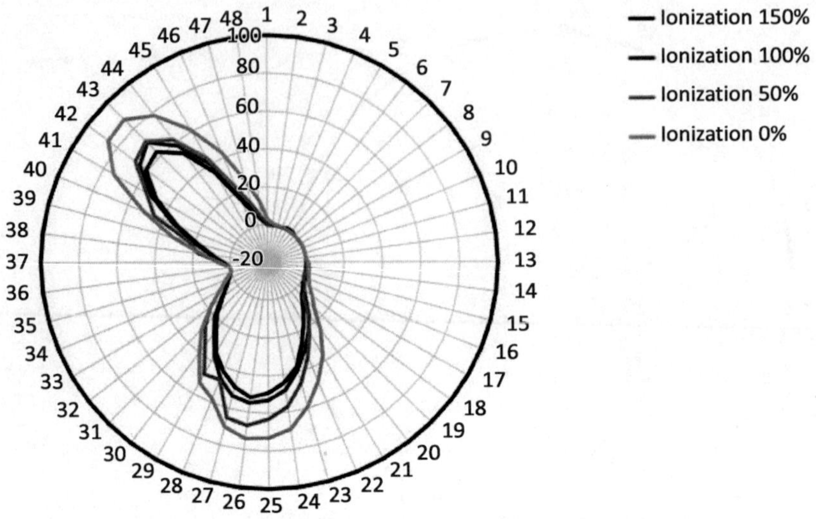

Fig. 17.25 Odour patterns at different levels of ionisation

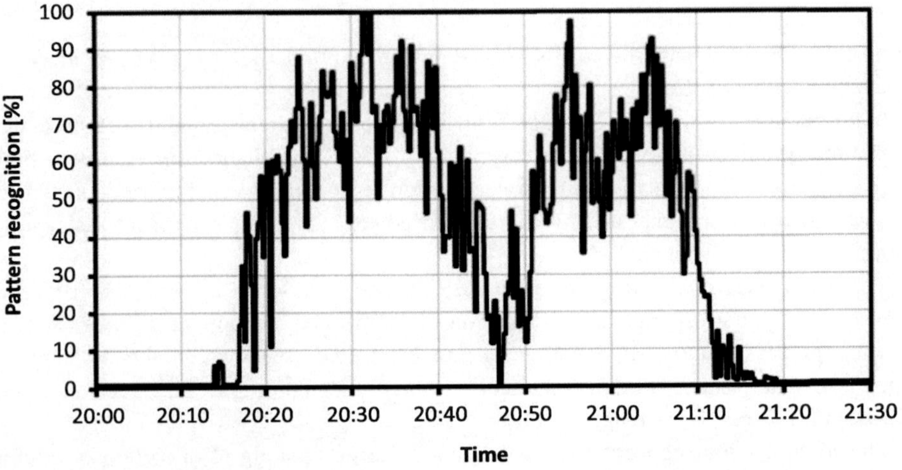

Fig. 17.26 Time course of an odour pattern recognition without ionization

Correlation of the Measurement Signal to the Subjective Sense of Smell

A correlation between the subjective sense of smell and the measurement signal is shown in Fig. 17.28. Here, the gas at the outlet of the analysis system is made available to a test person as an odour sample. A statement about the odour is then made. This can be freely defined using a scale (no perceptible odour to extremely strong odour pollution). The odour sample with the help of the test person is a non-representative individual test.

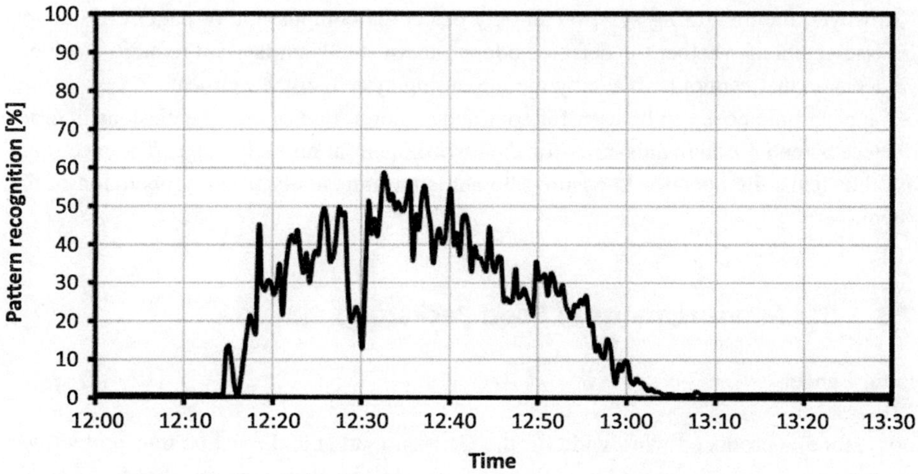

Fig. 17.27 Time course of an odour pattern recognition with ionization

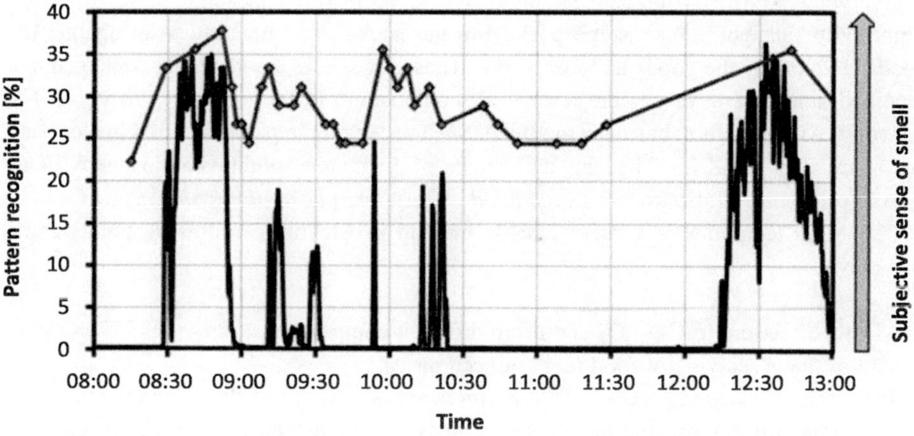

Fig. 17.28 Time course of an odour pattern recognition with subjective odour perception

Limits of Odour Measurement Technology

The described practical example shows that the measurement of odours in the exhaust air of breweries is possible with different sensors. However, this requires a complex gas sampling technique and extensive evaluation software. The performance of the ionization system can also be detected after appropriate calibrations.

The entire analysis system is largely adaptable and can also be used for other applications for monitoring air purification plants or for the recognition of odour sources.

However, the measuring system can only detect the composition or pattern of odours. The assessment of whether the detected odour pattern smells unpleasant remains a subjective decision that cannot be made by the measuring system at the moment.

The electronic nose can be used for continuous monitoring of odour emissions in order to detect potential odour nuisances for the environment at an early stage. The electronic nose thus helps the operator to ensure safe and environmentally friendly operation of the system.

17.3 Gas Measurements in Food Packaging

Martin Bender

Today, food is produced worldwide for the German market and must be transported over long distances to the end consumer. During transport and storage in the shops, the goods should meet the customer requirements in terms of quality for as long as possible. The packaging plays a decisive role in this.

Everybody knows it from daily shopping: the goods are often packed in plastic foil/trays supposedly "airtight". With some packaging the packaging film nestles up against the goods, with others the goods lie loosely in it. These foodstuffs are generally packaged in a modified atmosphere, that is the normal air composition (20.9 vol.% O_2, 78 vol.% N_2, 0.9 vol.% Ar + moisture) has been modified. Such packaging must be declared in Germany and many other countries. The label then reads "packed under modified atmosphere". This type of protection is also known as *modified atmosphere packaging* (MAP[2]).

The MAP technique is a very versatile method in which the following points come into play:

- Mostly a mixture of CO_2, O_2 and N_2 in different compositions.
- Part of the processing of food (e.g., minced meat).
- Part of the packaging process of food – prepared as a rinse and directly in the package.
- Now standard in more and more product areas – and indispensable for perishables.

The advantages of this method are above all:

- Longer shelf life and product freshness.
- Longer storage, further transport routes possible: development of further markets.
- Protection against microbes, chemical and physical changes.
- Ideal for convenience products.

[2]Modified atmosphere packaging.

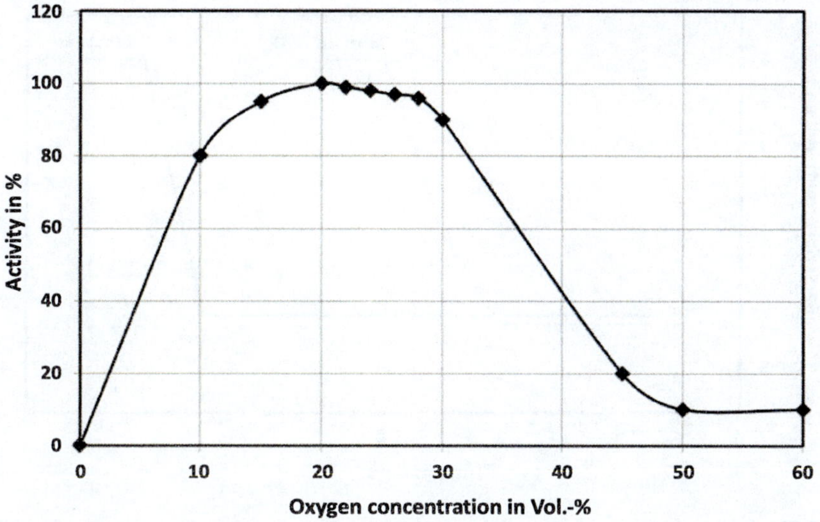

Fig. 17.29 Influence of oxygen (O_2) on the activity of microorganisms

The protective properties of various gases in the food sector result primarily from their effect on microorganisms (bacteria, fungi, etc.) and from the possibility of ensuring a stable protective gas atmosphere (Fig. 17.29) over a longer period of time. The protective effect (shelf life) for different food groups can be optimised by selecting the type of gas (Fig. 17.30). Many years of experience have shown that the following gases are suitable:

Oxygen (O_2)

O_2 basically causes oxidation-related spoilage of food and is the prerequisite for the growth of aerobic microorganisms. Oxygen is therefore often excluded from MAP. In some cases, for example with red meat, high oxygen concentrations are deliberately used to prevent the red colour from *fading* and to inhibit the growth of anaerobic organisms.

Carbon Dioxide (CO_2)

Is colorless, odorless and tasteless. It has an antioxidant and growth-inhibiting effect on most aerobic bacteria and moulds. The gas is often used to increase the shelf life of food. Usually, the higher the CO_2 content, the longer the shelf life of packaged or stored food. However, some products can become sour more quickly if the dosage is too high. In addition, the gas can diffuse out of the packaging or be absorbed by the product – the package collapses. The use of supporting or filling gases can slow down this effect.

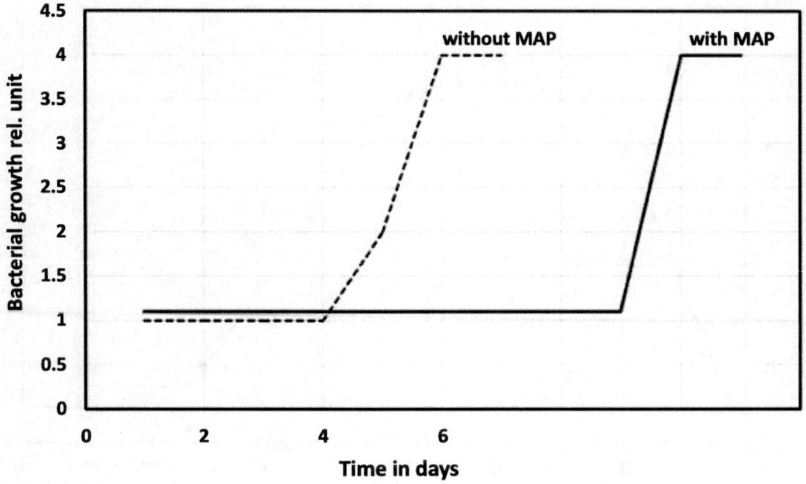

Fig. 17.30 Schematic representation of the shelf life of food, based on bacterial growth with and without MAP

Nitrogen (N_2)

Is an inert gas and has a high purity due to the manufacturing process. It is generally used to displace air, that is especially atmospheric oxygen, in food packaging. This prevents the oxidation of food and inhibits the growth of aerobic microorganisms. It is often used as a supporting or filling gas because it diffuses very slowly through plastic films and thus remains in the packaging for a long time.

Carbon Monoxide (CO)

Is colorless, odorless and tasteless. Similar to oxygen, carbon monoxide is partly used to preserve the red colour mainly of meat. The necessary concentrations are very low. However, in some countries, including the EU, the use of carbon monoxide for protective atmospheres in food packaging is prohibited.

Hydrogen (H_2)

Is colorless, odorless and tasteless. It is used exclusively for leak testing. The concentrations used are usually <5 vol.%, as otherwise flammable mixtures may be formed.

Helium (He)

Is colorless, odorless and tasteless. It is used exclusively for leak testing. The concentrations used are usually low, the disadvantage of helium is its high price.

Table 17.1 Examples of optimum gas mixtures for different foodstuffs (all figures in vol.%)

Product	O_2	CO_2	N_2
Raw red meat	70	20–30	0–10
Raw offal	80	20	0
Raw poultry with skin	0	30	70
Raw poultry without skin	70	20–30	0–10
Cooked meat and sausages	0	20–30	70–80
Hard cheese	0	30–100	0–70
Soft cheese	0	10–40	60–90
Cheese, sliced	0	30–40	60–70
Cream cheese	0	100	0
Yoghurt	0	0–30	70–100
Milk powder	0	0–20	80–100
Raw low-fat fish	20–30	40–60	20–40
Raw fatty fish	0	40	60
Shellfish and crustaceans	30	40	30
Cooked/smoked fish	0	30–60	40–70
Baked bread/bread rolls	0	50–100	0–50
Cakes, pastries	0	50	50
Fruit	3–10	3–10	80–90
Vegetables	0	30	70
Prepared dishes	0	30–60	40–70
Pasta/pizza	0	30–60	40–70
Sandwiches	0	30	70

Further gases like for example Ar (N_2-replacement) etc. naturally are also used, but this will not be discussed here. Gases are food additives with their own approval number:

- E290-carbon dioxide.
- E938-argone.
- E939-helium.
- E941-nitrogen.
- E942-laughing gas.
- E948-oxygen.
- E949-hydrogen.

Furthermore, the individual gases can also be mixed to further increase the protective effect for the respective food group. Here too, extensive experience is available, which is summarised in Table 17.1.

Applications

Ambient Air Monitoring

The best known application for gas sensor technology in food processing plants is ambient air monitoring. Wherever gases are used, leaks can occur that endanger the safety of the people in the room. For this reason, indoor air monitoring systems that monitor the ambient atmosphere are used there. This does not apply specifically to the food industry, but to practically all branches of industry. The food industry works with N_2, CO_2 and O_2. Depending on the application, the lower limit value for O_2 or CO_2 is monitored if CO_2 and N_2 are used, or the upper limit value for O_2 if O_2 is used.

Form Fill and Seal Machine

Form fill and seal machines are available in horizontal and vertical form. The operating principle is essential for both types: flat plastic films from the roll are formed into a tube and welded at the longitudinal seam. The tube is then sealed at the bottom and top by the flow pack machine. Between the seams, a cut is made, the newly started bag is filled and sealed crosswise towards the successor.

With the vertical form fill and seal machine, the product (e.g., potato chips, coffee or milk powder) falls into the bag before it is sealed.

If the machine is equipped with a gassing device, the still open hose runs over a gassing lance, through which gas flows in to displace the ambient air. The bag is only finally sealed behind the lance. By means of analysis, small quantities of gas can be sucked out again parallel to the gassing lance in order to measure the O_2 and CO_2 concentration in the bag again. This is a continuous measurement.

The following points are critical for the sensor technology here:

- Speed: due to the high speed of the machines, the measurement result should be available as quickly as possible, otherwise many defective products will have passed through in the meantime (Fig. 17.31).
- Temperature influence (Fig. 17.32).

Thermoforming Packaging Machines

This type of machine is characterized by the fact that a vacuum is often applied to the chamber before the packaging process to remove the ambient air. The chamber is then back-gassed with mixtures of N_2, CO_2 and O_2 to obtain the desired packaging atmosphere. Gas sensors are used to measure whether the process was successful.

The following points are critical for the sensor technology here:

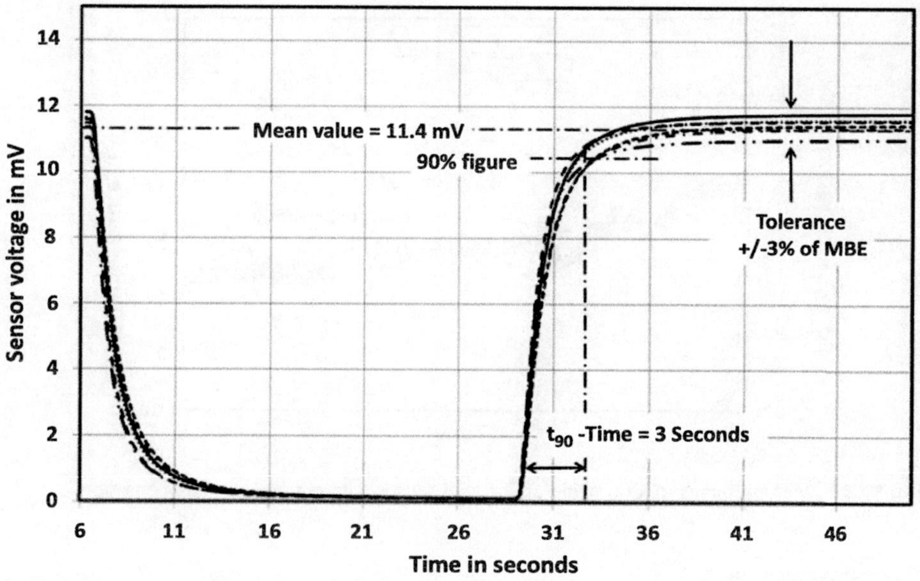

Fig. 17.31 Time response (t_{90} time) of several electrochemical oxygen sensors, for a gas jump from nitrogen to 20.9 vol.% oxygen

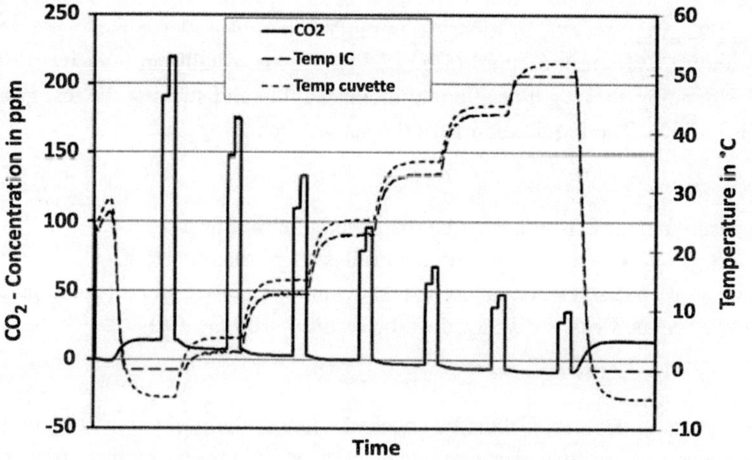

Fig. 17.32 Measurements of CO_2 by infrared sensor without temperature compensation

- Speed: the process of filling the chamber, evacuation, re-gassing, emptying the chamber takes about 4 s. After the re-gassing, an analysis must be made within the shortest possible time.

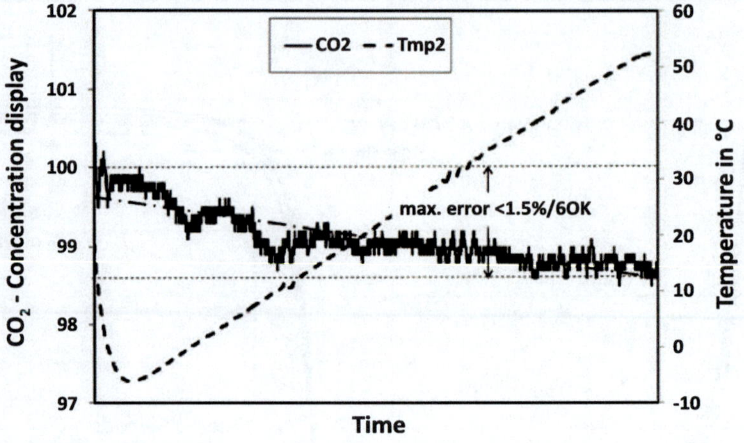

Fig. 17.33 Measurements of CO_2 using an infrared sensor with temperature compensation

- Vacuum resistance.
- Temperature fluctuation (Fig. 17.33).

Sampling

After the packaging process, in most cases random samples of the packaged goods are taken. O_2 and/or CO_2 are measured (Fig. 17.34). This is usually an invasive, destructive test. This is done by sticking a needle into the packaging and sucking the test gas into the sensor (Fig. 17.35). The requirements for the sensor technology are:

- Fast measurement: the result must be displayed after a few seconds.
- Small test gas volume (e.g., for coffee capsules, mini salamis etc.)
- Vacuum: some packages are vacuumed. Pressure compensation must take place here.
- N_2 as an associated gas is displayed as the result of 100-O_2–CO_2.

Leak Test

After the packaging process, it must be checked whether the packaging is really tight and *retains* the gas mixture that was previously injected. A proven method is to place the packaging in a vacuum chamber, create a vacuum and measure the increase in CO_2 concentration in the chamber (Fig. 17.36). Of course, this method only works if the packaging has been previously gassed with CO_2. The normal ambient air contains approximately 300 ppm CO_2. If this concentration increases significantly, the additional CO_2 can only come from the packaging.

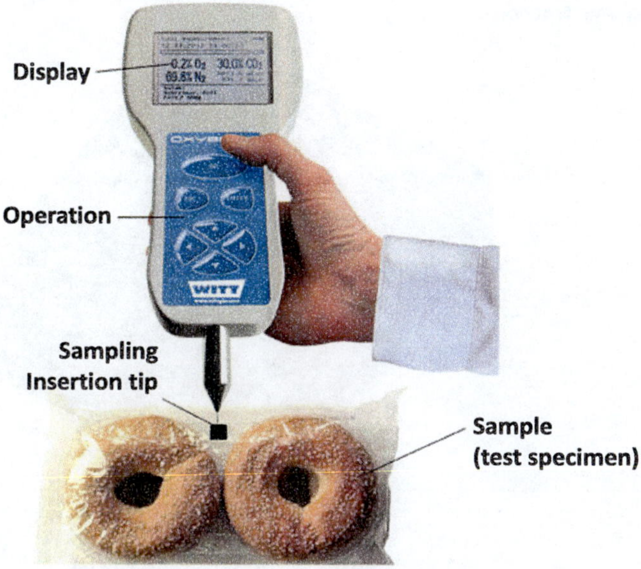

Fig. 17.34 Sampling device (*Oxybaby*)

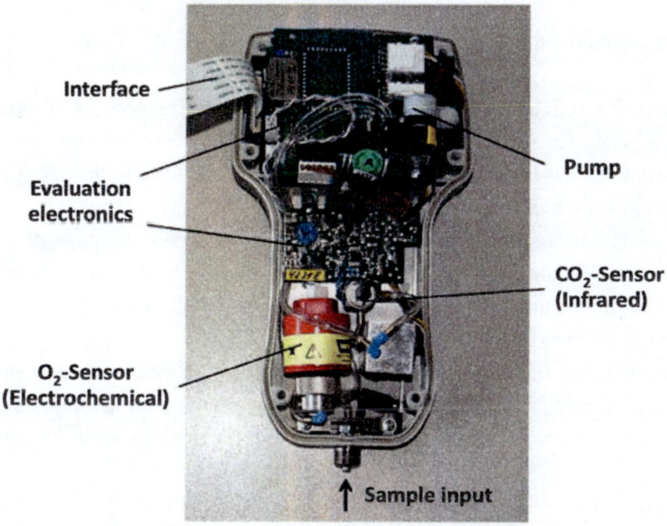

Fig. 17.35 Opened gas detector (*Oxybaby*) (Source: WITT-Gas Technology)

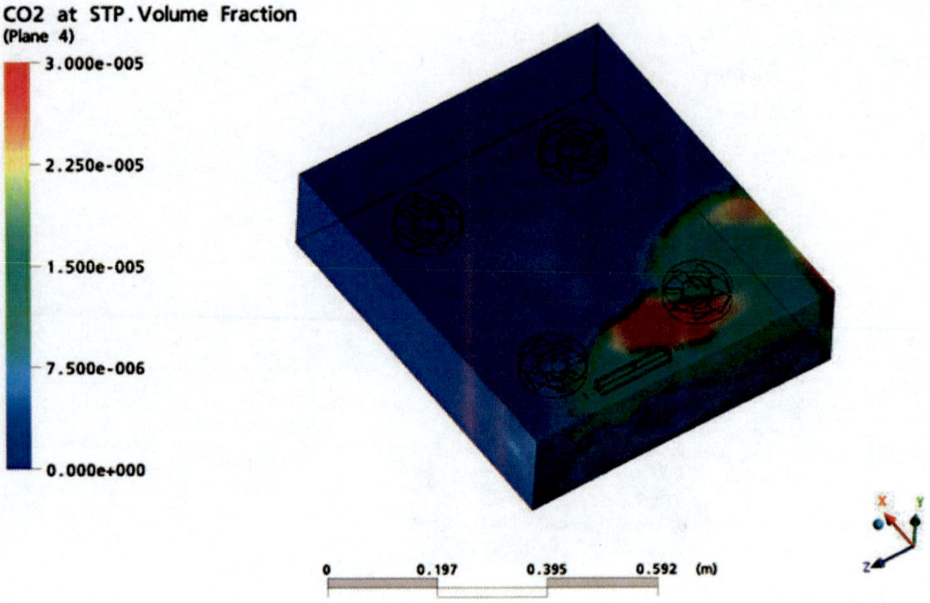

Fig. 17.36 Flow simulation of a CO_2 leak in a vacuum chamber

The requirements for the sensor technology are:

- Small measuring range for CO_2 with approximately 5000 ppm, at 1 ppm resolution https://www.witec-sensorik.de/en/products/infra-sens/.
- Compensation of CO_2 fluctuations in the ambient air (e.g., due to the fumigation process or operators simply breathing out)
- Alternatively H_2 or He can be used for leak detection. The advantage of these gases is their low density. Leaks can be detected relatively easily by the volatile gas. However, since both H_2 and He are not required for the shelf life process, an additional (expensive) gas must be introduced.
- Uniform distribution of CO_2 in the measuring chamber. Since only very small amounts of CO_2 escape locally from a leak, the gas must be distributed as optimally as possible (e.g., by means of fans) in order to be able to measure it at a specific location.

Ripening Chambers

The targeted metering of ethylene into air-conditioned storage rooms allows the ripening process of fruits – especially bananas – to be timed and guided to a desired point. The same applies to numerous types of fruit and vegetables and can also be used for storage of potatoes. Here, etyhlen has a sprout inhibiting effect. A similar effect is known to us all from the domestic area. If, for example, apples are placed next to bananas, the bananas ripen much faster due to the ethylene released by the apples.

Outlook
What are the future challenges for gas sensor technology in the field of food packaging? The keyword is *100% testing*. Random sampling is still a proven method, but the industrialised food industry and its customers demand a continuous 100% inspection. This requires new measuring methods that enable fast and, above all, non-destructive measurement. Approaches to this already exist, for example with optical measuring methods.

Also conceivable are electronic noses that continuously monitor the shelf life and composition of food.

17.4 NDIR Gas Analysis at High Ambient Temperatures

André Magi and Steffen Biermann

IR Gas Sensor Technology
NDIR[3]-based gas sensors are used in a wide range of applications for detection and concentration measurement of infrared active gases and fluids (https://www.witec-sensorik.de/en/products/infra-sens/). The fields of application, for example in environmental measurement technology, the medical sector, biotechnology and the chemical industry, are very broad in terms of the gases to be measured and the concentration ranges (from ppb to vol.%). The reasons for this are the high measuring accuracy, the long-term and system stability as well as the fundamental robustness of such systems due to their comparatively simple construction. However, many of these gas sensors are subject to restrictions with regard to environmental stability and especially temperature resistance. The maximum storage and working temperature of common NDIR gas measurement systems is about +60 °C and thus limits special applications where the sensor is exposed to higher temperatures.

This fact was the motivation to develop a NDIR gas sensor based on own components for a significantly extended temperature range (up to 190 °C) and harsh environmental conditions, such as high water vapour partial pressures, solvent atmospheres and high ambient pressures. This development was only possible based on the know-how available at MICRO-HYBRID Electronic GmbH for the temperature-stable and hermetic sealing of the IR components required for this application. The technological basis is formed by temperature-resistant circuit carriers, such as LTCC,[4] and their assembly and connection technology. In particular, adhesive-free manufacturing technologies are to be mentioned

[3]Non-dispersive infrared.

[4]Low Temperature Cofired Ceramics.

here, which enable a completely hermetic encapsulation without permeation of air/humidity into the sealed IR components. These patented process technologies are based on welding and soldering processes and form the basis for the system stability of the IR components (Biermann et al. 2014). This is an important basis for long-term stable and almost drift-free NDIR gas sensors.

Applications

The fields of application of the environmentally resistant NDIR gas sensor result from its resistance to harsh environmental conditions. An example of this is the use of the sensor for controlling the CO_2 atmosphere and for controlling or measuring the room humidity in CO_2 incubators. For optimal growth of cell cultures, specific and long-term stable environmental conditions are required. These include the temperature, humidity and especially the pH value of the indoor atmosphere. This pH value is adjusted by means of the CO_2 content in the incubator. Since the cells to be grown are usually highly sensitive to pH value deviations, exact and stable CO_2 control is absolutely necessary. Only in this way can the requirements for optimal and reproducible growth conditions of the cell cultures be guaranteed. Usual atmospheric conditions for ideal cell growth are a temperature of 37 °C, an air humidity of 95% rH and a CO_2 concentration of 5 Vol.-%. The most extreme humidity and temperature conditions can occur at 60 °C and 95% rH (Binder GmbH 2015). The resulting water vapour partial pressure of 190 mbar, together with the temperature, places the highest demands on the system stability of the hermetically encapsulated IR components of the gas sensor. Classical compounds based on organic joining aids, such as epoxy adhesives, cannot withstand these requirements with sufficient long-term stability. Due to the highly critical and sensitive fields of application such as:

- the *Bio Tissue Engineering*,
- in vitro fertilisation (IVF),
- of human and veterinary medicine,
- the pharmaceutical industry and
- of biotechnology.

However, extremely high demands are placed on sample safety (Binder GmbH 2015) and thus on the system stability of the gas sensor.

However, the decisive unique selling point of the MHE gas sensor is its high temperature stability up to 190 °C, which is necessary for hot air sterilization of the incubator at 180 °C, according to DIN 58947. This method of sterilization is the safest way to reliably eliminate deficient bacteria and spores and to decontaminate the interior of the incubator. This exposure to temperature represents an extraordinary thermomechanical load for the entire construction of the gas sensor, including the integrated infrared optical components. The system reliably withstands this load only on the basis of the special *packaging* technologies developed.

Further applications are for example in the concentration measurement of solvent-containing atmospheres at higher temperature conditions. The accelerated reaction rates prevailing here in accordance with the Arrhenius function represent a significantly increased chemical load on the hermetic joints and especially on the infrared-optical filters. Due to its design, the gas sensor can optimally withstand these harsh environmental stresses. As the resistance of the optical filter layers (with bandpass filters up to 100 levels) to various chemically active substances is not given in the long term, or a complex individual test of the special filter-media compatibility would be necessary, the functional IR filters were integrated into the interior of the detector. An optical window soldered on the outside protects the internal structure of the detectors including the IR filters from the deficient measuring atmosphere.

Requirements of the Sensor Elements

In addition to an adapted thermo-mechanical design, the design of the sensor element must also take greater account of temperature-dependent effects of the thermoelectric components and transmission elements. In the following, selected effects are described and the effect on the sensor signal is shown.

Thermopile Element

The signal voltage U_S of the thermopile detector is strongly dependent on the intrinsic temperature of the element. This is due to the temperature dependence of the thermoelectric coefficient (Jäckle 1998) and the effective radiation flux $\Delta\Phi_S$, which can be converted into heat on the membrane (Budzier and Gerlach 2010):

$$U_S = R_V \cdot \Delta\Phi_S \sim R_V\left(\Phi_{in} - \sigma \cdot T_{Det}^4\right) \tag{17.15}$$

Equation 17.15 describes this relationship schematically. Where $\Phi_{in}$is the total incident radiation flux, σ is the Stefan–Bolzmann constant, R_V *is* the voltage sensitivity and T_{Det} is the absolute detector temperature. Most of these input variables are generally dependent on temperature and wavelength.

Figure 17.37 shows the signal voltage as a function of the ambient temperature. As can be seen, with an extended operating temperature range up to 190 °C, the signal strength can drop over 40%. This effect must be taken into account when planning high-temperature IR sensors based on thermopile technology in order to achieve a sufficiently high signal-to-noise ratio in the application.

Another disturbing effect of thermopile-based measuring systems is the signal behaviour after rapid temperature changes, also known as thermal shock. This causes the sensor system, consisting of base, thermopile chip, cap and interference filter, to go out of thermal equilibrium. As a result, strong signal fluctuations occur, which can exceed the actual measuring effect of infrared radiation measurement many times over. During the

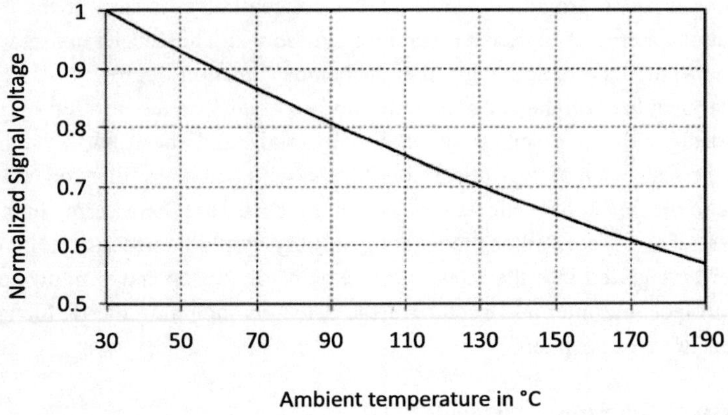

Fig. 17.37 Dependence of the signal amplitude on the ambient temperature

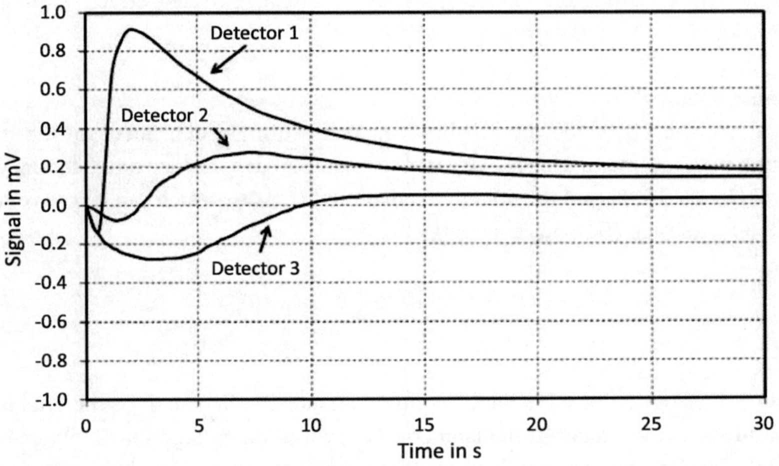

Fig. 17.38 Signal voltage curve of thermopile detectors after a rapid temperature change from $\Delta T \approx 30$ K

subsequent thermal equilibrium process, the signal cannot be evaluated. The strength of the effect is detector-specific and can vary greatly even with identical detectors. Figure 17.38 shows different signal responses of different detectors of the same design after a rapid ambient temperature change of 25 on 50 °C.

Interference Filter

In order to obtain a wavelength-selective sensitivity of thermopile detectors, narrow-band interference filters are introduced to the broadband absorbing sensor element. The central

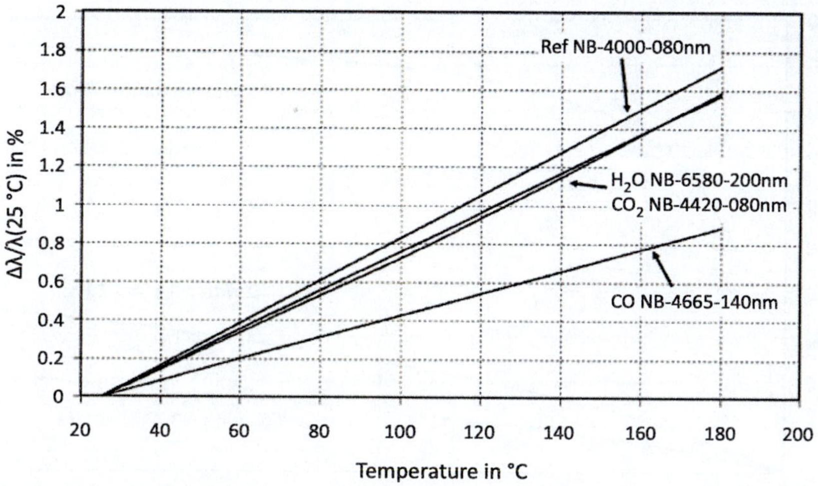

Fig. 17.39 Percentage change of the center wavelength λ, as a function of temperature T for different interference filters at 4.665 μm, 4.42 μm, 6.58 μm and 4.0 μm. *NB* narrow bandpass

wavelength (λ_{max}) and half-width (HW) of the filters are specially matched to the measurement task.

Due to the dielectric coating, the transmission properties change depending on the ambient temperature T. The reason for this effect is the change in the optical thickness ($n \cdot d$) of the dielectric materials. Both the refractive index n and the geometric layer thickness d change with temperature (Morelli 2006). The temperature dependence of the central wavelength $\lambda(T)$ can be calculated approximately according to Eq. 17.16 (Spectrogon 1015), whereby the temperature coefficient α can lie between $0.3 \cdot 10^{-4}$/K and $1 \cdot 10^{-4}$/K according to own measurements.

$$\lambda(T) = \lambda_0(1 + \Delta T \cdot \alpha) \tag{17.16}$$

If exact values are required, the filters must be measured spectrometrically at the respective temperature T_x. Figure 17.39 shows measured temperature curves of different interference filters.

When selecting the specific interference filters, a compromise must always be found between high selectivity (=narrow-band filters) and good signal resolution of the thermal detector (=broadband filters). However, high signal selectivity over a wide operating temperature range can also have a negative effect on the linearization function. Compromises must also be made in filter selection if, for example several gases with different spectral absorption coefficients are to be measured simultaneously in one sensor. This will be explained using an infrared gas sensor for a CO_2 incubator.

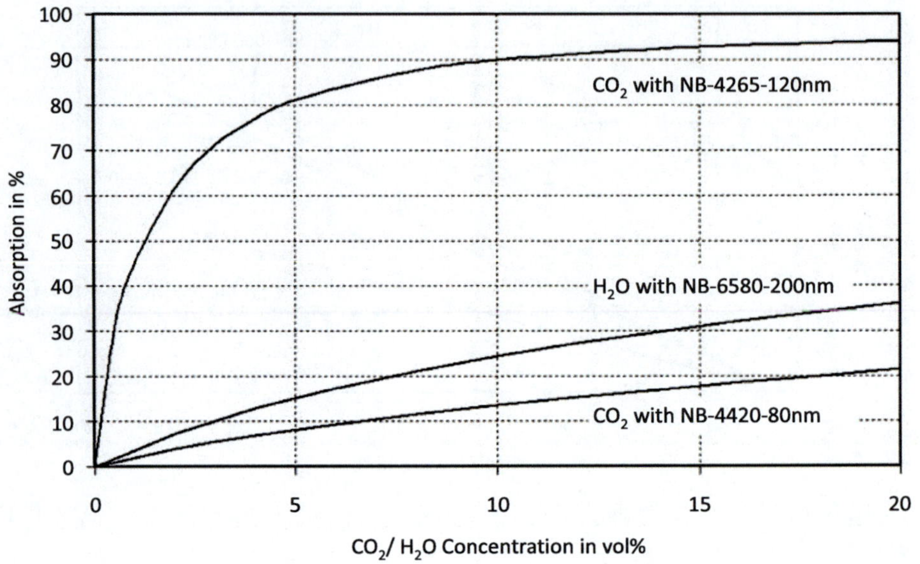

Fig. 17.40 Absorption curves with different interference filters at $T = 60\ °C$ and $L = 50$ mm measuring section

In a CO_2 incubator, the pH value of the nutrient solution should be adjusted via the CO_2 content in the gas atmosphere. Typical CO_2 concentration values are between 1 and 10 Vol.-%. At the same time, the humidity should be monitored to prevent the cells from drying out. The humidity concentrations of >90% rH are close to the saturation limit. The infrared gas sensor should have a CO_2 and a humidity measuring range up to 20 Vol.-%. The measuring distance L is determined by the weaker absorbing gas and in this example is set to $L = 50$ mm. At a maximum moisture concentration of 20 Vol.-%, the signal absorption is approximately 36%. For the CO_2 measurement, no filter can be selected where the central wavelength overlaps the CO_2 main band at 4.265 μm. In connection with the measuring distance of $L = 50$ mm and the high CO_2 concentration to be measured, the signal absorption would quickly go into saturation. An alternative in this case is a filter with a central wavelength on the $^{13}CO_2$ absorption band at 4.42 μm. The weak rotation lines cause only a moderate radiation absorption of approximately 22% at 20 Vol.-% CO_2 and 60 °C ambient temperature in this filter measuring path constellation. Figure 17.40 shows the absorption curves for different interference filters for a measurement path of $L = 50$ mm.

This measuring cell arrangement is fully acceptable for a limited temperature range. However, the CO_2 measurement collapses with increasing ambient temperatures. As shown in Fig. 17.41, the signal absorption decreases by more than 70% with increasing ambient temperatures. The reason for the strong shift in the characteristic curve is the shift of the filter characteristic to long wavelengths. This temperature effect depends on several

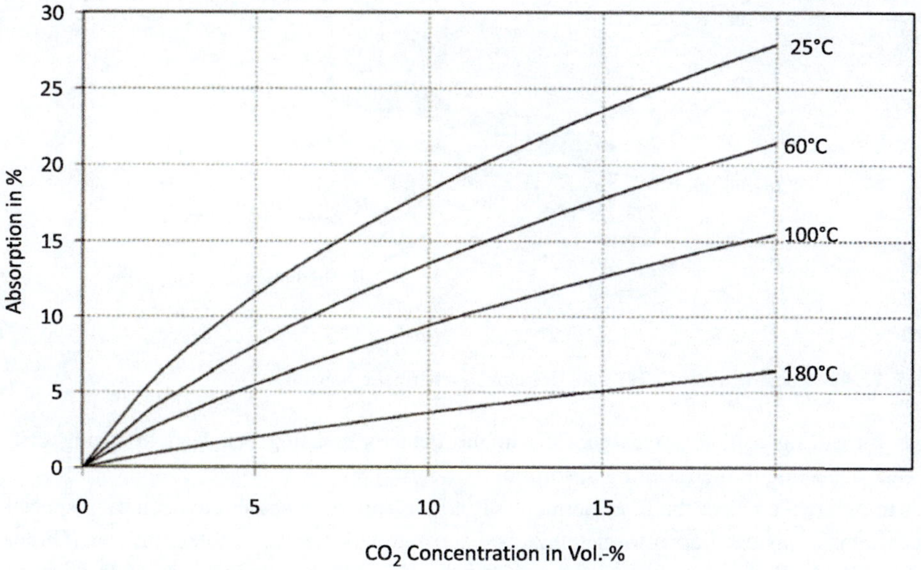

Fig. 17.41 Dependence of the linearization function on the ambient temperature *T*, for CO_2 measurement with an interference filter NB-4420-080 nm and 50 mm measuring distance

parameters, but can be simulated well with the help of infrared spectroscopy databases and known filter parameters (Staab 1994).

Furthermore, it is technically possible to optimize the filter design with regard to the temperature stability of the center wavelength or the angular stability of the center wavelength by selecting suitable dielectric materials. This makes it possible to design a suitable sensor configuration for the desired application temperature range in advance. It should be noted that the above parameters are contrary to each other with regard to their optimization.

For high temperature resistance and long-term stability, it is absolutely necessary to hermetically encapsulate the sensory components. However, this goal can only be achieved by an inorganic material connection, such as soldering or welding, of all joining elements. In this context, adhesive joints are to be regarded as only quasi-hermetic. The IR detectors are filled with an inert gas, which has a lower thermal conductivity than N_2, such as Krypton, to increase sensitivity. This gas filling would escape at higher temperatures through organic joining compounds via permeation processes and thus inevitably lead to a signal drift of the system. But even without a special inert gas filling, the sensor would not withstand the environmental conditions. Oxygen and other gases can diffuse into the housing during operation and destroy the infrared thermopile chip at high temperatures. A further disadvantage of adhesives is that even after long annealing times they still outgas and release substances such as CO_2, other hydrocarbons and water. Especially with a CO_2

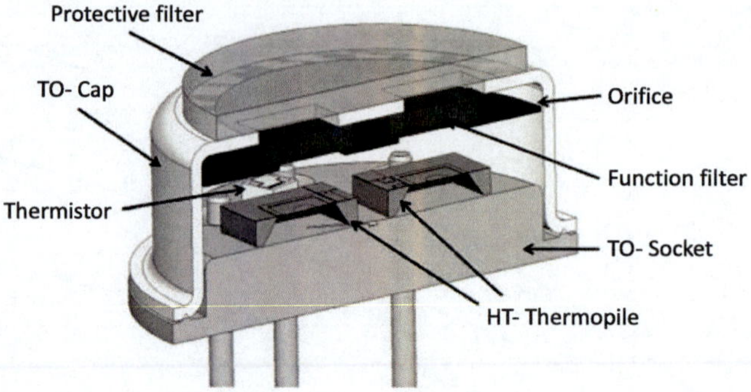

Fig. 17.42 Construction of a HT dual detector in a hermetic housing

gas measuring cell, the released CO_2 in the detector housing can lead to strong drift phenomena due to the resulting absorption.

In order to connect the filter hermetically to the cap via a solder connection, a special technology was developed to metallize and separate any free-form filter surfaces (Gittler et al. 2014). These filters are then preferably soldered onto the outside of the cap as protective windows. For the wavelength range up to 5 μm uncoated sapphire and above 5 μm anti-reflective silicon with an additional DLC[5] protective layer are used as filter material. The functional filters are then protected from environmental influences inside the detector housing. Figure 17.42 shows a classic dual detector setup for high temperature gas detection with one measurement and one reference channel together in a hermetic housing (Biermann et al. 2014).

The connection of the cap to the base is realized by a resistance welding process. In several test series, it could be shown that the glazing of the pins is temperature stable up to min. 250 °C and can withstand several 100 rapid temperature changes between $\Delta T = -40$ °C / +85 °C without losing its tightness. For the proof of hermeticity, a thermal conductivity effect is used for membrane-based microsystems. A heated membrane releases the excess thermal energy back to the housing in the form of radiation, heat conduction in the solid (membrane, silicon chip in the base) and via heat conduction through the gas atmosphere. In the manufacturing process, the thermal conductivity of the interior atmosphere can be controlled specifically via the filling gas. If inert gases with low thermal conductivity, such as krypton or xenon, are used, the membrane temperature is higher at the same irradiance than in inert gases with better thermal conductivity, such as nitrogen or helium. In Table 17.2 different filling gases for thermopile detectors with the respective thermal conductivities are summarized.

For thermopile detectors, this means that the sensitivity depends on the inert gas filling. The sensitivity of infrared detectors can be measured well with blackbody reference

[5]Diamond like carbon.

Table 17.2 Thermal conductivity coefficients λ of various filling gases (Stöcker 2004)

Gas	λ [W/m/K]	Ratio to N_2
He	0.143	6 : 1
Ne	0.046	1.9 : 1
N_2	0.024	1 : 1
Ar	0.016	1 : 1.5
Kr	0.0088	1 : 2.7
Xe	0.0051	1 : 4.7

sources and tempered detector mounts and apertures. To evaluate the tightness of the housing, a protective gas with poor thermal conductivity is first filled into the housing and the detector sensitivity is measured. Afterwards a helium overpressure storage is carried out for example 24 h at 3 bar. If the helium can diffuse into the housing due to a leak, the thermal conductivity of the interior gas atmosphere and thus the detector sensitivity changes significantly. The described method is many times more sensitive than standard helium leak test methods for these small detector volumes.

IR Emitters

Thermal emitters are usually used as infrared radiation sources. Depending on the application, different types with different construction variations are used. In contrast to incandescent bulbs, these emitters can also be operated openly at air with surface temperatures up to 800 °C, with long-term stability.

The filament radiator is a further development of the classic light bulb. Due to a special alloy with a high aluminium content, this material forms a passivation layer of aluminium oxide on the surface, which protects it from further oxidation processes. The porous aluminium oxide surface also has a positive effect on the emission coefficient. However, this requires a special burn in regime to obtain closed surfaces with long-term stability. Figure 17.43 shows the cross section of the Al_2O_3 surface of a burnt-in filament.

In order to reduce the thermal mass of the element and to increase the resistance of the heating element, the heating structure is cut out of a 30 µm thick sheet metal as a double spiral. The effective radiating area is approximately 2.5 mm^2. As shown in Fig. 17.44, the heating element is contacted by a gap welding process directly onto the socket pins. Socket and heating element thus form a high temperature resistant unit for ambient temperatures far above 200 °C.

Unfortunately, the still relatively large thermal mass of the filament has a negative effect on the direct electrical modulation capability. With a cut-off frequency of 0.3 Hz this emitter type is mainly used with direct measuring thermal detectors such as thermopiles.

MEMS emitters are the second type of emitter that can be used in high temperature gas sensors. These emitters have a similar structure to thermopiles, except that the resistive heating element is applied as a thin layer on the membrane. The heating layer can be

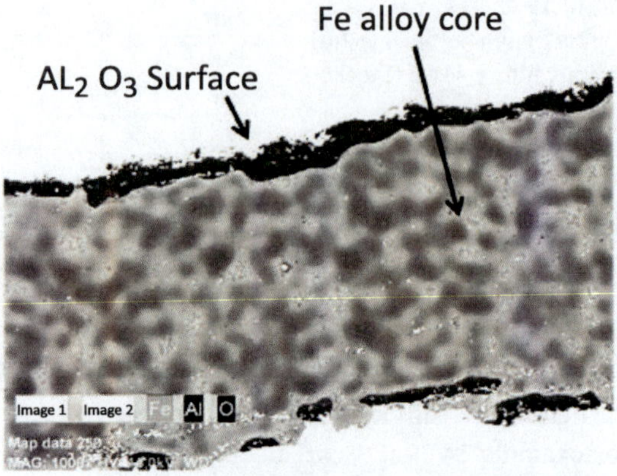

Fig. 17.43 REM analysis of a filament preparation

Fig. 17.44 Gap welding process of a filament on a TO socket

structured as for example platinum meander or full surface as for example NAC[6] material. In contrast to the flat heating layers made of NAC material, which usually already have a good emission coefficient of $\varepsilon \approx 0.9$, it is absolutely necessary to apply an additional emission layer in the case of platinum structures, otherwise interference phenomena will occur in the thin support membrane (Fig. 17.45).

One possibility to increase the emission coefficient of the radiating surface is shown in Fig. 17.46. In this example, the surface area is greatly increased by silicon needles and the emission characteristics are optimized by an additional platinum thin-film coating so that the radiation characteristic corresponds almost ideally to planks law.

[6]Nano-amorphous carbon.

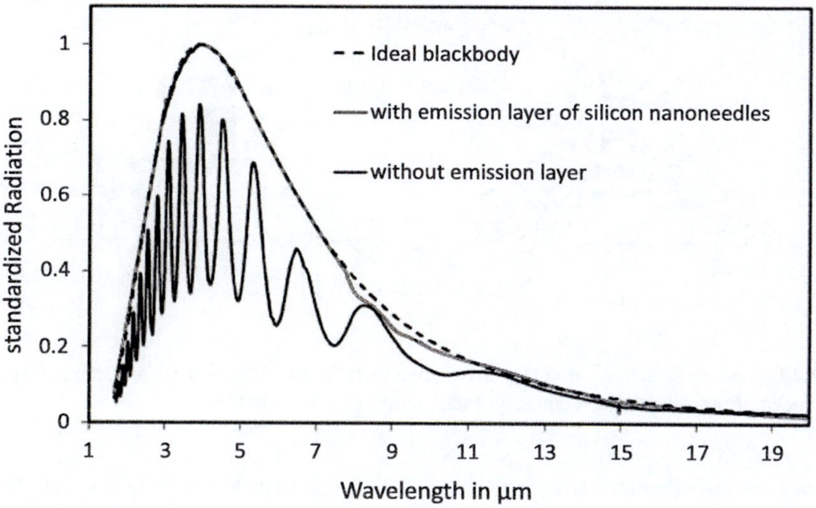

Fig. 17.45 MEMS emitters with and without emission layer of silicon nano-needles

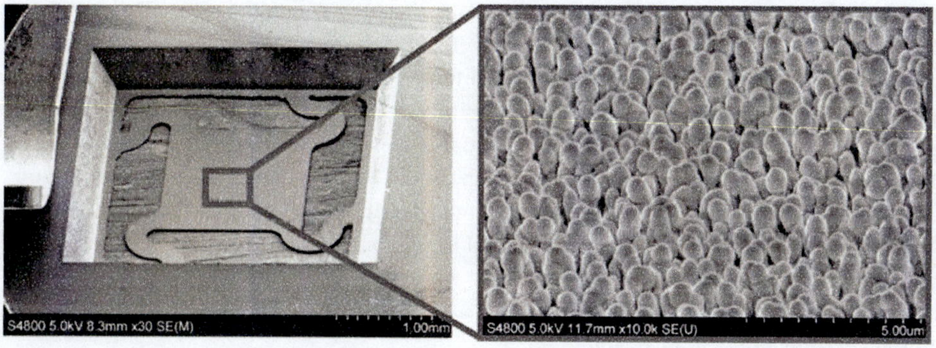

Fig. 17.46 MEMS emitter with additional emission layer of silicon nano-needle with Pt coating and SiO_2 passivation (Hoffmann et al. 2014)

Only a small part of the electrical energy is converted into infrared radiation. A large part of the energy is dissipated by heat conduction into the base and by heat conduction of the gas atmosphere onto the cap. This additional amount of heat puts additional strain on the measuring system. During design, care must be taken to ensure that the sum of self-heating and ambient temperature does not exceed the permissible total temperature of the components and the assembly and connection technology. By insulating the emitter chip from the socket, heat dissipation can be counteracted. Figure 17.47 shows two different AVT concepts of MEMS emitters. Heat conduction into the socket can be minimized by using suitable insulation measures. In closed systems, gas heat dissipation losses can also

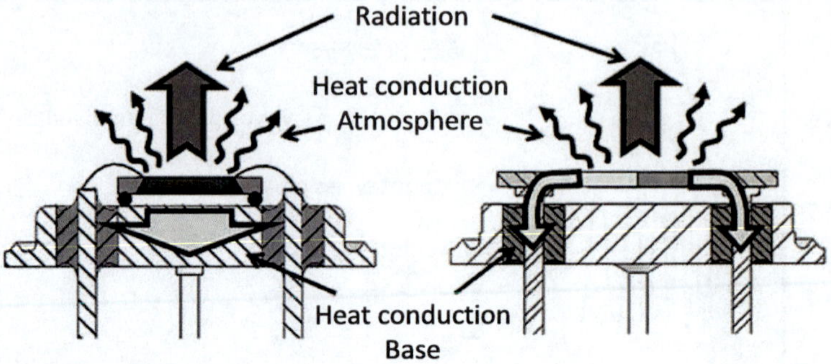

Fig. 17.47 Base construction variants with different thermal connections. *Left:* Mounting the emitter on the socket, *Right:* Insulated mounting of the emitter on the socket pins

be reduced by using a thermally insulating filling gas, such as Krypton. Due to the spectral intensity distribution, a wavelength-dependent factor for improved efficiency is established. In a broad band view, a factor of ~1.5 is detectable for the lower power loss of the emitter. When designing the system, care must be taken in this context to ensure that the turn-off time constant of the enclosed emitter increases due to the thermally insulating filling gas. The gas filling has practically no effect on the switch-on time constant.

All in all, less electrical energy is required to heat the membrane to the same temperature. In this way, highly effective infrared emitters can be constructed which have a high specific radiation and only place a small additional thermal load on the measuring system. Further advantages are the negligible distortion caused by the membrane being stretched on all sides, which leads to a long-term stable optical imaging quality, and the fast electrical modulation. Figure 17.48 summarizes several design variants of MEMS emitters in high-temperature-resistant packages.

Signal Processing/System Design

In many applications it is possible to connect the high temperature resistant sensor part with the temperature limited signal processing unit via a temperature resistant and shielded cable. However, the transmission distance is limited to a few decimetres due to the low output signal level, usually less than 100 µV, and the output impedance of the thermopile of around 20 kΩ. If longer transmission distances are required, pre-amplification is necessary. An integrated preamplifier on a multilayer ceramic in LTCC technology was developed for this purpose. The advantage is the high integration density of the components. Capacitors and resistors can be hermetically printed within the inner layers and are hermetically sealed after the sintering process to ensure long-term stability. At the same time, the LTCC circuit carrier serves as a base. The thermopile chip and operational amplifier are mounted on the top side. The filter or filter cap can be soldered on an external metallization rim at the LTCC substrate. The glass feed-throughs are not required. The connection to the connection cable

Fig. 17.48 Structure variants of different HT-MEMS emitters. (**a**) Insulated Soldered on. (**b**) Bonded to base. (**c**) HTCC SMD housing. (**d**) Hermetic housing

is made via solder pads on the back side. Figure 17.49 shows a high temperature dual detector setup with integrated preamplifier. The housing construction is only slightly larger than a standard TO39 version.

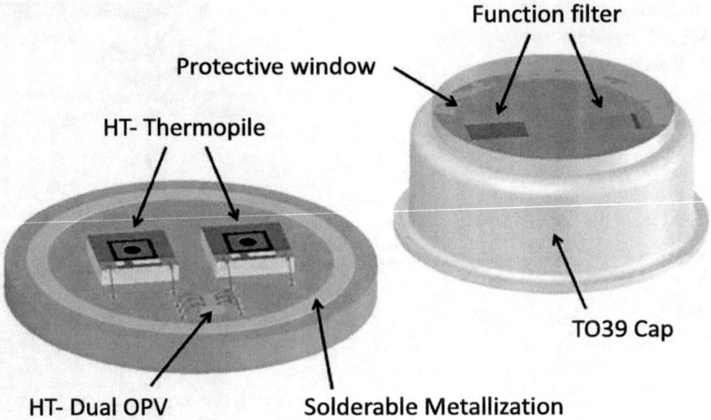

Fig. 17.49 High temperature dual detector with integrated LTCC-based amplifier

Temperature dependencies of the component parameters have a very negative effect on the sensor stability in a two-part sensor system. While the sensor head is usually at a certain stabilized temperature level, signal evaluation in the low temperature range is usually subject to room temperature fluctuations of several 10 K. In addition to gain fluctuations, changes in the emitter control in particular have a strong impact on the measured value. Depending on the emitter AVT, the emitter temperature changes by several 10 K with power fluctuations in the lower percentage range. Due to the nonlinear emission spectrum change, the energy ratios of the measuring and reference channel shift. This signal shift leads to a zero point error and thus to misalignment of the gas sensor. Figure 17.50 shows the zero point error of the gas sensor from the example CO_2 incubator for CO_2 and humidity measurement. If the measuring head and signal evaluation are at the same temperature level, these effects can be compensated relatively easily as temperature-dependent sum effects. At different temperature levels, this is only possible with considerably more effort. It is usually more effective to invest more in components with a very stable temperature coefficient.

Based on the knowledge gained, a demonstrator (Fig. 17.51) with an optimized sensor design for a high-temperature gas measuring cell with up to 20 vol.% CO_2 was built. The sensor should have a temperature resistant probe up to 190 °C. The signal processing is mounted in the low temperature range up to 70 °C, outside the measuring environment. Due to the short cable length, an integrated preamplifier is not required. A broadband interference filter NB-4260-180 nm is used as CO_2 measurement filter. This allows the measurement path to be shortened to $L = 3$ mm. Figure 17.52 shows the temperature dependence of the linearization function. The measuring effect is sufficiently strong over the entire operating temperature range.

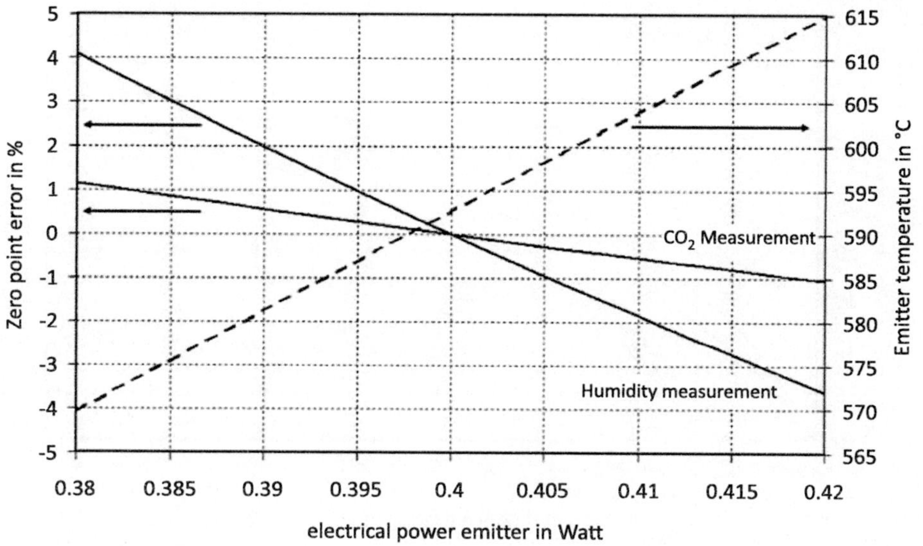

Fig. 17.50 Zero point error due to a power fluctuation of ±5%

Fig. 17.51 High temperature CO_2 gas sensor (demonstrator)

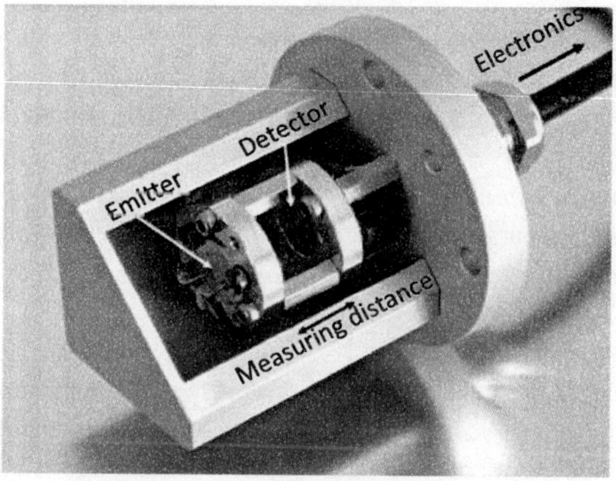

The dual detector as well as the MEMS emitter are each packaged in a hermetic housing with soldered sapphire protection windows. A thermistor for temperature compensation is additionally mounted in the housing of the two thermopiles. Both cases are filled with Krypton to increase efficiency. This increases the sensitivity of the thermopile detectors by 80% compared to a nitrogen filling. In terms of the emitters the Kr backfilling results in a membrane temperature of approximately 450 °C with only 300 mW electrical power. Measurement and reference channel are at this temperature level spectrally close to Planck's intensity maximum.

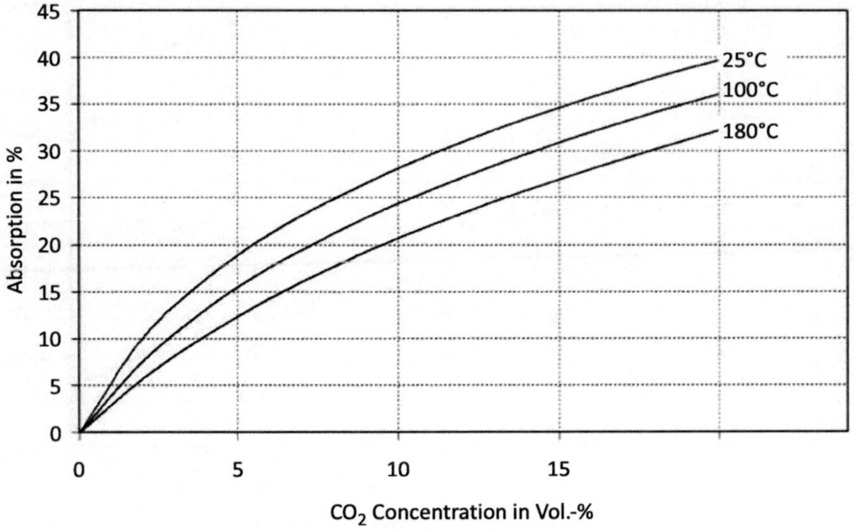

Fig. 17.52 Dependence of the linearization function on the ambient temperature for CO_2 measurement with the filter NB-4260-180 nm and 3 mm measuring section

The gas sensor was calibrated in a calibration process at different gas concentrations and temperatures. Using an adapted set of algorithms, the temperature error of less than 0.2 vol. % can be compensated over the entire operating temperature range. Figure 17.53 shows the signal curves of a 5 vol.% CO_2 measurement at different temperatures. The signal-to-noise ratio decreases from 42 to 38 dB due to the thermopile signal reduction and the flattening of the linearization function. The CO_2 resolution, at high temperatures, can however be considered sufficient for many applications (Table 17.3).

Conclusion/Outlook

With the development of the presented high temperature CO_2 gas sensor a platform for further projects in NDIR gas measurement technology was created. The basis for this and other projects are the special packaging technologies for hermetically sealed IR components and temperature-stable electronics at MICRO-HYBRID Electronic GmbH. On the basis of the modular design of the IR components on the one hand and the gas sensor on the other hand, new customer projects that address special measurement tasks under harsh environmental conditions or have high requirements for long-term stable results can be implemented in a shortest possible time. The optimal system design of the gas sensor is already made possible by an application-specific adaptation of the parameters of the core components, IR emitter and detector. The know-how available here, from more than 15 years of development and series production, helps users to design specific gas sensors. The mechanical, electronic and optical concept of the NDIR gas sensor can be

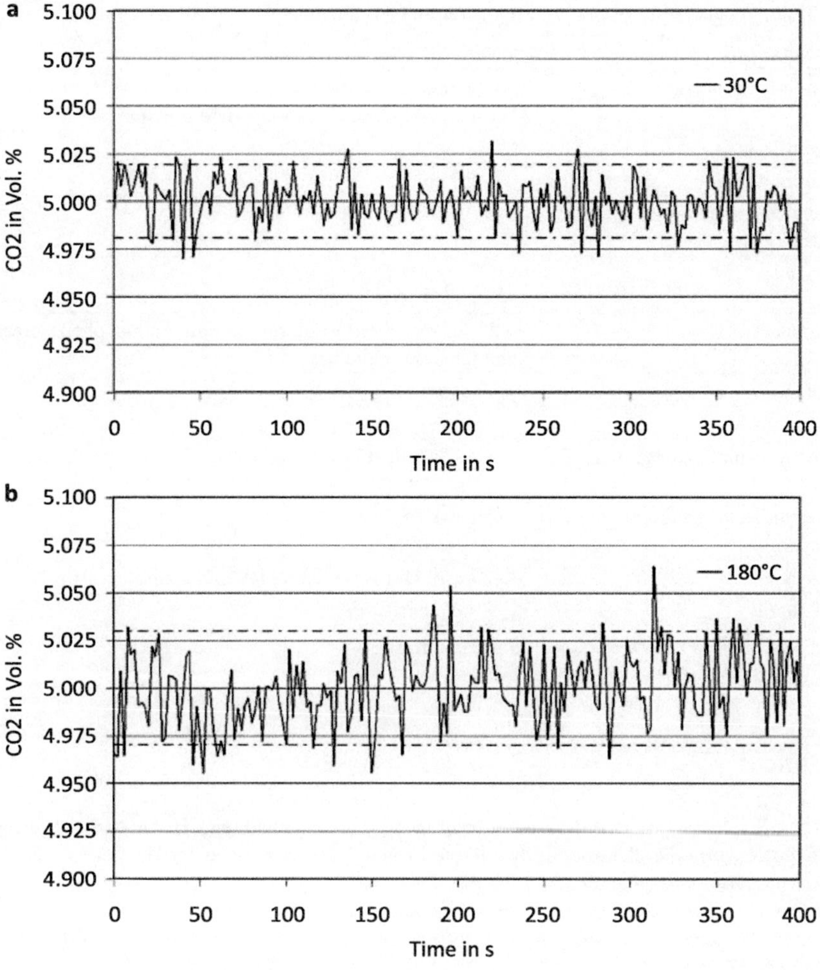

Fig. 17.53 Signal sequence of a 5 vol.% CO_2test gas measurement at (**a**) 30 °C and (**b**) 180 °C ambient temperature

newly assembled and adapted to the requirements of the specific application according to a modular principle. This flexibility, coupled with the technical possibilities of microsystem-based setups, up to calibration and the environmental simulation and long-term tests of the complete NDIR gas sensors, provide an optimal basis for the successful processing of customer-specific projects in the field of optical gas sensor technology.

Table 17.3 Specification of the high-temperature CO_2 gas sensor

Parameters	Value
Measurement procedure	NDIR dual beam method With compensation of atmosphere pressure
Measuring gas	CO_2
Measuring range	0 ... 20 vol.%
Temperature range of the sensor specification	0 . . . 190 °C
Accuracy	±0.2vol % + 2% of read
Additional errors	<2% of measured value for the entire atmospheric pressure and temperature range
Rise time	<20 s
Warm-up time	<5 min
Temperature application range of the sensor	−25. . . + 190 °C
Temperature application range of the electronics	−25. . . + 85 °C
Interfaces	4–20 mA, 0–1 V, 0–10 V, RS232, CAN
Power supply	12± 1 V max. 2 W

Source: MICRO-HYBRID Electronic GmbH Hermsdorf

References

Biermann, S., Magi, A., Sachse, P.: Europäische Patentanmeldung, Hermetisches gasdichtes optoelektrisches oder elektrooptisches Bauteil sowie Verfahren zu seiner Herstellung, EP 2 803 961. Europäisches Patentamt 19.11.2014

Binder GmbH: Produktkatalog CO_2 Inkubatoren Serie CB – NEU 60. 28.01.2015

Budzier, H., Gerlach, G.: Thermische Infrarotsensoren: Grundlagen für Anwender. WILEY-VCH, Weinheim (2010)

DWV: Jahresbericht 2014, Deutscher Wasserstoff- und Brennstoffzellen-Verband

FDA, Guidance for Industry, PAT – A Framework for Innovative Pharmaceutical Development, Manufacturing, and Quality Assurance, 2004. http://www.fda.gov/downloads/Drugs/Guidances/ucm070305.pdf. Zugegriffen: 21.02.2015

Födisch, H., Schengber, P.: Kontinuierliche Aufzeichnung von Geruchsemissionen. Freisetzung, Eintrag und Bewertung von Gerüchen in der Landwirtschaft, FAL/KTBL-Fachgespräch 14.–15. Oktober 1997, KTBL-Arbeitspapier 253, Landwirtschaftsverlag Münster-Hiltrup, 19–26 (1997)

Födisch, H., Schengber, P.: Applikationsbeispiele kontinuierlicher Geruchsemissionsmessungen [4], VDI Berichte; 1373; 99–112. Gerüche in der Umwelt: Innenraum- und Aussenluft, VDI-Tagung (1998)

Födisch, H., Schengber, P., Gennrich, R., Siebert, R.: Verfahren und Einrichtung zur sensorischen Erfassung und Reduzierung von Gerüchen. Deutsche Patentanmeldung DE19604600 A1 (1996)

Gesetzliche Grundlagen zur Ermittlung und Bewertung von Gerüchen: §3 (1) BImSchG, §3 (4) BImSchG

Gittler, E., Biermann, S., Brode, W., Störzner, F.: PCT Anmeldung, Verfahren zur Herstellung von in ein Gehäuse hermetisch dicht einlötbaren Fensterelementen und danach hergestellte Freiformfensterelemente, WO 2014/183746 A1. WIPO 20.11.2014

Hippler, M., Weingarten, R.: Unveröffentlichte Daten 2015. Westfälische Wilhelmsuniversität Münster, Institut für Biologie und Biotechnologie der Pflanzen (2015)

Hoffmann, M., Müller, L., Käpplinger, I., Biermann, S., Brode, W.: Highly efficient IR-emitters based on oriented nanostructures on silicon templates. In: Proceedings of 40th Micro and Nano Engineering (MNE 2014), Lausanne, Schweiz, 22.–26.09.2014

Höra, W.: Optimierung der Produktion und Aufarbeitung von Glucose-6-phosphat Dehydrogenase und Hexokinase aus Bäckerhefe. Diplomarbeit an der Fachhochschule Jena. diplom.de, 10–11 (2002)

Jäckle, J.: Über die Ursache der Thermospannung. http://www.uni-uonstanz.de/FuF/Physik/Jaeckle/papers/thermospannung/index.html (1998)

Melis, A., Zhang, L., Forestier, M., Ghirardi, M.L., Seibert, M.: Sustained photobiological hydrogen gas production upon reversible inactivation of oxygen evolution in the green alga Chlamydomonas reinhardtii. Plant Physiol. **122**, 127–136 (2000)

Morelli, D.M.: Interference Filter Handbook. 2. Aufl. JDS Uniphase Corporation (2006)

Müllner, K.C.: Optimierung und technische Umsetzung der photobiologischen Wasserstofferzeugung mit der Grünalge Chlamydomonas reinhardtii. Dissertation, Ruhr-Universität Bochum, Lehrstuhl Biochemie der Pflanzen in der Arbeitsgruppe Photobiotechnologie (2008)

Rohe, P.: Prozessnahe Hochdurchsatzoptimierung der heterologen Proteinproduktion in alternativen Wirtsorganismen. Forschungszentrum Jülich (2013)

Sahm, H., Antranikian, G., Stahmann, K., Takors, R.: Industrielle Mikrobiologie. Springer (2014)

Salomon, M.: Geschichte der Fermenter. (2010). http://www.fermenter.de/geschichte-der-fermenter.htm. Zugegriffen: 03.02.2015

Scherer, P., Neumann L.: „Methano-compost", a booster and restoring agent for thermophilic anaerobic digestion of energy crops. Biomass Bioenergy 56, 471–478 (2013)

Scherer, P., Müller, H., Schmale, U.: Gasmessung bei Biogasanlagen und Testreaktoren. Biogas J. **2**, 30–31 (2004)

Schön, M., Hübner, R.: Geruch-Messung und Beseitigung. Vogel, Würzburg (1996)

Staab, J.: Industrielle Gasanalyse. Oldenburg (1994)

Stöcker, H.: Taschenbuch der Physik. 5. Aufl. Harri Deutsch, Frankfurt am Main (2004)

Storhas, W.: Bioverfahrensentwicklung, pp. 752–753. John Wiley & Sons (2013)

VDI 2449 Bl. 2 und der Norm DIN 6879 (Messverfahren Olfaktometrie Richtlinie)

Wiegleb, G.: Industrielle Gassensorik. Expert, Esslingen (2001)

Further Reading

Lauper, I.: MAP-Qualitätskontrollen: online und Stichprobe. Lebensmittelindustrie. **1**(2), 14–15 (2002)

Safety Engineering

18

Safety equipment. (Source: fotolia)

18.1 Safety Engineering

Gaseous substances that spread in the ambient air can cause different effects that affect both people and buildings/facilities. Safety engineering is therefore concerned with these effects and the necessary protective measures to prevent damage. In order to be able to monitor the effects of these protective measures, appropriate gas measuring devices are required. This chapter describes the different safety areas and takes a close look at the requirements for the corresponding gas detection technology.

G. Wiegleb, *Gas Measurement Technology in Theory and Practice*,
https://doi.org/10.1007/978-3-658-37232-3_18

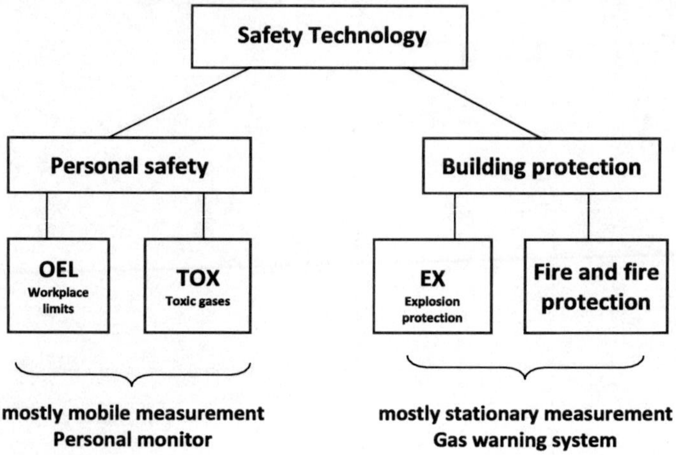

Fig. 18.1 Division of the safety areas

Basically, safety technology can be divided into two areas, which relate to personal protection and building protection. This division is shown graphically in Fig. 18.1.

Personal Safety

The possible effects of gaseous substances on the human organism can only be described by medical investigations. These investigations are evaluated by means of elaborate statistics, within the framework of epidemiological surveys. Whether a gaseous substance is toxic, harmful or even beneficial to health depends on various factors. One important criterion is the degree of exposure, also known as dose *(D)*. The dose results from the existing gas concentration *(c)* multiplied by the time period *(t)* over which the respective person was exposed to this concentration. It therefore applies:

$$\text{Dose } (D) = \text{Concentration } (c) \cdot \text{Period } (t) \tag{18.1}$$

The effects on the human organism are then signalled by certain symptoms, such as sudden headaches, nausea, vomiting, shortness of breath, etc. The last stage of these effects is then death due to an overdose. The statistical occurrence of certain diseases (e.g. Cancer) is also cited as an indicator of the harmful effects of gaseous substances on health. However, it cannot be concluded that a distinction between *generally* poisonous and *generally* non-toxic gases is useful, since only the dose counts (Fig. 18.2). *Paracelsus*[1] already recognized this:

[1] Philippus Theophratus Paracelsus von Hohenheim (1493–1541) German doctor, alchemist and philosopher.

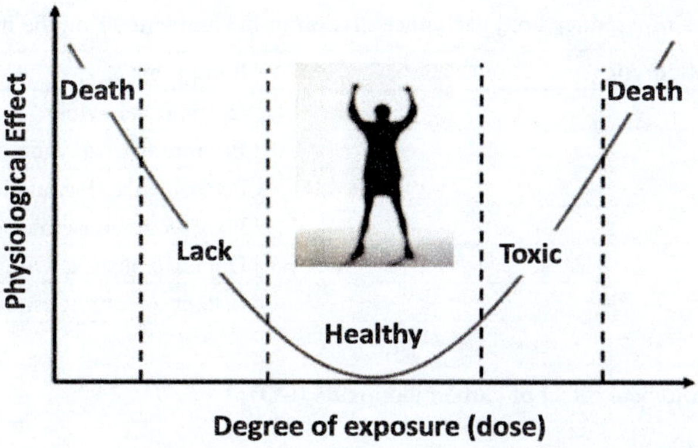

Fig. 18.2 Relationship between the degree of exposure (dose) and the physiological effect

Only the dose makes the poison
A well-known example of this is the so-called *Radon Therapy*, in which patients with rheumatic diseases come into contact with the radioactive gas up to 12 times 1 Std. during a stay at a health resort over a period of 4 weeks. Inhaling this radioactive gas triggers certain physiological processes in the body, which lead to a verifiable reduction in pain. If, on the other hand, one were to inhale radon permanently, this could lead to an increased risk of cancer.

Another example is *Nitric oxide therapy* in which patients inhale low NO concentrations to stimulate the metabolism and stabilize the cardiovascular system. Furthermore, this gas can also be used very effectively to test lung function.

The most important gas in the atmosphere for humans is oxygen. Just under 21% by volume of O_2 the air and the human body is optimally adjusted to this value. Increased Oxygen concentrations (100 vol.%) are not toxic, but can have a negative effect on the metabolism. Short-term and high-percentage oxygen surges, in the context of an oxygen therapy, however, show positive effects. However, if the Oxygen concentration is reduced compared to the natural level, the following gradations can be observed, which have a negative effect on the human organism (Table 18.1).

Foreign substances in the ambient air, such as Carbon monoxide (CO), which can also come from defective heating systems, on the other hand, leads to considerable damage as its concentration increases. Since CO is absorbed by the lungs much faster than Oxygen, Oxygen uptake is hindered and the lack of oxygen then leads to corresponding damage. Table 18.2 shows these effects. A concentration of $c > 1$ vol.% CO in the ambient air already results in a deadly atmosphere.

Toxic gases and vapours occur less frequently in the private sector. Only in heating systems or open fireplaces can low CO concentrations escape, which in rare cases can lead to death. In Germany, these accidents are particularly low, as the technical standard of heating technology is very high. In other countries (e.g. USA, Belgium) the number of

Table 18.1 Effects of a reduced oxygen concentration in the ambient air on the human organism

Oxygen concentration vol.%	Symptoms
≈21	Optimal behaviour
14 to 17	Beginning of oxygen deficiency
11 to 14	Performance degradation
8 to 11	Possible unconsciousness
6 to 8	Unconsciousness within a few minutes
Under 6	Instant unconsciousness

Table 18.2 Physiological effect of carbon monoxide (CO)

CO concentration in ppm	Physiological effects
30	OEL value for 8-hour work
200	Mild headache in 2–3 hours
400	Headache in the forehead area in 1–2 hours, spreads to the whole head in 2.3–3.5 hours
800	Dizziness, nausea +limb twitching in 45 minutes, unconscious in 2 hours
1600	Dizziness, nausea +headache in 20 minutes, death in 2 hours
3200	Dizziness, nausea +headache in 5–10 minutes, death in 30 minutes
6400	Dizziness +headache in 1–2 minutes, death in 10–15 minutes
12,800	Death in 1–3 minutes

fatalities is considerably higher both in percentage and absolute terms, so that appropriate gas measuring devices are recommended or prescribed here.

OEL Monitoring

In the operational environment, on the other hand, pollutants in the ambient air frequently occur, from which personnel must be adequately protected. The legislator has issued corresponding guidelines for these cases (*Ordinance on Hazardous Substances* (*Gefahrstoffverordnung), GefStoffV*), which must be observed by the respective employer. An essential part of these guidelines is the so-called *Occupational Exposure Limit (OEL)*, which was called *Maximum Workplace Concentration* or abbreviated *MAK* (Hartwig 2012) before 2006. For each substance or group of substances there is a maximum gas concentration which may be inhaled by personnel over a period of 8 hours per day (→shift) without medically proven damage occurring.

This limit value definition (TRGS 900 2006) also took into account that no delayed damage occurs and that the personnel may be exposed to this gas atmosphere for several years or decades. For this purpose, it is necessary to use appropriate measuring instruments to detect these substances. These personal monitors are usually worn on the body and record the gas concentrations during a shift. Figure 18.3 shows such a concentration curve.

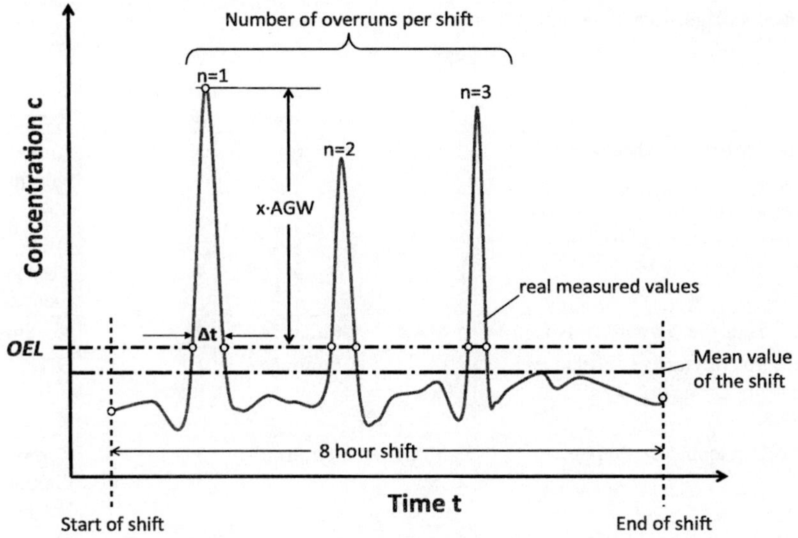

Fig. 18.3 Concentration curve of a pollutant to which personnel are exposed during a shift

The average value via 8 h must then be below the specified limit value. These measured values are stored over a period of 30 years and are also used to assess possible consequential damage.

Since in practice it is quite possible for the limit value to be exceeded even for short periods, the frequency (n) and duration (Δt) of exceedances per shift are determined for this case (Table 18.3). This definition distinguishes between five categories of gases, which have different effects on the human organism.

Section 18.3 (Mobile Gas Detection) describes in detail the instrumentation used for this application.

Explosion Protection

The combustion of gaseous substances is usually controlled by a locally limited flame, in which the fuel supply and the oxygen content are regulated. However, gaseous substances can also accumulate in the ambient air as a result of unnoticed leaks in gas pipes and gas tanks or through the evaporation of flammable liquids. Especially in closed rooms, this accumulation leads to a problem when an ignition source ignites this gas-air mixture. Combustion then takes place within a very short time (ms) and the combustion energy is thus released in the room almost abruptly. This process is called deflagration, explosion or detonation, depending on the speed (Table 18.4). The time period t_{EX} depends on the flame propagation speed v_F. For a mixture of petrol vapour and air, is $v_F \approx 25$ m/s. A gas explosion in a room of the size 5 m × 5 m × 2, 5 m therefore takes $t_{\mathrm{EX}} \approx 50 - 100$ ms.

Table 18.3 Categories and permitted limit value exceedances

	Category	Short term altitude	Duration Δt	Frequency n	Examples
I	Locally irritant substances	2 MAK·	5 min		Ammonia, chlorine
II	Resorptively active substances, onset of action within 2 h				
	II.1: Half-life <2 h	2 MAK·	30 min MW	4	Carbon monoxide
	II.1: Half-life 2 h until end of shift	5 MAK·	30 min MW	2	Acetone
III	Resorptively active substances, onset of action >2 h half-life >8 h	10 MAK·	30 min MW	1	DDT, hg
IV	Weak potential >500 ppm	2 MAK·	60 min	3	Carbon dioxide and butane
V	Odour-intensive substances	2 MAK·	10 min	4	Hydrogen sulphide

Table 18.4 Different forms of combustion processes

Event	Speed	Pressure increase
Flaming up	$v_F < 1$ m/s (40 cm/s[a])	0 bar
Deflagration	$v_F > 10$ m/s	$p < 1$ bar
Explosion	$v_F < 330$ m/s	$p < 10$ bar
Detonation	$v_F > 330$ m/s	$p > 10$ bar

Source: Scheuermann (2008)
[a]For natural gas (Cerbe 2008)

The resulting pressure increase or volume expansion can lead to the complete destruction of the room or building. However, such an explosion only occurs if three conditions are fulfilled simultaneously. These three conditions are often also shown graphically in the form of an explosion triangle (Fig. 18.4).

1. Sufficient fuel (gas, steam, mist, dust)
2. Sufficient oxygen available
3. Sufficient ignition energy (ignition source).

The ignition of an explosive gas mixture can be caused by various sources:

1. Hot surfaces
2. Flames or hot gases
3. Mechanically generated sparks

Fig. 18.4 Explosion triangle as prerequisite for an explosion

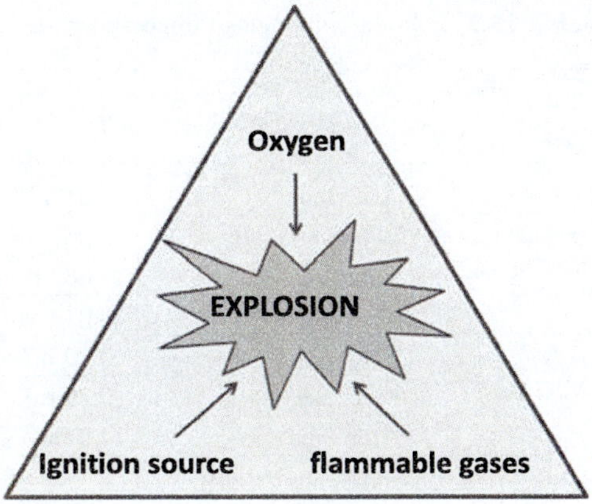

4. Electrical installations
5. Electrical compensation currents
6. Static electricity
7. Atmospheric discharges (lightning strike)
8. Electromagnetic fields from 9 kHz to 300 Ghz
9. Ionising radiation
10. Electromagnetic waves 0.1 to 1000 μm
11. Ultrasound >20 kHz
12. Adiabatic compression
13. Chemical or biological reactions.

However, ignition can only occur if the minimum ignition energy E_Z is exceeded. The individual substances differ considerably in terms of ignition energy. These substances are therefore divided into three different groups (A, B, C). Table 18.5 shows a selection of gases and their ignition properties. The data refer to the so-called *most ignitable mixture* c_{opt} where the lowest ignition energy $E_{Z\min}$ can be expected. The ignition temperature T_Z also leads to a classification in a corresponding temperature class TK (Table 18.6).

The minimum ignition energy E_Z can be generated, for example, by the discharge spark of a capacitor. For this purpose, the capacitor with known capacity is charged via a defined voltage source. The stored energy of a capacitor then results from

$$E = \frac{1}{2} \cdot C \cdot U^2 \tag{18.2}$$

Example

If a 1 μF capacitor is charged with a voltage of U = 6 V, the result is a stored energy of:

Table 18.5 Ignition properties of important gases

Ex-group	Substance	$E_{Z\,min}$	MIC	$c_{opt.\ Vol.\ -\ \%}$	T_Z	TK
II C MIC <0.45	Hydrogen sulphide	0.009 mJ	0,18	7,8	95 °C	T6
	Hydrogen	0.016 mJ	0,24	22	560 °C	T1
	Acetylene	0.019 mJ	0,26	7,7	305 °C	T2
II B MIC <0,8	Ethylene oxide	0.069 mJ	0,48	10,8	435 °C	T2
	Ethene	0.082 mJ	0,54	8,0	425 °C	T2
	Methanol	0.014 mJ	0,71	14,7	386 °C	T2
II A MIC >0,8	Benzene	0.20 mJ	0,85	4,7	560 °C	T1
	Methane	0.28 mJ	1,00	8,5	537 °C	T1
	Ammonia	14.0 mJ	7,10	20	630 °C	T1

Source: Redeker and Schön (1990)

Table 18.6 Classification of temperature classes

Temperature class	Ignition temperature range
T1	>450°C
T2	>300... < 450°C
T3	>200... < 300°C
T4	>135... < 200°C
T5	>100... < 135°C
T6	>85... < 100°C

$$E = \frac{1}{2} \cdot 1 \cdot 10^{-6}\,\mathrm{As/V} \cdot 6^2 \cdot \mathrm{V}^2 = 18 \cdot 10^{-6}\,\mathrm{VAs} = 0{,}018\,\mathrm{mJ}$$

If the capacitor terminals are connected together so that a spark is formed, this energy is suddenly released and is sufficient to cause a 22 vol% hydrogen-air mixture to explode.

If an inductance *L is* selected instead of a capacitor, the stored energy can be calculated as follows:

$$E = \frac{1}{2} \cdot L \cdot I^2 \tag{18.3}$$

The minimum ignition current I_Z is then obtained by transformation to:

$$I_Z = \sqrt{2 \cdot E_Z / L} \tag{18.4}$$

The minimum ignition current ratio *MIC*[2] is related to methane and can be calculated as follows:

[2]Minimum **Ignition** Current Ration.

$$\mathrm{MIC} = \frac{I_{Z,\mathrm{Gas}}}{I_{Z,\mathrm{Methane}}} = \sqrt{\frac{E_{Z,\mathrm{Gas}}}{E_{Z,\mathrm{Methane}}}} \tag{18.5}$$

By limiting current (I), voltage (U), capacitance (C) and inductance (L), it is thus possible to achieve effective explosion protection for electronic circuits, also known as *intrinsic safety (Ex-i)*.

A further prerequisite for a possible explosion is the gas concentration c of the flammable gas. If the concentration is too low, it cannot burn or explode. If the concentration is too high, the necessary oxygen is missing and combustion is also prevented. So there is a concentration range which is explosive and which is limited by the lower explosion limit (LEL) and the upper explosion limit (UEL). This range is shown graphically in Fig. 18.5. The LEL/UEL limits are different for the individual gases and are determined by the heat of combustion of the substances. This results in a pressure and temperature dependence of the limit ranges, which must be taken into account when using gas measuring instruments.

The relationship between the LEL and the heat of combustion Q_R can be derived from the energy rate (Jessel 2001). The result is proportional behaviour with LEL $\approx 1/Q_R$. Already in 1911 this relation was empirically found out by *Burgess* and *Wheeler* (Burgess 1911). Figure 18.6 shows the reciprocal value of the LEL as a function of the heat of combustion Q_R. A linear relationship can be seen ($y = 0.232x$). It therefore applies:

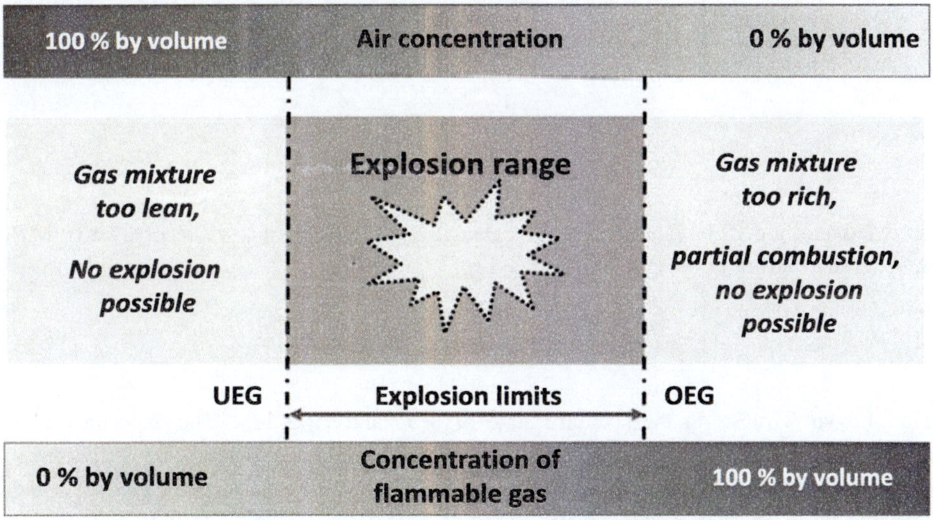

Fig. 18.5 Graphical representation of the explosion range defined by the lower explosion limit (LEL) and the upper explosion limit (UEL). No explosion is possible outside this range

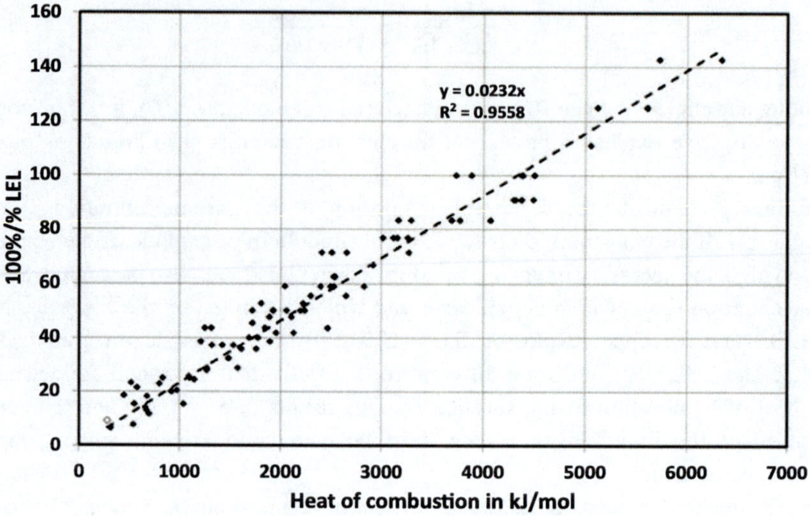

Fig. 18.6 Relationship between the combustion product Q_R and the reciprocal value of the LEL of 89 combustible substances. The course is described by the *Burgess-Wheeler law* (data from Jessel 2001)

$$\frac{100\%}{c_{\text{LEL}}} = 0.232 \cdot Q_R \tag{18.6}$$

With this expression, the LEL can be calculated by transformation if the unit kJ/mol is used for Q_R:

$$c_{\text{LEL}} \approx 100\% \cdot \frac{43.11}{Q_R} \tag{18.7}$$

For Methane, the LEL value would be calculated as follows, if a value of 39.831 MJ/ $m^3 \cdot 22.36\ m^3$/kmol = 890.6 kJ/mol is used for Q_R

$$\text{LEL}(CH_4) = 100\% \cdot \frac{43.11}{890.6} = 4.8\text{Vol.-\%}CH_4 \tag{18.8}$$

This 4.8 vol% refers to the standard state of 0 °C and 1013 hPa. The experimentally determined LEL value of methane (for 20 °C) is currently 4.4 vol.%. The deviation is therefore less than 10% of the LEL value. Table 18.7 shows the LELs and LELs for some selected gases. Further values can be found at Brandes and Möller (Safety Parameters 2003).

Table 18.7 LEL and UEL for some important gases

Gas type	LEL in % by volume	UEL in % by volume
Ethene	2.4	32.6
Ethanol	3.1	27.4
Methane	4.4	17.0
Ethane	2.4	14.3
Propane	1.7	10.8
Butane	1.4	9.40
Hydrogen	4.0	77.0

Data source: Brandes and Möller (2003)

The LEL values indicate a gas concentration in vol.%, which is referred to as 100% LEL. 4.4 vol.% CH_4 is therefore 10% LEL and 0.44 vol.% CH_4 then corresponds to 10% LEL.

In Eq. 18.7, the LEL for methane was calculated at an ambient temperature of 0 °C and a pressure of 1013 hPa. However, with increasing temperature T, the LEL decreases almost linearly according to the following equation:

$$c_{\mathrm{LEL}}(T) = c_{\mathrm{LEL}}(20^\circ\mathrm{C}) \cdot \left[1 - 0.00073/^\circ\mathrm{C}(T - 20^\circ\mathrm{C})\right] \tag{18.9}$$

From this equation the flame temperature ($T = 1/0.0073 = 1370^\circ$C) can be inferred if is$c_{\mathrm{UEG}}(T) = 0$. Figure 18.7 shows the LEL curve for methane from 20 to 400 °C. The explosion range increases considerably with rising temperature. A similar change in the explosion limits can be observed with increasing pressure. Figure 18.8 shows changes in the explosion limits for the pressure range from 1 bar to 100 bar This dependence corresponds to about 0.2% per bar pressure change. For atmospheric pressure changes (1013 hPa± 200 hPa) this dependence can therefore be completely neglected. This dependency only becomes problematic if, for example, Gases are compressed (filled) in compressed gas cylinders.

In contrast to the LEL, the UEL depends strongly on the Oxygen concentration c_{O_2}. In the LEL range, there is anyway an excess of Oxygen and a shortage of fuel, since this ratio is not stoichiometric. If the Oxygen ratio is increased, this has no effect on the LEL limit. The situation is different for the UEL. In this case, there is a fuel surplus and Oxygen deficiency, so that additional Oxygen shifts the LEL limit to higher values. Methane in air has an UEL of 16.8 vol%. In pure Oxygen, this limit shifts to 65 vol.%, and the Oxygen content is then only 35 vol.% (Molnarne et al. 2003).

Inerting

If, on the other hand, the Oxygen concentration is reduced, an explosion will no longer
$c_{O_2\,\mathrm{min}}$ occur below the minimum Oxygen concentration. A defined admixture of Oxygen-free gases can therefore prevent an explosive gas mixture (→ inerting). The minimum

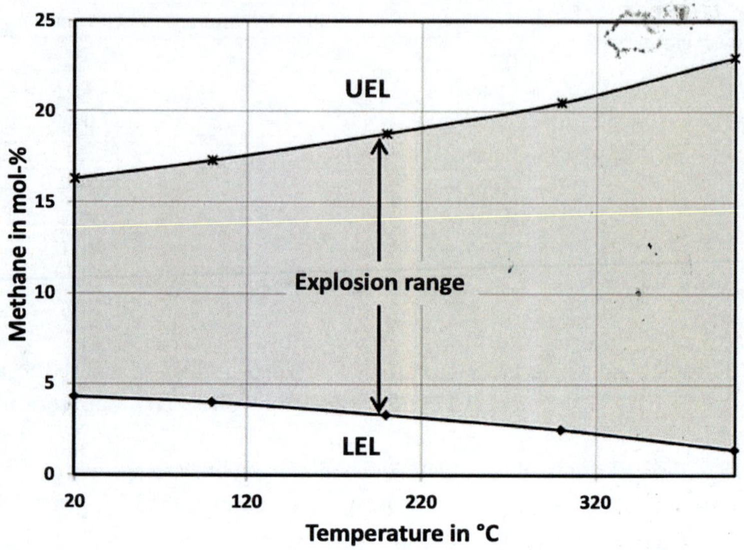

Fig. 18.7 Change of LEL and UEL as a function of temperature *T*. The increasing hazardous area is *highlighted in gray*. (Data source: Molnarne et al. 2003)

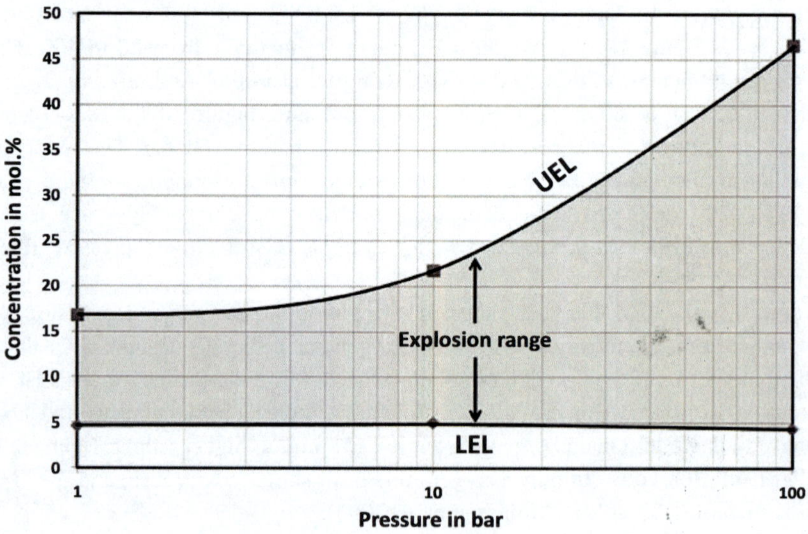

Fig. 18.8 Change of LEL and UEL as a function of pressure *p*. The increasing hazardous area is *highlighted in gray*. (Data source: Molnarne et al. 2003)

value of c_{O_2} mi[illegible]ygen concentration can be calculated with the stoichiometric coefficient ν_{O_2} as follows

$$c_{O_2\ \mathrm{min}} \cong \nu_{O_2} \cdot c_{LEL} \tag{18.10}$$

For methane is $\nu_{O_2} = 2$, so that the minimum Oxygen concentration is at ≈8, 8vol.% O_2. Experimentally determined concentrations have a value of ≈11, 6vol.% O_2 (Jessel 2001). The reduction of the Oxygen content is achieved by adding inert gases such as Nitrogen, Argon or Carbon dioxide. The Oxygen content can then be determined as follows:

$$c_{O_2} = 100 - c_{CH_4} - c_{N_2} \tag{18.11}$$

These relationships can be represented graphically in a three-substance diagram (Fig. 18.9). The three axes refer to the three components Methane, Nitrogen and Oxygen. The dotted lines (1, 2, 3), which run parallel to the Nitrogen axis, indicate areas with a constant O_2content. Accordingly, at point **A** there is an Oxygen content of 40% by volume and the Nitrogen content is also 40% by volume. The remaining part is then the Methane content of

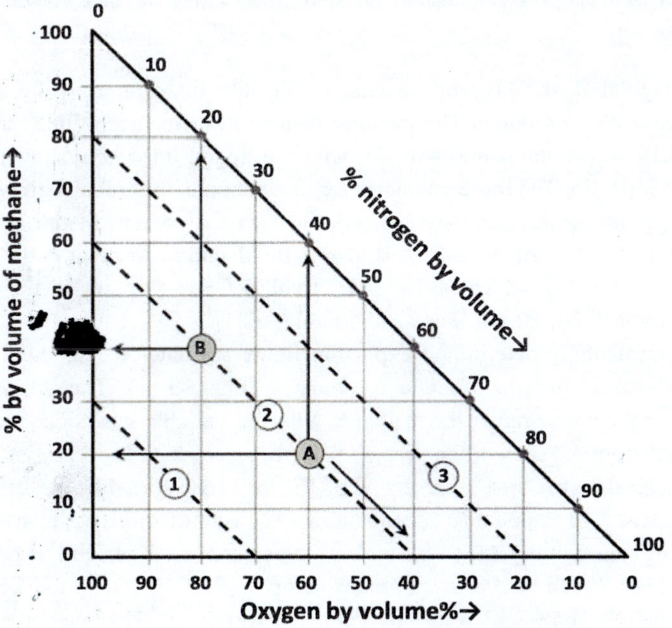

Fig. 18.9 Trivalent diagram for Methane with Nitrogen as inerting gas. Line 1 describes a constant O_2 content of 70% by volume

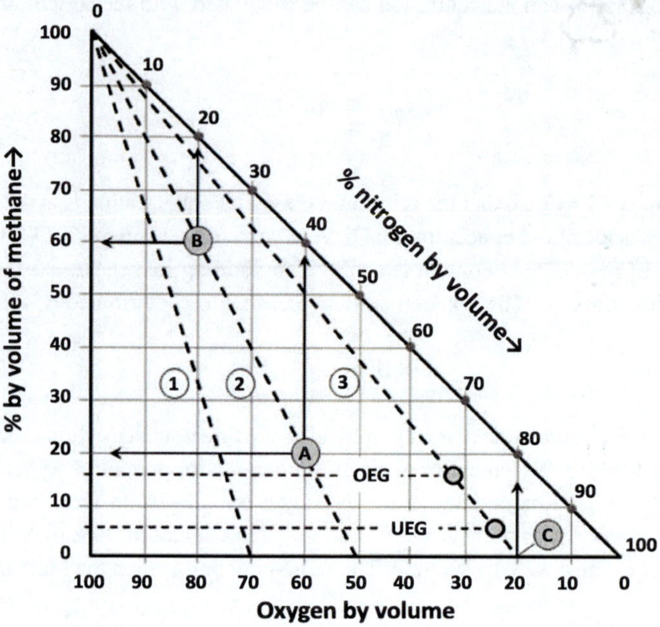

Fig. 18.10 Trivalent diagram for Methane with Nitrogen as inerting gas. Line 1 describes a constant O_2-N_2ratio of70 : 30

20 vol.%. At point **B,** the Oxygen content is still 40% by volume, while the Nitrogen content is only 20% by volume. The methane content has risen accordingly to 40 vol.%.

Figure 18.10 shows the same diagram, with the dotted lines now showing the same Oxygen/Nitrogen ratio. The line 3 characterizes, for example, the ambient air with a ratio of 21 : 79. The point of intersection C shows the Oxygen content in the air (≈21% by volume). The LEL and UEL for air are shown in the diagram. At point **A** the gas mixture consists of 40 vol.% O_2, 40 vol.% N_2 and 20 vol.% CH_4. At point **B**, the gas mixture consists of 20 vol.% O_2, 20 vol.% N_2 and 60 vol.% CH_4.

If the experimentally determined explosion limits are entered into this tri-substance diagram, a so-called inerting triangle is obtained (Fig. 18.11). The grey-shaded area characterizes the ignition range for methane, with the variable gas concentrations. Each line that comes from the apex of the triangle describes a constant O_2:N_2-ratio. The point of intersection with the grey area represents the UEL or LEL value. For the dashed line the ratio 50 : 50. The LEL value also remains almost constant for this ratio, while the UEL value increases considerably due to the higher oxygen content. The solid line describes the course in air (21 : 79) and the points of intersection for the LEL value and UEL value are the known values of 4.4% and 16.8% by volume respectively. The tangent at the edge of the inerting triangle intersects the O_2 axis at a point described $c_{O_2\ min}$ as the minimum oxygen concentration. Concentration values which are $< c_{O_2\ min}$, do not allow an ignitable

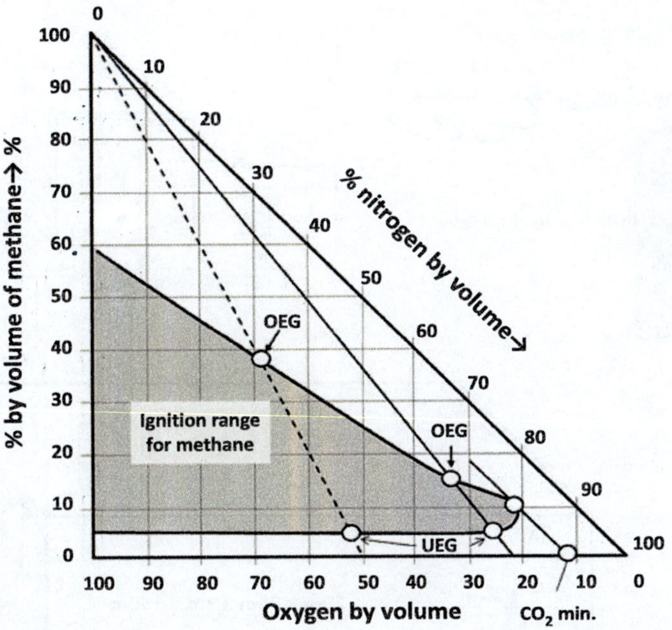

Fig. 18.11 Inerting triangle for determining the explosion limits. (According to Jessel 2001)

mixture. In order to achieve this goal by inerting, sufficient nitrogen must be added to keep the concentration below this value. This so-called *partial inerting* is achieved when the following condition is met for the inerting factor[3] f_i:

$$f_i = \frac{20.9\text{Vol.-}\%O_2}{c_{O_2\,\min}} - 1 \tag{18.12}$$

For methane, the following factor therefore results from the available values:

$$f_i = \frac{20.9}{11.6} - 1 = 0.8 \tag{18.13}$$

If the air is inerted with 0.8 times the amount of nitrogen, no explosion can occur, regardless of the methane content. In the event of a leak, however, pure air can be returned to the inert air, so that an ignitable atmosphere is created again.

[3]Dilution factor $=N_2$/air.

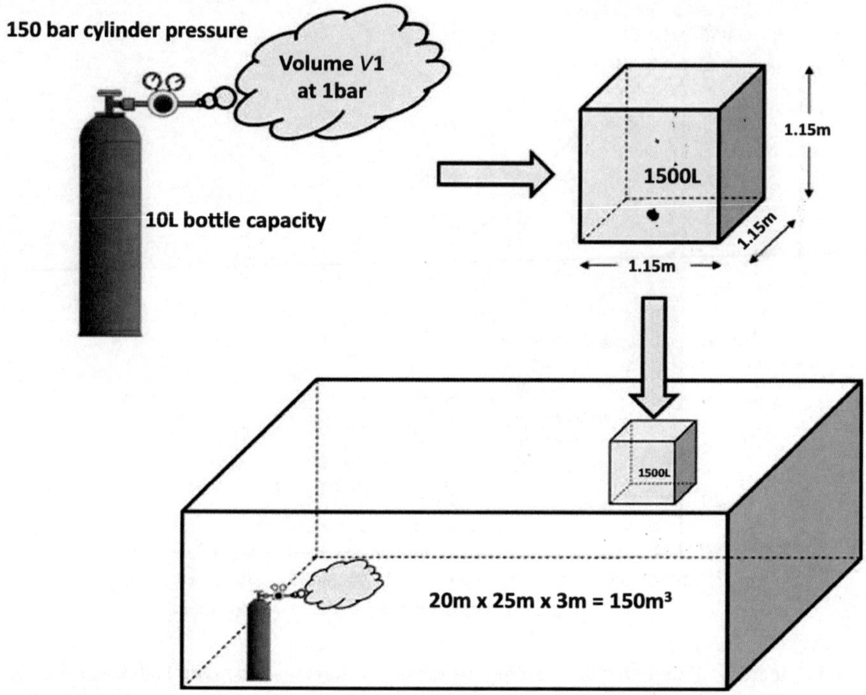

Fig. 18.12 Example of a possible damage situation caused by a leakage from a full methane gas cylinder

Total inerting also covers this situation by selecting a dilution ratio $_{fi}$ so high that even any admixture of pure air will not allow ignition. The inerting factor for this case can be determined with the following equation:

$$f_i = \frac{100}{\text{LEL}} \cdot \left[1 - \frac{c_{O_2\,\text{min}}}{20.9}\right] - 1 \tag{18.14}$$

For Methane one then gets a value of $f_i = 9.1$.

Formation of Explosive Atmospheres

Dangerous explosions that cause major damage or even death are comparatively rare in Germany. This is certainly also due to the fact that there are technically high-quality installations and regulations in this country, which are also observed and checked by expert personnel.

Explosive concentrations are still difficult to generate. Figure 18.12 shows a 10 L gas cylinder filled with Methane (CH_4) at an internal pressure of $p_F = 150$ bar. If the gas now escapes completely into the environment, the cylinder content V_F takes up one volume V_U according to the law of *Boyle-Mariotte* at $p_U = 1$ bar:

$$V_U = \frac{V_F \cdot p_F}{p_U} = \frac{10\,\mathrm{L} \cdot 150\,\mathrm{bar}}{1\,\mathrm{bar}} = 1500\,\mathrm{L} \qquad (18.15)$$

In a room with a total volume of 150 m^3 = 1500 L, this volume of methane gas of $1500\,\mathrm{L} = 1.5\,\mathrm{m}^3$only 1% by volume would make up only 1% by volume. However, the LEL value for methane is 4.4% by volume, so that this concentration value, after complete mixing, corresponds to only 23% LEL and is therefore not yet explosive.

The situation is different if, instead of the compressed gas cylinder, a camping gas cylinder filled with liquid gas (Propane C_3H_8) is opened. The filling weight of a full cylinder is 11 kg. The molar mass of Propane is 44 g / mol. If the total Propane quantity is relaxed at 1 bar, the 11 kg Propane takes up the volume V_{gas}:

$$V_{\mathrm{gas}} = \frac{11.000\,\mathrm{g} \cdot 22.4\,\mathrm{L/mol}}{44\,\mathrm{g/mol}} = 5600\,\mathrm{L} = 5.6\,\mathrm{m}^3 \qquad (18.16)$$

The volume concentration of Propane in the room ($V = 150\,\mathrm{m}^3$) is 3,7% by volume. The LEL of propane is only 1.7% by volume, so that the 100% LEL value was more than doubled. This means that a potentially explosive Propane gas atmosphere would have already been present in the room at half of the cylinder filling.

If a flammable liquid is spilt in an enclosed space and leaks, a potentially explosive atmosphere may also form (Fig. 18.13). We assume that 1 L of Methylated spirits (=Ethyl alcohol/ethanol) evaporates in a room and mixes with the air. Ethyl alcohol has a molecular weight of 46 g/mol. 1 L of Methylated spirits weigh 790 g, so that the total amount of spirit after evaporation takes up a volume of 384 L. In relation to the total volume of the room, this is only 0.64 vol.%. However, the LEL of ethyl alcohol is 3.1 vol.%. In order to create an explosive atmosphere in this room, at least 5 L alcohol would have to evaporate.

Zone Classification

Ex-hazardous areas are divided into different danger zones, which are classified according to the probability of the occurrence of an explosive atmosphere. Basically, a distinction is made in this classification between gaseous substances and dust. In practice, this classification is carried out by suitably trained personnel who are guided by the relevant standards and the explosion protection regulations of the Employers' Liability Insurance Association for the Chemical Industry. Figure 18.14 shows such a zoning for gases and vapours, using a filling station as an example. Due to the petrol vapour pressure, a gaseous atmosphere forms above the petrol tanks, which always contains an explosive gas mixture. The highest Ex level here is *Zone 0*. When refuelling the vehicles or filling the underground tanks, explosive gas mixtures can occasionally form at the connection points, which then justify classification in *Zone 1*. No explosive gas mixtures occur in the vicinity of the filling station

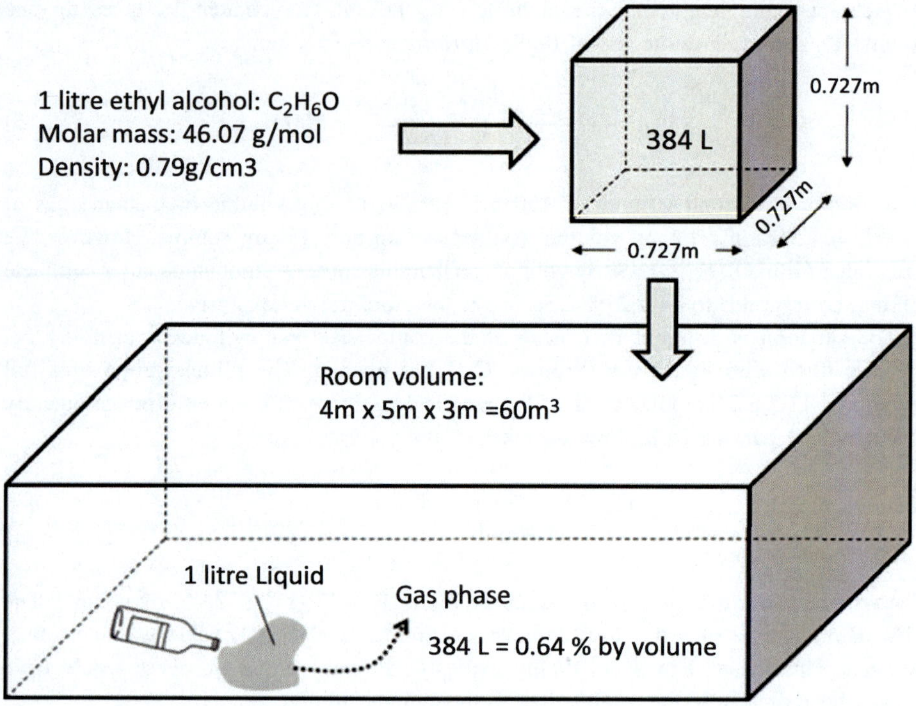

Fig. 18.13 Possible formation of an explosive gas atmosphere by leaking and evaporating a flammable liquid in a connected room

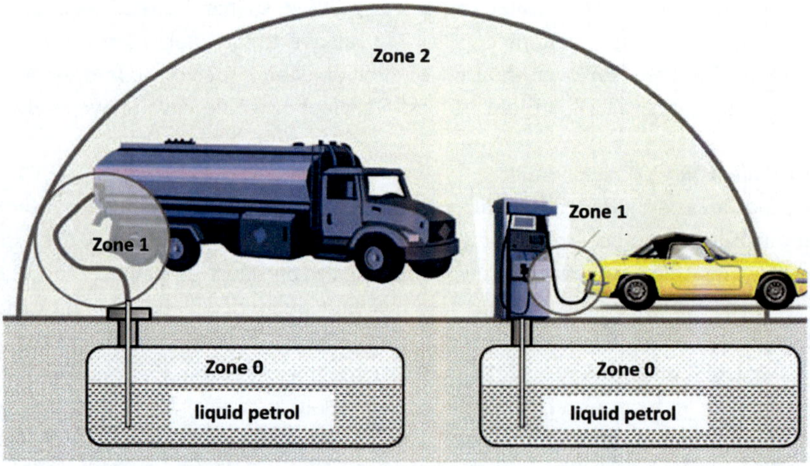

Fig. 18.14 Classification of Ex-hazard zones at a filling station

Table 18.8 Classification of Ex-zones

Type of fabric	Classification	Definition
Gases, vapours, mists	Zone 0	Areas in which a potentially explosive gas atmosphere is present continuously, for long periods or frequently
	Zone 1	Areas in which a potentially explosive gas atmosphere is occasionally present
	Zone 2	Areas in which a potentially explosive gas atmosphere is not normally to be expected, or only rarely and for a short time
Dust	Zone 20	Areas in which a potentially explosive dust atmosphere is present continuously, for long periods or frequently
	Zone 21	Areas in which a potentially explosive dust atmosphere is occasionally present
	Zone 22	Areas in which a potentially explosive atmosphere caused by whirled-up dust is not expected under normal circumstances or only rarely and for a short time

during normal operation and if so, only for a short time. These boundary conditions lead to a classification in *Zone 2*.

The overview of the Ex-zones can be seen in Table 18.8. A detailed description of the Ex-zones is given in EN 60079-10-2. Further information is also available in Sect. 18.2 (Gas warning systems in safety engineering).

Ignition Protection Types
The operation of electrical equipment, gas sensors or gas warning devices in hazardous areas (zones) is only permitted if these devices cannot trigger an explosion at any time. For this purpose it is necessary to separate the electrical equipment and the surrounding gas mixture. Another possibility is to design the electrical circuits in such a way that the possible ignition energy is always below the possible ignition energy. The technical protection options are therefore classified according to how they are implemented in practice. An exact definition of the individual protection types is given by EN/IEC 60079 part 0 ... 28. Table 18.9 contains a short description of the types of protection and a schematic diagram showing the design.

Protection Class Ex-p
In gas measurement technology, however, the measuring instruments must be brought into contact with the gas to be detected, otherwise no measurement is possible. For complex gas measuring instruments, protection type *Ex-p* is usually used, which ensures appropriate dilution by means of a permanent flow of purge gas and thus prevents an ignitable gas mixture (→ inerting). The housing must therefore have a certain tightness so that the purging gas flow can build up a corresponding overpressure inside the housing

Table 18.9 Description of the most important types of protection

Structure/symbol	Description
	*Pressurized enclosure Ex-p* Pressurized enclosure describes a type of protection in which the ingress of explosive gases is prevented. This is usually achieved by purging with inert gases (e.g. air or N_2) under positive pressure. The type of protection Ex requires a monitoring device which switches off the device (or system) when the overpressure (Δp) is no longer present.
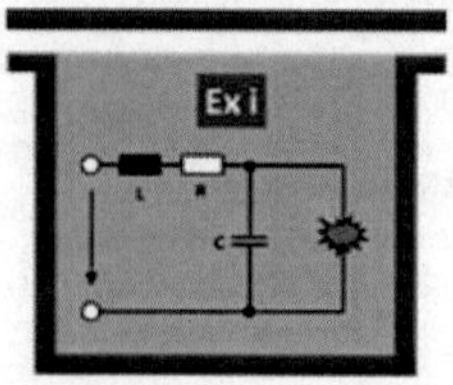	***Intrinsic safety Ex-i*** With this protection class, the electrical and thermal energy is limited to ensure safe operation of the electronics below the ignition energy. The ignition energies are between 280 ɰ for methane and 60 ɰ for hydrogen. This protection class is realized by circuit limitations, such as small capacitances/inductances and Zener diodes for voltage limitation.
	***Flameproof enclosure Ex-d*** Flameproof enclosure allows an explosion within an enclosure, with the design of the enclosures preventing the explosion from spreading. Stable cast housings are required for this purpose. Pressure equalisation or gas exchange is achieved via gaps or flame arresters which prevent propagation.
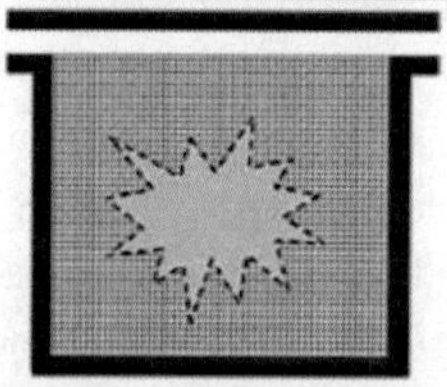	***Potting, sand or oil encapsulation Ex-m,-q,-o*** The encapsulation in the gas measurement technology is done with a 2-component casting resin, oil and sand are less common. This type of encapsulation prevents contact with the explosive gas mixture and thus prevents ignition.
	***Increased safety Ex-e*** The protection principle is based on design measures in which clearances and creepage distances are specified for the live parts. This prevents electrical sparks that could cause an explosion. Furthermore, at least the enclosure protection class IPS4 must be achieved.

($\Delta p \approx 10$ hPa). This overpressure prevents the explosive atmosphere from entering the device from the Ex-zone and causing an ignition by the electrical units inside. Figure 18.15 shows such a gas measuring device in connection with a purge gas monitoring system with a control unit (Fig. 18.16). Both the purging gas flow $\varnothing_{SG}$, which passes through the

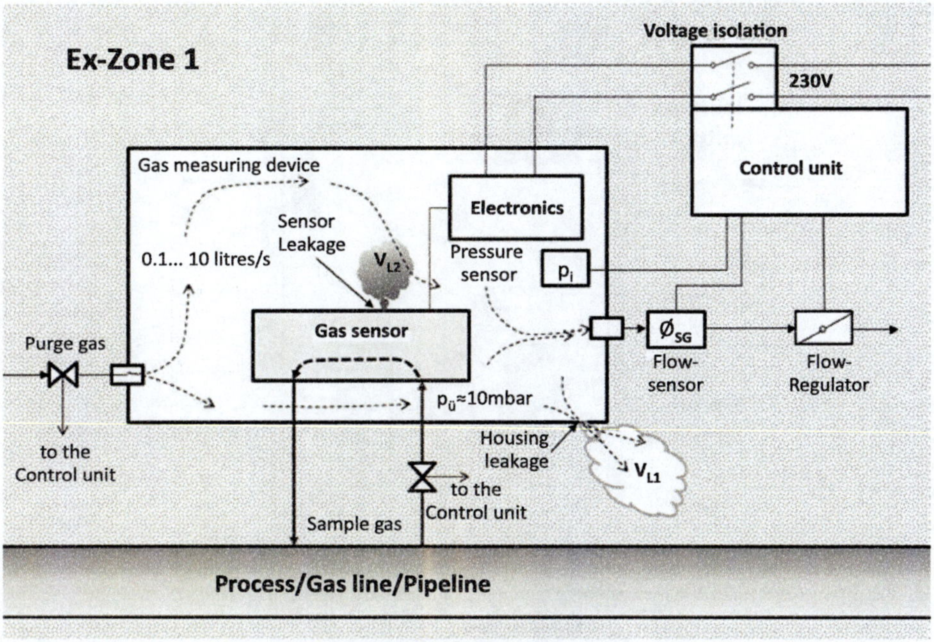

Fig. 18.15 Basic structure of an Ex-p protection device for gas detection

housing, and the pressure Δp are detected by corresponding sensors and fed to a control unit. If the pressure drops due to a malfunction (e.g. no more purge gas available) or gross leakage, the gas measuring device to be monitored is completely disconnected from the mains and switched off via a relay. Smaller leaks in the sample gas inlet within the measuring device or leaks in the housing are also neutralized by the rather high purging gas flow. Large leaks are also detected by the pressure sensor and can then lead to a safety shutdown. If necessary, the sample gas supply can then be additionally switched off with a further solenoid valve.

Protection Class Ex-i

The simplest possibility is to design the sensors and electronics in such a way that the minimum ignition energy E_Z *is* never exceeded in any situation. This type of protection is known as *intrinsically safe* (i). Since the ignition energies for the individual substances differ considerably, the gases are divided into substance groups according to EN 60079-11 (Table 18.10). Testing or certification is carried out within these substance groups.

Compliance with the maximum energy is achieved by appropriate dimensioning of the electronic components (capacitor, inductance, ...). Furthermore, the shortest connections between the potentials must have a minimum distance to prevent flashovers. In this context,

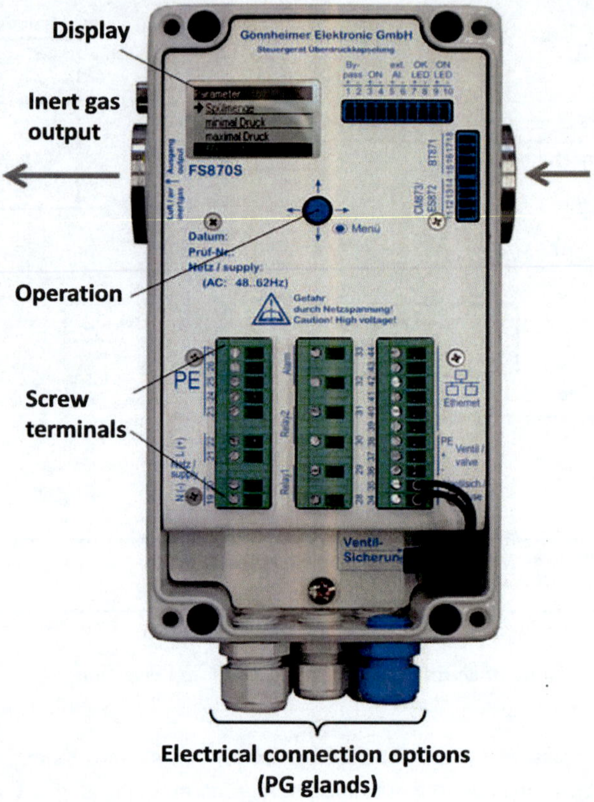

Fig. 18.16 Control unit for an Ex-p monitoring. (Source: Gönnheimer Elektronic GmbH Neustadt Germany)

Table 18.10 Ex-i group classification

Group	Maximum energy
II C	20 μJ
II B	80 μJ
II A	160 μJ

a distinction is made between the *clearances* (shortest path) and the *creepage distances* (shortest path over a surface). In Fig. 18.17 these relations are sketched.

If all boundary conditions for the intrinsically safe application area are fulfilled, the gas sensor can be used directly in the corresponding Ex zone for which it has been certified. The electrical supply and signal transmission, on the other hand, could also transmit ignition energy into the Ex-zone and thus cause an ignition. For this reason, *associated electrical equipment* is used in the Ex-free area to prevent such transmission. As a rule, this measure is achieved by using Zener barriers or electrical isolation. Figure 18.18 shows this

Fig. 18.17 Clearance and creepage distances in an electronic circuit

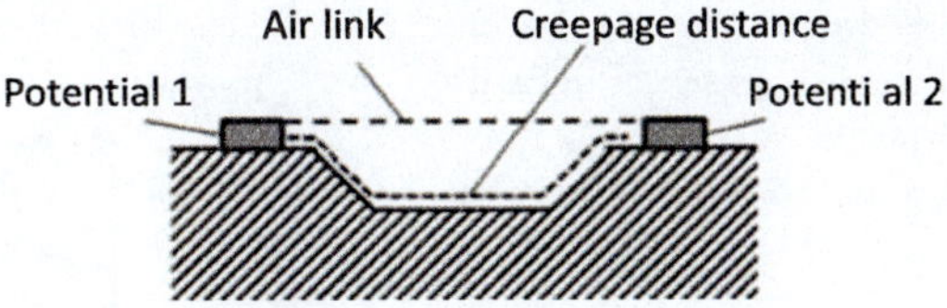

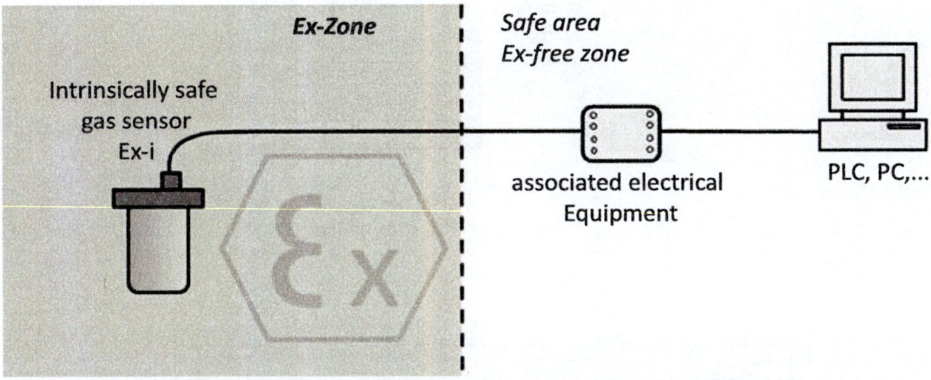

Fig. 18.18 Installation of an Ex-i gas sensor in the Ex-zone and signal and supply transmission with an appropriate device for decoupling the external circuits

decoupling between the gas sensor in the Ex zone and a data processing device outside the Ex zone.

Protection Class Ex-d

Another option that is frequently used in gas measurement technology is the so-called flameproof enclosure (Ex-d). With this type of protection, the gas sensor is installed in a solid metal housing that can withstand an internal explosion. The internal volume is kept as small as possible in order to obtain a limited explosion range which releases a correspondingly low amount of energy. The gas exchange takes place via so-called flame arresters, which in the simplest case consist of a sintered metal disc. The porosity and thickness of the disc are selected so that a flame resulting from an internal explosion never reaches the outside and causes a secondary explosion in the surrounding area. The damage in this situation is limited to the gas sensor, which in the worst case must be replaced. The sintered metal frit is installed in the sensor housing in such a way that a pressure shock resulting from an internal explosion cannot push the disk outwards. The electrical feedthroughs are cast from the rear side. The potting must also be designed in such a way that the potting plug cannot be pressed out by the explosion pressure (Fig. 18.19). These properties are examined in an Ex-test by carrying out a test ignition with the appropriate measuring gases in the sensor assembly.

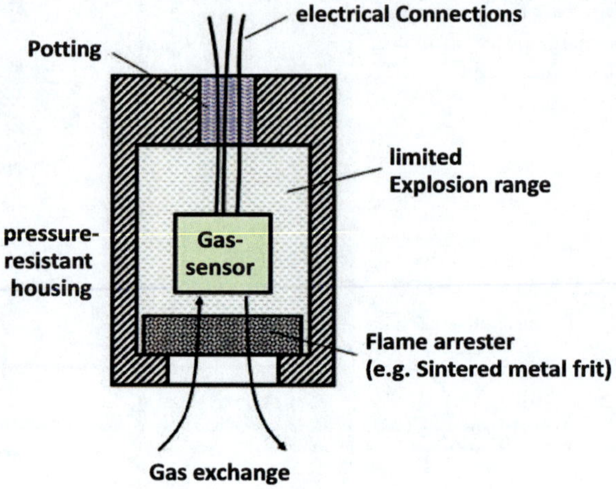

Fig. 18.19 Cross-section through a gas sensor in Ex-d version

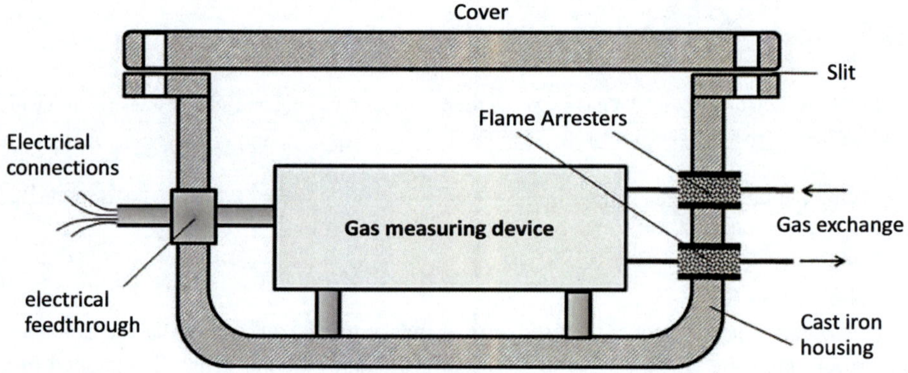

Fig. 18.20 Cross-section through an Ex-d housing in which a gas detector has been integrated. The arrangement must be tested according to the relevant Ex-directives and approved by a notified body

Larger sensor assemblies are installed in corresponding flameproof housings, which were specially designed for this application. Figure 18.20 shows a cross-sectional view of such a structure. The gas exchange also takes place with this arrangement by means of corresponding flame arresters, which are located in the inlet and outlet of the gas supply. The cover is connected directly to the cast housing by means of a corresponding number of screws. The gap between the cover and the cast housing is so small that no flame can penetrate. The electrical connections are also led to the outside via Ex-suitable bushings. Figure 18.21 shows a complete installation of a gas measuring device for checking the quality of natural gas (calorific value determination) in an Ex-d version. Figure 18.22 shows an oxygen analyser in Ex-d version for wall mounting.

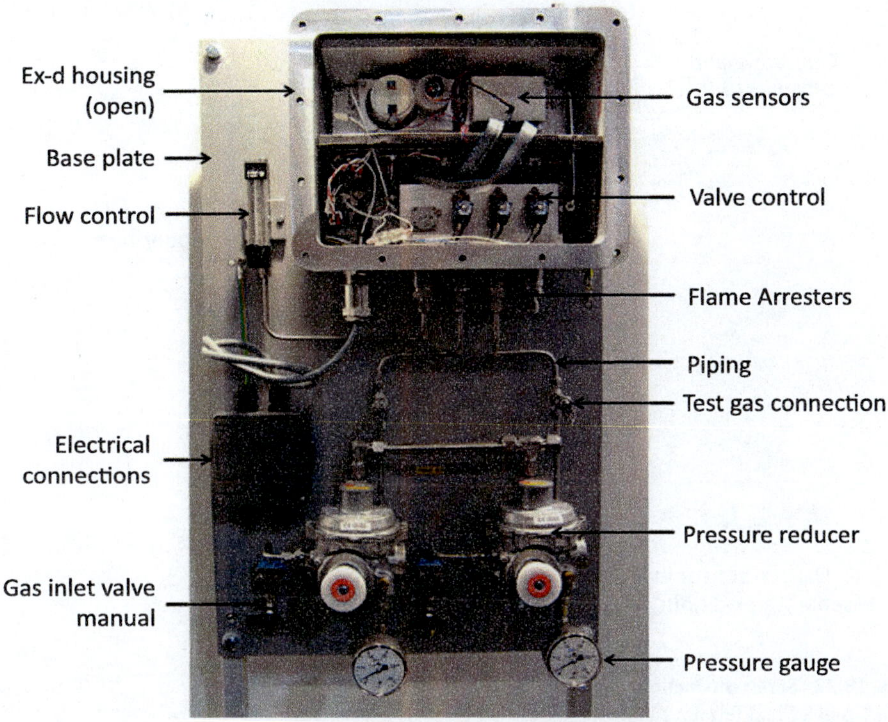

Fig. 18.21 Gas measuring device in an Ex-d housing for monitoring the calorific value of natural gas

Flame Arresters

An important component, which has already been mentioned several times, is the so-called flame trap. The propagation of a flame during an explosion can be described by the propagation velocity *vF*, which can be up to 100 m/s depending on the type of gas. If this flame front reaches a metallic barrier (see Fig. 18.23), which is permeable to gas and has constrictions (gaps), the flame is extinguished. The flame front also enters this gap, but the heat transfer dQ/dt, at the metallic narrow point reduces the flame temperature very quickly. Furthermore, the fuel (gas-air mixture) can only be replenished by diffusion, which, however, has a much lower speed. As a result of these effects, the flame within this barrier extinguishes after a short distance. The penetration depth of the flame must, however, be much smaller than the total thickness of the barrier, otherwise the flame would still penetrate to the outside. The total gap length (thickness) of the barrier is generally $L = 25\ mm$. The decisive factor for the effectiveness of a flame barrier is the gap width *d*. Technically, this task is usually carried out with sintered metal discs consisting of individual stainless steel particles which are pressed (sintered) into a disc under high pressure and temperature. Decisive for the effectiveness of the flame arresters are the porosity and the total thickness of the disc. Furthermore, in the event of an explosion, the disc must also

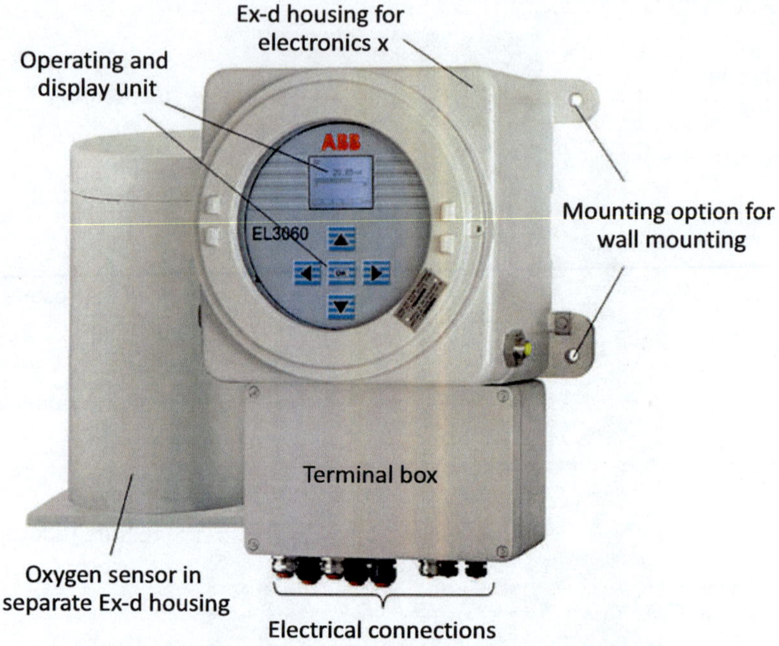

Fig. 18.22 Series production version of an oxygen analyser in Ex-d design. (Source: ABB Automation GmbH, Frankfurt am Main Germany)

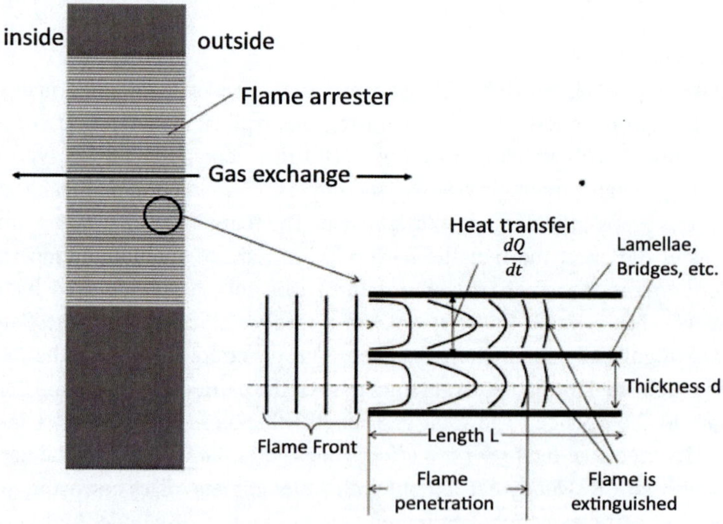

Fig. 18.23 How a flame arrester works, consisting of individual rips or slats

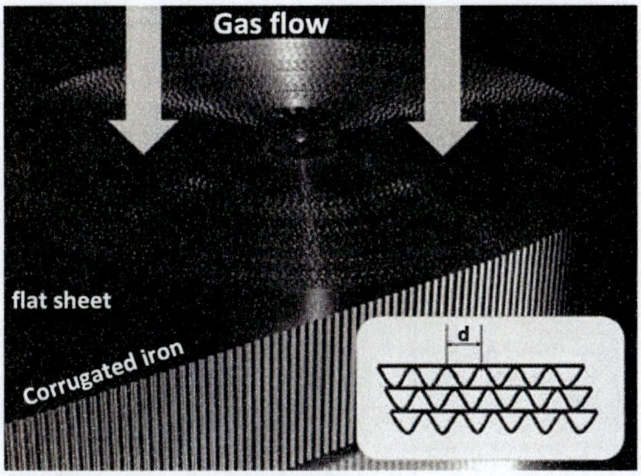

Fig. 18.24 Construction of a flame arrester consisting of a corrugated and a flat sheet metal. (Source: Kito Armaturen GmbH, Braunschweig Germany)

Table 18.11 Limiting gap widths of the explosion groups

Explosion group	Border column	MIC
II A	>0.9 mm	>0.8
II B	0.5–0.9 mm	0.45–0.8
II C	<0.5 mm	<0.45

withstand the resulting pressure without bursting. A disadvantage of these flame arresters is the high pressure drop when gas is pumped through these barriers. For diffusion-limited systems, however, this effect is not important.

Alternatively, corrugated sheets can also be used, which are wound up. Between two corrugated sheets, a flat sheet is always tied in, so that defined gap widths *d are* obtained for each corrugation. Since these individual gaps are connected in parallel for gas passage, the pressure drop is considerably reduced and thus large volume flows can also be secured with these flame arresters. Figure 18.24 shows this design.

The limiting gap width at which complete flame extinction occurs depends very much on the gas in question. The limiting gap widths for the individual explosion groups are shown in Table 18.11.

The limiting gap width at which an explosion can propagate also depends on the gas concentration *c* of the respective mixture. Figure 18.25 shows the results of a measurement carried out by PTB using a standard gap width measuring instrument. Carbon disulphide and acetylene therefore require gap widths which are <0.5 mm and are therefore assigned to explosion group II C.

Ex-marking

Every device that is used in an Ex-zone must be tested and certified by an independent body. In order to enable a clear assignment to the tested properties, the areas of application of the device must be clearly identified on the type plate. This marking has been defined in the relevant standards. Fig. 18.26 shows such a marking in accordance with the ATEX Directive and EN 60079–0.

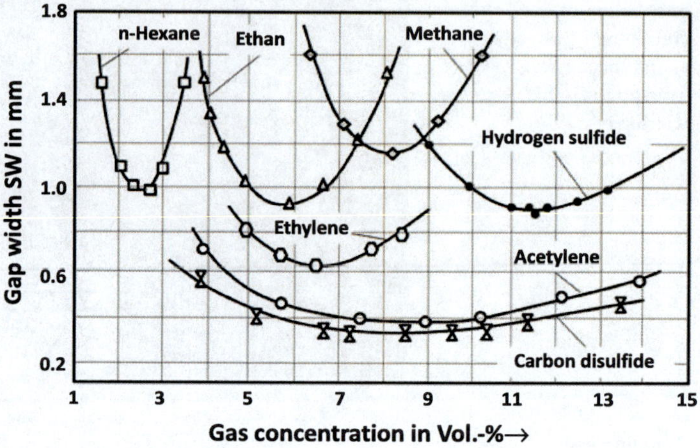

Fig. 18.25 Dependence of the minimum gap width for different gases and variable gas concentration. (Source: according to data from www.explosionsschutz.net)

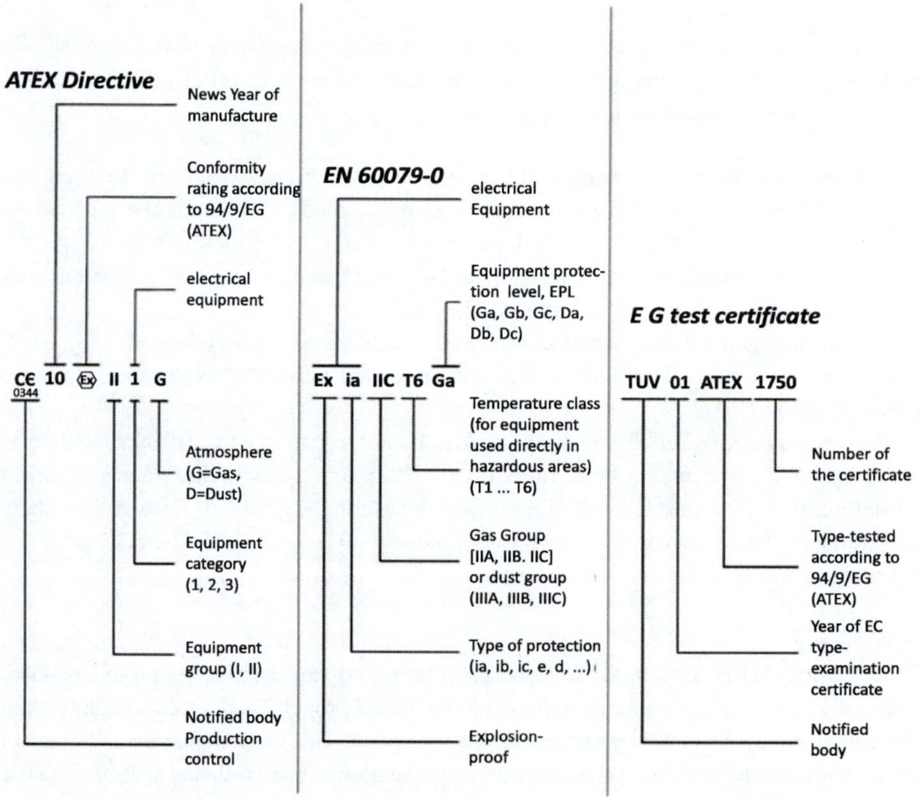

Fig. 18.26 Marking of Ex devices. (According to information from Phoenix Contact 2010)

18.2 Gas Detection Systems in Safety Engineering

Wolfgang Jessel

In industrial sectors, especially in the chemical and petrochemical industry, flammable gases or flammable liquids are often produced, processed, filled, stored and distributed. A release of such substances and their ignition can never be ruled out. Whether deflagration or explosion – the destruction of the plant by subsequent fires, personal injury, life-threatening injuries and even death, and possibly also a considerable impairment of our environment would be the consequence. Nevertheless, such risks can be largely controlled by measures of so-called explosion protection, which affects all equipment used in potentially explosive atmospheres.

Preventive Explosion Protection
Explosion protection is a legal matter, regulated throughout Europe by two binding directives (Atex directives), whose national implementation in Germany has led to the Explosion Protection Ordinance (ExVO) on the one hand and is contained in the Industrial Safety Ordinance (BetrSichV) and the Hazardous Substances Ordinance (GefStoffV) on the other (Table 18.12).

The ExVO regulates the quality requirements: Which requirements must equipment meet to ensure that it does not pose a hazard and therefore does not become a potential source of ignition? Depending on the required safety level, an accredited testing body (so-called *notified body*) must be involved in the relevant assessment, which not only confirms that the equipment under consideration meets the requirements of the directive (type examination), but also that its manufacturing process is designed in such a way that the equipment manufactured in series corresponds to the type and – accompanied by regular audits of the manufacturer – is just as safe as the tested equipment. These inspections are concluded by the issue of an EC type examination certificate and a quality certificate, on the basis of which the manufacturer (or the person placing the equipment on the market) certifies the conformity of the equipment with the directive in an EC declaration of conformity. This fulfils the requirements for affixing the CE mark and thus for placing the equipment on the European market. Of course, gas detection systems are also affected by this directive, because the transducers, that is sensors with associated

Table 18.12 Comparison of directives

Atex directive	is addressed to	Regulates	National implementation in Germany
94/9/EC	Manufacturer	Placing equipment on the market	Explosion protection ordinance - ExVO
1999/92/EC	Operator	Protection of The employee	Industrial safety regulation - BetrSichV

electronics, hereinafter referred to as (gas measuring) transmitters, must never become an ignition source. For such transmitters, the four types of protection that can be considered are preferably flameproof enclosure, intrinsic safety, potting and increased safety, which are based on the following simplified principles:

- Flameproof enclosure "d": Ignitable atmosphere can penetrate into the equipment and ignite if necessary. However, the enclosure is designed to withstand the explosion pressure and to prevent the flame from penetrating into the outside area.
- Intrinsic safety "i": It has been demonstrated that equipment whose electrical parameters voltage, current, energy and power remain below certain limits even in the event of a fault is not capable of igniting even the most ignitable gas mixture. Appropriately designed electronics and limitation of these electrical parameters ensure that the equipment produces neither hot surfaces nor electrical sparks – even in the event of a fault.
- Potting "m": Due to the embedding of the electronics in a plastic mass, the ignitable atmosphere has no direct contact with the electronics and therefore cannot be ignited. High demands are placed on the quality of the cured plastic, especially with regard to its resistance to ageing.
- Increased safety "e": This type of protection is limited to passive parts such as terminals, terminal boxes and cable glands. The design of such components reduces the probability of accidental heating or even sparking.

All these types of protection have in common that the presence of a potentially explosive atmosphere is assumed, that is, the lower explosion limit of a gas-air mixture is exceeded at least temporarily.

Prevent the Formation of Explosive Atmospheres: Concentration Limitation
However, explosion protection concepts which already prevent the formation of an explosive atmosphere have a higher, even the highest priority (hence often referred to as "primary explosion protection") and are also classified as such by the legislator:

1999/92/EC, Article 3, Prevention of and protection against explosions

- Prevention of the formation of explosive atmospheres, or where the nature of the activity does not allow that,
- Avoidance of the ignition of explosive atmospheres, and
- Mitigation of the detrimental effects of an explosion so as to ensure the health and safety of workers.

The first two measures fall under the so-called preventive explosion protection because they prevent explosions. The plant operator is thus required to take the "technical and/or organisational measures appropriate to the type of operation" according to this order of priority and also to review them regularly, in particular after significant changes to the plant.

What primary measures can the operator take?

- In its production process, for example, it can replace flammable liquids with non-flammable liquids (which means that it is no longer covered by the Atex directives),
- or at least replace them with liquids whose flash point is significantly above the processing temperature
- It can – which is usually only possible in closed apparatus – reduce the oxygen concentration in such a way that the oxygen concentration is significantly below the oxygen limit concentration (OLC): Inerting
- It can ensure that the gas or vapour concentration remains well below the lower explosion limit (LEL): Concentration limiting.

Ultimately, all these measures result from the well-known explosion triangle, a pictorial representation of the fact that if and only if three conditions are fulfilled simultaneously and at the same place, ignition, combustion, deflagration or explosion is possible:

- There is an ignition source of sufficiently high energy or temperature
- Sufficient oxygen is available (oxygen concentration OLC)
- Sufficient fuel is available (gas or vapour concentration LEL).

The use of gas detection systems for the detection of combustible gases and vapours is based on the last statement. Oxygen, namely air, is available in sufficient quantities. Also ignition sources cannot be excluded operationally at least. And yet it is even possible to allow a certain gas or vapour concentration in air – because as long as this remains below the lower explosion limit, the LEL, ignition can definitely not occur. With such gas detection systems, one operates active explosion protection to a certain extent by continuously monitoring the gas or vapour concentration in the air: The danger can be controlled by continuously measuring the concentration and switching off (or rendering ineffective) potential ignition sources well below the LEL, usually at 40% of the LEL.

Unpleasant for the operator, because this harsh measure (so-called *shutdown*) leads to a loss of production, the machines stand still! For this reason, gas detection systems have a pre-alarm (usually at 10 or 20% of LEL), which is intended to initiate countermeasures. If, for example, a leak occurs in an acetone filling plant, the acetone vapour concentration rises, switches on an effective fresh air ventilation at the pre-alarm concentration of 20% LEL and dilutes the vapour-air mixture to such an extent that the *shutdown concentration* of 40% LEL is not even reached. If it does: Switch off all electrical equipment that is not already *explosion-proof*.

Explosion Protection: No Chance Without Approvals

Ex-protected is jargon. – More correct would be the wording:

> ***Designed by one or more combined types of protection in an explosion-proof construction and approved for operation in the intended area, that is, type-tested as a type examination.***

Such type examination is carried out at the request of the manufacturer by the notified body which, after successful examination, assures the conformity of the type with the European Directive 94/9/EC and issues an EC type examination certificatc for this purpose. Conformity means in almost all cases that binding standards that have been recognized (harmonized) throughout the EU and implemented in country-specific standards are complied with. For the types of protection mentioned, for example, the following standards:

- EN 60079-0 Equipment – General requirements
- EN 60079-1 Equipment protection by flameproof enclosures "d"
- EN 60079-7 Equipment protection by increased safety "e"
- EN 60079-11 Equipment protection by intrinsic safety "i"
- EN 60079-18 Equipment protection by encapsulation "m"

However, if explosion protection is not ensured by ignition protection types but by a primary measure, namely the use of gas detection systems, then such a gas detection system must consequently be certified in the same way. And indeed, some standards have been harmonized for gas detection systems, and only those gas detection systems are approved for explosion protection which have been tested for suitability for the intended application in accordance with these standards and whose conformity with Directive 94/9/EC has been confirmed by an EC Type Examination Certificate, in particular the following standards

- EN 60079-29-1 Gas detectors – Performance requirements of detectors for flammable gases
- EN 50104 Electrical apparatus for the detection and measurement of oxygen – Performance requirements and test methods
- EN 50270 Electromagnetic compatibility – Electrical apparatus for the detection and measurement of combustible gases, toxic gases or oxygen
- EN 50271 Electrical apparatus for the detection and measurement of combustible gases, toxic gases or oxygen – Requirements and tests for apparatus using software and/or digital technologies

The fact that even the central units of a gas detection system also receive the Ex-marking known from ignition protection types may at first seem strange, but the Atex Directive 94/9/EC already expressly points out in its first article that safety, monitoring and control devices for use outside potentially explosive atmospheres also fall within the scope of this Directive if they are necessary for or contribute to the safe operation of equipment with regard to explosion hazards. Central units of gas detection systems are such safety devices. Although they are installed outside the hazardous area, their measuring function and the countermeasures activated by them have an effect inside the hazardous area, thus ultimately preventing explosions. It is clear that the measuring function of such a gas detection system, from the gas inlet of the sensors to the alarm contact, must meet certain standardized requirements. Of course, even with fluctuating environmental parameters (temperature, pressure, humidity, wind), electromagnetic interference, vibration and fluctuations in the supply voltage, the measurement result may only deviate from the nominal value within certain limits. The software and digital electronics of the devices are also subjected to thorough testing. Type tests according to EN 60079-29-1 or EN 50104 expressly include tests according to EN 50270 and EN 50271. Directive 94/9/EC states in point 1.5.7 of Annex II

> *The design of devices with a measuring function must be based on a safety factor which ensures that the alarm threshold is sufficiently far outside the explosion and/or ignition limits of the atmospheres to be detected, taking into account in particular the operating conditions of the device and any variations of the measuring system.*

But where do you find such explosion limits? In the meantime, a standard has been published, from the appendix of which not only the explosion limits but also other safety-related parameters of more than 300 gases and vapours, sorted by CAS No., are listed (EN 60079–20-1).

Potentially Explosive Areas Can Shrink

Back to preventive explosion protection through concentration limitation. Potentially explosive atmospheres are divided into zones according to the frequency of occurrence and duration of the presence of a hazardous explosive atmosphere:

Zone 0:	Area in which explosive atmospheres ... are present continuously, over long periods or frequently.
Zone 1:	Area in which a potentially explosive atmosphere ... May occasionally form during normal operation.
Zone 2:	Area in which a potentially explosive atmosphere ... Does not occur during normal operation or only occurs for a short time.

Experts interpret these definitions to mean that hazardous explosive atmospheres are present in Zone 0 for most of the time (more than 50% of the operating time), but in Zone 2 they are present only a few times a year, and no longer than about 30 minutes per incident. A Zone 1 is present if the frequency and duration lies between these two interpretations, for example, more frequently than a few times per year, and in each case for longer than approximately 30 minutes, but not for the most part. Preferably in these areas, i.e. Zone 1 areas, gas detection systems are used for explosion protection. And since a gas detection system counteracts the formation of explosive atmospheres by means of the switched countermeasure, this will not occur occasionally, and – if at all – only for a short time. Zone 1 becomes Zone 2! At the very least, Zone 1 "shrinks" considerably in favour of Zone 2 and the extent of the potentially explosive atmosphere is also reduced.

In room air monitoring, the so-called technical ventilation is one of the most frequently automatically initiated countermeasures. For this countermeasure to be effective, calculations must be made during the planning of the ventilation system which take into account the risk potential associated with the handling of flammable gases and vapours. This is also in the operator's interest, because otherwise the main alarm threatens to activate compulsory measures! *Shutdown*!

The standard EN 60079-10-1 offers help, also with calculation bases, ventilation models and practical examples:

> *The most important factor is that the ventilation quantity (degree of ventilation) is adapted to the type of release sources and their release rates. This is independent of the type of ventilation, whether it is wind speed or the number of air changes per unit of time. In this way, optimal ventilation conditions can be achieved in the hazardous area, and the higher the degree of ventilation in relation to the possible release rates, the smaller the extent of the zones created (hazardous areas) will be. In some cases it will even be negligible (non-hazardous area).*

In the best case – as also stated in Section E 1.3.4.2 of BGR 104 (Explosion Protection Rules) – technical ventilation can prevent potentially explosive atmospheres.

Risk of Explosion in Case of Failure!

It is very much in the operator's interest to make the countermeasures triggered by the pre-alarm as effective as possible so that the main alarm, which triggers forced measures, is not reached. This is because the main alarm forces a safe shutdown of the monitored system to a safe state, at the expense of availability – but in favour of safety. If the main alarm fails, therefore, there is no safety: there is a risk (of explosion)!

In the sense of the *Functional Safety* regulations, the gas detection system requests the execution of a specified safety function, in this case the activation of the main alarm, only rarely, once a year or even less frequently. One speaks of an operating mode with a *low demand* rate, the so-called *low demand mode*. In general, the question arises as to how reliable the monitoring system (or safety system, *Safety Instrumented System, SIS for* short) is. For example, after 100 uneventful days, will it meet the requirement and be able to perform its safety function? Or has an error crept in completely unnoticed on the tenth day, which is only detected on the 101st day because the main alarm is unfortunately not triggered at all despite the presence of a dangerously high gas concentration? One answer to this is a purely statistical statement, which becomes a measure of reliability by stating the so-called *Probability of Failure on Demand* (*PFD*).

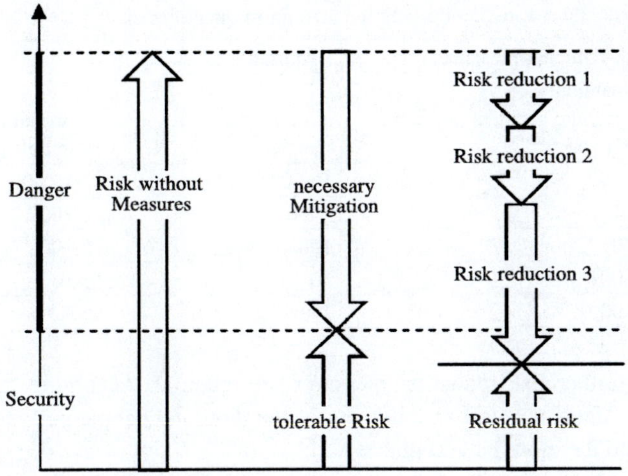

Fig. 18.27 Safety gained by risk reduction

Risk Reduction by Safety Systems

An installation is considered to require monitoring if the occurrence of a hazardous state cannot be excluded without risk-reducing measures. If the risk is applied upwards, as shown in Fig. 18.27, the safe area by definition extends to the limit of the tolerable risk. Above this limit, the hazardous area extends up to a line representing the risk present. This results in the following definitions:

- Safety: The risk is lower than the tolerable risk
- Danger: The risk is greater than the tolerable risk.

In order to make a plant safe, one or more risk-reducing measures must result in the risk falling below the tolerable risk. Then the actual risk, the residual risk, lies in the safe area. Risk reduction can be achieved in various ways, namely through *organisational measures* and/or *technical measures*.

In connection with safety systems, the technical measures are the main focus here. In order to be able to describe the necessary risk reduction measures also quantitatively, so-called *risk reduction factors* (*RRF*) are defined, the application of which leads to reaching or falling below the tolerable risk.

As can be seen from Table 18.13, if the risk is very high, a risk reduction by a factor of 10,000 must be achieved in order to just reach the safe area (i.e. the residual risk then corresponds precisely to the tolerable risk). In order to make such a plant safe, i.e. to fall below the limit of the tolerable risk, the RRF must actually be ≤10,000. The last line of this table states that for very low risks, an RRF of 1 is already sufficient to reach the tolerable risk, in other words: the risk is acceptable because it is already at the level of the tolerable risk. It can be concluded that installations with very low risk are safe and therefore

Table 18.13 Risk and risk reduction with the attempt to quantify risk

Risk without risk-reducing measures		Required RRF	
Qualitative	Quantitative e.g. Fatal accident per person and year	... to achieve thetolerable risk	... to fall below the tolerable risk
Very high	1	10,000	>10,000
High	0.1	1000	>1000
Medium	0.01	100	>100
Low	0.001	10	>10
Very low	0.0001	1	>1

practically no further risk reduction measures are required. According to Table 18.13, 0.0001 fatal accidents per year (i.e. one out of ten thousand people per year are killed in accidents) would therefore be acceptable.

A Fictitious Scenario

Consider a chemical production process in which, for example, acetone is produced. Let us assume that on average once a year a dangerous situation occurs where hot acetone vapour escapes due to a leak in the pipeline, which immediately ignites and causes a deflagration. In this case, people in the vicinity of this production process would be injured, and for 1% of the people affected, this incident would be fatal. A risk is described by the product of the probability of occurrence H of the damaging event and the damage severity S as a consequence of the event: $R = H \cdot S$.

With $S = 1$ person out of 100 people have a fatal accident $= 0.01$ and $H =$ once a year $= 1/\mathrm{a}$ then risk is $R = 0.01/\mathrm{a}$.

In order to reduce the risk, a gas detection system is now being used which alerts the operator to 10% LEL acetone and calls for safety measures (find the cause of the leakage with portable gas detectors, close off and seal off parts of the building, etc.). The safety function of the gas detection system, however, is to close the affected pipeline with a valve as soon as 40% LEL acetone occurs and to shut down the affected part of the process.

If, however, the gas detection system fails on average once a year and the safety function cannot be carried out, then there is nothing to be gained from the gas detection system. The risk remains 0.01/a. The situation is different with a gas detection system which, on average, only refuses to operate once every 10 years, that is, does not execute the safety function due to an undetected failure that has occurred within this year. In this case the probability of failure of the gas detection system is included 0.1/a and the probability of a deflagration with the consequences described above is reduced to one tenth because the gas detection system prevents the formation of an explosive acetone vapour-air mixture at an early stage. As shown in Fig. 18.28, this gas detection system reduces the risk by a tenth: $R = 0.001/\mathrm{a}$.

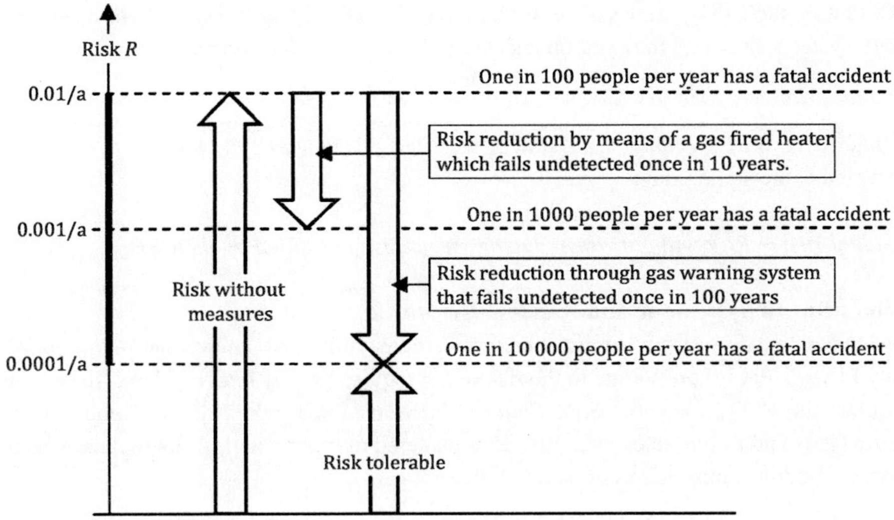

Fig. 18.28 Risk reduction by a gas detection system

Table 18.14 Risk reduction and probability of failure for a given SIL

Risk	RRF	PFD_{avg}	SIL
Very high	>10.000	<0.0001	4
High	>1000... ≤ 10.000	<0.001... ≥ 0.0001	3
Medium	>100... ≤ 1000	<0.01... ≥ 0.001	2
Low	>10... ≤ 100	<0.1... ≥ 0.01	1
Very low	>1... ≤ 10	<1... ≥ 0.1	No SIL

With a gas detection system, the probability of failure of which is even close to 0.01/a, one arrives at a risk of $R = 0.0001$/a, which can already be called tolerable risk. By using a gas detection system with a probability of failure 0.01/a, the risk has been reduced by 2 orders of magnitude. The scenario described thus illustrates the connection between risk reduction factors and the average probability of failure of the safety system. Table 18.14 shows that the risk reduction factors are closely related to the four so-called *Safety Integrity Levels (SIL)*. If risk reducing measures are implemented by electrical, electronic or programmable electronic systems (so-called E/E/PE systems), then $PFDs_{avg}$ are required of these, which correspond to the reciprocal value of the RRF. Their assignment to the four *Safety Integrity Levels is* based on orders of magnitude, that is, is obviously logarithmic.

Accordingly, very low risks require an RRF of at least 1, in other words – a *Safety Integrity Level* makes no sense. However, if, for example, a plant requiring monitoring has a medium risk, safety systems must ensure that the risk is reduced by more than a factor of

100, that is, the PFD_{avg} must be less than 0.01. Statistically speaking: Out of a hundred safety systems only one may fail undetected (within a specified time interval).

For the residual risk, the following relationship then results:

Residual risk = Risk without risk-reducing measures, divided by the RRF
or, which is the same thing:

Residual risk = Risk without risk-reducing measures, multiplied by the PFD_{avg}

Failure Due to Systematic and Random Errors
Risk-reducing safety systems must therefore demonstrably have an average failure probability PFD_{avg} that is appropriate to the risk or the required *safety integrity level.* In order to calculate the PFD_{avg} and the *Safe Failure Fraction (SFF)*, which is also required, the failure types and failure rates must first be considered in more detail. A distinction is made between random faults and systematic faults.

Systematic errors are planning and development errors which are already – undetected – present and reproducible at the time of delivery (e.g. software errors, incorrect dimensioning or operation of electronic components outside their specification). Of course they *cannot* be intercepted by redundancies, but they can be reduced to a minimum by organisational measures and deliberately safety-oriented development. Software can rarely be tested at 100%. To avoid errors, special methods for safe software design (e.g. V-model) are therefore used and meticulously tested and documented. While the incorrect dimensioning of components can be avoided through expert development and development processes, it is only possible to avoid improper operation through organizational measures on site. On site, the assigned safety engineer is responsible for compliance with the manufacturer's specifications, which are available in the form of a *safety manual.* Production errors (this would affect an entire series, for example) are also systematic errors. Precisely defined production processes, which are monitored, for example, by the manufacturer's quality management, are suitable for excluding such errors. A very common cause of systematic errors is the wrong specification during the planning phase, which of course can only lead to incorrect behaviour. It can be seen as an advantage of a systematic error that it is at least reproducible, so that in case of such an error the cause can usually be traced relatively easily.

The *random failure* is an unavoidable (!) characteristic property of components. It *does not* exist at the time of delivery, but occurs at any time during operation. The random failure is described by the so-called constant failure rate λ, that is, within the same time intervals always the same quantity of components fails.

The Random Failure: Failure Rates – Failure Causes – Failure Types
The manufacturer of electrical or electronic components usually states such failure rates in the associated data sheets. They are often taken from tables such as the US manual

Table 18.15 The four important failure modes in safety engineering

Failure mode	Riskiness	Detectability	Failure rate
DU	Dangerous	Undetectable	λ_{DU}
DD	Dangerous	Detectable	λ_{DD}
SD	Safe	Detectable	λ_{SD}
SU	Safe	Undetectable	λ_{SU}

MIL-STD-217 or the Siemens standard SN 29500 and are specified in *FIT* (*Failure in Time*, 1 FIT = 1.0E-09 Failure per hour = $10^{-9}h^{-1}$). For example, the failure rate for a silicon universal transistor is $\lambda = 5$ FIT, for a metal film resistor 2 FIT.

When considering a safety-related module such as signal-processing electronics of a switching amplifier, it is not only this failure rate that is important, but also the effect of the failure on the safety function. For this purpose, it must also be defined what 'failure' means. In the case of a metal film resistor, for example, experience has shown that the failure consists of 10% of a short circuit, 60% of an infinitely high resistance value ("*open circuit*"), 15% of a resistance value halved and 15% of a doubled resistance value. For these four causes of failure, the electronics must now be examined to determine whether they prevent the safety function from being triggered (if so, they are dangerous) and whether they can be detected by internal diagnostic functions (typically a fault alarm). These four types of faults (so-called *failure modes*) are identified by two indices (see Table 18.15), the first indicating the danger, the second the detectability.

In this, *safe* is to be interpreted as harmless, because the safety function is not impaired. Detectable failures are so important because the diagnostic electronics can always generate a signal that transfers the monitored system to the safe state (and allows it to remain there). This also applies to the dangerous failure if it is detected. In fact, the procedure just illustrated using the example of a metal film resistor, also known as FMEDA (*Failure Modes, Effects and Diagnostics Analysis*) must be carried out for all components of a safety system under consideration without exception. With the help of suitable computer-aided tools and experienced electronic engineers, very extensive tables are created in this way, but in the end, the results are just added up.

The result: The failure rates λ_{DU}, λ_{DD}, λ_{SD} and λ_{SU} for the entire safety system under consideration, four key figures that describe the effect of any random failure on the safety function and are therefore extremely important for characterizing the safety system in terms of the *safety integrity level.* Figure 18.29 illustrates these types of failures and their consequences. It is important to know that even harmless errors, i.e. those that do not affect the safety function at all (SD failures), bring the system into a safe state. So although there is no dangerous state at all, the system still goes into the safe state and remains unavailable. This is a typical false alarm for the operator! Such SD failures unfortunately accumulate with redundant systems, as will be shown later.

The λ_{DU}-value is the focus of the safety consideration, both as an absolute value and in relation to the sum of the four failure rates, it must be kept as low as possible.

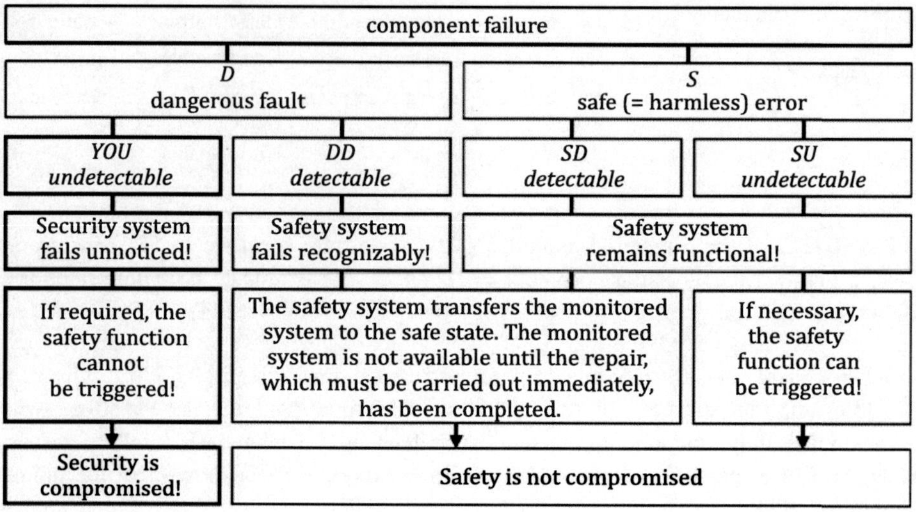

Fig. 18.29 Overview of the four safety-related failure modes

Probabilities of Failure

The definition of the random failure, namely that of n components always the same quantity Δn of components fails in equally large time intervals Δt, opens the way for mathematics, because it is

$$\frac{\Delta n}{n} = -\lambda \cdot \Delta t \tag{18.17}$$

where λ is again the constant failure rate. The corresponding differential equation:

$$\mathrm{d}n = -\lambda \cdot n \cdot \mathrm{d}t \tag{18.18}$$

has the solution:

$$n(t) = n_0 \cdot (1 - \exp(-\lambda \cdot t)) \tag{18.19}$$

Thus, one can calculate how many components n of originally n_0 components are still functional at time t. If you set this value $n_0 = 1$, you get the failure probability $F(t)$ of only one component:

$$F(t) = 1 - \exp(-\lambda \cdot t) \tag{18.20}$$

The failure probability of a component is therefore initially zero (for $t = 0$) and then increases with increasing time, that is, the longer a component is in operation, the greater its failure probability.

This also applies to a safety system: Immediately after a complete and successful test of the safety function, the so-called *proof test* (*proof* in the sense of evidence of functionality), the probability of failure is still zero, but increases over time. If the *proof test is* carried out again at a later date, the same applies: The probability of failure becomes zero again. This is a little confusing at first, but in fact the occurrence of a random failure ('random' means the failure rate λ is constant) is independent of its history (that is operating time) of the component under consideration, one speaks of a failure rate without history. Aging processes, that is, the increase of the failure rate λ at the end of the service life of a component, also known as wear, are not subject of the safety consideration. On the contrary, it is assumed that consumable parts are replaced at such an early stage in the course of regular maintenance that failures due to wear and tear cannot occur. This is the basis for the restoration of the quasi-new condition required for safe operation.

That one is allowed to sum up the failures of individual components at all is also a property of the random failure and its probability of failure, which can be represented so elegantly by an exponential function, because it is according to the laws of exponential calculation:

$$F(t) = 1 - \exp(-\lambda_1 \cdot t)\exp(-\lambda_2 \cdot t)\ldots\exp(-\lambda_n \cdot t) = 1 - \exp(-(\lambda_1 + \lambda_2 + \ldots + \lambda_n) \cdot t) \quad (18.21)$$

Back to *low demand mode*. The one that said that a safety system is left to itself for a longer period of time, but must perform the safety function in case of demand. After commissioning, the probability that a random failure will occur unnoticed, which will lead to the failure of the safety function when requested, slowly increases. In this case the failure rate is $\lambda_{DU.}$ And this *probability of failure on demand* (PFD) becomes zero again with the next *proof test.*

$$\mathrm{PFD}(t) = 1 - \exp(-\lambda_{DU} \cdot t) \quad (18.22)$$

Figure 18.30 shows the course of the failure probability. It is strongly exaggerated, the individual "shark fins" within a *Proof Test* Interval T_P are rather below 1% with their peaks and are there by far not so curved, rather straight, so that one is actually dealing with a sawtooth curve. In the safety assessment, however, one is not interested in the time-dependent, but in the average or mean probability of failure within the time interval T_P:

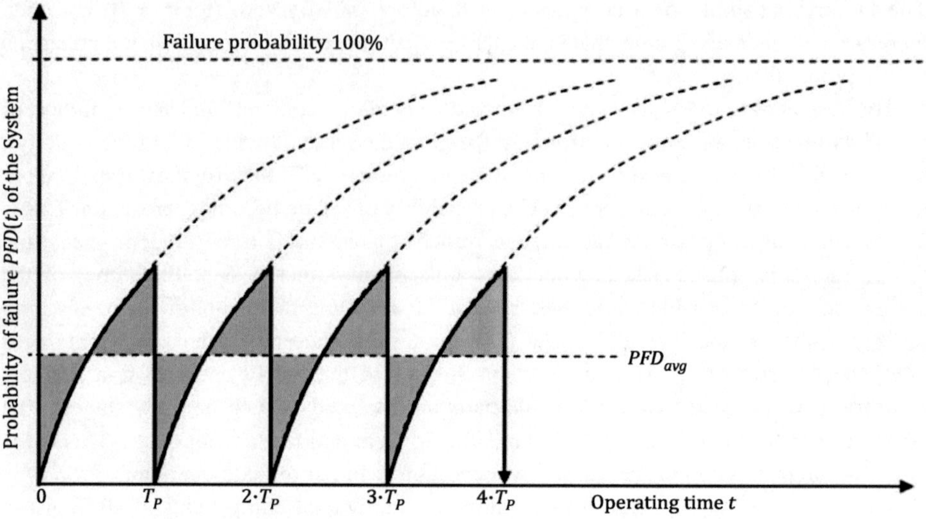

Fig. 18.30 Time course of the failure probability and its average value $\mathrm{PFD}_{\mathrm{avg}}$

$$\mathrm{PFD}_{\mathrm{avg}} = \frac{1}{T_P} \cdot \int_0^{T_P} 1 - \exp(-\lambda_{\mathrm{DU}} \cdot t) \cdot \mathrm{d}t \approx \frac{1}{2} \lambda_{\mathrm{DU}} \cdot T_P \tag{18.23}$$

This can already be seen from the figure, because for very small λ_{DU} the exponential function simply becomes a straight line, which increases linearly with time and has for $t = T_P$its maximum value:

$$\mathrm{PFD}(t) = \lambda_{\mathrm{DU}} \cdot t, \text{ and therefore } \mathrm{PFD}_{\mathrm{max}} = \mathrm{PFD}_{\mathrm{max}}(T_P) = \lambda_{\mathrm{DU}} \cdot T_P \tag{18.24}$$

The average value is then understandably half: $\mathrm{PFD}_{\mathrm{avg}} = \frac{1}{2} \lambda_{\mathrm{DU}} \cdot T_P$.

One might think that the sawtooth curve could be arbitrarily shifted together and thus the average probability of failure could be minimized. If one chooses the *Proof Test* interval T_P, that is, the time in which the safety system is left to itself, very small, for example, if one performs the *Proof Test* not annually but monthly, one reduces the average probability of failure to a twelfth! Correct in principle, but the *proof test is to* be understood as the conclusion of a procedure in which the quasi-new state of the safety system must first be established, accompanied by repairs, calibration work, etc., and during this period (according to the standard maximum 8 hours) the safety function is not available. Operating failure! The operator of the system will not really want a monthly repeated test. The repair time must be negligible compared to the interval of the proof test, only under this condition the simple formula for PFD_{avg} derived above applies.

Reach a Specified SIL with PFD and SFF?

What is the procedure if the proof test has to be carried out annually and the calculated PFD_{avg} is not sufficient to achieve a specified SIL?

As long as electronic components are concerned: The electronics engineer picks the thick chunks, i.e. the components that contribute particularly conspicuously to the DU failure rate, from the FMEDA list and uses qualitatively better ones, namely those with a significantly lower failure rate as indicated by the manufacturer. Or he supplements the electronics with diagnostic circuits and makes the failure detectable: In this way he converts a DU failure into a DD failure, which reduces the PFD_{avg} accordingly.

The system integrator, i.e. the person who assembles finished safety-related components, is worse off: the only thing left to do is to make his safety system fault-tolerant by redundantly combining such components. The calculation of the PFD_{avg} for redundant systems will be described later.

However, not only the absolute value of the DU failure, but also its relation to the total failure rate is subject to certain requirements, that is, its share must not exceed a certain percentage. Or, to put it in positive terms: the proportion of safe failures, the so-called *Safe Failure Fraction* (*SFF*) must be greater than a specified percentage, depending on the required SIL:

$$\mathrm{SFF} = 1 - \frac{\lambda_{\mathrm{DU}}}{\lambda_{\mathrm{total}}}, \text{ with } \lambda_{\mathrm{total}} = \lambda_{\mathrm{DU}} + \lambda_{\mathrm{DD}} + \lambda_{\mathrm{SD}} + \lambda_{\mathrm{SU}} \tag{18.25}$$

whereby the failure rates given here are the failure rates already summed up in the FMEDA. An FMEDA is carried out for so-called subsystems, whereby the previously considered assembly has become a subsystem. Depending on whether such a subsystem is simple (Type A) or complex (Type B), different achievable *Safety Integrity Levels* (SIL) result depending on the SFF:

In this context, simple subsystems are those containing, for example, relays, simple sensors, analog and to a certain extent digital circuits, so that the effect of a failure is predictable. However, this is not the case with complex subsystems, those containing software-based controls, microprocessors and ASICs.

Subsystems and HFT

Since the achievable SIL is not only dependent on the type, but also on the so-called *Hardware Failure Tolerance* (*HFT*), the architecture of a safety system must first be discussed in more detail.

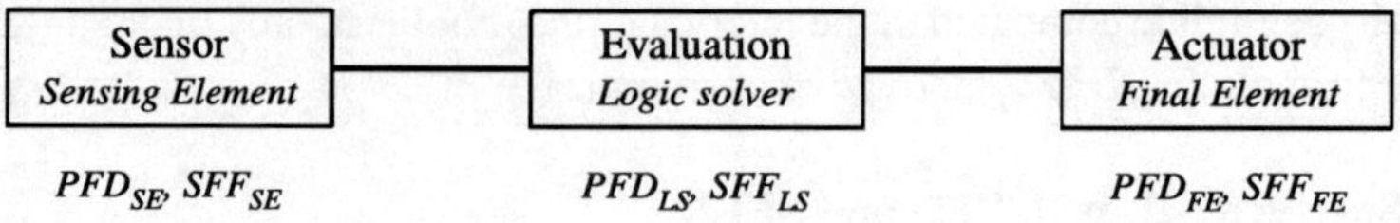

Fig. 18.31 A simple safety system consists of three subsystems

Safety systems are typically characterized by three subsystems:

- Subsystem Sensor (*SE, Sensing Element*), for detection of a possible hazardous situation
- Subsystem evaluation (*LS, Logic Solver*), which detects when predefined limits are exceeded
- Subsystem Actuator (*FE, Final Element*) that performs the required safety function.

Such subsystems form, as Fig. 18.31 shows, in the simplest case, a safety-related chain, whereby for each of these subsystems an average failure probability (the index *avg* is now omitted) and a *Safe Failure Fraction* are known:

The survival probability of this safety system is given by the product of the survival probabilities of the three subsystems, that is:

$$1 - \mathrm{PFD_{Sys}} = (1 - \mathrm{PFD_{SE}}) \cdot (1 - \mathrm{PFD_{LS}}) \cdot (1 - \mathrm{PFD_{FE}}) \quad (18.26)$$

Multiplying, sorting and neglecting all higher order products (which is allowed since the PFDs are considerably smaller than 1) results in

$$1 - \mathrm{PFD_{Sys}} \approx 1 - (\mathrm{PFD_{SE}} + \mathrm{PFD_{LS}} + \mathrm{PFD_{FE}}) \quad (18.27)$$

Thus, the probability of failure of the safety system is calculated simply by addition:

$$\mathrm{PFD_{Sys}} = \mathrm{PFD_{SE}} + \mathrm{PFD_{LS}} + \mathrm{PFD_{FE}} \quad (18.28)$$

This safety system with a so-called linear architecture has the disadvantage that it already fails when only one dangerous failure occurs, the HFT (see above) is zero.

Safety systems with dual redundancy or triple redundancy have an HFT of 1 or 2, that is, even if one or two DU failures occur, the safety of the system is not yet at risk.

Redundant Safety Systems and Architectures

System architectures are often represented by the abbreviation MooN (M ***out of*** N). The well-known safety system with a 2-out-of-3 voting therefore has a 2oo3 architecture. Table 18.16 lists the five most common architectures.

Table 18.16 The most common architectures in safety systems and their HFT

Architecture	Redundancy	HFT	PFD_{Sys}
1oo1	Linear system, no redundancy	0	$0,5\,\lambda_{DU} \cdot T_P$
2oo2	No redundancy, so-called pseudo-redundancy	0	$\lambda_{DU} \cdot T_P$
1oo2	Dual redundancy	1	$0,33 \cdot (\lambda_{DU} \cdot T_P)^2$
2oo3	Majority redundancy, two out of three decision	1	$(\lambda_{DU} \cdot T_P)^2$
1oo3	Triple redundancy	2	$0,25 \cdot (\lambda_{DU} \cdot T_P)^3$

Table 18.17 *Safe Failure Fraction* and achievable SIL

Safe Failure Fraction SFF	Type A simple subsystems	Type B complex subsystems
<60%	SIL 1	–
60 ... <90%	SIL 2	SIL 1
90 ... <99%	SIL 3	SIL 2
≥99%	SIL 4	SIL 3

Table 18.18 *Safe Failure Fraction* and achievable SIL as a function of the HFT

Safe Failure Fraction	Type A simple subsystems			Type B complex subsystems		
SFF	HFT = 0	HFT = 1	HFT = 2	HFT = 0	HFT = 1	HFT = 2
<60%	SIL 1	SIL 2	SIL 3	–	SIL 1	SIL 2
60... < 90%	SIL 2	SIL 3	SIL 4	SIL 1	SIL 2	SIL 3
90... < 99%	SIL 3	SIL 4	SIL 4	SIL 2	SIL 3	SIL 4
≥99%	SIL 4	SIL 4	SIL 4	SIL 3	SIL 4	SIL 4

Table 18.17, which shows the achievable SIL as a function of the SFF, can now be extended to include the HFT (see Table 18.18).

As expected, a higher HFT can achieve a higher SIL under certain conditions:

A "hard-wired" linear safety system (1oo1, HFT = 0) without software and microprocessors (type A), whose minimum SFF is caused by the *final element* and whose $SFF_{FE} = 78.3\%$, meets SIL 2 requirements with regard to random failures. In addition, the probability of failure PFD_{Sys} must also be less than 0.01 at an appropriate *proof test interval* of, for example, $T_P = 1$ year (8760 hours), as shown in Table 18.14.

It seems that redundant safety systems are the panacea for the combination of subsystems whose "SIL parameters" PFD and SFF are not suitable to achieve a required SIL.

However, one must keep in mind that multi-channel systems, apart from the multiple procurement costs, also contain the same electronics several times. The unpleasant SD failure (keyword false alarm!) also multiplies, that is, surprisingly, redundancies increase the probability of a false alarm occurring.

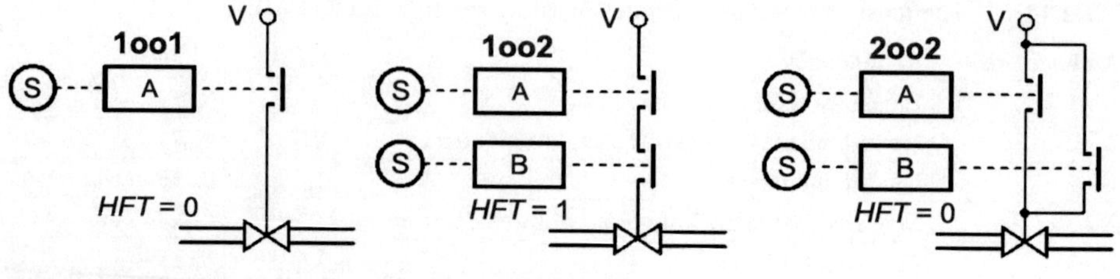

Fig. 18.32 The three simplest architectures and their HFT

Figure 18.32 shows the three simplest architectures, each with similarly structured channels. S (sensor) is a transmitter, the rectangles A and B represent the respective signal evaluation with subsequent switching output. The pipeline indicated below contains a flammable gas under high pressure, the valve is closed without current. The currentless state is shown, that is, the contacts only close and the valve opens after commissioning, provided that the voltage V is applied. Let the PFD_{Sys} of a single channel be 0.01.

The safety function is to de-energize the valve, that is, open the corresponding switching contacts again. However, if an undetected failure (DU failure) occurs, the contact does not open.

In a 1oo2 system, the probability of a DU failure occurring within the time interval T_P in both channels is given by the product of the individual probabilities and is therefore very low (about 0.00001). The disadvantage, however, is that even detected failures that do not impair the safety function at all (SD failure, contact opens) will de-energize the valve and thus reduce the system availability. This is different with the 2oo2 system: a detected failure does not de-energize the valve as long as it does not also occur in the second channel. This system is virtually immune to SD failures, but is no gain in terms of DU failure: Double electronics – double PFD. This is the reason why it is called pseudo-redundancy. The availability has been increased, but not the safety – the HFT is not greater than the HFT of a 1oo1 system.

As Fig. 18.33 shows, the 2oo3 system is a good compromise between availability and safety: As with the 1oo2 system, only the occurrence of two DU failures in two channels leads to the failure of the safety function. And as in the 2oo2 system, only the occurrence of two SD failures in two channels de-energizes the valve. The 1oo3 system is even better in terms of safety, because the safety function only fails if a DU failure occurs in each channel within the time interval T_P. Very safe (therefore also HFT = 2), but unfortunately also quite sensitive to SD failures.

It is not by chance that the HFT is called *Hardware Failure Tolerance.* Software errors as systematic errors cannot be intercepted by redundancies. Therefore the software must have been developed and tested according to the rules valid for the SIL achieved. For example, if SIL 3 is achieved by means of a higher HFT, the software involved must meet the SIL 3 requirements.

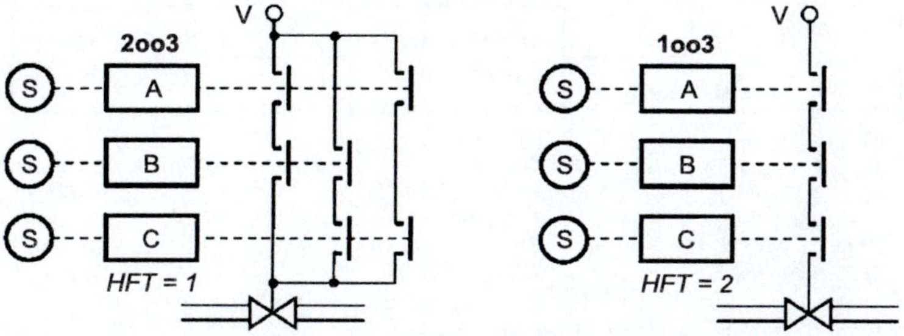

Fig. 18.33 Majority redundancy (HFT = 1) (**a**) and triple redundancy (HFT = 2) (**b**)

Not only software errors, but also other systematic errors that affect *all* redundant channels destroy the redundancy gained at great expense. These are the so-called *Common Cause Failures* (*CCF*), which can usually only be estimated as a proportion of $\beta = 1, 2, 5$ or 10% of the DU failure.

The supposedly best-known failure due to common cause is the failure of the power supply, although it is particularly easy to detect by a power failure relay. However, the energy supply is not the subject of the safety considerations of a gas detection system, since all considerations in this regard are based on a system in operation which goes into a safe state in the event of a power failure. In this sense, emergency power supply operation increases availability, but not safety.

If redundant architectures are composed of similarly structured channels, an additional additive term of size $0.5 \cdot \lambda_{DU} \cdot T_P$ must therefore be taken into account when determining thePFD_{Sys}, for example, for the 1oo2 architecture (see Table 18.16) we have

$$PFD_{Sys} = \frac{1}{3} \cdot (\lambda_{DU} \cdot T_P)^2 + \beta \cdot \frac{1}{2} \lambda_{DU} \cdot T_P \tag{18.29}$$

and for the 1oo3 architecture:

$$PFD_{Sys} = \frac{1}{4} \cdot (\lambda_{DU} \cdot T_P)^3 + \beta \cdot \frac{1}{2} \lambda_{DU} \cdot T_P. \tag{18.30}$$

This β term must not be neglected under any circumstances, it is usually much greater than the first term.

Real redundancy with almost no *common cause failures* can only be achieved by diversity, that is, by using different measuring methods (*Sensor Elements*) or products, if possible even from different manufacturers.

Optimum redundancy is also characterised by the fact that they have diagnostic circuits that switch off a channel that has been recognised as faulty and issue a message in the form of a repair request. The system therefore remains functional due to the intact channel. The repair must of course be carried out immediately. The system is only susceptible to a further

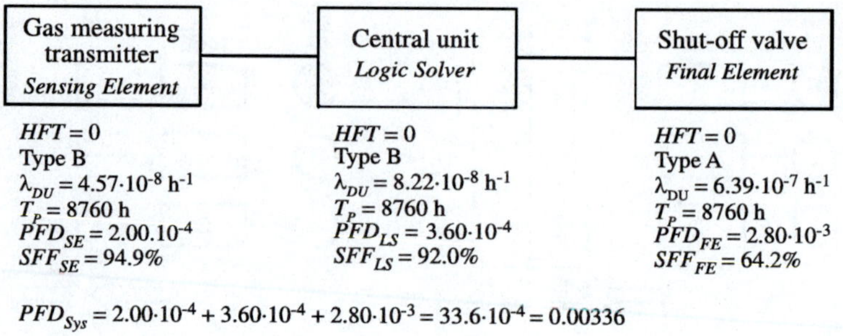

Fig. 18.34 Example: This 1oo1 architecture is suitable for SIL 2 applications

failure during the relatively short period of repair. Redundant architectures with such channel failure messages provided by diagnostic circuits are usually marked by a trailing D, e.g. 1oo2D.

The PFD_{Sys} given here deviate slightly from the formulae given in Part 6 of the EN 61508 standard. However, if one takes into account that the mean repair time *MTTR* is generally estimated at 8 hours and is negligible compared to T_P (a few 1000 hours) and that the factor β is also so small that an approximation of the form $1 - \beta \approx 1$ is justified, the calculation formulae with $MTTR = 0$given there merge into those used here.

Figure 18.34 illustrates the procedure for calculating the PFD_{Sys} using a typical example. The upper part shows the familiar linear 1oo1 system, the required data is listed under the respective subsystem. The *proof test* is performed annually ($T_P = 8760$ hours). It results in:

$$\mathrm{PFD_{Sys}} = \frac{1}{2} \cdot \lambda_{\mathrm{DU1}} \cdot T_P + \frac{1}{2} \cdot \lambda_{\mathrm{DU2}} \cdot T_P + \frac{1}{2} \cdot \lambda_{\mathrm{DU3}} \cdot T_P \tag{18.31}$$

Since the SFF of the subsystems (greater 90% for type B and greater 60% for type A) are sufficient for SIL 2 according to Table 18.17 and the average system PFD is less than 0.01, this system therefore meets the requirements for SIL 2 in this respect.

A redundant 1oo2 system consisting of two similar channels is shown in Fig. 18.35, whereby the two shut-off valves (closed when currentless) in the pipeline to be shut off are flowed through one after the other. The β box symbolizes the influence of the *common cause failure*, which was pessimistically assumed here to be 10%. It results:

$$\begin{aligned}\mathrm{PFD_{Sys}} &= \frac{1}{3} \cdot (\lambda_{\mathrm{DU1}} \cdot T_P)^2 + \frac{1}{3} \cdot (\lambda_{\mathrm{DU2}} \cdot T_P)^2 + \frac{1}{3} (\lambda_{\mathrm{DU3}} \cdot T_P)^2 + \beta \\ &\cdot (\mathrm{PFD_{SE}} + \mathrm{PFD_{LS}} + \mathrm{PFD_{FE}}) = 0.00034667\end{aligned} \tag{18.32}$$

The proportion of the first three squared terms is negligible, so that the simplification

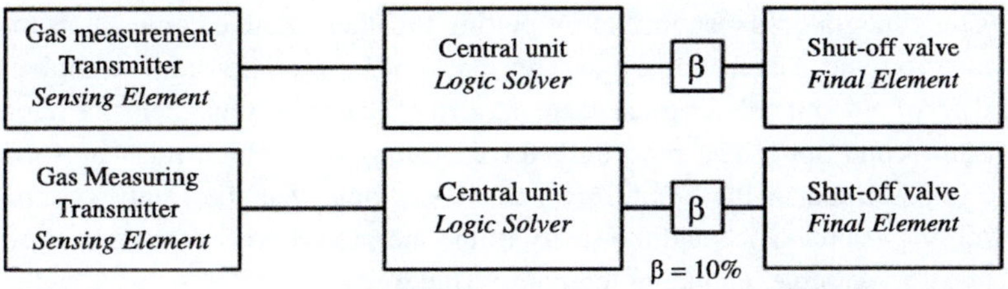

$HFT = 1$	$HFT = 1$	$HFT = 1$
Type B	Type B	Type A
$\lambda_{DU} = 4.57 \cdot 10^{-8}\ h^{-1}$	$\lambda_{DU} = 8.22 \cdot 10^{-8}\ h^{-1}$	$\lambda_{DU} = 6.39 \cdot 10^{-7}\ h^{-1}$
$T_P = 8760$ h	$T_P = 8760$ h	$T_P = 8760$ h
$PFD_{SE} \approx \beta \cdot 2.00 \cdot 10^{-4}$	$PFD_{LS} \approx \beta \cdot 3.60 \cdot 10^{-4}$	$PFD_{FE} = \beta \cdot 2.80 \cdot 10^{-3}$
$SFF_{SE} = 94.9$ %	$SFF_{LS} = 92.0$ %	$SFF_{FE} = 64.2$ %

$PFD_{Sys} \approx 0{,}10.(2.00 \cdot 10^{-4} + 3.60 \cdot 10^{-4} + 2.80 \cdot 10^{-3}) = 0.00336$

Fig. 18.35 Example: This 1oo2 architecture meets SIL 3 requirements, if applicable

$$\mathrm{PFD_{Sys}} \approx \beta \cdot (\mathrm{PFD_{SE}} + \mathrm{PFD_{LS}} + \mathrm{PFD_{FE}}) = 0.000336 \tag{18.33}$$

is justified. Since this average system PFD is smaller than 0.001 (see Table 18.14), this redundant system can be used to operate applications according to SIL 3, provided that the software used has been developed according to SIL 3 requirements. The fact that the *Safe Failure Fractions* of the subsystems are sufficient for SIL 3 applications can be seen for HFT = 1 directly from Table 18.18.

The calculable (and therefore very popular with engineers) SIL parameters such as PFD and SFF must by no means obscure the fact that qualitative requirements, namely requirements for safe operating behaviour, must also be met. For developers in particular, but also very informative for system integrators, there is even a product standard for gas detectors in which such requirements are listed very systematically (EN 50402). This standard also contains the important note that it can be assumed that gas detectors approved according to EN 60079-29-1 already meet requirements according to SIL 1. This standard deliberately does not refer to SIL, but only to *SIL capability* – because *Safety Integrity* always refers to the complete safety system. In this standard, however, the *final element* is not considered. Similarly, the suitability test according to EN 60079-29-1 only covers the gas detection system from gas inlet to alarm contact – without a *final element*.

The current approach to demonstrably meet the requirements for gas detectors with regard to SIL 2 is that the complete development process, including software development, is monitored by external authorized institutions. For this purpose, FMEDA reports with clearly defined operating conditions are prepared and the results are compiled in a *safety manual.* As is customary for type examinations, no changes may be made to the product without the agreement of the supervising institution. This is monitored by a *safety management* as an important part of quality control.

The system integrator is responsible for putting together suitable components to form a safety system on site. Here, too, the system may not be modified after completion and successful *proof test* and only original spare parts may be used for maintenance (creation of the quasi-new condition). The procedure for the *proof test,* which must be carried out regularly, is described in the *Safety Manual.* Please note that the safety system is not available during maintenance and the safety of the monitored process must be ensured by other means, for example, including temporary shutdown.

Concluding Remarks on Gas Detection Systems

A further prerequisite for the use of a gas detection system as a safety system is of course that the gas detection system from the gas inlet of the sensor to the switching relay is demonstrably suitable for the intended application. This is where the circle closes: Without the metrological suitability test confirmed by a notified body according to EN 60079-29-1 (or for oxygen according to EN 50104) no SIL! The use of a suitability-tested gas detection transmitter in combination with a switching amplifier, although SIL-certified, is not permitted without the corresponding EC type examination certificate.

No matter how safely a gas detection system is designed and no matter how reliably it measures – if the gas cannot reach the sensor, no alarm will be triggered even in the most dangerous situation! This statement shows quite clearly how important sensible planning and system design is. Operators (as best experts on their system and their measuring problem) and system engineers of the gas detection system manufacturer (as best experts on the various applications of their products) should consult together on the design of such gas detection systems.

A standard can also advise. The content of standard EN 60079-29-2, which can also be understood as a guide, is highly recommended in this respect. In particular, this standard also provides important information on the arrangement of sensors and on the commissioning and maintenance of gas detection systems. The correct calibration of sensors is also an important topic. Since a sensor is also differently sensitive for different substances, it must always be calibrated to the substance to which it is most insensitive. Then it will react more sensitively to all other substances, that is, the alarm thresholds are already exceeded at lower concentrations, but you are "on the safe side". Safety knows no compromises.

Such calibrations may have to be performed more frequently, especially if special influences change the measurement sensitivity of the sensor (sensor poisoning). At this point, the important note that a sensor calibration may well be necessary more frequently than – in the case of safety systems – specified by the time interval T_P of the proof test to be performed regularly. Infrared sensors do not suffer from such sensor poisoning, and the calibration intervals can be selected accordingly long, but possible contamination of the optical system must be taken into account. However, due to certain compensation procedures, nowadays an IR transmitter is more or less tolerant of contamination and issues a maintenance request long before it reports a fault. This is a classic self-diagnosis

Fig. 18.36 The Dräger PIR 7000 without dust protection cap and with open terminal box, on the *left* the infrared measuring section with front heated reflector

which, in addition to other self-checks, makes such an IR transmitter virtually predestined for use in safety systems.

Intelligent IR transmitters can output such *information about the well-being of the transmitter* not only via the industrially standardized *HART interface*, but the 4–20 mA signal, which is very common in the industrial sector, can also output status messages (fault messages and advance warnings) under 4 mA. A prerequisite, however, is that these status messages are also understood by a central controller. Such transmitters, such as the Dräger PIR 7000 (see Fig. 18.36) for the detection of combustible gases and vapors and also CO_2, have almost become the industry standard. Tested for more than 30 gases and vapors, whose characteristics and parameters can be retrieved from an implemented gas library of the transmitter, usable in a wide temperature range from −40 to +70°C, and with a SFF of 95% and a PFD_{avg} of only 0.0002 ($T_P = 1$ year), they can be used without any problems in plants which have to meet the requirements according to SIL 2. A product that has proven itself fifty thousand times over and is recognized worldwide. Its development with regard to *safety integrity* has been monitored and certified in accordance with EN 61508 by the TÜV SÜD Product Service, which is accredited for this purpose.

Gas measurement technology does not therefore mean that a usable concentration-dependent electrical signal can be elicited from a sensor during gassing under laboratory conditions! Rather, gas measurement technology stands for durable sensors and robust field transmitters that provide continuous service year in year out, even in the harshest industrial atmospheres, in arctic cold or desert climates, under heavy mechanical stress and vibration, under severe electromagnetic interference, especially the dreaded line-conducted *bursts,* and detect gases so early that their harmful effects are largely eliminated. Gas detection systems as a safety system can reduce a risk emanating from industrial plants by orders of magnitude if they have been demonstrably developed in such a way that possible dangerous failures are avoided or tolerated or – if this is not possible – at least detected.

18.3 Mobile Gas Measurement Technology

Rainer Krage

Time and again, the media report about devastating gas explosions caused by leaked natural gas from leaking pipes. Or about carbon monoxide poisoning by inhaling the exhaust gases of incomplete combustion. In the working environment there are countless other potential gas hazards to which people are exposed when entering rooms or which arise during the work process. Especially in closed rooms, basements or sinks there is a risk of accumulation of hazardous gases.

One of the first users of electrical gas detectors were the miners, for whom the aureole formation, a bluish glow of the miner's lamp, used to warn of methane gas. To protect them from toxic carbon monoxide, canaries were used, which fell off the pole already at such low gas concentrations that there was still enough time for the miners to get to safety.

Since the beginning of the nineteenth century, colorimetric measurement methods, catalytic, semiconductor and electrochemical sensors as well as PID and infrared sensors have been increasingly used in portable gas detection devices. Especially for personal and plant protection, mobile gas detectors are used today in many areas as useful aids for the detection of toxic or explosive gases and of oxygen deficiency. These devices are known as gas detectors, each of which is optimized for the specific hazards of the application. Leak detectors are used to detect leaks in tanks and pipelines and can detect even the smallest gas concentrations.

The list in Table 18.19 shows that in the majority of applications flammable and explosive gases can also occur and should be detected. The equipment used in these areas must therefore not pose any hazard that could trigger an explosion that would result in personal injury or damage to property.

For this reason, the vast majority of gas detectors established on the market are approved for use in potentially explosive atmospheres, which can be seen from the marking on the type plate and a corresponding certificate of conformity in the operating instructions. The extent to which the approval actually meets the requirements of the respective application must be checked in each individual case.

Device Structure

Modern portable multi-gas detectors are capable of selectively measuring up to 7 gases continuously and simultaneously. The linearized measured values are shown on graphic displays and can often be stored in data loggers. Sensors are used to measure flammable gases, especially for monitoring the explosion hazard in the range up to the lower explosion limit, but also for leak detection in the ppm measuring range. Oxygen sensors are mostly used for oxygen deficiency monitoring. This means that the oxygen content of fresh air of 20.95 vol.% is monitored, the measuring range is then usually 0–25 vol.%. Toxic gases should often be detected as selectively as possible with electrochemical sensors, alternatively as broadband as possible, for example, with photoionization sensors (PID).

Table 18.19 Applications of mobile gas detection devices

Field of application	Typical gases (extract)
Automotive industry	Carbon monoxide, carbon dioxide, nitrogen oxides
Chemical and petrochemical industry	Oxygen, combustible gases (nonane, methane), carbon monoxide, various other
Gases industry	Oxygen, combustible gases (nonane, methane), various other
Energy utilities	Natural gas
Food industry	Nitrogen, refrigerant, carbon dioxide
Steel industry	Hydrogen, natural gas, carbon monoxide, hydrogen sulphide
Mining	Methane, carbon monoxide
Canal construction	Hydrogen sulphide, carbon dioxide, nonane
Kohlen monoxid	Methane, carbon dioxide
Water and wastewater management	Ozone, chlorine, methane
Fire and rescue services	Solvents, acetylene, carbon monoxide
Hospitals	Oxygen, anaesthetic gases
Research institutes and universities	Oxygen, semiconductor process gases (arsine, phosphine, ...), combustible gases (nonane, methane)

Fig. 18.37 Device structure of the Microtector II G460

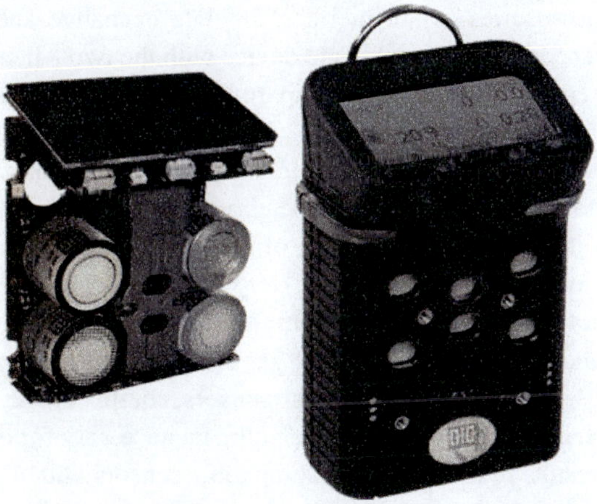

A view into the interior of such a portable gas detector is shown in Fig. 18.37. The right half of the picture shows the fully assembled device in a polycarbonate plastic housing moulded with thermoplastic elastomer (TPE). The elastic material protects the device from damage when it is dropped from a great height or subjected to strong shocks. On the back of the unit there is a *crocodile clip* (similar to a braces) for attaching to work clothing and a mounting bracket for attaching to a carabiner or similar. In the upper part of the device the graphic display for simultaneous visualization of all gas measurement values and the state of charge is visible. The device can be operated using the three keys.

Fig. 18.38 Smart gas sensors with EEPROM and standard interface

For visual alarm when limit values are exceeded, numerous red light-emitting diodes are located behind the areas recessed by the elastomer. The display illumination can change from green to orange to red, depending on the hazard potential. A horn with resonance body is located in the device for loud acoustic alarm.

The gas sensors are positioned behind the six oval recesses on the front of the device. These areas have been kept free of the elastomer, otherwise the measured values may be falsified for some gases. A calibration cap can be attached to the threaded inserts in the area of the diffusion openings to allow gas to flow through the sensors. The additional threaded inserts further down the device are required to connect a suction pump firmly to the G460.

The nine contacts in the lower part of the device serve in conjunction with the accessories as an interface for data exchange and as charging contacts for the battery. The battery is fixed in the device with the two Allen screws, but can be easily substituted or replaced by a battery compartment with alkaline, non-rechargeable batteries.

The left half of Fig. 18.37 shows the SMD PCBs[4] with five gas sensors mounted on them. The elongated sensor in the middle is an infrared sensor for simultaneous detection of carbon dioxide and hydrocarbons. To optimize the reaction time, there are two diffusion openings in the cuvette of the sensor. The remaining four sensors are immediately noticeable that they look very similar in geometry. This design with a diameter and a total height of about 20 mm each is, independent of the sensor manufacturer, one of the most common designs for gas sensors in mobile devices.

Figure 18.38 shows such sensors, equipped with electronics for analog signal processing and an EEPROM[5] in which all data necessary for digital signal processing are stored. This results in mechanically compatible sensors with a uniform electronic interface, which can be integrated interchangeably in one or even different devices. The device reads the data from the EEPROM and automatically adjusts the corresponding device parameters, so that the sensor can be used for measurement without any problems. These intelligent, so-called *smart sensors* considerably increase the flexibility of application, which in turn increases the economic efficiency.

[4] **S**urface-**m**ount **d**evice – **p**rinted **c**ircuit **b**oard.

[5] **E**lectrically **e**rasable **p**rogrammable **r**ead-**o**nly **m**emory.

Table 18.20 Energy consumption of gas sensors in mobile devices

Principle	Measuring component	Energy demand
Electrochemical 2 electrodes	Toxic gases	Negligible
	Oxygen	Negligible
Electrochemical 3 electrodes	Toxic gases selective	Negligible
	Oxygen unleaded	0.5 mW
Catalytic pellistor	Flammable gases	300 mW
Catalytic MEMS	Flammable gases	100 mW
Metal oxide	Miscellaneous	130 mW
Photoionization	Volatile organic compounds	100 mW
Infrared laser	Methane, carbon dioxide, ammonia	500 mW
Infrared thermal emitter	Methane, carbon dioxide, ammonia	100 mW
Infrared LED	Methane, carbon dioxide	3 mW

Power Supply, Charging Technology

The energy requirement of a mobile gas detector is essentially dependent on the sensor technology used. Devices with only one electrochemical sensor can be operated for months with a single alkaline mignon cell (AA cell or LR6), for example, as long as display lighting or alarms are not triggered frequently. Even disposable devices with a permanently installed lithium cell are offered, which are intended to be operational for up to three years. If the battery of such devices exhausts earlier in real field use, replacement or recharging is not possible. For some sensors, rough approximate values of the energy requirement as a function of the functional principle are listed in Table 18.20 as examples. The requirement for the analogue and digital signal processing is not taken into account.

In multi-gas appliances, in which pellistors are usually also used for universal monitoring of the explosion hazard, the use of rechargeable batteries is common. For devices with explosion protection approval, only those batteries that are part of the approval may be used. Rechargeable batteries usually have to be specially encapsulated, for example in epoxy resin, to be approved. Only chargers that meet the specifications of the device manufacturer may be used. Otherwise it cannot be ruled out that components relevant to ignition protection may be destroyed during the charging process. In order to achieve optimum equipment running times, chargers from the original accessories should be used anyway and the instructions for charging in the operating instructions should be observed. The most common way is to use NiMH batteries, for which some basic rules of conduct must be observed. In order to maintain the full capacity of the NiMH battery permanently, care should be taken to ensure that the battery is only charged with the charger when necessary, depending on its service life and frequency. The charger should not be used for weeks to store the gas detector. Table 18.21 shows recommendations for charging the battery depending on the use of the instrument.

Table 18.21 Recommended charging of NiMH batteries

Device usage	Recommendation for charging the battery
More than 3 h every day	Charge after use
Daily less than 3 h	Charge every second or third day
1× per week	1 day before the next use
1× per month more than 3 h	Charge after use and charge 1 day before the next use
1× per month less than 3 h	1 day before the next use
1× per quarter or less	Charge after use and charge 1 day before the next use

If a device is seldom used, the battery should be charged after use, as in many devices, parts of the electronics have to be further supplied with energy also in the switched off condition and the batteries slowly loose their capacity by a self-discharge. If a device has not been used for a very long time and the battery is completely discharged, it should be charged about one day before the next use. A normally discharged battery is usually recharged to well over 90% of its nominal capacity in a few hours in normal charging mode. The chargers then switch to a lower charging current, and in this trickle charge mode the battery is then charged and held at 100% of its nominal capacity. If, despite a fully charged battery, the normal equipment operating time is not achieved, this could be caused by the *lazy battery effect*. The discharge behaviour changes in such a way that the battery voltage drops after a short time in operation, even though the battery is fully charged.

With NiMH power supply units, the influence of temperature above 50 °C, after a long period of non-use, unfavorable use of the device or incorrect charging behavior can lead to the so-called *lazy battery effect* and thus to a reduction in the operating time of the device. This can happen if the battery is never completely discharged by the device or if the battery is charged too often or for too long. It should be avoided that charging processes are started several times a day and that devices remain in the charger for several days or weeks.

The *lazy battery effect* can usually be eliminated by completely discharging the NiMH battery. This discharge can optionally be carried out on some devices before a charging process, for example once or on set weekdays, so that the readiness for use of the device is always guaranteed. If the *lazy-battery effect is* strongly developed, this option should be switched on repeatedly after charging the battery.

Limit Values and Units

Mobile gas measuring devices are generally used for monitoring limit values in personal and plant protection. If defined alarm thresholds are exceeded or undercut, visual and acoustic alarms are triggered in addition to the pure measured value display.

To monitor the explosion hazard, the detected gas concentration is usually not converted and displayed as an absolute concentration in volume %, but as a relative concentration in % of the lower explosion limit, % LEL. However, there is no generally valid rule for the definition of alarm thresholds. In practice, limits of 20% of LELs with an advance warning at 10% of LELs have proved to be useful for portable gas detectors. Depending on the type of gas and the application, however, other limit values may be prescribed or may be appropriate.

In oxygen monitoring, an alarm due to lack of oxygen in the breathing air is the most common application. If the oxygen content falls from the 20.95% normally present in the earth's atmosphere, an alarm is established if the oxygen content falls below 17% with a pre-alarm at 19%. If oxygen enrichment is to be expected, the common alarm value is established at 23% oxygen. It is remarkable that the absolute oxygen concentration, i.e. the oxygen partial pressure, is relevant for most chemical and biological processes. Accordingly, the total air pressure must be taken into account when assessing whether the oxygen content in the air we breathe is harmless.

The concentration of toxic gases in the air we breathe can be subject to considerable fluctuations at workplaces. To prevent health hazards, various limit values must be monitored. In addition to limit values for the permissible maximum instantaneous values, limit values for average exposure over time are also set. For many substances that are harmful to health, a limit value for the average exposure over an entire working day is specified. Typically, the device calculates an average value over eight hours and triggers an alarm if the limit value is exceeded. In addition, maximum permissible short-term concentrations can be defined, for which mean values over a time of 15 minutes are taken as reference.

For Germany, the Federal Ministry of Labor and Social Affairs has published these limit values to be complied with in the Joint Ministerial Gazette and the Federal Gazette since 2007. Information on International Occupational Exposure Limits can be found, for example, in the GESTIS substance database of the Institute for Occupational Safety and Health of the German Social Accident Insurance, which can be viewed on the Internet.

Sampling

Mobile gas detectors usually work in diffusion mode. This means that only those gases to which the device is directly exposed can be detected. The diffusion openings are usually equipped with hydrophobic, gas-permeable membranes to protect the sensors behind them from dirt and moisture. These membranes must be kept clean so as not to hinder the entry of gas. During operation, care must also be taken that the devices are not worn on the body in such a way that no gas can reach the openings.

Fig. 18.39 Multi-gas detector *Microtector II G460* with attached pump and telescopic suction tube

If gas concentrations are to be detected in areas which cannot be entered for safety reasons or which are inaccessible, there are various options for the clearance measurement.

Before descending into sewer shafts, for example, the unit can be lowered by means of a rope. Especially for devices with peak value memory, even better with a data logger, the situation prevailing in the examined area can be assessed after the recall. This prevents entering areas without breathing protection where dangerous gas concentrations have accumulated. Special care must be taken during this procedure so that the device is not immersed in dirty water or other refuse and the diffusion openings become dirty or even the device becomes completely inoperative due to moisture entering. In unlit shafts, a flashlight integrated in the unit is helpful for this purpose, with which the area below the unit is illuminated.

Another possibility is the use of suction pumps. The spectrum ranges from simple bellows pumps to electrical accessories and pumps integrated into the gas detector. These pumps can be used to bring the gas to be detected to the sensors via hoses, float probes and so-called telescopic suction pipes, as shown in Fig. 18.39. However, a number of instructions must be observed in order to obtain precise measurement results.

- With long hoses, a delayed reaction time must be observed, the hose should not be longer or thicker than necessary
- The flow of the pump must be monitored to ensure a continuous gas flow
- Filters against dirt and water may be necessary to protect the pump and gas meter
- Avoid "dilution" of the gas due to leaking tubing or unsuitable diffusion cover of the sensors
- Reactive gases require suitable, chemically inert materials in the pump and in the hose
- Moisture in the sampling system can be problematic, for example, for ammonia measurement.

Maintenance, Calibration and Adjustment
Portable gas detectors are often used in a wide variety of locations and under changing, often harsh environmental conditions with extreme temperatures, humidity and dirt. They may have been dropped or subjected to severe impacts. Damage to the device caused by such dropping or impact is not necessarily obvious, so that regular checks of the error-free functioning of the device are strongly recommended. This is the only way to ensure that reliable measured values are determined and that an alarm is triggered if limits are exceeded. Depending on the intended use, there are also specifications for such checks, for example, by the Professional association. For example, the German Professional Association for Raw Materials and Chemical Industry's leaflets T 021 and T 023 require daily visual inspections and display tests with test gas for both portable gas detectors for toxic gases/vapors and oxygen and for explosion protection.[6]

In order to be able to carry out the necessary tests, gases from compressed gas containers (gas cylinders) are usually used. The zero point of most sensors can also be tested with fresh air, but it must be ensured that this is really gas-free. For carbon dioxide sensors synthetic air or nitrogen must be used, for oxygen sensors nitrogen. More difficult is the correct selection of the suitable test gas for checking and, if necessary, adjusting the display sensitivity. Suitable test gas concentrations and compositions depend on the sensors used and the measuring task. As a rule, the test gas should correspond to the target gas and the test gas concentration should be approximately in the middle of the measuring range, but in any case above any alarm limits. If the component to be measured is difficult to handle as a test gas, a substitute gas can also be used if cross-sensitivity is known. This option is particularly useful when measuring solvents with low vapor pressure or very toxic gases. The choice of the test gas can also be difficult for instruments with several sensors. Mixed gases are often used to test all sensors simultaneously. If the composition of mixed gases is incorrectly selected, the sensors can be influenced by cross-sensitivities or damage from sensor poisons. In the case of test gases for devices in which both catalytic sensors (pellistors) for the detection of combustible gases and hydrogen sulfide sensors are installed, the hydrogen sulfide concentration should therefore be selected as low as possible due to possible damage to the pellistor. When using a combination of oxygen sensor and pellistor, care must be taken that the measured values of the pellistor are falsified if there is too great a shortage of oxygen, because oxygen for catalytic combustion is missing. This effect can be observed below a concentration of 10% oxygen. In order to be able to test the functionality of a gas measuring instrument in a qualified manner, comprehensive basic knowledge is therefore necessary. In addition, special instructions in the operating manuals of the devices must be observed.

---

[6]BG RCI. 2012. T 021/T023. [http://www.bgrci.de/exinfode/dokumente/gaswarneinrichtungen-und-geraete/]. Heidelberg: Trade Association for Raw Materials and Chemical Industry, 2012.

Fig. 18.40 Docking station *DS404* with *Microtector II G460*

For the test gas feeding, the sensors must be supplied with gas in such a way that the measured values correspond to those in measuring mode. Numerous sensors provide different measured values depending on the flow velocity, temperature or pressure of the gas. For calibration and adjustment, suitable attachments are used for diffusion measuring instruments, with which the test gas task can be carried out at a defined volume flow.

Incorrect maintenance, especially incorrect adjustment, can also cause risks. In order to avoid this, it is state of the art for modern, processor-controlled mobile gas measuring devices to use so-called test or docking stations, which can be used to carry out automated functional tests, also known as bump tests, as well as an adjustment of the zero point and sensitivity of the sensors. A *docking station* with associated test gas cylinder, sampling fitting, CO_2 filter cartridge and power supply unit is shown in Fig. 18.40.

The display test typically covers testing the response behavior of the sensors with reaction times as well as the functionality of visual and acoustic alarm transducers. If the display test is negative or if the time interval for a function check has been reached, a zero-point adjustment and a sensitivity adjustment is also carried out.

The software-controlled docking stations enable the specifications to be met via corresponding user interfaces. All test results as well as information on the test gas used and other relevant parameters are also stored in databases or in CSV (*Comma Separated Value*) format for easy export to Excel®. This enables a permanent, complete documentation of the test results without any problems.

Another very helpful function can be the automated check of the cylinder pressure. A test gas cylinder that is running low can thus be replaced in good time without any downtime if the appropriate signal is given.

New Trends

The desire for wireless networking does not stop at gas detection devices. As far as safety-related gas detectors are concerned, however, the use of the popular W-LAN technology or the GSM standard is not suitable. With both technologies it cannot be guaranteed that sufficient capacities for on-line data transmission are available if required. W-LAN networks are operated in a largely unregulated manner, which is why failures can occur due to channel utilization or interference from high field strengths of individual transmitters. Mobile radio networks do not have complete area coverage and cannot be extended ad-hoc with base stations for optimization if necessary. Even in areas with good network coverage, mobile radio networks are occasionally already overloaded with the various private users.

When selecting appropriate radio technologies, not only generally applicable national legal requirements regarding spectrum use must be taken into account. In potentially explosive atmospheres, additional requirements are imposed to limit the maximum output radiated power or radiant energy,[7] so that the energy level does not cause an explosion hazard.

In radio networks, the available bandwidth usually increases at higher transmission frequencies, but the range and ability to overcome obstacles decreases. The use of long wavelengths, i.e. low transmission frequencies, thus enables better penetration of built-up areas or buildings due to the physical laws of wave propagation through diffraction and reflection. Correspondingly, wireless data transmission on frequencies from 863 to 870 MHz, which have been reserved exclusively for short-range radio communication with so-called Short Range Devices (SRD) throughout Europe, is well suited. In Germany, the use of frequencies by these devices is regulated by the Federal Network Agency,[8] with frequency usage parameters for maximum radiated power, frequency access and interference reduction techniques, and other usage restrictions with regard to the data to be transmitted.

The networks shown in Fig. 18.41 operate in this frequency range. Mobile gas detectors with low transmission power and correspondingly low range connect directly to a control station computer in system 1 or to permanently installed base stations (repeaters) in system 2. Several base stations are then wirelessly networked with each other with higher transmission power, alternatively via cable, and finally connected to a control station computer. The communication paths with the corresponding transmission power and range are described in the legend. In this way, not only measurement data and alarms of mobile gas detectors can be transmitted. Using the base stations within range, whose location is known, it is also possible to locate the mobile devices. Building and/or site plans can then be used to visualize the positions of the mobile devices with associated

[7] IEC 60079-0.

[8] Vfg 30/2014.

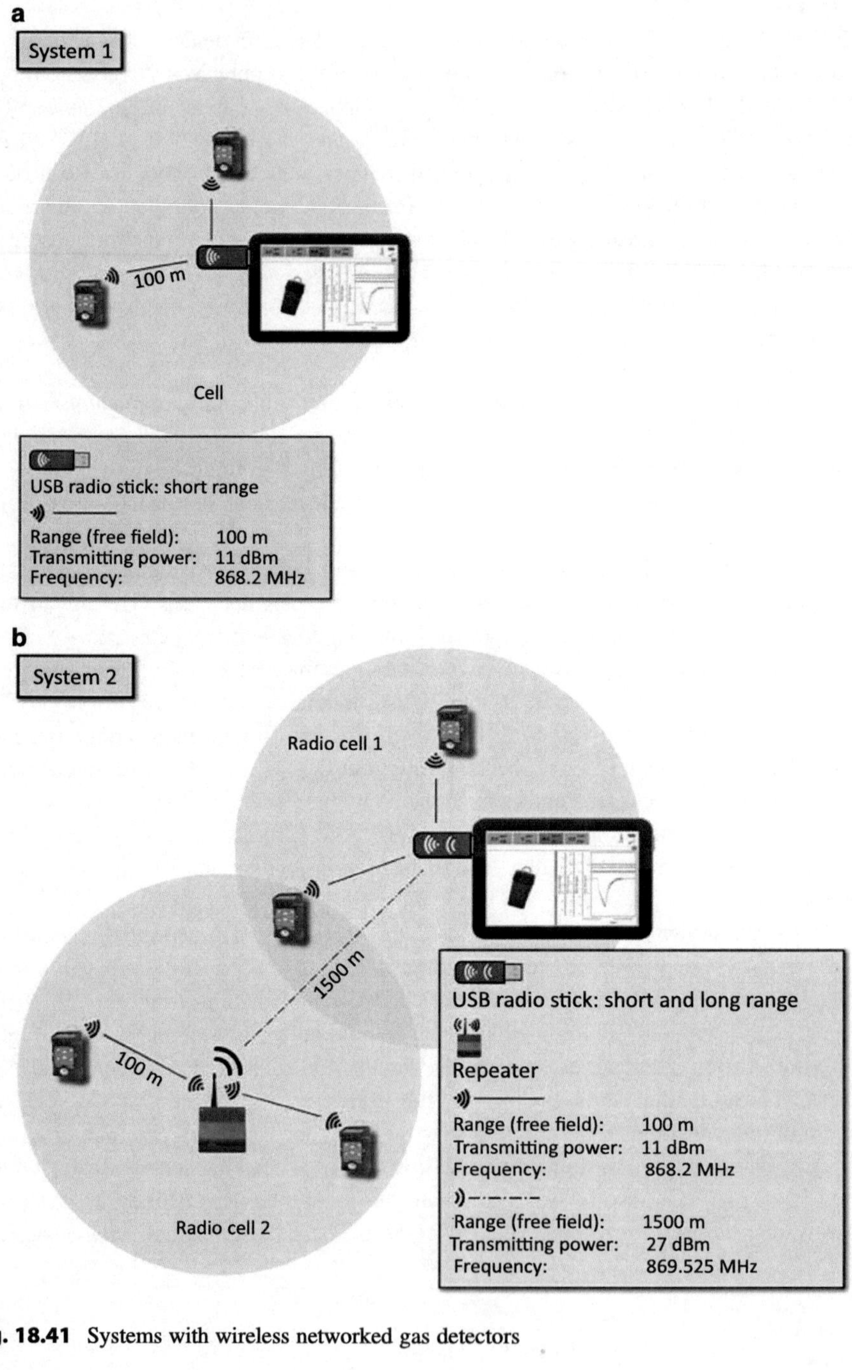

Fig. 18.41 Systems with wireless networked gas detectors

measured values by means of GUI (*graphical user interface*) and to link them clearly with other relevant information. Hazard warnings known in the control center can be sent to the mobile devices via a bidirectional radio link and can trigger alarms locally or be shown as text messages on the display to warn those affected in good time.

A useful addition to mobile gas detectors can be the so-called *dead man detection.* Here, the signal of an acceleration sensor is evaluated in order to be able to judge whether the wearer of the device has fallen or is motionless, i.e. possibly unconscious. In conjunction with wireless networking and localization options, this function can provide valuable services in the event of unobserved accidents.

18.4 Annex

Safety related material data of important gases. The table was developed by the Gesellschaft für Gerätebau mbH Dortmund (GfG) and is based on data from CHEMSAFE (The database CHEMSAFE is a joint project of the Physikalisch-Technische Bundesanstalt (PTB-Braunschweig) of the Federal Institute for Materials Research and Testing (BAM-Berlin) and the DECHEMA Society for Chemical Engineering and Biotechnology (Frankfurt am Main). The GESTIS substance database is created and maintained by the Institute for Statutory Accident Insurance and Prevention for occupational safety-BIA). Values from the Gestis database are marked with (WT = Heat tone, WL = Heat conductivity, EC = Electrochemical sensors, CS = Chemosensor, IR = Infrared, PID = Photoionization).

		TRGS 900 12.06.2008		Gas density	Ignition temperature	Flash point
Gas types (substances)	Chemical formula	ppm	mg/m^3	Luft = 1	°C	°C
Acetone	$CH_3-CO-CH_3$	500	1200	2	435	<−20
Acetonitrile	$CH_3-C\equiv N$	20	34	1.42	525	6
Acetylene (= Ethin)	$HC\equiv CH$	–	–	0.91	305	–
Ammonia	NH_3	20	14	0.6	630	–
Arsine (= hydrogen arsenic)	AsH_3	0.005	0.016	2.69	285	–
Petrol (petrol)	Mixture	–	–	–	–	−21
Benzene	C_6H_6	1	3.25	2.7	498	−11
Boron trichloride (HCI measurement)	BCI_3	–	–	4.06	–	–
Boron trifluoride (HF measurement)	Bf_3	0.35	1	2.37	–	–
Bromomethane (R40 B1)	CH_3Br	−5	−20	3.07	535	194

(continued)

Gas types (substances)	Chemical formula	TRGS 900 12.06.2008 ppm	mg/m³	Gas density Luft = 1	Ignition temperature °C	Flash point °C
Bromotrifluoromethane Halon 1301	$CBrF_3$	1000	6200	5.24	–	–
Butadiene	$CH_2 = -CH$ $CH{=}CH_2$	–	–	1.92	415	−85
n-Butane	C_4H_{10}	1000	2400	2.11	365	−60
Iso-butane	$(CH_3)_3CH$	1000	2400	2.11	460	−83
1-butanol	C_4H_9OH	100	310	2.55	359	29
2-butanol	C_2H_5- $CHOH-CH_3$	100	300	2.56	405	24
Isobutanol (= 2-methyl-1-propanol	$(CH_3)_2CH-$ CH_2OH	100	310	2.56	415	28
2-butanone (= Methylethylketon)	CH_3-CO- C_2H_5	200	600	2.49	404	−9
1-butene (= 1-butylene)	$CH_2 = -CH$ CH_2-CH_3	–	–	2	440	−112
n-butyl acetate	$CH_3-CO-O-$ C_4H_9	100	480	4.01	425	22
Iso-butyl acetate	CH_3-CO- $O-_2-CH\ CH$ $(CH_3)_2$	100	480	4.01	421	18
Chlorine	CI_2	0.5	1.5	2.49	–	–
Chlorofluoromethane (R22)	$CHCIF_2$	500	1800	2.99	–	–
Chlorine dioxide	CIO_2	0.1	0.28	2.33	–	–

Molar weight	Ex-limits in air (20 °C and 1013 hPa)		Possible measuring methods and measuring ranges					
			WT	WL	EC	CS	IR	PID
g/mol	UEG	OEG	%UEG	Percent by volume	ppm	ppm	Percent by volume	ppm
58.08	2.5	14.3	0 ... 100	–	–	20 ... 300	0 ... 2.5	0 ... 500
41.05	3	16	0 ... 100	–	–	–	0 ... 3	–
26.05	2.3	100	0 ... 100	–	–	20 ... 1000	0 ... 2.3	–
17.03	15.4	33.6	0 ... 50	–	0 ... 100	20 ... 10.000	–	0 ... 1500
77.95	3.9	77.8	–	–	0 ... 1	–	–	–
–	0.7	7.6	0 ... 100	–	–	–	0 ... 0.7	0 ... 500
78.11	1.2	8.6	–	–	–	20 ... 300	–	0 ... 250

(continued)

Molar weight	Ex-limits in air (20 °C and 1013 hPa)		Possible measuring methods and measuring ranges					
			WT	WL	EC	CS	IR	PID
g/mol	UEG	OEG	%UEG	Percent by volume	ppm	ppm	Percent by volume	ppm
117.17	–	–	–	–	0 . . . 3	–	–	–
67.81	–	–	–	–	0 . . . 3	–	–	–
94.94	8.6	20	–	–	–	–	–	0–700
148.91	–	–	–	–	–		–	–
54.03	1.4	16.3	0 . . . 100	–	–	20 . . . 300	0 . . . 1.4	–
58.12	1.4	9.4	0 . . . 100	–	–	30 . . . 2000	0 . . . 1.4	–
58.12	1.5*	9.4*	0 . . . 100	–	–	30 . . . 2000	0 . . . 1.5	–
74.12	1.7	11.3	0 . . . 100	–	–	20 . . . 1000	0 . . . 1.7	0 . . . 1500
74.12	1.6	11	0 . . . 100	–	–	20 . . . 1000	0 . . . 1.6	–
74.12	1.7	11	0 . . . 100	–	–	20 . . . 1000	0 . . . 1.7	–
72.11	1.5	12.6	0 . . . 100	–	–	20 . . . 1000	0 . . . 1.5	0 . . . 400
56.11	1.2	10.6	0 . . . 100	–	–	20 . . . 1000	0 . . . 1.2	0 . . . 500
116.16	1.2	7.5	0 . . . 100	–	–	20 . . . 1000	0 . . . 1.2	–
116.16	1.3	10.5	0 . . . 100	–	–	20 . . . 1000	0 . . . 1.3	–
70.91	–	–	–	–	0 . . . 5	–	–	–
86.47	–	–	–	–	–	30 . . . 1000	–	–
67.45	–	–	–	–	0 . . . 1	–	–	–

Gas types (substances)	Chemical formula	TRGS 900 12.06.2008		Gas density	Ignition temperature	Flash point
		ppm	mg/m³	Luft = 1	°C	°C
Chloromethane (= Methylchlorid)	CH_3Cl	50	100	1.78	625	–
Hydrogen chloride	HCl	2	3	1.27	–	–
Hydrogen cyanide	HCN	1.9	2.1	0.95	538	−18
Cyclohexane	C_6H_{12} (6-ring)	200	700	2.91	245	−20
Cyclopentane	C_5H_{12} (5-ring)	–	–	2.42	361	< – – 17
Landfill gas	$CH_4 + CO_2$	–	–	–	Mixture – values variable	
Diboran	B_2H_6	–	–	0.97	Approx. 40	−90
Dichlorodifluoromethane (R12)	CCl_2F_2	1000	5000	4.26	–	–

(continued)

Gas types (substances)	Chemical formula	TRGS 900 12.06.2008		Gas density	Ignition temperature	Flash point
		ppm	mg/m^3	Luft = 1	°C	°C
1,1-dichloroethane	$CHCl_2-CH_3$	100	410	3.42	660	−17
Dichloromethane (= methylene chloride)	CH_2Cl_2	75	260	2.93	556	–
Diethyl ether	$C_2H_5-O-C_2-H_5$	400	1200	2.55	175	−20
Difluoroethane (= 152a)	CHF_2-CH_3	–	–	2.28	455	–
Dimethyl ether	CH_3-O-CH_3	1000	1900	1.63	235	−42.2
1,4-dioxane	$(C_2H_4-O-C_2H_4)O$ (6-ring)	20	73	3.04	300	12
Ethane	C_2H_6	–	–	1.05	515	−135
Ethanol	C_2H_5OH	500	960	1.59	363	12
Ethyl acetone	$CH_3-co-O-C_2H_5$	400	1500	3.04	426	−4
Ethylene (ethylene)	$CH_2=CH_2$	–	–	0.98	425	–
Ethylene oxide	$(CH_2-CH)O_2$	–	–	1.52	440	−30
Fluorine	F_2	1	1.6	1.31	–	–
Hydrogen fluoride (= hydrofluoric acid)	HF	1	0.83	−0.71	–	–
Formaldehyde	$H_2C=O$	0.3	0.37	1.04	424	–
English (= germanium hydrogen)	GeH_4	–	–	2.65	< 85	–
Helium	He	–	–	0.14	–	–
n-Heptane	C_7H_{16}	500	2100	3.46	220	−4
n-Hexane	C_6H_{14}	50	180	2.98	225	−20
Kerosene (for example, petroleum 180/220)	Mixture (low in aromatics)	–	–	Approx. 4	355	60
Carbon dioxide	CO_2	5000	9100	1.53	–	–
Carbon monoxide	CO	30	35	0.97	605	−191
Methane	CH_4	–	–	0.55	537	–

References

Brandes, E., Möller, W.: Sicherheitstechnische Kenngrößen. Bd. 1: Brennbare Flüssigkeiten und Gase. Wirtschaftsverlag, Bremerhaven (2003)

Burgess, M.J., Wheeler, R.V.: The lower limit of inflammation of mixtures of the paraffin hydrocarbons with air. Trans. Chem. Society. **99**, 2013–2030 (1911)

Cerbe, G.: Grundlagen der Gastechnik. Hanser Verlag, München (2008)

EN 50402: Elektrische Geräte für die Detektion und Messung von brennbaren oder toxischen Gasen und Dämpfen oder Sauerstoff – Anforderungen an die funktionale Sicherheit von ortsfesten Gaswarnsystemen

EN 60079-10-1: Explosionsgefährdete Bereiche – Teil 10-1: Einteilung der Bereiche

EN 60079-20-1: Stoffliche Eigenschaften zur Klassifizierung von Gasen und Dämpfen – Prüfmethoden und Daten

EN 60079-29: Gasmessgeräte – Auswahl, Installation, Einsatz und Wartung von Geräten für die Messung von brennbaren Gasen oder Sauerstoff

Hartwig, A.: MAK- und BAT-Werte-Liste Senatskommission zur Prüfung gesundheitsschädlicher Arbeitsstoffe. Mitteilung 48. Wiley-VCH, Weinheim (2012)

Jessel, W.: Gase-Dämpfe-Gasmesstechnik. Dräger Safety AG, Lübeck (2001)

Molnarne, M., Schendler, T., Schröder, V.: Sicherheitstechnische Kenngrößen. Bd. 2: Explosionsbereiche von Gasgemischen. Wirtschaftsverlag, Bremerhaven (2003)

Phoenix Contact: Explosionsschutz Theorie und Praxis. MNR 5107652/14.01.2011–3 (2010)

Redeker, T., Schön, G.: 6, Nachtrag zu Sicherheitstechnische Kennzahlen brennbarer Gase und Dämpfe. Deutscher Eichverlag GmbH, Braunschweig (1990)

Scheuermann, K.: Praxishandbuch Brandschutz. Beuth-Verlag (2008)

TRGS 900: Technische Regeln für Gefahrstoffe, Arbeitsplatzgrenzwerte. BArBl H. 1/2006, 41–55 (2006)

Further Reading

BARTEC: Broschüre Grundlagen Explosionsschutz. 11. überarbeitete Aufl. BARTEC GmbH, Bad Mergentheim (2012)

Baumbach, G.: Luftreinhaltung. Springer, Heidelberg (1990)

Stahl, R.: Grundlagen Explosionsschutz. ID-NR. 00 006 26 77 0S-PB-ExGrundlagen-03-de-10/ (2006)